A TEXTBOOK OF
MANUFACTURING TECHNOLOGY
(Manufacturing Processes)

For B.E./ B.Tech., A.M.I.E.-Section B, and Competitive Examinations

CW01506542

By
Er. R.K. RAJPUT

M.E. (Hons.), *Gold Medallist*; Grad. (*Mech. Engg. & Elect. Engg.*);
M.I.E. (*India*); M.S.E.S.I.; M.I.S.T.E.; C.E. (*India*)

Recipient of:
"Best Teacher (Academic) Award"
"Distinguished Author Award"
"Jawahar Lal Nehru Memorial Gold Medal"
for an outstanding research paper
(Institution of Engineers–India)

Principal (*Formerly*):
● *Thapar Polytechnic College;*
● *Punjab College of Information Technology,*
PATIALA

LAXMI PUBLICATIONS (P) LTD

(An ISO 9001:2008 Company)

BENGALURU ● CHENNAI ● COCHIN ● GUWAHATI ● HYDERABAD
JALANDHAR ● KOLKATA ● LUCKNOW ● MUMBAI ● RANCHI ● NEW DELHI
BOSTON (USA) ● ACCRA (GHANA) ● NAIROBI (KENYA)

A TEXTBOOK OF MANUFACTURING TECHNOLOGY

Printed and bound in India
Typeset at Goswami Associates, Delhi
First Edition: 2007; Second Edition : 2015
ISBN 978-81-318-0244-1

Branches			
✆	Bengaluru	080-26 75 69 30	
✆	Chennai	044-24 34 47 26,	24 35 95 07
✆	Cochin	0484-237 70 04,	405 13 03
✆	Guwahati	0361-254 36 69,	251 38 81
✆	Hyderabad	040-27 55 53 83,	27 55 53 93
✆	Jalandhar	0181-222 12 72	
✆	Kolkata	033-22 27 43 84	
✆	Lucknow	0522-220 99 16	
✆	Mumbai	022-24 91 54 15,	24 92 78 69
✆	Ranchi	0651-220 44 64	

PUBLISHED IN INDIA BY

Laxmi Publications (P) Ltd.
(An ISO 9001:2008 Company)
113, GOLDEN HOUSE, DARYAGANJ,
NEW DELHI - 110002, INDIA
Telephone : 91-11-4353 2500, 4353 2501
Fax : 91-11-2325 2572, 4353 2528
www.laxmipublications.com   info@laxmipublications.com

C— 9791/015/02

Printed at: Ajit Printing Press, Delhi

A TEXTBOOK OF

MANUFACTURING TECHNOLOGY
(Manufacturing Processes)

By the Same Author:

- *Thermal Engineering*
- *Engineering Thermodynamics*
- *Applied Thermodynamics*
- *Internal Combustion Engines*
- *Automobile Engineering*
- *Power Plant Engineering*
- *Elements of Mechanical Engineering*
- *Steam Tables and Mollier Diagram (SI Units)*

Contents

RECENT TRENDS IN MANUFACTURING

ADDITIONAL TOPICS

Preface to the Second Edition

I take the pleasure in presenting the **"Second Edition"** of this book. The warm reception which the previous edition and reprints have received is a matter of much satisfaction to me.

In this edition, the book has been thoroughly revised, and a new **"Section"** on **"Short Answer Questions"** has been added to make the book still more useful to the students.

The constructive suggestions for improvement of this book are most welcome.

Er. R.K. Rajput
(Author)

Preface to the First Edition

This treatise on **"Manufacturing Technology"** (*Manufacturing Processes*) contains comprehensive treatment of the subject matter in simple, lucid and direct language. It also envelopes an large number of solved examples. It covers comprehensively the syllabii of various Indian Universities on the above mentioned subject for B.E./B.Tech. courses. The book will prove a boon to the students preparing for Engineering undergraduate, A.M.I.E. (Section B) and competitive examinations.

The book comprises 25 chapters, covering various topics systematically and exhaustively. All chapters are saturated with much needed text supported by simple and self explanatory figures. **"Questions with Answers"**, *Highlights, Objective Type Questions, Theoretical Questions* and *Unsolved Examples* have been added at the end of each chapter to make the book a comprehensive and a complete unit in all respects.

The author's thanks are due to his wife Ramesh Rajput for extending all cooperation during the preparation of manuscript and proofreading.

Although every care has been taken to make the book free of errors both in text as wells as in solved examples, yet the author shall feel obliged if any errors present are brought to his notice. Constructive criticism will be warmly received.

Er. R.K. Rajput
(Author)

1

Concept of Manufacturing

1.1. INTRODUCTION

In general, **manufacturing** *is the economic term for making goods and services available to satisfy human wants.* In fact, manufacturing involves a series of related activities and operations such as :

(*i*) Product design and development;

(*ii*) Material selection;

(*iii*) Process planning;

(*iv*) Inventory control;

(*v*) Quality assurance;

(*vi*) Marketing, etc.

In view of the above activities and operations, manufacturing is no longer a simple operation but has become a *system* where a number of sub-systems interact in a dynamic manner.

1.2. MANUFACTURING SYSTEM

1.2.1. General Aspects

- *Any system that produces useful products or services, in general, is called a* **Production system.**

- **Production** *may be considered as a process of transformation of a set of input elements whereby the utility of goods or services is increased.* For example, the input could be parts and the assembled product serves as output.

- *The manufacturing processes are collected together to form a* **manufacturing system.**

The manufacturing system takes inputs and produces products for the customer.

A manufacturing system is depicted as an input-output system, in Fig. 1.1. Here the input elements undergo technological transformation to yield a set of output elements. The technological

1

transformation must be optimised with reference to an objective function which could be cost, productivity or product.

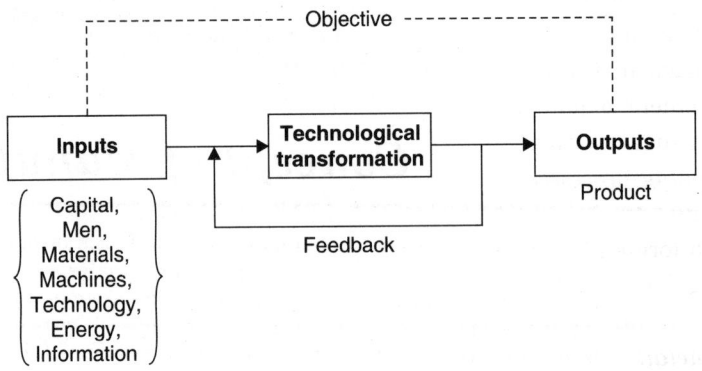

Fig. 1.1. Manufacturing as an input-output system.

— *"Manufacturing" is the heart of the system where material is converted from one form to another and value is added.*

— *The "manufacturing sub-system" is a collection of processes and operations that are used to obtain the desired product in the required quantity.*

● The present day concept of 'manufacturing system' has evolved through the following *distinct stages :*

  1. Manual manufacturing ;

  2. Mechanisation ;

  3. Hard automation ;

  4. Soft automation ;

     — NC/CNC/DNC

     — Industrial robots

     — FMC (Flexible Manufacturing Cell)/FMS (Flexible Manufacturing System)

The major components of the FMS are : (*i*) *Machine tools ;* (*ii*) *Control system ;*
(*iii*) *Handling system ;* (*iv*) *Operators.*

— Computer Aided Design (CAD)/Computer Aided Manufacturing (CAM).

1.2.2. Classification of Manufacturing Processes

The principal types of manufacturing are :

1. Process-type manufacturing. It involves *continuous flow of materials through a series of process steps to obtain a finished product like chemicals.*

2. Fabrication-type manufacturing. It involves *manufacturing of individual parts or components* by a series of operations, such as rolling, machining and welding.

Here, the following *basic manufacturing processes* are used :

Casting ; Forming ; Machining ; Grinding and Finishing ; Unconventional Machining ; Joining ; Heat treatment.

3. Assembly-type manufacturing. In this type of manufacturing the *parts or components are put together to get a complete product* such as a machine.

*The manufacturing processes are **classified** as follows :*

I. Constant mass processes :

1. *Casting :*

    (*i*) Sand casting   (*ii*) Shell mould casting

    (*iii*) Precision investment casting   (*iv*) Plaster mould casting

    (*v*) Permanent mould casting   (*vi*) Die casting

    (*vii*) Centrifugal casting.

2. *Metal forming processes :*

    (*i*) Rolling   (*ii*) Drop forging

    (*iii*) Press forging   (*iv*) Upset forging

    (*v*) Extrusion   (*vi*) Wire drawing

    (*vii*) Sheet metal operations.

3. *Powder metallurgy processing.*

4. *Heat treatment.*

II. Metal removing processes :

1. *Machining :*

    (*i*) Turning   (*ii*) Drilling

    (*iii*) Milling   (*iv*) Shaping and planning

    (*v*) Sawing   (*vi*) Broaching.

2. *Grinding and finishing*

3. *Unconventional machining.*

III. Material addition processes :

1. *Welding and allied processes :*

    (*i*) Gas welding   (*ii*) Electric arc welding

    (*iii*) Electric resistance welding   (*iv*) Thermit welding

    (*v*) Cold welding   (*vi*) Brazing

    (*vii*) Soldering.

2. *Mechanical joining :*

    (*i*) Bolting   (*ii*) Riveting, etc.

1.2.3. Selection of a Manufacturing Process

- In order to produce the product with least cost within reasonable time without compromising the quality of the product, it is imperative to *select the right type of manufacturing process.*

- The final product should satisfy both functional and physical objectives at a minimum cost that is acceptable to the ultimate user. For this purpose *production and process-product topology technique is used.*

For *selecting a manufacturing process*, the following points should be given due consideration :

    (*i*) Manufacturing cost ;

    (*ii*) Production volume and production rate ;

    (*iii*) Characteristics and properties of workpiece material ;

    (*iv*) Limitations on shape and size ;

(*v*) Surface finish and tolerance requirements ;

(*vi*) Functional requirements of the product.

1.2.4. Process Planning

- Process planning aims at planning method or series of methods for economic manufacturing of a product of the quality called for by the drawings or specifications laid down.
- It is a fundamental part of the industrial activity. An efficient and economic planning leads the firm towards success whereas faulty planning creates hindrances and bottlenecks at each stage of manufacturing.

Following are the *purposes of process planning :*

(*i*) To determine the most economical process to be followed to manufacture components of the product which are to be manufactured in the shops.

(*ii*) To determine what parts to be manufactured and what parts to be purchased from the market.

(*iii*) To determine the sequence of operations to be performed on each component in particular process.

(*iv*) To determine the blank sizes of raw materials in processes like forging, welding, processing and gross weight of material, for casting purpose.

(*v*) To prepare a list of materials for all components of the product in prepration to purchasing the raw materials.

(*vi*) To determine the machine tools to do the operations at required accuracies and prepare complete specifications of such machine tools.

(*vii*) To determine the need of any special equipment like tools, jigs and fixtures, and dies in the light of production quantities.

(*viii*) To determine the stages of inspection and also, the need of designing the inspection devices and limit gauges for different stages of manufacturing.

(*ix*) To determine the time standards for performance of the job and fix the rates of payments in piece payment system.

(*x*) To determine the type of labour required to do the job and the estimated product cost prior to the start of manufacturing.

1.2.5. Analysis of Manufacturing System

Methods employed for manufacturing system analysis are enumerated and briefly discussed as follows :

    1. Linear programming

    2. Waiting line model

    3. Simulation models

    4. Network models

    5. Statistical techniques.

1. Linear programming :

- This method has emerged from the combination of mathematics and economics and is an improved method of doing the job.
- It provides a general model to obtain the most economic way of allocating the scarce resources.
- It deals equally well when a group of limited resources must be shared amongst a number of competing demands and all decisions are interlinked because of the common set of

fixed limits. Fixed limits are set by machine tool capacity, plant capacity, raw materials storage space, etc.

Simplex method and graphical method, etc. may be used to solve such distribution problems.

2. Waiting line model :

- This method involves the arrival of units which require service at one or more service facilities; under situation, units may have to wait for certain period.

- *Waiting is always associated with costs.* Hence the problem is to organise the system in such a fashion that *waiting and costs are minimum.*

3. Simulation models :

- It has been observed that some industrial systems are very complicated and as such it is very difficult to express them in any mathematical form. However, the data has got some characteristics which can help to lead to simple solution of managerial problems. This approach to problems sets up a simulated experiment and then carries through the experiment completely on paper or computer to observe the effect of the variables on the measure of effectiveness.

- In general simulation models follow a conceptual structure and may not lead to optimum answer.

- It is *a good tool to compare the various alternatives.*

4. Network models. This method is commonly used to carryout the various projects where the number of operating parameters are involved and large number of decision points are there in the sequence. It is expected that the work required to be done can be easily handled with the help of network techniques.

5. Statistical techniques. In order to solve the problems, various statistical techniques can be employed such that effective decision can be approached related to manufacturing system analysis.

1.3. TYPES OF PRODUCTION

The various *types of production* are :

1. Job order production
2. Batch or quantity production
3. Mass production.

1. Job order production :

- It consists of *small scale production to meet the requirements of individual customers.*

- It is carried out in *small factories* and suits for various works.

- It has a lot of flexibility for operation and is capable of technical economics, but does not have commercial and financial economics.

- Continuous and careful thought must given to the development of cheaper manipulative process which involves a number of types of operations.

- The workers in the department should have skill, intelligence and very good ability.

2. Batch or quantity production :

- This is a common type of production.

- It requires very good managerial skill to achieve an economic plan in production.

- The *most economic size* is determined by sales demand, delivery and stock requirements.

- The jobs are produced in a *lot or in certain quantity* and this *varies between the job production and the mass production.*
- This can be had in a *medium size enterprise where equipment, etc. cannot* be purchased in a *large scale,* but to cater say a local market and for the local demand.

3. Mass production :

- It means a continuous production without loss of time.
- It requires a good site plan, factory layout and special machinery and expensive jigs and fixtures.
- Processing and assembling are carefully timed, sometimes to a fraction of a second.
- Inspection is to be very rigid.
- The cost of machining will be very low due to less stock involved, good control of production and higher output of the machines.
- The *disadvantages* in this type of production are : *It is not easily changeable to other types of production, costlier to changeover* and *loss of time involved in the period of changeover,* etc.

1.4. CONTROL SYSTEMS

1.4.1. Introduction

Automatic control has played a significant role in the advance of engineering science. Besides its extreme importance in space-vehicle systems, missile-guidance systems, etc., *automatic control has become an important and integral part of modern manufacturing and industrial processes.* Automatic control, for example, is essential in :

— Design of auto pilot systems in *aerospace industries.*

— Design of cars and trucks in the *automobile industries.*

— *Industrial operations as controlling pressure, temperature, humidity, viscosity and flow in the process industries.*

1.4.2. System

A system may be defined as follows :

- "A **system** *is an arrangement, set or collection of things connected or related in such a manner as to form an entirely or whole".*

Or

"A **system**, *is an arrangement of physical components connected or related in such a manner as to form and/or act as an entire unit."*

- A system consists of a sequence of components in which *each component has some cause as input and its effect will be its output. Broadly it is a sequential set of cause and effects.*

Each system may have a *large number of subsytems.* Examples :

(*i*) This universe is itself a system consisting of large number of subsystems.

(*ii*) Human body as a system has digestive system, respiratory system, etc.

1.4.3. Control System

A **control system** *is an arrangement of physical components connected or related in such a manner as to command, direct or regulate itself or another system.*

Elements of a control system :
The elements of a control system are enumerated and defined below :

Element	Definition
1. *Controlled variable*	The quantity or condition of the controlled system which can be directly measured and controlled is called *controlled variable*.
2. *Indirectly controlled variable*	The quantity or condition related to controlled variable, but cannot be directly measured is called *indirectly controlled variable*.
3. *Command*	The input which can be independently varied is called *command*.
4. *Reference input*	A standard signal used for comparison in the closed-loop system.
5. *Actuating signal*	The difference between the feedback signal and reference signal is called *actuating signal*.
6. *Disturbance*	Any signal other than the reference which affects the system performance is called *disturbance*.
7. *System error*	The difference between the actual value and ideal value is called *system error*. The negative value is called *deviation*.

Examples of control system applications :
Following are some examples of control system applications :
1. Steering control of automobile
2. Printwheel control system
3. Industrial sewing machine
4. Sun-tracking control of solar collectors
5. Speed control system
6. Temperature control of an electric furnace.

1.4.4. Classification of Control Systems
Control systems are *classified* into the following *two basic types* :
1. Open-loop control systems (Unmonitored or non-feedback control systems),
2. Closed-loop control systems (Monitored or feedback control systems).

Comparison between Open-loop and Closed-loop Systems

Open-loop	Closed-loop
1. Less accurate.	1. More accurate.
2. Generally build easily.	2. Generally complicated and costly.
3. Stability can be ensured.	3. May become unstable at times.
4. Presence of non-linearities cause malfunctioning.	4. It usually performs accurately even in the presence of non-linearities.
5. Any change in system component cannot be taken care of automatically.	5. Change in system component is automatically taken care of.
6. Input command is the sole factor responsible for providing the control action.	6. The control action is provided by the difference between the input command and the corresponding output.
7. The control adjustment depends upon human judgement and estimate.	7. The control adjustment depends on output and feedback element.
Examples :	*Examples :*
(*i*) Automatic washing machine	(*i*) Liquid level control system
(*ii*) The electric switch	(*ii*) Traffic signal system
(*iii*) An automatic toaster.	(*iii*) Human being reaching for an object.

Note. All control systems *operated by present timing mechanisms are open-loop.*

1.4.5. Open-loop Control Systems (Non-feedback Systems)

● An **open-loop control system** is *one in which the control action is independent of the desired output. The actuating signal depends only on the input command and output has no control over it.*

● The **elements** of an open-loop control system can usually be *divided* into the following two parts (See. Fig. 1.2) :

(*i*) Controller ;

(*ii*) Controlled process.

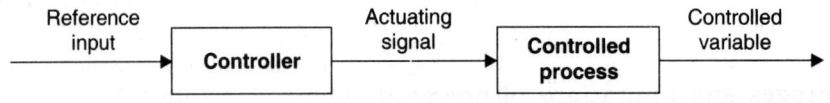

Fig. 1.2. Elements of an open-loop control system.

— An input signal or command is applied to the controller, whose output acts as the actuating signal ; the actuating signal then controls the controlled process so that the controlled variable will perform according to prescribed standards.

— In *"simple cases"*, the controller can be an *amplifier, mechanical linkage, filter or other control element,* depending on the nature of the system. In *"more sophisticated cases"*, the controller can be a computer such as a *microprocessor.*

— Because of the *simplicity and economy* of open-loop control systems we find this type of system in many *non-critical applications.*

Examples :

1. *Idle-speed control system :*

● The following are the main *objectives* of the idle-speed control system of automobile :

(*i*) To eliminate or minimize the speed drop when engine loading is applied.

(*ii*) To maintain the engine idle-speed at a desired value.

● Fig. 1.3 shows an idle-speed control system from the standpoint of inputs-system outputs. In this case the throttle angle and the load torque (due to the application of air-conditioning, power steering, transmission, power brake, etc.) are the inputs, and the engine speed is the output. The engine is the controlled process of the system.

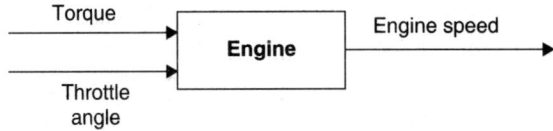

Fig. 1.3. Idle-speed control system.

2. *Printwheel control system :*

Fig. 1.4 shows an example of the printwheel control system of a word processor or electronic typewriter (and also shows a typical input-output set for the system).

— With a reference command input is given, the signal is represented as a step function. Since the electric windings of the motor have inductance and the mechanical load has inertia, the printwheel cannot respond to the input instantaneously. Typically it will follow the response and settle at the new position after sometime. Printing should not begin until the printwheel has come to complete stop; otherwise, the character will be smeared.

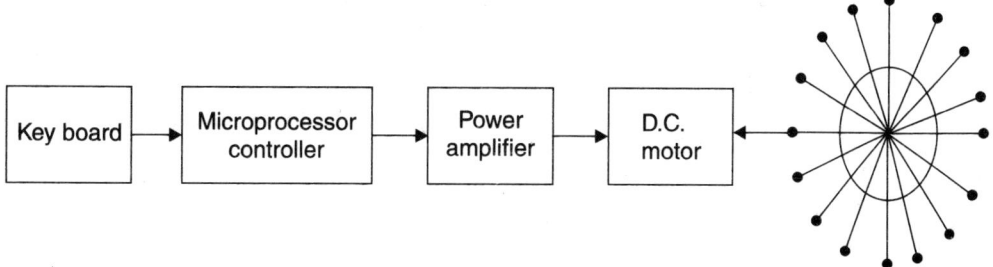

Fig. 1.4. Open-loop word processor control system.

Advantages and limitations of open-loop control system :

Advantages :

1. Simple construction.
2. Easy maintenance.
3. Less costly than a closed-loop system.
4. No stability problem.
5. Convenient when output is difficult to measure or measuring the output precisely is economically not feasible.

Limitations/Disadvantages :

1. Since the system is affected by internal and external disturbances, the *output may differ from the desired value.*
2. For getting accurate results, this system needs frequent and careful calibrations.
3. Any change in system component cannot be taken care of automatically.
4. Presence of non-linearities cause malfunctioning.

1.4.6 Closed-loop Control System (Feedback Control System)

- A **closed-loop system** is one in which *control action is somehow dependent on the output.* In this case the controlled output is fedback through a *feedback element* and compared with the reference input. Thus, the *actuating signal is the difference of desired output and reference input.*

- **Feedback** is that property of a closed-loop system which permits the output or some other controlled variable of the system, to be compared with the input to the system, so that the appropriate control action may be formed as some function of the output and input. A *feedback is said to exist in a system when a closed sequence of cause and effect relations exist between system variables.*

The **characteristics of feedback** *are as follows :*

(*i*) Increased bandwidth.

(*ii*) Increased accuracy.

(*iii*) Tendency towards oscillation or instability.

(*iv*) Reduced effects of non-linearities and distortion.

(*v*) Reduced sensitivity of the ratio of output to input to variations in system characteristics.

- A **closed-loop idle-speed control system** is shown in Fig. 1.5.

— The reference input (ω_r) sets the desired idling speed. The engine speed at idle should agree with reference value ω_r, and any difference such as load torque is sensed by the speed transducer and the error detector. The controller will operate on the difference and provide a signal to adjust the throttle angle to correct the error.

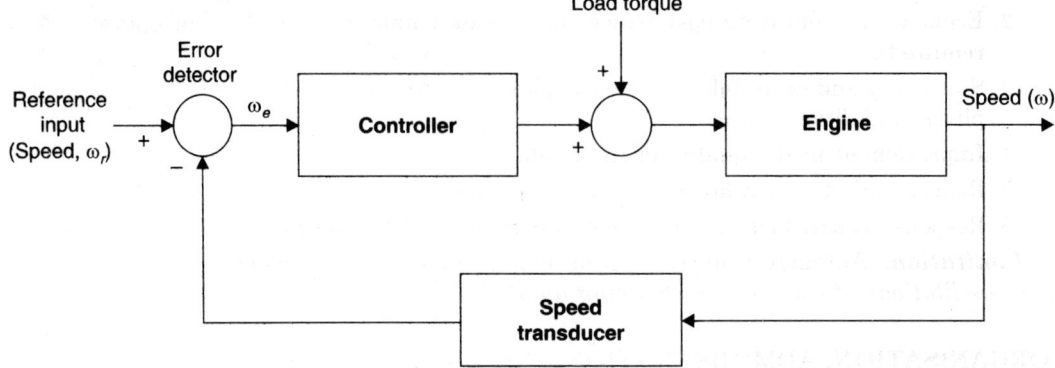

Fig. 1.5. Closed-loop idle-speed control system.

Advantages and Limitations :

Advantages :

1. More accurate comparatively.
2. Usually performs accurately even in the presence of non-linearities.
3. Change in system component is automatically taken care of.
4. The use of feedback system response is relatively insensitive to external disturbances and internal variations in system parameters. It is thus *possible to use relatively inaccurate and inexpensive components to obtain the accurate control of a given plant* (whereas doing so is impossible in the open-loop case).

Limitations/Disadvantages :

1. Generally complicated in construction.
2. Generally higher in cost and power.
3. May become unstable at times.

1.4.7. Automatic Control Systems

● A closed-loop control system operating without human operator is called an *automatic control system.*

Examples :

(*i*) *Centrifugal watt governer,* where the lift of the rotating balls is used as speed monitor. The supply of steam is automatically controlled as speed tends to increase or decrease beyond a set point.

(*ii*) A *pressure control system* where the pressure inside the furnace is automatically controlled by affecting changes in the position of the damper.

(*iii*) The *level control system* where the inflow of water to the tank is dependent on the water level in the tank. The automatic controller maintains the liquid level by comparing the actual level with a desired level and correcting any error by adjusting the opening of the control valve.

Advantages and Limitations :

Advantages :

1. Increased output.
2. Economy in operating cost (since continuous employment of human operator is not required).
3. Suitability and desirability in the complex and fast acting systems which are beyond the physical abilities of a man.
4. Improvement in the quality of the products.
5. Reduced effect of non-linearities and distortions.
6. Response is satisfactory over a wide range of input frequencies.

Limitation. Automatic control system has a tendency to *overcorrect errors* which may *result in oscillations of constant or changing amplitude.*

1.5. ORGANISATION, ADMINISTRATION AND MANAGEMENT

These terms may be defined as follows :

Organisation. *It is defined as a systematic, co-ordination and combination of the efforts with the aid of the money, men, machinery, materials and methods in a manner as would result in maximum manufacturing efficiency with minimum cost.* It must embrace within itself the formulation of sound financial and commercial policy for the business, efficient planning of works building and plant equipment, judicious buying, effective conduct of productive operations, proper direction, supervision and control of personnel and a co-ordination and co-relation of the productive as also the non-productive forces into a harmonious whole.

Administration. *It is the effectuation of the purpose by means of organisation and is concerned with "Carrying out" policies, rules, regulations and designated methods.*

Management. *It is the art of creating industrial relations of any kind, between people engaged in industry viz. relations between employers and employees, relation between individuals entering into commercial contracts, relation between investors and debtors, etc.*

1.6. PLANT ORGANISATION

Plant. *A plant is a place, where men, materials, money, equipments, machinery, etc. are brought together for manufacturing products.*

Organisation. Organisation *is the pattern of ways in which a large number of people engaged in a complexity of tasks relate themselves to each other in systematic establishment and accomplishment of mutually agreed purposes.*

A few *common principles of organisation are :*

(*i*) Consideration of objectives	(*ii*) Relationship of basic components
(*iii*) Responsibility and authority	(*iv*) Span of control
(*v*) Dividing and grouping work	(*vi*) Effective delegation
(*vii*) Communication	(*viii*) Line and staff relationships
(*ix*) Balance, stability and flexibility.	

The structure of an industrial organisation differs from that of another organisation and it depends upon :

(*i*) Size of the organisation,

(*ii*) Nature of product being manufactured, and

(*iii*) Complexity of the problem being faced.

A few commonly known forms of *organisation structures* or types of organisations are :

1. "Line" or "Departmental" type of organisation.

2. "Functional" type of organisation.

3. "Line and staff" type of organisation.

4. "Complex organisation" or "Line, functional and staff organisations".

1. "Line" or Departmental type of organisation. It is also called *"Scalar"* or *Military* type organisation. In this type of organisation the authority flows directly from the boss to various sub-executives in charge of particular phases of the business and from them to the other workers. Fig. 1.6 illustrates this type of organisation. It is evident from the figure that superintendent is under direct control of general manager and foremen of the departments receive orders and instructions straight way from superintendent whilst workmen are under direct authority of foremen of different departments.

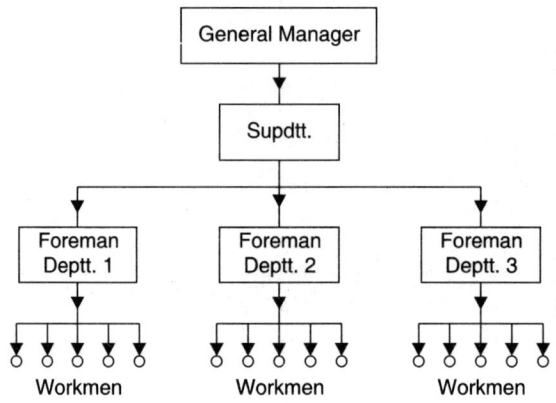

Fig 1.6. Line organisation.

The activities of such an organisation are limited to a fewer functions such as finance, production and distribution. Each of these functions are further subdivided into self-sufficient sections which are managed and supervised by departmental heads.

Advantages :

(*i*) Simplicity   (*ii*) Flexibility

(*iii*) Quick decisions   (*iv*) Communication

(*v*) Executive development   (*vi*) Unified control

(*vii*) Fixed responsibility   (*viii*) Effective discipline

(*ix*) Economy.

Demerits :

(*i*) Overburdening   (*ii*) Instability

(*iii*) Lack of specialisation   (*iv*) Autocratic control

(*v*) Difficulty in staffing   (*vi*) Inadequate communication.

2. "Functional" type of organisation. The limited scope of working to which the *"Line"* organisation is liable has given rise to another type of organisation known as *"Functional"* type of organisation. *It divides the responsibility according to the functions to be performed and not in* *departments or sections of the works.* Here each individual executive is placed in full control of a function and by virtue of his total efforts being exercised in one direction, becomes a highly skilled specialist in his own field. He is thus able to look after its working with great ability, confidence and technical skill. Furthermore, this scheme presents an adequate scope for the growth and development of business, for despite its expansion even to any considerable extent; each functional head will still be able to attend to and control the activity for which he is responsible.

Fig. 1.7 shows a functional type of organisation.

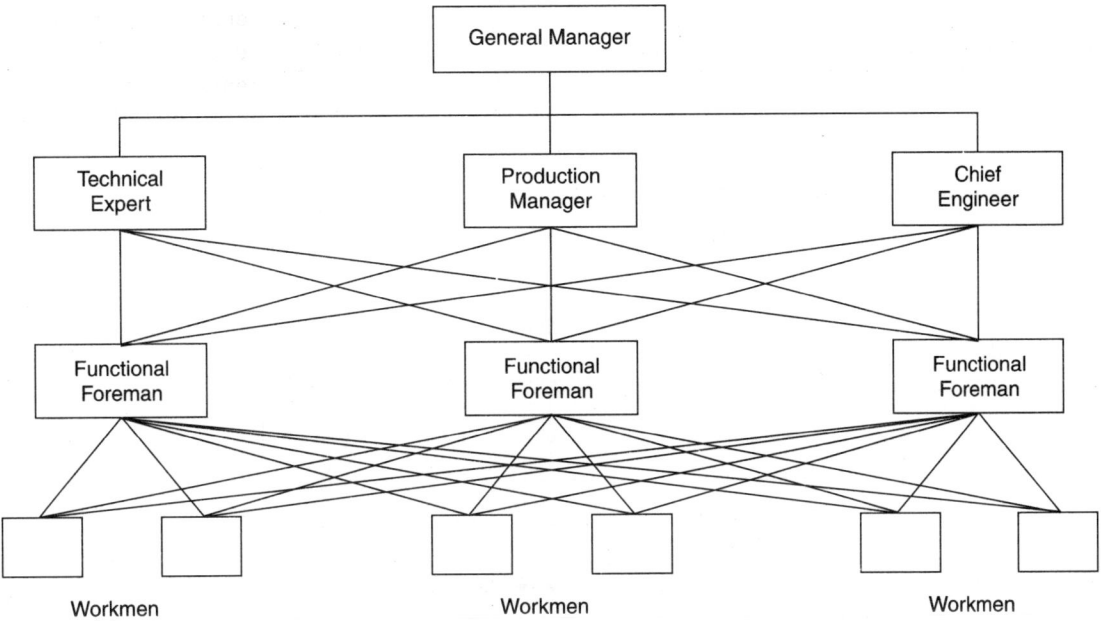

Fig. 1.7. "Functional" organisation.

Technical expert, production manager and chief engineer are under the authority of general manager. Each of the functional foremen is responsible to the above mentioned departmental heads. Similarly, all the workmen are to carryout the orders issued by every functional foreman.

Advantages :

(*i*) Specialisation

(*iii*) Simplified control

(*v*) Scope for expansion

(*ii*) Easier staffing

(*iv*) Better supervision

(*vi*) High efficiency.

Demerits :

(*i*) Lack of co-ordination

(*iii*) Poor discipline

(*v*) Lack of executive development

(*vii*) Divided responsibility.

(*ii*) Delayed decisions

(*iv*) Low morale

(*vi*) Uneconomical

3. Line and staff organisation. This type of organization assimilates the features of both the previously discussed types. Its main characteristic feature is that *it draws a rigid line between direction on one hand and the execution of work on the other. The* **"Staff"** *represents the directing,* standardising, analysing and advising part of the efforts whilst **"Line"** *corresponds to the actual performance of the tasks.* The "staff" points out the way to efficient and economic performance and shoulders the responsibility for efficiency of equipment and quality of material, maintaining cost and analytical records, conducting time study, fixing pieces, rates, etc. At times, the staff work is conducted through the medium of experts with the help of committees framed for multipurposes such as—Tools committee, Research committee, Production committee, etc. The foreman or executive head of each department is also taken as a member of such committees and the members meet at regular intervals to discuss most economical and efficient ways of carrying out the objectives effectively. These committees are simply advisory and supplementary to the line, and in no way should supercede the management which is entrusted mainly with the duty to manage the entire affairs.

Fig. 1.8 shows a line and staff organisation. Vertical line starting from General manager represents the 'line relations' while horizontal lines emanating from process and product specialists signify the 'staff relationship'.

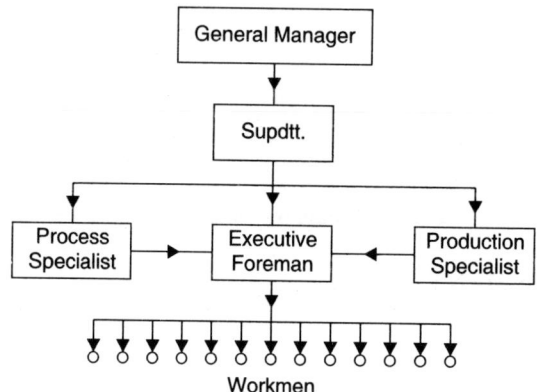

Fig. 1.8. Line and staff organisation.

Advantages :

(*i*) Discipline   (*ii*) Balanced decisions

(*iii*) Planned specialisation   (*iv*) Undivided responsibility

(*v*) Flexibility   (*vi*) Staffing and development.

Demerits :

(*i*) Ineffective staff   (*ii*) Conflicts

(*iii*) Expensive   (*iv*) Lack of co-ordination.

4. "Complex" or "Line, functional and staff" organisation. Fig. 1.9 represents the line, functional and staff relationship. It reveals that personnel manager is shown in a staff relationship to the remainder of the organisation ; though not a standard arrangement yet it recognizes the way in which most personnel managers conceive their role in the organisation, as being advisory to all departments on personnel matters. Further there is a line relationship in the personal department itself and some of the line coordinates of the personnel manager are entrusted with functional responsibility and authority.

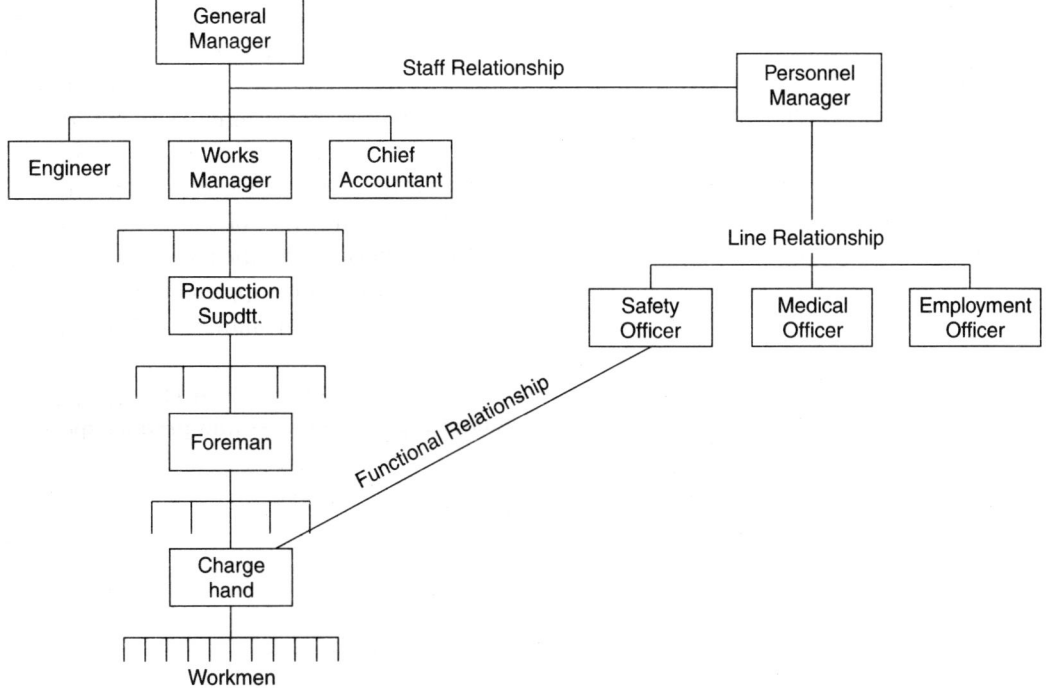

Fig. 1.9. Complex organisation.

1.7. SCIENTIFIC MANAGEMENT

1.7.1. Definition

Scientific management may be defined as a *systematic approach to manage the enterprise on the basis of observation, experimentation and rational decisions.* The methods based on guess work, trial and error should always be avoided.

1.7.2. Principles

In order to achieve the objectives of an organisation following *principles* constituting scientific management are observed :

(*i*) Scientific method of production

(*ii*) Standardisation

(*iii*) Time and motion study

(*iv*) Costing and cost control

(*v*) Production planning and control through functional foremanship

(*vi*) Scientific selection, training and remuneration of the workers.

1.7.3. Aims

The *aims of scientific management* are :

1. Placement of right person on the right job through scientific selection and training.

2. Reduction in cost of production by rational planning and regulation of cost control techniques.

3. Increase in rate of production by use of standardised tools, equipment and methods.

4. Elimination of wastage in the use of resources, time and methods of operation.

5. Relative wage payment according to the efficiency of the worker.

6. Improvement in the quality of the products by research, quality control and inspection devices.

7. Ensuring steady flow of standard goods to customers at fixed price.

1.8. FUNCTIONS OF MANAGEMENT

Function of management can be *classified* into the following *six* activities :

(*i*) Planning   (*ii*) Organising

(*ii*) Staffing   (*iv*) Directing

(*v*) Controlling   (*vi*) Co-ordinating.

Elements of communication. The process of communication consists of the following *components* :

(*i*) Communicator   (*ii*) Message

(*iii*) Communication symbol   (*iv*) Communication channel

(*v*) Receiver.

1.9. PLANT LOCATION

Some *important factors affecting plant location* are :

(*i*) Nearness to raw materials   (*ii*) Transport facilities

(*iii*) Nearness to markets   (*iv*) Availability of labour

(*v*) Availability of fuel and power   (*vi*) Availability of water

(*vii*) Climatic conditions   (*viii*) Financial and other aids

(*ix*) Land   (*x*) Community attitude.

1.10. PLANT LAYOUT

A few sound *principles of plant layout are :*

(*i*) Integration   (*ii*) Minimum movements and material handling

(*iii*) Smooth and continuous flow   (*iv*) Cubic space utilization

(*v*) Safe and improved environment   (*vi*) Flexibility.

Objectives of good layout :

Following are the important *objectives of a good layout* :

1. Reduce manufacturing costs.

2. Better quality of product.

3. Better service to the customer.

4. Increase flexibility.

5. Increase employee safety.

6. Reduce work-in-process to minimum.

7. Minimise materials handling and loss.

8. More effective utilization of floor space.

9. Better work methods and utilization of labour.

10. Improve control and supervision.

11. Better utilization of equipment and facilities.

12. Reduce manufacturing cycle.

Fundamental considerations in layout :

The following are among the major fundamentals most often cited :

1. Products manufactured
2. Operations sequence
3. Special requirements
4. Equipment
5. Maintenance and replacement
6. Minimum movement
7. Flow
8. Waiting and service areas
9. Plant climate
10. Flexibility.

Types of layout :

Plant layout may be *classified* in two fundamental types :

1. Process layout,
2. Product layout.

1. Process layout. It is sometimes also called *functional layout.* A plant is said to have process or functional layout *when all the machines of a particular class doing a particular type of work or process are arranged in a separate department.* In other words, performance of each type of work or operation in a distinct department on all types of products is the basic characteristic of process layout. Suppose 'B' has set-up a factory for the manufacture of taps, reamers and drills. To manufacture these tools three principal operations naming turning, heat treatment and grinding will be involved. It may be noted that only these three different tools have common feature and there will be established three sections or departments, *viz.,* turning department, heat treatment department and a grinding department. Thus, in this example all the three types of products would travel through the above mentioned principal operations without which the manufacture will not be possible.

It is evident from the above discussion that process grouping or layout creates such conditions in which *similar machines or operations are grouped functionally.* This grouping or set-up is carried out in the domain of the factory known as shops or departments.

This type of layout is suitable for *job order plants which involve non-repetitive processes.* A typical process or functional layout is shown in Fig. 1.10.

Advantages :

1. Greater specialisation.
2. Better utilization of high production equipment.
3. Greater flexibility of production process and higher degree of machine utilization.
4. Greater margin of safety of breakdowns.
5. Better supervision due to specialized process.
6. Less investments for equipments due to less duplication.
7. Higher level of individual operator performance.
8. Changes in the sequence of operations are not disruptive and can be incorporated with minimum inconvenience.

9. It provides better control of total manufacturing costs.

10. There is less interruption in work flow due to machine breakdown.

11. A lower proportion of fixed costs to total costs.

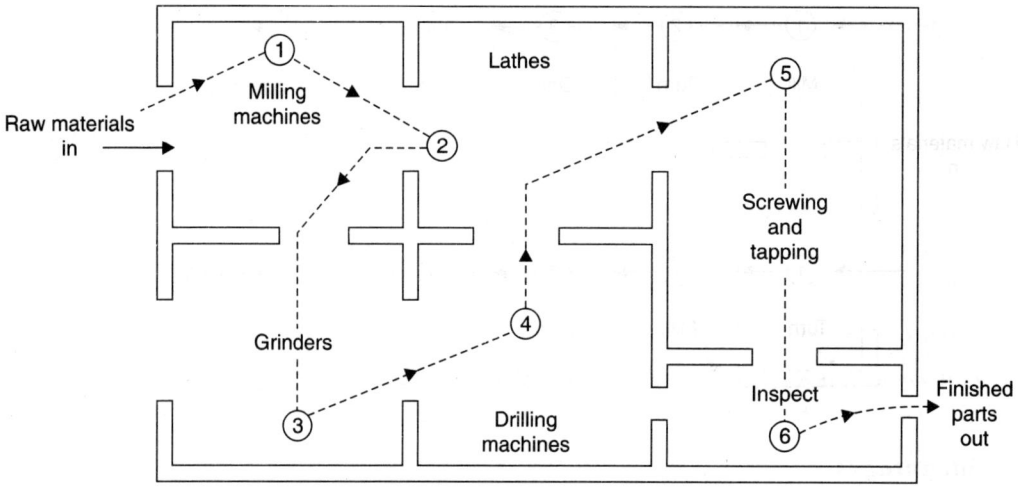

Fig. 1.10. Process layout.

Disadvantages :

1. It groups the machines functionally in such a way that the cost of material handling is increased.

2. It involves difficulties in routing, scheduling and controlling the manufacture of products because of the almost endless combinations of sequences that often can be used in processing similar items.

3. It entails more difficult control as well as more costly supervision inside and outside the 'shop' or 'department'.

4. Co-ordination and control are difficult because of manufacturing variations.

5. A plant breakdown in one department may sterilise the work of other departments, which means the wastage of time and effort.

6. It is difficult to trace the final responsibility for the finished product as the work has to move through different departments during its processing.

7. Requires more skilled labour and entails difficulty in labour procurement.

2. Product layout. In this layout, machines and other manufacturing facilities are located in the sequence required to manufacture the product. Thus, if the sequence of operations for a given part consists of drilling a hole, milling a slot, and forming the part, then the arrangement or layout of machines to manufacture this part would consist of, first, a drill press to drill the hole; next, a milling machine to mill the slot; and last a press to form the required part. In effect the operations necessary to fabricate the part are determined and the machines are then placed in the sequence needed to effect these operations. It is also called *straight line layout.*

This type of layout mostly and best suits those industries which manufacture a large volume of standard products involving repetitive processes.

Fig. 1.11 shows a typical product layout.

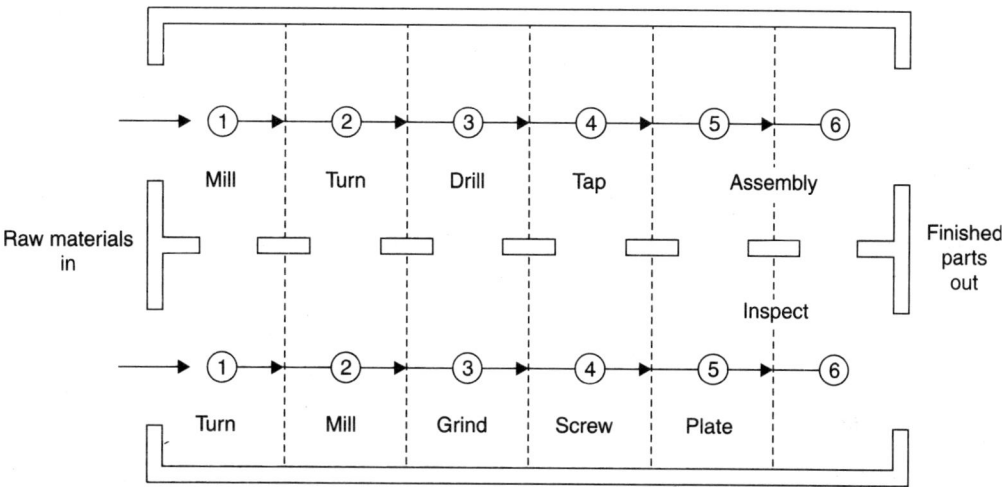

Fig. 1.11. Product layout.

Advantages :
1. It makes more automation applicable.
2. It allows for labour specialization.
3. It needs reduced manufacturing area.
4. Simpler production control and reduced total inspection.
5. More utilisation of unskilled persons.
6. Labour and material flow can be controlled easily.
7. Reduced materials handling costs.
8. It results in lower manufacturing costs.
9. Low cost labour and ease of procurement and training.
10. Requires less inspection.

Disadvantages :
1. Specialised facilities require a high initial investment.
2. It entails high aggregate overhead cost.
3. The arrangement leads to an inflexible layout and as such cannot be adopted to the manufacture of any other type of product.
4. Even a minor change in the machine arrangement means rather a complete change in layout.
5. This layout involves such grouping in a continuous line that machine breakdown at any point along the line will lead to stoppage of the entire line of machines.
6. Supervision is more difficult.

Combined Process and Product Layout. Product layout and process layout are the classic types of layout but they are more often found in combination than in their pure forms. For example, a plant may use process layout in the manufacture of the component parts for an item but assembling and testing of the item may be carried out on a product layout basis.

A proper layout is one which contributes most effectively the economical flow of material through the sequences of the manufacture.

1.11. PRODUCTION, PLANNING AND CONTROL (PPC)

1.11.1. Definitions

Production. *It involves sequence of operations that transform raw materials into the desired shape and size.* Transformation in an industry is done by carrying out a number of operations.

Planning. *It begins with the analysis of given data, on the basis of which a scheme of utilisation of firm's services can be outlined so that the desirable target may be achieved in an efficient manner.*

Control. *It involves supervising operations with the aid of control mechanisms and feed-back information about the progress of work.*

1.11.2. Functions of PPC

The various *functions* of PPC are :

(i) Forecasting	(ii) Order writing
(iii) Product design	(iv) Process planning and routing
(v) Material control	(vi) Tool control
(vii) Loading	(viii) Scheduling
(ix) Despatching	(x) Progress reporting
(xi) Corrective action.	

1.11.3. Advantages of Production Control

1. More effective use of man power and equipment.
2. Lost time of works cut to a minimum.
3. Overtime cut considerably.
4. More efficient purchasing of materials results in smaller raw materials inventory.
5. Good for morale of workers.
6. Definite delivery dates can be maintained.

1.11.4. Constituents of Production Control

The main *constituents or techniques of production control* are :

1. Production planning.	2. Routing.
3. Scheduling.	4. Despatching.
5. Follow-up.	

1. Production planning. Production planning covers a careful and exhaustive study of a pre-arranging of the technique involving a long and complicated series of separate operations so that the required product of the right quantity may be manufactured at the right time and at the most economic cost. It is aimed at achieving a manufacturing output that will achieve one or more of the following *objectives* :

(i) Capture a desired share of the market demand.

(ii) Operate the plant at a predetermined level of efficiency.

(iii) Bring a prescribed level of profit.

(iv) Create a specific number of jobs.

(v) Utilise available plant facilities.

A *balanced production planning* provides the following *advantages* :

(a) Increases operating efficiency by stabilising productive activities,

(b) Facilitates selling and customer service,

(c) Helps reduce production cost by providing reliable basis for investment in raw materials and tools, and

(d) Promotes fuller utilisation of plant, equipment and labour by controlling all time and efforts essential in manufacture.

Functions of planning department. Planning department *performs* the following *functions*:

(i) It lays down the production schedule in advance, drafts designs, specifications and instructions for each Works Order and also determines the routing of operations through the plant.

(ii) It decides the quantity and quality of materials required, specifies the appropriate tools and machines to be used, and in some cases, fixes the rate of wage payment.

(iii) It determines the grade of labour required, the quantity of output to be produced on each Works Order; and regulates the allocation and distribution of work amongst the different groups of men and machines so as to enable each task to be completed within its scheduled time.

(iv) It is always on the constant lookout for any improvement in machinery or tools, or variation in existing methods of operation, as would tend to give more efficient results.

(v) It decides what should be most economic size of each unit of production in execution of large or small orders.

2. Routing. *'Routing' means determination of path on which manufacturing operations will travel establishing the sequence of such operations and looking to the proper category of machines and personnel which these operations will require.* It is a technical function in the first instance and is originally performed by the methods or engineering department. Routing includes the following *steps* :

(i) The analysis of the finished article from the manufacturing standpoint, including the determination of components if it is an assembly product.

(ii) The fixing of the sequence of completion in manufacture that one part or piece of material bears to another in order that all may be brought together as needed in the process of manufacture.

(iii) Determination of manufacturing operations and establishing their sequence for their performance.

(iv) Determination of process time in respect of each individual operation and the class or number of machines required for manufacturing the articles.

(v) The division of total quantities required into proper manufacturing lots or batches. This should be done with particular reference to (a) length of operations, (b) space occupied by the material while moving through the shop, and (c) the requirements of the master schedule.

(vi) Determination of scrap factors leading to spoilage or shrinkage.

(vii) Preparation of the forms to be used by the plant departments.

3. Scheduling. *'Scheduling' means the determination of the relative time at which each operation or event in connection with manufacturing is to occur.* It furnishes a logical time-table which shows when work will be released to the plant in a prescribed order in the proper sequence.

Scheduling procedure. *'Master schedule'* and *'Production schedule'* usually provide the base for carrying out scheduling.

Master schedule (i) provides the production manager relevant statistics so that arrangements may be made to meet delivery and sales commitments; (ii) assures the utilisation of plant capacity in an effective and logical order, and; (iii) helps calculation of machine loading.

For preparing the master schedules a specified date of delivery according to the contract with the customer is fixed. Beginning at that date and measuring backward a series of dates is set fixing the latest time on which important features of the work must be completed if the delivery date is to be met. In the schedule, the heavy horizontal lines indicate the scheduled time; the heavy broken lines show whether the work is up to schedule, behind the schedule or ahead of schedule.

Production schedule is that schedule which provides a time-table showing when the detailed operations of manufacture will start and when the same will finish. Thus, it is a bridge spanning the gap existing between the 'starting' and 'finishing' time involved in the performance of the detailed operations of manufacture embracing two considerations : *order of priority* and *proper sequence.*

The *limitations* of production schedule are : (*i*) Required materials, and parts which will be purchased. Such materials and parts must be available to process the materials which find their way on the schedule. (*ii*) Availability of plant facilities required for processing. (*iii*) Availability of right type of personnel required for the performance of the work which is being scheduled.

The *objectives* of production schedule are : (*i*) Production schedule should be so designed as to dance to the tunes called by the master schedule; (*ii*) To set a constant supply of work in advance so that each machine may be kept engaged in relation to its production capacity; (*iii*) To complete manufacturing orders in the most economical time coupled with correct sequence and proper relation.

4. Despatching. When routing and scheduling of a product have been completed, there still remains the job of despatching. *The despatching function involves the actual granting of permission to proceed according to plans already laid down.* It is the contact between planning and operation. It carries out the physical work which has been planned by scheduling. Despatching actually means day to day control of the work in progress, to issue daily instructions or work orders, move orders, checkup receipt of materials, issue of materials, assigning of particular persons to particular machines everyday. So far 'routing' meant laying down operations and their sequence in advance, 'scheduling' meant the timing of different operations through different machines, men and materials whereas *dispatching actually controls the activity on the spot, co-ordinates what has been planned and sees to it everything goes on smoothly.*

5. Follow-up. The *final step* in production control is the checking or *follow-up stage.* This includes reporting production operations as they take place and investigating variance from the operation schedules in an effort to assure a close connection between planned and actual production. The expediter is ever on the alert for bottlenecks caused by breakdowns of equipment, lack of proper tools or material, labour difficulties or absences, lags in speed on some operations as compared with others, excessive production of defective parts, and other causes of description in the planned scheduled of factories activities.

Gantt charts. The Gantt chart is one of the fundamental charts used in a manufacturing concern. It is especially valuable in production control. It was originated by H.L. Gantt during World War I.

The *underlying principle of a Gantt chart is to schedule and then measure the progress of different jobs in the different periods scheduled to complete the same.* A schedule of work to be carried out in a department is prepared for a number of weeks in advance after taking into considerations the orders in hand, raw materials and on order, the number of workmen in the department, the capacity of the machines and various other factors that may be applicable in a given case in preparing such a schedule. Table 1.1 presents an example of such schedule and its data is incorporated in Fig. 1.12.

Table 1.1

	Jobs scheduled	Jobs completed
1st week	80	70
2nd week	60	80
3rd week	100	80
4th week	120	100
Total	**360**	**330**

Table 1.2

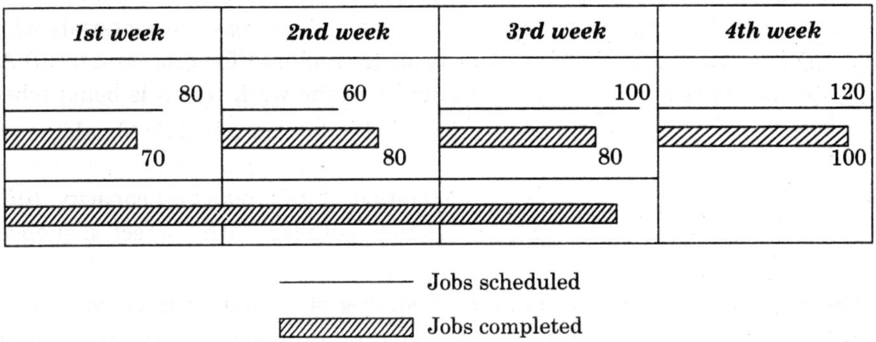

1st week	2nd week	3rd week	4th week

——————— Jobs scheduled

////////////// Jobs completed

Fig. 1.12. Progress chart.

As shown in the Fig. 1.12 the schedule of work in number of jobs for each week is indicated by fine horizontal line while the hatched bar (drawn parallel to fine line) represents the number of jobs completed during one week. When the number of jobs produced is more than the jobs shown in the schedule the length of the bar will be more than that of the fine line indicating the schedule and when completed jobs' number is less, the bar will fall short of the fine line showing the schedule. The continuous fine line and hatched bar represent the total number of jobs scheduled and completed for a period of four weeks. In such a chart, the width of the daily space usually represents the standard tasks or schedule of work rather than time because spaces on the chart representing time can be converted in the ultimate analysis into work units or represents amounts that can be produced in that time.

Gantt charts in actual practice are used in the following five typical forms :

1. Planning, 2. Load, 3. Machine idleness, 4. Man idleness, 5. Progress.

QUESTIONS WITH ANSWERS

Q. 1.1. (a) What is a process ?

(b) What are the general categories of manufacturing ?

Ans. (a) A **process** *is simply a method by which products can be manufactured from raw materials.* Any product that we use is processed in some form or the other.

(b) The *general categories of manufacturing* are :

1. Casting and moulding
2. Cutting
3. Forming
4. Assembly.

Besides these processes the final operation on a product could be a finishing process which includes :

- Heat treatment
- Plating
- Cleaning
- Painting.

Q. 1.2. What are the functions of a process engineer ?

Ans. The *functions of a process engineer are as follows :*

1. To determine the basic manufacturing process.
2. To determine the order of sequence of operations necessary to manufacture in terms of operation routing, process details and process pictures.
3. To specify production tolerances on blanks and on auxiliary surfaces.
4. To specify the process parameters for the various manufacturing operations selected.
5. To determine the order of tooling and inspection gauges required to manufacture the part.
6. To determine and select the equipment needed to manufacture the part.
7. To provide the necessary documentation to be used by the shop people.
8. To specify the methods and means of inspection depending upon the accuracy desired.

Q. 1.3. What is a 'Product Cycle' in manufacturing ?

Ans. The impetus of a new product or a revision of an existing product may come from sources such as sales and marketing people or maintenance and service people based on their contact with the consumers or the market demand. Then the product (design) engineering staff based on their expertise and constraints such as appearances, functions, durability, cost and ease of maintenance would design the product and provide the following details :

1. *Build-models for testing.*
2. *Provide part prints :*
   - Physical dimensions
   - Material
   - Special finishing required.
3. *Provide tool design and construction aids :*
   - Master layouts
   - Templates
   - Master models.
4. *Provide production rates and release dates.*

The process engineering takes place directly after product engineering has completed the design of product. From the information received it creates the plan of manufacture. Processing is then the function of determining exactly how a product will be made to satisfy the requirements specified at the most economical cost.

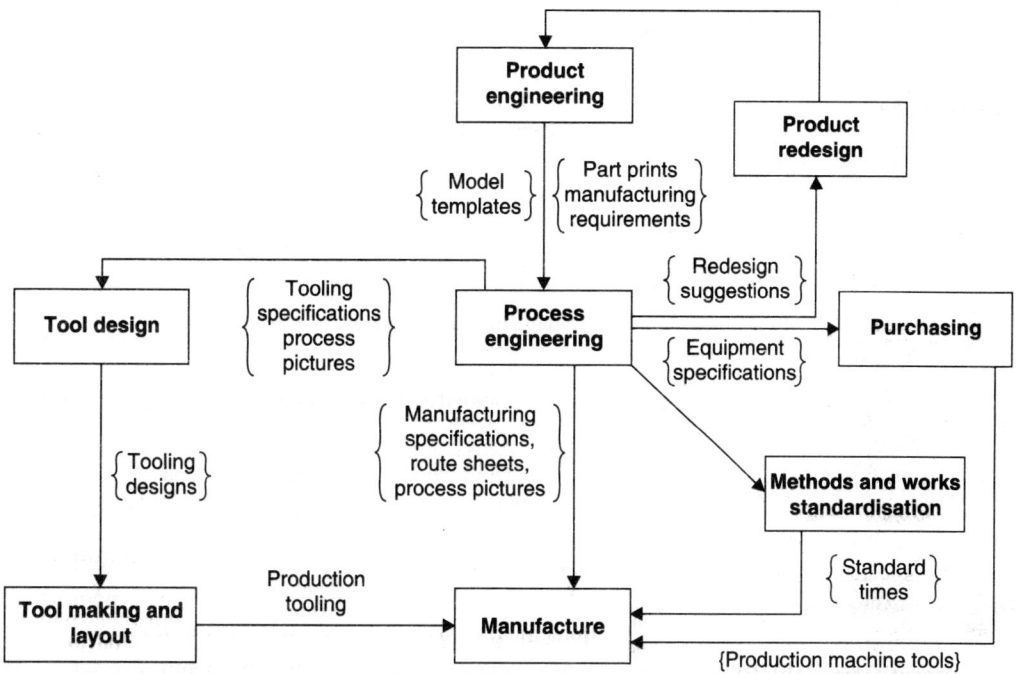

Fig. 1.13. Product cycle in manufacturing.

Fig. 1.13 shows 'Product cycle' in engineering.

Q. 1.4. Explain briefly "Process engineering function" ?

Ans.

- The *process engineering function* (see Fig. 1.13) *forms the heart of the total manufacturing operation.* It interacts with practically all areas of the total manufacturing situation.

- In *large companies,* the process engineering function is of a magnitude of a separate department. This department has many process engineers who specialise in this work. Each process or methods engineer may further specialise by dealing with only one of the products being manufactured.

In *smaller companies,* the process engineer may work in a department having other functions as well. In fact, in very small organisations, the title of a process engineer may not exist. The function of process engineering is carried out as a part time job by one or more individuals. Alternatively in large organisations each separate department may have its own group of process engineering staff to cater to their products only.

The function of process engineering, in whatever way it is organised, must be present in all manufacturing industries.

Q. 1.5. What is the difference between process planning and operation planning ?

Ans. Process planning :

The task of process planning consists of *determining the manufacturing operations required to transform a part form a rough to the finished status specified on engineering drawing.* Fig. 1.14. shows the *specific activities or steps* involved in determining the manufacturing plan for converting a part from a rough to a finished condition.

Process Planning

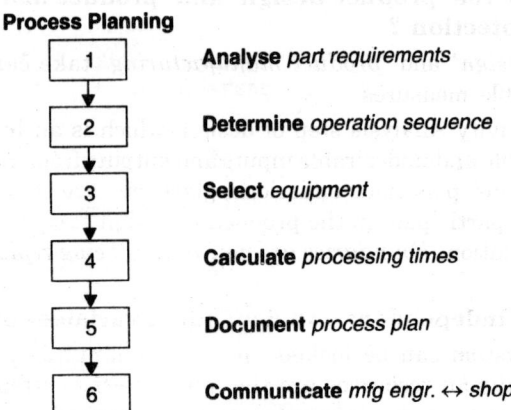

Fig. 1.14. Steps/activities involved in process planning.

1. *'Analyse' part requirements.* The part's requirements defining its features, dimensions and tolerance specifications will determine the corresponding processing requirements. These requirements will include not only the operations involved in generating part shape but likewise operations encompassing inspections, testing, heat treating, surface coating and packaging.

2. *'Determine' operation sequence.* This step involves determining the sequence of operations required to transform the features, dimensions and tolerances on the part from a rough to finished state. This helps in determining the type of processing operation that has the capability to generate the various types of features given the tolerance requirements.

3. *'Select' equipment.* In this step, a specific piece of equipment is selected to accomplish the required operations. Selection of the machine is determined based on size and tolerance capabilities, economics, availability and capacity considerations.

4. *'Calculate' processing times.* This step involves calculation of the specific operation set-up times on each machine. After this appropriate times for part loading, unloading, machine indexing and other factors involved in one cycle for processing a part are included.

Steps 5 and 6 then follow, as the need be.

Operation Planning :

- **Operation planning** is *that stage in planning which marks the completion of routing at process planning level.* The operation planning is concerned with planning the details of the method to be used to complete each operation at its chosen work centre and with designing the necessary tooling.

- Operations are divided into work elements. The record used to show the planned sequence of work elements is generally known as an *operation sheet.* It is in effect a record showing how an operation should be carried out.

- The *purpose of operation sheet is to record and communicate information that is essential for making each part.* This is the sole determinant and criterion for the design of the form that will be used. It is intended to achieve a level of specification that can be coated, evaluated and altered in specific rather than in abstract terms.

- Operation sheets are prepared for each part, sub-assembly and assembly. They *indicate the route of the parts* through the various departments, the sequence of required operations, the machines, special tools and gauge needed, the time required to do each operation, the details of speeds, feeds, etc.

Q. 1.6. Explain how the 'product design' and 'product manufacturing' take care of the environmental protection ?

Ans. The *'product design'* and *'product manufacturing'* take care of the environmental protection by adopting suitable measures.

Example. During Activity Analysis step of design, which is an Input-Output analysis, the designer estimates the desirable and undesirable inputs and outputs from the design. Any undersiable output such as noise, vibrations, poisonous emissions, glare, etc. are estimated at the design stage itself. Production people also participate in the production design stage to support factors which do not create environmental pollution. *Foundaries are now-a-days being replaced by fabrication plants to minimise pollution.*

Q. 1.7. Describe the independent and dependent variables of a production system.

Ans. A production system can be looked upon as an input-output system in which raw materials, equipments and human resources are the inputs to the system. The outputs are product or service to market. Apart from the *internal system variables* which **depend** wholly upon the management, there are **independent** or exogenous variables such as *market conditions, competitors, government laws, etc.* which are important and decide the success of any business.

Q. 1.8. On what considerations manufacturing processes are selected for a given product ?

Ans. Producing design of any product is a critical link in the chain of events that starts with creative ideas of design and ends with final and successful product in the market place. In modern technological age the function of production is not a routine activity. Rather we can say that *design, material selection and processing are closely related to each other* as shown in Fig. 1.15.

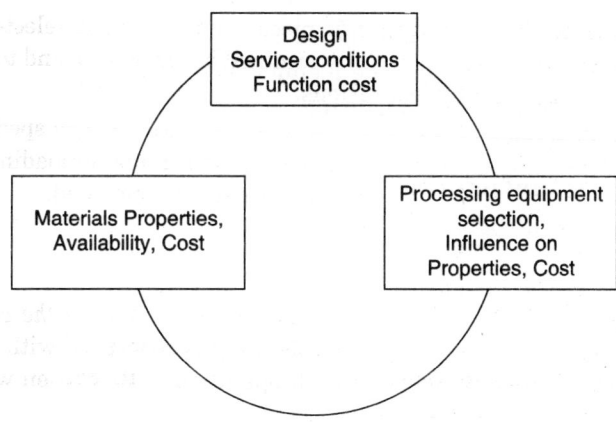

Fig. 1.15.

When form of any product is being developed, it becomes important to identify materials and production techniques and to be aware of their specific engineering requirements. An experienced designer already has in mind during design phase the short list of materials and processes. The important points to be considered for selection of manufacturing processes to be used are mentioned below :

1. The quantity of product to be developed.
2. Prior use knowledge for similar applications; this knowledge can be both blessing and a curse.
3. Knowledge and experience.
4. Availability of material and process.

The following list of factors also influence selection of manufacturing process :

(i) Cost of manufacture	(ii) Quantity or pieces required
(iii) Material	(iv) Geometric shape
(v) Surface finish	(vi) Tolerance
(vii) Gauges	(viii) Tooling jigs and fixtures
(ix) Available equipment	(x) Delivery date.

- The selection of a best possible manufacturing process is not an easy task. Rarely can a product be made by only one method, there are several competitive processes available. Since in all engineering design cost is very important factor, therefore, the selection of optimum manufacturing process depends largely on the costs of manufacture by the competing processes. The evaluation not only depends upon the cost of processing the material to the finished product but also the material utilisation factor and the effect of processing method on material properties and the subsequent performance of parts in service.

Q. 1.9 Discuss with examples the following types of manufacturing systems :
(i) Discrete manufacturing,
(ii) Continuous manufacturing.
Ans. (i) **Discrete manufacturing :**

"Discrete manufacturing system" is typified by the intermittent or interrupted flow of material through the plant.

- It makes use of general purpose machines and produces components different in nature and in small quantities.

  Machine-shops, repair and maintenance-shops, welding shops, etc. are some of the *examples of intermittent production.*

- Intermittent production or discrete manufacturing can be *classified* as :

  (a) Batch production or manufacturing,

  (b) Job manufacturing.

(ii) Continuous manufacturing :

"Continuous manufacturing" involves a continuous or almost continuous physical flow of material.

- It makes use of special purpose machines and produces standardized items in large quantities.
- Chemical processing, Cigarette manufacturing and Cement manufacturing are some of the industries engaged in continuous production or manufacturing.

Q. 1.10. What are the features of a production system ?
Ans. The *features of a production system* are :

1. A production system is goal oriented; goal being production of goods and services.
2. The goal is achieved through technological transformation of raw material, using energy.
3. Technological transformation is carried on by a suitable choice of technique, an optimal combination of capital (machinery) and labour from the broad spectrum of techniques ranging from complete automation (all capital, no labour) to completely manual labour (no capital, all labour).
4. Irrespective of choice of technique, specialisation or breaking down total quantum of transformation needed for goal to smaller simplified packets, so that the labour and machine repetitively perform then with reflexive (involuntary) and automatic efficiency.

5. Division of work entails corollary division of total team objectives for the subsystem, hence of authority and of responsibility.

Q. 1.11. Discuss briefly input-output model of a production system.

Ans. Owing to the complexity of the modern system it is necessary to analyse the production system on "system concept". The system concept provides a conceptual framework of production situation. It is worthwhile to put a production system in a model form which can represent all the parameters involved in the industry.

The model formulation process involves the listing of various input such as materials, tools parameters, energy requirements, etc., to the production process and getting the output from the production system.

Fig. 1.16 depicts the input-output model of a production system.

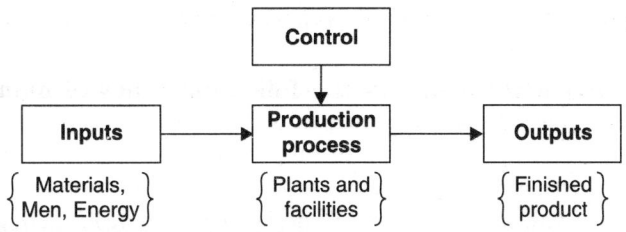

Fig. 1.16. Input-output model of a production system.

Q. 1.12. What are the advantages and disadvantages of a mass production system ?

Ans. Following are the advantages and disadvantages of a mass production system operating as a continuous flow line :

Advantages :

1. *Smooth flow of material* from one work-station to the next.
2. *Minimum material handling* (since the work-stations are so located as to cut down the distance between consecutive operations).
3. *Less space required* for the work transit and for temporary storage of material.
4. The operators *do not need special training* at the production line (as such the training is simple and inexpensive).
5. *Simple* production planning and control system can be followed.

Disadvantages :

1. Mostly it has been felt that *high investments* are needed due to the specialised nature of the machines.
2. In the event of breakdown of one machine, there will be *complete stoppage* of the remaining jobs on other machines.
3. The production pace is known by the *'slowest machine'*.
4. Maintenance and repair work becomes most difficult.
5. Originally the layout is established by the product but any changes incorporated in the product design *results in major change in the layout.*

Q. 1.13. Explain briefly a systematic approach for formulating model and solving practical problems.

Ans. Following **steps** are involved in a systematic approach for formulating model and solving practical problems :

Step 1 : The first and foremost phase is to *'Define the problem'.* It should point to the exact need and should be too easy to be understood by everyone. If the definition of problem is ambiguous, it may lead to loss of time, money and labour.

- After getting the exact solution, it is to be *analysed.* For its analysis, production system with its sub-systems can be represented into graphical and schematical model for better understanding.

  All inputs, outputs and information channels are clearly drawn by block diagrams.

Step 2 : After the identification of the exact definition of problem, *process of transformation starts.*

- To start with the *alternative solutions are thought-out.* These solutions are governed by various factors influencing the production system.

- Effectiveness of the alternatives is weighed by various criteria, *e.g.,* cost, profits, invest- ment, machine utilisation, men utilisation, etc. Any of these parameters may be altered depending on the nature of criteria.

- All the variables influencing system can be categorised as *controllable* and *uncontrolla- ble variables.* According to the nature of problem, these variables will be different. **Con- trollable variables** *are those that may be manipulated by management while* **un- controllable variables** *cannot be controlled by the management.*

  Based on these variables, a *model is developed for 'effectiveness' of alternatives.*

Step 3 :

- After finding *'effectiveness',* management has to find alternative courses of action. For this, existing science and technology, research market surveys, standards, handbooks, etc., in consultation with experts are consulted.

- Each of the possible alternative is evaluated and then it is accepted or rejected.

Step 4 :

- Finally, the problem is to be *analysed.* It is analysed by *evaluation of outputs of the system corresponding to given inputs in terms of effectiveness.*

In any production system, the target remains maximising the profits with minimum of additional investment. Possible methods for this aim may be the *exploitation of plant capacity to the maximum, increasing plant capacity by proper planning and also by overtime.* These three plans must be analysed. Breakeven analysis can prove very useful for the evaluation of appropriate level.

Q. 1.14. Explain briefly an Input-Output model for an automobile industry.

Ans.

- In automobile industries, the layout is *"line layout"* which is *preferred for mass manu- facturing.*

- Huge quantity of output can be attained only if inputs meet with proper standards and specifications. For maintaining proper control over the process *automation is preferred* which can be attained by a systematic arrangement of electronic instruments, scanners, automatic inspection and general purpose machines.

Robots find wide application in automobile industries at those spots where atmosphere is not suitable to human beings or high speed uniform operation is to be carried out.

Fig. 1.17 depicts an idea of Input-Output model for an automobile industry, producing cars. Inputs include a variety of raw materials which will be transformed to final products, some finished components which are to be used as such. All these finished components are assembled to give final shape of car.

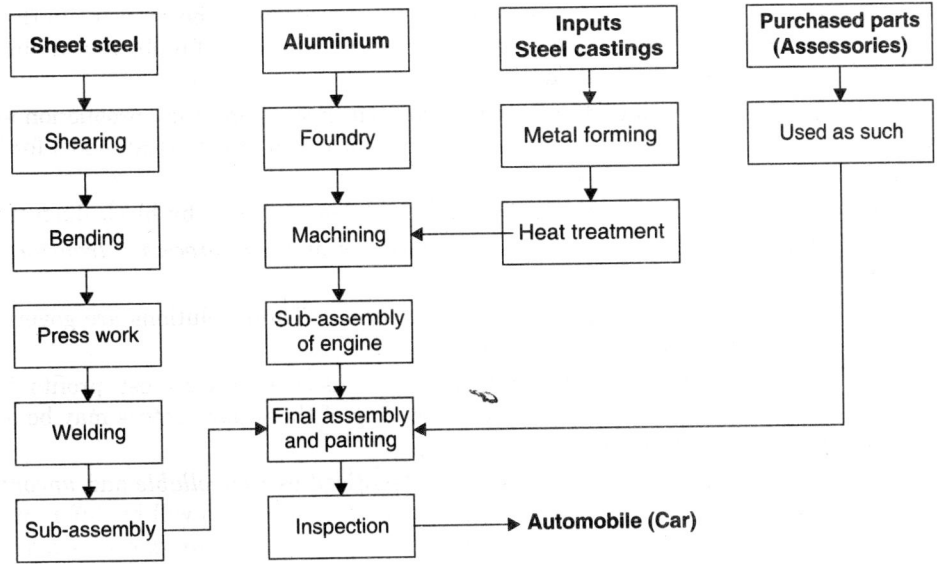

Fig. 1.17. Input-output model for an automobile industry.

Q. 1.15. Explain briefly various feedback loops which provide information for redesign.

Ans. Following are the various feedback loops which provide information for redesign.

1. The feedback loop from *market to function analysis.* This is owing to the market needs effecting a design change.

2. Feedback loop from *production to design, i.e.,* in case the design is difficult to produce in terms of accuracy and processing.

3. Feedback loop from *prototype shop to design*—Just in case the prototype testing indicates some need for design alteration due to insufficient performance of the prototype from the point of view of various standards of performance.

Q. 1.16. How will you classify the manufacturing processes ? Describe the important points to be observed while designing for heat treatment.

Ans. Manufacturing processes are broadly *classified* as follows :

1. *Primary processes :*
   - Casting
   - Welding
   - Rolling, etc.

2. *Secondary processes* .... These give final dimension.
   - Machining
   - Forging
   - Die casting
   - Grinding and abrasive processes, etc.

Design for heat treatment :

Parts which have to undergo heating and cooling cycle should be designed so that during cooling, deformation cracks, stress concentration should be avoided. Sharp corners, sudden change of cross-section, under-cut should also be avoided in the parts to be heat treated.

Q. 1.17. Explain what do you understand by 'Product Engineering' and 'Process Engineering' ? Outline the scope of each.

Ans. Product engineering. It refers to *various parts of a product and their specifications* and how they are inter-related to each other; size tolerances and other specifications also come under this.

The *scope of product engineering* include the following :

● Product life cycle ;

● Product profit planning ;

● Product improvement and development ;

● Design for competitive advantages.

Process engineering. It refers to the *analysis and control of processes* involved in manufacturing a product; how the process takes place step-by-step, its time schedule, machine working, all other aspects related to processing. Now-a-days it can be done through a computer also.

The *scope of process engineering* include the following :

● Process planning;

● Selection of machines and machine tools;

● Planning of operation sequence;

● Process sheets, etc.

Q. 1.18. Explain the meaning of 'designing for production'.

Ans. *'Design for production'* is establishing the shape of the components to allow for efficient high quality production for meeting with buyer's satisfaction.

The key concern of the design for production is in specifying the best manufacturing process for the component (or part of product) and ensuring that component form supports the production process.

Q. 1.19. Explain how designing for production rules help in speedy production, with minimum scrap of parts requiring :

(*i*) Cutting of external threads;

(*ii*) Drilling of holes on standing surfaces.

Ans. (*i*) Cutting of external threads :

1. External threads made by all processes *should not terminate too close to a shoulder* or other large diameter. Space must be provided for threat cutting tool.

2. The *length* of threads should be kept *as short as possible,* consistent with the function requirement of the part.

3. Slots, cross holes and flats *should not be placed* where they intersect screw threads.

4. Tubular parts must have a *wall heavy enough to withstand* the pressure of the cutting or forming action.

(*ii*) Drilling of holes :

1. The drill entry surface should be *perpendicular* to the *drill bit* to avoid starting problems and to help in the proper location.

2. The exit surface of the drill also should be *perpendicular to the axis of the drill* to avoid breakage problems if the drill leaves the work.

3. If the straightness of the hole is particularly critical it is best *to avoid interrupted cuts* unless a guide bushing can be placed at each re-entry surface.

4. It is *best to use standard drill sizes* when ever possible to avoid the added cost of special drill grinding.

5. *Through holes are preferable to build holes.*

6. When blind holes are specified, they should not have flat bottoms.

7. *Avoid deep holes* (over 3 times the diameter).

8. *Avoid designing parts, with very smaller holes.*

Q. 1.20. What are the factors to be considered for ease of manufacturing ? Explain with suitable examples.

Ans. Product development and design should consider whether geometrically a design has a good compatibility with material and production process for making a producible design. A typical *example* is that of a *pulley* in which curved arms are provided so that during production there are no casting problems and high tensile stresses being set-up in the arms during casting solidification are prevented. Therefore, it is always suggested to take care of production process involved in manufacturing the required product in designing the product.

Design for manufacturability. A product design should be such that it is easily producible by the production process. Design for production or manufacturability is a topic of importance in product design. Each production process and corresponding design configuration that can be easily produced is considered in brief.

1. Design for casting. Casting is a solidification process. There would have been no problem if every portion of a casting were solidified at the same time. Such a situation is only an ideal one. In practice products have cross-section so designed that there are heavy sections where loading in that portion is more and vice-versa. A pulley with straight arms has problem during solidification due to the heavy rim portion and boss portion putting tensile stresses on the straight arms during cooling. This causes 'hot tear'. Thus, pulley arms are made *curved for enhanced flexibility.*

Likewise, junctions such as T, V, Y create problems due to less volume ratio at the junction.

Sudden variation in thickness of material is harmful to castings. Gradual changes are provided by designer in such area. Sharp corners are avoided and radius is provided.

2. Design for forging. Closed die forgings are produced due to compression of material between upper and lower half of the dies. In order that material may flow freely in die cavities, there should be *no transitional surfaces* such as intersection of cylinder and rectangle, etc., in a properly designed part. Otherwise this part cannot be manufactured easily due to problem of filling of die cavity.

3. Design for machining. Component design for each of machining should be such that :

(*i*) The *ample overtravel* is available to the tools;

(*ii*) The tools should *not get unbalanced* in drilling of inclined surfaces;

(*iii*) As much as possible machining should be done *in one pass* to avoid time wastage;

(*iv*) A good design should *minimise machining to a minimum.*

4. Design for sheet metal parts. Shapes of sheet metal parts should ensure that minimum wastage of variable sheet metal should be in scrap form. 'Nesting technique' is very important in strip layouts of sheet metal parts design. *Operations such as deep drawing should be avoided.*

There is modern trend to combine welding along with bending operation to produce surfaces of revolution which were initially produced by tube sinking and deep drawing.

5. Design for powder metallurgy. Parts produced from powders and particulate processing should be purely *prismal, i.e.,* there is no need to provide a draft as in case of casting/forgings. The male and female dies have a nice sliding fit and produce a dense and strong component.

Q. 1.21. What are the design rules for manufacturing and ease of assembly ?

Ans. The *design rules for manufacturing and ease of assembly* are listed below :

1. Develop a modular design.
2. Design for minimum number of parts.
3. Minimise part variations.
4. Design parts to be multifunctional.
5. Design parts to be multiuse.
6. Design parts for ease of fabrication.
7. Avoid non-standard fasteners.
8. Minimise assembly directions; preferably top down.
9. Maximise compliance in parts; make assembly easy through chamfers, leads, tapers.
10. Evaluate assembly sequence for efficiency.
11. Improve component accessibility.
12. Avoid flexible components as they are difficult in automated assembly.
13. Allow for maximum intolerance of parts.
14. Use known vendors and suppliers.
15. Use parts at derated values of stress for increased reliability.
16. Minimise sub-assemblies.
17. Use new technology only when necessary.
18. Emphasize standardisation.
19. Design components for symmetry from end to end and about axis insertion.
20. Use the simplest operations with known capability.

Q. 1.22. What factors affect the quality of a product ?

Ans. The following *nine M's* directly affect the quality of products and services and so these should be identified and dealt with appropriately to obtain good results :

1. Material
2. Manpower
3. Money
4. Market for products, services
5. Motivation of employees
6. Modern information approaches
7. Management
8. Machines used
9. Mounting product needs.

HIGHLIGHTS

1. *Manufacturing* is the economic term for making goods and services available to satisfy human wants.
2. Any system that produces useful products or services, in general, is called a *Production system*.
3. *Production* may be considered as a process of transformation of a set of input elements whereby the utility of goods or services increased.
4. *Process type manufacturing* involves continuous flow of materials through a series of process steps to obtain a finished product like chemicals.

   Fabrication-type manufacturing involve manufacturing of individual parts or components by a series of operations, such as rolling, machining and welding.

   In *Assembly-type manufacturing* the parts or components are put together to get a complete product, such as a machine.

5. The manufacturing processes are *classified* as :
    (*i*) Constant mass processes
    (*ii*) Metal removing processes
    (*iii*) Metal addition processes.

6. Methods used for manufacturing system analysis :
    (*i*) Linear programming
    (*ii*) Waiting line model
    (*iii*) Network models
    (*iv*) Statistical techniques.

7. Types of production :
    (*i*) Job order production
    (*ii*) Batch or quantity production
    (*iii*) Mass production.

8. A *system is an arrangement of physical components* connected or related in such a manner as to form and/ or act as an entire unit.
A *control system* is an arrangement of physical components connected or related in such a manner as to command, direct or regulate itself or another system.

9. Control systems are classified as :
    (*i*) Open-loop control systems (Unmonitored or non-feedback control systems).
    (*ii*) Closed-loop control systems (Monitored or feedback control systems).
A closed-loop control system operating without human operator is called an *automatic control system.*

10. *Organisation* is defined as a systematic coordination and combination of the efforts with the aid of the money, men, machinery, materials and methods in a manner as would result in maximum manufacturing efficiency with minimum cost.
Administration is defined as the effectuation of the purpose by means of organisation and is concerned with "carrying out" policies, rules, regulations and designated methods.
Management is defined as the art of creating industrial relations of any kind, between people engaged in industry, *viz.,* relations between employers and employees, relation between individuals entering into commercial contracts, relation between investors and debtors etc.

11. *Scientific management* may be defined as a systematic approach to manage the enterprise on the basis of observation, experimentation and national decisions.

12. *Plant layout* may be classified as :
(*i*) Process layout : It is suitable for job order plants which involve non-repetition processes.
(*ii*) Product layout : This type of layout mostly and best suites those industries which manufacture a large volume of standard products involving repetitive processes.

13. *Production* involves sequences of operations that transform raw materials into the desired shape and size.
Routing means determination of path on which manufacturing operations will travel establishing the sequence of such operations and looking to the proper category of machines and personnel which these operations will require.
Scheduling means the determination of the relative time at which each operation or event in connection with manufacturing is to occur.

OBJECTIVE TYPE QUESTIONS

Fill in the blanks or say "Yes" or "No" :

1. .... is the economic term for making goods and services available to satisfy human wants.
2. Any system that produces useful products or service, in general, is called a .... system.
3. .... may be considered as a process of transformation of a set of input elements whereby the utility of goods or services is increased.
4. The manufacturing processes are collected together to form a .... system.
5. Manufacturing is the heart of the system where material is converted from one form to another and value is added.

6.   .... manufacturing involves manufacturing of individual parts or components by a series of operations such as rolling, machining and welding.

7.   In Assembly-type manufacturing the parts or components are put together to get a complete product, such as a machine.

8.   Casting is a metal removing process.

9.   Powder metallurgy processing is a constant mass process.

10. Wire drawing is a metal removing process.

11. Welding is a .... addition process.

12. Thermit welding is a metal forming process.

13. Process planning is a fundamental part of industrial activity.

14. .... planning aims at planning method or series of methods for economic manufacturing of a product of the quality called for by the drawings or specifications laid down.

15. Linear programming is one of the methods of .... system analysis.

16. Waiting is always associated with costs.

17. In general, simulation models follow a conceptual structure and may not lead to optimum answer.

18. .... production consists of small scale production to meet the attainment of individual customers.

19. Batch order or quantity production is a special type of production.

20. Mass production means a continuous production without loss of time.

21. Job order production is carried out .... factories and suits for .... works.

22. A .... is an arrangement of physical components connected or related in such a manner as to form and/ or act as an entire unit.

23. Each system may have a large number of sub-systems.

24. A .... system is an arrangement of physical components connected or related in such a manner as to command, direct or regulate itself or another system.

25. .... control system is one in which the control action is independent of the desired output.

26. .... control system is one in which the control action is somehow dependent on the output.

27. Liquid level control system is example of open-loop control system.

28. Traffic signal system is an example of closed-loop control system.

29. Automatic wasting machine is an example of .... control system.

30. Human being reaching for an object is an example of closed-loop system.

31. A closed-loop control system operating without human operator is called an .... control system.

32. Automatic control system has a tendency to overcorrect errors which may result in oscillations of constant or changing amplitude.

33. .... is defined as systematic coordination and combination of the efforts with the aid of the money, men, machinery, materials and methods in a manner as would result in maximum manufacturing efficiency with minimum cost.

34. .... is the evaluation of the purpose by means of organisation and is concerned with "Carrying out" policies, rules, regulation and designated methods.

35. .... is the art of creating industrial relations of any kind, between people engaged in industry, *viz.*, relations between employers and employees, relation between individuals entering into commercial contracts, relation between investors and debtors etc.

36. A plant is a place, where men, materials, money equipment, machinery etc. are brought together for manufacturing products.

37. .... organisation is also called scalar or military type organisation.

38. Functional type of organisation divides the responsibility according to the functions to be performed and not in departments or sections of the works.

39. .... management may be defined as a systematic approach to manage the enterprise on the basis of observation, experimentation and rational decisions.

40. Ensuring steady flow of standard goods to customers at fixed price is one of the aims of scientific management.

41. Organising is one of the activities of ....

42. Transport facilities is not an important factor affecting plant location.

43. Flexibility is one of the sound principles of plant layout.

44. Process layout is sometimes called .... layout.

45. .... layout is suitable for job order plants which involve non-repetitive processes.
46. In .... layout, machines and other manufacturing facilities are located in the sequence required to manufacture the product.
47. Product type of layout mostly and best suits those industries which manufacture a large volume of standard products involving repetitive processes.
48. A proper layout is one which contributes most effectively the economical flow of material through the sequences of the manufacture.
49. .... involves sequences of operations that transform raw materials into the desired shape and size.
50. .... begins with the analysis of given data, on the basis of which a scheme of utilisation of firm's services can be outlined so that the desirable target may be achieved in an efficient manner.
51. Control involves supervising operations with the aid of control mechanisms and feedback information about the progress of work.
52. .... means determination of path on which manufacturing operations will travel establishing the sequence of such operations and looking to the proper category of machines and personnel which these operations will require.
53. .... means the determination of the relative time at which each operation or event in connection with manufacturing is to occur.
54. 'Master schedule' and 'Production schedule' usually provide the base for carrying out scheduling.
55. The Gantt chart is one of the fundamental charts used in a manufacturing concern.

ANSWERS

1. Manufacturing	2. production	3. Production	4. manufacturing	5. Yes
6. Fabrication-type	7. Yes	8. No	9. Yes	10. No
11. material	12. No	13. Yes	14. Process	15. manufacturing
16. Yes	17. Yes	18. Job order	19. No	20. Yes
21. small, variety	22. system	23. Yes	24. control	25. Open-loop
26. Closed-loop	27. No	28. Yes	29. Open-loop	30. Yes
31. automatic	32. Yes	33. Organisation	34. Administration	35. Management
36. Yes	37. Line	38. Yes	39. Scientific	40. Yes
41. management	42. No	43. Yes	44. functional	45. Process
46. product	47. Yes	48. Yes	49. Production	50. Planning
51. Yes	52. Routing	53. Scheduling	54. Yes	55. Yes.

THEORETICAL QUESTIONS

1. List the activities and operation involved in manufacturing.
2. Distinguish between production and 'production system'.
3. Depict a manufacturing system as an input-output system.
4. Name the distinct stages through which the present day 'manufacturing system' has evolved.
5. Enumerate the principal types of manufacturing.
6. How are manufacturing processes classified ?
7. How is right type of manufacturing process selected ?
8. What points should be considered for selecting a manufacturing process ?
9. What is "Process planning" ?
10. What are the purposes of process planning ?
11. Explain briefly the methods employed for the analysis of a manufacturing system.
12. Enumerate the types of production.
13. Explain briefly the following types of production :
    (i) Job order production   (ii) Batch or quantity production
    (iii) Mass production.
14. How is 'system' defined ?

15. What is a 'control system' ?
16. Enumerate and define the elements of a control system.
17. Give five examples of control system applications.
18. How are control systems classified ?
19. Give the comparison between 'Open-loop' and 'Closed- loop' systems.
20. What is an 'open-loop' control system ? Enumerate and explain the elements of an open-loop control system.
21. State the advantages and limitations of open-loop control system.
22. Explain briefly the following :
     (i) Idle-speed control system ;
     (ii) Open-loop word processor control system.
23. What is a 'Closed-loop' control system ?
24. What is a feedback ? What are its characteristics ?
25. What are the advantages and limitations of a 'Close-loop' control system.
26. What is an automatic control system ? Explain with examples.
27. What are the advantages and limitations of automatic control systems ?
28. Define the following :
     (i) Organisation;   (ii) Administration;
     (iii) Management.
29. Explain briefly any two of the following organisation structures :
     (i) Line organisation   (ii) Line and staff organisation
     (iii) Functional type organisation   (iv) Complex organisation.
30. Define 'Scientific management'.
31. State the principles and aims of scientific management.
32. What are the functions of management ?
33. Enumerate elements of communication.
34. Name the factors which affect the plant location.
35. List the principles of plant layout.
36. What are the objectives of good layout ?
37. What are the fundamentals considerations in layout ?
38. How are layouts classified ?
39. Describe briefly the following types of layout :
     (i) Process layout   (ii) Product layout.
40. State the advantages and disadvantages of process and product layouts.
41. Explain briefly combined process and produce layout.
42. What are the functions PPC (Production Planning and Control).
43. What are the advantages of production control ?
44. Explain briefly the following constituents of production control :
     (i) Production planning   (ii) Routing
     (iii) Scheduling   (iv) Despatching
     (v) Follow up.
45. Write a short note on Gantt charts.

2

Casting Processes

2.1. INTRODUCTION

Casting is perhaps the oldest method of manufacturing and *invariably the first step in the sequence of manufacturing a product.*

In this process (casting) the *raw material is melted, heated to the desired temperature and poured into the mould cavity where it takes the desired shape. After the molten metal solidifies in the mould cavity the product is taken out to get the casting.*

- *Casting is preferred* because of the following **reasons :**

 (*i*) It is cheap and direct way of producing a shape with desired mechanical properties.

 (*ii*) Metals and alloys, *e.g.*, highly creep resistant alloys cannot be worked mechanically and can only be cast.

 (*iii*) Casting is best suited when different properties are required in different sections of product. These are made by incorporating pre-fabricated inserts in a casting.

 (*iv*) Cost associated in giving details by casting process is minimum whereas cost in mechanical working would be too high to produce them.

 (*v*) Casting is preferred when components are desired in low quantities.

Basic features. The basic features common to various casting processes can be summarized as :

 (*i*) *Pattern and mould.*

 (*ii*) *Melting and pouring.*

 (*iii*) *Solidification and cooling.*

 (*iv*) *Removal, cleaning, finishing and inspection.*

2.2. PATTERNS

2.2.1. Definition

- A **pattern** is *defined as a model of a casting, constructed in such a way that it can be used for forming an impression (mould) in damp sand.*
- In making a casting the first step is to prepare a model, known as *pattern*, which differs in a number of respects from the resulting casting. These differences, known as pattern *allowances, compensate for metal shrinkage, provide sufficient metal for machining the surfaces and facilitate moulding.*

2.2.2. Requirements of a Good Pattern

The *requirements of a good pattern* are :

1. Light in weight.
2. Convenient to handle.
3. Simple in design and ease of manufacture.
4. Smooth and wear resistant surface.
5. Retain its dimensions and rigidity during the definite service life.
6. High strength and long life.
7. Secure the desired shape and size of the casting.
8. Ability to withstand rough handling.
9. Cheap and readily repairable.

2.2.3. Pattern Materials

For pattern making the following materials are commonly used :

1. Wood   2. Metal
3. Plastic   4. Quick setting compounds.

1. **Wood :**

- The wood used for pattern making should be properly dried and seasoned.
- It should be straight grained.
- It should be free from knots.
- It should be free from insects and excessive sap wood.

Advantages :

  (i) Cheapness.
 (ii) Ease of availability.
(iii) Lightness.
 (iv) Ease of obtaining smooth surface and preserving surface ; by applying coating of shellac.
  (v) Ability to be worked on easily.
 (vi) Ease of joining.
(vii) Ease of fabricating to numerous shapes.

Limitations :

  (i) Easily affected by moisture, its shape changes by change in moisture content.
 (ii) It wears out quickly by sand abrasion.
(iii) It may warp during improper storing.
 (iv) It cannot stand rough usage.
  (v) Inherently non-uniform in structure.

The following types of wood are commonly used for pattern making :

(i) *White pine :*
— It is the most widely used wood.
— It is soft and easy to work.
— It is unlikely to warp.

(ii) *Mahogany :*
— It is less likely to warp comparatively.
— When straight grained, it can be worked easily.
— It is harder and more durable than white pine.

(iii) *Mapple, Kirch and Cherry :*
— These woods tend to warp in large sections ; as such these should be used for small patterns only.
— Since they pick up moisture readily, they should be treated carefully.
— They are heavier and harder than white pine.

(iv) *Teak, Shisham, Kail and Deodar.*

2. **Metal.** Where durability and strength are required, patterns are made from metals.
● A metal pattern can be either cast from master wooden pattern or may be machined by the usual methods of machining.
● *Metal patterns* are usually used in *machine moulding.*

Advantages :
(i) Resistant to wear, abrasion, corrosion and swelling.
(ii) Possess a smooth surface.
(iii) Ability to withstand rough handling.
(iv) Do not undergo deformation in storage.
(v) More durable and accurate in size than wooden patterns.

Limitations :
(i) Cannot be repaired easily.
(ii) More expensive than wooden patterns.
(iii) Ferrous patterns may get rusted.
(iv) Heavier than wooden patterns.

The following metals are commonly used for pattern making :

(i) *Aluminium :* It is the best pattern material.
— It is light in weight.
— It is corrosion resistant.
— It is easily worked.

However, it is subject to shrinkage and wear by abrasive action.

(ii) *Brass :*
— It has a smooth, closed pore structure.
— It may be easily worked and built-up by soldering or brazing.

However, it is expensive and, therefore, generally employed for small cast parts.

(iii) *White metal :*
— It has low melting point.
— It can be cast easily.
— It has low shrinkage.

— It is light in weight.

— It may be built-up by soldering.

However, it is subject to wear by abrasive action of sand.

(*iv*) **Cast iron :**

— Cast iron with fine grain can be used as a pattern material.

— It is cheaper and more durable than other metals.

— It has low corrosion resistance unless protected.

However, it is difficult to work.

3. **Plastic.** The plastics when used as pattern material entail the following *advantages* :

(*i*) Highly resistant to corrosion.

(*ii*) Lighter and stronger than wood pattern.

(*iii*) Strong and dimensionally stable.

(*iv*) The production process is facilitated.

(*v*) Moulding sand sticks less to plastics than to wood.

(*vi*) The surface of pattern is smooth.

(*vii*) No moisture absorption.

● The various plastics which are employed for the production of patterns are the compositions based on epoxy, phenol formaldehyde and polyester resins ; polyacrylates, polyethylene, polyvinyl chloride, etc.

4. **Quick setting compounds :**

● *Gypsum* patterns are capable of producing castings with intricate details and to very close tolerances.

● Gypsum can be easily formed, has plasticity and can be easily repaired.

Note. *Soft wood patterns* are used upto 50 pieces of medium size castings, *hard wood patterns* for 50 to 200 castings and *metal patterns with hardened steel wear plates* for 200 to 5000 castings.

2.2.4. Types of Patterns

Types of patterns depend upon the following *factors :*

(*i*) The shape and size of casting.

(*ii*) Number of castings required.

(*iii*) Method of moulding employed.

(*iv*) Anticipated difficulty of the moulding operation.

The following *types of patterns* are commonly used :

1. Solid or single piece pattern.	2. Split pattern.
3. Loose piece pattern.	4. Match plate pattern.
5. Gated pattern.	6. Skeleton pattern.
7. Sweep pattern.	8. Cope and drag pattern.
9. Follow board pattern.	10. Segmental pattern.
11. Shell pattern.	12. Lagged-up pattern.

1. Solid or single piece pattern : Refer to Fig. 2.1.

● It is simplest of all the patterns and the cheapest.

● It is made in one piece and carries no joint, partition or loose pieces.

● Depending upon the shape, it can be moulded in one or two boxes.

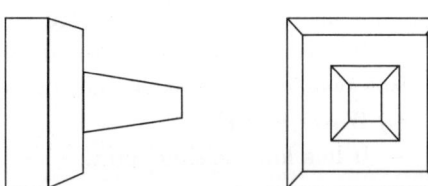

Fig. 2.1. Solid pattern.

● Its use can be made to a limited extent of production only since its moulding involves a large number of manual operations like gate cutting, providing runners and risers and the like.

2. Split pattern : Refer to Fig. 2.2.

● Most of the patterns are not made in a single piece because of the difficulties encountered in moulding them. In order to eliminate this difficulty, some patterns are made in two or more pieces.

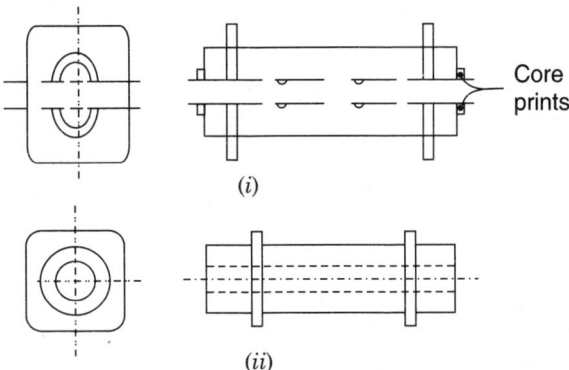

(i) Two halves of pattern. (ii) Prepared casting

Fig. 2.2. A split pattern.

● A pattern consisting of two pieces is called a *two piece split pattern.* One-half of the pattern rests in the lower part of the moulding box known as *drag* and the other half in the upper part of the moulding box known as *cope.* The line of separation of the parts is called *parting line* or *parting surface.*

● Sometimes a pattern is constructed in three or more parts for complicated castings. Such a pattern is called *multi-piece pattern* (Fig. 2.3).

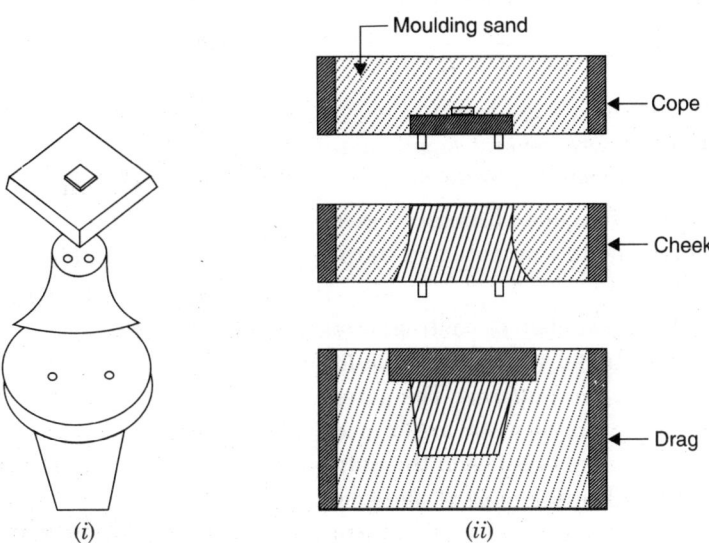

Fig. 2.3. Multipiece pattern.

3. Loose piece pattern : Refer to Fig. 2.4.

In some cases a pattern has to be made with projections or overhanging parts. These projections make the removal of the pattern difficult. Therefore such projections are made in loose pieces and are fastened loosely to the main pattern by means of wooden or wire dowel pins. These pins are taken out during moulding operation. After moulding the main pattern is withdrawn first and then the loose piece is removed by using a lifter.

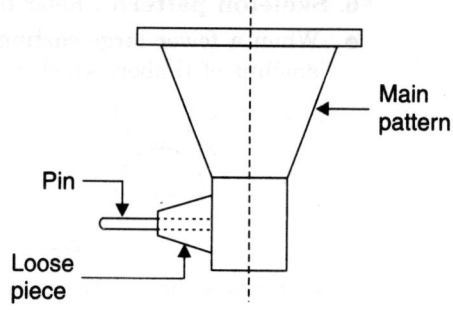

Fig. 2.4. Loose piece pattern.

4. Match plate pattern :

- A match plate pattern is made by fastening each half of a split pattern to the opposite side of one plate. The plate provides a substantial mounting for patterns and is *widely used in machine moulding.* The gates and runners are also attached in their correct position. On the drag side the plate is equipped with location holes which fit into the pins provided on the drag portion of the flask.

When the match plate is lifted off the mould all patterns are drawn, and the cope or upper half of the mould matches perfectly with the drag or lower half of the mould. *The gates and runners are also completed in the same operation.*

- Fig. 2.5 shows a match plate upon which the patterns of two small dumb bells are mounted.

- Match plate patterns are *expensive to construct* (but the initial cost is justified in mass production).

- These patterns are *suited for mass production of small castings in moulding machines.*

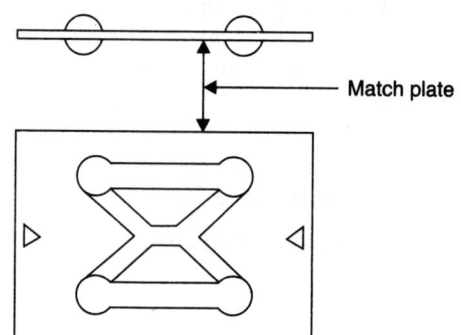

Fig. 2.5. Match plate pattern.

5. Gated pattern :

- In production where several small castings are required, gated patterns are used.

- Such patterns are made of metal to give them strength and to eliminate any warping tendency.

- To save time, a number of castings are produced in a single *multicavity mould by joining a group of patterns.* The gates or runners for the molten metal are formed by the connecting parts between the individual patterns. These groups of patterns with gate formers attached to them are called *gated patterns* (see Fig. 2.6).

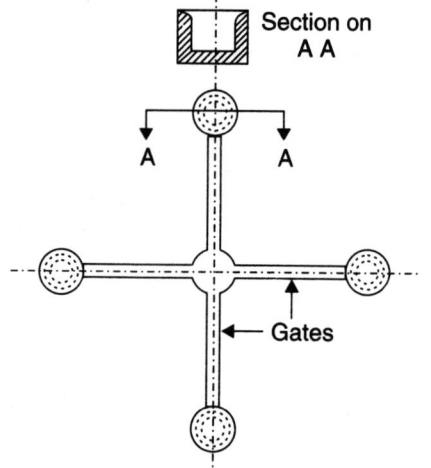

Fig. 2.6. Gated pattern.

6. Skeleton pattern : Refer to Fig. 2.7.

● When a fewer large castings are required, solid patterns would require a tremendous amount of timber, which may not be economical.

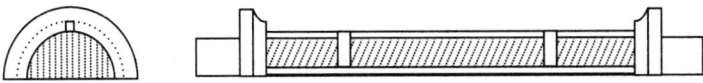

Fig. 2.7. A skeleton pattern for a flanged pipe.

In such cases the pattern is made of *wooden frame and rib construction* (skeleton) so that it will form a partially or interior outline of the casting and provide the general colour and size of the desired casting. The ribbed construction with a large number of square or rectangular openings between the ribs is filled and rammed with clay sand or foam. A strike off board known as *'strickle'* board is used to scrape the excess sand out of the spaces between the ribs so as to make the exterior surface even with the outside of the skeleton (see Fig. 2.7).

● Usually, skeleton pattern is built in two parts ; one for the cope and another for the drag.

7. Sweep pattern : Refer to Fig. 2.8.

● Such patterns are employed in making mould of *large symmetrical,* circular cross-section to effect saving in time, labour and material.

● A *sweep* is a template of wood or other material which has contour corresponding to the shape and size of casting. It is *rotated about a central spindle* (see Fig. 2.8).

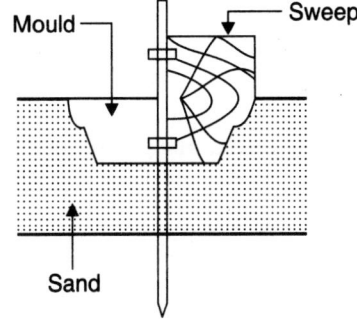

Fig. 2.8. Sweep pattern.

8. Cope and drag pattern :

When very large castings are to be made, the complete pattern becomes too heavy to be handled by a single operator. Such a pattern is made *in two parts which are separately moulded in different moulding boxes. After completion of the moulds, the two boxes are assembled to form the complete cavity, of which one part is contained by the drag and the other in cope.*

9. Follow board pattern :

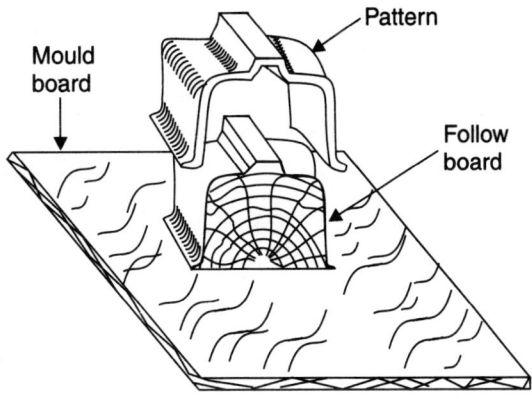

Fig. 2.9. A follow board pattern.

The patterns having thin sections (see Fig. 2.9) tend to get distorted or collapse during ramming. Sagging of this pattern due to ramming can be easily overcome by constructing a supporting block (follow board) which may fit inside the pattern to serve as a support during ramming.

10. Segmental pattern : Refer to Fig. 2.10.

- These patterns are generally applied to circular work, like rings, wheels, rims, gears etc.

- In principle they work like a sweep, but the difference is that a sweep is given a continuous revolving motion to generate the desired shape whereas a *segmental pattern is a portion of the solid pattern itself and the mould is prepared in parts by it.* It is mounted on a central pivot (see Fig. 2.10) and after preparing the part mould in one position, the segment is moved to next position. The operation is repeated till the complete mould in ready.

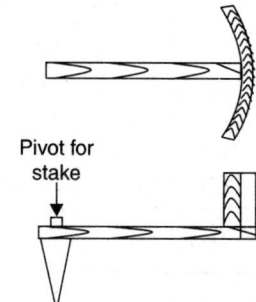

Fig. 2.10. Segmental pattern.

11. Shell pattern : Refer to Fig. 2.11.

- A shell pattern is largely used for *drainage fittings and pipe work.*

- This type of pattern is usually made of metal and parted along the centre line, the two sections being accurately dowelled together. The short bends are usually moulded and cast in pairs.

- The shell pattern is a hollow construction like a shell. *The outside shape is used as a pattern to make the mould, while the inside is used as a core box for making cores.*

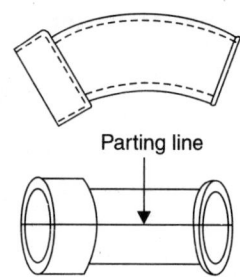

Fig. 2.11. Shell pattern.

12. Lagged-up pattern :

Refer to Fig. 2.12. Cylindrical patterns, *e.g.,* barrels, pipes or columns are built up with lag or stave construction to ensure proper shape. Longitudinal strips of wood, called *lags* or *staves* are bevelled on each side and glued to the wooden pieces called '*heads*' (see Fig. 2.12). Such a construction gives the maximum amount of strength and permits building close to the finished outline of the pattern so that there is comparatively little excess stock to be removed to bring it to the required form.

2.2.5. Pattern Allowances

While making patterns certain dimensional allowances must be given in the pattern so that the casting obtained is of the required specifications.

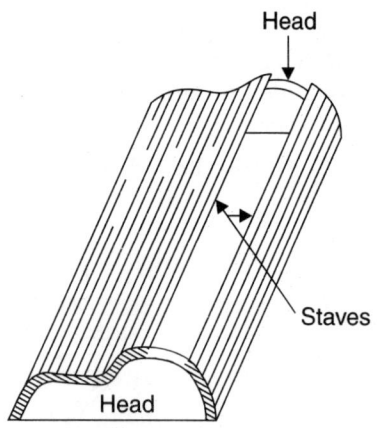

Fig. 2.12. Lagged-up pattern.

The following *allowances* are usually provided in a pattern :

1. Shrinkage allowance.

2. Draft or taper allowance.

3. Machining allowance.

4. Rapping or shaking allowance.

5. Distortion allowance.

1. Shrinkage allowance :

- Shrinkage allowance is an allowance added to the pattern *to compensate for the metal shrinkage that takes place while the metal solidifies.* All metals except Bismuth and Gallium shrink.

The following table shows the contraction allowances for castings of different metals in sand moulds :

Table 2.1. Contraction Allowance for Different Metals.

S. No.	Metals/Alloys	Contraction allowance mm/metre
1.	Grey cast iron	7 to 10.5
2.	White cast iron	21
3.	Malleable iron	15
4.	Steel	20
5.	Copper	16
6.	Brass	16
7.	Bronze	10.5 to 21
8.	Zinc	24
9.	Lead	24
10.	Aluminium	16
11.	Magnesium	18

Note. The contraction of metals/alloys is always *volumetric*, but the contraction allowances are always expressed in *linear measures.*

2. Draft or taper allowance :

- It is the taper provided on vertical surfaces of a pattern to *facilitate its removal from the mould without excessive rapping or breakage of cavity edges.* The amount of taper varies with the type of pattern.
- The taper on the inner surfaces should be greater than on the outside surface.
- The amount of taper varies from 0.5° to 1.5°. It may be reduced to less than 0.5° for larger castings.
- The wooden patterns require more taper than metal patterns because of the greater frictional resistance.

3. Machining allowance :

- Machining or finishing allowance is the *extra material provided on certain details of a casting so that the casting may be machined to exact dimensions.* The machining allowance depends on the following factors.

    (*i*) Casting process.

    (*ii*) Size of the casting.

    (*iii*) Degree of finish.

    (*iv*) Machining method.

    (*v*) Metallic alloy from which the casting is made.

- The amount of this allowance varies from 1.6 mm to 12.5 mm.
- The *ferrous metals require more machining allowance than non-ferrous metals.*

4. Rapping or shaking allowance :

- This allowance is provided to *compensate for enlargement of the mould cavity because of excessive rapping.*
- In small and medium-sized castings, this allowance can be neglected. But in larger casting this allowance is considered by making the part *slightly smaller than the casting* (*i.e.,* the allowance is a negative one as the pattern is made smaller to allow rapping operation).

5. Distortion allowance :

- This allowance is provided on the pattern *to compensate for possible distortion of the casting because of the unequal cooling rates of different sections of the casting and uneven internal stresses.*
- Such an allowance depends on the judgement and experience of the pattern maker, who understands the shrinkage characteristics of the metal.

2.3. MOULD MAKING

2.3.1. General Aspects

A **mould** *may be defined as the negative print of the part to be cast and is obtained by the pattern in the moulding sand container (boxes) into which molten metal is poured and allowed to solidify.* Sand moulds are destroyed as the casting is removed from the moulds.

Moulding is *an art of making sound mould out of sand by means of pattern and cores so that metal can be poured into the moulds to produce casting.*

— *Moulding is done both by hands and by machines.* Hand moulds are restored to *odd castings* generally less than 50 pieces at a time or so. Here ramming in done by hand which takes more time than machine moulding. However the quality is better for odd castings.

For mass manufacture, machine moulding is suitable. Moulding machines are prominently used in big foundries.

The **moulding machines** *perform the following basic operations :*

1. Ramming or sand in the mould,
2. Lifting or drawing of pattern from the mould, and
3. Rolling over mould section.

Following are the *two main classes of moulding machines :*

(*a*) Hand moulding machines.

(*b*) Power moulding machines :

    (*i*) Jolt-machine.

    (*ii*) Squeezing machine.

    (*iii*) Jolt-squeeze machine.

    (*iv*) Sand slinger.

    (*v*) Diaphragm moulding machine.

    (*vi*) Stripper-plate machine.

- Moulding is carried out in moulding boxes called *flasks*, which are open at the top and bottom ; the top part is called the *cope* and the lower part as the *drag*. The moulding boxes are usually either of fabricated mild steel or cast iron which can be clamped together. To avoid misfitting of two halves there are two pins on one side and one pin on other side in the top half which go into corresponding holes of the bottom half. This avoids the possible misfit. In case of very big castings, the moulds may be made on ground without moulding boxes. In some cases moulding boxes are put in 3 pieces (the intermediate part is called a *cheek*) to facilitate moulding. The section of moulding box is shown in Fig. 2.13.

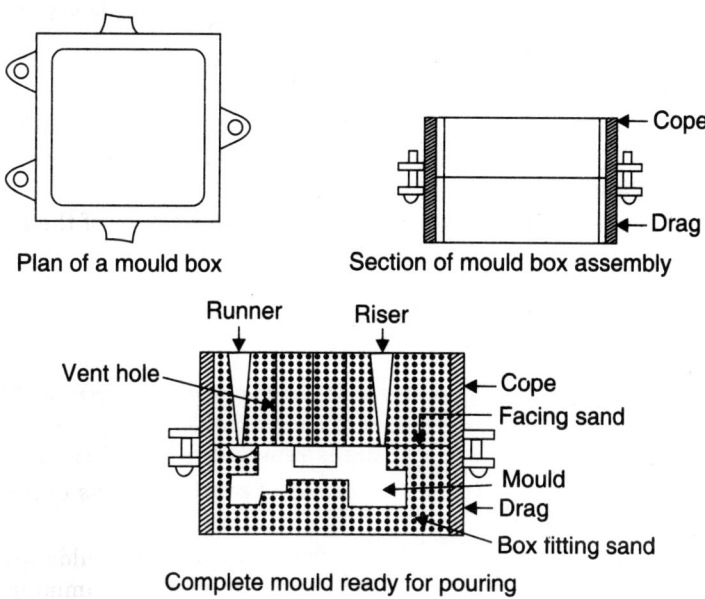

Plan of a mould box   Section of mould box assembly

Complete mould ready for pouring

Fig. 2.13. Mould making.

2.3.2. Types of Moulds

The moulds are of the following two types :

1. Temporary moulds :

These moulds are destroyed at the time of removing castings from them.

Example : Sand moulds.

2. Permanent moulds :

These moulds are used in die casting. These moulds are used time and again.

Example : Metallic moulds.

2.3.3. Moulding Processes

The moulding processes may be *classified* as follows :

1. Bench moulding :

● The moulding done on a bench of convenient height to the moulder is called *bench moulding.*

● It is used for small work.

2. Floor moulding :

● The moulding done on the foundry floor is called *floor moulding.*

● It is used for all medium sized and large castings.

3. Pit moulding :

- Very large moulds made in a pit or cavity cut in the floor to accommodate very large castings is called *pit moulding*. The pit acts as a drag.
- Since pit moulds can resist pressures developed by hot gases, therefore, it greatly saves pattern expenses.

4. Machine moulding :

- The mouldings done by a machine is called *machine moulding*.
- Small, medium and large moulds may be made with the help of a variety of machines.
- Machine moulding is usually faster and more uniform than bench moulding.
- Machine moulding generally requires mounted patterns.

2.3.4. Types of Sand Moulding

Sand moulding methods may be *classified* as follows :

1. Green sand moulds.
2. Dry-sand moulds.
3. Skin-dried moulds.
4. Loam moulds.
5. Metal moulds.

1. Green sand moulding (moulds) :

Among the sand-casting processes, moulding is often done with green sand. *Green moulding sand* may be defined as a *plastic mixture of sand grains, clay, water, and other materials, which can be used for moulding and casting processes.* The sand is called "*green*" because of moisture present and is thus distinguished from dry sand.

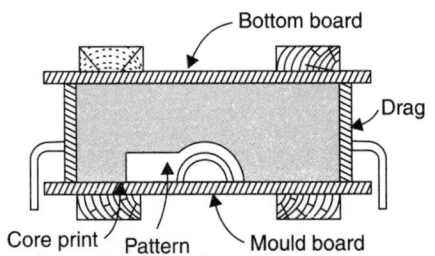

Fig. 2.14. Drag half of mould made by hand. Drag is ready to be rolled over in preparation for making the cope.

The basic steps in green-sand moulding are as follows :

(*i*) ***Preparation of the pattern.*** Most green-sand moulding is done with match plate or cope and drag pattern. Loose patterns are used when relatively few castings of a type are to be made. In simple hand moulding the loose pattern is placed on a mould board and surrounded with a suitable-sized flask as illustrated in Figs. 2.14 and 2.15 respectively.

(*ii*) ***Making the mould.*** Moulding requires ramming of sand around the pattern. As the sand is packed, it develops strength and becomes rigid within the flask. Ramming may be done by hand, as in simple set up illustrated in

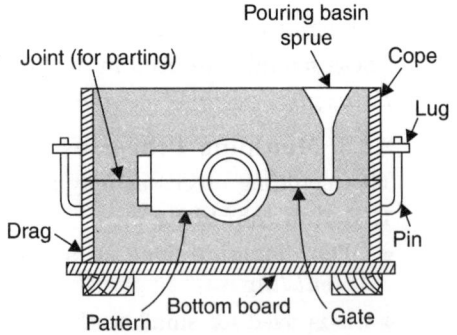

Fig. 2.15. Cope mould rammed up.

Fig. 2.14. Both cope and drag are moulded in the same way, but the *cope must provide for the sprue.* The gating-system parts of the mould cavity are simply channels for the entry of the molten metal, and can be moulded as illustrated in Fig. 2.15.

(*iii*) **Core setting.** With cope and drag halves of the mould made and the pattern withdrawn, cores are set into the mould cavity to *form the internal surfaces of the casting.*

(*iv*) **Closing and weighing.** With cores set, the cope and drag are closed. The cope must usually be weighted down or clamped to the drag to prevent it from floating when the metal is poured :

Advantages :

1. Great flexibility as a production process.
2. The least costly method of moulding.
3. Less time consuming since no backing operations or equipment is required.
4. Green sand moulds can be used for all types of the ferrous and non-ferrous alloys.

Disadvantages/Limitations :

1. Certain metals and some castings develop defects if poured into moulds containing moisture.
2. More intricate castings cannot be made.
3. The dimensional accuracy and surface finish of green-sand castings may not be adequate.
4. Green sand moulds are not very strong and may be damaged during handling or by metal corrosion.
5. Storage of green sand moulds for longer periods is not possible.

2. Dry-sand moulds :

- The sand mould made with a sand that does *not* require moisture to develop strength (the binder provides strength) are called *dry sand moulds.*
- These moulds are used for *steel castings.*

Advantages :

1. The dry-sand moulds are stronger and may be handled more easily with less damage.
2. Dry sand eliminates the possibilities of moisture related defects in casting.
3. Surface finish of castings is better, mainly because dry sand moulds are coated with a wash.

Disadvantages :

1. These are *more expensive* comparatively.
2. Castings are more prone to hot tearing.
3. Distortion is greater than for green sand moulds because of baking.
4. Production is slower than for green sand moulds.
5. More flash equipment is needed to produce the same number of finished pieces because processing cycles are longer than for green sand moulds.

3. Skin-dried moulds :

- The sand moulds with a dry sand facing and a green sand backing are called *skin-dried moulds.*
- They can be employed for casting all ferrous and non-ferrous alloys.

- They are more commonly used for *large moulds.*
- As compared to dry-sand moulds they are less expensive to construct, but more expensive than green sand moulds.
- They are less stronger than dry-sand moulds.

4. Loam moulds :
- These moulds are made with loam sand (a mixture of sand and clay). The loam sand also contains fire clay or gainster.
- They are used for large work.
- A loam mould is constructed of porous bricks cemented together with loam mortar. The inside of the brick structure forms the rough contour of the casting and it is faced with a 6 to 12 mm layer of loam sand to give the required shape.
- A loam mould requires enough area and space, difficult to give proper contour and shape and is *suitable only for a single casting.*

5. Metal moulds :
- The metal moulds are permanent type of moulds. These are used in die casting where molten metal is introduced into the metallic mould cavity by means of pressure. Sometimes even gravitational force is sufficient to feed the metal into the mould cavity.
- These moulds are used *in the casting of low-melting temperature alloys.*

Advantages :
(*i*) Improved surface furnish.
(*ii*) Since the castings produced by metal moulds have a smooth finish, therefore much of the machine work is eliminated.
(*iii*) High production rate.
(*iv*) Thin sections can be cast.
(*v*) Castings produced are less defective.

Disadvantages :
(*i*) High cost of moulds and equipment.
(*ii*) For maintenance of moulds/equipment, special skill is required.
Following points are worth noting :
1. In a foundry, upto 90 per cent of the moulding sand can be reprocessed to make new moulds.
2. Majority of castings are poured in green sand moulds.
3. More than 85 per cent of all metal castings are poured in sand moulds, the balance are made in ceramic shell or metal moulds.

2.4. CORE

A **core** *is a body made of refractory material (sand or metal, metal cores being less frequently used), which is set into the prepared mould before closing and pouring it, for forming the holes, recesses, projections, undercuts and internal cavities.*

- The cores are subject to much more severe thermal and mechanical effects than the moulds, because they are surrounded on all sides (except for the ends) by molten metal. Consequently core sands should meet more stringent requirements.
- *Refractoriness or thermal stability of core can be increased by giving a thin coating of graphite or similar material to the surface of the core.*

Characteristics of cores. A core should have the following *characteristics* :

(*i*) Core must have dry strength or be capable of being handled after drying. It must be able to withstand force of molten metal.

(*ii*) It should produce minimum amount of gas when in contact with molten metal.

(*iii*) It should be highly refractory, and able to withstand high temperatures of molten metal.

(*iv*) It should be more permeable than the mould itself and for this reason, coarse sand with large grain size mixed with *molasses* is used for core making.

(*v*) It should be capable of collapsing shortly after the molten metal has solidified around the core. Collapsibility provides freeness in the contraction of metal.

(*vi*) The surface of core should be smooth so as to provide a good finish to the casting.

2.4.1. Core Making

- Cores are made in simple wooden, metallic or plastic core boxes. These core boxes are part of the pattern equipment for the castings. The complicated shapes may require support of sand or metal formers until these are baked.

- The *simple method of core making is similar to that of mould making.* The sand mixture is rammed into the core box with a wooden rammer. Sometimes the cores may need reinforcement with wire or nails in order to provide internal support so that they may not collapse while handling. The core-sand mixture is rammed by hand or pneumatic rammers. Venting and other necessary operations are performed during construction of the core.

- For *production work, machines are used for core makings* where core-sand mixture is rammed by *jolting, squeezing or blowing by means of suitable machines.* The most common core making machine is the *core blower.* Venting, reinforcing and other necessary operations are performed by hand during core construction.

Sometimes cores may be made by *extruding a core-sand mixture through a suitable die opening and called "stock cores"* which are of symmetrical cross-section.

The cores are removed from the core box placed in metal trays and are baked in an oven at a suitable temperature varying from 150°C to 400°C for the required duration of time. The source of heat may be the burning of gas, oil, coke, or electric heating.

2.4.2. Types of Cores

A *core is a specially designed shape employed to take the place of metal in a mould.*
The cores may be *classified* as follows :

I. According to the state of the core :

(*i*) Green sand cores.

(*ii*) Dry sand cores.

(*iii*) Oil sand cores.

(*iv*) Loam cores.

(*v*) Metal cores.

(*i*) Green sand cores :

- These are made from ordinary moulding sand, mixed with floor-sand, thoroughly vented and not too damp.

- They are restricted to *simple shapes,* reinforced with substantial core irons for handling, and *not dried before using.*

(*ii*) **Dry sand cores :**

- These cores consist of moulding sand with controlled conditions of such opening materials as horse manure, sawdust or chopped straw, blended together in a mill with either water, clay water, molasses or any other suitable binding agent. Ample core irons are used for strengthening.
- The interiors of large cores are filled with coke to facilitate venting, and to overcome contraction strains.
- They are baked until perfectly dry.

(*iii*) **Oil sand cores :**

- They consist chiefly of sea shore or other silica sands, to which has been added a binding medium ; a few typical binders being linseed oil, resin, molasses and cereal flour.
- All cores of this type are baked before handling.

(*iv*) **Loam cores :**

- Loam cores are perforated cast-iron or steel barrels on to which has been wound straw or a hayband and then coated with loam. The whole is then baked perfectly dry.

(*v*) **Metal cores :**

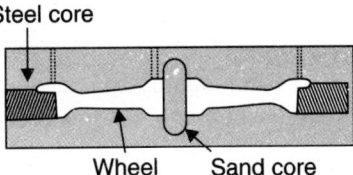

Fig. 2.16. Steel cores in a sand casting.

- These cores are usually made of steel (see Fig. 2.16) and are mostly used in the making of non-ferrous castings, acting as densers, and they also impart a fine finish.
- They are also used for the production of iron castings, they produce a *white hard skin*.

Note. Owing to contraction stresses, means should be provided whereby the cores can be released at a reasonable time after the metal in the casting has set.

II. According to the position of the core in the mould :

(*i*) Horizontal core.   (*ii*) Vertical core.

(*iii*) Balanced core.   (*iv*) Cover core.

(*v*) Hanging core.   (*vi*) Wing core.

(*i*) **Horizontal core :** Refer to Fig. 2.17 (*i*) :

- The core is usually cylindrical in form and is laid horizontally at the parting line of the mould.
- The ends of the core rest in the seats provided by the core prints on the pattern.

(*ii*) **Vertical core :** Refer to Fig. 2.17 (*ii*) :

- The core is placed vertically in the mould.
- Usually top and bottom of the core are provided with a taper, but the amount of taper on the top is greater than that at the bottom.

(*iii*) **Balanced core :** Refer to Fig. 2.17 (*iii*) :

- This core is similar to horizontal core, but it is supported at one end only.
- The core print in such cases should be large enough to give proper bearing to the core.

(*iv*), (*v*) **Cover core and hanging core :** Refer to Fig. 2.17 (*iv*) and (*v*) :

- The cover core, as shown in Fig. 2.17 (*iv*), is used when the entire pattern is rammed in the drag and the core is required to be supported from the top of the mould. This type of core usually requires a hole through the upper part to permit the metal to reach the mould.

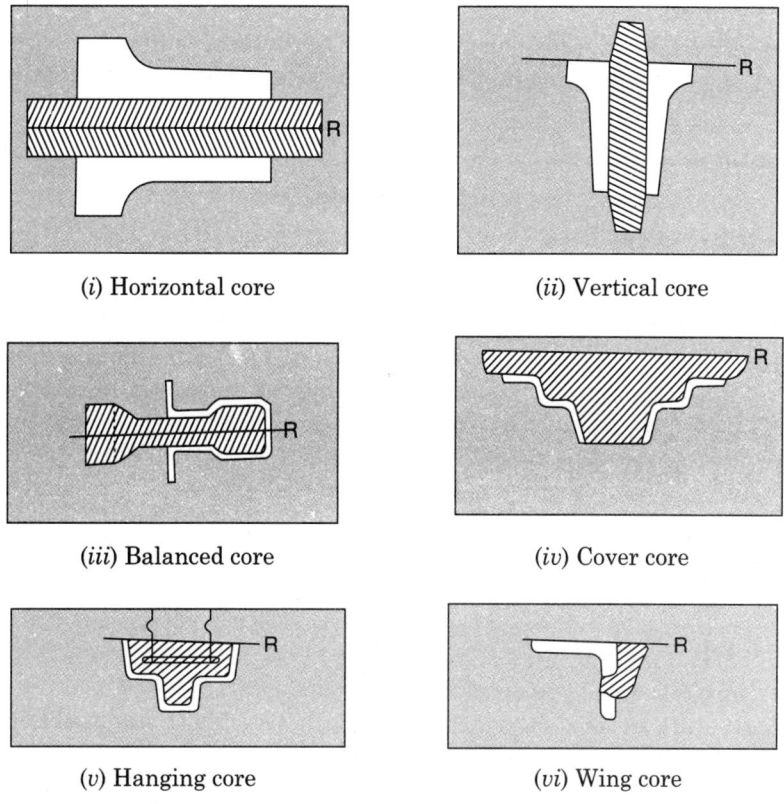

(i) Horizontal core   (ii) Vertical core

(iii) Balanced core   (iv) Cover core

(v) Hanging core   (vi) Wing core

Fig. 2.17. Types of cores.

- If the core hangs from the cope and does not have any support at the bottom in the drag, then it is called *hanging core* [Fig. 2.17 (v)].

(vi) **Wing core :** Refer to Fig. 2.17 (vi) :

- A wing core is used when a hole or recess is to be obtained in casting either above or below the parting line. In this case, the side of the core point is given sufficient amount of taper so that the core can be placed readily in the mould.
- The core is sometimes designated by other names such as *tail core, drop core, chair core* and *saddle core* according to its shape and position in the mould.

Core sand :

- Core sand is a variety of silica sand. Rock sand, river sand and sea shore sand, commonly known as sharp sand, are generally used for making of cores, chiefly because they are *capable of withstanding high temperatures, and resisting the penetrating action of the molten metal. They have in addition, high porosity, together with good permeability.*
- Having no natural bond, these sands are mixed with a suitable binder of which there are several kinds in the form of creams, oils and resins. These binders are burnt out by the time the casting is cold making the core friable and easy to remove.

2.4.3. Core Prints

For supporting the cores in the mould cavity an impression in the form of a recess is made in the mould with the help of a projection on pattern. This projection is known as **core print.** Core prints are of the following types :

1. Horizontal core print :

It produces seats for horizontal core in the mould.

2. Vertical core print :

It produces seats to support a vertical core in the mould.

3. Balanced core print :

It produces a single seat on one side of the mould and the core remains partly in this formed seat and partly in the mould cavities, the two portions balancing each other. The hanging portion of the core may be supported on chaplets.

4. Cover core print :

It forms seat to support a cover core.

5. Wing core print :

It is used to form a seat for a wing core.

2.4.4. Core Box

- **A core box** *is a type of box used for the production of sand cores.* Core boxes are used in foundry work to form shapes in sand, called cores, which are used in connection with moulds, when *holes or internal shapes are required.*

Various methods of construction are used depending on the shape or size of the core, and its removal from the box after it is made. The inside of a core box must be cut out to form the exact shape of the hollow part required in the casting, plus the extensions for locating in the prints.

- The core boxes are filled with sand which is made firm with ramming and removed by opening the box at a centre joint, or taking it away from the core in various directions by a number of joints.

- If a plain hole is required in a casting, the core is a cylindrical piece of sand, the length of the hole plus the print portion ; this core can be made in a **simple core box** cut from *two pieces of wood held together with dowels on the centre joint.* Fig. 2.18 (*i*).

- Another method of building up core boxes is to make the joints in such positions as will enable straight cuts to be made, *e.g.,* **piston core box,** Fig. 2.18 (*ii*). Each piece of the box is made to thickness of the sections at which the diameter of the piston alters, and screwed together. After making out of the joint on the box, the pieces are taken apart and cut through, glued and screwed back in position.

- **Stickles** *are frequently used with core boxes to obtain shapes of large works.* They work from a *guide held against a parallel side of the core box.* Fig. 2.18 (*iii*). A considerable amount of time in building up and shaping is saved by their use.

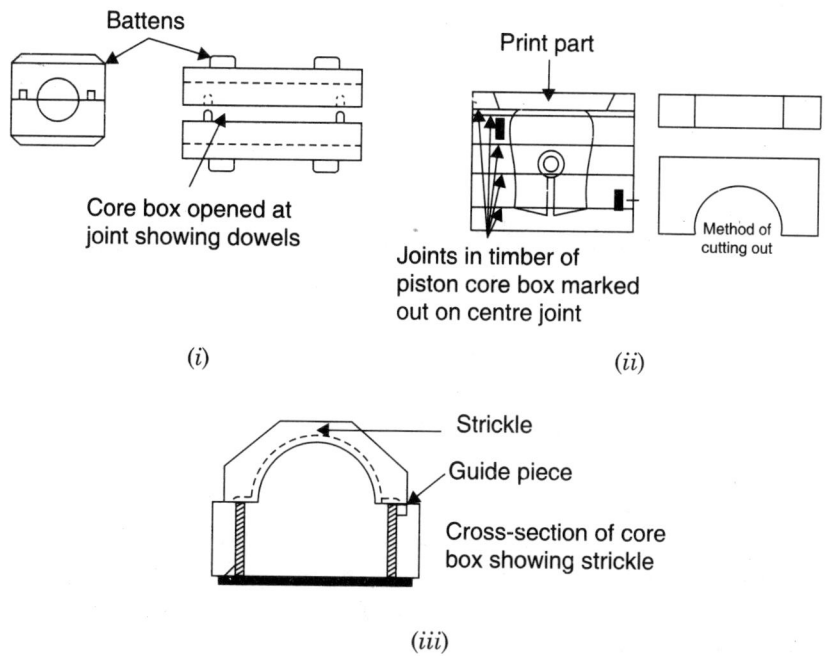

Fig. 2.18. Core boxes : (*i*) Simple box for cylindrical core ; (*ii*) Piston core box, arranged for straight cuts ; (*iii*) Strickle used to shape large cores.

2.5. MOULDING SAND

In a foundry shop sand is the principal moulding material and is used for all types of castings. The moulding sand possesses all the properties which are vital for foundry purpose and is used time and again.

2.5.1. Properties of Moulding Sand

1. Permeability :

- It is the property to allow gases to escape easily from the mould.
- Higher the silt content of sand, the lower is gas permeability. If the mould is rammed too hard, its permeability will decrease and vice versa.
- It is measured in number such as 60, 80, 100, 120, etc.

2. Strength or cohesiveness :

- It is defined as the *property of holding together of sand grains.*
- A moulding sand should have ample strength so that mould does not collapse or get partially destroyed during conveying, turning over or closing.
- The strength of the moulding sand *grows with density, clay content of the mix* and *decreased size of sand grains.* Thus as the *strength of the moulding sand increases, its porosity decreases.*

3. Refractiveness :

- *It is the ability of the moulding sand mixture to withstand the heat of melt without showing any sign of softening or fusion.*
- *It increases with the grain size of sand and its content and with the diminished amount of impurities and slit.*

4. Plasticity or flowability :

It should be of plastic nature so that it can easily take any desired shapes.

5. Collapsibility :

- *This is the ability of the moulding sand mixture to decrease in volume to some extent under the compressive forces developed by the shrinkage of metal during freezing and subsequent cooling.*
- This property permits the moulding sand to collapse easily during shake out and permits the core to collapse easily during its knock out from the cooled casting.
- Lack of collapsibility in the moulding sand and core may result in the formation of cracks in the casting.
- This property depends on the amount of quartz sand and binders and their type.

6. Adhesiveness :

- This is the property of sand mixture to *adhere* to another body.
- The moulding sand should cling to the sides of the moulding boxes so that it does not fall out when the flasks are lifted and turned over.
- This property depends on the type and amount of binder used in sand mix.

7. Coefficient of expansion :

The sand should have low co-efficient of expansion.

8. Chemical resistivity :

The sand should not chemically react or combine with molten metal.

2.5.2. Types of Moulding Sand

The moulding sands are *classified* as follows :

I. *According to their clay bonding material :*

1. Natural sand :

It contains sufficient amount of binding clay and, therefore, no more binder is required to be added.

2. Synthetic sand :

It is one which is artificially compounded by mixing sand, and selected type of clay binders, etc. These sands have the following *advantages :*

- Lower cost in large volume.
- Widespread availability.
- The possibility of sand reclamation and reuse.

II. *According to their use :*

1. Green sand :

- The sand in its natural or moist state is called *green* sand.
- It is a mixture of silica sand with 20 to 30% clay, having total amount of water from 6 to 10%.
- The green sand moulds are used for small size castings of ferrous and non-ferrous metals.

2. Dry sand :

- When the moisture from the green sand is evaporated by drying or baking, after the mould is made is called *dry sand* mould.
- The dry sand moulds have greater strength, rigidity and thermal stability. The dry sand moulds are used for large and heavy castings.

3. Loam sand :

- The loam sand consists of as high as 50% of clay contents.
- It is used for loam moulding of *large grey-iron castings.*

4. Facing sand :

- A sand used for facing of the mould is called *facing sand.*
- Since it comes in contact with molten metal when poured, therefore it must possess high strength and refractiveness.

5. Parting sand :

- Parting sand is purely clay-free silica sand which is sprinkled on the pattern and the parting surfaces of the mould so that the sand mass of cope and drag separate without clinging and do not stick to the pattern.

6. Backing or flour sand :

- A sand used to back up the facing sand not used next to the pattern, is called *backing sand.*
- Because of its black colour, it is sometimes called *black sand.*

7. Core sand :

- A sand used for the preparation of the cores is called *core sand.*
- It is sometimes called *oil sand.*

2.5.3. Composition of the Green Sand

(*i*) Silica	....	up to 75 per cent
(*ii*) Clay	....	8 to 15 per cent
(*iii*) Bentonite	....	2 to 5 per cent
(*iv*) Coal dust	....	5 to 10 per cent
(*v*) Water	....	7 to 8 per cent

- A large proportion of "*silica*" gives a refractory sand, but if in excess destroys its cohesiveness. This would be termed as open, weak or sharp sand, and indicates that it has good refractoriness and permeability but no cohesive or bonding qualities.
- The presence of "*alumina*" gives the sand its clay content which acts as a binder, but if in excess it destroys the permeability of the sand ; this condition is known as *close* or *strong.*
  - Alumina vitrifies at high temperature, and if a high percentage is in the sand it will lower its bonding efficiency for any subsequent moulding.
- "*Metallic oxides*" and "*lime*", if in excess, impair the refractory qualities of the sand, and increase its fusibility. This condition is to be avoided, as the castings would have a very rough surface, indicating poor stripping quality.

Following points are worth noting :

- Sands may be blended together. An open sand is added to a strong sand to make the resultant mixture more refractory or to increase its permeability, or a close sand may be mixed with a weak sand to give more cohesion.
- Natural sands used alone are not suited for all the metals and alloys with their varying characteristics. Admixtures must be made such as :
  - Coal-dust, for cast-iron in green-sand moulds ;
  - Cow-hair, horse-manure, for very heavy work in iron and bronze ;

— Sulphur and boric acid, for magnesium castings ;

— Bentonite, for general work in green sand ;

— Various oils and creams for core-making.

For mixing and blending sands, various types of machines are used, mainly of the pan-mill and centrifugal pattern.

2.5.4. Sand Testing

In order to meet the required level of accuracy and surface finish of castings the moulding sands should be of proper quality. Proper quality of the moulding sand results in sound castings that will decrease the cost per unit and increase the production.

Periodic tests are necessary to determine the essential qualities of foundry sand. The properties of the moulding sand depend upon shape, size, composition and distribution of sand grain. There are standard tests to be used which are given in relevant Indian Standards and that of other foundry societies.

Tests are conducted on a sample of standard sand. The moulding sand should be prepared exactly as is done in the shop on the standard equipment and then carefully enclosed in a closed container to safeguard its moisture content.

B.I.S. has recommended the following tests :

1. Fineness test or Sand grain size test.

2. Permeability test.

3. Strength test.

4. Moisture content test.

5. Clay content test.

6. Mould hardness test.

1. **Fineness test.** This test determines the size of grains and the distribution of grains of different sizes in the moulding sand. *This test is performed on completely dry and clay free sand.*
Procedure. To carry out this test the dried clay-free sand grains are placed on the top sieve of a sieve shaker which contains a series of sieves one upon the other with gradually decreasing mesh sizes. The sieves are shaken continuously for a period of 15 minutes. After this shaking operation, the sieves are taken apart and the sand left over on each of the sieve is carefully weighed. Each weight is converted to a percentage basis. Each percentage is multiplied by a weighting factor and these are added to get the sum of products. Then the grain fineness number (GFN) is expressed as :

$$\text{G.F.N.} = \frac{\text{Sum of products } (\Sigma M_i f_i)}{\text{Sum of percentages of sand retained on each sieve and pan } (\Sigma f_i)}$$

(M_i = Multiplying factor for i_{th} sieve ; f_i = Amount of sand retained on i_{th} sieve)

According to AFS (American's Foundrymen's Society), the various sieve mesh numbers and the corresponding multiplying (weightage) factors are as follows :

Sieve number :	6	12	20	30	40	50	70	100	140	200	270	Pan
Weightage factor :		3	5	10	20	30	40	50	70	100	140	200	300

Example. A typical example is given in the table on next page ; this type of sand is normally used in malleable and grey iron foundries for castings which weigh more than 20 kg.

Sieve number	Retained sample f_i : (g)	Retained percentage P_i	Multiplying factor M_i	$M_i \times f_i$	$M_i \times P_i$
6	—	—	—		
12	—	—	—		
20	—	—	—		
30	—	—	—		
40	2.495	5	30	74.85	150
50	13.972	28	40	558.88	1120
70	23.952	48	50	1197.60	2400
100	6.986	14	70	489.02	980
140	2.495	5	100	249.50	500
200	—	—	—		
270	—	—	—		
Pan	—	—	—		
Total	49.900	100		2569.85	5150

$$\therefore \qquad \text{GFN} = \frac{2569.85}{49.90} = \mathbf{51.50}$$

or, $$\qquad \text{GFN} = \frac{5150}{100} = \mathbf{51.50.}$$

- By the above definition the fineness number is the average grain size and corresponds to a *sieve number through which all the sand grains would pass through, if they were all of the same size.*
- This is a very convenient way of describing the grain size and its value can be *expected between 40 and 220 for those sands used by most of the foundries.*
- Although the properties of sand depend on both the *grain size* and *grain size distribution, GFN is a very convenient way of finding the sand properties since it takes both into account.*

2. **Permeability test.** *Permeability* (or porosity of the moulding sand) *is the measure of its ability to permit air to flow through it.* It is measured in terms of "Permeability number".

Procedure. To carry out this test, a test specimen of moulding sand (50.8 mm dia × 50.8 mm long) is placed in a specimen tube. Time taken for 2000 cm³ of air at a pressure of 980 Pa (10 g/cm²) to pass through the specimen is noted. Then, the permeability number, P is given as :

$$P = \frac{V \times H}{p \times A \times t} \qquad \qquad ...(2.1)$$

where, V = Volume of air = 2000 cm³

H = Height of the sand specimen = 50.8 mm = 5.08 cm

p = Air pressure = 10 g/cm²

A = Cross-sectional area of sand specimen = $\frac{\pi}{4} \times (5.08)^2 = 20.268 \text{ cm}^2$

t = Time in minutes for complete air to pass through.

Substituting the above standard values into the expression, we get,

$$P = \frac{2000 \times 5.08}{p \times 20.268 \times t} = \frac{501.28}{p.t}$$

- This permeability number is a relative number. It does not necessarily tell the permeability of a would made with the same sand which depends on the compactness of the sand.
- The permeability test is conducted for two types of sands :
  - Green permeability is the permeability of the *green sand.*
  - Dry permeability is the permeability of the *moulding sand,* dried at 105 to 110°C *to remove the moisture completely.*

3. **Strength test.** The strength of moulding sands can be carried out on the universal sand strength testing machine. The strength can be measured in compression, shear and tension.

The sands that could be tested are green sand, dry sand or core sand. The compression test and shear test involve *standard cylindrical specimen that was used for the permeability test.*

Green compression strength. It refers to the stress required to rupture the sand specimen under compressive loading. The sand specimen is taken out of the specimen tube and is immediately (any delay causes the drying of the sample which increases the strength) put on the strength testing machine and force required to cause the compressor failure is determined.

- The green strength of sands is generally in the range of 30 to 160 *kPa.*

Green shear strength. With a sand similar to the above test, a different adapter is fitted in the universal machine so that loading now be made for the shearing of the sand sample. The stress required to shear the specimen along the axis is then represented as the green shear strength.

- The green shear strength may vary from 10 to 50 *kPa.*

Dry strength. The test similar to the above can also be carried with the standard specimens dried between 105 and 110°C for 2 hours. Since the strength greatly increases with drying, it may be necessary to apply larger stresses than the previous tests.

- The range of dry compression strengths found in moulding sands is from 140 to 1800 *kPa,* depending on the sand sample.

4. **Moisture content test.** Moisture content may be determined *by loss of weight, after evaporation.*

Procedure :

- To test the moisture of a moulding sand a carefully weighted test sample of 50 g is dried at a temperature of 105°C to 110°C for 2 hours by which time all the moisture in the sand would have been evaporated. The sample is then weighted. The weight difference in grams divided by the weight of sample (50 g) and multiplied by 100 gives the percentage of moisture contained in the moulding sand.
- *Alternatively* a **'moisture teller'** can also be used for measuring the moisture content. In this arrangement the sand is dried by suspending the sample on a fine metallic screen and allowing hot air to flow through the sample. This method of drying completes the removal of moisture in a matter of minutes compared to 2 hours as in the earlier method.

Another **'moisture teller'** utilises calcium carbide to measure the moisture content. A measured amount of calcium carbide (a little more than that actually required for complete reaction) in a container along with a separate cap consisting of measured quantity of moulding sand is kept in the moisture teller (Sand and calcium carbide should not come into contact with each other). The apparatus is then shaken vigorously such that following reaction takes place :

$$CaC_2 + 2H_2O \longrightarrow C_2H_2 + Ca(OH)_2$$

(Calcium carbide) (Acetylene)

The amount of C_2H_2 produced is proportional to the content of the moisture in the sand. The moisture content is directly measured from a calibrated scale on the instrument.

5. **Clay content test.** The clay content of moulding sand is determined by dissolving or washing it off the sand.

Procedure :

(i) A 50 g sample of moulding sand is dried at 105 to 110°C. This dried sample is taken in a one litre glass flask and added with 475 ml of distilled water and 25 ml of a 1% NaOH solution (NaOH 25 g per litre). This sample is thoroughly stirred.

(ii) After the stirring, for a period of 5 minutes the sample is diluted with fresh water upto a 150 mm graduation mark and sample is left undisturbed for 10 minutes to settle. The sand settles at the bottom and the clay particles washed from the sand would be floating in water. 125 mm of this water is siphoned off the flask and it is again topped to the same level and allowed to settle for 5 minutes.

(iii) The above operation is repeated till the water above the sand becomes clear, which is an indication that all the clay in the moulding sand has been removed.

(iv) The sand is removed from the flask and dried by heating.

The difference in weight of dried sand gives the clay content.

6. **Mould hardness test.** The mould hardness is measured by a method similar to the Brinell hardness test.

Procedure. A spring loaded steel ball with a mass of 0.9 kg is indented into the standard sand specimen prepared. The depth of indentation can be directly measured on the scale which shows units 0 to 100. When no penetration occurs, then it is a mould hardness of 100 and when it sinks completely, the reading is zero indicating a very soft mould.

2.6. MELTING EQUIPMENT

The main types of furnaces used in foundries for melting various varieties of ferrous and non-ferrous metals and alloys are enumerated and described below :

1. Crucible furnace.
2. Reverberatory or air furnace.
3. Open hearth furnace.
4. Electric furnace.
5. Cupola furnace.

2.6.1. Crucible Furnace

Refer to Fig. 2.19.

A crucible furnace is most suited for *small foundries* and can be designed for melting any of the metals. It consists of the following two main types :

(i) Pit type furnace.

(ii) Tilting type furnace.

(i) Pit type crucible furnace : Refer to Fig. 2.19 (i).

● It is built to suit the type of metal to be melted.

● These are fixed wholly or party in the ground from which the crucible must be lifted when the metal is ready.

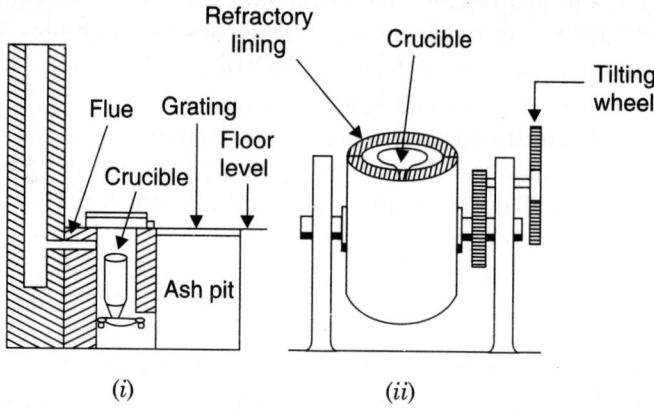

Fig. 2.19. Crucible furnace.

- Here a crucible (a heat resisting pot for metal melting and made of fire clay mixed with coke dust or graphite) is placed in a pit in the floor. The furnace is usually fired with sufficient coke being packed round and above the crucible pots to melt and superheat the charge without re-coking. The natural draught provided by tall chimney is controlled by means of loose brick or damper at the foot of the stack.

(ii) Tilting type crucible furnace : Refer to Fig. 2.19 (ii).

- This type of furnace is built above the ground level, and contains a firmly fixed crucible.
- The furnace is fired with coke, oil or gas and the forced draught is used.
- When the metal charge is ready for pouring, the whole furnace is tilted and the crucible emptied by operating a geared trunnion.
- For the metals of high melting points, clay or plumbago crucibles are used ; for the low-melting-point metals, such as zinc-base or aluminium, cast iron or steel cucibles are suitable.

2.6.2. Reverberatory or Air Furnace

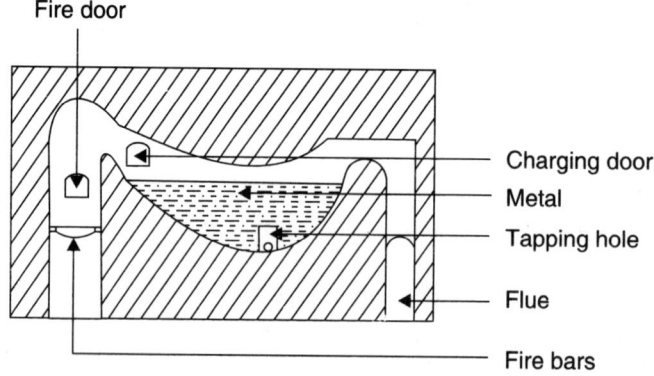

Fig. 2.20. Reverberatory or air furnace.

Refer to Fig. 2.20.

- This is used for *melting in one heat large quantities of metal,* those most suited being all grades of cast-iron and the alloys of brasses and bronzes.

This type of furnace is also used for the production of wrought iron, and is then situated not in a foundry, but near forge or rolling mill, and is known as a *puddling furnace.*

- It may have either a sloping roof, or a double arched roof which forms a dip in the centre. A chimney is provided at one end and a fire grate or burners at the other end. A hearth or well is provided in the centre for holding the metal.
- It employs natural draught, which is controlled by dampers.
- The fuel can be either small lumpy coal, which is used on the fire grate, or powdered fuel, which is supplied through burners. The object is to create a long flame which reverberates or strikes back from the furnace roof on to the metal to be melted in the hearth.

2.6.3. Open Hearth Furnace

Refer to Fig. 2.21.

- An open hearth furnace is used chiefly for the *production of steel and for refining purposes.*

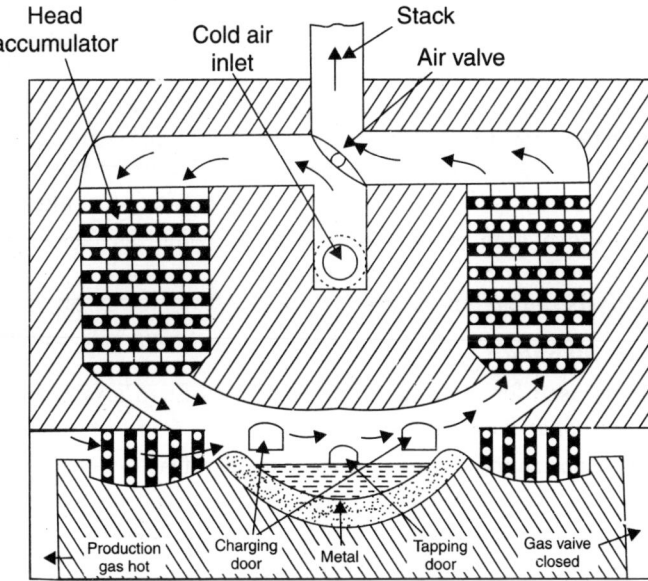

Fig. 2.21. Open hearth furnace.

- Gas and heated air, admitted through the ports on the left, burn above the hearth. The hot, spent gases heat a brickwork chamber, before reaching the chimney stuck. After about twenty minutes the direction of air and gas flow is reversed, so that the cold air passes through the newly-heated brick chamber, while that on the left is reheated in preparation for the next cycle.

2.6.4. Electric Furnace

This is used especially where rigid control over temperature and analysis is required. It is *suitable for all types of metals and alloys.*

The electric furnaces are classified as follows :

   I. *Arc type furnaces :*
      (*i*) Direct arc.
      (*ii*) Indirect arc.
  II. *Induction furnaces.*

I. *Arc type furnace :*

(*i*) **Direct-arc furnace** : Refer to Fig. 2.22.

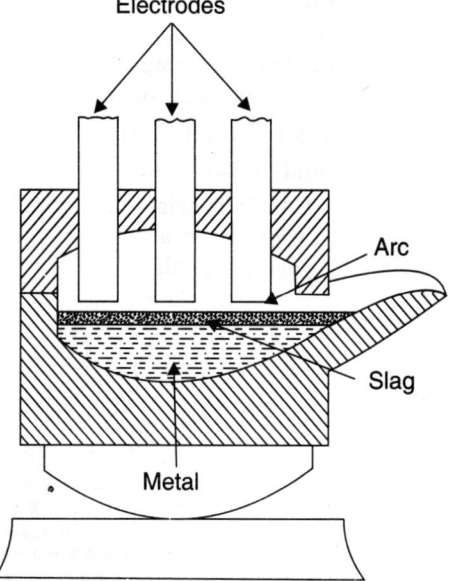

- It consists of a round, bowl-shaped carbon hearth with a domeshaped roof supporting one or more carbon electrodes through which passes the current which strikes arcs with the metal in the hearth, thus giving heat direct to the metal.

- This type of furnace can be either stationary or made to tilt. The roof is usually so made that it can be removed for charging purposes.

- The capacity of these furnaces for production work varies from 3 to 10 tonnes. These are best suited for laboratory work where very small quantity of a few kg is needed for research work.

- These furnaces give *high melting rate, high pouring temperature and excellent-control of metal analysis and temperature.*

Fig. 2.22. Direct arc furnace.

(*ii*) **Indirect-arc furnace :**

- This furnace is used for melting *all types of metallic alloys but especially useful in the production of copper-base alloys.*

- It consists of a horizontal cylinder lined with a refractory material with two electrodes on the horizontal axis. An arc is struck between the electrodes in the centre of the furnace. The arc does not come in contact with the metal to be melted, the heat being given to the charge by radiation from the arc and reflection from the walls of the furnace. The furnace is designed to give a rocking motion as the melting proceeds, thus quickening up the melt by distributing the heat more rapidly. The charging, tapping and slagging are done through an opening in the side of the furnace.

II. *Induction furnace :*

- An induction furnace is a tilting furnace used chiefly *for the melting of non-ferrous metals. Heat is generated by the resistance offered to an induced current set up within the metal in the furnace.*

- The design of the furnace is such that a small channel is formed inside at its base ; this channel is filled with metal, which should never be allowed to solidify owing to the amount of damage it would do to the lining.

- *When working, an alternating current is supplied to the primary coil of a transformer which is built with the furnace. This induces a current to pass through the liquid metal in the channel which acts as a secondary coil of the transformer. The forces set up by the current in the secondary circuit induce the metal in the channel to heat up and circulate through the bath of metal.*

2.6.5. Cupola Furnace

This furnace (obtained in different sizes) is most commonly used for *melting and refining pig iron along with cast iron and steel scraps.* Besides iron castings, cupola can be used for melting some copper base alloys also.

Fuel for cupola is generally a *good grade low-sulphur coke, anthracite coal or carbon briquettes.*

Cupola can be employed to duplexing and triplexing operations for making steels. Duplexing and triplexing melting operations employ two and three furnaces respectively.

2.6.5.1. Construction and working

Construction : Refer to Fig. 2.23.

— Cupola consists of a cylindrical steel shell with its interior lined with heat resisting fire bricks. It is a vertical shaft furnace (similar to a blast furnace in principle), into which raw materials and fuel are charged at the top.

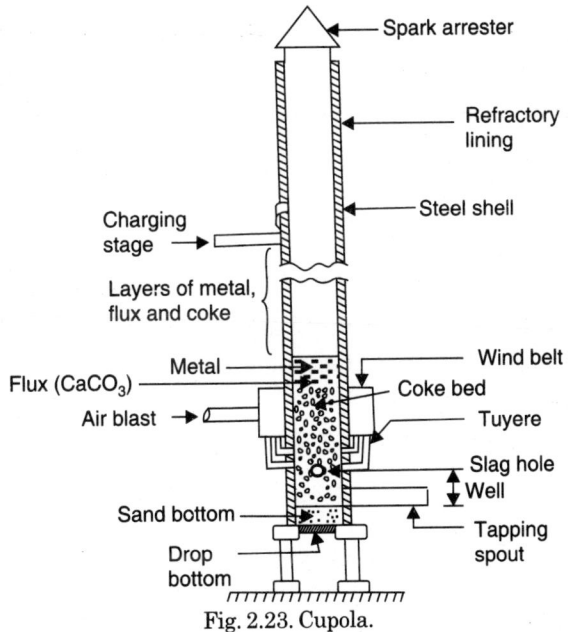

Fig. 2.23. Cupola.

— Air for combustion of fuel is introduced through one or more rows of tuyeres a short distance above the bottom. Since the cupola is only concerned with the *melting of metal and not with the reduction of ores as in the blast furnace it is considerably smaller than a blast-furnace of the same output.*

Its diameter varies from 1 to 2 metres with a height of 4 to 5 times the diameter.

Working :

— In a cupola, the first operation is to light the fire at the bottom. When the fire is burning strongly, coke is added gradually till the level above tuyeres is about 0.6 metres. This coke serves as a bed for the alternate charges of metal, flux and coke which follow.

— When the shaft of the cupola is filled level with the charging door the blast is put on and the combustion of the coke near the tuyeres increases rapidly until a very intense heat is attained. The gases of combustion move upwards and pass on a portion of the heat to the metal and coke waiting to descend.

— In 5 to 10 minutes the first charge of metal starts melting and trickles down through the coke and finally collects at the bottom of the cupola. When an adequate quantity (say 1 or 2 tonnes) has accumulated the plug of clay called *'bout'* is removed from the tap hole and metal allowed to run into the ladle. The temperature of tapping metal is 1200–1400°C.

— After melting a number of charges as per requirements the bed coke is removed through a drop-bottom door and quenched with water so as to be available for use the next day.

The **"*fluxes*"** are added in the charge to *remove the oxides and other impurities present in the metal*. The flux most commonly used is lime ($CaCO_3$) in proportion of about 2 to 4% of the metal charge. Some of the other fluxes that may be used are dolomite, sodium carbonate and calcium carbide. The flux is expected to react with the oxides and form compounds which have low melting point and are also *lighter*. As a result, the *molten slag tends to float on the metal pool and thus, can very easily be separated.*

2.6.5.2. Hot blast cupola

A hot blast cupola differs from the ordinary cupola in the respect that it utilises a *preheated blast of air* for combustion purposes. The air supply is preheated to a temperature of 200 to 400°C with the help of hot gases coming out of the stock or by a separate heat input.

- The use of preheated or air blast offers the following **advantages :**

    (*i*) Higher temperatures can be obtained.

    (*ii*) Melting rate is fast.

    (*iii*) Improved combustion.

    (*iv*) Coke is saved by 20 to 25 percent.

    (*v*) Cupola operation is improved.

    (*vi*) Saving in flux.

    (*vii*) Reduced oxidation losses of Fe, Si and Mn.

    (*viii*) The composition of molten iron is more uniform.

    (*ix*) Sulphur pick-up is less.

    (*x*) A higher proportion of steel scrap can be added in the cupola charge.

    (*xi*) Even a low grade coke may be used.

Limitations :

(*i*) The maintenance problems are increased.

(*ii*) Additional cost of equipment required for preheating the air blast.

Uses. Owing to the additional equipment and extra care needed for operation, the hot blast cupolas are *used only in shops that require large amounts of metal to be melted on a continuous basis.*

2.6.5.3. Advantages and limitations of cupola

Advantages :

1. Simplicity of operation.

2. Low initial cost as compared to other furnace of same capacity.

3. Continuity of production.

4. Economy of working.

5. High degree of efficiency.

6. Increased output.

7. Less floor space requirements as compared to those of other furnaces of same capacity.

Limitations:

1. It is difficult to maintain close temperature control.

2. Since molten iron and coke come in contact with each other, certain elements such as Si, Mn are *lost* while others are *picked up*. Consequently the final analysis of molten iron changes.

2.6.5.4. Efficiency of cupola

The percentage efficiency of cupola

$$= \frac{\text{Heat utilised in preheating, melting and superheating}}{\text{Heat input due to heat in the coke + heat evolved due to oxidation of C, Fe, Si and Mn + Heat in the air blast}} \times 100$$

- 'Preheating zone' starts from above the melting zone and extends upto the bottom of the charging door.
- 'Melting zone' starts from the first layer of metal charge above the coke bed and extends upto a height of 90 cm or less.
- 'Superheating zone' is situated normally 15 cm to 30 cm above the top of the tuyeres.

The term heat in the air blast is more purposeful with regard to a hot blast cupola.

The *efficiency of a cupola varies from 30 to 50 percent.*

— When a cupola is examined after dumping the remains through the bottom door, a groove can be seen in the cupola lining. This groove shows the location of high temperature zone. The coke bed charge should reach the height of this groove above the sand bottom.

Example 2.1. *A cupola of 80 cm diameter has melting ratio of 8 : 1. Calculate :*

(i) *Air required for complete combustion.*

(ii) *Air required to melt 600 kg of iron at this ratio.*

(iii) *The coke required to melt 600 kg of iron if melting ratio is 10 : 1 and 8 : 1.*

(iv) *Air required to melt 600 kg of iron if the melting ratio is 10 : 1.*

Assume the weight of regular charge of coke as 35 kg.

Solution.

(i) **Air required for complete combustion :**

With a melting ratio of 8 : 1, 35 kg of coke will melt $35 \times 8 = 280$ kg of iron.

Since coke is 88 percent carbon, 35 kg of coke contains

$$\frac{35 \times 88}{100} = 30.8 \text{ kg of carbon.}$$

Now,
$$C \quad + \quad O_2 \quad \longrightarrow \quad CO_2$$
$$12 \qquad\qquad 32 \qquad\qquad\quad 44$$
$$1 \qquad\qquad \frac{8}{3} \qquad\qquad\quad \frac{11}{3}$$

i.e.,   To burn 1 kg of carbon $\dfrac{8}{3}$ kg (*i.e.,* 2.67 kg) of oxygen is required.

Further since air contains about 23.2% of oxygen by weight, the weight of air needed to produce 2.67 kg of oxygen is $= \dfrac{2.67}{0.232} = 11.5$ kg of air.

If air weighs 1.28 kg/m^3, the volume of air required for *every one kg of carbon* $= \dfrac{11.5}{1.28}$ m^3 of air = 8.98 m^3 of air.

Hence, volume of air required *for complete combustion* $= 30.8 \times 8.98 =$ **276.58 m^3.**   (**Ans.**)

(*ii*) **Air required to melt 600 kg of iron (at 8 : 1 melting ratio) :**

Air required to melt 600 kg of iron at melting ratio of 8 : 1

$$= \dfrac{276.58}{280} \times 600 = \textbf{592.5 m}^3.   (\textbf{Ans.})$$

(*iii*) **Coke required to melt 600 kg of iron :**

Coke required to melt 600 kg of iron at 10 : 1 and 8 : 1 melting ratio $= \dfrac{600}{10} =$ **60 kg** and

$\dfrac{600}{8} =$ **75 kg** respectively. (**Ans.**)

(*iv*) **Air required to melt 600 kg of iron at 10 : 1 melting ratio :**

Air required to melt 600 kg of iron at 10 : 1 melting ratio

$$= \dfrac{592.5}{75} \times 60 = \textbf{474 m}^3.   (\textbf{Ans.})$$

2.6.5.5. Cupola charge calculations

In a foundry it is very important to know the final composition of the metal being obtained, so as to control it properly. In the final analysis, out of the various elements, carbon, silicon, manganese and sulphur are the relevant ones ; these are discussed below :

- *Carbon.* As the charge comes through the coke bed, some amount of carbon is *picked up by the metal* depending on the temperature and the time when the metal is in contact with the coke. It may be reasonable to assume a ***pick up of 0.15% carbon.***
- *Silicon.* Silicon is likely to get oxidised in the cupola, and therefore, a **loss of 10% of the** *total silicon contained in the charge is normal. Under the worst conditions, it may go as high as 30%.*
  — If the silicon content in the charge is not high, *extra silicon can be added by inocu-lating the metal in the ladle with ferrosilicon.*
- *Manganese.* This element is also likely to be lost in the melting process. The *loss could be of the order of* **15 to 20%.**
  — *Loss of manganese in the final analysis, can be made up by the addition of ferromanganese.*
- *Sulphur.* This element is also likely to be *picked up* from coke during melting. The pick-up depends on the sulphur content of the coke, but a *reasonable estimate* could be **0.03 to 0.05%.**

The following **steps** *are involved in the cupola charge calculations :*

1. Fix arbitrarily the compositions of various constituents of the cupola charge.
2. Knowing the percentage of elements like carbon, silicon, manganese, phosphorus and sulphur present in each constituent of the charge estimate the final analysis of iron.
3. Determine whether it has come out to be the same as desired in the iron or not.
4. If negative, adjust the percentage of each constituent of the charge and recalculate the final analysis of iron.
5. Repeat this trial and error procedure till the calculated final analysis of iron conforms to the analysis desired.

Example 2.2. *A cupola charge weighs 1000 kg and is made up of the constituents of the following proportions :*

Constituents	Charge %	Carbon %	Silicon %	Manganese %	Sulphur %	Phosphorous %
Pig iron 1	15	3.5	2.50	0.7	0.016	0.17
Pig iron 2	20	3.5	3.0	0.65	0.018	0.11
New scrap	30	3.4	2.3	0.5	0.030	0.20
Returns (gates, rises, etc.)	35	3.3	2.5	0.65	0.035	0.16

Assume carbon pick-up as 0.15%, sulphur pick-up as 0.05%, silicon loss as 10% and manganese loss as 15%.

(i) Work out the final iron analysis for cupola melting.

(ii) If the desired analysis requires a silicon of 2.4%, adjust the charge and calculate the final proportions of charge constituents.

Solution. (*i*) **Final iron analysis for cupola melting :**

Constituents	Charge mass (kg)	Carbon %	Carbon kg	Silicon %	Silicon kg	Manganese %	Manganese kg	Sulphur %	Sulphur kg	Phosphorus %	Phosphorus kg
Pig iron 1	150	3.5	5.25	2.5	3.75	0.7	1.050	0.016	0.024	0.17	0.255
Pig iron 2	200	3.5	7.00	3.0	6.00	0.65	1.300	0.018	0.036	0.11	0.220
New scrap	300	3.4	10.20	2.3	6.90	0.5	1.500	0.030	0.090	0.20	0.600
Returns	350	3.3	11.55	2.5	8.75	0.65	2.275	0.035	0.1225	0.16	0.560
Total	1000		34.00		25.40		6.125		0.2725		1.635
Charge %			3.4 $\left(\dfrac{34 \times 100}{1000} = 3.4\right)$		2.54		0.6125		0.02726		0.1635
Change in cupola			+ 0.15		Loss − 0.254 (10% of total silicon)		− 0.09187 (20% loss of total manganese)		+ 0.050		—
Final iron analysis :			**3.55%**		**2.286%**		**0.5206%**		**0.07725%**		**0.1635%**

(Ans.)

(*ii*) Desired silicon content = 2.4%

Calculated silicon content = 2.286%

Additional silicon required = 2.4 − 2.286 = 0.114%

or,   Silicon required for 1000 kg of charge = $\dfrac{0.114}{100}$ × 1000 = 1.14 kg

In order to have net additional silicon of 1.14 kg, total silicon required to be added

$$= 1.14 + \frac{10}{100} \times 1.14 \ (10\% \ \text{silicon loss}) = 1.254 \ \text{kg.}$$

Silicon can be increased by adding ferro-silicon. Ferrosilicon contains 50% silicon. If 1 kg of ferrosilicon is added, 0.5 kg of silicon goes into the charge, but at the same time to keep the total charge as 1000 kg, one kg of foundry returns may be reduced. Since 1 kg of foundry return (gates, risers etc.) contain 2.5% *i.e.*, 0.025 kg of silicon this means addition of 1 kg of ferrosilicon and in turn reduction of 1 kg of foundry returns will increase silicon content by 0.500 − 0.025 = 0.475 kg.

Thus to compensate 1.254 kg of silicon, $\dfrac{1.254}{0.475}$ = 2.64 kg of ferrosilicon will be added and the foundry returns will be reduced by the same amount.

Hence, the final proportions of the charge constituents are :

- *Pig iron 1*   ....   **150 kg**
- *Pig iron 2*   ....   **200 kg**
- *New scrap*   ....   **300 kg**
- *Returns*   ....   350 − 2.64 = **347.36 kg**
- *Ferrosilicon*   ....   **2.64 kg**

       Total   1000 kg   **(Ans.)**

2.7. MELTING AND POURING

2.7.1. Melting

In order to get good, defect free casting proper melting of the metal is essential. *Factors controlling the proper melting includes* :

— Gases in metals ;

— Selection and control of scrap ;

— Flux ;

— Approximate furnace with temperature and atmosphere control.

- The *melting technique* should not only provide the molten metal at the required temperature, but should also provide the material of good quality and in the required quantity.

- The gases produced during casting may lead to poor quality of casting. In metal casting, the gases may be mechanically trapped, may be developed due to variation in their solubility at different temperatures and phases, and may be produced due to chemical reaction. Therefore, *vacuum melting* of metal is increasingly used for preventing the solution of gases in the molten metals and combination of reactive elements is the melt. Vacuum melting has been found more effective for controlling the dissolving gases and chemical composition.

Molten metal often reacts with oxygen and the metal oxides (slag) produced may get into the mould during pouring, and produce an inferior casting. This can be controlled by *covering the molten metal with fluxes or by carrying out melting and pouring in vacuum.* Alternately, ladles which pour the molten metal from beneath the surface can be used.

- Furnaces used for melting metals differ widely in operation, design from one another. A furnace is mainly selected on the basis of *metal chemistry, maximum temperature required and molten metal delivery rate and size and shape of available raw materials.*

— Cupola is extensively used for melting cast iron, primarily because of lower initial cost and cost of melting.

— A number of other furnaces such as induction furnace and side-blow converter are available for melting foundry alloys, but their choice essentially depends upon the type of alloy being melted, maximum temperature required and rate and mode of delivery.

2.7.2. Pouring

Molten metal is poured or injected into the mould-cavity. Many difficulties are encountered during pouring and this can be overcome by using an appropriate gating system, which ensures distribution of molten metal in the mould cavity at proper rate, without much temperature drop, without much turbulence and with minimum entrapping of gases and slags.

During pouring, the two main considerations are :

(i) *Pouring temperature :*

Selection of pouring temperature is marked by the factors such as hot tearing and the lower limit by the ability of molten metal to flow in thin sections and avoidance of skulls in ladles. It also influences *metallographic structure, particularly grain size and structure.*

— The fluidity of the molten metal, which affects its ability to flow and fill the mould cavity, depends on the casting material and the mould configuration. In general, lower is the viscosity, the higher is the fluidity. Therefore, at higher pouring temperatures, the fluidity will increase since the viscosity decreases. If pouring is carried out at a temperature higher than that required, increased fluidity may cause the molten metal to even get into small voids between the sand particles and cause surface defects. However, at low pouring temperatures, the fluidity of molten metal may not be adequate to fill the entire mould cavity.

— With the pouring temperature, another problem associated is the amount of trapped gases in the liquid metal. Since the *solubility of gas increases significantly with an increase in temperature,* therefore, in order to minimise the defects in casting due to trapped gases, the pouring temperature or extent of superheat should be *kept low.* In ferrous metals, the removal of gases like hydrogen and nitrogen is accomplished by passing CO gas through the molten metal. For non-ferrous metals, chlorine, helium or argon gases are invariably used.

— As the fluidity considerations and minimisation of gas solubility are two requirements, the "*optimum temperature*" is, however, decided on the basis of *fluidity requirements.* The temperature should be such that the *molten metal fills all sections of the mould before beginning to freeze and at the same time there should be no penetration into voids between the sand particles in the mould.*

(ii) *Pouring time or Pouring rate :*

"*Pouring time*" is the *time taken in pouring the molten metal in sprue for casting.* It plays an important role as this time is to be adjusted according to the requirements of a good casting.

The "*pouring rate*", in general, depends upon the *casting material and the casting configuration.* If the pouring rate of liquid metal is *low*, then the time taken to fill the mould cavity may be sufficiently long and the molten metal may begin to solidity before the mould is completely filled. This could be avoided by using higher pouring temperature, but gas solubility and fluidity may cause problems. Alternatively, *higher* pouring rate may be used, but this may cause erosion of mould surface and turbulence. *The "optimum pouring rate" is, therefore, a compromise taken to ensure that the entire mould cavity is filled before the start of solidification, without causing erosion of mould surfaces and undue turbulence.*

Following are some empirical relationships for estimations pouring time (T_{pour}) :

1. *Grey iron castings :*

$$T_{pour} = K \left(1.41 + \frac{t}{14.59} \right) \sqrt{W} \quad ...... \text{ for W} < 450 \text{ kg} \qquad ...(2.2)$$

$$T_{pour} = K \left(1.236 + \frac{t}{16.65} \right) \sqrt[3]{W} \quad ...... \text{ for W} > 450 \text{ kg} \qquad ...(2.3)$$

2. *Steel castings :*

$$T_{pour} = K \, (2.4335 - 0.3953 \log W) \, \sqrt{W} \, s$$

where, T_{pour} = Pouring time (s),

   K = Fluidity factor, which depends on iron composition and pouring temperature,

   t = Average thickness of casting (mm), and

   W = Mass of the casting (kg).

2.8. GATING SYSTEM

The passage way in the mould meant for carrying molten metal to the mould cavity is known as **gating system.**

Fig. 2.24. shows a gating system.

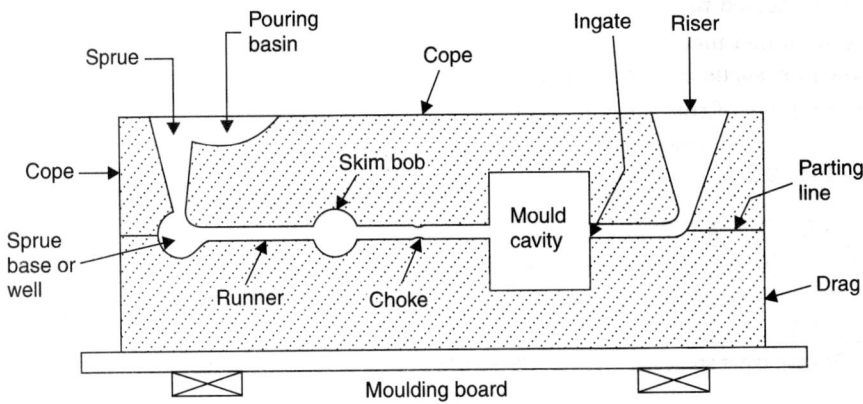

Fig. 2.24. Gating system.

Design requirements of the gating system. Any gating system should *aim* at *providing a defect free casting.* This can be achieved by making provision for certain requirements while designing the gating system. There are as follows :

   (*i*) The metal should flow smoothly into the mould without any turbulence.

   (*ii*) The mould should be completely filled in the smallest time possible without having to raise metal temperatures nor use higher metal heads.

   (*iii*) The gating system should ensure that enough molten metal reaches the mould cavity.

(*iv*) The metal entry into the mould cavity should be properly controlled in such a way that aspiration of the atmospheric air is prevented.

(*v*) The metal flow should be maintained in such a way that no gating or mould erosion takes place.

(*vi*) Unwanted material such as slag, dross and other mould material should not be allowed to enter the mould cavity.

(*vii*) A proper thermal gradient be maintained so that the casting is cooled without any shrinkage of cavities or distortions.

(*viii*) The design of the gating should be economical and easy to implement and remove after casting solidification.

Basic elements of the gating system :

Refer to Fig. 2.24.

The basic elements of the gating system are :

1. Pouring basin.	2. Sprue.
3. Sprue base or well.	4. Runner.
5. Skim bob.	6. Gates and foringates.
7. Choke.	8. Riser.

1. **Pouring basin.** The molten metal from the ladle is poured into the *pouring basin* from where it moves into the sprue and through the runner to other areas. The basin maintains a constant pouring head through weir (not shown) and holds back slag and dirt which float on the surface of the molten metal.

2. **Sprue.** *The vertical passageway through which the molten metal flows down from a parting plane is called the* ***sprue.***

- It is connected to the mould in cavity by a gate or series of gates.

- The solidified metal that occupies the sprue passage after the mould has been poured, is known as *sprue of the casting.*

- The function of the sprue is to provide an entrance to mould cavity for the molten metal. Sprues may be designed with either a positive taper, a reverse taper or no taper at all (see Fig. 2.25).

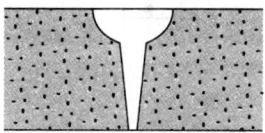 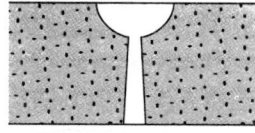 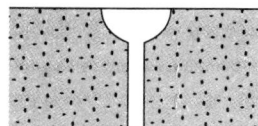

| Positive taper sprue | Reverse taper sprue | Straight or no taper sprue |

Fig. 2.25. Sprue designs.

3. **Sprue base or well :**

- At the bottom of the sprue is a reservoir for molten metal called *sprue well*. It serves to dissipate the kinetic energy of the falling stream of molten metal. The molten metal then changes direction and flows into the runner. Ceramic or wiremesh filters are sometimes placed at the sprue base to filter out dross and other large inclusions.

- *Splash core* (not shown) is a piece of ceramic or baked sand core inserted in the mould directly beneath the sprue to prevent the erosion of the mould sand where the molten metal strikes it at the base of the sprue.

4. Runner :

- It is generally located in the horizontal plane (parting plane) which connects the sprue to its ingates, thus letting the metal enter the mould cavity. The runners are normally made *trapezoidal in cross-section.*

- It is a general practice for ferrous metals to cut the runners in the cope and the ingates in the drag. The main reason for this is to trap the slag and dross which are lighter and thus trapped in the upper portion of the runners. For effective trapping of the slag, runners should flow full. When the amount of molten metal coming from the down sprue is more than the amount flowing through the ingates, the runner would always be full and thus slag trapping would take place. But when the metal flowing through the ingates is more than that flowing through the runners, then the runner would be filled only partially and the slag would enter the mould cavity.

Runner extension. The runner is extended a little further after it encounters the ingate. This extension is provided *to trap the slag in the molten metal.* The metal initially comes along with the slag floating at the top of the ladle and this flows straight, going beyond the ingate and then trapped in the runner extension.

5. **Skim bob.** It is an enlargement along the runner whose function is to trap heavier and lighter impurities such as dross or eroded sand. It, thus, prevents these impurities from going into mould cavity.

6. **Gates.** Gates, also called the ingates, are the openings through which the molten metal enters the mould cavity. The shape and the cross-section of the ingate should be such that it *can readily be broken off after casting solidification and also allow the metal to enter quietly into the mould cavity.*

Gates, depending on their position may be *top, parting, bottom* type, and *step gate.*

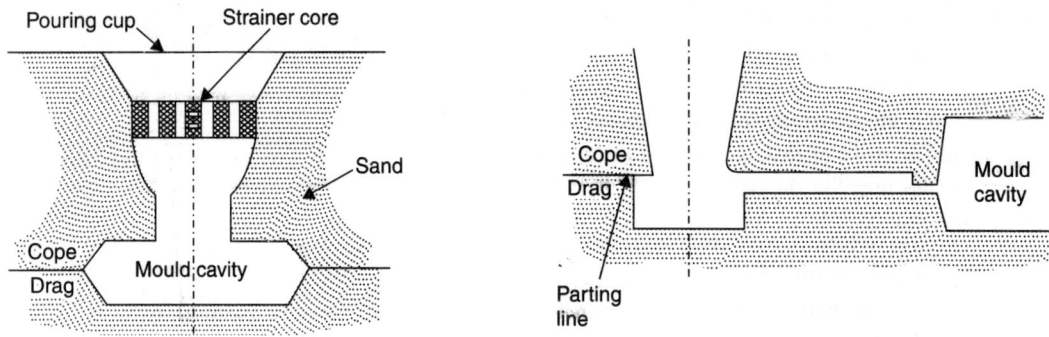

Fig. 2.26. Top gate.   Fig. 2.27. Parting gate.

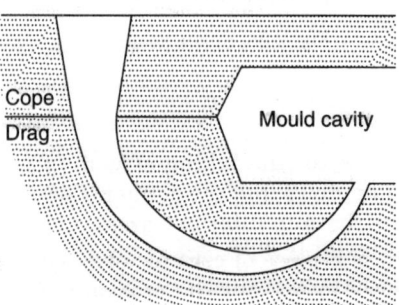

Fig. 2.28. Bottom gate.

(*i*) **Top gate :**

● Top gates (see Fig. 2.26) are usually limited to small and simple mould or larger castings made in moulds of erosion resistant material.

— Top gating is not advisable for light and oxidisable metals like aluminium and magnesium because of fear of entrapment due to turbulent pouring.

(*ii*) **Parting gate :**

— As the name implies the metal enters the mould at the parting plane when part of the casting is in the cope and part in the drag (see Fig. 2.27). For the mould cavity in the drag, it is a top gate and for the cavity in the cope it is a bottom gate. Thus, this type of gating tries to derive the best of both the types of gates, *viz* top and bottom gates.

— Of all the gates, this is *also the easiest and most economical in preparation.*

● This is the *most widely used gate in sand castings.*

(*iii*) **Bottom gate :**

— In the bottom gating system (Fig. 2.28) the molten metal flows down the bottom of the mould cavity in the drag and enters at the bottom of the casting and rises gently in the mould and around the cores.

— In this case the turbulence and mould erosion are the least. However time taken to fill up the mould is more.

— In bottom gating, directional solidification is difficult to achieve because the metal continue to lose its heat into the mould cavity and when it reaches the riser, metal becomes much cooler.

● These gates are best suited for *large sized steel castings.*

(*iv*) **Step gate.** In a step gate the metal enters (see Fig. 2.29) mould cavity through a number of ingates which are arranged in vertical steps. The sizes of ingates are normally increased from top to bottom such that metal enters the mould cavity from the bottom most gate and then progressively moves to the higher gates. This ensures a gradual filling of the mould without any mould erosion and produces a sound casting.

● These gates are used for *heavy and large castings.*

While designing a casting, it is essential to choose a suitable gate, considering the casting material, casting shape and size so as to produce a sound casting.

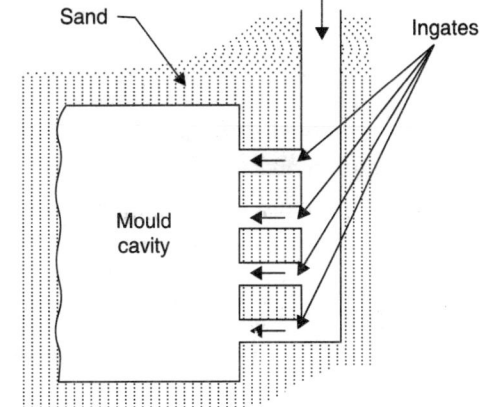

Fig. 2.29. Step gate.

Gate ratio. It is defined as the *ratio of sprue area to total runner area to total gate area* *i.e.,* Sprue area : Runner area : Gate area.

7. **Choke.** It is that part of the gating system which has the smallest cross-sectional area. It performs the following *functions* :

— To control the rate of metal flow to help lower the flow velocity in the runner ;

— To hold back slag and foreign material and float these in the cope side of the runner ;

— To minimise sand erosion in the runner.

According to its location, the gating system may be *classified as :*

(*i*) *Pressurised (or choked) gating system.*

— In this system, the ingates serve as the choke. This system maintains a back pressure and causes the entire gating system to become pressurised.

— In this system the molten metal enters the mould cavity uniformly.

— A typical gating ratio in this system can be 4 : 3 : 2.

— Pressurised systems are generally smaller in volume for a given metal flow rate than unpressurised ones.

— This system is adopted for metals like iron, steel, brass, etc.

(*ii*) *Unpressurised gating system :*

— In this system, the sprue base serves as the choke.

— The typical gating ratios in this system can be 1 : 2 : 2, 1 : 2 : 4, 1 : 3 : 3, or 1 : 4 : 4.

— In such systems careful design is required to ensure them being kept filled during pouring. Drag runners and cope gates aid in maintaining a full runner, but careful streaming is essential to eliminate the separation effects and consequent air separation.

— Such a system is adopted for light, oxidisable metals like aluminium and magnesium where the turbulence is to be minimised by slowing down the rate of metal flow.

8. Riser :

— *"Risers" are reservoirs designed and located to feed molten metal to the solidifying casting to compensate for solidification shrinkage. Riser* is a hole cut or moulded in the cope to permit the molten metal to rise above the highest point in the casting. It provides a visual check to ensure filling up of mould cavity.

— The riser should be designed for minimum possible volume while maintaining a solidification time longer than that of casting. The flow of liquid metal from the riser to the solidifying casting occurs only during the early part of the solidification process. This means that the volume of riser should be much more than the shrinkage volume ; a value of three times the shrinkage volume is generally considered to be adequate.

Requirements of a riser : The following are the *requirements of a riser :*

(*i*) It should be effective in establishing a pronounced temperature gradient with the casting so that the casting will solidify directionally towards the riser.

(*ii*) Its volume must be sufficient to compensate for metal shrinkage within the casting.

(*iii*) It must be the last part of the casting to solidify, such that all the shrinkages that are likely to occur should be in the riser.

(*iv*) It must cover completely the casting section that requires feeding.

(*v*) Inside the riser, the fluidity of the metal must be maintained so that the metal can flow from it and penetrate to the last contraction cavity.

Location of the riser. In order to ensure that the liquid metal from the riser is supplied to the desired locations within the casting, the location of the riser is an important consideration. To achieve this, a riser should be located in such a way that *directional solidification is obtained.* Since the heaviest section of the casting solidifies last, the riser should be located to feed this section. The heaviest section will now act as a riser for other sections which are not so heavy or thick.

For small castings, a single riser can feed the entire casting, but more than one riser is required for large castings. The number of risers and their locations will depend upon the casting configuration.

● The risers are generally located at a short distance away from the casting since they are ultimately separated from the final casting. The connecting channel between the casting and the riser should be large enough to ensure that this link does not freeze before the casting.

Types of risers. Various types of risers in use are : Refer to Fig. 2.30.

(i) *Top riser (or dead or cold riser).* It is located on top of the casting and has the advantage of additional pressure head and small feeding distance over the side riser which is placed adjacent to the casting.

(ii) *Side riser (live or hot riser).* It is filled last and contains the hottest metal. It receives the molten metal directly from the runner before it enters the mould cavity and is more effective than the top riser.

(iii) *Open riser.* These risers are open to the atmosphere at the top surface of the mould.

Advantages :
● Can be easily moulded.
● Since it is open to atmosphere, it will not draw metal from the casting as a result of partial vacuum in the riser.
● Such risers serve as collectors of non-metallic inclusions floating upto the surface.

Limitations :
● Their height should commensurate with the height of the cope ; this reduces the yield of the casting.
● These are the holes through which foreign matter may get into the mould cavity.

(iv) *Blind riser.* The riser which does not break to the top of the cope and is entirely surrounded by moulding sand is known as *'Blind riser'* (see Fig. 2.30). These risers are often made rounded at the top to effect metal savings, because the hemispherical shape has the smallest-surface area to volume ratio.

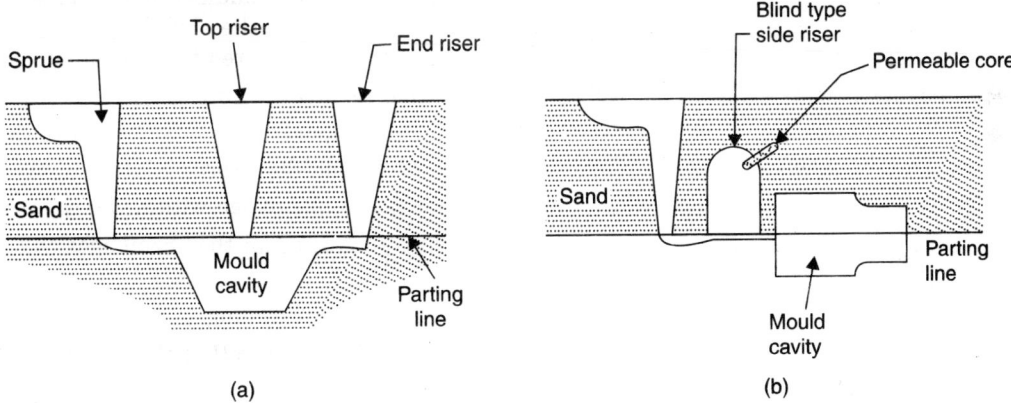

Fig. 2.30. Types of risers.

Advantages :
● Can be removed more easily from the casting than an open riser.
● Since a blind riser is surrounded on all sides by moulding sand, therefore, it looses heat slowly which helps in better directional solidification of the casting.
● It can be smaller than a comparable open riser, therefore, more yield is obtained.

Limitation. As the metal in it cools, metal skins may quickly form on its walls. This results in a vacuum in the riser and the riser will not actually draw metal from the casting. This may be avoided by inserting a permeable dry sand core into the riser cavity, connecting it to the mould sand layers. Through these sand layers air passes into the riser interior and thus the riser operates under atmospheric pressure.

Riser design :

If during casting no riser is provided, the solidification will start from walls and liquid metal in the centre will be surrounded by a solidified shell and the contracting liquid will produce voids towards the centre of the casting. Further cooling of the solid in centre sets up undesirable stresses in the casting.

The above problems are overcome by the provision of risers as these supply molten metal for a solidifying casting. To accomplish this, the risers must be large enough to remain liquid after the casting has solidifed and must contain sufficient metal to provide for the contraction losses. Further, these should be so positioned that they continue to supply metal throughout the solidification period.

The risers should *loose heat at a slower rate,* since they are designed to solidify last as to feed enough metal to heavy sections of the casting to make up for shrinkage before and during solidification. For a given size the risers should be designed with a high $\dfrac{V \text{ (volume)}}{A \text{ (surface area)}}$ ratio.

This condition can be met when the riser is *spherical in shape* so that its surface area (A) is minimum, for a given volume (V). The cylinder is next best practical shape; the rectangular sections being very inefficient and must be avoided as far as possible. Since risers of spherical shapes are difficult to mould, therefore, the *cylinder* is probably the best shape to be used for general run of casting. The *height of the riser should be tall enough so that any pipe formed in it may not penetrate the casting. The ratio of height to diameter usually varies from* **1 : 1 to 3 : 2.**

The *riser size* for a given casting can be obtained form the following relations :

1. *Chvorinov's rule :*

Chvorinov's rule for metal casting states the postulation that *total freezing (solidification) time for a casting is a function of the ratio of volume to surface area.*

$$\text{Solidification time, } t = C \left[\frac{\text{Volume (V)}}{\text{Surface area (A)}} \right]^2 \quad i.e., \quad t = C \left(\frac{V}{A} \right)^2 \qquad \qquad ...(2.4)$$

where, C = A constant, that reflects mould material, metal properties like latent heat, and temperature.

Best riser is one whose $\left(\dfrac{V}{A} \right)^2$ is *10 to 15% larger than that of the casting.*

Since *V* and *A* for the casting are known, $\left(\dfrac{V}{A} \right)_{\text{riser}}$ can be determined. Assuming the height to diameter ratio for the cylindrical riser, the riser size can be determined.

This rule helps in determining the solidification time of casting. Accordingly we can select the method of casting.

- Chvorinov's rule/formula is not *very accurate,* since it does not take into account the solidific⬚ ⬚ contraction or shrinkage. This method is *valid for calculating proper riser size for short freezing-range alloys such as steel and pure metals.*
- There is no *satisfactory relationship for determining riser size for non-ferrous alloys.*

2. *Caine's method :*

This method of determining riser size is based on *experimentally determined hyperbolic relationship between relative freezing times and volumes of the casting and the riser.*

According to Caine: If the casting solidifies *infinitely rapidly,* the feeder (riser) volume should be *equal to the solidification shrinkage of the casting,* and if the feeder and casting solidify at the *same rate,* the feeder should be *infinitely large.*

Relative *freezing time or freezing ratio* (R_F) is defined as :

$$R_F = \frac{(A/V)_{casting}}{(A/V)_{riser}}$$

Volume ratio (R_V) is given as :

$$R_V = \frac{V_{riser}}{V_{casting}}$$

Then, **Caine's formula** is given as :

$$R_F = \frac{a}{R_V - b} + c \qquad \qquad ...(2.5)$$

where, a = Freezing characteristic constant for the metal,
  b = Contraction ratio from liquid to solid, and
  c = Relative freezing rate of riser and casting.

Typical values of a, b and c for commonly used metals are given below :

S. No.	Cast-metals	a	b	c
1.	Grey cast-iron	0.33	0.03	1.00
2.	Cast-iron, brass	0.04	0.17	1.00
3.	Steel	0.12	0.05	1.00
4.	Aluminium	0.10	0.06	1.08
5.	Aluminium bronze	0.24	0.17	1.00
6.	Silicon bronze	0.24	0.17	1.00

Fig. 2.31 shows a typical hyperbolic curve. Such curves for various cast metals are available in Handbooks.

For determining the riser size for a given casting, the diameter and height of the riser are *assumed.* Then knowing the values of *a, b,* and *c,* the values of R_F and R_V are calculated and plotted on hyperbolic curve figure. In case the *values of R_F and R_V meet the above curve, the assumed riser size is satisfactory, otherwise a new assumption is made.*

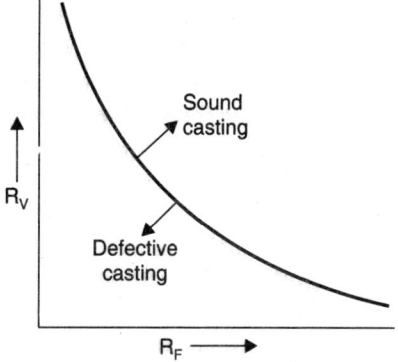

Fig. 2.31

2.9   COOLING AND SOLIDIFICATION

- After the mould cavity is filled with the molten metal, the metal is allowed to solidify into the desired shape. The *solidification process decides the structural features of cast material, and controls the properties of casting.*

- The liquid metal is poured into the mould cavity at a temperature *much higher than the freezing temperature* (it is the temperature above which pure metals are completely liquid and below it completely solid, also called equilibrium melting temperature) to allow sufficient time for the liquid metal to flow into all corners of the mould cavity before it begins to freeze (the difference between the pouring temperature and freezing temperature is known as *superheat*).

- Refer to Fig. 2.32. For proper understanding of *solidification mechanism* and rate of heat loss from the material to the mould etc. it is necessary to predict how the casting will solidify which avoids casting defects such as seams, gas holes and hot tears etc.

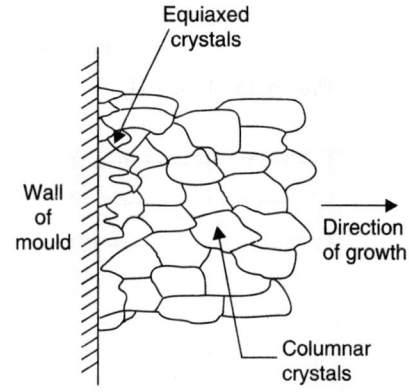

Fig. 2.32

We know, solidification requires energy to produce a crystalline structure, some supercooling (below freezing point) is required before the liquid metal starts solidifying. *This cooling is provided by the mould which provides sites around which crystals can grow initially and subsequently by the solidified particles, and the metal itself.*

In the case of casting, after pouring of the molten metal in the mould, the temperature falls steadily until freezing commences at a particular point. During solidification, the temperature more or less remains constant due to release of latent heat. In fact there may be a slight increase in temperature if supercooling has occurred. The temperature again starts falling steadily in case of pure metal. *In the case of alloy, commencement of crystallization is followed by a period of less steep temperature reduction while the metal is passing through the mushy state.*

Since an alloy does not have a sharply defined freezing temperature solidification takes place over a *range of temperature.* The solids separating out at different temperatures therefore possess different compositions. The direction of crystal growth is thus dependent upon the composition gradient within the casting variation of solids, temperature with composition and thermal gradient within the mould. *The crystal growth in the case of alloys is of dendritic structure.*

- During cooling many characteristics such as crystal structure and alloy composition at different parts of the casting are decided. Unless a proper care is taken during cooling, defects like shrinkage, cavity, cold shut, misrun and hot tear occur. Due to all these reasons, the direction of crystal growth depends on the various factors such as :

    (*i*) The variation of temperature with composition.

    (*ii*) The thermal gradient within the mould.

    (*iii*) Composition gradient within the casting.

The cooling patterns of an ordinary mould are shown in Fig. 2.33.

- *The cooling curves are useful to find out the too low pouring temperature which will cause partially filled cavities. Also these are used to find the too high pouring temperature which will create hard casting, etc.*

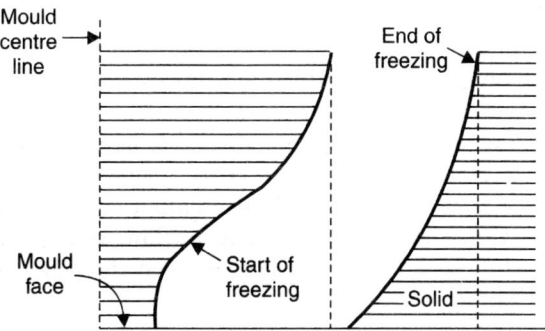

Fig. 2.33. The cooling patterns of an ordinary mould.

Fig. 2.34 shows the cast structure of metals.

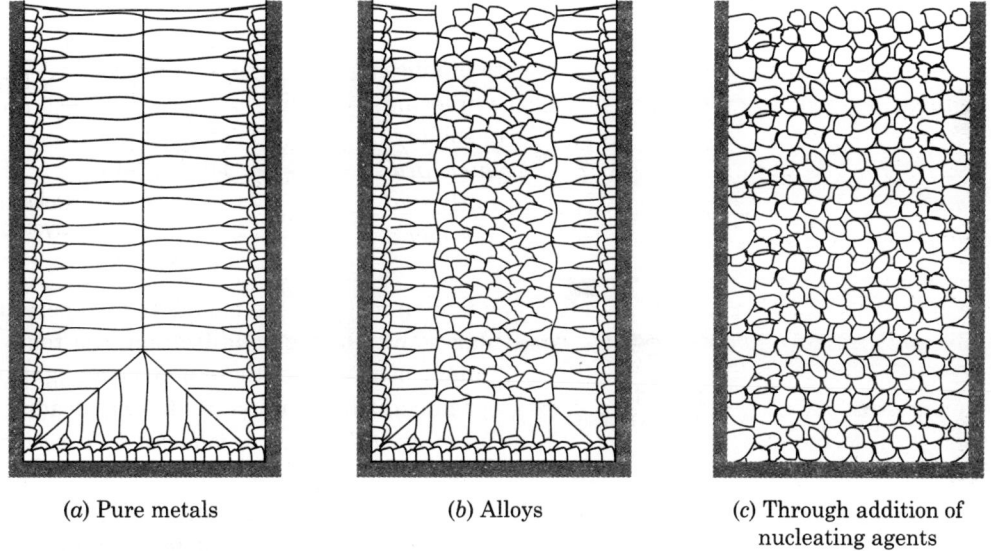

(a) Pure metals   (b) Alloys   (c) Through addition of nucleating agents

Fig. 2.34. Cast structure of metals.

- In **directional solidification** (aimed at producing sound castings) the solidification continues progressively from the thinnest section which solidifies first towards the risers which should be last to solidify.

Directional solidification can be ensured by :

(i) Designing and positioning the gating system and risers properly ;

(ii) Increasing the thickness of certain sections of the casting by the use of *padding* (the tapering of thinner section towards the thicker section) ;

(iii) Using exothermic materials in the risers or in the facing sand around certain portions of the casting ;

(iv) Using chills in the moulds.

Chills :

- **Chills** *are provided in the mould so as to increase the heat extraction capability of the sand mould. A chill provides a steeper temperature gradient so that directional solidification as required in a casting be obtained.*

- The *chills are metallic objects having a higher heat absorbing capability than the sand mould.* These can be of two types :

(*i*) External chills (placed in the mould cavity adjoining the mould cavity at any required position).

(*ii*) Internal chills (placed inside the mould cavity where an external chill cannot be provided).

- These are used when it may be *either impractical or impossible* to use riser on thick sections of the casting. However they can be used along with risers if possible.

Chills and risers placed on massive portions produce a joint effect which enables the production of castings completely free from contraction voids.

- *Chills* are metal objects in the form of rods, wire and nails, either cast or made from rolled stock.

Fig. 2.35 shows use of chills.

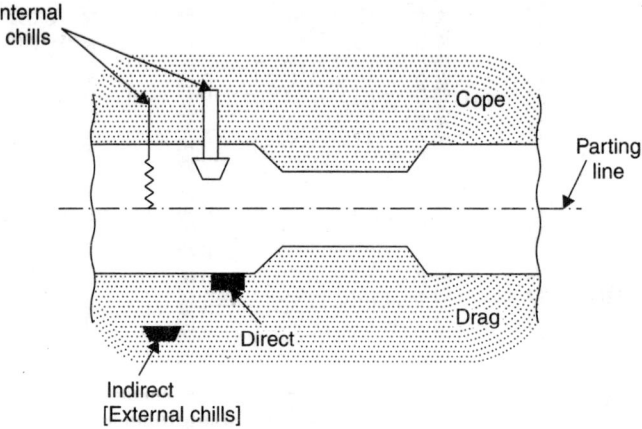

Fig. 2.35. Use of chills.

Example 2.3. *Assuming uniform cooling in all directions, determine the dimensions of a 90 mm cube casting after it cools down to room temperature. The solidification shrinkage for the cast metal is 5% and the solid contraction is 7.5%.*

Solution. *Given* : Side of the cube = 90 mm

Solidification shrinkage = 5%

Solid contraction = 7.5%

Dimension of each side of the cube after cooling :

Volume of casting, $V = (90)^3 = 7,29,000$ mm^3

Volume after solidification shrinkage $= 7,29,000 \left(1 - \dfrac{5}{100}\right) = 6,92,550$ mm^3

Volume at room temperature $= 6,92,550 \left(1 - \dfrac{7.5}{100}\right) = 6,40,609$ mm^3

∴   *Dimension of each side of the cube* $= (6,40,609)^{1/3}$

$$= \textbf{86.2 mm.} \textbf{(Ans.)}$$

Example 2.4. *What will be the solidification time for a 1100 mm diameter and 33 mm thick casting of aluminium if the mould constant is 2.2 sec/mm² ?*

Solution. Diameter of casting, d = 1100 mm

Height/thickness of casting, h = 33 mm

Mould constant, C = 2.2 sec/mm²

Solidification time, t :

Volume of the casting, $V = \dfrac{\pi}{4} d^2 \times h$

$$= \frac{\pi}{4} \times (1100)^2 \times 33 = 31{,}360{,}948 \text{ mm}^3$$

Surface area of the casting, $A = 2 \times \dfrac{\pi}{4} d^2$

$$= 2 \times \frac{\pi}{4} \times (1100)^2 = 1{,}900{,}663 \text{ mm}^2$$

$\therefore$ $t = C\left(\dfrac{V}{A}\right)^2$

$$= 2.2. \times \left(\frac{31{,}360{,}948}{1{,}900{,}663}\right)^2 = 598.95 \text{ s or } \textbf{9.98 min.} \quad \textbf{(Ans.)}$$

Example 2.5. *Two castings of the same metal have the same surface area. One casting is in the form of sphere and other is a cube. What is the ratio of the solidification time for the sphere to that of a cube.*   **(GATE)**

Solution. Let, V = Volume of casting, and

A = Surface area of casting

Also, $A_{sphere} = A_{cube}$ ...*(Given)*

According to Chvorinov's rule,

Solidification time, $t \propto \left(\dfrac{V}{A}\right)^2$

$\therefore$ Ratio of solidification time for the sphere to that of cube,

$$\frac{t_{sphere}}{t_{cube}} = \left(\frac{V_{sphere}}{A_{cube}}\right)^2$$

$$= \left(\frac{\frac{4}{3}\pi R^3}{a^3}\right)^2 = \frac{16\pi^2 R^6}{9a^6}. \quad \textbf{(Ans.)}$$

where, R = Radius of sphere, and

a = Side of the cube.

Example 2.6. *Two solid workpieces (i) Sphere with radius R, (ii) a cylinder with diameter equal to its height, have to be sand cast. Both workpieces have the same volume. Show that the cylindrical workpieces will solidify faster than the spherical workpieces.* **(GATE)**

Solution. Let, $\quad$ R = Radius of sphere, d = Dia. of cylinder, and h = Height of cylinder.

Then, volume of sphere, $V_{sphere} = \dfrac{4}{3}\pi R^3$

and, volume of cylinder, $\qquad V_{cyl.} = \dfrac{\pi}{4} d^2 \times h = \dfrac{\pi}{4} d^3 \qquad\qquad$ $(\because \quad h = d$ ... given$)$

Also, $\qquad\qquad V_{sphere} = V_{cyl.} \qquad\qquad\qquad\qquad$ ...*(Given)*

$\therefore \qquad\qquad\qquad \dfrac{4}{3}\pi R^3 = \dfrac{\pi}{4} d^3$

or $\qquad\qquad\qquad\qquad d^3 = \dfrac{16}{3} R^3 \qquad i.e., \qquad d = \left(\dfrac{16}{3}\right)^{1/3} R$

According to *Chvorinov's rule* :

Solidification time, $\qquad\qquad t \propto \left(\dfrac{V}{A}\right)^2$

Now, $\qquad\qquad\qquad \left(\dfrac{V}{A}\right)_{sphere} = \dfrac{\dfrac{4}{3}\pi R^3}{4\pi R^2} = \dfrac{R}{3} = 0.33\ R$

and, $\qquad\qquad \left(\dfrac{V}{A}\right)_{cyl.} = \dfrac{\dfrac{\pi}{4} d^2 \times h}{\pi d h + 2 \times \dfrac{\pi}{4} d^2} = \dfrac{\dfrac{\pi}{4} d^2 \times d}{\pi d \times d + 2 \times \dfrac{\pi}{4} d^2} = \dfrac{d}{6}$

$\qquad\qquad\qquad\qquad = \dfrac{1}{6} \times \left(\dfrac{16}{3}\right)^{1/3} R = 0.29\ R$

From the above results, it is evident that the **cylindrical** *workpiece will solidify faster than the spherical workpiece* (it is known that *risers with a higher value of* $\left(\dfrac{V}{A}\right)$ loose heat at a slower rate). **(Ans.)**

Example 2.7. *Which one of the following casting shapes would have least solidification time ?*

$\quad$ *(i) A sphere of diameter D = 25 mm ;*

(ii) A cylinder with both diameter d and height h = 25 mm ;

(iii) A cube with a length of side l = 25 mm.

Solution. *Given :* D = 25 mm ; $d = h$ = 25 mm ; l = 25 mm

Least solidification time :

According to Chvorinov's rule, solidification time,

$$t = C \left(\dfrac{V}{A}\right)^2$$

where, V and A are the volume and surface area of the casting respectively.

$$\therefore \quad (i) \qquad t_{\text{sphere}} = C\left(\frac{\frac{\pi}{6}D^3}{\pi D^2}\right)^2 = \frac{CD^2}{36} = \frac{C \times (25)^2}{36} = \textbf{17.36 C.}$$

$$(ii) \qquad t_{\text{cyl.}} = C\left(\frac{\frac{\pi}{4}d^2h}{2 \times \frac{\pi}{4}d^2 + \pi dh}\right)^2$$

$$= C\left(\frac{\frac{\pi}{4}d^2 \times d}{2 \times \frac{\pi}{4}d^2 + \pi d \times d}\right)^2 = C\left(\frac{\frac{d^3}{4}}{\frac{d^2}{2} + d^2}\right)^2$$

$$= C\left(\frac{d}{6}\right)^2 = C\left(\frac{25}{6}\right)^2 = \textbf{17.36 C.}$$

$$(iii) \qquad t_{cube} = C\left(\frac{l^3}{6l^2}\right)^2 = C\left(\frac{l}{6}\right)^2 = C\left(\frac{25}{6}\right)^2 = \textbf{17.36 C.}$$

From above results we find that *all three castings will have the same solidification time.* **(Ans.)**

Example 2.8. *Compare the solidification times for castings of three different shapes of same volume : Cubic, cylindrical (with height equal to its diameter) and spherical.*

Solution. Let,   A = Surface area of a casting, and

         V = Volume of each casting = unity (say)   ...(*Given*)

According to Chvorinov's rule, we have

Solidification time, $t = C\left(\dfrac{V}{A}\right)^2,$ where C is a constant.

(*i*) **Cubic casting :**

Let *l* be the side of the cube.

Then, $V = l^3 = 1 \quad \therefore \quad l = 1$

Surface area, $A = 6l^2 = 6 \times 1^2 = 6$ units,

$\therefore$ $t = C\left(\dfrac{1}{6}\right)^2 = \textbf{0.0278 C.}$

(*ii*) **Cylindrical casting :**

Let *r* and *h* be the radius and height of the cylinder.

$\therefore$ $h = 2r$ ...Given

Then, $V = \pi r^2 h$ or $\pi r^2 \times (2r) = 1$

or, $2\pi r^3 = 1 \quad \therefore \quad r = \left(\dfrac{1}{2\pi}\right)^{1/3} = 0.5419$ unit.

Surface area, $A = 2\pi r^2 + 2\pi rh$

$$= 2\pi r^2 + 2\pi r \times 2r = 6\pi r^2 = 6\pi \times (0.5419)^2 = 5.53 \text{ units}$$

$\therefore$ $t = C\left(\dfrac{1}{5.53}\right)^2 = \textbf{0.0327 C.}$

(iii) Spherical casting :

Let R be radius of the sphere,

Then, $V = \dfrac{4}{3}\pi R^3 = 1$ or $R = \left(\dfrac{3}{4\pi}\right)^{1/3} = 0.62$ unit

$\therefore$ Surface area, $A = 4\pi R^2 = 4\pi \times (0.62)^2 = 4.83$ units

$\therefore$ $t = C\left(\dfrac{1}{4.83}\right)^2 = \textbf{0.0429 C.}$

Thus, 'cubic casting' has the least solidification time and as such it will *solidify the fastest* ; the 'spherical casting' has the maximum solidification time and therefore will *solidify the slowest.* **(Ans.)**

Example 2.9. *With cylindrical riser, prove that for a longer solidification time, diameter of riser = height of riser.*

Solution. According to Chvorinov's rule,

Solidification time, $t \propto \left(\dfrac{V}{A}\right)^2$ ...(i)

where, V and A are volume and surface area of the casting respectively.

From the expression (i), it is evident that for longer solidification time, the ratio $\left(\dfrac{V}{A}\right)$ should be maximum or the ratio $\left(\dfrac{A}{V}\right)$ should be minimum.

Now, $V = \dfrac{\pi}{4}d^2 \times h,$

where, d and h are the diameter and height of the cylinder respectively.

or, $h = \dfrac{4V}{\pi d^2}$ ...(ii)

Also, $A = \pi d \times h + 2 \times \dfrac{\pi}{4}d^2$

$$= \pi d \times \dfrac{4V}{\pi d^2} + \dfrac{\pi}{2}d^2$$

$$= \dfrac{4V}{d} + \dfrac{\pi}{2}d^2$$

For A to be minimum for a given V,

$$\dfrac{dA}{dd} = 0$$

i.e.
$$\frac{d}{dd}\left(\frac{4V}{d} + \frac{\pi}{2}d^2\right) = 0$$

or,
$$-\frac{4V}{d^2} + \pi d = 0 \quad \text{or} \quad -4V + \pi d^3 = 0$$

or,
$$d^3 = \frac{4V}{\pi}$$

Also,
$$\frac{4V}{\pi} = hd^2 \qquad\qquad \text{... From expression } (ii)$$

$\therefore$
$$d^3 = hd^2$$

or,
$$\mathbf{d = h} \quad \textbf{Proved.}$$

This optimum is true only for a *side riser.*

Note. For a *top riser*, $A = \pi dh + \dfrac{\pi}{4}d^2$

Following the above procedure, it will be seen that

$$d = 2h \quad \text{or} \quad \frac{h}{d} = \frac{1}{2}.$$

Example 2.10. *Calculate the ratio of solidification times of two steel cylindrical risers of sizes 36 cm in diameter by 72 cm height and 72 cm in diameter by 36 cm in height subjected to identical conditions of cooling.*   **(GATE)**

Solution. *Given :* $d_1 = 36$ cm ; $h_1 = 72$ cm ; $d_2 = 72$ cm ; $h_2 = 36$ cm.

Ratio of solidification times of the two cylindrical risers :

According to Chvorinov's rule, solidification time,

$$t \propto \left(\frac{V}{A}\right)^2 \qquad\qquad ...(i)$$

where V and A are the volume and surface area of the casting respectively.

Riser-1 :
$$\frac{V}{A} = \frac{\frac{\pi}{4}d_1^2 \times h_1}{\pi d_1 h_1 + 2 \times \frac{\pi}{4}d_1^2} = \frac{\frac{d_1^2}{4}h_1}{d_1 h_1 + \frac{d_1^2}{2}}$$

$$= \frac{\frac{36^2}{4} \times 72}{36 \times 72 + \left(\frac{36^2}{2}\right)} = \frac{23328}{3240} = 7.2$$

Riser-2 :
$$\frac{V}{A} = \frac{\frac{72^2}{4} \times 36}{72 \times 36 + \frac{72^2}{2}} = \frac{46656}{5184} = 9$$

From expression (i), we have

$$\frac{t_{\text{riser-1}}}{t_{\text{riser-2}}} = \left(\frac{7.2}{9}\right) = 0.64. \quad \textbf{(Ans.)}$$

Example 2.11. *An aluminium cube of 12 cm side has to be cast along a cylindrical riser of height equal to its diameter. The riser is not insulated on any surface. If the volume shrinkage of aluminium during solidification is 6 per cent ; calculate :*

(i) *Shrinkage volume of cube on solidification.*

(ii) *Minimum size of the riser so that it can provide the shrinkage volume.*   **(GATE)**

Solution. *Given :* Side of the aluminium cube, $a = 12$ cm

Diameter of cylindrical riser, $d =$ height of the riser (h)

Volume shrinkage of aluminium during solidification = 6%.

(i) **Shrinkage volume of cube on solidification :**

Volume of casting $= a^3 = (12)^3 = 1728$ cm^3

Shrinkage volume $= 6\% = \dfrac{6}{100} \times 1728 = \textbf{103.68 cm}^3. \quad \textbf{(Ans.)}$

(Normally this shrinkage, depending upon metal, varies from 2.5 to 7.5%)

(ii) **Minimum size of the riser :**

A riser should be designed with minimum possible volume while having a longer solidification time than the casting.

Now from practice, *minimum volume of riser is approximately three times the shrinkage volume.*

$\therefore$ Minimum volume of riser $= 3 \times 103.68 = 311.04$ cm^3

$\therefore$ $\dfrac{\pi}{4} d_r^{\,2} \times h_r = 311.04$

where suffix 'r' stands for riser.

Now, $h_r = d_r$ ...(Given)

$\therefore$ $\dfrac{\pi}{4} d_r^{\,2} \times d_r = 311.04$ or $d_r = 7.34$ cm

In order to have a sound casting, the metal in the riser should be the last to cool, that is, the riser should have a longer solidification time than the casting, so

$$\left(\frac{A}{V}\right)_r \leq \left(\frac{A}{V}\right)_c \quad \text{or} \quad \left(\frac{V}{A}\right)_r \geq \left(\frac{V}{A}\right)_c, \text{ where suffix } c \text{ stands for casting.}$$

$$\text{Now, } \left(\frac{A}{V}\right)_r = \left(\frac{\pi d h + 2 \times \dfrac{\pi}{4} d^2}{\dfrac{\pi}{4} d^2 \times h}\right)_r = \left(\frac{\pi d \times d + 2 \times \dfrac{\pi}{4} d^2}{\dfrac{\pi}{4} d^2 \times d}\right)_r = \left(\frac{\pi d^2 + \dfrac{\pi}{2} d^2}{\dfrac{\pi}{4} d^3}\right)_r = \frac{6}{d_r} = \frac{6}{7.34}$$

$\phantom{\text{Now, }} = 0.817$

and, $\left(\dfrac{A}{V}\right)_c = \dfrac{6 \times 12 \times 12}{12 \times 12 \times 12} = \dfrac{1}{2} = 0.5$

As is clear $\left(\dfrac{A}{V}\right)_r$ is $> \left(\dfrac{A}{V}\right)_c$, which is not desirable

$\therefore \qquad\qquad \left(\dfrac{A}{V}\right)_r \le \left(\dfrac{A}{V}\right)_c$

$$\le 0.5$$

or $\qquad\qquad \dfrac{6}{d_r} \le 0.5, \quad \therefore \quad \dfrac{6}{0.5} \le d_r \quad \text{or} \quad d_r \ge \dfrac{6}{0.5} \text{ or } 12$

$\therefore$ *Minimum size of riser* = **12 cm diameter × 12 cm height.**

i.e., $\qquad\qquad V_r = \dfrac{\pi}{4} \times (12)^2 \times 12 = \textbf{1357.2 cm}^3. \quad \textbf{(Ans.)}$

Example 2.12. *In a sand mould, a sprue of 210 mm height and 1152 mm² top area is provided to maintain the flow rate of liquid at 1,728,000 mm³/s. What should be the area at the base of down sprue to prevent aspiration of molten metal ? Take g = 9815 mm/s².*

Solution. *Given :* h = 210 mm ; Top area = 1152 mm ; Flow rate, Q = 1,728,000 mm³/s, g = 9815 mm/s²

Area at the base of down sprue :

Velocity is down sprue, $v = \sqrt{2gh}$

$\qquad\qquad\qquad = \sqrt{2 \times 9815 \times 210} = 2030.3 \text{ mm/s}$

$\therefore$ Area at the base of down sprue to maintain the flow rate,

$$A = \dfrac{Q}{v} = \dfrac{1,728,000}{2030.3} = \textbf{851.1 mm}^2. \quad \textbf{(Ans.)}$$

2.10. CASTING

- **Casting** *means the pouring of molten metal into a mould, where solidification occurs.*

Metal casting may also be *defined as a process of production of objects of desired shape and sizes by introduction of molten metal into a predesigned mould cavity created commonly in a compact sand mass, with the help of a pattern, or in a metallic mould (as in die casting) and allowing it to solidify.*

- Almost every finished metal product has been cast at some stage of its manufacture. For example, all rolled and forged steels are initially in the form of cast ingots, and even after extensive hot working, evidence of cast structure may still remain in the form of solids, chemical segregation, or surface defects.
- The main advantage of the foundry process is its *flexibility* and the *possibility of making all sorts and types of casting for a wide range of applications.*

2.11. ADVANTAGES AND DISADVANTAGES OF CASTING PROCESS

Advantages :

Casting process entails the following *advantages and disadvantages :*

1. *Cheapest* method of fabrication.
2. Objects of large size can be *produced easily.*
3. The objects having *complex and complicated shapes*, which cannot be produced by any other method of production, can usually be cast.
4. Castings with wide *range of properties* can be produced by adding various alloying elements.
5. By proper selection of type of moulding and casting process, *required dimensional accuracy in casting can be achieved.*

6. *Almost all the metals and alloys and some plastics can be cast.*
7. The number of castings can vary from *very few to several thousands.*

Disadvantages :
1. The *time* required for the process of making casting is *quite long.*
2. Metal casting involves melting of metal which is a high *energy consuming process.*
3. The *working conditions* in foundaries are *quite bad* due to heat, dust, fumes, slag etc., compared to other processes.
4. Metal casting is still *highly labour-intensive compared to other processes.*
5. The *productivity is less* than other automatic processes, *e.g.,* like rolling.

2.12. PREPARATION OF A CASTING

Preparation of a casting involves the following *steps :*
1. Preparation of a pattern.
2. Preparation of moulding sand.
3. Preparation of mould and core(s).
4. Melting the metal.
5. Pouring of metal into the mould.
6. Cooling and solidification.
7. Removing the casting from the mould.
8. Fettling (*i.e.,* cutting off the unwanted projection in the form of gates, risers, etc.)
9. Heat treatment.
10. Testing and inspection.

2.13. BASIC RULES FOR GOOD CASTING DESIGN

Following are some important *basic rules for good casting design :*
1. Design all sections as nearly uniform in thickness as possible.
2. All parts should be of sufficient thickness for proper running of metal in the mould.
3. Avoid abrupt section changes. Make transition gradual, blending heavier sections into light ones.
4. Minimum number of sections should be brought together, member functions to be staggered.
5. Use gentle corners, avoid sharp angles. A cooling surface to be always prevented.
6. Reduce bosses, lugs or pad and other projections.
7. Reentrant angles should be avoided.
8. The openings in casting walls should be large radii of curvature and gradual changes.
9. For adjoining sections, replace corners with radii and avoid heat and stress concentration.
10. The use of ribs meeting at an acute angle should be avoided.
11. Ribs should be employed only where necessary to increase strength or reduce weight or avoid wastage.
12. Avoid intersecting ribs by staggering.
13. Since the ribs are provided to increase stiffness and reduce weight the thickness of the rib should be equal to 80 percent of casting thickness and should be rounded at edge and correctly filleted.
14. Multiplicity of cores should be avoided.
15. Avoid intersecting ribs by staggering.

2.14. CASTING PROCESSES

The various casting processes are as follows :

1. Sand casting
2. Shell-mould casting
3. Plaster mould casting   ... **Expendable mould,**
4. Ceramic mould casting   **permanent pattern**
5. Vacuum casting

6. Evaporative-pattern casting (lost foam)   ... **Expendable mould,**
7. Investment casting   **expendable pattern**

8. Slush casting
9. Pressure casting
10. Die casting   ... **Permanent mould**
11. Centrifugal casting
12. Squeeze casting
13. Semisolid metal forming
14. Continuous casting

2.14.1. Sand Casting

- A commonly used method involves pouring molten metal into a cavity in a mass of packed sand.
- Fig. 2.36 (*i*) shows a typical mould in cross-section. It illustrates the use of *chills* to produce a local hard surface, a **core** to form a shaft opening, and a **sprue** for running the molten metal into the cavity. A wood or metal **pattern** approximately the shape of the final casting is used to produce the cavity in the sand mould. So that the pattern may be removed from either the cope (upper) or the drag (lower) section of the mould without disturbing the sand that has been packed around it, a taper or draft of a few degrees must be allowed on the metal faces of the pattern.

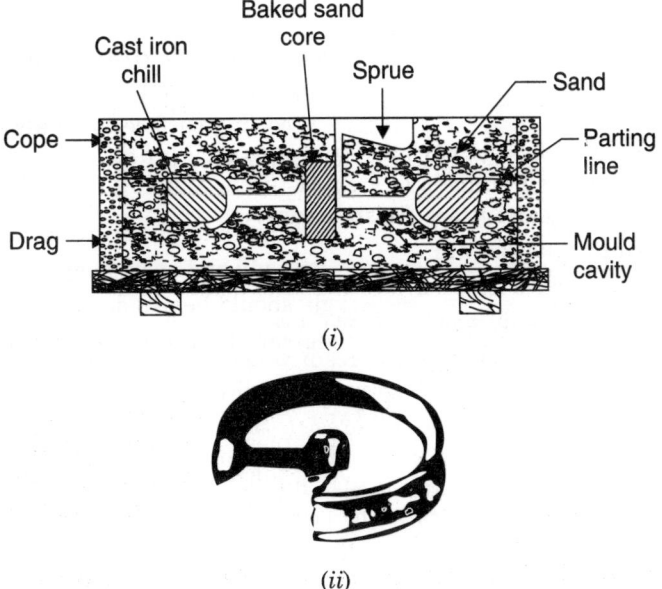

Fig. 2.36. (*i*) and (*ii*). Typical mould in cross-section.

- Since casting alloys decrease in volume as they solidify and cool to room temperature, it is necessary to make the pattern larger than the final casting by an amount known as the shrinkage allowance. Shrinkage depends on such factors as the kind of alloy being cast, the design of the casting, the pouring temperature, and the size of the casting. In making castings such as those of a U-shape it is necessary to "*rake*" or distort the pattern in order to obtain the desired form in the final casting. This is called *distortion allowance*. An additional *machine finish allowance* of 1.6 mm or more must be allowed on surfaces that are to machined. Finally even with the use of the best available information on shrinkage allowance, it is unlikely that final dimensions of the casting can be corrected exactly. Therefore, size tolerance equal to half the shrinkage allowance are suggested for use with castings of new design.

Advantages :

1. No limit on size and shape.
2. Almost any metal can be cast.
3. Low equipment cost.
4. Economical for low-volume production.
5. Low production rate.
6. Extreme complexity possible.
7. Low tool cost.
8. Most direct route from pattern to casting.

Limitations :

(*i*) Product gives rough surface.

(*ii*) Dimensional accuracy difficult.

(*iii*) Thin projections not practical.

(*iv*) Machining always necessary.

2.14.2. Shell-mould Casting

Shell moulding is modification of sand casting in which a *relatively thin shell forms the mould cavity into which the molten metal is poured.*

Shell-mould casting is capable of producing *semiprecise castings* which may not require subsequent machining. However, grinding on casting can be required, if the dimensional accuracy demands.

The *mould material consist of phenolic-resin mixed with fine, dry silica in the presence of alcohol (No water).* Pattern used is made of grey cast-iron, aluminium or brass and is accurately machined (good dimensional and surface finish accuracy).

Normally, the **steps** followed are :

(*i*) Machined pattern with good surface finish is heated to about 230–260°C.

(*ii*) Sand-resin mixture is dumped or blown over its surface.

(*iii*) Hot pattern melts and hardens the resin which binds the sand-grains with resin very closely and firmly within 20–30 seconds (dwell time) over the pattern.

(*iv*) A layer of sand-resin mixture adheres to the pattern. Thus a shell of about 6 mm thickness is prepared. Rest of the sand grains are cleaned off. The thickness of shell is controlled by dwelling-time and it also decides the required strength and rigidity to hold the weight of molten metal to be poured into the shell cavity.

(*v*) The shell-mould with pattern is then heated in an oven to about 300°C for 15–60 seconds (curing time) to make the shell rigid.

(*vi*) Now the pattern is ejected by means of ejector pins mounted on the pattern.

(*vii*) The shell thus formed constitutes of one half of the mould. The second half is also prepared in the similar way. *Two such halves are clamped, and make a complete shell-mould-cavity in which molten metal can be poured.*

The smoothness of the shell-cavity wall is independent to the skill of moulder.

● Castings produced by this process have *better surface finish and closer dimensional tolerances than sand castings.*

● This *process is economical when a large number of similar but shell items with reasonable sharp corners, small projections and thin sections are required.*

Advantages :

1. A very smooth surface is generally obtained.
2. The shell cast parts can be produced with dimensional tolerance of ± 0.2 mm.
3. Reduced cleaning and machining costs.
4. Gives rapid production rate.
5. Uniform grain structure.
6. Minimum finishing operations.
7. The moulds can be stored until required.
8. Complex shapes can be produced with less labour.
9. The process can be automated fairly easily.
10. Less skilled labour is required.

Disadvantages :

1. The resin binder is more expensive than other binders.
2. The initial cost of metal patterns and other specialised equipment is high.
3. Dimensional limitations.
4. Limited to some specific metals.
5. The minimum thickness of section that can be cast is 4 mm.

Applications :

1. Shell-mould casting applications include small mechanical parts that require high levels of precision, such as *gear housings, cylinder heads,* and *connecting rods.*
2. Shell moulding is also widely used in producing high-precision moulding cores, such as *engine block water jackets.*

2.14.3. Plaster-mould Casting

— In this precision casting process, the mould is made of *plaster of paris* (gypsum, or calcium sulphate), with the addition of talc and silica flour to improve strength and to control the time required for the plaster to set. These components are mixed with water, and the resulting slurry is poured over the pattern. After the plaster sets, usually within 15 minutes, the pattern is removed, and the mould is dried to remove the moisture.

— The moulds are then assembled to form the mould cavity and preheated to about 120°C for 16 hours.

— Next, the molten metal is poured into the mould. Because plaster moulds have very low permeability, gas evolved during solidification of the metal cannot escape. Consequently, the molten metal is poured either in a vacuum or under pressure. Plaster-mould permeability can be increased substantially by the *Antioch process.* In this process, the moulds are dehydrated in an autoclave (pressurised oven) for 6 to 12 hours and are then rehydrated in air for 14 hours. Another method of increasing permeability is to use foamed plaster, which contains trapped air bubbles.

- Patterns for plaster moulding are generally made of aluminium alloys, thermosetting plastics, brass or zinc alloys ; wood patterns are not suitable for making a large number of moulds.

- Since plaster moulds have *lower thermal conductivity* than other types of moulds, the *casting cools slowly, yielding more uniform grain structure with less warpage and better mechanical properties.*

Advantages :

1. High dimensional accuracy (of the order of 0.008 to 0.01 mm per mm).
2. Smooth surface.
3. Low porosity.
4. Mould easily repairable.
5. Because of low thermal conductivity of plaster the metal does not chill rapidly and thus *very thin sections can be cast.*

Disadvantages :

1. Limited to non-ferrous metallic castings.
2. Dimensional limitations.
3. Time consuming.
4. The moulds are not permanent and are destroyed when the castings are removed.

Applications :

Used for casting silver, gold, aluminium, magnesium, copper and alloys of those metals (particularly brass and bronze).

2.14.4. Ceramic-mould Casting

The ceramic-mould casting (a precision casting process) is similar to the plaster mould process, with the exception it uses *refractory mould materials suitable for high-temperature applications.* This process is also called **cope-and-drag investment casting.**

— The slurry is a mixture of fine grained zircon (Z_rSiO_4), aluminium oxide, and fused silica, which are mixed with bonding agents and poured over the pattern, which has been placed in a flask. The pattern may be made of wood or metal.

— After setting, the moulds (ceramic facings) are removed, dried, burned off to remove volatile matter, and baked. The moulds are clamped firmly and used as all ceramic moulds. In *shaw process,* the ceramic facings are backed by fireclay to give the moulds strengths. The facings are then assembled into a complete mould, ready to be poured.

- The high-temperature resistance of the refractory moulding materials allows these moulds to be used *in casting ferrous and other high-temperature alloys, stainless steels, and tool steels.*

- The castings have good dimensional accuracy and surface finish over a wide range of sizes and intricate shapes, but the process is *some what expensive.*

Typical parts made are :

— Impellers ;

— Dies for metal working ;

— Clutters for machining ;

— Moulds for making plastic or rubber components (some parts weighing as heavy as 700 kg).

2.14.5. Vacuum Casting

A schematic illustration of *vacuum-casting process,* or *counter-gravity low-pressure* (CL) *process* is shown in Fig. 2.37.

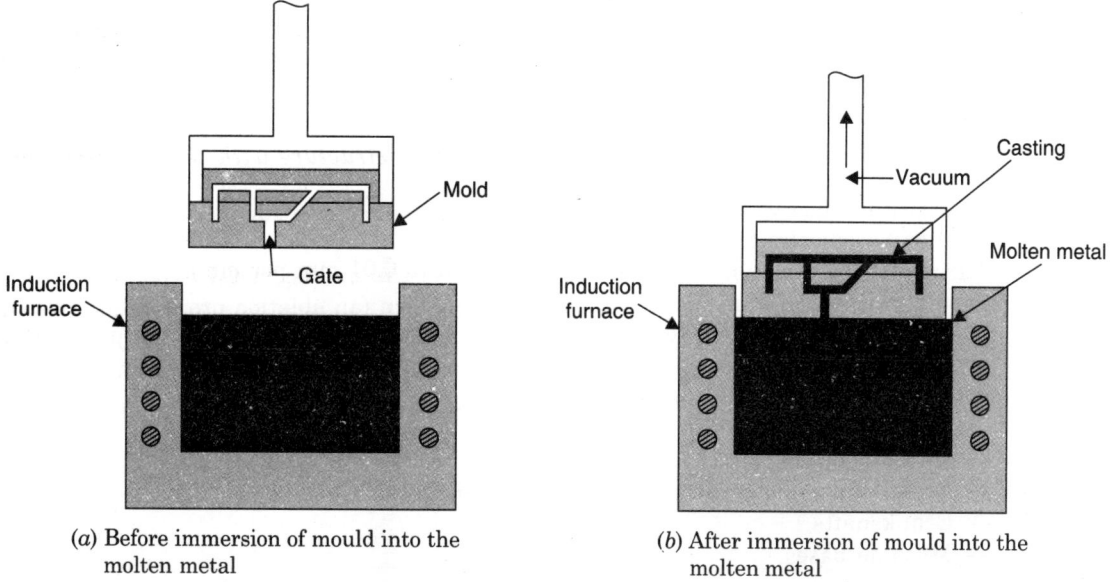

(a) Before immersion of mould into the molten metal

(b) After immersion of mould into the molten metal

Fig. 2.37. Vacuum casting process.

— A mixture of fine sand and urethane is moulded over metal dies and cured with amine vapour. Then the mould is held with a robot arm and partially immersed into molten metal held in an induction furnace. The metal may be melted in air (*CLA process*) or in a vacuum (*CLV process*).

— The vacuum reduces the air pressure inside the mould to about two thirds cf atmospheric pressure, *drawing the molten metal into the mould cavities through a gate in the bottom of the mould.* The molten metal in the furnace is usually at a temperature 55°C above the liquidus temperature ; consequently, the molten metal begins to solidify within a fraction of a second.

— After the mould is filled, it is withdrawn from the molten metal.

  • This process is an alternative to investment, shell-mould, and green-sand casting and is *particularly suitable for thin-walled (0.75 mm) complex shapes with uniform properties.*

— Carbon and low-and high-alloy steel and stainless-steel parts, weighing as much as 70 kg, have been vacuum cast by this method.

— CLA parts are easily made at high volume and relatively low cost. CLV parts usually contain reactive metals, such as aluminium, titanium, zirconium and hafnium.

  • The process can be automated, and production costs are similar to those for green-sand casting.

2.14.6. Evaporative-pattern Casting (Lost foam)

This process is also known as *lost-pattern casting* and under the trade name *"Full-mould process".* It uses a *polystyrene pattern, which evaporates upon contact with molten metal to form a cavity for the casting.*

The evaporative-pattern casting process is *carried out* as follows :

— Raw expendable polystyrene (EPS) beads, containing 5% to 8% pentane (a volatile hydro-carbon), are placed in a preheated die, usually made of aluminium. The polystyrene expands and takes the shape of the die cavity. Additional heat is applied to fuse and bond the beads together.

— The die is then cooled and opened, and the polystyrene pattern is removed. Complex patterns may also be made by bonding various individual sections of the pattern, using hot-melt adhesive.

— The pattern is then coated with a water-base refractory slurry, dried, and placed in a flask.

— The flask is filled with loose fine sand, which surrounds and supports the pattern. The sand is periodically compacted by various means.

— Then, without removing the polystyrene pattern, the molten metal is poured into the mould. This action immediately vaporizes the pattern (an ablation process) and fills the mould cavity, completely replacing the space previously occupied by the polystyrene pattern. The heat degrades (depolymerizes) the polystyrene, and the degradation products are vented into the surrounding sand.

● Typical applications of this process are :
   — Cylinder heads ;
   — Brake components and manifolds for automobiles ;
   — Crankshafts ;
   — Machine bases.

2.14.7. Investment Casting (Precision casting process or Lost Wax casting process)

This is the process where the *mould is prepared around an expendable pattern.* The term investment refers to special covering apparel, in this case a refractory mould, surrounding a refractory-covered wax pattern.

This process involves the following **steps :**

   (i) The pattern is made by injecting semisolid or liquid wax or plastic into a metal die in the shape of a pattern.

   (ii) The pattern is then removed and dipped into a slurry of refractory material, such as very fine silica and binders, ethyl silicate, and acid. After this initial coating has dried, the pattern is coated repeatedly to increase its thickness. Wax patterns require careful handling, because they are not strong enough to withstand the forces involved during mould making.

   (iii) The one-piece mould is dried in air and heated to a temperature of 90°C to 175°C for about four hours, depending on the metal to be cast, to drive off the water of crystallisation (chemically combined water).

   (iv) The molten metal is then poured into the mould. After the metal has solidified, the mould is broken up and casting is removed.

A number of patterns can be joined to make one mould called a **tree,** thus increasing process productivity.

Fig. 2.38 shows the sequences involved in investment casting.

● Investment castings have *excellent surfaces* and *dimensional accuracy* and for this reason, they are *used for parts made of non-machinable and non-forgeable alloys. All extremely complex sections can be produced by this method, since there are no problems of draft, parting lines* and so on (as in sand casting).

Advantages :

1. In average work the close tolerances (± 0.05 mm) are easily maintained.

2. Extremely smooth surfaces are produced.

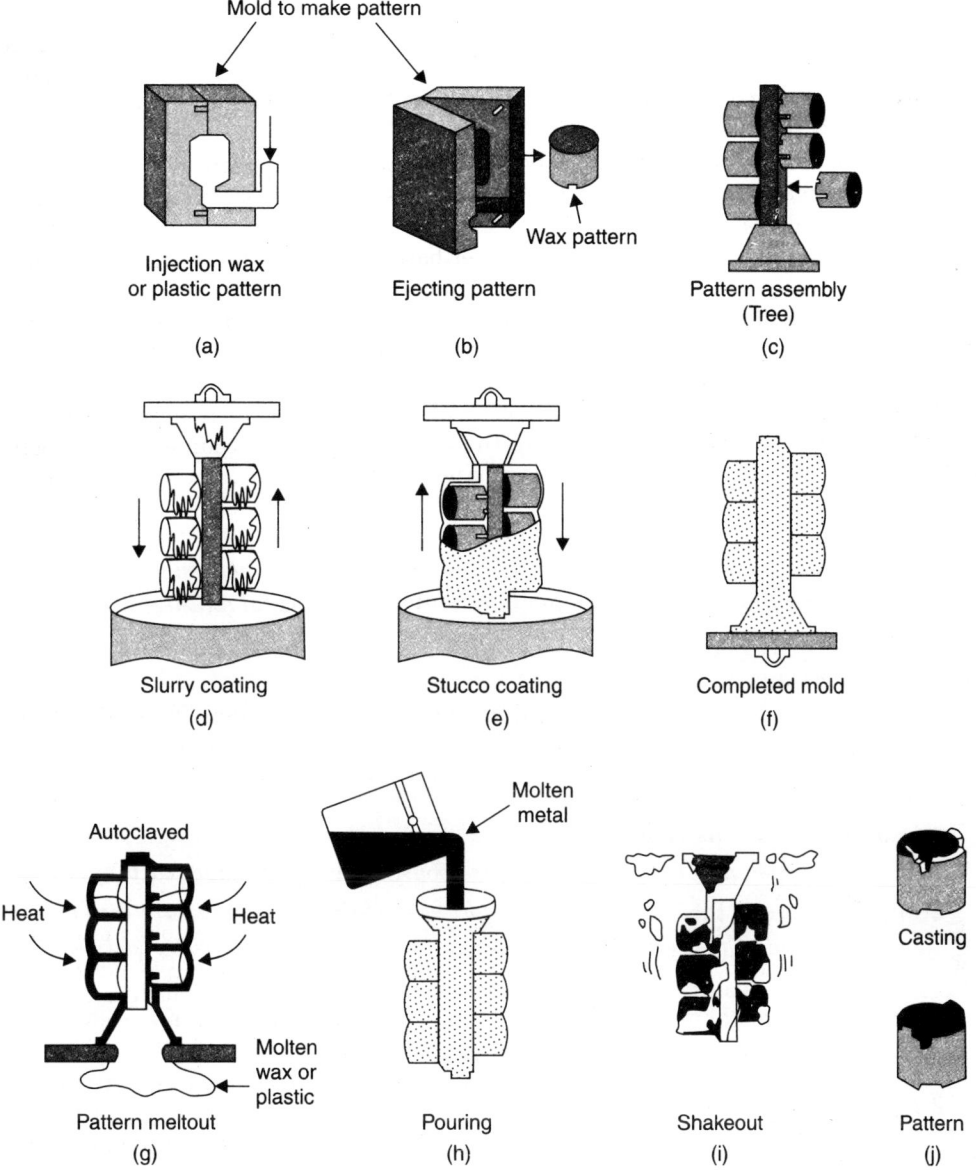

Fig. 2.38. Investment casting—sequences involved.

3. Most machining operations including thread cutting and gear tooth forming are eliminated.

4. Adaptable to metallic alloys.

5. Intricate details can be cast.

6. More than one casting can be made at a time.

7. Undercuts and other shapes, which would not allow the withdrawal of a normal pattern, are easily obtained.

Disadvantages :

1. The large size objects are impractical for investment casting due to equipment size limits.
2. The investment moulds as well as the materials from which they are made are single purpose, therefore they cannot be reused ; this increases the production cost.
3. The process has the limitations in use of and location of holes.
4. The process is expensive.

Applications :

● Parts for computers and data processing equipment, aerospace industry, machine tools and accessories.
● Costume jewellery.
● Nozzles, buckets, vanes blades for gas turbines.
● Radar wave guides etc.

2.14.8. Slush Casting

Slush casting generally involves the processing of *low-temperature-melting alloys.*

The *principle involves pouring the molten metal into a permanent mould. After the skin has frozen, the mould is turned upside down or slung to remove the metal still liquid.*

A thin-walled casting results, the thickness depending on the chilling effect from the mould and the time of operation.

● The process is suitable for *small production runs* and is generally used for making *ornamental and decorative objects and toys from low-melting point metals*, such as zinc, tin and lead alloys.

2.14.9. Pressure Casting

This casting process is also called *pressure pouring* or *low-pressure casting.*

● In this process, the *molten metal is forced upward by gas pressure into the graphite or metal mould. The pressure is maintained until the metal has completely solidified in the mould.*

● The molten metal may also be forced upward by *vacuum,* which removes dissolved gases and produces a casting with lower porosity as well.

2.14.10. Die Casting

Die casting *(or Pressure die casting) is essentially permanent mould casting in which pressure forces the molten metal into the mould cavity. However, the mould used is much more expensive (it is called a "die") and a complex machine is employed to produce casting at a very high rate.* If the molten metal is forced into a metallic die under a gravity head (as done in sand cavity) the process is known as *"Gravity die casting"* or *"Permanent mould casting".*

The mould is normally called a metallic die with two halves. One half is *fixed* and the other is *movable.*

Die materials like medium carbon, low alloys, hot steels are used. *The die carries all details required on the casting* and then molten metal is forced into the cavity. Die is cooled by circulating water to increase the life (of die).

Fig. 2.39 shows the schematic arrangement of 'Die casting'.

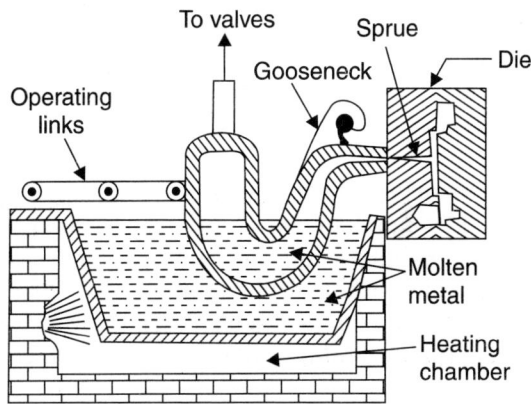

Fig. 2.39. Die casting.

In die casting the following **steps** are involved :

(*i*) Close and lock the two halves of the die.

(*ii*) Force the molten metal into the die cavity under pressure.

(*iii*) Maintain the pressure for a short time, and permitting the metal to solidify.

(*iv*) Open the die halves.

(*v*) Eject the casting with its assembly of sprue, runners and gates, by pins. The above cycle is repeated.

In order to obtain uniformity of die castings and maximum speed of operation it is imperative to employ a predetermined and automatically controlled cycle.

● The weight of most castings range from 90 g to about 25 kg.

● Die casting machines are normally rated by the magnitude of clamping force. Another method specifies the shot-weight capacity of the *injection system*. Machine capacities usually ranges from 50 kN to 25 MN.

Following are the two types of die-casting processes used in practice :

1. Hot-chamber process.

2. Cold-chamber process.

1. *Hot-chamber process :*

● This process involves the use of piston, which traps a certain volume of molten metal and forces it into the die cavity through a goose neck and nozzle. The pressure ranges upto 35 MPa, with an average of about 15 MPa. The metal is held under pressure until it solidifies in the die.

● *Low-melting-point alloys* such as zinc, tin, and lead are commonly cast by this process.

2. *Cold-chamber process :*

● In this process, molten metal is introduced into the injection cylinder (*shot chamber*). The shot chamber is not heated, hence the term *cold chamber*. The metal is forced into the die cavity at pressures usually ranging from 20 to 70 MPa, although it may be as high as 150 MPa. The machines may be horizontal or vertical.

● *High-melting-point alloys* of aluminium, magnesium, and copper are normally cast by this method, although other metals (including ferrous metals) can also be cast in this manner.

Advantages of die casting :

1. Large quantities of identical parts can be produced rapidly and economically.
2. Very little machining is required on the parts produced.
3. The parts having thin and complex shapes can be casted accurately and easily.
4. The die casting requires less floor area than is required by other casting processes.
5. The castings produced by die-casting process are less defective, owing to increased casting soundness.
6. The rapid cooling rate produces high strength and quality in many alloys.
7. Cored holes down to 0.75 mm diameter at accurate locations are possible.
8. Inserts of any metal can be successfully embedded into a die casting, which largely eliminates secondary operations such as drilling and certain types of threading.
9. The sprue, runners and gates can be remelted, resulting in low scrap loss.
10. Die casting dies retain their usefulness and accuracy over a very long time of production.

Disadvantages :

1. The cost of equipment and die is high.
2. There is a limited range of non-ferrous alloys which can be used for die castings.
3. The die castings are limited in size.
4. It requires special skill in maintenance.
5. Die castings usually contain some porosity due to the entrapping of air.
6. The minimum economic quality for die casting is around 20,000.

Applications :

Typical parts made by die casting include :
— Transmission housings ;
— Valve bodies ;
— Carburettors ;
— Motors ;
— Business machine and appliance components ;
— Hand tools ;
— Toys, etc.

2.14.11. Centrifugal Casting

The centrifugal casting is normally carried out in a permanent mould which is rotated during the solidification of casting. For developing the hollowness at the central-axis of rotation, the die is rotated around the same axis (*no core is used in this process*). For developing a solid casting, the axis of rotation is shifted at the end of the casting. Thus, this process is useful to *produce dense casting whether hollow or solid castings. Centrifugal action segregates the less dense non-metallic inclusions and slag particles near the centre of rotation which are removed in a later machining operation.*

— *Cylindrical parts and pipes are most adaptable to this process.*
— The castings are produced with promoted directional solidification as the cold metal is thrown to outer side of the castings and the hotter metal nearer the axis of rotation which further acts as a feeder during solidification of metal.

Following are the *three* types of centrifugal casting :

1. True centrifugal casting.
2. Semicentrifugal casting.
3. Centrifuging or pressure casting.

1. True centrifugal casting :

- In true centrifugal casting, hollow cylindrical parts, such as pipes, gun barrels, and lamp posts, are produced by the technique shown in Fig. 2.40, in which molten metal is poured into a rotating mould. The axis of rotation is usually horizontal, but can be vertical for short-workpieces. Moulds are made of *steel, iron* or *graphite* and may be coated with a refractory lining to increase mould life. Mould surfaces can be shaped so that pipes with various outer shapes, including square or polygonal, can be cast. The inner surface of the casting remains cylindrical, because the molten metal is uniformly distributed by centrifugal forces.

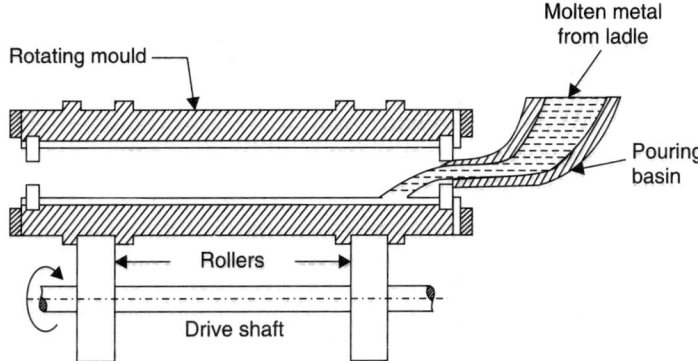

Fig. 2.40. True centrifugal casting process.

- *Castings of good quality, dimensional accuracy, and external surface detail are obtained by this process. In addition to pipes, typical parts made include bushings, engine cylinder liners, and bearing rings with or without flanges.*

2. Semicentrifugal casting :

- It is also known as *profited centrifugal casting*. It is nearly similar to true centrifugal casting with the only difference that a *central core is used to form the inner surface. The particular shape of the casting is produced by the mould and core shapes and not by centrifugal force. However centrifugal force aids in proper feeding of mould cavities.* In this process the axis of *spin is always vertical.*

    This method is used to cast parts with rotational symmetry such as a *wheel with spokes.*

- Rotational speeds for this form of casting are not so great as for the true centrifugal process.

- This process is used only for symmetrical objects and its yield is not as high as that of true centrifugal casting.

- Parts produced by this process include *gears, flywheels,* and *track wheels.*

3. Centrifuging or pressure casting :

In this type of casting, several casting cavities are located around the outer portion of the mould, and the metal is fed to these cavities by radial gates from the centre, either single or stack can be used. The mould cavities are filled under pressure from the centrifugal force of the metal as the mould is rotated.

- This method of casting, *not-limited to symmetrical objects,* can produce castings of irregular shapes such as *bearing caps or small brackets.* The *dental profession uses this process for casting gold inlays.*

- This type of casting is possible only in *vertical direction.*

Advantages of centrifugal casting :

1. Quick and economical than other methods.
2. In this process the use of risers, feed heads, cores etc. is eliminated.
3. The ferrous as well as non-ferrous metals can be casted.
4. The castings produced have dense and fine grained structure with all impurities forced back to the centre where they can be frequently machined out.
5. Good surface finish.
6. Gates and risers can be kept to a minimum.

Disadvantages :

1. Metallic composition of alloys is not uniform throughout the casting.
2. Casting must be symmetrical.
3. Limited to small intricate castings.
- The mechanical properties of centrifugal castings are superior to those of sand castings, but gravity segregation is encountered in some alloys.

2.14.12. Squeeze Casting

This process involves solidification of the molten metal under high pressure ; hence, the process is combination of casting and forging.

The machinery includes a *die, punch,* and *ejector pin.* The pressure applied by the punch keeps the entrapped gases in solution (especially hydrogen in aluminium alloys), and high pressure contact at the die-metal interface promotes heat transfer, resulting in a fine microstructure with good mechanical properties and limited microporosity.

In this casting process, parts can be made to *near-net shape,* with complex shapes and fine surface detail, from both non-ferrous and ferrous alloys.

The pressures required in squeeze casting are typically *higher than those used in pressure die casting, but lower than those for hot or cold forging.*

- Typical products made include :
  - Automotive wheel ;
  - Motor bodies (a short-barreled canon).

2.14.13. Semisolid Metal Forming

This process, also called *semisolid metal working* was developed in 1970s.

- When the metal or alloy enters the die or mould, it has nondendritic, roughly spherical, and fine grained structure. The alloy exhibits its *thixotropic* behaviour, that is, its viscosity decreases when agitated ; hence this process is known as **thixoforming** or **thixocasting.**

Semisolid-metal-forming technology was in commercial production by 1981 and is also used in making *cast-matrix composites.*

  - Another technique for forming in a semisolid state is **rheocasting,** in *which slurry is produced in a mixer and delivered to the mould or die.* However, this process is yet to get commercial success.

2.14.14. Continuous Casting

A *"continuous casting"* or *"strand casting"* produces higher quality steels for less cost. It involves the following *procedure :*

  (*i*) The molten metal in the ladle is cleaned and equalised in temperature by blowing nitrogen gas through it for 5 to 10 minutes.

(*ii*) The metal is then poured into a re-fractory-lined intermediate pouring vessel (tundish), where impurities are skimmed off. The tundish holds as much as three tonnes of metal. The molten metal travels through water-cooled copper moulds and begins to solidify as it travels downward along a path supported by rollers (*pinch rolls*).

(*iii*) Before the casting process is started, a *solid starter,* or *dummy bar,* is inserted into the bottom of the mould. The molten metal is then poured and solidifies on the starter bar. The bar is withdrawn at the same rate the metal is poured. The cooling rate is such that the metal develops a solidified skin (shell) to support itself during its travel downward, at a speed typically of 25 mm/s. The shell thickness at the exit end of the mould is about 12 to 18 mm. Additional cooling is provided by water sprays along the travel path of the solidifying metal. The moulds are generally coated with graphite or similar solid lubricants to reduce friction and ad-hesion at the mould-metal interfaces. The moulds are vibrated to further reduce friction and sticking.

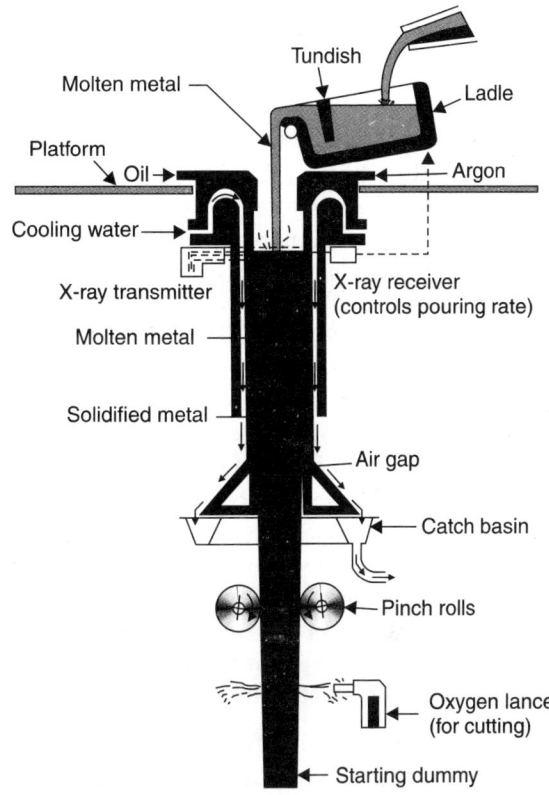

Fig. 2.41. Continuous-casting process for steel.

(*iv*) The continuously cast metal may be cut into desired lengths by shearing or torch cutting, or it may be fed directly into a rolling mill for further reductions in thickness and for shape rolling of products such as channels and *I*-beams.

● In addition to costing less, *continuously cast metals have more uniform compositions and properties than metals obtained by ingot casting.*

Fig. 2.41 shows the continuous-casting process for steel.

● In **strip casting,** thin slabs or strips are produced from molten metal, the metal solidifies in similar fashion to strand casting, but the hot solid is then *rolled to form the final shape.*

2.15. DEFECTS IN CASTINGS

A large number of defects occur in sand castings produced through various methods. The *factors* which are normally responsible for the *production of these defects are* :

(*i*) Design of casting.

(*ii*) Design of pattern equipment.

(*iii*) Moulding and core-making equipment.

(*iv*) Mould and core materials.

(*v*) Gating and risering.

(*vi*) Melting and core-making techniques.

(*vii*) Melting and pouring.

(*viii*) Composition of the metal.

Some of the common defects in casting are described below :

1. Blow holes : Refer to Fig. 2.42.

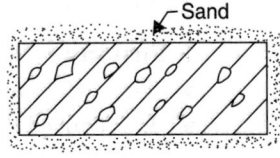

Fig. 2.42. Blow or gas holes.

- They appear as *cavities in a casting*. When they are visible on the upper surface of the casting, they are called "*open blows*". When they are concealed in casting and are not visible from outside, they are known as **blow holes**. They are due to the *entrapped bubbles of gases* in the metal and are exposed only after machining.

- They are caused mainly by *hard ramming, excessive moisture, low permeability, excessive fine grains and incomplete or improper venting.*

2. Misrun : Refer to Fig. 2.43.

- This defect is *incomplete cavity filling.*

- It is caused mainly by *inadequate* metal supply, too low mould or melt temperature and improperly designed gates.

- This defect determines the minimum thickness that can be cast for a given metal, superheat, and type of mould.

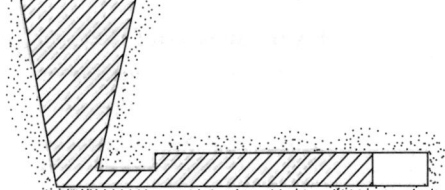

Fig. 2.43. Misrun.

3. Cold shut : Refer to Fig. 2.44.

A *cold shut is an interface within a casting that is formed when two metal streams meet without complete fusion.* The causes are the same as for misrun.

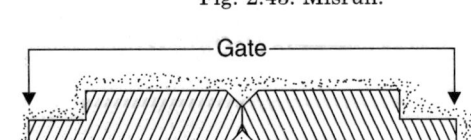

Fig. 2.44. Cold shut.

4. Mismatch : Refer to Fig. 2.45.

- It is *shift of the individual parts of a casting with respect to each other.*

- It is caused by an inexpert assembling of the two halves of the mould and dimensional discrepancy between the core prints of the pattern and the core prints of the core.

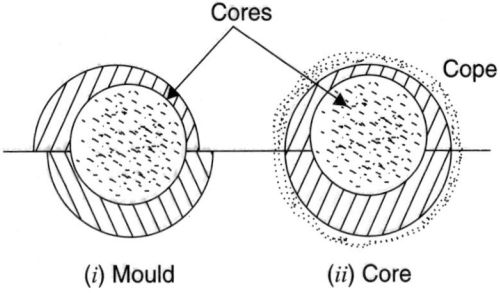

(*i*) Mould   (*ii*) Core

Fig. 2.45. Mismatch.

5. Drop : Refer to Fig. 2.46.

- This defect appears as *an irregular deformation of a casting.*

- It occurs on account of a portion of the sand breaking away from the mould and dropping into the molten metal.

- *Increase in green strength of the sand by suitable modification in its composition, hard ramming and adequate reinforcing of cope and other sand projections by means of bars, nails and gaggers, etc. are the principal remedies of this defect.*

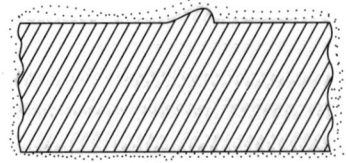

Fig. 2.46. Drop.

6. Flashes or fins :

● These are *thin projections of metal not intended as a part of casting.* These usually occur at the parting line of the mould or core sections.

● These are caused by loose clamping of the mould, insufficient weight on the top part of the mould and excessive rapping of the pattern before it is withdrawn from the mould.

7. Fusion :

● This defect appears as a rough glassy surface over the casting.

● It is caused due to lack of enough refractoriness in sand, faulty gating, too high pouring temperature of the metal and poor facing sand.

8. Metal penetration :

● This defects occurs as a *rough and uneven external surface on the casting.*

● The principal causes for the promotion of this defect are the use of coarse sand, having high permeability and low strength, and soft ramming.

9. Cut or wash : Refer to Fig. 2.47.

● It is a low projection on the drag face of a casting that extends along the surface, decreasing in height as it extends from one side of the casting to the other end.

● It usually occurs in bottom gating castings in which the moulding sand has insufficient hot strength, and when too much metal is made to flow through one gate into the mould cavity.

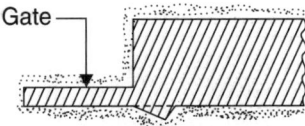

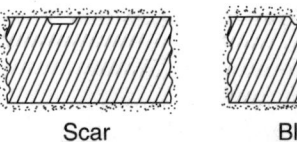

Scar   Blister

Fig. 2.47. Wash.   Fig. 2.48. Scar and blister.

10. Scars and blisters : Refer to Fig. 2.48.

● A *scar* is a shallow blow. It generally occurs on a flat surface, whereas a blow occurs on a convex casting surface.

● A *blister* is a shallow blow like a scar with a thin layer of metal covering it.

11. Hot tears : Refer to Fig. 2.49.

● These are the cracks having *ragged edges due to tensile stresses during solidification.* It is due to the discontinuity in the metal casting resulting from hindered contraction, occurring just after the metal has solidified.

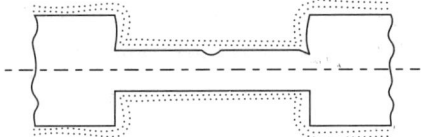

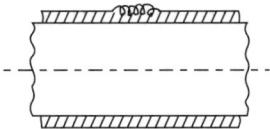

Fig. 2.49. Hot tears.   Fig. 2.50. Sponginess.

● This defect is caused by excessive mould hardness by ramming, high dry and hot strength and improper metallurgical and pouring temperature controls.

12. Sponginess : Refer to Fig. 2.50.

- Sponginess or honeycombing is an external defect, *consisting of a number of small cavities in close proximity.*
- It is caused by *'dirt'* or *'inclusions'* held mechanically in suspension in molten metal.

13. Scab : Refer to Fig. 2.51.

- This defect occurs when a portion of the face of a mould lifts or breaks down and the recess thus made is filled by metal.

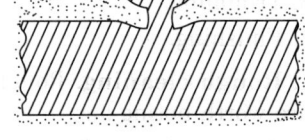

Fig. 2.51. Scab.

- It is caused by too fine a sand, low permeability of sand and uneven ramming of the mould.

14. Swell : Refer to Fig. 2.52.

- A swell is a slight, smooth bulge usually found on vertical faces of castings, resulting from liquid metal pressure.
- It is caused due to low strength of mould because of too high water content or when the mould is not rammed sufficiently.

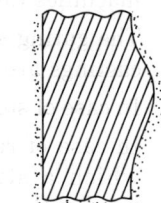

Fig. 2.52. Swell.

15. Buckle : Refer to Fig. 2.53.

- A buckle is a *long, fairly shallow, broad, vee depression that occurs in the surface of flat casting.*
- It occurs due to the sand expansion caused by the heat of the metal, when the sand has insufficient hot deformation. It is also caused due to poor casting design.

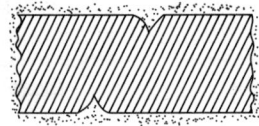

Fig. 2.53. Buckle.

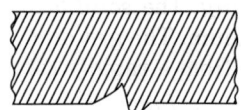

Fig. 2.54. Rat tail.

16. Rat tail : Refer to Fig. 2.54.

A rat tail is a *long, shallow, angular depression in the surface of a flat casting and resembles a buckle except that it is not shaped like broad vee.* The reasons for this defect are the same as for buckle.

17. Slag holes :

- These are smooth depressions on the upper surfaces of the casting. These usually occur near the ingates.
- This defect is due to imperfect skimming of the metal or due to poor metal.

18. Pour short :

- It occurs when the mould cavity is incompletely filled because of insufficient metal.
- This defect occurs due to interruptions during pouring operation, and insufficient metal in the ladles being used to pour the mould.

2.16. CLEANING OF CASTINGS

Generally, the cleaning of casting refers to all operations related to the *removal of adhering sand, gates, risers or other metal not a part of the casting.* The cleaning operations may also include a certain amount of metal finishing or machining for obtaining the required casting dimensions.

The various cleaning operations usually performed on a casting are enumerated and discussed below :

1. Rough cleaning
2. Surface cleaning.
3. Trimming.
4. Finishing.

1. Rough cleaning. Rough cleaning includes the *removal of gates of risers*. The following points are worth-noting :

- In case of a ductile material casting, rough cleaning may be done with mechanical cut-off machines (using abrasive cut-off wheels, band saws and metal shears).
- The gating system of a brittle material casting may be broken off by impact when the castings are dumped and vibrated in shake-out or knock-out devices.
- In case of steel castings, very large risers and sprues may be removed by cutting torches.
- In case of risers being large and cast of oxidation-resisting alloys, *powder cutting* (in which a stream of iron powder in introduced into the oxygen torch flame) is employed.

2. Surface cleaning. Surface cleaning includes cleaning of interior and exterior surfaces when sand, scale and other adhering materials are involved. This type of cleaning involves the following procedures :

(*i*) *Tumbling.* This operation is carried out with a barrel-like machine called *tumbling mill*, which removes sand, scale and some fins and wires.

(*ii*) *Blasting.* The *sand blasting* is performed by using coarse sand as abrasive and air as the carrying medium. The grit or sand blasting is carried out by throwing the metallic particles by centrifugal force from a rapidly rotating wheel.

(*iii*) *Other surface cleaning methods* :

The following methods aid in surface cleaning :

— Wire brushing ;
— Buffing ;
— Pickling ;
— Various polishing procedures.

3. Trimming. Trimming involves the removal of fins, gate and riser pads, chaplets, wires and other similar *unwanted* appendages to the casting which are not a part of its final dimensions.

It involves the following *procedures* :

(*i*) *Chipping.* It is used to remove pins, gates and riser pads, wires etc. It may be carried out by hammer and chisel or by pneumatic chipping hammers.

(*ii*) *Grinding.* It is employed to remove excess metal and is carried out, through portable grinders, stand grinders and swing-frame grinders.

4. Finishing. It is the later stage of cleaning. In certain cases cleaning is complete after trimming operations, but others may require additional surface finishing, *e.g.,* machining, polishing, buffing, etc.

Note. The complete process of cleaning of castings, involving the removal of the cores, gates and risers, cleaning of the casting surface and chipping of any of the unnecessary projections on the surfaces is known as **Fettling.**

2.17. INSPECTION OF CASTINGS

In order to determine the presence of any defects (not readily visible) it becomes necessary to inspect the casting. Following methods are employed to *inspect the casting*.

1. Destructive inspection method. In this type of inspection the casting sample is destroyed during inspection. This method is used to test mechanical properties, *e.g.,* tensile strength, hardness etc. These tests are performed on the test bars or pieces cut from the casting sample.

2. Non-destructive inspection method. Following are the various methods of non-destructive inspection :

(*i*) **Visual inspection.** The main aim of this type of inspection is to ensure that the outward appearance of the casting looks good. Through this inspection the defects like cracks, tears, run outs, swells etc. may be detected.

(*ii*) **Dimensional inspection.** The dimensional inspection may be carried out by surface plates, height and depth gauges, and plug gauges, etc. Through this inspection it can be ascertained whether certain details are within tolerances or not.

(*iii*) **Pressure testing.** It is employed to locate leaks in a casting or to check the overall strength of a casting in resistance to bursting under hydraulic pressure. It is carried out on tubes and pipes.

(*iv*) **Radiographic inspection.** This type of inspection is employed to inspect *internal defects of a casting*, by the use of X-ray or gamma ray technique.

(*v*) **Magnetic particle inspection.** This inspection method is employed on magnetic ferrous castings for detecting invisible surface or slightly subsurface defects.

(*vi*) **Fluorescent penetrant :**

- This type of inspection is employed to find minute pores and cracks on the surface of castings that may be missed even under magnification.

- In this method a fluorescent penetrating oil mixed with whiting powder is applied to the casting surface by dipping, spraying or brushing. The cracks or other defects become visible after the surface has been wiped dry (the oil creeping out of cracks).

(*vii*) **Eddy current inspection.** In this method, the material of the casting need not be ferromagnetic. The test includes a probe which is supplied with a high frequency current. It induces an electric field in the casting. The field changes in the presence of surface or near surface defects. These changes show up on instruments.

QUESTIONS WITH ANSWERS

Q. 2.1. What are the requirements of a good moulding sand ?

Ans. Requirements of good moulding sand :

The requirements of a good moulding sand are :

1. It must allow the free passage of air and gases generated when in contact with molten metal. This is the *permeability* of the sand.

2. When rammed it must retain the shape given to it and resist the pressure of the molten metal. This is known as its *"cohesive"* quality.

3. It must be able to withstand high temperature without fusing. This is called *"refractory"* quality.

4. It should easily come away from the cold casting, and leave a clean, smooth surfaces. This is known as it *"stripping"* quality.

Q. 2.2. Discuss briefly the materials which are added to moulding sand to improve their moulding properties.

Ans. The following materials are added to moulding sand to improve their mould properties :

1. **Coal dust and silica flour.** *These provide surface finish* and resistance to metal.

2. **Saw dust or wood flour :**

- *The addition of wood flour reduces (about 1.5 percent max) the expansion defects while improving the flowability of the moulding sand and help maintain the uniform mould density.*

- *Too high a wood flour makes the moulding sand brittle.*

3. **Starch and dextrin.** These materials *increase resistance to deformation, skin hardness and expansion defects such as scale.*

4. **Iron oxide :**

- Iron oxide (about 3 percent max.) in moulding sand improves surface finish, decreases metal penetration, reduces burn on, increases the chilling effect of the mould and increases glazing.

- With enough iron oxide in combination with silica flour, mould washes could be avoided.

- Iron oxide decreases green strength and permeability while improving the hot strength. It reduces collapsibility and makes the shake out of the mould difficult.

Q. 2.3. Discuss briefly the influence of water-content on moulding sand properties.

Ans.

- Normally, mould is prepared by a mixture of refractory sand, water and same organic additives. In a typical moulding sand, the approximate percentage of different ingredients is used on weight basis *i.e.,* 70–85% sand, 10–20% clay, 3–6% water and 1–6% additives. Percentages may vary depending upon ferrous casting or non-ferrous casting.

- The success of a casting process depends greatly on the properties of the moulding sand *i.e., Strength, permeability deformation, flowability, refractoriness.* These properties should be evaluated to decide upon the quantity of different constituents.

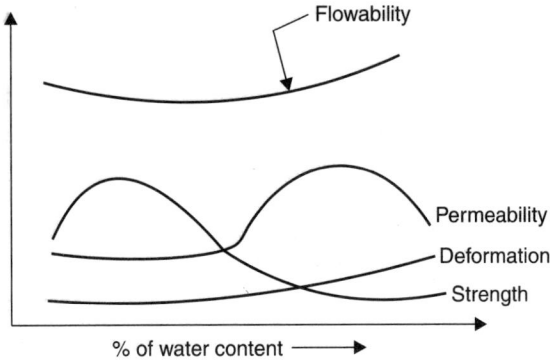

Fig. 2.55

Q. 2.4. What is "Stop-off" ? Explain briefly.

Ans. Stop-off is *the portion of the pattern which is added for its strength only if the pattern is fragile.*

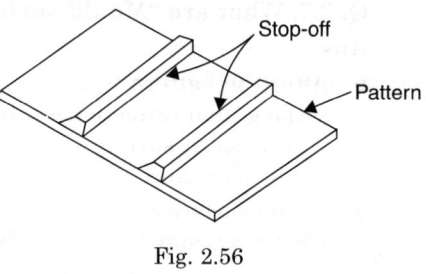

Fig. 2.56

- It forms a cavity in the mould when the moulder withdraws the pattern, which is *refilled* with sand before pouring.
- Stop-offs are wooden pieces used to reinforce some portions of the pattern which are structurally weak, especially from the stand point of repeated handling. They have no connection with the completed casting.

Q. 2.5. Explain briefly the following :

1. Parting sand ; 2. Facing sand ; 3. Backing sand.

Ans. 1. Parting sand :

- This type of sand is sprinkled on the pattern and to the parting surfaces of the mould halves before they are prepared, *to prevent the adherence of the moulding sand.* This helps in easy withdrawal of the pattern and easier separation of the cope and drag flasks at parting surface.
- It is essentially a non-sticky material such as washed silica grains.

2. Facing sand :

- This sand is used next to the pattern to obtain cleaner and smoother casting surfaces.
- Usually, sea coal or coal dust is mixed with the system sand to improve the mouldability and surface finish. The sea coal being carbonaceous, will slowly burn due to the heat from the molten metal and give off small amounts of reducing gases. This creates a small gas pressure in the surroundings of the cavity such that molten metal is prevented from entering into the silica grains or fuse with them. This *helps in generating good casting surface and also lets the moulding sand peel off from the casting during shake out.*

3. Backing sand :

- This sand is normally the *reconditioned foundary sand* and is *used for ramming the bulk of the moulding flask.* The moulding flask is completely filled with backing sand after the pattern is covered with a thin layer of facing sand.
- As the casting is not affected to any great extent by backing sand, it usually contains the burnt facing sand, moulding sand and clay.

Q. 2.6. List some specific properties which moulds should possess to produce sound castings.

Ans. Following are some specific properties which moulds are required to possess to produce sound castings :

1. It should generate minimum amount of gases as a result of the temperature of the molten metal.

2. A mould should have *good venting capacity* to allow the generated gases to completely escape from it.

3. It must resist the erosive action of the flowing hot metal.

4. It must be strong enough to withstand the temperature and weight of the molten metal.

Q. 2.7. What are "Mould surface coatings" ? Explain.

Ans.

● After the pattern is drawn, the mould surfaces are coated with certain materials possessing high refractioness. It eliminates the possibility of burn-on and enables castings with smooth surface to be obtained. However, the permeability of the mould gets reduced. Therefore, the *coatings should not contain gas forming materials.*

● Mould surface coatings which are also known as *facings, dressings, washes, blackings* or *whitening,* may be applied dry (by dusting) or wet in the form of thin cream.

— The various mould surface coating materials include : Coal dust, pitch, graphite, China clay, Zircon flour or French chalk (calcium oxide).

Q. 2.8. Write short note on "Moulding sand for non-ferrous castings".

Ans. The moulding sands for non-ferrous castings may be less refractory and permeable since the melting point of non-ferrous metals is much lower than that of ferrous metals. Further, a smooth surface is desirable on non-ferrous castings. In view of this, the moulding sand for non-ferrous castings contain a *considerable amount of clay and are fine grained.*

Q. 2.9. What is "Stack moulding" ?

Ans. In an upright stack moulding process from 10 to 12 flask sections are arranged one above, having a common sprue through which all the moulds are poured.

● An advantage of this process is that it *requires much less floor space in the foundry.*

● Stack moulding is used to make *small castings.*

Q. 2.10. (*a*) Define the terms "Casting" and "Foundry".

(*b*) What are the types of foundries ?

Ans. (*a*) **Casting** *is the solidified piece of metal which is taken out of the mould.*

Foundry *is a plant where the castings are made.*

(*b*) All the foundries are basically of the following two types :

(*i*) *Jobbing foundries :*

● These foundries are mostly independently owned.

● They produce castings on contract, within their capacity.

(*ii*) *Captive foundries :*

● The foundries are usually a department of a big manufacturing company. They produce castings exclusively for the parent company.

● Some captive foundries which achieve high production, sell a part of their output.

Q. 2.11. What are "Cushion materials" ? Explain.

Ans. The cushion materials (*e.g., wood flour, cereals, cellulose, etc.*) when added to the moulding sand burn and form gases when the molten metal is poured into the mould cavity. This gives rise to the *space for accommodating the expansion of the sand at the mould cavity surfaces.*

Q. 2.12. (*a*) How does machine moulding differ from hand moulding ?

(*b*) What are the advantages and limitations of moulding machines ?

Ans. (*a*) ● **Hand moulding** is a slow and laborious process and does not yield good results as it *does not impart uniform hardness to the rammed mould.*

● In **machine moulding :**

— Production becomes faster.

— Labour is minimised.

— Less skill is required.

— The castings of good quality are produced as the moulds of greater uniformity can be produced by machine moulding.

The moulding machines, for mass production, justify their initial cost and at the same time the quality of work will also be good. However, the use of moulding machine is limited due to size and complexity of the job.

(b) **Advantages and limitations of moulding machines :**

Advantages :

(i) Semi-skilled persons can be employed whereas in hand moulding skilled artisans are needed.

(ii) Prove to be economical by reducing the slow hand operation and fatigue on workers.

(iii) Best suited to mass production work.

(iv) A higher quality of product is maintained.

Limitations :

(i) Cannot be employed for bigger and complex jobs.

(ii) Suitable only for mass production work since for different types of moulds, the machines can't justify their work.

(iii) The moulding machines like flexibility for more specialised mould-making procedure.

Q. 2.13. Discuss briefly "Moulding machines".

Ans. A *moulding machine consists of a large number of interconnected parts and mechanisms which transmit and guide various motions in order to prepare mould.*

According to the method in which the sand is compacted around the pattern to make a mould, the moulding machines are classified as :

1. Jolt moulding machines.

2. Squeeze moulding machines.

3. Sand slingers.

1. **Jolt moulding machines :**

In *Jolt moulding* the sand is first filled into the flask from an overhead hopper, and then it is raised to a certain height before it is allowed to free fall onto a solid bed plate. The resulting impact forces the sand to get compacted uniformly into the mould. This lifting and dropping process continues repeatedly till the required mould hardness is achieved.

● This type of ramming is *suitable for horizontal surfaces.*

Limitations :

(i) Noisy operation.

(ii) Non-uniform density of sand around the pattern. The sand at the bottom experiences highest force and consequently is packed well compared to the sand in the layers.

(iii) Considerable load on the foundation.

(iv) The moulding flask should be of sufficient strength because it has to be lifted and dropped repeatedly.

2. **Squeeze moulding machines :**

In this type of machine sand is compressed through the application of compressed air or other suitable force transmitted through a piston-table arrangement which squeezes the sand against a platen. The ramming is best at the sand platen interface from where pressure is applied and not near the platen.

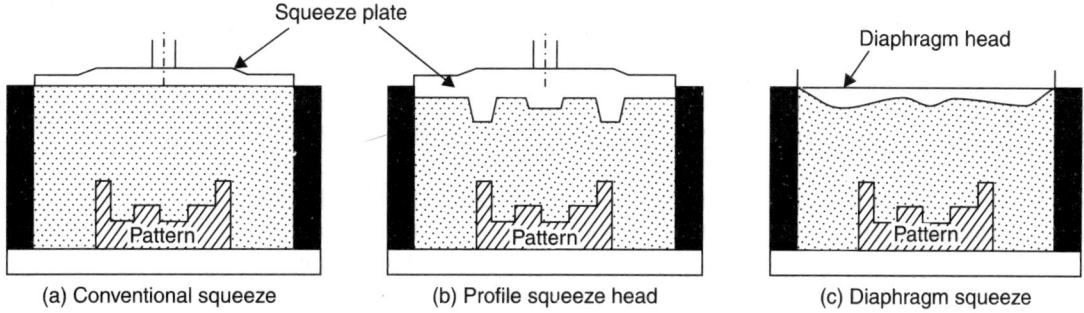

Fig. 2.57. Types of squeeze heads used for machine moulding.

— Fig. 2.57 (*a*) shows a conventional squeeze comprising a plate slightly smaller than the inside dimensions of the moulding flask fitted into the flask already filled with the moulding sand.

— Fig. 2.57 (*b*) shows a profile squeeze head which is provided with contour to match the pattern to achieve better uniform hardness of the mould.

— Fig. 2.57 (*c*) shows a diaphragm squeeze which may be used to provide differential ramming force required for the contour of the pattern.

● This machine is *suitable for small castings and for small flasks.*

Limitations :

(*i*) Plain squeezing is limited only to shallow work.

(*ii*) In case the pattern contains cavities for formation of green sand cores, squeezing does not make sand flow into cavities effectively and get it packed properly.

(*iii*) The results are not as good as in jolting.

Note. For most of the cases, it is preferable to use *combined squeezing and jolting.*

3. **Sand slingers :**

In these moulding machines the *sand is thrown out by centrifugal force* from a rapidly rotating single bladed impeller and directed over pattern in the flask. The moulding sand should be mixed with suitable binders, etc., so that it can flow into place readily and afterwards binder should become hard.

● In this method the *sand density is more uniform* and at all the levels and it is, therefore, *best suited for any type of mould.* However, it is best adapted to work ranging from medium to large size.

— The density can be controlled by changing the impeller speed.

Advantages :

(*i*) The process is *very fast.*

(*ii*) The ramming is *very uniform.*

Limitation :

Initial cost of equipment is *high* comparatively.

Q. 2.14. **What is mould wash ? Explain briefly.**

Ans.

● After the pattern is withdrawn purely carbonaceous materials such as sea coal, finely powdered graphite or proprietary compounds are applied to the mould cavity. This is called *"mould wash"* and is done by spraying, swabbing or painting in the form of a wet paste. These are used essentially for the following *reasons :*

(*i*) To avoid mould-metal interaction and prevent sand fusion.

(*ii*) To prevent metal penetration into the sand grains and thus ensure a good casting finish.

- To deposit the mould wash, either water or alcohol can be used as a carrier. But because of the problem of getting the water out of mould, alcohol is preferred as a carrier. The proprietary washes are available in powder, paste or liquid form. The powder needs to be first prepared and applied whereas the paste and liquid can be applied straight away.

Q. 2.15. How can the casting defect 'Hot tears' be eliminated ?

Ans. In a casting, *'hot tears'* result from temperature gradients, establishing different rates of contraction during solidification and thereby inducing stresses due to resistance of the sand of a magnitude sufficient to cause fracture. These can be *minimised by adopting good design i.e., avoiding abrupt changes in section, sharp angles and non-uniform webs connected to flanges.*

Q. 2.16. Explain briefly "CO_2 moulding process".

Ans. CO_2 moulding process is a sand moulding process in which sodium silicate ($Na_2O \cdot x \cdot SiO_2$), that is, water glass is used as a binder, 2 to 6 per cent, rather than clay. After the mould is prepared, CO_2 gas is made to flow through the mould and the sand mixture hardens through a very rapid reaction (of duration about 1 minute, which is very less than several hours needed to produce a dry sand mould) yielding a stiff gel.

- CO_2 moulds can be *used for producing very smooth and intricate castings, because the sand mix has a very high flowability to fill up corners and intricate contours.*

Advantages :

(*i*) CO_2 moulds can be made without flasks.

(*ii*) Sands are free flowing, therefore ramming is eliminated or reduced.

(*iii*) Tensile strength of moulds are higher than those of conventional moulds. This permits reduction of mould weight and easier handling of large moulds.

Disadvantage :

Sands must be used *immediately.*

Q. 2.17. What are composite moulds ? Explain briefly.

Ans. The composite moulds are made of two or more different materials, such as shells, plaster, sand with binder and graphite. These moulds combine the advantages of each material.

- These moulds are used in shell moulding and other casting processes, generally for casting *complex shapes,* such as turbine impellers.

Advantages :

(*i*) Increased mould strength.

(*ii*) Improved dimensional accuracy, and surface finish of castings.

(*iii*) Reduced overall costs and processing times.

Q. 2.18. What is dry sand moulding ?

Ans. Invariably the sand mould is *dried* after pattern is removed in order to increase the strength of moulding sand so that it can withstand the higher static pressures of the liquid in case of *big castings.*

- Sometimes clay-bonded dried mould castings are used to obtain *greater dimensional tolerances* and obtain *better surface finish.*

Q. 2.19. Explain briefly CO_2 process of making cores.

Ans. CO_2 process of making cores is briefly discussed below :

● Clean and dry sand is mixed with a solution of sodium silicate and rammed or blown into the core box.

● The mixture is then gassed with CO_2 gas for several seconds, and CO_2 forms a silica gel which binds the sand grains into a strong solid form. If required, additional hardening of resulting core can be carried out by baking.

— Sodium silicate and sand mixture gel cannot be reclaimed after use.

Q. 2.20. List the various moulding defects.

Ans. Various moulding defects are :

● Sand spots.
● Internal air pockets.
● Sand holes.
● Blow holes.
● Honey combing or sponginess.
● Oversize castings.
● Poured short.
● Swelling.
● Mismatch, lifts and shifts.

Q. 2.21. What are the characteristics of sand castings ?

Ans. Following are the *characteristics of sand castings :*

1. Poor ductility.

2. Porous in nature.

3. Lower density and poor strength.

4. Less stronger than wrought products.

5. Poor finish.

6. Less costly.

7. Castings obtained by moulding process have good hardness.

8. The sand castings are susceptible to cooling cracks if proper care in design is not taken.

9. The suitability of sand castings lies with high melting point of molten metal.

10. The moulding method is suitable for moderate and particularly large castings and unsuitable for thinner sections.

11. The sand castings can be made internally sound by minimising gas evaluation during solidification by avoiding turbulence while pouring.

Q. 2.22. Explain briefly the following :

(i) Core dressing ;   (ii) Core chaplet.

Ans. (*i*) **Core dressing.** It is operation of applying a compound to the surface of a core, either in the green state or after baking, for the purpose of *providing protection against the scouring action of flowing molten metal and to assist the formation of a smooth surface in a cored hole or cored form.*

(*ii*) **Core chaplet :**

● Core chaplet is a metal location piece inserted in a mould either to prevent a core shifting its position or to give extra support to a core.

● The molten metal melts the chaplet which then forms the part of the cast material.

Q. 2.23. List some important design rules which must be considered in sand casting.

Ans. Following are some important design rules which need be considered in sand casting to ensure maximum dispersal of stress, and minimum stress concentration :

1. In order to increase resilience of ductile metals to fatigue rupture external corners should be *rounded* with radii of 10–20 percent of the section thickness.

2. Complex sections like *X, Y, V* and *K* should be staggard to staggard *T*.

3. In joining sections of dissimilar size or at *L* or *T* junction radii equal to the thickness of small section should be provided.

— In case of *L* junctions, largest possible radii should be used.

4. The load bearing capability with complex loads is increased by adopting tubular or reinforced *C* sections rather than *I, H* or channel sections.

Q. 2.24. Explain briefly "Casting yield".

Ans. All the metal that is used while pouring does not end up as a casting. There are some losses in the melting. Also there is a possibility that some castings may be rejected because of the presence of various defects. On completion of the casting process, the gating system used is removed from the solidified casting and remelted to be used again as raw material. Hence the casting yield is the proportion of actual casting mass.

● Casting yield depends to a great extent on the *casting materials and complexity of the shape.* Generally those materials which shrink heavily have lower casting yields. Also massive and simple shapes have higher casting yield compared to small and complex parts.

● *Higher the casting yield higher is the economics of the foundry practice.* It is therefore desirable to give consideration to *maximising the casting yield, at the design stage itself.*

Q. 2.25. What are the technical aspects of making sound castings ?

Ans. Following are some important technical aspects of making sound castings :

1. In order to facilitate removal of pattern from moulds, or separate solidified casting from permanent moulds tapers must be incorporated in all sections.

— Taper also *induces directional solidification.*

2. As far as possible all corners should be rounded and re-entrant sections must be avoided to *permit free flow of liquid metal.*

3. In order to allow for shrinkage of metal from freezing point to room temperature pattern maker's shrinkage allowance should be provided.

4. For forming cavities and re-entrant sections cores must be used.

5. The number and complexity of cores melted should be minimised.

Q. 2.26. How are gases formed in casting and how can these be eliminated ?

Ans.

● The gases in castings may appear as *gas holes* and *pin holes.* These defects can be avoided by proper riser design and adequate venting of permeable moulds.

● Another source of gases is from the dissolved gases in the liquid metal at high temperature, which on cooling are given off. The gases in melts can be reduced by *vacuum melting and vacuum degassing* (placing liquid metal in low pressure chamber to remove dissolved gases).

Q. 2.27. Write short note on "Residual stresses in castings".

Ans. In case of non-uniform cross-section casting different sections solidify at different rates depending on their cross-sectional areas. This results in varying amount of contraction in different parts, producing high internal stresses which may cause *tearing or cracking of casting.*

High residual internal stresses can be avoided by *placing chills over large cross-sectional areas so that whole of casting cools at uniform rate.*

Such stresses can also be controlled by taking out casting at an average temperature of around 750°C and putting it in an *insulated pit and allowing to cool at 5.5°C/hour.*

Following points are worth mentioning :

- Any temperature gradient above 540°C does not give rise to elastic strain because same is relaxed to plastic strain due to high rate of creep.
- In case high residual stresses exist in casting it has to be relieved by a suitable heat treatment or by other methods of stress relieving.

Q. 2.28. What is chill casting ? Explain briefly.

Ans.

- Chill casting process is nearly similar to sand casting and is *used where very hard outer surfaces and wear resistant castings are required.*
- Moulds are mode of sand or cast iron and for the purpose of *chilling the cast iron, steel blocks are used.*
- Metallic chills are used at outer surfaces so that the rate of cooling increases and hence hardness increases. Where hardness is of ample importance, *metallic moulds* are used as in the case of railway brake shoe. The excessive chilling effect is reduced by preheating the moulds.
- In case of bushes and bearings, the inner surfaces of the holes should be hard and wear resistant and to meet this requirement *core chills* are used in the moulds. Extensive chills are used to reduce the possibility of the defect called 'hot-tear'.
- The cooling rate has a considerable effect upon the hardness of the surface ; the greater the rate of cooling the lesser amount of carbon will come out in graphite state. In other words, carbon will be in *combined form* and hence casting will be hard.
- Following are the *examples of chill casting* :
- Wheel tread.
- Tram-car wheels.
- Chilled rolls used in rolling mills and machine tools ways.
- Railway brake shoe.
- Crusher jaw.

Q. 2.29. What are the relative advantages and disadvantages of the following types of furnaces used in foundry shops ?

Crucible furnace ; Induction furnace ; Arc furnace, Open health furnace ; Converters ; Cupola.

Ans. 1. **Crucible furnace :**

- The melting pot is of graphite.
- It could be fired by gas, oil or coke.
- Its *cost is more than that of cupola but quality of iron is better.*

2. Induction furnace :
- It is *easier to melt* and obtain different grades of iron.
- Its *cost is high* and power requirement is also *more*.

3. Arc furnace :
- These furnaces are *good for melting steel*.
- The additives can be *easily* added.

4. Open health furnace :
- The grade of iron produced is *good*.
- In this furnace hot gases are circulated.
- Lining needs to be renewed every year.

5. Cupola :
- Its material cost is *low*.
- It is a *high production unit*.
- The quality of melt is average.
- A water jacket is necessary at the melting zone.

Q. 2.30. What do you mean by the term "Fluidity" ? Explain briefly.
Ans.
- The term *"fluidity"* is normally used in a foundry to *designate the casting material's* ability to fill the *mould cavity*.

Fluidity depends on the *casting material* as well as the *mould*.

The *casting material's properties* which affect the fluidity to a great extent are :
— Viscosity of the melt.
— Heat content of the melt.
— Surface tension.
— Freezing range.
— Specific weight of the liquid metal.

The *mould properties* that affect the fluidity are :
— Thermal characteristics.
— Permeability.
— The mould cavity surface.
- The most commonly used fluid test is the *spiral fluidity test*.

Q. 2.31. What is "System sand" ?
Ans.
- In mechanised foundaries, where machine moulding is employed a so called "system sand" is used to fill the whole flask. Since the whole mould is made up of this system sand, the strength, the permeability and refractoriness of the sand must be higher than that of backing sand.
- Whereas facing sand is always used to make dry sand moulds, the *system sand is frequently used for green sand moulding*.

Q. 2.32. Explain briefly the following :
(a) Process of crystallisation of pure metals.
(b) Dendritic solidification.

Ans. (a) Process of crystallisation of pure metals :

- A crystal may be visualised as forming from a *centre of freezing, or nucleus,* which is *composed of a small group of atoms* oriented into one of the common crystal patterns, (Fig. 2.58). During the process of solidification many of these nuclei spring up, each nucleus being a potential crystal and able to grow to form a crystal large enough to be seen with the unaided eye. As each nucleus is a growing crystal, and the atoms within it are all similarly oriented, no nucleus within the freezing melt may form with its planes or groups of atoms the same as those of any other nucleus. Thus, when the individual crystals have grown to the point where they have absorbed all of the liquid atoms and, therefore, come in contact with each other along their boundaries, *they do no line up, i.e., their planes of atoms change direction in going from one crystal to another. This results in the solid states being composed of a number of crystals of different orientation, and we have a crystal aggregate or mixed crystals.* Each crystal, therefore, is composed of a group of similarly oriented atoms, but on going from one crystal to the neighbouring crystals, the orientation changes.

Melt ——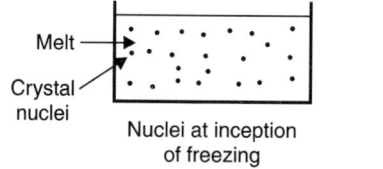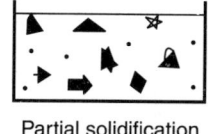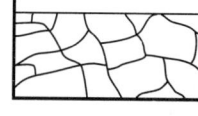

Crystal
nuclei

Nuclei at inception   Partial solidification   Completion of crystallisation
of freezing   dendritic grains nuclei   growing crystals meet

Fig. 2.58. Progressive freezing of a uniformly cooled metal.

- So far the nature of the *crystal border* is still not known completely but we may assume that *it is an interlocking border line where the atoms of one crystal change orientation from the atoms of another crystal.* It may be that these are some left-over atoms along this border line separating the differently oriented crystals. Such atoms not knowing to which crystal to attach themselves, and these act as *non-crystalline cement between the various crystals.* This *condition may account for the greater strength of the crystal* boundaries as compared to the strength of the individual crystals, and for many of actions that take place at the crystal boundaries.

Fig. 2.59 shows the *structure of a polycrystalline metal.*

Primary axis ⟍   ⟋ Secondary axis

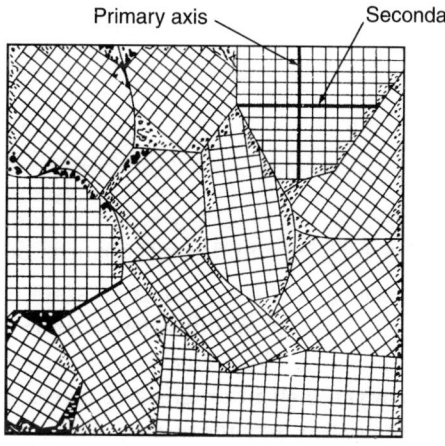

Fig. 2.59. Structure of a polycrystalline metal.

(*b*) **Dendritic solidification :**

The crystals which form in the process of solidification of a metal may have many different structures (dendritic, lamellar, needle-type or acicular etc.) depending on the *rate of cooling, and the type and amount of admixtures or impurities in the melt.*

Perfect crystals of proper external shape can be obtained only if crystallisation develops under conditions when *the degree of supercooling is very slight and the metal has a very high purity.*

In the great majority of cases, branched or tree-like crystals are obtained, which are called *dendrities* (Fig. 2.60).

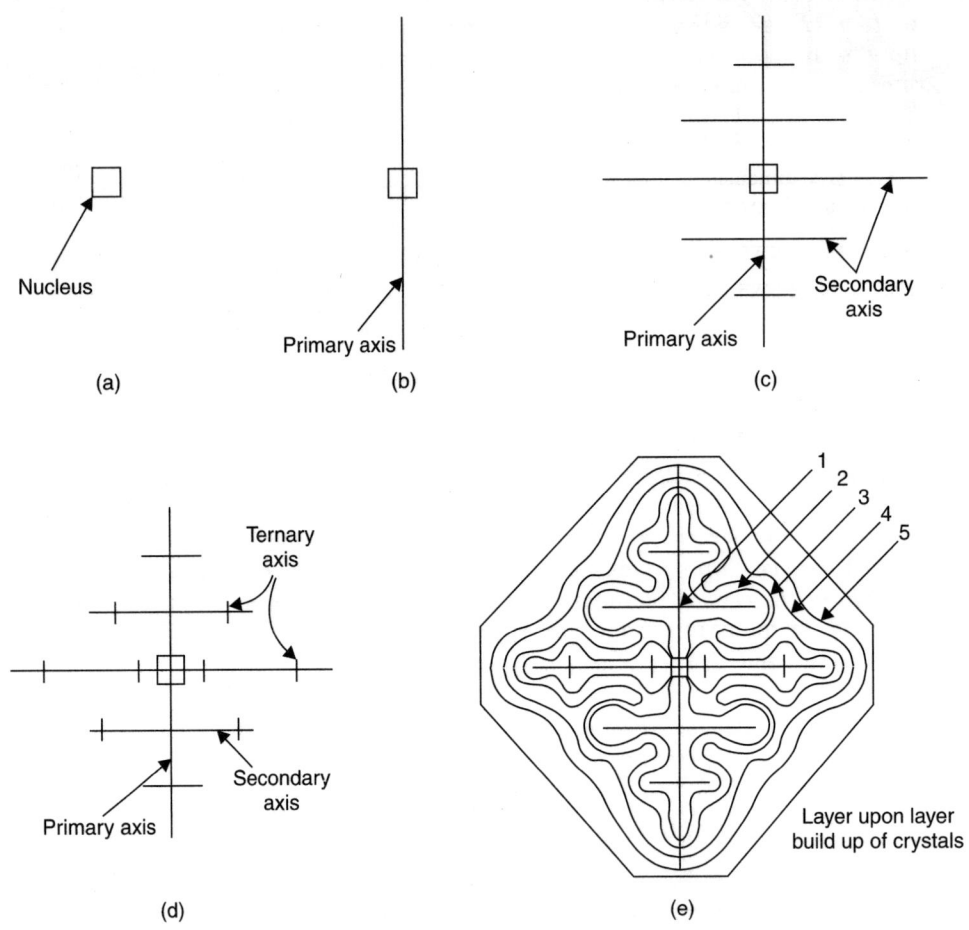

Fig. 2.60. Formation of a crystal.

Fig. 2.60 shows steps in the formation of a crystal. A crystal nucleus forms as shown in (*a*) and then proceeds to send out shoots or axes of solidification as shown in (*b*), (*c*), (*d*), forming the skeleton of a crystal in much the same way as frost patterns form. Atoms then attach themselves to the axes of the growing crystal from the melt in progressive layers, finally filling up these axes, and thus forming a completed solid or crystal, as shown in (*e*). Three dimensional view of *dendritic growth* is shown in Fig. 2.61.

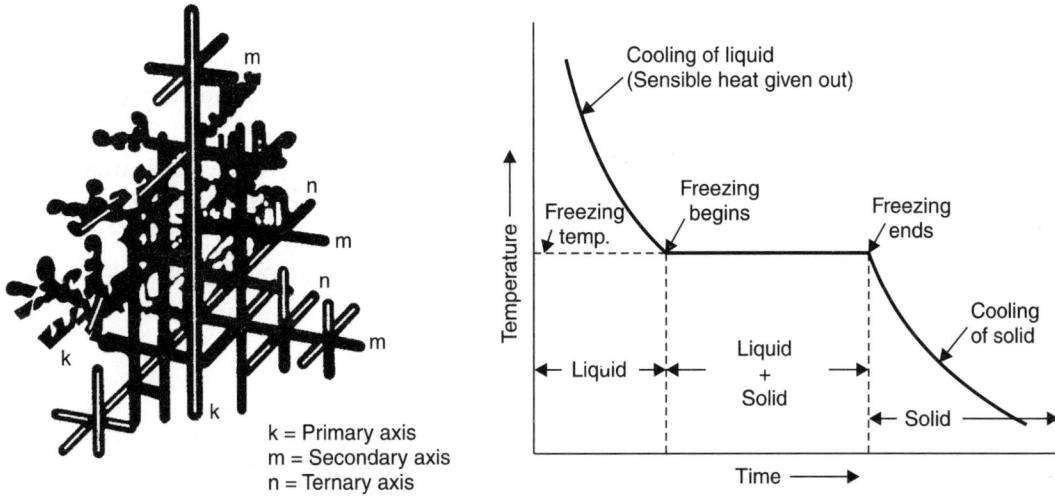

Fig. 2.61. Three dimensional view of dendritic growth.

Fig. 2.62. Cooling curve for a pure metal.

- Fig. 2.62 shows a cooling curve for a pure metal.
- Fig. 2.63 shows a cooling curve for a Cu-Ni alloy.

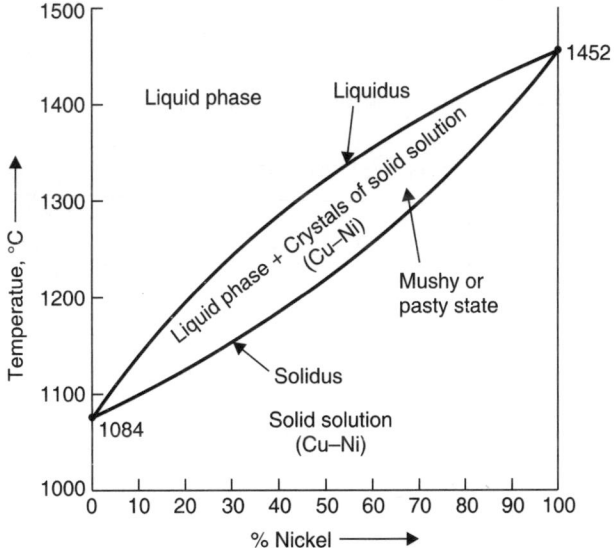

Fig. 2.63. Cooling curve for an alloy.

Q. 2.33. What is critical size of nucleation that nucleates from melt ? Deduce the expression for free energy concept.

Critical radius of nucleus :

Nucleation of supercooled grains depends upon the following two factors :

1. *Free energy available from the solidification process :*

It depends upon the volume of the particle formed. The replacement of old phase (molten state) by the new phase (solid) accompanies a *free energy decrease,* ΔF per unit volume and this contributes to the stability of the new phase.

The free energy for a spherical particle

$$= -\frac{4}{3}\pi R^3 \Delta F \qquad \text{(–ve sign indicates decrease in energy)}$$

where, R = Radius of sphere.

2. *Energy required to form a liquid-solid interface :*

Particles formed in the melt have some surface area. Solid-liquid phases possess a surface in between the two. Such a surface has a +ve free energy σ per unit area, associated with it.

Creation of new interface results in free energy increase which is

$$= \sigma \times 4\pi R^2$$

∴   Net free energy, $\Delta f = -\frac{4}{3}\pi R^3 \Delta F + \sigma \times 4\pi R^2$ ...(*i*)

In order to get critical radius, R_c, let us proceed as follows :

$$\frac{d(\Delta f)}{dR} = 0$$

i.e.,

$$\frac{d}{dR}\left(-\frac{4}{3}\pi R^3 \cdot \Delta F + \sigma \times 4\pi R^2\right) = 0$$

or

$$-\frac{4}{3}\pi \times 3R^3 \Delta F + \sigma \times 4\pi \times 2R = 0$$

∴

$$R_c = \frac{2\sigma}{\Delta F}$$

Substituting this value of R_c in equation (*i*), we get

$$\Delta f_c = -\frac{4}{3}\pi\left(\frac{2\sigma}{\Delta F}\right)^3 \Delta F + \sigma \times 4\pi \times \left(\frac{2\sigma}{\Delta F}\right)^2$$

$$= -\frac{4}{3}\pi \times \frac{8\sigma^3}{(\Delta F)^3} \times \Delta F + \sigma \times 4\pi \times \frac{4\sigma^2}{(\Delta F)^2}$$

$$= -\frac{32}{3}\pi \times \frac{\sigma^3}{(\Delta F)^2} + 16\pi \times \frac{\sigma^3}{(\Delta F)^2}$$

$$= \frac{\pi\sigma^3}{(\Delta F)^2}\left(-\frac{32}{3} + 16\right) = \frac{16}{3} \cdot \frac{\pi\sigma^3}{(\Delta F)^2}$$

i.e.,

$$\Delta f_c = \frac{16}{3} \cdot \frac{\pi\sigma^3}{(\Delta F)^2} \cdot$$

Q. 2.34. Explain the following :

Normal segregation, Gravity segregation, Micro-segregation and Macro-segregation.

Ans. Segregation *is the separating out of the constituent elements with different freezing temperatures when a liquid alloy metal cools and solidifies.*

(*i*) **Normal segregation :**

- Normal segregation occurs, when the solidification moves away from the mould walls *as a plane front* (dendrites are not formed).

- The constituents of alloy having lower freezing point are driven towards the centre of the resultant casting.

(ii) **Gravity segregation :**

- This type of segregation takes place due to *gravity*.
- The elements having higher density sink to the bottom of the mould while higher elements will float to the surface. This results in a casting richer in higher density elements at the lower part and the upper part will be richer in higher elements.

(iii) **Micro-segregation :**

- The "cored dendrites" formed (see Fig. 2.64) have higher concentration of alloying elements at the surface than at the core of the dendrite ; this is due to solute rejection from the core towards the surface during solidification of dendrite. This is known as *"Micro-segregation"*.

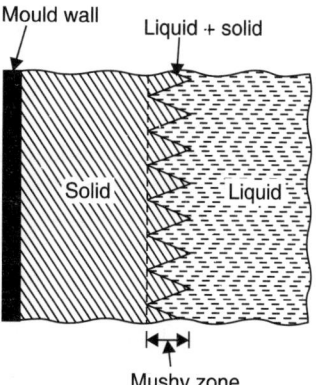

Fig. 2.64. Solidification of a liquid metal.

(iv) **Macro-segregation :**

- An alloy consists of elements which freeze at different temperatures. When a liquid is poured into a mould, the higher freezing point metal will form the first crystal and the metals having lower freezing points will start nucleating.
- In a casting mould, the cooling and solidification starts from the mould walls and progresses inwards towards the centre of the mould. Consequently, the outer surface of the casting will be richer in higher freezing point elements and the central zone of the casting will be richer in lower freezing point constituents. Owing to this, the casting will be non-homogeneous in properties and will contain different composition from crust to core. This is known as *"Macro-segregation"*.

Q. 2.35. Discuss the role of moisture content (water percent) on the performance of sand casting moulds.

Ans. Moisture content affects the various properties of moulds as under :

1. It increases the permeability first and then it starts decreasing permeability when water content is more.

2. It increases the compressive strength first at faster rate and subsequently at slower rate.

3. It decreases the green strength.

Q. 2.36. Why risers are not used in die casting ?

Ans. Die castings are made by forcing molten metal at high pressure into a split steel die cavity. Within a fraction of second, the fluid alloy fills the entire die. The die is water cooled ; therefore low temperature is being maintained. *Because of the low temperature of the die, the casting solidifies quickly.* Therefore *risers are not required in die casting.*

Q. 2.37. What purpose is served by risers in sand casting ?

Ans. Riser *is a hole cut or moulded in the cope to permit the molten metal to rise above the highest point in the casting.*

- Riser serves as feeder to feed the molten metal into the main casting cavity to *compensate for shrinkage*. The design of the riser should be such that it establishes temperature gradients within casting so that the casting solidifies directionally towards the riser. It *also helps in easy ejection of steam, gas and air from the mould cavity while filling the mould with the molten metal.*

Risers act as reservoir and heat gradient regulator and provide the necessary fluid metal to *compensate solidification.*

- *In case no riser is provided during casting the solidification will start from walls and liquid metal in the centre will be surrounded by a solidified shell and the contracting liquid will produce the voids towards the centre of casting.*

Q. 2.38. What are the causes of blow holes in castings and how they can be minimised during casting operations ?

Ans. Normally gases exist in a molecular form but at higher temperature and in contact with metal a significant portion of gases may dissociate in the atomic form and enter into the metal, these gases can be accommodated in the relatively loose and disordered structure of melts. This causes high solubility of gases above the melting point. *Solubility of gases falls steeply as the melt solidifies, some gas may be trapped in the solid in the atomic form but much is rejected at the solid liquid interface to combine into molecules.* These molecules come together and form a group into gas bubbles which rise in the melt or if trapped during solidification cause *gas porosity (pinholes or larger blow holes)* in the structure. These gas pores are generally round and if they contain a neutral or reducing gas, they have a clear, bright surface. They too can be regarded as inclusions of zero strength but their large radius makes them less damaging to mechanical properties. The *blow holes can also cause blistering.*

All the gases in all the metals are not equally soluble. Hydrogen is soluble in practically all metals because of small size of its atoms. Hydrogen may be introduced into the melt by dissociation of water from air, the charge or combustion products. In constrat to hydrogen, nitrogen is soluble in iron but not in non-ferrous metals. Noble gases (of which argon is technically most significant) are completely insoluble.

The solute S of any gas in melt increases or decreases with square root of the partial vapour pressure (p_g) of the gas over the melt (which is known as *Sivert's law*) and is given by :

$$S = K \sqrt{p_g}$$

where, K = The equilibrium constant.

It follows that the *concentration of any gas in the melt can be reduced by either reducing the overall pressure (by drawing a vacuum) or by bubbling a non-soluble scavenging gas through the melt just before pouring.* Because the partial pressure of the offending gas is zero in the scavenging gas bubbles, the offending gas is drawn out of the solution into the rising scavenging gas and is removed.

Q. 2.39. What is a centrifugal casting ? For what type of jobs would you recommend this casting process ?

Ans. The centrifugal casting process is carried out in a permanent mould which is *rotated during the solidification of the casting.*

For producing a *"hollow part"* the axis of rotation is placed at the centre of the desired casting. The speed of rotation produces a centripetal acceleration which segregates less dense non-metallic inclusions near the centre of rotation.

"Solid parts" can be made by a variation of this process by placing the entire mould cavity *on outside of the axis of rotation.*

- The castings produced by this method are *very dense and are used for such critical parts as cylinder liners, etc.*

Q. 2.40. Explain briefly the following :

(*i*) True centrifugal casting ;

(*ii*) Semi-centrifugal casting ;

(*iii*) Centrifugal casting.

Ans. (*i*) **True centrifugal casting.** In this process the metal is held against the wall of the mould by centrifugal force, and *no core is required to form a cylindrical cavity on the inside.*

True centrifugal casting is used for *pipes, liners* and *symmetrical objects* that *are cast by rotating the mould about its horizontal or vertical axis.*

Vertical castings are much smaller in size and weight because of the instability of a spinning vertical cylinder.

(*ii*) **Semi-centrifugal casting.** In this type of casting, the mould is completely free of metal as it is spun about its vertical axis, and *riser and cores may be employed.* The centre of the casting is usually solid, but *because the pressure is less there, the structure is not so dense and inclusions and entrapped air are often present.*

Rotational speeds for this form of centrifugal casting are *not so great as for the true centrifugal process.*

- This method is *normally used for parts in which the centre of the casting will be removed by machining.*

(*iii*) **Centrifugal casting.** In this case, several casting cavities are located around the outer portion of a mould, and metal is fed to these cavities by radial gates from the centre, either single or stack can be used. The mould cavities are filled under pressure from the centrifugal force of the metal as the mould is rotated.

- The centrifuge method, *not limited to symmetrical objects,* can produce castings of irregular shape such as *bearing caps or small brackets.* The *dental profession uses this process for casting gold inlays.*

Q. 2.41. Describe the need of investment casting. Explain the investment casting process.

Ans. Need of investment casting :

- Investment casting is used *when intricate shapes, good dimensional accuracy and a very good surface finish are required.*
- Investment casting is suitable for *high melting point alloys* as well as difficult to machine metals. It is also suitable for *processing small size castings having intricate shapes.*

Investment casting process :

The various **steps** in investment casting are as under :

1. Preparation of heat-disposable pattern, together with its gating system is done by injecting wax or thermoplastic into the die cavity.

2. A *'tree'* is prepared from number of such pattern fixed to a wax or plastic runner bar with a suitable ceramic cup to act as pouring basin.

3. The tree is then dipped into a ceramic slurry (containing silica flour in ethyl acetate). Sufficient fine silica sand is sprinkled on the tree dipped in ceramic slurry. This enables the formations of a self-supporting ceramic shell mould to be formed around the wax assembly.

4. The ceramic shell mould is then baked so that the wax melts and flows out leaving a precise mould cavity.

5. The shell is fired between 850°C and 1000°C to eliminate all the wax and give more strength to the mould.

6. The molten metal is poured into the mould while it is still hot and cluster of castings is obtained.

- These days, *lost wax process* is used for manufacturing larger objects like *cylinder heads, crankshafts,* etc. In these processes *'styrofoam'* is used *instead of wax.*

Q. 2.42. Explain briefly 'slush casting'.

Ans.

- Slush casting generally *involves the process of low-temperature-melting alloys.*
- In this process hollow castings can be produced without the use of cores. The *principle involves pouring the molten metal into a permanent mould. After the skin has frozen the mould is turned upside down or slung to remove the metal still liquid. A thin-walled casting results, the thickness depending on the chilling effect from the mould and the time of operation. Toys and ornaments* are made by this process from zinc, lead or tin alloys.

Q. 2.43. Explain very briefly the following casting processes :

(i) Electroslag casting ; (ii) Precision casting ; (iii) Pressed casting.

Ans. (*i*) **Electroslag casting.** This process is unusual in that it does not employ a furnace. Instead, consumable electrodes melting or striking beneath a slag layer furnish molten metal to fill a water-cooled permanent mould.

(*ii*) **Precision casting.** It employs techniques that enable very smooth, highly accurate castings to be made from both ferrous and non-ferrous alloys.

(*iii*) **Pressed casting.** This method of casting resembles both the gravity and slush processes but differs somewhat in procedure. This is also called *corthias casting.*

Q. 2.44. Describe the pre-design consideration in the design of castings.

Ans. A product designer who selects casting as the primary manufacturing process should make a design not only to serve the function (by being capable of withstanding the loads and the environmental conditions to which it is going to be subjected during its service life) but also to facilitate or favour the casting process.

Following are some *design considerations and guidelines :*

(*i*) *Promote directional solidification.* When designing the mould, the riser should be properly dimensioned and located to *promote solidification of the casting towards risers.* In other words the *presence of large sections or heat masses in locations distant from the risers should be avoided* and good rising practice should be followed. *Failure to do so may result in shrinkage cavities, porosity or cracks in those large sections distant from the risers.*

(*ii*) *Avoid the shortcomings of columnar solidification.* Dendrites often start to form on the cold surface of the mould and then grow to form a columnar casting structure. This almost always results in planes of weakness at sharp corners. Therefore *rounding of the edges is necessary to eliminate the development of planes of weakness.*

(*iii*) *Avoid hot spots.* The rate of solidification (and the rate of heat dissipation to start with) is slower at locations having a low ratio of surface area to volume. Such location are usually referred to as *hot spots* in the foundry practice. Unless precautions are taken during the design phase, hot spots and consequently shrinkage cavities are likely to occur at the L, T, V, Y and $+$ junctions. Shrinkage cavities can be avoided by modifying the design. Also, it is *always advisable to avoid abrupt changes in sections and to use taper together with generous radii, to join thin to heavy section.*

(*iv*) *Avoid the causes of hot tears.* Hot tears in casting are defects caused by tensile stresses as a *result of restraining a part of the casting.* Efforts should be made to avoid the causes of hot tears.

(v) *Ensure easy pattern withdrawal.* It is important that the pattern can be easily withdrawn from the mould. Undercuts, protruding bosses (especially if their axes do not fall within the parting plane) and the like should be avoided.

Q. 2.45. What is a continuous casting ?

Ans. The *"continuous casting process" consists of continuously pouring molten metal into a mould that has the facilities for rapidly chilling the metal to the point of solidification and then withdrawing it from the mould.*

- It has been proved through research and experimental work that there are many opportunities for cost economies in the continuous casting having a degree of soundness and uniformity not possessed by other methods of producing bars and billets.

Q. 2.46. List the advantages claimed by permanent mould casting over sand casting.

Ans. The *permanent mould casting claims the following advantages over sand casting :*
1. Smoother surfaces.
2. Freedom from gas porosity.
3. Freedom from sand and dirt.
4. Rapid rate of production.
5. Low cost of unskilled labour needed.
6. Closer tolerance.
7. Free from internal shrinks.
8. High percentage of good castings.

Q. 2.47. Explain briefly the following :
(i) Composite moulds ; (ii) Rammed graphite moulding.

Ans. (*i*) **Composite moulds :**

— These moulds are made of *two or more different materials* and are used in shell moulding and other casting processes. Commonly used moulding materials include shells, plaster, sand with binder, metal and graphite.

— These moulds may also include cores and chills to control the rate of solidification in critical areas of castings.

— Composite moulds have *increased strengths, improve the dimensional accuracy* and *surface finish of castings,* and *may help reduce overall costs and processing time.*

- They are generally used for *casting complex* shapes, such as *impellers for turbines.*

(*ii*) **Rammed graphite moulding :**

— *Rammed graphite,* instead of sand, is used to make moulds for *casting reactive metals,* such as titanium and zirconium. Sand cannot be used because these metals react vigorously with silica.

— In this method, the moulds are packed rather like sand moulds and are then air dried, baked at 175°C, fired at 870°C, and stored under controlled humidity and temperature.

- The casting procedures are similar to those for sand moulds.

Q. 2.48. What are 'Bars' and 'Gaggars' ? Explain.

Ans. Added support is normally required with large cope flasks to *keep the moulding sand from sagging or falling out when the cope is raised to remove the pattern.*

This support is provided by two elements, namely *'bars'* and *'gaggars'* (see Fig. 2.65).

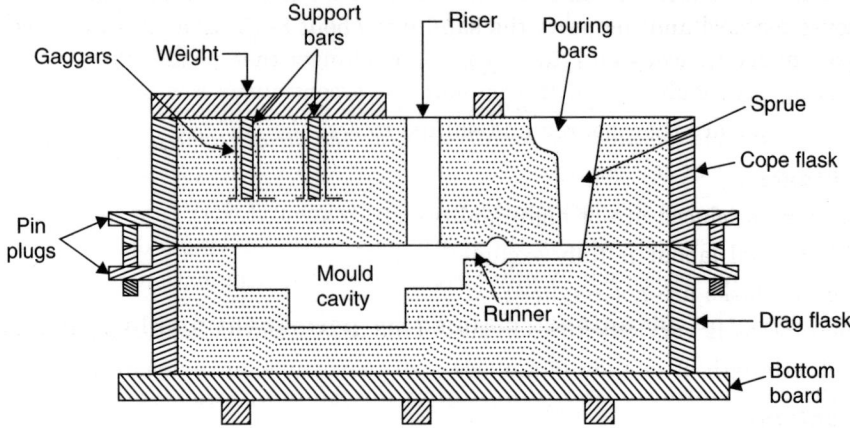

Fig. 2.65. Bars and gaggars.

- The **bars** subdivide the large cope flask areas into smaller areas which are able to support themselves. The bars should extend downward to within 25 to 50 mm of the pattern or parting surface.

- When added support is needed, **gaggars** should also be used. A gaggar is a *L*-shaped steel rod (6.35 to 12.5 mm in diameter).

Gaggars extend vertically downward from the top of the bar within 25 to 50 mm of the pattern or parting surface. The gaggars are generally dipped in a clay slurry to improve the bonding of the mould sand to the gaggar. Or their surfaces should be rough to help them obtain a better grip in the sand.

Q. 2.49. Explain briefly "Flaskless moulding process".

Ans. The flaskless mould process employs the principle that *when properly bonded and high strength moulding sands of not less than 185 kPa in green state, are used and the ramming pressures are high enough (around 8 to 15 MPa), the sand acquires adequate strength to maintain the integrity of the mould.*

This moulding process uses a vertically parted flaskless moulding machine and two pattern plates, one fastened to the pressure plate of the machine and the other to the counter pressure plate (Fig. 2.66).

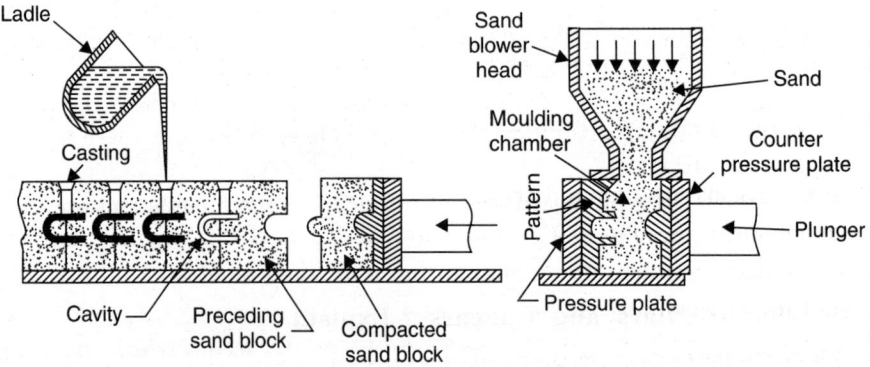

Fig. 2.66. Automatic flaskless moulding.

— Initially, sand is blown from blower head into the moulding chamber under the action of compressed air. The entrance to the chamber is chopped off and a squeeze plunger moves forward and compacts the sand into about 8–15 MPa. The pressure plate of the machine moves away and turns aside. The plunger then pushes the sand block, thereby forming the mould cavity. In this manner, a continuous row of moulds is produced.

— The moulds are then poured and shifted to a shake out grid.

Advantages :

(i) Castings produced are of high quality.

(ii) The manual labour is fully excluded.

(iii) Casting flasks are not required.

(iv) The process is very effective for mass production, as well as for batch production.

(v) High strength of moulds minimises wall movement during casting.

Applications :

(i) Malleable iron pipe fittings.

(ii) Brass valve bodies.

(iii) Gas stove grills, etc.

Q. 2.50. What is Electro-magnetic casting ? Explain.

Ans. This recent casting process is a variant of conventional continuous casting method. The molten *metal is contained and solidified in an electro-magnetic field instead of in a conventional mould. There is no sliding contact of the molten metal with the mould walls, and the metal can be solidified by direct water impingement.*

The surface finish of the casting is very good and the process can be automated.

The method is being *used for obtaining continuous strands of metals.*

Q. 2.51. List the factors which influence the choice of the proper casting method.

Ans. The following factors influence the choice of the proper casting method :

1. Volume of castings required.

2. Weight.

3. Complexity of the shapes, that is, external and internal shape, minimum wall thickness.

4. Types of cores needed.

5. Cost of the pattern or die.

6. Working characteristics (strength rigidity, stiffners, etc.) required.

7. Surface quality required.

Q. 2.52. Enumerate the methods used for repairing and salvaging defective castings.

Ans. The methods used for repairing minor defects of the castings are :

1. Cold welding.

2. Hot welding.

3. Liquid metal welding.

4. Metal spraying.

5. Luting and impregnation—In these methods, the defects in the castings are repaired with the help of sealing agents.

Q. 2.53. What is a 'Aspiration effect' in a casting process ? Explain.

Ans. The metal velocity in a faulty design may be high and thus *pressure may fall below atmosphere and the gases originating from baking of organic compounds may alter molten metal stream, producing porous castings.*

Following are the two possible cases where *negative pressure* may be experienced :

(*i*) *Sprue design.* Refer to Fig. 2.67. It will be seen that pressure at points 1 and 3 is atmospheric and at 2 (By Bernoullie's theorem) is negative (sprue shown by dotted lines). This problem may be overcome by making sprue tapered, as shown in firm lines between 2 and 3.

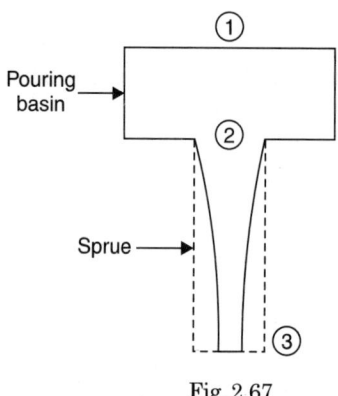

Fig. 2.67

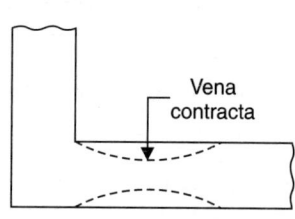

Fig. 2.68

(*ii*) *Sudden change in flow direction.* Refer to Fig. 2.68. In this case, due to change in direction of flow of metal, vena contracta effect is experienced. The negative pressure in this region can be avoided by making the mould shape as per profile of vena contracta.

Q. 2.54. Explain briefly hand tools commonly used in foundry.

Ans. The various hand tools commonly used in foundry are described below :

1. Shovel : Refer to Fig. 2.69.

- It consists of a square pan fitted with a wooden handle.

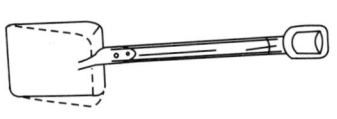

Fig. 2.69. Shovel.

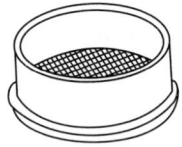

Fig. 2.70. Riddle.

- It is used for *mixing and for moving sand from one place to another in the foundry.*

2. Riddle : Refer to Fig. 2.70.

- It consists of a wooden frame fitted with a screen of standard wire mesh at its bottom.
- It is used for *hand riddling of sand to remove foreign material from it.*

3. Rammers :

- A rammer is a tool used by foundry workers for consolidating sand. Unless the sand is rounded evenly, swelling may occur in the metal in the neighbourhood of any soft spots, so that the resulting casting will not be true.

- Fig. 2.71 shows floor rammer and bench rammer, the only difference being the size of the shaft and weight of head.

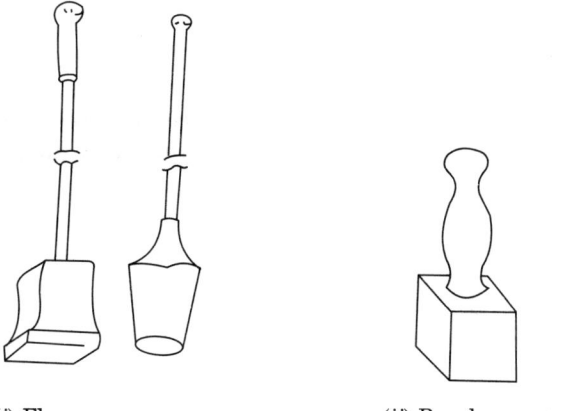

(*i*) Floor rammers   (*ii*) Bench rammers

Fig. 2.71. Rammers.

4. Strike off bar : Refer to Fig. 2.72.

- It is a flat bar, made of wood or iron, to *strike off the excess sand from the top of a box after ramming.*

- Its one edge is made bevelled and the surface perfectly smooth and plane.

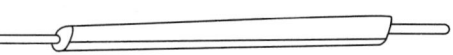

Fig. 2.72. A strike off bar.

5. Vent wire : Refer to Fig. 2.73.

After ramming and striking off the excess sand, vent wire is *used to make small holes in the sand mould to allow the exit of gases and steam during casting.*

Fig. 2.73. Vent wire.

6. Slick : Refer to Fig. 2.74.

- It is a small double ended tool having a flat on one end and a spoon on the other. It is also made in a variety of other shapes.

- A slick is used for *repairing and finishing the mould.*

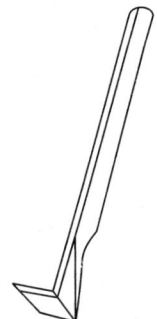

Fig. 2.74. Slick.   Fig. 2.75. Lifter.

7. Lifter : Refer to Fig. 2.75.

● It is made of thin sections of steel of various widths and lengths with one end bent at right angles.

● Lifters are used for *removing loose sand from inside the mould cavity.*

8. Swab :

● A simple swab is a small brush having long hemp fibres.

● A *bulb swab* (Fig. 2.76) has a rubber bulb to hold the water and a soft hair brush at the open end.

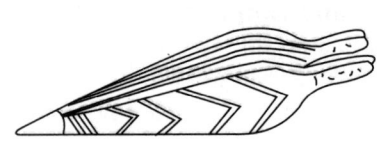

Fig. 2.76. Swab bulb.   Fig. 2.77. Bellow.

● It is used for *moistening the sand around the edge before the pattern is removed.*

9. Bellow : Refer to Fig. 2.77.

It is *used to blow out the loose or unwanted sand from the surface and cavity of the mould.*

10. Trowels : Refer to Fig. 2.78.

● Trowels are used for *finishing flat surfaces and joints in a mould.*

● They are *made of iron and are provided with a wooden handle.*

Fig. 2.78. Trowels.

11. Gate cutter : Refer to Fig. 2.79.

● A gate cutter is a "*U-shaped*" piece of thin sheet.

Fig. 2.79. Gate cutter.

- It is *used for cutting shallow trough in the mould to act as a passage for the hot metal.*

12. Draw screws and rapping plate : Refer to Fig. 2.80.

- Draw screws are straight mild steel rods carrying a loop or ring at one end and a wood or machine screw at the other. They are always used in conjunction with a *rapping plate* for rapping and withdrawing the pattern from sand.

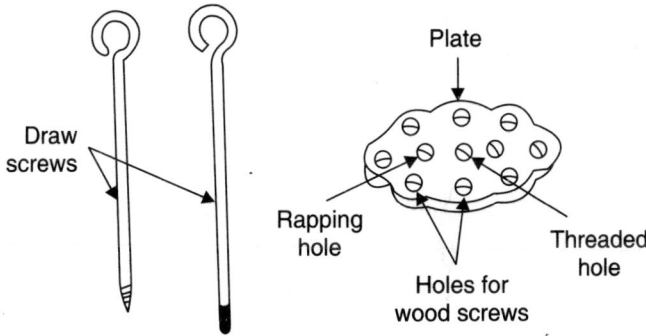

Fig. 2.80. Draw screws and rapping plate.

- The rapping plate is provided with several holes (see Fig. 2.80) to accommodate either a wood screw type or machine screw type draw rod. Rapping holes are provided to accommodate separate rapping rods so that the threaded holes are not spoiled.

13. Sprue pin :

Sprue pin is embedded in the sand mould and later withdrawn to produce a hole called runner, through which the molten metal is poured into the mould.

14. Mallet : Refer to Fig. 2.81.

- It is similar to a wooden mallet as that used in carpentry work.

- In foundry work a mallet is used for *driving the draw spike into the pattern and then rapping it.*

Fig. 2.81. Mallet.

15. Gaggars :

These are *bent pieces of wires and rods and are used for reinforcing the downward projecting sand mass in the cope.*

16. Clamps, cotters and wedges :

They are *made of steel* and are used *for clamping the moulding boxes firmly together during pouring.*

HIGHLIGHTS

1. *Casting* is perhaps the oldest method of manufacturing and invariably the first step in the sequence of manufacturing a product.

2. The *basic features* common to various casting process are :

   (*i*) Pattern and mould.

   (*ii*) Melting and pouring.

   (*iii*) Solidification and cooling.

   (*iv*) Removal, cleaning, finising and inspection.

3. A *pattern* is defined as a model of a casting, constructed in such a way that it can be used for forming an impression (mould) in damp sand.

4. The following *allowances* are usually provided in a pattern :

   (*i*) Shrinkage allowance.

   (*ii*) Draft or taper allowance.

   (*iii*) Machining allowance.

   (*iv*) Rapping or shaping allowance.

   (*v*) Distortion allowance.

5. A *mould* may be defined as the negative print of the part to be cast and is obtained by the pattern in the moulding sand container (boxes) into which molten metal is poured and allowed to solidify.

6. *Sand moulding methods* may be classified as follows :

   (*i*) Green sand moulds.

   (*ii*) Dry sand moulds.

   (*iii*) Skin-dried moulds.

   (*iv*) Loam moulds.

   (*v*) Metal moulds.

7. A *core* is a specially designed shape employed to take the place of metal in a mould.

8. *Casting* means the pouring of molten metal into a mould, where solidification occurs.

9. The various important casting processes are :

   (*i*) Sand casting.

   (*ii*) Shell moulding.

   (*iii*) Permanent mould casting.

   (*iv*) Die casting.

   (*v*) Centrifugal casting.

   (*vi*) Investment casting.

   (*vii*) Plaster casting.

   (*viii*) Slush casting.

   (*ix*) Continuous casting.

OBJECTIVE TYPE QUESTIONS

Fill in the blanks or say 'Yes' or 'No :

1. .... is perhaps the oldest method of manufacturing and invariably the first step in sequence of manufacturing a product.

2. A .... is defined as a model of casting, constructed in such a way that it can be used for forming an impression (mould) in damp sand.

3. A pattern consisting of two pieces is called a two piece...................... pattern.

4. Match plate patterns are suited for mass production of small castings in moulding machines.

5. .... patterns are generally applied to circular work, like rings, wheels, rims, gears, etc.

6. The construction of metals/alloys is always ...., but the contraction allowances are always expressed in .... measures.

7. .... allowance is the extra material provided on certain details of a casting so that the casting may be machined to exact dimensions.

8. .... allowance is provided to compensate for enlargement of mould cavity because of excessive rapping.

9. For manufacture .... moulding is suitable.

10. Moulding is carried out in moulding boxes called ....

11. Sand mould is an example of permanent mould.

12. Bench moulding is used for large castings.

13. In pit moulding, the pit acts a drag.

14. .... moulds are used in the casting of low-melting temperature alloys.

15. A .... is a specially designed shape employed to take the place of metal in a mould.

16. A core box is a type of box used for production of sand cores.

17. Stickles are frequently used with core boxes to obtain shapes of large works.

18. An open hearth furnace is used chiefly for the production of steel and for refining purposes.

19. .... means the pouring of molten metal into a mould, where solidification occurs.

20. Permanent mould casting is a casting process in which .... moulds and .... are employed.

21. Investment casting is a process also known as the lost wax process or precision casting.

22. A .... is an interface within a casting that is formed when two metal streams meet without complete fusion.

23. Mismatch is shift of the individual parts of a casting with respect to each other.

24. A scar is a shallow blow.

25. .... are the cracks having ragged edges due to tensile stresses during solidification.

26. .... defect occurs when a portion of the face of a mould lifts or breaks down and the recess thus made is filled by metal.

27. Rough cleaning of castings does not include the removal of gates or risers.

ANSWERS

1. Casting	2. Pattern	3. Split	4. Yes	5. Segmental
6. Volumetric, linear	7. Machining	8. Rapping	9. machine	10. Flasks
11. No	12. No	13. Yes	14. Metal	15. core
16. Yes	17. Yes	18. Yes	19. Casting	20. steel, cores
21. Yes	22. Cold shut	23. Yes	24. Yes	25. Hot tears
26. Scab	27. No			

THEORETICAL QUESTIONS

1. Define the term casting.
2. Why casting is preferred over other methods of manufacturing ? Discuss.
3. What are the basic features common to various casting processes ?
4. What is a pattern ?
5. Discuss briefly various pattern materials.
6. List the factors on which the types of patterns depend.
7. Enumerate the various types of commonly used patterns.
8. Explain briefly with neat sketches any *three* of the following patterns :
   (i) Solid pattern.   (ii) Loose piece pattern.
   (iii) Skeleton pattern.   (iv) Segmental pattern.
   (v) Cope and drag pattern.   (vi) Sweep pattern.
9. List the various types of allowances which are usually provided in a pattern.
10. Explain briefly the following pattern allowances :
    (i) Shrinkage allowance.   (ii) Machining allowance.
    (iii) Rapping or shaking allowance. (iv) Distortion allowance.
11. Define the following terms :
    (i) Mould.   (ii) Moulding.
12. Name the basic operations which are performed on the moulding machines.
13. What are the two main classes of moulding machines ?
14. Explain briefly with neat sketches the 'mould making' process.
15. What is the difference between a 'temporary mould' and a 'permanent mould' ?
16. How are moulding processes classified ?
17. Explain briefly the following moulding processes :
    (i) Floor moulding.
    (ii) Machine moulding.
18. How are sand moulding methods classified ?
19. Explain briefly any two of the following moulding methods :
    (i) Green sand moulding (moulds).   (ii) Dry-sand moulds.
    (iii) Loam moulds.   (iv) Metal moulds.
20. What are the advantages and limitations of green sand moulding ?
21. What is a core ? How are 'cores' made ?
22. How are cores classified ?
23. Explain briefly any two of the following cores :
    (i) Green sand cores.   (ii) Dry sand cores.
    (iii) Oil sand cores.   (iv) Loam cores.
    (v) Metal cores.
24. What are core prints ?
25. What is a core box ? Explain with a neat sketch ?
26. Explain briefly the following properties of moulding sand :
    (i) Permeability.   (ii) Strength or cohesiveness.
    (iii) Refractoriness.   (iv) Plasticity or flowability.
    (v) Collapsibility.   (vi) Adhesiveness.
    (vii) Coefficient of expansion.   (viii) Chemical resistivity.
27. How are moulding sands classified ?
28. Give the composition of the green sand.

29. Name the main type of furnaces used in foundaries for melting various varieties of ferrous and non-ferrous metals and alloys.

30. Explain briefly any two of the following furnaces :
    (*i*) Crucible furnace.   (*ii*) Reveberatory or air furnace.
    (*iii*) Open hearth furnace.   (*iv*) Electric furnace.
    (*v*) Cupola furnace.

31. What do you mean by the term 'casting' ?

32. What are the advantages and disadvantages of casting processes ?

33. What steps are involved in the preparation of a casting ?

34. What points should the casting designer consider while designing a casting ?

35. Name the various casting processes in use.

36. Explain briefly the following casting processes :
    (*i*) Centrifugal casting.   (*ii*) Die casting.
    (*iii*) Investment casting.   (*iv*) Slush casting.

37. State the advantages and limitations of sand casting.

38. Discuss permanent mould casting, stating its advantages and disadvantages.

39. Explain with a neat sketch die casting. State its advantages and disadvantages.

40. What is plaster casting ? What are its advantages and disadvantages ?

41. What are the factors which are normally responsible for the production of defects in castings.

42. Name the common defects found in castings.

43. Explain briefly the following defects in casting :
    (*i*) Blow holes.   (*ii*) Misrun.
    (*iii*) Cold shut.   (*iv*) Mismatch.
    (*v*) Fins.   (*vi*) Hot tear.
    (*vii*) Scab.   (*viii*) Swell.

44. Explain briefly the following cleaning operations usually performed on a casting :
    (*i*) Rough cleaning.   (*ii*) Surface cleaning.
    (*iii*) Trimming.   (*iv*) Finishing.

45. Discuss briefly the various methods employed to inspect the castings.

Metal Forming Processes

3.1. INTRODUCTION

The materials which are covered under the scope of material science are available either from nature or industry. However, these materials cannot be used in raw form (whatever the source may be) for useful purposes. They have to be shaped and formed into articles through difference manufacturing processes. Besides there are some processes which improve material properties. In some processes the materials are changed into their primary forms for some selected parts. In some cases the materials are suitably finished for commercial uses. In other cases, neither surface finish nor the dimensions are satisfactory for the final product, and further work is necessary. However, the selection of the best process for a given product requires a knowledge of all possible production methods.

Some important processes are listed below :

1. **Cold working :**
   (i) Drawing
   (ii) Squeezing
   (iii) Bending
   (iv) Shearing
   (v) Hobbing
   (vi) Shot peening
   (vii) Cold extruding.

2. **Hot working :**
   (i) Rolling
   (ii) Forging
   (iii) Pipe welding
   (iv) Hot piercing

(v) Hot drawing   (vi) Hot spinning

(vii) Hot extruding.

3. **Forging :**

(i) Hand forging   (ii) Machine forging.

4. **Casting :**

(i) Sand casting   (ii) Shell moulding

(iii) Permanent mould casting   (iv) Die casting

(v) Centrifugal casting   (vi) Investment casting

(vii) Plaster casting   (viii) Slush casting.

- *Mechanical working processes are used to achieve optimum mechanical properties in the metal. Metal working reduces any internal voids or cavities present and thus makes the metal dense.* The impurities in the metal also get elongated with the grains and in the process get broken and *dispersed throughout the metal.* This *decreases the harmful effect of impurities and improves the mechanical strength.*

- When materials are subjected to external loads, they get deformed. The deformation may be *elastic, plastic* or *fracture,* depending upon the load and the properties of the material. Elastic deformation is said to have occurred when the material returns to its original shape on removal of load, and when it *does not return to its original shape* on removal of load but retains its new configuration, the material is said to have *deformed plastically.* Further deformation causes the material to fracture resulting in separation of a part of material from the body of material.

- *"Plastic deformation"* is the deformation which is permanent and beyond the elastic range of the material. Often, metals are worked by plastic deformation because of the beneficial effect that is imparted to the mechanical properties by it. *The necessary deformation in the metal can be achieved by application of large amount of mechanical force only or by heating the metal and then applying a small force.*

- The deformation of metals which is caused by the displacement of atoms is achieved by one or both of the processes called *slip* and *twinning.* When plastic deformation occurs, on macroscopic scale the metal appears to flow in the solid state along specific directions which are dependent on the type of processing and direction of applied force. The crystals or grains of the metal get elongated in the direction of metal flow. This flow of metal can be seen under microscope after polishing and suitable etching of the metal surface. These visible lines are called *'fibre flow lines'.* Since it is possible to control these flow lines in any specific direction by careful manipulation of the applied forces, it is possible to achieve optimum mechanical properties.

- In metal working processes the wastage of material is either *negligible* or very small, and the production rate is in general *very high.* These two factors give rise to the economy in production.

3.2. COLD AND HOT WORKING

3.2.1. Cold Working

When plastic deformation of metal is carried out at temperature below the recrystallisation temperature the processes performed on metals are termed as **cold working.**

The various cold working processes are :

- Drawing
- Bending
- Hobbing
- Cold extruding, etc.
- Squeezing
- Shearing
- Shot peening

Advantages of cold working :

1. Handling of material is easy.
2. Good surface finish and better dimensional accuracy.
3. Energy saving since heating is not required.
4. Strength, fatigue and wear properties are improved.
5. Minimum contamination because of low working temperature.
6. No possibility of decarburisation of the surface.
7. Economical for smaller sizes.
8. Highly suitable for mass production and automation, because of low working temperatures.
9. The physical properties of metals that do not respond to heat treatment can be improved by cold working.
10. Thin gauge sheet can be produced.

Disadvantages :

1. Ductility of metal is reduced.
2. Deformation energy required is high, so rugged and more powerful equipment is required, thus equipment cost is high.
3. Severe stresses are set up, this requires stress relieving, which increases the cost.
4. Owing to limited ductility at room temperature, the complexity of shapes that can be readily produced is limited.
5. Cold working, for large deformation, requires several stages with interstage annealing, which increases the production cost.

- Following points are *worthnoting* :

    (i) Cold working produces an *improved surface finish and closer dimensional tolerance* and because of this characteristic cold working processes are generally used in making *end-use products.*

    (ii) Since recrystallization does *not* take place in cold working, the grains are *permanently distorted.*

    (iii) *During cold working residual stresses are set up. As their presence is undesirable a suitable heat treatment is generally necessary to neutralise these stresses and restore the metal to its original structure.*

    (iv) Cold forming is most suitable for axisymmetric components such as *shaft components, flanged components, finished gears and bearing races, etc.*

3.2.2. Hot Working

When plastic deformation of metal is carried out at temperature above the recrystallisation temperature the processes performed on metals are termed as **hot working**.

The various hot working processes are :

- Rolling
- Pipe welding
- Hot extruding, etc.
- Forging
- Hot spinning

Advantages of hot working :

1. High production rate (since the process is faster).
2. Very high reduction is possible without fear of fracture.
3. Metal is made tougher because pores get closed and impurities are segregated.
4. Deformation energy required is low, hence, less powerful equipments are required.
5. Structure can be altered to improve the final properties.
6. The process does not change hardness or ductility of the metal since distorted grains soon change into new undeformed grains.
7. Interstage annealing and stress relieving are not required.
8. Since hot working promotes diffusion of constituents, segregation can be reduced or eliminated.

Disadvantages :

1. Handling of material is not so easy.
2. Heat resistant tools are required which are expensive.
3. High temperature may promote undesirable reactions.
4. Close tolerances cannot be held because of non-uniform cooling and thermal contraction.
5. Surface finish is poor because of scale formation.
6. Metalligical structure may be non-uniform because of cooling history after deformation.

Following points are *worth noting* :

(i) Hot working process can be considered as simultaneous *combination of cold working and annealing.* Any work hardening effect caused by plastic deformation is neutralised immediately by the effect of high temperature.

(ii) Hot working process *facilitates metal shaping with low power requirements though it is expensive to handle hot materials.*

(iii) In hot working there is *loss of metal by scaling and fine dimensional tolerance cannot be achieved.*

(iv) Hot working *increases the density* since any pores or cavities in the cast metal disappear.

Warm working or Semi-hot working :

It is defined as *plastic deforming of a metal or alloy under conditions of temperature and strain rate, such that the drawbacks of both cold working and hot working are eliminated and their advantages are combined together.*

In this type of working, the selection of proper temperature is very important ; it depends upon the following *factors* :

(i) Dimensional tolerance on the components.

(ii) The ductility of material.

(iii) Yield or flow strength of metal or alloy.

(iv) Losses due to oxidation and scaling.

3.2.3. Comparison of Cold Working and Hot Working Processes

The comparison of cold working and hot working processes is given below :

S. No.	Cold working	Hot working
1.	Cold working is done at a temperature below the value required for recrystallization, so no appreciable recovery takes place during deformation.	Hot working is done at a temperature above recrystallization temperature, so it can be regarded as a simultaneous occurrence of deformation and recovery process.
2.	Hardening is *not eliminated* as working is done at a temperature below recrystallization, so this is always accompanied by strain hardening.	Hardening due to plastic deformation is *completely eliminated* by recovery and recrystallization. This is true, however, only if the rate of crystallization is higher than rate of deformation.
3.	Cold working *decreases* the value of elogation, reduction of area and impact values.	Mechanical properties like elongation, reduction of area and impact values are *improved*.
4.	Crystallization does *not* take place, so refinement of crystals is out of question. Grains are elongated.	Refinement of crystals occurs.
5.	Uniformity of material is *lost* and properties are affected a lot.	Promotes uniformity of material by facilitating diffusion of alloys, constitutes and breaks brittle film of hard constituents or impurity, *e.g.,* cementite in steel.
6.	Chances of crack propagation is more.	Cracks and unoxidised below holes are sometimes *welded up* ; alternatively, serious cracks or faults are usually shown up at an early stage.
7.	Cold working *increases* ultimate tensile strength, yield point, hardness and fatigue strength while resistance to corrosion is *decreased*. If severely worked, yield point may coincide with ultimate strength value.	Ultimately tensile strength, yield point, fatigue strength, hardness and resistance to corrosion, etc. are *not affected* if hot working is done properly.
8.	Internal and residual stresses *are produced*.	Internal and residual stresses are *not produced*.
9.	Energy required for plastic deformation is *more*.	Energy required for plastic deformation is *less* because at high temperatures, metals become soft and ductile.
10.	*More stress* is required for deformation.	*Less stress* is required for deformation.
11.	No oxidation of metal occurs during working and hence pickling is *not required*.	Heavy oxidation occurs during working and pickling *is required* to remove the oxide.
12.	*Embrittlement does not occur* due to less diffusion and no reaction of oxygen at lower temperature.	Reactive metals get *severely embrittled* by oxygen and hence must be protected from the action of oxygen by using inert atmosphere.
13.	Surface decarburisation in steels does *not* occur.	Surface decarburisation in steels is *likely to occur* at higher temperatures unless the steel is protected by a proper atmosphere.

14.	Surface finish is *good*.	Surface finish is *not so good* due to oxidation at high temperatures.
15.	It is *easy to control* the dimensions within the tolerance limit.	It is *difficult to control* the dimensions because of contraction occurring during cooling.
16.	Ordinary steels can be used for shaping and hence the cost of the cold working plant is *less*.	Alloy steels are necessary for shaping and hence the cost of the hot working plant is *high*.
17.	Handling of materials is *easy*.	Handling of materials is *difficult*.

3.3. ROLLING

Rolling *is a forming operation on cylindrical rolls wherein cross-sectional area of a bar or plate is reduced with a corresponding increase in length.* The metal is thinned and elongated by compression and shear forces but increased in width only slightly. Because of the high surface finish maintained on the rolls, the surface of stock is burnished by the rolling action and attains a smooth bright finish.

This process is one of the most widely used of all the metal working processes, because of its *high productivity and low cost*. Rolling would be able to produce components having *constant cross-section* throughout its length. Many shapes such as *I, T, L and channel sections are possible, but not very complex shapes*. It is also possible to produce special sections such as railway wagon wheels by rolling individual pieces.

Rolling is normally a hot working process unless specifically mentioned as cold rolling.

3.3.1. Principle and Mechanism of Rolling

— The process is illustrated in Fig. 3.1. The rolls are in contact with the passing metal piece over a sufficient distance, represented by the arc *LM*. The angle *LOM* subtended at the centre of the roll by the arc *LM* is called the *'angle of contact'* or *the 'maximum angle of bite'*. It is the friction between the surfaces of the metal piece and the rolls which provides the required grip of rolls over the metal piece to draw the latter through them. *The greater the coefficient of friction more the possible reduction.*

— The pressure exerted over the metal by the roller is not uniform throughout, it is minimum at both the extremities *L* and *M* and maximum at a point, known as *no-slip point or the point of maximum pressure*. At this point the surfaces of the metal and the roll move at the *same speed*. Before reaching this point, *i.e.*, from *L* to *S* the metal moves *slower* than the roll and the frictional force acts in the direction to draw the metal piece into the rolls. After crossing the neutral point *S*, *i.e.*, from *S* to *M*, the metal moves *faster* than the roll surface, as if it is being extruded, and the friction opposes the travel tending to hold the metal track. This results in setting up of stresses within the metal to obstruct its reduction.

Refer to Fig. 3.1. Let t_i, l_i, b_i and t_f, l_f and b_f be the *initial* and *final* thickness, lengths and breadths of the metal piece respectively.

Then, *Absolute draft,* $\qquad \delta t = (t_i - t_f)$ mm

$\qquad$ *Absolute elongation,* $\qquad \delta l = (l_f - l_i)$ mm

$\qquad$ *Absolute spread,* $\qquad \delta b = (b_f - b_i)$ mm

— Spread is *proportional to the draft and depends upon the thickness and width of the job. Spread increases with increase in roll diameter and co-efficient of friction, as well as with a fall in temperature of the metal in course of hot rolling.*

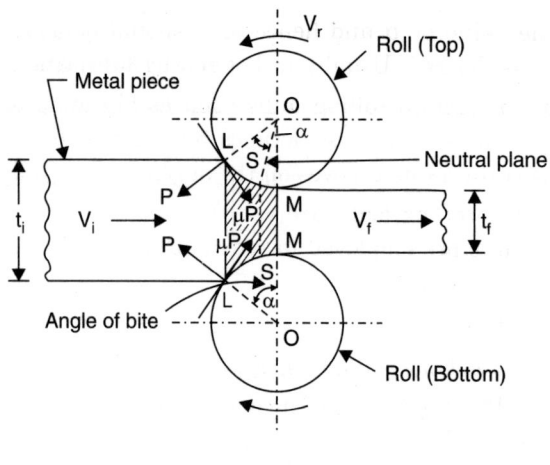

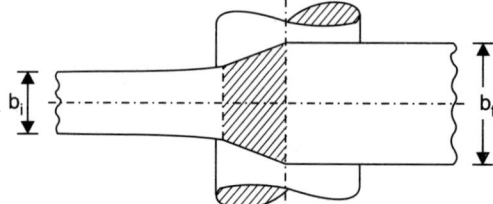

Fig. 3.1. Metal rolling process.

Relative draft, $$R_t = \frac{\delta t}{t_i} = \left(\frac{t_i - t_f}{t_i}\right) \times 100$$

Elongation co-efficient, $\gamma = \dfrac{l_f}{l_i}$.

- At the moment of bite, two forces act on the metal from the side of each roll, *radial or normal force P* and the *tangential forced frictional force* μP where μ is the co-efficient of friction between the metal and roll surfaces. The part would be dragged in if the resultant of horizontal component of the normal force *P* and tangential force μP is directed in that direction. In the limiting case,

$$P \sin \alpha = \mu P \cos \alpha \qquad \qquad \text{...(3.1)}$$

or, $\mu = \tan \alpha$ ...(3.1 (*a*))

or, $\alpha = \tan^{-1} \mu$ ...(3.1 (*b*))

When $\alpha > \tan^{-1} \mu$, the metal would not enter the space between the rolls automatically, that is unaided.

The *maximum permissible angle of bite (or contact) depends upon the value of 'μ' which in turn depends upon* :

— Materials of rolls ;

— Job being rolled ;

— Roughness of their surfaces ;

— Rolling temperature ;

— Speed.

- In *hot rolling,* the value of α and hence of μ should be *greater* since the maximum possible reduction is desired. Usually in hot rolling lubrication is not necessary.

— In case of primary reduction rolling mills such as blooming or rough rolling mills for structural elements the rolls may sometimes be *"ragged"* to increase μ (*Ragging* is the process of making certain fine grooves on the surface of the roll to increase the friction).

In *cold rolling,* since the rolling loads are very high μ should not be much. Rolls for cold rolling are ground and lubricants are employed to reduce μ.

The usual values of biting angles are :

$2°-10°$ ...... For cold rolling of oiled sheet and strip ;

$15°-20°$ ...... For hot rolling of sheet and strip ;

$24°-30°$ ...... For hot rolling of heavy billets and blooms.

- *The volume of metal that enters the rolling stand should be the same as that leaving it* except in initial passes when there might be some loss due to filling of voids and cavities in the ingots. Since the area of the cross-section gets decreased, the metal leaving the rolls would be at a higher velocity than when it entered. Initially when the metal enters the rolls, the surface speed of rolls is higher than that of the incoming metal, whereas the metal velocity at the exit is higher than that of surface speed of the rolls. *Between the entrance and exit, the velocity of the metal is continuously changing, whereas the roll velocity remains constant.*

Now, $Forward\ slip = \dfrac{V_f - V_r}{V_r} \times 100$

$Backward\ slip = \dfrac{V_r - V_i}{V_r} \times 100$

where, V_i = Initial metal speed,

V_f = Final metal speed, and

V_r = Surface/peripheral speed of rolls.

$$\%\ age\ of\ cold\ work\ \ = \dfrac{A_i - A_f}{A_i} \times 100$$

$$= \dfrac{t_i b_i - t_f b_f}{t_i b_i} \times 100$$

$$= \dfrac{t_i - t_f}{t_i} \times 100 \qquad\qquad (\because\ \ b_i \cong b_f)$$

where A_i and A_f are the initial and final areas of cross-section.

Process variables in rolling process :

The main process variables in rolling process are :

(*i*) Diameter of roll.   (*ii*) Angle of bite.

(*iii*) Speed of rolling.   (*iv*) Strength of work material.

(*v*) Temperature.   (*vi*) Roll gap or draft.

(*vii*) Co-efficient of friction.   (*viii*) Dimensions of sheet.

The rolling load (*P*) can be calculated as :

$$P = l \cdot b \cdot p_m$$

where, l = Roll-strip contact length,

   b = Breadth of sheet, and

   p_m = Mean specific pressure.

Since l depends on roll diameter and angle of bite, it is approximately given as :

$$l \cong \sqrt{R \cdot \delta t} \qquad\qquad \text{...(3.2)}$$

where, R = Roll radius

and, δt = draft = $t_i - t_f$

p_m depends on R, t_i, t_f and yield strength of work material.

- *Hot rolling* is carried out to roll ingots into slabs, blooms or billets. On further hot rolling, plates, bars, rounds, structural shapes and rails are obtained. Because of limitations in equipment and workability of metals, rolling is done in *progressive steps,* that is, a number of passes through the rolls may be required to get the required configuration. For example, ten roll passes are required to get a 100 mm × 100 mm billet reduced to a 12 mm rod (Fig. 3.2). The *initial few passes* are designed to merely reduce the cross-sectional area, while the *intermediate passes* not only reduce the area but also try to bring the shape close to the final shape. *Final or finishing passes* bring the material to the required shape and size.

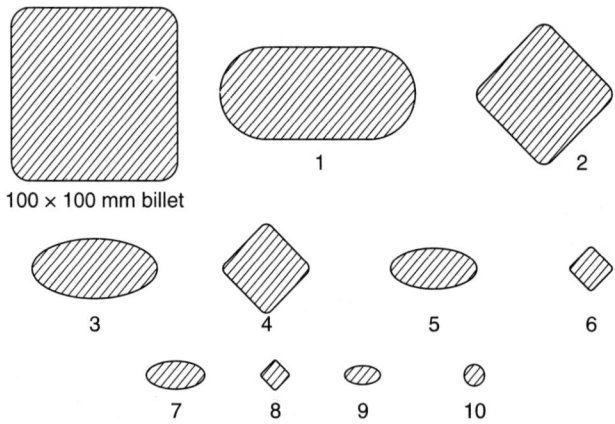

Fig. 3.2. Roll passes to get a 12 mm rod from 100 × 100 mm billet.

- *Cold rolling* is employed to finish bars, rods, sheets and strips of most common metals because this process provides *better dimensional accuracy, good surface finish and improved physical properties.*

  Roll materials commonly used are *cast iron, cast steel and/or forged steel because of high strength and high wear resistance requirements.*

- Fig. 3.3 shows the effect of both cold working and hot working on the microstructure of cast metals.

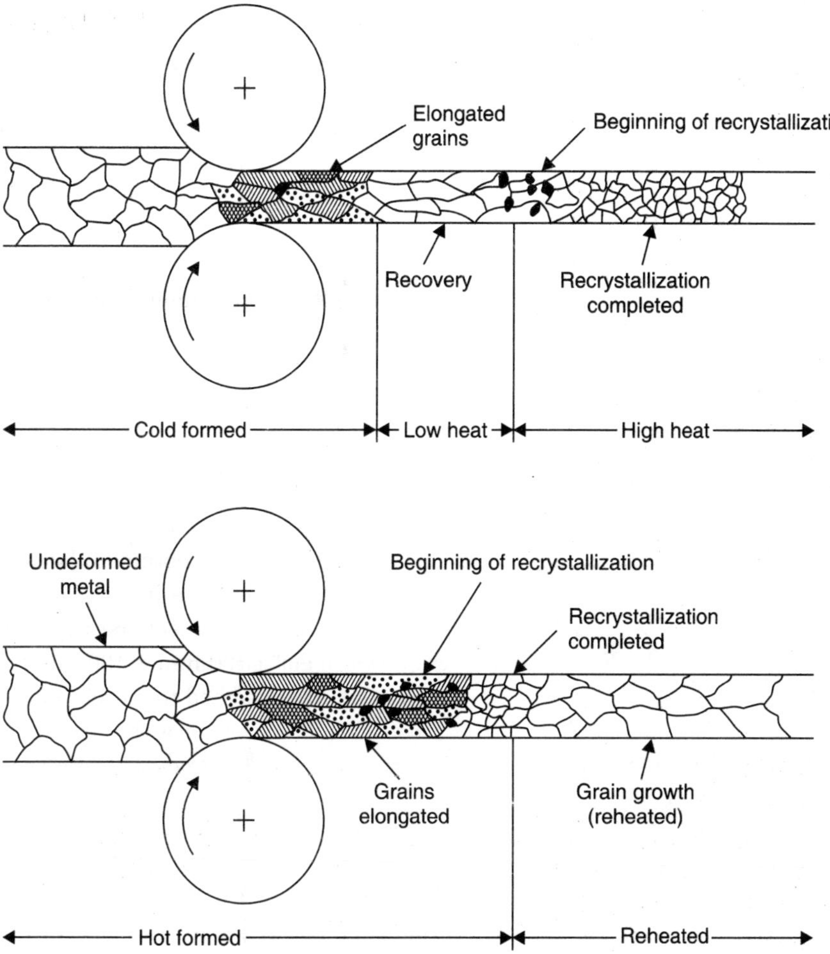

Elongated
grains

Beginning of recrystallization

Recovery

Recrystallization
completed

Cold formed — Low heat — High heat

Undeformed
metal

Beginning of recrystallization

Recrystallization
completed

Grains
elongated

Grain growth
(reheated)

Hot formed — Reheated

Fig. 3.3. Effect of both cold working and hot working on the microstructure of cast metals.

3.3.2. Rolling Stand Arrangement

The arrangement of rolls in a rolling mill, also called *rolling stand,* varies depending on the application. The names of the rolling stand arrangements are given by the number of rolls employed. The various possible configurations are presented in Fig. 3.4 to Fig. 3.8.

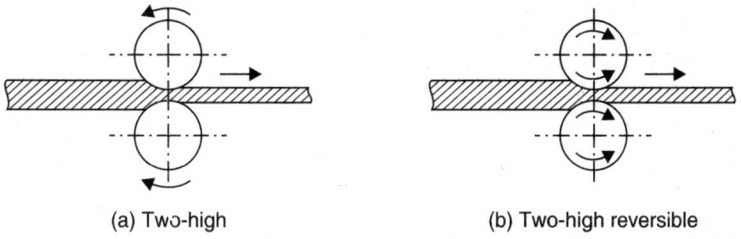

(a) Two-high   (b) Two-high reversible

Fig. 3.4. Two-high rolls.

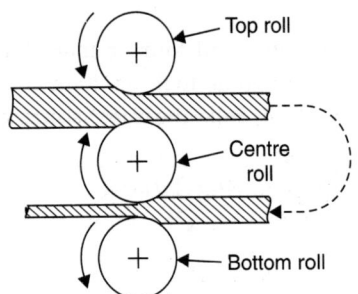

Fig. 3.5. Three-high rolls.

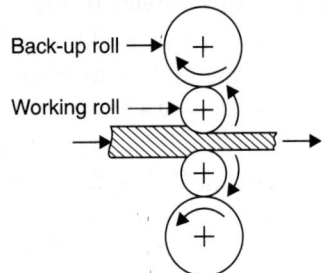

Fig. 3.6. Four-high rolls.

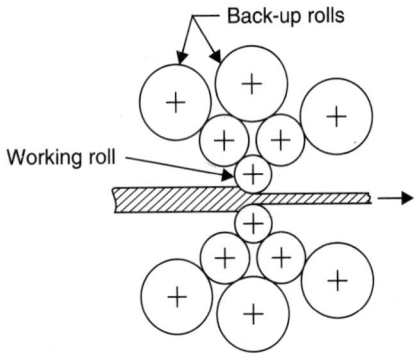

Fig. 3.7. Cluster roll.

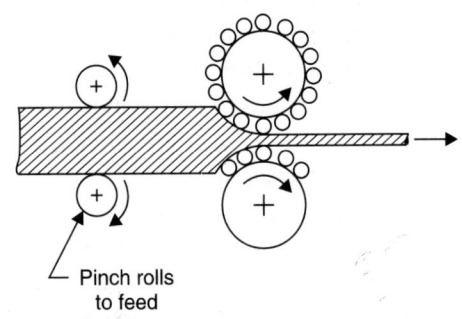

Fig. 3.8. Planetary mill.

1. **Two-high rolls :**

● Both the rolls rotate in opposite directions to one another as shown in Fig. [3.4 (a)]. Their direction of rotation is fixed and cannot be reversed. Thus, the work can be rolled by feeding from one direction only.

— The space between the rolls can be adjusted by raising or lowering the upper roll. The position of the lower roll is fixed.

● There is another type of two-high mill [Fig. 3.4(b)] which incorporates a drive mechanism that can *reverse* the direction of rotation of the rolls. This facilitates rolling of the workpiece continuously through back-and-forth passes between the rolls.

This type of rolling mill is known as *two-high reversing mill*. These stands are more expensive compared to the non-reversible type because of the reversible drive needed.

2. **Three-high rolls :** Refer to Fig. 3.5.

● This stand arrangement is used for rolling of two continuous passes in a rolling sequence without reversing the drives. After all the metal has passed through the bottom roll set, the end of the metal is entered into the other set of the rolls for the next pass. For this purpose a table-tilting arrangement is required to bring the metal to the level with the rolls.

● This arrangement may be used for *blooming, billet rolling* or *finish rolling*.

3. **Four-high rolls :** Refer to Fig. 3.6.

● This rolling stand is essentially a two-high rolling mill, but with small-sized rolls. The other two rolls are the back-up rolls for providing the necessary rigidity to the small rolls.

● These mills are generally employed for subsequent rolling of slabs. The common products of these mills are *hot or cold rolled sheets and plates*.

4. **Cluster rolls :** Refer to Fig. 3.7.

● It consists of two working rolls of smaller diameter and four or more back-up rolls of larger diameter. The number of back-up rolls may go as high as 20 or more, depending upon the amount of support needed for the working rolls during the operation.

● This type of mill is generally used for *cold rolling.*

5. **Planetary mill.** For the rolling arrangements requiring *large reduction,* a number of free rotating wheels instead of a single small roll, are fixed to a large back-up roll in the planetary rolling mill arrangement shown in Fig. 3.8.

3.3.3. Defects in Rolling

The various *defects in rolling process* are enumerated and discussed below :

1. Surface defects.

2. Structional defects.

1. **Surface defects :**

These defects may result from :

— Inclusions and impurities in the material ;

— Scale, rust, dirt ;

— Roll marks ;

— Other causes related to the prior treatment and working of the material.

In hot rolling blooms, billets, and slabs, the surface is usually preconditioned by various means, such as by torch (**scarfing**), to remove scale.

2. **Structural defects :**

These defects *distort or affect the integrity of the rolled product.*

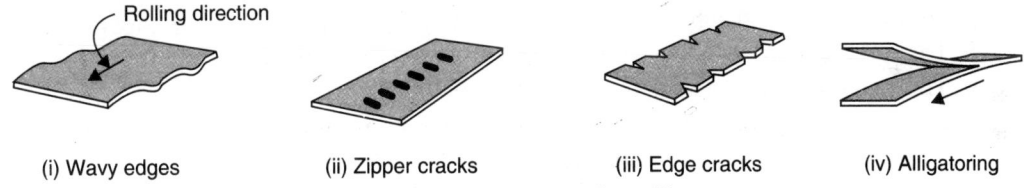

| (i) Wavy edges | (ii) Zipper cracks | (iii) Edge cracks | (iv) Alligatoring |

Fig. 3.9. Typical defects in flat rolling.

(*i*) **Wavy edges.** Refer to Fig. 3.9 (*i*). These are caused by *bending of the rolls ;* the edges of the strip are thinner than the centre. Because the edges elongate more than the centre and are restrained from expanding freely, they buckle.

(*ii*), (*iii*) **Zipper cracks and edge cracks :** Refer to Fig. 3.9 (*ii*), (*iii*). Zipper cracks in the centre of strip and edge cracks are usually caused by *low ductility and barreling.*

(*iv*) **Alligatoring.** Refer to Fig. 3.9 (*iv*). Alligatoring is a complex phenomenon that results from *inhomogeneous deformation of the material during rolling or from defects in the original cast ingot, such as piping.*

Note : *"Residual stresses"* can be generated in rolled sheets and plates because of *inhomogeneous plastic deformation in the roll gap.*

— *Small-diameter rolls or small reductions* tend to work the metal plastically at its surfaces. This working generates compressive residual stresses on the surfaces and tensile stresses in the bulk.

— *Large-diameter rolls and high reductions,* however, tend to deform the bulk to a greater extent than the surfaces, because of frictional constraint at the surfaces along the arc of contact. This situation generates residual stresses that are opposite to those of the previous case.

3.4. FORGING

3.4.1. Introduction

Forging *is the process by which heated metal is shaped by the application of sudden blows or steady pressure and characteristics of plasticity of material are made use of.*

- Forging can be carried out at room temperature (*cold working*), or at elevated temperatures, a process called *warm or hot forging*, depending on the temperature.
  - Simple forgings can be made with a heavy hammer and an anvil by techniques used by blacksmiths for centuries. Usually, though, a set of dies and a press are required. The three basic categories of forging are open die, impression die, and closed die.
- *Typical parts made by forging* today are :
  - Crankshafts and connecting rods for engines ;
  - Turbine discs, gears, wheels, bolt heads, hand tools ;
  - Many types of structural components for machinery and transformation equipment.

3.4.2. Advantages and Disadvantages of Forging

Following are the advantages and disadvantages of forging process :

Advantages :

1. Forging improves the structure of metal and hence its mechanical properties.
2. Forging distorts the previously created uni-directional fibre in such a way as to strengthen the component.
3. Owing to intense working, flaws are seldom found and the workpiece has a high reliability.
4. Forgings are easily welded.
5. Rapid duplication of components.
6. Ability of the forging to withstand unpredictable loads.
7. Metal removing in machining is minimum.
8. The forging can withstand unpredictable loads.
9. The surface of the forging is relatively smooth.
10. Superior machining qualities.
11. Minimum weight per unit strength and better resistance to shock.
12. Forgings can be held to within fairly close tolerances.

Disadvantages :

1. The initial cost of dies and the cost of their maintenance is high.
2. In hot forging, due to high temperature of metal, there is rapid oxidation or scaling of the surface resulting in poor surface finish.
3. Forging operation is limited to simple shapes and has limitations for parts having undercuts, re-entrant surfaces, etc.
4. Forgings are usually costlier than castings.

3.4.3. Classification of Forging

Forging can be *classified* in two ways :

1. Hand forging.
2. Machine forging.

Hand Forging :

- Hand forging or blacksmithing is employed for small quantity production and for special work.

Generally speaking, the *accuracy obtained is less than that of drop forging.*

In hand forging the metal is heated in a Smith's forge or hearth (Fig. 3.10). It consists of a hearth for holding the fuel, a cast iron tuyere for supplying air blast to the fire, a centrifugal blower driven by a power preferably electric motor, to produce the blast, a chimney to carry the smoke and poisonous gases to air, a water tank behind the hearth to water cool the tuyere, a cool bunker to stock coal or coke, a water trough in front for quenching cutting tools and an air valve to control the blast. In operation, the work is paced in the fire pot and heated to the proper temperature for forging.

Fig. 3.11 shows the various tools used in smithy. The list of important smithy tools and their uses are given below :

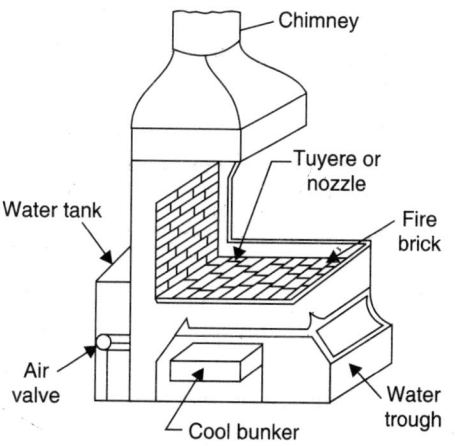

Fig. 3.10. Smith's forge.

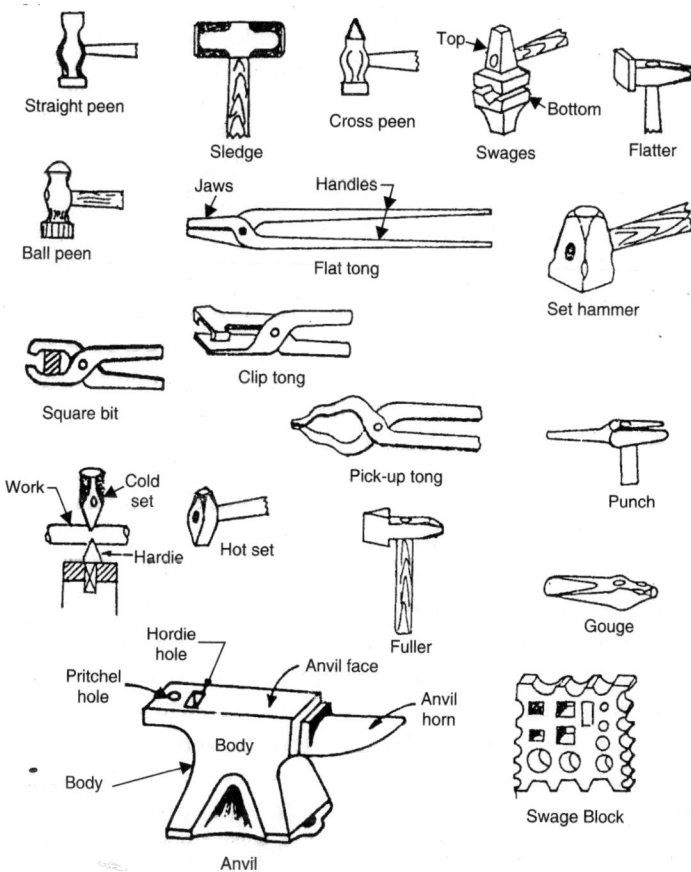

Fig. 3.11. Tools used in smithy.

Tools used in smithy

Name of tool	Use
1. *Sledge hammers, straight, flat and cross peen*	— To forge big jobs (heavy work).
2. *Smith's ball peen hammer*	— To forge light and medium work.
3. *Tongs, flat or square bit pick up tong.*	— To hold the hot work.
4. *Chisel long cold set*	— To cut cold metal.
5. *Hot set*	— To cut hot metal.
6. *Fullers, top and bottom*	— To shape inside curves. To form corrugations for elongating metal.
7. *Swages, top and bottom*	— To shape convex surfaces and to give finish to round, square, hexagonal or octagonal shaped sections.
8. *Flatter or flattener*	— To give smooth finish to flat surfaces.
9. *Set hammer*	— To form square shoulders and to clean the rounding in corners.
10. *Punches*	— To make recesses of any shape in hot metal.
11. *Hardie*	— To nick the bar and to shape the cold work.
12. *Anvil*	— To forge art, bend and shape the work.
13. *Swage block*	— To shape or bend the work to any form and to knock heads of bolts, etc.
14. *Gouge*	— To cut plates to curves.

Forging on anvil is usually done with : (*i*) one man or (*ii*) two men-two handed working—the smith and his striker. The former uses a small hammer, the latter the sledge. To indicate, where he requires his mate to strike a blow the smith lightly taps the work with the small hammer ; the striker's job is to hit the spot with the sledge. If working three handed the same procedure is followed, a ligh tap from the smith preceding a heavy one from the striker. To indicate when to finish, the smith taps the anvil with his hammer.

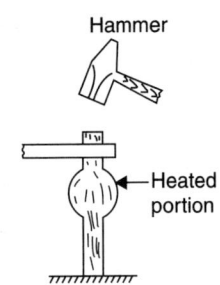

Fig. 3.12. Upsetting.

Upsetting. It (Fig. 3.12) is the process of increasing cross-sectional dimensions when forging. The process implies that *cross-section is increased and the length decreases.* It may be done in a number of ways, each varying according to the details of the article required and the equipment in the shop. The simplest is to place the heated article on the anvil and hammer directly on the upper end. This increases the cross-section and reduces the length of the metal being worked.

Drawing down. It is the process of increasing the length of a bar at the expense of its cross-sectional area. It is illustrated in Fig. 3.13.

Setting down is a localised drawing down or swaging operation. **Punching** is the process of removing a slug of metal,

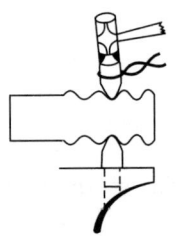

Fig. 3.13. Drawing down.

generally cylindrical, by using a hot punch over the pritchel hole of the anvil, over a hole of correct size in the swage block.

Cutting out. It is the process of cutting large holes of various shapes by using a hot chisel over a hole in the sewage block.

Forging Machines :

A forging machine is one which is designed to shape a metal article while the material is in hot plastic state.

The term forging machine in its widest sense includes :

1. Drop stamp (whether of rope, belt or board type).
2. Steam hammer.
3. Pneumatic hammer.
4. Hydraulic hammer.

1. Drop stamp. The *drop stamp of board type* (Fig. 3.14) is widely employed when shaping hot bars, and finally to bring the work to size and shape between a set of drop stamping dies. In board hammer the tup is attached to a board which passes between two rollers. The latter run in an over head attachment, are belt-driven and run in opposite directions. The tup is lifted by means of eccentric (foot, or hand operated or self acting) and they (eccentrics) cause the rollers to grip or release the board, when the board is gripped by the rollers their direction of rotation is such as to lift it (board) and the attached tup, when the board is released the tup falls with it. The height of lift depends upon the timing of release, which is instantaneous.

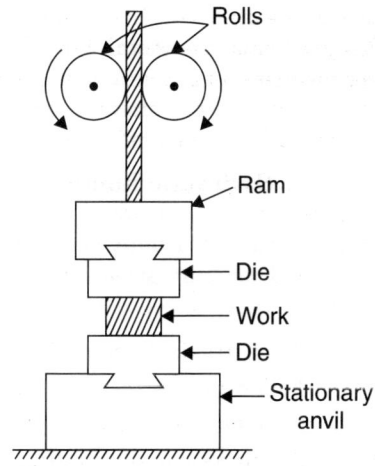

Fig. 3.14. Drop stamp of board type.

When producing small drop forgings or hot pressings the drop stamps in its various forms is a very effective method of obtaining the desired results. For shallow sheet metal work drop stamp is first class production machine as it permits a solid blow to be struck without any fear of bending a crank or breaking a press frame.

2. Steam hammer. A *steam hammer* (Fig. 3.15) operates on the principle of the steam engine. The main parts are frame, a steam chest or cylinder, piston, piston rod and the anvil. The hammer head is attached to the piston rod and is raised by admitting steam in the cylinder through the valve beneath the piston. The downward stroke of the hammer is obtained by exhausting the steam from beneath the piston and admitting from above the piston. The hammer descends by gravity and steam pressure is 5.5 to 8.5 bar. For varying the intensity of the hammer blows, light to heavy, steam is admitted below the piston while the hammer is descending to create cushioning to the falling hammer. The steam inlet and outlet are controlled by a special slide valve. For generating steam a boiler is required. A wide range of work is done on this class of forging machine.

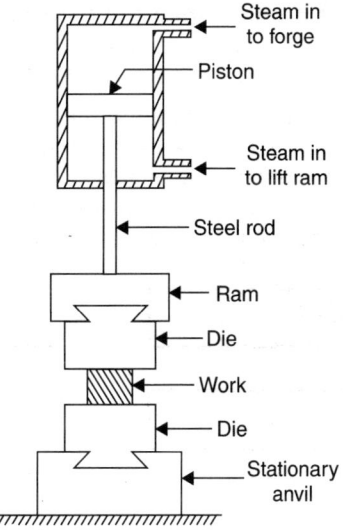

Fig. 3.15. Steam hammer.

3. Pneumatic hammer. In *pneumatic hammer* (Fig. 3.16) air is compressed on both upward and downward strokes of the piston which is worked by the electric motor. This compressed air is supplied to the ram cylinder by the long valve kept between the two cylinders which is moved by the control lever. By lowering and raising of control lever, the strokes and the speeds of the blows per minute can be varied from 50 to 200.

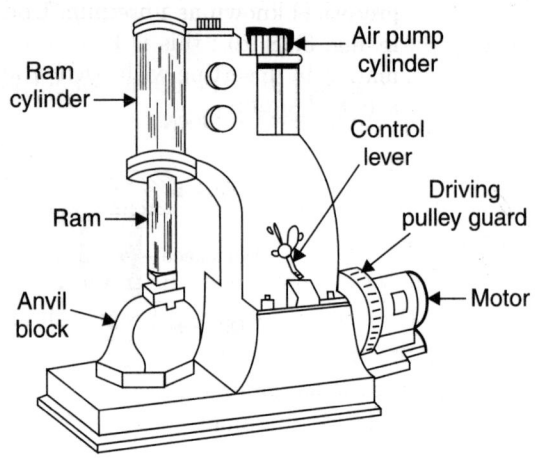

Fig. 3.16. Pneumatic hammer.

The steam and air hammers are designed to give sharp and fast blows, reproducing to a marked extent the action of the smith and his hammer. They may be used with a standard pair of anvils or with a set of dies, the latter often being so designed that the metal can be drawn out to the approximate length and width, and then placed in the dies for the final shaping stage. The flash, which is formed, is clipped off as the last operation.

4. Hydraulic hammer. For the large castings and in cases where a heavy pressure is required the use of *hydraulic hammer* is restored to. The hydraulic forging machine being sluggish in action cannot usually compare with the steam or air hammer which operate more quickly, for small and medium sized forgings. The main advantage of the hydraulic forging press is that it gives a definite squeeze and the time element permits the material to flow.

3.4.4. Basic Categories of Forging

Following are the three basic categories of forging :

1. Open-die forging.
2. Impression-die forging.
3. Closed-die forging.

3.4.4.1. Open-die forging

This type of forging is distinguished by the fact that the metal is *never completely confined as it is shaped by various dies.*

Most open-die forgings are produced on flat-V, or swaging dies (Fig. 3.17). Round swaging dies and V dies are used in pairs or with a flat die. The top die is attached to the ram of the press, and the bottom die is attached to the hammer anvil or, in the case of press open-die forging, to the press bed. As the workpiece is hammered or pressed, it is repeatedly manipulated between the dies until hot working forces the metal to the final dimensions.

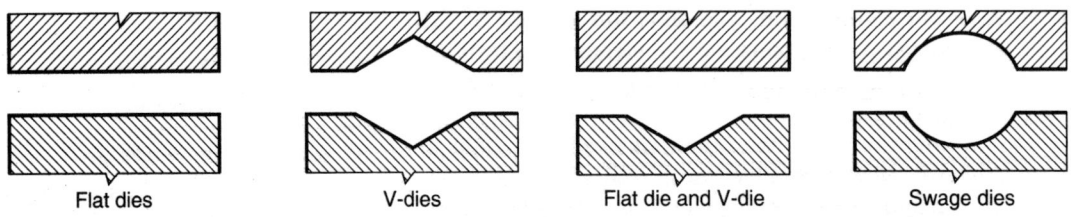

Flat dies   V-dies   Flat die and V-die   Swage dies

Fig. 3.17. Types of dies used in open-die forging.

- Open-die forging, in its simplest form generally involves placing a solid cylindrical workpiece between two flat dies (platens) and reducing its height by compressing it. This process is known as *upsetting.* Under ideal conditions, a solid cylinder deforms as shown in Fig. 3.18 (*a*) ; this is known as *homogeneous deformation.* Fig. 3.18 (*b*) shows deformation in upsetting with *friction* as the die-workpiece interfaces ; the specimen develops a *barrel shape.*

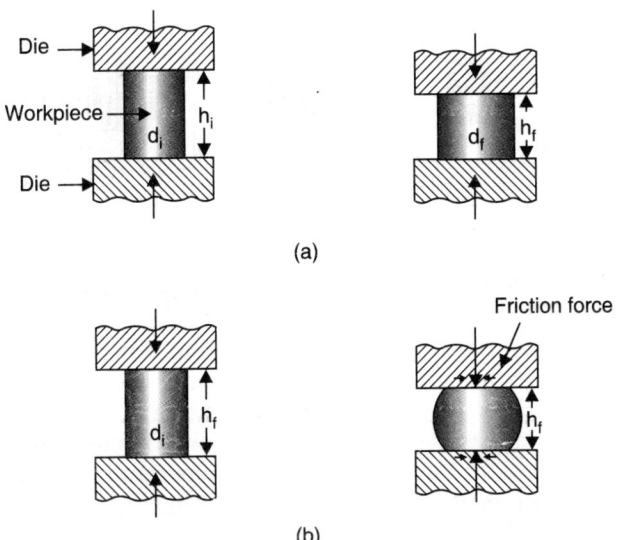

Fig. 3.18

— *Barreling* caused by friction can be minimised by an *effective lubricant* or *ultrasonic vibration* of the platens. The use of *heated platens or thermal barrier at interfaces* will also reduce barrel is hot working.

Advantages of open-die forging :

(*i*) Simple to operate.

(*ii*) Simple for low production volume.

(*iii*) Inexpensive tooling and equipment.

(*iv*) Wide range of workpiece sizes can be used.

Limitations :

(*i*) Suitable for simple shapes only.

(*ii*) Can be employed for short run production only.

(*iii*) It is difficult to maintain moderately close tolerances.

(*iv*) Material utilisation is poor.

(*v*) Less control in determining grain flow, mechanical properties and dimensions.

(*vi*) Fairly skilled workers are required.

(*vii*) Since machining is often required, final cost of production may be higher than other forging methods.

3.4.4.2. Impression-die forging

● In impression-die forging, the workpiece acquires the shape of the die cavities (impressions) while it is being upset between the closing dies.

● In the simplest example of this type of forging, two dies are brought together, and the workpiece undergoes plastic deformation until its enlarged sides touch the side walls of the die (Fig. 3.19). A small amount of material is forced outside the die impression, forming *flash* that is gradually thinned. The flash cools rapidly and presents increased resistance to deformation, effectively becoming a part of the tool, and helps build up pressure inside the bulk of the workpiece that *aids material flow into unfilled impressions.*

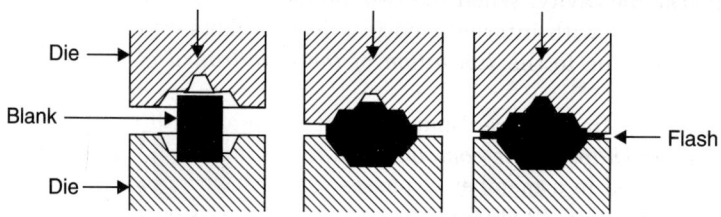

Fig. 3.19. Impression-die forging.

3.4.4.3. Closed-die forging

● Though not quite correct, the example shown in Fig. 3.19 is also referred to as *closed-die forging.* In *true* closed-die forging, *no flash is formed,* and the workpiece is completely surrounded by dies, while in impression-die forging, any excess metal in the die cavity is formed into a flash. Since no flash can be formed in closed-die forging, proper control of the volume of material is essential to obtain a forging of desired dimensions (and to avoid generating extreme pressures in the dies from overfilling). One approach to getting the right amount of metal for the die cavity and reduce forging time is the use of roll-formed shapes or extruded preform shapes.

Advantages :

 (*i*) Can be used for production of complex shapes.

 (*ii*) Good dimensional accuracy and reproducibility.

 (*iii*) Suitable for high production rate.

 (*iv*) Less time consuming than open-die forging.

 (*v*) Workpiece materials are utilised effectively.

 (*vi*) The grain flow of the metal can be controlled ensuring high mechanical properties.

 (*vii*) Since the forgings are made with smaller machining allowances, therefore, there is a considerable reduction in the machining time and consumption of metal required for the forging.

Limitations :

 (*i*) More than one step required for each forging.

 (*ii*) Finishing required for achieving final shape.

 (*iii*) High equipment and tooling cost.

 (*iv*) Appropriate die set for production of each component.

● Both open-die and closed-die forgings can be carried out in *hot or cold state.*

● **Cold forging** obviously requires *higher deformation energy* and is usually carried out *for only those materials which are sufficiently ductile at room-temperature.* Cold forged parts have *better dimensional accuracy and have good surface finish.*

'Hot forged parts' although require lower forces but give inferior finish and dimensional accuracy.

3.4.5. Methods of Forging

3.4.5.1. Drop forging

This method of forging uses a closed impression die to obtain the desired shape of the component. The *shaping is done by the repeated hammering given to the material in the die cavity.* The equipment employed for delivering the blows are called *drop hammers.*

The die used in drop forging consists of two halves. The lower half of the die is fixed to the anvil of the machine, while the upper half is fixed to the ram. The heated stock is kept in the lower die while the ram delivers four to five blows on the metal in quick succession so that the metal spreads and completely fills the die cavity. When the two die halves close, the complete cavity is formed.

Too complex shapes with internal cavities, deep pockets, re-entrant shapes, etc. cannot be obtained in drop forging due to the *limitation of the withdrawal of the finished forging from the die.*

In drop forging, the final desired shape cannot be obtained directly from the stock in a single pass. *Depending on the shape of the component, and the desired grain flow direction,* the material should be manipulated in a member of passes ; the various **passes** used are :

- *Fullering impression* (Reducing the stock to the desired size).
- *Edging impression* (Preform).
- *Bending impression* (Required for those parts which have a bent shape).
- *Blocking impression* (Semi-finishing impression).
- *Finishing impression* (Final impression).
- *Trimming* (Removal of extra flash present around the forging).

Typical products obtained in drop forging are :

— Wrench

— Crane hook

— Crank

— Crankshaft

— Connecting rod, etc. (For *forging sequence of connecting rod refer to* **Q. 3.17**).

Note : The difference between drop forging and smith forging is that in drop forging closed-impression dies are used and there is drastic flow of metal in the dies due to repeated blows the impact of which compels the plastic metal to conform to the shape of the dies ; whereas in the smith forging open face dies are used and the hammering of the heated metal is done by hand tools to get the desired shape by judgement.

3.4.5.2 Press forging

- In *press forging*, the metal is shaped not by means of a series of blows as in drop forging, but by *means of a single continuous squeezing action.* This squeezing is obtained by means of presses. Owing to the continuous action of the press, the material gets uniformly deformed throughout its entire depth. The impressions obtained in press forging are clear compared to that of the likely jarred impressions which are likely in the drop forged components.

- The press is generally of vertical type and the squeezing action is carried *completely to the centre of the part being pressed.*
  Forging presses are of two types : *Mechanical* and *hydraulic*. Mechanical presses may be either of screw type used for brass forging only ; or crank type. Presses can be readily automated.

— *"Hydraulic presses"* are used for *heavy work* and *"mechanical press"* for *light work.* Mechanical presses operate faster than the hydraulic presses, but hydraulic presses are designed to provide greater squeezing force.

- For press forging operation, the drive should be capable of giving huge force which is needed at the end of the stroke when the metal is forced into desired shape. For this purpose, *copper alloys* are well suited as these flow easily in the die and are readily extruded.

Advantages of press forging :

Following are the *advantages of press forging over drop forging* :

1. Presses provide a faster rate of production because the die in press forging is filled in a single stroke.

2. Superior structural quality of the product.

3. Quicker operation comparatively.

4. High output even with unskilled operators.

5. Low susceptibility to failure and simple maintenance.

6. Uniform forgings with exacting tolerances and low machining allowances.

7. Alignment of the two die halves can be move easily maintained than with hammering.

3.4.5.3 Machine forging

The machine forging, as it involves the upsetting operation, sometimes it is simply called *upset forging.*

Like press forging, in machine forging also, the material is plastically deformed by squeeze pressure into the shape provided by the dies in the forging machine ; but unlike press forging, it operates in *horizontal direction.* It is the forging method which is often selected when certain parts are required with an increased volume of metal at the centre or only at one end.

- Upsetting machines called *up setters* are generally horizontal acting. The die set consists of a die and a corresponding punch or a heading tool. The die consists of two parts, one called stationary gripper die which is fixed to the machine frame and the other, movable gripper die, which moves along with the die slide of the upsetter. The stock is held between these two gripper dies by friction. The *upset forging cycle starts with the movable die sliding against the stationary die to grip the stock.* The two dies when in closed position, form the necessary die cavity. Then the *heading tool advances against the stock and upsets it to completely fill the die cavity.* Having completed the upsetting, the *heading tool* moves back to its back position. Then the *movable gripper die releases the stock by sliding backwards.*

In machine forging, similar to drop forging, the operation is carried out in a number of stages. The die cavities required for the various operations are all arranged vertically on the gripper dies. The stock is then moved from one stage to the other in a proper sequence till the final forging is ready.

Fig. 3.20 shows the operation of upsetting the end of a bar.

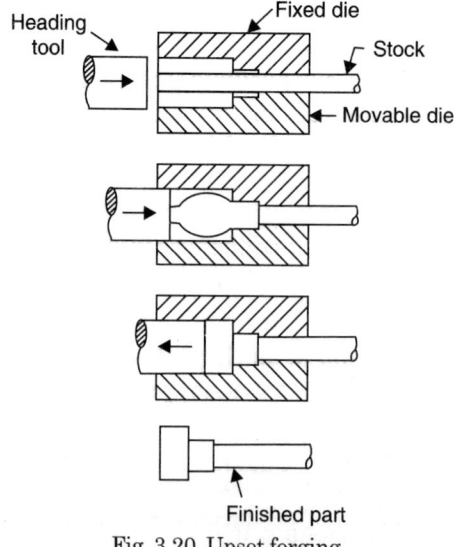

Fig. 3.20. Upset forging.

● The automatic forging machines are employed for the mass production of :

— Nuts   — Bolts
— Rivets   — Wood screws
— Roller and ball bearing balls — Gear blanks
— Valve stems   — Axles
— Couplings, etc.

Advantages :

1. Better quality of forging.
2. Since there is no or little draft is needed on forging made by upsetters, therefore, there is saving in material and also machining expenses.
3. The upsetting process can be automated.
4. As compared to drop forging hammers, forging machines have a higher productivity and their maintenance is much cheaper.
5. In forging machine the forging is accompanied by little or no flash (whereas in drop forging the flash is quite large).

Disadvantages :

1. High tooling cost.
2. It is difficult to forge intricate, non-symmetric and heavy jobs on a forging machine.
3. Owing to the material handling difficulties it is not convenient to forge heavier jobs.
4. The maximum diameter of the stock which can be upset is limited (about 250 mm maximum).

3.4.6 Other Forging Processes

Under this heading the following processes will be discussed :

1. Roll forging
2. Rotary forging or swaging
3. High velocity forging (HVF)
4. Orbital forging
5. Incremental forging
6. Liquid metal forging
7. Gatorizing.

3.4.6.1 Roll forging

The primary function of the forging rolls is to reduce the cross-section of a bar over a designed length or to produce taper on its surface over a certain length.

The machine used is called a ***roll forging machine,*** and consists of two horizontal *rolls* (Fig. 3.21) arranged directly over each other. These rolls are not completely circular ; about half or more portion of these rolls is cut away to allow the stock to enter through them. One or more sets of grooves, according to the shape required on the job surface, are arranged on the circular portion of the rolls. The *total reduction is accomplished in several stages.*

When the rolls open, the bar is fed into the groove and rolled. During the next opening it is shifted to the next groove and rolled. This continues

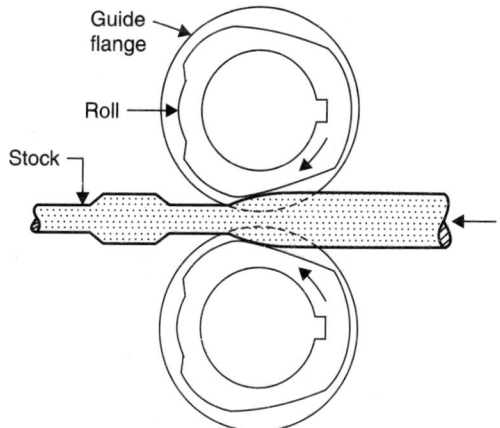

Fig. 3.21. Roll forging.

till the total reduction is achieved. After each rolling the bar is turned through 90°, before feeding into the next groove to prevent the flash formation.

Pressure on rolls may be as high as 1 MN.

- Roll forging is employed for producing *long slender forged components,* such as *axles* and *leaf springs.*

3.4.6.2. Rotary forging or swaging

- This process can be used to *reduce diameters of round bars or tubings.* The other applications include *fabrication of stepped and tapered shafts, pipes with forged out ends, etc.*

- The process can be done both in hot and cold state. However, cold working is preferable because of the greater use of handling and better surface finish obtained.

- In swaging, also known as **rotary swaging** or **radial forging,** a solid rod or tube is reduced in diameter by the reciprocating radial movement of two or four dies (Fig. 3.22).

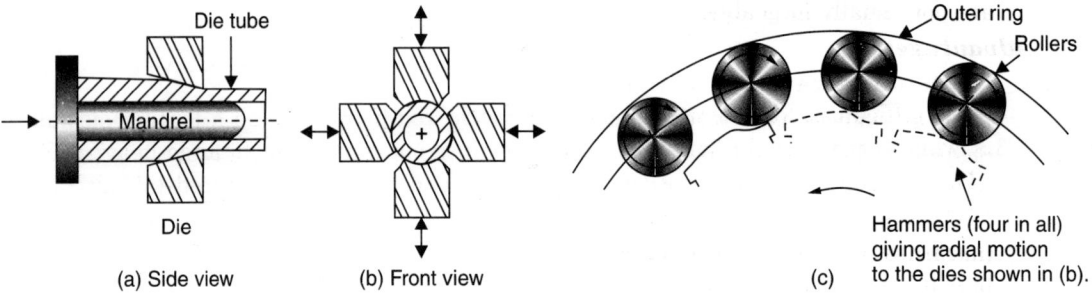

Fig. 3.22. Swaging process.

The die movements are generally obtained by means of a set of rollers in cage. The internal diameter and thickness of the tube can be controlled with or without mandrels. Mandrels can also be made with longitudinal groove (similar in appearance to a splined shaft) ; thus, internally shaped tubes can be swaged.

- The swaging process is usually limited to workpiece diameters of about 50 mm, although special machinery has been built-up to swage gun barrels of large diameter.

- Die angles are usually only a few degrees and may be compound, that is, the die may have more than one angle, for more favourable material flow during swaging.

- Lubricants are used for improved surface finish and longer die life.

- The process is generally carried out at room temperature.

- Parts produced by swaging have *improved mechanical properties* and *good dimensional accuracy.*

Advantages :

(*i*) Tooling cost is low.

(*ii*) Low initial investment.

(*iii*) Consistency of the product.

(*iv*) Labour cost is low.

(*v*) Rapid production.

(*vi*) Maintenance is easy.

Limitation :

The process is limited to parts of *symmetrical cross-section only.*

3.4.6.3. High velocity forging (HVF)

High velocity forging is also known as High Energy Rate Forming (HERF) or Pneumatic-Mechanical High Velocity Forging (PMHVF).

HVF machines provide a *very high impact rate at a very low operating cost.* These machines are based on essentially the same principles as are drop hammers and impact forging machines. They provide greater energies for a given ram weight by using ram velocities of the order of 2 to 10 times those of hammers. These are based upon the following methods of releasing energy (*i*) *Chemicals*—High explosives, propellants, gas mixtures ; (*ii*) *Electrical*—Exploding wires, spark discharge, magnetic field.

These machines can be effectively used for forging, extruding, compacting and many other metal forging operations. The most efficiently performed operation, however, is the *closed die forging of metals,* although even ceramics can be worked on these machines.

- A wide variety of materials can be formed or forged including exotic and refractory metals, stainless steels, non-ferrous alloys, and high-strength materials, some of which are not usually forgeable.

Advantages :

1. Complex parts can be made is one blow.
2. Draft allowances are reduced and is some cases eliminated.
3. Scaling and decarbonisation of metal surface during the operation are almost eliminated.
4. Extrusion of even such materials, which either cannot be extruded on the conventional machines or will need conventional machines of giant sizes for this purpose, is also possible on these machines.
5. Tolerance and surface finish are improved over those obtained with conventional forging techniques.
6. Because strength and fatigue resistance are improved, parts can be made smaller.
7. Repeatability is excellent.
8. Overall production cost is very low.

Limitations :

1. Part configuration is usually limited to one-piece dies. If two-piece or split dies are necessary, the economics of process become debatable.
2. There is a limitation of size and weight of the product these machines can handle. Extremely large and very heavy forgings (say above 25 kg) cannot be easily produced on these machines.

Applications :

A few examples of the type of products made on these machines are :

— Valve bodies ;

— Gears ;

— Engine housings ;

— Rocket components ;

— Missile components, etc.

3.4.6.4. Orbital forging (or Rota forming)

It is a *cold forming process.*

In this process, the pressure of the upper die of the workpiece is *concentrated on a small area at any time,* and not on the total workpiece area as in conventional forging. The *upper die is slightly inclined to the vertical axis of the machine and it imparts a high frequency circular rocking motion across the top surface of the workpiece.* At the same time, the workpiece is slowly moved hydraulically upward and *pressed against* the *orbiting upper die.* The forging operation is completed when the hydraulic ram touches a preset stop. The ram is then lowered and the hydraulically operated ejector ejects the forging from the lower die.

Advantages :

1. Low equipment cost.
2. Press capacity requirements are only 5 to 10 per cent of that needed for conventional forging (since only a small portion of die actually contacts the workpiece).
3. Better surface finish.

Limitation :

The limitation of this process is that the forging is obtained by filling the lower die and so the bottom surface of the upper die should be flat and smooth.

- Examples of parts produced by this process are : *Parts flanged with indented or crown shapes and discs, etc.*

3.4.6.5 Incremental forging

- In this forging method, very big forgings are made by working different areas of the forging into shape, one at a time.
- As only a limited area is worked at any time, the forging equipment can be *much smaller in capacity* as compared to conventional forging. This makes it possible to forge huge parts on presses of modest capacity.

Limitation :

The workpiece tends to cool below the forging temperature as it moves from incremental step to step. Reheating the part to its original forging temperature may destroy the thermomechanical work already done on it in the case of many alloys. It may be prevented as follows :

(i) By reheating the workpiece progressively to a lower temperature after its temperature falls below the forging temperature.

(ii) By covering the billet using insulating blankets and other coatings so as to reduce the heat loss.

3.4.6.6. Liquid metal forging

- Liquid metal forging process also called *squeeze casting* is hybrid between conventional casting and forging methods and produces *complex shaped components from molten metal in a single step.*
- In this process of forging, the molten metal is poured into the bottom forging die and allowed to solidify *partially.* Then the upper and bottom dies are closed and pressure is applied and maintained for a fixed time until solidification is completed. The dies are then opened and the forged component is ejected from the bottom die.

Advantages :

This forging process claims the following *advantages over conventional casting and forging processes* :

1. Low capacity presses are required.
2. Economical from material point of view.
3. Yield is higher in comparison to sand castings, since gates and risers are not needed.
4. The tooling and equipment needed are basically simple, low cost and readily available.
5. No gas and shrinkage porosities.
6. Thinner and more complex components can be produced.
7. The cast as well wrought alloys can be used.
8. The mechanical properties of the product are comparable to those by conventional forging ; they are considerably better than sand castings.

Limitations :

1. This process cannot be employed to produce all the component shapes.

2. Liquid forged parts necessitate different heat treatments (cooling rates being *faster,* the effect on microstructure of the component will be different as compared to forging or casting).

3.4.6.7 Gatorizing

In this forging technique the forging stock is *preconditioned in inert atmosphere* to obtain a temporary condition of fine grained microstructure resulting in low strength and high ductility (This is accomplished by mechanically working the workpiece at temperature slightly below the recrystallisation temperature). After preconditioning, the isothermal forging operation is carried out at slightly below the recrystallisation temperature. Finally, certain heat treatment operations are done on the forgings to get normal high strength and hardness.

● This forging method is employed for making *aircraft components of nickel and titanium based alloys.*

3.4.7. Defects in Forging

Although the forging process generally gives superior quality products compared to other manufacturing processes, there are some defects that are likely to come if proper care is not taken in forging process design. A brief description of such defects is given below :

1. **Cold shut.** This usually occurs at the corners and at right angles to the surface.

   — This is *caused* mainly by the *improper design of the die* wherein the corner and fillet radii are small as a result of which the metal does not flow properly into the corner and ends up as a cold shut.

2. **Unfilled section.** It is similar to misrun in casting and occurs when metal does not completely fill the die cavity.

   — It is usually caused by using insufficient metal or insufficient heating of the metal.

3. **Flakes.** Basically, these are internal ruptures.

   — These are caused by the *improper cooling of large forging* and can be remedied by following proper cooling practice.

4. **Scale pits.** These are irregular depressions on the surface of the forging.

   — These are primarily caused because of the *improper cleaning of the stock* used for forging.

5. **Improper grain flow :**

   — This is caused by the *improper design of the die* which makes the flow of metal not following the final intended directions.

6. **Internal cracks :**

   — These can result from *too drastic a change* in the shape of the raw stock at too fast a rate.

7. **Die shift :**

   — This defect is caused by the *misalignment* of the two die halves, making the two halves of the forging to be of improper shape.

8. **Burnt and overheated metal :**

   — This defect is caused by *improper heating conditions* and *soaking the metal too long.*

3.4.8. Cleaning and Finishing of Forgings

Cleaning and finishing of forgings include the following ·

1. **Removal of oxide scale :**
   - Due to the contact of heated steel with air a thin layer of scale (iron oxide) is formed on the surface of steel forging. The amount of scale depends upon the *forging temperature* and *length of time of the operation.*
   - The scale can be removed by employing steam or compressed air.

2. **Cleaning by pickling :**
   - The hard scale from the surface of the forgings can be removed by *pickling process.*
   - The pickling process consists of immersing the forgings in a tank filled with an acid solution, which is 12 to 15 percent concentrate of H_2SO_4 in water. The solution acts to loosen the hard scale from the forging surface and remove it.

3. **Tumbling process :**
   - This process is employed to remove scale and for general cleaning of the forgings.
   - In this process, the forgings along with some abrasive materials such as coarse sand or small metallic particles are placed in barrel ; the tilted barrel is rotated at low speeds. This action loosens the scale from the surface of the forgings and results in general cleaning of the forgings.

4. **Blast cleaning :**
   - The process consists of directing a jet of sand, grit or metallic shots against the forgings.
   - By this process the scale is removed and a smooth surface finish is imparted to the forging.

3.4.9. Heat Treatment of Forgings

The forged parts are generally heat treated for the following *reasons* :

1. To relieve internal stresses set up during working and cooling.
2. To normalise the internal structure of the metal.
3. To improve machinability.
4. To improve hardness, strength and other mechanical properties.
   - The common heat treatments given to forged components are *annealing, normalising* and *tempering.*

3.4.10. Design Considerations

While designing a forging, the following *considerations* need be given :

1. The parting line of a forging, as far as possible, should lie in one plane.
2. The forged component should ultimately be able to achieve a radial flow of grains or fibres.
3. In order to facilitate easy removal of forgings from the dies, sufficient draft on surfaces should be provided. Generally, a 1° to 5° draft is provided on press forgings and 3° to 10° on drop forgings.
4. As far as possible, sharp corners should always be avoided, to prevent concentration of stresses leading to fatigue failures and to facilitate ease in forging.
5. In order to facilitate an easy flow of metal too thin sections should be avoided.
6. The presence of pockets and recesses in forgings should be avoided.
7. While deciding the forging and finishing temperatures metal shrinkage and forging method should be duly taken into account.

8. Adequate allowances to compensate for metal shrinkage, machining, die wear, trimming and mis-match of dies should be provided.

3.5. EXTRUSION

3.5.1. Introduction

Extrusion *is the process in which metal is caused to flow through a restricted orifice so creating an extremely elongated strip of uniform, but comparatively small cross-section.* The operation is identical to the squeezing of tooth paste out of the tooth paste tube. It is also similar to cold drawing except that the material is pushed ; not pulled, through the hole in the die, and the operation is often carried out at high temperature.

- An important feature of the extrusion process in general is that the *grain flow of the material lies in the direction most suited to resist imposed stresses in service.*

- Metals that can be extruded include aluminium and its alloys, copper and its alloys, lead and its alloys, magnesium and magnesium alloys, steel ; tin and zinc and their alloys, with few exceptions.

- By the extrusion process, it is possible to make components which have a constant cross-section over any length as can be had by the rolling process. Some extrusion shapes are shown in Fig. 3.23.

Fig. 3.23. Some extrusion shapes.

- The complexity of parts that can be obtained by extrusion is more than that of rolling, because the *die required being very simple and easy to make.* Also extrusion is a *single pass* process unlike rolling.

- The amount of *reduction* that is possible in extrusion is *large.*

- Generally, *brittle materials can also be very easily produced.*

- Large diameter, thin walled tubular products with excellent concentricity and tolerance characteristics can be produced.

- The *extrusion pressure for a given material depends on the extrusion temperature, the reduction in area and the extrusion speed.* The extrusion speed depends on the work material. Too high an extrusion speed would cause excessive heat generation in the extruded metal causing lateral cracks.

Extrusion ratio. *It is defined as the ratio of cross-sectional area of the billet to the cross-sectional area of the product.* Its value for hot extrusion for steel is about 40 : 1 and for aluminium as high as 400 : 1.

3.5.2. Advantages, Limitations, and Applications of Extrusion Process

Advantages :

Extrusion process entails the following *advantages* :

1. Shapes can be extruded, those if produced by other methods shall entail more cost comparatively.
2. Thinner walls can be obtained by increasing the forming pressure.
3. Due to high reduction ratio (the cross-sectional area of the billet to the cross-sectional area of the shape produced) the metal has excellent transverse flow lines, consequently the part has an added *strength and secondary operations are made easier.*
4. The extrusion dies are less expensive comparatively and so, moderately short runs are practical.
5. Extrusion process allows low cost in process redesign.
6. The *dimensional tolerances* of extrusions are very good.
7. There is more flexibility in design for adjacent thin and heavy sections as well as for difficult re-entrant angles. Sharp corners, not practicable in other processes can readily be obtained by extrusion.
8. It is also possible to get shapes with internal cavities in extrusion by the use of spider dies.
9. Extruded shapes can often replace weldments and members previously machined from bar stock.
10. Extrusion is an ideal process for obtaining *rods from metal having poor ductility.*

Limitations :

The following are *limitations* of extrusion process :

1. As compared to roll forming extruding *speed is slow.*
2. Though the accuracy is good and entirely adequate for most applications yet it is not as close as a machine part would be.
3. The size of dies and presses that can be economically built is a *limiting factor.*
4. Costs of extrusion are generally greater as compared to other techniques.
5. In productivity, extrusion is much inferior to rolling, particularly to its continuous varieties.
6. Process waste is higher in extrusion than in rolling.
7. High tooling costs.

Applications :

The main *fields of application of extrusion processes* are :

(*i*) Manufacture of parts of high dimensional accuracy.

(*ii*) Manufacture of sections and pipes of complex configuration.

(*iii*) Medium and small batch production.

(*iv*) Working of poorly plastic and non-ferrous metals and alloys.

3.5.3. Classification of Extrusion Processes

Extrusion processes can be *classified* as follows :

1. **Hot extrusion :**

(*i*) Forward (or direct) extrusion.

(*ii*) Backward (or indirect) extrusion.

2. **Cold extrusion :**

(*i*) Forward extrusion :

- Hydrostatic extrusion.

(*ii*) Backward extrusion :

- Impact extrusion
- Cold extrusion forging.

3.5.4. **Hot Extrusion Processes**

3.5.4.1. **Forward or direct extrusion process**

Refer to Fig. 3.24. In this process a billet (piece of metal worked to a suitable shape) of the material to be extruded is placed in a container. At one end of the container is fixed a die while from the other end the metal is forced to flow through the die by hydraulically driven ram through a follower pad.

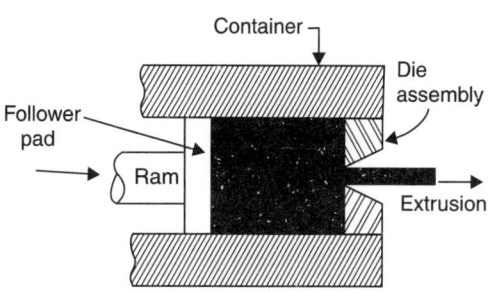

Fig. 3.24. Forward or direct extrusion.

Much work must be supplied to overcome the resistance and high frictional forces between the billet and wall of the container and to produce the required rate of deformation.

- The problem of friction is particularly severe in case of steels because of their higher extrusion temperatures. To reduce this friction lubricants are to be used.
- At lower temperatures, a mixture of oil and graphite is generally used. Molten glass is generally used for extruding steels.
- To reduce the damage to equipment, extrusion is finished quickly and the cylinder is cooled before further extrusion.

3.5.4.2. **Backward or indirect extrusion**

Refer to Fig. 3.25. In this process the metal is confined fully by the cylinder. The ram which houses the die is the hollow plunger or ram. It is termed *backward* or *indirect* or *inverted* because of the opposite direction of the flow of metal to that of ram movement. Thus, the billet in the container remains stationary and hence no friction. Also the extrusion pressure is not affected by the length of the billet in the extrusion press since friction is not involved.

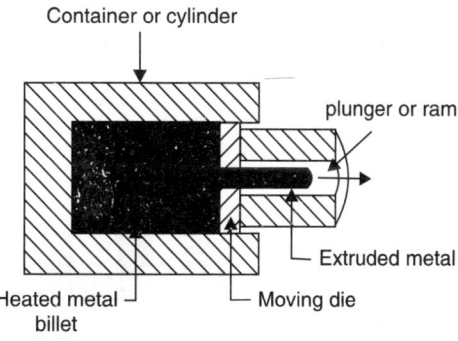

Fig. 3.25. Backward or indirect extrusion.

The *surface quality achieved in generally good* since there is no heat cracking due to the friction between the billet and the extrusion cylinder interface.

The *disadvantage* of backward extrusion is that the *surface defects of the billet would end up in the final product* unlike direct or forward extrusion where these are discarded in the extrusion container.

- The *use of backward extrusion is restricted due to the fact that the ram must be hollow and that the extrusion product must be passed back through the ram.* The process of direct extrusion is generally mechanically more convenient.

Comparison between Forward and Backward Extrusion :

S. No.	Forward or Direct extrusion	Backward or indirect or inverted extrusion
1.	It is the *simplest,* but is limited by the fact that as the ram moves the billet must slide or shear at the interface between billet and container.	The billet proper does not move relative to container, instead the *die moves.*
2.	*High friction forces* involved.	*Low friction forces ;* the friction involved is only between the die and the container and this is independent of billet length.
3.	High extruding forces required ; however, *mechanically more convenient.*	Extruding force is 25 to 30% less than in forward extrusion. However, a long hollow ram is required and this limits the loads which can be applied. Due to this, and complex design of tools, this type of extrusion finds *limited application.*
4.	Scrap or process *waste* is about *18 to 20%* of the billet weight.	Scrap or process *waste* is only *5 to 6%* of billet weight.

Advantages of hot extrusion :

(*i*) A very rapid process.

(*ii*) Involves low tool costs.

(*iii*) Owing to high pressures used and compressive nature of the process a very dense structure of metal is obtained.

(*iv*) In several cases it proves cheaper than pressure die casting with comparable finish and tolerances.

(*v*) It is an ideal process for producing parts of uniform cross-section in large quantities.

3.5.5. Cold Extrusion Processes

3.5.5.1. Hooker extrusion

In the Hooker or extrusion down method, a cup is first formed by a suitable working operation ; extrusion then consists of elongating and thinning the walls of cup, using a shouldered punch and die as shown in Fig. 3.26.

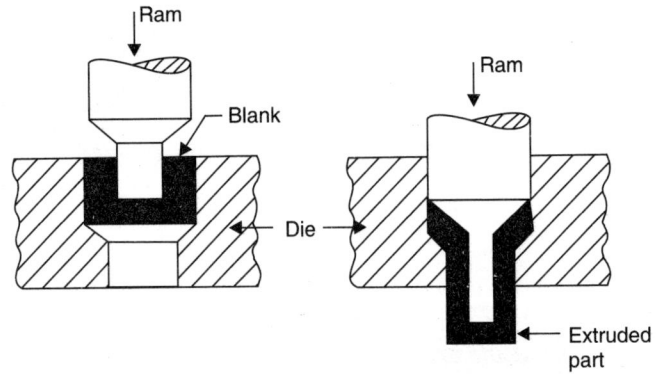

Fig. 3.26. Hooker extrusion.

● This cold working process is commercially applied mainly for the production of *small, thin walled copper and aluminium seamless tubes and cartridge cases.*

3.5.5.2. Hydrostatic extrusion

Hydrostatic extrusion Fig. 3.27 is an extrusion method in which the required pressure is applied through a *fluid medium* surrounding the billet. Although the presence of fluid inside the extrusion chamber eliminates the container wall friction, the process finds limited industrial application because of need for specialised equipment and tooling and low production rate (high set-up time). It is usually carried at room temperature.

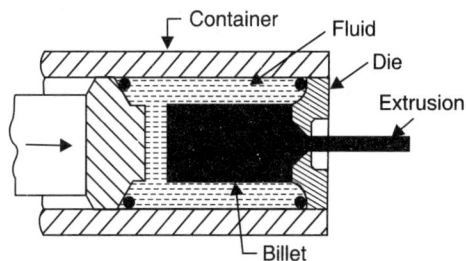

Fig. 3.27. Hydrostatic extrusion.

— The pressure transmitting fluids commonly used for hydrostatic extrusion are : Glycerine, Ethylgol, SAE 30 mineral lubricating oil, castor oil with 10% alcohol and isopentane. The hydrostatic pressure ranges from 1100 to 3150 N/mm^2.

● Because the hydrostatic pressure increases the ductility of the material, brittle materials can be extruded successively by this method.

The *main commercial applications* of this process are :

— Cladding of metals ;

— Making wires of less ductile materials ;

— Extrusion of nuclear reactor fuel rods.

3.5.5.3. Impact extrusion

This process is quite similar to Hooker process, but the flow of metal is in *opposite direction.*

In impact extrusion (Fig. 3.28) the punch descends at a high speed and strikes the *blank (slug)*, extruding it upward. The thickness of the extruded tubular section is a function of the clearance between the punch and the die cavity.

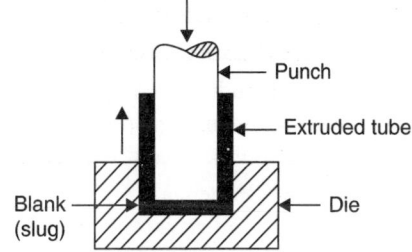

Fig. 3.28. Impact extrusion.

● This process is applied primarily to lead, aluminium, manganese, tin, zinc, and their alloys.

About 95 percent of die products of impact extrusion are *collapsible paste tubes.*

3.5.5.4. Cold extrusion forging

It is similar to impact extrusion but with the *main difference that the side walls are much thicker and their height is smaller.* Fig. 3.29 shows the process schematically.

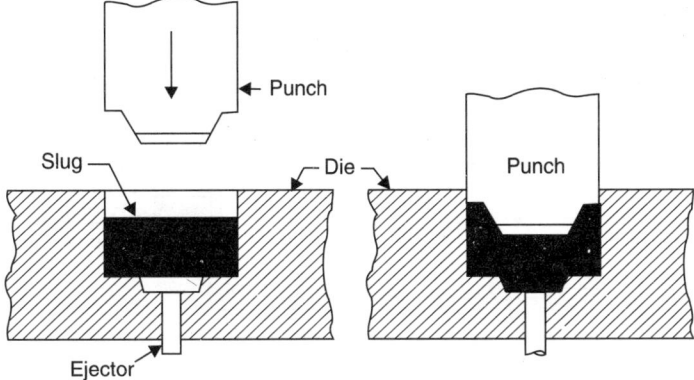

Fig. 3.29. Cold extrusion forging.

The punch slowly descends over the slug kept on the die, thus forging some metal between the punch and the die and the rest being extruded through the clearance between the punch and die side walls. The side walls thus generated are short and thick with any profile in the end unlike the impact extension. Afterwards, the component is ejected by means of the ejector pin provided in the die.

Note : The **backward cold extrusion** processes are different from other extrusion processes in that *each stroke of the punch prepares a directly usable single component which may not necessarily have a uniform cross-section over its entire length.* Also, these are *limited to smaller sizes and for non-ferrous alloys only.*

Advantages and limitations of cold extrusion :

Advantages :

Cold extrusion has gained wide acceptance in industry because of the following *advantages* over hot extrusion :

(*i*) Improved surface finish, provided that lubrication is effective.

(*ii*) High production rates and relatively low costs.

(*iii*) Lack of oxide layers.

(*iv*) Good control of dimensional tolerances, thus requiring a minimum of machining operations.

(*v*) Improved mechanical properties resulting from strain hardening, provided that the heat generated by plastic deformation and friction does not recrystallise the extruded metal.

Limitations/Drawbacks :

(*i*) In cold extrusion the *stresses on tooling are very high,* especially with steel workpieces. The design of tooling and selection of appropriate tool materials are therefore crucial to success in cold extrusion.

— Punches are a critical component ; they must have sufficient strength, toughness and resistance to wear and fatigue.

(*ii*) *Lubrication* also is crucial, especially with steels, because of the generation of new surfaces and the possibility of seizure between the metal and the tooling, caused by the breakdown of lubrication.

(*iii*) *Temperature* rise in cold extrusion is an important factor, especially at high extrusion ratios. The temperature may be sufficiently high to initiate the complete recrystallisation process of the cold worked metal, thus reducing the advantages of cold working.

3.5.6. Extrusion Force

— The expression for determining the force required to produce an extrusion as shown in Fig. 3.30 is as follows :

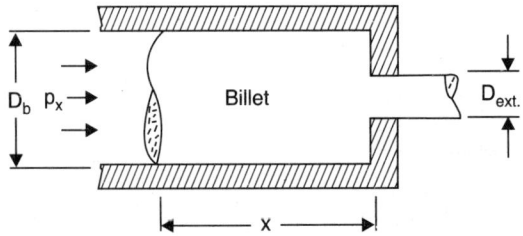

Fig. 3.30. Extrusion force.

$$\frac{p_x}{\sigma_y} = \left[1.7 \ln \frac{D_b}{D_{\text{ext.}}} + \frac{2X}{D_b}\right] \qquad\qquad ...(3.3)$$

where, p_x = Pressure,

σ_y = Yield strength of the material,

D_b = Billet diameter,

$D_{\text{ext.}}$ = Extruded component diameter, and

X = Working length of the billet/stock.

— A typical relationship developed by PERA (Production Engineering Research Association, England) for calculating the maximum pressure for *backward or indirect extrusion of carbon steels* (0.1, 0.2 and 0.3% carbon) is given below :

$$p = \tau \left[3.45 \ln \frac{A_{\text{ext.}}}{A_b} + 1.15\right] \text{kN/mm}^2 \qquad\qquad ...(3.4)$$

where, p = Extrusion pressure,

τ = The upper yield point, kN/mm^2,

(τ = 0.29, 0.31, 0.36 kN/mm^2 for 0.1%, 0.2% and 0.3% carbon steels respectively),

$A_{\text{ext.}}$ = Cross-sectional area of the extruded component, and

A_b = Cross-sectional area of the billet.

This expression is valid for extrusion ratios ranging from 1.65 to 4.25 using billets with 0.6 length to diameter ratio.

● In extrusion process the *main process variables* are :

1. Strength of billet material.
2. Size of billet.
3. Extrusion ratio.
4. Die angle.
5. Co-efficient of friction.
6. Speed of extrusion.
7. Temperature.

3.5.7. Extrusion Equipment

● The *presses* for extrusion processes may be either *hydraulic* or *mechanical,* according to the pressure required, and either vertical or horizontal.

— The *hydraulic press* is more usually vertical since it has great flexibility, and compactness, and less liable to injury resulting from improper use.

— The *mechanical* vertical extrusion press given a high output and each piece spends only a second or two in actual contact with the dies. This shortness of contact means that little heat is lost as a result of radiation and condition, so that, if required little the extended part can be extruded and can be transferred straight to reducing mill.

● The **dies** employed are *often of high speed steel* and are *always of high quality special alloys tool steels.*

● The **punches** may also be of high quality steel. Lubrication is plentifully employed in most instances.

3.5.8. Extrusion Defects

Following are the three principal defects in extrusion :

1. Extrusion defect.
2. Surface cracking.
3. Internal cracking.

1. **Extrusion defect :**

- The most common defect in extrusion known as *extrusion defect,* **pipe, tailpipe,** and *fish failing* arises from the *back flow of material,* pushing the end face of the bullet into the core of the product. Such a defect *weakens the product* since the *surface layer is normally contaminated by oxides.*
- This defect can be reduced/avoided :
— By modifying the flow pattern to a less inhomogeneous one, such as by controlling friction and minimising temperature gradients ;
— By machining the surface of the billet prior to extrusion to eliminate scale and impurities.
— By using a dummy block that is smaller in diameter than the container, thus leaving a thin shell along the container wall as extrusion progresses.

2. **Surface cracking :**

- Sometimes the heat generated due to extrusion may raise the temperature of the job, resulting in the *development of surface cracks.* These cracks are intergranular and are usually the result of hot shortness ; they occur especially with aluminium, magnesium, and zinc alloys, but are also observed with other metals, such as molybdenum alloys.
  The surface cracks can be *avoided by using lower temperatures and speed.*
— The surface cracking may also occur at low temperatures and has been attributed to periodic sticking of the extruded product along the die land **(stick-slip)** during extrusion.

3. **Internal cracking :**

- The centre of an extruded product can develop *cracks* variously known as *centre burst, centre cracking, arrowhead fracture.* These cracks are attributed to a state of *hydrostatic tensile* stress at the centre line of the deformation zone in the die.
- The tendency for centre cracking increases with increasing die angles and level of impurities and decreases with increasing extrusion ratios.

3.6. WIRE DRAWING

Wire drawing *is the process of reducing diameter of metal rods by drawing them through conical openings in die blocks.*
— Wire drawing is fundamentally a simple process. Steel, iron or non-ferrous rod is converted into wire by drawing it through a conical hole having an included angle of 8–24 degrees. *In continuous wire-drawing the wire passes through a succession of holes of decreasing size in dies made of steel, tungsten carbide, ruby or diamond, the reduction in cross-sectional area usually being about 30 per cent.*
— The rods used for wire-drawing are first pickled in acid to remove any scale and then electrically butt welded. The end of the rod is tapered sufficiently to fit the first dies by passing it through a pointing machine, which generally takes the form of two motor-driven rollers having a number of grooves of decreasing size between which the rod is rolled. The rod may be coated with iron hydroxide, copper or tin, applied during or after pickling. The rod is then fed into the wire drawing machine, which may be fitted with six or more dies, through which the wire is drawn by means of number of power-driven pulleys or rotating drums. Although the principle is similar in each case, the machines differ in design, and for practical purposes are broadly divided into *"non slip"* and *"slip types".*
  Fig. 3.31 shows some important features of wire-drawing.

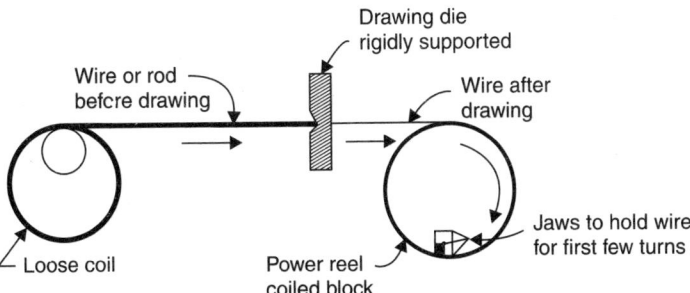

Fig. 3.31. Wire drawing.

— Due to the great heat generated by the friction of the wire in the dies, both the dies and drums are continuously cooled by circulating water through them. Lubrication is often ensured by passing the rod through dry soap on its way to the die. On some designs, however a synthetic lubricant is forced under pressure into dies, each of which is thus continuously lubricated over the whole of its surface. Separate cooling equipment is provided for the lubricant.

— A *different method of overcoming the cooling and lubrication problem is to employ a wet-drawing machine operating on the slip principle.*

— Modern alloy-steel **dies** have a long life, but maximum efficiency *tungsten-carbide dies are replacing steel.*

— Ruby or diamond dies are also employed but their use generally being confined to the drawing of fine wires of less than 3 mm in diameter. The jewels are first trimmed and then drilled with small drills which are fed with diamond powder and oscillated for several hours.

When the hole has been pierced it is finished and polished to within 0.00025 mm. The die is then mounted in a metal holder.

— When very fine wires are to be produced, a composite wire is employed for drawing. A platinum wire only 1/30,000 inch in diameter, for instance, many thousand times finer than a human hair, has been drawn by encasing the platinum wire in silver, thus increasing the overall diameter to a practicable figure. After drawing, the silver was dissolved by nitric acid, leaving a platinum wire which was visible only under a microscope.

Mechanics of wire and rod drawing :

The major variables in the drawing process are : (*i*) Reduction in cross-sectional area (*ii*) Die angle, and (*iii*) Friction ; these are illustrated in Fig. 3.32.

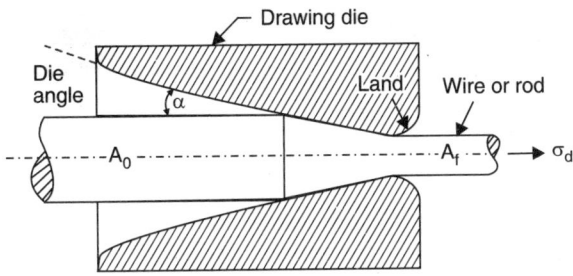

Fig. 3.32. Variables in drawing round rod or wire.

The following expressions can be obtained for the *drawing stress* σ_d :

(i) $\qquad \sigma_d = \sigma_y \ln\left(\dfrac{A_o}{A_f}\right) \qquad\qquad$ ... *Ideal deformation* ...(3.5)

(ii) $\qquad \sigma_d = \sigma_y \left(1 + \dfrac{\tan\alpha}{\mu}\right)\left[1 - \left(\dfrac{A_f}{A_o}\right)^{\mu\cot\alpha}\right]$ ...*Ideal deformation and friction* ...(3.6)

where, σ_y = Yield stress,

$\quad\alpha$ = Die angle,

$\quad\mu$ = Co-efficient of friction,

$\quad A_o$ = Original cross-sectional area of the wire or rod, and

$\quad A_f$ = Final cross-sectional area after the drawing operation.

- *Wire drawing improves the mechanical properties because of the colds working.* The material loses its ductility during the wire drawing process and when it is to be repeatedly drawn to bring it to the final size, *intermediate annealing is required to restore the ductility.*

The drawing machines can be arranged in *tandem* so that wire coming from one die is coiled upto a sufficient length before it is re-entered into the subsequent die and so on. This coiling of sufficient wire helps for any discrepency in the speed of wire drawing in any die. *Since there is no change in volume successive drawings have to be done at higher speeds.*

Defects in drawing :

- Defects in drawing are similar to those observed in extrusion, especially **center cracking.** The tendency for cracking increases with increasing die angle, with decreasing reduction per pass, with friction, and with the presence of inclusions in the material.
- Another type of defect is the *formation of* **seams,** which are longitudinal scratches or folds in the material. Such defects can open up during subsequent forming operations by upsetting, heading, thread rolling, or bending of the rod or wire.

Various *surface defects can also result from improper selection of process parameters and lubrication.*

3.7. TUBE DRAWING

Tube drawing is accomplished in most cases with the use of a *draw bench.* Following are the different methods used for drawing tubes : Refer to Fig. 3.33.

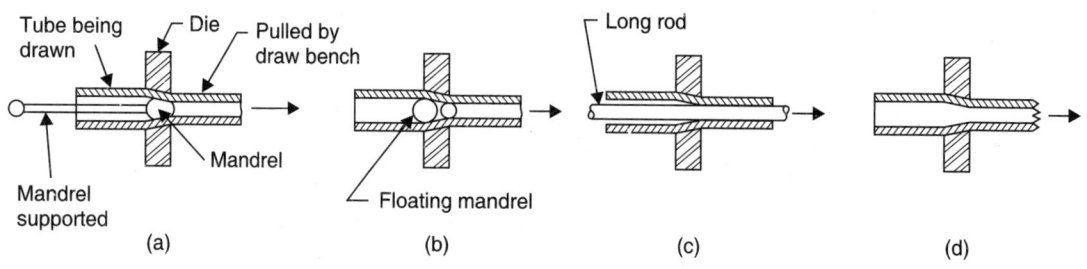

Fig. 3.33. Tube drawing.

- Method (*a*) is most commonly used.
- Method (*b*) uses a floating mandrel which adjusts itself to the correct position because of its stepped contour.

- Method (*c*) is usually used for small-sized tubing.
- Method (*d*) uses no mandrel or rod and has little or no control over inside diameter.

By repeated cold drawing and annealing, when necessary, it is possible to reduce a 50 mm diameter tube to a size of a hypodermic needle, which has an outside diameter of about 0.2 mm.

3.8 TUBE MAKING (By Rotary Piercing)

Here the material used is a *hot steel billet.*

The billet is gripped between two rolls which spin at high speed (Fig. 3.34). The spinning action tends to open a hole or cavity in the axis of the billet. A plug or mandrel bar is pierced through the centre and a round hole is easily produced. Rotary piercing is an important method, the rough pierced blank being finished by plug rolling over a mandrel between a pair of grooved rolls. For making tubes of large diameters the blanks are initially rough pierced and then given a second larger piercing before being finally rolled.

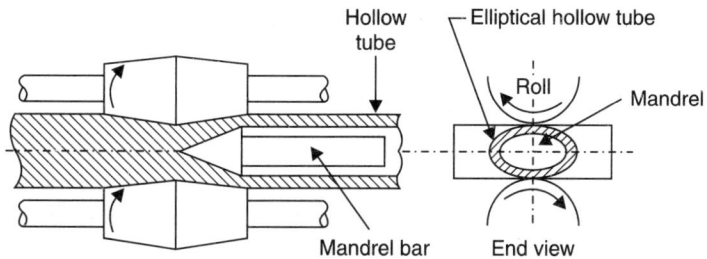

Fig. 3.34. Tube making.

3.9. METAL STAMPING AND FORMING

3.9.1. Introduction

Stamping *is a general term for a number of operations such as punching, blanking, shearing, bending and forming that are performed on a* **press** *with the use of dies.*

— Stamping dies consist of punches, usually held in the upper half of the die, and matching dies, which are normally located in the lower half.

Press working *may be defined as a chipless manufacturing process by which various components are made from sheet metal.* This process entails the following *advantages* :

(*i*) The weight of fabricated parts is small.

(*ii*) Items of diversified shapes, both simple and complex, such as washers, bushings, retainers of ball bearings, etc. can be made easily.

(*iii*) The parts made by cold sheet metal working have narrow tolerances with a high surface finish (In several cases, they require no subsequent machining and are delivered to the assembly shop directly).

(*iv*) The productivity of labour is quite high.

In cold press working, the initial material is : Low carbon steels, ductile alloy steels, copper and its alloys, aluminium and its alloys, as well as, other ductile materials from tenth of a mm to about 6 mm or 8 mm thick.

Die sets. The punches are set in a punch holder and the dies in a die shoe as shown in Fig. 3.35. The two together are called a *die set*. The die set has guide pins (not shown) that keep the upper and lower half in alignment, which facilitates die changing when changing from one product to another.

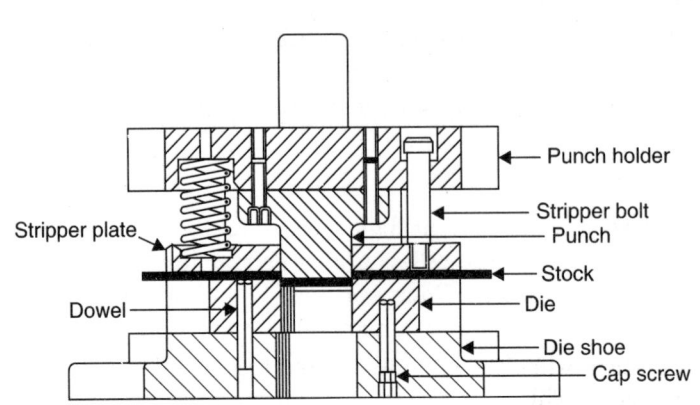

Fig. 3.35. Standard die set with a punch
and die mounted in place.

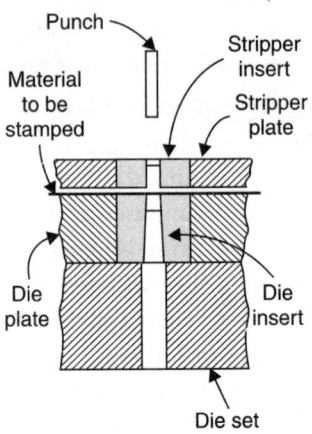

Fig. 3.36. Punch and die showing a
solid stripper plate and carbide
inserts (punch holder not shown).

Fig. 3.36 shows a punch and die for making small holes in sheet metal (punch holder not shown). A solid stripper is used to pull the part off the punch on the upstroke of the press.

3.9.2. Bending

Bending is the operation of deforming a flat sheet around a straight axis where the neutral plane lies. It is a very common forming process for changing sheet and plate into channel, drums, tanks, etc.

The types of bending dies/methods are shown in Fig. 3.37.

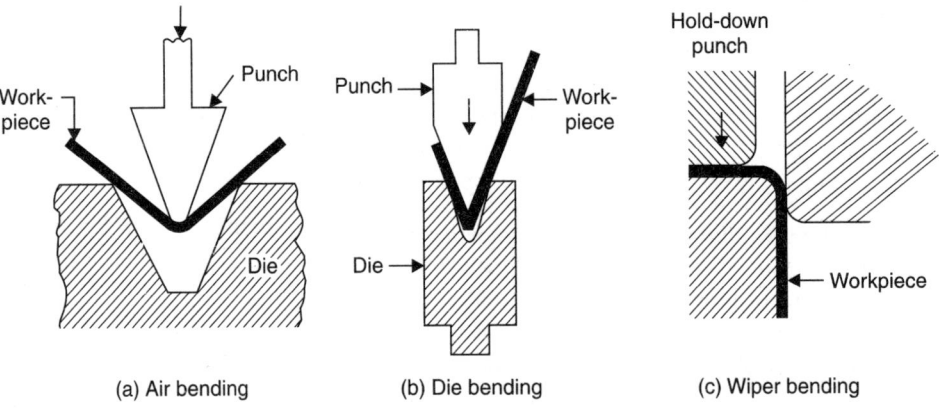

(a) Air bending   (b) Die bending   (c) Wiper bending

Fig. 3.37. Types of bending dies.

— In *V*–bending (Fig. 3.37 *a, b*) a wedge-shaped punch forces the metal sheet or strip into a wedge shaped die cavity. The bend angle may be acute, 90° or obtuse.

  V–dies are the ones that are most generally used.

— Wiper bending is used for 90° bends only. Here the work is held firmly to the die, and the punch bends the extended portion of the blank (Fig. 3.37*c*).

● Presses specifically designed for these operations are called *press brakes*.

● The bending load may be calculated from the knowledge of material properties and die characteristics as follows :

$$P_b = \frac{Kl\sigma_u t^2}{b} \qquad\qquad ...(3.7)$$

where, P_b = Bending force, N,

$\quad$ K = 1.33 for die opening of $8t$; 1.20 for die opening of $16t$; 0.67 for U bending ; 0.33 for a wiper die,

$\quad$ l = Length of bent part, mm

$\quad$ σ_u = Ultimate tensile strength, *MPa,*

$\quad$ t = Blank thickness, mm, and

$\quad$ b = Width between contact points, mm.

Spring back :

- At the end of the bending operation, when the pressure on the metal is released, there is an elastic recovery by the material. This causes a decrease in the bend angle and this phenomenon is termed as *"spring back"*. It is the extent to which the metals tends to return to its original shape or position after undergoing bending or forming operation.

- The *methods* used to overcome or prevent spring back are :

    (*i*) Stretch forming

    (*ii*) Overbending

    (*iii*) Bottoming

    (*iv*) Ironing.

3.9.3. Deep Drawing

In sheet metal, *drawing* is a process of forming flat sheet metal into hollow shapes by means of a punch that causes the metal to flow into the die cavity. *If the depth is one or more times the diameter, the process is called* ***deep drawing*** (See Fig. 3.38). The forming of shallow shapes is sometimes referred to as *stamping*, but the distribution is not clear since stamping is also used to describe cutting flat figures or patterns.

In deep-drawing, also called *"cup or radial drawing"*, a parallel-walled cup is created from a flat blank (Fig. 3.38). The blank may be circular, rectangular, or of a more complex outline. The blank is drawn into the die cavity by action of the punch. Deformation is restricted to the flange and draw radius. No deformation occurs under the bottom of the punch. As punch forms the cup, the amount of material in flange decreases.

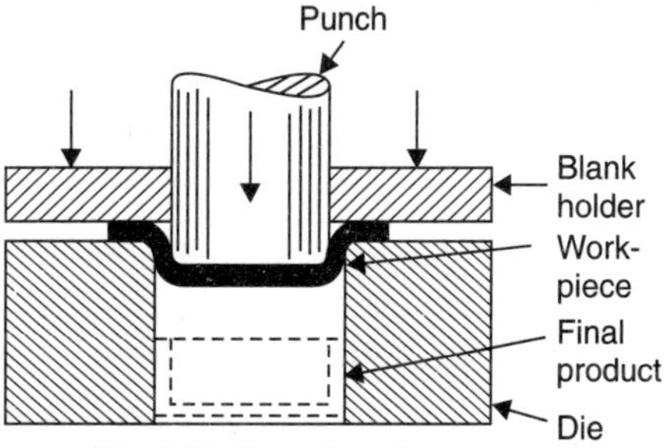

Fig. 3.38. Deep drawing.

- There should be no appreciable change in the thickness of the material between the blank and finished part. When it is desired to reduce the metal thickness, it may be done in secondary operations as in ironing.

- Important variables of the drawing operation are *friction, squeezing force,* and *formability of the metal.*

- The following materials are being used for deep drawing operations : Soft steel, copper, brass, aluminium and their alloys, magnesium and their alloys, zinc and a variety of other non-ferrous alloys.

Some *common products of deep drawing* are :

- Automobile bodies ;

- Lamp reflectors ;

- Cylinders and various small components required in the automobile industry.

— Metals and alloys suitable for deep drawing should be of special deep drawing quality and should be tested, so that the required properties conform to the specifications specified.

Variation in grain size and the presence of harmful impurities will impart bad finish to the finished articles. In case of materials which workharden, stress relieving anneal may be necessary during one or more intermediate stages.

3.9.4. Stretch Forming

It is a method of producing contours in sheet metal. Thinning and strain hardening are inherent in the process.

Refer to Fig. 3.39(a). The grips are stationary and the form block moves upwards to provide the necessary tension and motion.

— The jaws and form blocks are hydraulically actuated to provide control of pressure adjustment to meet the varying conditions characteristic of the material to be formed.

Refer to Fig. 3.39(b). Stretch wrapping consists of first stretching the metal beyond *its yield point* while it is straight and then wrapping it around a form block. It is particularly suited for *long sweep bends on tubes and extruded shapes.* The main *advantage* of this type of forming is that only *one die* is needed ; the die may be made out of *inexpensive material.*

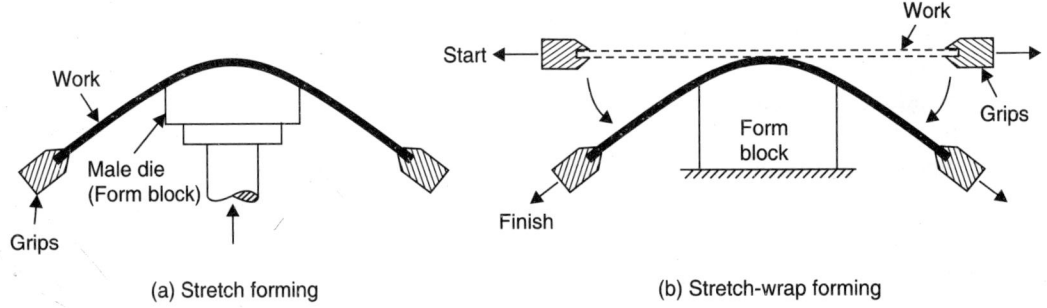

(a) Stretch forming   (b) Stretch-wrap forming

Fig. 3.39. Methods of stretch forming a metal.

— The thickness reduction should not exceed 5 percent of the original thickness.

● In stretch forming process the *spring back is completely eliminated* because the forming is done by introducing uniform tensile stresses in the fine increments of sufficient magnitude to exceed the elastic limits of the metal, thus causing the material to take a permanent set.

● The process is applicable to a wider range of materials, because the ductility is the least important factor in stretch forming. The *desirable qualities in the metal for maximum stretchability are toughness, fine grain structure and a large spread between the tensile yield and ultimate strength.*

— It has been found that an annealed material which has high strain hardening rate is best suited to stretch forming.

3.9.5. Metal Spinning

Spinning *means shaping a metal blank as it revolves at a high speed in a lathe.*

— It is one of the oldest methods of producing a wide range of goods which have an *axis of rotation.* For small-batch production, spinning, because of the simplicity of the equipment needed and the ease with which a chuck can be prepared, often gives *lower overall production costs than any alternative method.*

— The process lends itself to the production of goods *direct from a flat bank*. It is also a *very useful adjunct to the press and drop stamp*. In the latter direction it is used to finish articles that could not readily be handled by any other means. Hence, the spinning lathe is often used along with the press or drop stamp to trim, bead, burnish or stroke over, neck down or spin a raising, cupping, or a shell to the desired size and shape.

In practice metal spinning may be roughly grouped under two headings :

1. Hand spinning.

2. Machine spinning.

1. Hand spinning :

— Refer to Fig. 3.40. With hand spinning the blank, as it revolves at a high speed, is subjected by a steel tool to pressure exerted by the operator, who has the tool handle tucked under his right arm. The tool is levered off a tree rest and is prevented from slipping by means of a peg. The leverage obtained is adjusted to suit the class of work being hundled.

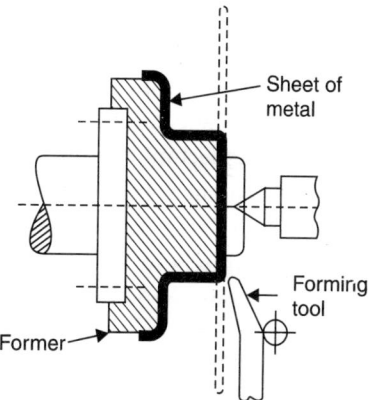

Fig. 3.40. Hand metal spinning.

— Skill in the use of the tools and experience in handling the varying classes of material is essential. If the metal is worked only in one place, it becomes work-hardened, and the result may be cracking, buckling or even tearing of the metal. The art in metal spinning lies in working the tool over the whole surface and coaxing the metal towards the desired cross-section. As the work proceeds and the surface becomes work-hardened, it is necessary to resort to interstage annealing.

2. Machine spinning :

— Machine spinning is very similar to manual spinning, but is *always used in conjunction with press to perform some operation that cannot be readily done on the latter type of machine*. Rarely is machine spinning used to draught blank straight down. The method of operating is also different. In place of the tee rest and hand spinning tool, a compound slide is used in conjunction with a roller mounted in a fork. By constantly adjusting the positions of the slides the roller is made to give the desired contour.

— For machine spinning chucks are made from either cast iron or steel suitably hardened. The rollers are best made from a good alloy steel that will take a hard surface.

— For some jobs a rocking slide is also fitted to the machine and used along with the compound rest. This permits such operations as trimming and beading to be done at the same setting as, say, spinning or burnishing.

Hot spinning :

— Hot spinning of metals is used commercially to dish or form thick circular plates to some shape over a resolving form block.

— A *hot spinning machine is a combination of a vertical press and a large vertical spinning lathe which rotates the blank about a vertical axis*. Power actuated rollers do the forming, as the part is rotated.

● Metals upto 150 mm thick is routinely hot spun into *dished pressure vessels and tank shapes*. Hot spinning is also used to shape thinner plates of hard to form metals like titanium.

Advantage of spinning :

(*i*) Some complex parts of re-entrant shapes such as kettles, pitchers, etc. can be produced economically.

(*ii*) Equipment cost is low.

(*iii*) Low tool costs (The form block may be of plaster, wood or metal).

Limitations/Drawbacks :

(*i*) Finished parts are not always uniform ; close tolerance cannot be maintained.

(*ii*) The spinning method depends to a large extent on the operator's skill.

Applications :

Product applications of spinning are :

— Funnels and large processing kettles ;

— Rocket motor cases ;

— Reflectors ;

— Kitchen ware ;

— Bells on musical instruments, etc.

Comparison of spinning and drawing :

● Spinning is comparable to drawing for making cylindrical shaped parts. Because of the simple tools used in *spinning*, it is *economical for smaller lots*. But the time required for making a cup is more in spinning and also more skill is required in the process. Thus, it is *not suitable for large-scale production.*

● Complicated shapes and re-entrant shapes are not feasible by drawing, but can be made by spinning using the sectional and collapsible form blocks.

● Large parts are much more easily made in spinning than by drawing.

● When *sheet thickness is more*, for example, in making the dished ends of pressure vessels, *cold spinning* is *not sufficient*. Then the *blank is heated to the forging temperature and so the process is called hot spinning*. Also, in hot spinning the tools are mechanically manipulated because of the higher magnitudes of forces required.

3.9.6. Shear Forming and Flow Forming

Shear forming :

Refer to Fig. 3.41. Unlike spinning (where the thickness of the finished part and that the starting blank are essentially the same), the *shear forming produces a deliberate reduction in wall thickness of the final component*. The setup is similar to that of the spinning lathe, but a *much higher force level is required.*

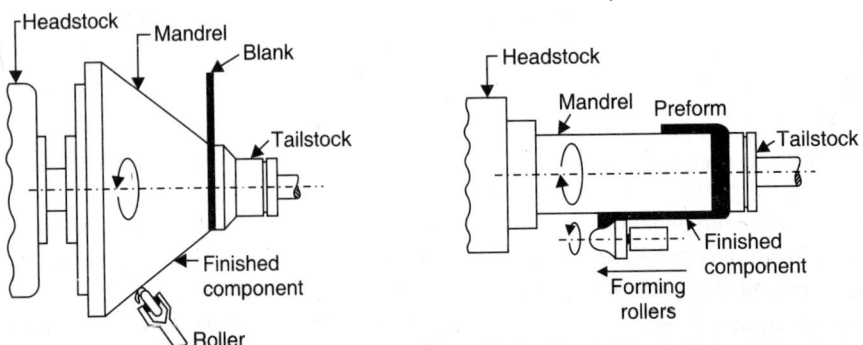

Fig. 3.41. Shear forming.   Fig. 3.42. Flow forming.

In some cases, both spinning and shear forming will be used on different portions of the same workpiece.

- The main *advantage* of shear forming is the cold working of the workpiece which adds significantly to its strength. Thus, wall sections thinner than those normally acceptable may be used with *considerable savings in material costs.*
- This technique can be employed to produce angles between 12° and 80° with the center line of rotation. Thinning is excessive below 12°, and above 80° the amount of working is insufficient to ensure metallurgical stability.

Flow forming :

Refer to Fig. 3.42. This forming process is similar to shear forming, but the process is used to produce *cylindrical items that usually start from a preformed blank.* Most modern flow-forming machines employ three rollers of a more complex design.

- Applications for flow forming range from kitchenware to truck wheels and high-pressure gas cylinder bodies.

Note : Newer hydraulic spinning machines are computer controlled with full program storage and editing capabilities. CNC has reduced set up time by 80 percent, making it possible to produce smaller lot sizes for flexible manufacture.

3.9.7. Blanking

***Blanking** is a process of cutting or shearing a blank from sheet or strip material.*

- In the production of many articles from sheet metal under a press or by spinning, the process of blanking is often the first operation in the production cycle. Simply expressed, *blanking is the process of shearing from sheet or strip metal a blank of a given shape.* The term normally implies two things :

(*i*) that the process is carried out under some type of press, and

(*ii*) that the entire edge of the resultant blank has been subjected to a shearing operation.

Reduced to simplest terms, *a set of blanking tools consists of a bed and punch.* The latter is attached to the ram of the machine and the former is clamped to the bed to the press. When the tools are set, the punch should first enter the bed without touching the sides in such a manner as to damage the cutting edge. The clearance between the cutting edges is regulated to suit the thickness of the metal to be cut with the tools in position, the punch should be adjusted to enter the bed for the minimum distance necessary to give complete shear. If it is permitted to enter too far, the *punch is subjected to unnecessary wear.* With tools in position and properly set, the tail end of the downward stroke of the ram forces the punch to carry the material it covers into the bed. The surrounding metal cannot enter, hence is sheared off. *In order to obtain a blank free from burr, that is, with a clean sheared edge, particular attention should be given to the clearance and condition of the cutting edges.*

Fig. 3.43 shows a typical layout for a small set of blanking tools. The punch is of the built-up type having a low-carbon steel shank, which is fitted in the press ram, and a hardened-steel cutting edge held by screws on to the face. The aim of having a built-up punch of this type is two fold. It gives a strong edge that will withstand the abrasive action for a considerable period without losing its size, and it permits a great economy in the use of expensive alloy steel. In this particular design the cutting edge is held in position by means of a spigot and hollow-head screws. To assist the clamping and make up the "shunt height" of the press, the bed is mounted on a bolster plate. It is

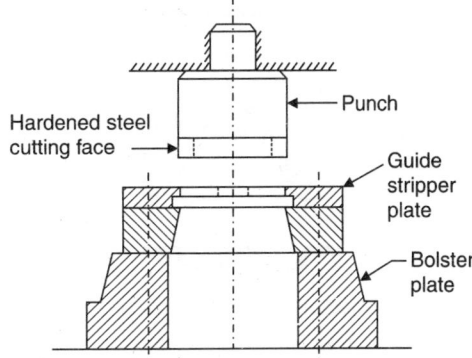

Fig. 3.43. Typical layout for a small set of blanking tools.

located by means of strong dowels and clamped in position with hollow-head screws. For simplicity neither the dowels nor the screws are shown in the figure. *In order to guide the material a guide plate fitted with a stop is attached to the face of the blank bed. The guide plate also acts as a stripper and takes the scrap webbing off the punch.* It aids in getting a good blank, and a reasonable output per hour, as time is not wasted in having to pull the scrap clear by hand.

- For economic reasons the blanking operations is often combined with others which can be effected during the same work stroke of die press (*e.g.*, blanking and cupping).

3.9.8. Piercing

The main difference between piercing and blanking is that the metal portions that are cut out in the former process are scrap. Holes of various shapes are pierced. When two or more piercing punches are employed together in a die, their lengths should differ slightly in order to reduce the force and impact required at one time.

— The diameters of holes which are to be pierced should be *at least twice the metal thickness, in order to avoid excessive punch breakage.* Pierced holes should not be located too close to adjacent holes.

— Drilling should be used for smaller holes and for holes which must be quite close together.

3.9.9. Embossing and Coining

Both the embossing and coining are cold press working operations in which the starting material is in the form of a blank of sheet metal. The aim of both the operations is to force impressions into the surface or surfaces of the metal. However, *whereas "embossing" is a forming operation, "coining" is a pressing operation.*

Embossing. Refer to Fig. 3.44. It is the operation used in *making raised figures on sheets with its corresponding relief on the other side.* The die set consists of a die and punch with the desired contours, so that when the punch and die meet the clearance between them is *same* as that of the sheet thickness. The metal flow is in the direction of the applied force. The forces needed are much less than in the coining process.

- Embossing operation is generally used for *providing dimples on sheets to increase their rigidity and for decorative sheet work used for panels in houses and religious places.*

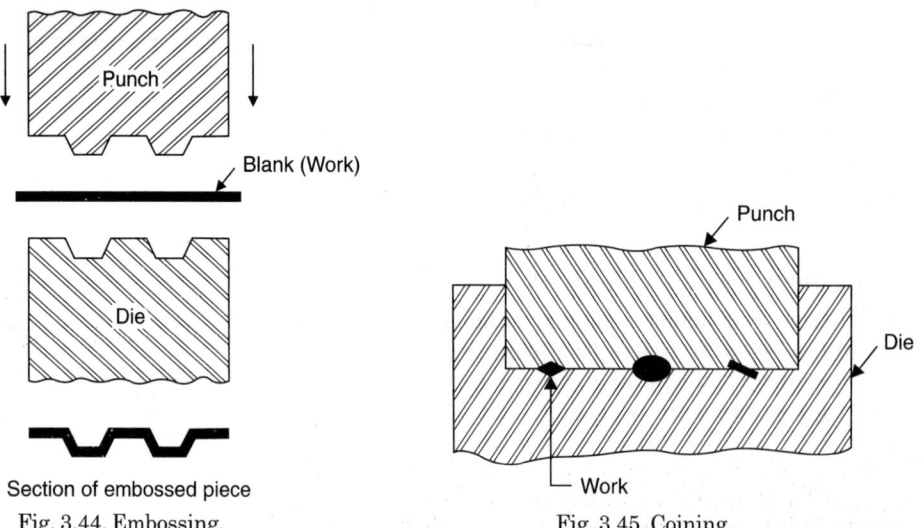

Section of embossed piece	
Fig. 3.44. Embossing.	Fig. 3.45. Coining.

Coining. It is essentially a cold forging operation excepts for the fact that the flow of metal occurs only at the top layers and *not* the entire volume. The coining die consists of the punch and die which are engraved with the necessary details required on *both* sides of the final object. A blank, kept on the die, is compressed by it as shown in Fig. 3.45.

Very large pressures are exerted (1200 to 3000 MPa) to cause the metal to flow to all portions of the die cavity. The metal is caused to flow in directions *perpendicular* to the compressive force along the die surfaces.

— Owing to the difficulty of forcing the metal of flow, the depth of impression is never very great, rarely exceeding 0.8 mm.

● Probably, the *hard currency* is the best known product of coining operation. Other applications of coining are the manufacture of *insignia, medals, badges, piece of art* and *household hardware.*

3.9.10. Roll Forming

Refer to Fig. 3.46. **Roll forming** *may be defined as a process used to change the shape of coiled stock into desired contours without altering the cross-sectional area.*

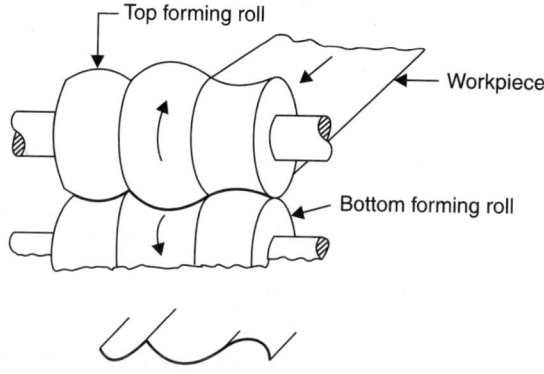

Fig. 3.46. Roll forming.

— The process utilizes a series of rolls to gradually change the shape of metal as it passes between them.

— Most ductile sheets or strips (0.1 to 20 mm) are used for roll forming.

— The main advantage of this process over other production methods is speed.

— This process can handle prepainted or electroplated surfaces without damage.

● This process is particularly suitable for producing *large quantities of long strips of desired shape, with minimum handling operations.*

3.9.11. Rubber Press Forming

Fig. 3.47 shows a schematic view of rubber-press forming operation. Rubber is restrained from sideward motion and acts as a deformable female die to exert essentially hydrostatic pressure on the workpiece.

Advantages:

(*i*) Tool setting time is amply reduced.

(*ii*) Greater depth of draw in one stroke.

(*iii*) Thinning of work material is almost nil.

(*iv*) Tools have fewer components and are made of cheap and easy to machine materials.

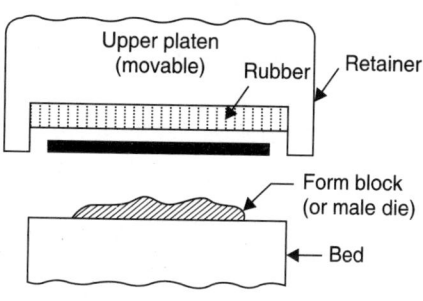

Fig. 3.47. Rubber press forming.

(*v*) Since only one tool is used in this process, the variation in sheet thickness is not a limitation as in conventional drawing.

(*vi*) One rubber pad takes the place of several die shapes.

● This process is quite suitable for short run production because of its *economy* since it eliminates the need for the more expensive mating steel dies.

Limitations :

(*i*) The formed components have *less sharp details* as compared to conventionally formed components.

(*ii*) Rubber pads wear out rapidly.

Rubber hydroform process :

● In this process, a *pressurised liquid* behind the rubber pad is used to exert the force required for forming.

● Since hydrostatic pressure acts equally all over the surface ; *deeper cups and complex shapes with sharper details* can be obtained.

3.9.12. Hydromechanical Forming

In this type of forming process (a modification of rubber hydroform process) the container containing the liquid (oil) is fixed on the bed of the press and rubber pad/diaphragm acting as a seal is not used. The blank which is directly placed on the draw ring acts as the sealing element. Blank holder descends on the blank and applies the blank holding pressure. As the punch travels into the container, the cup is formed *against the hydrostatic pressure developed inside the container.* This enables the *formation of deeper cups and complex shapes.*

3.9.13. Defects in Sheet Metal Formed Parts

Some of the defects encountered in sheet metal formed parts are :

1. *Burr and bend*—Caused due to improper and excessive clearance between punch and die.

2. *Wrinkling*—A wavy condition on metal parts, due to buckling under compressive stresses.

3. *Earing*—It occurs due to more easy deformation in some directions than the other, thus forming *ears* on the deep drawn parts.

4. *Sinking*—A small depression on the surface of the formed part which is not desirable on the finished part.

5. *Strain hardening*—This phenomenon reduces ductility (formability) and plasticity.

6. *Stretcher strain marks.*

7. *Orange peel effect.*

8. *Part failure.*

9. *Secondary tearing.*

3.10. SHOT PEENING

Shot peening *is mainly employed to improve the fatigue resistance of metal by setting up compressive stresses in the surface.*

— This process is also employed to prevent the cracking of workpieces in corrosive media to improve the oil retaining properties of the processed surfaces.

— This process is adopted to *remove stress concentration* on parts of irregular shapes or at local areas.

The process is carried out by *blasting or hurling a rain of small shots pneumatically or mechanically at high velocity against the worked surface.* Small indentations are produced due to striking of shots, which *causes the metal to flow plastically* to a depth of few tenths of mm.

3.11. TYPES OF DIES

3.11.1. Classification of Dies

The dies may be *classified* as follows :

1. Simple dies
2. Compound dies
3. Progressive dies
4. Transfer dies
5. Fluid-activated diaphragm dies
6. Multi-action dies.

1. Simple dies :

● The die sets which are made to perform a single press-working operation, such as punching or bending, that is accomplished in one stroke of the press, are called *simple dies*.

● A single operation die may be a bending die, curlie die, wiring die and buldging die.

2. Compound dies :

● Compound dies make close tolerance and concentric parts, as all work is done in one stroke.

● Fig. 3.48 shows a compound blank and draw die. The metal enters as a flat sheet, is cut to the right length, and is then formed over a reverse-type punch. Spring action on a pressure plate strips the part off the punch. The knockout pin, shown at the top, pushes the part out of the upper die if it stays on that side when the die opens.

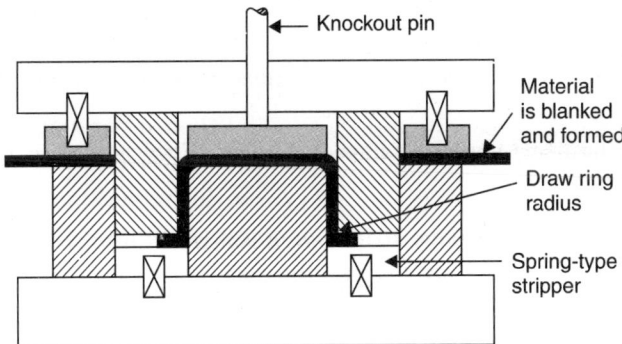

Fig. 3.48. Compound blank and draw die.

3. Progressive dies :

— These dies are made to *cut and form a part in successive stages or stations of the die.* The parts are held together by the strip skelton or tabs until the last station or cut off. Force or movement comes from the strip feeder attached to the press.

— A progressive die can perform very complex work, doing piercing, blanking, forming, lancing and notching.

● The *advantage* of this type of die is that the tooling can be spread out.

● The *cost* of progressive dies is *high* and therefore they are usually limited to high production operations.

4. Transfer dies :

● Transfer dies, like progressive dies, are also multistation dies and are constructed on the same principle, but unlike progressive dies, the blanks are cut out first from the strip stock and then mechanically moved through successive stages, sometimes arranged in a circle. An example of the type of part that would require a transfer die is shown in Fig. 3.49.

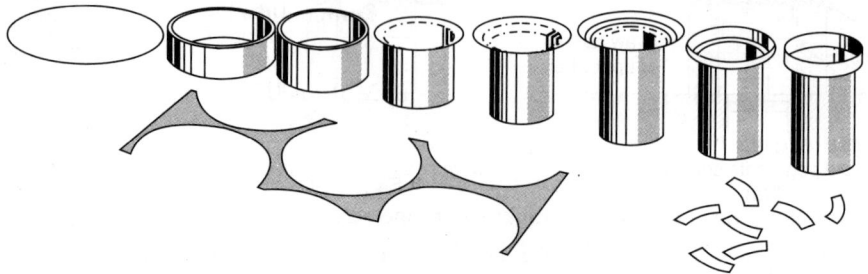

Fig. 3.49. Deep drawn part formed in a transfer die.

Comparison between Transfer dies and Progressive dies :

Transfer dies :

- Both are *multiple station operation dies.*

(*i*) The main advantage of transfer die over progressive die is a reduced liability to damage due to scrap metal getting into the dies as already cut blanks are fed to the dies.

(*ii*) The elimination of a great deal of scrap loss is the greatest advantage of transfer operations.

(*iii*) Transfer blanks can also be cut to take advantage of the grain direction of the metal stock.

(*iv*) Transfer mechanisms generally have a *higher initial cost than progressive die setups.*

Progressive dies :

(*i*) In progressive dies continuous strip from a coil is fed into the dies whereas in transfer dies previously cut blanks are fed into the dies.

(*ii*) Progressive operations can usually be run at higher speeds and with shorter press strokes than comparable transfer operations.

5. Fluid-activated diaphragm dies :

- Drawing of sheet-metal parts may be accomplished by the use of fluids and a diaphragm.
- The process has the *advantage of closer control of the drawing operation so that parts that usually require two drawing operations can be done in one.*

6. Multi-action dies :

- Multi-action dies are used to perform multistep forming in one operation. The *punch and die segments are operated in a sequence that is coordinated with the press stroke.*
- In this setup (Fig 3.50) the entire sheet is held between tooling plates at all times. Unsupported metal does not flow with the punch as it does in conventional tooling. The multiple, matched dies are made so that segments can be timed to move for an optimum forming sequence. The tools are mounted in a conventional hydraulic press, which is programmed to activate independent, external hydraulic circuits.

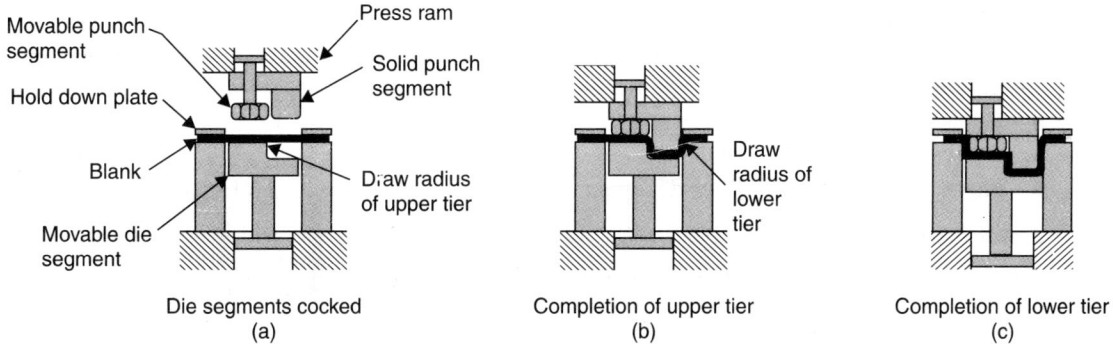

Fig. 3.50. Multi-action dies.

Note : Though both the fluid-activated diaphragm and the multi-action dies have good control over metal flow, which is necessary to avoid excessive thinning and buckling, it appears that multi-action dies provides *more versatility.*

3.11.2. Causes of Failure of Dies in Metal Working Operations

In metal working operations, failure of dies generally results from one or more of the following *causes :*

1. Excessive wear.
2. Improper installation, assembly, and alignment.
3. Overloading, misuse and improper handling.
4. Improper dies design.
5. Defective die materials.
6. Improper heat treatment and finishing operations.
7. Overheating and heat checking *i.e.,* cracking caused by thermal cycling.

3.12. PRESSES

3.12.1. Classification of Presses

Presses are employed to provide power to operate dies for blanking, forming, and shearing. They may be *classified as* follows :

1. According to type of frame :

(*i*) Open frame presses.

(*ii*) Closed frame presses.

Open frame press :

● This press is also known as 'Gap frame' or 'C' frame press.

— The most common type press is 'open back inclinable C-type frame press', commonly known as OBI press. Its frame is inclined backward which facilitates the removal of scrap or parts by gravity through the open back. It ranges from small 1 tonne bench press to floor presses rated upto 150 tonnes.

● Owing to their construction, open frame presses are *less rigid and strong* and are useful mainly for operations on *smaller work.*

● These presses are available upto 200-tonne capacity with strokes of 90 to 120 per minute.

Closed frame press :

- These presses have two upright columns on each side of the die. They are stronger, more rigid and balanced than C-type frame presses.
- These are suitable for *heavier work*. Work is fed either from the front or from the back of the press into the die area.
- These presses can be of two designs : Arch frame press and straight side frame press.

2. According to type of drive :

Press drives refer to the means of supplying power to the ram, which be belts, gears, or hydraulic.

— If *belts* are used, it is called non-geared and is *flywheel driven.*

— If *geared,* it may have several arrangements, from single gear drives to multiple-reduction gear drives that are used on the very large presses. Fig. 3.51 shows a single-geared, two-end drive.

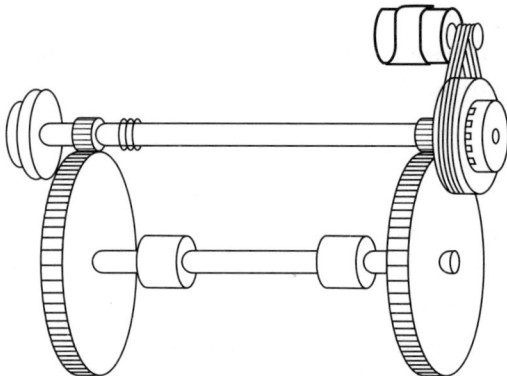

Fig. 3.51. Single-geared, two-end drive.

- *Advantages of mechanical presses (over hydraulic presses) :*

   (*i*) Lower capital cost.

   (*ii*) Lower maintenance cost.

   (*iii*) Faster operation.

   (*iv*) Higher punch-slide speeds.

- *Advantages of hydraulic presses :*

   (*i*) Constant pressure can be maintained throughout the stroke.

   (*ii*) Force and speed can be adjusted throughout the stroke.

   (*iii*) More versatile and easier to operate.

   (*iv*) More powerful than mechanical presses.

   (*v*) Tonnage adjustable from zero to maximum.

   (*vi*) It is safe, since it stops at a pressure setting.

   — However, it is *slower than mechanical press.*

Note : A press is rated in tonnes of force, it is able to apply without undue strain. To keep the deflections small, it is usual practice to choose a press rated 50 to 100 per cent *higher* than the force required for an operation.

3. According to slide actuation :

- The most common method of actuating the press slide is with a *'crankshaft'*.

- *'Double cranks'* are employed for *wider presses,* and some *large presses have 'multiple cranks'.*

- *'Eccentrics'* are used where only *short strokes* of the slide are needed.

- *'Knuckle joints'* (Fig. 3.52) are used where *short, powerful strokes are needed.*

4. According to action :

- *Action* refers to the number of slides used, such as single, double, or triple.

— A *double-action* press has two slides moving in the same direction ; the outer one is used for blank holding.

— A *triple-action* press has an inner and outer slide as in the double-action press, but in addition has a third slide moving upward through the fixed bed. This action can perform draws against the inner slide while both the upper slides are in a dwell position.

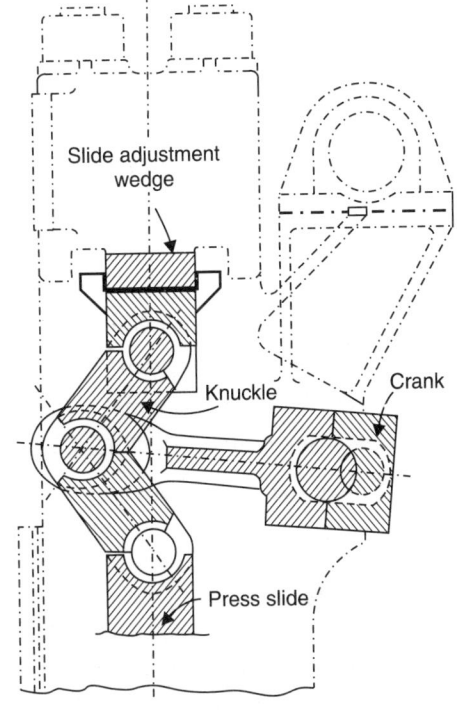

Fig. 3.52. Press slide actuated by a knuckle-joint.

5. According to suspension. *'Suspension'* refers to the number of points of attachment from the drive to the press slide. On small presses there is only one from the crankshaft to the slide. On large presses, there may be as many as four.

CNC turret presses :

- Numerically controlled **(NC)** and Computerised Numerical Control **(CNC)** *punching machinery differs from the conventional mechanical, hydraulic or pneumatic presses in that the workpiece is automatically positioned under a basic punch that is gripped by a turret bushing or adapter in a single station.*

- CNC manipulates the punch-controlling numbers electronically, performing such tasks as optimisation of punching instructions and other kinds of number juggling that earlier technology could not do.

- The operation of a turret press **FMS** (Flexible Machining Systems) consists of the following *steps :*

(*i*) Selecting the tooling for the job;

(*ii*) Computerized nesting of the parts;

(*iii*) Loading material;

(*iv*) Loading tools into storage carousel.

The automatic tool changer has a capacity of upto 22 different tools of any size or shape. Tool return and retrieval are accomplished simultaneously with the punching sequence. After parts are punched and cut, a trap door in the machine table unloads all parts upto 45 cm × 67 cm into bins for convenient handling and sorting.

3.12.2. Cutting Forces and Energy in Press Work

Cutting forces. The formulae for calculating the force required to cut sheet metal in a die are as follows :

General formula :

Cutting force (P) = Shear stress (τ) × perimeter (p) × sheet metal thickness (t)

For *rectangular cuts :* $P = \tau \times 2(l + b) \times t$ ...(3.8)

For *circular cuts :* $P = \tau \times \pi d \times t$ ...(3.9)

where, P = Cutting force, N

      τ = Shear stress, N/mm^2

      l = length of cut, mm

      b = Width of cut, mm

      t = Sheet metal thickness, mm.

Energy in press work :

Energy in press work or the work done to make a cut is given ideally as,

$$E = P_{max} \times \text{travel of punch}$$

or, $E = P_{max} \times K \times t$ ...(3.10)

where K is the percentage of penetration required to cause rupture.

After allowing for energy lost in machine friction and in pushing slugs through the die etc., the above equation get modified as :

$$E = P_{max} \times K \times t \times Z$$

                                              ...(3.11)

where Z is the factor that accounts for the amount of extra energy required (it depends upon the circumstances in each case). For *general purpose* the value of Z as 1.16 is recommended.

3.13. COMPARISON OF METAL FORMING PROCESSES

The comparison of metal forming processes is given below :

S. No.	Metal forming processes	Advantages	Limitations
1.	Cold rolling	(i) Suitable for production of plates, sheets and foils. (ii) High production rate. (iii) Good dimensional accuracy and finish.	(i) Deformation limited to small reductions. (ii) High equipment cost.
2.	Hot rolling	(i) Suitable for large reduction. (ii) High production rate. (iii) Wide range of shapes (Billets, blooms, slabs, sheets, bars, tubes and structural sections) can be produced.	(i) Suitable for production of large sections. (ii) High equipment cost. (iii) Poor dimensional accuracy and finish.

3.	Drawing	(i) Good surface finish and dimensional accuracy. (ii) Low equipment and tooling cost. (iii) High production rate. (iv) Long lengths of rounds, tubings, square and angles can be produced.	(i) Production of constant cross-sections only. (ii) Deformation limited to small reductions. (iii) Lubrication is necessary.
4.	Deep drawing	(i) Moderate equipment and tooling cost. (ii) High production rate. (iii) Good surface finish.	(i) Forming of shallow or deep parts of simple shapes only. (ii) Limited to forming of thin sheets. (iii) Finishing required.
5.	Hot extrusion	(i) Suitable for large reduction. (ii) Moderate cost of equipment and toolings. (iii) Complex sections and long products can be produced.	(i) Components with thin walls are difficult to produce. (ii) Only constant cross-section can be produced. (iii) Lubrication is necessary. (iv) Dimensional accuracy and finish achieved are not good.
6.	Impact extrusion	(i) Good finish and dimensional accuracy. (ii) High production rate. (iii) Generally no finishing is required. (iv) Suitable for production of thin sections.	(i) Deformation limited to small reductions. (ii) Suitable for production of light components from softer materials.
7.	Punching and blanking	(i) Low cost of labour. (ii) High production rate. (iii) Almost any shape can be obtained. (iv) Moderate equipment cost.	(i) Cost of tooling can be high. (ii) Limited to thin sheet applications.
8.	Open-die forging	(i) Simple to operate. (ii) Inexpensive tooling and equipment. (iii) Wide range of workpiece sizes can be used. (iv) Suitable for low production volume.	(i) Fairly skilled operators are required. (ii) Can be used for simple shapes only. (iii) Production rate is low. (iv) Dimensional accuracy and surface finish achieved are poorer. (v) Finishing required for achieving final shape.
9.	Closed-die forging	(i) Can be used for production of complex shapes. (ii) Suitable for high production rate. (iii) Good dimensional accuracy and reproducibility.	(i) Appropriate die set for production of each component. (ii) High equipment and tooling cost. (iii) More than one step required for each forging. (iv) Finishing required for achieving final shape.

| QUESTIONS WITH ANSWERS |

Q. 3.1. What is work (or strain) hardening ?

Ans :

- Work (or strain) hardening is a *phenomenon* which results in an *increase in hardening and strength of a metal* (specimen) subjected to plastic deformation at temperature lower than the recrystallization range (*cold working*).

- When a material is subject to plastic deformation, a certain amount of work done on it is *stored internally as stain energy. This additional energy in a crystal results in a strengthening or work hardening of solids.*

Thus *work (or strain) hardening may be defined* as *increased hardness accompanying plastic deformation.* This increase in hardening is accompanied by an *increase in both tensile and yield strength.*

- Work hardening *reduces ductility and plasticity.*

- Work hardening is used in many manufacturing processes such as rolling of *bars and drawing of tubes.* It is also used to improve the elastic strength in the manufacture of many parts such as : (*i*) Prestretching of hoisting chains and cables, (*ii*) Initial pressurisation of pressure vessles, cylinders of hydraulic press and guns.

Mechanism of work hardening. *Work hardening is caused by dislocations interacting with each other and with barriers which impede their motion through the crystal lattice.* Hardening due to dislocation interaction is a complicated problem because it involves large groups of dislocations, it is difficult to specify group behaviour in simple mathematical way. One of the easiest dislocation concepts to explain strain hardening was the idea that *dislocations pile up on slip planes at barriers in crystal. The pile ups produce back stress which opposes the applied stress on slip plane.* Another mechanism of work hardening in addition to that due to back stress resulting from dislocation pile up at barriers, is believed to occur when *dislocations moving in the slip plane cut through other dislocations intersecting the slip plane.* The dislocations threading through the active slip plane are often called a *dislocation forest* and this work hardening process is referred to the intersection of a forest of dislocations.

Theory of work hardening. According to all theories (of work hardening) work hardening is due to the increased resistance the dislocations experience in moving through the lattice when the metal has been subjected to cold working/deformation. The basic idea put forward by Taylor in 1934 was that work hardening is due to the *dislocations getting in each other's way.* Thus the stress (τ) necessary to move a dislocation in the stress field of other dislocations surrounding it, will have to have a higher value.

The value of τ is given by the relation :

$$\tau = K \left(\frac{b}{L}\right)^{1/2} \gamma^{1/2}$$

where, K = A constant,

   b = Burger's vector,

   L = A side of the dislocation loop if it is assumed that square loops are emitted ($L = R$ in case the dislocation loops are circular rings, where R is the radius of dislocation ring), and

   γ = Plastic strain.

Q. 3.2. Explain briefly strain hardening and normal anisotropy co-efficients.

Ans. The limits of forming relate primarily to the exhaustion of ductility. For obtaining better formability, a material must be capable of *straining uniformly.*

Through research, it has been seen that the following two properties strongly influence formability :

(*i*) Strain hardening co-efficient, or *n* value;

(*ii*) Anisotropy co-efficient, or *r* value.

- The *n* value determines the *ability of the material to be uniformly stretched.*
- The *r* value represents the *thickness-to-width strain.*

Strain-hardening co-efficient (*n* value)

— Strain hardening refers to the fact that, as a metal deforms in some area, dislocations occur in the microstructure. As these dislocations pile up, they tend to strengthen the metal against further deformation in the area. Thus the strain is spread throughout the sheet. However, at some point in the deformations, the strain suddenly localizes and necking, or localised thinning, begins. When this occurs, little further overall deformation of the sheet can be obtained without its fracturing in the necked region. Thus the *strain-hardening co-efficient reflects how well the metal distributes the strain throughout the sheet, avoiding or delaying localised necking.* The *higher the strain-hardening co-efficient, the more the material will harden as it is being stretched and the greater will be the resistance to localised necking* (Necks in the metal may affect structural integrity or harm surface appearance).

- The *stampings that entail much drawing should be made with metals having high average strain-hardening co-efficients.*

Normal anisotropy co-efficient (*r* value) :

— The *r* value (the ratio of the plastic width strain to the longitudinal strain) can be measured in a tensile test.

— A material that has a high plastic anisotropy also has a greater thinning resistance. In general, *the higher* the *anisotropy co-efficient,* the better the metal deforms in *drawing operations.* Although *r* values may vary in relations to the rolling direction, they may still average out at unity. This condition is termed *planar anisotropy* and is one that would lead to *earing* (it refers to a localised elongation that takes place when deep drawing, as in deep drawing a cup).

Q. 3.3. Explain the terms 'Recovery', 'Recrystallisation' and Grain growth. Give the differences between recovery and recrystallisation.

Ans. Recovery, Recrystallisation and Grain growth :

The deformed metal, in comparison with its underformed state, is in a non-equilibrium and thermodynamically unstable state. Subsequently, spontaneous processes occur in strain-hardened metal, even at room temperature. If the temperature is raised sufficiently, the metal attempts to approach equilibrium through the processes namely :

(*i*) Recovery,

(*ii*) Recrystallisation, and

(*iii*) Grain growth.

Fig. 3.53 shows a schematic diagram indicating recovery, recrystallisation and grain growth and changes in important properties in each region.

Recovery :

- *It is a low temperature phenomenon which results in the restoration of the physical properties without any observable change in microstructure.*
- During recovery, there is *negligible effect on hardness* whereas *electrical resistance decreases rapidly towards the annealed value.*

- The process of recovery is important for *releasing internal stresses in forging, moulded and fabricated equipments, cartridge cases and boiler tubes without decreasing the strength acquired during cold working.*

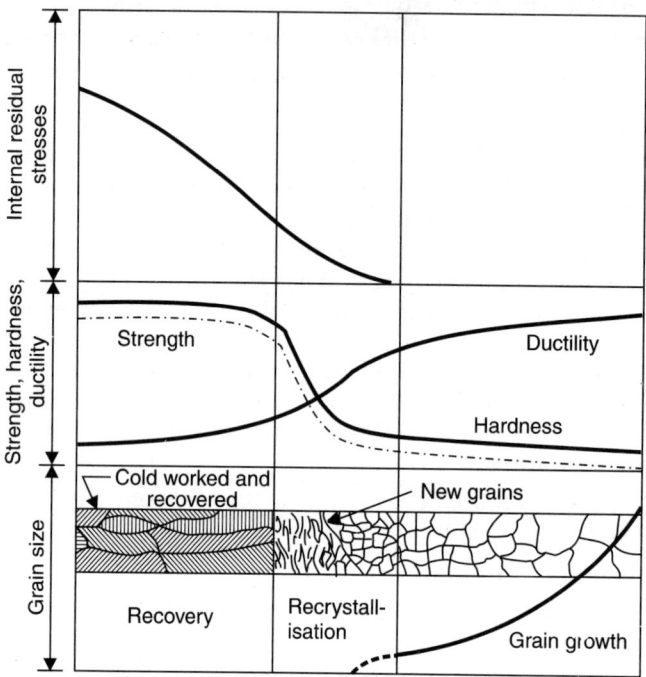

Fig. 3.53. Recovery, recrystallisation and grain growth.

Recrystallisation :

- *It is a process by which distorted grains of cold worked metal are replaced by new strain free grains during heating above a specific minimum temperature called **recrystallisation temperature.***
- Recrystallisation temperature is a function of :
  (*i*) *Particular metal.*
  (*ii*) *Purity of metal :* Soluble impurities raise the recrystallisation temperature.
  (*iii*) *Metal/Alloys :* Recrystallisation usually occurs at a temperature of about $0.3\ T_m$ in pure metals and about $0.5\ T_m$, in alloys, where T_m is the melting temperature.
  (*iv*) *Amount of prior deformation :* The greater the degree of cold work the lower the recrystallisation temperature and smaller the grain size.
  (*v*) *Annealing time :* The longer annealing time decreases the temperature necessary for recrystallization.
  (*vi*) *Grain size :* The finer the grain size of cold worked metal, the lower is the recrystallisation temperature.
- The process of recrystallisation can be divided into *three states :*
  (*i*) Nucleation.
  (*ii*) Primary grain growth.
  (*iii*) Secondary grain growth.

- During **nucleation** *small strain free nuclei form at points at crystal grain boundaries in the heated structure.*

- In the **primary grain growth** the nuclei grow into grains until they first meet, replacing the old grains by new ones, that are now strain free and ultimately **secondary grain growth** which is accompanied with new grain growth at the expense of others and are small in size. During prolonged heating at high temperature the grains grow rapidly and produce locally exaggerated grain growth.

Grain growth :

— *Grain growth is an increase in grain size.*

— When the material is held for longer times at temperature above crystallisation temperature, or when it is healed to a higher temperature the *grain size increases* and there is *decrease in hardness and strength and gain in ductility. The decrease in hardness is not as sharp as during recrystallisation.*

— At a given temperature the grain size D at a given time is given by the following relation known as the *law of grain size :*

$$D^2 - D_0^2 = C.t \qquad\qquad ...(3.12)$$

where, D = Grain size at a given time,

$\quad\quad D_0$ = Initial grain size,

$\quad\quad C$ = Constant of proportionality, and

$\quad\quad t$ = Time.

— The process of grain growth depends largely on the following factors :

(*i*) Annealing temperature.

(*ii*) Annealing time.

(*iii*) Rate of heating.

(*iv*) Degree of prior deformation.

(*v*) Insoluble properties.

(*vi*) Alloying elements.

Differences between Recovery and Recrystallisation

S. No.	Aspects	Recovery	Recrystallisation
1.	*Process*	A cold worked metal consists of networks of dense dislocations. Rearrangement of these dislocations to reduce lattice energy takes place due to polygonisation during recovery stage and is assisted by thermal activation. Polygonisation is a mechanism in which dislocations of the same sign align themselves into walls to form small angle subgrain boundaries.	As the temperature is increased, the dislocation networks in a cold worked metal tend to contract and regions of initially low dislocation density begin to grow. The driving force for recrystallisation comes from the stored energy of cold work. The elimination of subgrain boundaries is a basic part of recrystallisation process.
2.	*Temperature*	It involves low *temperature* for the same metal.	It involves *higher temperature.*

3.	*Effect on properties of metal*	Internal stresses are released without decreasing the strength acquired by cold working. Electrical conductivity and ductility *increase* rapidly towards the annealed value during recovery.	Strain hardening is removed and with that is removed the increased strength at rapid rate. Internal stress is completely eliminated. There is slow further improvement in electrical conductivity and ductility during this stage. Grain size gets refined.
4.	*Changes in micro-structure*	There is restoration of strain-free state without any change in micro-structure of the cold worked metal.	There is replacement of cold worked structure by a new set of strain-free grains.
5.	*Change in grain size*	No change	Slight increase.

Q. 3.4. What limits the maximum deformation in metal forming ?

Ans. In metal forming process the material plastically flows while the total volume of workpiece remains substantially constant but resulting in a corresponding change in the properties of the material. As a rule, the plasticity of metal increases with temperatures whereas its resistance to deformation decreases. The higher the temperature, the higher the plasticity and lower the yield point. Moreover no work hardening occurs at temperatures above recrystallisation temperature. This should be expected since recrystallisation denotes the formation and growth of new grains of metal from the fragments of the deformed grains, together with restoring any distortion in the crystal lattice.

In case of hot working the temperature at which deformation takes place is higher than recrystallisation. While in case of cold working it occurs at a temperature below the recrystallisation temperature of metal. So, we may say that *limit of metal forming is recrystallisation temperature of the metal.*

Q. 3.5. What do you mean by the following terms ?

(*i*) Coining; (*ii*) High-energy rate forming; (*iii*) Progressive piercing.

Ans. (*i*) **Coining.** The operation of coining is performed in dies that confine the metal and restrict its flow in a lateral direction.

(*ii*) **High-energy rate forming.** It includes a number of processes in which parts are formed at a rapid rate *by extremely high pressures.*

(*iii*) **Progressive piercing.** It is the method frequently employed on upset forging machines for producing parts such as *artillery shells* and *radial engine cylinder forgings.*

Q. 3.6. What is warm working ?

Ans. *"Warm working"* is deformation under the conditions of *transition, i.e.,* a working temperature between 0.3 and 0.6 times the melting point.

Q. 3.7. Explain briefly stretch forming process.

Ans. In order to form large sheets of thin metal involving symmetrical shapes or double-curve bends, a *metal stretch press* can be used effectively. In a simpler hydraulically operated press a single die mounted on a ram is placed between two slides that grip the metal sheet. The die moves in a vertical direction and the slides move horizontal. Large forces of 0.5 to 1.3 MN are provided for the dies and slides. The process is a stretching one and causes the sheet to be stressed above its elastic limit while confirming to the die shape. This is accompanied by a slight thinning of the sheet, and the action is such that there is little spring back to the metal once it is formed.

- The process *can be used with many hard-to-form alloys, there is little severe localised cold working, and the problem of unequal metal thin out is minimized.*
- *Scrap loss is fairly high* because material must be left at the ends and sides for trimming and there is a *limitation to the shapes that can be formed.*

Q. 3.8. What is electrohydraulic forming ?

Ans. Electrohydraulic forming, also known as *electrospark forming,* is a process whereby *electrical energy is directly converted into work.*

This process is *safe to operate and has low die and equipment cost. The energy rates can also be closely controlled.*

Q. 3.9. What is magnetic forming ?

Ans. Magnetic forming is another example of the direct conversion of electrical energy into useful work. At first it served primarily for swaging-type operations such as fastening fittings on the ends of tubes and crimping terminal ends of cables. More recent applications are embossing, blanking, focusing and drawing, all using the same power source but differently designed work coils.

The process has the *limitations :* Complex shapes may be impossible to form, pressure cannot be varied over the workpiece and present units are limited to 400 MPa pressure.

Q. 3.10. Explain briefly the following terms for rolled products :

Bloom, Billet, Slab, Plate, Sheet, Strip, Foil, Bar and Wire.

Ans. (*i*) **Bloom :**

- It is the product of the first breakdown of ingot.
- It has square or slight rectangular section, ranging in size from 150 mm × 150 mm to 250 mm × 300 mm.
- A bloom is used to make structural shapes *e.g.,* I beams, channels, etc. by hot rolling.

(*ii*) **Billet :**

- A reduction of bloom by hot rolling results in a *billet.*
- Its size ranges from 50 mm × 50 mm to 125 mm × 125 mm.
- It is rolled to make *rounds, wires* and *bars.*

(*iii*) **Slab :**

- It is a product obtained by hot rolling, either from ingot or from bloom.
- It has a rectangular cross-section, with thickness = 50 mm to 150 mm and width = 0.6 m to 1.5 m.
- Slabs are further *rolled to get plates, sheets, strips, wires and skelp.*

(*iv*) **Plate :**

- It is a finished or semi finished product with a minimum thickness of 6.35 mm.
- Its width will be equal to the width of the roll and the length equal to the maximum which can be handled or stripped.

(*v*) **Sheet :**

- It is a thin partner of plate with a maximum thickness of 6.35 mm.

(*vi*) **Strip :**

- It is a narrow sheet and has a maximum width of 600 mm with a maximum thickness of 6.35 mm.
- As it is normally handled in coil form, its length can be considerable and is limited only by the manufacturing and handling facilities.

(*vii*) **Foil :**

- It is thin strip with a maximum width of 300 mm and maximum thickness of 1.5 mm.
- It is available in coil form.

(*viii*) **Bar :**

- It is a long, straight, symmetrical piece of uniform cross-section. It may be round, square, or of another configuration.
- A circular bar is called a rod.

(*ix*) **Wire :**

- It is a thin variety of bar, available in coil form and not normally so identified over 9.75 mm cross-section.

Q. 3.11. Explain briefly the following rolling operations :

(*i*) **Ring rolling ;**   (*ii*) **Thread and gear rolling.**

Ans. (*i*) **Ring rolling :**

- In this type of rolling, a small-diameter, thick ring is expanded into a large-diameter, thinner ring. The ring is placed between two rolls, one of which is driven (Fig. 3.54). The thickness is reduced by bringing the rolls closer as they rotate. The reduction in thickness is compensated for by an increase in the diameter of the ring. A great variety of cross-sections can be ring rolled with shaped rolls.

- This process can be carried out at room or elevated temperatures, depending on the size and the strength of the product.

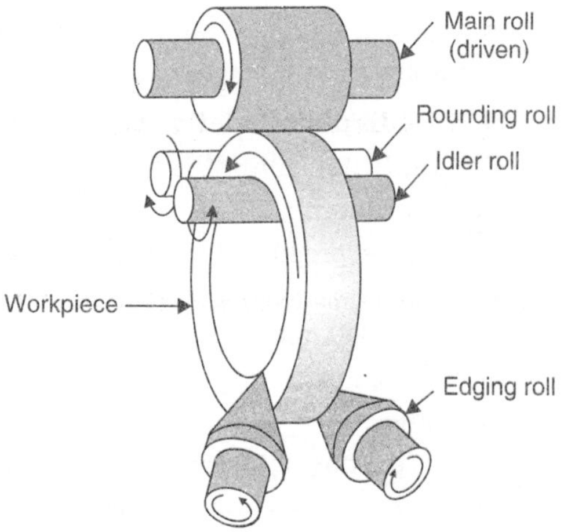

Fig. 3.54 Ring-rolling operation.

Advantages :

(*i*) Short production runs ;

(*ii*) Material savings ;

(*iii*) Close dimensional tolerances ;

(*iv*) Favourable grain-flow directions.

Applications. Typical *applications* are :

— Gearwheel rims ;

— Large rings for rocket and turbines ;

— Ball- and roller-bearing races ;

— Flanges and reinforcing rings for pipes and pressure vessels.

(*ii*) **Thread and gear rolling :**

Thread rolling. The threads are formed on round rods or workpieces by passing them between reciprocating or rotating dies (Fig. 3.55). Typical products made by this process include *screws, bolts,* and *similar threaded parts.*

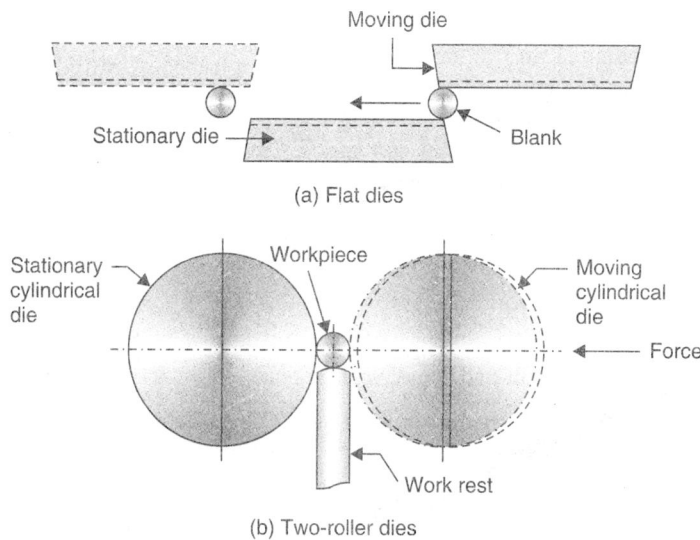

(a) Flat dies

(b) Two-roller dies

Fig. 3.55. Thread rolling processes.

The thread rolling process *generates threads without any loss of metal and with greater strength, because of the cold working involved.* The product is superior to that made by thread cutting and is used in the production of almost all externally threaded fasteners.

Gear rolling. Spur and helical gears are also produced by cold rolling processes similar to thread rolling. The process may be carried out on cylindrical blanks or on precut gears. Helical gears can also be made by a direct extrusion process, using specially shaped dies.

● *Cold rolling of gears has many applications in automatic transmission and power tools.*

Q. 3.12. Explain briefly the following extrusion processes :

(*i*) **Side or lateral extrusion ;**   (*ii*) **Helical extrusion ;**

(*iii*) **Continuous extrusion.**

Ans. (*i*) **Side or lateral extrusion :**

● In this process, the material flows at right angle to the direction of motion of the ram. Side extrusion may be *solid* or *hollow*. The tooling opening is determined by the split die and the mandrel.

● Since very high extrusion force is required, this process is mainly employed for non-ferrous metals and highly plastic materials like lead.

(*ii*) **Helical extrusion :**

● In this process a conical punch is driven slowly into the end of a copper billet, to form a short tube with an annular face. This is then steadily deformed by a tool that rotates about the axes of the billet and slowly advances in a helical path. The swart so produced is not worked but is trapped in a small chamber, where it is forced to escape through an orifice, as a wire.

This process has the following *advantages* :

(*i*) Very large reductions in cross-section are possible, and (*ii*) Extrusion forces are low.

● It is a novel technique for the production of *wire or tube.*

(*iii*) **Continuous extrusion :**

● In this process, *continuous feed stock (not a billet of finite length) is converted into continuous product.*

- In the *"conform process"* (the most successful technique of continuous extrusion) the continuous feed stock in the form of rod is inserted between the grooved extrusion wheel and the mating die shoe. With the rotation of the wheel, the die is carried round the wheel by friction and is finally pressed against a stationary block, known as abutment. It gets upset to conform to the container. Enough pressure is built up to force the material to extrude through a die opening.

Advantages :
(i) Low capital and operating costs.

(ii) Truly continuous process.

(iii) Metallic and non-metallic powders can be intimately mixed and extruded.

(iv) Raw stock in the form of red, powder or machined swarf can be used.

Limitation. The use of the process is limited to the extrusion of non-ferrous metals, mainly copper alloys and aluminium.

Q. 3.13. What is "forgeability" ? Explain.
Ans. Forgeability *may be defined as the tolerance of a metal or alloy for deformation without failure, regardless of forging-pressure requirements.*

Foregeability of a metal or alloy is influenced by the following *mechanical factors* :

(i) Strain rate.

(ii) Strain distribution.

Metals exhibiting low ductility at cold working temperatures show reduced forgeability at increasing strain rates and metals exhibiting high ductility at cold working temperatures are not noticeably affected by increasing strain rates.

Forgeability of a metal or alloy is influenced by the following *metallurgical variables :*

(i) Composition and purity.

(ii) Number of phases present.

(iii) Grain size.

— Alloys containing elements that form insoluble compounds exhibit poor forgeability.

— Alloys containing two phases are generally less forgeable than those containing one phase.

— The forgeability of an alloy improves with decreasing grain size.

Forgeability may be evaluated on the basis of the following **tests :**

(i) *Hot-twist test :*

- In this test, metal bar is twisted at elevated temperatures and number of twists to failure are counted.

- A larger number of twists before failure indicate better forgeability.

(ii) *Upset set :*

- In this, a number of cylindrical billets are upset-forged to various thicknesses.

- The limit for upset setting without failure by cracking (radial or peripheral) is considered as a measure of forgeability.

Q. 3.14. Explain briefly the following forging processes :
(i) **Precision forging.**

(ii) **Isothermal forging.**

Ans. (*i*) **Precision forging :**

- *Precision forging* or *flashless forging*, and similar operations where the part formed is close to the final dimensions of the desired component are also known as ***near-net-shape production.*** Any excess material is subsequently removed by various machining processes.
- In precision forging, special dies are made and machined to greater accuracy than in ordinary impression-die forging.
- The choice between combinational forging, and precision forging requires an economic analysis. *Precision forging requires special dies.* However, much less machining is involved, because the part is closer to the desired final shape.

Advantages :

(*i*) Parts can be made with minimum draft, 0° to 1°, permitting weight reduction.

(*ii*) High strength-to-weight ratio.

(*iii*) Surface finish is smooth.

(*iv*) Lateral protrusions and undercuts can be produced without machining.

(*v*) Small edge radii (3.30 mm) can be produced in the as-forged parts.

(*vi*) The control of parting-line placement provides the most desirable grain flow and metallurgical properties. Some "seamless" forgings are produced.

Limitations :

(*i*) The dies must be considerably sturdier.

(*ii*) The process is slower, requiring more press time.

(*ii*) Isothermal forging :

- In the isothermal forging process, also known as ***hot-die forging,*** the dies are heated to the same temperature as the hot blank. In this way, cooling of the workpiece is eliminated, the *low flow stress of the material is maintained, and material flow within the die cavities is improved.*
- The dies are generally made of nickel alloys, and complex parts with good dimensional accuracy can be forged is one stroke in hydraulic presses.
- Isothermal forging generally is expensive ; however, it can be economical for intricate forgings of expensive materials, provided that the quantity required is large enough to justify die costs.

Q. 3.15. Explain briefly various "Cold forging techniques".

Ans. The various *cold forging techniques* are :

(*i*) Sizing ;   (*ii*) Coining ;   (*iii*) Cold heading, etc.

(*i*) **Sizing.** The primary object of sizing operation is to obtain *closer dimensional tolerances on portions of a forging.*

— In "*plane sizing*", the forging is pressed between two flat dies. The operation *straightens* the forging, enables *accurate dimensions* to be obtained and *improves the surface finish.*

— Three dimensional sizing imparts the final size to the whole operation in impression dies with the formation of flash which is subsequently trimmed.

(*ii*) **Coining.** When the sizing operation is such that the metal flow is restricted in a space confined by the die impression (that is no *flash is formed*), the sizing operation is called "*coining*".

— Sizing and coining operations are commonly done on knuckle joint presses.

(*iii*) **Cold heading.** Fig. 3.56 shows the principle of operation of cold head machine :

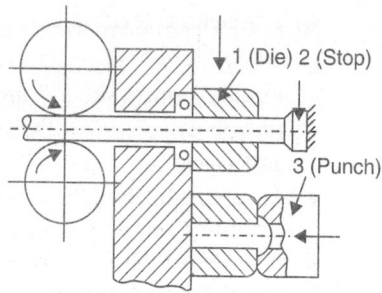

— The stock is fed through die '1' to stop '2'.

— Next, the die is shifted to cut off the blank and to carry it over to the heading line where required head is formed by the blow of punch '3'.

— The *cold heading process* is limited to axisymmetric components formed by a combination of extrusion and upsetting techniques. Small parts such as nails, rivets, pins, screws, bolts etc. can be produced in large quantities from the continuous wire stock.

Fig. 3.56. Cold heading.

● The *cold forging process* is most suitable for axisymmmetric components such as :

— Bearing races ;

— Shaft components ;

— Flanged components, etc.

Q. 3.16. What do you understand by 'draft' on forgings and why is it provided ?
Ans :

● *Draft* refers to the intentional taper of about 7° to provide convex contour on the forged parts.

● It is provided :

(*i*) To allow the metal to flow in the closing dies ;

(*ii*) To facilitate the extraction of the forging from the dies.

Q. 3.17. Explain briefly the forging sequence for a connecting rod.
Ans. Forging sequence for a connecting rod (Fig 3.57) involves the following **steps** :

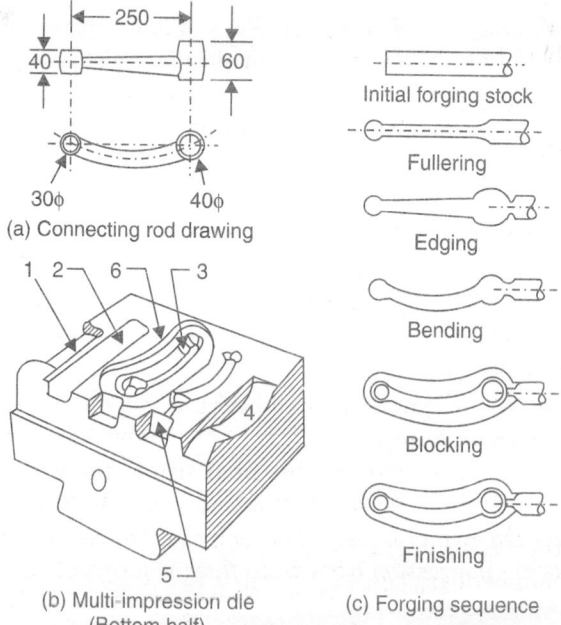

Fig. 3.57. Multi-impression die—forging sequence of a connecting rod.

(*i*) The heated forging stock is first placed in fullering impression '1' where it is reduced by hammer blows.

(*ii*) Edging impression '2' redistributes the metal.

(*iii*) Before being forged in finishing impression '3', the stock is processed in the bender '4' and the semifinish (blocker) impression '5'.

(*iv*) Flush gutter '6' is provided around the finishing impression '3'.

Q. 3.18. Explain briefly a 'Cored forging'.

Ans. Fig 3.58 shows a *'cored forging'*.

— The split die opens horizontally to remove the forging.

— The pressure to induce metal flow is applied by the top core pin, which comes down and extrudes metal into the die cavity. One or more movable side cores are made to advance and withdraw. This type of tooling *permits zero draft and undercuts as desired.*

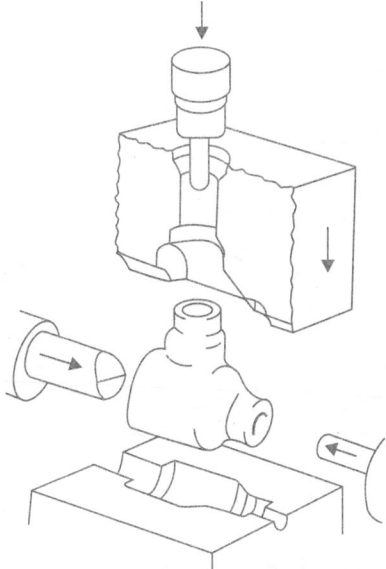

Fig. 3.58. Cored forging.

Q. 3.19. Explain briefly the various types of "forging equipment".

Ans. The various types of forging equipments are enumerated and discussed briefly below :

1. Hammers
2. Counterblow hammers
3. Mechanical presses
4. Screw presses
5. Hydraulic presses.

1. **Hammers :**

Hammers derive their energy from the *potential energy* of the ram, which is then converted to *kinetic energy* ; thus hammers are energy limited. The speeds of hammers are high ; therefore, the low forming times minimise cooling of the hot forging, *allowing the forging of complex shapes, particularly with thin and deep recesses.*

In power hammers, the ram is accelerated in the down stroke by steam or air, in addition to gravity.

— In power hammers the highest energy available is about 1200 kJ.

2. Counterblow hammers :

— These hammers have two rams that simultaneously approach each other to forge the part. They are generally of mechanical-pneumatic or mechanical-hydraulic type.

— These machines transmit less vibrations to the foundation than other hammers.

— The capacity of the largest counterblow hammer is around 1250 kJ.

3. Mechanical presses :

— These are stroke-limited presses.

— These are of either the crank or the eccentric type, with speeds varying from a maximum at the centre of the stroke to zero at the bottom. The force available depends on the stroke position and becomes extremely large at the bottom-dead-centre position ; thus, *proper setup is essential to avoide breaking the dies or other equipment.*

— The capacity of the largest mechanical press is around 110 MN.

Fig. 3.59 shows a mechanical forging press.

4. Screw presses :

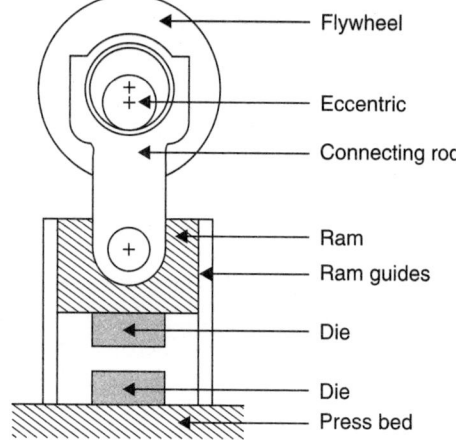

— These presses derive energy from a flywheel. The forging load is transmitted through a vertical screw.

— These are *energy limited* and are particularly suitable for producing small quantities, for parts requiring precision such as turbine blades.

5. Hydraulic presses :

— These presses have a constant low speed and are *load limited.* Ram speed can be varied during the stroke.

— These are used for both open-die and closed-die forging operations.

— The capacity of the largest hydraulic press is about 700 MN.

Fig. 3.59. A mechanical forging press.

● Generally *"presses"* are preferred for aluminium, magnesium, beryllium bronze, and brass ; *"hammers"* are preferred for copper, steels, titanium, and refractory alloys.

Q. 3.20. Explain briefly the following processes :

(*i*) **Swaging**   (*ii*) **Creep forming**   (*iii*) **Hot Isostatic Pressing (HIP).**

Ans. (*i*) **Swaging :**

● It is a metal working process in which the cross-sectional area of bars, rods, or tubes in the desired area is reduced by repeated blows or radially actuated shaped hammers.

● It is a batch process and not easily adaptable to continuous process.

(*ii*) **Creep forming :**

● This operation is the deformation of metal at a stress level below its yield point and the application of heat to the metal for time sufficient to permit metallurgical creep to cause relaxation of the induced elastic stresses to such a level that plastic deformation to the desired shape is achieved.

- This process is used almost exclusively in *aerospace industry of large comparatively thin and shallow contour components.*

(*iii*) Hot Isostatic Pressing (HIP) :

This process involves the following **steps** :

- The metal powder is seated in a metal or ceramic container that has the shape of the desired part.
- The container is placed in a special pressure vessel that has the capability of simultaneously heating the container and subjecting it to a hydrostatic pressure (gas).
- The powder is compacted, densified and sintered in one step.
- Upon removal of the container, a finished part close to the final shape is obtained.

This method is being tried *to forge exotic materials for aircraft industry.*

Q. 3.21. Explain briefly the following :

(*i*) **Staking ;**   (*ii*) **Precision blanking ;**

(*iii*) **Perforating ;**   (*iv*) **Buldging.**

Ans. (*i*) **Staking :**

- It is a method of fastening two parts permanently. It is a substitute to drilling/punching and riveting.
- In this method, a shaped punch is forced into the top of a projection of one of the parts to be fastened. This indentation action of punch deforms the metal sufficiently to squeeze it tightly against the second component so that they are firmly locked together.

(*ii*) **Precision blanking :**

- This process produces high dimensional accuracy of edge of the part along the entire punched surface.
- Fine blanking presses require completely ground dies with zero clearances between punch and die, and triple action presses offer following actions : (*i*) Holding of metal firmly against the face of die plate by embedding a V-ring (stinger) provided on face of stripper or pressure plate. This also prevents lateral movement and prevents metal from flowing away from punch as it enters ; (*ii*) Blanking pressure which remains constant throughout the shearing stroke ; (*iii*) Counter punch working against blanking punch is used for flattening, coining or bending as may be the case.

(*iii*) **Perforating :**

- It is the operation of *punching a pattern of holes in similar or continuous rows.*

(*iv*) **Buldging :**

- Buldging is a metal forming process to *expand a tubular or cylindrical part.*
- Relatively symmetrical shapes can be provided by *mechanical means, i.e.,* using segmented buldging die in which various segments, held together by springs, are pushed apart by the punch in order to expand a superimposed shell. Other method is *hydrostaic buldging* in which a fluid or rubber is used to apply internal pressure to the inside of the workpiece as the die is closed.

Q. 3.22. Explain various defects relating to forming processes.

Ans. 1. Defects in rolling :

(*i*) *Irregularies.* It leads to a trapping of scales which remain inside as laps.

(*ii*) *Non-metallic inclusions.* These defects are produced, especially during the hot rolling of a thick slab, crocodile cracks, separate the product into two halves.

(*iii*) *Internal blow holes in the stock.* It results in an elongation of the blow holes and the product becomes weaker.

- All the above defects can be minimised by a *careful inspection of the billet and by keeping the roll smooth and clean.* To avoid the internal cracks, it is necessary to design the roll pass properly.

2. Defects in forging :

(*i*) *Scale pits.* These are shallow depressions caused by hot removing scale from the dies. The scale is subsequently worked into the surface of the forging.

(*ii*) *Mismatch.* A mismatch occurs in drop forging when the dies are incorrectly aligned, and results in a lateral displacement between portions of the forging.

(*iii*) *Unfilled section.* An unfilled section in similar to misrun in casting and ocurs when metal does not completely fill the die cavity. It is usually caused by using insufficient metal or insufficient heating of metal.

(*iv*) Defects resulting from improper forging such as *seams, cracks, laps* etc.

(*v*) *Ingot defects such* as pipes, cracks, scabs and segregation.

(*vi*) Defects resulting from the melting practice, such as *dirt, slag* and *blow holes.*

(*vii*) Defects resulting from *improper heating* and *cooling of the forging such as burnt metal, decarburised steel and flakes.*

- These defects are found in all metals which are heated to plastic stage and then shaped.

3. Extrusion defects :

- The most common defect in extrusion (known as extrusion defect) arises from the *back flow of the material,* pushing the end face of the billet into the core of the product. Such a defect *weakens the product* since the surface layer is normally contaminated by oxides.

- Sometimes the heat generated due to extrusion may raise the temperature of the job, resulting in the *developement of surface cracks.*

4. Defect in drawing :

- The typical surface defects in rod and wire drawing are *due to a polughing by hard particles and local breakdown of the lubricating film.*

- The other kinds of defects include *formation of a buldge,* ahead of the die, with low reduction and high die angle, and the *development of a centre burst* with too large a deformation gradient along the cross-section.

5. Deep drawing defects :

- An *insufficient* blank holder pressure causes *to develop wrinkles on the flange, which may also extend to the wall of cup. Too much* of a blank holder pressure and friction may cause a *thinning of the walls* and a *fracture at the flange, bottom and corners* (if any).

- Due to the misplacement of the stock, *unsymmetrical flanges may result ;* this defect is commonly known as **miss strike.**

- The effect of a large grain size is to produce a *dull surface* (orange feel effect) ; this defect is also common in bending operations.

- While drawing a rolled stock, ear lobes tend to occur because of the anisotropy induced by rolling operation.

Q. 3.23. Explain briefly various "Sheet metal operations".

Ans. Several operations need to be carried out in sheet metal work. Some are done by hand, some by using machines and some can be performed either manually or by using machines. The main operations are enumerated and described as follows :

1. Measuring and marking ;	11. Joint making ;
2. Cleaning ;	12. Bending ;
3. Laying out ;	13. Notching ;
4. Hand cutting and shearing ;	14. Drawing ;
5. Hand forming ;	15. Soldering ;
6. Machine shearing ;	16. Sinking ;
7. Nibbling ;	17. Raising ;
8. Circle cutting ;	18. Planishing ;
9. Piercing and blanking ;	19. Hollowing blocking.
10. Edge forming and wiring ;	

1. Measuring and marking :

- The standard sizes of the metal sheet available in the market are quite large. The sheet size required for making a component may be much smaller. Therefore, it may be necessary to cut the required size for the component from the large sized sheet. These small sizes are first decided and then these sizes are marked on the large sheet to cut the latter into small pieces along the marked lines.

- While cutting the large sheet into small size(s) a little *allowance* for cutting is always added to the required overall sizes so that the cut pieces are not undersize.

- Overall dimensions, length and breadth, of the required smaller pieces are marked on the large sheet with the help of marking tools, including a steel straight edge, a steel square etc. ; and a scriber. The sheets may have to be coated with a colouring media, such as cellulose lacquer, so that the scribed lines are clearly visible. For producing circular pieces a divider or trammel may have to be used to mark the circles.

In mass production of identical small items *a template and a scriber may be used to mark the blank.*

2. Cleaning :

Invariably the blank surfaces require proper cleaning before being processed. This requirement is more predominant in case of non-ferrous metal sheets, like those of copper, brass and silver. The surfaces are cleaned by a *pickling process* which is described below :

- A *pickle bath,* consisting of 1 part dilute sulphuric acid and 20 parts water, is heated.
- The blanks are then immersed in the hot bath.
- After allowing adequate time for pickling the blanks are thoroughly washed in a stream of water and then allowed to dry.
- Sometimes *cold pickling* is used in such cases where cleaning operation is not required very frequently. In such cases either pickling period is to be increased or a smaller acid to water ratio of the pickle has to be used to ensure an effective cleaning of the surface.

3. Laying out :

- The *operation of scribing the development* of the surface of the component on the sheet, sheet blank, together with added allowances for overlapping, bending, *hammering,* etc. which when cut out of the blank and folded and joined will give the required component is known as laying *out.*

- Such a layout when made on the sheet is called a *pattern* and the process as *pattern layout.*

- In case the jobs to be made are small in size and in large number it is better to use a *template* for repetitive marking of the development and then cutting it along the marked contours. The required allowances are included in the template size.

4. Cutting and shearing :

- The cutting and shearing operations are used to separate a sheet piece into two parts along a premarked line or contour.
- The *cutting* is normally used when the sheet metal is cut by means of a *chisel* and *hammer* manually. The term *shearing* stands for cutting of sheet metal by two parallel cutting edges moving in opposite directions. This can be done either manually by using hand shear or snips (Fig. 3.60) or by means of machines called *shears*.

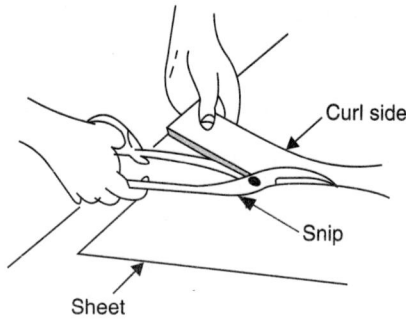

Fig. 3.60. Shearing by a snip.

- The selection of a particular method and means of cutting depend on the following factors :
— Thickness of sheet metal ;
— Sizes of blanks to be cut ;
— Amount of cutting required to be done ;
— Number of blanking to be cut ;
— Type of production—Jobbing, batch or mass production ;
— Available means of cutting, etc.

5. Hand forming :

- The process of shaping and/or bending of sheet-metal in three dimensions to give it the desired shape and size of the final product is termed as *metal forming.*
- Best examples of this operation can be seen if one observes the traditional metal workers manufacturing cooking utensils of brass out of blanks by hand hammering and shaping them into different shapes and sizes without any joint.

6. Machine shearing :

- Machine shearing is carried out by means of *shearing machines.*
- The shearing machines can be *hand operated* or *Guillotine shear.*
The former is used for smaller and thinner sheets while the latter for large and thicker ones.
- For continuous cutting along a straight line the *rotary shears* are used.

7. Nibbling :

- The *process of continuous cutting along a contour, which may be a straight line or an irregular profile*, is known as **Nibbling.**
- Nibbling operation is carried out by the machines known as *Nibblers*. These machines are portable shearing machines and can be either *electrically* or *pneumatically* operated.

8. Circle cutting :

- *Circle cutting* is the operation of cutting circular blanks or curved contours with the help of a *circle cutting machine.*
- It is *a continuous cutting operation.*

9. Piercing and blanking :

- *Piercing* is basically a hole punching operation while ***blanking*** is an *operation of cutting out a blank.*

- In both the cases blanks will be produced but in case of piercing operation obtaining a blank is not the objective. However, in case of blanking operation the production of a blank of the desired size is the main objective, which is a useful part for further processing.

10. Edge forming and wiring :

- The edges of sheet metal products are formed or folded to ensure *safety* of hands, while handling these products, and to provide *stiffness* to the products in order that they will retain their shapes during handling.

- When still stronger edges are required they are reinforced by inserting a metal wire or rod and then forming the edge by circling the edge of sheet metal around it. By doing so, not only the stiffness of the joint is increased but its appearence is also improved.

11. Joint making :

- For joining sheet metal parts together or securing them to other metallic or non-metallic bodies, several means are used which include the following :

— Screwed fastening ;

— Riveting ;

— Welding, brazing, and adhesives.

- The most commonly used methods involves joining the sheet metal parts by means of *folded joints* or *self securing joints,* followed by *soldering* and *adhesive jointing.*

12. Bending :

- The bending operation involves stretching of metal on the outer surface and compressing it on the inner surface along a *neutral* line, which remains unchanged in length.

- Sheet metal can be bent by hammering over a base by hand or by means of bending or rolling machines.

13. Notching :

- In bent sections that have folded edges, there should be some provision so that there is no overlapping of metal where the corners come together. In order to prevent bulging at such a place, it is necessary to slit or clip the metal or provide small openings. The openings left at the corners of seams and edges are known as notches and the operation is called ***notching.***

- The size, location and types of notches depend upon the shape of the object.

- The following are the different types of notches commonly used in sheet metals :

    (*i*) Straight notch ;

    (*ii*) Square notch ;

    (*iii*) V-notch ;

    (*iv*) Slant notch ;

    (*v*) Wire notch.

Fig. 3.61 shows laying out a straight notch.

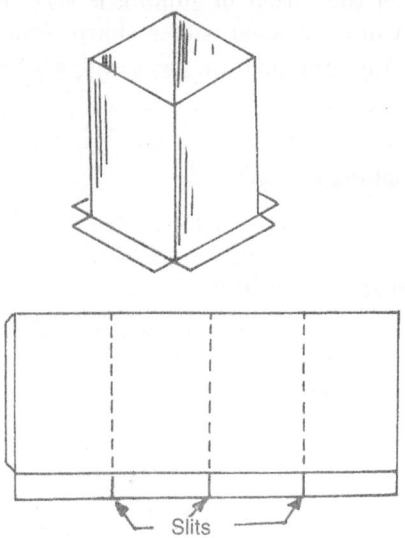

Fig. 3.61. Laying out a straight notch.

14. Drawing :

- In sheet metal, drawing operation is used *to produce thin walled hollow shapes.*
- The drawing operation is carried out with the help of a *die* and a *punch* on a suitable press.
- — If the drawn length of the component is *less than its width or diameter* it is called **box drawing** or **shallow drawing.**
- — When the length drawn is *more than the width* the operation is known as **deep drawing.**

15. Soldering :

- It is a very common method of joining sheet metal parts.
- It involves spreading of a low melting point alloy (solder) in molten condition between the surfaces or edges to be joined and allowing it to solidify there.

16. Sinking : Refer to Fig. 3.62

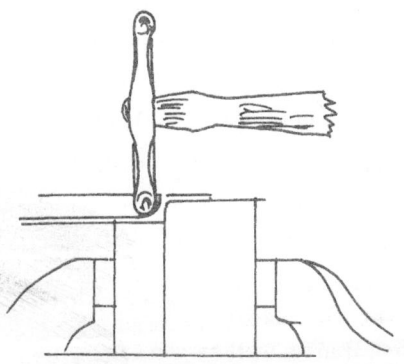

Fig. 3.62. Sinking a tray on wooden blocks.

- *Sinking* is a process *used to sink the bottom for forming a tray with a flat rim.*
- Firstly, the position of the corner of sinking is scribed and then the depth of tray is scribed inside that. A piece of wood with a sharp corner is placed vertically in the vice and over the edge of this, the metal is beaten down with a tray hammer to shape the tray.

17. Raising :

- It is the process of *denting the metal down to shape over a tool with a raising hammer or mallet.*
- Firstly, the concentric circles are drawn with pencil compass as a guide to keep the work true. The hammering is started from the centre and worked around in circles till the outer edge is reached. The metal must be annealed after each completed course and the process is repeated.

18. Planishing : Refer to Fig. 3.63

It is a process which is applied to sheet with the *main object of stretching the metal with an improved surface.* The process brings the article to its final shape and imparts some degree of hardness to the metal.

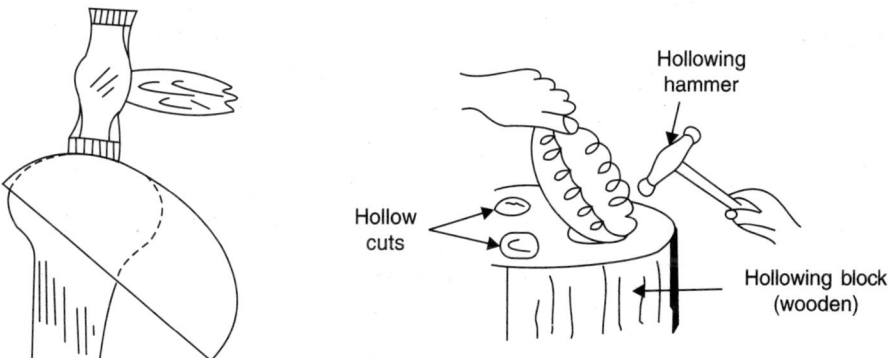

Fig. 3.63. Position of hammer   Fig. 3.64. Method of hollowing.
        when planishing.

19. Hollowing or blocking : Refer to Fig. 3.64.

- It is the process of *beating the sheet metal into a particular shape such as sauce pan lid or bowl.* It is usually done on the hollowing block which is a wooden block with hollow cuts into it.
- The hollowing process may also be done on a sand bag. The main advantage of sand bag is that there are no indentation on the top.

Q. 3.24. Enumerate and describe very briefly the commonly used machines in sheet metal shop.

Ans. Sheet metal working machines :

When a large number of jobs are to be made, particularly in the heavier type of steels, it is obvious that hand operation like shearing, bending, etc. will be difficult and uneconomical. Hence, to carry out the job many hand operated as well as power operated machines have been developed.

Some of the most commonly used machines in sheet metal shop are enumerated and described below :

    1. Shearing machine ;

    2. Folding machine ;

    3. Grooving machine ;

    4. Swaging machine ;

    5. Beading machine ;

    6. Universal cutting machine ;

    7. Burring machine ;

    8. Turning machine ;

    9. Wiring machine.

1. *Shearing machine* :

- A simple hand operated or lever shearing machine is shown in Fig. 3.65. It can perform the following number of operations :

— Sheet cutting ;

— Flat shearing ;

— Round bar shearing ;

— Square bar shearing ;

— Angle iron cutting.

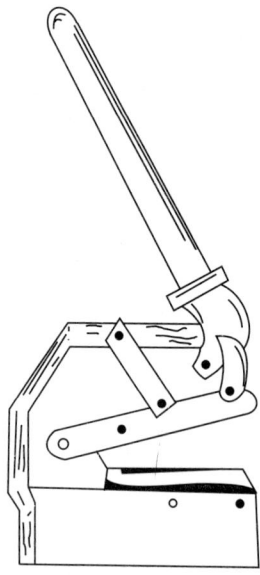

Fig. 3.65. Lever shearing machine.

- These machines are available in many sizes according to their capacity.

2. *Folding machine* :

- Folding machines are used for *bending* and *folding* the edges of the plates to form the joint at the seam.

- *Bending machines* or *bending rollers*, as they are better known, are used for shaping metal sheets into cylindrical objects.

3. Grooving machine :

A grooving machine is used for *making grooves in the sheet.* The depth of the groove as well as its position can be conveniently adjusted.

4. Swaging machine :

A swaging machine is used for *pressing i.e., locking the grooved joints or seam joints.*

It consists of a sharp teeth blade that can be moved by means of a handle. While moving on the joint, the joint gets pressed and locked.

5. Beading machine :

A beading machine is used for *beading operations.*

6. Universal cutting machine :

- It is used for *cutting the metallic sheet into any desired shape and size.*
- Its straight as well as circular cutting capacity is 22 S.W.G. sheets and thinner.
- The machine has two circular tools used for circle cutting.

7. Burring machine :

A burring machine is used for *preparing edges for seam joints of cylindrical shaped articles.*

8. Turning machines :

It is used for *making round edges for wiring operations.*

9. Wiring machine :

A wiring machine is used for making *wired edges, after the wire seat is prepared on turning machine.*

Q. 3.25. Discuss briefly some sheet metal joints commonly used in sheet metal work.

Ans. In sheet metal work, according to the requirements, many different types of joints are used. Some commonly used forms of these are shown in Fig. 3.66.

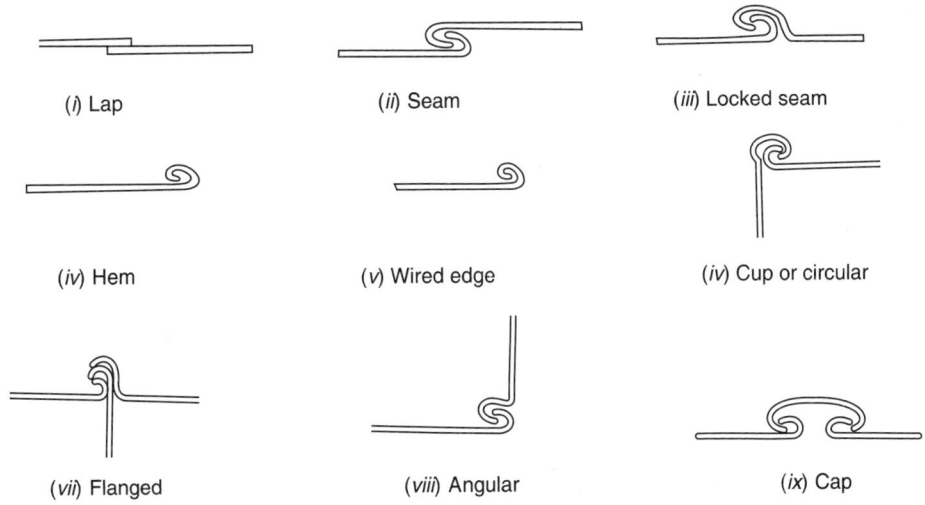

Fig. 3.66. Sheet metal joints.

- A *lap joint* is very frequently used and can be *prepared by means of soldering or riveting.*
- *Seam joint* is a very commonly used one. It is *locked* so as to *ensure a positive grip and also to make the joint flush with the surface.*

- **Hem** (single and double), **wired edge, cup** and **angular** joints enable the edges to *join the pieces along them.*
- **Flanged joint** is used normally in *making pipe connections.*
- **Cap joint** provides another *useful form of locked seam joint.*

HIGHLIGHTS

1. *Rolling* is a forming operation on cylindrical rolls wherein cross-sectional area of a bar or plate is reduced with a corresponding increase in length.

2. *Forging* is the process by which heated metal is shaped by the application of sudden blows of steady pressure and characteristics of plasticity of material is made use of.

3. *Extrusion* is the process in which metal is caused to flow through a restricted orifice so creating an extremely elongated strip of uniform, but comparatively small cross-section.

4. *Wire drawing* is the process of reducing diameter of metal rods by drawing them through conical openings in die blocks.

5. *Blanking* is a process of cutting or shearing a blank from sheet or strip material.

6. *Spinning* means shaping a metal blank as it revolves at a high speed in a lathe.

7. *Embossing* is an operation of raising a design or form above the surface of a component by means of a pressing or squeezing action.

8. *Casting* means the pouring of molten metal into a mould, where solidification occours.

9. *Machining* is the process of cold working the metal into different shapes by using different type of machine tools.

OBJECTIVE TYPE QUESTIONS

Fill in the Blanks/Choose the Correct Answer :

1. The most important property of metal is elastic/plastic deformation and it is extensively used for fabrication by mechanical working.

2. A metal is said to be cold/hot worked, if it is mechanically processed below the crystallisation temperature of the metal.

3. Residual stresses are set up during hot/cold working.

4. Shot peening is a cold /hot working operation.

5. When plastic deformation of metal is carried out at temperature above the recrystallisation temperature the processes performed on metal are termed as cold/hot working.

6. .... working process can be considered as simultaneous combination of cold working and annealing.

7. Pipe welding is a cold/hot working operation.

8. Hardening is not eliminated in cold/hot working process.

9. Cold/hot working decreases the value of elongation.

10. Refinement of crystals occurs in hot/cold working processes.

11. In cold/hot working chances of crack propagation is more.

12. In hot/cold working internal and residual stresses are not produced.

13. Rolling/casting is a forming operation which is carried on cylindrical rolls.

14. The two/three/four/six high rolls being least expensive are most common for both hot and cold rolling.

15. .... rolling is widely employed to produce a finish for hot rolled metals.

16. Cold/hot rolling gives improved surface finish.

17. Cold/hot rolling produces thickness dimensions.

18. For rolling strips, coils or sheets that may be thousands of metre in length .... mills are used.

19. Steel is nearly always rolled hot/cold for finishing passes on sheet.
20. Brass, nickel and silver are usually cold/hot rolled with many intermediate annealings.
21. If the temperature is uniform/non-uniform work hardening and crack may result.
22. In cold/hot rolling crystal structure is refined.
23. Forging/casting is the process by which heated metal is shaped by the action of sudden blows or steady pressure.
24. Flattener/punch is used to give smooth finish of flat surfaces.
25. Gouge/hardie is used to cut plates to curves.
26. Drawing down/setting down is the process of increasing the length of the bar at the expense of its cross-sectional area.
27. A drop stamp/steam hammer works on the principle of the steam engine.
28. Hooker extrusion is cold/hot extrusion process.
29. About 95% of the products of impact extrusion are .... paste tubes.
30. Indirect extrusion requires less/more force than direct extrusion.
31. The dimensional tolerances of extrusions are very good/good.
32. Wide drawing/tube drawing is the process of reducing diameter of metal rods by drawing them through conical openings.
33. Blanking/piercing is a process of cutting or shearing a blank from sheet or strip material.
34. .... holes should not be located too close to adjacent holes.
35. Spinning/blanking means shaping a metal blank as it revolves at a high speed in a lathe.
36. Casting/forging means the processing of molten metal into a mould, where solidification occurs.
37. Die/centrifugal casting has excellent surface finish and dimensional accuracy.

ANSWERS

1. Plastic	2. Cold	3. Cold	4. Cold	5. Hot
6. Hot	7. Hot	8. Cold	9. Cold	10. Hot
11. Cold	12. Hot	13. Rolling	14. Two	15. Cold
16. Cold	17. Cold	18. continuous	19. Hot	20. Cold
21. Non-uniform	22. Hot	23. Forging	24. Flattener	25. Gouge
26. Drawing down	27. Steam hammer	28. Cold	29. collapsible	30. Less
31. Very good	32. Wire drawing	33. Blanking	34. Pierced	35. Spinning
36. Casting	37. Die.			

THEORETICAL QUESTIONS

1. Describe what occurs in metal when it is rolled.
2. What is difference between bloom and a billet ?
3. What are the advantages and disadvantages of cold working ?
4. List the advantages and disadvantages of hot working.
5. Give the comparison between cold working and hot working processes.
6. What do you mean by warm working or semi-hot working ? Explain briefly.
7. Define 'Rolling'. How does cold rolling differ from hot rolling ?
8. Discuss briefly principle and mechanism of rolling.
9. List the main process variables in a rolling process.
10. Describe briefly "Cluster rolls" and "Planetary mill".
11. Explain briefly various defects in rolling.
12. What do you understand by the term ' forging' ? How does 'hand forging' differ from 'machine forging' ?

13. State the advantages and disadvantages of forging.
14. Write short notes on :
    (*i*) Drop stamp   (*ii*) Steam hammer
    (*iii*) Pneumatic hammer   (*iv*) Hydraulic hammer.
15. Explain briefly the following three basic categories of forging :
    (*i*) Open-die forging ;   (*ii*) Impression-die forging ;   (*iii*) Closed-die forging.
16. Describe briefly the following method of forging :
    (*i*) Drop forging ;   (*ii*) Press forging ;   (*iii*) Machine forging.
17. Explain briefly any two of the following forging process :
    (*i*) Roll forging ;   (*ii*) Rotary forging or swaging ;   (*iii*) High velocity forging ;
    (*iv*) Orbital forging ;   (*v*) Incremental forging.
18. Expalin briefly various defects in forging.
19. How are forgings cleaned and finished ? Explain briefly.
20. What is 'Extrusion' ?
21. List the advantages, limitations and applications of extrusion process.
22. How are extrusion processes classified ?
23. Explain briefly the following extrusion processes :
    (*i*) Hydrostatic extrusion
    (*ii*) Hooker extrusion
    (*iii*) Impact extrusion.
24. Describe briefly the 'extrusion defects'.
25. Explain with a neat sketch 'wire drawing' process.
26. Explain briefly the following processes :
    (*i*) Tube drawing   (*ii*) Tube making (By rotary piercing).
27. Explain briefly any three of the following metal forming processes :
    (*i*) Bending ;   (*ii*) Deep drawing ;   (*iii*) Stretch forming ;
    (*iv*) Metal spinning ;   (*v*) Embossing.
28. Discuss the following forming processes :
    (*i*) Roll forming ;   (*ii*) Rubber press forming ;   (*iii*) Hydromechanical forming.
29. What is 'shot peening' ? Explain briefly.
30. List some important defects, prevalent in sheet metal formed parts.
31. How are dies classified ? Explain them briefly.
32. What are the causes of failure of dies in metal working operations ?
33. How are presses classified ? Explain them briefly.

4

Powder Metallurgy

4.1. DEFINITION

Powder metallurgy *is defined as the art of making objects by the heat treatment of compressed metallic powders.*

"Powder metallurgy" includes the blending and mixing of powders, pressing or compacting powder into an appropriate shape, sintering the pressed-powder compact, and perhaps final sizing or finishing of the product to meet specified dimensional tolerances.

The process is applicable to a single metal powder, to mixtures of metals and non-metals. The operation of pressing may be carried out at ordinary or elevated temperatures depending upon the composition and properties desired in the product.

4.2. ADVANTAGES, DISADVANTAGES/LIMITATIONS OF POWDER METALLURGY

Advantages :

The advantages of powder metallurgy are :

1. Such parts which have special properties can be produced which otherwise cannot be obtained.
2. Machining operations are eliminated.
3. Scrap losses are reduced and often results in lower unit cost for a given part in comparison to any other production method.
4. Metals and alloys can be mixed together in any proportion which is difficult and sometimes not possible by melting.
5. Metals and non-metals can be mixed together in any proportion.
6. There is better control of composition and structure of the component by this process.
7. Articles of any desired porosity can be manufactured.
8. Super-hard cutting bits, which can never be manufactured by any other methods are made by powder metallurgy, *e.g.,* sintered carbides, satellites.
9. This process is suitable for mass production because the stroke of the pressing or compacting consists of a press at a speed of 60 strokes/minute.
10. Antifriction alloy strips made by powder metallurgy can be made to adhere on a strong alloy backing piece.
11. The process is very economical and the loss of material is lesser as compared to other processes.
12. Diamond impregnated tools for cutting porcelain, glass and tungsten carbides are made possible only by powder metallurgy.

219

Disadvantages/Limitations :

The process has the following *disadvantages/limitations* :

1. Owing to the fairly high compacting pressures required to press the powder, the wear on the dies is high.
2. Due to high rate of wear of dies, high costs for dies and presses the method is rendered uneconomical particularly for small runs.
3. Since the compacted parts must be ejected from the die without fracture, therefore, the shapes that may be made by this method are limited.
4. Equipments required are very costly.
5. A completely dense product is not possible without heating the product after pressing operation.
6. The physical properties obtained by this process are lower than those obtained by other processes.
7. In the low melting powders like tin, zinc and cadmium, sometimes certain thermal difficulties appear.
8. Pressed and sintered powder can approximately achieve the properties of the wrought alloy but at the cost of increased production cost.
9. The intricate shapes cannot be made by compacting since metal powders cannot flow like fluid under compact load.
10. The products are of small size because for large size bigger equipments and tools would be necessary involving very heavy investment.
11. Many metal powders are explosive at room temperature.
12. A few metals cannot be compressed because they have a tendency to cold-weld to the walls of the die causing wear on the die.

4.3. APPLICATIONS OF POWDER METALLURGY

Powder metallurgy has the following present day *applications* :

1. Porous and graphite containing metal bearings.
2. Electrical contacts consisting of a current and heat-conducting matrix in which are embedded wear resisting particles.
3. Tungsten wires.
4. Rotors of gear pump.
5. Diamond impregnated tools.
6. Magnetic materials.
7. Refractory metal composites.
8. Metal to glass seals.
9. Motor brushes.
10. Metallic filters.
11. Metallic coatings.
12. Babitted bearings for automobiles.
13. Cemented carbides.
14. Friction materials.

4.4. MANUFACTURE OF PARTS BY POWDER METALLURGY

The manufacture of parts by powder metallurgy involve the following *steps* :

1. Production of metal powders.
2. Blending powders.
3. Pressing or compacting of metal powders.
4. Sintering.
5. Finishing operations.

4.4.1. Production of Metal Powders

The methods of powder production are :

1. Atomising
2. Gaseous reduction
3. Electrolysis process
4. Carbonyl process
5. Stamp and ball mills
6. Granulation process
7. Mechanical alloying
8. Other methods.

1. **Atomising process.** In this process the molten metal is forced through an orifice into a stream of high-velocity air, steam or inert gas. This causes extremely rapid cooling and disintegration into a very fine powder.

   ● The use of this process is usually *limited to metals* with low melting point.

2. **Gaseous reduction.** This process consists of grinding the metallic oxide to a finely divided state and then *reducing* it by hydrogen or carbonmonoxide.

   ● It is employed for metals such as *iron, tungsten* and *copper* (whose melting points are near or above 1100°C).

3. **Electrolysis process.** In this process of producing powders the conditions of electrode position are controlled in such a way that a soft spongy deposit is formed ; which is then pulverised to form the powder. The particle size can be varied over a wide range by varying the electrolyte composition and various electrical parameters.

   ● The powders of copper, iron and other metals are made by this process.

4. **Carbonyl process.** This process is based upon the fact that a number of metals can react with carbon monoxide to form what are known as *carbonyls.* For example, the iron carbonyl is made from iron reduced from ferric oxide. Carbon monoxide at a pressure of 48–200 bar is then passed over heated iron. The resulting carbonyl is decomposed by heating to a temperature of 200°C to 300°C.

   ● This process yields powders of high purity but entails a heavy cost.

5. **Stamp and ball mills.** These are mechanical methods which produce a relatively coarse powder. The ball mill is employed for brittle materials while stamp mill for more ductile materials.

   ● The cost is usually high, and the powders produced by these methods are usually treated to remove the cold work-hardening received in the process.

6. **Granulation process.** This process consists in the formation of an oxide film on individual particles when a bath of metal is stirred in contact with air.

   ● This process produces a relatively coarse powder with a high percentage of oxide.

7. **Mechanical alloying.** In this method, powders of two or more pure metals are mixed in a ball mill. Under the impact of the hard balls, the powders repeatedly fracture and weld together by diffusion, forming alloy powders.

8. **Other methods.** Other, less commonly used methods include :

   (*i*) **Precipitation** *from a chemical solution.*

   (*ii*) *Production of fine metals by* **machining.**

   (*iii*) **Vapour condensation.**

   ● New developments include techniques based on high temperature *extractive metallurgical processes*. Metal powders are being produced using high temperature processing techniques based on :

   — The reaction of volatile halides with liquid metals ;

   — The controlled   reduction and reduction/carburization of solid oxides.

   ● Recent developments include the production of **nanopowders** of various metals such as copper, aluminium, iron and titanium. When the metals are subjected to large plastic deformation at stress levels of 5500 MN/m^2, their particle size is reduced, and the material becomes pore free and thus possesses enhanced properties.

4.4.2. Blending of Metal Powders

The process of blending (mixing) powders is carried out for the following *purposes* :

(*i*) To obtain *uniformity* (since the powders made by various processes may have different sizes and shapes).

(*ii*) To *impart special physical and mechanical properties and characteristics* to the powder metallurgy product.

(*iii*) Addition of lubricants (*e.g.*, stearic acid, zinc stearate in proportion of 0.25 to 0.5% by weight) to the powders improve the flow characteristics of the powders. Such blends result in reduced friction between the metal particles, improved flow of powder metals into the dies, and longer die life.

● In order to avoid contamination and deterioration, powder mixing must be carried out under controlled conditions.

● The metal powders like aluminium, magnesium, titanium, zirconium, and thorium powders are *explosive* owing to their *high surface area to volume ratio* consequently, a great case must be exercised during their blending, storage and handling.

4.4.3. Pressing or Compaction of Metal Powders

The principal object of pressing or compacting is to effect cold-pressure welds between the particles so that some cohesion is conferred, this is usually measured by the strength of the green compact and is termed the *green strength.*

Compacting exercises the following *effects* :

(*i*) Reduces voids between powder particles and increases density of the compact.

(*ii*) Produces adhesion and cold welding of the powder and sufficient green strength.

(*iii*) Plastically deforms the powder to allow recrystallisation during subsequent heating.

(*iv*) Plastically deforms the powder to increase the contact areas between the powder particles, increasing green strength and facilitating subsequent sintering.

● *Pressing or compacting* is carried out by pouring a measured amount of the metal powder into the die cavity and then compacting the metal powder into coherent mass by means of one or more plungers (Fig. 4.1). Compressing from both top and bottom (Fig. 4.2) of the compact is better than compressing from the top only (Fig. 4.1), as pressure distribution and porosity distribution are more uniform.

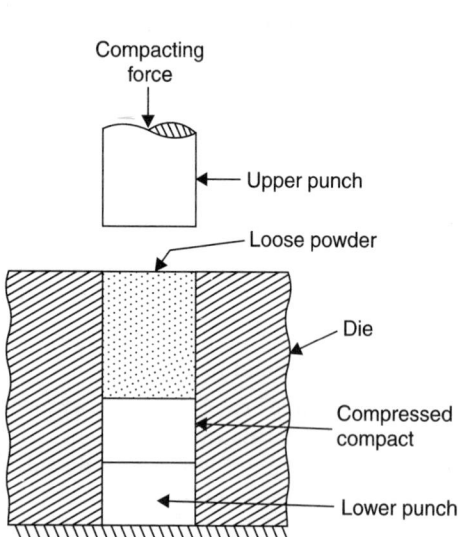

Fig. 4.1. Pressing or compacting by one or more plungers.

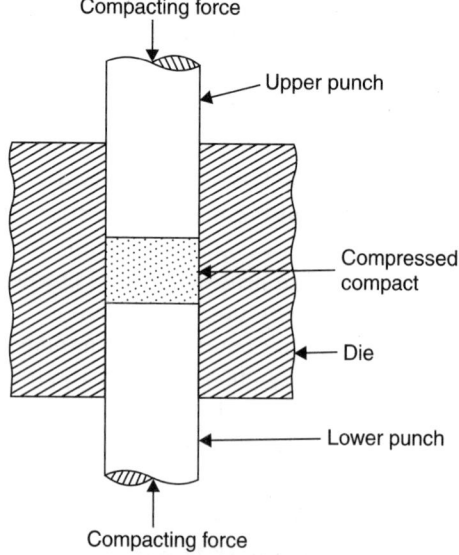

Fig. 4.2. Pressing or compacting from top and bottom.

— To improve uniformity of pressure and porosity through the piece the use of lubricants graphite, strearatex and zinc, aluminium and lithium strearate is made. Bond strength between particles is affected by the area of contact and by the cleanliness of particles (oxide layers being common sources of difficulty).

— The *compacting pressure required depends on the characteristics and shape of the particles, the methods of blending, and the lubrication.* The pressure required for pressing metal powders range from 70 MN/m^2 for aluminium to 800 MN/m^2 for high-density iron parts.

● In compacting powders in steel dies at room temperatures pressures from 7.5 to 37.5 (sometimes 150) kN/cm^2 are employed. Mechanical presses are employed for 500 kN and hydraulic presses for higher pressures. The moulding of small parts at great speeds and at relatively low pressures can be best accomplished in the mechanical press. However, large parts and parts to be moulded at higher pressures are best moulded in hydraulic presses.

— Extremely hard powders are slower and more difficult to press ; some organic binder is usually required to hold the hard particles together after pressing until the heat of sintering creates atomic bonds and promotes welding.

— In hot pressing if the powder is heated to proper temperature the pressure for complete densification is only 150 to 300 bar. Hot pressing requires a die material having appreciable high strength. Since even sturdy dies (made of graphite) usually sustain only one operation, therefore, *hot pressing is limited to the manufacture of articles of costly metals.*

4.4.4. Sintering

Sintering *means the heating of pressed compact to below the melting temperature of any constituent of the compact, or atleast below the melting temperature of all principal constituents of the compact.* Such heating facilitates *bonding action* between the individual powder particles and *increases the strength of the compact.* Heating is carried out *in a controlled, inert or reducing atmosphere, or in vacuum to prevent oxidation.*

— Prior to sintering, the compact is brittle, and its strength, known *as green strength* is low. The nature and strength of the bond between the particles, and hence of sintered product, depend on the *mechanism of diffusion plastic flow, evaporation of volatile materials in the compact, recrystallization, grain growth, and pure shrinkage.*

During the sintering process bonding of the individual powder particles takes places in any of the following three ways :

(*i*) Melting of a minor constituent ;

(*ii*) Diffusion ;

(*iii*) Mechanical bonding.

The important factors which control sintering are : (*a*) *temperature*, (*b*) *time, and* (*c*) *furnace atmosphere.* The sintering temperatures used vary with the compressive loads used, the type of powders, and strength required of the finished part.

— Aluminium and aluminium alloys can be sintered at temperatures from 350° to 500°C for periods upto 24 hours.

— Copper and copper alloys can be sintered at temperatures ranging from 700°C to temperatures that may melt one of the constituent metals.

— Compacts of iron powders are usually sintered at temperatures from 1000°C to 1200°C for approximately half an hour.

● During sintering, ordinarily, the density of the compact is increased and shrinkage results. Shrinkage fills the holes within the compact, increases the areas of contact and reduces the size of the holes in the compact. Shrinkage can be controlled by careful selection of the metal powder and determination of the correct pressure for cold forming. Instances are also met when pressed compacts grow during sintering and this is caused by the gas being trapped within the compact during pressing or produced by chemical reaction within the compact.

Fig. 4.3 indicates the qualitative effect of sintering temperature on the density of pressed metal powder.

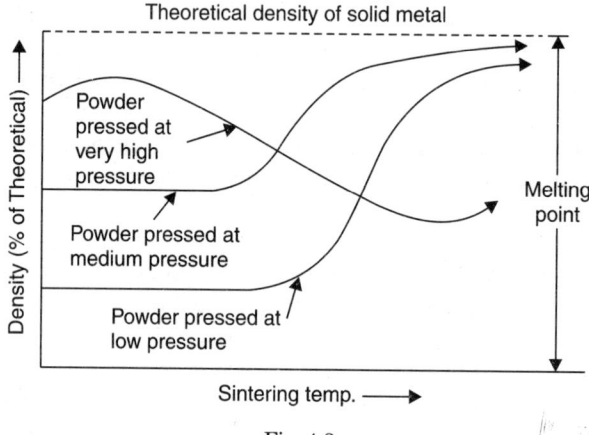

Fig. 4.3

- Sintering may be carried out in batches or continuously in batch type and continuous furnaces respectively.
- Batch type furnaces prove useful for laboratory or experimental work while continuous furnaces are more suitable for production work.
- A continuous furnace usually comprises three zones. The *first zone* warms the pressed parts and the protective atmosphere used in the furnace purges the work of any air or oxygen that may be carried into the furnace by work or trays. In the *second zone* the work is heated to the proper sintering temperature. The *third zone* cools the work to a temperature that allows handling. Fig. 4.4. shows a continuous-conveyor-type sintering furnace.

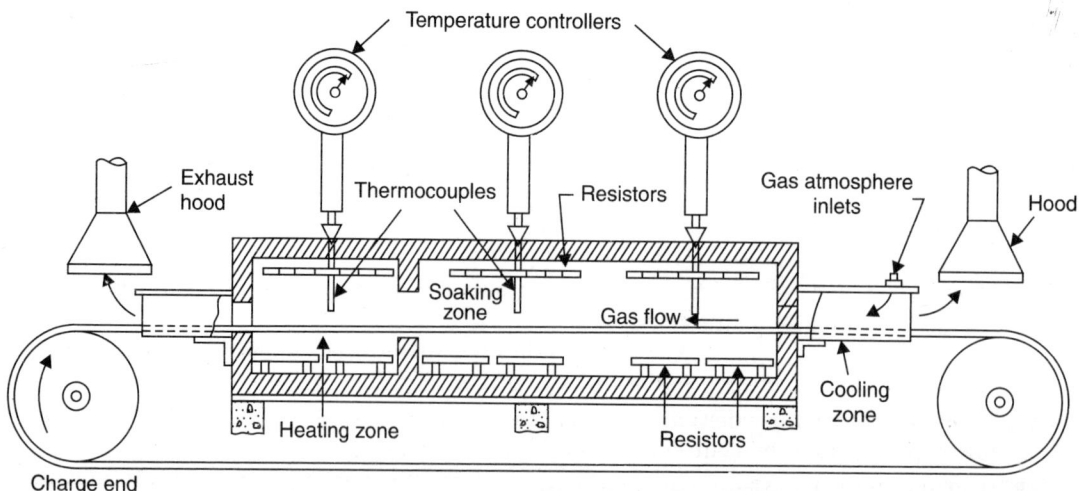

Fig. 4.4. Sintering furnace.

Note : (*i*) The strength, ductility and density of powder metallurgy parts is much lower than for correspond-ing cast parts. Density is usually less than 90% of the density of conventional metal. Ductility varies from perhaps less than 10 to 70% of the values for conventional metal.

(*ii*) Powder metallurgy parts generally speaking, are known for poor corrosion resistance probably be-cause of large area exposed by the pores of the object. This shortcoming can be overcome by impregna-tion with low-melting alloys or non-metallics.

(*iii*) Powder metallurgy structural parts compete with castings, particularly precision and die castings. They also compete with parts machined from wrought stock.

4.4.5. Finishing Operations

After sintering, the following additional operations may be carried out to further improve the properties of powder metallurgy products :

(*i*) **Re-pressing.** The purpose of this operation (also called *coining* and *sizing*) is to impart dimensional accuracy to the sintered part and to improve the part's strength and sur-face finish by *additional densification.*

(*ii*) **Forging.** This process involves the use of unsintered or sintered alloy-powder preforms that are subsequently hot forged in heated, confined dies to the desired final shapes ; the preforms may also be shaped by impact forging. The process is usually referred to as *powder-metallurgy forging,* and when the preform is sintered, the process is usually referred to as *sinter forging.*

(*iii*) **Infiltration.** In this process a metal slug with a lower melting point than that of the part is placed against the sintered part, and the assembly is heated to a temperature sufficient to melt the slug. The molten metal infiltrates the pores by *capillary action,* resulting in a relatively pore-free part with good density and strength.

(*iv*) Powdered-metal parts may be subjected to other operations including *heat treating, machining* and *finishing.*

4.5. DESIGN CONSIDERATIONS FOR POWDER METALLURGY

While using metal powders, because of their unique properties, the following *design principles* should be followed :

1. In order to facilitate easy ejection, narrow and deep sections should be avoided.

2. As far as possible, abrupt changes in section thickness should be avoided.

3. In order to increase tool and die life and reduce production costs, the powder metallurgy parts should be made with the widest dimensional tolerances, consistent with their in-tended applications.

4. In designing flat sections of high density, enough section thickness should be provided, otherwise the punch will break under pressure.

5. The design should be consistent with the capabilities and limitations of the available equipment.

6. Very close dimensional tolerances in the direction of pressing should be avoided.

7. The designed dimensions of parts should carry adequate allowances to compensate for the likely changes in dimensions due to shrinkage during sintering.

8. Holes should not be designed in the direction of pressing.

9. Production of very small holes through pressing should preferably be avoided.

10. Walls should not be less than 1.5 mm thick, although walls as thin as 0.34 mm have been successfully provided on components 1 mm in length.

    — Walls with length-to-thickness ratio greater than 8 : 1 are difficult to press, and density variations are virtually unavoidable.

11. The meeting plane between moulding punches should be on a flat or cylindrical surface and never on a spherical surface.

- Dimensional tolerances of sintered powder metallurgy parts are usually of the order of ± 0.05–0.1 mm. Tolerances improve significantly with additional operations such as sizing, machining, and grinding.

- Die and punch surfaces must be tapped or polished in the direction of tool movements for improved die life and overall performance.

QUESTIONS WITH ANSWERS

Q. 4.1. Compare the features of powder metallurgy with casting, forging (hot), extrusion (hot) and machining processes.

Ans. The comparison of the features of powder metallurgy with the other above given processes is given below :

S. No.	Process	Advantages over powder metallurgy	Limitations as compared with powder metallurgy
1.	Casting	• Wide range of part shapes and sizes produced. • Generally low mould and set up cost.	• Some waste of material in processing. • Some finishing required. • May not be feasible for high-temperature alloys.
2.	Forging (hot)	• High production rate of a wide range of part sizes and shapes. • High mechanical properties through control of grain flow.	• Some finishing required. • Some waste of material in processing. • Relatively poor surface finish and dimensional control.
3.	Extrusion (hot)	• High production rate of long parts. • Complex cross-sections may be produced.	• Only a constant cross-sectional shape can be produced. • Poor dimensional control.
4.	Machining	• Wide range of part shapes and sizes. • Short lead time. • Flexibility. • Good dimensional control and surface finish. • Simple tooling.	• Waste of material in the form of chips. • Relatively low productivity.

Q. 4.2. Explain briefly 'Particle size, distribution, and shape' in powder metallurgy technique.

Ans.

- **Particle size.** Usually, particle size is measured by *screening*, that is, by passing the metal powder through screens (sieves) of various mesh sizes.

Besides screen analysis, several other methods used for particle size analysis, particularly for powders finer than 45 µm, are :

(*i*) Sedimentation.   (*ii*) Microscopic analysis.

(*iii*) Light scattering.   (*iv*) Optical means.

(*v*) Suspension of particles-electrical sensors.

- **Size distribution of particles.** The distribution of particle size affects the processing characteristics of the powders. It is given in terms of a *frequency-distribution plot.*
- **Shape of particles.** The shape of particles immensely influences particles' processing characteristics. The shape is usually described in *terms of aspect ratio or shape index.*
- *Aspect ratio* is the ratio of largest dimension to the smallest dimension of the particle. This ratio ranges from 1 for a spherical particle to about 10 for flakelike or needlelike particles.
- *Shape index* or *shape factor* (SF), is a measure of the surface area to the volume of the particle with reference to a spherical particle of equivalent diameter ; thus the shape factor for a flake is higher than that for a sphere.

Q. 4.3. What is "Presintering" ?

Ans :

- *Presintering means heating the green compact to a temperature below the sintering temperature.* It is done to increase strength of green compact and remove the lubricants and binders added during blending.
- Some metals like tungsten carbide are easily machined in presintered state as they become too hard after sintering. However, if machining is not required, this operation can be avoided for them.

Q. 4.4. What are the objects of 'Sintering' ?

Ans. The *main objects of sintering* are :

1. To achieve good bonding of powder particles.
2. To produce a dense and compact structure.
3. To achieve high strength.
4. To produce parts free of oxides.
5. To obtain desired structure and improved mechanical properties.

Q. 4.5. Explain briefly the following compacting and shaping processes used in powder metallurgy :

1. Isostatic pressing.	**2. Metal injection moulding.**
3. Rolling.	**4. Extrusion.**

5. Pressureless compaction.

Ans. 1. Isostatic pressing. In *cold isostatic pressing (CIP)*, the metal powder is placed in a flexible rubber mould made of neoprene rubber, urethene, polyvinyl chloride or other elastomers. The assembly is then pressurised hydrostatically in a chamber, usually with water. The most common pressure is 400 MN/m^2, although pressures of upto 1000 MN/m^2 have been employed.

In *hot isostatic pressing (HIP)*, the container is usually made of a high-melting point sheet metal, and the pressurizing medium is inert gas or vitreous fluid. A common condition for HIP in 100 MN/m^2 at 1100°C, although the trend is towards higher pressures and temperatures.

- The main *advantage* of *HIP is its ability to produce compacts with essentially 100% density, good metallurgical bonding among the particles, and good mechanical properties.*
- The *limitations* of process are :
  (*i*) Wider dimensional tolerances produced ;
  (*ii*) High cost and time required ;
  (*iii*) Relatively small production quantities output.

Applications. (*i*) This process is routinely used as a final densification step for tungsten-carbide cutting tools and powder metallurgy tool steels.

(*ii*) It is also employed to close internal porosity and improve properties in super alloy and titanium-alloy castings for aerospace industry.

2. **Metal Injection Moulding (MIM).** It is also known as Powder Injection Moulding (*PIM*). It is carried out as follows :

- Very fine metal powders, generally < 45 μm, and often > 10 μm, are blended with a polymer or wax-based binder. The mixture then undergoes a process similar to injection moulding of plastics.
- The moulded green parts are placed in a low-temperature oven to burn off the plastic, or else the binder in partially removed by solvent extraction, and then the green parts are sintered in a furnace.

Advantages :

(*i*) Close tolerance can be maintained.

(*ii*) Highly complex components, with densities of 96 per cent of the theoretical and above, can be produced.

(*iii*) Ductility of as-sintered parts is usually high, with common elongation of 25 to 30 per cent.

Limitations :

(*i*) Due to the process to remove binder, the cross-sectional thickness of the part is limited (usually 6.35 mm or less).

(*ii*) Parts are usually limited in overall size and less than about 80 g ; however, larger parts are under development.

3. **Rolling :**

- In *powder rolling* (also called *roll compaction*), the powder is fed to the roll gap in a two-high rolling mill and is compacted into a continuous strip at a speed of 0.5 m/s. The process can be carried out at room or elevated temperatures.
- *Sheet metal for electrical and electronic components and for coins* can be made by this process.

4. **Extrusion :**

- Powders may be compacted by *hot extrusion* ; in this process, the powder is encased in a metal container and extruded. For example, superalloy powders are hot extruded for improved properties.
- Performs made from the extrusions may be reheated and forged in a closed die to their final shape.

5. **Pressureless compaction :**

- In this method, the die is filled with metal powder by gravity, and the powder is sintered directly in the die.
- This method (because of the resulting low density) is used principally for porous part, such as *filters*.

Q. 4.6. Discuss briefly the following processes for obtaining full density products :

1. **Ceracon process ;**

2. **Osprey process.**

Ans. 1. Ceracon process. In this process the hot preform is completely surrounded by a granular material capable of transmitting pressure in the pseudo uniform manner. The complete assembly is then pressed and compacted in a conventional hydraulic press.

Advantages :

(*i*) The granular material can be heated and reused.

(*ii*) No need of canning and decanning the parts.

(*iii*) The compacted part and the pressurizing medium can be separated conveniently.

2. **Osprey process.** In this process (also known as "In situ consolidation") the molten metal is atomized into powder. The powder is then sprayed by streams of inert gas into a shaped collector mould. Cooling of the droplets is controlled is such a way that they strike the surface of the mould in semi-solid state and quickly cool. This process can produce 98 to 99% dense parts. The preform, so obtained, is immediately hot-forged or otherwise hot-worked, to get the final shape.

Advantages :

(*i*) Low collector mould costs.

(*ii*) The product has uniform fine grain size and uniform chemistry.

(*iii*) The preforms have better workability as compared to conventional powder metallurgy performs.

(*iv*) No sintering step is required.

Q. 4.7. What do you understand by 'Spark discharge sintering' ? Explain briefly.

Ans. Spark discharge sintering :

- In this type of sintering, a high energy electric discharge spark is produced during compaction which results in instantaneous diffusion bonding. This method thus combines compacting and sintering the metal powders to a dense metal part in 12 to 15 seconds.

Advantages :

(*i*) Dimensional accuracy of the parts is maintained.

(*ii*) No separate sintering furnace is required for the process.

Q. 4.8. Explain briefly "Self lubricating bearings".

Ans. *Self lubricating or porous metallic bearing is that bearing in which the metal is porous, the pores containing the oil necessary for lubrication.* These bearings are produced by *powder metallurgy.* Powders of copper, tin and graphite are sintered, and are then passed through a sizing die to the correct dimensions. The metal thus produced is of a porous nature, capable of holding upto one third of its volume of oil. A pressure on the bearing surface, or a temperature rise will cause the oil to exude, so that lubrication is ensured continuously and automatically where it is needed.

- These bearings *help eliminating much routine oiling.* Dirt cannot enter the bearings since there are no oil holes or grooves. There is no leakage of oil hence no contamination of any products which are being handled, a point of vital importance in machinery for dealing with foodstuffs or delicate textiles.

- The bearing contains, in most cases, adequate oil to outlast the machine itself. Any loss of oil is made up by capillary action, from the source. When needed, further oil can be added to the outside of the porous bearing bush. It will be absorbed, filtered and fed uniformly over the running surface.

- Self lubricating bearings have sufficient strength to support the load applied by the shaft and to be installed in the bearing house. The permissible load for self lubricating bearings depends upon the operating conditions, housing conditions and construction of the assembly. Although *iron base* self lubricating bearings will stand higher loads, they are not nearly as frequently used because their resistance to seizure and to corrosion is much lower than for bronze (copper tin) type, self lubricating bearings.

Some of the self lubricated bearing are shown in the Fig. 4.5.

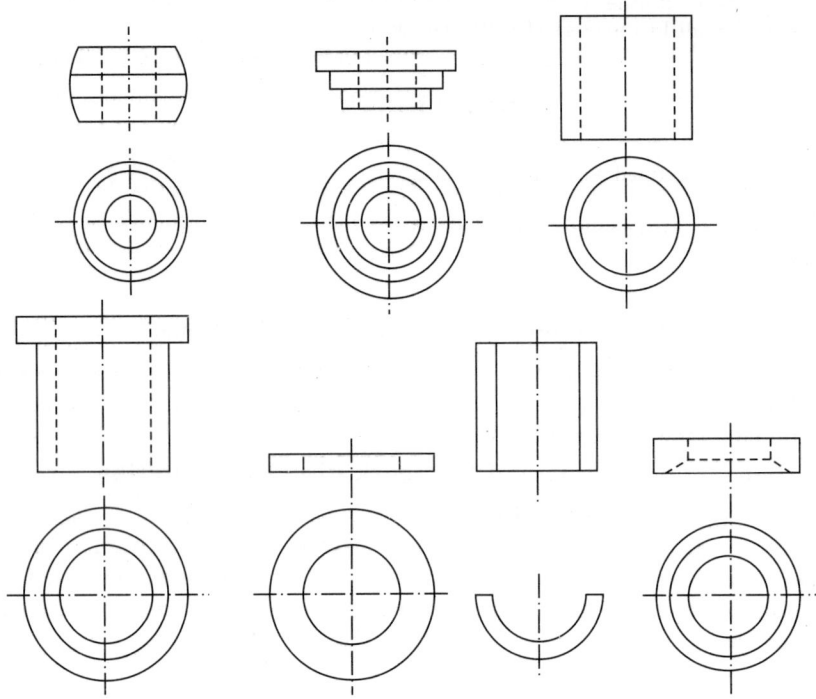

Fig. 4.5. Self lubricating bearings.

Advantages of self lubricating bearings over standard sleeve bearings :

Self lubricating bearings produced by powder metallurgy claim the following *advantages* over standard sleeve bearings.

1. No machining is required.

2. Cost is low.

3. Processing time from raw material to finished part is not more than ten minutes including heat treatment.

4. The powdered metal required is not very expensive comparatively.

Q. 4.9. Explain briefly how is tungsten carbide (for making tools and dies) is produced by powder metallurgy process.

Ans. The following powder-metallurgy procedure is used to make tungsten carbide ; the *steps* are :

1. Powders of tungsten and carbon are blended together in a ball mill or rotating mixer. The mixture (typically 94% tungsten and 6% carbon, by weight) is heat treated to approximately 1500°C in a vacuum-induction furnace. As a result of this process, the tungsten is carburized, forming *tungsten carbide in fine-powder form.*

2. A binding agent is then added to the tungsten carbide (with an organic fluid such as hexane), and the mixture is ball milled to produce a uniform and homogeneous mix, a process that can take several hours, or even days.

3. The mixture is then dried and consolidated, usually by cold compaction, at pressures in the range of 200 MN/m^2.

4. Finally, the mixture is sintered in a hydrogen-atmosphere or vacuum furnace at a temperature of 1350°C to 1600°C depending on its composition.

Note : A combination of other carbides, such as titanium carbide and tantalum carbide, can also be produced, using mixtures made by the method described above.

HIGHLIGHTS

1. *Powder metallurgy* is defined as the art of making objects by the heat treatment of compressed metallic powders.
2. The chief methods of producing metal powders are :
   (*i*) Atomising ;   (*ii*) Gaseous reduction ;
   (*iii*) Electrolysis ;   (*iv*) Carbonyl process ;
   (*v*) Stamp and ball mills ;   (*vi*) Granulation.
3. *Sintering* means the heating of pressed compact to below the melting temperature of any constituent of the compact, or at least below the melting temperature of all principal constituents of the compact.
4. *Self lubricating or porous metallic bearing* is that bearing in which the metal is porous, the pores containing the oil necessary for lubrication. These bearings are produced by powder metallurgy.

OBJECTIVE TYPE QUESTIONS

Fill in the Blanks :
1. ...... is defined as the art of making objects by the heat treatment of compressed metallic powders.
2. In .... process the molten metal is forced through an orifice into a stream of high-velocity air, stream or inert gas.
3. The process of .... consists of grinding the metallic oxide to a finally divided state and then reducing it by hydrogen or carbon monoxide.
4. .... process is based upon the fact that a number of metals can react with carbon monoxide to form what are known as carbonyls.
5. .... process produces a relatively coarse powder with a high percentage of oxide.
6. .... means the heating of pressed compact to below the melting temperature of any constituent of the product, or at least below the melting temperature of all principal constituents of the compact.
7. The important factors which control sintering are :
   (*a*) ....   (*b*) ...., and   (*c*) ....
8. During sintering, ordinarily, the density of impact is .... and .... results.
9. .... type furnaces prove useful for laboratory or experimental work.
10. .... bearing is that bearing in which the metal is porous, the pores containing the oil necessary for lubrication.

ANSWERS

1. Powder metallurgy   2. Atomising   3. Gaseous reduction
4. Carbonyl   5. Granulation   6. Sintering
7. Temperature, time, furnace atmosphere   8. Increased, shrinkage
9. Batch   10. Self lubricating.

THEORETICAL QUESTIONS

1. What is meant by powder metallurgy ?
2. State the advantages and limitations of powder metallurgy.
3. Mention the metals which are commonly made in powder form.
4. Explain the procedure of manufacturing parts by powder metallurgy.

5. What are self lubricated bearings ? How are they produced ?

6. Mention the advantages and disadvantages of self lubricated bearings.

7. State briefly, the process of making a powder metallurgy product having improved properties and discuss the advantages of powder metallurgy.

8. List the applications of powder metallurgy.

9. Explain briefly any three methods of production of metal powders :

    (*i*) Atomising   (*ii*) Gaseous reduction

    (*iii*) Electrolysis   (*iv*) Carbonyl process

    (*v*) Granulation   (*vi*) Mechanical alloying.

10. What do you mean by 'Blending metal powders' ? Explain briefly.

11. What is the object of pressing or compaction of metal powders ? Explain.

12. What do you mean by 'Sintering' ? Explain briefly.

13. List the factors which control sintering.

14. Explain briefly, a continuous-conveyor-type sintering furnace.

15. List the design considerations for powder metallurgy.

16. Discuss briefly the following processes :

    (*i*) Ceracon process ;   (*ii*) Osprey process.

17. Compare the features of powder metallurgy with the following processes :

    (*i*) Casting ;   (*ii*) Hot forging ;

    (*iii*) Hot extrusion ;   (*iv*) Machining.

18. Explain briefly the following compacting and shaping processes.

    (*i*) Isostatic pressing ;   (*ii*) Metal injection moulding ;

    (*iii*) Rolling ;   (*iv*) Extrusion.

19. What is presintering ?

20. What are the objects of 'sintering' ?

21. What is 'Spark discharge sintering' ? Explain briefly.

5

Processing of Plastics

5.1. Introduction to plastics— definition of plastic—classification of plastics—characteristics/properties of plastics—compounding materials—fiber glass reinforceu plastics—trade names and typical applications of some important plastics. 5.2. Processing of plastics—general aspects—plastic processing methods—compression moulding—transfer moulding—injection moulding—expandable bead moulding—rotomoulding—blow moulding—extrusion—thermoforming—calendering—casting. 5.3. Machining of plastics. 5.4. Joining of thermoplastics. 5.5. Plastic design rules. **Questions with Answers**—Highlights—Objective type Questions—Theoretical Questions.

5.1. INTRODUCTION TO PLASTICS

5.1.1. Definition of Plastic

A **plastic,** in broadest sense, is *defined as any non-metallic material that can be moulded to shape.* The most common definition for plastics is that they are *natural or synthetic resins, or their compounds, which can be moulded, extended, cast or used as films or coatings.*

- Most of the plastics are of organic nature composed of *hydrogen, oxygen, carbon and nitrogen.*

5.1.2. Classification of Plastics

Plastics, most commonly, are *classified* as :

1. Thermoplastic, and

2. Thermosetting.

1. **Thermoplastic materials.** *The plastics which soften on the application of heat with or without pressure, but require cooling to set them to shape are called **thermoplastic materials.***

These can be heated and cooled any number of times, only they should not be heated above their decomposition temperatures.

- They are main long chain straight or slightly branched molecules and the chains are held close to each other by secondary weak forces of type Vander Waals' forces. During heating, as the temperature increases the secondary forces are *reduced* and the sliding of these long chain molecules can *easily occur* one over the other at a reduced stress level.

- They are highly plastic and are easy for moulding or shaping.

- They have *low* melting temperatures and *not so strong* as the thermosetting plastics.

- Since they can be repeatedly used, they have a *resale value.*

Important Thermoplastic materials are :

   (i) Polyethylene or polythene $(C_2H_4)_n$ (ii) Polyvinyl chloride (PVC)

 (iii) Polypropylene (PP)   (iv) Teflon or polytetrafluoroethylene (PTFE)

  (v) Polystyrene   (vi) Acrylics (Polymethyl methacrylate PMMA)

(vii) ABS (Acrylonitric butadiene styrene)

233

(*viii*) Silicones   (*ix*) Polyvinylidene chloride

(*x*) Polyamides   (*xi*) Bitumen.

2. Thermosetting materials. *The plastics which require heat and pressure to mould them into shape are called* **thermosetting materials.**

- They cannot be resoftened once they have set and hardened.
- They are ideal for moulding into components which require rigidity, strength and some resistance to heat.
- In general, resins formed by condensation are thermosetting.
- Thermosetting resins have *three-dimensional molecular structure and have very high molecular weights.*
- Due to *cross-linking* thermosetting resins are hard, tough, non-swelling and brittle. Hence, they cannot be softened or remoulded as in the case of thermoplastic resins. Moulding and casting are the processes often used with such materials.

Important thermosetting materials are :

(*i*) Phenol formaldehyde (PF)—"Bakelite"

(*ii*) Amine formaldehyde (Urea and melamine formaldehyde)

(*iii*) Polysters (unsaturated)

(*iv*) Epoxy resins (epoxies).

Difference between thermoplastic and thermosetting materials :

The *difference between thermoplastic and thermosetting materials* may be explained *in terms of molecular structure* :

— The *"thermoplastics"* are essentially long chain macromolecules with a limited number of cross links. When heated and compressed, the chains glide over each other and fluid materials take the shape of any mould in which they are placed.

— The *"thermosetting plastics"* are characterised by strong cross-links between the chains; once these are formed by heat and pressure, the plastics set to a rigid infusible solid.

Fig 5.1 represents the two structures where *R* stands for monomer unit.

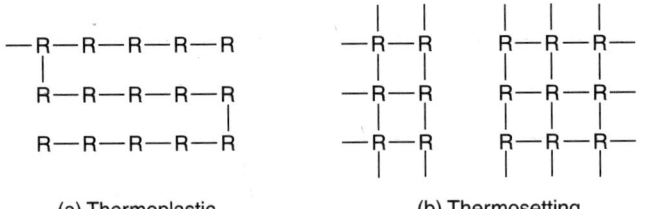

(a) Thermoplastic   (b) Thermosetting

Fig. 5.1. Structure of plastics.

Differences between thermosetting and thermoplastic materials

S. No.	Thermosetting materials	Thermoplastic materials
1.	They have three dimensional network of primary covalent bonds with cross-linking between chains.	They are linear polymers without cross-linking and branching.
2.	They are more stronger and harder than thermoplastic resins.	They are comparatively softer and less strong.
3.	Once hardened and set they do not soften with the application of heat.	They can be repeatedly softened by heat and hardened by cooling.

4.	Objects made by thermosetting resins can be used at comparatively higher temperature without damage.	Objects made by thermoplastic resins cannot be used at comparatively higher temperatures as they will tend to soften under heat.
5.	They are usually supplied in a monomeric or partially polymerized form in which they are either liquid or partially thermoplastic solids.	They are usually supplied as granular materials.
6.	It is difficult to fill an intricate mould with such plastics.	They can fill the complicated mould quite easily.
7.	They cannot be recycled.	The scrap of these plastics can be recycled again and thus they are economical.
	Uses : *Telephone receivers, electric plugs, radio and TV cabinets, camera bodies, automobile parts, circuit breaker switch panels, etc.*	**Uses :** *Toys, combs, toilet goods, photographic films, insulating tapes, hoses, electric insulation, etc.*

5.1.3. Characteristics/Properties of Plastics

The properties of plastics vary with the molecular structure, degree of polymerisation, etc., some of the properties are described below :

1. Specific gravity :

- Plastics are generally light having specific gravity between 0.9 to 3.0 compared to 3.0 to 12.00 for metals. However, their strength/weight ratio compares favourably with many light alloys.

2. Specific heat :

- Specific heat is the heat necessary to raise the temperature of 1 kg substance by 1 K.
- The specific heat of plastics varies between 200 to 800 J/kg/K as against 400 J/kg/K for steel.

3. Thermal conductivity :

Plastics have comparatively low thermal conductivity, hence they are good thermal insulating materials particularly expanded or cellular polymers.

4. Thermal expansion :

Thermal expansion of plastics is very high (approx. five times thermal expansion of aluminium and other metals) which is the *main disadvantages associated with the plastics.*

5. Electrical properties :

- Plastics are good electrical insulators. However, their usefulness is limited by their low heat resistance and softness.
- Also they can acquire electrostatic charges and may cause sparking which is potential fire hazard.

6. Corrosion resistance :

- Plastics are generally resistant to most inorganic chemicals, weathering and soil. However, most plastics become brittle and yellow when exposed to sunlight for long duration.
- Plastics are resistant to attack by oils and greases, hence superior to rubber.

7. Combustibility :

Plastics are combustible because of presence of carbon. The maximum service temperature is about 100°C.

8. Rigidity :

- Plastics have low rigidity compared to other materials, since they have low modulus of rigidity. However, this property can be improved by addition of fillers such as glass fibres.

5.1.4. Compounding Materials

In order to obtain the desired plastics certain materials are added with polymers (resins) to reduce the cost or to enhance mechanical properties. These materials are described below :

1. Binders :

- The main purpose of binders is to *hold other constituents of the plastic together,* as the name implies.
- A binder may compose of 30–100 percent of the plastic.
- The binders may be synthetic or natural resins. Thus, resins are the basic binding materials in the plastic.
- On the basis of the type of resin used in preparation, the plastic itself is called *thermoplastic or thermosetting plastic.*

2. Fillers :

- *Fillers* are added to reduce the *cost* and *enhance the strength and hardness of plastics.*
- These may be fibrous type like asbestos, glass fibre, mica etc., or non-fibrous type, *e.g.,* china clay, carbon black, talc, zinc oxide, calcium carbonate. Quartz and mica are used to improve hardness.
- Inorganic fillers like asbestos are added to improve heat and corrosion resistance, shredded textiles are used to increase impact strength and so on.
- The proportion of the filler added can be as high as 50 percent of the plastic.

3. Plasticisers :

- Low molecular weight (approx. 300) materials blended with polymers are called ***plasticisers.***
- These are added to *improve flexibility* and *processing of plastic articles,* and *to reduce the temperature and pressure required for moulding.* However, a plasticiser may reduce the tensile strength and chemical resistance of plastics.
- Some of the plastics are : ***Polyesters, epoxies, nitrile rubbers.*** These are normally *chemically inert, non-volatile and non-toxic.*
- The percentage of plasticisers can be as high as 60 percent of the plastic.
- If the plasticiser is added in excessive quantity, the bond will be mainly between the small molecules and we will have a liquid. In a common paint, excess plasticiser is added to make the paint liquid and allow brushing. After the paint is applied the plasticiser evaporates and the paint dries. This evaporation is usually accompanied by some polymerisation and cross-linking with oxygen. The residual plasticiser makes the film of paint tough and flexible with time, however, the combination of further oxidation and loss of plasticiser can result in brittleness and flaking of the paint.

4. Blending :

"Blending" or alloying" is *combining of two or more distinct polymer molecules to form a new product with different characteristics.*

5. Stabilisers :

- The stabilisers are added to minimise the effect of heat, sunlight, ozone.
- White lead, barium, cadmium laurate are examples of stabilisers.

6. Colouring agents (pigments) :

- These are added to give desired colour to plastics.
- Some of the colouring agents are : *Zinc oxide, white lead, titanium dioxide, dyes.*

5.1.5 Fiber Glass Reinforced Plastics

- The fiber glass reinforced plastic (or FRP) is formed by using two materials in conjunction with each other to form a composite material of altogether different properties.
- In FRP, the glass *fibers provide stiffness and strength while resin provides a matrix to transfer load to the fibers.*

Properties of FRP :

Following are the properties which have made the FRP the most commercially successful composite material of construction :

(*i*) Aesthetic appeal.   (*ii*) Dimensional stability.

(*iii*) Light weight.   (*iv*) Easy to repair.

(*v*) Durable.   (*vi*) Corrosion resistant.

(*vii*) Requires less energy for production. (*viii*) Least maintenance required.

(*ix*) The tooling is inexpensive and fast.   (*x*) FRP products transmit a great deal of light.

Applications :

Following are the *applications* of FRP :

(*i*) Water storage tanks.   (*ii*) Roof sheets.

(*iii*) Domes.   (*iv*) Structural sections.

(*v*) Doors and window frames.   (*vi*) Concrete shuttering.

(*vii*) Internal partitions and wall panelling. (*viii*) Temporary shutters.

5.1.6. Trade Names and Typical Applications of Some Important Plastics

The trade names, and typical applications of some important plastics are given in the table below :

S. No.	Material type	Trade name	Typical applications
A. Thermoplastics :			
1.	*Acrylics*	Lucite, Plexiglas	• Outdoor signs • Lenses • Transparent aircraft enclosures
2.	*Nylons*	Zytel, Plaskon	• Handles • Bearings • Cams • Gears
3.	*Polyethylene*	Alathon, Petrothene, Hi-fax	• Tumblers • Ice trays • Toys • Flexible bottles
4.	*Polystrene*	Styron, Lustrex, Rexolite	• Appliance housings • Battery cases • Wall tiles
5.	*Polystere (PET)*	Mylar, Colanar, Dacron	• Automatic tyre cords • Magnetic recording tapes • Clothing

6.	*Vinyls*	PVC, Tygon, Saran	• Phonograph records • Floor coverings • Garden hose
B. *Thermosetting :*			
7.	*Phenolics*	Bakelite, Resinox, Durez	• Auto distributors • Telephones • Motor housings
8.	*Epoxies*	Araldite, Epon	• Adhesives • Electrical mouldings • Sinks
9.	*Polysters*	Laminac, Selectron, Paraplex	• Fans • Chairs • Helmets.

5.2. PROCESSING OF PLASTICS

5.2.1 General Aspects

Properties of plastics materials substantially differ from those of metals and hence their processing also differs. Plastic components *are generally fabricated by a single operation* rather than a sequence of operations as in the case of metals.

- With plastics, *large and complex shapes can be formed as a single unit, thus eliminating various complicated assembly operations.*

- *Plastic fabrication processes often provide the required surface finish and precision eliminating the need for additional finishing operations.*

- Plastics are, however, *inferior to metals* from the point of view of their mechanical properties.

— Most of the plastics are *prone to dimensional instability with substantial creep and cold flow at all temperatures.*

— Some plastics get affected by moisture.

— Operating temperature of most of the plastic materials is low, generally ranging from 65°C to 315°C, which is considerably lower than the operating temperature of metals.

— Rigidity and fatigue strength of plastics are also *substantially lower* than most of metals.

- All plastic forming and shaping processes require heating plastic so that it flows. This liquid plastic or *polymer melt* has two important properties namely 'Viscosity' and 'Viscoelasticity' which are important for forming and shaping of plastics.

— "Viscosity" is defined as resistance to flow and is a measure of internal friction that arises when velocity gradient is present in fluid. Thus, the more viscous the fluid is, the higher is the internal friction and the greater is the resistance to flow.

— "Viscoelasticity" is a combination of viscosity and elasticity and *this property determines the strain experienced by a material when subjected to a combination of stress and temperature over time.*

Following points are *worthnoting* :

(*i*) Plastics, compared with metals, have a low elastic modulus and deflect easily under the cutting forces; therefore they must be supported carefully.

(*ii*) In order to avoid sticking or excessive expansion or deflection, the work should be kept cool during machining.

— Cooling may be by air jet, vapour mist, or water and soluble oil, 5 : 1.

(*iii*) Cutting tools should be of high speed steel, but for some thermosets, for filled plastics and for long runs, carbide tipped tools are preferred.

(*iv*) The finish on some plastic articles may be improved by polishing or buffing using only light pressure.

(*v*) *Twist drills* for drilling plastics should have wide, polished flutes, a low (< 30°) helix angle and 60°-90° point angle, particularly for the softer plastics.

(*vi*) *Cutting tools* should be of high speed steel, but for some thermosets, for filled plastics and for long runs, carbide-tipped-tools are preferred.

(*vii*) *Laminated plastics* can be milled with standard tools at speeds and feeds similar to those used with bronze and soft steel. The feed rate is determined by the desired finish.

(*viii*) The finish on some plastic articles can be improved by polishing or buffing using only light pressure.

5.2.2. Plastic Processing Methods

The common production processes are :

1. Moulding processes :

   (*i*) Compression moulding   (*ii*) Transfer moulding

(*iii*) Injection moulding   (*iv*) Rotomoulding

  (*v*) Expandable-bead moulding   (*vi*) Blow moulding.

2. Extrusion

3. Thermoforming

4. Calendering

5. Casting.

5.2.3. Compression Moulding

Most of the thermosetting plastics and a few thermoplastic plastics can be moulded through this process; the pressure used is in the range of 7 to 140 MN/m^2.

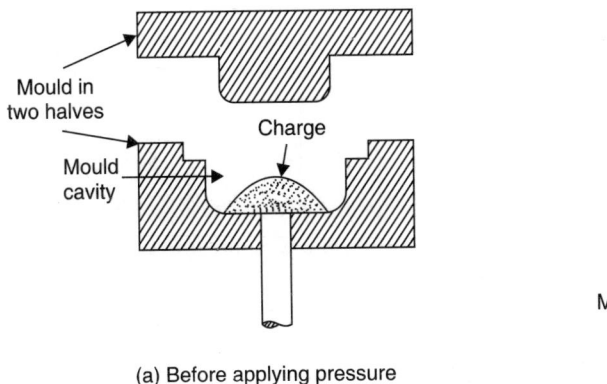
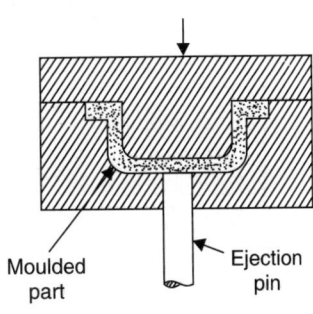

(a) Before applying pressure   (b) After applying pressure

Fig. 5.2. Compression moulding.

This process is carried out as follows :

(*i*) The raw plastic material (*i.e.*, charge) in the form of solid granules is placed in the cavity of an open mould which has been preheated to a temperature of 120 to 240°C.

(*ii*) A punch, also heated upto the same temperature, squeezes the material into the mould cavity. The plastic material melts at this temperature and under the pressure of the punch it flows into all portions of the mould cavity.

(*iii*) The material is kept inside the mould for sometime to allow it to cure under continued exposure of heat and then removed (with aid of ejection pin) to obtain the final product.

Sometimes the *raw plastic material is preheated to reduce the curing time.*

Compression moulding may be of the *positive type, semi-positive-type*, or the *flash type*. In the latter, some of the material is allowed to escape, usually along the moving-die perimeter, over a land or cut-off area, in the form of a thin *flash* or *pin* which is finally trimmed off. A mould of this type gives closer tolerances and is usually cheapest to make.

- Parts produced are *simple with uniform wall thickness* (usually not more than 3 mm).
- This moulding process, owing to non-uniform pressure, often results in turbulence and uneven flow of liquid plastic in the mould.

Cold moulding. In this process, a powder or fibres (often of refractory materials) are mixed with a binder and compacted in a cold die. These procedures are followed by *curing* in a separate oven.

- This method is *not* suitable where close tolerance and good surface finish is required.

Advantages of compression moulding :

1. Compression-moulded parts *do not have high stresses*, since the material is not forced through gates.

2. The process is comparatively *simple,* which makes *initial mould costs low.*

3. Simple setting.

Limitations :

1. Used mainly for thermosetting plastics.

2. Medium production rate.

3. Dimensional accuracy not so good.

5.2.4. Transfer Moulding

Transfer moulding (also known as *gate moulding* or *extrusion moulding*) is modified form of compression moulding.

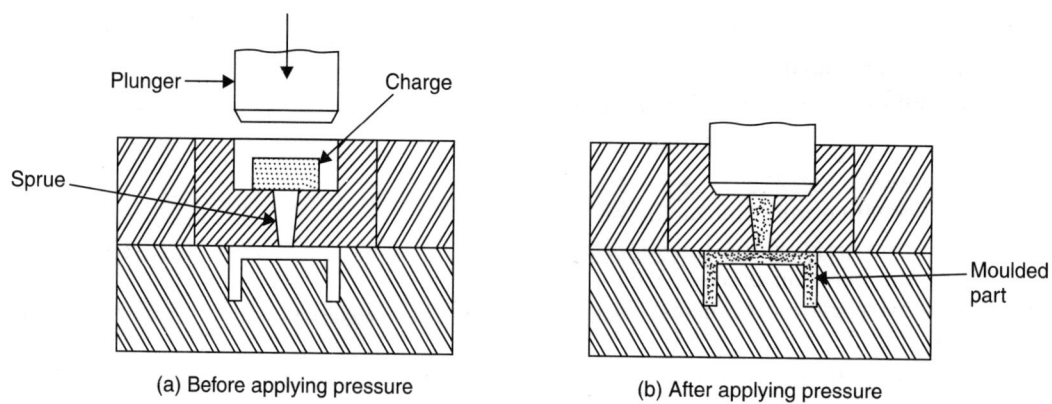

(a) Before applying pressure   (b) After applying pressure

Fig. 5.3. Transfer moulding.

In this method the moulding powder (charge) is heated to the plastic stage and then forced into mould proper when the curing is completed. Pressures in the range of 30–100 MN/m^2, depending upon the polymer properties, are generally used.

Advantages :

1. This process provides relatively close tolerances and fairly uniform density.

2. The operation is easier and enables trouble free production of intricate parts with thin sections as the mould is not directly subjected to the compression force.

3. High production rate.

4. Medium tooling cost.

Limitations :

1. The transfer moulds are more complex and more costly to build.

2. Loss of material as scrap.

5.2.5. Injection Moulding

This process is very commonly used for *thermoplastic plastics.*

Fig. 5.4 shows the schematic diagram of injection moulding.

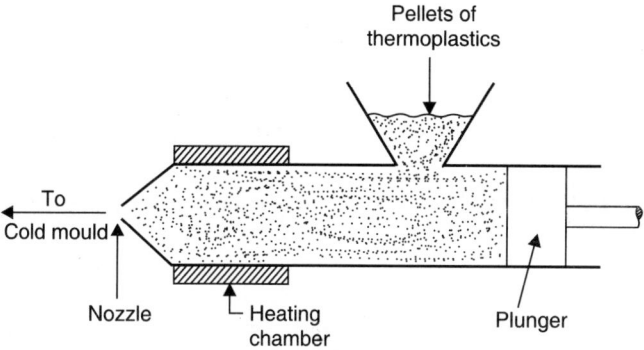

Fig. 5.4. Injection moulding.

— The pellets of thermoplastics are first compressed in the pressure chamber and then pushed into a heating chamber.

— The softened material in the flowing state is then forced with pressure (about $960 \, MN/m^2$) through a nozzle into a cold mould having cavity of the desired shape. Plastic articles of intricate shapes can be formed in these cavity moulds. The articles can be removed by opening the moulds.

— The following methods are used to inject the molten plastic into the mould :

   (*i*) Reciprocating screw injection moulding.

   (*ii*) Plunger injection moulding.

   (*iii*) Two-stage injection moulding.

An important part of the injection-moulding process is that the *mould can be securely clamped* during the injection process. High-pressure, fast action clamping may be provided either hydraulically or by toggle mechanisms. Toggle clamps are usually used on machines that require a clamping pressure 200 tonnes or less.

● Injection moulding is a *faster process* and *suits best for large-quantity production.* The greatest quantity of plastic parts are made by injection moulding.

Advantages :

1. Complex shapes of various sizes can be obtained.

2. Good dimensional accuracy.

Limitations :

1. High tooling tool.

2. High volume production is required.

Coinjection moulding. This type of moulding makes it *possible to mould articles with a solid skin of one thermoplastic and a core of another thermoplastic.* The skin material is usually solid, while the core material contains blowing agents.

The basic process may be one-, two-, or three-channel technology :

- In *'one-channel technology'*, the two melts are injected into the mould, *one after the other* (by shifting a valve). The skin material cools and adheres to the cooler surface ; a dense skin is formed under proper parameter settings. The thickness of the skin can be controlled by adjustment of injection speed, stock temperature, mould temperature, and flow compatibility of the two melts.

- In the *'two- and three-channel techniques'*, both plastic melts *may be introduced simultaneously.* This allows for better control of wall thickness of the skin, especially in gate areas on both sides of the part.

Reaction injection moulding (RIM). In this process *reactive liquid chemicals are mixed at high pressures and instantly dispensed into a closed mould.* When the gas-resin mixture is shot under pressure into the mould cavity, the gas expands within the plasticized material as it fills the mould, producing on *internal cellular structure* as well as *tough external skin* on the mould face. This structure is usually referred to as *integral skin.*

Advantages and limitations :

- Parts can be produced in a single shot that weights upto 45 kg with a cycle time of 2 or 3 min.

- The processing temperatures are low, raw materials are normally kept at 30 to 38°C and mould temperatures normally do not exceed 66°C.

- Clamping pressures are much less than those for conventional injection moulding since the mould pressure is only 0.105 to 0.525 MN/m^2.

- The major disadvantage of RIM for high production has been the long cycle time, 3 to 4 min.

Applications :

- At present RIM, solid and micro cellular, is used extensively in automative fascia, bumpers, etc.

5.2.6. Expandable Bead Moulding

This process consists of placing small beads of polystyrene along with a small amount of blowing agent in a tumbling container. The polystyrene beads soften under heat, which allows a blowing agent to expand them. When the beads reach a given size, depending on the density required, they are quickly cooled. This solidifies the polystyrene in its larger, formed size. The expanded beads are then placed in a mould (usually aluminium) until it is completely filled. The entrance pot is then closed and steam is injected, resoftening the beads and fusing them. After cooling, the finished, expanded part is removed from the mould.

Advantages :

1. Comparatively simple moulds can be used to make rather large parts such as ice chests, water jugs, water float toys, shipping containers, and display figures.

2. Expanded-bead products have an excellent strength-to-weight ratio and provide good insulation and shock-absorbing qualities.

5.2.7. Rotomoulding

In this process, the product is formed inside a closed mould that is rotated about two axes as heat is applied.

— Liquid or powdered thermoplastic or thermosetting plastic is poured into the mould, either manually or automatically. The mould halves are then clamped shut.

— The loaded mould is then rolled into an oven where it spins on both axes. Heat causes the powdered materials to become semi liquid or the liquid materials to gel (Permissible temperature ranges are considerably greater than for injection moulding). As the mould rotates, the material is distributed on mould-cavity walls solely by gravitational force; centrifugal force is not used.

— After the parts have been properly formed, the moulds are cooled by a combination of cold-water spray, forced cold air, and/or cool liquid circulating inside the mould.

Advantages :

1. Rotomoulding can be used to make parts in sizes and shapes that would be difficult by any other process.

2. Rotational moulds are relatively inexpensive when compared to moulds for injection and blow moulding.

3. In short runs, rotomoulding is generally less expensive than most other moulding processes.

4. Changing colours during a production run is comparatively easy, whereas in injection and blow moulding it is costly and time consuming.

Limitations :

1. The part must be hollow and typical tolerances are ± 5%.

2. The cycle time is slow because of the loading and cumbersome part removal unless the process is highly automated.

3. It is difficult to get varying wall thicknesses within a part, so parts are often designed with reinforcing ribs to stiffen flat areas.

4. The range of mouldable materials is not as broad as it is for other moulding processes.

5.2.8. Blow Moulding

This process is used for producing narrow neck plastic containers, like bottles and similar other articles.

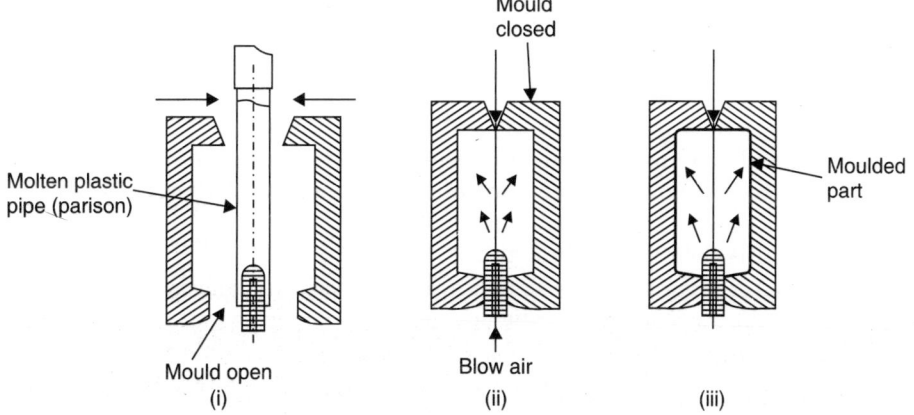

Fig. 5.5. Steps in blow moulding.

In blow moulding (Fig. 5.5), a hot tube of plastic material is placed between two halves of the mould. The mould is closed and air or non-reactive gas like argon is blown under of pressure of 20–40 MPa which expands the hot tube outwards to fill the mould cavity. The mould is then opened and the product, hollow in shape, is removed from the mould.

- It is used for fast and cheap production.
- This can be done manually or by automatic and semi-automatic machines.

Advantages :

1. High production rate.

2. Hollow shapes of various sizes with thin walls can be produced.

3. Low cost for making hollow shapes.

Limitation :

Limited to production of hollow shapes.

5.2.9. Extrusion

It is also known as *extrusion moulding*. Practically all thermoplastic materials can be extruded into various shapes like tubes, rods, sheets, films, pipes, ropes and other profiles. Thermosetting plastics, generally, are not suitable for extrusion.

Extrusions are produced by forcing a material through the opening of a die to produce the desired shape.

Common types of extruders are :

(*i*) Single screw; (*ii*) Cast film; (*iii*) Pipe and profile.

Fig. 5.6 shows a typical *single screw extruder*.

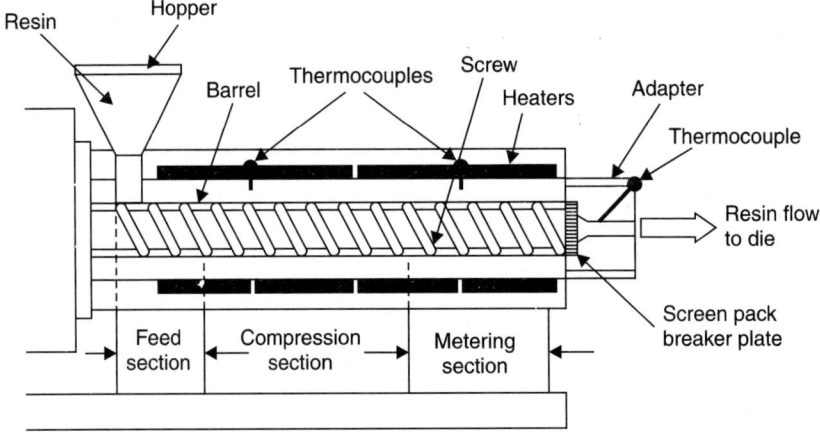

Fig. 5.6. Screw extruder.

— The powdered raw material, usually thermoplastic, is fed into a hopper and carried along by a screw conveyor through the heating chamber, where it becomes a viscous fluid.

— The material is then forced through the heated die.

— The material leaving the die rests on a moving conveyor and is cooled by air or water spray to retain the shape of the die opening.

— Rigid plastic materials are cut to desired lengths while flexible plastics are coiled.

● Owing to continuous nature of the process, it is extensively used for producing long product of uniform cross-section such as rods, tubes and channels from thermoplastics. Thermosetting plastics, generally, are not suitable for extrusion.

● Plastic coated wires and cables are also produced using this technique. For this purpose, the wire is fed into the die opening at a controlled rate.

Advantages :

1. Relatively inexpensive tooling.

2. High production rate.

Limitations :
1. Suitable for forming of long uniform sections.
2. High production volume is required.
3. Dimensions accuracy not very good.

5.2.10. Thermoforming

"Thermoforming" refers to *heating a sheet of plastic material until it becomes soft and pliable and then forming it either by vacuum, by air pressure, or between matching mould halves.* This process is depicted schematically in Fig. 5.7.

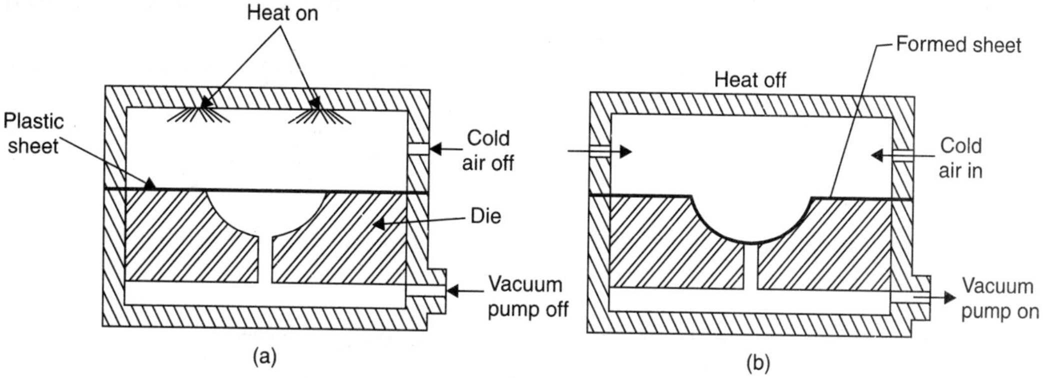

Fig. 5.7. Thermoforming.

— A sheet of thermoplastic material is placed over a die and heated until it becomes soft.
— A vacuum is then created inside the die cavity which draws down the heated plastic sheet into the shape of the die.
— The material is then cooled, the vacuum is released and the final product is taken out.

Heating systems used include quartz lamps, small ceramic modules, gas-fired ovens, emitter strip panels, and calrod-type resistance heaters. More recent developments include a microprocessor with computer interfacing.

Advantages :
1. Relatively inexpensive tooling.
2. Shallow and cheap cavities can be produced.
3. High production rate.
4. Quite economical.

Limitation :

Dimensional accuracy not so good.

Applications :

The process is used in making of liners, panels, housings, advertising signs, coffee-cream containers, meat trays, egg cartons, etc.

5.2.11. Calendering

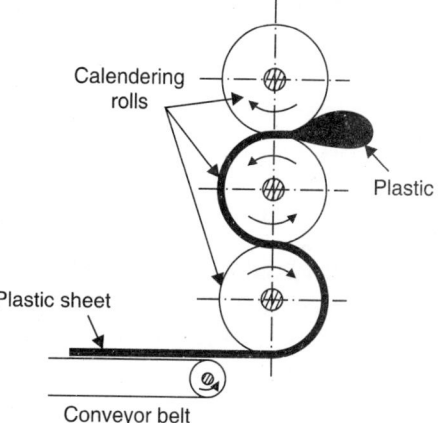

Fig. 5.8. Calendering.

"Calendering" is *the production of sheet of materials by rolling the plastics between multiple rollers.* This process can be used only for thermoplastics and not for thermosetting plastics.

In this process a heated doughy paste of plastic compound is passed through a series of hot rollers, where it is squeezed into the form of thin sheet of uniform thickness (see Fig. 5.8).

● This process is widely used for making plastic *films and sheets.*

5.2.12. Casting

This process is similar to the metal casting process, and involves only heating and no pressure is used.

Usually liquid resins, such as phenolics, polyesters, silicones, acrylics and expoxies, and also some cellulose derivatives, which can be considerably softened for pouring, are used for casting.

The casting process is carried out as follows :

— The plastics are heated to a suitable temperature and poured into the moulds (Moulds are made of lead, rubber, glass, plaster, wood or metal.

— A *catalyst* is added to unpolymerised plastics to help polymerisation.

— After due curing and hardening, the moulds are opened and castings knocked out.

— A few examples of such cast products include : sheets, films, jewellery, dies, jigs; punches.

In order to obtain different colour shades two or more plastics can be poured together in the same mould.

● The process is cheap, but slow; it is therefore quite suitable for small production.

● Plastics of commercial significance that can be cast are nylon 6, epoxies, and acrylics. The basic cast nylon formulation is used for heavy-duty applications requiring high strength, resilience, and wear resistance and where lubrication is difficult if not impossible.

Advantages :

1. Accurately contoured product, such as the gears, that can be made several centimeters thick and weighing upto about 200 kg, yet free of voids.

Cast nylon weighs about $\frac{1}{7}$ th the weight of steel and has excellent sound and vibration damping qualities.

2. Owing to the slower cooling cycle in the casting process than in moulding, surfaces are harder (more crystalline) and are more wear resistant.

3. Mould costs for casting are quite low compared to those for injection moulding.

4. Very economical.

Limitation :

Low production race.

5.3. MACHINING OF PLASTICS

Two type of plastics are being used for fabrication. These are plain plastics in cast conditions or laminated plastics. Plastics present such problems as *heating and edge fretting*. During machining material is removed in fine particles and in some case gases may also evolve in chips and gases are harmful and hence must be removed immediately. Plastics may be subjected to aimost all machining operations with the difference in cutting speed, feed, tool parameters, etc. Brief suggestions are given below :

1. Filing : Both plastics can be filed and surface finishes of Rough (V) and Fine (VV) types are obtainable.

● For rough filing use rasp or bastard file.

● For fine filing use smooth or dead smooth file.

2. Sawing : Both plastics can be cut by a circular saw using cutting speed of 50 m/min and hand feed. Pitch of the saw should be close to 10 mm. However, the thickness of saw will vary with thickness of plastic to be cut.

● For thickness 0.5 to 4 mm use *3 mm thick saw.*

● For thickness 5 to 8 mm use *4 mm thick saw.*

● For thickness 19 mm and above use *5 mm thick saw.*

3. Drilling : Drilling a plastic may pose such problems as damaged hole periphery on surfaces where drill enters and exists and also sticking of drill in the hole.

By using sharp cutting edges and by carefully setting them on the surface before starting the operation, by putting a wooden support directly below the hole where drill will exist, and by frequently lifting the drill from hole the above problems are overcome. The drill may also be dipped in mineral oil for obtaining a smooth surface finish of hole.

- For plain plastic use : point angle–90° to 116°.
- However, if section is thin this angle may be 50°
    : flute angle—60°
    : cutting speed—70 m/min
    : feed—0.2 to 0.3 mm/rev.
- For laminated plastics : use cutting speed 60 to 90 m/min and feed 0.2 to 0.5 mm/rev.
- However, drilling may be done parallel or perpendicular to laminated plastics. If parallel use point angle of 70° and rake angle of 0–20°.

If perpendicular use point angle 100–110°, rake angle 10 to 12°.

4. Planing : For both plain and laminated plastics use well sharpened and lapped carbide tipped tool at cutting speed of 10 to 20 m/min and feed of 0.2 to 0.8 mm/stroke.

5. Turning : Carefully the tool must be set at the centre of workpiece, and the tool tip must be rounded to a radius of 1.5 to 2 mm.

- *For rough surface (∇) :*

	Plain plastic	Laminated plastic
Cutting speed :	80 to 200 m/min	100 to 200 m/min
Feed r :	0.3 to 0.5 mm/rev	0.2 to 0.5 mm/rev
Depth of cut :	3 to 5 mm	3 to 5 mm

- *For fine surface (∇∇) :*

Cutting speed :	200 to 300 m/min	200 to 300 m/min
Feed :	0.1 to 0.3 mm/rev	0.1 to 0.3 mm/rev
Depth of cut :	2 mm	2 mm

For both materials use clearance angle of 3° and rake angle of 15°. Chips should be removed by suction.

6. Milling : Number of teeth on milling cutter must be 1/2 to 2/3 of the number of a corresponding steel job. Clearance angle of 20° and rake angle of 20 to 25° is recommended on carbide tipped tool. While feed for both types of plastic as 0.5 to 0.8 mm/rev is recommended the plain plastic should be cut with a cutting speed of 300 m/min and laminated plastic with a cutting speed of 120 to 250 m/min.

7. Grinding : Grinding of plastics is possible with belt, sand paper and grinding wheels. For rough grinding grit 20 wheels and for finish grinding grit 60 silicon carbide wheels are used. For belt grinding V = 6 m/min, sand paper V = 25 m/min.

For surface grinding water may be used as coolant.

8. Threading : Tapping, turning or milling can be used. In both turning and milling use HSS cutter with cutting speed 20 to 40 m/min.

Milling will result in better surface.

When using taps use wax and grease to lubricate with taps of wide flute.

Laminated plastics can be sheared upto thickness of 3 mm. However, if thickness greater than 3 mm is to be sheared the material is recommended to be warmed.

Note : The cutting tool angles for machining plastics are made somewhat different than those of tools for ferrous and non-ferrous metals. The rake angles are positive and relatively larger. Because of the visco-elastic behaviour of thermoplastics, some of the local elastic deformation is regained when the load is off. Therefore tools must be made with large relief angles (20° to 30°).

Laser machining of plastics :

Recent advances in laser systems have significantly enhanced the capabilities to machine plastics to tolerances normally associated with conventional machining and stamping of plastics.

- The laser machining process is characterised by a small heat-affected zone along the narrow kerf. The width of the transformed zone is dependent on the physical properties of the material. High cutting speeds can be used (*e.g.*, 12.7 mm thick acetal polymer can be cut at 508 mm/min. without crazing, cracking, or degradation of the edge).

5.4. JOINING OF THERMOPLASTICS

Thermoplastics can be *joined* by the following methods :

1. Fusion bonding.
2. Solvent bonding.
3. Friction welding.
4. Induction welding.
5. Ultrasonic welding.
6. Vibration welding.

1. **Fusion bonding.** Conventional fusion bonding is carried out as follows :

— The parts are placed in fixtures and held by chucks, gripping fingers, vacuum cups, or other mechanical devices.

— The fixtures then press the parts against a heating platen that is usually made of aluminium, aluminium-copper, or hot-rolled steel coated with Teflon or Teflon-impregnated fiber glass.

— The platen is heated to about 260°C to soften the plastic to a predetermined depth.

— Holding fixtures are then opened and the heating platen removed ; the fixtures are closed again until the melted material is cooled and fused.

2. **Solvent bonding.** This is the most widely used plastics-joining process.

— The solvent may be applied to the joint by *dipping, spraying, and brushing.*

— The most common method is to *hold the solvent-soaked pad against surfaces.* The edges dissolve, producing tacky surfaces that are then brought together, moved slightly to obtain mixing, and held in contact, in correct relationship, to dry and harden. Dry time ranges from a few minutes to hours.

3. **Friction welding.** Plastics can be friction welded in much the same way as metals. Mechanical energy is directly converted to heat on the surfaces being welded.

Limitations :

(*i*) One of the pieces must be a body of revolution and its section at the joint must be a circle or an annulus.

(*ii*) When pressure is applied to the pieces to complete the joint splash is formed at the connection.

4. **Induction welding.** Originally thermoplastics were induction welded by pressing two parts together around a metal insert. This combination was then subjected to an A.C. magnetic

field, which melted the surfaces of the plastics around the insert. Compression of the parts resulted in a fusion weld. Improved methods employ impregnated plastics with either metallic particles or tiny, permanent magnets.

Advantages :

(i) This method ensures a strong bond.

(ii) Several simple components can be joined to form complex shapes that would be expensive to mould.

(iii) Work coils need not touch the part and minimal compression is required to form the weld.

(iv) Control is easily maintained by adjusting the power supply.

(v) Parts with multiple bond lines can be welded in seconds.

(vi) Skilled personnel are not required.

(vii) Bonds can be hidden inside the part. Hemetic seals are possible.

(viii) Heat damage and distortion are reduced since they are localised to the bond. Parts need not be cooled but can be handled immediately.

(ix) Both stiff and soft materials can be joined, and the bonds can be flexible or rigid.

Limitations :

(i) In bonds between metals and plastics, the substrate may overheat.

(ii) A bonding agent is required; thus spin or vibration welding, which fuse at the bond line, may be more cost effective.

(iii) The bonding agent is opaque, so transparent seals are not possible.

(iv) Strong welds are sometimes difficult to make in reinforced thermoplastics because less plastic material is available at the joint, and direct application of heat may disrupt the fiber orientation of the material, reducing strength.

5. **Ultrasonic welding.** This type of welding requires the conversion of low-frequency electrical power to high-frequency mechanical vibration that are applied to thermoplastic components. When the plastic melts, the components are joined in a fusion bond.

- Ultrasonic welding is applicable to acrylics, PVC, polystyrene and synthetic textile.

- Lap and tee spot joints are made best of all. Satisfactory joints are also made in the case of lap welds in static jigs. In all cases, neither edge preparation nor filler material is needed.

- Ultrasonic welds can also be made in dissimilar plastics.

- The process can be used to *fabricate* delicate parts such as *electrical switches and relays.*

6. **Vibration welding.** In this method frictional heat is generated by clamping together two plastic parts under pressure and vibrating one part at a set amplitude and frequency.

5.5. PLASTIC DESIGN RULES

The following important *rules* need be followed for designing plastic moulded parts :

1. Provide due allowance for shrinkage after moulding.

2. Try to locate the parting surface of the mould in one plane, as far as possible.

3. Allow at least a minimum draft of 0.5° to 1° for easy withdrawal of the parts from the mould.

4. Avoid undercuts wherever possible.

5. As far as possible, long cored holes should be avoided.

6. Back tapered features in parts should be avoided since they require complicated moulds and improper moulding conditions.

7. In case any lettering or other details are to be produced on the surface, they should be located on a surface at right angles to the direction of opening or closing of mould.

8. Ribs should be provided to increase strength and rigidity and to reduce distortion.

9. Try to design projections to have circular cross-sections, as far as possible.

10. Thick sections should be preferably kept as nearly uniform in thickness as possible.

QUESTIONS WITH ANSWERS

Q. 5.1. What is "slush moulding" ? Explain briefly.

Ans. Slush moulding resembles casting in that no pressure is applied.

The process consists of preparing a slurry of thermoplastic resin and then pouring the same into a preheated mould. On account of the heat of the mould a uniform layer of resin sets all along the surface of the mould cavity. Excess slurry, if any, is then poured out. Additional heat is provided for curing the resin set in the mould, followed by hardening of the product by cutting and removing the same from the mould.

● Slush moulding is used to produce *flexible toys, artificial flowers,* etc.

Q. 5.2. Explain briefly the following methods of forming plastic sheets :

(*i*) **Pressure forming ;**

(*ii*) **Draw forming ;**

(*iii*) **Plug and ring forming.**

Ans. (*i*) **Pressure forming.** In this process the heated plastic sheet is formed into the required shape (under pressure) between a pair of male and female die. No vacuum is used in this process.

(*ii*) **Draw forming.** This process is similar to the deep drawing process for metals.

Here, a heated blank of plastic sheet is placed over a die and held firmly by means of a hold down plate. A punch is pressed down into the die cavity to draw the material into the die and around its own body. Flange, if remaining, may be trimmed separately or within the same operation.

(*iii*) **Plug and ring forming.** This process is a combination of stretch and draw forming methods. The heated plastic sheet is clamped over a draw ring which forces the sheet on the form block. Finally the mould mounted at the bottom of a ram is forced into the clamped sheet to complete the forming to final shape.

Q. 5.3. What do you understand by "Rigidized vacuum forming" ? Explain briefly.

Ans. *"Rigidized vacuum forming" is an intermediate process between basic hand layup and costly matched die moulding.*

A part is first made by thermoforming and then it is reinforced and rigidized by glass reinforcement. The thermoformed parts are placed on a holding fixture to ensure that during the curing of the reinforcing laminate the parts will not change dimensionally.

This process is employed to produce :

— Bathroom tubs and sinks and shower stalls;

— Automotive body and recreational vertical panels;

— Motor shrouds;

— Other items where production and uniformity are desired.

Q. 5.4. Explain briefly "Pultrusion".

Ans.

- Pultrusion, as the name implies, *is like pulling an extrusion.* Constant cross-sectional shapes are produced by pulling resin-impregnated, reinforcing material through a heated steel shaping die in such a way as to promote adequate polymerisation of the resin.

- Pultrusion consists of pulling continuous-strand woven roving, surfacing mat, woven fabrics, and reinforcing mat through a resin bath to wet out the fibers and then through a heated steel die. The entrance section of the die is water cooled to prevent premature curing or binding. The pultrusion moves through the die at speeds ranging from 50 to 1540 mm/min, depending on the cure time, which in turn depends on thickness of the section.

- Pultrusion is used mainly with thermosetting plastics, but it may also be used with thermoplastics.

- The main use of pultrusion has been for electrical, recreational, and construction purposes and where corrosion resistance is necessary.

Q. 5.5. What do you understand by "Filament winding" ? Explain briefly.

Ans. *"Filament winding" is a process of wrapping reinforcements consisting of resin-impregnated, tensioned, continuous filaments around a form called a mandrel.* After the resin mix is cured, the mandrel may be removed, leaving a hollow, monolithic shell.

The most common filament used is a continuous glass filament, but graphite is also used. The matrix resin is generally epoxy or polyester.

- The outstanding property of filament-wound structures is their *high strength-to-weight ratio,* which is better than that of steel or even titanium.

Applications :

- Rail road freight car;
- Solid rocket cases;
- Metal-lined compressed gas cylinder, etc.

Q. 5.6. Discuss briefly "Solid phase forming".

Ans. The *forging of plastic material* (solid phase forming) is *relatively a new process.* It was developed to shape materials that are difficult or *impossible to mould* and is used as a *low-cost solution for small production runs.*

The process is carried out as follows :

- The forging operations start with a blank or billet of the required shape and volume for the finished part. The blank is heated to a preselected temperature and transferred to the forging dies, which are closed to deform the work material and fill the die cavity.

- The dies are kept in the closed position for a definite period of time, usually 15 to 60 s.

- When the dies are opened, the finished forging is removed.

The process is applicable to thermoplastics, since forging involves deformation of the work material in heated and softened condition.

As yet the number of materials used for forging is limited. Materials used commercially are polypropylene, high-density polyethylene, and ultra high molecular weight polyethylene.

- The forging process is generally suited to small-quantity production, but it can be automated for high production rates.

Advantages :

1. Capability of producing thick parts with relatively abrupt changes in section.

2. Only simple tooling is required. The tooling pressure required less than 7 MPa as compared to 55 to 205 MPa for injection moulding. Forging can be done on a small hydraulic press.

3. Parts that are deformed significantly in the forging process have improved mechanically properties due to molecular attraction.

Note : Besides forging and injection moulding, machining is the only process that is capable of producing complex, solid parts. Since no special tooling is involved, machining is more cost effective when the quantity is on the order of a few hundred; however, machining produces scrap that may not be reusable. Therefore, with an increase in production, volume forging is favoured.

Q. 5.7. Explain briefly "Testing of plastics".

Ans. Plastics are tested for electrical and mechanical properties. The various characteristics measured are defined below :

1. Insulation resistance after immersion in water :

- A specimen immersed in distilled water for 24 hours is cleaned and dried.
- The specimen is then placed between pairs of electrodes and its resistance is measured at a potential difference of 500 V D.C., the temperature and humidity being under control.

2. Dielectric strength :

- This property measures the electrical breakdown through the material.
- Dielectric strength is measured as *voltage gradient as volts/mm to cause electrical breakdown.*

3. Vieat softening point :

It is the temperature at which a penetrator of circular section of 1 mm diameter will penetrate a specimen of thermoplastic to a depth of 1 mm under specified load and uniform rate of temperature rise.

4. Melt flow index :

- Plastic is extruded through a die by a loaded piston under controlled conditions of temperature.

Number of grams extruded per 10 minutes is determined as melt flow index.

5. Cross breaking strength :

Refer to Fig. 5.9.

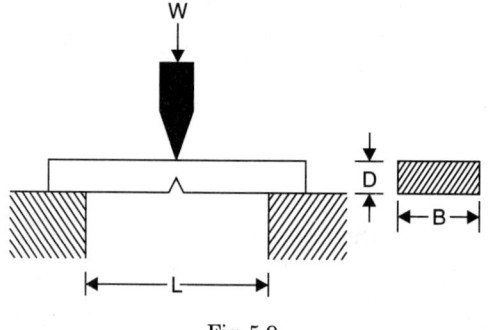

Fig. 5.9

- A plastic specimen of rectangular section (B = width, D = thickness) is supported like a simply supported beam over a span L and a load (W) in the centre of the span applied to cause fracture.

- The cross breaking strength is measured as $\dfrac{1.5WL}{BD^2}$.

6. Impact strength :

Refer to Fig. 5.10.

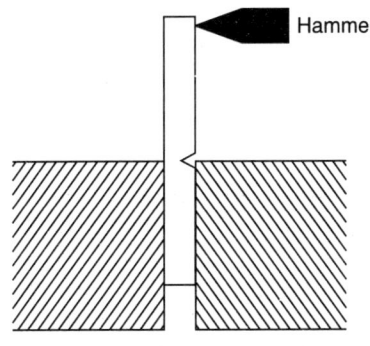

Fig. 5.10

- Izod impact test in which a matched specimen of specified dimension $\left(\frac{1}{88}'' \text{ thick} \times \frac{1}{2}'' \text{ wide} \times 2\frac{1}{2}'' \text{ long}\right)$ held like a cantilever is struck by a swinging hammer at free end.

- The energy absorbed in fracture is the measure of impact strength.

- The impact strength is measured in kgf-cm or Nm. This figure is calculated on the basis of 1 m match length.

Q. 5.8. Explain briefly the following :

1. Polymers   2. Polymerisation   3. Depolymerisation.

Ans. 1. Polymers. *Polymers are organic materials having carbon as the common element in their make-up.*

- These are composed of a large number of *repeating* units (small molecules) called **monomers.** A polymer is, therefore, made up of thousands of monomers joined together to form a large molecule of colloidal dimension, called *"macromolecule".*

- The unique feature of a polymer is that each molecule is either a *long chain or network of repeating units all covalently bonded* together. In some cases, molecules are hold together by *secondary bonds.*

- The size of a *molecule is determined by dividing the molecular weight by the mer weight.* The number is called *degree of polymerisation,* DP.

$$DP = \frac{\text{Molecular weight}}{\text{Mer weight}}$$

At DP values of above 10 to 20 mers per molecule, the substance formed is light oil. As the DP *increases,* the substance becomes *greasy,* then *waxy,* and finally at a value of D.P. of about 1000 the substance becomes a solid and is then, a true polymer. Natural DP is almost unlimited—it may increase to around 100,300 or so on.

2. Polymerisation. *The process of linking together of monomers is called* **polymerisation.** The need to start with the process of polymerisation lies on the necessity of breaking the double bond ($C = C$) of the monomers. This requires considerable energy.

Polymerisation mechanisms may be of the following two types :

(*i*) Addition polymerisation.

(*ii*) Condensation polymerisation.

(i) Addition polymerisation :

This polymerisation process is of simplest form. When a large number of simple molecules are chemically added together to increase the average molecule size without wastage, process of addition polymerisation takes place. Such a polymerisation takes place by three steps namely. (i) *Initiation,* (ii) *Chain propagation, and* (iii) *Termination.*

Example : Addition polymerisation of ethylene (Fig. 5.11).

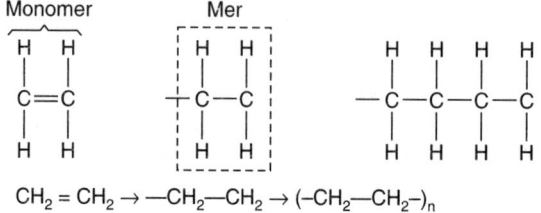

$$CH_2 = CH_2 \rightarrow -CH_2-CH_2 \rightarrow (-CH_2-CH_2-)_n$$

Fig. 5.11. Addition polymerisation.

Once polymerisation process is initiated, it does not continue indefinitely since it is impossible to link all the monomers in a plastic one long continuous chain. The polymerisation is terminated by a collosion between the active ends of two chains or by addition of a terminator, such as free radicals from the catalyst.

Copolymerisation : It is another type of addition polymerisation.

Copolymerisation is the addition polymerisation of two or more chemically different monomers. Many monomers will not polymerise with themselves but will copolymerise with other compounds.

Copolymerisation of vinyl chloride and vinyl acetate is shown in Fig. 5.12. This is comparable to a solid solution in metallic ceramic crystals.

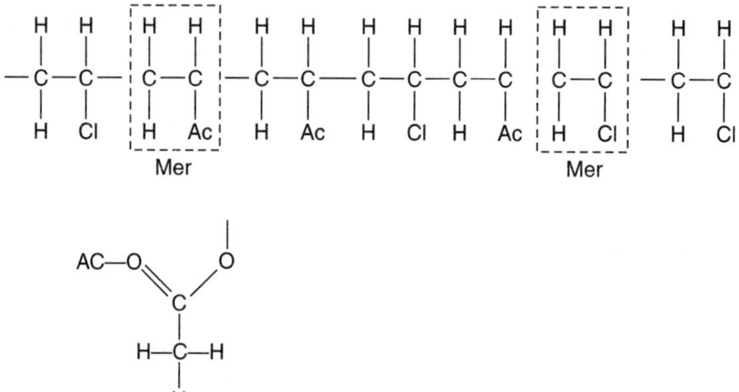

Fig. 5.12. Copolymerisation.

Copolymerisation has been applied extensively, in the *artificial rubbers.*

Copolymers may take several forms, *alternating* (as described), *block and graft* types.

Block polymers. Refer to Fig. 5.13 (a), (b). They contain relatively long chains of a particular chemical composition separated either by a chain of different composition (a) or a low molecular weight composition (b).

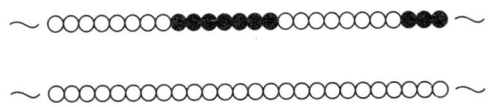

Fig. 5.13. (a), (b) Block polymers.

Graft polymers. The graft polymers consist of a main stem of one polymer with monomer stems off the sides. Grafting can be accomplished by the ionizing radiation that produces free radicals which initiate polymerisation of branches on the main chains (see Fig. 5.14).

Fig. 5.14. Graft polymers.

(ii) Condensation polymerisation :

"Condensation polymerisation" is defined as the process of linking together of unlike monomers accompanied by splitting off a small molecule. This process usually requires a **catalyst.**

In comparison to *addition* reaction in which a simple molecular summation occurs, *condensation* reactions result in *splitting out* of simple monopolymerisable molecules, *e.g.,* water which are considered to be *by-products* of the process. Thus, when a phenol and formaldehyde monomers are polymerised, water is released, and the resulting product is polymerised phenol formaldehyde, more commonly known as *Bakelite* (Fig. 5.15).

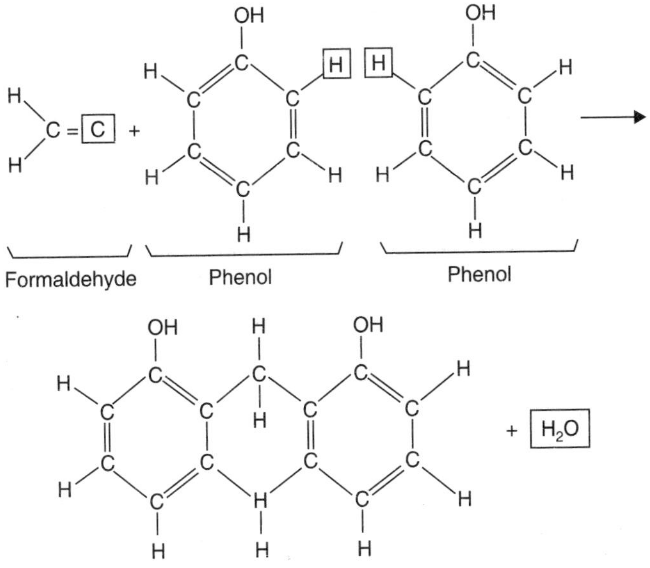

Fig. 5.15. Condensation polymerisation.

Comparison between Addition and Condensation Polymerisation

S. No.	Addition polymerisation	Condensation polymerisation
1.	It requires molecules which are unsaturated.	It requires two unlike molecules.
2.	It does not yield a by-product.	It yields a by-product.
3.	Reaction is very fast and may take 10^{-2} to 10^{-6} sec.	Reaction normally takes hours and days to complete.
4.	It is kinetic chain reaction.	It involves inter-molecular reaction.
5.	Polymer formed is a thermoplastic type.	Polymer formed is thermosetting type.

3. Depolymerisation :

- During polymerisation process the reaction proceeds *only in one direction under controlled conditions.* A series of reactions may occur which cause depolymerisation.

- A *reversal of the polymerisation reaction is known as* **depolymerisation.** Depolymerisation may also take place in plastics formed at high temperature due to thermal vibrations which may disrupt the inter-molecular bonds within the molecules. This can be represented as below :

$$nR \leftarrow (-R-)\, n$$

Examples of depolymerisation :

(i) Depolymerisation may occur with the urea formaldehyde plastic if it is used for extended periods of time with steam.

(ii) Depolymerisation may also occur in any plastic being formed at high temperatures, since thermal vibrations may disrupt the inter-molecular bonds within the molecules.

It may be noted that depolymerisation is *not* always harmful, it is useful as well.

Examples :

- Depolymerisation is commercially used for cracking petroleum into more combustible, light molecules.

- Charring of carbohydrates (toast) and cellulose (charcoal) are similar examples of depolymerisation.

Q. 5.9. What are laminated plastics ? Explain briefly.

Ans. Laminated plastics are also called *plastic laminates* and are formed by impregnating sheets of fibrous materials such as paper, linen, canvas or silk with a synthetic resin and then compressing the sheets together with application of heat. The synthetic resins may be phenolic resin, urea formaldehyde or a vinyl resin. The resin is usually dissolved in alcohol.

The material in roll form is immersed in the resin solution at atmospheric pressure at room temperature and then run through a drier at 150°C. The rolls are next cut into sheets of given size, which are arranged into stacks. These stacks finally are compressed in a hydraulic press at about 170°C under a pressure of 200 bar. The sheets are thus bonded to one another.

Properties. The laminated plastics have the following *properties* :

(i) Light and strong.

(ii) Machinable.

(iii) Resistant to wear, acids and alkalis.

(iv) Impervious to water and oil.

(v) Have a high dielectric constant.

Uses. The laminated plastics are used for :

(*i*) Electric insulation

(*ii*) Making *silent gears*

(*iii*) Water lubricated bearings

(*iv*) Pulley wheels

(*v*) Pump parts

(*vi*) Press tools

(*vii*) Decorative purposes in wall panelling, transluscent panelling and table and counter parts.

Q. 5.10. Explain briefly the following advanced polymeric materials.

(*i*) Ultrahigh molecular weight polyethylene (UHMWPE).

(*ii*) Liquid crystal polymers (LCPs).

Ans. **(*i*) Ultrahigh molecular weight polyethylene :**

It is a linear polyethylene and its molecular weight is extremely high (nearly ten times greater than that of high density polyethylene).

Properties :

(*i*) It is electrically insulating and possesses excellent dielectric properties.

(*ii*) It possesses very good chemical resistance.

(*iii*) It offers outstanding resistance to wear and abrasion.

(*iv*) Its low-temperature properties are excellent.

(*v*) Its impact resistance is extremely high.

(*vi*) It provides a self lubricating and non-stick surface.

(*vii*) It offers a very low coefficient of friction.

(*viii*) It assimilates outstanding damping and energy absorption characteristics.

(*ix*) Due to its relatively low melting temperature, its mechanical properties diminish rapidly with increasing temperature.

Uses :

(*i*) Blood filters

(*ii*) Marking pen nibs

(*iii*) Pump impellers

(*iv*) Valve gaskets

(*v*) Bullet proof vests

(*vi*) Golf ball cores

(*vii*) Ice skating rink surfaces.

(*ii*) Liquid crystal polymers :

● These polymers belong to a group of chemically complex and structurally distinct materials and possess unique properties.

● They are composed of extended, rod shaped, and rigid molecules.

● In the melt; unlike other polymers (which are randomly oriented) LCP molecules become *aligned in highly ordered configurations.*

● Some of the liquid crystal polymers are rigid solids at room temperature and exhibit the following properties/behaviour :

(*i*) High impact strengths, which are retained upon cooling to relatively low temperatures.

(*ii*) Excellent thermal stability.

(*iii*) Stiff and strong.

(*iv*) Chemically inert to wide variety of acids, solvents, etc.

(*v*) Inherent flame resistance.

(*vi*) Combustion products are relatively non-toxic.

Uses :
1. These polymers are mainly used in liquid crystal displays (LCDs), on digital watches, laptop computers and other digital displays.
2. Interconnect devices, relays and capacitor housings, etc.
3. Photocopiers and fibroptic components.

HIGHLIGHTS

1. *Plastics* are natural or synthetic resins, or their compounds which can be moulded, extruded, cast or used as films or castings.
2. Plastics are classified as '*Thermoplastic*' and '*Thermosetting*'.
3. The common plastic production processes are :
    (*i*) Compression moulding;   (*ii*) Transfer moulding;
    (*iii*) Injection moulding;   (*iv*) Expandable bead moulding;
    (*v*) Roto moulding;   (*vi*) Blow moulding;
    (*vii*) Extrusion   (*viii*) Thermoforming;
    (*ix*) Calendering   (*x*) Casting.
4. Plastics are *joined* by the following methods :
    (*i*) Fusion bonding;   (*ii*) Solvent bonding;
    (*iii*) Friction welding;   (*iv*) Induction welding;
    (*v*) Ultrasonic welding;   (*vi*) Vibration welding.

OBJECTIVE TYPE QUESTIONS

"Fill in the Blanks" or say "yes or no" :
1. A ...... is defined as any non-metallic material that can be moulded to shape.
2. The plastics which soften on the application of heat with or without pressure, but require cooling to set them to shape are called thermosetting materials.
3. ...... materials are highly plastic and are easy for moulding and shaping.
4. Thermoplastic materials have a resale value.
5. The plastics which require heat and pressure to mould them into shape are called ...... materials.
6. In general, resins formed by condensation are thermosetting.
7. ...... resins have three-dimensional molecular structure and have very high molecular weights.
8. Thermosetting materials can be recycled.
9. Thermoplastic resins are more stronger and harder than thermosetting resins.
10. Plastics have comparatively high thermal conductivity.
11. Plastics are good electrical ......
12. Plastics are combustible because of presence of carbon.
13. Fillers are added to reduce the cost and enhance the strength end hardness of plastics.
14. Low molecular weight (approx. 300) materials blended with polymers are called ......
15. ...... moulding process, owing to non-uniform pressure, often results in turbulence and uneven flow of liquid plastic in the mould.
16. Cold moulding method is not suitable where close tolerance and good surface finish is required.
17. One of the limitations of transfer moulding process is the loss of material as scrap.
18. The greatest quantity of plastic parts are made by ...... moulding.
19. Expandable-bead products have an excellent strength-to-weight ratio and provide good insulation and shock-absorbing qualities.
20. ...... moulding process is used for producing narrow neck plastic containers, like bottles and similar other articles.

21. ......refers to heating a sheet of plastic material until it becomes soft and pliable and then forming it either by vacuum, by air pressure, or between matching mould halves.
22. ...... is the production of sheet of materials by rolling the plastics between multiple rollers.
23. Solvent welding is the least used plastic-joining process.

ANSWERS

1. plastic	2. No	3. Thermoplastic	4. Yes	5. Thermosetting
6. Yes	7. Thermosetting	8. No	9. No	10. No
11. insulators	12. Yes	13. Yes	14. plasticizers	15. Compression
16. Yes	17. Yes	18. injection	19. Yes	20. Blow
21. Thermoforming	22. Calandering	23. No.		

THEORETICAL QUESTIONS

1. Define the term 'Plastic'.
2. How are plastics classified ?
3. What are thermoplastic materials ? How do they differ from thermosetting materials ?
4. List six important thermoplastic materials.
5. What are thermosetting materials ? What are their characteristics ?
6. Name three important thermosetting materials.
7. Compare thermosetting materials with thermoplastic materials.
8. What are the characteristics/properties of plastics ?
9. What are compounding materials ? Explain briefly.
10. What are "fibre glass rainforced plastics" ? State their properties and applications.
11. Name the methods/processes which are commonly used for processing of plastics.
12. Explain briefly any two of the following plastic processing methods.
   (i) Compression moulding;   (ii) Transfer moulding;
   (iii) Injection moulding ;   (iv) Expandable bead moulding.
13. What is 'Roto moulding' ? State its advantages and limitations.
14. What is 'Blow moulding' ? Explain briefly.
15. What is 'Thermoforming' ? What are its advantages and limitations ?
16. Explain briefly the following plastic production processes :
   (i) Calendering;   (ii) Casting.
17. Discuss briefly "Machining of plastics".
18. Write short note on "Laser machining of plastics".
19. Explain briefly any three of the following methods of joining thermoplastics :
   (i) Fusion bonding;   (ii) Solvent bonding;
   (iii) Friction welding;   (iv) Induction welding;
   (iv) Ultrasonic welding;   (vi) Vibration welding.
20. Mention some important rules which should be followed for designing plastic moulded parts.

6

Ceramic and Composite Materials—Their Structure, Properties and Processing

6.1. Introduction to ceramic materials. 6.2. Classification of ceramics. 6.3. Advantages of ceramic materials. 6.4. Applications of ceramics. 6.5. Properties of ceramic materials. 6.6. Structure of crystalline ceramics. 6.7. Silicate structures—types of silicate structures. 6.8. Polymorphism. 6.9. Glass—definition and structure—constituents of glass and their functions—properties of glass—glass furnaces—fabrication of glass—classification of glass—uses of glass—the glass industry in India. 610. Advanced ceramics. 6.11. Processing of ceramics—general aspects—shaping processes—casting—plastic forming—pressing—drying and firing—finishing operations. 6.12. Design considerations for ceramics. 6.13. Introduction to composite materials. 6.14. Classification—particle-reinforced composites—fibre-reinforced composites—structural composites. 6.15. Production of composite structures. **Questions with Answers**—Highligths—Objective Type Questions—Theoretical Questions.

I. CERAMIC MATERIALS

6.1. INTRODUCTION TO CERAMIC MATERIALS

Ceramic materials *are defined as those containing phases that are compounds of metallic and non-metallic elements.*

The *science of ceramics*, nearly as old as mankind, is the *processing of earthly materials by heat.* The crude cooking utensils of early man were the first application of the materials now used in jet engines and atomic reactors. All the early ceramic products were made from *clay* because the ware could be easily formed. It was then dried and fired to develop the permanent structure. Because the other ceramic materials lacking plasticity also have desirable properties, other methods of forming and processing have been developed. Other forming methods used for ceramic materials are injection moulding, sintering and hot pressing. In other cases the formed materials are allowed to harden on the job by the addition of water, as in case of cements.

6.2. CLASSIFICATION OF CERAMICS

A. Classification of Ceramic Materials. Ceramic materials are *classified* as follows :

1. Functional classification :

(*i*) Abrasives :	Alumina, carborundum
(*ii*) Pure oxide ceramics :	MgO, Al_2O_3, SiO_2
(*iii*) Fire-clay products :	Bricks, tiles, porcelain, etc.
(*iv*) Inorganic glasses :	Window glass, lead glass, etc.
(*v*) Cementing materials :	Portland cement, lime, etc.
(*vi*) Rocks :	Granites, sandstone, etc.

(*vii*) Minerals :   Quartz, calcite, etc.

(*viii*) Refractories :   Silica bricks, magnesite, etc.

2. Structural classification :

(*i*) *Crystalline ceramics* : Single-phase like MgO or multi-phase from the MgO to Al_2O_3 binary system.

(*ii*) *Non-crystalline ceramics* : Natural and synthetic inorganic glasses.

(*iii*) *"Glass-bonded" ceramics* : Fire clay products-crystalline phases are held in glassy matrix.

(*iv*) *Cements* : Crystalline or crystalline and non-crystalline phases.

B. Classification of Ceramic Products. A general classification of 'ceramic products' is difficult to make because of the great versatility of these materials, but the following list includes the major groups :

1. Whitewares.	2. Bricks and tiles.
3. Chemical stonewares.	4. Cements and concretes.
5. Abrasives.	6. Glass.
7. Insulators.	8. Porcelain enamel.
9. Refractories.	10. Electrical porcelain.
11. Mineral ores.	12. Slags and fluxes.

● Ceramics are generally divided into the following *two categories* :

(*i*) *Traditional ceramics :*

   — Whiteware ;

   — Tiles ;

   — Bricks ;

   — Pottery ;

   — Abrasive wheels ;

(*ii*) *Engineering or high tech. ceramics :*

   — Heat exchangers

   — Cutting tool

   — Semiconductors

   — Prosthetics.

6.3. ADVANTAGES OF CERAMIC MATERIALS

The ceramic materials entail the following *advantages* :

1. The ceramics are hard, strong and dense.

2. They have high resistance to the action of chemicals and to the weathering.

3. Possess a high compression strength compared with tension.

4. They have high fusion points.

5. They offer excellent dielectric properties.

6. They are good thermal insulators.

7. They are resistant to high temperature creep.

8. Availability is good.

9. Good sanitation.

10. Better economy.

6.4. APPLICATIONS OF CERAMICS

The *applications* of ceramics are listed below :

1. The **whitewares (older ceramics)** are largely used as/in :
- Tiles ;
- Sanitary wares ;
- Low and high voltage insulators ;
- High frequency applications ;
- *Chemical industry*—as crucibles, jars and components of chemical reactors ;
- *Heat resistant applications*—as parameters, burners, burner tips, and radiant heater supports.

2. **Newer ceramics** (*e.g.*, borides, carbides, nitrides, single oxides, mixed oxides, silicates, metalloid and intermetallic compounds) which have the *high hardness values* and *heat and oxidation values* are largely used as/in :
- *Refractories* for industrial furnaces.
- *Electrical and electronic industries*—as insulators, semiconductors, dielectrics, ferro-electric crystals, piezo-electric crystals, glass, porcelain alumina, quartz and mica, etc.
- *Nuclear applications*—as fuel elements, fuel containers, moderators, control rods and structural parts. Ceramics such as UO_2, UC, UC_2 are employed for all these purposes.
- *Ceramic metal cutting tools*—made from glass free Al_2O_3.
- *Optical applications*—Ytralox, a comparative newcomer in the ceramic material field, is useful since it is as transparent as window glass and can resist very high temperature.

3. **Advanced ceramics** (*e.g.*, SiC, Si_3N_4, ZrO_2, B_4C, SiC, TiB_2, etc.)

The advanced ceramics are utilized as/in :
- Internal combustion engines and turbines, as armour plate ;
- Electronic packaging ;
- Cutting tools ;
- Energy conversion, storage and generation.

6.5. PROPERTIES OF CERAMIC MATERIALS

1. **Mechanical Properties.** The ceramic materials possess the following *mechanical properties* :

(*i*) The compressive strength is several times more than the tensile strength.

(*ii*) *Non-ductile/brittle*. Stress concentration has little or no effect on compressive strength.

(*iii*) The ceramic materials possess ionic and covalent bonds which impart high modulus of elasticity. The modulus decreases with increase in temperature (due to increase in interatomic distance at elevated temperature).

(*iv*) As compared to pure metals, more force is required to cause slip in diatomic ceramic materials, because diatomic material consists of a mixture of positively and negatively charged ions which have strong forces of attraction between them.

(*v*) Below recrystallisation temperature, non-crystalline ceramics are fully brittle. The cleavage failure occurs along crystallographic planes and propagation of the crack takes place at high speed.

(*vi*) At high temperature rigidity is high.

(*vii*) In case of alloy consisting of two or more metals, each phase may have appreciable difference of coefficient of thermal expansion which generates stress. This stress may then cause the metal to fail.

·**2. Electrical Properties.** The electrical properties of ceramic products vary from the low loss, high frequency dielectrics to semiconductors. Electrical insulators fall into two general classifications, the classical electrical porcelain for both high and low tension service and the special bodies such as steatite, rutile, cordierite, high alumina, and clinoestatite for high frequency insulation.

Dielectric constant :

- *Dielectric constant* is the *ratio of the capacitance of a dielectric compared to the capacitance of air under the same conditions.*
- A low dielectric constant contributes to low power loss and low loss factor ; a high dielectric constant permits small physical size.
- The dielectric constant for electrical porcelain varies between 4.1 and 11.0. Some special bodies have reported values of several thousands.
- Porcelain has large positive temperature coefficient.
- Rutile bodies have large negative coefficients.
- By combining capacitor dielectrics having different temperature coefficients it is possible to reduce effect of the temperature change.

Dielectric strength :

- The *dielectric strength* of a material is defined as the *ability of a material to withstand electrical breakdown.*
- The specific values of dielectric strength vary from 100 V per mil for low-tension electrical porcelain to 500 V per mil for some special bodies.
- Rutile bodies show higher breakdown strength at higher frequencies.

Volume and surface resistivity :

- A *volume resistivity* of 10^6 ohms/cm^3 is considered the lower limit for an insulating material. At room temperature practically all ceramic materials exceed this lower limit. As the temperature of ceramic materials is raised, the volume resistivity decreases ; the volume resistivity of side-lime glasses decreases rapidly with temperature, whereas some special bodies are good insulators (above 10^6 ohms/cm^3) at 700°C. Crystallised alumina has a volume resistivity of 500 ohms/cm^3 at 1600°C.
- *Surface resistivity* for dry, clean surface is 10^{12} ohms/cm^2. At 98% humidity, the surface resistivity may be 10^{11} ohms/cm^2 for a glazed piece or 10^9 ohm/cm^2 for an unglazed piece. The presence of dissolved gases and other deposits also tends to decrease the surface resistivity of ceramic materials.

3. Thermal properties. Since the ceramic materials contain relatively few electrons, and ceramic phases are transparent to radiant type energy, their thermal properties differ amply from that of metals. The following are the most important thermal properties of ceramic materials (which vary from material to material and from condition to condition) :

(*i*) Thermal capacity

(*ii*) Thermal conductivity

(*iii*) Thermal shock resistance.

(*i*) Thermal capacity :

- The specific heats of fine clay bricks are 0.25 and 0.297 and 1000°C and 1400°C respectively.
- Carbon bricks possess specific heats of about 0.812 at 200°C and 0.412 at 1000°C.

(*ii*) **Thermal conductivity :**

● The ceramic materials possess a very low thermal conductivity since they do not have enough electrons (for bringing about thermal conductivity). The conduction of heat takes place by phonon conductivity and the interaction of lattice vibration, while at elevated temperatures conduction takes place by the transfer of radiant energy.

● The impurity content, porosity and temperature decrease the thermal conductivity.

● In order to have maximum thermal conductivity, it is imperative to have maximum density which most of the ceramic materials do not possess.

(*iii*) **Thermal shock :**

Thermal shock resistance is the *ability of a material to resist cracking or disintegration of the material under abrupt or sudden changes in temperature.*

Thermal shock is developed primarily because of thermal expansion or contraction, which is largely a function of internal structure particularly the inter-atomic bonding. Loosely packed structures can provide *internal expansion.* Thus, the coefficient of expansion is low.

● *Lithium compounds* are used in many ceramic compounds to *reduce thermal expansion* and to *provide excellent thermal shock resistance.*

● Common ceramic materials graded in order of decreasing thermal shock resistance are given below :

(*i*) Silicon nitride	(*vi*) Beryllia
(*ii*) Fused silica	(*vii*) Alumina
(*iii*) Cordierite	(*viii*) Porcelain
(*iv*) Zircon	(*ix*) Steatite.
(*v*) Silicon carbide	

4. Chemical, Optical and Nuclear Properties :

Chemical properties :

● Several ceramic products are highly resistant to all chemicals except hydrofluoric acid and to some extent, hot caustic solutions. They are not affected by the organic solvents.

● Oxidic ceramics are completely resistant to oxidation, even at very high temperatures.

● Ziconia, magnesia, alumina, graphite, etc. are resistant to certain molten metals and are thus employed for making crucibles and furnace linings.

● Where resistance to attack from acids, bases and salt solutions is required, ceramics like glass are employed.

Optical properties :

● Several types of glasses have been employed for the production of windows, subjected to high temperatures and optical lenses.

● Special glasses, in large number, have also been used for selective transmission or absorption of particular wavelengths such as infrared and ultraviolet.

Nuclear properties. As ceramics are refractory, chemically resistant and because different compositions offer a wide range of neutron capture and scatter characteristics, they are finding nuclear applications and are being used as : *Fuel elements, moderators, controls and shielding.*

6.6. STRUCTURE OF CRYSTALLINE CERAMICS

Most ceramic phases, like metals, have *crystalline structure.* Ceramic crystals are formed by either a pure ionic bond, a pure covalent bond or by bonds that possess the ionic as well as covalent characteristics.

- **Ionic bonds** *give ceramic materials of relatively high stability*. As a class, they have a much higher melting point, on the average, than do metals or organic materials. Generally speaking, they are also *harder and more resistant to chemical reaction.*
- **Covalent crystals** *usually also possess high hardness, high melting point and low electrical conductivity at room temperature.*

The ceramics crystals structures are, however, invariably more complex as compared to those of metals, since atoms of different sizes and electronic configurations are assembled together.

Common crystal structures found in crystalline ceramics particularly those of *oxide type* include the following :

    (*i*) Rock salt structure.

    (*ii*) Cerium chloride structure.

    (*iii*) Zinc blende structure.

    (*iv*) Wurzite structure.

    (*v*) Spinel structure.

    (*vi*) Fluorite structure.

    (*vii*) Ilmenite structure.

6.7. SILICATE STRUCTURES

The silicates are *co-ordinate structures based upon large anions arranged about small cations.* The dimensions of the lattice in general are controlled by the anions rather than cations because of the larger sizes of the former. Most important are the Si^{+4} and O^{-2} ions. In all silicates the basic unit is the SiO_4 *tetrahedron*. This appears to remain essentially unaltered regardless of the other materials present.

Silicates are important constituents of most of the ceramic materials since they are *plentiful, cheap and have certain distinct properties, necessary for certain engineering applications.*

- *Portland cement is the most widely known silicate.* It has the very advantage of forming a hydraulic bond.
- Certain other construction materials made of silicates are brick, tile, glass, various enamels, etc.
- Silicates are also used as *reinforcing glass fibres, chemical wares and electrical insulators*

6.7.1. Types of Silicate Structures

The various silicate structures are :

1. Silicon-oxygen tetrahedron $(SiO_4)^{4-}$ structure.

2. Double and poly-tetrahedral structures.

3. Chain structures.

4. Sheet structures.

5. Framework structures.

6. Vitreous structures.

1. Silicon-oxygen tetrahedron $(SiO_4)^{4-}$ structure : Refer to Fig. 6.1.

In this structure (primary structural unit of silicates) one silicon atom fits interstitially among four oxygen atoms.

Example : Forsterite (Mg_2SiO_4), a mineral which is a high temperature refractory.

2. Double and poly-tetrahedral structures : Refer to Fig. 6.2.

● This type of structure results when three or more tetrahedral units link together (a ring type structure is produced). One of the oxygens is a member of two units.

● The composition of a three poly-hedral unit is Si_3O_9, which produces $(Si_3O_9)^{6-}$ ions.

Example : Polysilicates (double tetrahedral structure).

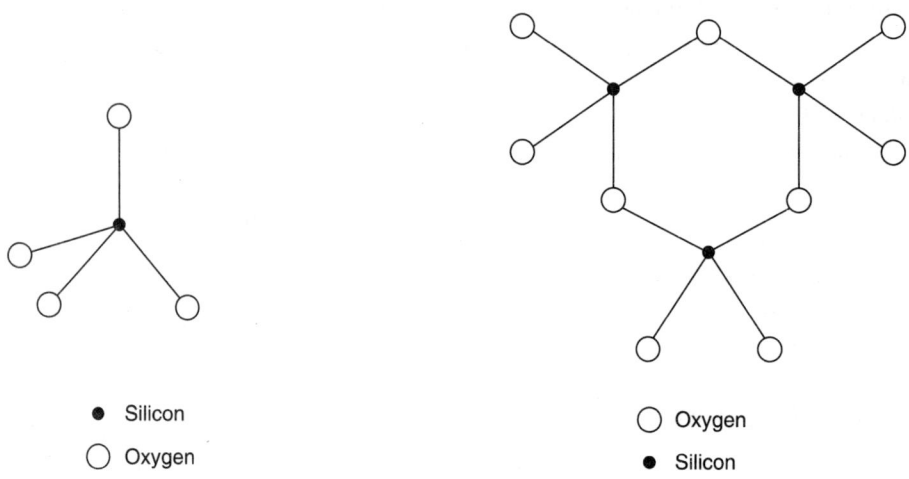

● Silicon

○ Oxygen

○ Oxygen

● Silicon

Fig. 6.1. Silicon-oxygen tetrahedron $(SiO_4)^{4-}$.   Fig. 6.2. Poly-tetrahedral silicate $(Si_3O_9)^{6-}$.

3. Chain structure :

● A chain structure is formed when two corners of each tetrahedra are linked.

● A *single chain structure* (Fig. 6.3) can be noticed in **proxenes.**

● A *double chain structure* (Fig. 6.4) results when two parallel identical chains are polymerized by sharing oxygen to every alternate tetrahedranol. *Example* : Amphiboles.

● Theoretically, the length of these chain structures can be almost infinite.

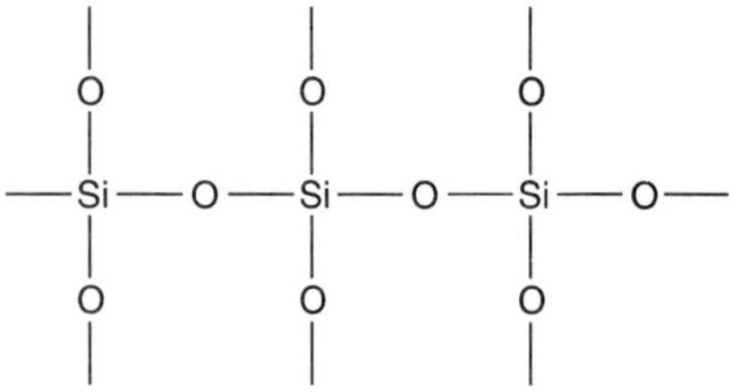

Fig. 6.3. Single chain structure. 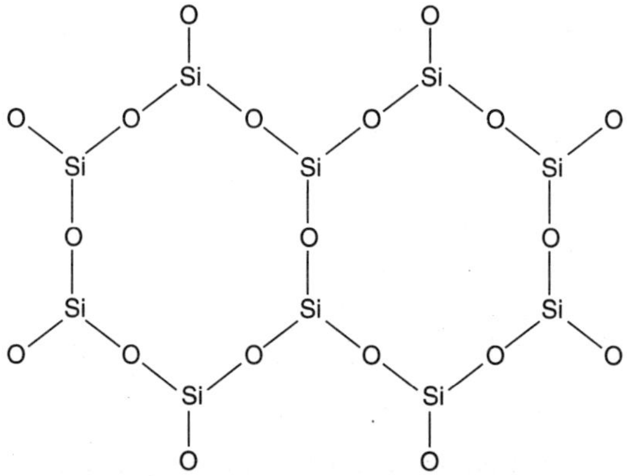   Fig. 6.4. Double chain structure (Si-silicon, O-oxygen).

4. Sheet structure : Refer to Fig. 6.5.

● When the double chain structure extends infinitely in a two dimensional plane, a sheet structure results.

● This structural arrangement provides certain important properties, *e.g.,* the lubricating characteristics of tale and plasticity of clay, the cleavage of mica, etc.

● The sheet structure is found in ceramic materials which as clays, micas and talc.

5. Framework structure :

● This type of structure is an extension of silicate tetrahedral unit into *three dimensions.*

● A framework structure is generally hard, has low atomic packing factors and possesses relatively low densities.

Examples : Quartz, feldspar, cristobalite, etc.

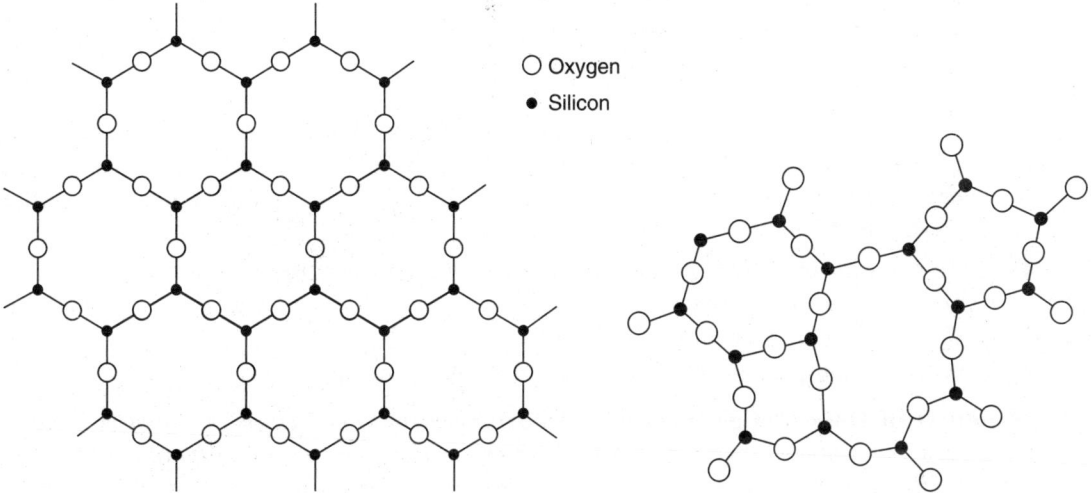

○ Oxygen

● Silicon

Fig. 6.5. Sheet structure.   Fig. 6.6. Vitreous structure.

6. Vitreous structure :

● Fig. 6.6 shows a viterous form of silica (glass) in two dimensions.

● Glass is a vitreous silicate, having a vitreous structure. Glass has a three-dimensional framework structure containing covalent bonds.

6.8. POLYMORPHISM

Depending on the conditions one and the same substance may crystallize in different forms, this is called **polymorphism** usually referred to as *allotropy.* The two or more types of crystals which have the same composition are called **polymorphs.** Polymorphism is found in pure elements as well as among chemical compounds both organic and inorganic. According to the number of distinct polymorphic forms a substance may be known as di-or trimorphic.

● In nature, the phenomonon of polymorphism is widespread. *Silica is an important ceramic and it undergoes polymorphic changes.* The tetrahedral units of silica, $(SiO_4)^{4-}$, are arranged in a hexagonal pattern and this is stable at room temperature. At higher temperatures different transformations takes place and the bonds which excited earlier are broken to form new ones. Three modifications of silica generally observed are : (*i*) *Quartz,* (*ii*) *Tridymite,* (*iii*) *Cristobalite.*

- In aluminosilicates (Al_2SiO_5) three different crystalline forms *viz.*, *sillimanite, andalusite, and kyanite* are formed from the compound mullite ($3Al_2O_3 2SiO_2$) which is stable at the liquidus.
- Similarly, alpha and gamma forms of alumina are obtained from fused alumina.
- Among metals, the most important example of polymorphism is *iron.*
- Among *non-metallic elements, polymorphism is found in phosphorus and carbon.*

Polymorphic transformations are usually brought about by temperature changes but several have been induced by high pressure or drastic cold work treatment also.

6.9. GLASS

6.9.1. Definition and Structure

- **Glass** is *any substance or mixture of substances that has solidified from the liquid state without crystallization.* Elements, compounds and mixture of wide varying composition can exist in the glass state, but the term *"glass"* as ordinarily used *refers to material which is made by the fusion of mixture of silica, basic oxides and a few other compounds that react either with silica or with the basic oxides.* No definite chemical compounds can be identified in glass. Many of its properties correspond to those of a supercooled liquid whose ingredients cannot be identified because they have not separated from the solution in crystalline form.
- *Glass may also be defined as a hard, brittle, transparent or transluscent material chiefly compound of silica, combined with varying proportions of oxides of sodium, potassium, calcium, magnesia, iron and other minerals.*
- Glass is an amorphous substance having a homogeneous texure.

Structure of glass. The *glass is a random arrangement of molecules, the great majority of which are oxygen ions bounded together with the network forming ions of silicon, boron or phosphorus.* A glass made of silica alone has many desirable characteristics but unfortunately the high temperatures involved make it expensive, and difficult to prepare. In order to reduce the temperature, required network-modifying ions are added. *Sodium, potassium, and calcium* are the most common. The network-modifying ions increase the competition for the oxygen ions, thus loosening the Si-O bonds. Certain other ions may substitute for either the network-forming ions or network-modifying ions ; *aluminium, zinc, beryllium, lead and ion* are a few of these intermediate ions.

6.9.2. Constituents of Glass and their Functions

The various *constituents of glass* and their *functions* are described below :

1. Silica :

- It is the principal constituent of glass.
- If silica alone is used in the manufacture of glass, it could be fused only at a very high temperature but it would give a good glass on cooling. However, it is imperative to add some alkaline materials (sodium or potassium carbonate) and lime in suitable proportions to make the molten silica glass sufficiently viscous to make it amply workable and resistant against weathering agencies.

2. Sodium or potassium carbonate :

- It is an alkaline material and forms an essential component of glass.
- It is added in suitable proportion to *reduce the melting point* of silica and to *impart viscosity to the molten glass.*

3. Lime :

● It is added in the form of chalk.

● It imparts durability to the glass.

In place of lime, sometimes, *lead oxide* is also added ; it makes the glass *bright and shining.*

4. Manganese dioxide :

● It is added in suitable proportion to *correct the colour of glass due to the presence of iron in raw.*

● It is also called 'Glass maker' soap.

5. Cullet :

● It is the old broken glass of the same type as that which is intended to be prepared.

● It is added in small quantity to provide body to the glass.

6. Colouring substance. While manufacturing a coloured glass, a suitable colouring substance is added at fusion stage to provide the desired colour to the glass.

The various colouring substances for manufacturing glass of different colours are given below :

Colour	Colouring substance	Colour	Colouring substance
1. **Black**	Cobalt, nickel and manganese oxide	4. **Violet**	Manganese dioxide
2. **Green**	Chromic oxide	5. **White**	Cryolite, tin oxide
3. **Red**	Cuprous oxide, selenium	6. **Yellow**	Cadmium sulphate.

6.9.3. Properties of Glass

Following are the *properties* of glass :

1. No definite crystalline structure.

2. No sharp melting point.

3. Absorbs, refracts or transmits light.

4. Affected by alkalies.

5. An excellent electrical insulator at elevated temperatures.

6. Extremely brittle.

7. Available in beautiful colours.

8. Not affected by air or water.

9. Not easily attacked by ordinary chemical reagents.

10. Capable of being worked in several ways.

11. Can take up a high polish (and may be used as substitute for very costly gems).

12. Possible to weld pieces of glass by fusion.

13. As a result of advancement made in the science of glass production, it is possible to make glass lighter than cork or softer than cotton or stronger than steel.

14. Glass can be cleaned easily by any of the following methods :

　(*i*) Applying methylated spirit.

　(*ii*) Rubbing finely powdered chalk.

　(*iii*) Rubbing damp salt for cleaning paint spots.

　(*iv*) Painting the glass panes with lime-wash and leaving it to dry and then washing with clean water.

Commercial glass must meet the following requirements :

1. The material must melt at commercially obtainable temperature. Fused silica cools to a glass that is superior to ordinary glass, but the temperature required to melt it is so high that its production is expensive and its use restricted.

2. The molten mixture must remain in the amorphous or non-crystalline condition after cooling.

3. The fluidity of molten glass must persist to a sufficient extent to permit the formation of *desired shapes while the glass is cooling.*

4. The glass must be reasonably permanent in the use for which it is intended. Glass with a high proportion to sodium oxide is more readily attacked by water and acids than glass that contains less sodium oxide and more lime and magnesia. Glasses low in basic oxides are less readily attacked.

Mechanical properties :

Ranges of *physical properties* of glasses :

Refractive index	1.46–2.179
Specific index	2.215–8.120
Compressive strength	600–1200 MN/m^2
Tensile strength	27–700 MN/m^2
Thermal conductivity	0.0018– 0.0028 cal/cm °C/sec.
Expansion coefficient	8×10^{-7}–140×10^{-7} cm/cm/°C
Softening point	500–1510°C
Annealing point	350–890°C
Volume resistivity	10^8–10^{18} ohms/cm^3
Dielectric constant	3.7–16.5

Strength. Glass is weak in tension and it always fails in tension no matter how stress is applied. Nevertheless, glass is four to six times as strong in compression as it is in tension and it should, therefore, be used under compressive loads. Strength depends upon the factors such as surface conditions, cross-section and amount of addition, etc.

Hardness and brittleness. The hardness of glass cannot be measured by the Brinell or Rockwell machines because the test specimen fails when the surface is ruptured. Yet some glasses will scratch steel. The relative value of hardness can be evaluated by scratch or abrasion test. *For ordinary purpose the hardness of glass is a function of its tensile strength.*

The concept of brittleness has not been clearly defined. When glass is fixed it bends only a little, then shatters. There is no plastic deformation, and the curve of the *stress-strain diagram is straight until the test specimen breaks.*

Thermal endurance. The lower the coefficient of expansion and the thinner the piece the less likely it is to break under thermal shock. Glass does not break from thermal causes but from the tensile stresses setup by the temperature gradient.

Electrical Properties :

Conductivity. Conductivity depends on composition, temperature and surface conditions. Sodium and potassium ions are largely responsible for imparting electrical conduction, but the function of the calcium ions or the mechanism of conduction in an alkali-free glass is not known. Na_2O and K_2O increase the conductivity. Al_2O_3 increases the conductivity slightly ; ZnO, PbO, MgO, Fe_2O, BaO, B_2O_2 and CaO decrease the conductivity.

The presence of a film on the surface glass affects the surface resistivity to a marked degree. In high alkali glasses the surface conductivity may exceed the volume conductivity. The presence of CO_2, SO_2, H_2S solids and mineral soils also affects the surface conductivity.

In making conductivity tests on glass a precise value is seldom obtained because of the influence of the surrounding medium. Even oil immersion will not give the true value because it has been shown that such a test is a measure of the oil, which fails first, bringing about the failure of the glass. Care in eliminating the edge effects and other conditions show that glass has puncture values several times higher than that obtained in service. The practical limit must take the operating conditions into account.

In the usual method of testing glass for its electrical conductivity the glass specimen is placed between electrodes. Such a system consists of a capacitor with glass as the dielectric medium. Upon charging such a capacitor, the initial charging current is high and then drops off. The charging current rapidly diminishes and straightens out, approaching a low but finite value. The discharge current behaves in like manner. If the charged capacitor is short circuited, the initial discharge current is high and drops off rapidly, approaching a low but finite value. The residual charge may be removed by short-circuiting the capacitor. It may be necessary to repeat this process several times. Because of this behaviour, the measurement of electrical resistance may be in error, and the time interval should be stated.

Dielectric constant, power factor, dielectric loss. The dielectric constant K of glass varies from 3.7 to 16.5 and remains relatively constant over a wide frequency, the square of the voltage gradient, and to the product of dielectric constant and the power factor. For conditions of high frequency or high voltage or both, the value of K should be kept as low as possible. Both the dielectric constant and the power factor can be varied by changes in the glass composition by a ratio of more than 200 to 1. Two general glasses are available : those with high dielectric constant and low power factor, and those with low dielectric constant and low power factor.

Dielectric strength. Electrical breakdown is influenced by the characteristics of the dielectric, the thickness, the duration of the stress, temperature and edge effects, the surrounding medium and characteristics of the voltage, whether direct or alternating ; thermal failure of dielectric is a result of internal heating, which causes it to become a conductor. Alternating currents causes failure at a lower voltage than direct currents do.

6.9.4. Glass Furnaces

For the manufacture of glass two types of furnaces are used :

1. Pot furnaces.

2. Tank furnaces.

1. Pot furnaces. This type of furnace is used to make small amounts of glass such as optical glass that require careful control of ingredients and of melting and fining operations. The pots are fire clay crucibles ranging up to 0.9 metres in height and up to 1.5 metres in diameter. They are arranged on edges in a circular or oblong furnace where they are exposed to a steadily increasing temperature.

2. Tank furnaces. In a continuous tank a constant level of molten glass is maintained by the addition of raw material or "batch" at the proper rates. The furnace is divided into two compartments : the large melting compartment and the smaller working compartment from which the molten glass is withdrawn. Molten glass flows from the melting to the working compartment through a small opening at the bottom of the dividing wall. The "day tank" is a smaller tank, designed to complete melting of a small charge in one day. It is a particularly useful type of tank in cases where frequent changes in glass composition are required.

Melting. When the well-mixed raw materials are heated in the glass furnace, those substances with the lower melting points liquify and act as solvents for other substances, at the same time reacting with them and enabling them to react with each other. The system is very complicated one and the exact sequence of events cannot be determined. The temperature during melting is kept as high as practicable (1400 to 1500°C) in order to reduce the viscosity. Bubbles of gas (carbon dioxide, steam, etc.) that are formed during the heating must escape, and their escape will be slower from a more viscous liquid. In order to remove the gases entirely, the glass must be held for some-time at the highest possible temperature. This is called **plaining or fining.** Additions are often made to the batch to lower its viscosity and to generate gas in large volumes at later stages in the melting process, thus sweeping out the small bubbles along with large ones. Salt cake is often used for this purpose. The under composed portion forms a liquid layer on the surface of the glass and reacts with the scum of siliceous material that tends to collect there. The working temperature, *i.e.,* the temperature at which the glass has the proper viscosity for the shaping operations, is much lower than the highest temperature during melting, the actual working temperature being depend-ent on the kind of glass and the type of machine used. The range is roughly from 1000 to 1300°C or higher.

6.9.5. Fabrication of Glass

The various processes involved in the *fabrication of glass* are enumerated and described below :

   1. Blowing   2. Flat drawing

   3. Polling   4. Pressing into moulds

   5. Casting   6. Spinning.

1. Blowing. In this process, a blow-pipe, 1.5 m to 2 m long and 12 mm diameter is used. One end of the blow-pipe is dipped in a molten mass of glass and a lump (of about 50 N weight) is taken out. The operator then blows vigorously from other end of the blow pipe (it can also be done with the help of a compressor) ; this blowing causes the molten mass of assume a cylindrical shape. It is then heated for ten seconds and is blown again. The blowing and heating are continued till the cylinder of required size is formed. It is then kept on an iron plate disconnected from the blow pipe and a cut is made on its top surface longitudinally by a diamond glass cutter so that by gravity the two ends of the cut cylinder will fall out, forming a rectangular thin sheet of glass.

2. Flat drawing. In this process, the molten glass in viscous form is drawn in the form of a plate by moving an iron bar side way through it. The plate is then passed over a large rotating roller which helps it in spreading out in a thin sheet.

3. Rolling. Rolling is carried out in two ways :

   (*i*) In one method the molten glass is poured on a flat iron casting table and it is then turned flat with the aid of a heavy iron roller.

   (*ii*) In another method, the molten mass of glass is passed between heavy iron rollers and flat glass plate of uniform thickness is obtained.

4. Pressing into moulds. In this process, the molten glass is pressed into moulds by ma-chines. Such a glass is stronger and more durable than unpressed glass. This process is used for *hollow glass articles, ornamental articles,* etc.

5. Casting. In this process, molten glass is poured in moulds and then allowed to cool, down slowly. This method is used to prepare large pieces of glass of simple design, mirrors, lenses, etc.

6. Spinning. In this process, the molten glass is drawn into threads or is spun on a high speed wheel operated by power, to produce even finer than cotton threads.

The spun glass has *tensile strength equal to that of mild steel. It is very soft and flexible and does not fade, decay or shrink.* It is *not* attacked by acids, fire and vermins.

It is used for *producing insulation against heat, sound and electricity.*

6.9.6. Classification of Glass

As per composition and properties glass may be *classified* as :

1. Soda-lime or crown glass

2. Flint glass

3. Pyrex or heat-resistant glass.

1. Soda-lime or crown glass :

- It is the cheapest quality of glass.
- Its composition, like that of most glass, is not rigidly fixed, but can be varied both as to the amount of ingredients and chemical compounds used.

Composition by weight : Sand 75 parts ; lime 12.5 parts ; soda 12.5 parts ; alumina 1 part and waste glass 50 to 100 parts.

- It is easily fusible at comparatively low temperatures.
- It is available in clean and clear state.
- It is possible to blow or to melt articles made from this glass with the help of simple sources of heat.

Uses : Its principal uses are for *window glass, plate glass and container glass* (bottles, glass, etc.)

2. Flint glass :

- It contains varying proportion of lead oxide to make it suitable for various purposes. Lead provides *brilliance and high polish* which makes the glass available for special purposes.

Composition by weight : Sand 100 parts ; red lead 70 parts ; potash 32 parts ; waste glass 10 parts.

- It liquifies at lower temperature than soda-lime glass and has better lustre.
- Owing to the ease with which lead compounds are reduced, the glass must be melted in an oxidizing atmosphere.

Uses : This is potash-lead glass used for better quality of *table-wares* and for *optical glass.* It is also used for *electric lamps, thermometers, electron tubes, laboratory apparatuses, containers for foods,* etc.

3. Pyrex or heat resistant glass. Both soda-lime and flint glasses are unable to withstand sudden temperature changes because of their large coefficients of thermal expansion. The basic oxides that they contain make them susceptible to chemical attack by water and acids. Elimination of the basic oxides and inclusion of boron oxide produce a glass that is very resistant to thermal shock and to attack by water and acids. The temperature required to melt and fine such a glass is so high that it has to be heated in the electric arc. The familiar Pyrex glasses, which are used extensively for cooking utensils and laboratory wares, are **borosilicate glasses.**

Composition by weight : Silica 80 parts ; boron oxide 14 parts ; sodium oxide 4 parts ; alumina 2 parts, with traces of a postassium oxide, calcium oxide and magnesium oxide.

Special Types of Glasses :

1. Annealing glass. To prevent glass articles becoming too brittle and falling into pieces at the slightest shock, they are kept while still hot in an annealing furnace to cool very slowly. *The longer the annealing period, the better, the quality of the glass.*

2. Sheet glass :

- It is roughly composed of 100 parts sand, 35 parts limestones or chalk, 40 parts soda and 50 parts waste glass.
- It is made by blowing glass into hollow cylinder, splitting the cylinder and finally flattening it over a plane surface. It is manufactured in thickness varying from 1.5 to 5 mm and sizes upto 1.5 × 1 metre.

Uses : It is generally used for *doors and windows.*

3. Plate glass :

- Its composition is : White sand 100 parts ; soda carbonate 33 parts ; slaked lime 14 parts ; manganese peroxide 0.15 part and waste glass 100 parts.
- It is *made by pouring white hot glass over an iron table and rolling it to a uniform thickness under heavy roller.*
- It is thicker than sheet glass, and its thickness varies from 5 to 25 mm and sizes upto 275 cm × 90 cm.
- It is stronger and more transparent to sheet glass.
- In modern glass-fabrication, rolled plate glass is annealed because glass cooled normally is brittle.
- Plate glass includes *transparent, transluscent, opaque and structural glasses.*

Uses :

(*i*) It is used for making looking-glass, wind screens of motors, car skylights and glass houses.

(*ii*) It is also used for *sales counter and table tops* after being laminated with plywood or metal sheet.

4. Fluted glass. When there are *corrugations* on one side of the plate glass then it is known as *fluted glass.* The other side is wavy but smooth. The light is admitted without glare of the sun.

Uses : It is used in situations where it is desirable to *secure privacy without obstruction of light.* Horizontal ribs give more light in the middle and less at the sides, while upright ribs give more light at the sides and less in the middle. It is thus more *used for skylight roofts and for windows of railway stations and factories.*

5. Ground glass :

- It is made either by *grinding* (usually by sand blasting) one side, or by melting powdered glass upon it.
- This glass is mostly transluscent.
- This type of glass is also known as *frosted glass or obscured glass.*

Uses : This glass is used in situations where light is required without transparency. It is normally used for *window panes* and *bath room ventilators,* etc.

6. Wired glass :

- It is glass with *wire netting or similar strengthening material embedded in it during manufacture.* This is why this glass is known as **reinforced glass.**
- It resists fire better than ordinary plate glass.
- In case the glass is fractured it does not fall into pieces.

Uses : It is used for *skylight and roofs,* also for *fire-resisting doors.*

7. Safety glass :

- This type of glass is produced by sandwitching sheets of celluloid or other transparent plastic between two sheets of glass and sticking the whole combination together by means of colourless and heat resisting glue.
- Wired glass falls into the category of safety glass.

8. Bullet-proof glass :

- This glass is made of several layers of plate glass and alternate layers consist of vinyl-resin plastic. The outer layers of plate glass are made thinner than the inner layers. The special care is to be taken for heating and cooling of layers during manufacture.
- The thickness of this type of glass may vary from 15 mm to 75 mm or more.
- This glass will not allow bullet to fierce through it.

Uses : It is extensively used for *glazing bank teller, cages, cashier booths, jewellery stores, display cases,* etc.

9. Insulating glass.

It is transparent glass unit composed of two or more plates of glass separated by 6 to 13 mm of dehydrated captive air, hermetically sealed inside, is scientifically cleaned and dried.

Insulating glass provides a high resistance to heat-flow. *The sealed air makes the coefficients of heat transmission of the glass low and hence keeps the apartment cool in summer and warm in winter.*

10. Foam glass :

- It is prepared from powdered glass to which is added the desired quantity of carbon or any gas which makes the mass porous and light in weight.
- This glass is water-proof also.
- It can be easily cut and worked with common masonry tools.

Uses : It is used for *sound and heat insulation purposes. It is specially recommended for use in air-conditioning of buildings.*

11. Glass blocks :

- These consist of two halves so fused together as to *form a hollow inside.*
- They *provide insulation against heat, cold and noise and are easy to clean.*

Uses : They are widely used for construction of *wall partitions.*

12. Soluble glass :

- It is prepared by melting quartz sand, grinding and thoroughly mixing it with soda ash, sodium sulphate or potassium carbonate. The melting is carried out in glass tank at a temperature between 1300°C to 1400°C and it takes about 7 to 10 hours. The resultant glass mass flows out from the furnace and it cools rapidly and breaks up into pieces, known as **silica lumps.**
- It is soluble in water under normal conditions.
- The soluble glass in the form of silicate lumps is transported in containers and in the form of liquid, it is transported in barrels or glass bottles.

Uses : It is used for preparing *acid-resistant* cement.

13. Ultra-violet glass :

- It transmits ultra-violet rays *effectively* even though it is not in the direction of the rays of sun.
- It is made from the raw mixture with minimum admixtures of iron, titanium and chrome.

Uses : It is used in the windows of schools, hospitals, etc.

14. Structural glass :

- This type of glass is available in the form of *glass-crete square blocks, tiles or lenses in thicknesses varying from 5 mm to 30 mm.*
- These glass products are *hollow, light and transparent.*
- This type of glass can be *sawn, placed and drilled* like wood work, inspite of having general properties of glass.

Uses : Widely used for *payment lights, partitions, lantern lights ; also used for roof covering material in industrial buildings, factories,* etc.

Glass-fibre or glass-wool

- The usual *composition* of glass-fibres is that of a *soda-lime glass* but it may be varied for different purposes.
- The *glass-fibres are made by letting the molten glass drop through tiny orifices and blowing with air or steam to attenuate the fibres.*
- They have *very high tensile strengths,* upto about 2750 N/mm^2.
- Glass-fibre or glass-wool differs from mineral wool in that it is a glass made to a definite formulation with a uniformity not found in mineral wool.

6.9.7. Uses of Glass

Besides other uses, some of the important *uses of glass,* based on the recent development in the glass industry, are as follows :

1. The fibre glass reinforced with plastics can be used in the construction of furniture, cars, trucks, lampshades, bath room fittings, etc.
2. Glass is used to form a rifle barrel which is lighter and stronger than conventional type.
3. Thousands of items in the body of a guided missile are made of glass.
4. Glass is used in the construction of noses of deep-diving vehicles.
5. Optical glass is finding wide application for the development and advancement of sciences of astronomy and bacteriology.
6. The glass linings are applied on equipments likely to be affected by the chemical corrosion such as valves, pipes, pumps, etc.
7. Hollow glass blocks can be used for the construction of the walls and ceilings of the modern homes.
8. These days, it is possible to prepare the *colour-changing glass* ; a window with such a glass will be transparent during the day and it will be a source of light at night.

6.9.8. The Glass Industry in India

Names/addresses of some important glass manufacturing concerns in India are given below :

S. No.	Name	Trade mark	Address
1.	*Hindustan Pilkington Glass Works Ltd.*	"HPG"	Hindustan Building, 4, Chitranjan Avenue, CALCUTTA.
2.	*Borosil Glass Work Ltd.*	"CORNING" Glass	44, Khanna Construction House, Khan Abdul Gaffar Khan Road, Worli, MUMBAI.
3.	*The Associated Glass Industries Ltd.*	"AGI" Make	Veradanagar, Kukatpally, HYDERABAD.
4.	*Alembic Glass Industries (India) Ltd.*		Alembic Road, BARODA-3.
5.	*Vallabh Glass Works Ltd.*		Vallabh Vidyanagar ANAND (Gujarat)

6.10. ADVANCED CERAMICS

1. Glass Ceramics :

- These are special glass compositions that are thermally treated prior to forming operations to *devitrify or precipitate a crystalline phase from the material* ; this phase gives that material special properties such as *zero thermal expansion* for applications involving *high thermal-shock application.*

- The compositions (typical of glasses) in which nucleation and crystallization have been commerically produced are : $MgO\text{-}Al_2O_3\text{-}SiO_2$; $LiO_2\text{-}Al_2O_3\text{-}SiO_2$; $LiO\text{-}MgO\text{-}SiO_2$.

Characteristics :

(*i*) Very low coefficient of thermal expansion.

(*ii*) Relatively high mechanical strengths.

(*iii*) High thermal conductivities.

(*iv*) Can be easily fabricated (conventional glass-forming techniques may be employed conveniently in the mass production of nearly pore-free ware).

Uses :

(*i*) Owing to their excellent resistance to thermal shock and their high conductivity, glass ceramics are used as *ovenware* and *tableware.*

(*ii*) As insulators.

(*iii*) As substrates for printed circuit boards.

2. Dielectric Ceramics :

- The use of ceramic materials is made both as *electrical insulators* and as *functional parts* of an electrical circuit. Since the electrical insulators can breakdown under high electrical voltages, the insulators are *designed with lengthened surface paths to decrease the possibility of surface shorting.* Since internal pores and cracks provide opportunity for additional *surface* failure, the insulators are *glazed* to make them non-absorbent.

- *Non-linear dielectric ceramics* are suitable in the miniaturization of electronic parts which have led to the development of increasingly sophisticated electrical circuitry.

- These ceramics are also used in capacitors.

- Some typical non-linear dielectric ceramics are : Lead zirconate-titanate, lead niobates, barium titanate, etc.

3. Electronic Ceramics. Ferrites ferro-electric ceramics, etc. are the ceramic materials with unusual properties that are of specific use in electronic circuits.

- *Ferrites* are mixed-metal-oxide ceramics (almost completely crystalline). They assimilate high electric resistivity and strong magnetic properties. *Soft ferrites* can be used for specific uses such as *memory cores for computers* and cores for radio and television loop antennas. *Barium and lead ferrites* are widely used in permanent-magnet motors in automobiles, portable electrical tools and small appliances.

- *Ferro-electric ceramics* can convert electrical signal into mechanical energy (such as sound) ; and can also change sound, pressure or motion into electrical signals. Thus, they function as *transducers.*

Examples : *Barium titanate* (most common), *tantalates, zirconates, niobates,* etc.

4. Cermets :

- Cermets are ceramic-metal composites.

- Cermets contain alumina (Al_2O_3) and chromium in varying proportions.

- These are used in *brake shoe linings, oxidation-resistant parts and inject engines.*

- The most common cermet is *cemented carbide* and such like composites are extensively used as cutting tools for hardened steels.

6.11. PROCESSING OF CERAMICS

6.11.1. General Aspects

The procedure of processing ceramics involves the following *steps* :

1. Crushing or grinding the raw materials into very fine particles.

— Crushing (also called *comminution or milling*) of raw materials is generally done in a ball mill, either dry or wet. Wet crushing is more effective, because it keeps the particles together and prevents the suspension of fine particles in air.

2. Mixing the particles with *additives* to impart certain desirable characteristics.

3. Shaping, drying and firing the material.

6.11.2. Shaping Processes

For ceramics, following are the three basic *shaping processes* :

(*i*) Casting ;

(*ii*) Plastic forming ;

(*iii*) Pressing.

6.11.2.1. Casting

(*i*) **Slip casting.** Also called *drain casting,* it is the most common casting process. A *slip* is a suspension of ceramic particles in a liquid, generally water. This process is carried out as follows :

— The slip is poured into a porous mould made of plaster of paris. The slip must have sufficient fluidity and low viscosity to flow easily into the mould, much like the fluidity of molten metals.

— After the mould has absorbed some of the water from the outer layer of suspension, it is inverted, and the remaining suspension is poured out (for making hollow objects, as in slush casting of metals).

— The top of the part is then trimmed, the mould is opened, and the part is removed.

By this process large and complex parts such as *plumbing ware*, *art objects*, and *dinner ware* can be made.

● Although dimensional control is limited and product rate is low, mould and equipment costs are also low.

(*ii*) **Doctor-blade process.** This casting technique can be used to make thin sheets of ceramics, less than 1.5 mm thick. In this operation, the slip is cast over a moving plastic belt, and its thickness is controlled by a blade. Other processes include *rolling* the slip between pair of rolls and *casting* the slip over a paper tape, which is then burned off during firing.

6.11.2.2. Plastic forming

Also called *soft, wet, or hydroplastic forming* can be done by various methods such as *extrusion, injection moulding,* or *moulding* and *jiggering* (as done as a potter's wheel).

This process tends to orient the layered structure of clays along the direction of material flow. This orientation leads to *anisotropic behaviour* of the material, both in subsequent processing and in the final properties of the ceramic product.

● In **extrusion** the clay mixture, containing 20% to 30% water, is forced through a die opening by screw-type equipment.

— The cross-section of the extended product is constant, but there are limitations on the wall thickness for hollow extrusions.

The extruded products may be subjected to additional shaping operations.

— Tooling costs are low.

— Production rates are high.

6.11.2.3. Pressing

The various methods of *pressing* are :

1. Dry pressing.

2. Wet pressing.

3. Isostatic pressing.

4. Jiggering—an operation in which the clay bat is formed with templates or rollers ; the part is then dried and fired.

5. Injection moulding.

6. Hot pressing.

6.11.3. Drying and Firing

Drying. In drying, control of atmospheric humidity and temperature is important in order to reduce working and cracking. Loss of moisture results in shrinkage of the part by as much as 15% to 20% of the original moist size.

The dried part (called *green,* as in powder metallurgy) can be machined relatively easily to bring it closer to its final shape, although it must be handled carefully.

Firing. Also called *sintering*, involves heating the part to an elevated temperature in a controlled environment, similar to sintering in powder metallurgy.

Firing gives the ceramic part its strength and hardness.

6.11.4. Finishing Operations

After the completion of firing operation, the following additional operations may performed to give the part its final shape, remove surface flaws, and improve surface finish and dimensional tolerances :

(*i*) Grinding ;   (*ii*) Lapping ;

(*iii*) Ultrasonic machining ;   (*iv*) Electric-discharge machining ;

(*v*) Laser-beam machining ;   (*vi*) Abrasive water-jet machining ;

(*vii*) Tumbling (to remove sharp edges and grinding marks.

● In order to improve appearance and strength, and to make them *impermeable*, ceramic products are often coated with a **glaze** material, which forms a glassy coating after firing.

6.12. DESIGN CONSIDERATIONS FOR CERAMICS

Some important *design considerations* for ceramics are :

1. To avoid large flat surfaces—for eliminating warping of the product.

2. To avoid large changes in thickness—for eliminating non-uniform design and cracking.

3. To provide generous dimensional tolerance—for avoiding the need for machining which is very difficult and expensive.

4. To avoid features like sharp corners, notches and unstrengthened holes (since ceramics are sensitive to stress concentration, being brittle).

5. To make an effort to load ceramics in compression and avoid tensile loading (since the ceramics are about 10 times stronger in compression than in tension).

● Ceramics can be successfully attached to steel by press fits and shrink fits. This permits prestressing the ceramic part in compression, which increases its load carrying capacity.

II. COMPOSITE MATERIALS

6.13. INTRODUCTION TO COMPOSITE MATERIALS

- **A composite material** is *a combination of two or more materials having compositional variations and depicting properties distinctively different from those of the individual materials of the composite.* The composite material is generally better than any of the individual components as regards their strength, heat resistance or stiffness.
- Composites include the following:

(*i*) Multiphase metal alloys;   (*ii*) Ceramics;

(*iii*) Polymers.

Examples of composites are:

(*i*) **Pearlitic steels.** The *pearlitic steels* have a microstructure consisting of alternating layers of α ferrite and cementite; the *ferrite phase is soft and ductile*, whereas *cementite is hard and very brittle.* The combined mechanical characteristics of the pearlite, reasonably high ductility and strength, are superior to those of either constituent phases.

(*ii*) **Wood.** It is one of the examples of the composites that occur in nature. It consists of strong and flexible cellulose fibres surrounded and held together by a stiffer material called *lignin.*

Plywood is the composite of thin sheets of wood with grains of alternate sheets perpendicular to each other and bonded together by a polymer in between them.

(*iii*) **R.C.C.** It has steel rods embedded in the concrete mix, which, itself, is a composite of cement, sand aggregate and water. The resulting R.C.C. structure can take loads which, otherwise, cannot be carried by the concrete alone. Steel rods can be of different shape, size and provided in various directions.

(*iv*) **Vehicle tyres.** The vehicle tyres are rubber reinforced with woven cords.

- Each of the materials in composites serves one or more specific functions. The properties of composites are affected by the following :

(*i*) The size and distribution of the constituents in relation to each other ;

(*ii*) The bond strength between them ;

(*iii*) The shape, size, amount and properties of each material.

- The base material surrounding other materials is normally present in higher percentage and is called *matrix.* Other materials which reinforce the properties of base material are called *reinforcements.* Cohesion between the matrix and reinforcement is essential and may take place in any or combination of the following ways :

(*i*) Chemical reaction at the interfaces of the constituents.

(*ii*) Mechanical keying between the matrix and the reinforcement.

(*iii*) Mechanical bonding between the matrix and reinforcement by vander Walls' forces acting between the surface molecules of the various constituents.

- The selection of matrix and reinforcement is made in such a way that their *mechanical properties complement each other, while deficiencies are neutralised.* In particular cases there may be more than one type of reinforcement present at the same time. The *effect of reinforcement is to increase both the tensile strength and tensile modulus of the composites.*

Advantages :

1. Improved appearance.

2. Reduction in structural mass.

3. Elimination of corrosion and stress corrosion problems.

4. High stiffness-to-weight and strength-to-weight ratios.

5. Fatigue problems are reduced significantly.

6. Improved control of surface contour and smoothness.

6.14. CLASSIFICATION

The composite materials may be *classified* as follows : Refer to Fig. 6.7.

1. Particle reinforced :

(*a*) Large particle ;

(*b*) Dispersion strengthened.

2. Fibre reinforced :

(*a*) Continuous (aligned) ;

(*b*) Discontinuous (short) ;

    (*i*) Aligned ;

    (*ii*) Randomly oriented.

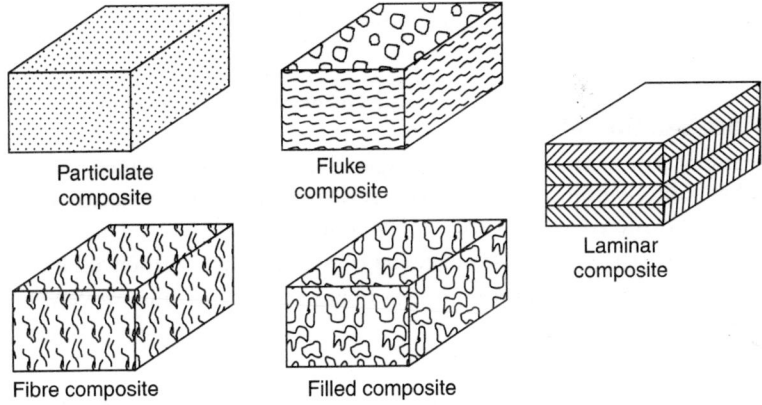

Particulate composite

Fluke composite

Laminar composite

Fibre composite

Filled composite

Fig. 6.7

3. Structural :

(*a*) Laminates

(*b*) Sandwich panels.

6.14.1. Particle-reinforced Composite

(*a*) Large-particle composites :

● These composites are utilized with all three types of materials, *viz., metals, polymers and ceramics.*

Examples :

    (*i*) *Automobile tyres*—containing 15 to 30% volume of spherical particles of carbon with 20–50 μm diameter in vulcanized rubber which improves the tensile strength, toughness and resistance against corrosion.

    (*ii*) *Concrete*—consisting of cement (the matrix), and sand and gravel (the particulates).

    (*iii*) **Cements**, *e.g., cementied carbide*—consisting of extremely hard particles of a refractory carbide ceramic (such as tungsten carbide or titanium carbide) embedded in a matrix of a metal such as cobalt or nickel (load is shared between ceramics and metals). These composites are widely used for *cutting tools* for hardened steels, *drills, electrical contacts, magnets* and *rocket nozzles,* etc. Other applications of cemented carbides are *burner nozzles, gauges and plugs used for inspection of materials, grinding balls and liners in grinding mills. The cermets are prepared by powder metallurgy techniques.*

● In the case of *two-phase composites,* the following two mathematical expressions have been formulated for the dependence of the elastic modulus on the volume fraction of the constituent phases :

$$E_c \text{ (upper bound)} = E_m \, V_m + E_p \, V_p \qquad \qquad ...(i)$$

$$E_c \text{ (lower bound)} = \frac{E_m \, E_p}{V_m \, E_p + V_p \, E_m} \qquad \qquad ...(ii)$$

where, E and V denote elastic modulus and volume fractions respectively ; and

c, m, p represent composite, matrix and particulate phases respectively.

(b) Dispersion-strengthened composites :

● The metals and metal alloys may be strengthened by the *uniform dispersion of several volume percent of fine particles of a very hard and inert material.* The dispersed phase may be metallic or non-metallic ; oxide materials are often used.

● The strength of nickel alloys at elevated temperature may be enhanced considerably by the addition of about 3 volume percent of thoria (ThO_2) as finely dispersed particles, this material is known as *thoria dispersed (or TD) nickel.*

6.14.2. Fibre-reinforced Composites

The most important composites, technologically, are those in which the disperse phase is in the form of the fibre. Fibre-reinforced composites with high specific strengths and moduli have been produced that utilize low-density fibre and matrix materials.

The strength and other properties of these composites are influenced by the following :

(*i*) The arrangement or orientation of the fibres relative to one another :

(*ii*) The fibre concentration;

(*iii*) The fibre distribution.

When the *fibre distribution is uniform, better overall composite properties are realized.*

● **The fibre phase.** The fibres, on the basis of diameter and character, are grouped into three different classifications : *whiskers, fibres and wires.*

Whiskers :

— These are very thin crystals having extremely large length-to-diameter ratios (crystal size ranges from 0.5 to 2.0 microns in diameter to 20 mm length). As a consequence of their small size, they have a high degree of crystalline perfection and are virtually free from flaws, which accounts for their exceptionally high strengths (tensile strength of whisker is approx. 21 GN/m^2 compared to 3 GN/m^2 for carbon fibre) ; they are strongest known materials.

— They are *difficult and costly to manufacture* (as such whiskers are not employed extensively as a reinforcement medium).

— Whisker materials include graphite, silicon carbide, silicon nitride and aluminium oxide.

— Boron and carbon whiskers are used in polymeric matrix. Alumina whiskers are used to reinforce the metal nickel.

Fibres :

— Fibres are either *polycrystalline or amorphous and have small dismaters.*

— The fibrous materials are generally either polymers or ceramics.

Wires :

— Wires are used as a radial steel reinforcement in automobile tyres, in filament-wound rocket castings, and in wire-wound high-pressure hoses.

— Fine wires have relatively large diameter ; typical materials include steel, tungsten and molybdenum.

● **The matrix phase.** In fibrous composites, the matrix phase may be a metal, polymer, or ceramic. The metals and polymers, in general, are employed as matrix materials because some *ductility is desirable*. In case of ceramic-matrix composites, the reinforcing component is added to *improve fracture toughness*.

The matrix phase, in case of fibre-reinforced composites, serves the following purposes/functions :

(*i*) *It binds the fibres together* and acts as a medium by which an externally applied stress is transmitted and distributed to the fibres ; only a very small proportion of an applied load is sustained by the matrix phase.

(*ii*) *It protects* the individual fibres *from surface damage* as a result of mechanical abrasion or chemical reactions with the environment.

(*iii*) The matrix separates the fibres and, by virtue of its relative softness and plasticity *prevents the propagation of brittle cracks from fibre to fibre*, which could result in catastropic failure.

In order to minimize fibre pull-out it is essential that *adhesive forces between fibre and matrix be high*.

Some important fibre-reinforced composites are :

● *Polymer-matrix composites*

— Glass fibre-reinforced polymer (GFRP) composites

— Carbon fibre-reinforced polymer (CFRP) composites

— Aramid fibre-reinforced polymer composites.

● *Metal-matrix composites*

● *Ceramic-matrix composites*

● *Carbon-carbon composites*

● *Hybrid composites.*

6.14.3 Structural Composites

(*a*) Laminar composites :

● These composites are made from the *two dimensional* sheets or panels that have a *preferred high strength direction* (such as found in wood and continuous and aligned fibre reinforced plastics).

● The layers are sacked and subsequently cemented together such that the orientation of the high strength direction varies with each successive layer (*e.g.*, in *plywood*, adjacent wood sheets are aligned with the *grain direction at right angles to each other*).

(*b*) Sandwich panels :

● These composites are composed of *two strong outer sheets or faces* (typical face materials include aluminium alloys, fibre-reinforced plastics, titanium steel and plywood), separated by a *layer of less-dense material* or *"core"* (typical core materials include foamed polymers, synthetic rubbers, inorganic cements and wood) which has lower stiffness and lower strength. The core, structurally perform the following *two functions :*

(*i*) It separates the face and resists deformation perpendicular to the face plane.

(*ii*) It provides a certain degree of shear rigidity along planes which are perpendicular to the faces.

● Sandwich panels find wide applications in the following :

(i) Roofs, floors and walls of building ;

(ii) In aircraft for wings, fuselage and tailplane skins.

Tensile Modulus of Composite :

Consider a composite cylinder made up of continuous fibres, all parallel to the axis of the cylinder in a matrix. Each component in the composite shares the applied force. Thus,

$$\text{Total force} = \text{Force on matrix} + \text{force on fibre}$$

$$= \text{Stress on matrix} \times \text{area of matrix} + \text{stress on fibre} \times \text{area of fibre}$$

$$[\because \quad \text{Force} = \text{stress} \times \text{area}]$$

$$\therefore \quad \frac{\text{Total force}}{\text{Total area}} = \left[\text{stress on matrix} \times \frac{\text{area of matrix}}{\text{total area}} \right] + \left[\text{stress on fibre} \times \frac{\text{area of fibre}}{\text{total area}} \right]$$

∴   Stress on composite = (stress on matrix × % area of matrix)

+ (stress on fibre × % area of fibre)   ...(1)

Strain on composite = strain on matrix = strain on fibre   ...(2)

From eqns. (1) and (2), we get

Tensile modulus of composite = Modulus of matrix × % area of matrix

+ modulus of fibre × % area of fibre   ... (3)

Example 6.1. *Determine the Young's modulus of a composite containing 65 vol % of glass fibre* $(E_f = 70 \text{ GN}/m^2)$ *in a matrix of epoxy resin* $(E_m = 3 \text{ GN}/m^2)$ *under isostress condition.*

Solution. Volume fraction of glass, $V_f = 65\%$ or 0.65

∴   Volume fraction of matrix, $V_m = 1 - 0.65 = 0.35$

We know that Young's modulus of composite is given by

$$E_c = E_f V_f + E_m V_m$$

∴ $E_c = 70 \times 0.65 + 3 \times 0.35 =$ **46.55 GN/m².   (Ans.)**

Example 6.2. *Calculate the volume ratio of aluminium and boron in Al-Boron composite which can have the Young's modulus equal to that of iron. The Young's modulus of Al, iron and boron are 71, 210, 440 GN/m² respectively.*

Solution. Given : $E_{al} = 71$ GN/m² ; $E_i = 210$ GN/m² ; $E_b = 440$ GN/m².

Volume ratios : V_{al} ; V_b :

We know, $E_i = E_{al} V_{al} + E_b V_b$

$$210 = 71 \times V_{al} + 440 \ V_b \qquad\qquad\qquad ...(i)$$

Also, $V_{al} + V_b = 1$

∴ $V_{al} = 1 - V_b$

Substituting the values of V_{al} in eqn. (i), we get

$$210 = 71(1 - V_b) + 440 \ V_b = 71 - 71 \ V_b + 440 \ V_b$$

∴ $V_b =$ **0.377** and $V_{al} = 1 - 0.377 =$ **0.623.   (Ans.)**

6.15. PRODUCTION OF COMPOSITE STRUCTURES

Particulate composites *are normally* made *via* "Powder metallurgy" route. However, a few of them are made by dispersing the particles in the matrix materials through introduction into a slurry.

Fibre-reinforced composites are fabricated by the following *processes* :

(i) **Open mould process.** In this process only one mould (die) is used to fabricate the reinforced part. The following techniques are used :

- Hand lay-up technique
- Bag moulding
— Vacuum-bag moulding.
— Pressure-bag moulding
— Spray up.

The above techniques are used for fabricating part such as ducts, truck bodies, tanks, etc.

(ii) *Matched-die moulding.*
— Compression moulding (employed for moulding bulk moulding compounds)
— Resin injection moulding.

(iii) *Pultrusion.* Process of extrusion of resin-impregnated rooing (a bundle of fibres) to manufacture rods, tubes and structural shapes.

In "Pulmoulding", the process begins with pultruding ; then the part is placed in a compression mould.

(iv) *Filament winding.* In this process resin-impregnated strands are applied over a rotating mandrel, to produce high strength, reinforced cylindrical shapes. Fibres or tapes are drawn through a resin bath and wound on to a rotating mandrel.

(v) *Laminating.* In this process, composite parts are produced by combining layers of resin-impregnated material in a press under heat and pressure.

- **Honeycomb structures and Sandwich panels.** The first step in fabricating honeycomb structures is to fabricate the core (honeycomb panel or the sandwich panel), for which the following *two methods* are used :

(i) Corrugated process.
(ii) Expansion process.

The honeycomb structure is fabricated by attaching the face sheets with adhesives or by brazing to the top and bottom surfaces of the honeycomb block.

QUESTIONS WITH ANSWERS

Q. 6.1. Give briefly the types and general characteristics of ceramics and glasses.

Ans. The types and general characteristics of ceramics and glasses are given below :

1. **Oxide ceramics.**

(i) *Alumina :*
- High hot hardness and abrasion resistance.
- Moderate strength and toughness.
- Most widely used ceramic ; used for cutting tools, abrasives and electrical and thermal insulation.

(ii) *Zirconia :*
- High strength and toughness.
- Resistance to thermal shock, wear, and corrosion.
- Partially-stabilized zirconia and transformation-toughened zerconia have better properties suitable for heat-engine components.

2. **Carbides :**

(i) *Tungsten carbide :*
- High hardness, strength, toughness and wear resistance, depending on cobalt binder content.
- Commonly used for dies and cutting tools.

(*ii*) *Titanium carbide :*
- Not as tough as tungsten carbide, but has a higher wear resistance.
- Has nickel and molybdenum as the binder.
- Used as cutting tools.

(*iii*) *Silicon carbide :*
- High-temperature strength and wear resistance.
- Used for heat engines and as abrasives.

3. **Nitrides :**

(*i*) *Cubic boron nitride :*
- Second hardest substance known, after diamond.
- High resistance to oxidation.
- Used as abrasive and cutting tools.

(*ii*) *Titanium nitride :*
- Used as coatings on tools, because of its low frictional characteristics.

(*iii*) *Silicon nitride :*
- High resistance to creep and thermal shock.
- High toughness and hot hardness.
- Used in heat engines.

4. **Sialon :**
- Consists of silicon nitrides and other oxides and carbides.
- Used as cutting tools.

5. **Cermets :**
- Consist of oxides, carbides, and nitrides.
- High chemical resistance but is somewhat brittle and costly.
- Used in high-temperature applications.

6. **Nanophase ceramics :**
- Stronger and easier to fabricate and machine than conventional ceramics.
- Used in automobile and jet engine applications.

7. **Silica :**
- High temperature resistance.
- Quartz exhibits prezoelectric effects.
- Silicates containing various oxides are used in high-temperature, non-structural applications.

8. **Glasses :**
- Contain at least 50% silica.
- Amorphous structure.
- Several types available, with a wide range of mechanical, physical, and optical properties.

9. **Glass ceramics :**
- High crystalline components to their structure.
- Stronger than glass.
- Good thermal shock resistance.
- Used for cookware, heat exchangers, and electronics.

10. **Graphite :**
- Crystalline form of carbon.
- High electrical and thermal conductivity.
- Good thermal-shock resistance.
- Also available as fibres, foam, and bucky balls for solid lubrication.
- Used for moulds and high temperature components.

11. **Diamond :**
- Hardest substance known.
- Available as single crystal or polycrystalline form.
- Used as cutting tools and abrasives and as die insert for fine wire drawing ; also used as coatings.

Q. 6.2. Explain briefly 'Metal-matrix composits'.

Ans. In composite materials, new developments are continually taking place, with a wide range and form of polymeric, metallic, and ceramic materials being used as fibres and as matrix materials.

— The *advantage* of a metal matrix over a polymer matrix is the former's higher resistance to elevated temperatures and higher ductility and toughness.

— The *limitations* of the metal matrix are higher density and greater difficulty in the processing components.

In metal-matrix composites the *matrix materials* are usually aluminium, aluminium-lithium, magnesium and titanium, although other metals are also being investigated for this purpose. The *fibre materials* are graphite, aluminium oxide, silicon carbide, and boron, with beryllium and tungsten as other possibilities.

Applications. Current applications of metal-matrix composites are in :

— Electrical components ;
— Gas turbines ;
— Various structural components.

Methods of manufacture. The following *methods* are used to manufacture these composites into near-net-shape parts :

(*i*) *Liquid-phase processing.* It consists of casting the liquid matrix and the solid reinforcement, using either conventional casting processes or pressure infiltration casting.

(*ii*) *Solid-phase processing.* It consists of powder-metallurgy techniques, including cold and hot isostatic pressing .

(*iii*) *Two-phase processing.* It consists of rheocasting, and spray atomisation and deposition.

Q. 6.3. What are ceramic-matrix composites ? Explain briefly.

Ans. Ceramic-matrix composites are another development in engineered materials. Ceramics are strong and stiff and resist high temperatures, but generally lack toughness. Silicon carbide, silicon nitride, aluminium oxide, and mullite (a compound of aluminium, silicon and oxygen) are new matrix materials that retain their strength to 1700°C.

Applications. Ceramic-matrix composites are used in :

— Jet and automotive engines ;
— Equipment for deep sea mining ;
— Pressure vessels ;
— Various structural components.

Processes of manufacture. The common processes for manufacture are :

(*i*) *Slurry infiltration.* It involves the preparation of a fibre preform that is hot pressed and then impregnated with a slurry that contains the matrix powder, a carrier liquid, and an organic binder.

(*ii*) *Chemical-synthesis processes.*

● *Solegel process.* In this process, a sol (a colloidal fluid with the liquid as the continuous phase) containing fibres is converted to a gel, which is then subjected to heat treatment to produce a ceramic-matrix composite.

● *Polymer-precursor method.* It is analogous to the process used in making ceramic fibres.

(*iii*) *Chemical-vapour infiltration.* In this process, a porous fibre preform is infiltrated with the matrix phase, using the chemical vapour deposition technique.

— The product has very good high-temperature properties.

— The process, however, is time consuming and costly.

Q. 6.4. How are ceramic superconductors processed ? Explain briefly.

Ans. Ceramic superconducting materials are available in powder form. The fundamental difficulty in manufacturing them is their inherent brittleness and anisotropy, which make it difficult to align the grains in the proper direction for high efficiency ; the smaller the grain size, the more difficult is to align the grains.

The following *steps* are involved in the processing of superconductors :

(*i*) Preparing the powder, mixing it, and grinding it in a ball mill to a grain size of 0.5 mm to 10 mm.

(*ii*) Forming it into shape.

(*iii*) Heat treating to improve grain alignment.

'Oxide powder in tube' (OPIT) is the most common forming process. In this process, the powder is packed into silver tubes (since silver has the highest electrical conductivity of all metals) and sealed at both ends. The tubes are then mechanically worked, by such deformation processes as swaging, drawing, extrusion, isostatic pressing, and rolling, into final shapes, which may be wire, tape, coil, or bulk.

Q. 6.5. What are 'laminates' ? Give examples.

Ans. Laminates (or *laminar composites) are those structures which have alternate layers of materials bonded together in some manner.*

Common *examples* of laminar composite :

(*i*) Plywood   (*ii*) Bimetallic strips

(*iii*) Safety glass   (*iv*) Sandwich material

(*v*) Roll cladding (bonding) and explosive cladding (welding)

(*vi*) Laminated plastic sheet

(*vii*) Tufnol   (*viii*) Laminated carbides

(*ix*) Laminated wood.

Q. 6.6. Discuss briefly surface coatings.

Ans. The surface coatings are applied to materials for the following *purposes* :

(*i*) To protect the material against corrosion.

(*ii*) To improve visibility through luminescence and better reflectivity.

(*iii*) To provide electrical insulation.

(*iv*) To improve the appearance.

(*v*) For decorative, wear resistance and processing purposes.

The various types of surface coatings are :

1. **Metallic coatings.** Metallic coatings of copper, chromium, nickel, zinc, lead and tin, etc. are applied by hot dipping, electroplating or spraying techniques to protect the base metal from corrosion and for other purposes.

2. **Inorganic chemical coatings :**

- *Oxide and phosphate coatings* are done to make iron or steel surfaces free from rust and this is done by chemical action. These coatings also provide protection against corrosion.
- Vitreous coatings are commonly applied to steel in the form of a powder or frit and are then fused to the steel surface by heat.
  — *Enamel* is an example of a ceramic coating on metal and *glaze or tiles* is an example of a glassy ceramic or a crystalline ceramic base.
  — Coatings of TiC and TiN on HSS base are examples of ceramics on steel.

3. **Organic coatings :**

- These coatings include *paints, varnishes, enamels* and *lacquers.*
- They serve to protect the base metal and to improve its appearance.
- Polymer coated metals are used for making beverage cans.

HIGHLIGHTS

1. *Ceramic materials* are defined as those containing phases that are compounds of metallic and non-metallic elements.
2. *Glass* is any substance or mixture of substance that has solidified from the liquid state without crystallization.
3. The various processes involved in the fabrication of glass are :
   (*i*) Blowing ;   (*ii*) Flat drawing ;
   (*iii*) Rolling ;   (*iv*) Pressing into molds ;
   (*v*) Casting ;   (*vi*) Spinning.
4. *Advanced ceramics* include : Glass ceramics, dielectric ceramics, electronic ceramics, cermets.
5. The three basic shaping processes of ceramics are :
   (*i*) Casting,   (*ii*) Plastic forming ;
   (*iii*) Pressing.
6. A *composite material* is a combination of two or more materials having constitutional variations and depicting properties distinctively different from those of the individual materials of the composite.

OBJECTIVE TYPE QUESTIONS

Fill in the blanks or say "Yes" or "No" :

1. The compressive strength of ceramic materials is several times more than the tensile strength.
2. In ceramic materials, stress concentration has significant effect on compressive strength.
3. At high temperature, the rigidity of ceramics is ......
4. Porcelain has ...... positive temperature coefficient.
5. Rutile bodies have large negative coefficients.
6. The silicates are co-ordinate structures based upon large anions arranged about small cations.

7. Among metals, the most important example of polymorphism is ....
8. ...... is any substance or mixture of substances that has solidified from the liquid state without crystal-lisation.
9. Silica is the principal constituent of glass.
10. Lime imports durability to glass.
11. Glass has a sharp melting point.
12. It is possible to weld pieces of glass by fusion
13. Glass is strong in tension
14. The hardness of glass can be measured by Brinell or Rockwell machines.
15. The spun glass has tensile strength equal to that of mild steel.
16. Soda-lime or crown glass is the cheapest quality of glass.
17. Glass ceramics have very ...... coefficient of thermal expansion.
18. Barium titanate is a ...... dielectric ceramic.
19. Ferrites are mixed-metal-oxide ceramics.
20. ...... are ceramic-metal composites.
21. Slip casting is also called drain casting.
22. The vehicle types are rubber reinforced with woven cords.

ANSWERS

1. Yes	2. No	3. high	4. large	5. Yes
6. Yes	7. iron	8. Glass	9. Yes	10. Yes
11. No	12. Yes	13. No	14. No	15. Yes
16. Yes	17. low	18. non-linear	19. Yes	20. Cermets
21. Yes	22. Yes			

THEORETICAL QUESTIONS

1. Define the term 'Ceramics'.
2. How are ceramics classified ?
3. What are advantages of ceramic materials ?
4. List the applications of ceramics.
5. Discuss briefly properties of ceramic materials.
6. Write a short note on 'structure of crystalline ceramics.'
7. Explain briefly various types of silicate structures.
8. What is 'polymorphism' ? Explain briefly.
9. Define 'Glass'.
10. What is the structure of glass ?
11. Explain briefly the constituents of glass.
12. List the properties of glass.
13. State the requirements which a commercial glass must meet.
14. Discuss briefly 'Glass furnaces'.
15. Explain briefly the following processes involved in the fabrication of glass :
    (i) Blowing   (ii) Flat drawing
    (iii) Rolling   (iv) Pressing into moulds
    (v) Casting   (vi) Spinning.
16. Give the classification of glass.
17. Write a short note on 'Glass-wool'.
18. What are uses of glass ?

19. Explain briefly the following advanced ceramics.
    (*i*) Glass ceramics   (*ii*) Dielectric ceramics
    (*iii*) Electronic ceramics   (*iv*) Cermets.
20. Explain briefly 'Processing of ceramics'.
21. Mention some important design considerations for ceramics.
22. What is a composite material ?
23. What do composites include ?
24. How are composite materials classified ?
25. Explain briefly the following composites :
    (*i*) Particle-reinforced composites.
    (*ii*) Fibre-reinforced composites.
    (*iii*) Structural composites.
26. Explain briefly the production of composite structures.

7

Welding and Allied Processes

7.1. Introduction. 7.2. Advantages, disadvantages and applications of welding. 7.3. Classification of welding processes. 7.4. Forge welding. 7.5. Resistance electric welding—general aspects—resistance spot welding—resistance seam welding—resistance projection welding—resistance butt welding—upset welding—flash welding. 7.6. Gas welding—fusion welding—general aspects—advantages and disadvantages of gas welding—applications of gas welding—oxy-acetylene welding. 7.7. Electric arc welding—introduction—advantages and limitations—metallic arc welding—carbon arc welding—atomic hydrogen welding—shielded arc welding—arc blow—comparison between A.C. and D.C. arc welding—types of welded joints. 7.8. Thermit welding. 7.9. Tungsten inert-gas (TIG) welding. 7.10. Metal inert-gas (MIG) welding—difference between TIG and MIG welding processes. 7.11. Submerged arc welding. 7.12. Electro-slag and electro-gas welding—electro-slag welding—electro-gas welding. 7.13. Electron-beam welding. 7.14. Ultrasonic welding. 7.15. Plasma arc welding. 7.16. Laser beam welding. 7.17. Friction welding. 7.18. Explosive welding. 7.19. Diffusion welding. 7.20. Induction welding. 7.21. Cold welding. 7.22. Stud-arc welding. 7.23. Hydrodynamic welding. 7.24. Under-water welding. 7.25. Oxy-acetylene torch cutting. 7.26. Solid/liquid-state bonding-soldering and brazing. 7.27. Soldering—definition—classification of soldering methods—types of solder—selection of solder—flux or soldering fluid—soldering equipment—soldering procedure—characteristics of a good joint—important tips for effective soldering operation—advantages of soldering—applications of soldering—types of soldered joints. 7.28. Brazing—introduction—fluxes—brazing equipment—brazing methods—brazing procedure—advantages and limitations of brazing—applications of brazing—silver soldering (or silver brazing)—comparison between soldering and brazing. 7.29. Electrodes—electrode materials—electrode coatings—electrode's designation—typical data on use of "mild steel electrodes". 7.30. Welding of various metals. 7.31. Rebuilding. 7.32. Hard facing. 7.33. Characteristics of good weld. 7.34. Defects in welds. 7.35. Weldability. 7.36. Testing of welded joints. 7.37. Effect of welding on the grain size of the metal. 7.38. Principles of welding design. 7.39. Comparison of welding and allied process—Worked Examples—**Questions with Answers**—Highlights—Objective Type Questions—Theoretical Questions.

7.1. INTRODUCTION

I. Welding :

It is method of joining metals by applications of heat, without the use of solder or any other metal or alloy having a lower melting point than the metals being joined.

Or

Welding is defined as "a *localised coalescence of metals, wherein coalescence is obtained by heating to suitable temperature, with or without the applications of pressure and with or without the use of filler metal*". The filler metal has a melting point approximately the same as the base metal.

— The large bulk of materials that are welded are metals and their alloys, although the term welding is also applied to the joining of other materials such as thermoplastics. Welding joins different metals/alloys with the help of a number of processes in which heat is supplied either electrically or by means of a gas torch. In order to join two or more pieces of metals together by one of the welding processes, the *most essential requirement is heat. Pressure may also be employed, but this is not, in many processes essential.*

— A good welded joint is as strong as the parent metal. The product is known as *"weldment"*.

II. Soldering and brazing :

Soldering. *It is a process of joining two pieces of metal with a different fusible metal applied in a molten state.* The fusible metal is called '*solder*'.

Or

It is a process of joining two metals with low melting point metal.

Brazing. *It is a process of joining two metal pieces in which a non-ferrous alloy is introduced in the liquid state between the pieces to be joined and allowed to solidify.*

☞ • Soldering and brazing are two common *solid/liquid-state bonding processes*. These are *different from welding* as bonding here requires *capillary action* and that some degree of *alloying action* between the filler and the base metal always occurs. Also the *composition of filler metal is significantly different and its strength and melting point are substantially lower than that of the base metal.*

— In '*soldering*' (very similar to brazing) the filler material is usually a *lead-tin based alloy* which has much lower strength and melting temperature (about 250°C). Also, less alloying action between the base and filler metals gives *lower joint strength*. Since in this process much lower temperatures are involved, it is usually carried out with *electric resistance heating*.

— In '*brazing*' the joint is made by heating the base metal red hot and filling the gap with molten filler metal whose melting temperature is above 427°C but below the temperature of base metal. The filler metals, generally used for brazing are *copper alloys*. This process is usually carried out with a *gas flame*.

Joining processes. Base on the composition of the joint, the joining processes may be classified as following :

(*i*) *Autogeneous process.* In this type of joining process, no filler material is added during the joining process as in the case of resistance welding, cold welding, friction welding, diffusion welding and hot forge welding.

(*ii*) *Homogeneous process.* This process make use of filler metal but of same composition as the parent metal as in the case of arc, gas and thermit welding.

(*iii*) *Heterogeneous process.* In this process the filler material is soluble in both the parent metals which themselves are insoluble in each other.

7.2. ADVANTAGES, DISADVANTAGES AND APPLICATIONS OF WELDING

Following are the *advantages and disadvantages* of welding :

Advantages :

1. A large number of metals/alloys both similar and dissimilar can be joined by welding.

2. Welding can join workpieces through spots, as continuous pressure tight seams, end-to-end and in a number of other configurations.

3. A good weld is as strong as the base metal.

4. Welding permits considerable freedom in design.

5. General welding equipment is not very costly.

6. Portable welding equipments are available.

7. Welding results in a good saving of material and reduced labour content of production.

8. Low manufacturing costs.

9. Welding is also used as a method for repairing broken, worn or defective metal parts. Due to this, the cost of reinvestment can be avoided.

Disadvantages :

1. Welding results in residual stresses and distortion of the workpieces.
2. Welding heat produces metallurgical changes. The structure of the welded joint is not same as that of parent metal.
3. Jigs and fixtures are generally required to hold and position the parts to be welded.
4. A welded joint, for many reasons, needs stress-relief heat treatment.
5. Welding results in residual stresses and distortion of the workpieces.
6. Welding gives out harmful radiations (light), fumes and spatter.
7. For producing a good welding job, a skilled worker is a must.

Applications :

The welding process finds wide applications in almost all branches of industry and construction.
- Extensively employed in the *fabrication* of :
    — Structural members of bridges and buildings, etc ;
    — Vessels of welded-plate construction *e.g.,* steel reservoirs, boilers, pressure vessel tanks and pipelines, etc.
    — Concrete reinforcement.
- Chief means of *fastening* panels and members together into automobile bodies and in aviation industry.

7.3. CLASSIFICATION OF WELDING PROCESSES

A. The welding processes may be *classified* as follows :

I. Pressure Welding :

1. Forge welding.
2. Resistance electric welding.
 (*i*) Butt welding
 (*ii*) Flash welding
(*iii*) Spot welding
 (*iv*) Seam welding
  (*v*) Projection welding
 (*vi*) Percussion welding.
- The characteristics of a pressure weld is that the metal joined is *never brought to a molten stage*, it is heated to a welding temperature and the actual union is brought about by *application of pressure.*

II. Fusion Welding :

1. Gas welding.
2. Electric arc welding.
   (*i*) Metallic arc welding
  (*ii*) Carbon arc welding
 (*iii*) Atomic hydrogen welding
  (*iv*) Shielded arc welding.
3. Thermit welding.
- The characteristic of a fusion weld is that the material being joined is a *actually melted* and the union is produced on subsequent solidification.

III. Modern/Miscellaneous Welding Techniques :

(*i*) Tungsten inert-gas (TIG) welding or GTAW (Gas tungsten arc welding)

(*ii*) Metal inert-gas (MIG) welding

(*iii*) Submerged arc welding

(*iv*) Electro-slag and electro-gas welding

(*v*) Electron-beam welding

(*vi*) Ultrasonic welding

(*vii*) Plasma arc welding

(*viii*) Laser beam welding

(*ix*) Friction welding

(*x*) Explosive welding

(*xi*) Diffusion welding

(*xii*) Induction welding

(*xiii*) Cold welding

(*xiv*) Stud-arc welding

(*xv*) Hydrodynamic welding.

Allied Processes :

1. Soldering

2. Brazing

B. The welding processes may also be *classified* as follows :

1. Solid-state welding processes :

(*i*) Forge welding

(*ii*) Friction welding

(*iii*) Explosive welding

(*iv*) Ultrasonic

(*v*) Diffusion.

2. Liquid-state (or fusion) welding processes :

(*i*) Gas welding

(*ii*) Electric arc welding

(*iii*) TIG welding

(*iv*) Resistance welding

(*v*) Thermit welding, etc.

3. Solid/liquid-state bonding processes :

(*i*) Soldering

(*ii*) Brazing.

Note : For ISO 4063 classification of welding processes, Refer to Q. 7.28.

7.4. FORGE WELDING

- In this method of welding the surfaces to be joined are heated in an open hearth until they reach the welding temperature of metal, *which is below its melting point*. The blacksmith will judge this temperature by the colour of the metal, which may be between red-hot and white-hot. The parts are then placed on an anvil and hammered together.

- In this welding process there is a risk of oxide and other inclusion when the metal is heated in an open fire and accurate judgement of the temperature is called for if the structure of metal is not to be changed. Modern alloy steels can be ruined by injudicious heating.

- When the wide range of light alloys is considered, it becomes imperative to use a more scientific method. However, this process is still widely used for heavier classes of work, such as *manufacture of anchor chains, while controlled heating furnaces and automatic forging machines have been designed to replace the* open-forge fire and the blacksmith's anvil.

  — A modern version of welding in this category is *manufacture* of *butt-welded pipes.* In this process, the skulp heated upto the required welding temperature is pulled through a die which forces the two edges of a heated skulp to come into contact under pressure and get welded.

Advantages :

1. Inexpensive equipment.
2. Semi-skilled operation.

Limitations :

1. Poor joint strength
2. Labour intensive process (low production rate).
3. Weld quality dependent on operator's skill.
4. Can be used only when hammering is possible.

7.5. RESISTANCE ELECTRIC WELDING

7.5.1. General Aspects

It is the method of uniting two pieces of metal by the passage of a heavy electric current while the surfaces are pressed together. The fusing temperature is obtained by placing the surfaces to be joined in contact with one another, and passing a *current of two to eight volts, at a high amperage* through them. The *heat is developed around the point to which they touch, forcing them together* (by pressure mechanically applied), and at the same time *switching off the current, completes the weld.*

A special feature of resistance welding is the *rapid heating* of the surface being welded (in hundredths of a second) due to application of currents of high amperage.

- Successful operation of a resistance welding process depends upon correct application and proper control of the following *factors* :

  (*i*) Welding current.

  (*ii*) Welding pressure.

  (*iii*) Time of application (cycle time) :

  — Weld time

  — Squeeze or forge time

  — Hold time

  — Off-time

  (*iv*) Contact area of electrodes.

Electrodes. The electrodes in resistance welding should have *higher electrical conductivity as well as higher hardness.*

  — Steel though strong, do not have conductivity required for electrodes. Hence, *copper in alloyed form is generally used for making electrodes* (Pure copper is poor in mechanical properties).

  — *Copper cadmium (0.5 to 1.0%) alloys* have the highest electrical conductivity with moderate strengths and are used for welding *non-ferrous materials* such as aluminium and magnesium alloys.

— *Copper chromium (0.5 to 0.8%) alloys* have slightly lower electrical conductivities than the above but better mechanical strength. These are used for resistance welding of *low strength steels* such as wild steel and low alloy steels.

Advantages and disadvantages/Limitations of resistance welding process :

Advantages :

1. The heat is localised, action is rapid and *no* filler metal is needed.

2. The operation requires *little skill* and can be easily mechanised and automated.

3. A *high degree of reliability* and *reproducibility* can be achieved.

4. Very well suited for *mass production* (owing to high production rate).

5. Very *economical* process.

6. It is possible to weld *dissimilar* metals as well as metal plates of different thicknesses.

7. Heating of workpiece is confined to a very small part which results in *less distortion*.

Disadvantages/Limitations :

1. High cost of equipment.

2. Certain resistance welding processes are limited only to lap joints.

The various resistance welding processes are :

- Resistance spot welding.
- Resistance seam welding. ⎫
- Resistance projection welding. ⎬ *Lap* joints used
- Resistance butt welding : ⎭

   (*i*) Upset welding ⎫
   (*ii*) Flash butt welding ⎬ *Butt* joints used

These process are discussed in the following articles.

7.5.2. Resistance Spot Welding

Spot welding is the simplest and most commonly used resistance-welding process.

Refer to Fig.7.1. Spot welding, as the name implies, is carried out by overlapping the edges of two sheets of metal and fusing them together between copper electrode tips at suitably spaced intervals by means of a heavy electrical current. The resistance offered to current as it passes through the metal raises the temperature of the metal between the electrodes to welding heat. The current is cut-off and mechanical pressure is then applied by the electrodes to forge the welds. Finally the electrodes open.

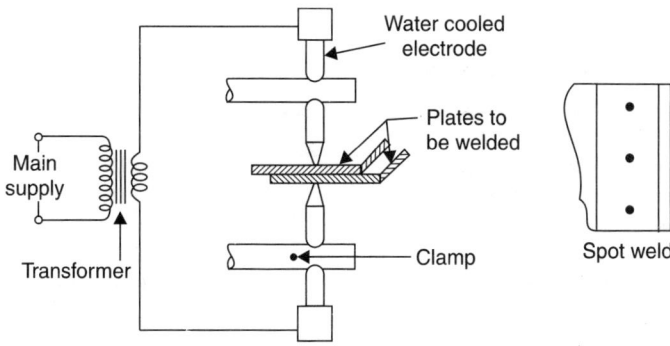

Fig. 7.1. Spot welding.

When sheets of unequal thickness are joined, the current and pressure setting for the thinner sheets are used. Similarly four thickness may be welded, using the same settings as for two thickness.

- Currents usually range from 3000 A to 40,000 A, depending on the materials being welded and their thickness. *Modern equipment for spot welding is computer controlled for optimum timing of current and pressure, and the spot-welding guns are manipulated by programmable robots.*

- *Steel, brass, copper and light alloys can be joined by this method,* which forms a *cheap and satisfactory substitute for riveting.* The area of fusion at each spot weld, in fact, is approximately equal to the cross-sectional area of the rivet which would be employed for a similar gauge of material.

Applications :

Spot welding is widely used for fabricating *sheet-metal products.* Examples of its applications range from attaching handles to stainless-steel cookware to rapid spot welding of automobile bodies, using multiple electrodes.

Advantages :

1. High production rate.
2. Very economical process.
3. High skill not required.
4. Most suitable for welding sheet metals.
5. Dissimilar metals can be welded.
6. No edge preparation is needed.
7. Operation may be made automatic or semi-automatic.
8. Dependability.
9. Small heat affected area.
10. More general elimination of warping or distortion of parts.

Limitations :

1. Suitable for thin sheets only.
2. High equipment cost.

Spot welding machines :

The following three types of spot welding machines are in common use :
1. Standard machines.
    — Rocker arm type
    — Press type spot or projection welders
2. Special multiple-electrode machines.
3. Portable welders.

7.5.3. Resistance Seam Welding

Refer to Fig. 7.2. Seam welding is analogous to spot welding with the difference the electrodes are in the form of rollers ; and the *work moves in direction perpendicular to roller axis.* The current is interrupted 300 to 1500 times a minute to give a series of overlapping spot welds. The welding is usually done under water to keep the heating of the welding rollers and the work to a minimum, and thus to give lower roller maintenance and less distortion of the work.

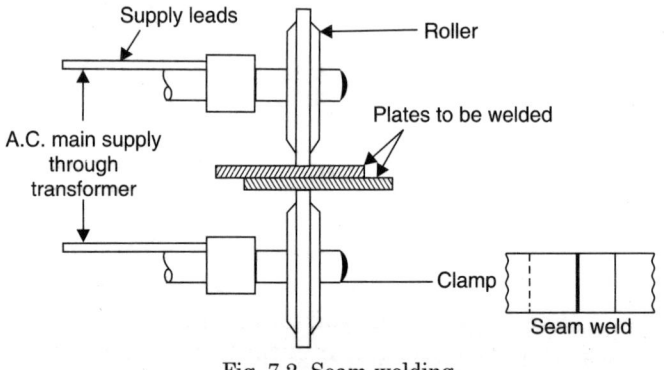

Fig. 7.2. Seam welding.

- Welding currents range from 2000 A to 5000 A while the force applied to the rollers may be as high as 5 kN to 6 kN.
- The typical welding speed is 1.5 m/min for thin sheet.
- With intermittent application of current to the rollers a series of spot welds at various intervals can be made along the length of the seam, a procedure called *roll spot welding*.

Applications :

It is employed on many types of pressure (light or leak proof) tanks, for oil switches, transformers, refrigerators, evaporators and condensers, aircraft tanks, paint and varnish containers, etc.

7.5.4. Resistance Projection Welding

Refer to Fig. 7.3. It is in effect, a form of multi-spot welding in which a number of welds are made simultaneously.

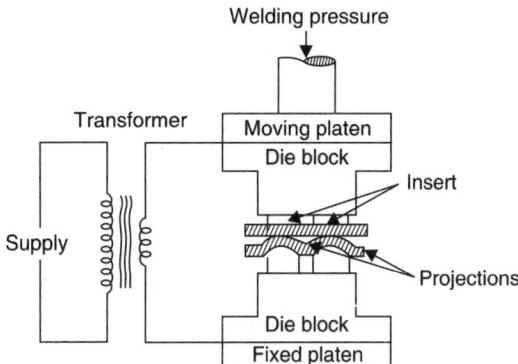

Fig. 7.3. Projection welding.

- — The pieces to be welded are arranged between two flat electrodes which exert pressure as the current flows.
- — The projections, and the areas with which they make contact, are raised to welding heat and are joined by the pressure exerted by the electrodes.
- — The projections are flattened during the welding.

Applications :

The process is used chiefly to join pressings together since it is relatively simple to make the press-tools so that the projections are produced during the main forming operation in the press.

- — The materials like brass and aluminium *cannot* be projection welded satisfactorily.
- The same principle is used in the cross welding of a number of wires or rods to make a mesh.

7.5.5. Resistance Butt Welding

There are two types of butt welding : Upset and Flash.

7.5.5.1. Upset welding

Refer to Fig. 7.4. In this type of welding which is employed to join bars and plates together end-to-end, one bar is held in a fixed clamp in the butt welding machine ; and the other bar in a movable clamp, the clamp being electrically insulated, the one from the other, and being connected

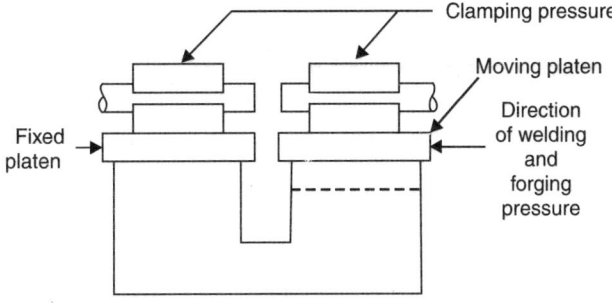

Fig. 7.4. Butt welding.

to a source of current. When the two ends to be joined are brought into contact and current is switched on, the resistance at the joint causes the ends to heat up to welding temperature. Current is then switched off and the movable clamp *forced* up, so that pressure applied upsets or forges the parts together. The voltage applied across the clamps is a low one, from 2 to 6 volts, and the current is usually alternating. If the bars being joined are different in cross-section the amounts they project from their clamps may have to be adjusted so as to modify the heat losses and ensure both bars being brought to the welding temperature simultaneously.

Applications :

- Upset welding is used principally on non-ferrous materials for welding bars, rods, wire.
- *This process is being used for welding such things as steel rails whose cross-sectional area is as much as 6.25 cm².*

7.5.5.2. Flash welding

Refer to Fig 7.5. In this process, the parts to welded are clamped to the electrode fixtures, as in butt welding *but the voltage is applied before the parts are butted together.* As the *parts touch each other, an arc is established which continues as long as the parts advance at the correct speed.* This arc bursts away a portion of the material from each piece. When the welding temperature is reached, the speed of travel is increased, the power switched off and weld is upset.

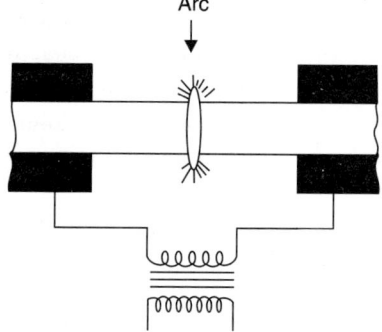

Fig. 7.5. Flash welding.

- The upsetting action forces out the impurities caused by flashing. *The forced-out metal is called flash.* The inner weld is then sound and free of oxides and cast metal.
- Many different materials and combination can be flash butt welded ; steels and the ferrous alloys other than cast iron are probably the most easily welded. Those materials which cannot be flash butt welded are lead, tin, zinc, antimony, bismuth and their alloys, and the copper alloys in which these metals are present in large percentage.

Flash welding claims the following *advantages* over upset method of welding :

(*i*) Power consumed is less once the arc creates more heat with a given current.

(*ii*) The weld is made in clean virgin metals as the surfaces are burned away.

(*iii*) More quicker.

Applications :

- *It is widely used in automobile construction on the body, axles, wheels, frames and other parts.*
- *It is also employed in welding motor frames, transformer tanks and many types of sheet steel containers such as at barrels and floats.*

Percussion welding :

- It is a very *fast* method of welding.
- It consists of holding the parts at a small distance with their end faces opposite to each other, bringing them closer at a fast speed after switching on the current, thus creating an arc between their end faces just before they come in contact and completing the weld under *impact.* Some of the metal may squeeze out of the joint, but it is very small.
- The *use* of this process is limited to *very thin* wires, with their diameters ranging between 0.05 mm and 0.38 mm. It can also be used for joining wire of *dissimilar* metals, such as copper to nichrome and copper to stainless steel.

7.6. GAS WELDING

7.6.1. Fusion Welding—General Aspects

In *fusion (or liquid state) welding* the material around the joint is melted in both the parts to be joined. If necessary a molten filler metal is added from a filler rod (or otherwise). The important *zones* in fusion welding are :

(*i*) Fusion zone ;

(*ii*) Heat affected unmelted zone around the fusion zone.

(*iii*) The unaffected original part.

— The characteristics of a fusion weld is that the metal being joined is *actually melted* and the union is produced on subsequent solidification.

Factors affecting fusion welding process :

(*i*) Nature of weld pool.

(*ii*) Chemical reaction in the fusion zone.

(*iii*) Characteristics of heat source.

(*iv*) Contraction, residual stresses and metallurgical changes.

(*v*) Heat flow from the joint.

● The fusion welding group includes :

1. Gas welding ;

2. Electric arc welding ;

3. Thermit welding.

Gas welding :

It is a method of fusion welding in which a flame produced by a combustion of gases is employed to heat and melt the parent metal and filler rod of a joint. It can weld most common materials.

7.6.2. Advantages and Disadvantages of Gas Welding :

Advantages :

1. The oxy-acetylene torch is *versatile*. It can be used for brazing, bronze welding, soldering, heating, heat treatment, metal cutting, metal cleaning, etc.

2. It is portable and can be moved almost everywhere for repair of fabrication work.

3. The oxy-acetylene flame is easily controlled and not as piercing as metallic arc welding, hence, extensively used for sheet metal fabrication work.

4. Welder has considerable control over the temperature of the metal in the weld zone. When the rate of heat input from the flame is properly coordinated with the speed of welding, the size, viscosity and surface tension of the weld puddle can be controlled, *permitting the pressure of the flame to be used to aid in positioning and shaping the weld.*

5. The cost and maintenance of the gas welding equipment is low when compared to that of some other welding processes.

6. The rate of heating and cooling is relatively low. In some cases, this is an advantage.

7. Good weld quality.

Disadvantages :

1. As compared to arc welding, it takes considerably longer time for the metal to heat up.

2. Owing to prolonged heating harmful thermal effects are aggravated which results is a larger heat affected area, increased grain growth, distortion and less of corrosion resistance.

3. Oxygen and acetylene gases are expensive.

4. Flux applications and the shielding provided by the oxy-acetylene flame are not so positive as those supplied by the inert gas in TIG, MIG or CO_2 welding.

5. The handling and storing of gas necessitate lot of safety precautions.

6. Heavy sections cannot be joined economically.

7. Flame temperature is less than the temperature of the arc.

8. Skilled operator required.

9. Difficult to prevent contamination.

10. Large heat affected zone.

7.6.3. Applications of Gas Welding

Following are the *applications of gas welding :*

1. To join most ferrous and non-ferrous metals, *e.g.,* carbon steels, alloy steels, cast iron, aluminium, copper, nickel, magnesium and its alloys, etc.

2. To join thin materials.

3. To join materials in whose case excessively high temperatures would cause certain elements in the metal to escape into the atmosphere.

4. To join materials in whose case excessively high temperatures or rapid heating and cooling of the job would produce unwanted or harmful changes in the metal.

5. Automative and Aircraft industries.

6. Sheet metal fabricating plants, etc.

7.6.4. Oxy-acetylene Welding

The oxy-acetylene welding process can be used for welding almost all metals and alloys used in engineering practice. The advantage of using acetylene, instead of other fuels, with oxygen is that it *produces a comparatively higher temperature and also an inert gas envelop, consisting of CO_2 and water vapours, which presents the molten metal from oxidation.*

Refer to Fig. 7.6. The *principle* of oxy-acetylene welding is the ignition of oxygen and acetylene gases, mixed in a blow pipe fitted with a nozzle of suitable diameter ; this flame is applied to the edges of the joint and to a wire filler of the appropriate metal, which is thereby melted and run into the joint. When the acetylene is burned in an atmosphere of oxygen an intensely hot flame with a temperature of about 3300°C is produced. As the melting point of steel is approximately 1300°C, the metal is fused very rapidly at the point at which the flame is applied.

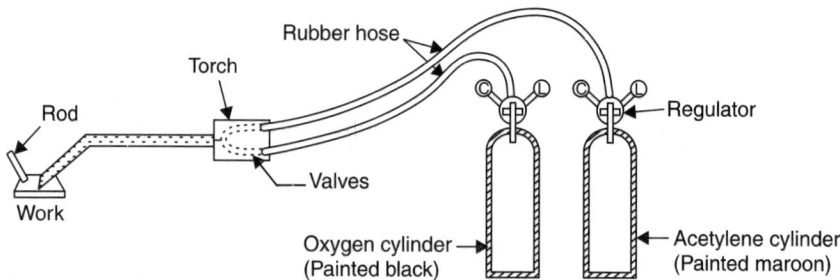

Fig. 7.6. Oxy-acetylene welding.

There are *two systems* of oxygen acetylene welding :

(*i*) **High pressure system.** In this method both oxygen and acetylene are derived for use from high pressure cylinders.

(*ii*) **Low pressure system.** In this system oxygen is taken as usual form a high pressure cylinder but *acetylene is generated,* by action of water on carbide (usually calcium carbide), in a low pressure acetylene generator.

- The use of an oxy-acetylene flame is the most widely employed method of *welding iron, steel, aluminium, cast-iron* and *copper*, the equipment required (Fig 7.6) being *considerably cheaper and simpler than that needed for electric welding.* For a certain class of mass production work, however, electric welding will always prove superior both in quickness and cheapness.

Methods of welding :

There are two methods of welding by means of the oxy-acetylene blow pipe :

(*i*) Leftward or forward welding.

(*ii*) Rightward or backward or backhand welding.

(*i*) *Leftward (or forward or forehand) welding :*

In *leftward welding* after suitable preparation of the joint the weld is commenced at the right-hand side of the joint and blow pipe is given a steady forward movement, with a slight sideways motion, zigzagging along the weld towards the left as shown in Fig. 7.7. The blow pipe is kept at an angle of 60° to 70° to the surface of the work so that the flame plays ahead of it, and the filler rod held at an angle of 30° to 40°, is held just ahead of the flame and progressively fed into it.

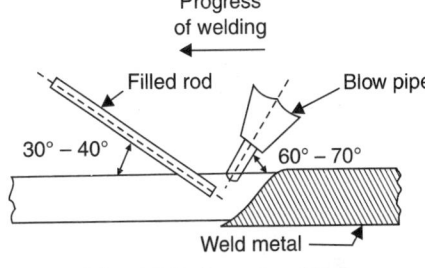

Fig. 7.7. Leftward welding.

— *Vertical joints are welded by this technique.*

- This technique is restricted to welding of *mild steel plates upto 5 mm thick, cast iron and non-ferrous metals.*

(*ii*) *Rightward (or backward or backhand) welding :*

In this welding technique the flame is directed towards the completed part of the joint and welding proceeds from left to right as shown in Fig. 7.8. The filler rod is given a circular movement as it is fed into the flame.

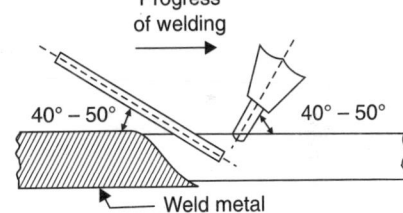

Fig. 7.8. Rightward welding.

— *Horizontal and overhead welding are usually done by the backhand technique.*

- The technique is used for *thicker materials, chiefly steel.*

Advantages :

(*i*) Rightward welding is faster by 20 to 25% from and 15 to 25% less acetylene is needed in comparison to leftward welding.

(*ii*) The mechanical properties of the weld are better due to the annealing effect of the flame which is directed on the completed weld.

(*iii*) The amount of distortion in the work is minimum.

Note : The angle at which the torch is inclined to the surface being welded depends upon the thickness of the metal. *Thicker metals require a higher concentration of heat and cousequently a larger torch angle.*

Types of flames :

Following are the *three types of flames of oxygen and acetylene mixture :*

1. Neutral flame
2. Carburising flame
3. Oxidising flame.

The brief description of these flames is given below :

1. *Neutral flame.* Refer to Fig. 7.9.

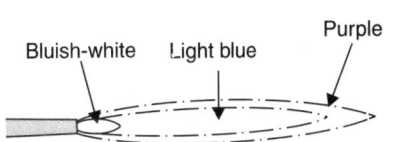

Fig. 7.9. Neutral flame (3250°C).

- When the *ratio of oxygen and acetylene is equal,* a neutral flame is obtained.

- This type of flame has a temperature of about 3250°C, is white in colour and has a sharply defined central cone with a reddish purple envelope.
- It does not react chemically with the parent metal and protects it (the metal) from oxidation.
- The neutral flame is *used to weld carbon steels, cast iron, copper, aluminium, etc.*

2. *Carburising flame.* Refer to Fig. 7.10.

- *The ratio of oxygen to acetylene is 0.9 to 1.* It consists of the following three zones :
  — Luminous zone,
  — Feather or intermediate cone of white colour, and
  — Outer envelope.

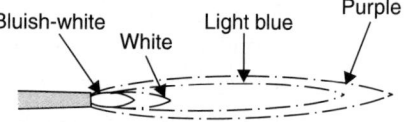

Fig. 7.10. Carburising flame (3150°C) (*Excess acetylene*).

- It is also called as *reducing flame* and has a temperature of 3150°C.
- *The carburising flame is used* for the following purposes :
  — To join those materials which are *readily oxidised*. Thus, it is used to weld *aluminium* since it prevents the formation of aluminium oxide at the time of welding.
  — To weld *monel metal, high carbon steel* and *alloy* steel.
  — To give a *hard facing* material in some cases.

3. *Oxidising flame.* Refer to Fig. 7.11.

- The *ratio of oxygen to acetylene varies from about 1.2 to 1.5.*
- It is used in the following cases :
  — To weld *copper, brass* and *bronze* and *zinc-bearing alloys.*
  — For *gas cutting.*

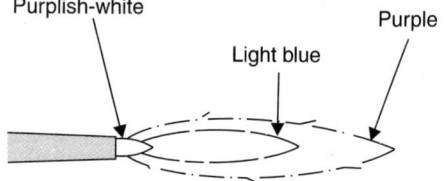

Fig. 7.11. Oxidising flame (3480°C) (*Excess oxygen*).

Other fuel gases. It may be noted that although, in gas welding, oxygen and acetylene mixture is popular, other fuel *gases like propane, hydrogen and coal gas may also be used, along oxygen to produce gas flames for welding.*

   — *Methyl acetylene propadiene (MAPP) gas is replacing acetylene gas particularly when portability is important,* because :

(*i*) It is more *dense,* thus providing *more energy* for a given volume.

(*ii*) It can be stored *safely in ordinary pressure tanks.*

Qualities of welding flame :

The welding flame should possess the following *qualities* :

(*i*) Must not burn the metal (oxidise it).

(*ii*) High temperature to melt the metals.

(*iii*) Products of combination should not be toxic.

(*iv*) Very intense concentrated flame so that a spot under the flame becomes molten and forms a liquid puddle.

(*v*) Must not add dirt or foreign material to the metal.

Equipment :

For gas welding following equipments are used :

1. Gas cylinders.

2. Pressure regulators.

3. Pressure gauges.

4. Welding torch.

5. Hoses and hose fittings.

6. Safety devices, etc.

The brief description of the above equipments is given below :

1. Gas cylinders :

A. *Oxygen cylinder :*

— For safety purposes oxygen cylinders are filled at a pressure 12500 to 14000 kN/m^2 and cylinder capacity is 6.23 m^3.

— The cylinder is provided with a right *hand thread valve* and is *painted* **black.**

— The cylinders are usually provided with fragile disc and fusible plug to relieve the cylinder of its contents if subjected to overheating or excessive pressure.

B. *Acetylene cylinder :*

— The cylinder is usually filled to pressure of 1600 to 2100 kN/m^2.

— The cylinder is provided with *left hand threads* for accommodating pressure regulator and is painted **maroon.**

— Acetylene gas above one atmospheric pressure is highly explosive. Hence, acetylene is stored with calcium silicate saturated with acetone. Acetone can absorb 25 times its own volume of acetylene for each atmosphere pressure.

2. Pressure regulators. The cylinders are provided with pressure regulators to control the working pressure of oxygen and acetylene to the welding torch. The pressure of oxygen and acetylene depends on the thickness of the metal to be welded/cut.

3. Pressure gauges. Two pressure gauges are fitted on each pressure regulator. While one pressure gauge shows the pressure inside the cylinder, the other one shows the working pressure of the fuel gas and oxygen.

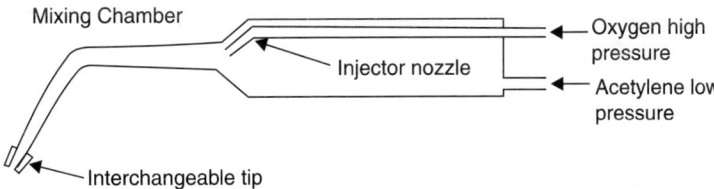

Fig. 7.12. Low pressure blow pipe.

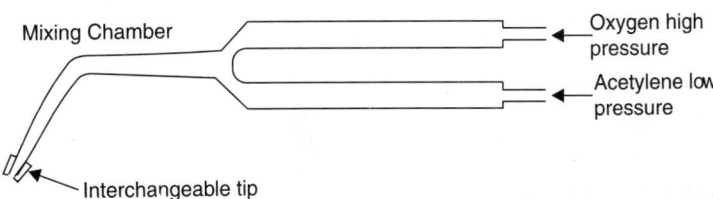

Fig. 7.13. High pressure blow pipe.

4. Welding torch/blow pipe. It is a device ; for moving oxygen and acetylene in the required volume and igniting it at the mouth of its tip. Generally, following two types of torches are available :

1. Low pressure blow pipe (Injector type).

2. High pressure blow pipe.

5. Hose and hose fittings :
- Hoses are the rubber and fabric pipes used to connect gas cylinder to blow pipe and are painted black or green for oxygen and red or maroon for acetylene. It should be strong, durable, non-porous and light.
- Special fittings are used for connecting hoses to equipment.

6. Safety devices :
- *Goggles* fitted with coloured glasses should be used to protect the eyes from harmful heat ultraviolet rays.
- *Gloves* made of leather, canvas and asbestos should be worn to protect hands from any injury. Gloves should be light so that the manipulation of the torch may be done easily.

Other requirements include *spark-lighter, apron, trolley, wire brush, spindle key, spanner set, filler rods and fluxes and welding tips.*

Welding rods (Filler materials) for gas welding :

The welding wire or rod used as filler material in gas welding *should have a chemical composition similar to that of the base metal.* The welding rod diameter, $d = \dfrac{t}{2} + 1$ mm (app.), where t is the thickness of the base metal, mm.

- Gas welding *"fluxes"* (composing of borates or boric acid, soda ash and small amount of other compounds *e.g.,* sodium chloride, ammonium sulphate and iron oxide) must melt at a lower temperature than the metals being welded so that surface oxides will be dissolved before the metal melts.

7.7. ELECTRIC ARC WELDING

7.7.1. Introduction

Arc welding *is the system in which the metal is melted by the heat of an electric arc.* It can be done with the following methods :
- (*i*) Metallic arc welding.
- (*ii*) Carbon arc welding.
- (*iii*) Atomic hydrogen welding.
- (*iv*) Shielded arc welding.

7.7.2. Advantages and Limitations

Following are the advantages and limitations of electric arc welding :
Advantages :
1. Portable and relatively inexpensive equipment.
2. Very versatile process.
Limitations :
1. Large heat affected zone.
2. Weld quality depends upon operator's skill in normal operations.
3. Not suitable for thin sections.

7.7.3. Metallic Arc Welding

Refer to Fig. 7.14. In metallic arc welding an arc is established between work and the filler metal electrode. The intense heat of the arc forms a molten pool in the metal being welded, and at the same time melts the tip of the electrode. As the arc is maintained, molten filler metal from the electrode tip is transferred across the arc, where it fuses with the molten base metal. Arc may be formed with direct or alternating current. Petrol or diesel driven generators are widely used for welding in open, where a normal electricity supply may not be available. D.C. may also be obtained from electricity mains through the instrumentality of a transformer and rectifier. A simple transformer

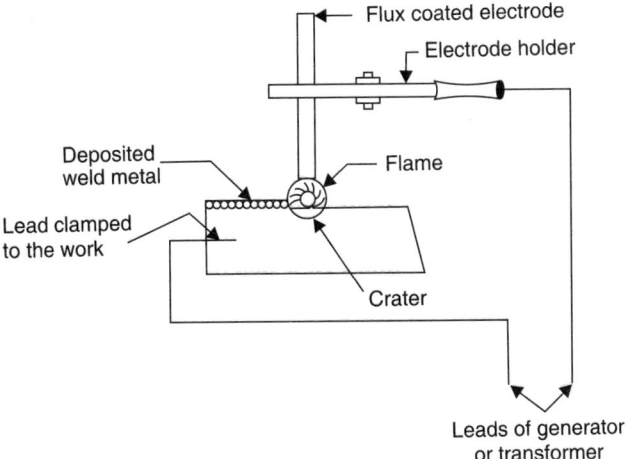

Fig. 7.14. Metallic arc welding.

is, however widely employed for A.C. arc welding. *The transformer sets are cheaper and simple having no maintenance cost as there are no moving parts.*

- With Arc system, the *covered or coated electrodes are* used, whereas with D.C. system for cast iron and non-ferrous metals, bare electrodes can be used.
- In order to strike the arc an open circuit voltage of between 60 to 70 volts is required. For maintaining the short arc 17 to 25 volts are necessary ; the current required for welding, however, varies from 10 amp. to 500 amp. depending upon the class of work to be welded.
- The great *disadvantage* entailed by D.C. welding is the presence of *arc blow* (distortion of arc stream from the intended path owing to magnetic forces of a non-uniform magnetic field). With A.C. arc blow is considerably reduced and use of higher currents and large electrodes may be restored to enhance the rate of weld production.

Applications :
- *The field of application of metallic arc welding includes mainly low carbon steel and the high-alloy austenitic stainless steel.*
- Other steels like low and medium-alloy steels can however be welded by this system but many precautions need be taken to produce ductile joints.

7.7.4. Carbon Arc Welding

Refer to Fig. 7.15. Here the work is connected to negative and the carbon rod or electrode connected to the positive of the electric circuit. Arc is formed in the gap, filling metal is supplied by fusing a rod or wire into the arc by allowing the current to jump over it and it produces a porous and brittle weld because of inclusion of carbon particles in the molten metal. It is therefore *used for filling blow holes in the castings which are not subjected to any of the stresses.*

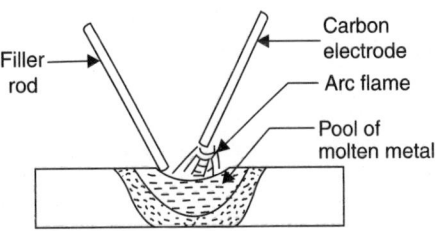

Fig. 7.15. Carbon arc welding.

- The voltage required for striking an arc with carbon electrodes is about 30 volts (A.C.) and 40 volts (D.C.).
- A *disadvantage* of carbon arc welding is that *approximately twice the current* is required to raise the work to welding temperature as compared with a metal electrode, while a carbon electrode can only be used economically on D.C. supply.

7.7.5. Atomic Hydrogen Welding

Refer to Fig. 7.16. In this system heat is obtained from an alternating current arc drawn between two *tungsten electrodes in an atmosphere of hydrogen.* As the hydrogen gas passes through the arc, the hydrogen molecules are broken up into atoms and they recombine on contact with the cooler base metal generating intense heat sufficient to melt the surfaces to be welded, together with the filler rod, if used. The envelop of hydrogen gas also shields the molten metal from oxygen and nitrogen and thus prevents weld metal from deterioration.

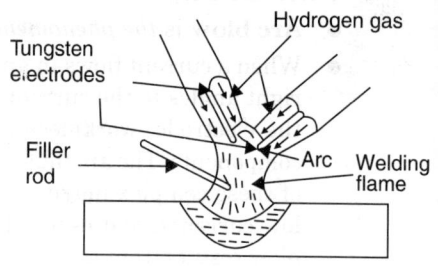

Fig. 7.16. Atomic hydrogen welding.

- The welds obtained are homogeneous and smooth in appearance because the hydrogen keeps the molten pool.

Advantages :

1. No flux or separate shielding gas is used ; hydrogen itself acts as a shielding gas and avoids weld metal oxidation.

2. Due to high concentration of heat, welding can be carried out at fast rates (specially when filler metal is not needed) and with less distortion of the workpiece.

3. Welding of thin materials is also possible which otherwise may not be successfully carried out by metallic arc welding.

4. The job does not form a part of the electrical circuit. The arc remains between two tungsten electrodes and can be moved to other places easily without getting extinguished.

Limitations :

1. For certain applications, the process becomes uneconomical because of higher operating cost as compared to that of other welding processes.

2. The process cannot be used for depositing large quantities of metals.

3. Welding speed is less as compared to that of metallic arc or MIG welding.

Applications :

- *Atomic hydrogen welding being expensive is used mainly for high grade work on stainless steel and most non-ferrous metals.*

7.7.6. Shielded Arc Welding

In this system molten weld metal is protected from the action of atmosphere by an envelope of chemically reducing or inert gas.

As molten steel has an affinity for oxygen and nitrogen, it will, if exposed to the atmosphere, enter into combination with these gases forming oxides and nitrides. Due to this injurious chemical combination metal becomes weak, brittle and corrosion resistant. Thus, several methods of shielding have been

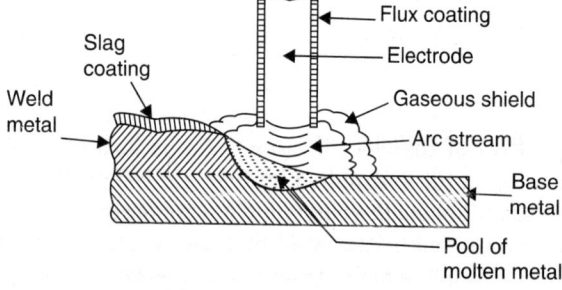

Fig. 7.17. Shielded arc welding.

developed. The simplest (Fig. 7.17) is the *use of a flux coating on the electrode* which in addition to producing a slag which floats on the top of the molten metal and protects it from atmosphere, has organic constituents which turn away and produce an envelope of inert gas around the arc and the weld.

- Welds made with a completely shielded arc are more *superior* to those deposited by an ordinary arc.

7.7.7. Arc Blow

- **Arc blow** is *the phenomenon of wandering of arc and it occurs in D.C. welding.*
- When a current flows in any conductor, a magnetic field is formed around the conductor at right angles to the current. Since in the case of D.C. arc welding, there is current through the electrode, workpiece and ground clamp, magnetic field exists around each of these components. The arc thus lacks control as though it were being blown to and by the influence of these complex magnetic fields. This is more common in welding with very high or very low currents, and especially in welding in corners or other confined spaces. Usually arc blow results from the interaction of magnetic fields of the electrode workpiece with that of the arc. *The movement of arc blow causes atmospheric gases to be pulled into the arc, resulting in porosity or other defects.*

The severity of arc blow problem can be reduced by taking the following corrective measures :

1. *Change to A.C. welding*, if possible (since due to change in the polarity, the effect of magnetic field is nullified).
2. *Reduce the current used* so that the strength of magnetic field is reduced.
3. *Use a short arc length* so that filler metal would not be deflected but carried easily to the arc crater.
4. Place more than one ground lead from the base metal (preferably on each from the ends of the base metal plate).
5. The ground cable may be wrapped around the workpieces such that the current flowing in it sets up a magnetic field in a direction which will counteract the arc blow.

7.7.8. Comparison between A.C. and D.C. Arc Welding

The Comparison between A.C. and D.C. arc Welding is given below :

S. No.	Aspects	A.C. Welding	D.C. Welding
1.	Power consumption	Low	High
2.	Arc stability	Arc unstable	Arc stable
3.	Cost	Less	More
4.	Weight	Light	Heavy
5.	Efficiency	High	Low
6.	Operation	Noiseless	Noisy
7.	Suitability	Non-ferrous metals cannot be joined	Suitable for both ferrous and non-ferrous metals
8.	Electrode used	Only coated	Bare electrodes are also used
9.	Welding of thin sections	Not preferred	Preferred
10.	Miscellaneous	Work can act as cathode while electrode acts as anode and *vice versa.*	Electrode is always negative and the work is positive.

Specifications of A.C. Transformer/D.C. generator :

A.C. transformer : Step down, oil cooled = 3 phase, 50 Hz ; Current range = 50 to 400 A ; Open circuit voltage = 50 to 90 V ; Energy consumption = 4 kWh per kg of metal deposit ; Power factor = 0.4 ; Efficiency = 85%.

D.C. generator : Motor generator—3 phase, 50 Hz ; Current range = 125 to 600 A ; Open circuit voltage = 30 to 80 V ; Arc voltage = 20 to 40 V ; Energy consumption = 6 to 10 kWh/kg of deposit ; Power factor = 0.4 ; Efficiency = 60%.

Electrodes :

The electrodes may be of the following two types :

1. *Consumable electrode :*

 (*i*) Base electrode

(*ii*) Flux coated electrode.

2. *Non-consumable electrode :*

1. *Consumable electrode :*

 (*i*) *Bare electrode :*

● These electrodes do not prevent oxidation of the weld and hence the joint is weak. They are used for minor repairs where strength of the joint is weak.

● Employed in automatic and semi-automatic welding.

(*ii*) *Flux-coated electrode :*

● The flux is provided to serve the following *purposes* :

   — To prevent oxidation of the weld bead by creating a gaseous shield around the arc.

   — To make the formation of the slag easy.

   — To facilitate the stability of the arc.

2. *Non-consumable electrode :*

● These electrodes are 12 mm in diameter and 450 mm long.

● These are not consumed during the welding process.

Examples of these electrodes are : *Carbon, graphite and tungsten.*

7.7.9. Types of Welded Joints

The type of joint is determined by the relative positions of the two pieces being joined.

The following are the *five* basic types of commonly used joints :

1. Lap joint

2. Butt joint

3. Corner joint

4. Edge joint

5. T-joint.

1. *Lap joint.* Refer to Fig. 7.18.

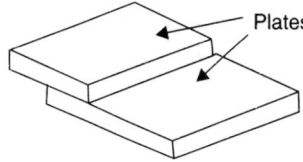

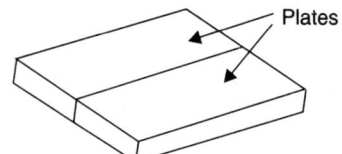

Fig. 7.18. Lap joint.   Fig. 7.19. Butt joint.

● The lap joint is obtained by overlapping the plates and then welding the edges of the plates.

● The lap joints may be *single traverse, double traverse and parallel* lap joints.

● These joints are employed on plates having thickness *less than 3 mm.*

2. *Butt joint :*

● The butt joint is obtained by placing the plates edge to edge as shown in Fig. 7.19.

● In this type of joints, if the plate thickness is *less than 5 mm,* bevelling is *not* required. When the thickness of the plates ranges *between 5 mm to 12.5 mm,* the edge is required to be bevelled to V or U-groove, while the plates having thickness *above 12.5 mm* should have a V or U-groove on both sides.

● The various types of butt joints are shown in Fig. 7.20.

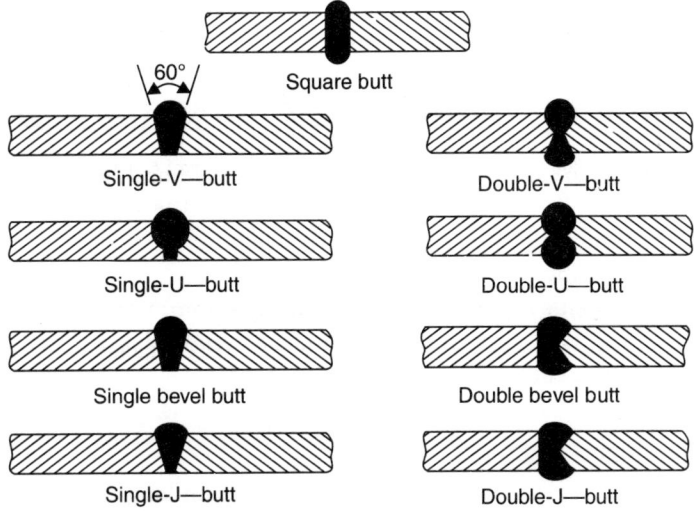

Fig. 7.20. Various types of butt joints.

3. Corner joint. Refer to Fig. 7.21.

● A corner joint is obtained by joining the edges of two plates whose surfaces are at an angle of 90° to each other.

● In some cases corner joint can be welded, without any filler metal, by melting off the edges of the parent metal.

● This joint is *used for both light and heavy gauge sheet metal.*

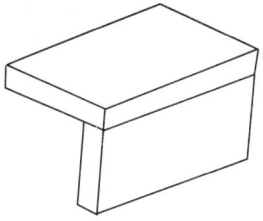

Fig. 7.21. Corner joint.

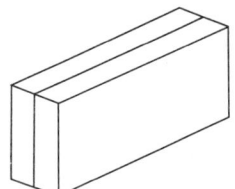

Fig. 7.22. Edge joint.

4. Edge joint. Refer to Fig. 7.22.

● This joint is obtained by joining two parallel plates.

● It is *economical for plates having thickness less than 6 mm.*

● It is *unsuitable for members subjected to direct tension or bending.*

5. T-joint. Refer to Fig. 7.23.

● It is obtained by joining two plates whose surfaces are approximately at right angles to each other.

● These joints are suitable up to 3 mm thickness.

● *T*-joint is widely *used to weld siffeners in aircraft and other thin walled structures.*

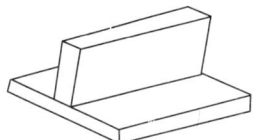

Fig. 7.23. *T*-joint.

Note : The lap joints, corner joints and T-joints are known as **fillet weld joints.** The fillet cross-section is approximately triangular. Fig. 7.24 shows the three types of fillet welds.

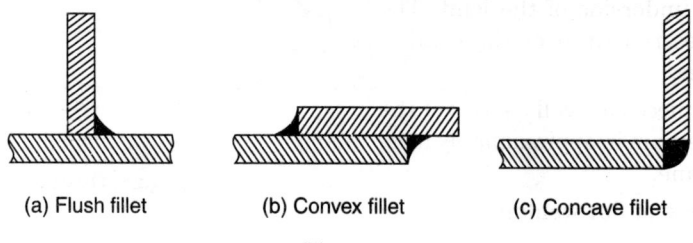

(a) Flush fillet   (b) Convex fillet   (c) Concave fillet

Fig. 7.24

Welding positions :

It is easiest to make welds in *flat positions, i.e.,* both the parent metal pieces lying in horizontal plane over a flat surface. But, several times it becomes unavoidable to weld the workpieces in some other positions also. The common welding positions are :

1. Flat position
2. Horizontal position
3. Vertical position
4. Overhead position.

1. *Flat position.* Refer to Fig. 7.25.

- In this welding position, the welding is done from the upper side of the joint and the welding material is normally applied in the downward direction.
- On account of the downward direction of application of welding material this position is also sometimes called as *downward position.*

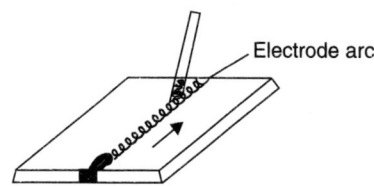

Fig. 7.25. Flat position.

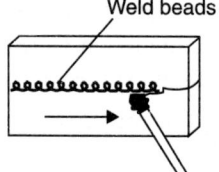

Fig. 7.26. Horizontal position.

2. *Horizontal position.* Refer to Fig. 7.26.

In this case, the weld is deposited upon the side of a horizontal and against a vertical surface.

3. *Vertical position.* Refer to Fig. 7.27.

- In this position, the axis of the weld remains either vertical or at an inclination of less than 45° with the vertical plane.
- The welding commences at the bottom and proceeds upwards.
- The tip of the torch is kept pointing upwards so that the pressure of the outcoming gas mixture forces the molten metal towards the base metal and prevents it from falling down.

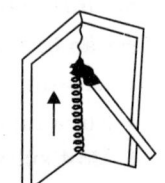

Fig. 7.27. Vertical position.

4. Overhead position. Refer to Fig. 7.28.

- In this case, the welding is performed from the underside of the joint. The workpieces remain over the head of the welder.
- The workpieces as well as axis of the weld all remain in approximately horizontal plane.
- It is reverse of flat welding.

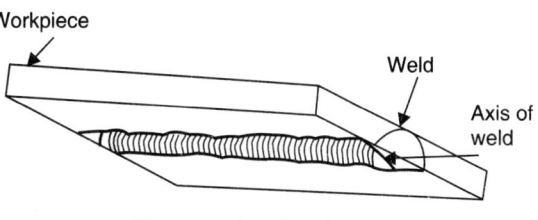

Fig. 7.28. Overhead position.

7.8. THERMIT WELDING

Refer to Fig. 7.29. *It is the method of uniting iron or steel parts by surrounding the joint with steel at a sufficient high temperature to fuse the adjacent surfaces of the parts together.*

- Here a wax pattern of desired size and shape is prepared around the joint or region where the weld is to be affected.

- The wax pattern is then surrounded by sheet iron box and the space between box and pattern is filled and rammed with sand.

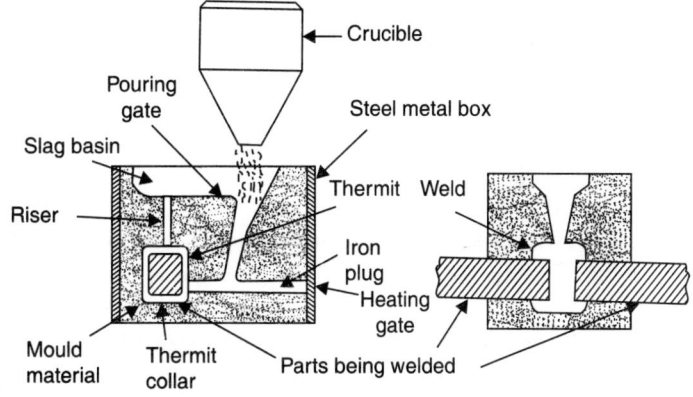

Fig. 7.29. Thermit welding.

- After cutting, pouring and heating gates and risers a flame is directed into the heating oven due to which the wax pattern melts and drains out, the heating is continued to raise the temperature of the parts to be welded.

- The thermit mixture (finely divided aluminium iron oxide) is packed in the crucible of conical shape formed from a sheet-iron casting lined with heat resisting cement and is ignited with magnesium or torch yielding a highly superheated (nearly 3000°C) molten-iron and a slag of aluminium oxide (the reaction is : $8Al + 3 Fe_3O_4 = 4 Al_2O_3 + 9Fe$ + heat).

- The molten iron is then run into the mould which fuses with the parts to be welded and forms a thermit collar at the joint. *The welds thus obtained are metallurgically very sound and strong.*

Advantages :

1. Can be used anywhere.
2. Low set-up cost.
3. Not a highly skilled operation.
4. Most suitable for welding of thick sections.

Limitations :

1. Only thick sections can be welded.
2. High set-up and cycle time.

Applications :

- The process is widely employed in the *shipping, steel and railroad industries.*
- It can also be used for welding non-ferrous parts by selection of a mixture of oxides which on reduction with aluminium will provide an alloy approximating the material to be welded.

7.9. TUNGSTEN INERT-GAS (TIG) WELDING

This welding process is also called Gas Tungsten Arc Welding (GTAW)

Refer to Fig. 7.30. In this process the heat necessary to melt the metal is provided by a very intense electric arc which is struck between a virtually *non-consumable tungsten electrode and metal workpiece. The electrode does not melt* and become a part of the weld. On joints where filler metal is required, a welding rod is fed into the weld zone and melted with base metal in the same manner as that used with oxyacetylene welding. The weld zone is shielded from the atmosphere by an inert-gas (a gas which does not combine chemically with the metal being welded) which is ducted directly to the weld zone where it surrounds the tungsten. The major inert gases that are used are *argon* and *helium.*

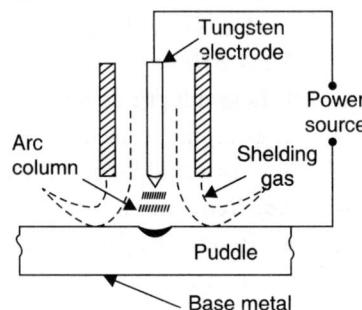

Fig. 7.30. Tungsten inert-gas (TIG) welding.

TIG process offers the following *advantages :*

1. TIG welds are stronger, more ductile and more corrosion resistant than welds made with ordinary shield arc welding.

2. Since no granular flux is required, it is possible to use a wide variety of joint designs than in conventional shield arc welding or stick electrode welding.

3. There is little weld metal splatter or weld sparks that damage the surface of the base metal as in traditional shield arc welding.

Applications :

(*i*) The TIG process lends itself ably to the fusion welding of *aluminium and its alloys, stainless steel, magnesium alloys, nickel base alloys, copper base alloys, carbon steel and low alloy steels.*

(*ii*) TIG welding can also be used for the combining of dissimilar metals, hard facing, and the surfacing of metals.

7.10. METAL INERT-GAS (MIG) WELDING

This welding process is also called Gas Metal Arc Welding (GMAW).

Refer to Fig. 7.31. The inert-gas *consumable electrode process,* or the MIG process is a refinement of the TIG process, however, in this process, the *tungsten electrode has been replaced with a consumable electrode.* The electrode is driven through the same type of collet that holds a tungsten electrode by a set of drive wheels. *The consumable electrode in MIG process acts as a source for the arc column as well as the supply for the filler material.*

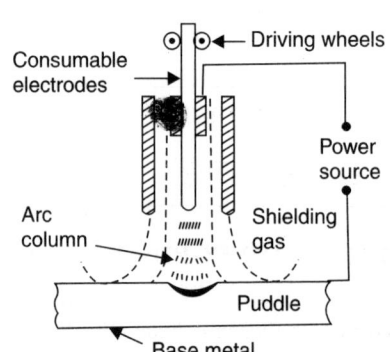

Fig. 7.31. Metal inert-gas welding (MIG).

MIG welding employs the following three basic processes.

1. Bare-wire electrode process

2. Magnetic flux process

3. Flux-cored electrode process.

Advantages :

1. It provides higher deposition rate.
2. It is faster than shielded metal-arc welding due to continuous feeding of filler metal.
3. Welds produced arc of better quality.
4. There is no slag formation.
5. Deeper penetration is possible.
6. The weld metal carries low hydrogen content.
7. More suitable for welding of thin sheets.

Limitations :

1. Less adaptable for welding in difficult to reach portions.
2. Equipment used is costlier and less portable.
3. Less suitable for outdoor work because strong wind may blow away the gas shield.

Applications :

● Practically all commercially available metals can be welded by this method.
● It can be used for deep groove welding of plates and castings, just as the submerged arc process can, but it is more advantageous on light gauge metals where high speeds are possible.

7.10.1. Difference between TIG and MIG Welding Processes

The difference between TIG and MIG welding processes is given in tabular form below :

S. No.	Aspects	TIG welding	MIG welding
1.	Name of the process	Tungsten inert-gas welding.	Metal inert-gas welding.
2.	Type of electrode used	Non-consumable tungsten electrode.	Consumable metallic electrode.
3.	Electrode feed	Electrode feed not required.	Electrode need to be fed at a constant speed from a wire reel.
4.	Electrode holder	It is called welding torch and has got a cap filled on the back to cover the tungsten electrode. It has also got connections for shielding gas, cooling water and control cable. It may be air-cooled also.	It is called welding gun or torch. It has facility to continuously feed wire electrodes ; shielding inert-gas, cooling water and control table.
5.	Welding current	Both A.C. and D.C. can be used.	D.C. with reverse polarity is used.
6.	Feed metal	Filler metal may or may not be used.	Filler metal in the form of fire wire is used.
7.	Bases metal thickness	Metal thickness which can be welded is limited to about 5 mm.	Thickness limited to about 40 mm.
8.	Welding speed	Slow.	Fast.

7.11. SUBMERGED ARC WELDING

The *submerged arc process* (which may be done manually or automatically) *creates an arc column between a base metallic electrode and the workpiece.*

— The arc, the end of the electrode, and the molten weld pool are *submerged in a finely divided granulated powder that contains appropriate deoxidizers, cleansers and any other fluxing elements.*

— The fluxing powder is fed from a hopper that is carried on the welding head. The tube from the hopper spreads the powder in continuous mount in front of the electrode along the line of the weld.

— This *flux mound is of sufficient depth to submerge completely the arc column so that there is no splatter or smoke,* and the weld is shielded from all effects at atmospheric gases. As a *result of this unique protection, the weld beads are exceptionally smooth.*

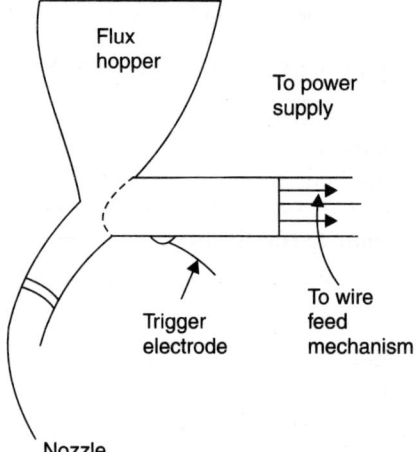

— The flux adjacent to the arc column melts and floats to the surface of the molten pool ; then it solidifies to form a slag on the top of the welded metal. The rest of the flux is simply an insulator that can be reclaimed easily.

— The slag that is formed by the molten flux solidifies and is easy to remove. In fact, in many applications, the slag will crack off by itself as it cools.

— The unused flux is removed and placed back into the original hopper for use for the next time.

— Granulated flux is a complex, metallic for silicate that can be used over a wide range of metals.

● The process is characterised by high welding currents. The current density in the electrode is 5 to 6 times that used in ordinary manual stick electrode arc welding, consequently the melting rate of the electrode as well as the speed of

Fig. 7.32. Apparatus used in manual submerged arc welding.

welding is much higher than in the manual stick electrode process.

Fig. 7.32 shows an apparatus used in manual submerged arc welding.

● *Welds made by the submerged arc welding process have high strength and ductility with low hydrogen or nitrogen content.*

Advantages :

1. Higher welding speeds can be employed, effecting saving in welding time.
2. Very high deposition rate.
3. Flux acts as a deoxidiser to purify the weld metal.
4. Shallow grooves can be used for making joints, requiring less consumption of filler metal. In some cases no edge preparation is at all needed.
5. No chance of weld spatter (since the arc is always covered under flux blanket).
6. If required, the flux may contain alloying elements and transfer them to the weld metal.
7. Can be employed with equal success for both indoor and outdoor welding work.
8. Less distortion.
9. Few passes are required due to deep penetration.
10. It is often used in automatic mode.

Limitations :

1. This process can be performed only in flat and horizontal welding positions.
2. In order to obtain good weld the base metal has to be cleaned and made free of dirt, grease, oil, rust and scale.
3. Flux may get contaminated and lead to porosity in weld.
4. Normally unsuitable for welding of metal thickness less than 4.8 mm.
5. Removal of slag is an additional follow-up operation. In multiplass beads it has to be done after every pass.

Applications :

● This process is suitable for welding low-alloy, high tensile steels as well as the mild, low-carbon steels.

● This process is also capable of joining medium carbon steels, heat resistance steels, and many of high-strength steels.

● Also the process is adaptable to nickel, monel and many other non-ferrous metals.

● The submerged arc process is also capable of welding *fairly thin gauge materials.*

7.12. ELECTRO-SLAG AND ELECTRO-GAS WELDING

These methods are employed to *fuse two sections of thick metal,* forming a seam in a *single pass.* Elimination of the need for making multiple passes and special joint preparations make these methods commonly used welding processes when heavy ferrous metals are to be joined. These processes have reduced costly time in fabrication of *large vessels and tanks.* There is theoretically no limit to the thickness of the weld bead.

7.12.1. Electro-slag Welding

Refer to Fig. 7.33. This process is a vertical and uphill ; two copper shoes, dams, or moulds must be placed on either side of the joint that is to be welded in order to keep the molten metal in the joint area.

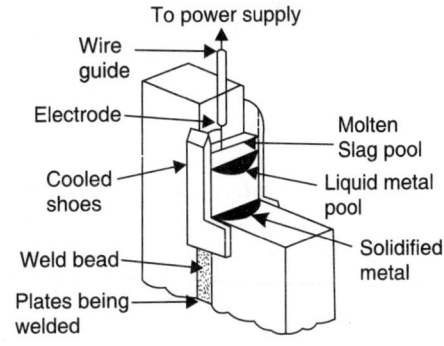

— One or more electrodes may be used to weld a joint, depending upon the thick-ness of the metal. The electrodes are fed into the weld joint almost vertically from special wire guides. Electrodes need not be of a special deoxidized nature but they may contain a flux, if it is needed.

— A mechanism for raising the equipment as the weld is completed and A.C. power source that has approximately 100 amperes output and a 100 per cent duty cycle are needed.

Fig. 7.33. Electro-slag welding.

● *Electro-slag welding depends upon the generation of heat that is produced by passing an electric current through molten slag.*

Applications :

Welding of *heavy steel forgings, large steel castings, thick steel plates* and *heavy structural members.*

7.12.2. Electro-gas Welding

Refer to Fig. 7.34. Electro-gas welding works on the same general principle as electro-slag welding, with *the addition of some of the principles of submerged arc welding.*

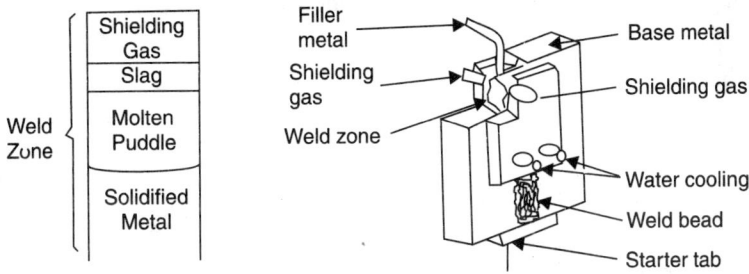

Fig. 7.34. Electro-gas welding.

- The major difference between electro-slag and electro-gas welding is that an inert gas, such as CO_2, is used to shield the weld from oxidation, and there is continuous arc, such as in submerged arc welding, to heat the weld pool. The joints and the use of flux to cleanse the weld are the same as in electro-slag process. The shoes that are used to form the weld, as in electro-slag process, are also used in the electro-gas process to control the weld zone through water cooling. However, the flux, instead of being issued to the weld zone through a hopper mechanism, is *incorporated within the electrode itself in the form of cored wires.*

Applications :

Welding of low carbon and medium carbon steels, and with specific precautions for alloy steels and stainless steels as well.

— Thickness ranges commonly welded are from 12 mm to 75 mm ; for thinner sections other processes prove more economical and for thickness above 75 mm electro-slag process proves superior.

7.13. ELECTRON-BEAM WELDING

Electron-beam welding fusion joins metal by bombarding a specific confined area of the base metal with high velocity electrons. The operation is performed in a vacuum to prevent the reduction of electron velocity. If a vacuum were not used, the electrons would strike the small particles in the atmosphere, reducing their velocity and decreasing their heading ability.

- The electron beam welding process allows fusion welds of great depth with a minimum width because the beam can be focused and magnified (Fig. 7.35). The depth of the weld bead can exceed the width of the weld bead by as much as 15 times.

- The *process joins separate pieces of base metal by fusing of molten metals. The melting is achieved by a concentrated bombardment of a dense stream of electrons, which are accelerated at high velocities,* sometimes as high as the speed of light. Under most circumstances the entire process is done inside a vacuum chamber.

- Most chambers house not only the workpiece but also the cathode, the focusing device and the remainder of the gun, preventing contamination of the weldment and the electron-beam gun itself (Fig. 7.36).

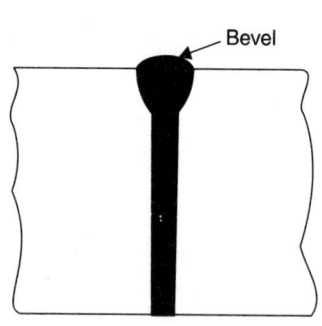

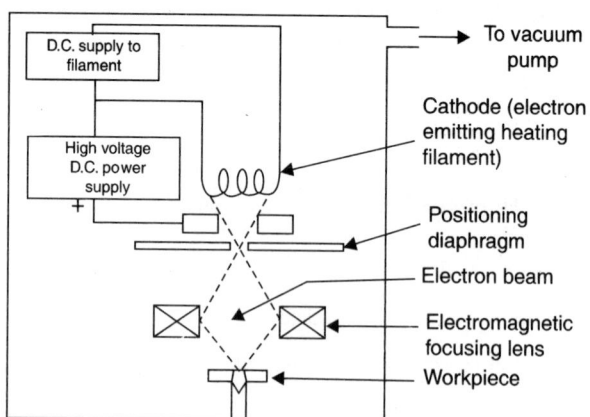

Fig. 7.35. Electro-beam welding.   Fig. 7.36. Electron-beam gun.

Advantages :

1. The greatest advantage of electron-beam welding is that *it eliminates contamination of both the weld zone and the weld head* because of the vacuum in which the weld is done because of the electrons doing the heating.

2. Even though initial costs are high, *operating costs are low* due to the low power usage. Many of the more costly fabrication methods could be replaced by electron beam process.

3. The narrow beam *reduces the distortion of the workpiece, making the replacement of costly jigs and fixtures less necessary* than when using other types of welding processes.

4. The speed may be as fast as 2500 mm/min and it will weld or cut any metal or ceramic, diamond, sometimes as thick as 150 mm.

5. Clean and sound welds.

6. Energy conversion efficiency is high, about 65%.

Limitations :

1. High operating costs.

2. Expensive equipment.

3. Limitations of the vacuum chamber. Work size is limited by the size of the chamber.

4. High cost of precision joint preparation and precision tooling.

Applications :

Welding of automobile, airplane, aerospace, form and other types of equipment including ball bearing over 100 mm.

7.14. ULTRASONIC WELDING

A schematic diagram of a typical ultrasonic welding is shown in Fig. 7.37. The welding equipment consists of two units :

    (*i*) *A power source of frequency converter which converts 50 cycle line power into high frequency electric power.*

    (*ii*) *A transducer which changes the high frequency electric power into vibratory energy.*

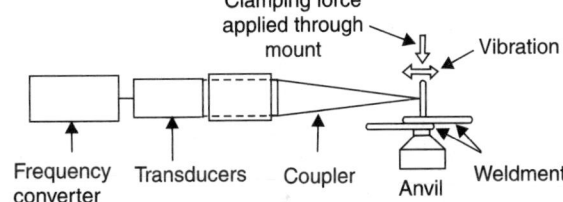

Fig. 7.37. Ultrasonic welding.

The components to be joined are simply clamped between a welding tip and supporting anvil with just enough pressure to hold them in close contact. The high frequency vibratory energy is then transmitted to the joint for the required period of time. The bonding is accomplished without applying external heat, filler rod or melting metal.

• Either spot-type welds or continuous-seam welds can be made on a variety of metals ranging of thickness from 0.000425 mm (aluminium foil) to 0.25 mm.

• Thicker sheet and plate can be welded if the machine is specifically designed for them.

• High strength bonds are possible both in similar and dissimilar metal combinations.

Advantages :

1. High productivity.

2. Thin pieces can be welded to thicker pieces.

3. Welds are free from foreign inclusions.

4. Post cleaning of welds is usually not necessary.

5. Very little preparation is required for the weld ; usually it involves degreasing.

Applications :

Ultrasonic welding is particularly adaptable for :

1. Joining electrical and electronic components.

2. Thermatic sealing of materials and devices.

3. Splicing metallic foil.

4. Welding aluminium wire and sheet.

5. Fabricating nuclear fuel elements.

7.15. PLASMA ARC WELDING

- *Plasma* is often considered the *fourth state of matter.* The other three are gas, liquid and solid. Plasma results when a gas is heated to high temperature and changes into positive ions, neutral atoms and negative electrons. When a matter passes from one state to another latent heat is required to change into steam, and similarly, the plasma torch supplies energy to a gas to change it into plasma. When the *plasma changes back to a gas, the heat is released. Any high current arc is composed of plasma, which is nothing more than an ionized conducting gas.* The plasma gas is forced through the torch, surrounding the cathode. The main function of the plasma gas is shielding the body of the torch from the extreme heat of the cathode. Argon and argon mixtures are most commonly used (since they do not attack tungsten or copper cathode).

- *Plasma arc consists of an electronic arc plasma gas, and gases used to shield the jet column. The equipment necessary for plasma arc welding includes a conventional D.C. power supply with a drooping volt ampere output and with 70 open line volts.*

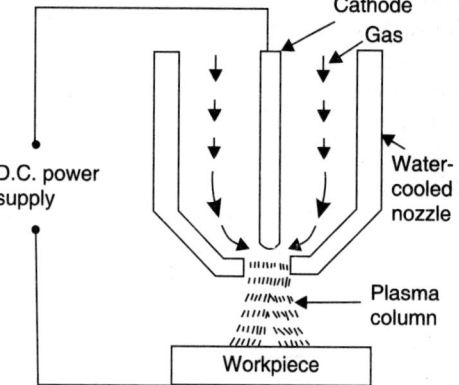

Fig. 7.38. Transferred arc plasma jet torch.

- The two main types of torches for welding and cutting with plasma arc are :

  (*i*) Transferred arc, and

  (*ii*) Non-transferred arc.

- The **'transferred arc'** plasma jet torch (Fig. 7.38) is similar to TIG torch, except that it has the water-cooled nozzle between the electrode and the work. This nozzle constricts the arc, increasing its pressure.

The plasma, caused by the collision of gas molecules with high-energy electron, is then swept out through the nozzle, forming the main current path between the electrode and the workpiece. *The plasma arc and transferred arc are generated between the tungsten electrode or cathode and the workpiece, or anode.*

- The **'Non-transferred arc'** torch extends the arc from the electrode, or the cathode, to the end of the nozzle. *The nozzle acts as the anode. This type of plasma jet is completely independent of the workpiece, with the power supply contained within the equipment.*

Key-hole method is used in actual process of welding with plasma jet. Jet column burns a small hole through the materials to be welded. As the torch progresses along the material the hole progresses also. However, it is filled by molten metal as the torch passes. Hundred per cent penetration is ensured by this method.

7.16. LASER BEAM WELDING

- The laser welding process is the *focusing of a monochromatic light into extremely concentrated beams.* It employs a carefully *focused beam of light that concentrates tremendous amount of energy on a small area to produce fusion.*

Refer to Fig. 7.39. The laser welding system comprises the following :

1. Electrical storage unit.
2. Capacitor bank.
3. Triggering device.
4. Flash tube that is wrapped with a wire.
5. Lasing material.
6. Focusing lens mechanism.
7. Work-table (operatable in three axes X, Y and Z).

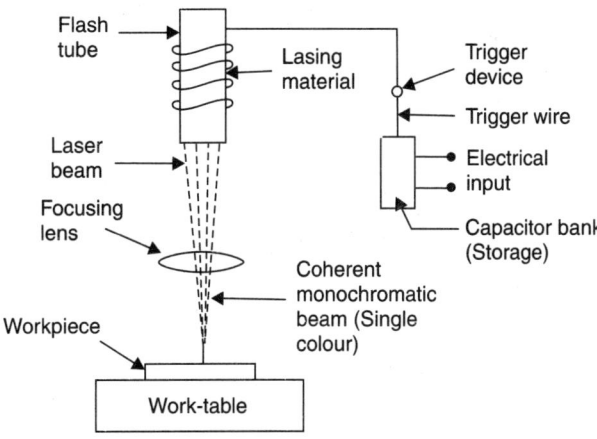

Fig. 7.39. Laser beam welding.

When capacitor bank is triggered energy is injected into the wire that surrounds the flash tube. This wire establishes an imbalance in the material inside the flash tube. Thick xenon often is used in the material for the flash tube, producing high power levels for very short period of time. The flash tubes or lamps are designed for operation at a rate of thousands of flashes per second. By operating in this manner, the lamps become an efficient device for converting *electrical energy into light energy,* the process of pumping the laser. The laser is then activated. The beam is emitted through the coated end of the lasing material. It goes through a focusing device where it is pin-pointed on the workpiece. Fusion takes place and the weld is accomplished.

- Both the Nd : YAG and CO_2 lasers may be used for welding.
  - Since *Nd : YAG* is laser is pulsed, it is ideal for producing *spot and seam welds.*
  - CO_2 laser can produce deeper welds at higher rates of speed than possible with the Nd : YAG laser.
  - Butt joints make the most efficient use of laser energy.

Advantages :

1. This process can be used to weld *dissimilar metals with widely varying physical properties.*
2. *Metals with relatively high electrical resistance and parts of considerably different sizes and mass can be welded.*
3. Because the laser is simply a beam, *no electrode is required,* so that any part in a particular position can be welded if there is a direct line of sight from beam to the workpiece.
4. Welds can be made with a *high degree of precision* and on material that is only a few thousands of a centimetre thick.
5. Laser welding holds thermal distortion and shrinkage to a minimum.
6. High depth-to-width ratio of weld.
7. Welding can be done in inaccessible locations.
8. High production rate.
9. Work cells can be made to cut and trim parts in addition to welding.
10. Welds can be made in air or with shielding gas.

Limitations :

1. High energy losses.
2. Highly skilled operation.
3. High equipment cost (but not as high as for electron-beam welding)
4. Eye protection required.

5. Suitable for narrow and deep joints.

6. Weld joints made for lasers must be prepared and fixtured to close tolerances.

7. Welts are normally limited to materials 0.3 mm thick or less.

7.17. FRICTION WELDING

In this welding process, often termed as *"inertia welding"*, (Fig. 7.40) the two surfaces to be welded are rotated relative to each other under light normal pressure. When the interface temperature increases due to frictional rubbing and when it reaches the required welding temperature, sufficient normal pressure is applied and maintained until the two pieces get welded.

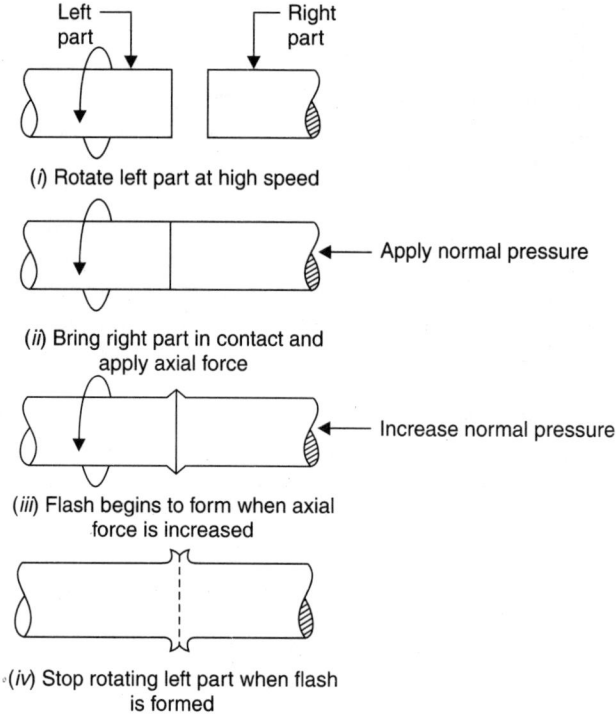

Fig. 7.40. Friction welding.

- The shape of the welded joint depends on the rotational speed and the axial force applied ; these factors must be controlled to obtain a uniformly strong joint. The *radially outward movement of the hot metal at the interface pushes oxides and other contaminants out of the interface.*

- A wide variety of metals and metal combinations can be welded by this process. Filler metals, fluxes, or shielding gases are not required, and welds can be made with a minimum of joint preparation.

- The method is *most suitable for circular parts, that is, butt welding of round bars or tubes.*

Advantages :

1. High quality welds.

2. The process is clean.

3. Low initial capital cost.

4. Low-cost power requirements.

5. Very little loss of material through exclusions.

6. Annealing of weld zone is not necessary.

7. The heating zone being very thin, therefore, dissimilar metals are easily joined.

Limitations :

● The method is limited to smaller components.

● The parts to be welded must be essentially round and must be able to withstand the high torque developed during welding.

Applications :

● H.S.S. twist drills ;

● Gas turbine shafts ;

● Aero-engine drive shafts and valves ;

● Refrigerator tubes of dissimilar metals ;

● Steering columns ;

● Welding of sintered products, etc.

Note : Refer to also Q. 7.6.

7.18. EXPLOSIVE WELDING

In this process welding is achieved through very high contact of pressure developed by detonating a thin layer of explosive placed over one of the pieces to be joined (Fig. 7.41). The detonation imparts high kinetic energy to the piece which on striking the other piece causes plastic deformation and squeezes the contaminated surface layers out of interface resulting in a high quality welded joint.

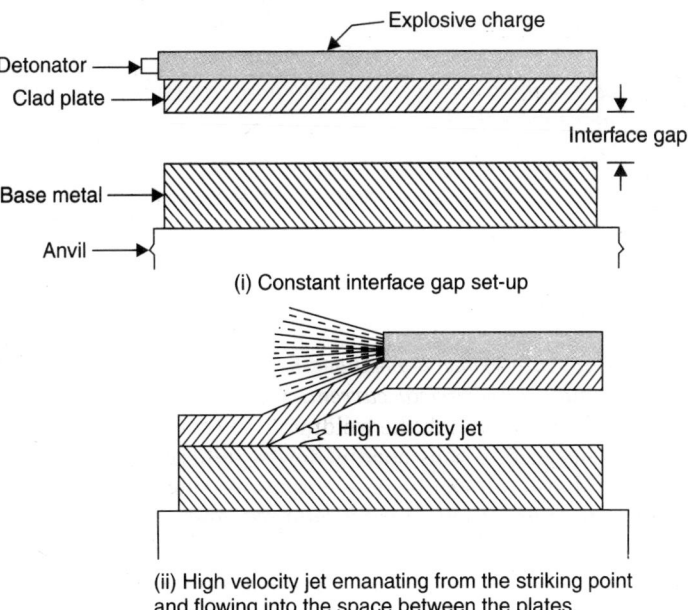

(i) Constant interface gap set-up

(ii) High velocity jet emanating from the striking point and flowing into the space between the plates.

Fig. 7.41. Explosive welding process.

Advantages :

(i) High joint strength.

(ii) Dissimilar metals can be welded.

(iii) Inexpensive equipment.

(*iv*) Large size plates can be welded.

(*v*) Good for plate cladding.

Limitations :

(*i*) Trained operators are required.

(*ii*) Inherently dangerous processes.

(*iii*) High set-up time.

Applications :

● This process is particularly suitable for cladding plates and slab with dissimilar metals, such as for the chemical industry.

● Explosive cladding is also finding use in the die-casting industry for nozzles, die-casting biscuits and other components.

● Tubes and pipe are often joined to the holes in the header plates of boilers and tubular heat exchangers by placing the explosive inside the tube ; the explosion expands the tube and seals it against the plate.

7.19. DIFFUSION WELDING

Diffusion Welding (DFW), *also called a diffusion bonding (a solid-state joining process), is the process of joining two parts purely by the diffusion. The diffusion can be achieved by keeping the two pieces in intimate contact under pressure.* The pressures used are in the range of 35 to 70 MPa, because of the large contact areas used. The diffusion being a rate process, can be accelerated by the use of heat, though is not essential. By the application of heat, the bonding time can be reduced from hours to minutes. A *filler metal is generally used and kept in between the two plates to be joined.*

● *Diffusion welding can be used for joining metals to metals as well as metals to non-metals.* The weld is very neat requiring no further processing on the joint.

● *Diffusion welding is more expensive and, therefore, can only be justified for closer tolerance work or for expensive materials.*

Applications :

● This process is used for fabricating complex parts in *small quantities* for the aerospace, nuclear, and electronics industries. It has also been automated to make it suitable and economical for *moderate* volume production.

7.20. INDUCTION WELDING

In **induction welding** *coalescence is produced by the heat obtained from the resistance of the weldment to the flow of an induced electrical current.* Pressure is frequently used to complete the weld. The inductor coil is not in contact with the weldment ; the current is induced into the conductive material. Resistance of the material to this current flow results in the rapid generation of heat. In operation a high current is induced into both edges of the work close to where the weld is to be made. *Heating to welding temperature is extremely rapid and the joint in completed by pressure rolls or contacts.*

Vacuum tube oscillators are the source of power for most high frequency welding.

● Induction welding can be used for most metals and has been successfully employed for some dissimilar metals.

Applications :

● Butt and seam welding of pipe, sealing containers ;

● Welding expanded metal ;

● Fabricating various structural shapes.

7.21. COLD WELDING

Cold welding is *defined as "A solid-state welding process wherein coalescence is produced by the external application of the mechanical force alone".* The process is quite similar to electric resistance welding with the exception that no heat is used to remove the oxide and gas layers but rather severe plastic deformation is relied upon to get metal-to-metal contact.

- In this process, pressure is applied to the surfaces of the parts, through either *dies* or *rolls.* Because of the resulting deformation, it is necessary that at least one, but preferably both, of the mating parts be sufficiently *ductile.* The interface is usually cleaned by wire brushing prior to welding.

The best bond strength and ductility are obtained with two *similar materials.* When two dissimilar metals, that are mutually soluble, brittle intermetallic compounds may form, resulting in a weak and brittle joint (*e.g.,* bonding of aluminium and steel).

Applications :
- The closing of aluminium cable sheaths ;
- Manufacture of kitchen utensils ;
- Lap and butt welding of wires and brushers of electrolytic cells, etc.

7.22. STUD-ARC WELDING

In this welding process, an arc is generated between the stud or similar part and the base metal.

- The operation sequence begins with the stud being placed in the stud gun.
- The gun is then positioned over the spot where the stud is to be placed.
- When the trigger, or switch, is depressed, current flows through the stud, which at the same time is lifted back slightly from the work, creating an arc.
- After a short arcing period, the stud is plunged into the molten pool created on the base plate and the gun is withdrawn.

Stud welding may be done with the conventional D.C. power supply or with a capacitor-discharge power supply.

Heat supplied by an electric current can be adjusted and varies depending on the size of the stud and kind of metal. The duration of the arc is also adjustable on control unit.

Advantages :
1. This provides a quick and easy way for securing fasteners to plate or round stock.
2. Very economical.

Limitations :
1. Stud must be of a size and shape that permits chucking, and the cross-sectional area of the weld base must be within the range of the welding-equipment power supply.
2. Areas to be welded must be clean and free from paint, scale, rust, grease, oil, dirt, zinc, or cadmium plating. Aluminium surfaces may also need oxide removal.

Applications :
- Joining of bolts and other projections onto workpieces without the need for drilling and tapping holes
- Used for attachment of handles and feet to appliances and electrical panel construction.
- Extensively employed in ship building, automotive, railroad and building construction industries.

7.23. HYDRODYNAMIC WELDING

- The process utilises both thermal energy and shock waves discharged through a common weld head. The *two important steps involved are :*

  (*i*) Induction heating ;

  (*ii*) Magnetic pulse pressurizing.

After welding the object can be cooled at a controlled rate to produce the desired microstructure in the joint region.

Advantages :

Hydrodynamic welding technique offers many *advantages* such as speed reproducibility of results and an automatic operation and it *requires no weld metal.*

Applications :

- It can be *used to join steel to many ferrous and non-ferrous metals and alloys.*

7.24. UNDER-WATER WELDING

Occasionally, welding has to be carried out under water and this is known as **"submerged welding".** Obviously gas welding is not possible as the heat necessary for fusion cannot be generated under water by gas welding. However, submerged welding is possible by **electric arc.** *Heat generated by an electric arc in water is much lower than in open air welding hence special electrodes are used which have low fusion points and are not affected under water.*

- Under electric welding, the welding may be by '*arc*' method or it may by '*resistance*' method.
  - In the *arc method* the arc passes between the electrode held in the holder with the welder and the job which is connected to the second terminal of the electric supply. The arc drawn between an electrode and the workpiece completes the electrical circuit.
  - In *resistance welding,* current is passed through the two pieces held together under pressure. The *electric current causes heat in the joint due to eddy currents and loss of energy in the resistance of the circuit. This is also known as* **Thomson process.**

7.25. OXY-ACETYLENE TORCH CUTTING

Cutting steel with a torch is an important production process. *A simple hand torch for flame cutting* differs from a welding torch. *It has several small holes for preheating flames surrounding a central hole through which pure oxygen passes. The preheating flames are exactly like the welding flames and are intended to preheat the steel before cutting.*

The **principle of flame cutting** *is that oxygen has affinity for iron and steel. At ordinary temperatures this action is slow, but eventually an oxide in the form of rust materializes.* As the temperature of the steel is increased this action becomes much more rapid. If the steel is heated to a red colour, about *870°C and a jet of pure oxygen is blown on the surface, the action is almost instantaneous and the steel is actually burnt into an iron oxide.*

- Metal upto 760 mm can be cut by this process.
- Flame-cutting machines which replace many machining operations where accuracy is not paramount, are widely used in the *ship-building industry, structural fabrication, maintenance work and the production of numerous items made from steel sheets and plates.*
- *Cast iron, non-ferrous alloys and high-manganese alloys are not readily cut by this process.*

7.26. SOLID/LIQUID-STATE BONDING—SOLDERING AND BRAZING

In solid/ liquid-state bonding, the *joint is made by distributing the molten filler metal between the closely fitted surfaces of the parts, without melting the base metal.* The two processes which fall under this category are : *Soldering* and *Brazing*. By AWs, definition the melting point of the filler metal for brazing will be above 427°C ; below this temperatures are the solders.

- *'Brazing' makes stronger joints as compared to 'soldering'.*

Advantages :

1. Parts with greatly unequal wall thickness can be joined.
2. Dissimilar materials can be joined.
3. In these processes, the molten filler, material is drawn to the various points of joint by surface tension (capillary attraction), so the access to all the parts of the joint is not required. Hence, complicated assemblies and assemblies consisting of many parts can be *simultaneously joined.*
4. In these processes no post joining heat treatment is normally necessary and joints are virtually free of internal stresses. The processes lend them to automation.

Limitations :

1. Cost of joint could be high.
2. The joint inspection could be difficult.
3. Relatively lower strength, particularly in case of the lower melting-point alloys.
4. Temperature resistance is limited by the melting point of the filler metal.

Applications :

- Joining of wires.
- Manufacture of automobile radiators, plate and tube heat exchangers, impellers, etc.

Soldering and Brazing are discussed in detail in the following articles.

7.27. SOLDERING

7.27.1. Definition

Soldering *is an operation of joining two or more parts together by molten metal.*

- *It is a quick method of making joints in light articles* made from steel, copper and brass and for wire joints such as occur in electrical work.
- *Soldering should not be used where much strength is required,* or in case *where the joint will be subjected to vibration or heat, as solder is comparatively weak and has a low melting point.* When joints must stand these treatments they should be riveted, welded or brazed.

7.27.2. Classification of Soldering Methods

Soldering methods are best classified by the method of heat application. The heat should be applied in such a manner that the solder melts while the surface is heated to permit the molten solder to wet and flow over the surface.

The various soldering methods are :

1. Soldering iron methods.
2. Torch method.
3. Dip and wave methods.
4. Induction method.
5. Resistance method.
6. Furnace and hot plate method.

7. Spray method.

8. Ultrasonic method.

9. Condensation method.

7.27.3. Types of Solder

Roughly there are two types of solder :

1. Soft solder, which is usually a *lead-tin mixture.*

2. Hard solders, which may be roughly grouped into the following sub-division :

(*i*) Brass solders (copper-zinc alloy).

(*ii*) Silver solders (copper-silver alloy).

(*iii*) Copper solders.

(*iv*) Nickel-silver solders.

7.27.4. Selection of Solder

When choosing any solder containing zinc the question of the solder setting up a *galvanic action should be examined.* With some *brass articles* this is very likely. The solder must be chosen in relation to the material of which the articles to be joined are made. In each instance the melting point of the solder should, where possible, be several hundred degrees below that of the metals to be united. Articles made in brass, copper, phosphor bronze or nickel-silver are both hard soldered and soft soldered. *If an assembly is subjected to vibration, then a silver solder should be chosen.* This is particularly necessary on pipe work for engines, and experience over many years indicates that silver solder of the correct grade cannot be improved upon for such articles.

7.27.5. Flux or Soldering Fluid

During the operation of a soft soldering a *flux is necessary to cover the surface of the components and solder with a film so that the formation of an oxide is prevented and the solder aided in its flow around or along the joint.*

Fluxes are of two kinds :

1. *Those which not only protect the surface, but play an active chemical part in cleaning it.* These are *most efficient.*

Examples : *Zinc chloride* ("killed spirits"), *ammonium chloride and zinc ammonium chloride,* a combination of the two.

Killed spirits gives the best all-round results and is extensively used, but care is necessary to ensure that the articles are thoroughly *washed afterwards* so that the acid does not attack the metal.

Killed spirits is muriatic (hydrochloric) acid in which pure zinc has been placed. The acid attacks the zinc, and much gas is liberated. When the acid cannot take up any more of the zinc it is said to be killed.

2. *Those which merely protect a previously cleaned surface.*

Examples : Tallow, resin, vaseline, olive oil, etc. and in addition to these there are various patent fluxes of which "Fluxite" is a well known example.

When the nature of the work is light and continuous, the problem of fluxing is greatly facilitated by using one of the strip solders with the flux incorporated. One such solder is *resin cored,* being in the form of a small thin tube having a core of resin which melts and fluxes the work as the solder is consumed.

7.27.6. Soldering Equipment

Soldering equipment broadly includes soldering iron and soldering gas stove/heater.

(*a*) **Soldering iron :** It is a *tool used during a soldering operation to heat the solder and the parts to be jointed.*

Soldering irons are of two main types :

(*i*) Those heated by either solid or gaseous fuel ;

(*ii*) Those *heated electrically.*

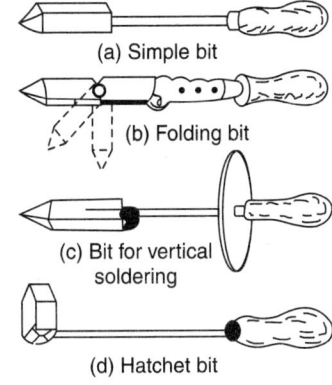

(a) Simple bit

(b) Folding bit

(c) Bit for vertical soldering

(d) Hatchet bit

Fig. 7.42. Soldering bits.

An ordinary soldering iron consists of a copper point or "*bit*", usually of *square or rectangular cross-section-a stem, usually of mild steel and a wooden handle.* The point is of a size chosen to suit the class of work, *its purpose is to absorb heat while in the fire or muffle, and to give this out when applied to the job,* thus heating the two parts and melting the solder. The point is tinned that is covered with solder, and should be kept clean.

When applying the iron it must be held in position for a sufficiently long time to ensure that the work is heated to the required temperature, it is then drawn slowly over the surface so that it will heat the adjoining area and melt the solder. The stem forms the connection between the point with its heat and high conductivity and the handle with its low conductivity which permits the tool to be easily manipulated.

When small assemblies have to be soft-soldered, an electrically heated iron proves exceedingly useful and is often far better than the type requiring solid or gaseous fuel for heating.

Fig. 7.42 shows various types of soldering bits.

(*b*) **Soldering gas stove/heater :** Fig. 7.43 shows a soldering gas stove/heater used for heating the soldering bit for accomplishment of soldering operation.

7.27.7. Soldering Procedure

The process of soldering is carried out as follows :

1. The surfaces of the pieces to be joined must be thoroughly cleaned and made free from rust, grease, oil and dirt by scraping with dull knife or emery paper.

2. Then coat the surfaces with flux.

3. Take a blob of solder on the bit of the hot soldering iron and allow it run down hill filling the recess of the joints for light work. For heavy work the hot iron should be held against the soldering stick and molten solder be allowed to fill the longer length of the joint.

4. Wipe off excess of solder with a piece of felt or cotton waste.

5. Wash the joint thoroughly with warm water to remove the traces of acid flux.

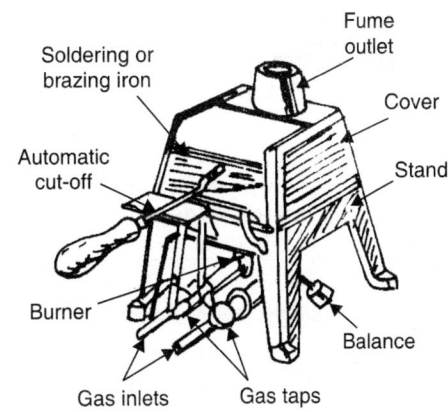

Fig. 7.43. Soldering gas stove/ heater.

7.27.8. Characteristics of a Good Joint

A *small amount of solder and perfect adhesion* are characteristics of a good joint. While carrying out tests on joints it has been established that the best joints are those in which thickness of solder varies between 0.0075 and 0.05 mm. It has been observed from experiments that the *thinner the layer of solder, the higher the soldering temperature necessary for maximum strength.*

7.27.9. Important Tips for Effective Soldering Operation

1. Always use an iron as large as can be handled and in the direction of having it too hot rather than not hot enough : it should not, of course, be red hot.

2. A better joint can be made if the work is warm than cold.

3. Iron tinning is facilitated by having some blobs of solder in a tin lid with a little spirits and touching both the spirits and the solder at the same time.

4. Quenching the hot joint in spirits, or painting on spirits whilst hot, will often effect remarkably thorough cleaning.

7.27.10. Advantages of Soldering

The advantages of soldering are :

1. Low cost.

2. Simplicity and cheapness of the equipment.

3. The properties of base metal are not affected due to low operating temperature.

4. Good and effective sealing in fabrication as compared to other process like riveting, spot welding and bolts.

7.27.11. Applications of Soldering

The soldering, in practice, is done for the following work :

1. Connections in wireless set (radio), T.V. sets, etc.

2. Wiring joints in electrical connections, battery and other terminals.

3. Drain water gutters and pipes.

4. Radiator brass tubes for motor car.

5. Copper tubing carrying liquid fuel, gas or air used on engines.

6. Brass halved bearings are joined with solder when relined with white metal and bored on lathe.

7. When halved bearings are to be lined with babbit metal.

8. It is sometimes used to repair utensils.

Table 7.1 gives the details of soft solders.

Table 7.1. Soft solders

S. No.	Purpose for which solder is used	Tin %	Lead %	Anti-mony %	Melting range °C	Remarks
1.	Plumber's piped joints	30	69	1	180–250	Prolonged pasty stage when melting or solidifying.
2.	Tinsmith's general work and hand soldering	45	52.5	2.5	180–210	Solidifies fairly rapidly.
3.	Tinsmith's and copper smith's fine work. Hand soldering	50	47.5	2.5	180–210	Solidifies fairly repidly.
4.	Steel tube joints.	65	35	—	180	Solidifies quickly. No pasty stage.

7.27.12. Types of Soldered Joints

In *soft soldering*, the only types of joint used is the *'lap joint'*. The lap should be 3 to 60 mm depending on the thickness of the metal and the working conditions of the joint. On thickness from 2 to 5 mm and a pressure of upto 5 bar (gauge), the lap should be at least 40 mm and the joint should be designed to work in shear. *In brazing, the most commonly used type of joint is also the 'lap joint'.* The other types of joints used in soldering practice are shown in Fig. 7.44.

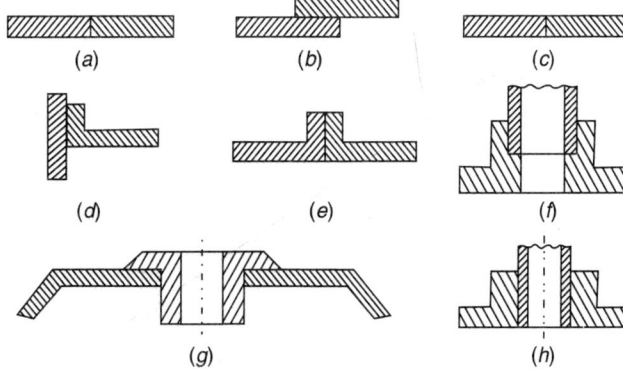

Fig. 7.44. Soldering joints.

The factors which govern the strength of a joint are the area of the joint (or lap), the lift-up of members to be joined and the clearance between them. The clearance though, is of importance to the strength of the soldered joints in high-strength metals, while in low-strength metal it is of minor consequence.

The clearance should be kept to a minimum to facilitate capillary attraction of the molten solder and to produce the strongest joint. When a clearance is sufficiently small, the metal of the joint is an alloy of the 0.2–0.25 mm.

7.28. BRAZING

7.28.1. Introduction

Brazing *is a soldering operation using brass as the joining medium.* The brazing operation is simply a form of *hard soldering using a copper-zinc alloy,* that is, brass, as the uniting medium (the term hard soldering is used because the welding alloy used in the joint is harder than solder, naturally the joint is much more stronger than soldering).

The brass used for making the joint in brazing is generally called *"spelter"* and its composition depends upon the metal being brazed because it is essential that the spelter shall have a lower melting point than the material being jointed.

Three brazing alloys are :

1. *Copper = 70%, Zinc = 30%* ; Melting point = 960°C

2. *Copper = 60%, Zinc = 40%* ; Melting point = 910°C

3. *Copper = 50%, Zinc = 50%* ; Melting point = 870°C.

When brazing or hard soldering using a brass mixture, the heating may be by means of :

 (*i*) Coal-gas and a mouth blow pipe for very small work.

 (*ii*) Coal-gas and compressed air using the normal blow pipe.

(*iii*) Oxy-acetylene torch.

(*iv*) Oxy-hydrogen torch.

 (*v*) Coal-gas and oxygen with a suitable torch.

(*vi*) Electrical resistance as on a spot welder.

7.28.2. Fluxes

When hard soldering, the chief flux is *borax,* which may be obtained in powder, granulated or stick form. It may be dissolved in hot water to form a paste and is then applied to the joint by means of a brush. Slick *borax* is supplied in the form of a cone, and is rubbed down on a rough slate with water to form a paste. In this condition it is applied to the joint, preferably with a brush.

The action of the flux is *three fold :*

(*i*) *It is used to prevent an oxide forming along the joint faces as they are heated.*

(*ii*) *A cleansing medium to remove the dirt.*

(*iii*) *It aids the capillarity of the molten metal and facilitates its flow around and through the joint.*

7.28.3. Brazing Equipment

The brazing equipment mainly comprises a *blow pipe* (Fig. 7.45) and *brazing hearth* (Fig. 7.46). The heat for brazing is obtained by a blow pipe fed with coal gas, and air at a slight pressure. When heating of the work is taking place it is advisable to make the most of the heat given out from the blow pipe flame, and for this reason a small sheet-metal hearth containing fire brick or coke should be used. If the work is placed upon a substance heat is reflected back on its underside and this conserves the heat during the operation.

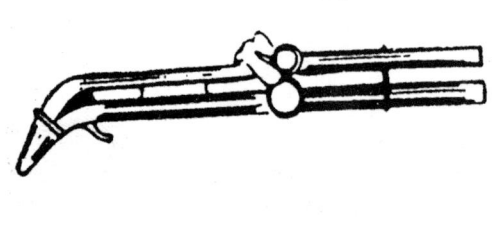

Fig. 7.45. Blow pipe.   Fig. 7.46. Brazing hearth.

7.28.4. Brazing Methods

Depending upon the heat source used to melt the brazing metal, the various brazing methods are :

1. Torch brazing
2. Furnace brazing
3. Induction brazing
4. Dip brazing
5. Resistance brazing
6. Laser brazing and electron beam brazing.

- **Braze welding** differs from the conventional brazing in that *much under gap is filled with brazing brass with the help of torch. Here, capillary action plays no part in making the joint.* The technique is also used for the repair of iron and steel casting.

7.28.5. Brazing Procedure

The process of brazing is carried out as follows :

1. The surfaces to be joined are thoroughly cleaned.
2. Then a paste made of flux and spelter is kept in the joint, the joint being held in position by suitable clamps or tongs.

3. The flame is directed over the joint held on a fire brick piece. The flux and spelter will soon melt and fill the recess between the joint.

The liquid is spread uniformly over the joint either with a pointed wire piece or by moving the jet of flame circular over the joint.

4. When the joint is hot common salt is put to soften the glossy hard flux that sets over the joint.

5. The work is removed from the clamp after it is cooled.

7.28.6. Advantages and Limitations of Brazing

Following are the *advantages* and *limitations of brazing* :

Advantages :

1. Cast and wrought metals can be joined.
2. Metallurgical properties of the base materials are not seriously disturbed.
3. Assemblies can be brazed in a stress free condition.
4. Dissimilar metals can be joined.
5. Non-metals can be joined to metals, when the non-metal is coated.
6. Materials of different thickness can be joined easily.
7. Complex assemblies can be brazed in several steps by using filler metals with progressively lower melting temperatures.
8. Little or no finishing is required by the brazed joints.
9. Several operations can be mechanised.
10. Almost all the common engineering materials may be satisfactorily brazed or braze welded.

Limitations :

1. High degree of skill required.
2. Limited size of parts.
3. Machining of the joint edges forgetting the desired fit is costly.
4. Joint design is somewhat limited if strength is a factor.

7.28.7. Applications of Brazing

The applications of brazing include the following :

1. Parts of bicycle such as frames and rims.
2. Pipe joints subjected to vibrations.
3. Exhaust pipes in motor engines.
4. Band saws.
5. Tipped tools.
6. Nipples and unions to M.S. and copper tubing.

7.28.8. Silver Soldering (or Silver Brazing)

The melting temperature of most silver solders is little lower than that of spelter used for brazing and so the *operation of silver soldering is somewhat easier to carry out than brazing.* A joint made by silver solder is not so strong as a brazed one but is stronger than the soldered joint. For different work, silver solders with varied composition of silver, copper and zinc are used.

Calcined borax and powdered glass are the suitable *fluxes.* Before silver soldering, the joint is pickled in the dilute sulphuric acid. The process of silver soldering is carried out with a blow-pipe and spirit lamp or oil lamp directing the flame to the joint only for melting the flux and silver solder. Clamps of various types are used to hold the work in position while joint is being made. Small thin pieces or filings of silver solder are mixed with flux and water. This mixture formed in a paste is kept in between the joint before the flame is blown through the blow pipe. Work is held either over a charcoal piece or fire brick for retaining heat at the joint. After the joint is done by silver soldering, the flux starts setting hard and glossy over the joint which is softened by putting the common salt. *Unlike and like metals can be joined together.*

Table 7.2. gives the composition, melting point and uses of some of the important silver solders.

Table 7.2. Silver solders

S. No.	Composition %	Melting point	Uses
1.	Ag = 20, Cu = 52, Zn = 38	820°C	Used for practically for all
2.	Ag = 20, Cu = 45, Zn = 35	775°C	non-ferrous alloy (except aluminium
3.	Ag = 80, Cu = 16, Zn = 4	740°C	and steels and iron).
4.	Ag = 45, Cu = 30, Zn = 25	675°C	

7.28.9. Comparison between Soldering and Brazing

The comparison between soldering and brazing is given below :

S. No.	Aspects	Soldering	Brazing
1.	Melting point of filler metal.	Below 400°C.	Above 400°C
2.	Stability of joints made.	Less.	More.
3.	Effect of high pressure and temperature on the joint.	Joints are affected.	Joints are not affected.
4.	Cost of the equipment.	Very low.	Comparatively high.

7.29. ELECTRODES

- Proper choice of electrodes is a vital point in getting good welds. Good preliminary precautions are of no use by improper choice of electrodes.

- The electrodes have two parts. One is *core wire* which contains the metal to be deposited. The second one is *coating. The heat causes the coating to emit gases which cover the weld and prevent bare metal from oxidation.* Contamination with water impares the effectiveness of coating and hence the *electrodes must be kept in fairly dry places.* Some electrodes do not have any coating.

- *Hand operated electrodes* are normally "extruded wire" with coating over it. The coating may contain ingredients like SiO_2, TiO_2, FeO, MgO, Al_2O_3 and cellulose in various proportions.

7.29.1. Electrode Materials

Depending upon job material, the following electrode materials are in use :

(*i*) Mild steel	(*ii*) Low alloy steel
(*iii*) Nickel steel	(*iv*) Chrome-moly steel
(*v*) Manganese-moly steel	(*vi*) Nickel manganese-moly steel
(*vii*) Nickel-moly-vanadium steel	(*viii*) Aluminium
(*ix*) Copper-aluminium	(*x*) Lead-bronze
(*xi*) Phosphor bronze.	

7.29.2. Electrode Coatings

The electrode coatings are made to serve the following *purposes* :

(*i*) To provide a gaseous shield to prevent atmospheric contamination.

(*ii*) To help stabilise and direct the arc for effective penetration.

(*iii*) To act as scavengers to reduce oxides.

(*iv*) To control surface tension in the pool to influence the shape of the bead formed when the metal freezes.

(*v*) To add alloying elements to the weld.

(*vi*) To insulate the electrode electrically.

(*vii*) To minimise splatter of the weld metal.

(*viii*) To form a plasma to conduct current across the arc.

(*ix*) To form a slag to carry off impurities, protect the hot metal and slow the cooling rate.

7.29.3. Electrode's Designation

● The *electrodes are designated by numbers which indicate grade and by sizes of the core wire.*

● The sizes are given by absolute sizes (*e.g.*, 1/8″, 5/16″, 7/16″, etc.) or by standard wire gauge (S.W.G.) sizes. Most common sizes are 4, 6, 8, 10 and 12. The electrodes with higher SWG size number are thinner.

● Although any size electrode can be used for the job, but *for quick welding, the electrode size should be as large as permissible. Larger size electrodes permit heavier currents to be used hence, the metal deposition is faster and the job is done quicker. It must be, however, emphasised that choice of the electrode size does not determine the speed of the work. Good welds come out rapidly if consistent higher currents are used.*

— Normally 12 SWG electrodes for welding can be used upto 3 mm plates or sections, 10 SWG upto 5 mm sections, 8 SWG upto 10 mm, 6 SWG upto 10/15 mm and 4 SWG for sections beyond 20 mm thickness.

● As the jobs get heated during welding, they should not be handled by naked hands. After the welding is over, the part may be put in *water for quick cooling* ; otherwise it may be allowed to cool in open air. After this, the scum at the surface may be knocked out by welding hammer.

— If *rebuilding* is being done, it is good practice to *beat the weld so that oversize grains are broken and this increases the strength of the joint.*

7.29.4. Typical Data on use of "Mild Steel Electrodes" :

(*a*) *Recommended currents :*

Electrode size		Current range in amps.	
SWG	*mm*	*Min.*	*Max.*
10	3.25	95	125
8	4.06	140	170
6	4.88	160	220
4	5.89	200	275

(*b*) *Composition (Chemical) :*

Carbon	......	0.09%
Manganese	......	0.47%
Silicon	......	0.20%
Sulphur	......	0.016%
Phosphorus	......	0.34%.

7.30. WELDING OF VARIOUS METALS

Some welding methods/techniques as applied to different metals/alloys are briefly given below :

Metals/Alloys and Welding methods/Techniques :
1. Low carbon steels (or mild steel) :
- *Welded by :*
- — Forge welding ;
- — Resistance welding ;
- — Arc welding ;
- — Gas welding.
- No flux required.
- Welding rods are made of pure iron or mild steel.
- No preheating is required even for welding thicker sections.
- In gas welding, a *neutral* flame is used in order to minimise oxidation of steel.

2. Medium carbon steels :
(Carbon content between 0.3 and 0.5%)
- *Welded by*
- — Arc welding ;
- — Resistance welding ;
- — Gas welding ;
- — Thermit welding.
- The preheating temperature (varying from 100°C to 400°C) depends upon the carbon content in the steel.
- As this steel is harder and brittle than mild steel, therefore, it is necessary to *normalise* the components after the welding in order to relieve the residual stresses present in them.
- In gas welding, a slight *carburising flame* is used.

3. High carbon steels :
(High %age of carbon)
- *Welded by :* Same as at (2).
- Preheating to about 400°C is essential.
- Sudden cooling should be avoided to avoid cracking along weld metal.
- Heat treatment of these steels after welding is necessary to a relieve residual stresses set up during welding.

4. Alloy steels :
(In addition to carbon, these steels contain small amount of nickel, chromium, molybdenum, manganese, silicon, copper. These steels may be low, *medium* and *high alloy steels*).
- The welding of low alloy steels is similar to the welding of medium or high carbon steels, *Shielded metal arc and submerged arc welding* are oftenly used.
- Preheating and slow cooling after welding is essential to obtain crack free welds.
Welding of "stainless steel" :
- — Can be welded by oxy-acetylene and metal arc welding methods ; best method being *electric butt welding* followed by prompt annealing between 730° to 800°C.

— A suitable electrode should be selected as per manufacturer's advice.

— For obtaining a sound weld, cleaning of edges to be welded and removal of slag after each run is necessary.

— Flux may or may not be used.

5. Cast steel :

● *The welding operation, for the repair of defective casting is termed as 'welding of cast steel'.*

● Gas welding with a *neutral flame* is usually used.

● Preheating of heavy casting is essential to minimise straining of metal due to local heating at the time of welding.

6. Cast-iron :

● *Welded by :*

— Metal arc welding ;

— Oxy-acetylene welding ;

— Braze welding.

● The weldability of cast-iron usually decreases as the amount of free carbon in cast-iron increases.

● The cast-iron parts are generally preheated to a dull red heat and then welded.

● After welding, it is very important to anneal the casting.

7. Aluminium and its alloys :

(Cast and wrought forms)

● *Welded by :*

— Welding of all types including inert gas and atomic hydrogen welding.

— The inert gas-tungsten arc welding is extensively used than other arc welding processes.

● In welding cast aluminium, preheating is necessary and after welding it should be cooled slowly.

8. Copper and its alloys :

● *Welded by :*

— Metal-arc welding

— Carbon-arc welding

● The direct current with straight polarity is usually employed for welding these metals.

● In gas welding of copper, a neutral flame and a filler rod of copper and silver alloy is used.

9. Nickel and its alloys :

● *Welded by :*

— May be easily welded by most of the major welding processes like metal—arc welding, resistance welding and oxy-acetylene welding.

● Pre-heat treatment and post-heat treatment are rarely used except for special cases.

● For welding nickel flux is not required.

7.31. REBUILDING

● Welding is used not only for fabrication of new parts but is used considerably for rebuilding of worn out parts, repair of *broken or cracked parts also.*

- *The worn out parts can be build up by use of proper electrodes. Where the high temperature required for welding might change the composition of the parent metal, low temperature or 'Eutectic' electrodes may be used.* After building up sufficient material the part can be re-machined to desired dimensions.

- This is done for *shafts, housings, etc.*

- *Repair of cracked parts is also possible.* The cracked castings, bearing housings, broken gears, etc. can be repaired this way. There is no difficulty of welding cracked mild steel, cast steel, manganese steel spares. However, cast iron and aluminium spares are difficult to weld. These require preheating and post heating so that the welding stresses do not weaken the repaired parts.

Thermit welding requires a special mould to be built up. A wax pattern is made around the gap and sand/clay mould is rammed outside it. A magnesite crucible contains charge and is mounted above the weld. After igniting crucible, the reaction takes about 30 seconds for completion. Metal is poured into the joint by removal of the tapping pin at the bottom of the crucible.

7.32. HARD FACING

- Welding is considerably used to reduce the wear on parts subjected to heavy abrasion such as *jaws of stone crushers, digging teeth of shovels, surfaces of rollers, etc.* If austenitic steels are used for manufacturing, then the difficulty arises when any machining is required. Further these steels may not have sufficient shock absorbing capacity. To obtain it, the usual practice is to *hard the face of mild steel with harder deposits.* These may be machinable or may be completely unmachinable.

- This hard facing may be even used for austenitic manganese steel wearing parts *used in cement or stone crushing industries. These layers should not be thick and must not be used for building up.*

- *If the hard facing is to be done on worn out parts, these may have to be built up by ordinary electrodes, before hard facing is done.* This is carried out in layers and a good pining (hammering out) should be carried out before second layer is built up. Many electrodes are available in the market, by which longitudinal joints are welded in uniform fashion. This is called seam welding and is valuable in pipe manufacturing. Water cooling is used to keep the electrodes cool.

- Hard facing gives 2 to 3 times the life in terms of wear as compared to plain wearing parts. This is very economical and requires less idle time for machines.

7.33. CHARACTERISTICS OF GOOD WELD

A weld not properly welded is a *defective weld.* A properly made weld should have the following *characteristics :*

1. The weld *should not crack in the bend test.*
2. It should not contain scum or slag imbedded in the well.
3. Its appearance should be ripple like and not spongy.
4. It should not have cavities, and the grain size should be uniform.
5. The contour of the weld should be even.
6. The weld should have even width.

- *Overcurrent tries to dissolve scum in the weld while undercurrent tries to give cracks in the weld.*
- If electrode distance from the weld is varying this will cause the unevenness of the weld.

7.34. DEFECTS IN WELDS

Some of the defects encountered in the weld metal are :

1. Porosity and blow holes.
2. Inclusions.
3. Cracks.
4. Fish eyes.

1. Porosity and blow holes :

- These defects are produced when during solidification and cooling, the gases like hydrogen, oxygen and nitrogen are evolved as a result of sudden drop in solubility. The gases are absorbed by the molten weld metal from various sources including the atmosphere, fluxes and electrode coatings.
- Gas porosity is detrimental to the quality of welds and represents discontinuity in the metal structure. Supersaturation of the molten metal with gases and an excessively high cooling rate of the weld metal lead to the formation of blow holes.

2. Inclusions :

- Inclusions are the slag or non-metallic particles and are derived from the environments around the weld metal. Basic slags are more viscous and are difficult to remove from the molten metal.
- Slow cooling of the molten weld metal pool helps in the elimination of inclusions.
- The weldability of the parent metal is normally affected by the size and distribution of inclusions. Such inclusions that are formed by sulphur and phosphorus must be kept at a minimum. Aluminium, zirconium and the rare earth elements are known to greatly modify the shape, size and distribution of inclusions and thereby enhance most of the engineering properties of the welds.

3. Cracks :

- In steel welds, the cracks are the most serious defects.
- Cracks may be formed due to various causes including *unequal physical properties of the parent and weld metals, high stress conditions and faulty welding.* Normally these can be attributed to the metal characteristics, particularly to the *hydrogen content in the metal.* Cracks caused by hydrogen can appear in the etched, pickled, plated, heat treated parts, vessels holding hydrogen under pressure and in the welds.

Hot cracks. The weld metal cracks are called *hot cracks* because these appear as a result of stress and lack of ductility of the deposited metal at the high temperature. Low melting point compounds such as FeS that get deposited on the grain boundaries during solidification are usually responsible for a low ductility and the appearance of the cracks.

Cold cracks. These cracks are formed near the weld area and are due to *excessive cooling rates and the absorbed hydrogen.* Since these appear a long time after the welding operation when the material is cold, these are termed *cold cracks.* They constitute a great danger in the low alloy and high carbon steel welds.

4. Fish eyes :

- Fish eyes usually appear on the *fractured surfaces of the welded sections in the form of white areas, circular in shape and varying in size between 1 to 10 mm diameter.* They always tend to form around inclusion and are frequently accompanied by microcracks.
- These are attributed to the presence of hydrogen in welds and the existence of stress conditions.
- The appearance of fish eyes in weld metal, after a small degree of low speed deformation, causes a drastic drop of weld ductility. But if the material is aged for a long time at room temperature or a few hours at higher temperature before testing, fish eyes are not formed and the ductility of the weld metal is high.

7.35. WELDABILITY

Weldability *is defined as the capacity of a metal to be welded under the fabrication conditions imposed in a specific suitably designed structure and to perform satisfactorily in the intended service.*

The real criteria in deciding on the weldability of a metal is the *weld quality* and the *ease with which it can be obtained.*

Weldability is of significant importance for fabrication of metals into various structures.

The weldability of a metal is affected by the following factors :

(*i*) Composition of metal.

(*ii*) Brittleness of metal.

(*iii*) Thermal properties.

(*iv*) Welding technique.

(*v*) Filler materials.

(*vi*) Flux material.

(*vii*) Strength of metal at high temperature.

(*viii*) Stability of micro-constituents upto welding temperature.

(*ix*) Affinity of oxygen and other gases before and at welding temperature.

(*x*) Shielding atmosphere.

(*xi*) Proper heat treatment before and after deposition of metal.

- *Weldability can be known by determining a metal's behaviour under fusion and cooling, by crack and notch sensitivity, or by comparison of the heating and cooling effects* which take place at the joint of the metal with the metal of known weldability.

Effects of alloying elements on weldability :

The alloying elements affect the weldability in the following ways :

(*i*) Improve mechanical properties.

(*ii*) Form carbides.

(*iii*) Increase or decrease hardenability in the heat affected zone.

(*iv*) Provide grain refinement.

(*v*) Provide deoxidation of molten metal.

(*vi*) Form age hardening precipitates.

(*vii*) Control ductile to brittle transformation temperature.

- To determine the suitability of iron and steel bar the following *welding tests* are carried out : (*i*) Bending test (on a welded joint), (*ii*) Bar welded into a link should stand being closed up without failure.

- The following materials have good weldability in the ascending order : (*i*) Strainless steel ; (*ii*) Low alloy steel ; (*iii*) Cast iron ; (*iv*) Carbon steel ; (*v*) Iron.

Service weldability. It refers to the ability of the process-material combination to turn out a weld that will stand upto *thermal and mechanical stresses, corrosion, and other service requirements.*

Fabrication weldability. It is the capacity of the combination of materials and processes to produce proper fusion, limited porosity, absence of cracks, and proper weld geometry. Fabrication weldability factors include *melting point of the base metals, thermal conductivity, thermal expansion and contraction, electrical resistance, and surface condition.*

7.36. TESTING OF WELDED JOINTS

- It is very essential to have careful examination of the component at each stage of manu-facture. Thus, by *inspection or quality control, the welding defects are located and preventive measures are devised to reduce or eliminate them.*

- It is very difficult to ascertain whether the finished weld is up to the expected standard or not. In order to locate the defects and hidden flaws the welded portion may be inspected visually or examined with instruments.

- The welded joints may be subjected to the following tests :

Destructive tests :
— Tensile test ;
— Impact test ;
— Bending test ;
— Hardness test, etc.

Non-destructive tests :
— Visual examination ;
— Liquid penetrant test ;
— Magnetic-particle inspection test ;
— Eddy-current testing ;
— Ultrasonic inspection ;
— Radiography ;
— Thermal testing ;
— Hydraulic air pressure tests for pressure vessels.

7.37. EFFECT OF WELDING ON THE GRAIN SIZE OF THE METAL

During welding, there is an ample variation of temperature from molten metal of the weld to the edge of the heat-affected zone. Some of the metals have been heated far above the critical temperature, some just at the critical temperature, some not upto critical, and all the way down to the unaffected base metal. Thus, the grain size of the weld will be rather large. The grain size will become gradually small till the recrystallisation temperature is reached. The grain size will be minimum here and then will advance gradually again until it blends with the unaffected parent metal. In case of *metals and alloys there is no transformation zone* and only a gradual reduction of grain size from weld metal to the parent metal occurs. In heavy welds where several passes are required, the *heat of each succeeding pass can be used to refine the grain of previous pass.*

Effect of preheating on microstructure of the weld area in high carbon steel :

The microstructure of the weld area is determined by the rate at which cooling takes place in the weld area. The preheating before welding to some temperature (*e.g.,*150° to 200°C) is thus carried out to reduce the rate cooling by lowering the temperature gradient (difference of temperature of molten metal and the original temperature of metal before welding). The temperature gradient produces the heat treatment of heated metal. Hard martensite may be formed in high carbon steels due to rapid cooling. This hardened steel at the heat-affected zone is less ductile, and thus is subjected to failure under impact loading.

7.38. PRINCIPLES OF WELDING DESIGN

While designing a weldment (*the optimum choice of a weld design is one that meets all design and service requirements at a minimum cost*) the following design principles should be kept in mind :

1. Welds should be so located that adequate strength will be provided at the proper places on a structure or part.
2. Weldments should be so designed as to require a minimum of weld metal.
3. Welds at the valuable cross-sections should be avoided.
4. Laps, straps and stiffening angles should be avoided except as required for strength.
5. The joint should have properly prepared grooves.
6. Avoid large flat falls, which tend to buldge and flux.
7. An important strength weld should not be located where much of it may be removed later by machining.
8. Welds should not be subjected to bending.
9. Weld bend size should be kept to a minimum to consume weld metal.
10. Components should fit properly before welding.
11. As far as possible, the use of welding fixtures should be avoided.
12. A weld should not be located at the point of maximum deformation.
13. Sharp discontinuities in metal should be kept at a minimum since these cause stress concentration.
14. Wherever possible use butt welds.
15. Provide for easy access to welds so that they are accessible for inspection.

7.39. COMPARISON OF WELDING AND ALLIED PROCESS

The comparison of welding and allied process is given in Table 7.3 below.

Table 7.3. Comparison of welding and allied process

S. No.	Process	Advantages	Limitations
1.	Forge welding	(*i*) Inexpensive equipment. (*ii*) Semi-skilled operation.	(*i*) Poor joint strength. (*ii*) Labour intensive process (low production rate). (*iii*) Weld quality dependent on operator's skill. (*iv*) Can be used only where hammering is possible.
2.	Resistance spot welding	(*i*) High production rate. (*ii*) Very economical process. (*iii*) High skill not required.	(*i*) Suitable for thin sheets only (*ii*) High equipment cost.

		(iv) Most suitable for welding sheet metals. (v) Dissimilar metals can be welded. (vi) Small heat affected area.	
3.	Gas welding	(i) Good weld quality. (ii) Portable low cost equipment. (iii) Suitable for repair work and low quantity production.	(i) Manual operation hence low production rate. (ii) Skilled operator required. (iii) Difficult to prevent contamination. (iv) Large heat affected zone.
4.	Electric arc welding	(i) Portable and relatively inexpensive equipment. (ii) Very versatile process.	(i) Large heat affected zone. (ii) Weld quality depends upon operator's skill in manual operations. (iii) Not suitable for thin sections.
5.	Thermit welding	(i) Can be used anywhere. (ii) Low set-up cost. (iii) Not a highly skilled operation. (iv) Most suitable for welding of thick sections.	(i) Only thick sections can be welded. (ii) High set-up and cycle time.
6.	Laser beam welding	(i) High depth-to-width ratio of weld. (ii) Good weld quality. (iii) Dissimilar metals can be welded. (iv) Suitable for thin sections. (v) Welding can be done in inaccessible locations. (vi) Minimum distortion. (vii) High production rate.	(i) Highly skilled operation. (ii) High equipment cost. (iii) Eye protection required. (iv) Suitable for welding narrow and deep joints.
7.	Soldering	(i) Suitable for making leak proof joints. (ii) No distortions. (iii) Low equipment cost.	(i) Joints suitable for low temperature applications. (ii) Low joint strength.
8.	Brazing	(i) Not much skill is required. (ii) Good joint strength. (iii) Dissimilar metals can be joined. (iv) Suitable for joining intricate and light weight shapes. (v) Very little distortion. (vi) Low equipment cost.	(i) Not suitable for thick sections. (ii) Joint strength not as high as in gas or arc welding. (iii) Joints not suitable for high temperature applications. (iv) Joint colour may not match with base metal.

$$\boxed{\textbf{WORKED EXAMPLES}}$$

Example 7.1. *Determine the melting efficiency in the case of arc welding of steel with a potential of 22 V and current of 230 A. The cross-sectional area of the joint is 25 mm^2 and the travel speed is 6 mm/s. Heat required to melt steel may be taken as 10 J/mm^3 and the heat transfer efficiency as 86 percent :*

Solution. *Given* : $V = 22$ volts, $I = 230$ amp. ; Area of cross-section of the joints, $a = 25$ mm^2 ; Travel speed, $v = 6$ mm/s ; Heat required to melt the steel = 10 J/mm^3 ; Heat transfer efficiency $\eta = 86\%$.

Melting efficiency :

Net heat supplied $\qquad = VI \times \eta\ = 22 \times 230 \times 0.86 = 4351.6$ W

Heat required for melting = Volume of base metal melted × heat required to melt the steel

$$= (a \times v) \times 10 \text{ J/s}$$

$$= 25 \times 6 \times 10 = 1500 \text{ J/S or W.}$$

Melting efficiency $\qquad = \dfrac{\text{Heat required for melting}}{\text{Net heat supplied}}$

$$= \frac{1500}{4351.6} = 0.3447 \quad \text{or} \quad \textbf{34.47\%} \quad \textbf{(Ans.)}$$

Example 7.2. *It is required to weld a low carbon steel plate by the manual metal arc welding process using a linear V.I. characteristic D.C. power source. The following data are available :*

Open circuit voltage of power source = 62 V

Short circuit current = 130 A

Arc length ; L = 4 mm

Transverse speed of welding = 15 cm/min

Voltage is given as, V = 20 + 1.5 L (L being arc length in mm)

Efficiency of heat input = 84 percent.

Calculate the heat input to the workpiece.

Solution. *Given* $\quad V_{oc} = 62$ V, $I_{sc} = 130$ A ; $L = 4$ mm, $v = 15$ cm/min ;

$$V = 20 + 1.5\ L,\ \eta = 84\%$$

Heat input to the workpiece :

Volt-amp. characteristics of source is given as :

$$\frac{V}{62} + \frac{I}{130} = 1 \qquad\qquad\qquad ...(i)$$

Here, $\qquad\qquad V = 20 + 1.5 \times 4 = 26$ volts

Substituting the value of V in eqn. (i), we get

$$\frac{26}{62} + \frac{I}{130} = 1$$

or, $$I = 130\left(1 - \frac{26}{62}\right) = 75.5\ A$$

Now, Power consumed $= VI$

$$= 26 \times 75.5 = 1963\ W$$

∴ Heat input into the workpiece

$$= 1963 \times \eta = 1963 \times 0.84 = \textbf{1648.9 W (Ans).}$$

Example 7.3. *The arc-length characteristic of a D.C. arc is given by the equation : V = 24 + 4L, where V is the voltage in volts and L is arc length in mm. The static volt-ampere characteristic of the power source is approximated by a straight line with a no load voltage of 80 V and a short circuit current of 600 A.*

Determine the optimum arc length for maximum power. **(GATE-2002)**

Solution. *Given :* $V = 24 + 4\,L$; $V_{nl} = 80\ V$; $I_{sc} = 600\ A$.

Optimum arc length for maximum power, L :

The static vol.-amp. characteristic of power source is given as :

$$\frac{V}{80} + \frac{I}{600} = 1$$

or,

$$I = \left(1 - \frac{V}{80}\right) \times 600$$

Now, Power,

$$P = V \times I$$

$$= (24 + 4\,L)\left(1 - \frac{V}{80}\right) \times 600$$

$$= (24 + 4\,L)\left[1 - \frac{(24 + 4\,L)}{80}\right] \times 600$$

$$= (24 + 4\,L)\left(\frac{80 - 24 - 4\,L}{80}\right) \times 600$$

$$= \frac{(6 + L)}{20}\,(50 - 4L) \times 600$$

$$= 120(6 + L)\,(14 - L)$$

$$= 120(84 + 8L - L^2)$$

For max. P,

$$\frac{dP}{dL} = 0$$

∴

$$\frac{dP}{dL} = \frac{d}{dL}\,[120(84 + 8L - L^2) = 0$$

or, $\qquad 120 \times 8 - 120 \times 2\,L = 0 \quad$ or $\quad 960 - 240\,L$

or, $\qquad\qquad\qquad$ **L = 4 mm (Ans.)**

QUESTIONS WITH ANSWERS

Q. 7.1. Differentiate between pressure welding and fusion welding.

Ans. Pressure welding. Pressure (or solid state) welding may be carried out at room temperature or at elevated temperature. The joint is formed *owing to plastic flow at the joint due to applied pressure*. In pressure welding at high temperature, heat is generated due to resistance at the joint due to flow of current being high. Other methods of achieving high temperature include friction, induction heating, impact energy as in the case of explosive welding. Seam welding, projection welding, butt welding are well known resistance welding techniques.

Fusion welding. In fusion (or liquid state) welding the material around the joint is melted in both the parts to be joined. If necessary a molten filler metal is added from a filler rod (or otherwise). The important zones in fusion welding are : (*i*) Fusion zone ; (*ii*) Heat affected unmelted zone around the fusion zone ; (*iii*) The unaffected original part.

Important factors affecting fusion welding process are :

(*i*) Nature of weld pool.

(*ii*) Chemical reaction in the fusion zone.

(*iii*) Characteristics of heat source.

(*iv*) Contraction, residual stresses and metallurgical changes.

(*v*) Heat flow from the joint.

Q. 7.2. What is the fundamental feature that distinguishes solid-state welding from liquid-state welding ?

Ans. The fundamental feature which distinguishes solid-state welding from liquid-state welding is the process by which two parts are joined together by heating them to fuse in *liquid form without fuse*. In liquid-state welding also called arc welding, an electric arc is used to melt and then join the work to be welded. While in solid-state welding the joint is made without heating the work to fuse ; the joint is made by explosion pressure or other methods.

The following methods of solid-state welding clearly give idea of doing solid-state welding and how it differs from liquid-state welding :

(*i*) *Friction welding.* In this process one of the parts to be welded is held in the headstock of a lathe like friction welding machine and the other is rotated and pressed against its surface.

(*ii*) *Explosive welding.* In this type of welding, strong metallurgical bonds are produced between solid metal surfaces by impacting one over another at a very high speed.

(*iii*) *Forge welding.* This method involves heating of the solids to their plastic state and then hammering the ends to be joined.

(*iv*) *Ultrasonic welding.* In the method two solids to be joined together are tightly pressed against each other and given relative movement (vibration) at frequency between 2 kHz to 6 kHz for a short interval.

Q. 7.3. State the effects of current and voltage on the quality of weld.

Ans. The *effects of current and voltage on the quality of weld* are as follows :

(*i*) *Too high current*	...	Gives deeper crater and penetration, flat bead, much spatter, electrode becomes red hot, etc.
(*ii*) *Too low current*	...	Imparts poor penetration, shallow crater, weld overlapping on the plate, etc.
(*iii*) *Too high voltage*	...	Produces a fierce wandering and noisy hissing arc, bead is often porous and flat, spattering of metal.
(*iv*) *Too low voltage*	...	Causes sticking of electrode with work and arc becomes difficult to maintain, weld is deposited in blobs with no penetration, etc.

Q. 7.4. Explain briefly "Principle of resistance welding".

Ans. In resistance welding the heat required for welding is produced by means of *electrical resistance* between the two members to be joined. The heat produced is given by the expression.

$$H = I^2 Rt\, K \qquad\qquad ...(7.1)$$

where,　H = The total heat generated in the work, J,

　　　　I = Electric current, A,

　　　　R = The resistance of the joints, Ω (ohms),

　　　　t = Time for which electric current is flowing through the joint, s, and

　　　　K = A factor that represents the energy losses through radiation and conduction ; its value is *less than unity*.

The resistance of the joint, R is composed of :

(*i*) The resistance of the electrodes ;

(*ii*) The electrode-workpiece contact resistances.

(*iii*) The resistances of the individual parts to be welded.

(*iv*) The workpiece-workpiece contact resistances.

- In resistance-welding operations, the magnitude of the current may be as high as 100,000 A, although the voltage is typically only 0.5 V to 10 V.

- The actual temperature rise at the joint depends on the specific heat and thermal conductivity of the materials to be joined ; consequently, because they have high thermal conductivity, *metals such as aluminium and copper require high heat concentrations.*

Q. 7.5. Explain briefly the following resistance welding processes :

(*i*) Metal fibre welding.

(*ii*) Metal foil welding.

(*iii*) Spike welding.

Ans. (*i*) **Metal fibre welding :**

- In this process (may be described as an extension of resistance spot welding), the equipment used is same as in spot welding.

- In metal fibre welding a small piece of *metal fibre sheet* is introduced between the workpieces to be joined. The workpieces are held as usual between the electrodes. The electrodes pressure employed is *very low* and, therefore, the indentation on the work surface is also very small.

- It is mainly used for making lap joints.

(*ii*) **Metal foil welding :**

- In this process the *metal sheets* to be joined are placed between electrode wheels (as for seam welding) and the abutting edges are *covered at top and bottom with very thin metal foils,* the foil thickness is approximately 0.25 mm and the same is made of the *same material as the workpiece metal.*

- A distinct feature of the weld formed is that it has a *raised beat.*

(*iii*) **Spike welding :**

- In this resistance welding process a *large amount of electricity is stored up in capacitors and then the same is released rapidly through the electrodes and the metal pieces to be joined.* This current flow is allowed for a very small interval and its time is controlled electronically. The welding is, therefore, *very rapid* and consequently *chances of warpage of workpieces and of contamination are altogether eliminated.*

Applications. Successfully used for welding of almost all metals and alloys, metals of different thicknesses together and dissimilar metals as well.

Q. 7.6. Explain briefly the following variations on the friction-welding process :

(*i*) Inertia friction welding ;   **(*ii*) Linear friction welding ;**

(*iii*) Friction stir welding.

Ans. (*i*) **Inertia friction welding :**

- The energy required in this process is supplied *through the K.E. of a flywheel.* The flywheel is accelerated to the proper speed, the two members are brought into contact, and an axial force is applied ; as friction at the interface begins to slow down the flywheel, the axial force is increased. The weld is completed when the flywheel comes to rest.

(*ii*) **Linear friction welding :**

- In this process, one part is moved across the face of the other part, using a balanced reciprocating mechanism.
- This welding process is capable of joining square or rectangular components, as well as round parts, made of metals or plastics.

(*iii*) **Friction stir welding :**

- Whereas in conventional friction welding heating of interfaces is achieved through friction by rubbing two contacting surfaces, in the *friction stir-welding* process, a third body is rubbed against the two surfaces to be joined. The rotating tool is a small (5 mm to 6 mm in diameter, 5 mm in height) rotating member that is plunged into the joint. The contact pressures cause frictional heating, raising the temperature to the range 230°C to 260°C. The probe at the tip of the rotating tool forces heating and mixing or stirring of the material in the joint.
- The welding equipment can be a conventional, vertical spindle milling machine, and the process is relatively easy to implement.

Q. 7.7. List the factors which affect the selection of electrodes for electric arc welding.

Ans. The following factors affect the *selection of electrodes* for electric arc welding :

1. Power source : A.C. or D.C.
2. Base metal composition.
3. Base metal thickness.
4. Welding current conditions recommended by the manufacturer of the electrodes.
5. Welding position.
6. Economic considerations.
7. Extent of penetration required in welding.
8. Mechanical properties desired in the joint.

Q. 7.8. Why is laser welding used only for microwelding applications ?

Ans. Laser is used only for microwelding, *i.e.,* welding very small wires to electronic devices as laser for generating high energy is *very costly* but *microwelding applications can be precisely controlled by laser.*

Q. 7.9. List the advantages and limitations of D.C. and A.C. power sources in arc welding.

Ans. D.C. power source :

Advantages :

1. Arc stability is much higher with D.C. than with A.C.
2. It can handle all situations and jobs.
3. It is known to most-operators.
4. It is preferred for difficult task like overhead welding.
5. Base electrodes are also suitable for welding.

Limitations :

1. Electric energy consumption per kg of metal deposited is higher.
2. More chances of "magnetic arc blow".
3. Machine efficiency is low (30 to 60%).
4. The equipment is heavier in weight, larger in size and operation and maintenance is difficult because of moving parts.

A.C. Power source :

Advantages :

1. Higher efficiency of A.C. welding transformers.

2. Little "magnetic arc blow".

3. A.C. welding equipment is considerably less expensive, light in weight, smaller in size and simpler in operation due to absence of moving parts.

4. Less electric energy consumption per kg of metal deposited (as compare to D.C. welding).

Limitations :

1. Power factor is low.

2. Unstable arc.

3. Only coated electrodes are suitable for welding.

Q. 7.10. What is "Arc cutting" ? Explain briefly.

Ans. *In " arc cutting," the metal is simply melted by the intense heat of the arc and is then blown away by the force of arc itself or by other gases such as air or shielding gases.*

Depending on the source of heat input, the various processes of arc cutting are :

— Carbon arc cutting ;

— Air carbon arc cutting ;

— Oxygen arc cutting ;

— Shielded metal arc cutting ;

— Gas metal arc cutting ;

— Gas tungsten arc cutting.

— Plasma arc cutting.

In all these processes, the equipment used is similar to that used for the corresponding welding process with the exception of the *torch which is different. The torch holds the electrodes as also has the provision for the supply of high pressure gas wherever needed.*

● Any metal can easily be melted and blown away by shielded gas cutting processes.

Q. 7.11. Explain briefly "Thermal machining".

Ans. Although *manual cutting is used extensively for salvage and repair work, flame machining provides greater speed, accuracy, and economy.*

With a cutting torch mounted on a variable-speed electric motor, referred to as a radiograph, straight cuts can be made with the addition of a track and circular cuts can be made with a compass attachment.

Other methods of guiding the cutting torch are :

— Templates ;

— Photoelectric tracters ;

— Numerical control.

In recent years the optical tracer used in thermal cutting in gradually being replaced by *"CNC flame cutters".*

NC and CNC entail the following *advantages* and *limitations* for thermal machining :

Advantages :

(*i*) Elimination of template preparation, storage, and maintenance.

(*ii*) Minimum scrap allowance due to optimum nesting of multiple cuts.

(*iii*) Simple, quick plate layout that can be checked on the *x–y* plotter.

(*iv*) Elimination of plate or template alignment for each row speeds-multiple part production.

(*v*) Keyboard entry and editing of basic shapes at the machine console.

(*vi*) Standard shapes can be cut without waiting for templates.

Limitations :

(*i*) To use CNC profitably, procedures must be exact. Each step must be thought through in a precise, disciplined manner.

(*ii*) High production is required to justify the cost of the equipment.

(*iii*) Although various programming languages and a minicomputer system may be used, much of the software (program to suit in-house machine) must be developed for maximum performance.

Q. 7.12. Name the metals welded by flash welding.

Ans. The *following metals can be welded by flash welding :*

(*i*) Low carbon steels.

(*ii*) Medium strength and high strength low alloy steels.

(*iii*) Tool steels.

(*iv*) Stainless steels.

(*v*) Aluminium alloys (with thickness greater than 1.25 mm).

(*vi*) Copper alloys (with high zinc content).

(*vii*) Magnesium alloys.

(*viii*) Molybdenum alloys.

(*ix*) Nickel alloys.

(*x*) Titanium alloys.

Q. 7.13. What are the advantages of Gas Metal Arc Welding (GMAW) or Metal Inert Gas (MIG) welding ?

Ans. Following are the advantages of GMAW or MIG :

1. Flux not required.

2. High welding speeds are possible.

3. Process can be automated.

4. High corrosion resistance.

5. The metals like aluminium and stainless steel, which are difficult to weld, can be welded.

Q. 7.14. Why is it difficult to start A.C. arc ? How is it simplified in practice ?

Ans. It is difficult to start A.C. arc *because of alternating current flow.* This difficulty is overcome by having hot start circuit which provides an extra flow of very high frequency current at the time of striking the arc. In some machines, capacitors are employed in arc (secondary) circuit to give high current surges for the arc striking.

Q. 7.15. What are the advantages of A.C. arc welding ?

Ans. The major advantage of A.C. arc welding is *complete absence of magnetic arc blow and thus quality welds are produced.*

● Once the arc is started, it is easy to control and maintain it.

● It is usually faster because large electrodes and more current can be used due to minimum magnetic blow conditions.

● It is very well suited to weld aluminium and is very popular for welding on heavy gauge steel.

Q. 7.16. Describe briefly the relative applications of A.C. and D.C. welding.

Ans. While D.C. welding is best suited for thinner sheel metal (below 6 mm) and also for welding non-ferrous metals, A.C. welding is used for most manual welding of 6 mm and thicker steel. As steel is the largest used structural material, A.C. welding finds maximum use, though D.C. welding has a greater variety of welding processes like GTAW and GMAW, straight polarity and reverse polarity, etc.

- Direct current straight polarity and reverse polarity welding can be used for overhead and vertical welds but A.C. welding is used for welding steel in the flat or horizontal position.

Q. 7.17. Name the commercially used gas welding and cutting flames.

Ans. The commercially used gas welding and cutting flames are :

- Oxygen — acetylene
- Oxygen — hydrogen
- Oxygen — natural gas or artificial gas
- Oxygen — liquefied petroleum gas.

Q. 7.18. Which gases are used for welding the following materials by GMAW or MIG processes ?

(i) Stainless steel

(ii) Steel

(iii) Copper or aluminium

(iv) Copper-nickel and high-nickel alloys

(v) Titanium.

Ans. Following gases are used for welding various materials by GMAW or MIG processes

(i) Stainless steel : *Argon-oxygen or helium-argon mixtures.*

(ii) Steel : CO_2.

(iii) Copper or aluminium : *Argon or argon-helium mixtures.*

(iv) Copper-nickel and high-nickel alloys : *Argon-helium mixtures.*

(v) Titanium : *Pure argon gas.*

Q. 7.19. Name the fluxes used for welding the following metals :

(i) Copper and its alloys

(ii) Aluminium

(iii) Cast iron.

Ans. (i) *Copper and its alloys :*

- Boric acid
- Borax
- Boric acid 50%, Borax 50%
- Boric acid 35%, Borax 50%, Sodium phosphate 15%
- Borax 56%, Potassium carbonate 22%, Sodium chloride 22%.

(ii) *Aluminium :*

- Sodium chloride 30%, Potassium chloride 45%, Lithium chloride 15%, Potassium fluoride 7%, Sodium bisulphate 3%.
- Sodium chloride 28%, Potassium chloride 50%, Lithium chloride 14%, Sodium fluoride 8%.
- Sodium chloride 19%, Potassium chloride 29%, Barium chloride 48%, Fluorspar 4%.

Aluminium fluxes must be stored in hermetically sealed cans.

(iii) *Cast Iron :*

- Borax.
- Borax 56%, Sodium carbonate 22%, Potassium carbonate 22%.
- Borax 23%, Sodium carbonate 27%, Sodium nitrate 50%.

Q. 7.20. Name the common welding troubles.

Ans. The common welding troubles are :

(*i*) Warping   (*ii*) Porous welds

(*iii*) Poor penetration   (*iv*) Distortion

(*v*) Undercutting   (*vi*) Cracked welds

(*vii*) Magnetic blow   (*viii*) Brittle welds

(*ix*) Spatter.   (*x*) Poor appearance

(*xi*) Poor fusion.

Q. 7.21. Explain briefly soft soldering and hard soldering.

Ans. Soft soldering. It is used extensively in *sheet-metal work* for joining parts that are *not exposed to the action of high temperatures and are not subjected to excessive loads and forces.* It is also employed for *joining wires and small parts.*

- The solder, which is mostly composed of lead and tin has a melting range of 150 to 350°C. A suitable *flux* is always used in soft soldering. *Its function is to prevent oxidation of the surfaces to be soldered or to dissolve oxides that settled on the metal surfaces during the heating process.* Although corrosive, *zinc chloride* is the most common soldering flux. Rosin is non-corrosive, but it does not have the cleaning properties of zinc chloride.

- A blow torch or soldering iron constitutes the equipment for heating the base metals and melting the solder and the flux.

Hard soldering. It employs solders which *melt at higher temperatures* and are *stronger* than those used in soft soldering.

- *Silver soldering is a hard soldering method, and silver alloyed with tin is used as solder.* The temperatures of the various hard solders vary from 600 to 900°C. The *fluxes are mostly in paste form and are applied to the joint with a brush before heating.*

Q. 7.22. Explain the process of joining in brazing and discuss the influence of gap on the strength of the joint.

Ans. Brazing is a joining process between parts 1 and 2 (say) with braze metal placed in the gap (a clearance) and heat applied to the work material. *Brazed joint is formed by capillary action.* It is essential to have *optimum gap* between two parts. *Too much gap is not desirable ;* at the same time *too little a gap will leave the joint unbrazed at some points and affect the strength.*

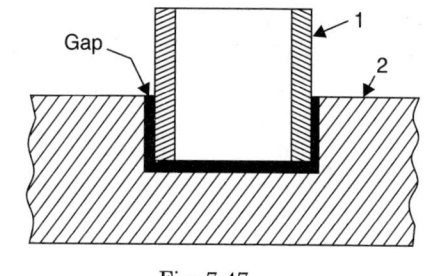

Fig. 7.47

Q. 7.23. What are the important design considerations in brazing parts ?

Ans. The following factors require due considerations in brazing parts :

(*i*) Base metals ;

(*ii*) Filler metals ;

(*iii*) Joint configuration ;

(*iv*) Service conditions like oxidation and corrosion resistance.

- Brazed joints may be either butt or lap or combination of two types. In designing joints, it should be remembered that the strength of the filler is *often less* than that of the base material. *Joint clearance* is important because it determines the *maximum joint strength* that can be developed by the particular filler metal.

Q. 7.24. What are the functions of 'Electrode coatings' ?

Ans. The *electrode coatings perform the following functions :*

1. Stabilize the arc.
2. Perform the metallurgical refining operations.
3. Provide a protecting atmosphere.
4. Facilitate overhead and position welding.
5. Slow down the cooling rate of the weld.
6. Increase the deposition efficiency.
7. Reduce spatter of weld metal.
8. Remove oxides and impurities.
9. Influence the depth of arc penetration.
10. Provide slag of suitable characteristics to protect the molten metal.

Q. 7.25. What points should be kept in view to avoid the weld defects ?

Ans. Following points should be kept in view *to ensure a correct weld :*

1. To maintain proper arc length.
2. To do properly the joint preparation of the workpiece to be welded.
3. To use correct welding.
4. To select the electrode of correct type and size.
5. To select the welding current according to the nature of job.

Q. 7.26. Explain briefly the following :

(*i*) **Furnace brazing ;**   (*ii*) **Induction brazing ;**   (*iii*) **Dip brazing.**

Ans. (*i*) **Furnace brazing :**

● The parts that are to be furnace brazed have the flux and brazing material preplaced in the joint. If the furnace has a neutral or shielding atmosphere, the flux may not be necessary.

● The furnace may be batch or conveyor type. Automatic controls regulate both time and temperature and, where applicable, atmosphere.

(*ii*) **Induction brazing :**

● The heat for induction brazing is furnished by A.C. coil placed in close proximity to the joint.

● The high-frequency current is usually provided by a *solid-state oscillator that produces a frequency of 200 Hz to 5 MHz.* These alternating high-frequency currents induce opposing currents in the work, which, by electrical resistance, develop heat.

Advantages :

(*i*) Accurate heat control.

(*ii*) Good heat distribution.

(*iii*) Speed.

(*iv*) Uniformity of results.

(*iii*) **Dip brazing :**

This process gets its name from the fact that parts are jigged (in some cases the parts are self jigging) and are placed in a chemical or molten-metal both maintained at the correct brazing temperature.

— In a chemical bath, the parts are first thoroughly cleaned and then assembled with a filler-metal preform placed on or in the joint. The bath, usually molten salts, is maintained at a higher temperature than the filler metal being used.

— After dipping, the parts are removed and immediately cleaned to remove the flux.

The number and size of the parts to be brazed are limited only by the capacity of the bath and the handling facilities.

Q. 7.27. What is "braze welding" ? Explain briefly.

Ans. Braze welding is similar to brazing in that the base metal is not melted but joined by an alloy of lower melting point. *The main difference is that in braze welding the alloy is not drawn into the joint by capillary action.*

A braze welded joint is prepared very much like a joint as prepared for welding except that an effort should be made to *avoid sharp corners*, because they are easily overheated and may also be points of stress concentration.

Applications :

- Braze welding is extensively *used* for *repair work*, as well as some *fabrication* on such metals as cast iron, malleable iron, wrought iron, and steel.
- It is used, but to a lesser extent, on copper, nickel, and high-melting point brasses.

Q. 7.28. Give the ISO 4063 classification of welding processes.

ISO 4063 classification of welding processes is given below :

No.	Process	No.	Process
1	**Arc welding**	3	**Gas welding**
11	Metal-arc welding without gas protection	31	Oxy-fuel gas welding
111	Metal-arc welding with covered electrode	311	Oxy-acetylene welding
112	Gravity arc welding with covered electrode	312	Oxy-propane welding
113	Bare wire metal-arc welding	313	Oxy-hydrogen welding
114	Flux cored metal-arc welding	32	Air fuel gas welding
115	Coated wire metal-arc welding	321	Air-acetylene welding
118	Firecracker welding	322	Air-propane welding
12	Submerged arc welding		
121	Submerged arc welding with wire electrode	4	**Solid phase welding ;**
122	Submerged arc welding with strip electrode		**pressure welding**
13	Gas shielded metal-arc welding	41	Ultrasonic welding
131	MIG welding	42	Friction welding
135	MAG welding metal-arc welding with non-inert gas shield	43	Force welding
		44	Welding by high mechanical
136	Flux cored metal-arc welding with non-inert gas shield		energy
		441	Explosive welding
14	Gas-shielded welding with non-consumable electrode	45	Diffusion welding
141	TIG welding	47	Gas pressure welding
149	Atomic-hydrogen welding	48	Cold welding
15	Plasma arc welding	7	**Other welding processes**
18	Other arc welding processes	71	Thermit welding
181	Carbon arc welding	72	Electroslag welding
185	Rotating arc welding	73	Electrogas welding
2	**Resistance welding**	74	Induction welding
21	Spot welding	75	Light radiation welding

22	Seam welding		751	Laser welding
221	Lap seam welding		752	Arc image welding
225	Seam welding with strip		753	Infrared welding
23	Projection welding		76	Electron beam welding
24	Flash welding		78	Stud welding
25	Resistance welding		781	Arc stud welding
26	Other resistance welding processes		782	Resistance stud welding
29	HF resistance welding			

HIGHLIGHTS

1. *Welding* is the method of joining metals by application of heat, without the use of solder or any other metal or alloy having a lower melting point than the metals being joined.

2. Welding processes may be broadly classified as :

   1. Pressure welding ;   2. Fusion welding.

3. *Seam welding* is employed on many types of pressure tanks, for oil swithces, transformers, refrigerators, evaporators and condensers, aircraft tanks, paint and varnish containers, etc.

4. The characteristic of a *fusion weld* is that material being joined is actually melted and the union is produced on subsequent solidification.

5. Fusion welding is classified as :

   (*i*) Gas welding ;   (*ii*) Electric arc welding ;

   (*iii*) Thermit welding.

6. Three types of welding flames are :

   (*i*) Neutral flame ;   (*ii*) Carburising flame ;   (*iii*) Oxidising flame.

7. *Arc welding* is the system in which the metal is melted by the heat of an electric arc.

8. *Thermit* welding is the method of uniting iron or steel parts by surrounding the joint with steel at a sufficient high temperature to fuse the adjacent surfaces of the parts together.

9. *Soldering* is an operation of joining two or more parts together by molten metal.

10. *Brazing* is a soldering operation using brass as the joining medium.

OBJECTIVE TYPE QUESTIONS

A. Choose the correct Answer :

1. Upon which of the following parameters does the current intensity in arc welding depend ?

   (*a*) Stability of arc

   (*b*) Electrode diameter

   (*c*) Gap between the electrode and parent metals

   (*d*) Thickness of parent metals

   (*e*) All of the above.

2. In .... welding two non-consumable electrodes are used.

   (*a*) MIG   (*b*) TIG

   (*c*) atomic hydrogen   (*d*) submerged arc

   (*e*) none of the above.

3. .... brazing process is good for mass scale.

   (*a*) Furnace   (*b*) Induction

   (*c*) Dip   (*d*) Torch.

4. For gray cast iron, which of the following welding methods is preferable ?
   (a) MIG   (b) Submerged arc
   (c) Gas flame   (d) Electric arc
   (e) Any of above.

5. Due to which of the following reasons, *no flux* is used in atomic hydrogen welding ?
   (a) The burning hydrogen shields the molten metal.
   (b) Two electrodes are coated which gradually release the flux.
   (c) The filler rod is coated with flux.
   (d) One of two electrodes is coated which releases the flux.
   (e) None of the above.

6. In resistance welding, between the electrodes, a current of .... voltage and .... ampere is passed.
   (a) high, high   (b) high, low
   (c) low, low   (d) low, high.

7. .... is the welding process in which heat is produced for welding by chemical reaction.
   (a) Resistance welding   (b) Thermit welding
   (c) Forge welding   (d) Gas welding
   (e) None of the above.

8. In case of submerged arc welding, the electrodes upto .... diameter may be used
   (a) 30 mm   (b) 20 mm
   (c) 15 mm   (d) 12 mm.

9. In arc welding, arc in created between the electrode and work by
   (a) contact resistance   (b) flow of voltage
   (c) flow of current   (d) electrical energy.

10. Material used for coating the electrode is called
    (a) flux   (b) slag
    (c) protective layer   (d) deoxidiser.

11. .... is the welding process in which two pieces to be joined are overlapped and placed between two pointed electrodes.
    (a) Seam welding   (b) Resistance welding
    (c) Projection welding   (d) Spot welding.

12. Which of the following gases are used in Tungsten inert gas welding ?
    (a) Helium and neon   (b) Hydrogen and oxygen
    (c) Argon and helium   (d) Carbon dioxide and hydrogen.

13. Preheating is essential in welding
    (a) copper   (b) aluminium
    (c) cast iron   (d) stainless steel.

14. The temperature, in arc welding, is of the order of
    (a) 2000°C   (b) 3000°C
    (c) 5500°C   (d) 7000°C

15. Acetylene gas is generated from
    (a) calcium   (b) carbon
    (c) calcium carbonate   (d) calcium carbide
    (e) none of the above.

16. Striking voltage as compared to voltage during arc welding is
    (a) less   (b) same
    (c) more   (d) unpredictable

17. Carburising flame has ...... zones.
    (a) one   (b) two
    (c) three   (d) four.

18. Due to which of the following reasons distortion in welding occurs ?
    (a) Oxidation of weld pool   (b) Use of high voltage
    (c) Improper clamping methods   (d) Use of high current
    (e) All of the above.

19. In reverse polarity welding
    (a) work is negative and holder is earthed
    (b) electrode holder is connected to negative and work to positive
    (c) electrode holder is connected to positive and work to negative
    (d) any of the above.

20. Where does maximum flame temperature occur ?
    (a) At the inner cone   (b) Next to inner cone
    (c) At the tip of the flame   (d) At the outer cone
    (e) Any of the above.

21. In which of the following welding processes, electrode gets consumed ?
    (a) TIG welding   (b) Resistance welding
    (c) Thermit welding   (d) Arc welding
    (e) None of the above.

22. Which of the following statements about welding is *incorrect* ?
    (a) Increased corrosion resistance
    (b) Even materials like stainless steel and aluminium can be welded
    (c) No flux required
    (d) High welding speed
    (e) None of the above.

23. Where is half corner weld used ?
    (a) where efficiency of the joint should be 50 percent
    (b) where longitudinal shear is present
    (c) where severe loading in encountered and the upper surfaces of both pieces must be in the same plane
    (d) all of the above
    (e) none of the above.

24. .... percent carbon steel is must weldable
    (a) 0.15   (b) 0.25
    (c) 0.35   (d) 0.8.

25. In which of the following metals does the phenomenon of 'weld decay' occurs ?
    (a) Stainless steel   (b) Cast iron
    (c) Carbon steel   (d) Bronze
    (e) None of the above.

26. On which of the following principles does the 'positive pressure type torch' work ?
    (a) Equal volume   (b) Positive pressure
    (c) Equal pressure   (d) Equal flow
    (e) None of the above.

27. Why is post cleaning necessary at brazed joint ?
    (a) To avoid corrosion   (b) To avoid slagging
    (c) To avoid oxidation   (d) To avoid scaling
    (e) All of the above.

28. Neutral flame is used to weld
    (a) Cast iron   (b) steel
    (c) copper   (d) all of the above.

29. Which of the following statements about 'Projection welding' is *correct* ?
    (a) It is a multi-spot welding process.
    (b) It is an arc welding process.
    (c) It is a continuous spot welding process.
    (d) It is a process used for joining round bars.
    (e) None of the above.

30. By which of the following welding processes is gray cast iron usually welded ?
    (a) TIG welding   (b) MIG welding
    (c) Resistance welding   (d) Gas welding
    (e) None of the above

31. Which of the following statements about copper is *correct* ?
    (a) It is very difficult to be spot welded.
    (b) It is preferred to be welded by spot welding.
    (c) It is easily spot welded.
    (d) It is as good for spot welding as any other material.

32. Which of the following statements about 'Submerged arc welding is *correct* ?
    (a) Arc is submerged is molten metal bath.
    (b) Arc is maintained under a blanket of flux.
    (c) There is no arc in actual.
    (d) None of the above.

33. Due to which of the following reasons, welding of stainless steel is difficult ?
    (a) Formation of oxide film.   (b) Melting point of stainless steel is very high.
    (c) Fear of cracking.   (d) Formation of rust.
    (e) None of the above.

34. In arc welding the length of arc should be equal to
    (a) half the electrode diameter   (b) electrode diameter
    (c) twice the electrode diameter   (d) none of the above.

35. Compared to oxyacetylene flame, temperature of oxy-hydrogen flame is
    (a) less   (b) same
    (c) more   (d) unpredictable.

36. Compared to oxidising flame, carburising flame is
    (a) less luminous   (b) equal luminous
    (c) more luminous   (d) any of the above.

37. In gas welding, more commonly used flame is
    (a) carburising flame   (b) neutral flame
    (c) oridising flame   (d) mixture of the three.

38. .... welding will be best suited for joining two stainless steel foils of thickness 0.1 mm.
    (a) MIG   (b) TIG
    (c) Plasma arc   (d) Gas.

39. Oxidising flame is used to weld metals/materials like
    (a) copper and brass   (b) aluminium, stainless steel, nickel, etc.
    (c) abrasive   (d) any of the above.

40. In thermit welding, iron oxide and aluminium oxide are mixed in the proportion of
    (a) 1 : 1   (b) 2 : 1
    (c) 3 : 1   (d) 1 : 3

41. Weld spatter is a
    (a) catalyst   (b) welding defect
    (c) flux   (d) none of these.

42. TIG welding is best suited for welding

(*a*) silver

(*b*) mild steel

(*c*) aluminium

(*d*) stainless steel.

43. Projection welding refers to

(*a*) pressure welding

(*b*) TIG welding

(*c*) submerged welding

(*d*) resistance welding.

44. In MIG welding, metal is transformed in the form of

(*a*) molecules

(*b*) molten drops

(*c*) weld pool

(*d*) a fine spray of metal.

45. In case of neutral flame, oxygen to acetylene ratio is

(*a*) 0.6 : 1.0

(*b*) 1 : 1

(*c*) 2 : 1

(*d*) 3 : 1.

46. Neutral flame has .... zones.

(*a*) two

(*b*) three

(*c*) four

(*d*) unpredictable.

47. In lost wax casting, tolerance is of the order of

(*a*) + 2 mm

(*b*) + 0.2 mm

(*c*) + 0.02 mm

(*d*) 0.005 mm.

48. In resistance welding, pressure is released

(*a*) during heating period

(*b*) after the weld cools

(*c*) no pressure is applied

(*d*) none of the above.

49. In which of the following welding processes the non-consumable electrode is used ?

(*a*) TIG welding

(*b*) LASER welding

(*c*) MIG welding

(*d*) Plasma arc welding.

ANSWERS

1. (*b*)	**2.** (*c*)	**3.** (*a*)	**4.** (*a*)	**5.** (*a*)	**6.** (*d*)	**7.** (*b*)
8. (*d*)	**9.** (*d*)	**10.** (*a*)	**11.** (*d*)	**12.** (*c*)	**13.** (*c*)	**14.** (*c*)
15. (*d*)	**16.** (*c*)	**17.** (*c*)	**18.** (*c*)	**19.** (*c*)	**20.** (*b*)	**21.** (*d*)
22. (*c*)	**23.** (*e*)	**24.** (*a*)	**25.** (*a*)	**26.** (*c*)	**27.** (*a*)	**28.** (*d*)
29. (*a*)	**30.** (*d*)	**31.** (*a*)	**32.** (*b*)	**33.** (*a*)	**34.** (*b*)	**35.** (*a*)
36. (*c*)	**37.** (*b*)	**38.** (*c*)	**39.** (*a*)	**40.** (*c*)	**41.** (*b*)	**42.** (*c*)
43. (*d*)	**44.** (*d*)	**45.** (*b*)	**46.** (*a*)	**47.** (*d*)	**48.** (*b*)	**49.** (*a*)

B. Fill in the blanks or say 'Yes' or No' :

1. .... is a process of joining two materials with the help of heat or pressure or by some other means.

2. .... is a process of joining two pieces of metal with a different fusible metal applied in a molten state.

3. A good weld is as strong as the base metal.

4. Welding does not result in residual stresses and distortion of the workpieces.

5. Welding does not permit any freedom in design.

6. Welding heat produces .... changes.

7. The characteristics of .... weld is that the metal joined is never brought to molten stage, it is heated to a welding temperature and the actual union is brought about by application of pressure.

8. The materials like brass and aluminium can be projection welded satisfactorily.

9. When the ratio of oxygen and acetylene is equal a .... flame is obtained.

10. In oxidising flame the ratio of oxygen to acetylene varies from about 1.2 to 1.5.

11. Heavy sections can be joined economically by gas welding.
12. The field of application of metallic arc welding includes mainly low carbon steel and the high-alloy austenitic stainless steel.
13. Atomic hydrogen welding being expensive is used mainly for high grade work or stainless steel and most non-ferrous metals.
14. .... welding is the method of uniting iron or steel parts by surrounding the joint with steel at a sufficient high temperature to fuse the adjacent surfaces of the parts together.
15. The consumable electrode in MIG process acts as the source for arc column as well as the supply for the filler material.
16. The .... arc process creates an arc column between a base metallic electrode and the workpiece.
17. Welds made by the submerged arc welding process have low strength.
18. The submerged arc process is capable of welding fairly thin gauge materials.
19. .... welding depends upon the generation of heat that is produced by passing an electric current through molten slag.
20. .... welding fusion joins metal by bombarding a specified confined area of the base metal with high velocity electrons.
21. Plasma is often considered the fourth state of matter.
22. The .... welding process is the focusing of a monochromatic light into extremely concentrated beams.
23. The laser welding process can be used to weld dissimilar metal with widely varying physical properties.
24. Hydrodynamic welding can be used to join steel to many ferrous and non-ferrous metals and alloys.
25. The .... are designated by numbers which indicate grade and by sizes of the core wire.
26. In steel welds, the .... are the most serious defects.
27. .... welding process uses only the pressure to produce the joints at room temperature and no heat is applied at any stage.
28. The welding operation for the repair of defective casting is termed as 'welding of cast steel'.
29. The brass used for making the joint in brazing is generally called ....
30. When hard soldering, the chief flux is borax.

ANSWERS

1. Welding
2. Soldering
3. Yes
4. No
5. No
6. metallurgical
7. Pressure
8. No
9. neutral
10. Yes
11. No
12. Yes
13. Yes
14. Thermit
15. Yes
16. submerged
17. No
18. Yes
19. Electro-slag
20. Electron-beam
21. Yes
22. Laser
23. Yes
24. Yes
25. electrodes
26. cracks
27. Cold
28. Yes
29. speller
30. Yes.

THEORETICAL QUESTIONS

1. Define the term 'welding' and name the various welding techniques.
2. Explain briefly the following types of flames :
   Neutral flame, Carburising flame and Oxidising flame.
3. Name and briefly explain the various equipment used in gas welding.
4. List the advantages and disadvantages of gas welding.
5. Give the comparison between A.C. and D.C. arc welding.
6. Write short notes on any three of the following :
   (a) Gas shielded arc welding
   (b) Submerged arc welding
   (c) Thermit welding
   (d) Plasma arc welding
   (e) Laser beam welding.

7. Explain briefly any two of the following :
   (*i*) Metallic arc welding   (*ii*) Carbon arc welding
   (*iii*) Atomic hydrogen welding   (*iv*) Shielded arc welding.
8. Explain briefly defects in welding.
9. Write short note on under-water welding.
10. How are welded joint tested ?
11. What do you mean by "Hard facing" ?
12. What is "Rebuilding" ?
13. What is thermit welding ?
14. Explain briefly cold welding.
15. What is the difference between shielded and unshielded arc welding processes ?
16. What are the advantages and disadvantages of D.C. and A.C. welding ?
17. Explain the inert-gas metal-arc welding ? How does it differ from other metal arc welding processes ?
18. Explain the principle of atomic hydrogen welding and the role of hydrogen in this welding.
19. What are advantages and disadvantages of spot welding processes ?
20. Differentiate between soldering and brazing.
21. Why a flux is used in soldering and brazing ?
22. What is the difference between welding, brazing and soldering processes ?
23. Explain briefly the following welding processes :
    (*i*) Friction welding   (*ii*) Diffusion welding
    (*iii*) Explosive welding.
24. Give a comparison between TIG and MIG welding processes.
25. (*a*) Define '*Soldering*'. Name types of solder.
    (*b*) Explain briefly the term '*Flux*' or '*Soldering fluid*'. Enumerate the fluxes commonly used in soldering process.
26. Describe briefly the equipment used in soldering.
27. How is soldering process carried out ?
28. What are the characteristics of a good soldered joint ?
29. State the applications of soldering.
30. Explain the term 'Brazing'.
31. What is the action of a flux in brazing process. Name the chief flux used in 'Brazing'.
32. Describe step-wise the procedure of 'Brazing'.
33. Enumerate the applications of brazing.
34. Write short note on 'silver brazing'.

8

Metal Cutting

8.1. Introduction. 8.2. Chip formation. 8.3. Types of chips. 8.4. Cutting tools—classification—single point cutting tool—tool elements and tool angles—tool signature (or tool designation). 8.5. Orthogonal and oblique cutting. 8.6. Chip control. 8.7. Forces of a single-point tool. 8.8. Mechanics of metal cutting—shear zone, shear plane and shear angle—chip thickness ratio—velocity relationship in orthogonal cutting—forces on the chip (Merchant's analysis)—stress and strain on the chip—work done during metal cutting and specific cutting energy—theories on mechanics of metal cutting. 8.9. Friction in metal cutting. 8.10. Thermal aspects of metal cutting—general aspects—factors affecting temperature—temperature distribution in metal cutting—measurement of chip-tool interface temperature. 8.11. Tool wear and failure—tool wear—tool failure. 8.12. Tool life. 8.13. Cutting speed, feed and depth of cut. 8.14. Machinability. 8.15. Cutting fluids—functions of cutting fluids—requirements of a cutting fluid—types of cutting fluids. 8.16. Cutting tool materials—characteristics of an ideal cutting-tool material—types of tool materials. **Questions with Answers**—Highlights—Objective Type Questions—Theoretical Questions—Unsolved Examples.

8.1. INTRODUCTION

The **metal cutting** (*machining*, a generic term, refers to all material removal processes) *refers to only those processes where material removal is affected by the relative motion between tool made of harder material and the workpiece.* The tool would be *single-point cutting tool* as used in operations like turning or shaping, or a *multi-point tool* as used in milling or drilling operation.

- The *conditions* which have an important influence on metal cutting are :
  (*i*) Work material ;
  (*ii*) Cutting tool material ;
  (*iii*) Cutting tool geometry ;
  (*iv*) Cutting speed ;
  (*v*) Feed rate ;
  (*vi*) Depth of cut ;
  (*vii*) Cutting fluid used.

- Metal cutting processes are performed on metal cutting machines, more commonly termed as *"Machine tools"* by means of various types of *"cutting tools"*.

- One major drawback of metal cutting or machining process is the loss of material in the form of *chips*.

8.2. CHIP FORMATION

The cutting tool removes the metal from the workpiece in the form of *"chips"*. *As the tool advances into the workpiece, the metal infront of the tool is compressed and when the compression*

362

limit of the metal has been exceeded, it is separated from the workpiece and flows plastically in the form of chip. The plastic flow of the metal takes place in a localised region called *shear plane,* which extends from the cutting edge obliquely upto the uncut surface infront of the tool. The cutting tool causes shearing action bearing the metal along the plane (Fig. 8.1).

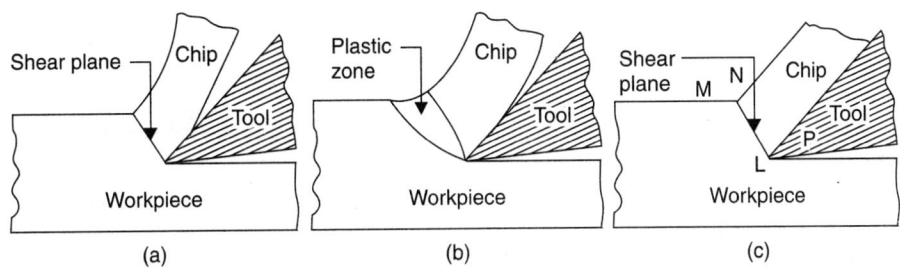

Fig. 8.1. Chip formation.

The shearing of the metal in the process of chip formation does not however, occur sharply across a straight line. The grains of the metal infront of the cutting edge of the tool start elongating along the line *LM* and continues to do so until they are completely deformed along the line *NP*. The region between the lines *LM* and *NP* is called *shear zone.* After passing out the shear zone, the deformed metal in the form of chip, slides along the tool face due to the velocity of the tool. This shear zone is treated as a shear plane for the mathematical analysis.

- Every machining operation involves the formation of chips, the nature of which depends upon the *operation, properties of the workpiece material and cutting conditions.*

8.3. TYPES OF CHIPS

The chips produced, whatever the cutting conditions be, may belong to one of the following *three types :*

1. Continuous chip
2. Discontinuous chip
3. Built-up chip.

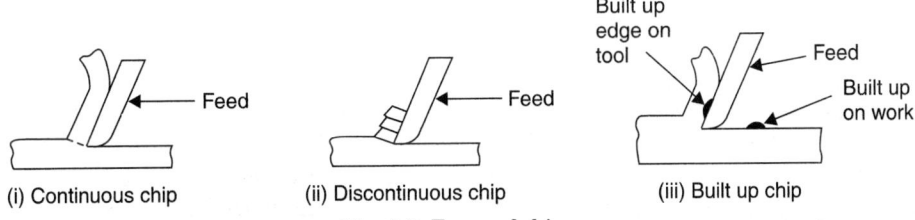

Fig. 8.2. Types of chips.

1. Continuous chip : Refer to Fig. 8.2 (*i*).

- These chips are produced while machining *more ductile materials.* This type of chip is *most desirable.*
- The continuous chip which is like a ribbon flows along the rake face. Production of continuous chips is possible *because of ductility of metal.*
- *About 95% of the power expended for metal removal is used in the deformation taking place in the shear zone.* This is the work required to form and remove the chip and incidental plastic deformation of the surface layer of the finished workpiece. The remaining power consumed, *about 5% of the total, is expended in stored elastic energy or residual stresses in the workpiece and friction.*

Note : The chips are formed largely by *shearing action* and *compressive stresses* on the metal infront of the tool. The compressive stresses are the greatest farthest from the cutting tool and are *balanced by tensile stresses* in the zone nearest the tool ; hence the *chip curbs outwardly or away from the cut surface.*

- Some ideal conditions that promote continuous chips in metal cutting are :
  — Small chip thickness (fine feed) ;
  — Small cutting edge ;
  — Large rake angle ;
  — High cutting speed ;
  — Less friction between the chip tool interface through efficient lubrication ;
  — Ductile work materials.
- These chips are *most useful chips* since the *surface finish obtained is good and the cutting is smooth.* It also helps in *having higher tool life and lower power consumption.*

However, because of the large coils of chips, *chip disposal is a problem.* For this purpose various forms of *chip breakers* have been developed which are in the form of a step or groove in the tool rake face. The chip breakers allow the chips to be broken into small pieces so that they can be easily disposed off.

2. Discontinuous chip : Refer to Fig. 8.2 (*ii*).

- These chips are usually produced while cutting *more brittle materials* like grey cast-iron, bronze and hard brass.
- In this type the chip produced is in the form of *discontinuous segments* (deformed material instead of flowing continuously) gets ruptured periodically.
- Discontinuous chips are easier from the view point of chip disposal. However, the cutting force becomes unstable with the variation coinciding with the fracturing cycle. Also they generally provide better surface finish. However, in case of ductile materials they cause poor surface finish and low tool life.
- Discontinuous chips are likely to be produced under the following conditions :
  — Low cutting speeds ;
  — Small rake angles ;
  — Higher depths of cut (large chip thickness).

3. Built-up chip : Refer to Fig 8.2 (*iii*).

When machining ductile materials, conditions of high local temperature and extreme pressure in the cutting zone and also high friction in the tool-chip interface may cause the work material to *adhere or weld to the cutting edge of the tool forming the built-up edge (BUE).* This *causes the finished surface to be rough.* However, since the cutting is being carried by the BUE and not the actual tool tip, the *life of the cutting tool increases* while cutting with BUE. That way BUE is not harmful while rough machining.

- In general low cutting speed, high feed and small rake angle are conducive to BUE formation.
- Presence of BUE increases power consumption.

8.4. CUTTING TOOLS

8.4.1. Classification

Cutting tools are *classified* as follows :

1. ***Single point cutting tools :***

- These tools have only one cutting edge ; such as lathe tools, shaper tools, planer tools, boring tools, etc.

2. *Multi-point cutting tools :*

(*i*) Solid tool.

(*ii*) Brazed tool.

(*iii*) Inserted bit tool.

● These tools have more than one cutting edges ; such as milling cutters, drills, broaches, grinding wheels, etc.

8.4.2. Single Point Cutting Tool :

● Fig. 8.3 (*a*) shows a single point right-hand cutting tool. Although these tools have traditionally been produced from solid tool-steel bars, they have been so largely replaced by carbide or other inserts of various shapes and sizes (*b*).

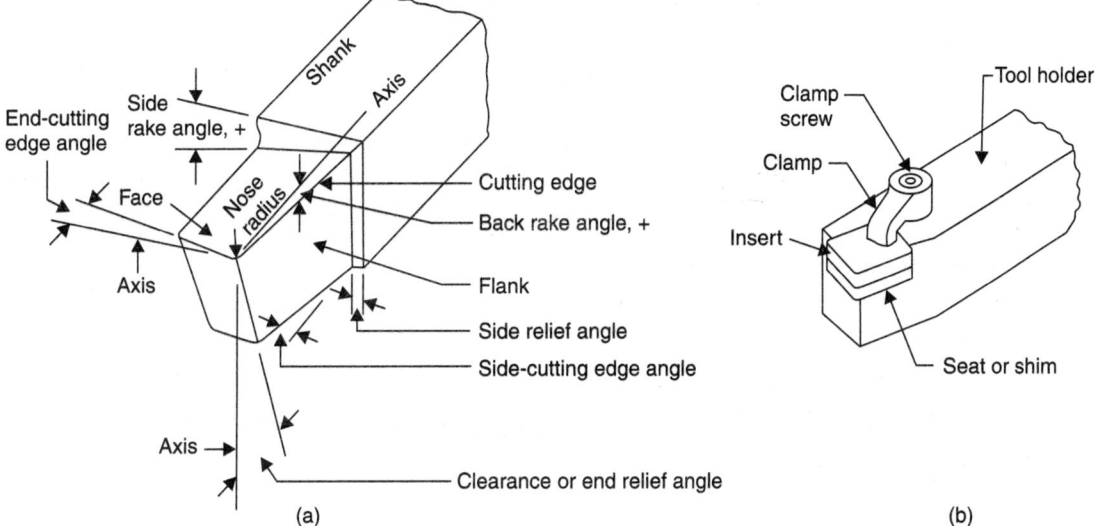

Fig. 8.3. Single point right-hand cutting tool.

● Fig. 8.4 shows the various angles of a single point cutting tool.

8.4.3. Tool Elements and Tool Angles :

Refer to Figs. 8.3 and 8.4.

Tool elements. The definitions of various tool elements are :

(*i*) **Shank.** It is the main body of the tool at one end of which the cutting portion is formed.

(*ii*) **Flank.** The surface (or surfaces) below and adjacent to the cutting edge is called the *flank* of the tool.

(*iii*) **Face.** The surface on which the chip slides is called the *face* of the tool.

(*iv*) **Heel.** It is the intersection of the flank and base of the tool.

(*v*) **Nose.** It is the *point* where the side cutting edge and end cutting edge intersect.

(*vi*) **Neck.** The portion which is reduced in section to form necessary cutting edges and angles is called *neck*.

(*vii*) **Cutting edge.** It is the edge on the face of the tool which removes the material from the workpiece.

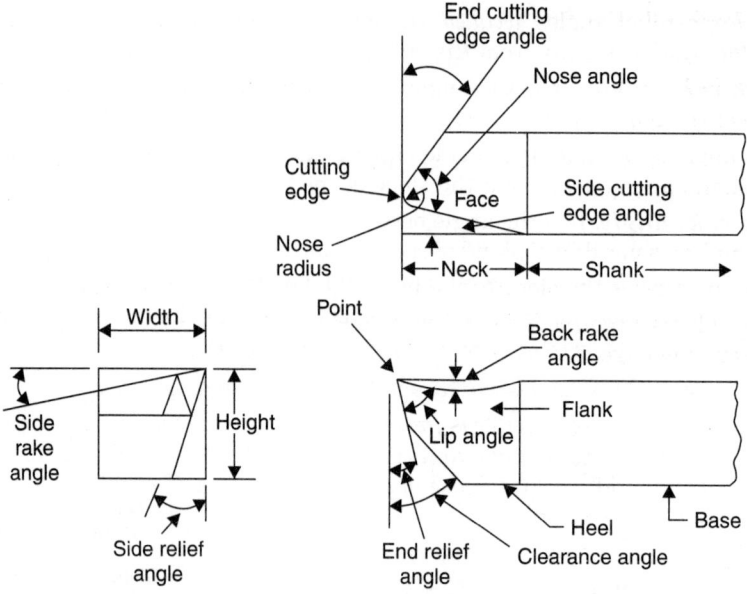

Fig. 8.4. Various angles of a single point tool.

Tool angles :

(*i*) *Side cutting edge angle.* It is angle between the side cutting edge and the side of the tool shank.

— It is also known as *'lead angle'*.

— Its complementary angle is called *'Approach angle'*.

● This angle *prevents interference as the tool enters the work material.*

● Its satisfactory values vary from 15° to 30° for general machining.

(*ii*) *End cutting edge angle.* This is the angle between the end cutting edge and a line normal to the tool shank.

● This angle provides a clearance or relief to the trailing end of the cutting edge to *prevent rubbing or drag* between the machined surface and the trailing part of the cutting edge. Only a small angle is sufficient for the purpose.

● An angle of 8° to 15° has been found satisfactory in most cases on side cutting tools, like boring and turning tools.

● End cutting tools, like cut off and necking tools often have no end cutting-edge angle.

(*iii*) *Side relief angle.* It is the angle between the portion of the side flank immediately below the side cutting edge and a line perpendicular to the base of the tool, and measured at right angle to the side flank.

(*iv*) *End relief angle.* It is the angle between the portion of the end flank immediately below the end cutting edge and a line perpendicular to the base of the tool, and measured at right angle to the end flank.

● The *side and relief angles* are provided so that the flank of the tool clears the workpiece surface and there is no rubbing action between the two.

— These angles range from 5° to 15° for general turning.

— *Small relief angles* are necessary to give strength to the cutting edge when machining *hard and strong materials.*

— Tools with *increased values* of relief angles penetrate and cut the workpiece material more efficiently and this *reduces* the cutting forces.

— *Too large* relief angles *weaken* the cutting edge and there is less mass to absorb and conduct the heat away from the cutting edge.

(*v*) **Back rake angle.** It is the angle between the face of the tool and a line parallel to the base of the tool and measured in a plane (perpendicular) through the side cutting edge.

— This angle is *positive,* if the side cutting edge slopes downwards from the point towards the shank and is *negative* if the slope of the side cutting edge is *reverse.*

(*vi*) **Side rake angle.** It is the angle between the tool face and a line parallel to the base of the tool and measured in a plane perpendicular to the base and the side cutting edge.

— This angle gives the slope of the face of the tool from the cutting edge.

The side rake is *negative* if the slope is *towards* the cutting edge and *positive* if the slope is *away* from the cutting edge.

● The *"rake angle" specifies the ease with which a metal is cut.*

— *Higher the rake angle, better is the cutting and less are the cutting forces.* There is a maximum limit to the rake angle and this is generally of the order of 15° for high speed steel tools cutting mild steel (*increase in rake angle reduces the strength of the tool chip as well as the heat dissipation*).

— It is possible to have rake angle as *zero* or *negative.* These are generally used in case of highly brittle tool materials such as carbides or diamonds *for giving extra strength to the tool tip.*

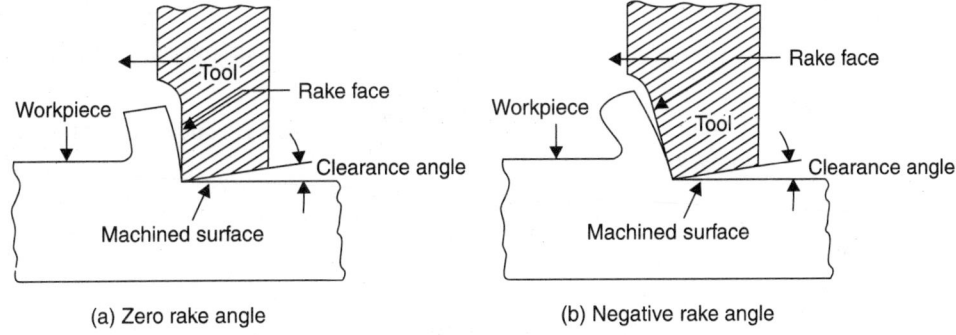

Fig. 8.5. Tool cutting at different rake angles.

(*vii*) **Clearance angle.** This is the angle between the machined surface and underside of the tool called the flank face.

● The clearance angle is provided such that the tool will not rub the machined surface thus spoiling the surface and increasing the cutting forces.

● A very large clearance angle reduces the strength of the tool lip, and hence normally an angle of the order of 5°–6° is used.

(*viii*) **Nose angle.** It is the angle between the side cutting edge and end cutting edge.

● *Nose radius* is provided to remove the fragile corner of the tool ; it increases the tool life and improves surface finish. Too large a nose radius will induce chatter.

8.4.4 Tool Signature (or Tool Designation)

The *seven important elements* comprise the signature of the cutting tool and are always stated in the following order :

(*i*) Back rake angle ;

(*ii*) Side rake angle ;

(*iii*) End relief angle ;

(*iv*) Side relief angle ;

(*v*) End cutting edge angle ;

(*vi*) Side cutting edge angle ;

(*vii*) Nose radius.

It is usual to omit the symbols for degrees and mm, simply listing the numerical value of each component :

- *A typical tool designation (signature) is :*

  0—10—6—6—8—90—1 mm.

8.5. ORTHOGONAL AND OBLIQUE CUTTING

In the metal cutting operation (Fig 8.1), the tool is wedge-shaped and has a straight cutting edge. Basically, there are *two* methods of metal cutting, depending upon the arrangement of the cutting edge with respect to the direction of relative work-tool motion :

1. Orthogonal cutting or two dimensional cutting.

2. Oblique cutting or three dimensional cutting.

1. Orthogonal cutting : Refer to Fig. 8.6.

- When the tool is pushed into the workpiece, a layer of material is removed from the workpiece and it slides over the front face of the tool called rake face. *When the cutting edge of wedge is perpendicular to the cutting velocity, the process is called* **orthogonal cutting.**

- In this case, the material gets deformed under plane strain conditions ; the *chip slides directly up the tool face.*

— Rarely in practice, however, is the cutting edge at right angles to the direction of cutting (*i.e.*, orthogonal cutting).

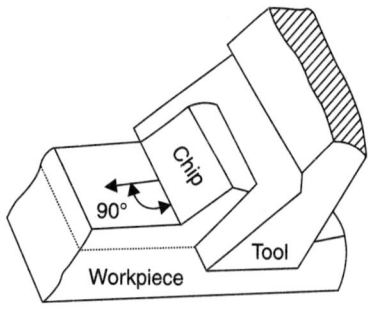

Fig. 8.6. Orthogonal cutting.

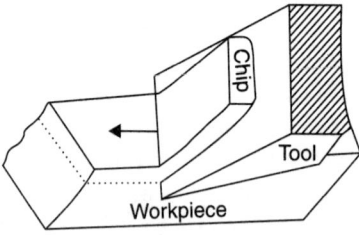

Fig. 8.7. Oblique cutting.

2. Oblique cutting : Refer to Fig 8.7.

- In most practical metal-cutting processes, the cutting edge of the tool is *not perpendicular* to the cutting velocity but set at angle with the normal to the cutting velocity.

- *Cutting in this case takes place in three-dimensions (turning or milling) and represents the general case of oblique cutting.*

- In oblique cutting a *lateral direction of chip movement is obtained.*

Comparison between 'Orthogonal cutting' and 'Oblique cutting'

S. No.	Aspects	Orthogonal cutting	Oblique cutting
1.	Inclination of the cutting edge of the tool.	Perpendicular to the direction of tool travel.	Inclined at an angle with the normal to the direction of tool travel.
2.	Clearance of the workpiece width by the cutting edge.	The cutting edge clears the width of the workpiece on either ends.	The cutting edge may or may not clear the width of the work-piece.
3.	The chip movement.	The chip flows over the tool face and direction of chip flow velocity is *normal* to the cutting edge. The chip coils in a tight flat spiral.	The chip flows on the tool face *making an angle* with the normal on the cutting edge. The chip flows sideways in a long curl.
4.	Number of components of cutting force acting on the tool.	Only *two* components of the cutting force act on the tool. These two components are perpendicular to each other and can be represented in a plane.	*Three* components of the forces (mutually perpendicular) act at the cutting edge.
5.	Maximum chip thickness occurrence.	Maximum chip thickness occurs at its middle.	The maximum chip thickness may not occur at middle.
6.	Tool Life.	Less	More.

8.6. CHIP CONTROL

- The control and disposal of chips in high speed production turning, is important to protect both the operator and the tools. The long and ribbon-like continuous chip that curls round the cutting tool has sharp edges and can inflict deep, painful and dangerous cuts. It should never be handled with the bare hands. A swarf rake should be used to drag it away from the working zone of the machine.

- The usual procedure to avoid the formation of continuous chips is to break the chip intermittently with a *chip breaker*.

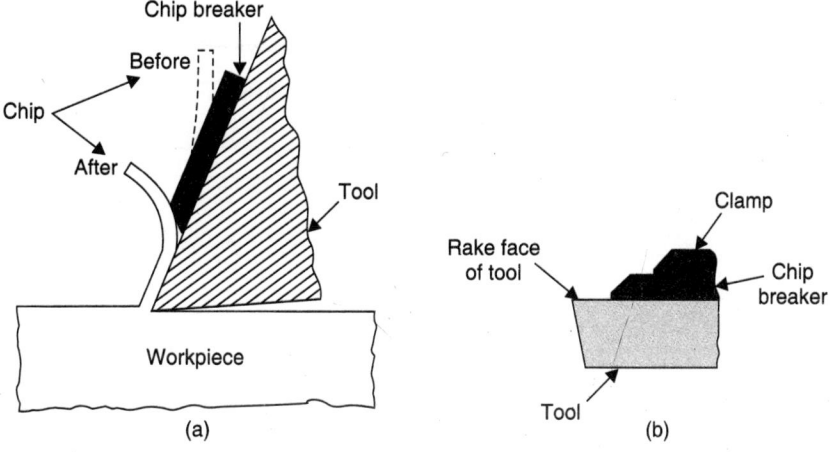

Fig. 8.8. Chip breaker.

— Fig. 8.8(*a*) shows the schematic illustration of the action of a chip breaker ; the chip breaker *decreases* the radius of the curvature of this chip.

— Fig. 8.8(*b*) shows the chip breaker clamped on the rake face of a cutting tool.

● A wide variety of cutting tools and inserts with chip-breaker features are available. However, with soft workpiece materials, such as commercially pure aluminium and copper, chip breakers are generally *not as effective* ; the remedy is usually to change the process parameters.

— In interrupted cutting operations, such as milling, chip breakers are *generally not necessary,* since the chips already have finite lengths, due to intermittent nature of operation.

8.7. FORCE OF A SINGLE-POINT TOOL

The work material offers resistance to the cutting tool, during metal cutting. This resistance is overcome by the cutting force applied to the tool face. The *work done by this force in cutting is expended in shearing the chip from the work, deforming the chip and overcoming the friction of the chip on the tool face and tool flank on the cutting surface.*

The *magnitude of the cutting force* depends upon following *factors* :

— Material being machined ;

— Rate of feed ;

— Depth of cut ;

— Tool angles ;

— Cutting speed ;

— Coolant used, etc.

Fig. 8.9 shows forces acting on a single-point cutting tool.

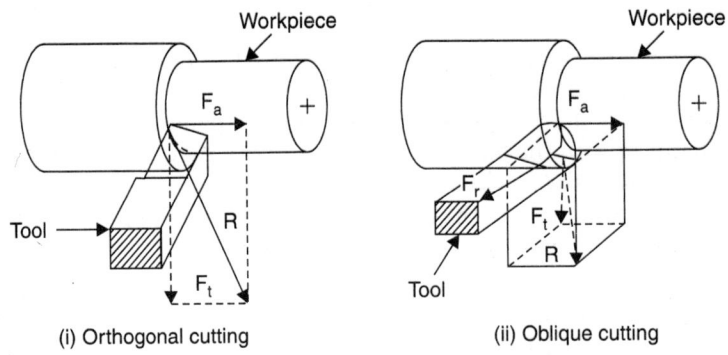

(i) Orthogonal cutting   (ii) Oblique cutting

Fig. 8.9. Forces acting on a cutting tool.

Orthogonal cutting : Resultant, $R = \sqrt{F_a^2 + F_t^2}$ ...(8.1)

Oblique cutting : Resultant, $R = \sqrt{F_a^2 + F_r^2 + F_t^2}$ ...(8.2)

Where F_a, F_r and F_t are *axial* (feed) *radial* and *tangential* forces respectively.

— Force, $\boldsymbol{F_a}$ acts in horizontal plane parallel to the work axis.

— Force, $\boldsymbol{F_r}$ acts in horizontal plane along a radius of the work.

— Force, $\boldsymbol{F_t}$ acts in a vertical plane tangent to the cutting surface.

● F_t is always the largest of the three components. It develops torque on the workpiece.

● F_a due to feed motion is about 35 to 55% of F_t.

● F_r which tends to push the tool back out of the work is about 25 to 30% of F_t.

Torque developed on the workpiece,

$$T = \frac{F_t \times D}{2 \times 1000} \text{ Nm (neglecting the components } F_a \text{ and } F_r) \qquad ...(8.3)$$

where, D = Diameter of the workpiece in mm.

Heat produced (= work done in metal cutting)

$$= \frac{F_t \times V}{60 \times 1000} \text{ kN m/s} \quad \text{or} \quad \text{kJ/s or kW} \qquad ...(8.4)$$

where, V = Cutting speed in m/min.

$$\textit{Power required} = \frac{F_t \times V}{60 \times 1000 \times \eta} \text{ kW} \qquad ...(8.5)$$

where, η = Efficiency of the machine.

The approximate values of efficiencies of the different machines when working at full loads are :

1. Lathes .... 80 to 90%
2. Drilling machines .... 85 to 90%
3. Milling machines .... 80 to 90%
4. Shapers and planers .... 65 to 75%
5. Grinding machines .... 80 to 85%.

The following points about the component forces (F_a, F_r and F_t) are worth noting :

● The forces are *not changed significantly by a change in cutting speed.*
● The *greater the 'feed',* of the tool, the *larger* the forces.
● The *greater the 'depth' of the cut,* the *larger* the forces.
● Tangential force increases with chip size.

Measurement of cutting forces :

Although an indirect method of measuring cutting forces acting on the tool is with the aid of a *"wattmeter"*, yet a more *exact method* is with the aid of a **tool dynamometer.**

The total force during metal cutting, in most metal cutting dynamometers, is determined by measuring the *deflections* or *strains* in the elements supporting the cutting tool. The design of the dynamometer should be such as to *give strains or displacements large enough to be measured accurately.*

The commonly used tool dynamometer are :

1. Mechanical dial gauge type.
2. Strain gauge dynamometer.
   ● A strain gauge dynamometer is *more accurate* than a mechanical dial gauge.
3. Pneumatic and hydraulic dynamometers.
4. Electrical dynamometers.
5. Piezoelectrical dynamometers.

8.8. MECHANICS OF METAL CUTTING

The *basic mechanism* by which chips are formed during the process of metal cutting is that of deformation of the material, lying ahead of cutting edge of the tool, because of *shearing action.*

8.8.1. Shear Zone, Shear Plane and Shear Angle

When cutting tool is introduced into the work material, *plastic deformation takes place in a narrow region in the vicinity of the cutting edge.* This region is called **shear zone** (see Fig. 8.10). The width of this zone is small and therefore chip formation is often described as a process of *successive* *shears of thin layers of the work material along particular surfaces.* At high speeds, this zone can be assumed to be restricted to a plane called **shear plane** (see Fig. 8.11) inclined at an angle ϕ *(shear angles).*

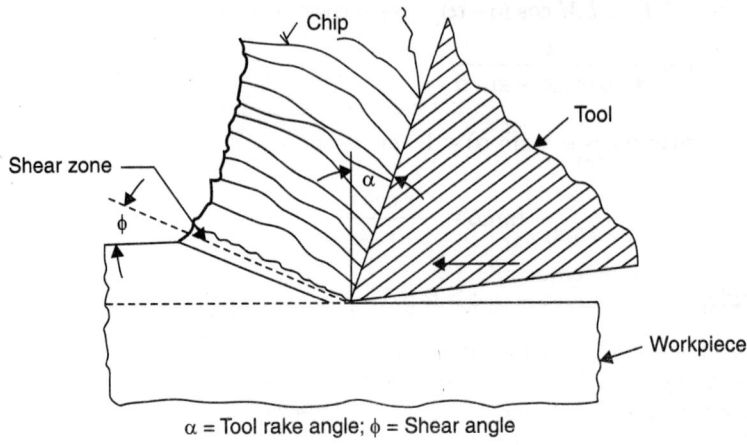

α = Tool rake angle; ϕ = Shear angle

Fig. 8.10. Basic mechanism of chip formation.

The sharp line *LM* separates the deformed and undeformed work material and indicates the projection of the shear plane.

- The *value of shear angle depends upon* :
  (*i*) Workpiece materials ;
  (*ii*) Cutting conditions ;
  (*iii*) Material of tool ;
  (*iv*) Geometry of tool.
  — When the shear angle is *small*, the plane of the shear will be larger, chip is thicker and therefore *higher force* is required to remove the chip.
  — When the shear angle is *large*, the plane of shear will be shorter, the chip is thinner and hence *less force* is required to remove the chip.
- The *shear angle is determined from chip thickness ratio (r).*

8.8.2. Chip Thickness Ratio

The machinability of a metal is expressed by *chip increases ratio* (also termed as *cutting ratio*).

In order to experimentally *determine the shear angle* we will have to stop the cutting process and study the zone with the help of a microscope or a photograph. Alternatively we can also derive a relationship from the geometry of chip formation (Fig. 8.11).

From Fig. 8.11,

Depth of cut,

$$t = LM \sin \phi \qquad ...(8.6)$$

Chip thickness,

$$t_c = LM \cos (\phi - \alpha) \qquad ...(8.7)$$

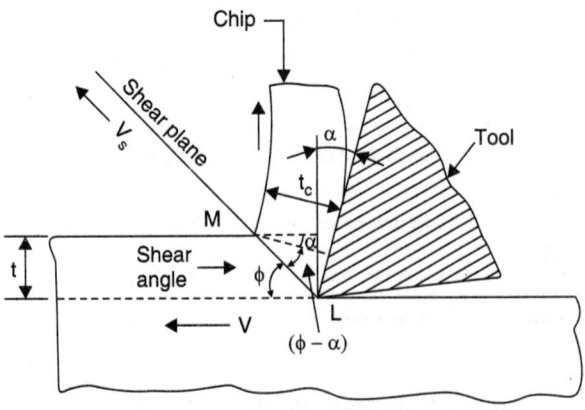

t = Original depth of cut ;
t_c = Thickness of chip.

Fig. 8.11. Shear angle (ϕ), shear plane and rake angle of the tool (α).

(The depth of cut is actually the feed in machining operation)

Then, chip thickness ratio, r would be :

$$r = \frac{t}{t_c} = \frac{LM \sin \phi}{LM \cos (\phi - \alpha)} = \frac{\sin \phi}{\cos \phi \cos \alpha + \sin \phi \sin \alpha}$$

or,
$$r = \frac{1}{\cot \phi \cos \alpha + \sin \alpha} \qquad \text{......Dividing numerator and denominator by } \sin \phi$$

or,
$$r(\cot \phi \cos \alpha + \sin \alpha) = 1 \quad \text{or,} \quad \cot \phi \cot \alpha = \frac{1 - r \sin \alpha}{r}$$

or,
$$\tan \phi = \frac{r \cos \alpha}{1 - r \sin \alpha} \qquad \qquad ...(8.8)$$

or, Shear angle,
$$\phi = \tan^{-1} \left(\frac{r \cos \alpha}{1 - r \sin \alpha} \right) \qquad \qquad ...[8.8(a)]$$

The value of r could be determined experimentally by measuring the average thickness of the chips produced under given conditions of feed and speed. From this it is possible to *evaluate the shear angle* using the above equation.

- The *cutting ratio* or *chip thickness ratio (r)* is *always less than unity* and can be evaluated by measuring chip thickness and depth of cut. But actually it is very difficult to measure chip thickness precisely due to the roughness of the back surface of chip.

Now, volume of metal removed = Volume of chip

$\therefore \qquad \qquad b \cdot t \cdot l \cdot \rho = b_c \cdot t_c \cdot l_c \cdot \rho_c$

(b, t, l, ρ being width, thickness or depth, length, and density of metal cut and c standing suffix for chip)

It is found that width of chip is same as of workpiece and also density of both is same.

$\therefore \qquad \qquad r = \frac{t}{t_c} = \frac{l_c \text{ (length of chip)}}{l \text{ (length of uncut chip)}}$

It is easier to measure the length of chip than thickness of work.

- *Cutting ratio* may also be defined *as the ratio of the chip velocity V_c to the cutting speed V.*

This ratio $\left(i.e., \dfrac{V_c}{V} \right)$ can be determined mathematically by finding the kinetic forces acting on the chip.

8.8.3. Velocity Relationship in Orthogonal Cutting

In an orthogonal cutting processes, there are three velocities ; these are :

1. *Cutting velocity (V)*—Velocity of tool relative to the workpiece.
2. *Velocity of chip (V_c)*—Velocity with which the chip moves over the rake face of the cutting tool.
3. *Velocity of shear (V_s)*—Velocity with which metal of the workpiece shears along the shear plane.

The cutting velocity V_c and rake angle α are always known ; the values of V_f and V_s can be calculated as follows :

Refer to Fig. 8.12 (b), which shows the velocity diagram $\left(\overrightarrow{V} = \overrightarrow{V_c} + \overrightarrow{V_s} \right)$.

Using sine rule, we get

$$\frac{V_c}{\sin \angle MSL} = \frac{V_s}{\sin \angle MLS} = \frac{V}{\sin \angle LMS}$$

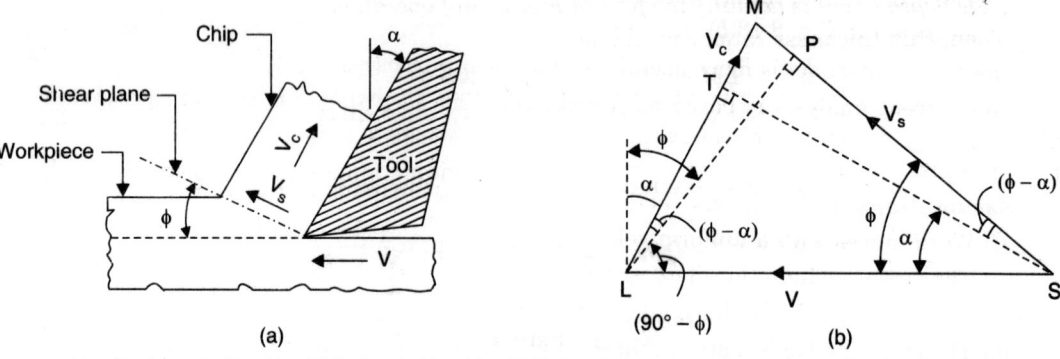

V = Cutting velocity ; V_c = Velocity of chip ; V_s = Velocity of shear ; α = Tool rake angle ; φ = Shear angle

Fig. 8.12. Velocity relationships in orthogonal cutting.

or,
$$\frac{V_c}{\sin \phi} = \frac{V_s}{\sin [(90° - \phi) + (\phi - \alpha)]} = \frac{V}{180 - [(90° - \phi + \phi - \alpha) + \phi]}$$

or,
$$\frac{V_c}{\sin \phi} = \frac{V_s}{\sin (90° - \alpha)} = \frac{V}{\sin \{90° - (\phi - \alpha)\}}$$

or,
$$\frac{V_c}{\sin \phi} = \frac{V_s}{\cos \alpha} = \frac{V}{\cos (\phi - \alpha)}$$

∴
$$V_c = \frac{V \sin \phi}{\cos (\phi - \alpha)} \qquad ...(8.9)$$

and,
$$V_s = \frac{V \cos \alpha}{\cos (\phi - \alpha)} \qquad ...(8.10)$$

8.8.4. Forces on the Chip (Merchant's Analysis)

There are usually *two* schools of thought in the analysis of the metal removal process :

— Our school of thought is that the deformation zone is very *thin and planar* as shown in Fig. 8.13(a).

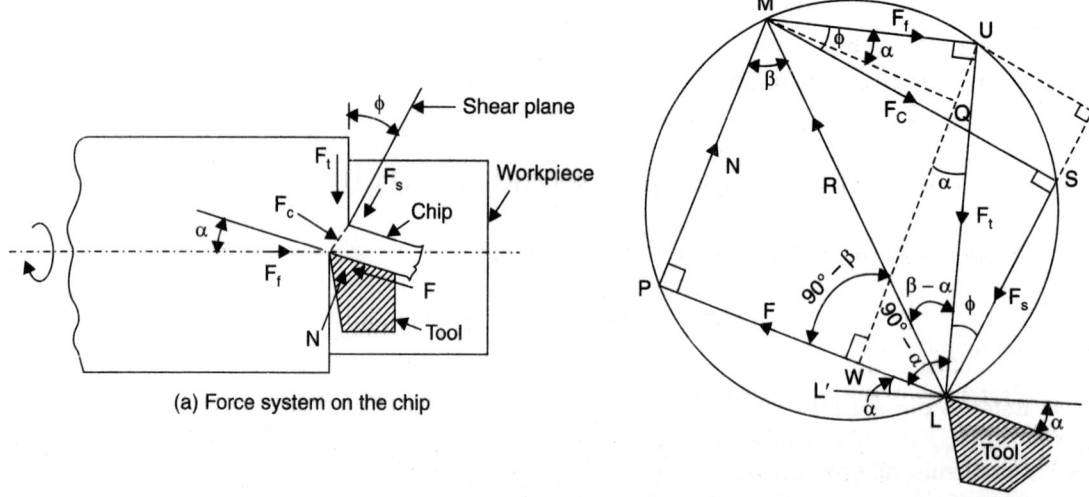

(a) Force system on the chip

(b) Merchant's circle diagram

Fig. 8.13. Merchant's analysis.

— The other school thinks that the actual deformation zone is a *thick one with a fan shape* as shown in Fig. 8.13(*b*).

The *thin zone model* is more useful for analytical purposes.

The current analysis is based on Merchant's *thin shear plane model* which considers the minimum energy principle. This *metal is applicable at very high cutting speeds which are generally practised in production.*

Assumptions. Following are the *assumptions* with regard to this model :

(*i*) Work moves with a uniform velocity.

(*ii*) The surface where the shear occurs is a plane.

(*iii*) The tool is perfectly sharp and there is no contact along the clearance face.

(*iv*) The cutting edge is a straightline which extends perpendicular to the direction of motion and generates plane surface as the work moves past it.

(*v*) Width of the tool is greater than the width of work.

(*vi*) The stresses on the shear plane are uniformly distributed.

(*vii*) Uncut chip thickness is constant.

(*viii*) A continuous chip is produced without any built up edge.

(*ix*) The chip does not flow to either side, or there is no side spread.

Refer to Fig. 8.13 (a) :

F_t = Tangential or cutting force ;   } *Forces acting on the tool, and*
F_f = Feed force ;   } *measured by the dynamometer.*

F_c = Compressive force on the shear plane ;   } *Forces exerted by the workpiece*
F_s = Shear force on the shear plane ;   } *on the chip.*

F = Frictional force along the rake force of tool ;   } *Forces exerted by the tool on*
N = Normal force at the rake face of tool.   } *the chip.*

Refer to Fig. 8.13(b) :

$$\alpha = \text{Tool rake angle ;}$$
$$\phi = \text{Shear plane angle ;}$$
$$\beta = \text{angle of friction.}$$

Now, $F = PW + WL$

or, $F = F_f \cos \alpha + F_t \sin \alpha$ ...(8.11)

$N = MP = UW - UQ$
$= F_t \cos \alpha - F_f \sin \alpha$ ...(8.12)

$\therefore$ $$\frac{F}{N} = \frac{F_f \cos \alpha + F_t \sin \alpha}{F_t \cos \alpha - F_f \sin \alpha}$$

Dividing numerator and denominator R.H.S. by cos α, we get

$$\frac{F}{N} = \frac{F_f + F_t \tan \alpha}{F_t - F_f \tan \alpha}$$ ...(8.13)

But, $\dfrac{F}{N} = \tan \beta = \mu$ (coefficient of friction)

[where F (frictional force) and N (normal reaction) are the components of resultant tool force R]

$\therefore$ $$\mu = \frac{F}{N} = \tan \beta = \frac{F_f + F_t \cdot \tan \alpha}{F_t - F_f \cdot \tan \alpha}$$ ...(8.14)

Now,

$$F_s = F_t \cos \phi - F_f \sin \phi \qquad \qquad \text{...(8.15)}$$
$$F_c = F_f \cos \phi + F_f \sin \phi \qquad \qquad \text{...(8.16)}$$
$$F_t = R \cos (\beta - \alpha) \qquad \qquad \text{...(8.17)}$$
$$F_f = R \sin (\beta - \alpha) \qquad \qquad \text{...(8.18)}$$

Also,

$$F_s = R \cos (\beta - \alpha + \phi) \quad \text{or} \quad R \cos (\phi + \beta - \alpha) \qquad \text{...(8.19)}$$

From eqns. (8.17) and (8. 19), we get

$$\frac{F_t}{F_s} = \frac{R \cos (\beta - \alpha)}{R \cos (\phi + \beta - \alpha)}$$

or,

$$F_t = F_s \left[\frac{\cos (\beta - \alpha)}{\cos (\phi + \beta - \alpha)} \right] \qquad \qquad \text{...(8.20)}$$

8.8.5. Stress and Strain on the Chip

Since the chips are formed (during machining) due to the plastic deformation of the workpiece material, they experience stress and strain. At the shear plane [see Fig. 8.13 (a)] two forces F_c and F_s (perpendicular to each other) exist.

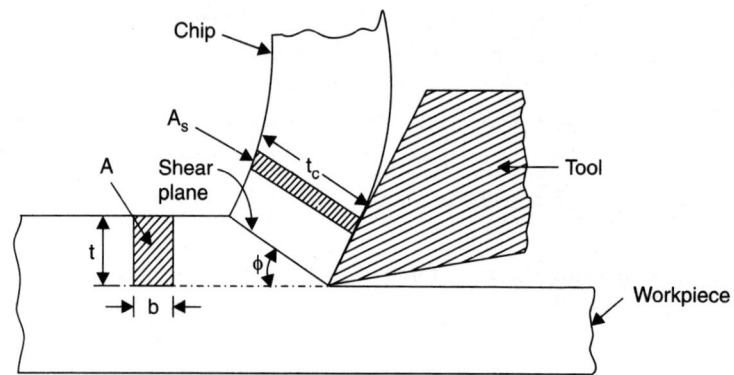

Fig. 8.14. Geometry of chip formation.

Refer to Fig. 8.14 :

$$A_s = \text{Area of shear plane,}$$
$$A (= b \times t) = \text{Cross-sectional area of uncut chip (i.e., before cutting)}$$
$$= A_s \sin \phi$$

where,

$$b = \text{Width of cut, and}$$
$$t = \text{Uncut chip thickness.}$$

Mean normal stress (σ) :

$$\sigma = \frac{F_c}{A_s} = \frac{F_c}{(A/\sin \phi)} = \frac{F_c \cdot \sin \phi}{A} \qquad \qquad \text{...(8.21)}$$

or,

$$\sigma = \frac{(F_f \cos \phi + F_t \sin \phi) \sin \phi}{A} \qquad \qquad \text{...(8.22)}$$

.... using eqn. (8.16)

Mean shear stress (τ) :

$$\tau = \frac{F_s}{A_s} \qquad \qquad \text{...(8.23)}$$

or, $\qquad \tau = \dfrac{F_s \sin \phi}{A}$ $\qquad\qquad$ ...(8.24)

$$F_s = \dfrac{\tau . b . t}{\sin \phi} \qquad\qquad ...[8.24(a)]$$

Using eqns. (8.15), (8.23) and (8.24), we have

$$\tau = \dfrac{F_t \cos \phi - F_f \sin \phi}{A_s} = \dfrac{(F_t \cos \phi - F_f \sin \phi) \sin \phi}{A} \qquad\qquad ...(8.25)$$

Shear strain. To evaluate the shear strains, let us take the help of Piispanen's model as shown in Fig. 8.15. (He considered the *undeformed metal as a stack of cards which would slide over one another as wedge shaped tools moved under these cards*. Though this idea is an over-simplified one, it accounts for a number of features that are found in practice. A practical example is when *paraffin is cut, a blockwise slip is clearly evident*).

Shear strain ($\in$) can be calculated as follows : Refer to Fig. 8.15

$$\in = \dfrac{\Delta s}{\Delta y} = \dfrac{BA}{CD} = \dfrac{BD + DA}{CD} = \dfrac{BD}{CD} + \dfrac{DA}{CD}$$

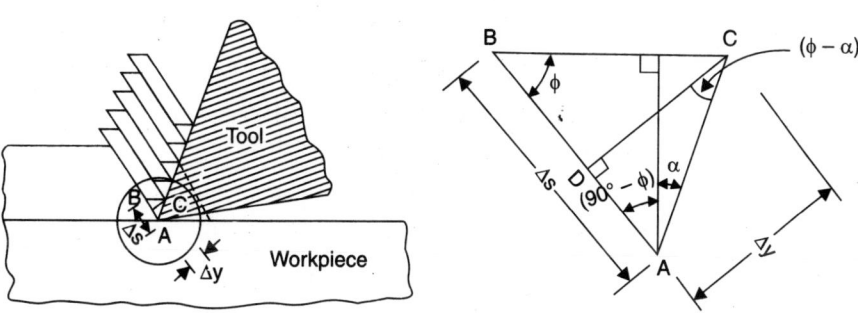

Fig. 8.15. Strain and strain rate in orthogonal cutting.

or, $\qquad \in = \cot \phi + \tan (\phi - \alpha) \qquad\qquad ...(8.26) \quad \begin{bmatrix} \because \ \angle DCA = 90° - (90° - \phi + \alpha) \\ = \phi - \alpha \end{bmatrix}$

$$= \dfrac{\cos \phi}{\sin \phi} + \dfrac{\sin (\phi - \alpha)}{\cos (\phi - \alpha)}$$

$$= \dfrac{\cos \phi . \cos (\phi - \alpha) + \sin \phi . \sin (\phi - \alpha)}{\sin \phi . \cos (\phi - \alpha)}$$

$$= \dfrac{\cos [\phi - (\phi - \alpha)]}{\sin \phi . \cos (\phi - \alpha)}$$

or, $\qquad \in = \dfrac{\cos \alpha}{\sin \phi \cos (\phi - \alpha)} \qquad\qquad ...(8.27)$

Strain rate is given by :

$$\dot{\in} = \dfrac{\Delta s}{\Delta y . \Delta t} = \dfrac{V_s}{\Delta y} = \dfrac{\cos \alpha}{\cos (\phi - \alpha)} . \dfrac{V}{\Delta y} \qquad\qquad ...(8.28)$$

$$\qquad\qquad\qquad\qquad\qquad\qquad\qquad\qquad .... \text{ using eqn. (8.10)}$$

where, Δy = Thickness of deformation zone, and

$\qquad t$ = Time to achieve the final value of strain.

8.8.6. Work Done During Metal Cutting and Specific Cutting Energy

● Most of the energy consumed in metal cutting is utilised in the plastic deformation.

Total work done in cutting, $W = F_t \times V$

Workdone in shear, $W_s = F_s \times V_s$

Work done in friction, $W_f = F \times V_c$

Thus, $W = F_t \times V = F_s V_s + FV_c$ ...(8.29)

where, F_t = Cutting force,

F_s = Shear force on the shear plane,

F = Frictional force along the rake face of tool,

V = Cutting velocity, (Velocity of tool relative to the workpiece)

V_s = Velocity of shear, and

V_c = Velocity with chip moves over the rake face of the cutting tool.

● In order to get a better picture of the efficiency of the metal cutting operation it is necessary to have a new parameter which does not depend upon the cutting process parameters. The *"specific cutting energy"* E_{sp} is such a parameter which can be obtained by dividing the total work done with the material remove rate (*MRR*).

$$MRR = V.b.t$$

$$E_{sp} = \frac{F_t . V}{V.b.t} = \frac{F_t}{b.t}$$

$$= \frac{1}{b.t} . F_s \left[\frac{\cos (\beta - \alpha)}{\cos (\phi + \beta - \alpha)} \right] \qquad \text{...using eqn. (8.20)}$$

$$= \frac{1}{b.t} . \tau . A_s \left[\frac{\cos (\beta - \alpha)}{\cos (\phi + \beta - \alpha)} \right]$$

$$= \frac{1}{b.t} . \tau . \left(\frac{bt}{\sin \phi} \right) \left[\frac{\cos (\beta - \alpha)}{\cos (\phi + \beta - \alpha)} \right] \qquad \left(\because A_s = \frac{A}{\sin \phi} = \frac{bt}{\sin \phi} \right)$$

or, $$E_{sp} = \frac{\tau \cos (\beta - \alpha)}{\sin \phi . \cos (\phi + \beta - \alpha)} \qquad \text{...(8.30)}$$

8.8.7. Theories on Mechanics of Metal Cutting

From the various relationships derived above, we find that they involve α (rake angle), β (friction angle), and ϕ (shear angle). The angle α on the tool can be easily measured but values of β and ϕ can be obtained by computation. Several investigators such as Ernst and Merchant, Merchant, Stabler, Lee and Shaffer, Palmer and Oxley have carried out lot of work to establish relationship between α, β and ϕ, and have proposed their own theories.

1. *Ernst-Merchant theory.* This theory is based on the following *assumptions* :

(*i*) The shear stress is maximum at the shear plane and it remains constant.

(*ii*) The expenditure is minimum in this process, *i.e.*, shear will take place in a direction in which energy required for shearing is minimum.

In Merchant's theory, it was found that ;

$$F_t = R \cos (\beta - \alpha) \qquad \text{...[Eqn. (8.17)]}$$

$$F_s = R \cos (\phi + \beta - \alpha) \qquad \text{...[Eqn. (8.18)]}$$

or, $$\frac{F_t}{F_s} = \frac{\cos (\beta - \alpha)}{\cos (\phi + \beta - \alpha)}$$

or, $$F_t = F_s \times \left[\frac{\cos (\beta - \alpha)}{\cos (\phi + \beta - \alpha)} \right]$$

$$\left(\text{where,} \quad F_s = \frac{\tau . b . t}{\sin \phi} \right) \quad ...[\text{Eqn. } 8.24(a)]$$

∴ $$F_t = \frac{\tau . b . t}{\sin \phi} \left[\frac{\cos (\beta - \alpha)}{\cos (\phi + \beta - \alpha)} \right] \qquad ...(1)$$

Differentiating eqn. (1) w.r.t. ϕ and equating to zero to find the value of shear angle, ϕ, for which F_t is a *minimum*, we get

$$\frac{dF_t}{d\phi} = - \tau . b . t \cos (\beta - \alpha) \left[\frac{\cos \phi . \cos (\phi + \beta - \alpha) - \sin \phi . \sin (\phi + \beta - \alpha)}{\{\sin \phi . \cos (\phi + \beta - \alpha)\}^2} \right] = 0$$

or, $\cos \phi . \cos (\phi + \beta - \alpha) - \sin \phi . \sin (\phi + \beta - \alpha) = 0$

or, $\cos (\phi + \phi + \beta - \alpha) = 0$

or, $$\cos (2\phi + \beta - \alpha) = \cos \frac{\pi}{2}$$

or, $$2\phi + \beta - \alpha = \frac{\pi}{2}$$

or, $$\phi = \frac{\pi}{4} - \frac{\beta}{2} + \frac{\alpha}{2}$$

or, $$\textit{Shear angle, } \phi = \frac{\pi}{4} - \frac{1}{2} (\beta - \alpha) \qquad ...(8.31)$$

2. Merchant theory. Merchant found that the above theory agreed *well* with experimental results obtained when *cutting synthetic plastics* but agreed *poorly* with experimental results obtained for *steel* machined with a sintered carbide tool.

Merchant then modified his theory by assuming that shear stress τ along the shear plane varies *linearly* with normal stress, *i.e.*,

$\tau = \tau_0 + k \sigma$, where k is constant. (τ_0 is the value of τ when $\sigma = 0$)

This *assumption* agreed with the work of *Bridgman*, where, in experiments on poly crystalline metals, the shear strength was shown to be dependent on the normal stress on the plane of shear. Merchant then derived,

$$2\phi + \beta - \alpha = C \qquad ...(8.32)$$

where, C = machining constant (= arc cot k) ; its value varies from 70° to 80° for various steels.

Most recent experimental work indicates that τ remains constant for a given material over a wide range of cutting conditions and therefore k would be expected to be zero.

3. Stabler theory. He modified the Ernst-Merchant equation as :

$$\phi = \frac{\pi}{4} - \beta + \frac{\alpha}{2} \qquad ...(8.33)$$

4. Lee and Shaffer theory. Lee and Shaffer obtained a solution to the problem of the mechanics of orthogonal cutting by using the *slip-line field theory*.

This theory is based on ideal theory of plasticity according to which shear occurs on a single plane.

The solution has been derived from the pattern of construction of a slip line field using a shear plane model (Fig. 8.16). They assume that there must be a stress field within the chip to transmit the cutting forces from the shear plane to the tool face. They represent this by a slip line field in which *no* deformation occurs although it is stressed upto the yield point. This shows the Mohr's circle for the stresses at the boundaries of the stressed zone, which results in the equation :

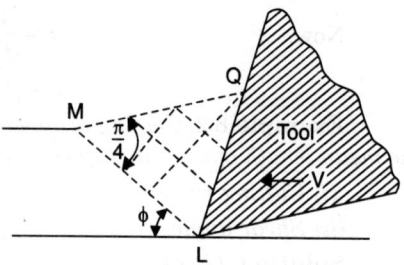

Fig. 8.16. Shear plane model of Lee and Shaffer.

$$\phi = \frac{\pi}{4} + (\alpha - \beta) \qquad\qquad ...(8.34)$$

Also,
$$\phi = \frac{\pi}{4} - \beta + \alpha - \theta \qquad\qquad ...[8.34(a)]$$

... considering built-up edge formation.

5. *Christopherson, Oxley and Palmer's theory*. During detailed study they found that the grains flow showed a *streamlined pattern and that the boundary between elastic and plastic zones was not a straight line but a narrow wedge shaped zone.*

6. *Empirical approach.* One classical example of empirical approach is *Dimensional analysis* which is extensively used in fluid mechanics and has been successfully applied to some problems in metal cutting by *"Kronenberg"*.

The above approach has been tried to get a suitable relationship for the shear angle, again of course assuming a thin shear plane as the shear zone corresponds to most of the practical high speed systems.

● Based on Dimensional analysis, the following two equations have been derived :

(*i*) $\dfrac{\phi}{\beta} = k \left(\dfrac{\alpha}{\beta}\right)^a \left(\dfrac{f}{d}\right)^b$ $\qquad\qquad ...(8.35)$

(*ii*) $\dfrac{\phi}{\beta} = k_1 \left(\cos \alpha\right)^{a_1} \left(\dfrac{f}{d}\right)^{b_1}$ $\qquad\qquad ...(8.36)$

where k, a, b, k_1, a_1 and b_1 are emiprical constants to be established from the data of experiments.

Example 8.1. *While doing orthogonal machining of a mild steel part, a depth of cut of 0.75 mm is used at 60 r.p.m. If the chip thickness is 1.5 mm and it is of continuous type then determine :*

(*i*) *Chip thickness ratio.*

(*ii*) *The length of the chip removed in one minute if work diameter is 60 mm before the cut is taken.*

Solution. *Given :* $t = 0.75$ mm ; $N = 60$ r.p.m. ; $t_c = 1.5$ mm ; $D = 60$ mm

(*i*) **Chip thickness ratio, *r* :**

$$r = \frac{t}{t_c} = \frac{0.75}{1.5} = 0.5 \quad \textbf{(Ans.)}$$

(*ii*) **Length of chip removed, t_c :**

Length of chip before cutting (*i.e.,* uncut chip length),

$$l = \pi DN = \pi \times 60 \times 60 = 11309.7 \text{ mm/min}$$

Now, $r = \dfrac{t}{t_c} = \dfrac{l_c}{l}$

$\therefore$ $l_c = r \times l = 0.5 \times 11309.7 = \textbf{5654.8 mm}$ **(Ans.)**

Example 8.2. *During orthogonal cutting a bar of 90 mm diameter is reduced to 87.6 mm. If the mean length of the cut chip is 88.2 mm and rake angle is 15°, calculate :*

(i) *Cutting ratio.*

(ii) *Shear angle.*

Solution. *Given :* $t = l_c = 88.2$ mm ; $\alpha = 15°$

(i) **Cutting ratio, r :**

Length of uncut chip, $t_c \, (= l) = \dfrac{\pi(90 + 87.6)}{2} = 278.97$ mm

$\therefore$ Cutting ratio, $r = \dfrac{\text{Cut chip length (depth of cut)}}{\text{Uncut chip length}} = \dfrac{t}{t_c} = \dfrac{l_c}{l} = \dfrac{88.2}{278.97} = \textbf{0.3162}$ **(Ans.)**

(ii) **Shear angle** ϕ **:**

We know that, $\phi = \tan^{-1}\left(\dfrac{r\cos\alpha}{1 - r\sin\alpha}\right)$ ...[Eqn. 8.8(a)]

$= \tan^{-1}\left[\dfrac{0.3162 \times \cos 15°}{1 - 0.3162 \times \sin 15°}\right] = \textbf{18.4°}$ **(Ans.)**

Example 8.3. *During orthogonal machining with a cutting tool having a 12° rake angle, the chip thickness is measured to be 0.44 mm, the uncut thickness being 0.18 mm. Determine :*

(i) *Shear plane angle.*

(ii) *Shear strain.*

Solution. *Given :* $\alpha = 12°$; $t_c = 0.44$ mm ; $t = 0.18$ mm

(i) **Shear plane angle,** ϕ **:**

Chip thickness ratio, $r = \dfrac{t}{t_c} = \dfrac{0.18}{0.44} = 0.41$

Now, $\phi = \tan^{-1}\left(\dfrac{r\cos\alpha}{1 - r\sin\alpha}\right)$

$= \tan^{-1}\left(\dfrac{0.41 \times \cos 12°}{1 - 0.41\sin 12°}\right) = \textbf{23.67°}$ **(Ans.)**

(ii) **Shear strain,** $\in$ **:**

$\in = \cot\phi + \tan(\phi - \alpha)$ ...[Eqn. (8.26)]

$= \cot 23.67° + \tan(23.67° - 12°) = \textbf{2.488}$ **(Ans.)**

Example 8.4. *The following data relate to the orthogonal cutting of a component :*

Feed force	.... *900 N*
Cutting force	.... *1800 N*
Chip thickness ratio	.... *0.26*
Tool rake angle	.... *12°.*

Determine : (i) *Compression and shear forces*

(ii) *Coefficient of friction of the chip on the tool face.*

Solution. *Given :* $F_f = 900$ N ; $F_t = 1800$ N ; $r = 0.26$; $\alpha = 12°$

(i) Compression and shear forces ; F_c, F_s :

We know that, $\phi = \tan^{-1}\left(\dfrac{r\cos\alpha}{1 - r\sin\alpha}\right)$

or, $\phi = \tan^{-1}\left(\dfrac{0.26\cos 12°}{1 - 0.26\sin 12°}\right) = 15.05°$

Now, $F_c = F_f\cos\phi + F_t\sin\phi$ ...[Eqn. (8.16)]

$= 900\cos 15.05° + 1800\sin 15.05° = \textbf{1336.5 N}$ **(Ans.)**

and, $F_s = F_t\cos\phi - F_f\sin\phi$ ...[Eqn. (8.15)]

$= 1800\cos 15.05° - 900\sin 15.05° = \textbf{1504.5 N}$ **(Ans.)**

(ii) Co-efficient of friction, μ :

$$\mu = \frac{F_f + F_t \cdot \tan\alpha}{F_t - F_f\tan\alpha} \qquad \text{...[Eqn. (8.14)]}$$

or, $\mu = \dfrac{900 + 1800 \times \tan 12°}{1800 - 900 \times \tan 12°} = \textbf{0.797}$ **(Ans.)**

Example 8.5. *Following data relate to an orthogonal cutting process :*

Chip length obtained	*= 96 mm*
Uncut chip length	*= 240 mm*
Rake angle used	*= 20°*
Depth of cut	*= 0.6 mm*

Horizontal and vertical components of cutting force = 2400 N and 240 N respectively.
Determine the following :

 (i) Shear plane angle   *(ii) Chip thickness*

 (iii) Friction angle   *(iv) Resultant cutting force.*

Solution. *Given :* $l_c = 96$ mm ; $l = 240$ mm ; $\alpha = 20°$; $t = 0.6$ mm ; $F_H (= F_t) = 2400$ N ; $F_V (= F_f)$ = 240 N

(i) Shear plane angle, ϕ :

Chip thickness ratio, $r = \dfrac{\text{Chip length }(l_c)}{\text{Uncut chip length }(l)} = \dfrac{96}{240} = 0.4$

$\therefore$ $\phi = \tan^{-1}\left(\dfrac{r\cos\alpha}{1 - r\sin\alpha}\right) = \tan^{-1}\left(\dfrac{0.4\cos 20°}{1 - 0.4\sin 20°}\right) = \textbf{23.5°}$ **(Ans.)**

(ii) Chip thickness, t_c :

Now, $r = \dfrac{t(\text{depth of cut})}{t_c(\text{chip thickness})}$,   or, $t_c = \dfrac{t}{r} = \dfrac{0.6}{0.4} = \textbf{1.5 mm}$ **(Ans.)**

(iii) Friction angle, β :

$$\mu = \tan\beta = \frac{F_f + F_t\tan\alpha}{F_t - F_f\tan\alpha} \qquad \text{...[Eqn. (8.14)]}$$

$$= \frac{240 + 2400\tan 20°}{2400 - 240\tan 20°} = 0.4815$$

$\therefore$ $\beta = \tan^{-1}(0.4815) = \textbf{25.7°}$ **(Ans.)**

(*iv*) **Resultant cutting force, R :**

$$R = \sqrt{F_t^2 + F_f^2}$$

$$= \sqrt{2400^2 + 240^2} \simeq \textbf{2412 N} \quad \textbf{(Ans.)}$$

Example 8.6. *The following data relate to orthogonal cutting of mild steel part :*

Cutting speed	.... *195 m/min*
Tool rake angle	.... *12°*
Width of cut	.... *1.75 mm*
Uncut thickness	.... *0.25 mm*
Average coefficient of	
friction between the tool and chip	.... *0.52*
Shear stress of work material	.... *385 N/mm²*

Calculate : (i) Shear angle.

(ii) Cutting and thrust components of the machining force.

Solution. *Given :* V = 195 m/min ; α = 12° ; b = 1.75 mm ; t = 0.25 mm ; μ = 0.52 ;
τ = 385 N/mm²

(*i*) **Shear angle,** ϕ :

We know that, $\mu = \tan \beta$

or, $0.52 = \tan \beta$,   or $\beta = \tan^{-1}(0.52) = 27.5°$

Using Merchant equation : $\phi = \dfrac{\pi}{4} - \dfrac{1}{2}(\beta - \alpha)$ ...[Eqn. (8.31)]

or, $\phi = 45° - \dfrac{1}{2}(27.5° - 12°) = \textbf{37.25°}$ **(Ans.)**

(*ii*) **Cutting and thrust components :** F_t, F_f :

$$\tau = \frac{F_s \sin \phi}{A} \qquad ...[\text{Eqn. (8.24)}]$$

or, $F_s = \dfrac{\tau \cdot A}{\sin \phi} = \dfrac{\tau \times (b \cdot t)}{\sin \phi} = \dfrac{385 \times 1.75 \times 0.25}{\sin 37.25°} = 278.3$ N

Also, $F_s = R \cos(\phi + \beta - \alpha)$ ...[Eqn. (8.19)]

or, $R = \dfrac{F_s}{\cos(\phi + \beta - \alpha)} = \dfrac{278.3}{\cos(37.25° + 27.5° - 12°)} = 459.78$ N

$\therefore$ $F_t = R \cos(\beta - \alpha)$ ...[Eqn. (8.17)]

         $= 459.78 \cos(27.5° - 12°) = \textbf{443.06 N}$ **(Ans.)**

and, $F_f = R \sin(\beta - \alpha)$ ...[Eqn. (8.18)]

         $= 459.78 \sin(27.5° - 12°) = \textbf{122.87 N}$ **(Ans.)**

Example 8.7. *Following data relate to an orthogonal cutting process :*

Depth of cut	.... *0.35 mm*
Chip thickness	.... *0.95 mm*
Width of cut	.... *2.8 mm*
Tool rake angle	.... *12°*
Tangential force	.... *1000 N*
Feed force	.... *500 N*

Calculate the following :

(i) Coefficient of friction between the tool and the chip.

(ii) Ultimate shear stress of the work material.

Solution. *Given :* $t = 0.35$ mm ; $t_c = 0.95$ mm, $b = 2.8$ mm ; $\alpha = 12°$, $F_t = 1000$ N ; $F_f = 500$ N.

(i) Co-efficient of friction, μ :

$$\mu = \frac{F_f + F_t \tan \alpha}{F_t - F_f \tan \alpha} \qquad \qquad ...\text{[Eqn. (8.14)]}$$

or,

$$\mu = \frac{500 + 1000 \times \tan 12°}{1000 - 500 \times \tan 12°} = \textbf{0.797} \quad \textbf{(Ans.)}$$

(ii) Ultimate shear stress, τ :

We know that, $\phi = \tan^{-1}\left(\dfrac{r \cos \alpha}{1 - r \sin \alpha}\right)$ $\qquad \qquad ...\text{[Eqn. 8.8(a)]}$

where, $r = \dfrac{t}{t_c} = \dfrac{0.35}{0.95} = 0.368$

$\therefore$ $\phi = \tan^{-1}\left(\dfrac{0.368 \cos 12°}{1 - 0.368 \sin 12°}\right) = 21.3°$

Now, $F_s = F_t \cos \phi - F_f \sin \phi$ $\qquad \qquad ...\text{[Eqn. (8.15)]}$

or, $F_s = 1000 \cos 21.3° - 500 \sin 21.3° = 750$ N

$\therefore$ $\tau = \dfrac{F_s \sin \phi}{A}$ $\qquad \qquad ...\text{[Eqn. (8. 24)]} \quad (\text{where, } A = b \times t)$

or, $\tau = \dfrac{750 \times \sin 21.3°}{2.8 \times 0.35} = \textbf{278 N/mm}^2 \quad \textbf{(Ans.)}$

Example 8.8. *The following observations were made during an orthogonal cutting operation :*

Depth of cut	.... *0.3 mm*
Chip thickness	.... *0.6 mm*
Rake angle	.... *20°*
Cutting velocity	.... *102 m/min*
Cutting force	.... *300 N*
Feed force	.... *120 N*

Determine :

(i) Shear angle.

(ii) Shear strain.

(iii) Velocity of chip along the tool force.

(iv) Work done in shear.

Solution. *Given :* $t = 0.3$ mm ; $t_c = 0.6$ mm ; $\alpha = 20°$, $V = 102$ m/min ; $F_t = 300$ N ; $F_f = 120$ N.

(i) Shear angle, ϕ :

$$\phi = \tan^{-1}\left(\frac{r \cos \alpha}{1 - r \sin \alpha}\right)$$

where, $r = \dfrac{t}{t_c} = \dfrac{0.3}{0.6} = 0.5$

$\therefore$ $\phi = \tan^{-1}\left(\dfrac{0.5 \cos 20°}{1 - 0.5 \sin 20°}\right) = 29.54°$

(ii) **Shear strain, $\in$:**

$$\in = \cot \phi + \tan (\phi - \alpha)$$
$$= \cot 29.54° + \tan (29.54° - 20°) = 1.932 \quad \textbf{(Ans.)}$$

(iii) **Work done in shear, W_s :**

$$W_s = F_s \times V_s$$

Now, $\qquad F_s = F_t \cos \phi - F_f \sin \phi$ ...[Eqn. (8.15)]

$$= 300 \cos 29.54° - 120 \sin 29.54° = 201.84 \text{ N}$$

and, $\qquad V_s = \dfrac{V \cos \alpha}{\cos (\phi - \alpha)}$ ...[Eqn. (8.10)]

or, $\qquad V_s = \dfrac{102 \times \cos 20°}{\cos (29.54° - 20°)} = 97.19 \text{ m/min}$

$\therefore$ Work done in shear, $W_s = F_s \times V_s$

$$= 201.84 \times 97.19 = \textbf{19616.8 NM/min} \quad \textbf{(Ans.)}$$

(iii) **Velocity of chip along the tool face, V_c :**

We know that, $\qquad V_c = \dfrac{V \sin \phi}{\cos (\phi - \alpha)}$ ...[Eqn. (8.9)]

or, $\qquad V_c = \dfrac{102 \times \sin 29.54°}{\cos (29.54° - 20°)} = \textbf{50.99 m/min} \quad \textbf{(Ans.)}$

Example 8.9. *The following observations were made during turning (orthogonally) of a mild steel tubing of 60 mm diameter on a lathe :*

Cutting speed	.... *24 m/min*
Toot rake angle	.... *32°*
Feed rate	.... *0.12 mm/rev.*
Tangential/cutting force	.... *3000 N*
Feed force	.... *1200 N*
Length of continuous chip in one revolution	.... *96 mm.*

Tangential/cutting force .... *3000 N* ⎫
Feed force .... *1200 N* ⎬ *Found by dynamometer*
⎭

Determine :

(i) Co-efficient of friction.

(ii) Shear plane angle.

(iii) Velocity of chip along tool face.

(iv) Chip thickness.

Solution. *Given :* $D = 60$ mm ; V (cutting speed) = 24 m/min ; $\alpha = 32°$; $f = 0.12$ mm/rev. ; $F_t = 3000$ N ; $F_f = 1200$ N ; $l_c = 96$ mm/rev.

(i) **Co-efficient of friction, μ :**

We know that, $\qquad \mu = \dfrac{F_f + F_t . \tan \alpha}{F_t - F_f \tan \alpha}$ ...[Eqn. (8.14)]

$$= \dfrac{1200 + 3000 \tan 32°}{3000 - 1200 \tan 32°} = \textbf{1.366} \quad \textbf{(Ans.)}$$

(ii) **Shear plane angle, ϕ :**

Using the relation : $\qquad \phi = \tan^{-1} \left(\dfrac{r \cos \alpha}{1 - r \sin \alpha} \right)$

where, r = chip thickness ratio = $\dfrac{t}{t_c} = \dfrac{l_c}{l} = \dfrac{96}{\pi \times 60} = 0.509$

$\therefore$ $\phi = \tan^{-1}\left(\dfrac{0.509 \times \cos 32°}{1 - 0.509 \times \sin 32°}\right) = \mathbf{30.6°}$ **(Ans.)**

(iii) **Velocity of chip along tool face, V_c :**

We know that, $r = \dfrac{V_c}{V}$ $\left(\because\ r = \dfrac{t}{t_c} = \dfrac{l_c}{l} = \dfrac{V_c}{V}\right)$

(where r = chip thickness ratio or cutting ratio)

or, $0.509 = \dfrac{V_c}{24}$

$\therefore$ $V_c = 0.509 \times 24 = \mathbf{12.22\ m/min}$ **(Ans.)**

(iv) **Chip thickness, t_c :**

$$r = \dfrac{t^*}{t_c} = \dfrac{Feed}{t_c}\quad \text{or}\quad t_c = \dfrac{Feed}{r} = \dfrac{0.12}{0.509} = \mathbf{0.236\ mm}\quad \textbf{(Ans.)}$$

* *The depth of cut (t) is actually the feed in the machining operation.*

Example 8.10. *The power required while turning mild steel rod is found to be 0.1 kW/cm³/min. The maximum power available at the machine spindle is 4 kW. Assuming a cutting speed of 38 m/min and feed rate 0.32 mm/rev. Calculate :*

(i) *Maximum metal removal rate.*

(ii) *Depth of cut.*

(iii) *Cutting force.*

(iv) *Normal pressure on the chip.*

Solution. *Given :* Power required during turning = 0.1 kW/cm³/min. ;
Maximum power available = 4 kW ; Cutting speed , V = 38 m/min ; Feed rate, f = 0.32 mm/rev.

(i) **Maximum metal removal rate, $(MRR)_{max}$**

$$(MRR)_{max} = \dfrac{\text{Max. power available at machine spindle}}{\text{Power required/cm}^3/\text{min}}$$

$$= \dfrac{4}{0.1} = \mathbf{40\ cm^3/min}\quad \textbf{(Ans.)}$$

(ii) **Depth of cut, t :**

$$(MRR)_{max} = V.t.f$$

or, $40 = (38 \times 100\ \text{cm/min}) \times t \times \left(\dfrac{0.32}{10}\ \text{cm/rev.}\right)$

or, $t = \dfrac{40}{38 \times 100 \times 0.032} = 0.329\ \text{cm}\quad \text{or}\quad \mathbf{3.29\ mm}\quad \textbf{(Ans.)}$

(iii) **Cutting force, F_t :**

Power available, $P = \dfrac{F_t \times V}{1000 \times 60}$ kW   (where F_t is in newtons, and V in m/min.)

or, $4 = \dfrac{F_t \times 38}{1000 \times 60}$

$\therefore$ $F_t = \dfrac{4 \times 1000 \times 60}{38} = \mathbf{6315.8\ N}$ **(Ans.)**

(*iv*) **Normal pressure on the chip, *p* :**

$$p = \frac{F_t}{\text{Chip area}} = \frac{6315.8}{3.29 \times 0.32} = \textbf{5999 N/mm}^2 \quad \textbf{(Ans.)}$$

Example 8.11. *The following observations were made during an orthogonal cutting operation :*

Tool rake angle	.... *10°*
Co-efficient of friction	.... *0.85*
Chip thickness	.... *2.5 mm*
Width of cut	.... *15 mm*
Cutting speed	.... *40 m/min*
Feed	.... *1.5 mm/rev.*
Shear strength	.... *650 N/mm²*

Determine the following :

  (*i*) *Chip thickness ratio.*

 (*ii*) *Shear angle.*

(*iii*) *Shearing force.*

(*iv*) *Friction angle.*

 (*v*) *Cutting force.*

(*vi*) *Power consumed at the cutting tool.*

Solution. *Given :* $\alpha = 10°$; $\mu = 0.85$; $t_c = 2.5$ mm ; $b = 15$ mm, $V = 40$ m/min. ; Feed, $f(= t) = 1.5$ mm/rev. ; $\tau = 6.5$ N/mm²

  (*i*) **Chip thickness ratio, *r* :**

$$r = \frac{t}{t_c} = \frac{1.5}{2.5} = \textbf{0.6} \quad \textbf{(Ans.)}$$

 (*ii*) **Shear angle, ϕ :**

$$\phi = \tan^{-1}\left(\frac{r\cos\alpha}{1 - r\sin\alpha}\right)$$

or,
$$\phi = \tan^{-1}\left(\frac{0.6 \times \cos 10°}{1 - 0.6 \times \sin 10°}\right) = \textbf{33.4°} \quad \textbf{(Ans.)}$$

(*iii*) **Shearing force, F_s :**

$$\tau = \frac{F_s \sin\phi}{A} \qquad\qquad\qquad ...[\text{Eqn. (8.24)}]$$

or,
$$F_s = \frac{\tau . A}{\sin\phi} = \frac{\tau . b . t}{\sin\phi} = \frac{650 \times 15 \times 1.5}{\sin 33.4°} = 26567.7 \text{ N} \quad \text{or} \quad \textbf{26.567 kN} \quad \textbf{(Ans.)}$$

(*iv*) **Friction angle, β :**

$$\mu = \tan\beta = 0.85$$

$\therefore$
$$\beta = \tan^{-1}(0.85) = \textbf{40.36° (Ans.)}$$

 (*v*) **Cutting force, F_t :**

$$F_t = F_s\left[\frac{\cos(\beta - \alpha)}{\cos(\phi + \beta - \alpha)}\right] \qquad\qquad ...[\text{Eqn. (8.20)}]$$

or,
$$F_t = 26.567\left[\frac{\cos(40.36° - 10°)}{\cos(33.4° + 40.36° - 10)}\right] = 51.85 \text{ kN}$$

(*vi*) **Power consumed at the cutting tool, *P* :**

$$P = \frac{F_t \times V}{60} \; kW$$

(where F_t is in *kN* and *V* is in *m/min*)

or, $$P = \frac{51.85 \times 40}{60} = \textbf{34.56 kW} \quad \textbf{(Ans.)}$$

Example 8.12. *The following observations were made during orthogonal cutting of an aluminium alloy :*

$t = 0.18$ *mm ; b = 4.0 mm ; l = 165 mm ; l_c = 45 mm ; b_c = 4.5 mm ; μ = 0.75, τ = 245 N/mm²,*
V = 35 m/min ; α = 20°

   Calculate : (*i*) *Cutting force.*

   (*ii*) *Feed force.*

   (*iii*) *Power consumed.*

Solution. (*i*) **Cutting force, F_t :**

Assuming that the volume of chip before cut remains the same as that after the cut,

   Then, $l \times b \times t = l_c \times b_c \times t_c$

or, $\dfrac{t}{t_c} = \dfrac{l_c}{l} \times \dfrac{b_c}{b} = \dfrac{45 \times 4.5}{165 \times 4.0} = 0.307$

i.e., $\dfrac{t}{t_c} = r = 0.3068$

   Shear angle $\phi = \tan^{-1}\left(\dfrac{r \cos \alpha}{1 - r \sin \alpha} \right)$

or, $\phi = \tan^{-1}\left(\dfrac{0.307 \times \cos 20°}{1 - 0.307 \times \sin 20°} \right) = 17.86°$

   Also, $\mu = \tan \beta = 0.75$

or, $\beta = \tan^{-1}(0.75) = 36.87°$

   Now, $F_t = F_s \left[\dfrac{\cos(\beta - \alpha)}{\cos(\phi + \beta - \alpha)} \right]$ ...[Eqn. (8.20)]

   But, $F_s = \dfrac{\tau A}{\sin \phi} = \dfrac{\tau \times (b.t)}{\sin \phi}$ ...[Eqn. (8.24)]

∴ $F_t = \left[\dfrac{\tau \times (b.t)}{\sin \phi} \right] \times \left[\dfrac{\cos(\beta - \alpha)}{\cos(\phi + \beta - \alpha)} \right]$

or, $F_t = \left[\dfrac{245 \times 4.0 \times 0.18}{\sin 17.86°} \right] \times \left[\dfrac{\cos(36.87° - 20°)}{\cos(17.86° + 36.87° - 20°)} \right] = \textbf{669.7 N} \quad \textbf{(Ans.)}$

(*ii*) **Feed force, F_f :**

From eqns. (8.17) and (8.18), we have

$$\frac{F_f}{F_t} = \frac{R \sin(\beta - \alpha)}{R \cos(\beta - \alpha)} = \tan(\beta - \alpha)$$

∴ $F_f = F_t \times \tan(\beta - \alpha) = 669.7 \tan(36.87° - 20°) = \textbf{203.09 N} \quad \textbf{(Ans.)}$

(iii) **Power consumption, *P* :**

$$P = \frac{F_t \times V}{1000 \times 60} \text{ kW}$$ 　　　　　　(where F_t is in newtons and V in m/min.)

or, 　　　　　　　　$$P = \frac{669.7 \times 35}{1000 \times 60} = \textbf{0.39 kW} \quad \textbf{(Ans.)}$$

8.9. FRICTION IN METAL CUTTING

In chip formation, the *friction* occurring between the chip and the tool face is a controlling influence.

- The actual contact of two *sliding* surfaces through the high spots can be seen under a microscope ; this is called *asperities.*

When a *normal load* is applied, yielding occurs at the tips of the contacting asperties (Fig. 8.17), and the real area of contact (A_r) increases until it is capable of supporting the applied load. This real area of contact A_r is only a small fraction of apparent contact area (A_a) for the vast majority of engineering applications.

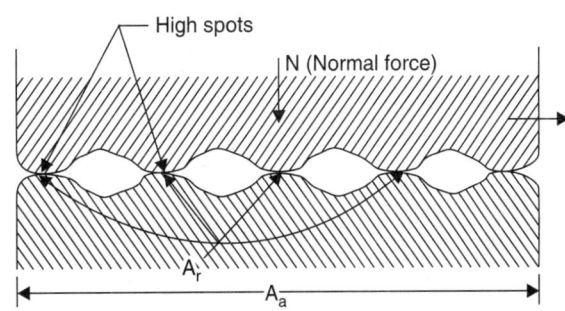

Fig. 8.17. Microview of asperities in contact.

$$A_r = \frac{N}{\sigma_y}$$ 　　　　　　　　　　　　　...(8.37)

where, N = Normal force, and

　　　　σ_y = Mean yield stress of the asperities.

- Real area changes by first the *elastic deformation* and when load increases by *plastic deformation.*

It has been seen that under the influence of normal and tangential load *very high temperatures are developed at the contacting asperities and the metallic bonding of the contacting high spots can occur.* Thus, sliding of one surface relating to the other must be accompanied by *shearing* of the welding asperities.

- When plastic deformation takes place at the contacting surfaces, then the mechanism of friction is *different* because the real area of contact *approaches* that of apparent area of contact. Under these conditions the *friction force is independent of normal force.*

Further, it has been observed that *coefficient of friction increases with an increase in rake angle.* Normally, it is expected that with an increase in the rake angle, the metal cutting forces decrease and should normally be associated with a decrease in friction. However, in actual practice the *coefficient of friction increases.* This happens *because the influence of the rake angle is not the same on the different components of the cutting force.* The normal force on the rake face *decreases* a great extent compared to the friction force. Thus, although there is an overall decrease in the forces friction coefficient increases. That is why it is called as *apparent coefficient of friction.*

- In metal cutting we have sliding situations of high normal load and with a metal surface which is *chemically clean* ; the clean metal surface explains the high value of friction coefficient (μ) and the high normal load and the departure from the usual laws of friction. Thus, the friction along the rake face of a cutting tool can be considered as *partially sticking* and *partially sliding* (Fig. 8.18).

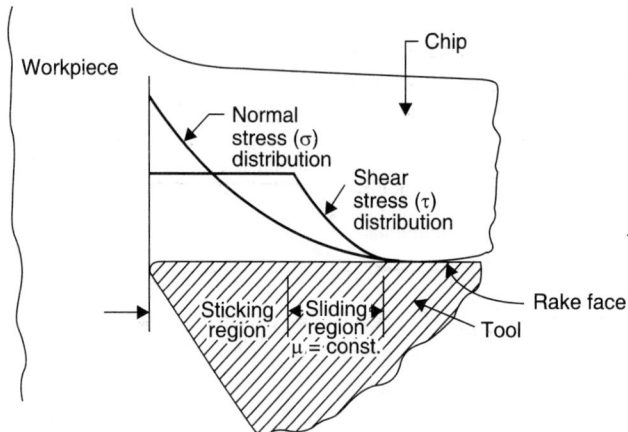

Fig. 8.18. Stress distribution expected along rake face—chip-tool friction.

— In the 'sticking zone', the *shear stress constantly approaches the yield stress of the work material*, while in the 'sliding zone', *the normal Coulomb's laws of friction hold good.*

   Note : During metal cutting, it has generally been observed that the mean coefficient of friction between the chip and the tool can vary considerably and is affected by changes in cutting speed, rake angle, etc. This variance of the mean coefficient of friction results from the very high normal pressures that exist at the chip-tool interface.

8.10. THERMAL ASPECTS OF METAL CUTTING

8.10.1. General Aspects

   During metal cutting, the energy dissipated gets converted into heat. Consequently, high temperatures are generated in the region of the tool cutting edge, and these temperatures have a controlling influence on the *rate of wear of the cutting tool and on the friction between the chip and tool.*

   — When a material is deformed *elastically*, the energy required for the operation is stored in the material as *strain energy*, and no heat is generated. However, when the material is deformed plastically, most of the energy used is converted into heat. In metal cutting the material is subjected to extremely high strains and the elastic deformation forms a *very small proportion* of the total deformation ; therefore it may be assumed that all the energy is converted into heat.

   In fact, heat is generated in three distinct regions (Fig. 8.19), these are :

   (i) **The shear zone :** Here the *energy needed to shear* the chip is the source of heat. In this region about 80–85% of the heat is generated.

   (ii) **The chip-tool interface region.** Here the *energy needed to overcome friction* is the source of heat. Some plastic deformation also occurs in this region. About 15–20% heat is generated in this region.

   (iii) **The tool-work interface region.** Here *energy needed to overcome frictional rubbing between flank face of the tool and workpiece* is the source of heat. In this region only 1–3% of heat is generated.

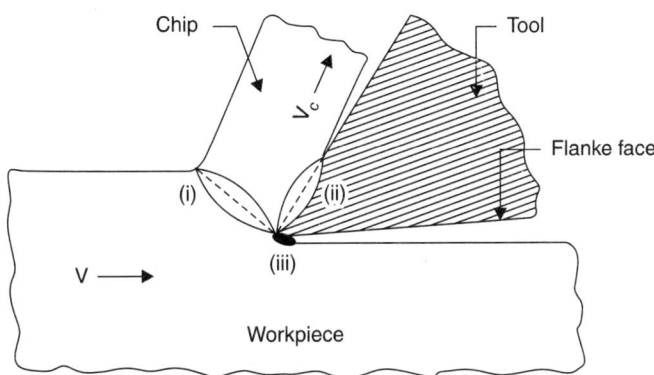

Fig. 8.19. Regions of heat generation in metal cutting.

It will be noted that each of these three zones lead to rise of temperature at the tool-chip interface and it is found that the maximum temperature occurs slightly away from the cutting edge, and not at the cutting edge. This temperature plays a major role in the formation of crater on the tool face and *leads to failure of tool by softening and thermal stresses*.

The various factors which lead to maximum tool temperature are cutting speed, feed , properties of materials, etc. These machining variables affect the size of shear zone and chip tool contact length and thereby, the area over which heat is distributed. *Shorter length* of contact of chip with tool results in *severe temperature rise*.

8.10.2. Factors Affecting Temperature

The following factors influence the cutting temperature :

(*i*) *Workpiece and tool material :*

— Materials with higher thermal conductivity produce lower temperatures than tools with lower conductivity.

(*ii*) *Tool geometry :*

— While rake angle has only a slight influence on the temperature, it increases considerably with increase in approach angle.

(*iii*) *Cutting conditions :*

"*Cutting speed* " has *predominant* effect on the cutting temperature (see Fig. 8.20).

— "*Feed*" has *little* effect.

— "*Depth of cut*" has the *least* effect.

(*iv*) *Cutting fluid :*

— At high speeds (such as employed for carbides), cutting fluid has *negligible effect* on tool-chip interface temperature. The fluid is carried away by the outward flowing chip more rapidly than it could be forced between the tool and the chip.

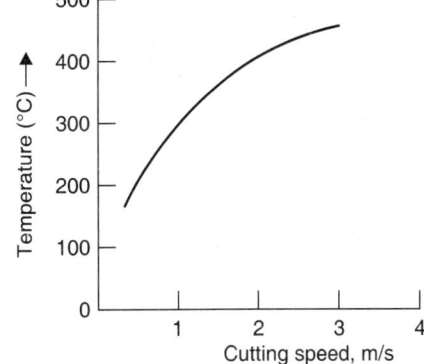

Fig. 8.20. Effect of cutting speed on temperature.

8.10.3. Temperature Distribution in Metal Cutting

Fig. 8.21 shows temperature distribution in workpiece and chip during orthogonal cutting (obtained from an infrared photograph, for free-cutting mild steel where the cutting speed is 0.38 m/s, the width of cut is 6.35 mm, the working normal rake is 30°, and the workpiece temperature is 611°C).

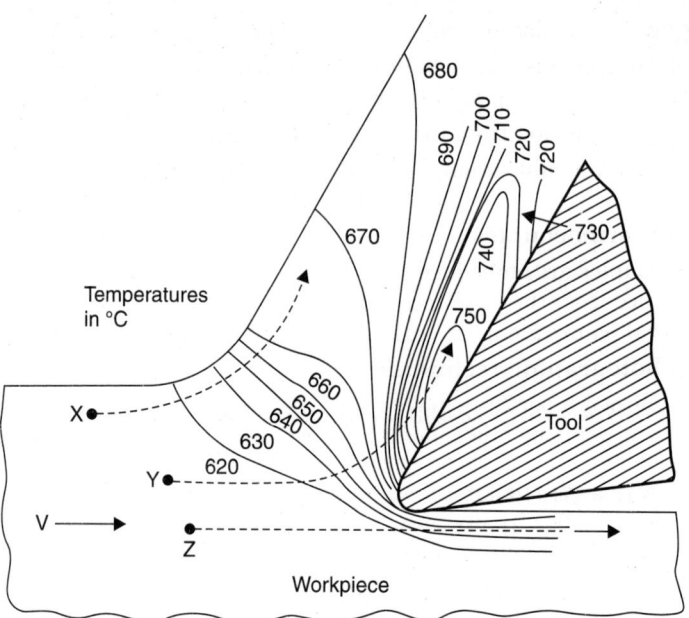

Fig. 8.21. Temperature distribution in workpiece and chip during orthogonal cutting (After Boothroyd).

The maximum temperature in the cutting zone occurs not at the tool tip but at some distance further up the rake face.

- Material at a point such as X gets heated as it passes through the shear zone and finally leaves as a chip.
- For points such as Y, heating continues beyond the shear plane into the frictional heat region. These points, however, loose shear zone heat to the chip while moving up but gain frictional heat.
- Points such as Z remain in the workspiece and their temperature rises merely due to conduction of heat into the workpiece.

The above factors cause maximum tool temperature to occur some distance away from the cutting edge.

8.10.4. Measurement of Chip-Tool Interface Temperature

Follow are the various methods of measuring chip-tool interface temperature :

1. Tool work thermocouple.
2. Embedded thermocouples.
3. Infra-red photographic technique.
4. Temperature sensitive techniques.
5. Temper colours.
6. Indirect calorimetric technique.

The *tool work thermocouple technique is the most widely used technique for the measurement of average chip-tool interface temperature*. The other methods suffer from various disadvantages such as slow response, indirectness, and complications in measurement.

Some of the above methods are discussed briefly below :

1. Tool work thermocouple :

In this technique, the hot end of the tool and workpiece and their cold ends act as thermocouple and e.m.f. proportional to temperature difference is produced.

- The workpiece is insulated from the chuck and tailstock centre.
- The end of workpiece is connected to a copper wire which dipped in mercury cup enables further connection serving as cold end.

This point and a connection from tool provide output for connection to a milli voltmeter.

It is possible to obtain calibration curve between tool temperature and e.m.f. by laboratory method.

2. **Embedded thermocouples :**

- The thermocouples are embedded in fine holes eroded in H.S.S. tool from bottom face upto a fixed distance from the rake face. This arrangement enables measurement of temperature at several points along the rake face of tool.

3. **Intra-red photographic technique :**

- This technique of temperature measurement is based on taking photographs of the sideface of tool-chip while cutting and comparing them with strips whose temperatures are known.

8.11. TOOL WEAR AND FAILURE

The usefulness of the tool cutting edges is lost through

— wear ;
— breakage ;
— chipping ;
— deformation.

Wear, which is responsible for most tool failures, is very complex and involves chemical, physical, and mechanical processes, often in combination.

'Tool failure' implies that the tool has reached a point beyond which it will not function satisfactory until it is reground.

8.11.1. Tool Wear

Wear *can be defined as the loss of weight or mass that accompanies the contact of sliding surfaces.* Wear seldom involves in a single unique mechanism.

The wear mechanism associated with gradual or progressive wear include :

(*i*) Abrasion wear.
(*ii*) Adhesion wear.
(*iii*) Diffusion wear.

(*i*) *Abrasion wear :*

- Abrasion wear occurs when *hard constituents of one surface plough through the material of the other surface.*
- This is basically a cutting process and, as a consequence, the amount of wear depends on the relative hardness of the contacting surfaces, as well as their elastic and plastic properties and mating geometries.
- Many steels, cast irons, and nickel-based alloys contain hard carbides, oxides and nitrides that may contribute to abrasive wear.
- The requirement for abrasive wear to occur in high-speed tools is that the constituents causing wear be harder than the martensitic matrix of the cutting tool.
- As with high-speed steel tools, but to a lesser extent, forging scale and surface sand on castings cause abrasive wear on carbide tools.

(*ii*) *Adhesion wear :*

- This form of wear takes place when *two surfaces are brought into intimate contact under normal loads and form welded junctions, which, when subjected to shearing loads, are subsequently destroyed.*

- The temperature at which adhesion occurs is influenced by the characteristics of the tool and work materials, as well as by the force acting between the tool and workpiece, which in turn is established by the cutting conditions.
- This type of wear is of *primary importance at relatively low cutting speeds*, and since it is a time-dependent mechanism, it tends to disappear at high cutting speeds.

(iii) *Diffusion wear :*

- The diffusion wear occurs by a *'solid-state diffusion mechanism' in which atoms in a metallic crystal lattice move from one lattice point to another in the direction of high atomic concentration to one of low atomic concentration.*
- The diffusion mechanism is dependent on the ambient temperature, and increases in temperature cause *exponential* increases in the rate of diffusion.
- In metal cutting the diffusion wear will occur if the mechanical process involved with adhesion raises the local localized interfacial-temperature sufficiently.

The amount of diffusion depends on :

— The bonding affinity of tool and the workpiece pair ;

— The level of atomic agitation (strongly dependent on temperature) ;

— The length of contact time at the elevated temperature.

- In machining with high-speed tools, there can be diffusion of carbon atoms from the tool into the stream of work material flowing by.
- *The wear of carbide cutting tools used in cutting steels is a well-known phenomenon in which diffusion plays an important role.*

8.11.2. Tool Failure

The tool failure may occur due to the following *factors* :

1. Excessive temperature.

2. Excessive stress.

3. Flank wear.

4. Crater wear.

1. **Excessive temperature :**

The high temperatures that occur in tool chip-workpiece contact zones may cause an initially sharp cutting tool to *lose* some of its strength and *flow plastically* under the pressures developed by the cutting force. The flow of the tool material along the flank surfaces causes the cutting tool to assume a configuration resembling that shown in Fig. 8.22.

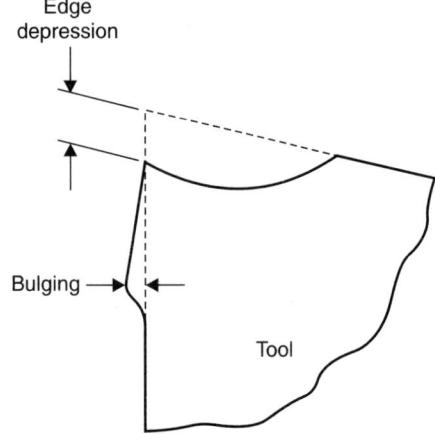

Fig. 8.22

- This type of failure is not to limited to high-speed steel cutting tools ; for, even though cemented carbide cutting tools are relatively brittle, they possess a certain amount of ductility under the high compressive loads and elevated temperatures present during cutting.

2. **Excessive stress :**

- When a cutting tool is acted upon by an excessive force its cutting edge may undergo immediate failure due to a lack of tool strength. Alternatively, the mechanical failure of the cutting tool may result from a *fatigue-type failure*.

- The *chipping* of a tool and the development of *cracks* along its cutting edge can be contributed to :
— Faulty tool design ;
— Material selection ;
— Reconditioning techniques ;
— Machining conditions conducive to chatter.

This type of failure can be *minimised* as follows :
— Use of small or negative tool rake angles on brittle materials ;
— Employing large side-cutting-edge angles to protect the tool type ;
— Honing a narrow chamfer along the cutting edge.

3. Flank wear :

The gradual or progressive wear that develops on the *flank surface* of a cutting tool is called *"flank wear"*.

- Flank wear (Fig. 8.23) occurs as a result of friction between the progressively increasing contact area on the tool flank and the newly generated workpiece surface.

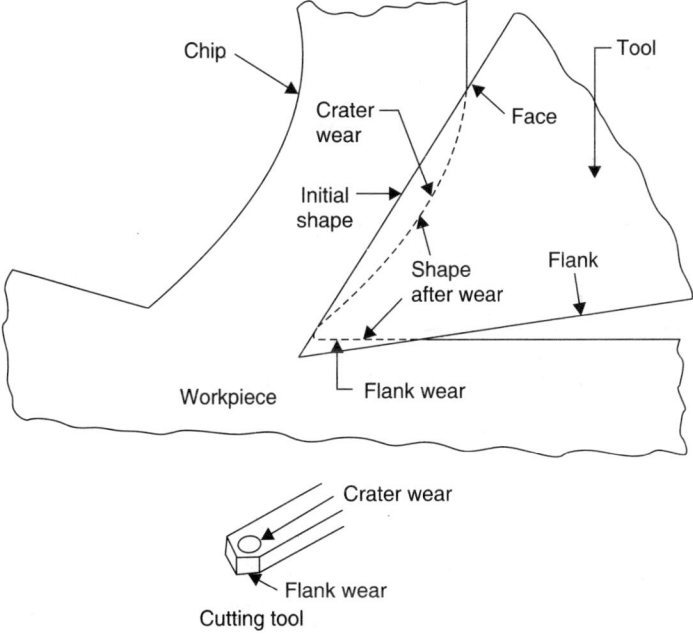

Fig. 8.23. Tool wear regions in metal cuttings.

Flank wear or wear land is on the *clearance surface of the tool.*

- The wear land can be characterised by the length of wear land, w. It modifies the tool geometry and changes the cutting parameters (depth of cut).
- With a high clearance angle, more flank wear is permissible before the critical wear land is reached ; however, excessive clearance weakens the cutting tool.

4. Crater wear :

The gradual or progressive wear that develops on the *rake surface* of a cutting tool is called *"crater wear"*.

- Crater wear (Fig. 8.23) occurs as a result of the friction developed as the chip flows over the rake surface of the cutting tool. It is largely *a temperature-dependent phenomenon.*

- The crater is on the rake face and is more or less circular. The crater does not always extend to the tool tip, but may end at a distance from the tool tip. *It increases the cutting forces, modifies the tool geometry and softens the tool tip.*

8.12. TOOL LIFE

Tool life *is defined as the time interval between two successive regrinds.*

Tool life represents the useful life of the tool expressed generally in time units from a start of a cut to some end point defined by a failure criterion. A tool that no longer performs the desired function is said to have failed and hence reached the end of its useful life. At such an end point the tool is not necessarily unable to cut the workpiece but is merely unsatisfactory for the purpose required. The tool may be resharpened and used again.

Factors affecting tool life :

Tool life depends upon the following *factors* :

(*i*) Tool material.

(*ii*) Hardness of the material.

(*iii*) Type of material being cut.

(*iv*) Type of the surface on the metal (scaly or smooth).

(*v*) Profile of the cutting tool.

(*vi*) Type of the machining operation being performed.

(*vii*) Microstructure of the material.

(*viii*) Finish required on the workpiece.

(*ix*) **Cutting speed.** As the cutting speed is reduced, the tool life increases.

(*x*) **Feed and depth of cut.** An increase in feed and depth of cut will shorten tool life but not nearly as much as an increase in cutting speed.

(*xi*) **Cutting temperature.** In general, higher temperatures cause shorter tool life.

- The *tool life* can be **specified** by any of the following *measurable quantities* :

(*i*) Actual cutting time to failure.

(*ii*) Length of work cut to failure.

(*iii*) Volume of metal removed to failure.

(*iv*) Cutting speed for a given time to failure.

(*v*) Number of components produced.

All these various parameters suggested are related and are used depending upon the final function of a given operation.

Tool failure criteria :

The following are some of the *possible tool failure criteria that could be used for limiting tool life* :

Based on tool wear :

(*i*) Wear land size.

(*ii*) Crater depth, width or other parameters.

(*iii*) A combination of the above two.

(*iv*) Chipping or fine cracks developing at the cutting edge.

(*v*) Volume or weight of material worn off the tool.

(*vi*) Total destruction of the tool.

Based on consequences of worn tool :
(i) Limiting value of change in component size.
(ii) Limiting value of surface finish.
(iii) Fixed increase in cutting force or power required to perform a cut.

Tool life equaticn :

Tool life of a cutting tool may be calculated by using the following relation :

$$VT^n = C \qquad \text{... This equation is known as } \textit{Taylor's tool equation.}$$

where, V = Cutting speed in m/min.,

T = Tool life in min.,

C = A constant (which is numerically equal to cutting speed that gives the tool life of one min.), and

n = Another constant (depending upon finish, workpiece material and tool material)
= 0.1 for H.S.S. steel tools : 0.2 to 0.25 for carbide tools and 0.4 to 0.55 for ceramic tools.

Although this is a fairly good formula, yet it does *not* take all the affecting parameters into account. As a result the applicability of the above formula is *restricted to very narrow regions of cutting process parameters.*

This formula was extended by a number of researchers to reduce this deficiency.

Example 8.13. *A cutting tool, cutting at 22 m/min, gave a life of 1 hour between re-grinds when operating on roughening cuts with mild steel. What will be its probable life when engaged on light finishing cuts ?*

Take : $n = \dfrac{1}{8}$ for roughening and $\dfrac{1}{10}$ for finishing cuts in the Taylor's equation $VT^n = C$.

Solution. *Given :* Cutting speed, V = 22 m/min ; T_r = 1 hour = 60 min ;

$$n = \frac{1}{8} \text{ and } \frac{1}{10} \text{ for roughening and finishing cuts respectively.}$$

Probable life—Light finishing cuts, T_f :

Taylor's equation : $\qquad\qquad VT^n = C$

For *roughening* cut : $\qquad 22(T_r)^{1/8} = C \qquad\qquad\qquad$ ...(i)

For *finishing* cut : $\qquad 22(T_f)^{1/10} = C \qquad\qquad\qquad$ ...(ii)

From (i) and (ii), we get

$$(T_r)^{1/8} = (T_f)^{1/10}$$
$$(60)^{1/8} = (T_f)^{1/10}$$

$\therefore \qquad\qquad\qquad T_f = (60)^{10/8} = \textbf{167 min.} \quad \textbf{(Ans.)}$

Example 8.14. *Calculate the cutting speed for a tool to have a tool life of 160 min. The same tool had a life of 9 min when cutting at 250 m/min.*

Take n = 0.22 in the Taylor's tool life equation.

Solution. *Given :* V_1 = 250 m/min ; T_1 = 9 min ; T_2 = 160 min ; n = 0.22.

Cutting speed, V_2 :

Taylor's tool life equation is given as : $VT^n = C$

$\therefore \qquad\qquad\qquad V_1 T_1^{\,n} = V_2 T_2^{\,n}$

or,
$$V_2 = \left(\frac{T_1}{T_2}\right)^n \times V_1$$

$$= \left(\frac{9}{160}\right)^{0.22} \times 250 = \mathbf{132.73 \ m/min} \quad \textbf{(Ans.)}$$

Example 8.15. *Calculate the percentage change in cutting speed required to give 60 percent reduction in tool life. Take n = 0.22.*

Solution. Taylor's tool life equation is given as : $VT^n = C$

or,
$$V_1 T_1^{\ n} = V_2 T_2^{\ n}$$

$$\frac{V_2}{V_1} = \left(\frac{T_1}{T_2}\right)^n$$

$$T_2 = (1 - 0.6)T_1 = 0.4 \ T_1 \qquad \text{... since the reduction in tool life in 60\%.}$$

$\therefore$
$$\frac{V_2}{V_1} = \left(\frac{T_1}{0.4 \ T_1}\right)^{0.22} = 1.223$$

This means that V_2 should be 22.3 percent *higher* than V_1 for the required percent reduction in tool life.

Example 8.16. *The relationship for H.S.S tools is $V(T)^{1/8} = C_1$, and for tungsten carbide tools $V(T)^{1/5} = C_2$.*

Assuming that at a speed of 24 m/min the tool life was 170 min. in each case, compare their cutting lives at 30 m/min.

Solution. For *H.S.S.* tools : $24 \times (170)^{1/8} = C_1$;

For *W.C.* tools : $24 \times (170)^{1/5} = C_2$.

Now, at cutting speed 30 m/min :

For *H.S.S.* tools : $24 \times (170)^{1/8} = 30 \times (T_{H.S.S.})^{1/8}$

$\therefore$
$$T_{H.S.S.} = \left(\frac{24}{30}\right)^8 \times 170 = 28.52 \ min$$

For *W.C.* tools : $24 \times (170)^{1/5} = 30 \times (T_{W.C.})^{1/5}$

$\therefore$
$$T_{W.C.} = \left(\frac{24}{30}\right)^5 \times 170 = 55.7 \ min$$

$\therefore$
$$\frac{T_{W.C.}}{T_{H.S.S}} = \frac{55.7}{25.52} = \mathbf{1.953} \quad \textbf{(Ans).}$$

Example 8.17. *A carbide tool with mild steel workpiece was found to give life of 2 hours while cutting at 48 m/min. If Taylor's exponent n = 0.27, determine :*

(i) The tool life if the same tool is used at a speed of 20 per cent higher than the previous one.

(ii) The value of cutting speed if the tool is required to have tool life of 3 hours.

Solution. *Given :* T_1 = 2 hours = 120 min ; V_1 = 48 m/min *;* n = 0.27 ; V_2 = 1.2 × 48 = 57.6 m/min ; T_2 = ? ; T_3 = 3 hours = 180 min ; V_3 = ?

(i) Tool life at higher speed, T_2 :

Taylor's tool life equation is given as : $VT^n = C$

Now, $V_1(T_1)^n = V_2 (T_2)^n$

$\therefore$
$$T_2 = \left(\frac{V_1}{V_2}\right)^{1/n} \times T_1 = \left(\frac{48}{57.6}\right)^{(1/0.27)} \times 120 = \mathbf{61.08 \ min} \quad \textbf{(Ans).}$$

(ii) **Cutting speed, V_3** :

$$V_1(T_1)^n = V_3(T_3)^n$$

or,

$$48 \times (120)^{0.27} = V_3 \times (180)^{0.27}$$

or,

$$V_3 = 48 \times \left(\frac{120}{180}\right)^{0.27} = \textbf{43.02 m/min}\quad\textbf{(Ans).}$$

Example 8.18. *During straight turning of a 24 mm diameter steel bar at 300 r.p.m. with an H.S.S. tool, a tool life of 9 min. was obtained. When the same bar was turned at 250 r.p.m., the tool life increased to 48.5 min. What will be the tool life at a speed of 280 r.p.m. ?*

Solution. *Given* : $D = 24$ mm ; $N_1 = 300$ r.p.m. ; $T_1 = 9$ min ; $N_2 = 250$ r.p.m. ;

$$T_2 = 48.5 \text{ min} ; N_3 = 280 \text{ r.p.m.} ; T_3 = ?$$

$$V_1 = \frac{\pi D N_1}{1000} = \frac{\pi \times 24 \times 300}{1000} = 22.62 \text{ m/min}$$

$$V_2 = \frac{\pi D N_2}{1000} = \frac{\pi \times 24 \times 250}{1000} = 18.85 \text{ m/min}$$

$$V_3 = \frac{\pi D N_3}{1000} = \frac{\pi \times 24 \times 280}{1000}\ 21.11 \text{ m/min}$$

Tool life at 280 r.p.m., T_3 :

Using Taylor's tool life equation : $V_1 T_1{}^n = V_2 T_2{}^n$, we have

$$n = \frac{\ln (V_2/V_1)}{\ln (T_1/T_2)}$$

or,

$$n = \frac{\ln (18.85/22.62)}{\ln (9/48.5)} = 0.108$$

Now,

$$V_1 T_1{}^n = V_3 T_3{}^n$$

or,

$$T_3 = \left(\frac{V_1}{V_3}\right)^{1/n} \times T_1$$

$$= \left(\frac{22.62}{21.11}\right)^{1/0.108} \times 9 = \textbf{17.06 min}\quad\textbf{(Ans).}$$

Example 8.19. *A mild steel workpiece is being machined by two different tools A and B under indentical machining conditions. The tool life equations for these tools are :*

For tool-A : $VT^{0.32} = 42.5$

For tool-B : $VT^{0.45} = 88.6$

where V and T are in m/s and s, respectively.

Determine the cutting speed above which tool-B will give better tool life.

Solution. Let, $V^* = $ *Break-even speed at which both the tools give the same tool life.*

Then, $T = \left(\dfrac{42.5}{V^*}\right)^{1/0.32}$ ...For tool-A

and, $T = \left(\dfrac{88.6}{V^*}\right)^{1/0.45}$ ...for tool-B

or, $\left(\dfrac{42.5}{V^*}\right)^{1/0.32} = \left(\dfrac{88.6}{V^*}\right)^{1/0.45}$

or,
$$\left(\frac{42.5}{V*}\right)^{3.125} = \left(\frac{88.6}{V*}\right)^{2.22}$$

or,
$$\frac{122663.5}{(V*)^{3.125}} = \frac{21052.5}{(V*)^{2.22}}$$

or,
$$\frac{(V*)^{3.125}}{(V*)^{2.22}} = \frac{122663.5}{21052.5} = 5.826$$

or,
$$(V*)^{0.905} = 5.826$$

or,
$$V* = (5.826)^{1/0.905} = \textbf{7.0 m/s} \quad \textbf{(Ans).}$$

Example 8.20. *The following equation for tool life is given for a turning operation :*
$$VT^{0.14} \times f^{0.78} \times d^{0.38} = C$$

One hour tool life was obtained while cutting at V = 28 m/min ; f = 0.3 mm/rev and d = 2.6 mm.

Calculate the tool life if the cutting speed, feed and depth of cut are increased by 25 percent individually and also taken together.

Solution. $\quad VT^{0.14} \times f^{0.78} \times d^{0.38} = C$ ... *Given*

Inserting the values given for V, f and d, we get
$$C = 28 \times (60)^{0.14} \times (0.3)^{0.78} \times (2.6)^{0.38} = 27.92$$

Change in tool life :

(*a*) *Increase taken individually :*

(*i*) Now, $V = 1.25 \times 28 = 35$ m/s

∴
$$T^{0.14} = \frac{27.92}{35 \times (0.3)^{0.78} \times (2.6)^{0.38}} = 1.419$$

or, $\mathbf{T} = (1.419)^{1/0.14} = \textbf{12.18 min.} \quad \textbf{(Ans).}$

(*ii*) Now, $f = 1.25 \times 0.3 = 0.375$ mm/rev.

∴
$$T^{0.14} = \frac{27.92}{28 \times (0.375)^{0.78} \times (2.6)^{0.38}} = 1.49$$

or, $\mathbf{T} = (1.49^{1/0.14} = \textbf{17.26 min.} \quad \textbf{(Ans).}$

(*iii*) Now, $d = 1.25 \times 2.6 = 3.25$ mm.

∴
$$T^{0.14} = \frac{27.92}{28 \times (0.3)^{0.78} \times (3.25)^{0.38}} = 1.629$$

or, $\mathbf{T} = (1.629)^{1/0.14} = \textbf{32.64 min.} \quad \textbf{(Ans).}$

(*b*) *Increased taken together :*

Now, $V = 35$ m/s ; $f = 0.375$ mm/rev ; $d = 3.25$ mm

$$T^{0.14} = \frac{27.92}{35 \times (0.375)^{0.78} \times (3.25)^{0.38}} = 1.095$$

∴ $T = (1.095)^{1/0.14} = \textbf{1.912 min.} \quad \textbf{(Ans).}$

8.13. CUTTING SPEED, FEED AND DEPTH OF CUT

Cutting speed. *It is defined as the relative surface speed between the tool and the job.*

It is expressed in metres per minute (m/min). It is thus the amount of length that will pass the cutting edge of the tool per unit of time.

The cutting speed to be used depends on the following *factors* :

(*i*) Work material ;

(*ii*) Cutting tool material ;

(*iii*) The depth of cut and feed ;

(*iv*) Desired cutting tool life ;

(*v*) Rigidity and conditions of the machine and tool and the rigidity of the work.

Feed. *It may be defined as the relatively small movement per cycle of the cutting tool, relative to the workpiece in a direction which is usually perpendicular to the cutting speed direction.*

It is expressed in millimetres per revolution (mm/rev.) or millimetres per stroke (mm/str).

— In *turning* and *drilling*, the feed is the *axial advance* of the tool along or through the job during each revolution of the tool or job ;

— In case of *shaper* and *planer*, it is *lateral offset* between the tool and work for each stroke ;

— For *multitooth milling cutters*, it is the advance of work or cutter between the cutting action of two successive teeth (expressed basically as mm per tooth).

● Feed to be used depends on the following *factors* :

(*i*) Smoothness of finish required ;

(*ii*) Power available, condition of machine and its drive ;

(*iii*) Type of cut ;

(*iv*) Tool life.

Depth of cut. *It is the thickness of the layer of metal removed in one cut, or pass, measured in a direction perpendicular to the machined surface.*

— The depth of cut is always, *perpendicular* to the direction of feed motion.

— *Depth of cut is usually taken 3 to 5 times the feed for rough operations.* The values for finishing operations are *usually small.*

● Depth of cut to be used depends on the following *factors* :

(*i*) Type of cut ;

(*ii*) Tool life ;

(*iii*) Power required.

8.14. MACHINABILITY

Machinability *is defined as the ease with which a given material can be cut permitting the removal of material with a satisfactory finish at lower cost.*

It is used to signify how well a material takes a good finish. It may also be called ***finishability.***

● Good machinability is associated with the following :

(*i*) High cutting speed.

(*ii*) Low power consumption.

(*iii*) Good surface finish.

(*iv*) Removal of material with moderate force.

(*v*) Medium degree of tool abrasion (longer tool life).

(*vi*) Formation of small chips.

● *Machinability depends on the following* **factors :**

(*i*) Chemical composition of the workpiece material.

(*ii*) Microstructure.

(*iii*) Mechanical properties.

(*iv*) Physical properties.

(*v*) Cutting conditions.

(*vi*) Coolant properties.

(*vii*) Feed and depth of cut.

(*viii*) Kind and shape of cutting tool.

(*ix*) Size and shape of cut.

(*x*) Coefficient of friction between chip and tool material.

(*xi*) Tool material.

(*xii*) Type of machine used.

(*xiii*) Type of machining operation.

- For judging the machinability the main factors to be chosen depend on the type of operation and production requirements. *The following criteria may be considered while **evaluating machinability** :*

(*i*) Value for cutting forces.

(*ii*) Tool life between two successive grinds.

(*iii*) Quality of surface finish.

(*iv*) Form and size of chips.

(*v*) Temperature of chips.

(*vi*) Metal removal rate.

(*vii*) Rate of cutting under standard force.

(*viii*) Cutting forces and power consumption.

- *The following factors increase machinability* :

(*i*) Small undistorted grains.

(*ii*) Uniform microstructure.

(*iii*) Lamellar structure in low and medium carbon steels.

(*iv*) Less hardness, less ductility and less tensile strength.

(*v*) Cold working of low carbon steel.

(*vi*) Annealing, normalising and tempering operations.

(*vii*) Addition of small amounts of sulphur, lead, phosphorus and manganese.

- *Machinability may be improved by the addition of small percentages of certain elements such as lead, selenium, sulphur, manganese etc.*

- *'Grey cast iron'* is much more machinable than *'white cast iron'* since the former processes *carbon in free form* as graphite flakes which assist the chips to break up easily during the machining (in addition graphite lubricates the tool during cutting operation) whereas the latter (white cast iron) possesses free carbides (carbon in combined form) which are very hard constituents and are difficult to be machined.

Machinability index : The machinability of different metals to be machined may be compared by using the *machinability index* of each material which may be defined as follows :

$$\text{Machinability index (\%)} = \frac{\text{Cutting speed of metal investigated for 20 min. tool life}}{\text{Cutting speed of a standard steel for 20 min. tool life}}$$

A *standard steel* has a carbon content of 0.13% max., manganese of 0.06 to 0.10% and sulphur of 0.80 to 0.03% and can be machined relatively easily ; its machinability index is arbitrarily fixed as 100 %.

Relative machinability of certain alloys is given below :

Excellent machinability :

(*i*) Magnesium alloys.

(*ii*) Aluminium alloys.

(*iii*) Zinc alloys.

Good machinability :

(*i*) Brass sheets and red brass

(*ii*) Gun metal

(*iii*) Malleable cast iron

(*iv*) Grey cast iron

(*v*) Free cutting steels

(*vi*) Copper aluminium alloys.

Poor machinability :

(*i*) Low carbon steel.

(*ii*) Annealed nickel.

(*iii*) Low alloy steel.

Fair machinability :

(*i*) Ingot iron and wrought iron.

(*ii*) High speed steel.

(*iii*) Monel metal.

(*iv*) Sintered carbide.

Not machinable :

(*i*) 18 : 8 stainless steel.

(*ii*) Stellite (alloy of W, Cr, C, CO).

(*iii*) White cast iron.

Machinability properties of material :

The material properties that affect machinability may be listed briefly as follows :

1. *Shear strength.*
2. *Strain hardenability*—the increase in strength and hardness with increasing plastic deformation.
3. *Hardness*—the characteristic of the material to resist indentation.
4. *Abrasiveness of the microstructure.*
5. *Thermal conductivity.*
6. *Coefficient of friction.*

8.15. CUTTING FLUIDS

A **cutting fluid** *may be defined as any substance (may be liquid, gas or solid) which is applied to a tool during a cutting operation to facilitate removal of chips.*

8.15.1. Functions of Cutting Fluids

1. To cool the cutting tool and the workpiece.
2. To lubricate the chip, tool, and workpiece.
3. To help carry away the chips.
4. To lubricate some of the moving parts of the machine tool.

5. To improve the surface finish.

6. To prevent the formation of built-up-ridge.

7. To protect the work against rusting.

However, the prime function of a cutting fluid in a metal cutting operation is to *control the total heat*. This can be done by dissipating the heat generated as well as reducing it. The mechanism by which a cutting fluid performs those functions are *cooling action and lubricating action*.

● The cutting fluid may be applied to the cutting tool in the following *ways* :

(*i*) By hand, using a brush.

(*ii*) By means of a drip tank attached to the machine body.

(*iii*) By means of a pump.

8.15.2. Requirements of a Cutting Fluid

A cutting fluid should possess the following *qualities* in order to be of practical value :

1. It should have long life, free of excessive oxide formation that might clog circulation system.

2. It should be suitable for a variety of cutting tools and materials and the cutting operations.

3. It should have *lubricating qualities, high thermal conductivity and low viscosity* to *permit easy flow* and *easy separation from impurities and chips,* and should *not stick to workpiece or machine.*

4. It should be *transparent* where high dimensional accuracy and fine finish are required in order to enable the operator to *have a clear view of tool and workpiece.*

5. It should present *no fire or accident hazards* or emit abnoxious odours or vapours harmful to operator or workpiece ; and should cause *no skin irritation.*

8.15.3. Types of Cutting Fluids

The broad *classification* of cutting and grinding fluids is as follows :

1. Straight or neat-oils.

2. Water-miscible cutting fluids.

3. Synthetics or semichemical cutting fluids.

1. **Straight or neat-oils :**

● These are derived from petroleum, animal, marine, or vegetable substances and may be used straight or in combination.

● Their main function is lubrication and rust prevention.

● They are chemically stable and lower in cost.

● They are usually restricted to *light-duty machining* on metals of high machinability, such as aluminium, magnesium, brass and leaded steels.

2. **Water miscible cutting fluids :**

Water miscible fluids form mixtures ranging from emulsions to solutions, which due to their high specific heat, high thermal conductivity, and high heat of vaporisation, are *used on about 90% of all metal-cutting and grinding operations.*

— The water blend is usually in the ratio of one part oil to 15 to 20 parts water for cutting and 40 to 60 parts water for grinding.

3. **Synthetics (chemical) or semichemical cutting fluids :**

● *Synthetic* coolants refer to any coolant-lubricant concentrate that does not contain petroleum oil.

- *Semichemical* coolants contain a small amount of mineral oils plus additives to further enhance the lubrication properties. These are gaining wide popularity in industry today because they incorporate the best qualities of both chemicals and normal water emulsions.

Advantages and disadvantages of chemical fluids :

Advantages :

(*i*) A very light residual film that is easy to remove.

(*ii*) Heat dissipation is rapid.

(*iii*) Good detergent properties.

(*iv*) An easy concentration to control with no interference from tramp oils.

Disadvantages :

(*i*) The lack of lubrication "oiliness" may cause some sticking in the moving parts of machine tools.

(*ii*) The high detergency may irrirate sensitive hands over a long period of time.

(*iii*) As compared to oils, there is less rust-control and lubrication, and there is some tendency to foam.

Note : *Pure water* is *by far the best cutting fluid available because of its highest heat carrying capacity.* Besides this it is *cheap* and *easily available.* Its *low viscosity* makes it to *flow at high rates* through the cutting fluid system and also *penetrates* the cutting zone.

However, water *corrodes the work material very quickly, particularly at high temperatures* prevalent in the cutting zone as well as the machine tool parts on which it is likely to spill. Hence *other materials would be added to water to improve its wetting characteristics, rust inhibitors and any other additives to improve lubrication characteristics.*

8.16. CUTTING TOOL MATERIALS

8.16.1. Characteristics of an Ideal Cutting-Tool Material

An ideal cutting-tool must possess the following *characteristics* :

1. The material must *remain harder than work material at elevated temperature.*
   (Hot hardness)

2. The material must *withstand excessive wear* even through the relative hardness of the tool-work materials changes.   *(Wear resistance)*

3. The material must have sufficient *strength and ductility* to withstand shocks and vibrations and to prevent breakage.   *(Toughness)*

4. The *coefficient of friction* at the chip tool interface must remain low for minimum wear and reasonable surface finish.

5. The cost and easeness of fabrication should be within reasonable limits.

8.16.2. Types of Tool Materials

- While selecting proper tool material the *type of service* to which the tool will be subjected should be given primary consideration. No one material is superior in all respects, but rather each has certain characteristics which limits its field of application.

- The *principal carbon tool materials* are :

1. Carbon steels
2. Medium alloy steels
3. High speed steels
4. Stellites
5. Cemented carbides
6. Ceramics
7. Diamonds
8. Abrasives.

1. Carbon steels :

- *Carbon tool steels are characterised by the low stability of the supercooled austenite.* Therefore, they have a high critical rate and a low hardenability. Through hardening can be achieved only in parts upto 12 to 15 mm in thickness or diameter. Consequently steel may be recommended for *small-sized tools*, which are quenched in oil or molten salts, and for comparatively large tools (15 to 30 mm diameter) in which the cutting section is only the surface layer (files, core drills, short reamers, etc.)
- When large tools are hardened (of diameter over 30 mm) the layer with a high hardness is so thin, even upon quenching in water, that the tools are not fit for cutting purposes.

Advantages :

  (*i*) *Cheapness.*

 (*ii*) *Low hardness* (BHN = 170 to 180).

(*iii*) Good machinability.

(*iv*) Formability in the annealed state.

 (*v*) Retain a tough unhardened core due to low hardenability.

This last factor improves their resistance to breakage under vibration and impacts.

Disadvantages :

  (*i*) *Narrow range of hardening temperature.*

 (*ii*) *Necessity for rapid quenching in water or aqueous alkali solutions (salt).*

This leads to *considerable deformation and warping and even to the formation of cracks.*

Therefore, tools of complex form with sharp changes in section and with a large length to-diameter ratio should not be made of carbon steels. *Warping and crack forming may be reduced somewhat by quenching in water only to 200-250°C with subsequent retarded cooling in oil.* Stepped quenching (martempering) is advisable for small-size tools. Good results may be obtained in applying induction hardening to certain types of tools.

Uses : Carbon steels are only applicable for tools operating at low cutting speeds (about 12 m/min) since their hardness is substantially reduced at temperatures above 190-200°C.

2. Medium alloy steels :

- The high carbon medium alloy steels have a carbon content akin to plain carbon steels, but in addition there is, say, upto *5 per cent alloy content consisting of tungsten, molybdenum, chromium and vanadium.* Small additions of one or more of these elements *improve* the performance of the carbon steels in respect of *hot hardness, wear resistance, shock and impact resistance and resistance to distortion during heat treatment.*
- The alloy carbon steels broadly occupy a midway *performance position between plain carbon and high speed steels.*
- They lose their required hardness at temperature from 250 to 350°C.

3. High speed steels :

- *High speed steels are distinguished for their high red-hardness, their* capability to retain their structure (martensite), hardness, and wear resistance at the high temperature generated on the cutting edges when machining at high cutting speeds. High-speed steels are designed for the manufacture of high production tools *with high wear resistance which must retain their cutting properties at temperatures upto 600-620°C.*
- High speed steels are obtained by alloying *tungsten, chromium, vanadium, cobalt* and *molybdenum with steel.* This alloying produces metals which remain hard at temperatures at which normal steel becomes quite soft.

(a) **18-4-1 high speed steel :** A common analysis is : *18 percent tungsten, 4 percent chromium, and 1 percent vanadium,* with a carbon content of 0.6-0.7 percent. This alloy is termed 18-4-1, while an increase of vanadium to 2 percent provides 18-4-2 steel.

(b) **Cobalt high speed steel :** This is sometimes called *super high-speed steel.* Cobalt is added from 2 to 15 percent to increase hot hardness and wear resistance. One analysis of this steel contains *20 percent tungsten, 4 percent chromium, 2 percent vanadium and 12 percent cobalt.*

(c) **Molybdenum high speed steels :** This class of high speed steels contains a *lower percentage of tungsten,* this being compensated by the addition of molybdenum.

This steel containing *6 percent molybdenum, 6 percent tungsten, 4 percent chromium and 2 percent vanadium have excellent toughness and cutting ability.*

4. Stellites :

- Stellite is a *non-ferrous alloy* with range of elements : *cobalt = 40 to 48 percent ; chromium = 30 to 35 percent and tungsten = 12 to 19 percent.* In addition to one or more carbide forming elements, carbon is added in amounts of 1.8 to 2.5 per cent.

- They *cannot be forced to shape,* but may be deposited directly on the tool shank in an oxyacetylene flame, alternately, small tips of cast stellites can be brazed into place.

- Stellites *preserve hardness upto 1000°C* and can be operated on steel at *cutting speed 2 times higher than for high speed steel.*

- These materials are not widely used for metal cutting, since they are very brittle, however, they are *used extensively in some non-metal-cutting application,* such as in rubbers, plastics where loads are gradually applied and the support is firm and where wear and abrasion are problems.

5. Cemented carbides :

- These are *so named because they are composed principally of carbon mixed with other elements.*

- The basic ingredient of most cemented carbide is *tungsten carbide which is* extremely hard. Pure tungsten powder is mixed under high heat at about 1500°C with pure carbon (lamp black) in the ratio of 94 percent and 6 percent by weight. The new compound, tungsten carbide, is then mixed with cobalt until the mass is entirely homogeneous. This homogeneous mass is pressed, at pressures from 100 to 420 MN/m^2, into suitable blocks and then heated in hydrogen. *The blocks are cut and ground into specified shapes and then sintered at high temperature heating in the presence of hydrogen.* Boron, titanium and tantalums are also used to form carbides.

- The amount of cobalt used regulates the toughness of the tool. A typical analysis of a carbide suitable for *steel machining is :*

  Tungsten carbide = 82 percent ;

  Titanium carbide = 10 percent ;

  Cobalt = 8 percent.

- The *carbide tools* are made by *brazing or silve-soldering* the formed inserts on the ends of the commercial steel holders.

- The most important properties of cemented carbides *are very high heat and wear resistance.* Cemented carbides tipped tools *can machine metals even when their cutting elements are heated to a temperature of 1000°C. They can withstand cutting speed 6 or more than 6 times higher than tools of high-speed steels.*

- Cemented carbide is the hardest manufactured material and has extremely high compressive strength. However, it is very brittle, has low resistance to shock, and must be very rigidly supported to prevent cracking.

Types of cemented carbides :

The two types are :

1. Tungsten-type cemented carbides.

2. Titanium-tungsten type cemented carbides.

"Tungsten type cemented carbides" are *less brittle* than titanium-tungsten type, they contain 92 to 98 per cent tungsten carbide and 2 to 8 per cent cobalt. These cemented carbides are *designed chiefly for machining brittle metals such as cast iron, bronze, but they may also be used for non-ferrous metals and alloy steels etc.*

"Titanium-tungsten type cemented carbides" are more wear resistant. They contain 66 to 85 per cent tungsten carbide, 5 to 30 percent titanium carbide and 4 to 10 percent cobalt. These cemented carbides are designed for *machining tougher materials chiefly for various steels.*

6. Ceramics :

- Ceramic tools are made by *compacting aluminium oxide powder in a mould at about 28 MN/m² or more*. The part is then sintered at 2200°C. This method is known as *cold pressing. Hot pressed ceramics are more expensive due to higher mould costs.*

- Ceramic tool materials are made in the form of the *tips* that are to be clamped as metal shanks.

- Ceramic toots have very *low heat conductivity and extremely high compressive strength, but they are quite brittle and have a low bending strength.* For this reason these materials *cannot be used for tools operating in interrupted cuts, with vibration as well as for removing a heavy chip. But they can withstand temperatures upto 1200°C* and can be used at cutting speeds 4 times that of cemented carbides.

- Ceramic tools are *chiefly used for single-point tools* in semi-finish turning of cast iron, plastics and other work, but only when they are *not subject to impact loads.*

- To give increased strength to ceramic tools often ceramic with a metal bond, known as *"cermets"* is used.

- Because of the high compressive strength and brittleness the tips are given a 5 to 8 degrees *negative rake for carbon steel and zero rake for cast iron and non-metallic materials* to strength their cutting edge and are *well supported by the tool holders.*

- Ceramic tools are generally used *without a coolant since they have very low heat conductivity.*

7. Diamonds :

- The diamond (hardest-known material) can be run at *cutting speeds about 50 times greater than that for high speed tools, and at a temperature upto 1650°C.*

- In addition to its hardness the diamond is incompressible, *is of a large grain structure, readily conducts heat, and has a low coefficient of friction.*

- Diamonds are suitable for *cutting very hard materials* such as glass, plastics, ceramics and other abrasive materials and *for producing fine finishes.*

- The maximum depth of cut recommended is 0.125 mm with feeds of, say, 0.05 mm.

8. Abrasives :

- *Abrasive is a class of mineral used to sharpen the edges of cutting tools, and to reduce or polish metallic or other surfaces.*

- Abrasive particles held together by a bonding material comprise the cutting edges in grinding wheels known as *abrasive wheels.*

- There is fairly small group of generally used abrasives, some *natural*, some *artificial*. Of the *natural abrasives* perhaps the best known are *emery, corrundum and diamond dust.*

- *'Emery' is rough and durable*, containing about 70 percent aluminium oxide, a valuable abrasive which does the actual cutting and is known as crystalline fused alumina. While aluminium oxide occurs in nature, it is also produced synthetically by fusing the mineral bauxite in the electric arc furnace. *Corrundum contains about 90 percent aluminium oxide.*

- *Diamond dust* is slowly forging ahead as abrasive. When diamond dust is used in a "loose" condition it is mixed with oil or grease into a paste-like consistency.

- *"Carborundum"* is a trade name for *silicon carbide*, one of the *most important modern artificial abrasives.*

QUESTIONS WITH ANSWERS

Q. 8.1. What are the advantages of providing side cutting edge angle (lead angle or principal edge angle) on the cutting tools ?

Ans :

- *Large side cutting edge angle* decreases the chip thickness, measured perpendicular to the cutting edge. Smaller chip thickness means less load on the tool and decreased wear. Or for keeping the same loading and wear conditions, *feed can be increased and thus production rate will be high.*

- Further if side cutting edge is more than 0, then the tool will first make contact at a point only which will gradually keep on increasing, thus load comes gradually on the tool. Also the first contact is at a position back of the point where the tool is quite strong. Due to gradual pick-up of load, it is specially advantageous when hard surfaces as of castings are to be machined. On the other hand, a tool with side cutting edge angle of 0 will pick-up the full load on the first contact, resulting in a shock or impact loading and reduced tool life. *Increasing side cutting edge angle too much is also not desirable as it would result in increase of radial force which can bend the work and cause chattering unless the work is stiff or well supported.*

Q. 8.2. Why can relief or clearance angles never be zero or negative ?

Ans. Relief (or clearance) angles are provided to prevent the end which is parallel to work and the side, which is at the cutting edge, from rubbing on the work. If these are made zero or negative or even very small, then these will *wear down and rubbing starts. This will lead to heating up of tool, chatter marks and marking up of smeared surfaces on the work.* Thus relief or clearance angle can never be made zero or negative.

Q. 8.3. Why a built-up edge on a tool is undesirable ?

Ans. A built up edge on a tool *increases the frictional resistance* to chip flow across the face of the tool. It *results in increased heat at chip-tool interface, absorption of more power and poor surface finish on the workpiece.*

Q. 8.4. What are the factors influencing in the selection of cutting speeds and feeds for a machining operation ?

Ans. *Important factors influencing speeds and feeds for metal cutting are* :

1. Machinability of workpiece material.

2. Material of cutting tool.

3. Objective criteria, *i.e.*, whether cost or time or surface finish/tolerance are of priority.

4. Cutting fluid used.

Q. 8.5. What is the effect of cutting speed, depth of cut and feed rate on the force on cutting tool ?

Ans. Forces on the cutting tool *increases only slightly with increase in speed though the* friction is less at higher speeds. With negative rake tools and carbides, the forces *actually* decrease at very high speeds.

Both depth of cut and feed rate also lead to *increase in forces on cutting tool,* but forces due to *increase in feed rate are less than due to increase in depth.*

Q. 8.6. Explain the process of crater formation on cutting tools. Why crater is formed at some distance above the tool tip ?

Ans. *Crater formation in metal cutting occurs due to rubbing action of continuous chip with the tool material at chip-tool interface.* The creater is formed ahead of the cutting edge because cratering effect is temperature dependent. Maximum interface temperature is mid-way of chip-tool contact length. Each mode, *i.e.*, adhesion wear, diffusion wear, abrasion wear, chemical wear is temperature dependent.

Q. 8.7. Why are chip breakers used ?

Ans. A continuous type of chip from a long cut is usually quite troublesome. Such chips foul the tools, clutter up the machine and workpiece, besides being extremely difficult to remove from the swarf tray. They should be broken into comparatively small pieces for ease of handling and to prevent it from being a work hazard. Hence chip breakers are used *to reduce the swarf into small pieces as they are formed.* The fact that the metal is already work hardened helps the chip breaker to perform effectively. Various types of chip breakers are made, but all of them consist mainly of a *step or groove ground into the leading edge of the tool or a piece of cutting tool material clamped on top of the cutting tool.*

Q. 8.8. What should be done to remove maximum material per minute with the same tool life and at the same time keeping good finish ?

Ans. The best method to increase material removal rate is *to increase depth of cut, as tool life is least affected by increase in depth of cut.* If depth of cut is increased, speed needs to be decreased for same tool life. However, depth of cut is also *restricted by the strength of the workpiece and amount of stock to be removed. Increase in feed rate has smallest decrease in tool life in relation to the increased metal removal rate.* Increasing feed rate beyond finish requirements is not possible. *Thus increase in feed is best method, of course, within limits of allowable finish.*

Q. 8.9. Explain why chatter and vibration are considerably reduced when a helix cutter is used instead of a plain milling cutter during milling.

Ans. A helical cutter operates more smoothly than a plain milling cutter. It is due to the fact that if the cutter teeth are made parallel to the axis, they strike the work simultaneously across the entire width. This *result in hammering action by the tool* on the work in a regular frequency and ultimately *results in chattering of the entire set-up.* This causes a poorer surface finish and shorter tool life. A cutter having helical teeth engages with work *progressively* and the cutting action is *continuous.* This eliminates chattering and smooth cutting action takes place.

Q. 8.10. State the conditions under which positive and negative rake angles are recommended.

Ans. Following are the *conditions* under which positive and negative rake angles are recommended :

Positive rake angles :   1. When cutting at low cutting speeds.

2. When machining long shafts of small diameters.

3. When machining low strength ferrous and non-ferrous materials and work-hardening materials.

4. When using low power machines.

5. When the set up lacks strength and rigidity.

Negative rake angles : 1. For rigid set-ups and when cutting at high speeds.

2. When machining high strength alloys.

3. When there are heavy impact loads such as in interrupted machining.

Q. 8.11. Give four examples each of orthogonal cutting and oblique cutting.

Ans. *Orthogonal cutting :*

— Sawing ;

— Broaching ;

— Slotting cutter ;

— Lathe cut-off tool.

Oblique cutting :

— Lathe tools ;

— Milling cutters ;

— Drills ;

— Planers.

Q. 8.12. Explain briefly the following new cutting materials :

  1. Coronite;   **2. Ucon.**

Ans. 1. **Coronite :**

● The properties of this tool, material lies between those of H.S.S. and cemented carbides. It combines the toughness of H.S.S. with hardness and wear resistance of cemented carbides. This *improves tool life, reliability and surface finish.*

● Majority of the tools are not produced from solid coronite but by compound and coating technology.

● Cutting tools made from this material are mainly endmills used for machining grooves, pockets and for profiling on majority of the workpiece materials.

2. **Ucon :**

● The constituents of Ucon (developed by Union Carbide, U.S.A.) are : Columbium = 50% ; Titanium = 30% ; Tungsten = 20%.

● It possess the following *properties* :

— High hardness ;

— High toughness ;

— Excellent shock resistance ;

— Excellent resistance to diffusion and adhesion wear.

● This is a *basically steel cutting material* and is not preferred for cutting cast iron, stainless steel and super alloys containing Ni, Co and Ti as base materials.

Q. 8.13. Describe briefly "Coated tools".

Ans. Since long, new alloys and engineered materials are being developed continuously. These materials have high strength and toughness, *but are generally abrasive and highly chemically reactive with tool materials.* The difficulty of machining these materials efficiently and the need to improve the performance in machining the more common engineering materials have led to important development in *coated tools.*

- The *coated tools,* because of their *unique properties can be used at high cutting speeds thus reducing the time required for machining operations and, hence, costs.*
- Commonly used *coating materials* include :
— Titanium nitride ;
— Titanium carbide ;
— Titanium carbonitride ;
— Aluminium oxide.
- These coatings are applied on tools and inserts (generally in the thickness range of 2-10 μm) by *chemical-vapour deposition (CVD)* and *physical-vapour deposition (PVD)* techniques.
— The *CVD* process is the most commonly used coating application method for carbide tools with multiphase and ceramic coatings.
— The *PVD*-coated carbides with T_iN coatings, on the other hand, have *higher cutting-edge strength, lower friction, lower tendency to form* a *built-up edge,* and are *smoother* and *more uniform in thickness,* which is generally in the range of 2-4 μm.
- Coatings should have the following general **characteristics** :
  (*i*) Low thermal conductivity.
 (*ii*) Little or no porosity.
(*iii*) High hardness at elevated temperature.
(*iv*) Chemical stability and inertness to the workpiece materials.
 (*v*) Good bonding to the substrate to prevent flaking or spalling.

Q. 8.14. What do you understand by boundary or extreme pressure (EP) lubrication mechanism in metal cutting ?

Ans. The boundary lubrication conditions can be achieved by adding some fluid additives (known as extreme-pressure or *EP* additives) which react chemically with metal to *form compounds on the metal surface.* These compounds may form layered structures which are easily sheared in sliding thus *reducing friction ; or these may inhibit the welding which would occur with base metal surface in contact.*

- The use of *EP* additives results in *improvement of surface finish, reduction in cutting forces and tool wear.*
- *Compound of sulphur and chlorine* (like chlorinated paraffins, elemental sulphur, or sulphurised fats) are often used as *EP* additives in many commercial cutting fluids used for more severe cutting operation.

Q. 8.15. What is "Cryogenic machining" ? Explain briefly.

Ans. In order to reduce or eliminate the adverse environmental impact of using metal working fluids, a recent technology is the use of *liquid nitrogen as a coolant* in machining, as well as in grinding.

— The liquid nitrogen at around 200°C is injected with small-diameter nozzles into the tool-workpiece interface, thus reducing its temperature. As a result, tool hardness and, hence,

*tool life is enhanced, thus allowing for higher cutting speeds ; further more, the chips be-
come more brittle and, thus, easier to remove from the cutting zone.*

— Since no fluids are involved and the liquid nitrogen simply *evaporates*, the chips can be
recycled more easily, thus also improving the economics of machining operations, without
any adverse effects on the environment.

Q. 8.16. State the means by which 'chip welding to the tool' can be prevented.

Ans. The chip welding can be prevented by the following means :

1. *Preventing metal-to-metal contact* by the use of a high pressure lubricant between chip and
tool interface.

2. *Reduce friction* by increasing the rake angle of the cutting tool and by using a lubricant
between the rake face and the chip. The rake face may also be polished.

3. *Reduce temperature* by reducing friction and by reducing cutting speed.

4. *Reduce pressure* between the chip and the tool by increasing the rake angle, reducing the
feed rate and using oblique instead of orthogonal cutting.

Q. 8.17. What is the most practical measure of machinability ?

Ans. The most practical measure of machinability is the comparison of the possible rate of
metal removal (cm³/min.) for different materials consistent with cost, tolerance and finish.

**Q. 8.18. Explain briefly the effect of cutting speed, feed and depth of cut on the
finish obtainable.**

Ans :

● In general, *increase in cutting speed tends to improve the finish.*

— With *carbide tools* particularly, *slow speed is not at all desirable* since it means wastage of
time and money and *tools wear out faster.*

● The *increase in depth of cut* influences the finish *slightly*, but greater depth makes the
finish poor, of course, slightly.

● As the *feed rate* increases, finish gets poorest because the tool marks show on the work.
However, its effect is modified by the nose radius of the tool bit.

**Q. 8.19. Give in summary form the factors influencing formation of various types of
chips.**

Ans. The factors influencing formation of various types of chips is given in summary form
below :

		Type of chip		
	Factors	*Continuous*	*Continuous with BUE*	*Discontinuous*
(i)	*Work material*	Ductile	Ductile	Brittle
(ii)	*Cutting speed*	High	Medium	Low
(iii)	*Feed*	High	Low	Low
(iv)	*Rake angle*	Large	Small	Small
(v)	*Cutting edge*	Sharp	Dull	—
(vi)	*Friction*	Low	High	—
(vii)	*Cutting fluid*	Efficient	Poor	—

Q. 8.20. List some extremely severe cutting conditions to which the cutting tools are subjected.

Ans. Cutting tools are subjected to extremely severe cuttings conditions such as :

1. Very high temperature.

2. Metal-to-metal contact with work and chip.

3. Very high stress.

4. Very high temperature gradients.

5. Very high stress gradients.

Q. 8.21. State the conditions under which use of positive and negative rake angles are recommended.

Ans :

● The use of *positive rake angles* is recommended under the following conditions :

(*i*) When cutting at low cutting speeds.

(*ii*) When using low power machines.

(*iii*) When machining low strength ferrous and non-ferous materials and work-hardening materials.

(*iv*) When machining long shafts of small diameters.

(*v*) When the set-up lacks strength and rigidity.

● The use of *negative rake angles* in recommended under the following conditions :

(*i*) For rigid set-ups and when cutting at high speeds.

(*ii*) When machining high strength alloys.

(*iii*) When there are heavy impact loads such as interrupted machining.

Q. 8.22. Why water based fluids and dilute emulsions find more use as cutting fluids than oils at higher speeds ?

Ans :

● Water has *high specific heat* (about twice that of organic fluids), *higher thermal conductivity* and *heat of vaporisation* than oils. Water also has the *better ability* of *wetting* the surface (due to its low surface tension) thereby increasing the effective heat transfer area in cooling.

● The advantage of oil to provide lubrication is *lost at higher temperature and speeds*.

Q. 8.23. Explain briefly the various methods by which cutting fluid can be applied.

Ans. The cutting fluid can be applied by the following *methods* :

1. Flooding ; 2. Jet application ; 3. Mist application.

1. **Flooding.** In this method, a high volume flow of the cutting fluid is generally applied *on the back of the chip*. This cutting fluid floods the entire machining area and thus takes away a large amount of heat generated in the process. The used fluid is then collected in the chip pan and returned by gravity to the sump of the pump.

2. **Jet application.** In this case the cutting fluid, which may be either a liquid or a gas is applied in the form of a fine jet under pressure (upto 4 MPa).

● Since this method allows the penetration of cutting fluid into the *chip-tool interface*, better than flooding, therefore, better cutting performance and surface finish are obtainable.

3. **Mist application.** In this method, very small droplets of the cutting fluid are dispersed in a gas medium, generally air, and the mixture is applied at the *cutting zone, through the clearance* crevice.

This method combines the attractive properties of gases and liquids.

- This method claims the following *advantages* :
  (*i*) The machine, operator and surroundings remain clean since there is no spillage of the cutting fluid.
  (*ii*) The penetrating ability of cutting fluids is improved due to the small size of the particles.
  (*iii*) The compressed air in the mist helps keep the chips away from the cutting zone, thus allowing the operator to follow the machining layouts easily.
  (*iv*) The consumption of the cutting fluid is much less of the order of 300 ml/hour.
  (*v*) Large surface to volume ratio for each drop provides the possibility of *rapid vapourisation,* which is an important step that must precede penetration of the chip-tool interface.
  (*vi*) There is no reclamation of the cutting fluid from the chips, nor frequent cleaning of the sumps.

HIGHLIGHTS

1. The *metal cutting* refers to only those processes where material removal is affected by the relative motion between tool made of harder material and the workpiece.
2. Chips are of three types :
   (*i*) Continuous chip   (*ii*) Discontinuous chip, and   (*iii*) Built-up chip.
3. The *seven* important elements comprise the signature of the cutting tool and are always stated in the following order :
   Back rake angle, Side rake angle, E relief angle, Side relief angle, End cutting edge angle ; Side cutting edge angle ; Nose radius.
4. When the cutting edge of the wedge in perpendicular to the cutting velocity, the process is called *orthogonal cutting.*
5. In *oblique cutting* a lateral direction of chip movement is obtained.
6. *Orthogonal cutting* : Resultant, $R = \sqrt{F_a^2 + F_t^2}$

   Oblique cutting : Resultant $R = \sqrt{F_a^2 + F_r^2 + F_t^2}$

   where F_a, F_r and F_t are axial (feed), radial and tangential forces.
7. The commonly used tool dynamometers are :
   (*i*) Mechanical dial gauge type.
   (*ii*) Strain gauge dynamometer.
8. The shear angle (ϕ) is determined from chip thickness ratio (r) ;

   $$\phi = \tan^{-1}\left(\frac{r \cos \alpha}{1 - r \sin \alpha}\right),$$

   where α is tool rake angle.
9. *Velocity relationships* in orthogonal cutting :

   $$V_c = \frac{V \sin \phi}{\cos (\phi - \alpha)} \; ; \; V_s = \frac{V \cos \alpha}{\cos (\phi - \alpha)}$$

   where V_c, V_s and V are chip velocity, shear velocity and cutting velocity respectively.
10. Forces on the chip.

    $$F_s = F_t \cos \phi - F_f \sin \phi$$
    $$F_c = F_f \cos \phi + F_t \sin \phi$$

    $$F_t = F_s \left[\frac{\cos (\beta - \alpha)}{\cos (\phi + \beta - \alpha)}\right]$$

    where F_s, F_c, F_f, F_t, are the shear force, compressive force on the shear plane, feed force and tangential (or cutting) force respectively ; β is the angle of friction.

11. Strain on the chip, $\varepsilon = \dfrac{\cos \alpha}{\sin \phi \cos (\phi - \alpha)}$.

12. *Theories on mechanics of metal cutting :*

   (*i*) Ernest and Merchant theory : $\phi = \dfrac{\pi}{4} - \dfrac{1}{2}\,(\beta - \alpha)$

   (*ii*) Merchant theory : $2\phi - \beta - \alpha = C_2$

   (*iii*) Stabler theory : $\phi = \dfrac{\pi}{4} - \beta + \dfrac{\alpha}{2}$

   (*iv*) Lee and Shaffer theory : $\phi = \dfrac{\pi}{4} + (\alpha - \beta)$

   Also, $\phi = \dfrac{\pi}{4} - \beta + \alpha - \theta$ ... considering built-up edge formation.

13. Heat is generated in the following three *distinct regions* :
   (*i*) The shear zone ;   (*ii*) The chip-tool interface zone ;
   (*iii*) The tool-work interface region.

14. *Tool failure* may occur due to the following factors :
   (*i*) Excessive temperature ;   (*ii*) Excessive stress ;
   (*iii*) Flank wear ;   (*iv*) Crater wear.

15. *Taylor's tool life equation.*
   $VT^n = C$
   where V is the cutting speed in m/min, T is the tool life in minutes, C and n are constants.

OBJECTIVE TYPE QUESTIONS

Fill in the Blanks or Say 'Yes' or 'No' :

1. The cutting tool removes the metal from the workpiece in the form of ... .
2. Continuous type of chips are least desirable.
3. Discontinuous chips are usually produced while cutting more brittle materials.
4. BUE causes the finished surface to be rough.
5. In general low cutting speed, high feed are not conducive to BUE formation.
6. The intersection of the flank and base of the tool is called ... .
7. Side cutting edge angle is also known as 'lead angle'.
8. The rake angle specifies the ease with which a metal is cut.
9. When the cutting edge of the wedge is perpendicular to the cutting velocity, the process is called ... cutting.
10. A strain gauge dynamometer is less accurate than a mechanical dial gauge
11. The shear angle is determined from chip thickness ratio.
12. Shorter length of contact of the chip with tool results is severe temperature rise.
13. Cutting speed has least effect on the cutting temperature.
14. ... can be defined as the loss of weight or mass that accompanies the contact of sliding surfaces.
15. ... wear occurs when hard constituents of one surface plough through the material of the other surface.
16. The diffusion wear occurs by a solid-state diffusion mechanism.
17. The crater increases the cutting forces, modifies the tool geometry and softens the tool tip.
18. ... is defined as the time interval between two successive regrinds.
19. As the cutting speed is reduced, the tool life increases.
20. $VT^n = C$ is known as ... tool life equation.

ANSWERS

1. Chips	**2.** No	**3.** Yes	**4.** Yes	**5.** No
6. heel	**7.** Yes	**8.** Yes	**9.** Orthogonal	**10.** No
11. Yes	**12.** Yes	**13.** No	**14.** Wear	**15.** Abrasive
16. Yes	**17.** Yes	**18.** Tool life	**19.** Yes	**20.** Taylor's.

THEORETICAL QUESTIONS

1. Define the term 'Metal cutting'.
2. State the conditions which have an important influence on metal cutting.
3. How are chips formed ? Explain briefly.
4. Explain briefly the following types of chips :
   (i) Continuous chip ;   (ii) Discontinuous chip ;
   (iii) Built-up chip.
5. How are cutting tools classified ?
6. With the help of neat sketches, explain briefly the tool elements and tool angles is case of a single-point tool.
7. Discuss briefly the following :
   Side cutting angle, Side relief angle, Back rake angle, Clearance angle.
8. What do you mean by tool signature ? Explain briefly.
9. Explain briefly with neat sketches the following :
   (i) Orthogonal cutting.
   (ii) Oblique cutting.
10. Give the comparison between 'Orthogonal cutting' and 'Oblique cutting.'
11. Discuss briefly 'chip control'.
12. What is the function of a chip breaker ?
13. How are cutting forces measured ?
14. Discuss briefly the basic mechanism of chip formation.
15. State the factors on which the value of shear angle depends.
16. What is 'chip thickness ratio' ?
17. Derive the following relation for the shear angle (ϕ) :

$$\phi = \tan^{-1}\left(\frac{r\cos\alpha}{1 - r\sin\alpha}\right)$$

   where, k = Chip thickness ratio, and
   α = Tool rake angle.
18. Derive the following velocity relationships in orthogonal cutting :

$$V_c = \frac{V\sin\phi}{\cos(\phi - \alpha)} ;$$

$$V_s = \frac{V\cos\alpha}{\cos(\phi - \alpha)}$$

   where, V_c = Velocity with which the chip moves over the rake face of the cutting tool,
   V_s = Velocity with which metal of the workpiece shears along the shear plane,
   V = Velocity of tool relative to the workpiece,
   ϕ = Shear angle, and
   α = Tool rake angle.
19. Discuss briefly 'Friction in metal cutting'.
20. Explain briefly the regions of heat generation in metal cutting.

21. Discuss briefly the factors which influence the cutting temperature.
22. Discuss briefly 'Temperature distribution in metal cutting'.
23. List the various methods of measuring chip-tool interface temperature, and explain any two of them.
24. Explain briefly how the chip-tool interface temperature is measured by a 'Tool work thermocouple'.
25. List the factors through which the usefulness of the tool cutting edges is lost.
26. What does 'Tool failure' mean ?
27. Explain briefly the factors which cause tool failure.
28. Define the term 'wear'.
29. Explain briefly the following :
     (i) Abrasion wear ; (ii) Adhesion wear ; (iii) Diffusion wear.
30. How is 'Tool life' defined ?
31. On what factors does the tool life depend ?
32. How is tool life specified ?
33. State the factors which affect tool life.
34. State the possible tool failure criteria that could be used for limiting tool life.
35. What is Taylor's tool life equation ?
36. Explain briefly the following terms.
     (i) Cutting speed ; (ii) Feed ; (iii) Depth of cut.
37. Define the term 'Machinability'.
38. List the factors with which machinability in associated.
39. State the factors on which machinability depends.
40. State the criteria which may be considered while evaluating machinability.
41. What is machinability index ?
42. What is a cutting fluid ?
43. What are the functions of cutting fluids ?
44. What are the requirements of a cutting fluid ?
45. Give the broad classification of fluids and explain them briefly ?
46. State the advantages and disadvantages of chemical fluids ?
47. What are the characteristics of an ideal cutting-tool material ?
48. Discuss briefly the following tool materials :
     Carbon steels ; High speed steels, Stellites ; Cemented carbides, Abrasives.

UNSOLVED EXAMPLES

1. During orthogonal cutting a bar of 75 mm diameter is reduced to 73 mm. If the mean length of the cut chip is 73.5 mm and rake angle is 15°, calculate :
    (i) Cutting ratio ;   (ii) Shear angle.   [**Ans.** (i) 0.317 ; (ii) 19°]
2. Following data relate to an orthogonal cutting process :
    Chip length obtained   ...... 80 mm
    Uncut chip length   ...... 240 mm
    Rake angle used   ...... 20°
    Depth of cut   ...... 0.5 mm
    Tangential force (F_t)   ...... 2000 N
    Feed force (F_f)   ...... 200 N
    Determine : (i) Shear plane angle.
    (ii) Chip thickness.
    (iii) Friction angle.
    (iv) Resultant cutting force.   [**Ans.** (i) 23°, (ii) 1.25 mm ; (iii) 25.8° ; (iv) 2008 N]

3. During orthogonal cutting with a cutting tool having a 10° rake angle, the chip thickness is measured to be 0.4 mm, the uncut thickness being 0.16 mm. Determine :
   (i) Shear plane angle ;   (ii) Shear strain.   [**Ans.** (i) 22.938° ; (ii) 2.59]

4. A 50 mm outside diameter mild steel tubing is turned on a lathe with cutting speed of 20 m/min., with a tool having rake angle of 35°. The tool is given a feed of 0.1 mm/rev. and it is found by dynamometer that the cutting force is 2500 N and the feed force in 1000 N. Length of continuous chip in one revolution is 80 mm.
   Calculate coefficient of friction, shear plane angle, velocity of chip along tool face and chip thickness.
                                            [**Ans.** 1.528 ; 30.5° ; 10.2 m/min ; 0.196 mm]

5. While doing orthogonal machining of a mild steel part, a depth of cut of 0.8 mm is used at 55 r.p.m. If the chip thickness is 1.6 mm and it is of continuous type determine the chip thickness ratio.
   Also, calculate the length of chip removed in one minute if work diameter is 50 mm before the cut is taken.   [**Ans.** 0.5 ; 4317.5 mm/min.]

6. Following data relate to an orthogonal cutting process :
   Depth of cut = 0.3 mm ; Chip thickness = 0.85 mm ; Breadth of cut = 2.5 mm ; Tool rake angle = 10° ;
   Tangential force = 900 N ; Feed force = 450 N.
   Determine : (i) Coefficient of friction between the tool and the chip ; (ii) Ultimate shear stress of the work material.   [**Ans.** (i) 0.74 ; (ii) 318.24 N/mm^2]

7. The following data relate to orthogonal cutting of mild steel part :
   Cutting speed = 200 m/min ; Tool rake angle = 12° ; Width of cut = 1.8 mm ; Uncut thickness = 0.2 mm ;
   Average value of the coefficient of friction between the tool and the chip = 0.55 ; Shear stress of work material = 390 N/mm^2.
   Calculate : (i) Shear angle ;   (ii) The cutting and thrust components.
                                            [**Ans.** (i) 36.6° ; (ii) F_t = 378.2 N ; F_f = 114.2 N]

8. A cutting tool, cutting at 24 m/min, gave a life of 1 hour between re-grinds when operating on roughening cuts with mild steel. What will be its probable life when engaged on light finishing cuts ? Take $n = \dfrac{1}{6}$
   for roughening and $\dfrac{1}{7.5}$ for finishing cuts in the Taylor's equation $VT^n = C$.   [**Ans.** 167 min.]

9. The relationship for H.S.S. tools is $V(T)^{1/8} = C_1$, and for tungsten carbide tools $V(T)^{1/5} = C_2$.
   Assuming that at a speed of 25 m/min the tool life was 180 min in each case, compare their cutting lives at 32 m/min.   [**Ans.** 1.582]

10. A carbide tool with mild steel workpiece was found to give life of 2 hours while cutting 50 m/min. If Taylor's exponent n = 0.27, determine :
    (i) The tool life if the same tool is used at a speed of 25 percent higher than the previous one.
    (ii) The value of cutting speed if the tool is required to have tool life 3 hours.
                                            [**Ans.** (i) 52.7 min. ; (ii) 44.8 m/min.]

11. During straight turning of a 25 mm diameter steel bar at 300 r.p.m with an H.S.S. tool, a tool life of 10 min. was obtained. When the same bar was turned at 250 r.p.m., the tool life increased to 52.5 min. What will be the tool life at a speed of 275 r.p.m.

                                            [**Ans.** 22 min.]

12. A mild steel piece is being machined by two different tools A and B under identical machining conditions. The tool life equations for these tools are :
    For Tool-A : $VT^{0.31} = 34.3$
    For Tool-B : $VT^{0.43} = 89.5$
    where V and T are in m/s and s, respectively.
    Determine the cutting speed above which tool-B will give better tool life.   [**Ans.** 4.71 m/s.]

9

Machine Tools and Machining Processes

9.1. Introduction—machining processes—machine tools. 9.2. **Centre/Engine lathe**—Introduction—working principle—parts of lathe—size and specifications of lathe—types of lathe—lathe tools—lathe operations—lathe accessories—lathe attachments—cutting speed, feed and depth of cut—metal removal rate (MRR)—machining time—power required in turning—thread cutting—eccentric turning. 9.3. **Turret and capstan lathes**—Introduction—limitations of a centre lathe—differences between a turret lathe and a centre lathe—comparison of turret lathe and capstan lathe—main parts of a turret or capstan lathe—types of turret lathes—size and specifications of turret lathe—common tools and attachments used on turret and capstan lathes—turret lathe operations—turret lathe tooling layout. 9.4. **Automatic lathes**—Introduction—classification of automatic Lathes—automatic vertical multi-station lathe—automatic screw machines. 9.5. **Shaping machine (Shaper)**—Introduction—principle of working—advantages, limitations and applications of shapers—classification of shapers—principal parts of a shaper—specification of a shaper—quick return mechanism in a shaper—work holding devices—operations performed—cutting tools and other tools used in shaper work—cutting speed, feed and depth of cut—machining time—material removal rate. 9.6. **Slotting machine (Slotter)**—Introduction—main parts of a slotter—size and specifications—slotter drive mechanisms—types of slotters—work holding devices—slotter cutting tools—slotter operations. 9.7. **Planing machine (Planer)**—Introduction—principle of working—planing operation—comparison between planer and shaper—types of planers—principal parts of a planer—size of a planer—planer driving and feed mechanisms—standard clamping devices—planer tools—planer operations—cutting speed, feed and depth of cut—machining time and material removal rate. 9.8. **Drilling machines**—Introduction—specifications of a drilling machine—types of drilling machines—classification of drills—twist drill nomenclature—work holding devices—drilling machine operations—cutting speed, feed and depth of cut—machining time in drilling—material removal rate. 9.9. **Boring machines**—Introduction—classification of boring machines—horizontal boring machines—vertical boring machines—jig boring machines—size of boring machines. 9.10. **Milling machines**—Introduction—milling processes—classification of milling machines—horizontal milling machine—vertical milling machine—universal milling machine—omniversal milling machine—planer type milling machine—bed type milling machine—drum type milling machine—planetary milling machine—specifications of a milling machine—types of milling cutters—nomenclature of a milling cutter—milling operations—cutting speed, feed and depth of cut—material removal rate—machining time—work holding devices—milling machine attachments—dividing or indexing head. 9.11. **Broaching machines**—Broaching—advantages, limitations and applications of broaching—broach—types of broaching machines—size of broaching machines. 9.12. **Sawing machines**—Introduction—classification of sawing machines—sawing machine blades—applications of sawing machines. 9.13. **Grinding and finishing processes**—Introduction—grinding process—advantages of grinding process over other cutting processes—special features of grinding process—grinding machines—grinding wheel—manufacture of grinding wheels—wheel shapes—mounting of wheels—wheel truing—material removal rate and machining time—Finishing processes—honing—lapping—super finishing—comparison among lapping, honing and superfinishing—polishing and buffing—burnishing—comparison of grinding and finishing processes. 9.14. **Coolants and lubricants**—**Questions with Answers**—Highlights—Objective Type Questions—Theoretical Questions—Unsolved Examples.

9.1. INTRODUCTION

9.1.1. Machining Processes

9.1.1.1. Machining

Machining *is the process of cold working the metals into different shapes by using different types of machine tools.* This process is mainly used to bring the metal objects produced by means of different fabrication techniques to final dimensions.

Machinability which is defined as the *ease of removing metal while maintaining dimensions and developing a satisfactory surface finish* is an important aspect affecting the metallurgical and properties stand-point of metals. Tool wear and power consumption are two factors which affect the metal removal rate. Greater effort and time is required to keep the tools sharp due to rapid tool wear and frequent machine stoppage for replacing the dull tools. Types of *metal chips* formed during machining operation also affect the different characteristics. Machinability of a metal is generally indicated by *machinability ratings* (which are dependent upon their techniques of determination as well as upon the particular metal cutting operation used for their measurement).

Machining is accomplished with the use of machines known as "**Machine tools**". For production of variety of machined surfaces different types of machine tools have been developed. *The kind of surface produced depends upon the shape of cutting, the path of the tool as it passes through the material or both.* Depending on them metal cutting processes are called either turning or planing or boring or other operations performed by machine tools like lathe, shaper, planer, drill, miller, grinder etc. as illustrated schematically in Fig. 9.1.

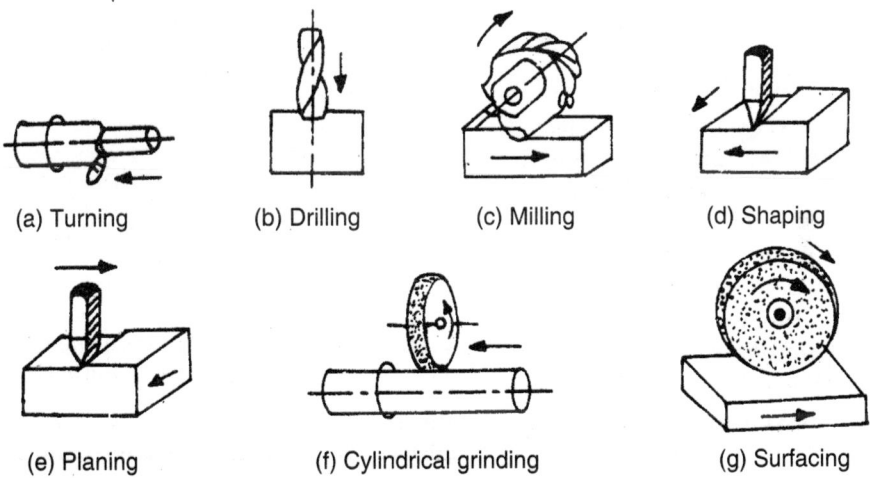

(a) Turning   (b) Drilling   (c) Milling   (d) Shaping

(e) Planing   (f) Cylindrical grinding   (g) Surfacing

Fig. 9.1. Principal machining methods–Tool work interaction.

9.1.1.2. Classification of machining processes

Machining process *are material removing operations in which the desired shape and surface finish on the finished product are obtained by removing surplus material.*

The machining processes are *classified* as follows :

1. *Metal Cutting :*

(*i*) *Single point cutting :*

- Turning
- Boring

- Shaping
- Planing.

(ii) Multi-point cutting :

- Milling
- Drilling
- Tapping
- Hobbing
- Broaching.

2. **Grinding and finishing :**

(i) Grinding :

- Surface grinding
- Cylindrical grinding
- Centreless grinding.

(ii) Finishing :

- Lapping
- Honing
- Superfinishing.

3. **Unconventional Machining :**

- Ultrasonic machining
- Electrodischarge machining
- Electro-chemical machining
- Laser beam machining.

- The **metal cutting** (machining, a generic term, refers to all material removal processes) *refers to only those processes where material removal is affected by the relative motion between tool made of harder material and the workpiece.* The tool would be single-point cutting tool as used in operations like turning or shaping, or a multi-point tool as used in milling or drilling operation.

- **Grinding and finishing** processes are those where metal is *removed by a large number of hard abrasive particles or grains which may be bonded as in grinding wheels, or be in loose form as in lapping.*

- **Unconventional machining processes** are those *which use electrical, chemical and other means of material removal for shaping high strength materials and for producing complicated shapes.*

9.1.1.3. Factor influencing the selection of a suitable process

Following are the major *factors* that influence the *selection* of a suitable process :

1. Cost consideration.

2. Material of workpiece.

3. Shape of workpiece.

4. Size of workpiece.

5. Degree of accuracy.

6. Surface finish requirement.

7. Number of products to be produced.

9.1.2. Machine Tools

9.1.2.1. Definition and functions

Machines tools *are the kind of *machines on which the metal cutting or metal forming processes are carried out.* They employ cutting tools to remove excess material from the given job.

The *functions* of a machine tool are :

(*i*) To hold the tool ;

(*ii*) To move the tool or the workpiece or both relative to each other ;

(*iii*) To supply energy required to cause the metal cutting.

9.1.2.2. Classification of machine tools

The machine tools are classified as follows :

1. *General purpose :*

(*i*) Lathe

(*ii*) Drilling machine

(*iii*) Shaping machine

(*iv*) Planing machine

(*v*) Milling machine

(*vi*) Sawing machine.

2. *Special purpose :*

(*i*) Special lathes like capstan, turret and copying lathes

(*ii*) Boring machine

(*iii*) Broaching machine

(*iv*) Production milling machine

(*v*) Production drilling machine.

3. *Automatic machine tools :*

These machine tools, also called Automatic screw cutting machines (or simply auto-mats), are used for mass production of essentially small parts using a set of pre-designed and job-specific cams.

4. *Computer numerical control (CNC) machine tools :*

Under CNC machine tools, we have *CNC turning centre,* which does all the work of a lathe and *CNC machining centre* which does milling, drilling etc., with provision for automatic tool changing and tool wear correction built into it.

- Characteristics of '*General purpose*' machine tools are :

(*i*) Less set-up and debugging time.

(*ii*) Less maintenance cost.

(*iii*) Usually less initial investment in equipment.

(*iv*) Less danger of obsolescence.

(*v*) Fewer machines may be required.

(*vi*) Greater machine flexibility.

- Characteristics of '*Special purpose*' machine tools are :

(*i*) Higher output.

(*ii*) Higher product quality.

(*iii*) Reduced skill requirements.

(*iv*) Reduced inspection cost.

---

* A *machine is* a device, which converts *one form of input into output*; *e.g.* washing machines, refrigerators, air cooler. Machine tools are also machines, but *not* all machines are machine tools. A *machine tool imparts the required shape to the workpiece with the desired accuracy by removing metal from the workpiece in the form of chips.*

(*v*) Uniform product flow.

(*vi*) Reduced manpower requirements.

(*vii*) Reduced factory floor space.

(*viii*) Reduced in-process inventory.

9.1.2.3. Elements of machine tools

Various elements of machine tools are :

1. *Structure*—formed by bed, column and frame.

2. *Slides and tool structure.*

3. *Spindles and spindle bearing.*

4. *Kinematics of machine tool drives.*

5. *Work holding, and tool holding elements.*

9.2. CENTRE/ENGINE LATHE

9.2.1. Introduction

● A centre/engine lathe (Fig. 9.2) is one of the oldest and perhaps most important machine tools ever developed.

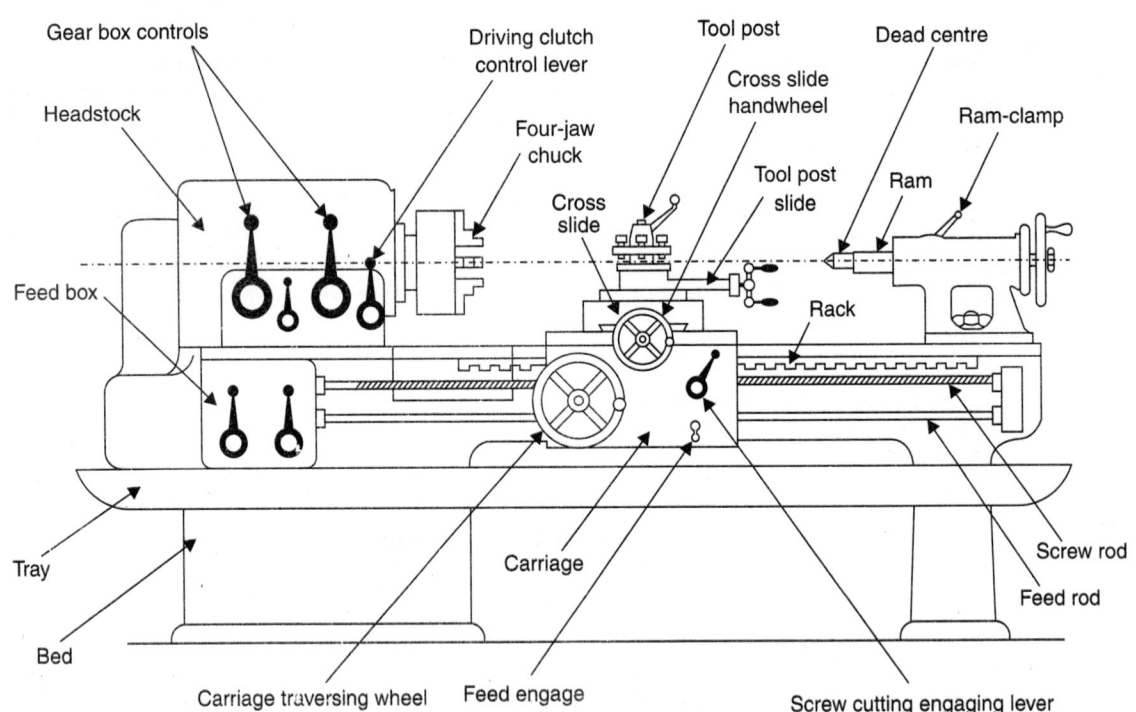

Fig. 9.2. Centre lathe.

● The job to be machined is rotated (turned) and the cutting tool is moved relative to the job. That is why, the lathes are also called *"Turning machines"*.

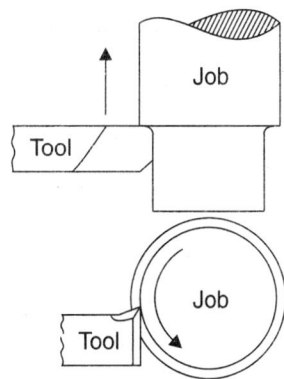

Fig. 9.3. Working principle of a lathe.

- If tool moves *parallel* to the axis of rotation of the workpiece, *cylindrical surface* is produced, while if it moves *perpendicular* to the axis, it produces a *flat surface.*

9.2.2. Working Principle

Fig. 9.3 shows the *working principle of a lathe.*

- In a lathe, the workpiece is held in a chuck or between centres and rotated about its axis at a uniform speed.
- The cutting tool held in tool post is fed into the workpiece for desired depth and in desired direction (*i.e.,* in linear, transverse or lateral direction).

Since their exists a relative motion between the workpiece and the cutting tool, therefore the material is removed in the form of chips and the desired shape is obtained.

9.2.3. Parts of Lathe

Fig. 9.2 shows a centre lathe. Its major parts are :

1. Bed.
2. Headstock.
3. Tailstock.
4. Carriage.
5. Feed mechanism.

1. Bed : Refer to Fig. 9.4.

- It is the base or foundation of the lathe.
- It is a heavy rigid casting made in one piece. In majority of cases the beds are made of grey cast iron-nodular cast iron, or high strength, wear resistance cast iron. The *cast iron* offers the following *advantages* over other materials :

(*i*) It is self lubricant : It can be hardened by induction hardening process.

(*ii*) It has better compressive strength.

(*iii*) It has excellent shock absorbing capacity.

(*iv*) It can easily be cast and machined.

- It holds or supports all other parts of the lathe. The top of the bed is planed to form *guide ways* for the carriage and tailstock.

   — The guide ways are of *two types* : (*a*) Flat guideways (Fig. 9.4(*a*)) or inverted Vee guideways. Generally, the combination of both the flat and inverted Vee guideways is used (Fig. 9.4(*b*)).

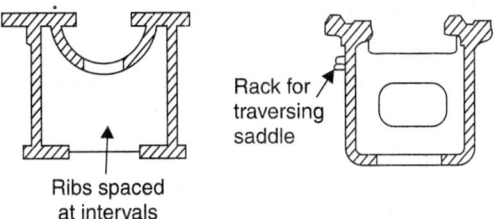

Ribs spaced at intervals

Rack for traversing saddle

(a) Flat guides   (b) Flat and inverted vee guides

Fig. 9.4. Bed.

2. Headstock :

- It is permanently fastened to the innerways at the left hand end of the bed.
- It serves to support the spindle and driving arrangements.
- All lathes receive their power through the headstock, which may be equipped with a step-cone pulleys or a gear head drive (the modern lathes are provided with all geared type head stock to get large variations of spindle speeds).
- In order to allow the long bar or work holding devices to pass through, the headstock spindle is made hollow. A tapered sleeve fits into the tapered spindle hole.

3. Tailstock : Refer to Fig. 9.5.

- It is situated at the right hand end of the bed.
- It is used for supporting the right end of the work.
- It is also used for holding and feeding the tools such as drills, reamers, taps etc.

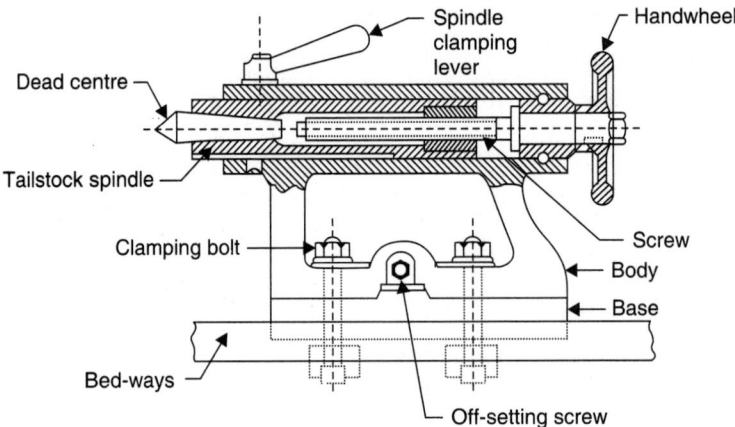

Fig. 9.5. Tailstock.

4. Carriage : Refer to Fig. 9.6.

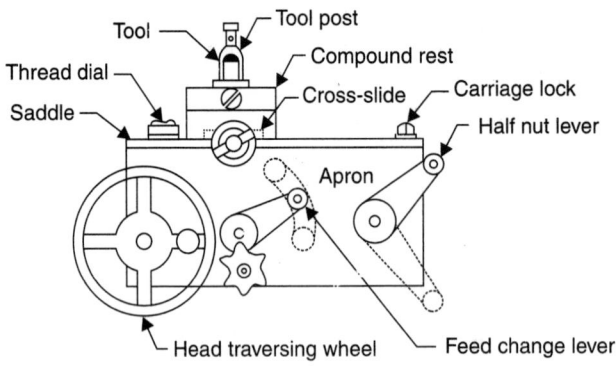

Fig. 9.6. Carriage.

- The carriage controls and supports the cutting tool.
- The carriage has the following five major parts :

(i) *Saddle*. It is a H-shaped casting fitted over the bed. It moves along the guideways.

(ii) *Cross-slide*. It carries the compound slide and tool post ; can be moved by power or by hand.

(iii) *Compound rest.* It is marked in degrees ; used during taper turning to set the tool for angular cuts.

(iv) *Tool post.* The tool is clamped on the tool post.

(v) *Apron.* It is attached to the saddle and hangs in front of the bed. It has gears, levers and clutches for moving the carriage with the lead screw for thread cutting.

5. Feed mechanism :

● It is employed for imparting various feeds (longitudinal, cross and angular) to the cutting tool.

● It consists of feed reverse lever, tumbler reversing mechanism, change gears, feed gear box, quick change gear box, lead screw, feed rod, apron mechanism and half nut mechanism.

9.2.4. Size and Specifications of Lathe

Size of a lathe is *specified* in any *one* of the following ways : Refer to Fig. 9.7.

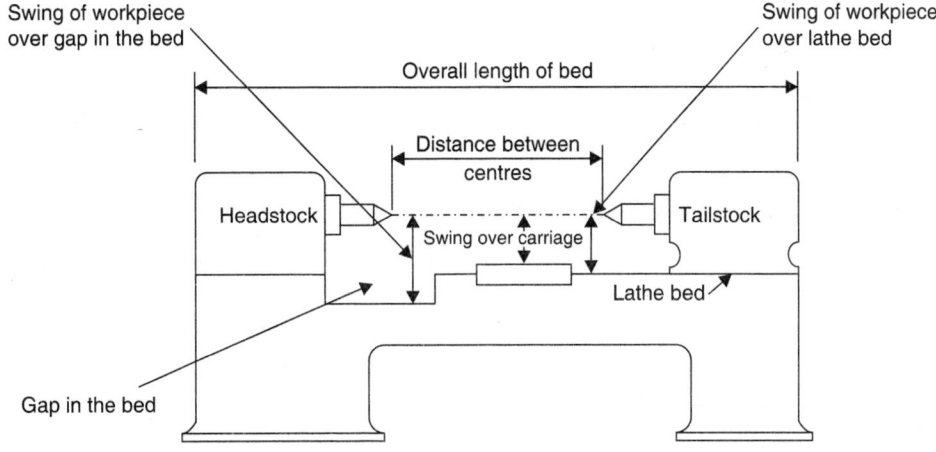

Fig. 9.7. Lathe specifications.

1. The *height of the centres* measured over the lathe bed.

2. *Swing or maximum diameter* that can be rotated over the bed ways.

3. *Swing or diameter over carriage.* This is the largest diameter of work that will revolve over the lathe saddle.

4. *Maximum job length* in mm that may be held between the centres (headstock and tail-stock centres).

5. *Bed length* in metres which may include the headstock length also.

6. *Diameter of the hole through lathe spindle* for turning bar material.

In addition to the above, the following specifications are necessary to provide while ordering a lathe :

(i) The length, width and depth of bed.

(ii) The depth and width of the gap, if it is a gap lathe.

(iii) The swing over gap.

(iv) The number and range of spindle speeds.

(v) The number of feeds.

(vi) The lead screw diameter.

(vii) The number and range of metric threads that can be cut.

(*viii*) The tailstock spindle travel.

(*ix*) The tailstock spindle set over.

(*x*) The back gear ratio.

(*xi*) The power rating of electric motor.

9.2.5. Types of Lathe

The following are the types of lathe :

1. Speed lathe :

● In this lathe spindle can rotate at a very high speed with the help of a variable speed motor built inside the headstock of the lathe.

● It is used mainly for *wood working, centering, metal spinning, polishing etc.*

2. Engine or centre lathe :

● It is the most common types of lathe and is widely used in *workshop.*

● The speed of the spindle can be widely varied as desired which is not possible in a speed lathe.

● The cutting tool may be fed both in cross and longitudinal directions with reference to the lathe axis with the help of a carriage.

3. Bench lathe :

● It is usually mounted on a bench.

● It is very similar to speed or centre lathe, the only difference being it is *smaller in size* which enables it *handle small work* (usually requiring considerable accuracy such as in the production of gauges, punches and beds for press tools).

4. Tool room lathe :

● It is similar to an engine lathe, designed for *obtaining accuracy.*

● It is used for manufacturing *precision* components, dies, tools, jigs etc ; and hence it is called as tool room lathe.

5. Turret and capstan lathes :

● These lathes have provision to *hold a number of tools* and can be used for performing wider range of operations.

● These are particularly suitable for *mass production of identical parts in minimum time.*

6. Automatic lathes :

● These lathes are so designed that the tools are automatically fed to the work and withdrawn after all operations, to finish the work, are complete.

● They require little attention of the operator, since the entire operation is automatic.

● These are used for mass production of identical parts.

7. Special purpose lathes :

These lathes are primarily designed for carrying out a particular operation with utmost efficiency. The lathes included in this category are :

(*i*) *Gap lathe :*

● It is used for machining extra large diameter pieces.

(*ii*) *Special purpose engine lathe :*

● These lathes are designed for machining special types of workpieces, *e.g.* wheel turning lathes for turning the tread on rail road-car and locomotive and so on.

(*iii*) *Instrument lathe :*

● These lathes are of smaller size than 'Bench lathes' and are used by instrument makers.

(iv) Facing lathes :
- These lathes are used to machine workpieces of large diameter but short in length in single piece production and in repair shops.
- Now these lathes have been replaced by more advanced boring and turning machines.

(v) Flow turning lathes :
- A flow turning lathe is used for roll flowing, as a method of cold-flowing metal. As heavy pressure is applied spirally with two hardened rollers against a metal block.

(vi) Heavy-duty lathe :
- A lathe, that has a swing of 500 mm or more and is used for roughing and finishing cuts, is often referred to as a heavy-duty lathe.

9.2.6. Lathe Tools

- In a lathe, for a general purpose work, the tool used is a single point tool (a tool having one cutting edge), but for special *operations multi-point tools may be used.*
- The commonly used materials are *high carbon steel, high speed steel, cemented carbides, diamond tips and ceramics.*

Depending upon the nature of operation done by the tool, the *lathe tools are classified* as follows :

(i) Turning tool (left hand or right hand)

(ii) Facing tool (left hand or right hand)

(iii) Chamfering tool (left hand or right hand)

(iv) Form or profile tool

(v) Parting or necking tool

(vi) External threading tool

(vii) Internal threading tool

(viii) Boring tool

(ix) Knurling tool.

The above mentioned tools are shown in Fig. 9.8.

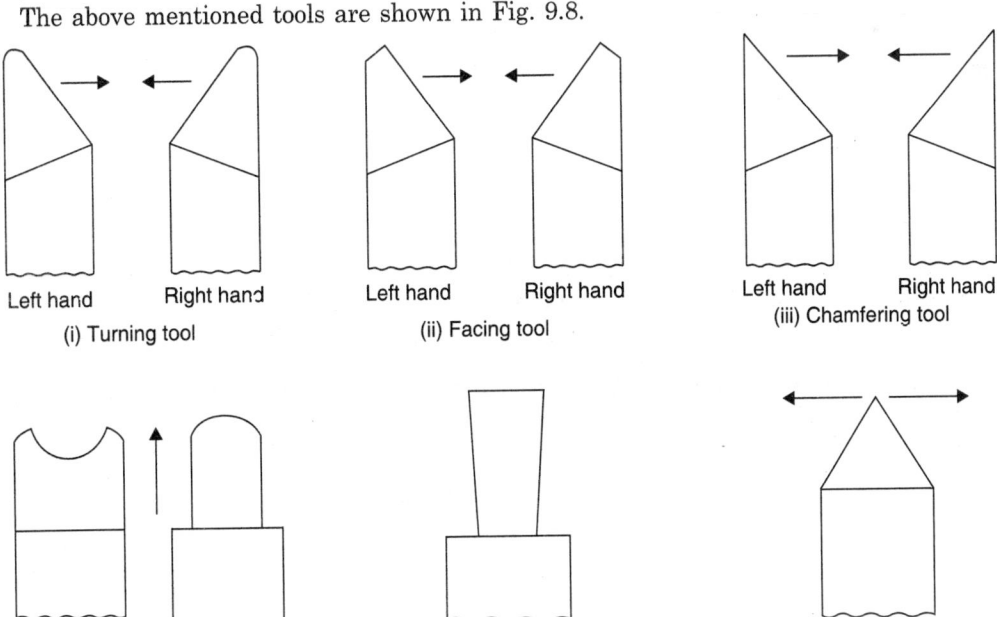

Left hand   Right hand
(i) Turning tool

Left hand   Right hand
(ii) Facing tool

Left hand   Right hand
(iii) Chamfering tool

(iv) Form or profile tools

(v) Parting or necking tool

(vi) External threading tool

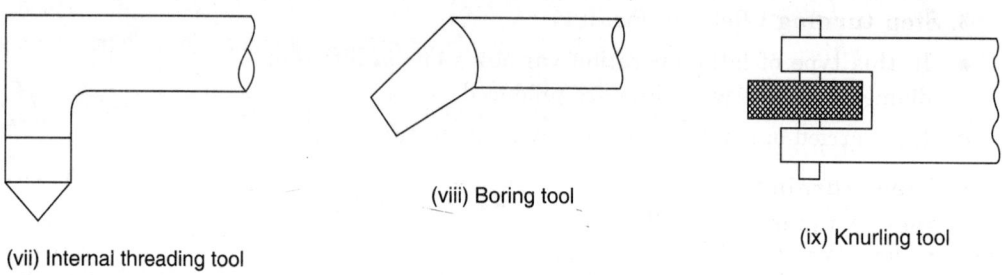

(vii) Internal threading tool   (viii) Boring tool   (ix) Knurling tool

Fig. 9.8. Lathe tools.

9.2.7. Lathe Operations

Common lathe operations which can be carried out on a lathe are enumerated and briefly discussed as follows :

1. Facing   2. Plain turning   3. Step turning

4. Taper turning   5. Drilling   6. Reaming

7. Boring   8. Undercutting or grooving   9. Threading

10. Knurling   11. Forming.

1. Facing : Refer to Fig. 9.9.

- *"Facing"* is an *operation of machining the ends of a workpiece to produce a flat surface square with the axis.* It is also used to cut the work to the required length.

- The operation involves feeding the tool *perpendicular* to the axis of rotation of the workpiece.

- A properly ground facing tool is mounted in the tool post. A regular turning tool may also be used for facing a large workpiece. The cutting edge should be set at the same height as the centre of the workpiece.

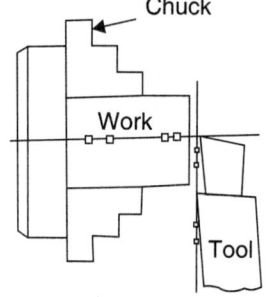

Fig. 9.9. Facing.

- The facing operation is usually performed in *two steps*.

In the first step a *rough facing* operation is done by using a heavy cross feed of the order of 0.5 to 0.7 mm and a deeper cut upto 5 mm (maximum). It is followed by a *finer cross feed* of 0.1 to 0.3 mm and a smaller depth of cut of about 0.5 mm.

2. Plain turning : Refer to Fig. 9.10.

- It is an operation of removing excess material from the surface of the cylindrical workpiece.

- In this operation, the work is held either in the chuck or between centres and the longitudinal feed is given to the tool either by hand or power.

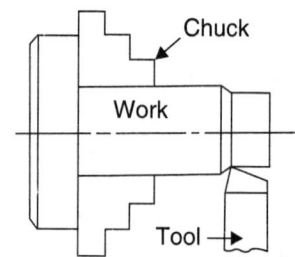

Fig. 9.10. Plain turning.

3. Step turning : Refer to Fig. 9.11.

- In this type of lathe operation various steps of different diameters in the workpiece are produced.

- It is carried out in the similar way as plain turning.

4. Taper turning :

Taper. A **taper** *may be defined as an uniform increase or decrease in diameter of a piece of work measured along its length.*

Refer to Fig. 9.12. The taper angle α can be found by using the following relationship :

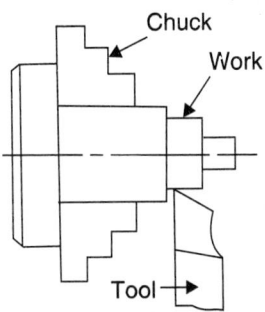

Fig. 9.11. Step turning.

$$\tan \alpha = \frac{D - d}{2L}$$

or,

$$\alpha = \tan^{-1}\left(\frac{D - d}{2L}\right) \qquad \qquad ...(9.1)$$

where, D = Large diameter of taper in mm,

d = Small diameter of taper in mm,

L = Length of tapered part in mm, and

α = Half of taper angle.

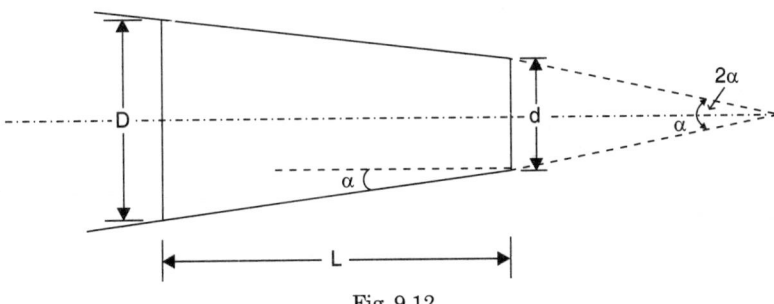

Fig. 9.12

The *conicity K* of a taper is defined as :

$$K (= 2 \tan \alpha) = \frac{D - d}{L} \qquad \qquad ...(9.1a)$$

Taper turning. *Taper turning means to produce a conical surface by gradual reduction in diameter from a cylindrical workpiece.*

— The tapering of a part has wide applications in the construction of machines. Almost all machine spindles have taper holes which receive taper shanks of various tools and work holding devices.

Taper turning methods. Taper turning can be carried out on lathes by the following methods :

   (*i*) By setting over the tailstock centre.

   (*ii*) By swivelling the compound rest.

   (*iii*) By using a taper turning attachment.

   (*iv*) By manipulating the transverse and longitudinal feeds of the slide tool simultaneously.

   (*v*) By using a broad nose form tool.

(*i*) **By setting over the tailstock centre :**

— This method is used for *small tapers only* (the amount of setover being limited).

— It is *based upon the principle of shifting the axis of rotation of the workpiece, at an angle to the axis, and feeding the tool parallel to the lathe axis*. The angle at which the axis of rotation of the workpiece is shifted is *equal to half angle of taper*. This is done when the body of the tailstock is made to slide on its base towards or away from the operator by a setover screw as shown in Fig. 9.13.

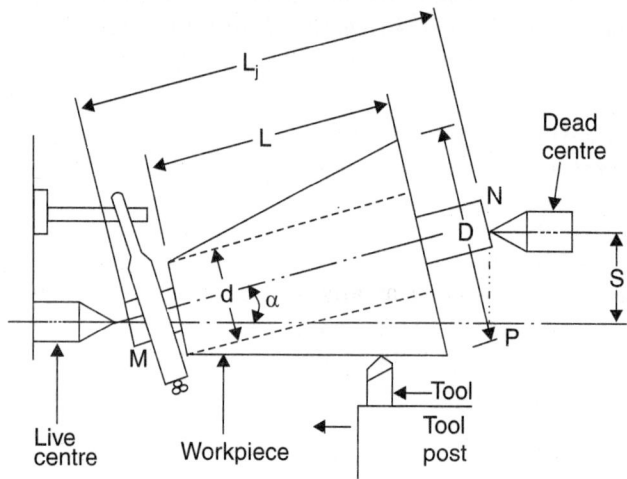

Fig. 9.13. Turnining taper by tailstock set-over method.

— By setting tailstock centre to the back (away from the operator) the taper will have bigger diameter towards the tailstock. If the tailstock centre is taken in the front, bigger diameter will be on the headstock side. The reduction in diameter will be twice the offset of tailstock centre if entire length is turned.

— The major **disadvantage** of this method is that the *live and dead centres are not equally stressed and the wear is non-uniform*. Also, the lathe carrier being set at an angle, the *angular velocity of the work is not constant*.

● It is useful for turning very long tapers up to about 5°. This method should be *avoided if possible*.

Calculation of setover (S) :

The *amount of setover required may be calculated as follows* :

From the right angled triangle *MNP* (Fig. 9.13), we have

$$NP = (\text{setover}),\ S = MN \sin (\alpha) = L_j \sin \alpha$$

For a very small angle, (α) it can be safely considered that :

$$\sin \alpha = \tan \alpha$$

i.e., setover, $\quad S = L_j \tan \alpha$

or, $\qquad\qquad\qquad S = L_j \times \dfrac{(D-d)}{2L}$ in mm $\qquad\qquad\qquad$ ...(9.2)

or, $\qquad\qquad S = \text{Total length of job/work} \times \dfrac{\text{Total taper}}{2 \times \text{taper length}}$ $\qquad$ ...(9.3)

where, $\quad\ S$ = The required setover in mm,

$\qquad\qquad D$ = Large diameter in mm,

$\qquad\qquad d$ = Small diameter in mm,

$\qquad\qquad L_j$ = Total length of job/work in mm, and

$\qquad\qquad L$ = Length of tapered portion in mm.

In case the job is to be tapered over its full length, L will be equal to L_j. Therefore, the setover will be given by :

$$S = \frac{D-d}{2} = \frac{\text{Total taper}}{2} \qquad \qquad ...(9.4)$$

— *The amount of the offset required may be quite accurately set by allowing the tool post to touch the tailstock barrel in the normal and in the offset position. This is accomplished by turning the cross-slide screw when the offset is measured directly by the difference of readings on the micrometer dial. A more accurate reading is obtained by using a dial indicator in conjunction with cross-slide.*

— For accurately setting of the tailstock, *slip gauges are sometimes used.*

(*ii*) **By swivelling the compound rest :**

— It is the *best method* as it does not affect the centering of the job or centres.

— In this method of taper turning the *workpiece is rotated on the lathe axis and the tool is fed at an angle to the axis of rotation of the workpiece.* The tool mounted on the compound rest is attached to the circular base, graduated in degrees, which may be swivelled and clamped at any desired angle as shown in Fig. 9.14. After the compound rest is set at the desired half taper angle, rotation of the compound slide screw will cause the tool to be fed at that angle and generate a corresponding taper.

— The *setting of compound rest* is done by swivelling the rest at the half taper angle, if this is already known. However, if the diameters of large (D) and small (d) ends are known, the half taper angle can be calculated as follows :

$$\tan \alpha \text{ or } \alpha = \tan^{-1}\left(\frac{D-d}{2L}\right).$$

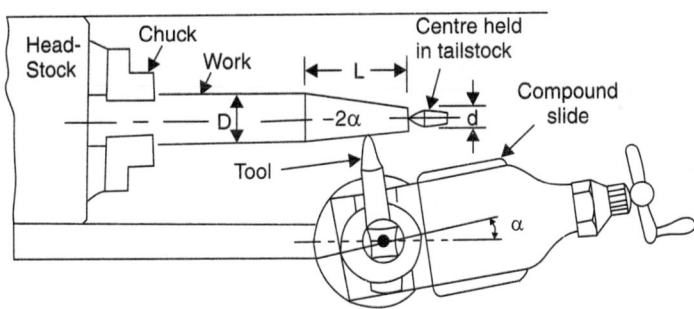

Fig. 9.14. Taper turning by swivelling the compound rest.

— Owing to the limited movement of the cross-slide, this method is limited to turn a *short taper* ; a *small taper* may also be turned.

● Short lengths of tapers not exceeding 45° included angle are usually turned by this method.

● This method gives a *low production capacity and poor surface finish because the movement of the tool is completely controlled by hand.* This method is tiring if the traverse is lengthy.

(*iii*) **By using a taper turning attachment :**

— This method *provides a very wide range of taper.*

— In this method of taper turning a *tool is guided in a straight path set at an angle to the axis of rotation of the workpiece, while the work is being revolved between centres or by a chuck aligned to the lathe axis.*

— As shown in Fig. 9.15, a taper turning attachment essentially consists of a bracket
or frame which is attached to the rear end of the lathe bed and supports a *guide bar*
pivoted at the centre. The bar is provided with graduations and may be swivelled on
either side of the zero graduation and is set at the desired angle with the lathe axis.

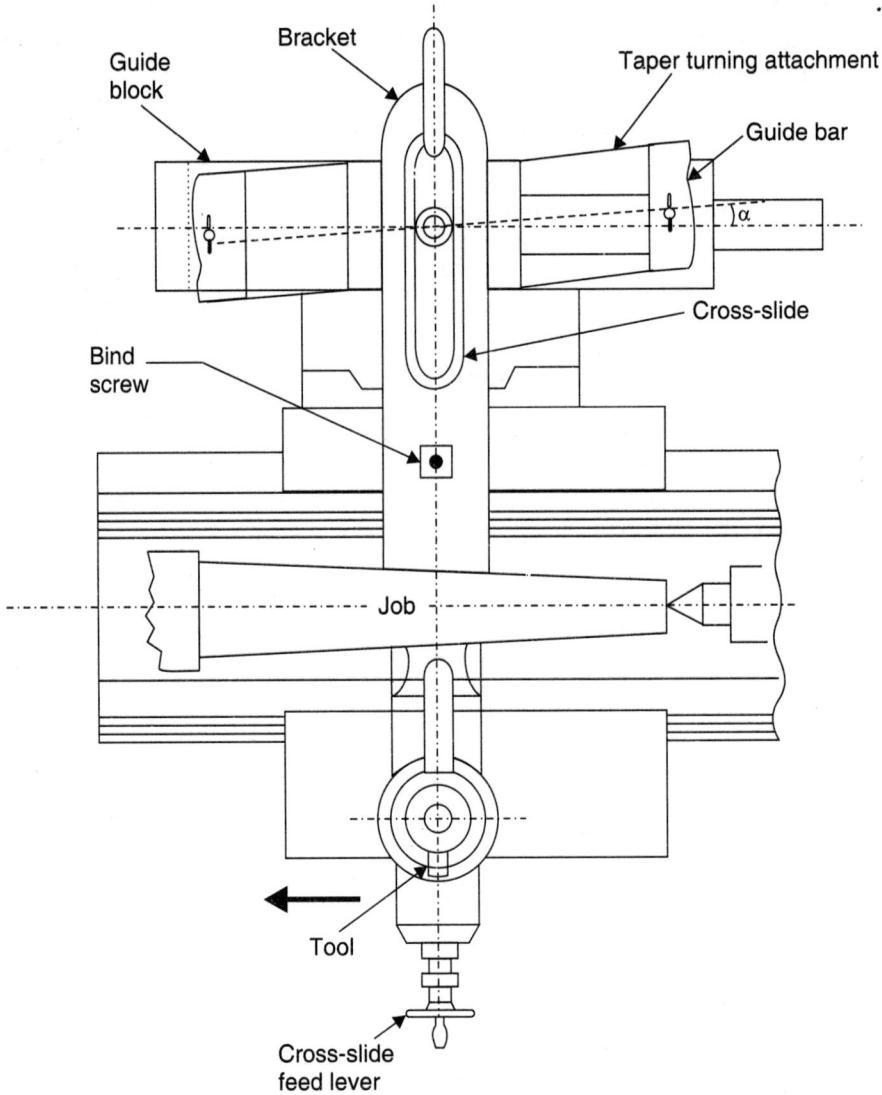

Fig. 9.15. Use of taper turning attachment.

The *taper turning attachment is used as follows :*

● The cross-slide is first made free of the lead screw by removing the *binder screw*. The
rear end of the cross-slide is then tightened with the guide block by means of a bolt.

● On the engagement of the longitudinal feed, the tool mounted on the cross-slide will
follow the angular path, as the guide block slides on the guide bar set at an angle to the
both axes.

● The required depth of cut is given by the compound slide which is placed at right angles
to the axis of the lathe.

- The guide bar must be set at half taper angle and the taper on the work must be converted in degrees. The maximum angle through which the guide bar may be swivelled is 10° to 12° on either side of the centre line.
- After every cut, the feed to the tool is given by moving the compound rest which is positioned parallel to the cross-slide (*i.e.*, at 90° to the axis of the job).
- The required angle (*i.e.*, angle of swivelling the guide bar) can be found out from the following relation :

$$\tan \alpha = \frac{D-d}{2L} \text{ (all dimensions in mm)}$$

or,
$$\alpha = \tan^{-1}\left(\frac{D-d}{2L}\right) \text{ degrees.}$$

where, D = Larger dia. in mm,
d = Smaller dia. in mm, and
L = Length of taper in mm.

Advantages of using a taper turning attachment :

1. Easy and quick setting.
2. The operator may not be highly skilled.
3. Accurate tapers can be easily obtained in a single setting.
4. Very steep taper on a long workpiece may be turned which is not possible with any other method
5. It is quite suitable for internal taper as well.
6. It provides a better finish.
7. It ensures an increased rate of production because it is possible to employ longitudinal power feeds easily.
8. During the operation, normal set-up and alignment of the lathe and main parts are not disturbed (as is the case with the other methods).

(*iv*) **By manipulating the transverse and longitudinal feeds of the slide tool simultaneously :**

- Taper turning by manipulation of both feeds is inaccurate and requires skill on the part of the operator.
- It is used for *sharp tapers only.*

(*v*) **By using a broad nose form tool :**

- In this method of taper turning (Fig. 9.16) a broad nose tool having straight cutting edge is set on to the work at half taper angle and is fed straight into the work to generate a tapered surface.
- With this method, tapers of short length only can be turned.

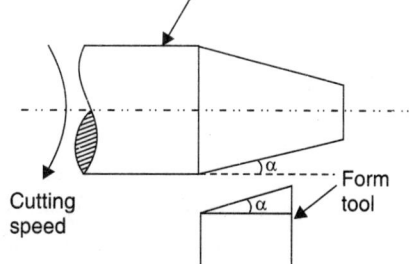

Fig. 9.16. Taper turning by a form tool.

Example 9.1. *Calculate the angle at which the compound rest would be swivelled for cutting a taper on a workpiece having a length of 180 mm and outside diameter 90 mm. The smallest diameter on the tapered end of the rod should be 60 mm and the required length of the tapered portion is 100 mm.*

Solution. *Given* : D = 90 mm ; d = 60 mm ; L = 100 mm

Let, α be the angle at which the compound rest would be swivelled.

Then, $\alpha = \tan^{-1}\left(\dfrac{D-d}{2L}\right)$ ...[Eqn. (9.1)]

$$= \tan^{-1}\left(\dfrac{90-60}{2\times 100}\right) = 8.53°. \text{ (Ans.)}$$

Note : The data on length of job (*i.e.* 180 mm) is a redundant data.

Example 9.2. *A taper pin of length 90 mm has a taper length of 50 mm. The larger diameter of taper is 95 mm and the smaller diameter is 85 mm. Determine : (i) Taper in mm/metre and in degrees.*

(ii) The angle to which the compound rest should be set up.

(iii) The tailstock setting over.

Solution. *Given :* $L_j = 90$ mm ; $L = 50$ mm ; $D = 95$ mm ; $d = 85$ mm.

(*i*) **Taper in mm/metre and in degrees :**

Taper, $T = \dfrac{D-d}{L} = \dfrac{95-85}{50} = \dfrac{10}{50}$

i.e. for a length of 50 mm, the taper is 10 mm.

$\therefore$ Taper in mm/metre $= \dfrac{10}{50} \times 1000 = $ **200 mm/metre length. (Ans.)**

$$\alpha = \tan^{-1}\left(\dfrac{D-d}{2L}\right) = \tan^{-1}\left(\dfrac{95-85}{2\times 50}\right) = 5.71°. \text{ (Ans.)}$$

(*ii*) **The angle to which the compound rest should be setup,** α :

$$\alpha = 5.71°. \text{ (Ans.)}$$

(*iii*) **Tailstock set over, S :**

$$S = L_j \times \dfrac{(D-d)}{2L}$$ ...[Eqn (9.2)]

$$= 90 \times \dfrac{(95-85)}{2\times 50} = 9 \text{ mm. (Ans.)}$$

5. Drilling : Refer to Fig. 9.17.

● It is an operation of producing a cylindrical hole in a workpiece by the rotating cutting edge of a cutter known as the drill.

● For this operation, the work is held in a suitable device, such as chuck or face plate, as usual, and the drill is held in the sleeve or barrel of the tailstock. The drill is fed by hand by rotating the handwheel of the tailstock.

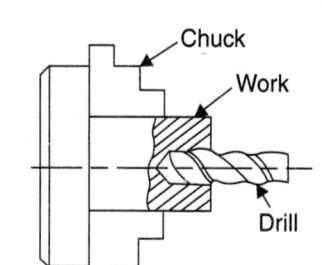

Fig. 9.17. Drilling.

6. Reaming : Refer to Fig. 9.18.

● *Reaming* is the operation which usually follows the earlier operation of drilling and boring in case of those holes in which a *very high grade of surface finish and dimensional accuracy is needed.*

● The tool used is called the *reamer*, which has multiple cutting edges. The reamer is held on the tailstock spindle, either direct or through a drill chuck and is held stationary while the work is revolved at very slow speed. The feed varies from 0.5 to 2 mm per revolution.

● For reaming tapered holes, taper reamers are used.

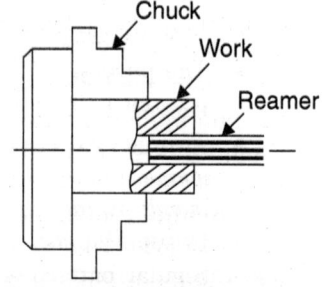

Fig. 9.18. Reaming.

7. Boring : Refer to Fig. 9.19.

● It is the operation of *enlarging and turning a hole produced by drilling, punching, casting or forging.*

● In this operation, as shown in Fig. 9.19, a *boring tool or a bit* mounted on a rigid bar is held in the tool post and fed into the work by hand or power in the similar way as for turning.

● *Boring cannot originate a hole.*

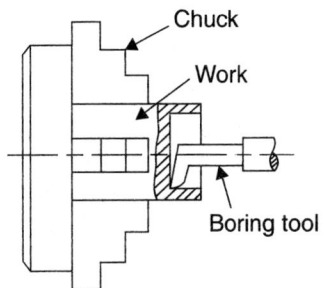

Fig. 9.19. Boring.

8. Undercutting/grooving : Refer to Fig. 9.20.

● It is the process of reducing the diameter of a workpiece over a very narrow surface. It is often done at the end of a thread or adjacent to a shoulder to leave a small margin.

● The work is revolved at half the speed of turning and a grooving tool of required shape is fed straight into the work by rotating the cross-slide screw.

9. Threading : Refer to Fig. 9.21.

● Threading is an operation of cutting helical grooves on the external cylindrical surface of the workpiece.

● In this operation, as shown in Fig. 9.21, the work is held in a chuck or between centres and the threading tool is fed longitudinally to the revolving work. The longitudinal feed is equal in the pitch of the thread to be cut.

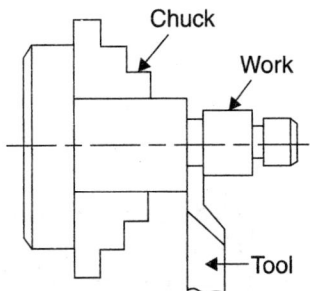

Fig. 9.20. Undercutting.

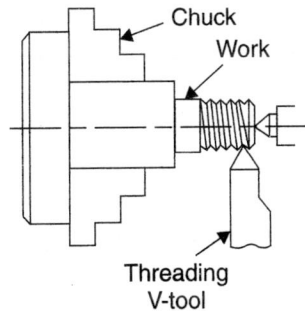

Fig. 9.21. Threading.

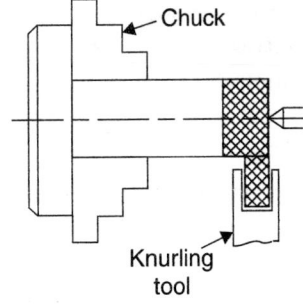

Fig. 9.22. Knurling.

10. Knurling : Refer to Fig. 9.22.

● It is an operation of *embossing a diamond shaped pattern on the surface of a workpiece.*

● The purpose of knurling is *to provide an effective gripping surface* on a workpiece to prevent it from slipping when operated by hand.

● The operation is performed by a special knurling tool which consists of 1 set of hardened steel rollers in a holder with the teeth cut on their surface in a definite pattern. *The tool is held rigidly on the tool post and the rollers are pressed* against the revolving workpiece to squeeze the metal against the multiple cutting edges, producing depressions in a regular pattern on the surface of the workpiece.

- Knurling is done at the *slowest speed* available in a lathe. Usually the speed is reduced to 1/4th of that of turning, and plenty of oil is flowed on the tool and workpiece.

11. Forming : Refer to Fig. 9.23.

- It is an operation of turning a convex, concave or any irregular shape.
- Form-turning may be accomplished by the following methods : (*i*) Using a forming tool, (*ii*) Combining cross and longitudinal feed, (*iii*) Tracing or copying a template.

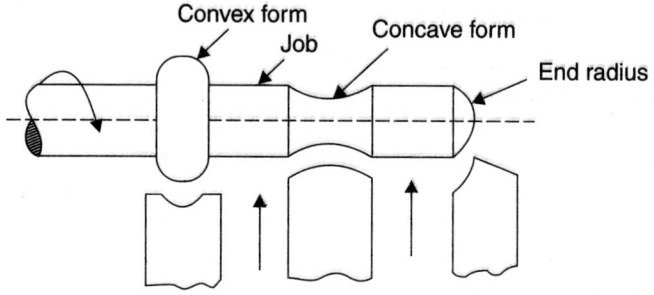

Fig. 9.23. Forming.

9.2.8. Lathe Accessories

*The devices employed for handling and supporting the work and the tool on the lathe are called its **accessories**.* The various accessories are enumerated below :

1. Chucks
2. Face plate
3. Angle plate
4. Driving plate
5. Lathe carriers or dogs
6. Lathe centres
7. Lathe mandrels
8. Rests
9. Jigs and fixtures.

The figures of some accessories are given below :

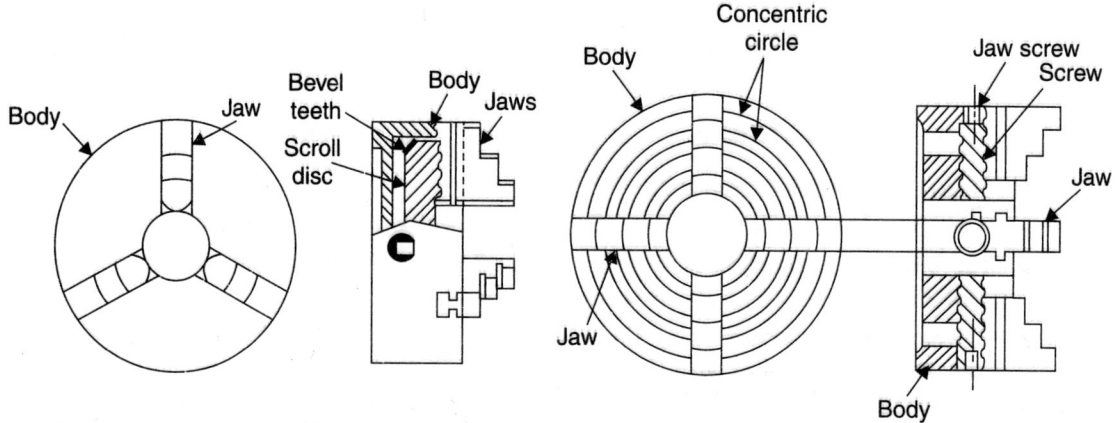

Fig. 9.24. Three jaw chuck.   Fig. 9.25. Four jaw chuck.

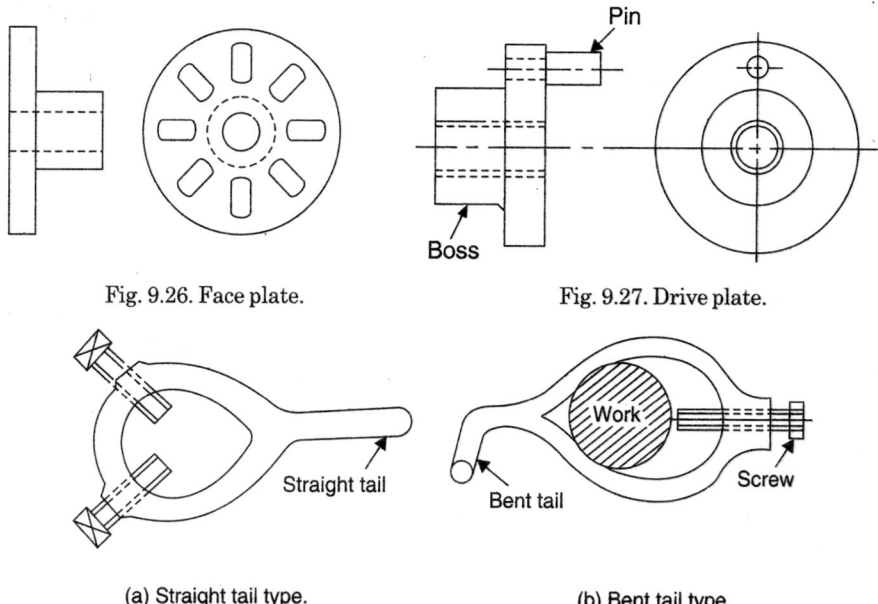

Fig. 9.26. Face plate.

Fig. 9.27. Drive plate.

(a) Straight tail type.

(b) Bent tail type.

Fig. 9.28. Lathe dog or carrier.

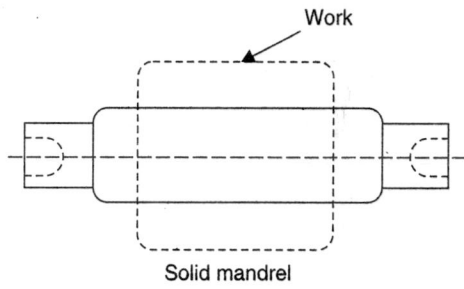

Fig. 9.29. Mandrel.

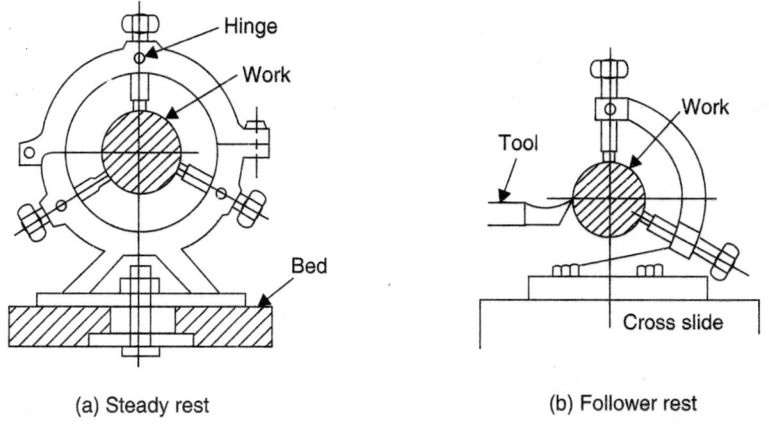

(a) Steady rest

(b) Follower rest

Fig. 9.30. Rests.

1. Chucks :

- The chucks provide a very efficient and true device of holding work on the lathe during the operation.
- Some of the commonly used *chucks* are :
- (*i*) Three jaw universal chucks (Fig. 9.20)
- (*ii*) Four jaw independent chuck (Fig. 9.21)
- (*iii*) Combination chuck
- (*iv*) Magnetic chuck
- (*v*) Air or hydraulic chucks
- (*vi*) Collet.

2. Face plate : Refer to Fig. 9.26.

- It is usually a circular cast iron disc, having a threaded hole at its centre so that it can be screwed to the threaded nose of the spindle.
- It consists of a number of holes and slots by means of which the work can be secured to it. A number of other things like bolts, nuts, washers, clamping plates and metallic packing pieces, etc., are required for holding the work properly on a face plate.

3. Angle plate :

- It is used for holding work in conjunction with a face plate.
- When the size or shape of the work is such that it is not possible to mount the work directly on the face plate the angle plate is secured to the face plate and the work mounted on it.

4. Driving plate : Refer to Fig. 9.27.

- It is a cast circular disc having a projected boss at its rear. The boss carries internal threads, so that it can be secrewed on to the spindle work. It also carries a hole to accommodate a *pin* which engages with the tail of a *lathe dog* or *carrier* when the job is held in the latter. When a bent tail dog is used, thin pin is taken out and bent portion of the tail inserted in the hole which serves the same purpose, or else the bent tail can be engaged in the slot made in the plate opposite to the pin hole :

5. Lathe carriers or dogs : Refer to Fig. 9.28.

- These are used in conjunction with the driving plate.
- The two common forms are straight tail and bent tail.
- The work to be held is inserted in the *'V' shaped hole* of the carrier and then finally secured in position by means of set screw.

6. Lathe centres : Refer to Fig. 9.2.

- They act as solid bearings to support the work during the operation.
- Cast steel or high grade tool steel is the common material used for their manufacture. They are then hardened and ground to correct angle. Sometimes, when very high speeds are to be employed, *tips* made of some other materials like cemented carbide or high speed steel are used which are fitted into usual types of shanks.
- The centre which is used in the headstock spindle revolves with it and it is known as *live centre,* whereas one fitted in tailstock remains stationary and is called *dead centre.*

7. Lathe mandrels :

- A *"mandrel"* can be described as a solid steel shaft or spindle which is used for holding *bored parts* for machining their outside surfaces on lathe. They are also known as *arbors.*

- Mandrels are usually employed for those jobs (relatively small) which have a finished hole which is concentric with the outer surface that is to be machined.
- The common types of mandrels are : Solid or plain (Fig. 9.29) : collar, stepped, expanding and double cone mandrels.

8. Rests :

- When a very long job is to be turned between centres on a lathe, due to its own weight it provides a springing action and carries a lot of bending moment. The result is that the turning tool is spoiled very soon and may even break sometimes. To avoid this, such jobs are always supported on an attachment known as *'steady rest* or *centre rest'* (Fig. 9.30a)
- Sometimes, when the job is too flexible, it becomes necessary to support the job very close to the cutting edge of the tool throughout the operation. In such cases a *'follower rest'* (Fig. 9.30 b), is used instead of the steady rest. It is attached to the saddle of the lathe carriage and thus travels along with the tool throughout the operation.

9. Jigs and fixtures :

- Jigs and fixtures are used in conjunction with the face plate on a lathe for supporting and holding odd shaped and eccentric jobs during the operation.
- This specific use is in the mass production of identical parts otherwise, if only a single item is to be made, the cost of production of the jigs or fixtures itself will be too high, preventing their use.

9.2.9. Lathe Attachments

A number of *attachments* are used on centre lathe *to increase production and efficiency and widen its scope of use for such works also which are normally not carried out on this machine.*

The following attachments are commonly used :

1. Stops
2. Grinding attachment
3. Milling attachment
4. Taper turning attachment
5. Copying attachment
6. Relieving attachment.

1. Stops :

- The stops are used in conjunction with both, the carriage as well as the cross-slide. *Their use enables a quick and accurate positioning of the carriage and the cross-slide.*
- Especially, when similar operation is to be done repeatedly, the use of stops effects a considerable saving in time and given more accuracy.

2. Grinding attachment :

- It is also known as 'Tool post grinder'.
- A typical form of this attachment (*for external grinding*) consists of a *bracket*, which is mounted on the cross-slide, a *grinding wheel* and a separate *motor*. Thus, the grinding wheel is driven separately by this motor.
- The job is held as usual in a chuck or between centres and the *rotating grinding wheel is fed against the job.*
- Some tool post grinders carry provisions such that the same attachment with a little change can be employed for internal as well external grinding.

3. Milling attachment :

- The milling attachment consists of a *vertical pillar,* to which is attached an individual motor and a device to mount the cutter on it. Base of the pillar is *fastened rigidly to the saddle in front of the lathe* and the unit carrying the motor and the cutter can be *moved vertically up and down* by means of a screw and the handwheel provided at the top of the pillar.

- The *job is held stationery between the centres* and the rotating cutter is moved longitudinally along it by operating the saddle. Depth of cut can be adjusted by moving the unit along the pillar by means of the handwheel.

- The *dividing head* is fitted in front of the pillar for indexing the job.

4. Tape turning attachment :
It is attached to the rear side of the lathe and is operated through the movements of the saddle and the cross-slide (Refer to Art. 9.2.7—taper turning).

5. Copying attachment :

- The common centre lathe, in the absence of a *copying lathe,* can be easily employed for repetition work by using copying attachment or the tracer attachment.

- The copying attachment consists of an *auxiliary slide* which is fitted on to the regular cross-slide of the lathe. At the rear of this slide is provided with a cylinder and piston which is operated by hydraulic pressure. A bent arm known as *overarm,* is attached to it and the free end of this arm carries the *tracer.* One end of the tracer moves along the profile of the *template* fitted at the back side of the lathe, of which the shape is to be copied, and the other operates an air nozzle which, in turn, through the air pressure regulates the oil supply and hence the hydraulic pressure. This controls the movement of the piston of the auxiliary slide and, therefore, that of the tool. The movement of the tool so controlled enables the *duplication or copying* of the profile of the template quite accurately and quickly.

6. Relieving attachment :

- Relieving is usually done on most of the multiple point tools such as milling cutters, taps and reamers etc. This operation can be easily performed on a centre lathe with the help of the relieving attachment.

- This attachment consists of an auxiliary slide mounted on the cross-slide in place of the compound rest. The tool post, carrying the tool, is mounted on the auxiliary slide. These slides are connected through gears to the lathe spindle. A cam rotates under the relieving slide which guides the tool in and out repeatedly as the work revolves with the spindle.

9.2.10. Cutting Speed, Feed and Depth of Cut
Refer to Fig. 9.31.

1. Cutting speed :

*The **cutting speed** (in a lathe for turning operation) is the peripheral speed of the workpiece past the cutting tool.*

Mathematically, $V = \dfrac{\pi DN}{1000}$ m/min   ...(9.5)

where, V = Cutting/peripheral speed, m/min,

D = Diameter of the job, mm, and

N = Job or spindle speed, r.p.m.

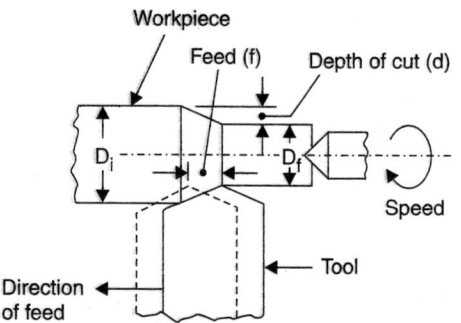

Fig. 9.31. Concept of cutting speed, feed and depth of cut.

- The *main factors* which influence the selection of a proper cutting speed are :
  (*i*) Material cf the cutting tool.
  (*ii*) Hardness and machinability of the metal to be machined.
  (*iii*) Quality of heat treatment, if it is a H.S.S. steel tool.
  (*iv*) Whether machining is to be done with or without the use of a coolant.
  (*v*) Rigidity of the tool and the work.
  (*vi*) Tool shape.
  (*vii*) Depth of cut.
  (*viii*) Feed to be given to the tool.
  (*ix*) Rigidity of the machine.

2. Feed (f) :

Feed may be defined as the distance that a tool advances into the work during one revolution of the headstock spindle.

- Feed is expressed in *mm/revolution.*
- The smaller the feed, the better the finish although a great deal depends on the type of lathe tool used, and a well sharpened tool is necessary.
- Larger feeds reduce machining time, but the tool life is reduced.
- Feed f may calculated as follows :

$$f = \frac{L}{N \times T_m} \qquad \qquad \ldots(9.6)$$

where, L = Length of cut, mm,
          N = r.p.m, and
          T_m = Machining/cutting time, min.

3. Depth of cut (d) :

The depth of cut 'd' is the perpendicular distance measured from the machined surface to the uncut (or previous cut) *surface of the workpiece.* For turning operations, the depth of cut is expressed as :

$$d = \frac{D_i - D_f}{2} \ \text{mm} \qquad \qquad \ldots(9.7)$$

where, D_i = Initial/original diameter of the workpiece, mm, and
          D_f = Final diameter of the workpiece, mm.

- For rough cutting, the depth of cut should be as large as possible, consistent with the size or capacity of the centre lathe and the material being turned.

The values of *speed, feed* and *depth of cut,* in general, depend upon the following factors :
— Type of workpiece material ;
— Type of tool material ;
— Type of surface finish required.

9.2.11. Material Removal Rate (MRR)

The **material removal rate** is *the volume of material removed per unit time.*

Volume of material removed is a function of speed, feed and depth of cut. Higher the values of these, more will be the material removal rate.

Let, D_i = Initial diameter of the workpiece, mm,
        d = Depth of cut, mm, and
        f = Feed, mm/revolution.

Then, material removed per revolution is the volume of chip whose length is πD_i and whose cross-sectional area is $d \times f$. That is,

Volume of material removed in one revolution = $\pi D_i \times d \times f$ mm^3 ...(9.8)

Since the job is making N r.p.m., the MRR in mm^3/min is given by

$$\text{MRR} = \pi D_i \times d \times f \times N \text{ mm}^3/\text{min} \qquad \qquad ...(9.9)$$

In terms of cutting speed V in m/min (see eqn. 9.5) is given by :

$$\text{MRR} = 1000 \times V \times d \times f \text{ mm}^3/\text{min} \qquad \qquad ...(9.10)$$

9.2.12. Machining Time

In a lathe, the machining time can be calculated as follows.

Refer to Fig. 9.32.

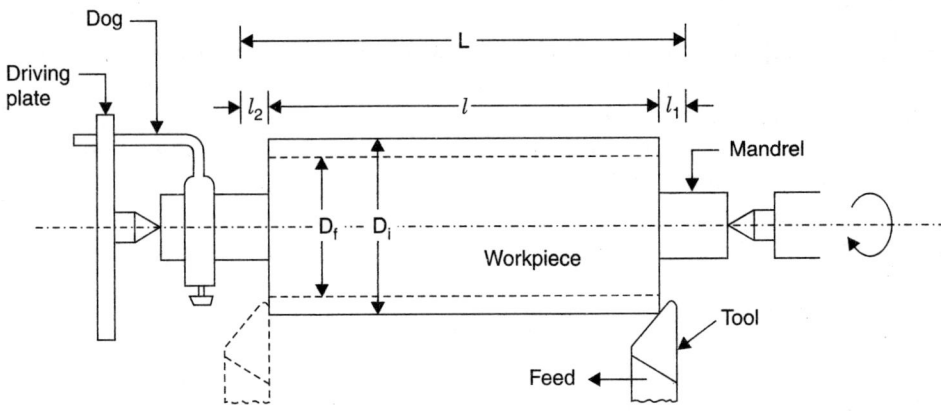

Fig. 9.32. Turning operation.

Let, l = Length of surface to be machined, mm,

l_1 = Distance required for feeding the tool cross-wise, to increase the depth of cut, mm (tool approach),

l_2 = Overtravel of the tool at the end of each cut, mm,

d = Depth of cut, mm,

f = Feed, mm/revolution,

N = Speed of work, r.p.m.,

n_p = Number of passes or cuts taken for obtaining the required diameter,

D_i = Initial diameter of the workpiece, and

D_f = Final diameter of the workpiece.

Now, the distance travelled by the tool in the direction of the feed in a *single cut*,

$$L = l + l_1 + l_2$$

(Approach of most of the single point cutting tools is negligible)

Then, machining time is given by :

$$T_{mp} = \frac{L}{fN}, \text{ per pass (cut)} \qquad \qquad ...(9.11)$$

Also, $V = \dfrac{\pi D N}{1000}$ m/min

or, $N = \dfrac{1000V}{\pi D}$

$$\therefore \quad T_{mp} \text{ (per pass or cut)} = \frac{L}{f \times \left(\dfrac{1000\,V}{\pi D}\right)} = \frac{\pi D L}{1000\,V_f} \quad \text{min.} \qquad\qquad \text{...(9.12)}$$

...Substituting for N

Total machining time, $T_{mt} = T_{mp} \times n$ ...(9.13)

$$\therefore \quad \text{Number of passes or cuts, } n_p = \frac{\text{Total machining allowance}}{\text{Material removed per cut}}$$

Now, total machining allowance $= \dfrac{D_i - D_f}{2}$

Material removed per cut = Depth of cut.

"Depth of cut" is *half the difference between the work diameter at the start of the cut and the diameter of machined surface obtained after the cut.*

— In practice, the depth of cut *per pass is not constant.* For roughing operations, it is much greater than for the finishing operations.

● **Total manufacturing time** = *Machining time + Set-up time + Moving and waiting time + Inspection time.*

9.2.13. Power Required in Turning

The power required at the spindle for *turning* depends upon the following factors :

— Cutting speed (V) ;
— Depth of cut (d) ;
— Feed rate (f) ;
— Hardness and machinability of workpiece material.

The power required depends upon the cutting force which is a power function of f and d. However, for the sake of gross estimation it can be safely assumed that

Cutting force, $\qquad\qquad F = K \times d \times f$ ...(9.14)

where K is a constant, which depends on the work material.

Then power, $\qquad\qquad P = F \times V$ ...(9.15)

or $\qquad\qquad\qquad\quad P = K \times d \times f \times V$ ...(9.16)

Example 9.3. *Calculate the time taken to face a workpiece of 80 mm diameter. The spindle speed is 90 r.p.m. and cross feed is 0.3 mm/revolution.*

Solution. *Given* : $D = 80$ mm ; $N = 90$ r.p.m. ; $f = 0.3$ mm/rev.

Time to face the workpiece, T_{face} :

$$T_{\text{face}} = \frac{L}{f \times N} = \frac{D/2}{f \times N} \qquad\qquad (\because \ L = D/2)$$

$$= \frac{(80/2)}{0.3 \times 90} = \textbf{1.48 min.} \quad \textbf{(Ans.)}$$

Example 9.4. *Calculate the time taken to turn a brass component 60 mm diameter by 84 mm long if the cutting speed is 50 m/min and the feed is 0.4 mm/revolution. Only one cut is to be considered. Neglect tool approach and tool travel.*

Solution. *Given* : $D = 60$ mm ; $L = 84$ mm ; $V = 50$ m/min ; $f = 0.4$ mm/rev.

$\mathbf{T_m}$:

$$T_m = \frac{\pi D L}{1000\,V_f} \quad \text{min.} \qquad\qquad \text{...[Eqn.(9.12)]}$$

$$= \frac{\pi \times 60 \times 84}{1000 \times 50 \times 0.4} \quad \text{min} = 0.792 \text{ min or } \textbf{47.5 s} \quad \textbf{(Ans.)}$$

Example 9.5. *A 160 mm long 15 mm diameter rod is reduced to 14 mm diameter in a single pass straight turning. If the spindle speed is 450 r.p.m. and feed rate is 225 mm/min, determine :*

(*i*) *Material removal rate.*

(*ii*) *Cutting time.*

Solution. *Given :* $L = 160$ mm ; $D_i = 15$ mm ; $D_f = 14$ mm ; $N = 450$ r.p.m. ; $f = 225$ mm/min.

(*i*) **Material removal rate (MRR) :**

We know that, MRR $= \pi D_i \times d \times f \times N$ mm^3/min   ...(Eqn. 9.9.)

where, d = depth of cut $= \dfrac{D_i - D_f}{2} = \dfrac{15 - 14}{2} = 0.5$ mm/rev.

          f = feed rate $= \dfrac{225}{450} = 0.5$ mm/rev.

$\therefore$ MRR $= \pi \times 15 \times 0.5 \times 0.5 \times 450$ mm^3/min = **5301.4 mm^3/min.   (Ans.)**

(*ii*) **Cutting time (T$_m$) :**

$$T_m = \frac{L}{fN} \text{ min.}$$

$$= \frac{160}{0.5 \times 450} \text{ min} = \textbf{0.711 min.   (Ans.)}$$

Example 9.6. *A workpiece of 300 mm diameter and 600 mm length is to be turned down to 282 mm for the entire length. The suggested feed is 1.2 mm/revolution and the cutting speed is 162 m/min. The maximum allowable depth of cut is 4.5 mm.*

Calculate the following :

(*i*) *Spindle r.p.m.*

(*ii*) *Feed speed.*

(*iii*) *Material removal rate.*

(*iv*) *Cutting time.*

Assume tool overtravel is 12.0 mm. Neglect tool approach.

Solution. *Given :* $D_i = 300$ mm ; $D_f = 282$ mm ; $f = 1.2$ mm/rev. ; $l = 600$ mm ; $l_2 = 12.0$ mm ; $V = 162$ m/min ; $d = 4.5$ mm.

(*i*) **Spindle r.p.m :**

$$V = \frac{\pi D N}{1000} \text{ m/min, where } D \text{ is in mm}$$

$\therefore$ Speed, $N = \dfrac{1000\,V}{\pi D} = \dfrac{1000 \times 162}{\pi \times 300} = \textbf{171.9 r.p.m.   (Ans.)}$

(*ii*) **Feed speed :**

$\therefore$ Feed speed $= f \times N = 1.2 \times 171.9 = \textbf{206.3 mm/min.   (Ans.)}$

(*iii*) **Material removal rate (MRR) :**

          MRR $= 1000 \times V \times d \times f$ mm^3/min

          $= 1000 \times 162 \times 4.5 \times 1.2 = \textbf{87.48} \times 10^4 \textbf{ mm}^3\textbf{/min.   (Ans.)}$

(*iv*) **Cutting time (T_{mt}) :**

Total cutting time, $T_{mt} = T_{mp} \times n_p = \dfrac{L}{fN} \times n_p$

where, T_{mp} = Cutting time per pass, and

          n_p = Number of passes.

Hence, $n_p = \dfrac{300 - 282}{4.5} = 4$

and $\qquad L = l + l_2 = 600 + 12 = 612$ mm

$\therefore \qquad T_{mt} = \dfrac{612}{1.2 \times 171.9} \times 4$ min. = **11.87 min.** **(Ans.)**

Example 9.7. *A hollow workpiece of 72 mm outside diameter and 180 mm length is held on a mandrel between centres and turned all over in 4 passes. If the approach length = 18 mm, over travel = 10 mm, average feed = 0.9 mm/rev. and cutting speed = 28 m/min, calculate the machining time.*

Solution. *Given* : $D = 72$ mm ; $l = 180$ mm ; $n_p = 4$; $l_1 = 18$ mm ; $l_2 = 10$ mm ; $f = 0.9$ mm/rev. ; $V = 28$ m/min.

Machining time (T_{mt}) :

Total machining time, $T_{mt} = T_{mp} \times n_p$

$$= \dfrac{L}{fN} \times n_p$$

where, $\qquad T_{mp}$ = Machining time *per pass,* min.

$\qquad n_p$ = No. of passes

$\qquad L$ = Total distance travelled by the tool in a single pass, mm,

$\qquad f$ = feed, mm/rev., and

$\qquad N$ = Speed, r.p.m.

Also, cutting speed, $V = 28 = \dfrac{\pi D N}{1000}$, where D is in mm.

or, $\qquad N = \dfrac{1000 \times 28}{\pi \times 72} = 123.8$ r.p.m.

and, $\qquad L = l + l_1 + l_2 = 180 + 18 + 10 = 208$ mm

$\therefore \qquad T_{mp} = \dfrac{208}{0.9 \times 123.8} \times 4 =$ **7.5 min.** **(Ans.)**

Example 9.8. *Calculate the time required to machine a workpiece 180 mm long 70 mm diameter to 175 mm long, 60 mm diameter. The workpiece rotates at 450 r.p.m., feed is 0.3 mm/rev and maximum depth of cut is 2 mm.*

Assume total approach and overall travel distance as 6 mm for turning operation.

Solution. *Given* : $l = 180$ mm ; $D_i = 70$ mm ; $D_f = 60$ mm ; $N = 450$ r.p.m. ; $f = 0.3$ mm/rev. ; $d = 2$ mm

Total time for machining :

The workpiece diameter is to be reduced from 70 mm to 60 mm and its length is to be reduced from 180 mm to 175 mm. It involves *both turning and facing operations.* Let us assume that first turning, and then facing is the sequence of operations.

Then, total time for machining = Time for turning + time for facing

Time for turning :

Total length of tool travel, $L = l + (l_1 + l_2)$

$$= 180 + 6 = 186 \text{ mm}$$

Required depth to be cut, $\quad d = \dfrac{D_i - D_f}{2} = \dfrac{70 - 60}{2} = 5$ mm

Since maximum depth of cut is 2 mm, 5 mm cannot be cut in one pass. Therefore, we calculate number of cuts or passes required.

Number of cuts/passes required, $n_p = \dfrac{5}{2} = 2.5$ or 3 (since cuts cannot be a fraction)

$\therefore$ Total turning time = Machining time for one cut × number of cuts

$$= \left(\frac{L}{f \times N}\right) n_p$$

$$= \frac{186}{0.3 \times 450} \times 3 = 4.13 \text{ min.}$$

Time for facing :

Now, the diameter of the job is reduced to 60 mm. In case of *facing operations length of tool travel is equal to half the diameter of the job.* That is $L = \dfrac{60}{2} = 30$ mm.

Time for facing (one pass) = $\dfrac{L}{f \times N}$ ...(Eqn. 9.6)

$$= \frac{30}{0.3 \times 450} = 0.22 \text{ min}$$

Number of passes required = $\dfrac{\text{Material to be removed}}{\text{Maximum depth of cut}} = \dfrac{5}{2} \approx 3$

$\therefore$ Total time for facing = 3 × 0.22 = 0.66 min

Hence, *total time for machining* = 4.13 + 0.66 = **4.79 min.**   **(Ans.)**

Note. The reader should find out the total machining time if facing is done first.

Example 9.9. *A component having length 120 mm and diameter 10 mm from a raw material of 120 mm length and 12 mm diameter, using a cutting speed of 32 m/min and a feed rate of 0.8 mm/rev. How many times we have to resharpen or regrind if 800 workpieces are to be produced ?*

In the Taylor's expression, use constants as n = 1.25 and C = 175.

Solution. *Given :* $L = 120$ mm, $D_i = 12$ mm ; $D_f = 10$ mm ; $V = 32$ m/min ; $f = 0.8$ mm rev ; $n = 1.25$; $C = 175$.

No. of resharpenings required :

From Taylor's tool life equation, we have

$$VT^n = C$$

$$32 \times (T)^{1.25} = 175$$

or, $T = \left(\dfrac{175}{32}\right)^{1/1.25} = 3.89$ min

Also, $V = \dfrac{\pi D N}{1000}$

or, $32 = \dfrac{\pi \times 12 \times N}{1000}$

or, $N \simeq 849$ r.p.m.

Now, Machining time/piece = $\dfrac{L}{f \times N} = \dfrac{120}{0.8 \times 849} = 0.177$ min.

$\therefore$ Machining time for 800 workpieces = 800 × 0.177 = 141.6 min.

Hence, *number of resharpenings required* = $\dfrac{141.6}{3.89} = 36.4$ or   **36 resharpenings**   **(Ans.)**

Example 9.10. *For turning a carbon steel cylinder bar of length 3.6 m and diameter 0.24 m at a feed rate of 0.7 mm/rev with a HSS tool, one of the two available cutting speeds is* *to be selected. These two cutting speeds are 120 m/min and 68 m/min. The tool life correspond-ing to the speed of 120 m/min is known to be 15 minutes with n = 0.6. The cost of machining time, set-up time unproductive time together is Rs. 1.50/sec. The cost of one tool resharpening is Rs. 25.*

Which of the above two cutting speeds should be selected from the point of view of the total cost of producing this part ? Prove your argument :

Solution. *Given :* $L = 3.6$ m $= 3600$ mm ; $D = 0.24$ m $= 240$ mm ; $f = 0.7$ mm/rev ; $V_1 = 120$ m/min ; $V_2 = 68$ m/min ; Tool life $= 15$ min ; $n = 0.6$.

When speed is 120 m/min :

$$N_1 = \frac{1000 \times V_1}{\pi D} = \frac{1000 \times 120}{\pi \times 240} = 159.1 \text{ r.p.m.}$$

Machining time, $T_{m_1} = \dfrac{L}{f \times N} = \dfrac{3600}{0.7 \times 159.1} = 32.32$ min.

Tool life corresponding to speed of 120 m/min is 15 min.

Number of resharpenings required $= \dfrac{32.32}{15} = 2.1$ or 2 resharpenings

∴   Total cost = Machining cost + cost of resharpening × Number of resharpenings

$$= 32.32 \times 60 \times 1.50 + 25 \times 2 = \text{Rs. } 2959$$

When speed is 68 m/min :

$$N_2 = \frac{1000 \times V_2}{\pi D} = \frac{1000 \times 68}{\pi \times 240} = 90.2 \text{ r.p.m.}$$

Using Taylor's tool life equation : $TV^n = C$, we have

$$V_1 T_1^n = V_2 T_2^n \quad \text{or} \quad \frac{T_2}{T_1} = \left(\frac{V_1}{V_2}\right)^{1/n} = \left(\frac{120}{68}\right)^{1/0.6} = 2.577$$

or,   Tool life, $T_2 = 15 \times 2.577 = 38.66$ min

Again,   Machining time, $T_{m2} = \dfrac{L}{f \times N_2} = \dfrac{3600}{0.7 \times 90.2} = 57$ min

Number of resharpenings required $= \dfrac{57}{38.66} = 1.4$ or 1 resharpening

∴   Total cost $= 57 \times 60 \times 1.5 + 25 \times 1 = \text{Rs. } 5155$

Thus the cost is *less when speed is* **120 m/min.** Hence *it should be selected.* **(Ans.)**

Example 9.11. *Tool life testing on a lathe under dry conditions gave n and C of Taylor's tool life equation as 0.12 and 130 m/min respectively. When a coolant was used, C increased by 10%. Find the percentage increase in tool life with the use of coolant at a cutting speed of 90 m/min.*   (GATE)

Solution. *Given :* $n = 0.12$, $C_1 = 130$ m/min ; $C_2 = 1.1 \times 130 = 143$ m/min ; $V = 90$ m/min.

Percentage increase in tool life :

Taylor's tool life equation is given as :

$$VT^n = C$$

Now, $$VT_1^n = C_1 \quad \text{or} \quad 90 \times (T_1)^{0.12} = 130$$

or, $$T_1 = \left(\frac{130}{90}\right)^{1/0.12} = 21.42 \text{ min}$$

and, $$VT_2^n = C_2 \quad \text{or} \quad 90 \times (T_2)^{0.12} = 143$$

or, $$T_2 = \left(\frac{143}{90}\right)^{1/0.12} = 47.4 \text{ min}$$

$\therefore$ Percentage increase in tool life

$$= \frac{47.4 - 21.42}{21.42} \times 100 = \textbf{121.29\%}. \quad \textbf{(Ans.)}$$

Example 9.12. *The following data relate to a metal cutting test conducted on a lathe :*

Vertical force on the tool = 2400 N

Depth of cut = 2.5 mm

Feed = 5 cuts/mm

Cutting speed = 35 m/min

Overall efficiency of the machine = 85 percent.

Calculate :

(i) Pressure (in MN/m²) of the chip cross-sectional area on the tool.

(ii) Power required for cutting the material.

Solution. Vertical force, $F_V = 2400$ N ; $d = 2.5$ mm ; $f = 5$ cuts/mm ; $V = 35$ m/min ; $\eta_0 = 85\%$.

(i) **Pressure of the chip cross-sectional area on the tool (MN/m²), p :**

Cross-sectional area of the chip, (A_c) = Depth of cut (d) × feed per revolution (f)

or $A_c = 2.5 \times \left(\dfrac{1}{\text{No. of cuts/mm}} \right) = 2.5 \times \dfrac{1}{5} = 0.5$ mm² $= 0.5 \times 10^{-6}$ m²

$\therefore$ $p = \dfrac{F_V}{A_c} = \dfrac{2400}{0.5 \times 10^{-6}} = 4800 \times 10^6$ N/m² or **4800 MN/m².** **(Ans.)**

(ii) **Power required for cutting the material, P_{motor} :**

Work done (joules) = Force (newton) × distance (metre)

and,   Power (watts) = Work done (joules) per second

Now, Power required at tool point, $P = \dfrac{\text{Force (newton)} \times \text{cutting speed (m/min)}}{60} = \dfrac{F_V \times V}{60}$

$$= \frac{2400 \times 35}{60} = 1400 \text{ watts}$$

Power required at motor, $P_{motor} = \dfrac{P}{\eta_0} = \dfrac{1400}{0.85} = \textbf{1647 W}$ or **1.647 kW.** **(Ans.)**

9.2.14. Thread cutting

Cutting screws is another important task carried out in lathes. Fig. 9.33 shows internal and external threads.

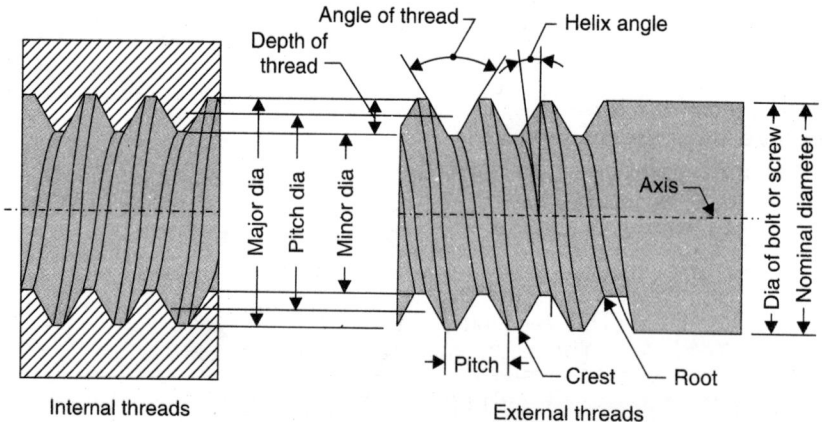

Fig. 9.33. Internal and external threads.

Thread cutting *is the operation of producing a helical groove of specific shape ; say V or square on a cylindrical surface.*

Threads can be cut on a cylindrical surface by forming or machining process. Thread production by forming operation is explained in chapter 10. *Thread cutting by the machining process is done using a lathe. Threads of any pitch, shape and size can be cut on a lathe.*

Fig. 9.34 shows the set-up of a lathe for thread cutting.

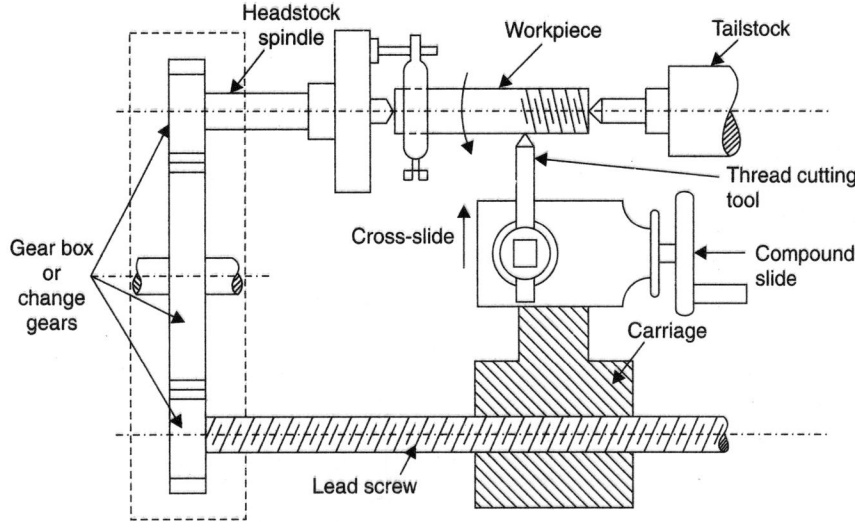

Fig. 9.34. Lathe set-up for thread cutting operation.

— *Thread cutting operation* is done on a lathe using a single-point tool called *thread cutting* tool.

— The workpiece is held between centres or in a chuck and the tool is held in tool post. For producing threads of pitch p mm, the tool must travel a distance equal to p mm as the workpiece makes one complete rotation.

— The definite relative rotary and linear motion between the workpiece and tool is achieved by locking or engaging the carriage with the lead screw through a screw and nut mechanism and fixing a gear ratio between the headstock spindle and lead screw. This is done by using change gear mechanism or gear box between the spindle and lead screw.

— To cut the threads, the tool is brought to the start of the workpiece and a small depth of cut is given to the tool using cross-slide. The carriage is engaged with the lead screw, the cut is made on the entire surface and at the end of the workpiece, carriage is disengaged. The tool is pulled out of the job and brought back to the starting position. The process is repeated until the full depth threads are obtained.

● To following relationship is used to determine the gears/wheels required to generate threads of definite pitch :

$$\text{Gearing ratio} = \frac{\text{Pitch of screw to be cut}}{\text{Pitch of lead screw}}$$

$$= \frac{\text{Lead of screw to be cut}}{\text{Lead of lead screw threads}}$$

(This result holds good for *multiple start threads also*)

$$= \frac{\text{No. of teeth of driver (stud gear)}}{\text{No. of teeth of driven (lead screw gear)}}$$

or,   simply = $\dfrac{\textbf{Driver}}{\textbf{Driven}}$ ...(9.17)

All the lathes are provided with a set of change gears, usually having 20 to 120 teeth, with a variation of 5 teeth. In addition, a gear of 127 teeth is also provided which is known as *translating gear. It is used in cutting metric threads.*

After the determination of above gearing ratio, the next step to follow is to multiply the numerator and denominator of the fraction by the same number in order to find out the number of teeth of the change gears. The following two types of gear trains are employed in cutting threads :

      1. Simple gear train   2. Compound gear train.

1. Simple gear train : Refer to Fig. 9.35.

● A simple gear train consists of a driving gear (mounted on the stud), a driven gear (mounted on the lead screw) and one or two intermediate gears.

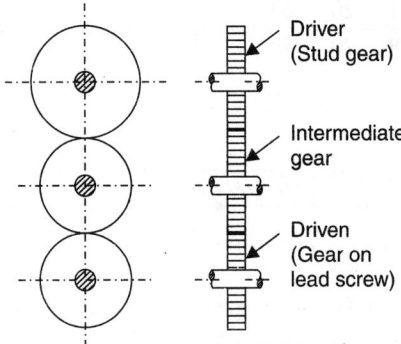

Fig. 9.35. Simple gear train.

● The intermediate gears are known as *idle gears,* have no effect on the speed ratio but are used only *to fill up the gap between the driver and driven gears,* and *to obtain a desired direction of rotation of the lead screw.*

2. Compound gear train : Refer to Fig. 9.36.

● A compound gear train consists of two studs instead of one. The second stud is suitably mounted on the bracket or quadrant carrying the change gears. Evidently there are two drivers and two driven gears.

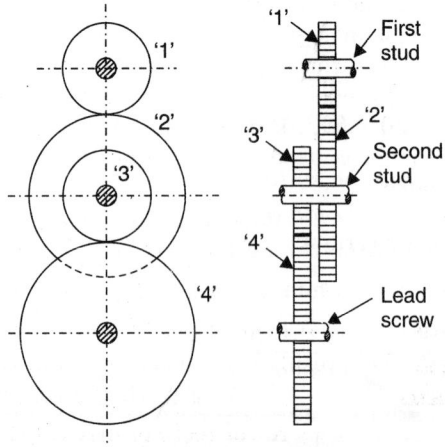

Fig. 9.36. Compound gear train.

- The *first driver* '1' is mounted on the first stud, which meshes with the first driver '2' on the second stud. The *second driver* '3' is also mounted on the second stud and it meshes with the second driven '4' mounted on the lead screw.

Such a gear train is employed when the desired gear ratio is such that it is *not* possible to arrange a simple gear train out of the given set of change gears.

Note. Try to use as small number of idle gears as possible.

Example 9.13. *Calculate the change gears to cut R.H. threads of 2 mm pitch on a lathe having lead screw of 6 mm pitch.*

Solution. *Given* : Pitch to be cut = 2 mm ; Pitch of lead screw = 6 mm

$$\frac{\text{Driver}}{\text{Driven}} = \frac{\text{Pitch to be cut}}{\text{Pitch of lead screw}} = \frac{2}{6} = \frac{1}{3}$$

$$= \frac{1}{3} \times \frac{20}{20} = \frac{20}{60} \text{ (simple train with one idler).} \quad \textbf{(Ans.)}$$

$$= \frac{1}{3} \times \frac{25}{25} = \frac{\textbf{25}}{\textbf{75}} \text{ (simple train with one idler).} \quad \textbf{(Ans.)}$$

Example 9.14. *Calculate the change wheels to cut threads of 1 mm pitch on a lathe having a lead screw of 8 mm pitch.*

Solution. Pitch to be cut = 1 mm ; Pitch of lead screw = 8 mm

$$\frac{\text{Driver}}{\text{Driven}} = \frac{\text{Pitch to be cut}}{\text{Pitch of lead screw}} = \frac{1}{8}$$

$$= \frac{1}{8} \times \frac{20}{20} = \frac{20}{160}$$

Since it is not possible is get a gear of 160 teeth, such a ratio will provide difficulty in adopting simple train. So a *compound train* will have to be used.

Now, $$\frac{\text{Driver}}{\text{Driven}} = \frac{1}{8} = \frac{1 \times 1}{4 \times 2}$$

$$= \frac{1 \times 20}{4 \times 20} \times \frac{1 \times 30}{2 \times 30}$$

$$= \frac{\textbf{20}}{\textbf{80}} \times \frac{\textbf{30}}{\textbf{60}} = \frac{\text{'1'} \times \text{'3'}}{\text{'2'} \times \text{'4'}} \qquad \text{...(see Fig. 9.36)} \quad \textbf{(Ans.)}$$

Example 9.15. *Calculate the suitable gear trains for cutting 10 mm pitch, 3 start on a lathe with a lead screw having 6.25 mm pitch.*

Solution. *Given* : Lead of the thread to cut = 3 × 10 = 30 mm

Lead of the lead screw = 6.25 mm

Now, $$\frac{\text{Driver}}{\text{Driven}} = \frac{\text{Lead of the threads to be cut}}{\text{Lead of lead screw}}$$

$$= \frac{30}{6.25} = 4.8 = \frac{48}{10}$$

$$= \frac{12 \times 4}{5 \times 2} = \frac{12 \times 5}{5 \times 5} \times \frac{4 \times 20}{2 \times 20}$$

$$= \frac{60}{25} \times \frac{80}{40} = \frac{\text{'1'} \times \text{'3'}}{\text{'2'} \times \text{'4'}} \qquad \qquad \text{...(see Fig. 9.36)}\quad \textbf{(Ans.)}$$

Example 9.16. *A two-start external square thread of 10 mm pitch and outside diameter of 62 mm is to be cut on a centre lathe which has a 6 mm pitch lead screw.*

Calculate :

(i) Depth of thread to give 0.12 mm clearance.

(ii) Lead of thread.

(iii) Core diameter.

(iv) Helix angle at a core diameter.

(v) Helix angle of thread.

(vi) Gear ratio between the headstock spindle and lathe lead screw.

Solution. *Given* : No. of starts = 2 ; Thread pitch = 10 mm ; Outside diameter = 62 mm
Pitch of lead screw = 6 mm

(i) Depth of thread to give 0.12 mm clearance :

$$\text{Depth of square thread} = \frac{\text{Pitch}}{2} = \frac{10}{2} = 5 \text{ mm}$$

Depth of thread to give a clearance of 0.12 mm

$$= 5 + 0.12 = \textbf{5.12 mm. (Ans.)}$$

(ii) Lead of thread :

Lead of thread = Pitch × No. of starts

$$= 10 \times 2 = \textbf{20 mm. (Ans.)}$$

(iii) Core diameter :

Core diameter = Outside diameter − 2 × depth of thread

$$= 62 - 2 \times 5.12 = \textbf{51.76 mm. (Ans.)}$$

(iv) Helix angle at a core diameter :

Helix angle at core diameter of thread

$$= \tan^{-1}\left(\frac{\text{lead}}{\text{core circumference}}\right)$$

$$= \tan^{-1}\left(\frac{20}{\pi \times 51.76}\right) = \textbf{7.01°. (Ans.)}$$

(v) Helix angle of thread :

$$\text{Helix angle of thread} = \tan^{-1}\left(\frac{\text{lead to be cut}}{\text{mean circumferance of work}}\right)$$

$$= \tan^{-1}\left[\frac{20}{\pi \times (62 - 5)}\right] = \textbf{6.37°. (Ans.)}$$

(*vi*) **Gear ratio :**

$$\text{Gear ratio} = \frac{\text{Driver}}{\text{Driven}} = \frac{\text{Lead of threads to be cut}}{\text{Lead of lead screw}}$$

$$= \frac{20}{6} = \frac{20 \times 5}{6 \times 5} = \frac{100}{30}.$$

100 teeth gear wheel may be keyed to the lathe spindle and 30 teeth gearwheel to the lathe lead screw. **(Ans.)**

9.2.15. Eccentric Turning

Although lathe does normal concentric turning, if some parts are required eccentric they can be turned so, on the lathe, as discussed below :

- One of the methods used, is to have *two centres* on the job faces countersunk. The central centre will turn out concentric turning. *Eccentric centre will produce eccentric turning.* The distance of the centres must be *half the eccentricity required for the job.*

- Another method is if the part is to be turned cylindrical around an axis other than central axis, the job can be turned centrally if it can be held in independent jaw chucks so the part to be turned remains central.

9.3. TURRET AND CAPSTAN LATHES

9.3.1. Introduction

Although centre/engine lathe is very useful and versatile yet it is not suitable for batch and mass production since the time taken for changing and setting the tools and time required for making measurements on the workpiece is quite large. Moreover, a skilled machinist is required to run to the centre lathe. Consequently. Turret, Capstan and Automatic lathes which reduce or eliminate the amount of skilled labour and reduce production time were developed and are widely used in producing the goods in large quantities.

9.3.2. Limitations of a Centre Lathe

The *main limitations* of a centre lathe are :

1. Only one tool can be used in the normal course (sometimes the conventional tool post can be replaced by a square tool post with four tools).
2. Large setting time of the job (in terms of holding the job)
3. The idle times involved in setting and movement of tools between the cuts is large.
4. If proper care is not taken by the operator, it is difficult to achieve precise movement of the tools to destined places.

The above limitations are taken care of in the various modified lathes such as turret and capstan lathes and semi-automatic and automatic machines by achieving improvements basically in the following areas :

(*i*) Multiple tool availability.

(*ii*) Work holding methods.

(*iii*) Automatic feeding of the tools.

(*iv*) Automatic stopping of the tools at precise locations.

(*v*) Automatic control of the proper sequence of operations.

9.3.3. Differences Between a Turret Lathe and a Centre Lathe

The differences between a turret lathe and a centre lathe are given in Table 9.1.

Table 9.1. Differences between turret lathe and centre lathe.

S. No.	Aspects	Turret lathe	Centre lathe
1.	Nature of production work	Adapted to quantity production work ; classed as a *production machine tool.*	Primarily used for *miscellaneous jobbing, tool room or single operation work.*
2.	Constructional difference (Turret head or tailstock)	*Hexagonal turret* (no tailstock), upon which are bolted various tool holders for knee turning, roller box turning, drilling, boring and recessing.	It has a *tailstock*
3.	No. of tools that can be handled at a time	Can hold a number of cutting tools at a time ; the tools can operate on the job simultaneously.	One tool cuts at a time.
4.	Set-up of tools	Tools may be permanently set-up in the turret in the sequence in which they need be used.	No such provision is available in a centre lathe.
5.	Provision of rigidity in the holding of work and tools	Extreme rigidity to permit multiple and combined cuts.	No special provision available.
6.	Machining time and handling time.	Use of a turret lathe can often cut machining time by 25 to 75% and handling time by 25% to 50%.	Consumes more time comparatively.
7.	Lead screw	Thread cutting is generally performed by tops and die heads. Hence no lead screw is provided for thread cutting.	Always provided on a centre lathe to enable thread cutting by a single point tool.
8.	Degree of automation	Semi-automatic.	Very normal.
9.	Rate of production	Higher.	Lower.
10.	Labour cost	Lower, because after tools and machine setting the operations can be performed by unskilled or semi-skilled operators only.	Higher, because of the requirement of highly skilled workers.
11.	Overhead charges	Higher than the centre lathe because of higher initial investment, more consumption of power and higher maintenance cost.	Lower comparatively.

- Summarily, the turret and capstan lathes *differ* from a general purpose centre lathe in the following aspects :

1. The headstock has more and heavier range of speeds due to which higher production rate is possible.

2. The tool post is indexable (four tools ; any one tool can be brought into the cutting position).

3. The tailstock is replaced by a tool turret with six tool positions.

4. Feed of each tool can be regulated by means of feed stops.

5. Two or more tools mounted on a single tool face can cut simultaneously.

6. These lathes are used for production operations involving better repeatability.

7. Semi-skilled operators are required.

9.3.4. Comparison of Turret lathe and Capstan Lathe

Turret lathe : Refer to Fig. 9.37.

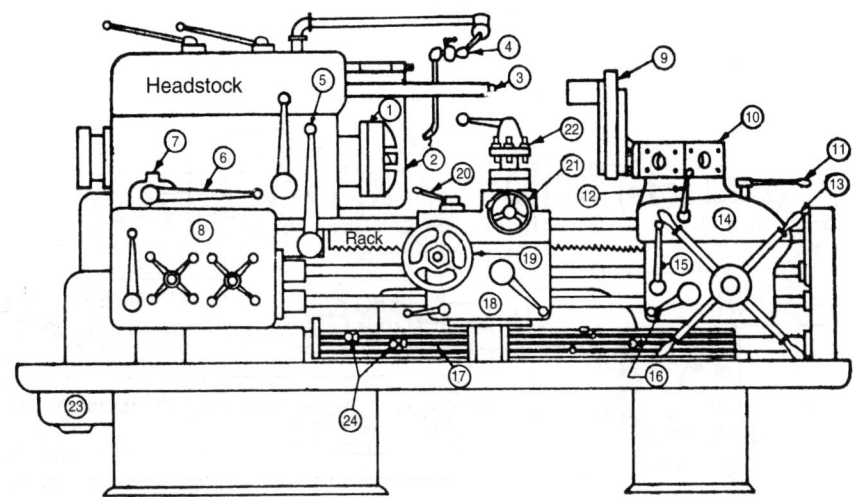

1. Chuck 2. Splash guard 3. Overarm support 4. Coolant pipe 5. Main driving clutch level
6. Lever for operating air chuck 7. Point for connecting air supply 8. Feed gear box 9. Knee tool holder
10. Turret head 11. Turret clamp 12. Turret lock and release lever 13. Star traversing wheel
14. Turret saddle 15. Feed engage lever 16. Engage lever for screw cutting 17. Stop bars 18. Chaser saddle
19. Handwheel for longitudinal feed 20. Carriage clamp 21. Cross slide handwheel 22. Turret tool post
23. Coolant pump assembly 24. Adjustable stops.

Fig. 9.37. Turret lathe.

Turret lathe is a machine *generally larger than a capstan lathe, but using similar tools.* The *main difference between them is that the turret saddle has longitudinal movement on the guides of the bed.* This feature ensures a longer traverse than is available on a similar sized capstan lathe, where the *turret saddle is clamped to the bed and the turret slide moves in the saddle.*

Turret lathes are available in wide range of sizes, the largest accommodating bars around 200 mm dia. through the spindle, so that very heavy work can be machined. In general, the tooling equipment is arranged for either bar or clutch work, high production being obtainable by use of multiple tooling on both the hexagon turret and square turret : Machines fitted with two traversing saddles are known as *combination turret lathe.*

The main tools for bar work are *roller steaay-turning tool-holders on the main hexagon turret and cutting-off tools on the square turret,* these being supplemented by *screwing drilling or forming tools to suit the work in hand.* For *chuck work it is necessary to provide equipment to cover comparatively large diameters.*

- **Features of turret lathe** *that make it a* **quantity production machine** :

  1. Rigidity in holding work and tools is built into the machine to permit multiple and combined cuts.

2. Tools may be set-up in the turret in the proper sequence for the operation.

3. Each station is provided with a feed stop or feed trip so that each cut of a tool is the same as its previous cut.

4. Multiple cuts can be taken from the same station at the same time, such as two or more turning and/or boring cuts.

5. Combined cuts can be made ; that is, tools on the cross slide can be used at the same time that tools on the turret are cutting.

6. Turret lathes may be fitted with attachments for taper turning, thread chasing, and duplicating, and can be tape controlled.

Capstan lathe : Refer to Fig. 9.38.

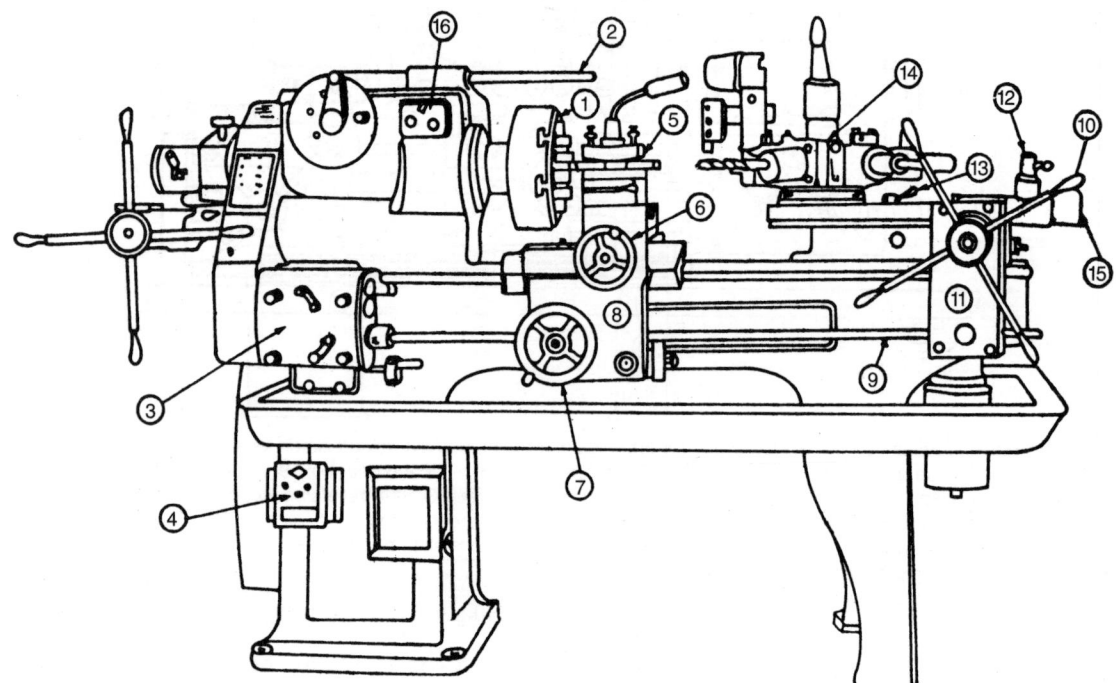

1. Chuck 2. Overarm support 3. Gear box 4. Push button starter 5. Tool post
6. Handwheel for cross feed 7. Handwheel for longitudinal feed to the carriage
8. Carriage 9. Feed bar 10. Star handwheel for slide 11. Saddle 12. Lever for locking the slide
13. Lever for free indexing of capstan head 14. Capstan head 15. Location for stop bars or screws.

Fig. 9.38. Capstan lathe.

A capstan lathe is a lathe designed *to use a number of cutting tools mounted in a rotating turret or capstan and arranged to perform turning operations successively.*

- This machine is similar both in appearance and operation to a turret lathe, but is used on *smaller work.*

- The main difference between the two types of machines is that *on a capstan lathe the turret saddle is clamped to the bed*, and the *turret slide has a limited amount of movement in the saddle,* whereas the *saddle on a turret lathe has movement on the slides of the bed.* A longitudinal power feed is provided in both cases as well as *hand movement by a pilot wheel.* In addition, a separate saddle is fitted at the front and a cut-off rest at the back, so that further operations can be performed either separately or simultaneously with the tools in the turret.

- To *ensure accurate length of work* being obtained, a set of *six adjustable stops* is fitted, each one corresponding to a face of the turret and coming into the correct position as the turret indexes trip the feed motion at any predetermined length.
- The tooling equipment varies not only on the type of work to be produced, but also on the material to be machined.

The differences between a turret lathe and a capstan lathe are given in Table 9.20

Table 9.2. Differences between a turret lathe and a capstan lathe

S. No.	Aspects	Turret lathe	Capstan lathe
1.	Turret position	Turret (head) is mounted *directly on the saddle.*	Turret is mounted on an auxiliary slide, which *moves on the guideways provided on the saddle.*
2.	Feeding of tools	For feeding the tools entire saddle is moved.	The saddle is fixed at a convenient distance from the work and the tools are fed by moving the slide.
3.	Extent of rigidity	Very high rigidity because all the cutting forces are transferred to the lathe bed.	Because of the overhung of the slide or ram, the tool support unit is subjected to bending and deflection, resulting in vibrations.
4.	Capability to handle jobs	Can handle *heavier jobs* (as a consequence of No. 3) involving heavy cutting forces and severe cutting conditions.	Since this type of lathe cannot withstand heavy cutting loads, therefore its use is confined to *relatively lighter and smaller jobs and precision work.*
5.	Maximum bar size that can be handled	Upto 200 mm diameter.	Upto 60 mm diameter.
6.	Tool travel	*Almost full length of the bed* (since the turret saddle directly rides over the bed way).	*Limited tool travel* (since the tool feeding is done by the traverse of the slide).
7.	Rate of tool feeding	Relatively slower and as such provides more fatigue to the operator's hands.	The tool traverse is faster and offers less fatigue to the operator's hands.
8.	Type of carriage	Reach-over type or side hung type.	Usually equipped with the reach-over type only since it is employed for relatively smaller jobs and therefore, does not require a large swing over bed; moreover this type of carriage provides better rigidity.
9.	Other provisions	<ul><li>Heavier designs are usually provided with pneumatic or hydraulic chucks to ensure a firmer grip over heavy jobs</li><li>Provision for cross feeding of the hexagonal turret (in some designs) to enable cross feeding of turret head tools.</li></ul>	These lathes do not have such provisions.

9.3.5. Main Parts of a Turret or Capstan Lathe

Refer to Figs. 9.39 and 9.40. The main parts of a Turret or Capstan lathe are :

1. Headstock   4. Bed

2. Carriage or chaser saddle   5. Legs.

3. Turret saddle

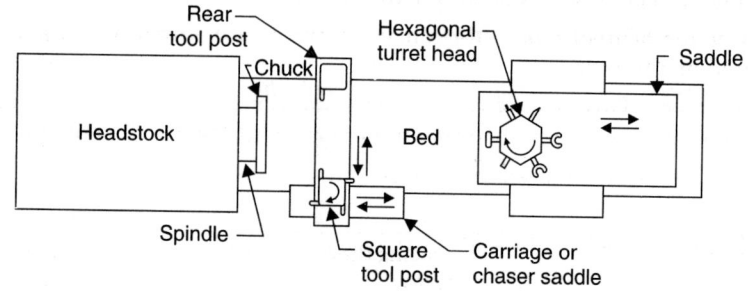

Fig. 9.39. Main parts of Turret lathe (Top view)

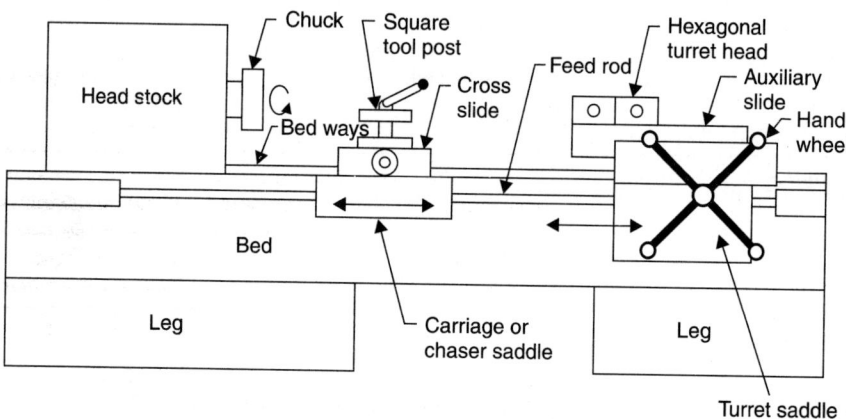

Fig. 9.40. Main parts of a capstan lathe (Front view).

1. **Headstock.** The *headstock* of a turret and capstan lathes houses a speed gear box similar in construction to the same unit as in a centre lathe but providing a narrower range of spindle speed variation and fewer speed steps.

The following types of headstocks are commonly used :

(i) Cone pulley type

(ii) Direct motor driven headstock

(iii) All geared headstock

(iv) Preoptive type headstock.

- *One of the chief characteristics of turret headstock is the provision for rapid stopping, starting and speed changing in order that the maximum advantage is taken by the operator of the most advantageous cutting speed for any job and at the same time to minimise the loss of time in speed changing, stopping and starting.*

2. **Carriage or chaser saddle.** It carries a *cross-slide* over it, on which are mounted *two tool posts*, one at the front and the other at the rear. Both these tool posts are usually square tool posts in which each is capable of holding *four* tools at a time. Tools in the rear tool post are mounted in an inverted position.

Both *hand* and *power* feeds can be employed to the saddle as well as the cross-slide, but the common practice is to use hand feed for the cross-slide until and unless a very heavy job is to be machined. *When power feeds are in operation, stops and trip dogs are used for controlling the longitudinal and cross feeds of the saddle and cross-slide respectively. These stops and trip dogs make the power feed to disengage as soon as the required tool travel is complete.*

The cross-slide carriage is of following two types ;

(*i*) *Reach over (or bridge) type :* Its construction is more rigid and allows a second tool holder to be mounted at the rear.

(*ii*) *Side hung type :* This type of carriage is generally fitted with heavy duty turret lathe where the saddle rides on the top and bottom guideways on the front of the lathe bed.

3. **Turret saddle.** It is mounted directly on the lathe bed on the same side as a tailstock in the centre lathe.

The turret head mounted on the slide or the saddle, as the case may be, is *usually hexagonal in turret lathes and circular or hexagonal in capstan lathes,* having six holes, one each on each flat face or equifaced along the periphery of the circular head.

The indexing of the tools is in a clockwise direction. After indexing, the automatic feed can be engaged.

4. **Bed.** The bed is a box shaped grey iron casting with a system of well developed internal stiffening ribs. The turret saddle and cross-slide travel along the ways on the top of the bed.

5. **Legs.** In each lathe there are two legs, one below each end of the bed. These legs are hollow castings which bear entire load of the bed, of the sliding and stationary parts mounted over the bed and also of the tooling and workholding devices or mechanisms.

9.3.6. Types of Turret Lathes

Turret lathes may be *classified* as follows :

1. Horizontal turret lathe

   (*i*) Ram type

   (*ii*) Saddle type

2. Vertical turret lathes

3. Numerically-Controlled (NC) turret lathes.

1. **Horizontal turret lathe.** These lathes are made in two general designs and are known as *ram* and *saddle.*

(*i*) **Ram-type turret lathe** : Refer to Fig. 9.41.

— It is so named because of the way the turret is mounted. The turret is placed on a slide or ram, which moves back and forth on a saddle clamped to the lathe bed. This arrangement permits *quick movement* of the turret.

— Speeds may be from 50 to 4000 r.p.m. depending on the size of lathe. Larger lathes have lower speeds.

— Trip stops are there to stop the feeding motion of the turret at any predetermined point.

— Ram type machines do not require the rigidity of chucking machines, because bar tools can be made to support the work.

● It is recommended for *bar* and *light-duty chucking work.*

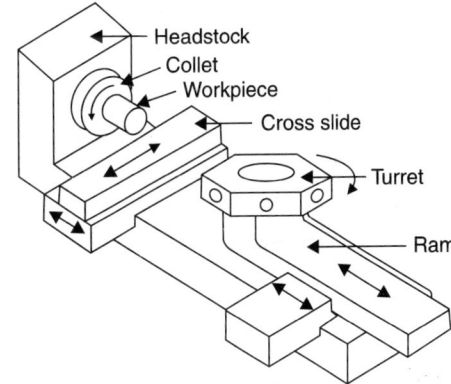

Fig. 9.41. Ram type turret lathe.

(*ii*) **Saddle-type turret lathe :** Refer to Fig. 9.42

— Here the turret is mounted directly on a saddle that moves back and forth with the turret. Since chucking tools overhang and are unconnected with the work through some sort of support ; greater strain on both work and tool support results. Chucking tools must have rigidity.

— The stroke is *longer*, which is an advantage in long turning and boring cuts, and saddle mounts assist the rigidity.

— Speed may be from 20 to 1500 r.p.m. depending upon the size of the machine.

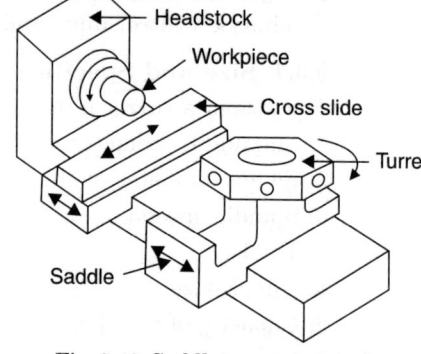

Fig. 9.42. Saddle type turret lathe.

● It is used for *chucking work.*

2. **Vertical turret lathe.** Vertical turret lathe resembles vertical boring mill, but it has the characteristic turret arrangement for holding the tools.

It consists of a rotating chuck or table in the horizontal position with the turret mounted above on a cross rail. In addition, there is at least one side head provided with a square turret for holding tools.

All tools mounted on the turret or side head have their respective stops set so that the length of cuts can be the same in successive machining cycles.

A vertical lathe, shown in Fig. 9.43, is provided with two cutter heads : The *swivelling main turret head* and the *side head.* Another side head is possible. The turret and side heads function in the same manner as the hexagonal and square turrets on a horizontal lathe.

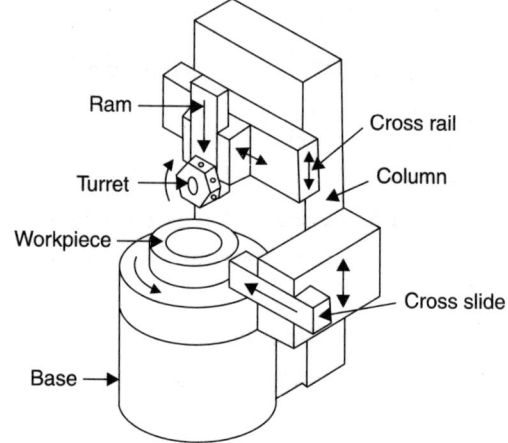

Fig. 9.43. Vertical turret lathe.

— To provide for angle cuts both the ram and turret heads may be swivelled 30° right or left of centre.

— The side head has rapid traverse and feed independent of the turret and, without interference, provides for simultaneous machining adjacent to operations performed by the turret.

— The ram provides another tool station on the machine, which can be operated separately or in conjunction with the other two.

● Vertical lathes are utilised solely for *complex chucking work, particularly* for *boring operations,* and are *not* adapted to bar work.

3. **Numerically-controlled (NC) turret lathes.** A numerically-controlled lathe is designed especially for *heavy-duty production.* Such lathes are now available in considerable variety and have proven to be for more productive than manually operated machines.

● The control in this type of lathe provides automatic functioning of spindle speed, slide movement, feeds, turret indexing, and other auxiliary functions. The slant bed, inclined rearward from the vertical, provides maximum rigidity and operator accessibility to the work area.

● This *machine can be set up quickly for small-lot jobs,* normally changing only jaw chucks, control tape, and possibly one or two cutters.

9.3.7. Size and Specifications of Turret Lathe

The size of a turret lathe can be *specified* by :

1. Maximum diameter of the bar that can be passed through the machine spindle.
2. Swing diameter of the workpiece.
3. Spindle speeds.
4. Feeds.
5. Chuck size.
6. Capacity of the drive motor.
7. Cost etc.

Specifications

| S.No. | Item | Turret lathe | |
		Ram type	Saddle type
1.	Bar stock capacity	25–64 mm	60–230 mm
2.	Chuck size	250–350 mm	up to 900 mm
3.	Swing over ways	550 mm	upto 900 mm
4.	Speed	50–4000 r.p.m.	20–1500 r.p.m.
5.	Capacity of drive motor	11–18.6 kW (15–25 H.P.)	15–56 kW (20–75 H.P.)

● *Specifications of capastan lathe :*

— *Either,* the total length of the bed, and height of the spindle centre above the bed.

Or

— The working length of bed, and the swing.

9.3.8. Common Tools and Attachments used on Turret and Capstan Lathes

Following tools are common to both the *turret* as well as *capstan* lathe :

1. Work stops or bar stops.
2. Centring and chamferring tools.
3. Drill and reamer holders.
4. Turning tools.
5. Tap and die holders.
6. Box tools.
7. Boring tools.
8. Reaming tools.
9. Knurling tools.
10. Centres and supports.
11. Attachments used on cross-slide.

● Following *special attachments* are mainly used on turret lathes :

1. Pilot bar
2. Multiple turning head

3. Cutter holders

4. Adjustable slide tool

5. Taper attachment

6. Screw cutting self-opening die head.

●   Commonly used *work holding devices are :*

1. Collet chucks.

2. Jaw chucks.

 (*i*) Three jaw self-centring chuck

(*ii*) Four jaw independent chuck

(*iii*) Two jaw box chuck

(*iv*) Power chucks.

3. Arbors.

4. Fixtures.

9.3.9. Turret Lathe Operations

Various operations performed on turret lathe are :

1. Cylindrical turning	7. Reaming
2. Taper turning	8. Threading
3. Form turning	9. Recessing
4. Facing	10. Chamferring
5. Boring	11. Knurling
6. Drilling	12. Parting off.

●   Operations performed on a *capstan lathe* are similar to those carried out on a turret lathe.

9.3.10. Turret Lathe Tooling Layout

The following factors should be considered while deciding turret lathe tooling layout.

1. Number of components to be made.

2. Set-up time.

3. Work handling time.

4. Machine controlling time.

5. Machining time.

6. Tool cost.

7. Set-up labour rate.

8. Lathe operator labour rate.

●   To illustrate the method of tooling and sequence of operations for a given job, a hexagonal turret set up *for making necessary internal cuts on a threaded adaptor* is shown in Fig. 9.44. This shows the details of the internal cuts required to machine the adaptor.

The sequence of operations for the hexagonal turret is as follows :

(*i*) The bar stock is *advanced against the combination stock stop and start drill and clamped in the collet.* The start drill is then advanced in the combination tool, and the *end of the work is centred.*

(*ii*) The hole through the solid stock is *drilled to the required length.*

(*iii*) The thread diameter is *bored to correct size for the threads specified.* A stub boring bar in a slide tool is used.

(*iv*) The drilled hole is *reamed to size with the reamer* supported in floating holder.

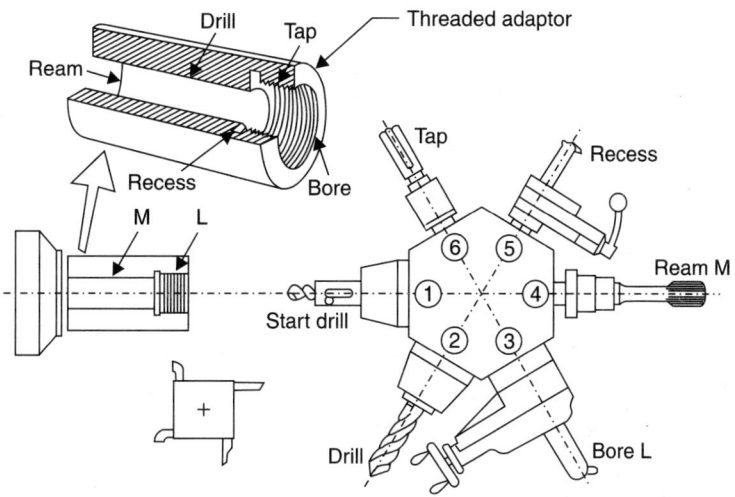

Fig. 9.44. Hexagonal turret set up illustrating the sequence of operations
to handle required internal cuts on thread adaptor.

(*v*) A *groove for thread clearance is recessed.* For this preparation a quick-acting slide tool is used with a recessing cutter mounted in a boring bar.

(*vi*) The *thread is cut with a tap* held in a clutch tap and die holder. This operation is followed by a cutting off operation (not shown in Fig. 9.44). Cut off uses the square turret on the carriage.

9.4. AUTOMATIC LATHES

9.4.1. Introduction

9.4.1.1. Automation

— ***Automation*** *is defined as any means of helping the workers perform their tasks more efficiently.*

— *Automation* is a technology of working in which handling methods, process and design of products are integrated to utilise economically justifiable mechanisation of thought and effort to achieve automatic and self regulating chain of process.

— The aim of automation is cost reduction in large scale production technology .

Automation entails the following "*advantages*" :

(*i*) Increased productivity.

(*ii*) Reduced unit cost (since a large number of components can be manufactured in shorter time).

(*iii*) Better utilisation of the resources.

(*iv*) Improvement in accuracy.

(*v*) Floor space, maintenance and inventory requirements are reduced.

● Highly automated machine tools especially of the lathe family are ordinarily classed as *automatic* and *semi-automatic.*

9.4.1.2. Automatic and semi-automatic lathes

*The lathes that have their tools automatically fed to work and withdrawn after the cycle is complete are known as **Automatic lathes***. The automatic machine, thus, performs a fully automatic cycle of operations which, when repeated, produces the desired number of identical parts without any participation of the operator. This is actually what we mean by *true automation* of the machine. The only operation required to be performed manually is the loading of bar stock or the individual casting or forged blanks and inspection of job during the operation.

In *semi-automatic lathes* (generally preferred in *chucking work*) all the machining operations are automatically performed by the machines, exactly in the same way as in fully automatic lathes, but loading or blank or bar stock, starting of machine for machining each new component, chucking the inspection of job during operation and unloading the finished articles are done by the operator *manually*.

— *Semi-automatic machines* are generally preferred in *lot-production work*.

● In all modern machine tools their operating cycles are automated through a *control system*, which may employ mechanical, electrical, electronic, pneumatic or hydraulic means or combination of some of these.

9.4.2. Classification of Automatic Lathes

Automatic lathes may be classified based on their *size, type of blank machined, processing capacity (operations performed), machining accuracy obtained, principle of operation design features, number of spindles and work positions, and type.*

A *typical classification* of automatic lathes is as follows :

Automatic lathes :

● *Horizontal spindle and vertical spindle :*

(*i*) *Semi automatic :*

    (*a*) Single spindle

        — Centre type

        — Chucking type

        — Turret

        — Special type.

    (*b*) Multiple spindle

        — Centre type

        — Chucking type

        — Turret

        — Special type.

(*ii*) *Automatics :*

    (*a*) Single spindle :

        — Cutting off

        — Swiss type

        — Automatic screw type

        — Special type.

    (*b*) Multiple spindle :

        — Cutting off bar

        — Drilling, forming, cutting off bar

        — Bar

        — Special type.

● As per the arrangement of the spindles, the automatics lathes are called *horizontal* or *vertical.*

— *Vertical machines* are *more rigid and powerful* than horizontal models and are designed for machining *large diameter work of comparatively short length.* These machines occupy *less floor space* in the shop but require *higher bays* than horizontal machines.

● *'Automatic bar machines'* are designed for producing workpieces of bar on pipe stock while magazine loaded automatic lathes process work from accurate separate blanks.

● *'Chucking machines'* are employed for machining separate blanks (hammer or die forgings, castings or pieces of previously cut off bar or pipe stock).

● *'Automatic bar machines'* are employed for manufacturing :

— High quality fastenings ;

— Bushing ;

— Shafting ;

— Rings ;

— Rollers ;

— Handles ;

— Other parts made of bar or pipe stock.

● *"Multiple spindle machines"* may have from 2 to 8 spindles. Their production capacity is higher than that of single spindle machines but their machining accuracy is somewhat lower.

Operations carried out on automatic lathes :

The following *typical operations* are carried out on automatic lathes :

(*i*) Centring

(*ii*) Turning cylindrical, tapered and formed surfaces

(*iii*) Drilling

(*iv*) Boring

(*v*) Reaming

(*vi*) Spot facing

(*vii*) Knurling

(*viii*) Thread cutting

(*ix*) Facing

(*x*) Cutting-off.

9.4.3. Automatic Vertical Multistation Lathe

Fig. 9.45 shows a multi spindle vertical chucking lathe.

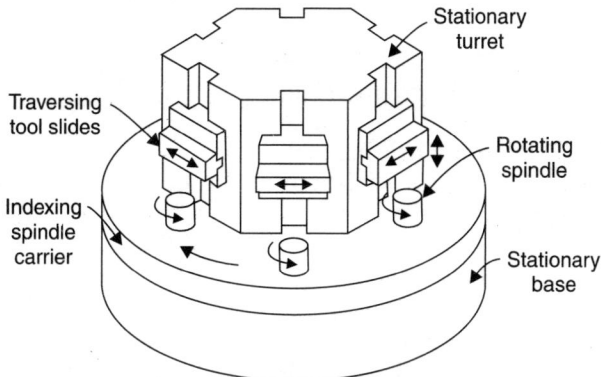

Fig. 9.45. Multiple spindle vertical chucking lathe.

— Such machines are designed for higher production and are usually provided with either 5 or 7 work stations and a loading position. In some machines two spindles are provided at each station, doubling the capacity for small diameter work.

— The work is mounted in chucks, the larger machines having chucking capacities upto 460 mm in diameter.

— All types of machining operations can be performed including milling, drilling, threading, tapping, reaming and boring.

● The *advantage* of the automatic vertical multistation lathe is that *all operations can be done simultaneously and in proper sequence.*

9.4.4. Automatic Screw Machines

The invention of an automatic screw machine includes the following *features* :

(*i*) A controlling movement for the turret so that tools can be fed into the work at desired speeds, withdrawn, and indexed to the next position. This was accomplished by a cylindrical or drum cam located beneath the turret.

(*ii*) Another feature, also cam controlled, is a mechanism for clamping the work in the collet, releasing it at the end of the cycle, and then feeding unworked bar stock against the stop.

9.4.4.1. Single-spindle automatic screw machine

Fig. 9.46 illustrates a single-spindle automatic screw machine designed for *bar work of small diameter.*

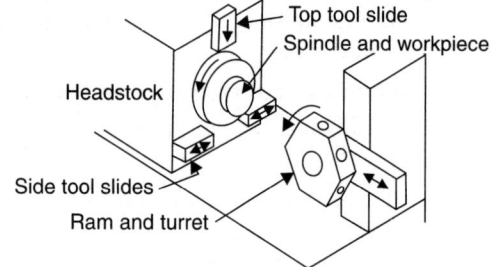

Fig. 9.46. Single-spindle automatic screw machine.

— This machine has a cross-slide capable of carrying tools both front and rear and a turret mounted in a vertical position on a slide with longitudinal movement.

— The tools used in the machine are mounted around the turret in a vertical plane in line with the spindle.

— Usual machining operations such as turning, drilling, boring and threading can be done on these machines.

9.4.4.2. Multi-spindle automatic screw machines (Refer to Fig. 9.47)

These machines, however, are not usually spoken of as screw machines, but rather as *multi spindle automatics.*

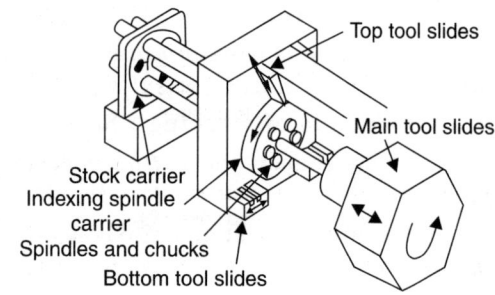

Fig. 9.47. Multi-spindle automatic screw machine.

— In these machines a large number of spindles usually 4, 6 or 8 are provided. These spindles are carried in a rotatable drum that is indexed in order to bring each spindle successively into a different working position. Each spindle carries a *bar* of material.

— A non-rotating tool slide holds the same number of tools as there are spindles in order to provide a cutting tool for each spindle. A cross-slide is provided at each spindle position so as to feed a tool from the side to carry out operations like facing, grooving, knurling and cutting off.

— The bars are loaded into hollow spindles. The tool slide is moved forward and cross-slides moved inward so that the various tools cut simultaneously. *All the motions are controlled automatically.*

— On completion of a particular operation the tool slide and cross-slides are moved backward and spindles indexed by one position by rotating the spindle drum so that the next operation can be carried out.

● *Owing to errors in indexing of the spindles and large number of fittings, these machines seldom maintain accuracy as good as single speed machines.*

● These machines are suitable when a very large number of components (say from 2000 to 5000 in a lot) are to be manufactured.

9.4.4.3. Swiss Automatic Screw Machine

These machines are called "Swiss" because their development began in Switzerland (a land noted for the production of watches).

This machine has been developed for *precision turning of small parts.*

— The single-point tools used on this machine (see Fig. 9.48) are placed radially around the carbon-lined guide bushing through which the stock is advanced during machining operations.

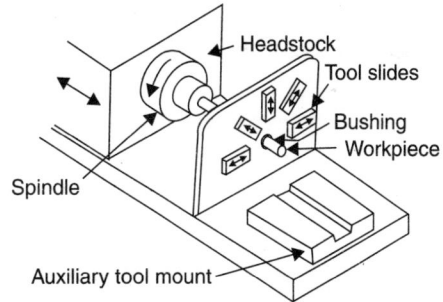

— Most diameter turning is done by the two horizontal slides, whereas the other three are used principally for knurling, chamfering, cutting off, and recessing.

— The stock is held by a rotating collet in the headstock back of tools, and all longitudinal feeds are accomplished by a cam that moves the headstock forward as a unit. This forward motion advances the stock through the guide bushing and to the single-point tools,

Fig. 9.48. Swiss automatic screw machine.

which are controlled and positioned by cams. By coordinating their movement with forward movement of the stock, small diameters on slender parts can be held to tolerance ranging from 0.005 mm to 0.013 mm.

9.5. SHAPING MACHINE (SHAPER)

9.5.1. Introduction—Principle of Working

A **shaper** *is a reciprocating type of machine tool intended primarily to produce horizontal, vertical or inclined flat surfaces (upto 1000 mm long).*

Principle of working : Refer to Figs. 9.49, 9.50 and 9.51.

In the shaper, the *cutting tool has a reciprocating motion, and it cuts only during the forward stroke only.*

The work is held in a vice bolted to the work table. The *regular feed is obtained by moving the worktable automatically at right angles to the direction of the cutting tool and the tool head gives downward feed at right angles to the regular feed or any other angle as desired.*

● Most shaping machines usually work with one cutting tool, and cutting is done in the forward stroke only, but the exceptions exist where more than one tool is used for simultaneous cutting or cutting on both (forward and return) strokes.

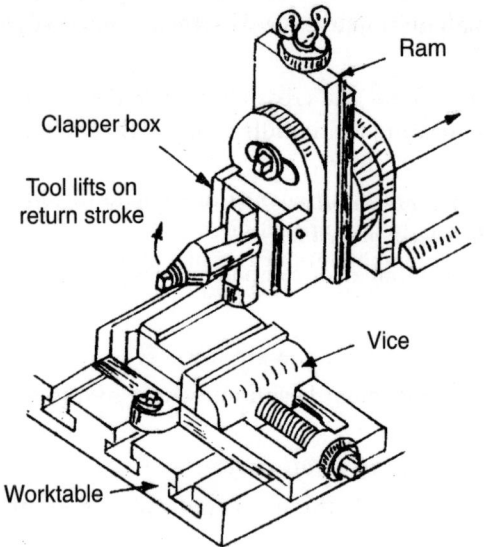

Fig. 9.49. Shaper.

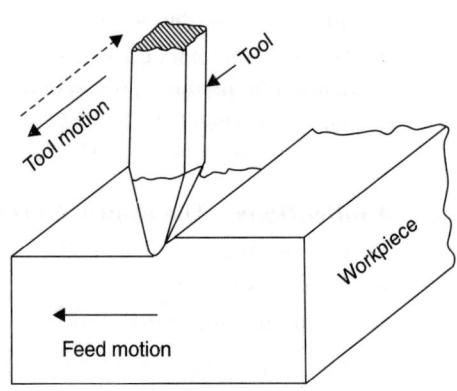

Fig. 9.50. Working principle of a shaper.

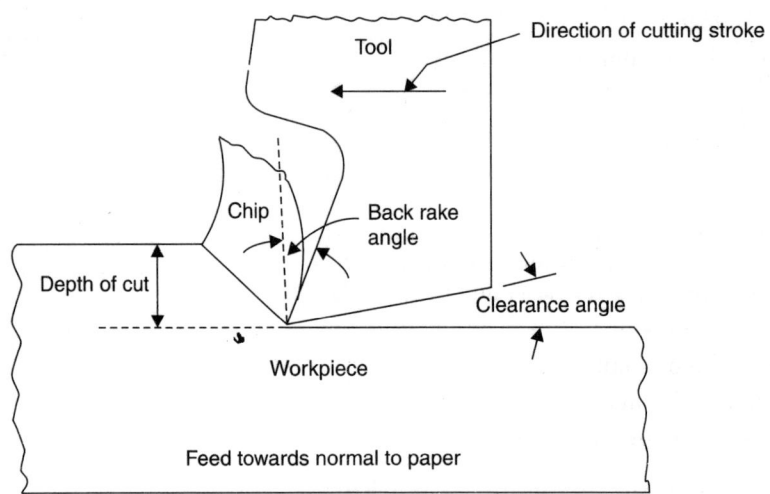

Fig. 9.51. Cutting operation of shaper.

9.5.2. Advantages, Limitations, and Applications of Shapers

The following are the advantages and disadvantages of shapers :

Advantages :

1. The set up of shaper is very quick and easy and can be readily changed from one job to another.

2. The work can be held easily.

3. The single point tools used are in inexpensive ; these tools can be easily grounded to any desired shape.

4. Lower first cost.

5. The cutting stroke has a definite stopping point.

6. Because of lower cutting forces, thin and fragile jobs can be conveniently machined on shapers.

Limitations :

1. A shaper, by nature, is a slow machine, because of its straight line, forward and return (idle) stroke. The single point tool requires *several strokes to complete a work.*

2. The cutting speeds are not usually very high since difficulties are encountered in designing machine tools with high speeds of reciprocating motion *due to high inertia forces developed* in the motion of the units and components of the machine. Owing to these reasons the shaper *does not find ready adaptability for assembly and production line.*

Applications : The shaper finds best use in (small and medium sized work) :

— Tool room ;

— Jobbing shops ;

— Die or jig making shops.

9.5.3. Classification of Shapers

The shapers are *classified* as follows :

1. According to the type of mechanism used for giving reciprocating motion to the ram :

(i) *Crank shaper :*

● In this type of shaper, a crank and a slotted lever quick return motion mechanism is used to give reciprocating motion to the ram.

● The crank arm is adjustable and is arranged inside the body of a bull gear (also called crank gear).

(ii) *Geared shaper :*

● In this shaping machine, the ram carries a rack below it, which is driven by a spur gear.

● This type of shaper is not widely used.

(iii) *Hydraulic shaper :*

● In this type of shaper, a hydraulic system is used to drive the ram.

● This shaper is *more efficient* than the crank and geared type shapers.

2. According to position and travel of ram :

(i) *Horizontal shaper :*

● In this shaping machine, the ram moves or reciprocates in a horizontal direction.

● This shaper is *mainly used for producing flat surfaces.*

(ii) *Vertical shaper :*

● In this shaper, the ram *reciprocates vertically in the downward as well as in upward motion.*

● This type of shaping machine is very convenient for machining *internal surfaces, keyways, slots or grooves.*

(iii) *Travelling head shaper :*

● A travelling head shaper has a reciprocating ram mounted on a saddle which travels sideways along the bed. The ram carries the tool slide.

● Heavy duty jobs which cannot be held on the standard shaper table, are kept stationary on the base travelling head shaper and machined as the ram reciprocates.

3. According to the type of cutting stroke :

(*i*) *Push-cut shaper :*

● In this shaper, the ram pushes the tool across the work during cutting operation. In other words, *forward stroke* is the *cutting stroke* and the *backward stroke* is an *idle stroke.*

● This is the most general type of shaper used in common practice.

(*ii*) *Draw-cut shaper :*

● In a draw-cut shaper, the ram draws or pulls the tool across the work during cutting operation. In other words, the backward stroke is the cutting stroke and forward stroke is an idle stroke.

4. According to the design of the table :

(*i*) *Standard or plain shaper* :

● In this type of shaper, the table has only two movements namely horizontal and vertical, to give the feed.

● It cannot be swivelled or tilted.

(*ii*) *Universal shaper :*

● In this shaper, in addition to the above two movements, the table can be swivelled about an horizontal axis parallel to the ram and the upper portion of the table can be tilted about a horizontal axis perpendicular to the first axis.

● A universal shaper is mostly used in *tool room.*

9.5.4. Principal Parts of a Shaper

Refer to Fig. 9.52. The principal parts of a shaper are described briefly below :

1. Base :

● It is made of cast iron to resist vibration and takes up high compressive load.

● It is so designed that it can take up the entire load of the machine and the forces set up by the cutting tool over the work.

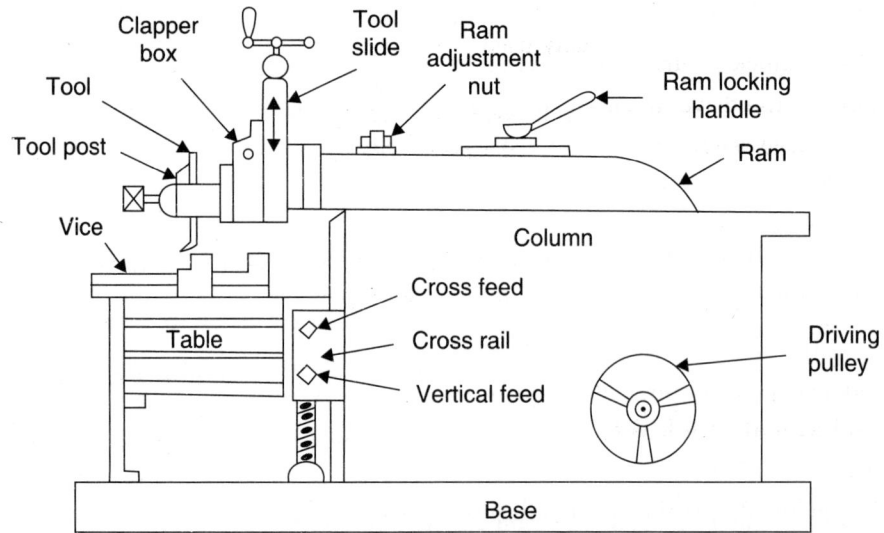

Fig. 9.52. Principal parts of a shaper.

2. **Column :**

- It is a box like casting mounted upon the base.
- It encloses ram driving mechanism.
- It is provided with guideways on its top to enable the ram to slide on it.

3. **Ram :**

- It is a reciprocating member which reciprocates on the guideways provided above the column.
- It carries a tool-slide on its head and a mechanism for adjusting the stroke length.

4. **Cross-rail :**

- It is mounted on the front vertical guideways of the column.
- It has two parallel guideways on its top in the vertical plane that are perpendicular to the ram axis. The table may be raised or lowered to accommodate different size of jobs by rotating an elevating screw which causes the cross-rail to slide up and down on the vertical face of the column.

5. **Table :**

- It is made of cast iron and is rectangular in shape.
- It has T-slots on its top surface.
- The table can be moved upward, downward or sideward with the help of elevating screws and other feed handle.

9.5.5. Specifications of a Shaper

The shaper is *specified* as follows :

1. Maximum length of the stroke (in mm).

2. Size of the table *i.e.,* length, width and depth of the table.

3. Maximum horizontal and vertical travel of the table.

4. Maximum number of strokes per minute.

5. Type of quick return mechanism.

6. Power of the drive motor.

7. Floor space required.

8. Weight.

- Horizontal shapers range in size from small bench models with *stroke* of 175 mm or 200 mm to heavy duty models with strokes as much as 900 mm. Shaping machines are commonly provided with *power feed* ranging from 0.2 to 0.5 mm/stroke.

9.5.6. Quick Return Mechanism in a Shaper

The shaper is driven by a *mechanical quick return mechanism* or by a *hydraulic system*.

1. **Mechanical quick return mechanism :**

Refer to Fig. 9.53.

— The motor drives the crank/bull gear which carries a pin (B centre) in a circular motion. The r.p.m. of the bull gear is controlled by the motor. This pin fits into the slot of the rocker arm and is free to slide in a straight line path.

— As the bull gear rotates, the rocker arm oscillates about its pivot point (*O*). The end of the rocker arm is connected to the ram of the shaper through a link arm.

— The length of the stroke is changed by changing the radius (*R*) of the circle is which the pin on the bull gear rotates (the larger the crank radius, the longer the ram stroke will be and vice versa). The length of travel should be a little longer than the actual length of the workpiece. This allows sufficient time for the tool block of the clapper box to swing back to its position for cutting.

Refer to Figs. 9.53. (*b*) and 9.53 (*c*) :

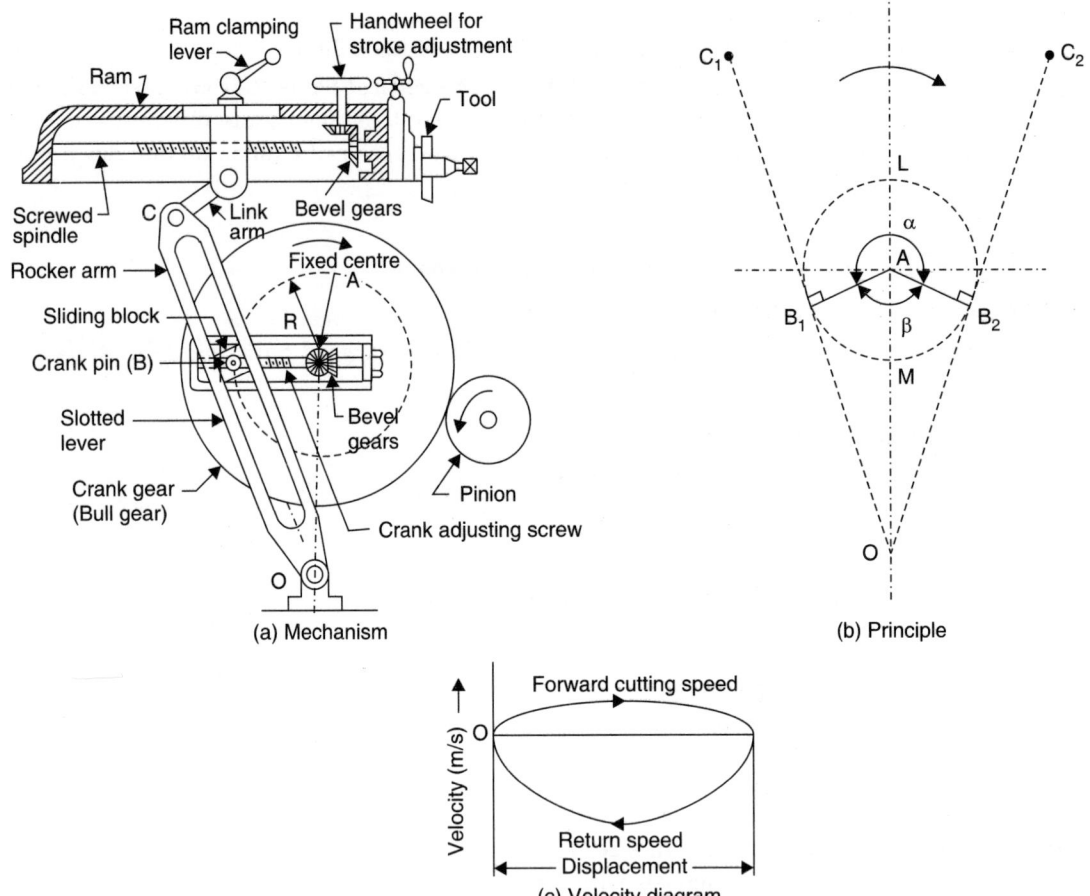

Fig. 9.53. Quick return mechanism of a shaper (Crank and slotted lever type).

— Since the crank rotates at a uniform speed, therefore

$$\frac{\text{Time of cutting stroke}}{\text{Time of return stroke}} = \frac{\angle B_1 \, LB_2}{\angle B_1 \, MB_2} \quad i.e., \quad \frac{\alpha}{\beta}$$

It is clear that the *return stroke is quick so as to minimise the return idle stroke time.*

— The ratio of return cutting speed is about 3 : 2

● The bull gear is massive in size so that it also performs the function of a flywheel. This makes the reciprocating movement of the ram smooth and without any jerks.

2. **Hydraulic drive :** Refer to Fig. 9.54.

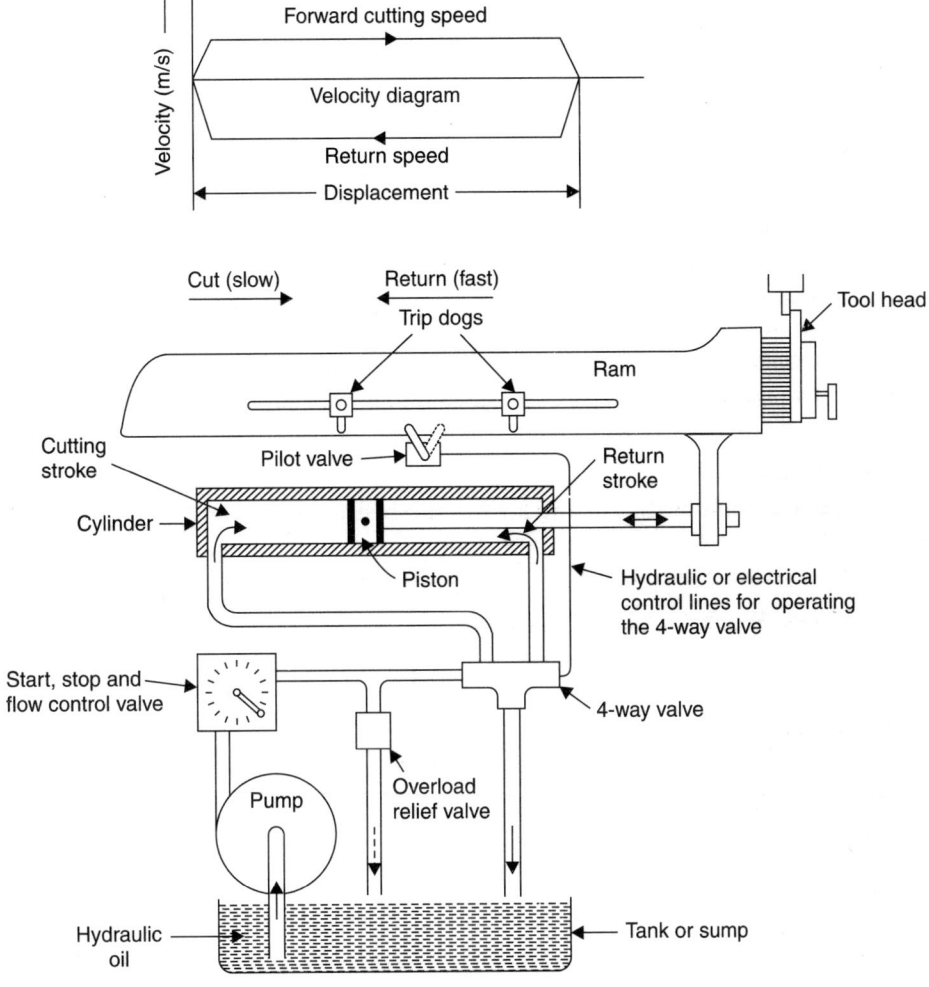

Fig. 9.54. Hydraulic drive.

Quick return in the hydraulic shaper is accomplished by increasing the flow of hydraulic oil during the return stroke.

— In the hydraulic shaper, the ram is connected to the hydraulic cylinder which is controlled by means of a 4-way valve.

— The hydraulic fluid is pumped to the hydraulic cylinder through 4-way valve ; this valve is also connected to the sump.

— The 4-way valve controls the direction of high pressure fluid into the cylinder and controls the direction of motion, either the cutting stroke or return stroke. The flow control valve controls the flow rate of the hydraulic fluid thereby controlling the speed at which the ram moves.

— The starting and stopping of the machine is achieved through a finger operated lever. An adjustable trip dog operated lever controls the operation of 4-way valve to control the ram reversal.

— *The return or idle stroke is faster than the cutting stroke because of the smaller area in the return side of the cylinder (due to presence of piston rod) if constant volume pump is used.*

— The maximum ratio of return to cutting speed is about 2 : 1.

Advantages of hydraulic drive :

1. The cutting stroke has a more constant velocity and less vibration is induced in the hydraulic shaper.
2. The cutting speed is generally shown on an indicator and does not require calculation.
3. Both the cutting stroke length and its position relative to the work may be changed quickly without stopping the machine.
4. Ram movement can be reversed instantly anywhere in either direction of travel.
5. The hydraulic feed operates while the tool is clear of work.
6. More strokes per minute can be achieved by consuming less time for reversal and return strokes.
7. Since the power available remains constant throughout, it is possible to utilise the full capacity of the cutting tool during the cutting stroke.

Disadvantages :

1. The stopping point of the cutting stroke in a hydraulic shaper can vary depending upon the resistance offered to cutting by the work material.

2. It is more expensive compared to the mechanical shaper.

9.5.7. Shaper Work Holding Devices

Shaper work holding devices include :

1. Vices
2. Parallel strips
3. Clamps
4. Jack
5. Angle plate
6. Vee blocks
7. Stop pins and toe dogs
8. Centres (for cutting of splines and gear teeth on shaper).

9.5.8. Operations Performed

On a standard shaper, the following operations can be performed : Refer to Figs. 9.55. to 9.59.

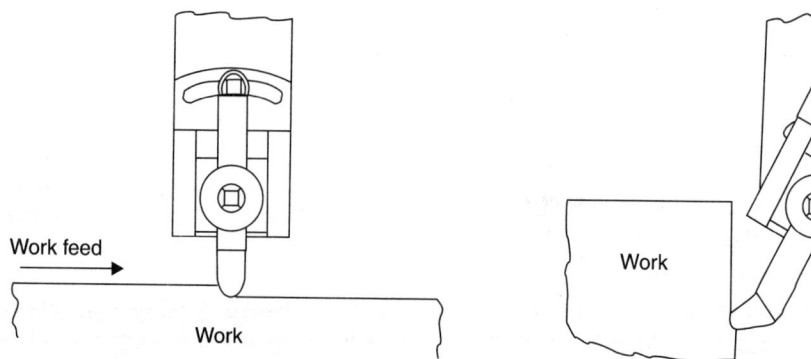

Fig. 9.55. Horizontal shaping.　　　　Fig. 9.56. Vertical shaping.

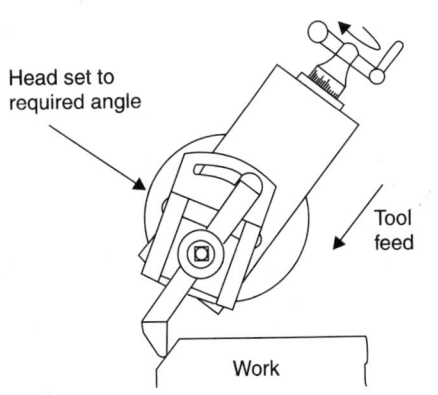

Fig. 9.57. Angular shaping.

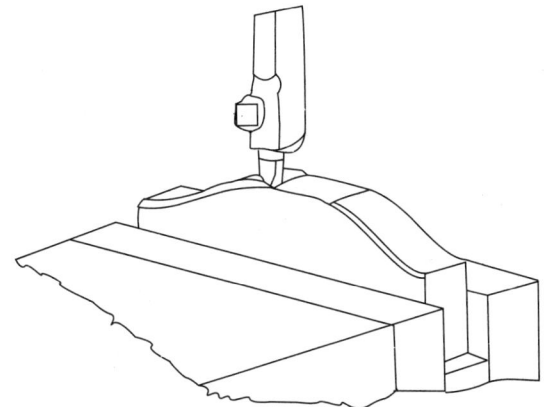

Fig. 9.58. Irregular (or contour) cutting.

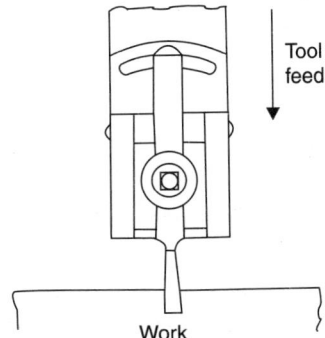

Fig. 9.59. Shaping grooves.

1. Machining of *horizontal surfaces*
2. Machining of *vertical surfaces*
3. Machining of *angular surfaces*
4. Machining of *curved surfaces*
5. Machining of *irregular surfaces*
6. Machining of *slots, grooves,* and *keyways* etc.

9.5.9. Cutting Tools and Other Tools Used in Shaper work

Cutting Tools :

— The cutting tools used in shapers are similar to those used in lathe work *except for side and front clearances.*

— Shaper tools have *less* side and front clearance because the work feeds into it on the return stroke, whereas the lathe tool is constantly feeding into the work. For best results these angles should not exceed 4° and be less than 2°.

The various types of cutting tools used in shaping work are shown in Fig. 9.60.

● High speed steel tools are commonly used in shaper work. However, in machining some very hard metals like chilled cast iron and die steel etc., where the life of HSS tools will be very short, carbide tipped tools can be effectively used for efficient and easy machining.

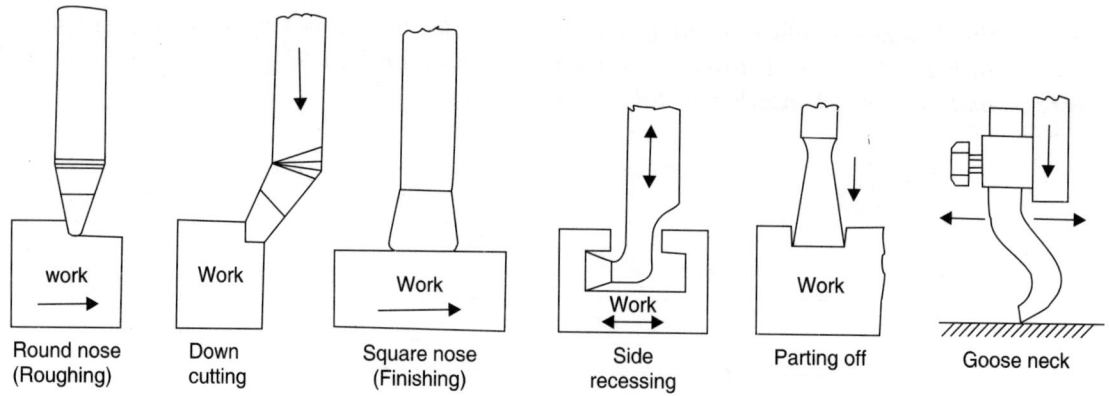

Fig. 9.60. Cutting tools used in shaping work.

Other tools :

1. Try square and square head of combination set.

2. Micrometer.

3. Surface gauge.

4. Sine bar.

5. Dial test indicator.

9.5.10. Cutting Speed, Feed and Depth of Cut

Cutting speed. *Cutting speed is defined as the average linear speed of the tool during the cutting stroke in m/min, which depends on the number of ram strokes (or ram cycles) per minute and the length of the stroke.*

● Cutting speed on shapers may be either constant or variable depending on the design of the machine tool. If the stroke length is changed the number of strokes per minute remains constant, the average cutting speed will be changed. To determine the proportion of time the cutting tool is working, the return stroke time to cutting stroke time ratio 'k' is used in calculation of the average cutting speed. The value of k as 2 : 3 implies that the tool is working $\frac{3}{5}$th of the time and the return stroke takes $\frac{2}{5}$th of the time.

This *return-to-cutting time ratio 'k' is a machine constant.*

The cutting speed V_c is determined by using the formula,

$$V_c = \frac{NL\,(1+k)}{1000} \quad \text{m/min} \qquad\qquad ...(9.18)$$

where, N = The *number of double strokes* or *cycles of the ram*/min(one double or full stroke comprises one cutting and one return stroke)

$\qquad$ L = Length of the ram stroke, mm, and

$\qquad$ k = Return stroke time/cutting stroke time $\left(\text{on an average, } k = \frac{2}{3} \text{ to } \frac{3}{4}\right)$.

● At each end of the stroke, the ram has to come to rest (zero speed) and start moving again. It takes definite time (and distance) to attain the desired cutting speed (and rest) ;

the distance is called *clearance* (*c*). Clearance, job length and stroke length are shown in Fig. 9.61 (*The clearance is a must in shaping and allied process unlike approach and over-travel in other metal cutting processes*).

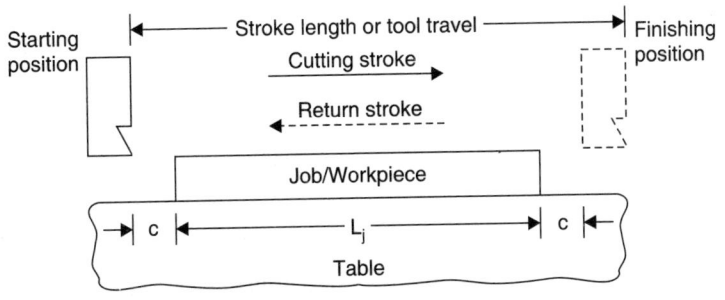

Fig. 9.61. Length of stroke and length of job/workpiece.

Thus, $\qquad$ $L = L_j + 2 \times c$ $\qquad$ ...(9.19)

where, $\qquad$ L = Length of stroke, and

$\qquad$ L_j = Length of job/workpiece.

Feed. The **feed** '*f*' *is the relative motion of the workpiece in a direction perpendicular to the axis of reciprocation of the ram.*

- In shaper, feed is *normally given to the workpiece* and can be automatic or manual.
- Feed is expressed in mm/double stroke or simply mm/stroke since no cutting is done in return stroke.

Depth of cut. Depth of cut '*d*' *is the thickness of the material removed in one cut, in mm.*

- The depth of cut may be given by the toolhead slide or by lifting the table.

9.5.11. Machining Time

The time required to *complete one double stroke*, from cutting speed (*V_c*), is given by :

$$t = \frac{L\,(1+k)}{1000\,V_c} \text{ min}$$ $\qquad$ ...(9.20)

With a feed of *f* mm/double stroke, number of double strokes required to machine a surface of width *B* will be

$$N_s = \frac{B}{f}$$ $\qquad$ ...(9.21)

Hence, total time for machining the surface will be

$$t_m = \frac{LB(1+k)}{1000\,V_c f}$$ $\qquad$ ...(9.22)

or, In terms of ram strokes *N*, the time for machining the surface is given by

$$t_m = \frac{B}{fN} \text{ min.}$$ $\qquad$ ...(9.23)

● Machining time (t_m) can also be calculated as follows :

$$t_m = \frac{B}{f}\left(\frac{L}{V_c \times 1000} + \frac{L}{V_r \times 1000}\right) \text{ min} \qquad ...(9.24)$$

where, B = Width of the job, mm,

f = Feed, mm/stroke,

L = Length of stroke, mm

V_c = Cutting speed, m/min, and

V_r = Return stroke speed, m/min.

If V_{av} is the average or mean speed, then

$$t_m = \frac{B}{f} \times \frac{2L}{V_{av} \times 1000} \text{ min} \qquad ...(9.25)$$

where, $$V_{av} = \frac{2V_c \times V_r}{V_c + V_r} \text{ m/min} \qquad ...(9.25(a))$$

9.5.12. Material Removal Rate

Material removal rate (MRR) in a shaping machine is given by :

MRR = $f.d.N.L.$ $(1 + k)$ mm^3/min $\qquad ...(9.26)$

where, f = Feed/stroke, mm,

d = Depth of cut, mm,

N = Number of double strokes/min, and

L_j = Length of stroke.

Example 9.17. *Calculate the machining time required for machining a surface 450 mm × 600 mm on a shaping machine, using the data given below :*

Cutting speed	*= 7.5 m/min*
Return-to-cutting time ratio	*= 2 : 3*
Feed	*= 2 mm/double stroke*
The clearance at each end	*= 50 mm.*

Solution. *Given :* L_j = 450 mm ; B = 600 mm ; V_c = 7.5 m/min ; k = 2/3 ; f = 2 mm/double stroke ; c = 50 mm.

Machining time t_m :

Now, Length of stroke, $L = L_j + 2c$ $\qquad ...[\text{Eqn. (9.19)}]$

= 450 + 2 × 50 = 550 mm

∴   Machining time required to machine the job,

$$t_m = \frac{LB\,(1+k)}{1000\,V_c\,f} \text{ min} \qquad ...[\text{Eqn. (19.22)}]$$

$$= \frac{550 \times 600\left(1 + \dfrac{2}{3}\right)}{1000 \times 7.5 \times 2} = \textbf{36.67 min.} \quad \textbf{(Ans.)}$$

Example 9.18. *A cast iron plate measuring 450 mm × 150 mm × 60 mm is to be rough shaped along its wider face. Calculate the machining time taking cutting speed = 10 m/min, return speed = 15 m/min, approach = 30 mm, over travel = 30 mm, allowance on either side of the plate width = 6 mm and feed per cycle = 1.5 mm.*

Solution. $L = 450 + 30 + 30 = 510$ mm ; Shaping width, $B = 150 + 6 + 6 = 162$ mm ; $V_c = 10$ m/min ; $V_r = 15$ m/min ;

Machining time, t_m :

Cutting stroke time $= \dfrac{L}{V_c} = \dfrac{510}{1000 \times 10} = 0.051$ min

Idle stroke time $= \dfrac{L}{V_r} = \dfrac{510}{1000 \times 15} = 0.034$ min

Time per cycle $= 0.051 + 0.034 = 0.085$ min

No. of cycles required $= \dfrac{B}{f} = \dfrac{162}{1.5} = 108$ cycles

$\therefore$ Machining time, t_m = No. of cycles × time per cycle

$= 108 \times 0.085 = $ **9.18 min. (Ans.)**

Example 9.19. *A cast-iron surface 300 mm long and 180 mm wide is to be machined on a shaper with cutting-to-return ratio of 3 : 2. Cutting speed, feed and clearance are 24.6 m/min, 2 mm/double stroke and 30 mm respectively. The available ram strokes on the shaper are : 28, 40, 60 and 90 strokes/min. If the depth of cut is 3.5 mm, determine :*

(i) Time required to machine the surface.

(ii) Material removal rate.

Solution. $L_j = 300$ mm ; $k = \dfrac{2}{3}$; $V_c = 24.6$ m/min ; $f = 2$ mm/double stroke ; $c = 30$ mm ; $d = 3.5$ mm.

(i) **Machining time, t_m :**

$$L = L_j + 2 \times c = 300 + 2 \times 30 = 360 \text{ mm}$$

Cutting speed, $V_c = \dfrac{NL(1 + k)}{1000}$

or, $N = \dfrac{V_c \times 1000}{L(1 + k)} = \dfrac{24.6 \times 1000}{360\left(1 + \dfrac{2}{3}\right)} \simeq 41$ strokes/min

Nearest available ram stroke is 40 strokes/min which is very near to the calculated value. Normally we should not exceed the specified cutting speed, as it will affect the tool life adversely. Hence select $N = 40$ strokes/min.

With a chosen value of N, we cannot use eqn. (9.22) for time calculation. Hence, substituting all the values in eqn. (9.23), we get.

$$t_m = \frac{B}{fN} = \frac{180}{2 \times 40} = \textbf{2.25. (Ans.)}$$

(ii) **Material removal rate (MRR) :**

We know that, MRR = $fdNL(1 + k)$ mm³/min [Eqn (9.26)]

$$= 2 \times 3.5 \times 40 \times 360 \left(1 + \frac{2}{3}\right) = \textbf{1,68,000 mm}^3\textbf{/min. (Ans.)}$$

9.6. SLOTTING MACHINE (SLOTTER)

9.6.1. Introduction

Basically the *slotting machine is a vertical axis shaper.*

— The chief difference between a shaper and a slotter is the direction of the cutting action ; *the tool moves vertically* rather than in horizontal direction.

— The slotter has a vertical ram and a hand or power operated rotary table. On some machines, the ram may be inclined as much as 10° to either side of the vertical position when cutting inclined surfaces.

— *The stroke of the ram is smaller in slotting machines than in shapers to account for the type of work that is handled in them.*

Uses of slotter. The uses of a slotter are as follows :

1. Internal machining of blind holes.

2. Work requiring machining on internal sections such as splines, keyways, various slots and grooves, and teeth.

3. Machining of die, punches, straight and curved slots.

4. Cutting of teeth on ratchet or gear rings which require primarily rotary feed.

5. By combining cross, longitudinal, and rotary feed movements of the table, even complex contours can be machined.

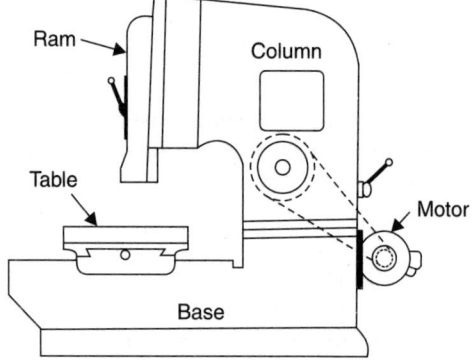

Fig. 9.62. Slotting machine.

9.6.2. Main Parts of a Slotter

The main parts of a slotter are shown in Fig. 9.62 ; These are briefly discussed below :

1. **Base :**

● It is also known as *bed* and is a heavy cast iorn construction.

● It acts as a support for the column, the driving mechanism ram, table and all other fittings.

● At its top it carries horizontal ways, along which the table can be traversed.

2. **Column :**

● It is also made of cast iron. It acts as housing for the complete driving mechanism.

● Its front face has guideways for the reciprocatory ram.

3. **Table :**

● On slotting machines, usually a circular table is provided. However, in some heavy duty slotters, such as a *puncher slotter*, either a rectangular or circular table can be mounted.

● On the top of the table are provided T-slots to clamp the work or facilitate the use of fixtures etc.

● The table can be moved is horizontal plane by two perpendicular cross-slides.

4. **Ram :**

● It moves in vertical direction between the vertical guideways provided in front of the column.

● The ram supports the tool head to which the tool is attached. The cutting action takes place during the downward movement of the ram.

9.6.3. Size and Specifications

The size and specifications/operating conditions are defined in the same way as that of shaping and planing machines.

The *complete specifications of a typical 300 mm stroke slotters are as under* :

Stroke maximum = 300 mm ; Stroke minimum = 0 mm ; Height between table and head = 450 mm ; Maximum diameter accommodated when machining at centre = 900 mm ; Diameter of the table = 500 mm ; Traverse of table, longitudinal = 450 mm ; Traverse of table, transverse = 350 mm ; Face of head = 575 mm × 250 mm ; Height overall = 2000 mm ; Length of bed = 1375 mm ; Width of bed = 412 mm ; Height of bed = 575 mm ; Height of head = 1275 mm ; Belt size = 75 mm ; H.P. required = 2 H.P.

9.6.4. Slotter Drive Mechanisms

The following drive mechanisms are used in slotters for driving the ram :

1. Slotted disc mechanism.

2. Variable speed reversible motor drive.

3. Hydraulic drive.

4. Slotted link drive.

The ram carrying the tool moves vertically up and down and cuts only in the downward stroke. The return stroke or the upward stroke is completed in a shorter period of time than the cutting stroke.

9.6.5. Types of Slotters

Various types of slotters are :

1. **Puncher slotter.** Used for *removing large amount of metal* from heavy work which has been forged, stamped or sawn roughly to shape.

2. **General production slotters.** Mostly used for *general production work.*

3. **Precision tool room slotters.** Primarily used for *tool room work* where accuracy, in most cases, is of paramount importance.

4. **Key seater.** Used exclusively for *machining keyways* on the inside of wheel and gear hubs.

9.6.6. Work Holding Devices

Following devices are used for holding workpieces on the worktable of the slotter :

1. Clamps   2. T-bolts

3. Vice   4. Parallel strips

5. Special fixtures for holding work.

9.6.7. Slotter Cutting Tools

The types of tools used in a slotter are very similar to that of shaper except that the cutting actually takes place in direction of cutting. However, in view of the types of surfaces that are possible in the case of a slotter, a large variety of boring bars or single point tools along with shanks are used.

9.6.8. Slotter Operations

The following operations can be performed on a slotter :

1. Cutting of internal grooves or keyways.

2. Cutting of internal gears.

3. Cutting of recesses.

4. Cutting of concave, circular and convex surfaces etc.

Fig. 9.63 shows a typical component machined in a slotter

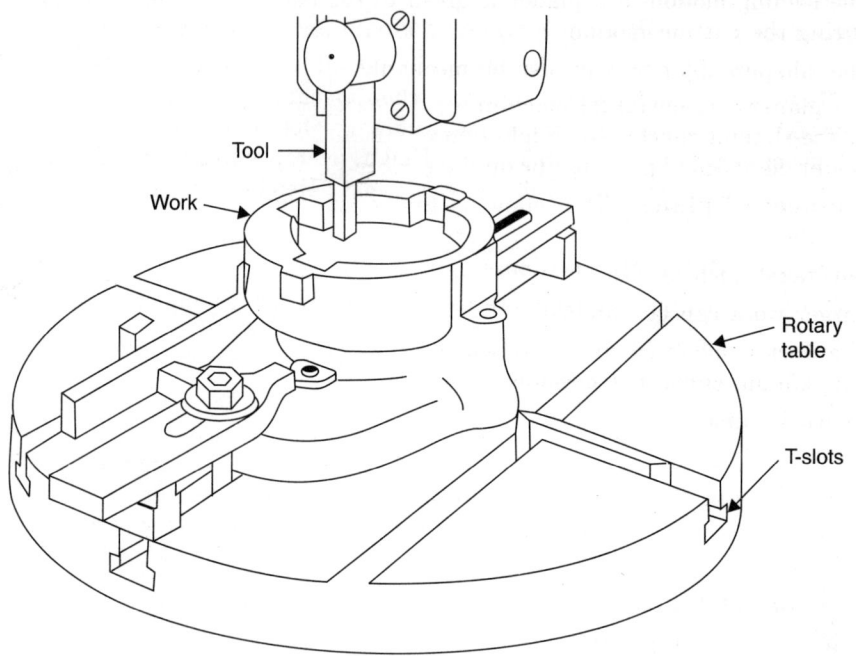

Fig. 9.63. A typical component machined in a slotter.

9.7. PLANING MACHINE (PLANER)

9.7.1. Introduction

Refer to Fig. 9.64.

The planing machine (*planer*) is a machine tool *used in the production of flat surfaces on workpieces too large or too heavy to hold in a shaper. In* this machine, the table called *PLATEN, on which the work is securely fastened, has a reciprocating motion.* The *tool head is automatically* fed horizontal in either direction along the heavily supported cross-rail over the work, and automatic downward feed is also provided.

    — The fundamental difference between a shaper and a planer is that the

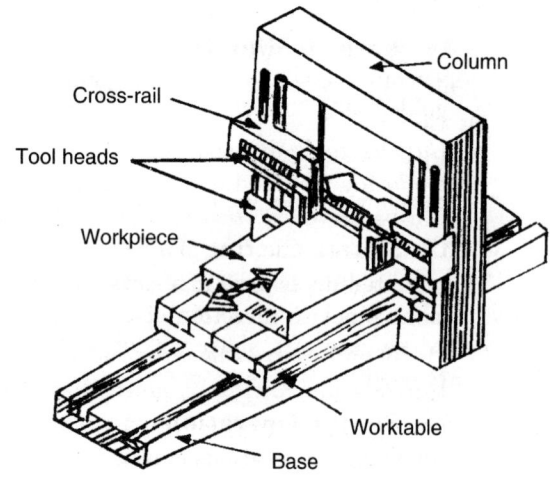

Fig. 9.64. Planer.

latter operates with an action opposite to the shaper, *i.e.*, the workpiece reciprocates past one or more stationary single point cutting tools.

— The feeding motion in a planer is given to the cutting tool, which remains stationary during the cutting motion.

— Like shapers, planers can also be mechanically or hydraulically driven.

● The planers are meant for machining large-sized workpieces, which cannot be machined by the shaping machines. In addition to machining the large parts it is also possible to mount many small parts in line on the table of planer and machine them simultaneously.

Advantages of planer. The following are the *advantages of a planer* as compared to shapers and millers :

1. Can take much heavier cuts.

2. Larger work can be handled.

3. The work is mounted on a table which is supported throughout its entire movement, so a maximum support is obtained.

4. No work or tool deflection or distortion (since there are no overhanging parts such as a ram).

Fig. 9.65 shows components manufactured by planning/shaping processes.

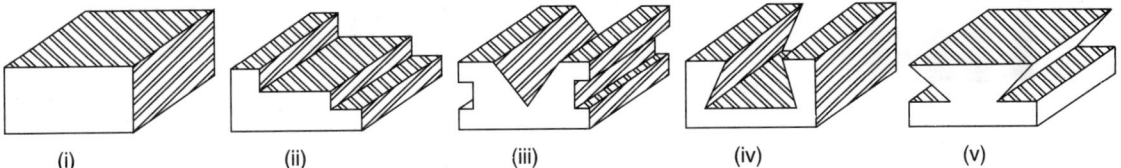

| (i) | (ii) | (iii) | (iv) | (v) |

Fig. 9.65. Components manufactured by planing/shaping processes.

9.7.2. Principle of Working—Planing operation

Fig. 9.66 shows the principle of operation of a planer.

● The workpiece is clamped on to a planer *table* with the help of special bolts that fit in the *T*-slots of the table, and provide a means for positioning and securely holding the workpiece. The table rides over *V*-groves on the bed of the planer and is accurately guided as it *travels back and forth.*

● *Cutting tools* are securely held in the *toolheads* mounted on the *housings* and can be moved *vertically or horizontally* from side to side. The *tool heads* are also mounted on a *horizontal cross-rail* that can be moved *up and down.*

— Use of multiple tool heads permits simultaneous machining of more than one surface of the workpiece and *increases productivity.*

— *With multiple tools cutting simultaneously, relatively slow strokes of the planer table are used.*

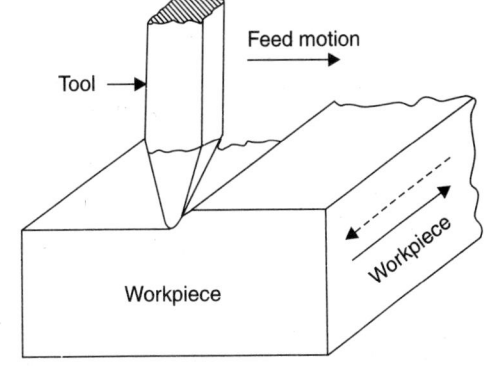

Fig. 9.66. Working principle of a planer.

9.7.3. Comparison between Planer and Shaper

Comparison between a planer and a shaper is given in the Table 9.3 below :

Table 9.3. Comparison between Planer and Shaper

S.No.	Planer	Shaper
1.	Heavier, more rigid and costlier machine.	A comparatively lighter and cheaper machine.
2.	Requires more floor area.	Requires less floor area.
3.	Work reciprocates horizontally.	Tool reciprocates horizontally.
4.	Tool is stationary during cutting.	Work is stationary during cutting.
5.	Heavier cuts and coarse feeds can be employed.	Very heavy cuts and coarse feeds cannot be employed.
6.	Work setting requires much of skill and takes a longer time.	Clamping of work is simple and easy.
7.	Several tools can be mounted and employed simultaneously, usually four as a maximum, facilitating a faster rate of production.	Usually one tool is used on a shaper.
8.	Used for machining large size workpieces.	Used for machining small size workpieces comparatively.

9.7.4. Types of Planers

The various types of planers commonly used are :

1. Double housing (standard) planer.

2. Open side planer.

3. Pit planer.

4. Edge or plate planer.

5. Divided table planer.

9.7.5. Principal Parts of a Planer

The principal parts of a double housing planer (see Fig. 9.67) are described below :

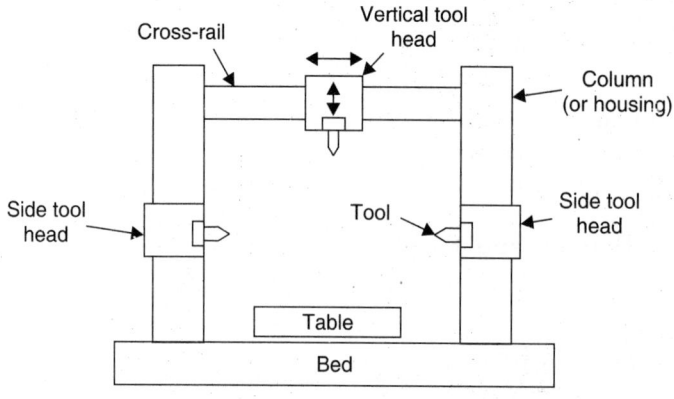

Fig. 9.67. Double housing or standard planer.

1. Bed :

- It is a big cast iron structure.
- The upper part of the bed is provided with precision Vee-type guideways on which the table slides.

2. Table :

- It is made of cast iron and its top surface (flat) is machined accurately.
- It reciprocates along the ways of the bed and supports the work.
- Its top surface is provided with slots to clamp the workpieces.
- It may be driven by rack and gear, by rack and double helical gear or by hydraulic system.

3. Column or Housing :

- The columns or housings are rigid column-like castings placed on each side of the bed.
- The front vertical surface of the column has guideways to enable movement of the cross-rail vertically up and down.

4. Cross-rail :

- It is mounted on the precision machined ways of the two housings.
- It may be raised or lowered on the housings to accommodate work of defferent heights on the table and to allow for the adjustment of the tools.

5. Tool heads :

These are mounted on the cross-rail or housings by means of a saddle which slides along the rail or housing ways. The saddle may be made to move transversely on the cross-rail to give cross feed.

9.7.6. Size of a Planer

- The size of a standard planer is specified by the *size of the largest rectangular solid that can reciprocate under the tool.*
- Double housing planers range from 750 mm × 750 mm × 2.5 m as the smallest and upto 3000 mm × 3000 mm × 18.25 m as the largest size.
- In addition to the basic dimensions, other particulars given below also need to be stated for specifying the planer completely :

(*i*) Number of speeds and feeds available

(*ii*) Power input

(*iii*) Floor space required

(*iv*) Net weight of the machine

(*v*) Type of drive etc.

9.7.7. Planer Driving and Feed Mechanisms

Driving mechanism. A planer driving mechanism provides the longitudinal to and fro motion of the planer worktable. The following methods are employed for the said purpose :

1. Open and cross belt drive.

2. Reversible motor drive.

3. Hydraulic drive.

— The worktable of the planer can be driven by one of the three methods :

(*i*) Rack and spur gears

(*ii*) Spiral rack and worm

(*iii*) Crank.

Feed mechanism. The feed may be given by *hand* or by *power*. The methods employed for power fed are :

1. Friction disc mechanism.

2. Electrical drive.

3. Hydraulic drive.

9.7.8. Standard Clamping Devices

The following are the standard clamping devices used for holding most of the work on a planer table :

1. Heavy duty vices

2. *T*-bolts and clamps

3. Step blocks, clamps, *T*-bolts

4. Poppets or stop pins and the dogs

5. Angle plates

6. Planer jacks

7. Planer centres

8. *V*-blocks

9. Stops.

Some of these clamping devices are shown in Fig. 9.68.

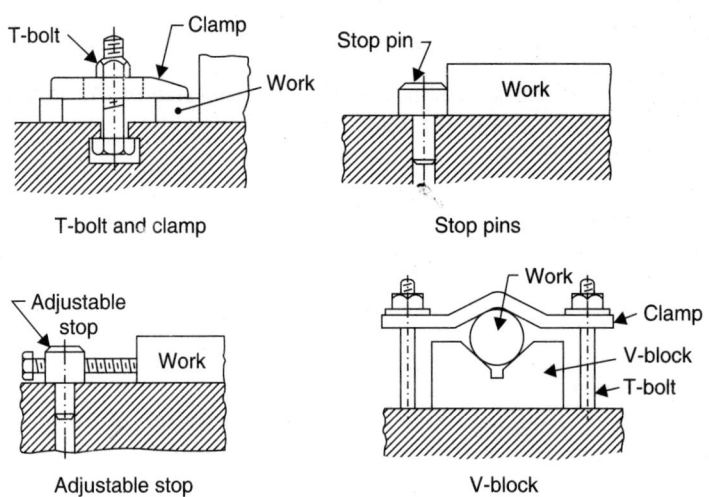

Fig. 9.68. Work holding devices for planer.

9.7.9. Planer Tools

- The planer tools are mostly made of H.S.S. The cemented carbide tipped tools are also used on planers for production work.

- The planer tools are similar to those used on shaper, but as the depth of cut are heavier and cutting edges are large, the planer tools are *large in size.*

- A planer tool may be *classified* as roughing or finishing and right hand or left hand type.

Fig. 9.69 shows the various types of tools used on planers.

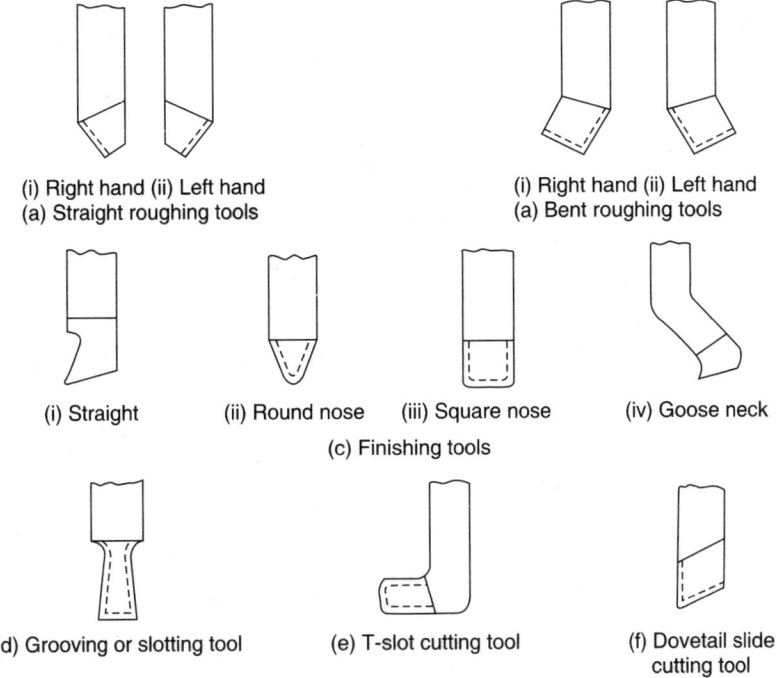

(i) Right hand (ii) Left hand
(a) Straight roughing tools

(i) Right hand (ii) Left hand
(a) Bent roughing tools

(i) Straight   (ii) Round nose   (iii) Square nose   (iv) Goose neck

(c) Finishing tools

(d) Grooving or slotting tool   (e) T-slot cutting tool   (f) Dovetail slide cutting tool

Fig. 9.69. Planer tools.

9.7.10. Planer Operations

The following operations can be performed on planer :

1. Planing horizontal surfaces.
2. Planing vertical surfaces.
3. Planing curved surfaces.
4. Planing slots and grooves.
5. Planing at an angle and machining dove-tails.
6. Planing a helix.
7. Gang or multiple planing.

9.7.11. Cutting Speed, Feed and Depth of Cut

Cutting speed :

- Similar to a shaper, the **cutting speed in planing** *is the average linear speed of the table during the cutting stroke (in m/min) because in planer the table reciprocates.*
- Formula for calculating the cutting speed is the same as that used in the shaping machines except that *N refers to the table strokes per minute and L is the length of table stroke.*

Feed :

- The **feed** *in a planing machine is the distance the tool travels at the beginning of each cutting stroke and is expressed in mm/double stroke.*

Depth of cut :

- *It is the thickness of metal removed in one cut and is measured by the perpendicular distance between the machined and non-machined surfaces and is expressed in mm.*

9.7.12. Machining Time and Material Removal Rate

The machining time and material removal rate (MRR) can be calculated by using eqn. (9.22), (9.23) and (9.26) with the changed meanings of the terms. *If multiple tools cut a workpiece simultaneously, the total cutting time will not cumulate but MRR of each tool should be added to get total MRR.*

Example 9.20. *Calculate the time required to machine the top surface of slab 450 mm (wide) × 3500 mm (long) on a planer, using the following data :*

Cutting speed = 19.2 m/min ; Return speed = 76.5 m/min ; Machining allowance = 10 mm ; Tool approach angle = 45° ; Cross feed of tool = 3 mm/full stroke ; Side overtravel of tool = 3 mm ; Table overtravel on both side = 300 mm.

Solution. *Given :* B = 450 mm ; L_j = 3500 mm ; V_c = 19.2 m/mm ; V_r = 76.5 m/min ; Machining allowance = 10 mm ; Tool approach angle λ = 45° ; f = 3 mm/full stroke.

Machining time, t_m :

Length of table stroke for planer (L)

$\qquad$ = Length of job (L_j) + table travel on both sides

i.e. $\qquad L$ = 3500 + 300 = 3800 mm

The cutting speed is given by :

$$V_c = \frac{N\,L\,(1+k)}{1000}\ \text{m/min} \qquad\qquad ...[\text{Eqn (9.18)}]$$

where, N = Number of full strokes/double strokes per min, and

$\qquad k$ = Return stroke time/cutting stroke time

$$= \frac{\text{Cutting speed}}{\text{Return speed}} = \frac{19.2}{76.5} = 0.251$$

$\therefore \qquad N = \dfrac{1000\,V_c}{L(1+k)} = \dfrac{1000 \times 19.2}{3800(1+0.251)} = 4.0$

Now, cutting time, $t_m = \dfrac{\text{Width of job } (B) + \text{side approach of tool} + \text{side over travel of tool}}{f\,N}$

where, Side approach of tool = 10 cot 45° = 10 mm

$\qquad$ Side overtravel of tool = 3 mm $\qquad\qquad\qquad\qquad$ ....(Given)

$\therefore \qquad t_m = \dfrac{450 + 10 + 3}{3 \times 4.0} = \textbf{38.58 min.}\ \ \textbf{(Ans.)}$

9.8. DRILLING MACHINES

9.8.1. Introduction

9.8.1.1. Drilling

Drilling is a *process of making hole or enlarging a hole in an object by forcing a rotating tool called "Drill".*

— The *drill* is generally called as '*twist drill*', since it has a sharp twisted edges formed around a cylindrical tool provided with a helical groove along its length to allow the cut material to escape through it. The sharp

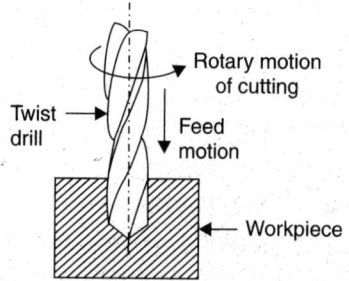

Fig. 9.70. Drilling operation.

edges of the conical surfaces ground at the lower end of the rotating twist drill cut the material by peeling it circularly layer by layer when forced against a workpiece. The removed material chips get curled and escape through the helical grooves provided in the drill. A liquid coolant is generally used while drilling to remove the heat of friction and obtain a better finish for the hole.

9.8.1.2. Drilling Machine

A power operated machine tool which holds the drill in its spindle rotating at high speeds and when manually actuated to move linearly simultaneously against the workpiece produces a hole is called **drilling machine.**

- Drilling machine is one of the simplest, moderate and accurate machine tool used in production shop and tool room. It consists of a ***spindle*** which imparts rotary motion to the drilling tool, or mechanism for feeding the tool into the work, a ***table*** on which the work rests and a ***frame.*** It is considered as *a single purpose machine tool since its chief function is to make holes.* However, it can and does perform operations *other than drilling also.*

9.8.2. Specifications of a Drilling Machine

A drilling machine is *specified* as follows (Refer to Fig. 9.71) :

1. Size of the drilling machine table.

2. Largest bit the machine can hold.

3. Maximum size of the hole that can be drilled.

4. Maximum size of the workpiece that can be held.

5. Power of the motor, spindle speed or feed.

Specifically, the various *types of drilling machines are specified as follows :*

- **Portable drilling machine**—*Maximum diameter of drill which can be held.*

- **Sensitive and upright drilling machines**—*The diameter of the largest workpiece that can be drilled.*

- **Radial drilling machine**—*The length of the arm and column diameter.*

- **Multiple spindle drilling machine**—*The drilling area, the size and number of holes a machine can drill.*

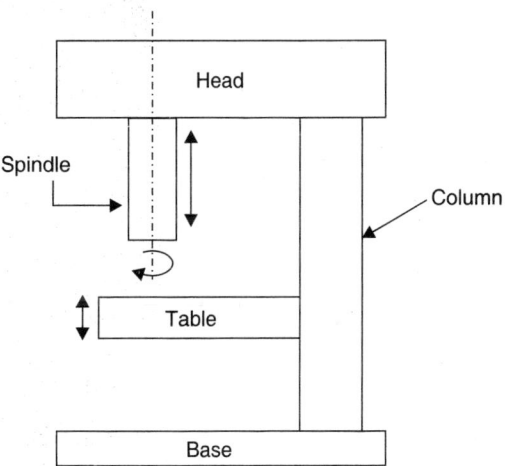

Fig. 9.71. Block diagram of a drill press.

9.8.3. Types of Drilling Machines

Drilling machines are manufactured in various sizes and varieties to suit the different types of work. They can, however, be broadly *classified* as follows :

1. Portable drilling machine.

2. Sensitive or bench drilling machine.

3. Upright drilling machine.

4. Radial drilling machine.

5. Gang drilling machine.

6. Turret machine.

7. Deep hole drilling machine.

8. Multiple spindle drilling machine.

9. Automatic drilling machines.

1. **Portable drilling machine :** Refer to Fig. 9.72.

Portable drilling machine is a very small, compact and self contained unit carrying a small *electric motor inside it.*

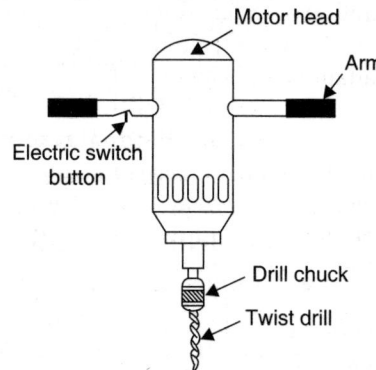

Fig. 9.72. Portable drilling machine.

- It is very commonly used to drill holes in the following *cases* : (*i*) when the component is bigger in size such that it can not be shifted to the shop floor ; (*ii*) when the space is restricted so that no other type of drilling machine can be used.

- Usually they are made to hold drills upto a maximum diameter of 12 mm. However, portable drills of upto 18 mm dia. capacity are available.

2. **Sensitive or bench drilling machine :** Refer to Fig. 9.73.

These are *light duty machines* used in workshops. They are normally mounted on work benches and hence the name. As the operator can feel the cutting operation while applying pressure using the feed lever, the machine is known as *sensitive drilling machine.*

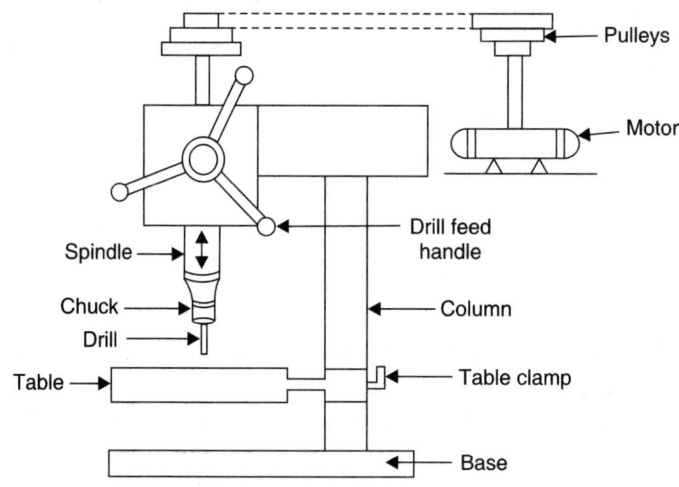

Fig. 9.73. Sensitive or bench drilling machine.

It consists of a cast iron *base* with a vertical *column* mounted over it. The vertical column is made of hollow steel pipe on which the *table* slides up and down. The *table* can be fixed to the required position by means of a table *clamp*. The table can also be swung radially at any desired position. The top of the column houses the *drive* consisting of endless belt running over the V-pulleys. Based on the speed of *spindle* required, V-belt can be shifted to different grooves of the *pulleys. To* drill small diameter holes, a twist-drill is fitted in the *drill chuck*, which in turn fits into the spindle of the machine. If the drill size is more, twist drill is directly fitted in the tapered portion of the spindle. The spindle can be moved up or down by means of drill feed handle or lever.

- This design is used to drill hole from 1.5 mm to 15 mm diameter. The controls are light and delicate speeds from 800 to 900 r.p.m. are a typical range.

3. Upright drilling machine :

Upright drills similar to sensitive drills have power-feed mechanisms for rotating drills and are designed for *heavier work.*

A box column machine is more rigid than a round column machine and consequently, is adapted to heavier work. These drilling machines *tap as well as drill.*

4. Radial drilling machine : Refer to Fig. 9.74.

A radial drilling machine is used to perform the drilling operations on the workpieces which are *too heavy and also may be too large to mount them on the worktable* of the vertical spindle drilling machine.

It consists of a heavy *base* and a *vertical column* with a long *horizontal/radial arm* extending from it and can be rapidly raised, lowered and swing in horizontal plane about the main column to any desired location. The *drilling head* can move to and fro along the arm and can be swivelled only in the universal radial drilling machines, to drill holes at an angle. The combinations of motions of the radial arm and drilling head offer a great deal of flexibility in moving the drill to any position.

The *main advantage* of the radial drilling machine is that the *drilling can be carried out on heavy workpieces in any position without moving them.*

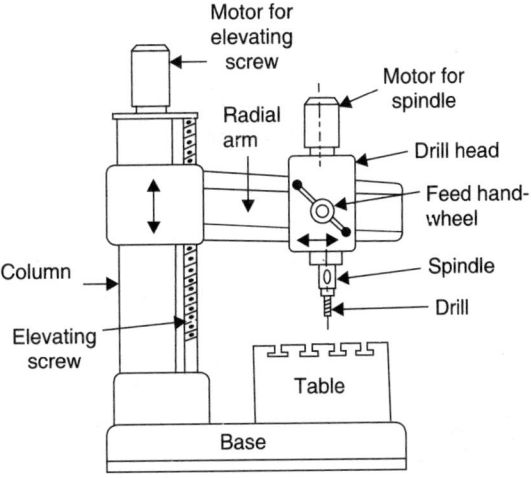

Fig. 9.74. Radial drilling machine.

- This type of drilling machine is used in *tool rooms and in large scale die manufacturing units.*

5. Gang drilling machine :

When several drilling spindles are mounted on a single table, it is known as a *gang drill* (Fig. 9.75 shows a three-spindle schematic).

In this type of drill, *each of these spindles can be independently set for different speed and depth of cut.*

- Such machines are useful when number of holes of different sizes are to be drilled on the same workpiece.
- Apart-from drilling, a number of other machining operations like *reaming, counter boring, tapping* etc ; can also be performed *at a time* on this machine.

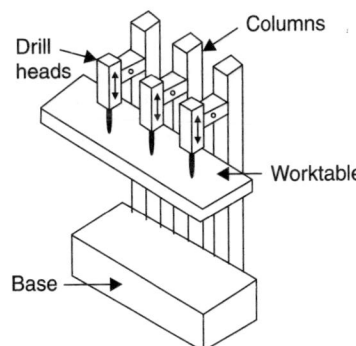

Fig. 9.75. Gang drilling machine.

6. Turret drilling machine :

A turret machine overcomes the floor space restriction caused by a gang drill press. A six-turret *NC* drill press is shown in Fig. 9.76.

— The stations are set up with a *variety of tools. Numerical control is also available.*

— Two fixtures can be located side by side on the worktable, thus permitting loading and unloading of one part while the other part is being machined ; this *reduces the machine cycle.*

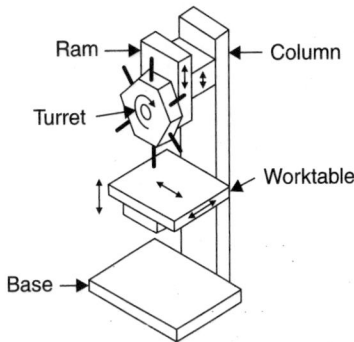

Fig. 9.76. Turret drilling machine.

7. Deep-hole drilling machines :

— These machines are used for drilling holes whose depth exceed normal drill size. These machines are operated at *high speed and low feed.*

— These machines are either horizontal or vertical. The work or the drill may revolve. Most machines are of horizontal construction using a center-cut gun drill, which has a single cutting edge with a straight flute running throughout its length (see Fig. 9.77). Oil under high pressure is forced to the cutting edge through a lengthwise hole in the drill. In gun drilling the feed must be light to avoid deflecting the drill and causing it to meander through it length.

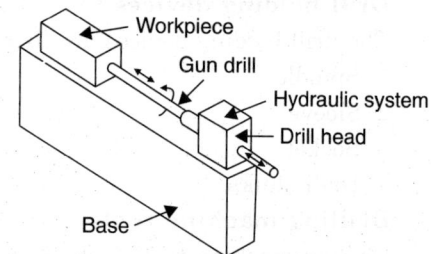

Fig. 9.77. Gun drilling machine.

● These machines are very useful for drilling deep holes in *rifle travels, crankshafts* etc.

8. Multispindle drilling machines :

These machines are vertical type machines. They *permit drilling of several holes of different diameters simultaneously.* Generally the spindles numbering 2 or 3 or even more are driven by only one gear in the head through universal joint linkages. Each spinde is mounted with a twist drill. A jig is used to guide the twist drills.

● These machines are mostly used in *continuous production shops* where several holes of same diameter or different diameters are to be drilled *simultaneously and accurately.*

9. Automatic drilling machines :

Automatic drilling machines are *production machines* arranged in series to perform a number of different operations in *sequence at successive work stations.*

— The workpieces, after completion of an operation at one station, are automatically transferred to the next station for another operation. Thus, it works as a *transfer line.*

— Several different operations like drilling, boring, tapping, milling, housing, etc. can be performed on a job in succession on these machines.

9.8.4. Classification of Drills

The tool used for drilling is called a *drill.* The commonly used drills may be *classified* in several ways, as follows :

1. *According to the type of shank* :
   (*i*) Parallel shank.
   (*ii*) Taper shank.
2. *According to the type of flutes* :
   (*i*) Flat or spade drills (parallel longitudinal flutes)
   (*ii*) Twist drills (spiral/helical fultes)
3. *According to length.*
   (*i*) Short series drills.
   (*ii*) Stub series drills.
   (*iii*) Long series drills.
4. *According to applications :*
   (*i*) Core drills.
   (*ii*) Drills for long hole drilling.
   (*iii*) Centre drills.
   (*iv*) Masonry drills.

5. *According to the tool material :*

    (*i*) High speed steel drills.

    (*ii*) Carbide tipped drills.

Drill holding devices :

The drill holding devices are listed below :

1. Spindle

2. Sleeve

3. Socket

4. Drill chuks.

Drilling machine tools :

Drilling machine tools include the following :

1. Flat drill	6. Centre drill
2. Straight drill	7. Reamer
3. Twist drill	8. Centre punch
4. Taper shank core drill	9. Drift
5. Oil tube drill	10. Hammer.

9.8.5. Twist Drill Nomenclature

Fig 9.78 shows the elements of a twist drill (dotted lines at the shank portion indicates the straight shank and firm line of the shank indicates the taper shanks.

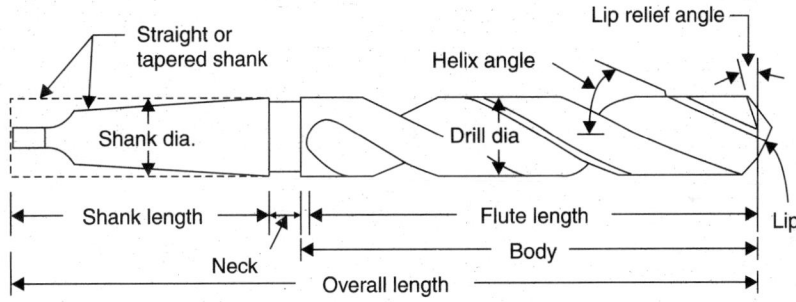

Fig. 9.78. Elements of a twist drill.

Fig. 9.79 shows the twist drill nomenclature.

The twist drill consists of mainly two parts **body** and **shank**. Both are separated by a neck. Two long and diametrically opposite helical grooves called **flutes** run throughout the length of the drill.

1. Body :

The body is the portion of the drill which extends from its extreme point upto the neck or shank of the drill. It consists of body clearance, chisel edge, face, flank, flutes, heel, land or margin, point, lip and web.

    (*i*) ***Body clearance.*** It is the portion of the body surface with reduced diameter which provides diametral clearance.

    (*ii*) ***Face.*** It is the portion of the flute adjacent to the lip on which chip flows as it is cut from the workpiece.

    (*iii*) ***Flank.*** It is the conical surface of a drill point which extends behind the lip to the following flute.

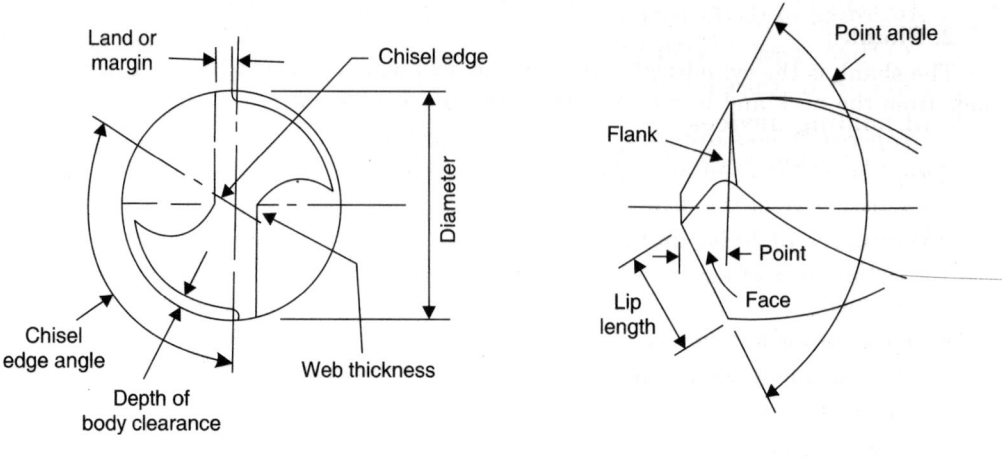

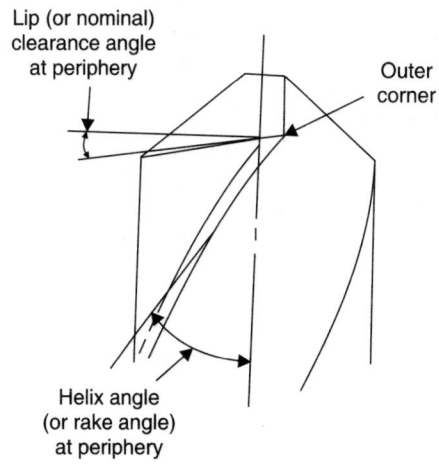

Fig. 9.79. Twist drill nomenclature.

(*iv*) **Flutes.** These are helical grooves cut on the cylindrical surface of the drill and provide the lip. They serve the following *purposes* :

● Ensure easy escape and flow of chips.

● Cause the chips to curl and provide passages for their flow.

● Form the lips and cutting edges on the point.

● Allow the cutting fluids to reach the cutting edges thus reducing their friction.

(*v*) **Heel.** It is the edge formed by the intersection of flute surface and the body clearance.

(*vi*) **Land.** It is the cylindrically ground narrow strip on the leading edges of drill flutes. It keeps the drill aligned. It is also known as "*margin*".

(*vii*) **Point.** It is the cone shaped sharpened end of the drill that produces lips, faces, flanks and chisel edge of the drill.

(*viii*) **Lips.** The lips, also known as "*cutting edges*", are the edges formed by the intersection of flanks and faces. They are two in number with identical length and angle.

(*ix*) **Web.** It is the thickness of the drill between the flutes which extends from point towards the shank. The point end of the web forms the chisel edge.

2. Shank :

The shank is the cylindrical portion of the drill which is used to hold and drive the drill. It extends from the neck and it may be either straight or papered.

— *Tapered shanks are used in drills of bigger sizes.*

Tang. It is flattened end of the taper shanks which fits into socket or drill holder. It ensures positive drive of the drill from the drill spindle.

Advantages of twist drills :

The *advantages* of using twist drills are :

1. For the same size and depth of the hole they need less power in comparison to other forms of drills.
2. Cutting edges are retained in good condition for a fairly long time, thus avoiding the frequent regrinding of the drill.
3. The chips and·cuttings of the metal are automatically driven out of the hole through the flute.
4. Heavier feeds and speeds can be employed quite safely, resulting in a considerable saving of time.

9.8.6. Work Holding Devices

The type of work holding device used on drilling machines depends upon the *shape and size of the workpiece, the required accuracy and the rate of production.* Some of the work holding devices are :

1. Machine vice.
2. V-blocks.
3. Strap clamps and T-bolts.
4. Drilling jigs.
5. Angle plate.

9.8.7. Drilling Machine Operations

In addition to drilling, the following operations are carried out on a drilling machine :

1. Reaming
2. Boring
3. Counter-boring
4. Counter-sinking
5. Spot facing
6. Tapping
7. Trepanning.

9.8.7.1. Reaming

Reaming is the operation of finishing an existing hole very smoothly and accurately in size. (See Fig. 9.80).

A drill will not produce a hole having sufficiently good qualities of finish and accuracy for many purposes. Therefore, when a very accurate, smooth hole is required the hole is first drilled a little undersize. Then it is reamed to the correct size.

— The accuracy to be expected is within ± 0.005 mm.

● A *reamer is a multi-tooth cutter which rotates and moves linearly into an already existing hole.*

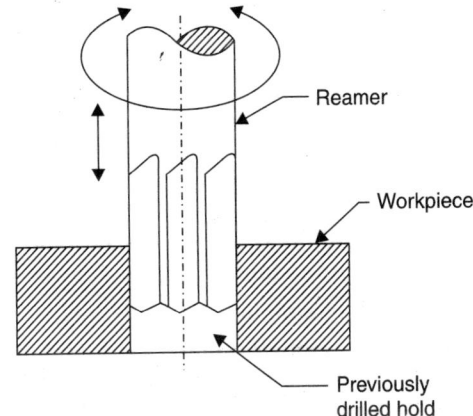

Fig. 9.80. Reaming for finishing to size a drilled hole.

The previous operation could be drilling or preferably boring. Reaming provides a smooth surface as well as close tolerance on the diameter of the hole. Generally the reamer follows the already existing hole and the, therefore, will not be able to correct the hole misalignment.

The reamers are of following types :

(*i*) Hand reamers [See Fig. 9.81 (*a*)].

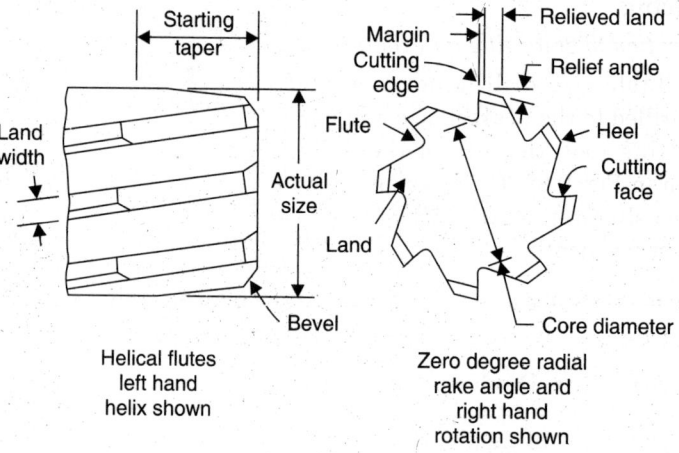

(a) Hand reamer

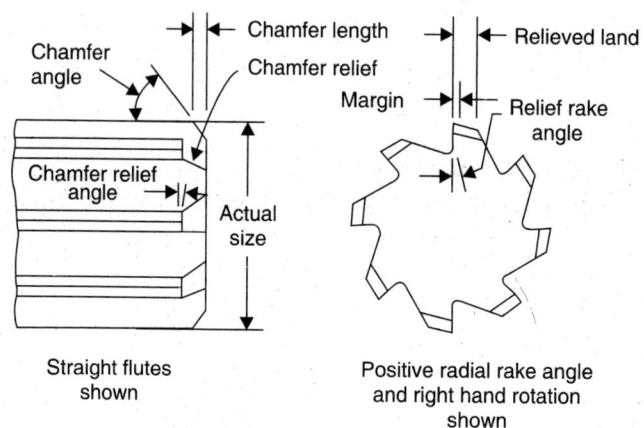

(b) Machine reamer

Fig. 9.81. Hand and machine reamers.

(*ii*) Chucking or machine reamers [see Fig. 9.81 (*b*)].

(*iii*) Adjustable reamers.

(*iv*) Expansion reamers.

(*v*) Taper reamers.

(*vi*) Taper pin reamers.

(*vii*) Shell reamers.

(*viii*) Carbide tipped reamers.

Some *other important forms of reamers are :*

 (*i*) Socket reamer for morse taper ;

 (*ii*) Machine bridge reamers ;

 (*iii*) Stub screw machine reamers ;

 (*iv*) Die maker's reamers.

9.8.7.2. Boring

It is an operation of enlarging an existing hole.

When a suitable size drill is not available, initially a hole is drilled to the nearest size and using a single point cutting tool, the size of the hole is increased as shown in Fig. 9.82. By lowering the tool while it is continuously rotating, the size of the hole is increased to its entire depth.

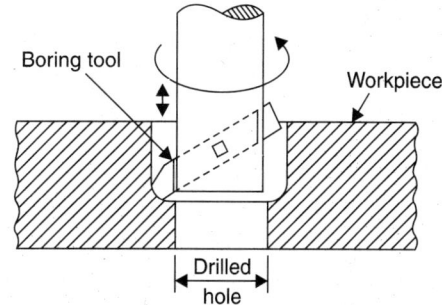

Fig. 9.82. Boring.

9.8.7.3. Counteboring

It is an operation of enlarging a drilled hole partially, that is for a specific length. (Fig. 9.83)

— The counterboring forms a large sized recess or a shoulder to the existing hole.

— The cutting tool will have a small cylindrical projection known as *pilot* to guide the tool while counterboring. The diameter of the pilot will always be equal to diameter of the previously drilled hole. Interchangeable pilots of different diameters are also used for counterboring holes of different diameters.

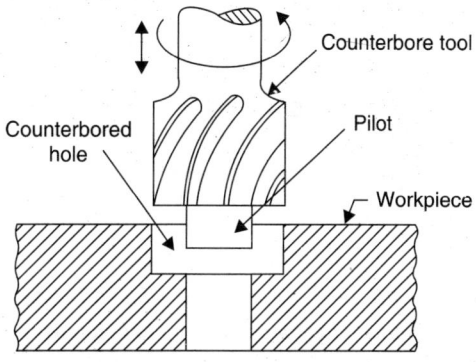

Fig. 9.83. Counterboring.

— *The speeds for counterboring must be two thirds of the drilling speed of the corresponding size of the drilled hole.*

● Generally the counterboring is done on the holes to accommodate the socket head screws, or grooved nuts, or round head bolts.

9.8.7.4. Countersinking

It is an operation of forming a conical shape at the end of a drilled hole. (Fig. 9.84). It is done using a countersink tool.

— The cutting speeds for countersinking must be about one-half of that used for similar size drill.

The countersunk holes are used when the countersunk screws are to be screwed into the holes so

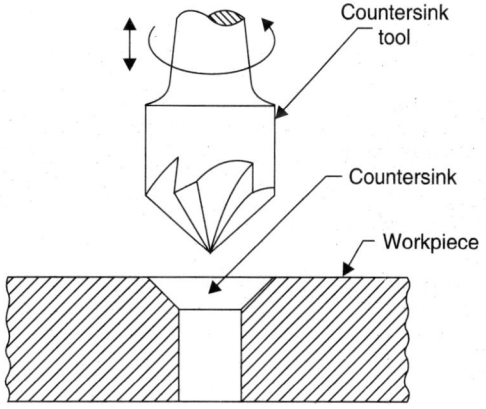

Fig. 9.84. Countersinking.

that their top faces have to be in flush with the top surface of the workpiece.

The countersinking process may also be employed for *deburring holes.*

9.8.7.5. Spot facing

It is the operation of smoothing and squaring the surface around a hole for the seat for the nut or the head of a screw. (Fig. 9.85).

Spot facing may be done with a counterboring tool or using a special spot facing tool.

9.8.7.6. Tapping

It is an operation in which external threads are cut in the existing hole.

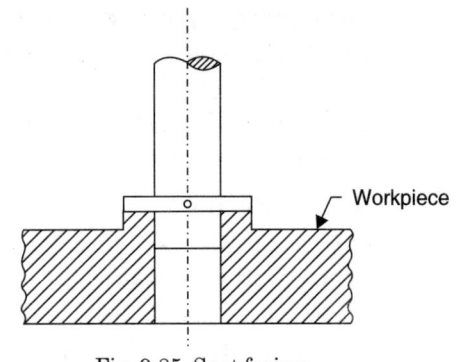

Fig. 9.85. Spot facing.

Fig. 9.86 shows tapping operation which uses a fluted threaded tool called *tap*. A tap is a cutting tool with threads cut accurately on its periphery. These threads are hardened and ground and act a cutting edges. The tap removes

metal when screwed into the hole and generates internal threads. A hole of required size in which internal threads are to be generated is drilled using a twist drill. The drill spindle is fitted with a tap and the feeding is done by operating the feed lever similar to conventional drilling operation. *During this operation spindle speed should be much lower than that used in conventional drilling.*

9.8.7.7. Trepanning

It is the operation of producing a hole by removing the metal along the circumference of a hollow cutting tool (Fig. 9.87).

This operation is performed for producing *large holes.*

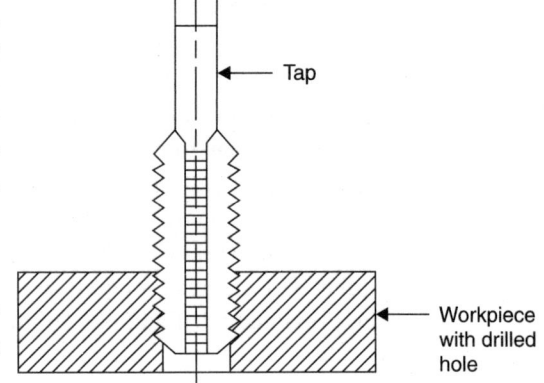

Fig. 9.86. Tapping operation and a tap.

9.8.8. Cutting Speed, Feed and Depth of Cut

Cutting speed. *It is the peripheral speed of a point on the surface of the drill in contact with the workpiece.* It is usually expressed in *metres per minute.*

Mathematically, $V_c = \dfrac{\pi DN}{1000}$

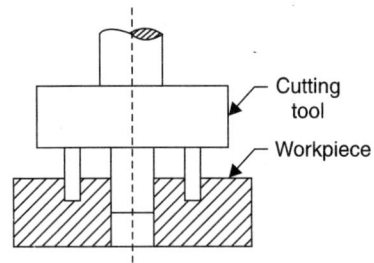

Fig. 9.87. Trepanning.

where, V_c = Cutting speed (surface), m/min,

   D = Diameter of the drill, mm, and

   N = Rotational speed of the drill, r.p.m.

The *cutting speed* depends upon the following factors :

(*i*) The type of material being drilled.

(*ii*) Cutting tool material.

(*iii*) The quality of hole desired.

(*iv*) The efficient use of cutting fluid.

(*v*) The way in which the work is set up or held.

(*vi*) The size and type of drilling machine.

Feed (f). *It is the distance the drill moves into the work at each revolution of the spindle.* It is expressed in mm/rev. It may also be expressed as *feed per minute.*

The correct feeds for different sizes of drill are given below :

Drill size, mm	Feed, mm/rev
3.2 and less	0.025—0.050
3.2 to 6.4	0.050—0.10
6.4 to 12.7	0.10—0.18
12.7 to 25.4	0.18—0.38
25.4 and large	0.38—0.64

— A twist drill gives satisfactory performance if it is run at correct cutting speed and feed. The following factors help in running the drill at correct cutting speed and feed :

(*i*) The drill is correctly selected and ground for the material being cut. The selection of drill depends upon the following factors :

(*a*) Size of drill hole ;

(*b*) Material of workpiece ;

(*c*) Point angle of drill.

(*ii*) The work is rigidly clamped

(*iii*) The machine is in good condition

(*iv*) A coolant is used if required.

— The rates of feed and cutting speed for twist drill are lower than most other machining operation because of the following *reasons* :

(*i*) The twist drill is weak compared with other cutting tools.

(*ii*) It is relatively difficult for the drill to eject chips.

(*iii*) It is difficult to keep the cutting edges cool when they are enclosed in the hole.

Depth of cut (d). *It may be defined as the distance from the machined surface to the drill axis. That is,*

$$d = \frac{D}{2} \qquad\qquad ...(9.27)$$

● The choice of operating conditions in drilling operations becomes more critical with increase in the hole depth. As the depth of hole increases, (*i*) the chip ejection becomes more difficult, and (*ii*) the fresh cutting fluid is not able to reach to the cutting zone. These factors lead to overheating of the drill and shortens its life. Hence, for *machining of lengthy holes, reduced feeds are used.*

— For machining holes of very large length, a special type drilling process, known as *gun drilling* is used. By this process, it is possible to machine the holes having length greater than 300 times of the diameter.

9.8.9. Machining Time in Drilling. Refer to Fig. 9.88.

Machining time, $t_m = \dfrac{L_j + l_1 + l_2}{fN}$, min

$$...(9.28)$$

where, L_j = Hole length or depth mm,

l_1 = Tool approach $\approx 0.29\,D$ (with point angle of 118°), and

l_2 = Tool overtravel, 1 to 2 mm.

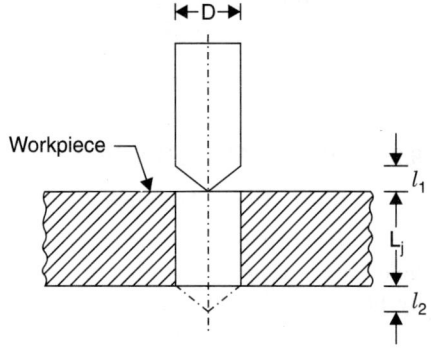

Fig. 9.88. Drilling operation.

9.8.10. Material Removal Rate

The material removal rate is indicated by the total volume of the material in the hole. In the case of a solid material without coring, the material removal rate *MRR* is given by the *area of cross-section of the hole times the tool travel rate through the material.* Thus

$$\text{MRR} = \frac{\pi D^2 fN}{4} \text{ mm}^3/\text{min} \qquad\qquad ...(9.29)$$

where, D = Drill diameter, mm,

f = Feed, mm/rev, and

N = Rotational speed of the drill, r.p.m.

Example 9.21. *A hole of 30 mm diameter and 75 mm depth is to be drilled. The suggested feed is 1.3 mm per rev. and the cutting speed is 62 m/min. Assuming tool approach and tool overtravel as 6 mm, calculate ;*

(*i*) Spindle r.p.m.

(*ii*) Feed speed.

(*iii*) Cutting time.

(*iv*) Material removal rate.

Solution. *Given* : D = 30 mm ; L_j = 75 mm ; f = 1.3 mm/rev ; V_c = 62 m/min ; $l_1 + l_2$ = 6 mm

(*i*) **Spindle r.p.m :**

Spindle r.p.m. $= \dfrac{1000\,V_c}{\pi D}$ $\left(\because\ V_c = \dfrac{\pi DN}{1000} \right)$

$$= \frac{1000 \times 62}{\pi \times 30} = \textbf{658 r.p.m.} \textbf{ (Ans).}$$

(*ii*) **Feed speed :**

Feed speed $= fN = 1.3 \times 658 = $ **855.4 mm/min.** **(Ans).**

(*iii*) **Cutting time :**

Cutting time, $t_m = \dfrac{L}{fN} = \dfrac{L_j + l_1 + l_2}{fN}$

$$= \frac{(75 + 6)}{1.3 \times 658} = \textbf{0.095 min.} \textbf{ (Ans).}$$

(iv) **Material removal rate :**

Material removal rate, MRR = $\dfrac{\pi D^2 \times f \times N}{4}$

$$= \frac{\pi \times (30)^2 \times 1.3 \times 658}{4} = \mathbf{60.4647 \times 10^4 \ mm^3/min.} \quad \textbf{(Ans).}$$

9.9. BORING MACHINES

9.9.1. Introduction

Boring is the process of using a single point tool to enlarge and locate a previously made hole. Drills tend to wander or drift, thus, where greater accuracy is required, drilling is followed by boring and reaming.

— Besides enlarging previously made holes, a boring machine can be used for drilling, facing, milling etc.

— Boring machines are one of the largest of the machine tools and are able to machine workpieces weighing upto 180 kN.

● The boring tool for a boring machine is usually a single point cutting tool made of HSS or carbide and is mounted on the tool head. It is capable of vertical movement and radial movement guided by the cross rail. The head can be swivelled to produce tapered internal surfaces or taper boring.

Comparison between boring and reaming

(i) Boring can correct hole location , size, or alignment and can produce a good finish if a fine feed and a correct tool are used.

The *reamer* follows the hole already in the workpiece and so *cannot correct location.*

(iii) Reaming involves the use of a tool of fixed size, which is different for each size of hole and a large hole would require an *expensive reamer,* while a *boring tool can make a hole of any size.*

(iii) Reaming is faster than boring but boring operation is often preferred because of location correction advantage.

9.9.2. Classification of Boring Machines

Boring machines are manufactured in various designs and sizes. They can be broadly be *classified* as follows ;

1. Horizontal boring machines (HBM) ⎤
    (i) Table type HBM
    (ii) Planer type HBM   *...Production machines*
    (iii) Floor type HBM
    (iv) Multiple spindle HBM.
2. Vertical boring machines ⎦
3. Jig boring machines   ... *Precision machine* used for precision boring operations such as jig boring.

9.9.3. Horizontal Boring Machines

The *table type* or *universal type* is the most versatile and commonly used horizontal boring machine. Fig. 9.89 shows the block diagram of a horizontal boring, drilling and milling machine. The principal features of this machine are,

1. **Bed.** A heavy and strong bed carries the entire load of different parts, workpiece and tooling over it.

2. **Columns.** Two vertical columns, one on each end of the table.

3. **Head stock.** The head stock can be moved vertically along the *main column.*

4. **Horizontal spindle.** It is suitably housed in the headstock, and can be rotated and fed forward and backward according to requirement.

5. **Load bearing end support.** It supports the end of a long boring bar and can be adjusted vertically along the *end support column.*

6. **Horizontal table.** It is mounted on a saddle and can be moved horizontally forward and backward and sideways by moving the saddle.

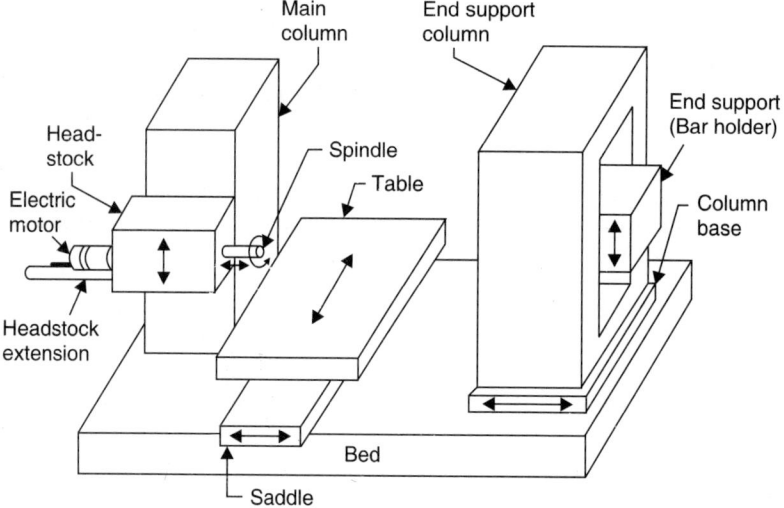

Fig. 9.89. Block diagram of a horizontal boring, drilling and milling machine.

- A horizontal boring machine nick named a *"bar"*, differs from the vertical boring mill in that the work is stationary and tool is revolved. It is adapted to the boring horizontal holes, and these holes and components are generally large. If these conditions of component size and hole size are not met, other smaller machines are used.

9.9.4. Vertical Boring Machines (VBM)

The parts whose length or height is less than diameter are machined, for convenience, on vertical boring machines.

On a VBM, the work is fastened on a horizontal revolving table, and the cutting tool or tools, which are stationary, advance vertically into it as the table revolves.

There are two designs of a vertical boring machine : (*i*) Single column VBM, and (*ii*) Double column VBM.

Fig. 9.90 shows the block diagram of a double column VBM.

— The work is accommodated on the horizontal revolving table at the front of the machine. The circular work can be clamped on to the table with the help of jaw chucks whereas the *T*-slots can be used with bolts and clamps for setting up and holding irregular work.

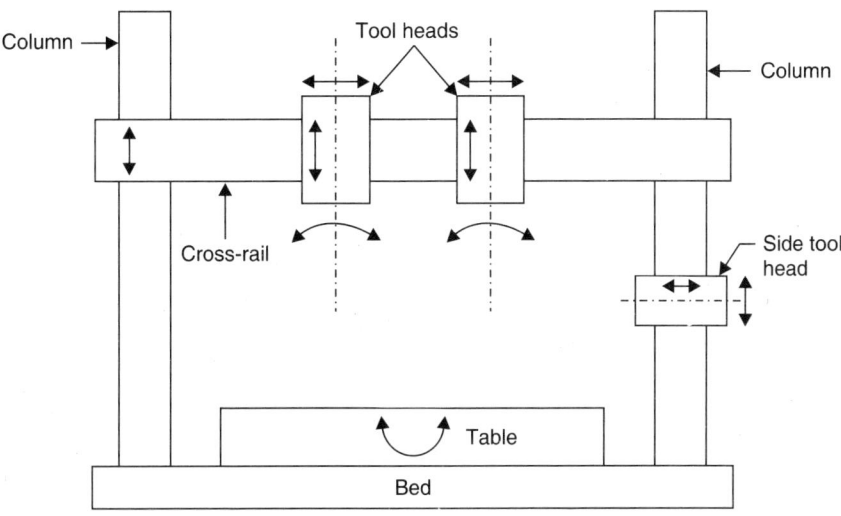

Fig. 9.90. Double column vertical boring machine.

— A horizontal cross-rail is carried on vertical slideways and carries the tool holder slide or slides.

On machines designed for working on large batches of similar articles, a single slide with turret may be employed.

— Most machines also have a slide tool head.

● A vertical boring machine is sometimes called a rotary *planer* and its cutting action on flat discs is identical with a planer. These machines, rated according to their table diameter, vary in size from 0.9 to 12 m.

9.9.5. Jig Boring Machine

A jig boring machine is a *very precise* vertical type boring machine.

Fig. 9.91 is a schematic of a machine *designed for locating and boring holes in jigs, fixtures, dies, gauges, and other precision parts.*

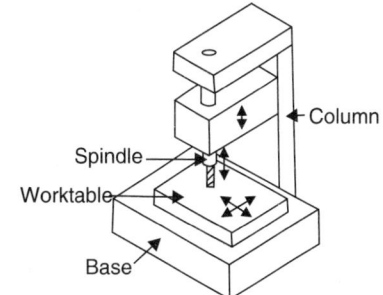

Fig. 9.91. Jig boring machine.

— Jig boring machines are constructed with greater precision and are equipped with accurate measuring devices for controlling table movements. On typical machines positioning to ± 0.003 mm can be dialled directly from a drawing.

— These machines are also operated by *numerical control.* By putting part design on tape, accurate repetition is ensured, jigs and fixtures are eliminated, and precision boring becomes practical for small-lot manufacturing.

— To prevent the influence of ambient temperature changes on machining, jig boring machines should be installed in special environmental enclosures with temperature maintained at a level of 20°C.

● *Jig-borers are used as coordinate measuring machines for inspection and precision layout operations.*

9.9.6. Size of Boring Machines

- The size of boring machine is given by the *diameter of the spindle* (75 mm to 350 mm).
- In case of vertical boring machines, the dimensions of column height and table size (diameter), 11200 mm to 3600 mm, also have to be given.

9.10. MILLING MACHINES

9.10.1. Introduction

Milling machine is a machine tool in which metal is removed by means of a revolving cutter with many teeth, each tooth having a cutting edge which removes metal from a workpiece.

- In a milling machine the work is supported by various methods on the workable, and may be fed to the cutter *longitudinally, transversely or vertically.*
- In the *milling process,* the workpiece is normally fed into a rotating cutting tool known as *milling cutter.* Equally spaced peripheral teeth on the cutter come in contact with the workpiece intermittently and machine the workpiece. This is called *intermittent cutting.* In some special milling machines, the workpiece remains stationary and the cutter is fed into the workpiece.
- Milling machines are used to produce parts having *flat as well as curved shapes. Intricate shapes,* which cannot be produced on the other machine tools, can be made on the milling machines.
- This machine in perhaps next to the lathe in importance.

9.10.2. Milling Processes—Principle of Milling

Generally there are *two types of milling processes*, namely :

1. Up milling (or conventional milling) process
2. Down milling (or climb milling) process.

1. Upmilling (or conventional milling) process. Refer to Fig. 9.92.

In *"upmilling process",* the *workpiece is fed opposite to the cutter's tangential velocity.* Each tooth of the cutter starts the cut with zero depth of cut, which gradually increases and reaches the maximum value as the tooth leaves the cut. *The chip thickness at the start is zero increases to maximum at the end of the cut.*

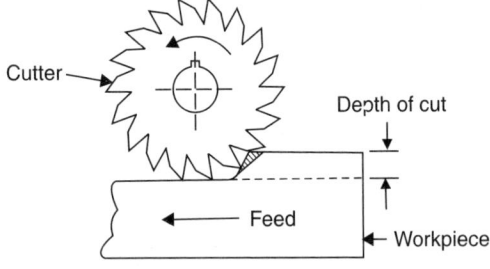

Fig. 9.92. Up milling process.

- The action of the cutter forces the workpiece and the table against the direction of table feed, thus each 'tooth' enters a clean metal gradually thus the *shock load on each tooth is minimised.*

Disadvantages :

1. When making deep cuts, such as in heavy slotting operations, the cutter tends to pull the workpiece out of the vice or the fixture since the cutting force is directed upward at an angle ; This requires secured clamping of workpiece.

2. Owing to typical nature of the cut, difficulty is experienced in pouring coolant on the cutting edge and as a result, chips accumulate at the cutting zone and may be carried over with the cutter, thus *spoiling the surface finish.* The surface becomes slightly wavy, as the cut does not begin as soon as the cutter touches the workpiece.

2. Down milling (or climb milling) process : Refer to Fig. 9.93.

In *"down milling process" the workpiece is fed in the same direction as that of the cutter's tangential velocity.* The cutter enters the top of the workpiece and removes the chip that gets *progressively thinner* as the cutter tooth rotates.

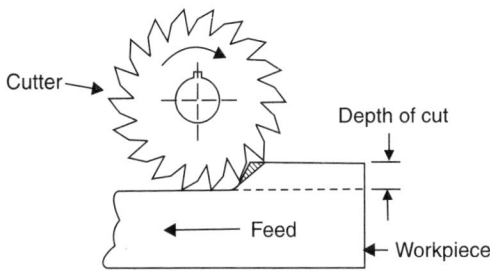

Fig. 9.93. Down milling process.

The **advantages** of this milling process are :

(*i*) The cutting force tends to hold the work against the machine table, permitting *lower clamping forces.*

(*ii*) This process produces *better finish and dimensional accuracy.*

(*iii*) The coolant can be fed easily. The chips are also disposed off conveniently and they do not interfere with the cutting. Thus the machined surface of the workpiece is *not* spoiled.

● *Down or climb milling is used only on materials that are free of scale and other surface imperfections that would damage the cutters.*

9.10.3. Classification of Milling Machines

The milling machines are broadly *classified* as follows :

1. Column and knee type :

(*i*) Horizontal milling machine.

(*ii*) Vertical milling machine.

(*iii*) Universal milling machine.

(*iv*) Omniversal milling machine.

2. Manufacturing or fixed bed type :

(*i*) Simplex milling machine.

(*ii*) Duplex milling machine.

(*iii*) Triplex milling machine.

3. Planer type :

4. Special type :

(*i*) Rotary table milling machine.

(*ii*) Drum milling machine.

(*iii*) Planetary milling machine.

(*iv*) Pantograph, profiling and tracer controlled milling machine.

9.10.4. Horizontal Milling Machine

Refer to Fig. 9.94. The main parts of a horizontal milling machine (*column and knee type*) are briefly described below :

1. Base :

● It is a heavy casting on which column and other parts are mounted.

● It may be bolted to the floor strongly.

2. Column :

● There are guideways on the front face of the column, on which the knee slides.

● It houses power transmission units such as gears, belt drives and pulleys to give rotary motion to the arbor. The drive mechanisms are also used to give automatic feed to the handle and table.

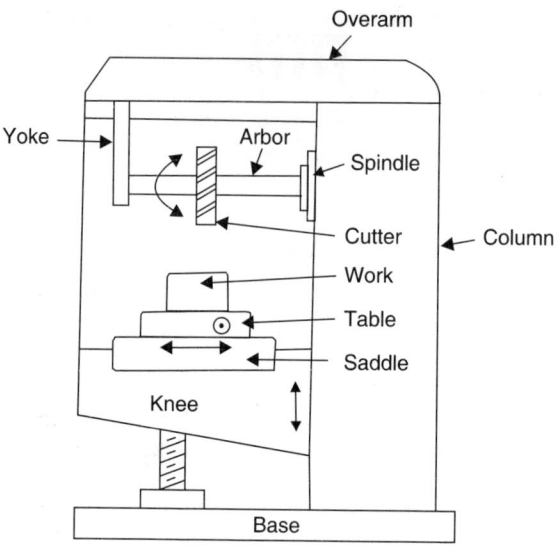

Fig. 9.94. Horizontal milling machine.

3. Knee :

● It supports the saddle, table, workpiece and other clamping devices.

● It moves on guideways of the column.

● It resists the deflection caused by the cutting forces on the workpiece.

4. Saddle :

● It is mounted on the knee and can be moved by a handwheel or by power.

● The direction of travel of the saddle is restricted to be towards or away from the column face.

5. Table :

● It is mounted on the saddle and can be moved by a handwheel or by power.

● Its top surface is machined accurately to hold the workpiece and other holding devices.

● It moves perpendicular to the direction of saddle movement.

6. Arbor :

● Its one end is attached to the column and the other end is supported by an overarm.

● It *holds and drives different types of milling cutters.*

7. Spindle :

● It gets power from gears, belt drives, to drive the motor.

● It has provision to add or remove milling cutters onto the arbor.

9.10.5. Vertical Milling Machine

In vertical milling machine, the position of the spindle head is vertical and the axis of the spindle is perpendicular to the work table as shown in Fig. 9.95.

— In this machine both the *'base'* and *'column'* are *integral castings.*

— The *spindle head* can be moved up and down over the guideways. The vertical milling machine may have fixed head, swivelling head or slidable swivelling head spindle.

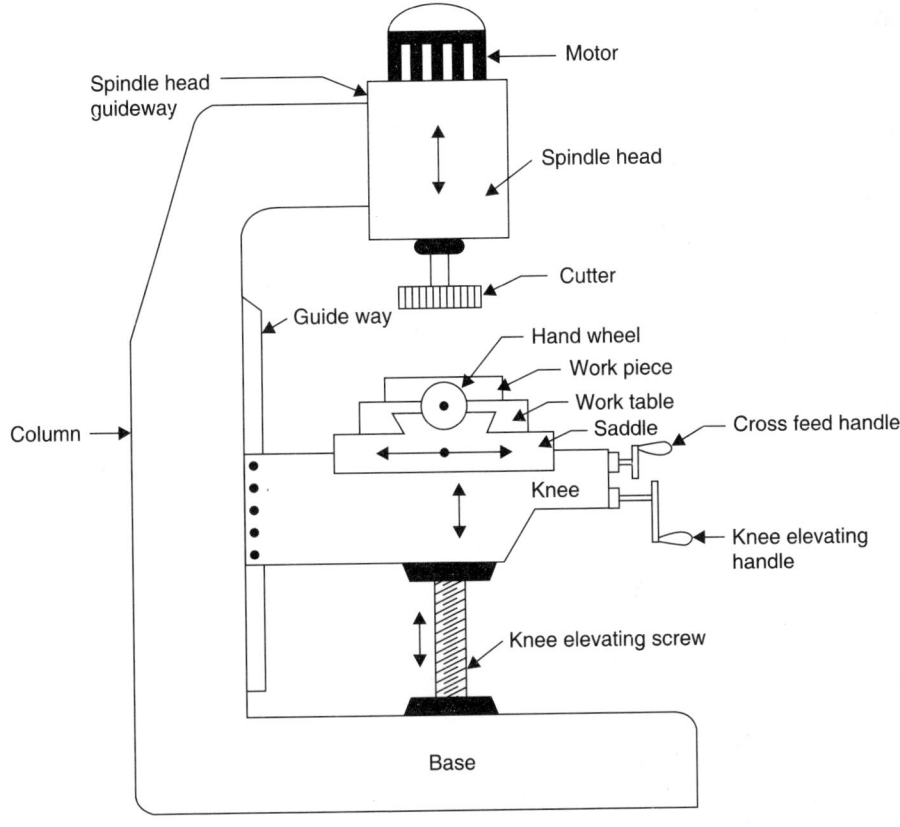

Fig. 9.95. Vertical milling machine.

— The '*saddle*' is mounted on a 'knee' which can be mould up and down over the guideways provided on the *column face*. The saddle can be moved horizontally either by hand or by power over the knee guideways.

— The *worktable* mounted on the saddle can be moved longitudinally over the guideways provided on the top of the saddle.

In this machine, the workpiece can be moved both in the vertical plane and on the horizontal plane.

● *The machine is used to machine grooves, slots and flat surfaces using end mill and face mill cutters.*

9.10.6. Universal Milling Machine

The universal horizontal milling machine differs from the plain horizontal type in that its *table can be swivelled to enable helical grooves to be milled* (*e.g.* The helical flutes of twist drills or the teeth of helical gears).

The *saddle is in two parts* so that the table can be horizontally rotated.

9.10.7. Omniversal Milling Machine

This milling machine has an *additional movement* as compared to universal milling machine. The *knee can be rotated about the column face on an axis perpendicular to it. This enables to machine tapered spiral grooves in reamers, teeth on bevel gears and angular holes etc.*

9.10.8. Planer Type Milling Machines

This type of machine looks like a double column planer, but has milling heads mounted in various planes, vertical milling heads on the cross-rail and horizontal heads at the sides (on column). This enables it to machine a workpiece on several sides simultaneously.

- These milling machines are primarily *intended for producing long straight surfaces on large and heavy machine parts.*

9.10.9. Bed Type Milling Machines

The smaller versions of planer type milling ma-chines, having one horizontal spindle on one side or two spindles, over each side, are called simplex, and duplex fixed bed milling machines (see Fig. 9.96).

- These machines are larger, heavier and have greater rigidity than the column and knee type and are *not* adapted to tool room.

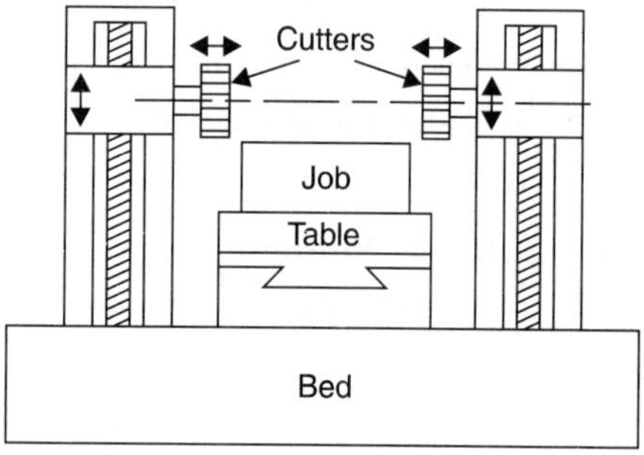

Fig. 9.96. Duplex bed type milling machine.

9.10.10. Drum Type Milling Machines

The drum type milling machines are one of continuous-operation type. They are mostly found in *large-lot and mass production shops for production of large parts such as motor blocks, gear cases, and clutch housings.*

— Two flat surfaces of the workpiece can be milled simultaneously.

— The output of such machine depends upon the number of simultaneously machined parts and the speed of rotation of the drum (feed rate).

9.10.11. Planetary Milling Machines

These machines are used for milling both internal and external short threads and surfaces.

— The work is held stationary and all movements necessary for the cutting are made by the milling cutters.

— At the start of a job, the rotating cutter is in centre or neutral position. It is first fed radially to the proper depth and then given a *planetary motion* either inside or outside the work.

- Typical applications of this machine include milling *internal and external threads on all kinds* of tapered surfaces, bearing surfaces, rear axle and holes, and shell and bomb ends.

9.10.12. Specifications of a Milling Machine

The following are the specifications of a *column and knee type milling machine :*

1. Width and length of the table.
2. Maximum distance the knee can travel.
3. Maximum longitudinal movement and cross feed of the table.
4. Number of spindle speeds.
5. Power of the main drive motor.

9.10.13. Types of Milling Cutters

Common types of milling cutters are :

1. Plain milling cutters.
2. Side milling cutters.

3. End milling cutters.

4. Face milling cutters.

5. Metal slitting cutters.

6. Angle milling cutters.

7. Formed milling cutters.

8. Wood ruff-key milling cutters.

9. T-slot milling cutter.

10. Fly cutter.

9.10.14. Nomenclature of a Milling Cutter

Fig. 9.97 shows the nomenclature of a milling cutter.

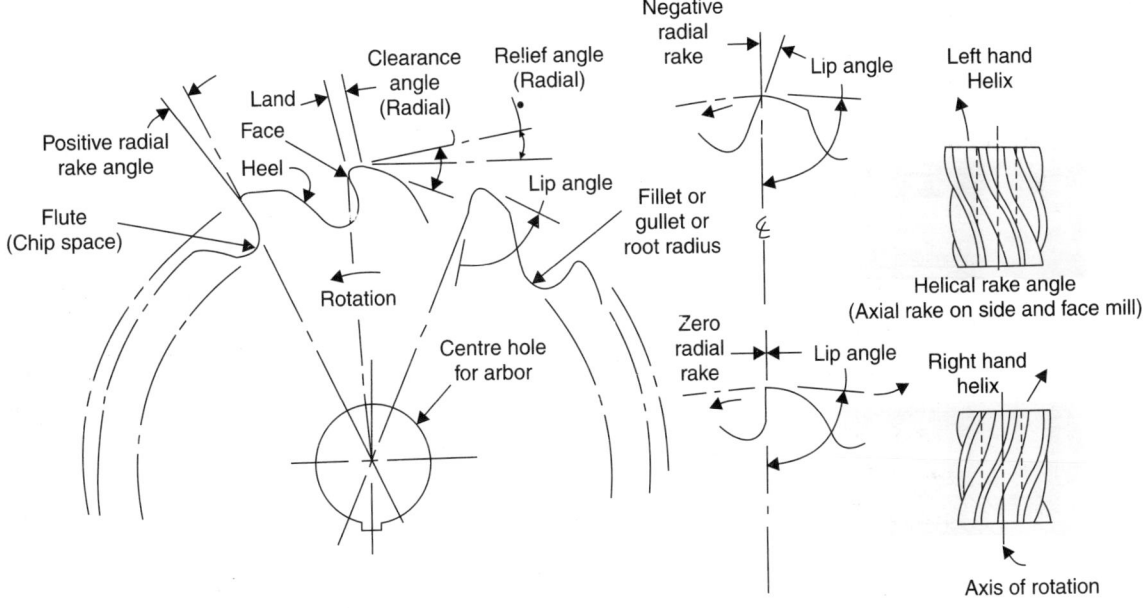

Fig. 9.97. Nomenclature of a milling cutter.

9.10.15. Milling Operations

The milling operations are *classified* as follows :

1. Plain or slab milling

2. Face milling

3. Angular milling

4. Form milling

5. Straddle milling

6. Gang milling

7. End milling

8. T-slot milling

9. Dove-tail milling

10. Saw milling

11. Involute gear cutting.

1. **Plain or slab milling.** Plain milling is used to *machine flat and horizontal surfaces* (Fig. 9.98). Here plain milling cutter is used, which is held in the arbor and rotated. The table is moved upwards to give the required depth of cut.

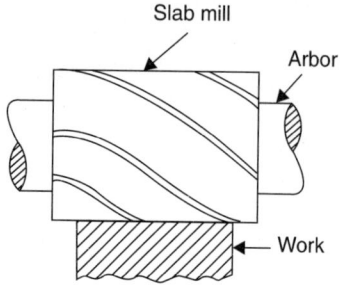

Fig. 9.98. Plain or slab milling.

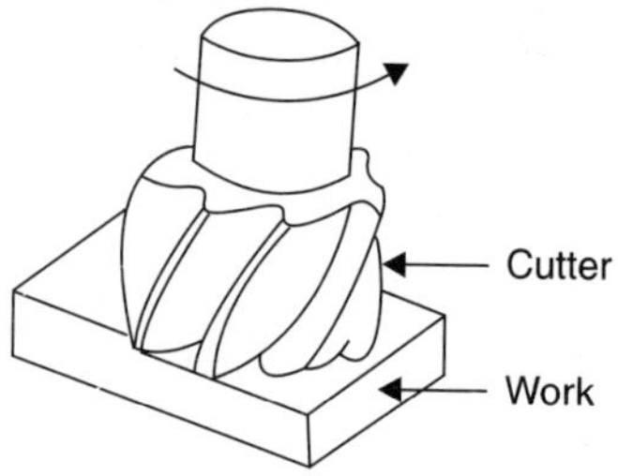

Fig. 9.99. Face milling.

2. **Face milling.** This milling process (Fig. 9.99) is used for machining a flat surface which is at right angles to the axis of the rotating cutter. The cutter used in this operation is the face *milling cutter.*

3. **Angular milling.** In angular milling, an angle milling cutter is used (Fig. 9.100). The cutter used may be a *single* or *double* angle cutter, depending upon whether a single surface is to be machined or two mutually inclined surfaces simultaneously.

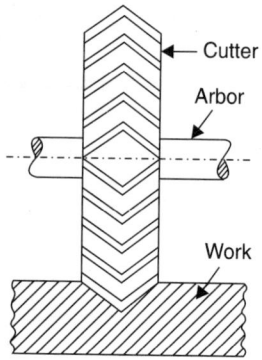

Fig. 9.100. Angular milling.

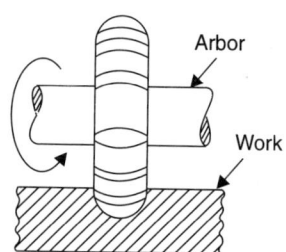

Fig. 9.101. Form milling.

4. **Form milling.** This milling process (Fig. 9.101) is used for machining those surfaces which are of *irregular shapes.* The form milling cutter used has the shape of its cutting teeth conforming to the profile of the surfaces to be produced.

5. **Straddle milling.** Refer to Fig. 9.102. Straddling milling is an operation in which a pair of *side milling cutters* is used for machining two parallel vertical surfaces of a workpiece simultaneously. The distance between the cutters is adjusted by the spacers. This process is used to mill *square and hexagonal surfaces.*

6. **Gang milling.** Gang milling (Fig. 9.103) is the name given to a milling operation which involves the use of a *combination of more than two cutters*, mounted on a common arbor, for milling a number of flat horizontal and vertical surfaces of a workpiece simultaneously. This method *saves much of machining time* and is widely used in *repetitive work*. The cutting speed of a gang of cutters is calculated from the cutter of the largest diameter.

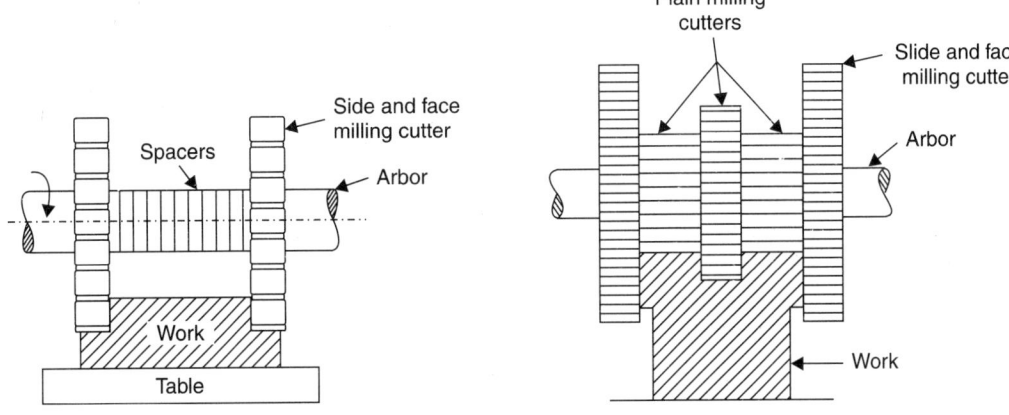

Fig. 9.102. Straddle milling.

Fig. 9.103. Gang milling.

7. **End milling.** Refer to Fig. 9.104. It is an operation of producing narrow slots, grooves and keyways using an end mill cutter. The mill tool may be attached to the vertical spindle for milling the slot. Depth of cut is given by raising the machine table.

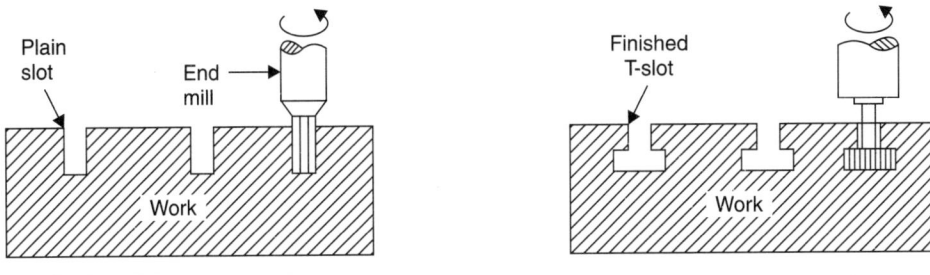

Fig. 9.104. End milling.

Fig. 9.105. T-slot milling.

8. **T-slot milling.** Refer to Fig. 9.105. In this milling operation, first a plain slot is cut on the workpiece by a side and face milling cutter. Then the T-slot cutter is fed from the end of the workpiece.

9. **Dove-tail milling.** Refer Fig. 9.106. In this milling operation, the end of the cutter is shaped to the required dove-tail angle. The cutter is passed from one end of the workpiece to the other end.

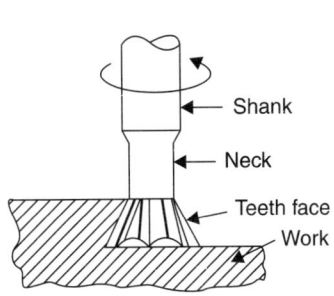

Fig. 9.106. Dove-tail milling.

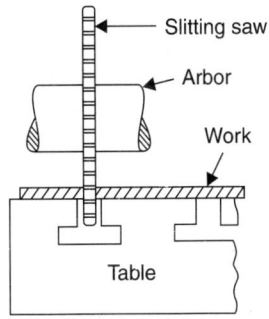

Fig. 9.107. Saw milling.

10. **Saw milling.** Refer to Fig. 9.107. It is an operation of producing narrow grooves and slots on the workpiece. A slitting saw is used for saw milling.

11. **Involute gear cutting.** Gear milling operation, often referred as *gear cutting*, involves cutting of different types of gears on a milling machine. For this, either an *end mill cutter or a form relieved cutter* is used, which carries the profile on its cutting teeth corresponding to the required profile of the gap between gear teeth.

Fig. 9.108 shows involute gear cutting operation. Shape of the cutter teeth resembles the involute profile. Gear blank is indexed after cutting each tooth.

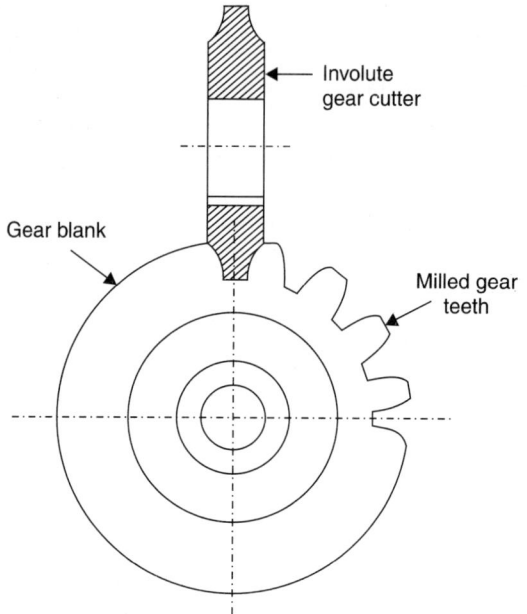

Fig. 9.108. Involute gear cutting.

9.10.16. Cutting Speed, Feed and Depth of Cut

Cutting speed. The '*cutting speed* (V_c)' for milling is *defined as the peripheral speed of the cutter.* The cutting speed (V_c) is given by :

$$V_c = \frac{\pi DN}{1000} \text{ m/min} \qquad \qquad ...(9.30)$$

where, D = Diameter of the cutter, mm, and

N = Cutter speed in r.p.m.

The selection of cutting speed depends on the following factors :

(*i*) The properties of the material being cut ;

(*ii*) Diameter and life of cutter ;

(*iii*) Number of cutter teeth ;

(*iv*) Feed ;

(*v*) Depth of cut as well as width of cut (or cutter) ;

(*vi*) Use of coolant.

Feed. The '*feed (f)' in a milling machine is defined as the movement of the workpiece relative to cutter axis.* It is the rate at which the workpiece is fed into the cutter.

The feed is expressed by the following *three* methods :

(*i*) Feed per tooth (mm per tooth of the cutter).

(*ii*) Feed per revolution (mm per revolution of the cutter).

(*iii*) Feed per minute (mm per minute).

Feed per min. (table feed) ϕ = feed per rev × cutter speed (r.p.m.)

or, $\qquad\qquad\qquad f \text{ (mm/min)} = f_t \times Z \times N$ ...(9.31)

where, f_t = feed rate per tooth, and

Z = Number of teeth on the cutter periphery.

— For rough milling, highest possible feed is employed while for finishing the feed is limited by the specified surface finish required.

— Feed for milling of *mild steel* varies from 0.03 mm/tooth to 0.25 mm/tooth. For milling *hard steel,* recommended feeds are one-third to one-half of these values.

Depth of cut. In the milling process, the *'depth of cut (d)' is defined as the thickness of the layer of material removed in one pass of the workpiece under the cutter.*

— A depth of cut from 3 mm to 8 mm is common for roughing cuts and is less than 1.5 mm for finishing cuts.

9.10.17. Material Removal Rate

'Material removal rate (MRR)' is the volume of metal removed in unit time. For milling, MRR is given by.

$\qquad\qquad \text{MRR} = B\ d\ f, \text{ mm}^3/\text{min}$ ...(9.32)

where, B = Width of cut,

d = Depth of cut, and

F = Rate of feed.

9.10.18. Machining Time

'Machining time (t_m) is defined as the *time required for one pass of width of cut B for milling or machining a surface and is expressed in minutes.*

Mathematically, $\qquad t_m = \dfrac{\text{Length of cut}}{\text{Feed rate}} = \dfrac{L}{f} = \dfrac{L}{f_t . Z . N} \text{ min}$ ...(9.33)

Above time is for making one pass across the length of surface. If the width of cutter is more than the width of surface to be milled, only one pass will machine the whole surface. However, if more than one pass is required, the above time is multiplied by the number of passes.

Figs. 9.109 and 9.110 show milling operations being performed by using the *plain* and *face* milling cutters respectively.

Plain milling : Refer to Fig 9.109.

Let, $\quad L_j$ = Length of the job to machined, mm,

l_1 = Approach, mm, and

l_2 = Overrun of the cutter, mm,

D = Diameter of the cutter, mm, and

d = Total depth of cut, mm.

Then, $\quad L = l_1 + L_j + l_2$ ...(9.34)

and, *Approach,* $l_1 = \sqrt{d\,(D-d)}$ mm ...(9.35)

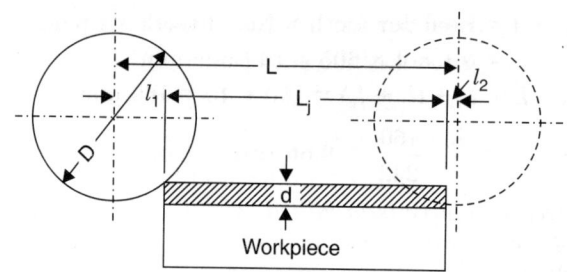

Fig. 9.109. Plain milling.

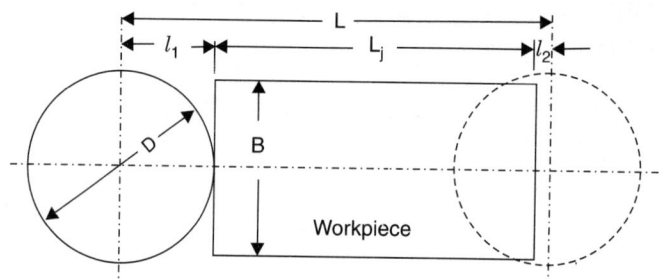

Fig. 9.110. Face milling.

Face milling : Refer to Fig. 9.110.

Approach, $\qquad l_1 = 0.5 \left(D - \sqrt{D^2 - B^2} \right)$ mm $\qquad\qquad$ ...(9.36)

where, $\qquad\qquad\qquad B$ = width of job, mm.

● The *overrun*/over travel of the cutter (l_2), depending upon the size of the machined surface, can be taken from 1 to 6 mm.

If n be the number of cuts and f the feed in mm per minute, then the machining time, t_m in minutes can be calculated as follows :

$$t_m = \frac{L \times n}{f} \text{ min} \qquad\qquad ...(9.37)$$

For finding out the *total floor time*, handling time should be added to the actual machining time calculated above.

Example 9.22. *Calculate the cutting time for cutting 150 mm long keyway using HSS end mill of 20 mm diameter having four cutting teeth. The depth of keyway is 4.2 mm. Feed per tooth is 0.1 mm and cutting speed is 38 m/min. Assume approach and overtravel distance as half of the diameter of the cutter and a depth of 4.2 mm can be cut in one pass.*

Solution. *Given :* L_j = 150 mm ; D = 20 mm ; $l_1 + l_2 = \dfrac{D}{2} = \dfrac{20}{2}$ = 10 mm ; No. of teeth, $Z = 4$; Feed per tooth = 0.1 mm ; V_c = 38 m/min ; d = 4.2 mm.

Machining/cutting time, t_m :

Speed (in r.p.m.) of the cutter,

$$N = \frac{1000\, V_c}{\pi D} \qquad\qquad\qquad \left(\because\ V_c = \frac{\pi D N}{1000} \right)$$

$$= \frac{1000 \times 38}{\pi \times 20} = 605 \text{ r.p.m.}$$

Then, feed per min, f = Feed per tooth × No. of teeth × r.p.m.

$$= 0.1 \times 4 \times 605 = 242 \text{ mm/min}$$

Now, Length of cut, $L = L_j + (l_1 + l_2) = 150 + 10 = 160 \text{ mm}$

∴ Machining time, $t_m = \dfrac{L}{f} = \dfrac{160}{242} = \textbf{0.66 min. (Ans).}$

Example 9.23. *A steel workpiece is to milled. Metal removal rate is 25 cm³/min. Depth of cut and width of cut are 4.5 mm and 90 mm respectively. Calculate the table feed.*

Solution. *Given :* MRR = 25 cm³/min = 25×10^3 mm³/min ; $d = 4.5$ mm ; $B = 90$ mm.

Table feed, f (mm/min) :

$$\text{MRR} = \text{Depth of cut } (d) \times \text{width of cut } (B) \times \text{Rate of feed } (f)$$

$$25 \times 10^3 = 4.5 \times 90 \times f$$

or, $f = \dfrac{25 \times 10^3}{4.5 \times 90} = \textbf{61.7 mm/min. (Ans).}$

Example 9.24. *Calculate the time required to mill a slot of 350 mm × 30 mm in a workpiece of 350 mm length with a side and face milling cutter of 120 mm diameter, 30 mm wide and having 20 teeth. The depth of cut is 6 mm, the feed per tooth is 0.1 mm and cutting speed is 34 m/min. Assume overtravel distance of 5 mm.*

Solution. *Given :* $L_j = 350$ mm ; $D = 120$ mm ; $Z = 20$ teeth ; $d = 6$ mm ; Feed per tooth = 0.1 mm ; $V_c = 34$ m/min, overtravel, $l_2 = 5$ mm.

Machining time, t_m :

We know that, $V_c = \dfrac{\pi D N}{1000}$ m/min

∴ Speed (in r.p.m.) of the cutter, $N = \dfrac{1000 \, V_c}{\pi D} = \dfrac{1000 \times 34}{\pi \times 120} \simeq 90$ r.p.m.

Feed rate, f = Feed per tooth × No. of teeth × r.p.m.

$$= 0.1 \times 20 \times 90 = 180 \text{ mm/min}$$

Also, Approach, $l_1 = \sqrt{d(D - d)}$ ...[Eqn. (9.35)]

$$= \sqrt{6(120 - 6)} \simeq 26 \text{ mm}$$

Now, Length of cut, $L = l_1 + L_j + \text{overtravel}$

$$= 26 + 350 + 5 = 381 \text{ mm}$$

∴ $t_m = \dfrac{L}{f} = \dfrac{381}{180} = \textbf{2.12 min. (Ans).}$

Example 9.25. *A plain surface 300 mm (long) × 100 mm (wide) is to be face-milled on a vertical milling machine. The cutter has 18 teeth and feed per tooth is 0.2 mm. The spindle speed is 120 r.p.m. Diameter of the cutter is 150 mm. The overtravel distance is 4 mm. Calculate the machining time.*

Solution. *Given :* $L_j = 300$ mm ; $B = 100$ mm ; No. of teeth in cutter, $Z = 18$; feed per tooth = 0.2 mm ; $N = 120$ r.p.m. ; $D = 150$ mm ; $l_2 = 4$ mm.

Machining time, t_m :

Cutter approach, $l_1 = 0.5 \left(D - \sqrt{D^2 - B^2} \right)$ mm ...[Eqn. 9.36]

$$= 0.5 \left(150 - \sqrt{150^2 - 100^2} \right) \simeq 19 \text{ mm}$$

Feed rate, f = Feed per tooth × No. of teeth in cutter × r.p.m.

$$= 0.2 \times 18 \times 120 = 432 \text{ mm/min}$$

$$\therefore \qquad t_m = \frac{L}{f} = \frac{l_1 + L_j + l_2}{f} = \frac{19 + 300 + 4}{432} \simeq \textbf{0.75 min. \quad (Ans.)}$$

Example 9.26. *When milling a slot 20 mm wide × 100 mm long in a rectangular plate 100 mm × 200 mm, the cutting speed = 60 m/min. ; Diameter of end drill = 20 mm ; Number of flutes = 8 ; Feed = 0.01 mm/flute ; Depth of cut = 3 mm.*

Determine the cutting time for the operation to be completed in a single pass. **(GATE)**

Solution. *Given :* $L_j = 100$ mm ; $V_c = 60$ mm/min ; $D = 20$ mm ; No. of flutes = 8, Feed = 0.01 mm/flute ; $d = 3$ mm.

Machining time, t_m :

Cutter approach, $\qquad l_1 = \dfrac{D}{2} = \dfrac{20}{2} = 10$ mm

Cutter over travel, $\qquad l_2 = 5$ mm (say)

Spindle speed (in r.p.m.), $N = \dfrac{1000 \, V_c}{\pi D}$

$$= \frac{1000 \times 60}{\pi \times 20} = 955 \text{ r.p.m.}$$

Table feed in mm/min $\qquad$ = Feed per tooth × No. of flutes × r.p.m.

$$= 0.01 \times 8 \times 955 = 76.4 \text{ mm/min}$$

$$\therefore \qquad t_m = \frac{L}{f} = \frac{l_1 + L_j + l_2}{f} = \frac{10 + 100 + 5}{76.4} = \textbf{1.505 min. \quad (Ans.)}$$

9.10.19. Work Holding Devices

The following devices are used for holding the workpiece to be machined :

1. *T*-bolts, strap clamps and pads.
2. Plain vice.
3. Swivel vice.
4. Universal vice.
5. Universal chuck.
6. Rotary table.
7. Dividing head.
8. Various types of milling fixtures.

9.10.20. Milling Machine Attachments

A wide variety of standard attachments, as listed below, are available to *increase the overall usefulness of the standard milling machine :*

1. Vertical milling attachment.
2. Universal milling attachment.
3. High speed milling attachment.
4. Rack milling attachment.
5. Slotting attachment.
6. Rotary table.
7. Universal spiral attachment.
8. Gear cutting attachment.
9. Dividing or indexing head.

9.10.21. Dividing or Indexing Head

9.10.21.1. Introduction

Indexing. *"Indexing" is the rotation of a work-part by small uniform amount, with or without the aid of change wheels, mainly in connection with milling operations.*

Dividing head. *"Dividing or indexing head"* is an attachment used on the milling machine table. *Its purpose is to accurately divide the circumferences of components for grooving or fluting, gear cutting, the cutting of splines, squares or hexagons.*

9.10.21.2. Types of dividing heads

The dividing heads are generally of the following *three* types :

1. *Plain dividing head* :
   - It is hand operated and used for simple indexing.
2. *Universal dividing head :*
   - It can be operated by hand, or gear drive, depending on the operation.
   - It is designed for simple, compound, differential or multiple indexing.
3. *Optical dividing head :*
   - It is the most precision attachment and is, therefore, used for very precision indexing work or for checking the indexing accuracy of the other types of dividing heads.

1. **Plain dividing head :** Refer to Fig. 9.111.
- The main spindle of the dividing head drives the workpiece by means of a 3-jaw universal chuck or a dog and live centre similar to a lathe.
- The index plate of a dividing head consists of a number of holes with a crank and pin. The index crank drives the spindle and the live centre through a worm gear, which generally has 40 teeth. Consequently, a full rotation of the workpiece is produced by 40 full revolutions of the index crank. Further indexing is made possible by having the index plates with equi-spaced holes around various circles. This would allow for indexing the periphery of the workpiece to any convenient number of divisions.

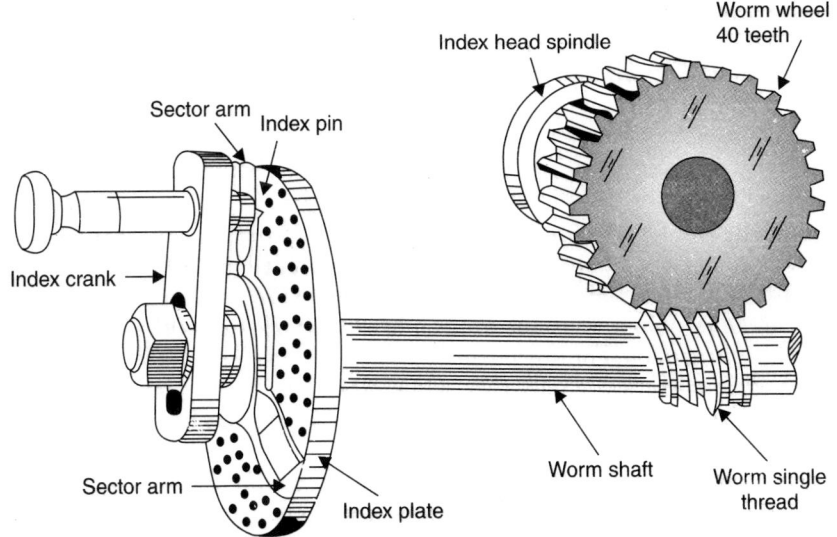

Fig. 9.111. Indexing mechanism of a dividing head.

- With the Brown and sharpe milling machines the following *index plates* are available.

> Plate no. 1 :   15, 16, 17, 18, 19, 20 holes
> Plate no. 2 :   21, 23, 27, 29, 31, 33, holes
> Plate no. 3 :   37, 39, 41, 43, 47, 49, holes

- The index plate used on Cincinnati and Parkinson dividing heads is
  > Plate 1 : Side 1   24, 25, 28, 30, 34, 37, 38, 39, 41, 42, and 43 holes
  >   Side 2   46, 47, 49, 51, 53, 57, 58, 59, 62, and 66 holes.

It is also possible to get additional plates from Cincinnati to increase the indexing capability as follows :

> Plate 2 : Side 1   34, 46, 79, 93, 109, 123, 139, 153, 167, 181, 197 holes
>   Side 2   32, 44, 77, 89, 107, 121, 137, 151, 163, 179, 193 holes
> Plate 3 : Side 1   26, 42, 73, 87, 103, 119, 133, 149, 161, 175, 191 holes
>   Side 2   28, 38, 71, 83, 101, 113, 131, 143, 159, 173, 187 holes

2. **Universal dividing head :** Refer to Figs. 9.112 and 9.113.

Many different makes of dividing heads are available, but although they may vary in construction, the principle on which they work is the same, and because of the limitation of the plain head the universal is much more extensively used. The attachment is comprised of the two parts :

(i) 'Head stock'– It carries the *dividing and driving mechanism.*

(ii) 'Tail stock'– It *holds the back centre,* for work that has to be supported between centres, when being milled.

- *The **headstock** consists of a main casting with the bearings, to accommodate the *main* and *auxiliary spindles.*

— The *main spindle* is hollow, the front end being bored to a standard taper ; the running or driving centre is made to fit this taper. The outside of the spindle at the front end is screwed to receive a chuck for use when the work cannot be held between centres. Fixed to the main spindle and inside the main casting is a *worm-wheel having 40 teeth cut on its circumference.*

— Set below the main spindle, and *at right angles to it, is another spindle passing through a flanged sleeve.* On the inner end of the spindle a *single-thread worm is fixed,* working in the worm-wheel, while on the outer end is a crank which carries a *spring-loaded plunger ;* The reduced end of the plunger engages into the hole of the division or hole plate. The flange sleeve has a bearing in the main casting, and can be resolved in either direction by gears connecting with another spindle running parallel with the main spindle and extending out at the back of the head.

— With a single-thread worm engaging into a 40–tooth worm-wheel it requires 40 turns of the worm-spindle to obtain one complete revolution of the main spindle. This is referred to as *40 to 1 ratio, and by this means the circle can be divided into any number of 40ths.* When divisions are required that are not 40ths of a circle, then other means must be applied. For this purpose dividing or hole plates are supplied (as discussed earlier) ; some makers use only two, while others use three plates. The hole plates are circular, having a central hole to pass over the worm-spindle, and are attached to the flange of the sleeve by two countersunk screws.

— Concentric circles of holes are drilled through the plates, each circle having a different number of holes ; they all start on a radial line which is vertical when the plate is fixed. *A sector with two adjustable arms that revolves on the arm-spindle fits against the plate and controls the travel of the crank arm. This also saves the counting of holes for each movement.*

— For 'spiral milling', the dividing head must be geared to the feed-screw of the table, in order to revolve the work in relation to the feed. This gives the lead or helix of spiral, and is obtained by placing change wheels on the warm-spindle and feed-screw with intermediate gears on studs.

● The **tailstock** casting is machined to fit a rectangular block which is adjustable for the milling of tapers, and carries the hand-wheel-operated barrel. The end of the barrel is bored to a standard taper to receive the hardened centre on which the work revolves.

— Work that is mounted between centres is revolved by a drive connected with a pin to the driving dog on the running centre and held by a set-screw.

Fig. 9.112 shows a universal dividing head.

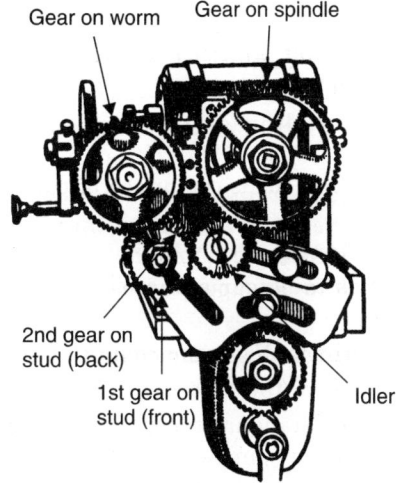
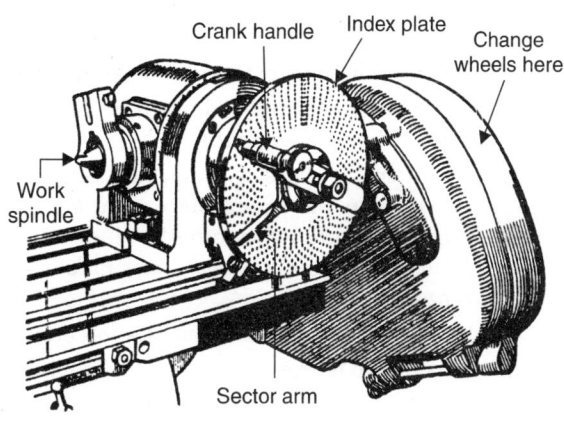

Fig. 9.112. Universal dividing head.　　　Fig. 9.113. The index plate of a universal dividing head.

Fig. 9.113 shows the *index plate* of a universal dividing head ; it carries centric circles of holes. A sector has two adjustable arms which act as stops for the movement of the crank arm, so that it is unnecessary to count individual holes in the plate for each indexing movement.

3. Optical dividing head :

This type of dividing head is used for *high precision operations,* when the error in the angular setting must not exceed 0.25 minute of arc. Such heads have a highly accurate dial which is read through the eyepiece of an optical system built into the head. *There is complete freedom from pitch and backlash errors.* The only gears employed are a single start worm and 40 teeth worm-wheel which are only used to rotate the spindle and not for any dividing purposes.

9.10.21.3. Methods of Indexing

Common *indexing methods* are :

1. Rapid or direct indexing.

2. Simple or plain indexing.

3. Compound indexing.

4. Differential indexing.

5. Angular indexing.

1. Rapid or direct indexing :

Rapid or direct indexing is the simplest method of indexing and is used only on work that requires a *small number divisions, such as square or hexagonal nuts etc.*

- In this indexing method, the spindle is turned through a given angle *without interposition of gearing.*

— The index plate has 24 holes ; crank may be rotated to divide the periphery of the workpiece into the divisions 2, 3, 4, 6, 8, 12 and 24 directly. Since the index plate is fastened directly to the spindle, the complete revolution of the index plate rotates the spindle also by one complete revolution.

— If indexing is to be carried out for square head screw, then the number of holes to move in the index plate.

$$= \frac{24}{N_{\text{div.}}} = \frac{24}{4} = 6 \text{ holes}$$

where, $N_{\text{div.}}$ is the required *number of divisions* to be made.

   2. **Simple or plain indexing :** Refer to Fig. 9.111.

   Simple or plain indexing is the name given to the indexing method which is carried out *using any of the indexing plates in conjunction with the worm.*

   Plain indexing is employed when it is required to divide a circle into more number of parts than is possible by rapid indexing.

- To index the work through any required angle, the index crankpin is withdrawn from the hole in the index plate. The spindle and hence the work is indexed through the required angle by turning the index crank through a calculated number of *whole revolutions and holes on one of the hole circles, after which the index pin is relocated is the required hole.*

   If the number of divisions on the job circumference needed is $N_{\text{div.}}$ then number of turns that the crank must be rotated for each indexing can be found from the following relationship :

$$n_{cr.} = \frac{40}{N_{\text{div.}}} \qquad \qquad \text{...(9.38)}$$

where $n_{cr.}$ is the number of crank rotations.

   Example 9.27. *It is required to divide the periphery of a job into 50 equal divisions. Find the crank movement.*

   Solution.   *Given :* $N_{\text{div.}} = 50$

   Required crank movement, $n_{cr.} = \dfrac{40}{N_{\text{div.}}} = \dfrac{40}{50} = \dfrac{4}{5}$

   Selecting 15 holes circle on plate No. 1, we get,

$$\frac{4}{5} = \frac{4 \times 3}{5 \times 3} = \frac{12}{15}$$

i.e.,   *12 holes on 15 holes circle.*   **(Ans.)**

   Example 9.28. *It is required to divide the periphery of a job i.:to 28 equal divisions. Find the indexing arrangement.*

   Solution. *Given :* $N = 28$

   Required crank movement $n_{cr.} = \dfrac{40}{28} = \dfrac{10}{7} = 1\dfrac{3}{7}$

   Now, $\dfrac{3}{7} = \dfrac{3 \times 3}{7 \times 3} = \dfrac{9}{21}$

   That is, for each indexing we need *one complete rotation of the crank plus 9 more holes on 21 hole circle* (Plate No. 2).   **(Ans.)**

3. **Compound indexing :**

The compound indexing method is employed when the number of divisions required is *outside the range that can be obtained by simple indexing.*

The compound indexing is achieved in *two stages* by using two *different hole circles of one index plate :*

1. By a movement of the crank in the usual way as in simple indexing, say, n_{h_1} holes in hole circle N_{h_1} with the lockpin engaged in circle N_2 of the index plate.

2. By *adding or subtracting* a further movement by rotating the *crank and index plate together forward or backward,* through n_{h_2} spaces in N_{h_2} circle (by disengaging the locking pin of the index plate so that it is free to turn).

Procedure :

Let, $N_{\text{div.}}$ = *No. of divisions needed on the work.*

$\therefore$ Crank movement for each indexing = $\dfrac{40}{N_{\text{div.}}}$

In order to obtain this movement (since it cannot be obtained by simple indexing), proceed as follows :

(*i*) Write $N_{\text{div.}}$ above and 40 below a straight line and factorize them.

(*ii*) Select the two numbers representing two hole circles in the *same plate.* Write these numbers below 40 and factorise them. Write their difference above $N_{\text{div.}}$ and factorise it, these hole numbers are to be chosen in such a manner *that all the factors above the line get cancelled out with factors below the line.*

Let these hole numbers be N_{h_1} and N_{h_2}

(*iii*) Let n_{h_1} be the number of holes to be indexed in N_{h_1} hole circle and n_{h_2} be the number of holes to be indexed in N_{h_2} hole circle.

Then, $\dfrac{n_{h_1}}{N_{h_1}} \pm \dfrac{n_{h_2}}{N_{h_2}} = \dfrac{40}{N_{div}}$ ...(9.39)

Now, find out n_{h1} and n_{h2} by trial and error.

Then, the total indexing will be : n_{h_1} holes in N_{h_1} hole circle $\pm$ n_{h_2} holes in N_{h_2} hole circle by rotating the crank and index plate together.

The procedure is discussed in detail in example 9.29.

Example 9.29. *Index for 87 divisions.*

Solution *Given :* N_{div} = 87

Steps : (*i*) $87 = 29 \times 3$

$40 = 2 \times 2 \times 2 \times 5$

(*ii*) Let $N_{h1} = 29$, and

$N_{h2} = 33$

$\therefore$ $(33 - 29)$ or $4 = 2 \times 2$ (*i.e.* Factors of defference of hole circles)

$87 = 29 \times 3$ (*i.e.* Factors of divisions required)

$40 = 2 \times 2 \times 2 \times 5$ (*i.e.* Factors of 40)

$29 = 29 \times 1$ (Factors of first hole circle N_{h_1})

$33 = 11 \times 3$ (Factors of second hole circle N_{h_2})

Now, $\dfrac{\text{Factors of divisions required} \times \text{factors of difference of hole circles}}{\text{Factors of } 40 \times \text{Factors of first hole circle } N_{h_1} \times \text{Factors of second hole } N_{h_2}}$

$$= \frac{(29 \times 3) \times (2 \times 2)}{(2 \times 2 \times 2 \times 5) \times (29 \times 1) \times (11 \times 3)}$$

Since all the factors above the line get cancelled out therefore, selection of N_{h_1} and N_{h_2} is
correct :

(*iii*) Now, indexing equation is given by :

$$\frac{n_{h1}}{29} \pm \frac{n_{h2}}{33} = \frac{40}{87} \qquad \text{.......using eqn. (9.39)}$$

$$33\, n_{h1} \pm 29\, n_{h2} = 440$$

By trial and error, $n_{h1} = 23$ and $n_{h2} = -11$

∴   Indexing equation will be :

$$\frac{23}{29} - \frac{11}{33} = \frac{40}{87}$$

i.e. Movement of crank by 23 holes in 29 hole circle *forward* and movement of crank and index
plate both by 11 holes in 33 hole circle *backwards.* **(Ans.)**

4. **Differential indexing :**

Although compound indexing is a convenient way to get any indexing required, it is *fairly*
cumbersome to use in practice. Hence differential indexing is used for that purpose which is an
automatic way to carry out the compound indexing method.

Differential indexing is also carried out is *two stages* :

(*i*) The crank is moved in a certain direction.

(*ii*) In the second stage, either some movement is added to the above crank movement or
subtracted from the same.

However, it should be noted that the said *loss or gain in movement* is accomplished by
moving the plate by means of a *train of gears, connecting the dividing head spindle to the worm*
spindle. The said motion is *gained* by rotating the index plate in the *same direction* as crank and
is *lost* by rotating the plate in the *opposite direction* to that of the crank. The index plate *locking*
pin, during differential indexing *should be taken out to make the plate free to rotate.*

The change gear set available (Brown and sharpe dividing heads) is :

24 (2 Nos.), 24, 28, 32, 40, 44, 48, 56, 64, 72, 86 and 100.

Both simple and compound gear trains are
used in differential indexing, these gear trains are
used as shown in Figs. 9.114 and 9.115 respectively.

● The direction of the movement of the index
plate depends upon the gear train
employed :

— If an *idle gear is added* between the spindle gear and the shaft gear in case of a
simple gear train, then the index plate
will move in the '*same direction*' to that
of the indexing crank movement.

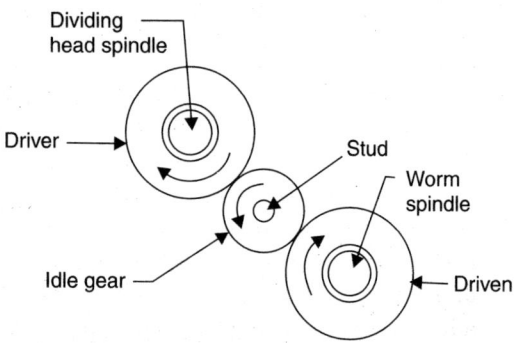

Fig. 9.114. Simple gear train.

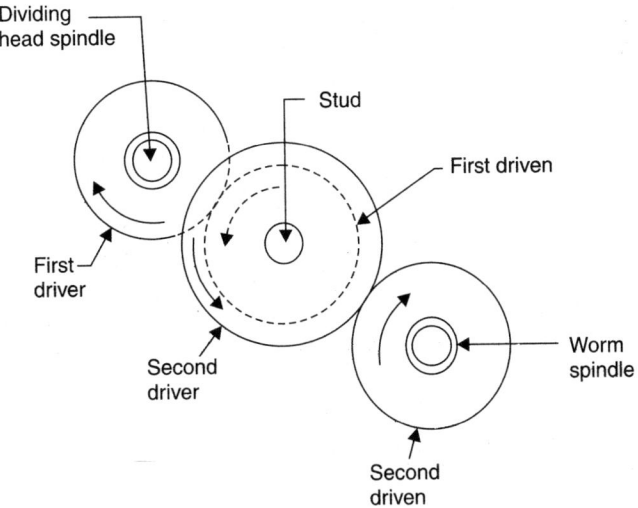

Fig. 9.115. Compound gear train.

— In the case of a *compound gear train an idler is used when the index plate is to move in the 'opposite direction'.*

Example 9.30. *Obtain the indexing for 137 divisions.*

Solution : *Given :* $N_{\text{div}} = 137$

The required indexing is $\dfrac{40}{N_{div}} = \dfrac{40}{137}$ which *cannot be obtained with any of the index plate.* Let us select a number *slightly greater or smaller* than the given number, such that the selected number can be easily indexed through simple indexing. Let us select the new number as 140.

Simple indexing for 140 divisions $= \dfrac{40}{140} = \dfrac{2}{7} = \dfrac{2 \times 3}{7 \times 3} = \dfrac{6}{21}$

i.e. 6 holes on 21 holes circle.

Now, if the index crank is turned $\dfrac{6}{21}$ of a revolution 137 times, it will take : $\dfrac{6}{21} \times 137$

$= \dfrac{822}{21} = 39\dfrac{3}{21}$ resolutions, whereas, for one complete turn of the job it should make 40 complete revolutions. Obviously, the job would not be, thus, indexed through exactly 137 equal divisions.

The total movement done by the crank is short of the required 40 turns by :

$$40 - 39\dfrac{3}{21} = \dfrac{18}{21} \text{ of a revolution.}$$

This fraction is to be *gained* by the movement of the plate. In order to gain the movement the plate will have to be turned in the *same direction as the crank.* Also, in order that the divisions are equal, this movement is to be gained gradually such that a certain amount of it is added equally to the crank movement in all the 137 movements of the latter, so as to make it complete 40 turns at the end of these movements. This will be done by employing a suitable gear train.

Now,   The *gearing ratio*, $\dfrac{\text{Driver}}{\text{Driven}} = \dfrac{18}{21} = \dfrac{6}{7} = \dfrac{6 \times 8}{7 \times 8} = \dfrac{\mathbf{48}}{\mathbf{56}}$

An *idle gear is to be used since the index plate has to move in the same direction* **(Ans).**

Example 9.31. *Obtain indexing for 51 divisions.*

Solution. *Given :* $N_{\text{div}} = 51$

The required indexing is $\dfrac{40}{N_{div}} = \dfrac{40}{51}$ which *cannot be obtained with any of the index plate.*

Let us select the new number = 50

Simple indexing $= \dfrac{40}{50} = \dfrac{4}{5} = \dfrac{4 \times 4}{5 \times 4} = \dfrac{16}{20}$

i.e. 16 holes on 20 holes circle.

Movement of crank for 51 division $= 51 \times \dfrac{4}{5} = \dfrac{204}{5} = 40\dfrac{4}{5}$

i.e. $\dfrac{4}{5}$ turn/rotation more than the reguired 40.

This is to be *lost* through plate movement.

∴   Gearing ratio, $\dfrac{\text{Drivers}}{\text{Driven}} = \dfrac{4}{5} = \dfrac{2 \times 2}{2.5 \times 2} = \dfrac{\mathbf{32 \times 24}}{\mathbf{40 \times 24}}$

i.e. A compound gear train with an *idler* (since the *index plate is to move in opposite direction*) is required. **(Ans.)**

Example 9.32. *Obtain indexing for 73 divisions.*

Solution. *Given :* $N_{div} = 73$

The required indexing is $\dfrac{40}{N_{div}}$ which *cannot be obtained with any of the index plate.*

Let us select the new number as 70.

Simple indexing $= \dfrac{40}{70} = \dfrac{4}{7} = \dfrac{4 \times 3}{7 \times 3} = \dfrac{12}{21}.$

i.e. 12 holes on 21 holes circle.

Movement of crank for 73 divisions

$$= 73 \times \dfrac{4}{7} = \dfrac{292}{7} = 41\dfrac{5}{7}$$

i.e. $1\dfrac{5}{7} = \dfrac{12}{7}$ crank rotation are to be *lost* through the plate movement.

∴   Gearing ratio, $\dfrac{\text{Driver}}{\text{Driven}} = \dfrac{12}{7} = \dfrac{12 \times 4}{7 \times 4} = \dfrac{\mathbf{48}}{\mathbf{28}}$

i.e. A simple gear train (with no idler or two idlers) is required.

5. **Angular indexing :**

Angular indexing is used when it is necessary to cut grooves or slots *subtending a given angle at the centre of the circle upon which they are spaced.*

In earlier discussions we have seen that 40 crank rotations make the work rotate through 360°. Therefore, for each rotation of the crank the work will rotate through $\dfrac{360}{40} = 9°$

Thus, crank movement $= \dfrac{\text{Angle required}}{9°}$.

Example 9.33. *Calculate the indexing for 42°.*

Solution. Indexing required $= \dfrac{42}{9} = 4\dfrac{2}{3}$

This is equivalent to *four full rotations of the crank followed by 12 holes in the 18 hole circle in plate No. 1.* **(Ans.)**

Example 9.34. *Calculate the indexing for 32° 20′.*

Solution. Indexing required $= \dfrac{32\dfrac{1}{3}}{9} = \dfrac{97}{27} = 3\dfrac{16}{27}$

i.e. 3 full rotations of the crank followed by 16 holes on 27 hole circle in plate No. 2. **(Ans.)**

Example 9.35. *Calculate indexing for 15° 40′.*

Solution. Indexing required $= \dfrac{15\dfrac{2}{3}}{9} = \dfrac{47}{27} = 1\dfrac{20}{27}$

i.e. 1 full rotation of the crank followed by 20 holes in the 27 hole circle in plate No.2. **(Ans.)**

9.11. BROACHING MACHINES

9.11.1. Broaching

Broaching *is a process of machining a surface with a special multipoint tool called a* **broach,** *whose teeth remove the whole machining allowance in a single stroke.*

- It is an operation designed to produce *high-precision forms* and the complex tools are expensive. The shapes produced may be flat surfaces but more often are holes of various forms or grooved components or the shaft of gears.
- In this operation *cutting speeds and feeds are low and adequate lubrication is essential.*
  - Although cutting speeds used in broaching is relatively low (2 to 15 m/min). The *production capacity is very high* since the length of the cutting edges that are simultaneously in operation is very great. The output in broaching can be raised still higher if broaching machines with continuous working motion are used, in conjunction with automatic workpiece loading and unloading.
- *Broaching machines* are designed for machining external and internal surfaces of various contours in *mass and large-lot production.* These machines are distinguished from their exceptionally high output. They produce surfaces that are highly accurate in shape and size. Fig. 9.116. shows some typical internally broached shapes.

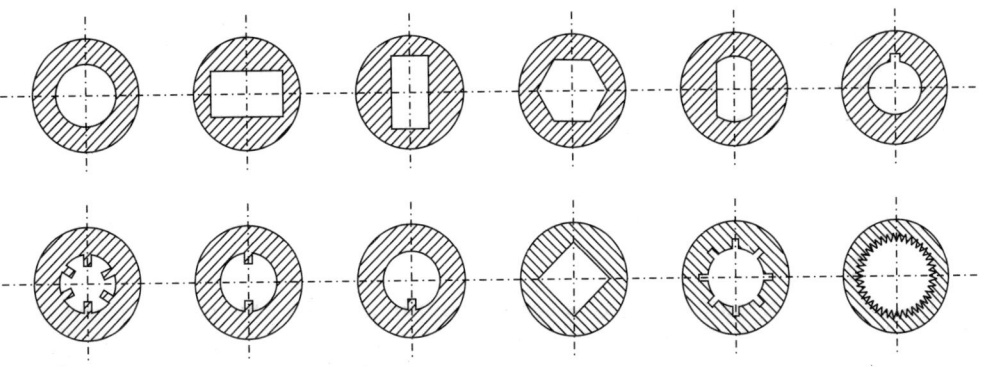

Fig. 9.116. Some typical internally broached shapes.

9.11.2. Advantages, Limitations and Applications of Broaching

Following are the *advantages and limitations of broaching.*

Advantages :

1. It is the fastest way of finishing an operation with a single stroke. The roughing as well as the finishing cuts are completed with one pass of tool.

2. A broaching machine is a simple machine since only a single reciprocating motion is required for cutting.

3. Internal or external surfaces can be broached.

4. Production rate is high because the actual cutting time is a matter of seconds. Rapid loading and unloading of fixtures minimizes total production time.

5. Since all the machining parameters are built into the broach, very little skill is required from the operator.

6. Any form that can be reproduced on a broaching tool can be machined into the part.

7. Final cost of machining operation is one of the lowest for mass production.

8. Production tolerances are suitable to interchangeable manufacture.

9. Finish comparable to milling work can be obtained.

10. Burnishing shells incorporated as the final teeth on the broach improve the surface finish.

Limitations :

1. Very high tool cost.

2. The workpiece must be rigidly held and broach firmly guided.

3. Large amounts of metals cannot be removed.

4. A broach has to be designed for a specific application and can be used only for that application. Hence the lead time for manufacture is more for custom designed broaches.

5. Broaching can only be carried out on a workpiece whose geometry is such that there is no interference for the broach movement for cutting.

Applications :

- Broaching is widely used in the manufacture of :
  - Special gears ;
  - Bushings and sleeves ;
  - Compressor wheels ;
  - Rotors ;
  - Chain sprocket teeth ;
  - Turbine blades etc. ;

- Broaching is also used to produce a wide variety of components in *rifle and gun manufacture, aircraft engines and structures and automobile manufacture.*

9.11.3. Broach

A broach or a broaching tool has a series of multiple teeth. The main elements of a broach are shown in Fig. 9.117 (a). The fixed teeth are designed to do the heaviest cutting and are called *roughing teeth* ; Next teeth are *semi-finishing teeth* which are followed by *finishing tooth.* The finishing teeth carry out finishing operations. Fig. 9.117 (b) shows the teeth details.

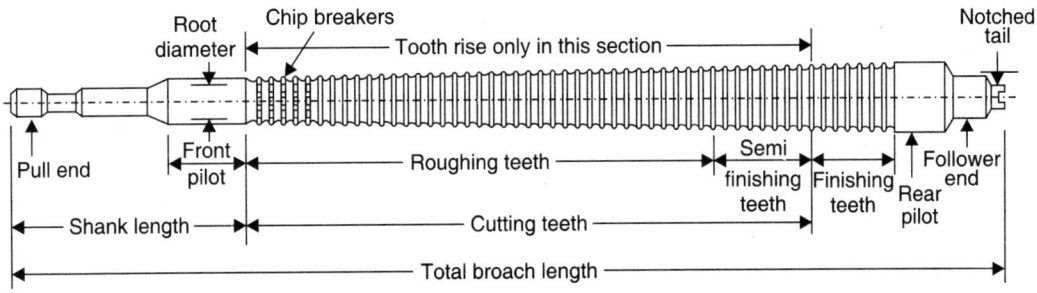

(a) Main elements of a broaching tool

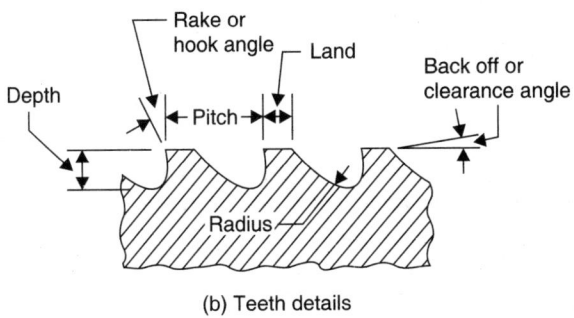

(b) Teeth details

Fig. 9.117. Standard broach parts and nomenclature.

— The *rake angle* (*face angle*) of the tooth depends on the material to be cut and its hardness, toughness and ductility.

— The *land* of the tooth determines its strength.

— The *pitch* determines the length of cut and chip thickness which a broach can handle.

● During broaching operation the broach is either *pushed* (Fig. 9.118 (*a*)) or *pulled* (Fig. 9.118 (*b*)) by the broaching machine past the surface of the workpiece. In doing so each tooth of the broach takes a small cut through the metal surface.

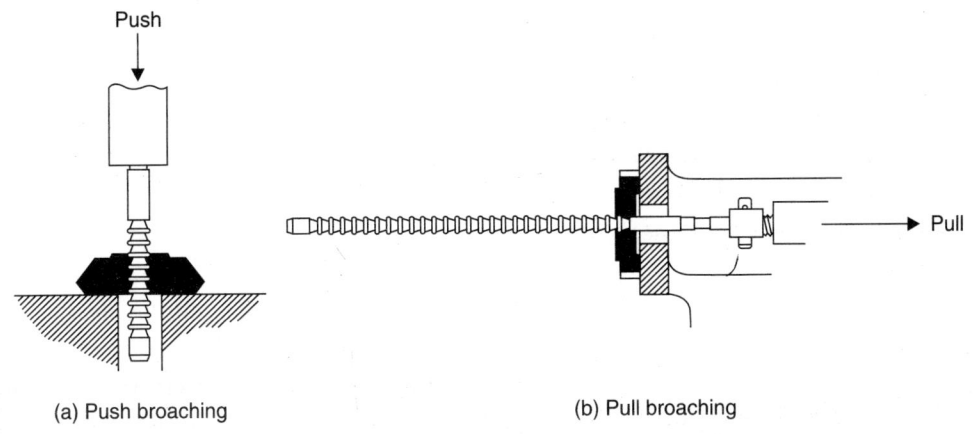

(a) Push broaching   (b) Pull broaching

Fig. 9.118. Round broaches for push-and pull-type broaching machines.

— *Push broaching* requires comparatively *short broaches of sufficient cross-section* to prevent column buckling resulting from the loads imposed during the operation.

— Large, heavy duty horizontal (*pull*) broaching machines are used in the high-speed production of cylinder blocks, intake manifolds, bearing cap clusters and aircraft turbine discs. The cutting speed approaches 1.0 m/s and the stock removal is upto 6.4 mm per stroke.

Broach materials :

— Most broaches are made from *18–4–1 high speed steel, ground after hardening.*

— Carbide broaches are used extensively in broaching cast iron.

9.11.4. Types of Broaching Machines

A broaching machine consists of the following :

(*i*) A broaching tool.

(*ii*) A work holding fixture.

(*iii*) A driving mechanism.

(*iv*) A suitable supporting frame.

Broaching machines are *classified* as follows :

1. *According to direction of broach travel :*

   (*i*) Horizontal broaching machines.

   (*ii*) Vertical broaching machines.

2. *According to method of operation :*

   (*i*) Push broaching machines.

   (*ii*) Pull broaching machines.

   (*iii*) Surface broaching machines.

   (*iv*) Continuous broaching machines.

3. *According to method of drive :*

   (*i*) Hydraulic.

   (*ii*) Pneumatic.

   (*iii*) Mechanical.

4. *According to type of operation :*

   (*i*) External (for surface broaching).

   (*ii*) Internal (for broaching a hole).

5. *According to special designs.*

9.11.5. Size of Broaching Machines

Broach machines are *specified* by :

● Length of stroke in mm ;

● Driving force in tonnes/kN.

9.12. SAWING MACHINES

9.12.1. Introduction

● *Sawing* is one of the most important cutting-off operations performed in a manufacturing/fabrication plant.

● Metal sawing is chiefly concerned with *cutting bar stock to convenient length or size for machining.*

- In sawing, the individual tooth of the saw *tracks* through the work, in each tooth deepening the cut made by the proceeding teeth in the direction of feed. Either the saw or work may be fed and, by controlling the direction of feed, either straight or curved cut can be produced. The *width of the cut is approximately equal to the width of the saw itself.*

9.12.2. Classification of Sawing Machines

The different types of metal cutting saws are broadly *classified* as follows :

1. Reciprocating saws (Power hacksaws).

2. Band saws.

3. Circular saws.

The above sawing machines are briefly described below :

1. **Reciprocating saws :**

- A reciprocating or power hacksaw carries a reciprocating frame, in which is mounted the cutting blade. This blade is in the form of a steel strip, on the edge of which the teeth are cut. This *blade moves forward and backward with the frame. Cutting takes place in one direction only.*

- The stock to be cut is held between the clamping jaws.

- Several pieces of barstock can be clamped together and cut at the same time.

- Both square and rectangular cuts can be made.

2. **Band saws :**

A band saw carries an *endless steel blade, having the teeth cut on its one edge.* This blade passes over two large wheels, one at the top and the other at the bottom. As the wheel rotates, the blade also rotates along with it and, thus *provides a continuous cutting action.* It also enables *sawing along a premarked contour.*

Fig. 9.119 shows a vertical metal band saw.

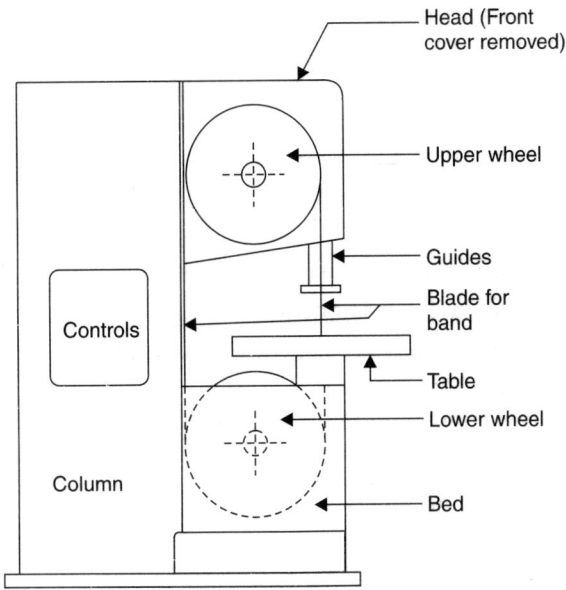

Fig. 9.119. Vertical metal band saw.

3. Circular saws :

A circular saw carries a disc type rotating blade which provides a continuous action. These machines are classified as :

(*i*) Cold sawing machine   (*ii*) High speed abrasive disc machine.

9.12.3. Sawing Machine Blades

The blades of a sawing machine are made of the following materials :

(*i*) Standard carbon steel   (*ii*) High speed steel

(*iii*) Bimetallic high speed steel.

● The main parts of a band tool are shown in Fig. 9.120.

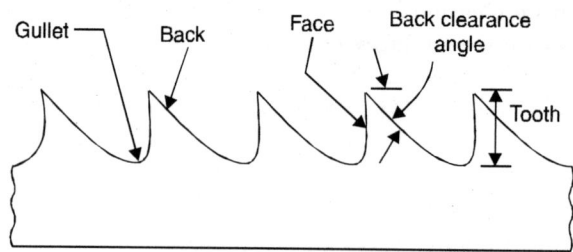

Fig. 9.120. Main parts of a band.

● Fig. 9.121 shows the setting of band saw teeth.

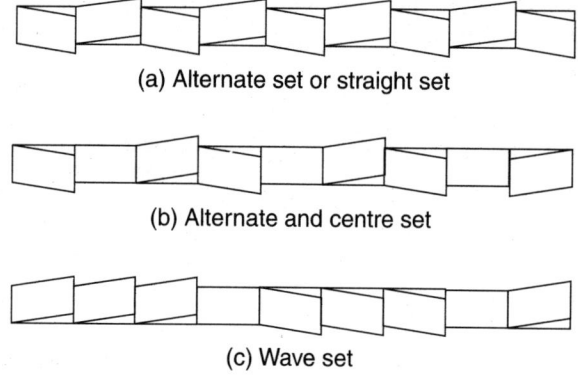

(a) Alternate set or straight set

(b) Alternate and centre set

(c) Wave set

Fig. 9.121. Setting of band saw teeth.

9.12.4. Applications of Sawing Machines

● *"Hacksaw machines"* are relatively slow and find usage in low-output production and tool room and maintenance work.

● where production is in the medium or high ranges, cut off and similar sawing operations are normally done on the *"cold saw"*.

— Manually operated cold saws are used for low production work and automatic-cycle and special units are used in the higher production ranges. Much larger sizes and cuts are possible than with hacksaws.

— Standard machines commonly available employ blades with diameters ranging from around 1000 mm to 1150 mm. Some machines use blades as much as 3 m in diameter.

— Capable of cuts other than straight, the *"band saw"* is useful for production operations other than cut off. With conventional plain contour band sawing it is practicable to produce blank parts in quantities upto about five hundred pieces, eliminating the usual dies costs for such parts.

9.13. GRINDING AND FINISHING PROCESSES

9.13.1. Introduction

● The *'grinding'* and *'finishing'* processes are used for *final finish* and *superfinish.* Out of various types of grinding, the most common one is *surface grinding.* This process is accompanied by a certain amount of metal removal. This process can give a surface finish in a range of 1.25 μm to 0.25 μm.

● For many applications, grinding cannot meet the accuracy and surface finish requirements. For such applications workpieces are subjected to final operations. Two such operations are **honing** and **lapping.**

A. GRINDING

9.13.2. Grinding Process

Grinding is a metal cutting operation performed by means of a rotating abrasive tool, called "grinding wheel". Such wheels are made of fine grains of abrasive materials held together by a bonding material, called a *"bond".* Each individual and irregularly shaped grain acts as a cutting element (a single point cutting tool).

— A magnified view of a grinding wheel and its cutting operation is shown in Fig. 9.122. The projecting grains of the abrasive material are held firmly by the board. The grains during rotation of the wheel remove very thin chips whose cross-section is similar to that obtained in milling. For this operation high wheel speeds are normally employed (upto 75 m/s). *As the section of the chip removed during the process is small, and high cutting speeds are involved, this operation results into very good finish and high accuracy.*

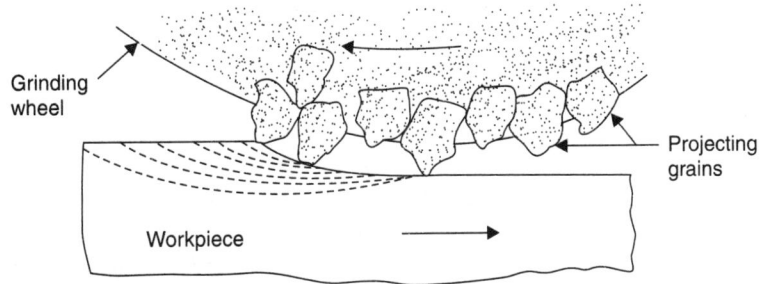

Fig. 9.122.

● Grinding is done on surfaces of almost all conceivable shapes and materials of all kinds. The grinding operation can be : (*i*) *Rough (or non-precision) grinding and* (*ii*) *Precision grinding.*

(*i*) **"Rough grinding"** is a commonly used method for removing excess material from castings, forgings and weldments etc.

(ii) **"Precision grinding"** is the principal production method of cutting materials that are too hard to be machined by other conventional tools or for producing surfaces on parts to higher dimensional accuracy and a finer finish as compared to other manufacturing methods.

Grinding, in *accordance with the type of surface to the ground, is classified as :*

(i) External cylindrical grinding.

(ii) Internal cylindrical grinding.

(iii) Surface grinding.

(iv) Form grinding.

(i) **External cylindrical grinding.** It produces a *straight* or *tapered surface* on a workpiece. The workpiece must be rotated about its own axis between centres as it passes lengthwise across the face of a revolving grinding wheel.

(ii) **Internal cylindrical grinding.** It produces *internal cylindrical holes and tapers.* The workpieces are chucked and precisely rotated about their own axes. The grinding wheel or, in the case of small bore holes, the cylinder wheel rotates against the sense of rotation of the workpiece.

(iii) **Surface grinding.** It produces *flat surface.* The work may be ground by either the periphery or by the end face of the grinding wheel. The *workpiece is reciprocated at a constant speed below or on the end face of the grinding wheel.*

(iv) **Form grinding.** This operation is done with *specially shaped grinding wheels* that grind the formed surface as in grinding gear teeth, threads, splined shafts, holes *etc.*

Fig. 9.123 shows *three basic kinds of precision grinding.*

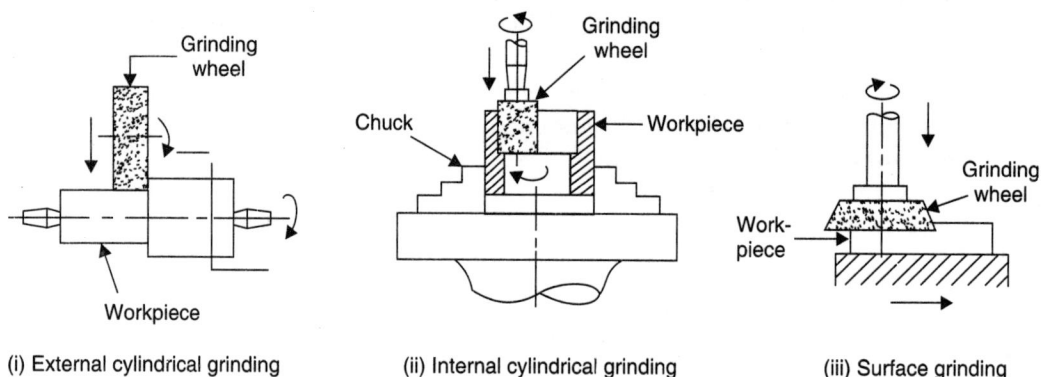

(i) External cylindrical grinding   (ii) Internal cylindrical grinding   (iii) Surface grinding

Fig. 9.123. Basic kinds of precision grinding.

9.13.3. Advantages of Grinding Process Over Other Cutting Processes

The grinding process claims the following *advantages over other cutting processes.*

1. It is possible to achieve very accurate dimensions and smoother surface finish in a very short time.

2. It is the only method of removing material from materials after hardening.

3. Owing to large number of cutting edges on the grinding wheel it is possible to produce extremely smooth surface desirable at contact and bearing surfaces by grinding operation.

4. Complex profiles can be produced accurately with relatively inexpensive turning templates.

5. Grinding unlike conventional machining need not cut through the hard skin of forgings etc.

6. Since the grinding wheel has considerable width therefore no marks as a result of feeding are there.

7. In this process little pressure is required, thus permitting its use on very light work that would otherwise tend to spring away from the tool. This characteristic permits the use of magnetic cluck for holding the work in many grinding operations.

9.13.4. Special Features of Grinding Process

Following are the *special features of grinding process* :

1. The grinding operation is intermittent in nature, and produces discontinuous chips.

2. The grinding wheel has a self sharpening character (*i.e.*, the dull or worn out grains of the grinding wheel during the operation are removed either by fracture or tearing of the bond, thus exposing fresh new grains).

3. The load acting on individual cutting grains is non-uniform.

4. The geometry of the grain is highly random and the time of contact between the chip and an abrasive grain is very small.

5. The grinding action depends strongly upon the characteristics of the grinding wheel.

6. High temperatures to the tune of 1000°C to 1400°C are usually encountered in grinding resulting into rapid grain wear and high induced surface in the workpiece.

7. The effective rake angle of abrasive grains is highly negative.

8. Grinding is associated with high specific cutting energy as compared to that encountered in conventional cutting operations.

9.13.5. Grinding Machines

The grinding machines are *classified* as follows :

I. According to the quality of surface finish :

1. *Roughing or non-precision grinders :*

   (*i*) Bench, pedestal or floor grinders.

   (*ii*) Swing frame grinders.

   (*iii*) Portable and flexible shaft grinders.

   (*iv*) Belt grinders.

2. *Precision grinders.*

II. According to the type of the surface generated or work done :

1. *Cylindrical grinders :*

   (*i*) Plain cylindrical grinders.   (*ii*) Universal cylindrical grinders.

   (*iii*) Centreless internal grinders.

2. *Internal grinders :*

   (*i*) Plain internal grinders.   (*ii*) Universal internal grinders.

   (*iii*) Chucking internal grinders.   (*iv*) Planetary internal grinders.

   (*v*) Centreless internal grinders.

3. *Surface grinders :*

   (*i*) *Reciprocating table :*

      (*a*) Horizontal spindle.   (*b*) Vertical spindle.

(ii) Rotating table :

    (a) Horizontal spindle   *(b)* Vertical spindle.

    4. *Tool and cutter grinders :*

        (a) Universal   *(b)* Special.

The various types of grinders are described below :

9.13.5.1. Plain cylindrical grinder

Fig. 9.124 shows a plain cylindrical grinder.

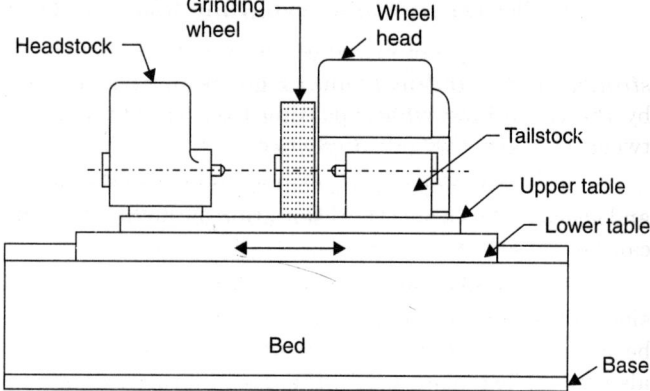

- In this type of a grinder, the workpiece is usually held between two centres. One of these centres is in the *headstock* and the other in the *tailstock*.

- In operation, the rotating work is traversed across the face of the rotating grinding wheel. At the end of *each* traverse, the wheel is fed into the *work* by an amount equal to the depth of cut. While mounting the work

Fig. 9.124. Block diagram of a plain cylindrical grinder.

between centres, the headstock centre is not disturbed.

    It is the tailstock centre which is moved in and out, manually or hydraulically, to insert and hold the work.

- The *table* is usually made in two parts. The upper table carries the headstock, tailstock and the workpiece and can be swelled in a horizontal plane, to a maximum of 10° on either side, along the circular ways provided on the *lower table*.

- The *wheel head* is usually mounted on the horizontal cross ways on the bed and travels along these to feed the wheel to the work. The movement is known as *infeed*.

9.13.5.2. Centreless grinder

Centreless grinding. Refer to Fig. 9.125.

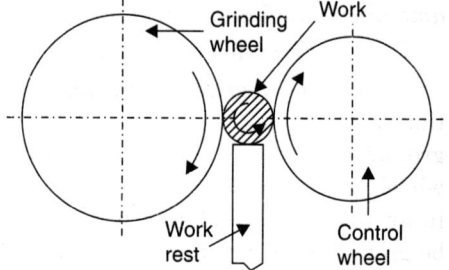

    It is the method of grinding metallic parts in which the piece to be ground is supported on a **work-rest***, and passed between a* **grinding wheel** *running at a high speed and a* **controlling wheel** *running at a slow speed.* The controlling wheel causes the work, which must be cylinder, to revolve in the opposite drection to that of the grinding wheel.

Fig. 9.125. Centreless grinding.

    The work-rest includes a number of guides that feed the work to the revolving wheels, and remove it as soon as the operation is over. The *pressure exerted by the grinding wheel drives the work into contact with controlling wheel and the work-rest.*

    It will be realised from this that a point on the surface of the work where it comes into contact with the grinding wheel revolves in the same direction as, but at a slower speed than a corresponding point on the grinding wheel. This has the *effect of producing a more accurately cylindrical surface.*

The controlling wheel is of the same composition as the grinding wheel, and its speed can be regulated to suit the requirements. *The work revolves at the same speed as the controlling wheel.*

Following are the three standard methods of *feeding* the work.

1. The through-feed grinding ;

2. The in-feed grinding ;

3. The end-feed grinding.

1. *The through-feed grinding.* Refer to Fig. 9.126.

In this type of grinding the part to be ground is *straight and cylindrical,* and is given an axial motion by the controlling wheel, passing from side to side between this and the grinding wheel.

Feed rate depends on controlling wheel diameter, and speed and its angle of presentation to the work, which can be controlled.

The number of times the work passes from one side to the next depends on the thickness of material to be ground off and on the degree of precision required, also on the extent to which the part is truly cylindrical at the start of operations.

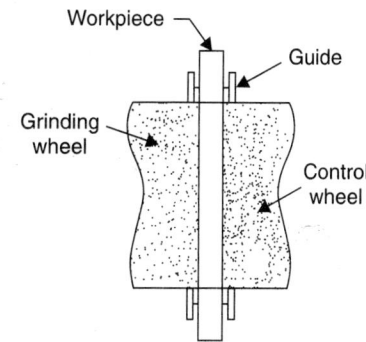

Fig. 9.126. Through feed grinding.

2. *The in-feed grinding.* Refer to Fig. 9.127.

Some parts change in cross-section forming *shoulders or heads.* These are centreless ground by the in-feed method, which resembles the method of form-grinding on centre grinder by taking a plunge cut. The wheel width governs the length of the portions capable of being ground in a single operation. *No axial feed is given,* but the controlling wheel is set so that its axis is roughly parallel to that of the grinding wheel. *The part is held firmly against the end stop by means of a small amount of inclination given to the controlling wheel.*

3. *The end-feed grinding.* Refer to Fig. 9.128.

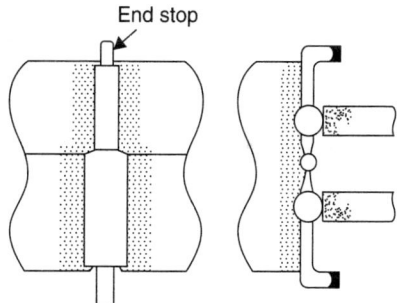

Fig. 9.127. The in-feed grinding.

While it has been said that centreless ground parts must be cylindrical, it should be noted that work having a *degree of taper* can be centreless ground by the end-feed process. In this, the grinding wheel, the controlling wheel and work rest are located in an unchanging relation to one another. The part to be ground is then automatically or by hand, fed in from the front towards a fixed end stop. Both the grinding and controlling wheels are tapered to suit the form of the work.

Fig. 9.129 shows a centreless grinding machine and also a selection of some typical examples of components that can be ground on this type of machine (on the right).

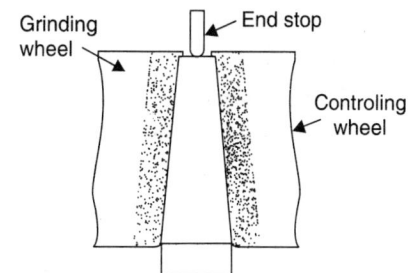

Fig. 9.128. The end-feed grinding.

Advantages and limitations of centreless grinding :

The following are the *advantages and limitations* of centreless grinding process :

Advantages :

1. There is *no need for centring* and *use of fixtures* etc.

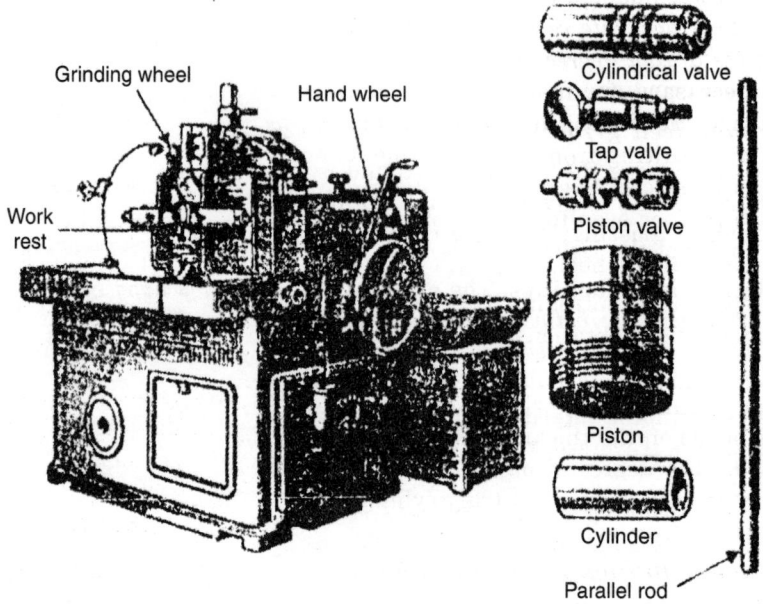

Fig. 9.129. Centreless grinding machine and example of components that can
be ground on this type of machine.

2. Since during the grinding process a true floating condition exists, therefore *less metal needs to be removed.*

3. It can be applied *equally to both external and internal grinding.*

4. Since the workpiece is supported throughout the entire length as grinding takes place, therefore, *small fragile or slender workpieces may be ground easily.*

5. The process is *continuous and adapted for production work.*

6. The *size of the work is easily controlled.*

7. The requirement of wheel adjustment is *minimum.*

8. For operating centreless grinders, a *low order of skill is needed.*

9. A *very little maintenance* is needed for the machine.

10. A *large variety of components can be ground.*

Limitations :

1. The set up time for a centreless grinding operation is usually large.

2. The process is useful only for large volume production. It may be necessary to have special equipments and additional set-up time for special profiles.

3. This process is *not suitable for large workpiece sizes.*

9.13.5.3. Internal grinders

Internal grinding is the mechanical grinding of the internal bores of gears, bushes and a wide variety of machine parts.

Fig. 9.130 shows the principle of internal grinding.

● Internal grinding is designed to complete the surfaces of holes, whether these have parallel sides, tapered bores or a combination of the two. It can also be adapted to holes of special form. It *produces accurate results, is not expensive and gives a high degree of surface finish*. It is often necessary, for instance, to remedy the slight distortion in long and slender hollow tools or parts resulting from heat treatment, and holes in these can be ground to an accuracy of 0.00635 mm, or even to 0.00254 mm. For this work choice of the right wheel is vital.

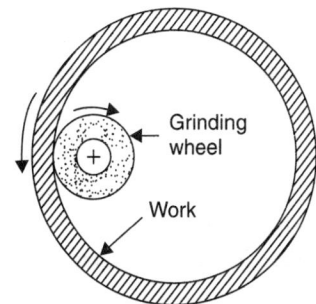

Fig. 9.130. Principle of internal grinding.

Internal grinding machines. The machines are *classified* in accordance with the *method of holding the work,* that is, *between centres or centreless* ; the *traversing or non-traversing of the work* ; and the *method* of *operation whether normal* or *automatic.*

In centreless grinding internally (Fig. 9.131) it is always the work that is traversed, the wheel being longitudinally fixed.

● The type of grinding wheel employed for internal grinding is *softer than that used for other types of grinding operations,* the reason being that the *area of contact between wheel and work is relatively great.*

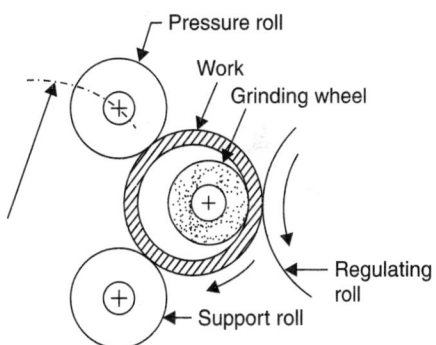

Fig. 9.131. Centreless internal grinding.

9.13.5.4. Surface grinders

Surface grinding is the method of grinding designed to carry out the removal of metal from a part or parts less expensively and with greater precision than could be achieved by machining processes with cutting tools of steel, or by hand or machine filing.

"Surface grinding" is particularly effective for *parts having hard spots that* would seriously blunt or impede a cutting tool, or where a hard superficial scale causes similar trouble in machining proper. The type of grinding finish resulting is often so good that a later polishing operation can be dispensed with, but for this to be so a well-designed machine is essential.

"*Surface-grinding machines*" differ according to the shape of the grinding wheel employed and the motion given to the worktable during operations. Same machines have reciprocating worktables and some have revolving worktables. They range in rigidity, size and weight from the relatively small, light machine used in the tool-room to the heavy and powerful machines which are used in the mass production of duplicate components.

1. Horizontal-spindle surface grinding machines. These machines use the circumference of straight grinding wheels, and are able to deal with a wide range of work needing superfinish and *extremely fine limits of accuracy.* They yield a *greater output and take off metal faster than similar machines* using cup-shaped, segmental or annular wheels. If a rotating worktable is used, a finish comprising concentric circles can be obtained and is often popular.

2. Vertical-spindle flat grinding machines. These grinders remove metal faster when using a cup, cylindrical or segmental wheel than when using a straight wheel. They have a *great precision,* and if strong and rigidly built can grind to extremely fine limits.

- **Disc grinding** is a form of surface grinding. In it one disc (or more) of abrasive type is mounted on a vertical spindle so that the plane of the disc is horizontal, the work resting on the surface of a flat, rotating carrier or worktable.

Disc grinding machines are employed where rough and semi-precision grinding is desired and where material must be ground off rapidly and effectively to tolerances somewhat greater than the most severe type. Their use in such cases is highly economical. Typical examples of work done by disc grinding are the sharpening of tools, forming the square ends of disc blanks, gear and crankcase covers etc. The abrasive discs are fairly thick and are usually reinforced with steel.

Fig. 9.132 shows various types of surface grinding machines :

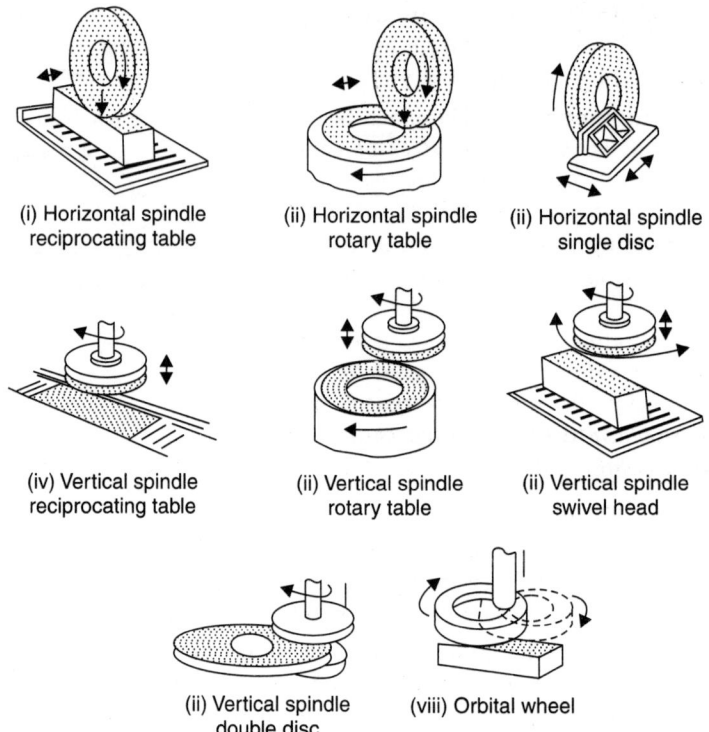

Fig. 9.132. Surface grinding machines.

9.13.5.5. Form grinding

Form grinding is grinding of tools, designed for machining and other operations, in such a way that they are provided with the precise form required for their work ; or regrinding them to restore the form after it has been lost as a result of service conditions. Form grinding must be done with great accuracy.

9.13.5.6. Creep feed grinding

It is a new form of grinding operation different from the conventional grinding process. *In creep feed grinding the entire depth of cut is completed in one pass only* using very small in feed rates.

As shown in Fig. 9.133 this process is characterised by high depth of cut of the order of 1 to 30 mm with *low work speeds* of the order of 1 to 0.25 m/min. The actual material removal rates calculated from these process parameters are generally in the same range as that of conventional

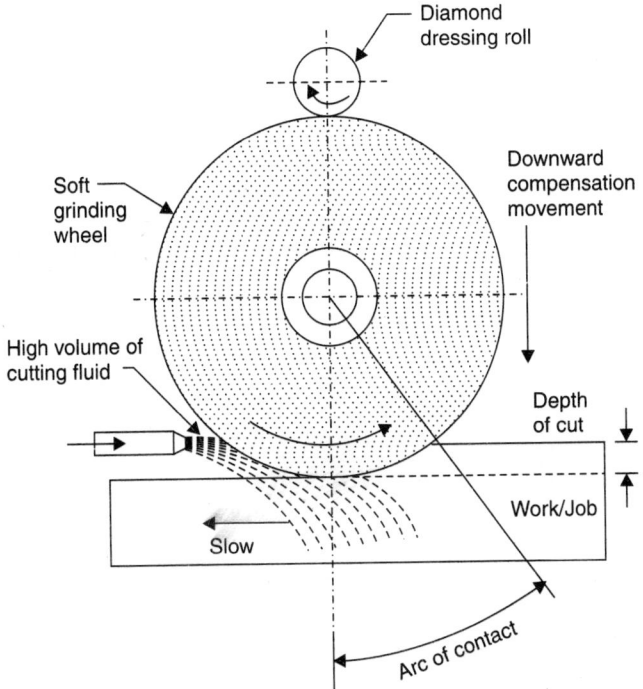

Fig. 9.133. Creep feed grinding.

grinding. However, the idle time (stopping and wheel/table reversal) gets reduced since the grinding operation is completed in one pass. *The cutting forces and consequently the power required increases in case of creep feed grinding but has a favourable G-ratio.*

It is necessary to continually dress the grinding wheel (to reduce the wheel dullness) for efficient operation. This however *causes wheel wear and makes it necessary to adjust the wheel head.*

— *Soft and open wheels* are used to take care of the wheel dressing and accommodate large volume of chips generated in the process.

— The grinding wheels speeds used are also *low* of the order of 18 m/s (compared to 30 m/s used in conventional grinding operations).

— The infeed rates used are low, of the order of 0.005 mm/pass.

— The grinding fluids used are *oil based* in view of the low grinding speeds employed. However, *the volume of grinding fluid is much more compared to conventional grinding* (in view of the *high heat* generated in the process).

9.13.5.7. Size and specifications of grinding machines

The size of a grinder is *specified* according to :

(*i*) The largest workpiece which can be held on a grinder.

(*ii*) The normal capacity (or power) of the grinder.

(*iii*) The width of the table.

(*iv*) The maximum traverse of the table.

(*v*) Wheel diameter.

(*vi*) Height of the grinding head etc.

9.13.6. Grinding Wheel

- A **grinding wheel** is a *multi-tooth cutter made up of many hard particles known as 'abrasive' which have been crushed to leave sharp edges which do the cutting.* The abrasive grains are mixed with a suitable bond, which acts as a matrix or holder when the wheel is in use.

- The wheel may consist of one piece or of segments of abrasive blocks built up into a solid wheel.

- The abrasive wheel is usually mounted on some form of machine adapted to a particular type of work.

The performance of a grinding wheel is usually evaluated in term of the grinding ratio (G) which is defined as :

$$G = \frac{\text{Volume of material required}}{\text{Volume of wheel wear}}$$

For fine grinding operations such as horizontal surface grinding, the value of G is usually in the range of 10 to 60, while for rough grinding operations it is much less than 10.

Characteristics of the grinding wheel :

Wheel parameters that influence the grinding performance are :

1. Abrasive material.
2. Abrasive size.
3. Bond
4. Grade.
5. Structure.

1. *Abrasive material :*

An *'abrasive' is a substance that is used for grinding and polishing operations.* Abrasives may be classified as follows :

1. *Natural :*

      (*i*) Sandstone   (*ii*) Emery

      (*iii*) Corundum   (*iv*) Diamonds.

2. *Artificial :*

      (*i*) Silicon carbide   (*ii*) Aluminium oxide.

Natural abrasives :

Almost all of the natural abrasives, except diamond are now considered obsolete.

- The "*sandstone*" is used only for the sharpening wood-working tools.

- "*Emery and corundum*" are the materials which were widely used formerly but now these have been replaced completely by artificial abrasives.

- "*Diamond*" is largely used for dressing the grinding wheels and as an abrasive for grinding hard materials.

Artificial abrasives :

(*i*) **Silicon carbide (SiC) :**

- Silicon carbide abrasive is manufactured from 56 parts of silicon sand, 34 parts of powered coke, 2 parts of salt, and 12 parts of saw dust.

- There are two types of silicon carbide abrasives :

    (*a*) *Green grit* which contains at least 97% silicon carbide.

    (*b*) *Black grit* which contains at least 95% silicon carbide.

- It follows the diamond in order of hardness, but it is not as tough as aluminium oxide.
- It is employed for grinding materials of *low tensile strength* such as *cemented carbides, stone* and *ceramic materials, grey cast iron etc.*

(ii) Aluminium oxide (Al_2O_3) :

- It is manufactured by heating mineral bauxite, a hydrated aluminium oxide clay containing silica, iron oxide, titanium oxide etc., mixed with ground coke and iron shavings in an arc-type electric furnace.
- As it is tough and is not easily fractured, it is better adopted to grinding materials of high *tensile strength*, such as carbon steels, high speed steels, tough bronzes etc.

2. *Abrasive size* :

Choice of the grain size depends upon the properties of the work material, surface finish, desired rate of metal removal etc.

- Coarse grains (grit size 10–24) give faster rate of metal removal, but yield a poor surface whereas fine grits (grit size 70–180) are used for finishing operation but the rate of *metal* removal is slow.
- *Coarse grain wheels are normally suitable for soft and ductile materials ; for hard and brittle materials finer grains are preferred.*

3. *Bond* :

To ensure an effective and continuous action, it is imperative that the grains of abrasive material should be held firmly together to form a series of cutting edges. The material used for holding them is known as **bond**. The principal bonds are enumerated and described below :

(*i*) Vitrified bond	(*ii*) Silicate bond
(*iii*) Oxychloride bond	(*iv*) Resinoid bond
(*v*) Shellac bond	(*vi*) Rubber bond.

(i) Vitrified bond :

- It is a clay bond, reddish brown in colour.
- The base material is *Felspar* which is a fusible clay.
- The bond itself is very hard and acts as an abrasive.
- It is not affected by water, oil, acids, temperature or climatic conditions.
- The structure of the wheel is uniform due to wet mixing of different components.
- Most of the grinding wheels possess this bond.
- Such bonds are abbreviated as *V.*

(ii) Silicate bond :

- The base material of this bond is *silicate of soda.*
- Wheels possessing this bond are light grey in colour.
- In this case, the cutting action of the wheel is smoother and cooler.
- Extra hard wheels *cannot* be produced with this bond.
- Such wheels are designated by the letter S.

(iii) Oxychloride bond :

- It is a *mixture* of *oxide* and *chloride of magnesium* and setting takes place in cold state.

- This bond provides a cool cutting action, but grinding is usually done dry.
- Such bonds are abbreviated as *O*.

(*iv*) **Resinoid bond :**

- It is made out of synthetic or organic resin.
- It is strong and flexible.
- Wheels made from such a bond can be used for high speed cutting at low temperatures.
- The bond is addressed by the letter B.

(*v*) **Shellac bond :**

- This bond is used for high finish work.
- It is designated by the letter-*E*.

(*vi*) **Rubber bond :**

- Rubber bonded wheels are composed of hard vulcanised rubber.
- Such wheels are hard flexible and can have very thin sections and are useful in cut-off operations.
- Use of cutting fluid is essential with such wheels.
- Rubber bond is abbreviated as-*R*.

4. *Grade* :

'Grinding wheel grade' refers to the strength with which the bond holds the grains together.

- The strength or hardness of the wheel depends upon the volume of the bonding material used. As the volume of the bonding material in a wheel increases its hardness improves.
- The wheel hardness is designated as soft, medium or hard. Wheel with hardness rating A to I are classified as *soft*, those having a rating of *J* to *P* are *medium* and wheels with hardness rating *Q* to *Z* are *hard*.

5. *Structure* :

'Structure' of a grinding wheel refers to the relationship between the volume of the abrasive material, volume of bond and the volume of voids present in a grinding wheel.

- A wheel would have a *dense structure when the percentage volume of the abrasive is large*.

Specification : The American Standard specifies the grinding wheels as follows :

1. Abrasive type : A for Alumina, S for Silicon carbide etc.
2. Grain size : mesh number.
3. Grade : Letters A for very soft, and Z for very hard.
4. Structure : 15 for very dense structure.
5. Bond type : V-vitrified, R-rubber etc.
6. Manufacturer's owned private mark.

Thus, a wheel marked as follows :

A-50   *Q*   8 *V*-30

represents an alumina wheel with *50 grit size, medium hardness, medium structure* and *vitrified bond.* The number 30 at the end is manufacturer's own identification number.

9.13.7. Manufacture of Grinding Wheels

The following *steps* are involved in the manufacture of grinding wheels :

1. Reduce the abrasive material to small size using roll and jaw crushes.
2. Remove the iron compounds with the help of a magnetic separator.
3. Wash the material to remove dust etc.
4. Grade the abrasive material grains by passing them over vibrating standard screen.
5. Select proper sized grains.
6. Mix grains with bonding material, mould or cut the wheel to proper shape and heat.
7. The wheels are then bushed, trued and tested.

9.13.8. Wheel Shapes. Refer Fig. 9.134 and 9.135.

- Grinding wheels are used in almost all shapes. *Most common is disc wheel.* It has a central hole for mounting and both edges or the sides of the wheels are used.
- For grinding piston rods or pins, *rod shaped wheel* is used.
- For grinding in restricted spaces like wheel teeth, etc. *tapered edges dish shape or wheel disc shape wheels are used.*
- Some grinding wheels have shapes like saucer, cylinder, flaring cup.
- Some of the grinding wheel shapes are shown in Fig. 9.134.

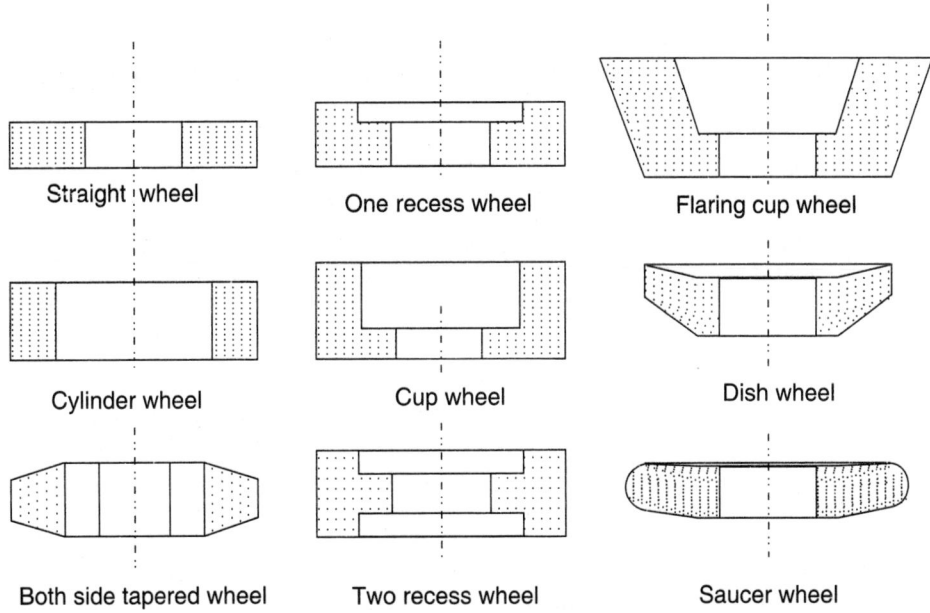

Fig. 9.134. Wheel shapes.

Maximum speeds for flexible shaft grinders :

3 × 100 mm shaft size	..... 40,000 r.p.m.
4 × 1250 mm shaft size	..... 30,000 r.p.m.
7 × 1500 mm shaft size	..... 18,000 r.p.m.
10 × 1500 mm shaft size	..... 15,000 r.p.m.
12 × 2000 mm shaft size	..... 8,000 r.p.m.
20 × 2500 mm shaft size	..... 5,000 r.p.m.

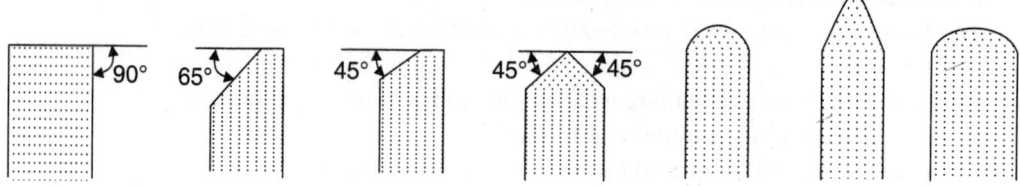

Fig. 9.135. Standard wheel edges.

Recommended wheel speeds for grinding :   *Metres/min.*

Internal grinding	.....	650—1950
Surface grinding	.....	1300—1630
Knife grinding	.....	1150—1450
Cutter grinding	.....	1630—1950
Tool grinding	.....	1630—1950

9.13.9. Mounting of Wheels

- Normally grinding wheels are supplied with a central hole fitted with a lead bush. This is then mounted on the spindle. The wheels are clamped axially with clamping collar. To distribute the clamping pressure, soft washers are put between wheels and collars. *In absence of the soft washers, there is a danger of breaking of wheels.* The soft washers can be of the *blotting paper or any other thick paper. Small washer increases the tightening pressure or if the lead bush is out of square, the pressure of tightening may not be equalised. Large washers are therefore essential.*

- To test the soundness of wheels, these are subjected to overspeeds for testing. A good wheel should not run out of centre and should not have hollows at the testing speed. Defective wheels will crack at overspeed test.

9.13.10. Wheel Truing

In the market, two types of wheel dressers are available. '*Hunting done*' type and '*Diamond dresser*'. Diamond dresser can be used as a hand dresser or can be mounted. It contains a diamond point. 'Tiam carbo' dresser is substitute for 'Diamond dresser'. The wheel has to be correctly mounted and *truer can traverse wheel backwards and forwards on the saddle. The purpose of the dressing is to expose newly made sharp wheel faces for grinding. Glazed wheels do not grind and require dressing. Dressing can be carried out by pressing abrasive sticks against the wheel and then moving all along the wheel face.*

9.13.11. Material Removal Rate and Machining Time

1. Cylindrical grinding :

This type of grinding includes the following two types of grinding operations :

(*i*) Traverse grinding.

(*ii*) Plunge cut grinding.

(*i*) ***Traverse grinding.*** In traverse grinding, the depth of cut is the layer of the metal removed by the grinding wheel in traverse stroke parallel to the axis of the job. It is also called "*in feed*".

Thus, depth of cut (d) is :

$$d = \frac{D_{ji} - D_{jf}}{2} \qquad ...(9.40)$$

where, $\qquad D_{ji}$ = Initial diameter of the job, and

$\qquad\qquad\qquad D_{jf}$ = Final diameter of the job.

The values of d : 0.01 mm to 0.025 mm for *rough grinding* ;

$\qquad\qquad\qquad\qquad$: 0.005 mm to 0.015 mm for *finish grinding*.

The traverse feed is the longitudinal feed parallel to the axis of the job, mm/rev.

For *rough grinding* : $\qquad f$ = (0.3 to 0.5) × width of wheel, for D_j < 20 mm

$\qquad\qquad\qquad\qquad\qquad$ = (0.7 to 0.85) × width of wheel, for $D_j \geq$ 20 mm

For *finish grinding* : $\qquad f$ = (0.2 to 0.4) × width of wheel

Table traverse will be, $\qquad f_t = f \cdot N_j$ mm/min

where, $\qquad\qquad\qquad N_j$ = Job speed, rev/min.

Peripheral speed of job, $\quad V_j = \dfrac{\pi D_j N_j}{1000}$ m/min $\qquad ...(9.41)$

Peripheral speed of grinding wheel cutting speed, $V_{gw} = \dfrac{\pi D_{gw} N_{gw}}{1000} \qquad ...(9.42)$

where, $\qquad\qquad D_{gw}$ = Diameter of grinding wheel, mm, and

$\qquad\qquad\qquad N_{gw}$ = Speed of grinding wheel, r.p.m.

● When the work and grinding wheel rotate in *opposite direction* to each other, the cutting speed (V_c) can be taken as :

$$V_c = V_{gw} + V_j \qquad ...(9.43)$$

Material removal rate (MRR). It can be taken *approximately* as :

$$MRR = \pi D_j \cdot d \, f \cdot N_{gw} \text{ mm}^3\text{/min} \qquad ...(9.44)$$

Machining time (t_m) :

$$t_m = \frac{L}{f \, N_j} \times n_p \times k$$

where, $\qquad\qquad n_p$ = Number of passes, and

$\qquad\qquad\qquad k$ = Sparking out factor.

● The sparking out factor takes into account the removal of metal even in passes in which there is *no in-feed*. This occurs due to :

— Non-uniform wear of wheel ;

— Springing of the work ;

— Breaking out of grains ;

— Absence of continuous edge.

All the above factors lead to an increase in machining time.

Values of k ; approximately are :

For *rough grinding* $\qquad\qquad$ 1.2

For finish grinding $\qquad\qquad$ 1.4

(*ii*) **Plunge-cut grinding.**

Material removal rate, $MRR = \pi \, D_j \, B_{gw} \cdot f_r \qquad ...(9.45)$

where, $\qquad\qquad D_j$ = Diameter of the job, mm,

B_{gw} = Width of the grinding, wheel, mm, and

f_r = Rate of radial feed, mm/min.

Machining time, $\qquad t_m = \dfrac{\text{Machining allowance}}{f_r}$ $\qquad\qquad$ ...(9.46)

where, $\qquad\qquad\qquad f_r = f \times N_{gw}$, f is in mm/rev.

2. Surface grinding :

Cutting speed, $\qquad\qquad V_c = \dfrac{\pi D_{gw} . N_{gw}}{1000}$ m/min

Material removal rate, MRR = $f_c . 2 . f_t$ $\qquad\qquad\qquad$ ...(9.47)

where, $\qquad\qquad\qquad f_c$ = Cross-feed per stroke, mm/stroke,

$\qquad\qquad\qquad\qquad f_t$ = Table traverse rate, mm/min, and

$\qquad\qquad\qquad\qquad d$ = Depth of cut per pass.

Machining time, $t_m = \dfrac{L_j \times B_j}{V_t \times 1000 \times f_c} \times n_p \times k$ $\qquad\qquad$ ...(9.48)

where, $\qquad\qquad\qquad L_j$ = Length of the job to be ground.

$\qquad\qquad\qquad\qquad B_j$ = Width of the job to be ground,

$\qquad\qquad\qquad\qquad V_t$ = Traverse velocity of table, m/min, and

$\qquad\qquad\qquad\qquad f_c$ = Cross feed of grinding wheel, mm/stoke.

B. FINISHING

9.13.12. Finishing Processes

For several applications, grinding cannot meet the accuracy and surface finish requirements. For such applications, workpieces are subjected to final finishing operations, *e.g.*, honing, lapping etc.

Some of the purpose of finishing surfaces of metal parts are :

(*i*) *To improve the surface appearance*—By polishing, buffing, burnishing.

(*ii*) *To improve dimensional accuracy and surface finish* (*smoothness*)—By lapping, honing, superfinishing etc.

(*iii*) *To provide a clean finish* to the surfaces of a machine part, *by buffing* etc. to enable them to be coated with other metal (aluminium and nickel plating)—By electro-depositing method.

(*iv*) *To improve the functional properties of the machine parts* (*e.g.,* wear resistance, fatigue strength, power losses in friction of motion, strength of interference fits of mating parts, corrosion resistance).

9.13.13. Honing

Honing is a grinding or abrading process in which very little material is removed. It is used primarily to *remove the marks on the surface left by previous operations.*

In honing, the material is removed by abrasive sticks (aluminium oxide or silicon carbide) mounted in a mandrel or fixture. By floating action between the work and tool, the pressure exerted in the tool is transmitted equally to all sides. The honing tool is given a slow reciprocating motion as it rotates (Fig. 9.136). This action results in rapid removal of stock and at the same time generation of a straight and round surface.

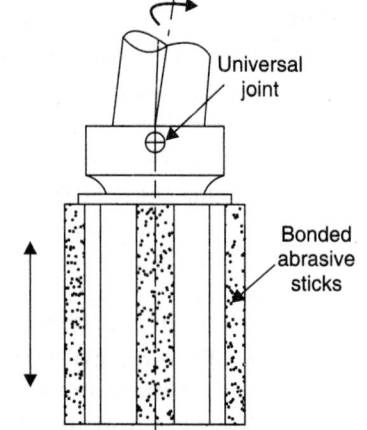

Fig. 136. Honing.

By honing, the surface defects, such as slight taper caused by previous operations, can be corrected.

Coolants are essentially required in this process to flood away small chips and to keep the temperature uniform.

A metal frame which holds the abrasive sticks during honing operation is known as a *hone* or a ***honing tool***. Hones are of two type :

(*i*) Internal hones.　　　　　　　　　　(*ii*) External hones.

Examples of honing work :

Following are the *examples of honing work :*

- Diesel engine cylinder bore.
- Roller bearing races.
- Bores of cannons.
- Automobile, aeroplane engine cylinder head bores.
- Bores in rocker arms.
- Hub holes in gears of gear boxes.

Advantages and disadvantages of honing :

Following are the *advantages* and *disadvantages* of honing :

Advantages :

1. Correction of geometrical accuracy :
   (*i*) Out of roundness ;
   (*ii*) Taper ;
   (*iii*) Axial distortions.
2. Dimensional accuracy.
3. Relatively high productivity as compared to lapping.
4. Holes of any diameter or length may be honed.
5. Several holes may be honed simultaneously on multiple spindle machines.

Disadvantages :

Tough non-ferrous metals cause glazing or clogging of the voides of the abrasive sticks and thus, they are difficult to hone.

Honing conditions :

- All materials can be honed. However, the material removal rate is affected by the hardness of the work material. The typical rates are :
   (*i*) *Soft material* 　　— 1.15 mm/min on diameter ;
   (*ii*) *Hard material* 　　— 0.30 mm/min on diameter :
- The maximum bore size that can be conveniently honed is about 1500 mm while the minimum size is 1.5 mm in diameter.
- The honing allowance should be small to be economical. However, the amount also depends upon the previous error to be corrected.
- The abrasive and grain size to be selected depend upon the work material and the resultant finish desired.

Vertical honing machines :

- These machines hold the work and honing tool with their axes in vertical position. These machines are available in both single spindle and multiple spindle types.
- These machines are *best suited for shorter jobs.* Part lengths up to 2000 mm are usually honed in vertical honing machines.

Horizontal honing machines :

● These machines hold the work and honing tool with their axes in horizontal position. These machines are available both in single spindle and multi-spindle types.

● These machines are employed for longer jobs. Jobs more than 2000 mm are usually honed in horizontal honing machines.

Note : According to new development in the field of honing, parts like cylinder blocks and liners, refrigerator compressors, gears and connecting rods are processed by *"diamond honing"*.

9.13.14. Lapping

Lapping is *a finishing process following after grinding, and designed to produce an exceptionally high degree of surface finish as well as a perfectly true surface accurate to size within extremely close limits.* In some work the finish is more important than the dimensional accuracy.

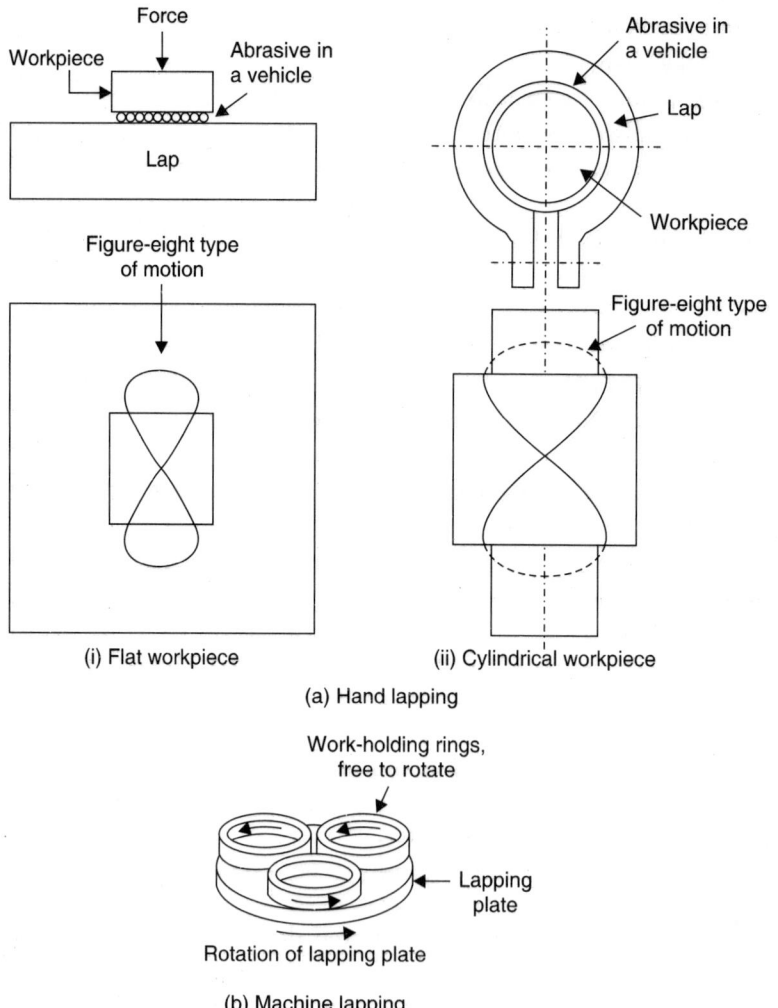

Fig. 9.137. Lapping process.

In *lapping process* (Fig. 9.137) a layer of fine abrasive particles, usually suspended in a liquid, is held between the workpiece and the lap. The lap material being softer than the workpiece (generally cloth, copper or cast iron) causes the abrasive grains to get embedded on to the lap surface when pressure is applied between the lap and the workpiece. These grains cut the work in the same way as in grinding when relative motion is provided between the workpiece and the lap. Because of variations in grain size from particle to particle, not all the grains get embedded into the lap and *these loose grains roll and slide between the workpiece and the lap and cause some material removal.* Embedded grains are however, responsible for bulk of material removed and abraded workpiece conforms to the shape of the lap.

Lapping is done manually or by specially designed machines (See Fig. 9.137). Soft materials are lapped with Al_2O_3 *and hard materials* with diamond or SiC grit.

- The lap materials generally used are :
  — Cast iron ;
  — Soft steel ;
  — Bronze ;
  — Brass.

Types of lapping operations :

1. *Equalizing lapping*—When work and lap mutually improve their shape and surface, for example ; when gears are run together with some abrasive, or tapered valves are seated in seats.

2. *Form lapping*—Shape of lap is imparted to the work.

Lapping Machines :

Lapping machines may be *classified* as follows :

1. **General purpose lapping machines :**

(*i*) Vertical axis lapping machine with metallic laps.

(*ii*) Vertical axis lapping machine with bonded abrasive wheels as laps.

(*iii*) Centreless lapping machine.

(*iv*) Abrasive belt lapping machine.

2. **Single purpose lapping machines :**

(*i*) Valve seat lapping machine.

(*ii*) Crankpins and journal lapping machine.

(*iii*) Crankshaft lapping machine.

(*iv*) Gear lapping machine etc.

Examples of lapping work :

1. *Hand lapping :*

- Surface plates
- Holes and pins
- Tappet valves and valve seats
- Jigs and fixture bushes
- Slip gauges
- Plug gauges etc.

2. *Machine lapping :*

- Diesel engine injectors
- Oil burner parts
- Gauges
- Ball and roller bearing faces
- Machine bearings
- Measuring instruments etc.

Advantages and Disadvantages of lapping :

Advantages :

(*i*) It can lap any type of material of any shape, if it is flat.

(*ii*) As the parts are not clamped and no heat is generated, there is no warping.

(*iii*) It produces no burrs, rather removes those left in earlier process.

Disadvantages :

The lapping is still an art because of large variables involved and thus past experience and skill in a new job go a long way. It is important to remember that flatness, surface finish and polished surface are not obtained all at the same time and in equal quantities.

9.13.15. Superfinishing

Superfinishing is an abrading process, efficient in surface refining of cylindrical, flat, spherical and cone shaped parts.

It is not primarily a dimension changing process but *mainly used for producing finished surface of fine quality on metals.* Only a slight amount of stock is removed (average 0.002 to 0.02 mm on a disc). The smoother finishes do not have scratch to exhibit any directional effect.

The honing process involves two motions whereas superfinishing requires three to five or even more. As a result of these motions the abrasive particle path is random and never repeats itself.

The operation is mainly concerned with *external work. Superfinishing is generally used for :*

    (*i*) Correcting inequalities in geometry.

    (*ii*) Removing surface fragmentation.

    (*iii*) Reducing surface stresses and burns and thus restoring surface integrity.

Superfinishing produces a high wear resistant surface on any object which is symmetrical.

Fig. 9.138 shows the superfinishing operation.

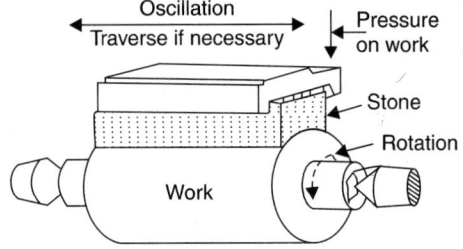

Fig. 9.138. Superfinishing operation.

— The contact surface in superfinishing is large and the tool maintains a rotary contact with the workpiece while oscillating as shown in Fig. 9.138.

— The typical stroke of the superfinishing stone is about 1 to 5 mm with an oscillating frequency of 2 kHz. Superfinishing speeds used are 10 to 40 m/min while the working pressure maintained is about 0.1 to 0.3 MPa. The heat generated under these conditions is appreciably small and hence *there is no metallurgical alteration of the work.*

— *The finish obtained on the surface depends upon the time for which the stone is in contact with the work.*

● The superfinishing operation basically differs from other abrasive finishing methods in the following *respects :*

    (*i*) It is primarily a finishing operation and not a dimensioning operation.

    (*ii*) The motion is multiple and random and very rapid.

    (*iii*) Strokes are 1 to 5 mm at 300 to 3000 reversals per minute as compared to 50 to 100 reversals of a grinding wheel.

    (*iv*) The reversal of stroke is short, as compared to the long stroke of the hone, which accumulates large amounts of chips that may scratch the surface.

    (*v*) A surface finish of 0.01 micron may take 30 seconds.

(*vi*) The abrasive pressure for external surfaces is 0.02 to 0.033 N/mm² against 1350 N/mm² for grinding with line contact.

(*vii*) The abrasive pressure for internal surfaces is 0.02 to 0.2 N/mm² against 3.5 to 7.0 N/mm² for average honing operations.

Examples of work :

- Crankshaft journals
- Automotive pistons
- Cylindrical shanks of valve tappets
- Roller bearing surfaces etc.

- Gun recoil machining
- Automobile valve systems

9.13.16. Comparison Among Lapping, Honing and Superfinishing

S.No.	Aspects	Lapping	Honing	Superfinishing
1.	Type of process	Abrading	Abrading	Abrading
2.	Type of surface produced	Used to produce geometrically true surface in addition to surface finish.	Used for finishing internal round holes.	Used to produce extremely light quality surface finish and not dimensional.
3.	Abrasive tools used	Lapping consists of the use of loose abrasive particles with some vehicle. Mesh size of abrasive particles ranges from 120-1200.	It makes use of bonded sticks called *hones*. Mesh size of abrasive particles ranges from 80–600.	It also makes use of bonded abrasive stones, but of finer mesh size ranging from 400–600.
4.	Types of surface/material for which used	It is normally used for hard surfaces, *e.g.*, steel and cast irons.	It can be used for both soft and hard materials.	It can also be used for both soft and hard materials.
5.	Type of process/operation	Finishing (metal removal is very small up to 0.025 mm)	Metal removal (metal removal is high up to 0.75 mm)	Superfinishing (Metal removal is very small up to 0.005 mm)
6.	Motion of the workpiece and tool	Both workpiece and laps are in motion.	Hone is rotating whereas the workpiece is held stationary.	Both workpiece and abrasive stone are in motion.

9.13.17. Polishing and Buffing

Both these processes are *used for making the surfaces smoother along with a glossy finish.*

Polishing and buffing wheels are *made of cloth, felt or such material which is soft and has a cushioning effect.*

Polishing. It is done with a very fine abrasive in *loose form* smeared on the polishing wheel with the work rubbing against the flexible wheel. *A very small amount of material is removed in polishing.*

Examples of work : The parts to be electroplated are usually polished before plating. All those which are to have a lustrous smooth appearance without plating such as stainless steel utensils, surgical instruments, bright finished hand tools like wrenches etc. are given a final finish by polishing alone or polishing followed by buffing.

Buffing. In this process the *abrasive grains in a suitable carrying medium such as grease are applied at suitable interval to the buffing wheel.* Negligible amount of material is removed in buffing while a very high lustre is generated on the buffed surface.

Examples of work : All parts to be electroplated and made of steel and harder materials are first finished by grinding, polishing and then buffing. Then they are plated. Die cast parts are mostly first polished and then buffed before being electroplated. Sheet aluminium, brass and copper usually require only the buffing operation before electrolating.

- The *dimensional accuracy of the parts is not affected by polishing and buffing operations.*

9.13.18. Burnishing

Burnishing *is an operation by which a bright, polished finish is produced on the surface of a metal by a rubbing action which smooths out small scratch marks.*

The finish is produced by the action of *burnishers,* which must be made of very hard material having a highly polished surface. When rubbed with pressure over the metal being treated, no metal is actually removed by the burnisher, but an action somewhat resembling a flow of metal takes place.

Uses :

- Burnishing is generally *used only to produce a decorative finish.*
- Gold leaf and brass can be burnished by rubbing an agate burnisher over the surface.
- Steel and brass can be burnished by rubbing with a number of interconnected hard steel rings mounted on a piece of leather and resembling a piece of "coat of mail".
- In some cases burnishing is used to produce a *finish which will give better wearing qualities than those left by a cutting tool. Thus, journals of railway axle* are sometimes burnished by the pressure of a plain hardened-steel roller mounted on a small shaft in the end of a tool held in a lathe slide-rest. The roller is slowly traversed along the rotating journal of the axle which is mounted on the lathe.

9.13.19. Comparison of Grinding and Finishing Processes

The comparison of grinding and finishing operations is given below :

Comparison of 'Grinding' and 'Finishing' processes

S.No.	Process	Advantages	Limitations
1.	*Horizontal surface grinding*	(*i*) Close tolerance (± 0.0025 mm) and good surface finish (0.2 – 2.5 μm). (*ii*) Suitable for grinding of flat surfaces. (*iii*) Medium labour skill.	(*i*) Suitable for improving tolerance and surface finish on machined surfaces. (*ii*) Low production rate. (*iii*) Thermal damage may occur due to high temperature during grinding.
2.	*Vertical surface grinding*	(*i*) High production rate. (*ii*) Suitable for rough grinding of flat surfaces. (*iii*) Medium labour skill.	(*i*) Suitable for stock removal only. (*ii*) Close tolerance cannot be obtained.

3.	Cylindrical grinding	(i) Tolerance and surface finish same as for horizontal surface grinding. (ii) Suitable for round shapes, stepped diameters, etc. (iii) Medium labour skill.	(i) Suitable for improving tolerance and surface finish on machined surfaces. (ii) Low production rate. (iii) Thermal damage may occur due to high temperature during grinding.
4.	Internal grinding	(i) Tolerance and surface finish same as for horizontal surface grinding. (ii) Suitable for bores. (iii) Medium labour skill.	(i) Suitable for improving tolerance and surface finish on machined surfaces. (ii) Low production rate. (iii) Thermal damage may occur due to high temperature during grinding.
5.	Centreless grinding	(i) High production rate. (ii) Suitable for long, round workpieces. (iii) Low labour skill.	(i) Not suitable for large diameter workpieces. (ii) Tolerances and surface finish not as good as those obtained in cylindrical grinding.
6.	Honing	(i) Very close tolerance (± 0.0015 mm) and extremely good surface finish (0.01 – 0.35 μm). (ii) Suitable for bores and holes. (iii) Low labour skill. (iv) No heat distortion.	(i) Expensive operation. (ii) Very low production rate. (iii) Limited amount of material can be removed.
7.	Lapping	(i) Tolerance and surface finish same as that for honing. (ii) Suitable for flat surfaces. (iii) High production rate. (iv) Low labour skill. (v) No heat distortion.	(i) Limited amount of material can be removed. (ii) Expensive operation.

9.14. COOLANTS AND LUBRICANTS

In all metal working operations in workshops, the use of metal working fluids is indispensable. These fluids are given various names *viz*, **Coolants**, **Lubricants** and **Cutting Compounds** or **Cutting fluids**, according to their use.

The *metal working fluid* usually performs the following *functions* :

1. Increases the tool life and produces better finish by carrying away the heat generated during metal working.

2. Minimises the friction between mating surfaces and, thus, prevents rise in temperature.

3. Protects the finished surface from corrosion.

4. Provides a cushioning effect between the job surface and the tool to prevent adhesion of the two, such as in stamping, extrusion etc.

5. Provides lubrication at high pressures called *boundary lubrication*.

6. Drives away the chips, scale and dirt, etc., from between the working or mating surfaces.

7. Prevents the work metal from a quick swelling on to the tool or into the die and the resulting wear on their surfaces.

Types of lubricants :

The various types of *metal working lubricants* are :

1. Mineral oils :

● These lubricants *do not* find much favour in boundary lubrication, as in deep drawing and extrusion processes. If at all they are to be used in such processes they are used in compounded form.

2. Fatty oils and acids :

These are *extensively used for boundary lubrication.*

● *Fatty oils* are used in heat-treatment process as a quenching medium for obtaining a high degree of hardness.

● *Fatty acid* is used as a flux in tinning work.

3. Waxes :

● Waxes (derived from petroleum) are used in various processes like *rolling, drawing, extrusion, tinning and wet coating on mould surfaces.*

● *Wax emulsions* and *compounded waxes*, are also used in *cutting* and *drawing.*

4. Graphite suspensions :

● These lubricants are widely used in *foundry* and *forging* work.

5. Compounded emulsions :

● The compounded emulsions are best suited for use in *heavy duty operations.*

— The water usually varies from 5 to 15 parts.

6. Conventional emulsions :

● These emulsions are prepared by mixing the neat soluble oils in water. The main constituents of these emulsions are *soap, fat, fatty acids* and *water.*

● Most of the *cutting* and *grinding work* done in the workshops involves the use of such emulsions.

7. Aqueous solutions :

● These solutions are principally used as *coolants.* However, some of them show reasonably good lubricating properties.

● Soda or borax in water is the cheapest and best solution mainly for cooling.

8. Compounded mineral oils :

● For drawing, cutting and forming operations the mineral oils are compounded with sulphurised fatty oils. The sulphurised minerals oils are commonly used under conditions of high pressures and excessive fraction.

● The sulphurised mineral-fatty oils are sometimes added with suitable amount of chlorine to give *chlorinated compounds*, which are widely used in different metal working operations.

9. Minerals :

● Various types of minerals (different types of salts, metals and refractory materials) are used as metal working lubricants or coolants.

Cutting Fluids :

The term *'cutting fluids'* is used to denote those *coolants* and *lubricants* which are used in metal machining and their allied operations like lapping, honing etc. The coolants and lubricants used in these processes perform the following *functions* :

(*i*) Cool the tool and the workpieces.

(*ii*) Provide adequate lubrication between the tool and workpiece and the tool and the chips.

(*iii*) Prevent adhesion of chips to the tool or workpiece or both.

● The sources of heat generation during metal cutting are :

— Friction ;

— Plastic deformation of metal ;

— Chip distortion.

Characteristics of a good cutting fluid :

Following are the *characteristics/qualities of a good cutting fluid* :

1. It should provide sufficient *lubrication* between the tool and work and the tool and chips so as to minimise tool wear and reduce power consumption.

2. It must carry away the heat generated during the process and, thus, cool the tool and workpiece both in order to minimise the tool wear and prevent distortion of the workpiece.

3. Its flash point should be amply high.

4. It should be able to impart antiwelding properties to the tool and the workpiece, otherwise very poor finish may result.

5. It should not discolour the finished work surface.

6. It should be non-poisonous and should not cause skin irritation.

7. It should carry such constituents which will prevent the finished work surface and the tool from being rusted or corroded.

Types of cutting fluids :

The cutting fluids are *classified* as follows :

1. *Cutting oils :*

(*i*) *Active cutting oils.*

(*ii*) *Inactive cutting oils.* These are straight mineral oils or straight mineral oils mixed with fatty oils, acids or sulphurised fatty oils.

● By *activeness* or *inactiveness* of the cutting oils we mean as to whether a particular cutting oil contains such constituents or not that can react chemically with work surface to help the machining operation.

2. *Water soluble oils or compounds.*

Cutting fluids used in different operations :

Operations	*Suggested fluids*
1. *Turning*	.... Emulsions or straight oils.
2. *Tapping and threading*	.... Active type mineral fatty oil. However, occasionally, emulsions of soluble oils are also used.
3. *Drilling and boring*	.... Soluble oils.
4. *Reaming*	.... Soluble oils.
5. *Planing and shaping*	.... Usually no cutting oil is used.
6. *Milling*	.... Sulphurised mineral fatty oils or emulsions in ample quantity.
7. *Broaching*	.... Heavy, active type cutting oils ; particularly in horizontal internal broaching.
8. *Thread rolling*	.... Straight mineral oil or emulsions.

9. *Gear cutting and shaping*	.... Active type mineral oils and compounds.
10. *Sawing*	.... Soluble oil.
11. *Automatic machining on single and multiple spindle automatic lathes, turret and capstan*	.... A high viscosity active cutting oil with some fatty oil.
12. *Grinding, lapping and honing*	.... Active type mineral oils and compounds ; emulsions of soluble oils or paste type.
13. *Machining plastic materials*	.... Brittle materials can be machined dry or with an air blast ; in case a fluid is required, a dilute solution of soluble oil should be used in ample quantity.

QUESTIONS WITH ANSWERS

Q. 9.1. Explain briefly various systems which provide single-point tool specifications.

Ans. The various *systems* providing *single-points tool specifications* are :

1. The American System (ASA)
2. The Orthoganal Rake System (ORS)
3. The Normal Rake System (NRS).

1. **The American System (ASA) :** Refer to Fig. 9.139.

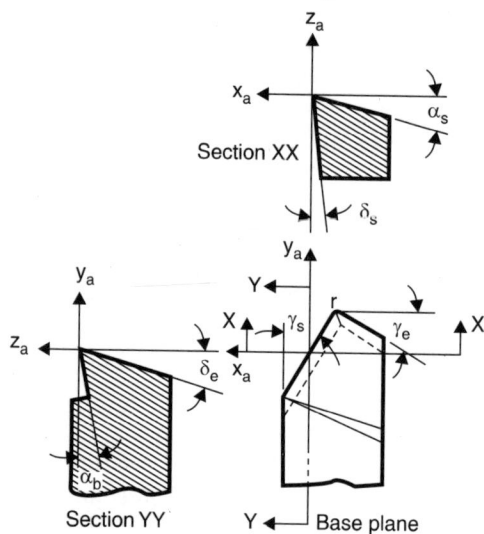

Fig. 9.139. Tool angles in ASA system.

The angles in this system are *indicated in a special* sequence as follows :

Back rake angle	....	α_b
Side rake angle	....	α_s
End flank angle	....	δ_e
Side flank angle	....	δ_s
End cutting-edge angle	....	γ_e
Side cutting-edge angle	....	γ_s
Nose radius	....	r

— Angle α_b is defined in the transverse plane w.r.t. y_a-axis, α_s in the longitudinal plane w.r.t. x_a-axis, δ_e in the transverse plane w.r.t. z_a-axis, δ_s in the longitudinal plane w.r.t. z_a-axis, and angles γ_e and γ_s on the base plane w.r.t. x_a and y_a axis respectively.

— The nose radius of the tool tip taken as r on the base plane.

● This system is followed in US and Canada.

2. **The Orthogonal Rake System (ORS)** : Refer to Fig. 9.140.

This system defines the rake face in terms of inclination angle i and orthogonal rake angle α_o, the principal and auxiliary flank faces by principal and auxiliary flank angles δ_p and δ_a, respectively, and the auxiliary and principal cutting edges by auxiliary and principal cutting-edge angles γ_a and γ_p, respectively. r as before is the nose radius of the tool.

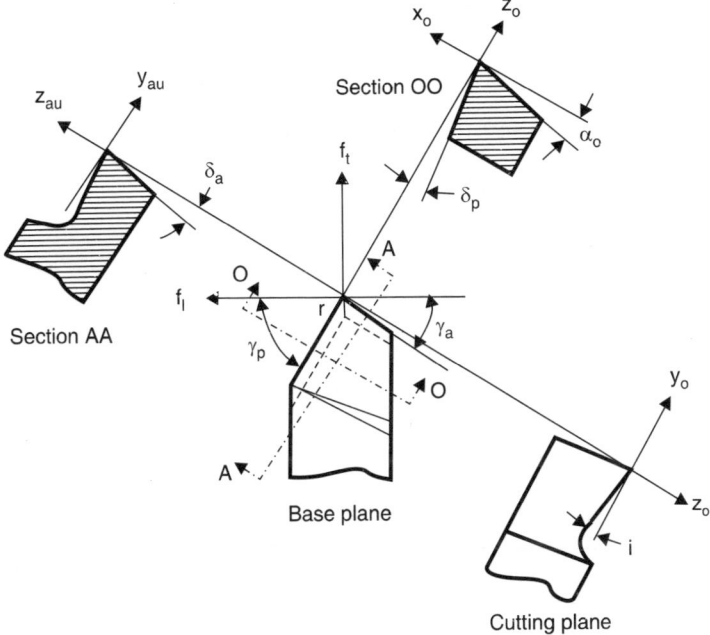

Fig. 9.140. Tool angles in ORS system.

The tool specification in ORS system is as follows :

Inclination angle	....	i
Orthogonal rake angle	....	α_o
Principal flank angle	....	δ_p
Auxiliary flank angle	....	δ_a
Auxiliary cutting edge angle	....	γ_a
Principal cutting-edge angle	....	γ_p
Nose radius	....	r

— The angle i is defined on the cutting plane with reference to y_o-axis while, α_o is defined on the orthogonal plane w.r.t. x_o-axis.

— The angle δ_p is defined on the orthogonal plane w.r.t. z_o-axis, but the angle δ_a is defined on a plane perpendicular to the auxiliary cutting edge, called auxiliary orthogonal plane, and is evaluated w.r.t. z_{au}-axis.

— The angles γ_a an γ_p are defined w.r.t. f_l-axis.

● This system is used in Germany, Russia and other Eropean countries.

3. **The Normal Rake System (NRS) :** Refer to Fig. 9.141.

Until recently, ORS was also followed in India but it has now adopted the system proposed by the International Standards Organisation (ISO) which is the NRS.

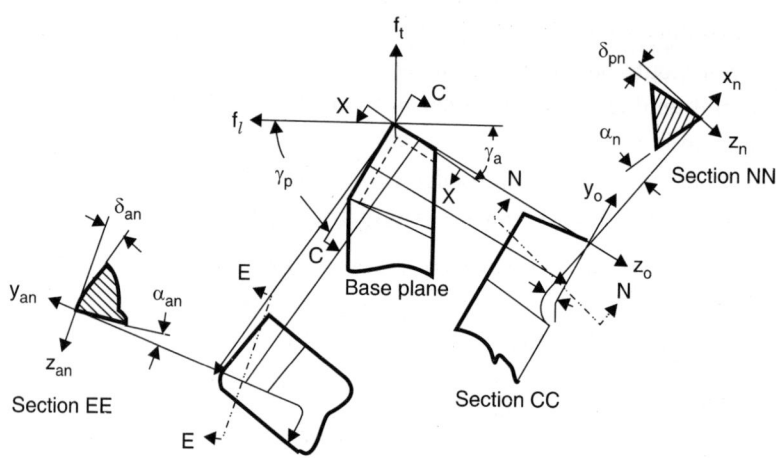

Fig. 9.141. Tool angles in NRS system.

The following is the tool specification in NRS system :

Inclination angle	 i
Normal rake angle	.... α_n
Principal normal flank angle	.... δ_{pn}
Auxiliary normal flank angle	.... δ_{an}
Auxiliary cutting-edge angle	.... γ_a
Principal cutting-edge angle	.... γ_p
Nose radius	.... r

— In this system the *reference planes and the axes are so chosen that every angle is a true angle.* Tool grinding, therefore becomes easy since *no angle corrections are required during grinding.*

Equations for conversion of tool angles in ASA and ORS systems into NRS systems :
ASA to NRS :

$$\tan i = \tan \alpha_b \cos \gamma_s - \tan \alpha_s \sin \gamma_s \qquad \qquad ...(9.49)$$

$$\tan \alpha_n = \cos i \,(\tan \alpha_s \cos \gamma_s + \tan \alpha_b \sin \gamma_s) \qquad \qquad ...(9.50)$$

$$\cot^2 \delta_p = \cot^2 \delta_e + \cot^2 \delta_s - \tan^2 i \qquad \qquad ...(9.51)$$

$$\cot \delta_{pn} = \cos i \cot \delta_p \qquad \qquad ...(9.52)$$

$$\gamma_p = 90 - \gamma_s \qquad \qquad ...(9.53)$$

ORS to NRS :

$$\tan \alpha_n = \cos i \tan \alpha_o \qquad \qquad ...(9.54)$$

$$\cot \delta_{pn} = \cos i \cot \delta_p \qquad \qquad ...(9.55)$$

Q. 9.2. Why should the cutting edge and top face of the tool be given a high finish ?

Ans. If the cutting edge and top face of the tool are *not* provided with high quality surface finish, then the following will happen :

- Friction between tool and chip would be high and more heat would be generated in the chip-tool contact zone.
- More power would be absorbed during the cutting process, and correspondingly greater forces would tend to dislodge the carbide tip from its shank. Under these conditions, the built-up edge is more likely to be formed.
- Roughness of the tool's cutting edge could result in a concentration of stresses which may cause surface cracks and eventual chipping of the tool.

Q. 9.3. What factors are likely to give rise to excessive heat during a metal-cutting operation ?

Ans. The main factors likely to give rise to *excessive heat during a metal cutting operation* are :

1. Formation of built-up edge on cutting face of tool.
2. Too high cutting speed.
3. Inadequate supply of cutting speed.
4. Poor surface finish on cutting face of tool.
5. Worn or incorrectly ground cutting face of tool.

Q. 9.4. Why are rake and clearance angles provided on cutting tools and on what factors the values of these angles depend ?

Ans. In order to remove a chip from the workpiece, it is essential to force a wedge type tool into the work material. When this tool is forced into the workpiece, the material of the workpiece is compressed and at certain load it fractures. With continuous cutting, chips are parted off from the workpiece and the leading tip of the tool keeps on cleaning the workpiece with a scraping action.

For getting better finish, it is also essential that only the leading tip of the tool should contact the workpiece and accordingly *clearance angle* of order of 5°–10° is provided on the tool. *Rake angle* is essential to provide wedge shape. It would be noted that the *sum of rake angle, wedge angle and clearance angle is equal to 90°.*

The value of rake angle depends upon the following *factors* :

(i) Tensile strength of material (rake angle being lower for more tensile materials and may be even negative for difficult-to-cut materials) ;

(ii) The desired tool life (being shorter for higher rake angle) ;

(iii) The surface finish (being poorer for greater rake angle).

Q. 9.5. List briefly the primary and secondary motions of the horizontal turret, capstan and vertical lathes.

Ans. The primary and secondary motions of horizontal, capstant and vertical lathes are given, in tabular form, as follows :

Lathe	*Primary motions*	*Secondary motions*
Horizontal and cap-stan lathes	(*i*) Rotation of head stock spindle to-gether with the workpiece. (*ii*) Longitudinal and cross-feeds of the carriage and cross-slide. (*iii*) Longitudinal feed of the turret saddle or ram.	(*i*) Indexing of front tool post to bring the required tool in position. (*ii*) Indexing of turret or capstan head to bring the required tool in cor-rect position. (*iii*) Rapid approach and withdrawal of cross slide and turret head, etc.
Vertical turret	(*i*) Rotation of table carrying the workpiece. (*ii*) Horizontal and vertical travels of turret slide or ram carrying the turret head (feed motions). (*iii*) Horizontal and vertical travels (feed motions) of the side heads, if used.	(*i*) Rapid vertical travels of the cross rail. (*ii*) Indexing of turret heads.

Q. 9.6. Explain briefly 'Bar feed mechanism' used in turret and capstan lathes.

Ans. Out of various methods employed for feeding the bar forward after each finished item is cut off the simplest method is to feed the bar by means of *a wire rope and weight.* (See Fig. 9.142). In this method a guide bar is fixed to the rear of the headstock. A rotating sleeve is mounted on the guide bar to carry the rear end of the bar stock. One end of the wire rope is tied to the sleeve and the other end carries a weight. When the chuck opens the bar is automatically pushed forward by weight and sleeve.

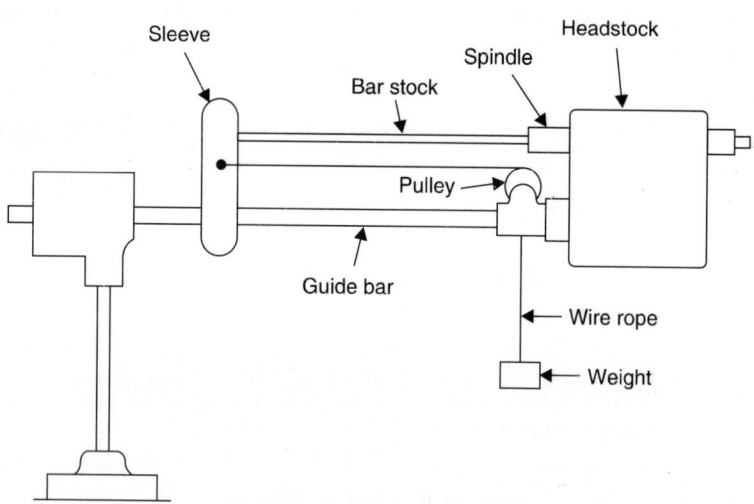

Fig. 9.142. Bar feed mechanism.

● This method however, is *confined to small machines.*

Q. 9.7. Discuss briefly an "Automatic cut off machine".

Ans. Automatic cut-off machines can perform operations like facing, form turning, chamfering and cutting off.

Fig. 9.143 shows the principle of an automatic cut-off machine.

— Two cross-slides are located on the bed at the front end of the spindle.

— A stock is clamped in the collet chuck of the rotating spindle. A stock stop is provided which is automatically advanced in time with the spindle axis at the end of the cycle.

— The stock is fed out by the bar feeding mechanism up to the stop.

— The various operations are carried out by feeding the tools on cross slides which are actuated by cams on a camshaft through a system of levers.

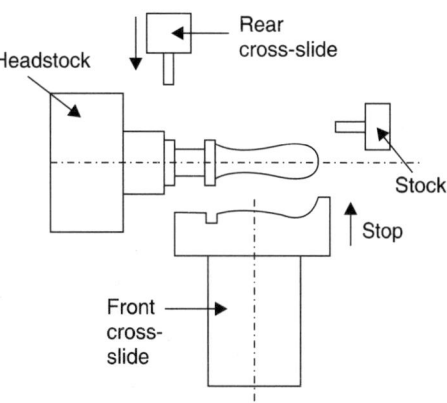

Fig. 9.143. Principal of an automatic cut-off machine.

● These machines are used to produce *short components of small diameter and simple shape from bar stock.*

Q. 9.8. Give the comparison of the following machining processes :
Turning ; Drilling ; Boring ; Milling ; Shaping ; Planning.

Ans. The comparison of various machining processes is given below :

Process	Advantages	Limitations
1. *Turning*	(*i*) Most versatile machine capable of producing external and internal circular profiles and flat surfaces. (*ii*) All types of materials can be turned. (*iii*) Low tooling cost. (*iv*) Large components can be turned.	(*i*) Low production rate. (*ii*) Requires skilled labour. (*iii*) Close tolerances and fine finish cannot be achieved.
2. *Drilling*	(*i*) Most suitable for producing round holes of various sizes. (*ii*) Inexpensive tooling and equipment. (*iii*) High production rate. (*iv*) Machine can be used for reaming and tapping.	(*i*) Basically a rough machining operation. (*ii*) Requires semi-skilled labour.
3. *Boring*	(*i*) Variety of internal circular profiles can be obtained. (*ii*) All types of materials can be bored. (*iii*) Low tooling cost. (*iv*) Large components can be bored. (*v*) Provides better dimensional control and surface finish.	(*i*) Low production rate. (*ii*) Requires skilled labour. (*iii*) Suitable for internal profiles only. (*iv*) Stiffness of boring bar is an important consideration.
4. *Milling*	(*i*) Versatile operation with wide variety of toolings and attachments. (*ii*) Variety of shapes including flats, slots and contours can be obtained. (*iii*) Suitable for low and medium production rate. (*iv*) Better dimensional control and surface finish.	(*i*) Tooling relatively more expensive. (*ii*) Requires skilled labour.

5. *Shaping*	(*i*) Suitable for low production rate. (*ii*) Suitable for producing flat and contour profiles on small size workpieces. (*iii*) Low tooling and equipment cost.	(*i*) Large size workpiece cannot be used. (*ii*) Requires skilled labour. (*iii*) Only simple profiles can be obtained. (*iv*) Close tolerance and fine finish cannot be obtained.
6. *Planning*	(*i*) Suitable for low production rate. (*ii*) Suitable for producing flat and contour profiles on large workpieces. (*iii*) Low tooling cost.	(*i*) Only simple profiles can be obtained. (*ii*) Requires skilled labour. (*iii*) Close tolerance and fine finish cannot be obtained.

Q. 9.9. What is grinding wheel and why better surface finish is obtained when this is used as a tool ? Explain.

Ans. A *"grinding wheel"* is a multi-tooth cutter made up of many hard particles called as abrasive which is crushed to leave sharp edges, these edges perform the cutting. The abrasive grains are mixed with a suitable bond. This bond acts as a matrix or holder when the wheel is in use. The wheel may consist of one piece or of segment of abrasive blocks built up into a solid wheel.

By *"bond"* we mean here an adhesive substance that is employed to hold abrasive grains together in the form of sharpening stones or grinding wheels.

9.10. Give general guidelines for selection of bond for the grinding of a range of materials.

Ans. Selection of bond mainly depends upon the following *factors* :

1. **Constant factors** :

(*i*) Work metal—It should be noted that for grinding a soft material, hard wheel should be used and vice versa.

(*ii*) Amount and rate of stock removal.

(*iii*) Area of contact between work and wheel.

(*iv*) Condition of grinding machine—A softer grade of wheel is used on robust and heavy machine.

(*v*) Finish and accuracy required on the job.

2. **Variable factors** :

(*i*) Wheel speed.

(*ii*) Work speed.

(*iii*) Condition of grinding machine (state of wheel spindle bearing).

(*iv*) Skill of operator (Personal factors).

Q. 9.11. Explain briefly a surface grinder.

Ans. A surface grinder uses a cylindrical grinding wheel or cup shaped wheel to *produce a flat surface.* Segmented abrasive stones fitted on the face of a cylindrical tool are also used for surface grinding. Two classes of surface grinding are called *peripheral grinding* and *face grinding.* The various arrangements showing relative arrangements of spindle and table are shown in Fig. 9.132.

Q. 9.12. What is surface grinding ? When do you recommend the use of this process ?

Ans. *'Surface grinding' produces flat or plane surfaces.* The work may be grounded by either the periphery or by the end face of the grinding wheel. The workpiece is reciprocated at a constant speed below or on the end face of the grinding wheel.

Surface grinding is done by two machines : (*i*) Planer type with a reciprocating table for work ; (*ii*) Machines with rotating table for continuous rapid grinding.

Modern surface grinding machine is provided with hydraulic control of table movements and wheel cross feed. Machine of this type is *adapted to reconditioning dies, grinding tool ways and other long surfaces.*

Q. 9.13. Give five applications of grinding process.

Ans. Following are the *five* applications of grinding process :

1. To sharpen the cutting tools.
2. To obtain better finish on the surface.
3. To grind threads in order to have close tolerances and better finish.
4. To remove a very small amount of metal from the workpiece to bring its dimensions within very close tolerances.
5. To machine hard surfaces which are otherwise difficult to be machined by the high speed steel tools or carbide cutters.

Q. 9.14. List the factors on which the performance of grinding wheel depends.

Ans. The performance of a grinding wheel depends upon the following *factors* :

1. *Type of abrasive*—SiC is more hard and brittle, so it is chosen for C.I., brass, hard alloy etc. Al_2O_3 is more suitable for grinding of steels and bronzes.
2. *The grade of the wheel*—The grade designates the force holding the grains. The grade is also called the hardness of the wheel. The grade of a wheel depends on the kind of bond, structure of wheel and amount of abrasive grains. It is designated by a letter, with 'A' representing the soft end and 'Z' the hard end of the scale.
3. *Wheel structure*—Structure relates to the spacing of abrasive grains. Open structures are used for high stock removal and consequently produce a rough finish. Dense structures are used for precision forms and profile grinding.
4. *Grain size.*
5. *The properties of the grains.*
6. *The geometry of the cutting edges of grains* (rake angles and cutting edge radius compared to depth of cut).
7. The process parameters (speeds, feeds, cutting fluids) and type of grinding (cylindrical, surface).

Q. 9.15. What is a lap ? Explain.

Ans. A **lap** *is a soft material disc, ring, plate or cylinder, charged with abrasive power or compound, and used for producing extremely accurate and finished surfaces.*

A lap is a cutting tool that is made by "charging" a metal body of lead, copper, soft cast iron or any other suitable soft material *with a fine abrasive.* Charging here means that the abrasive powder is embedded in the lap by rubbing or rolling.

- In general, copper and soft steel laps cut faster than cast iron laps but *cast iron laps retain their form better. Cast iron laps are widely used.* The faces of the cast iron laps are serrated, this makes it easier to remove the work after lapping and provides a storage space for oil and abrasives.

- A '*hand lap*' for finishing flat surfaces is *in the form* of a *flat plate.* The '*lapping machine' usually utilizes circular disc like laps.*

Q. 9.16. What is tumbling ? Explain.

Ans. *Tumbling (or liquid honing) is the process of revolving small workpiece in a barrel with abrasive and water for the purpose of producing a high lustre or removing burrs.*

The following are the *purposes of tumbling* :

- To debur the parts.
- To produce a high finish.
- To improve micro finish.
- To finish gears or threaded parts without damage.
- To remove paint or plating from parts, or descale the parts.
- To form uniform radii around the workpieces, *i.e.,* generating controlled radii.
- To finish high precision work to a high lustre.

Q. 9.17. What is diamond machining ? Explain.

Ans. Diamond is the hardest cutting tool material. It has *low friction, high thermal conductivity and low coefficient of expansion.* It finds applications for special jobs like *turning outside surface of aluminium alloy pistons, boring of white metal bearing liner to produce highly polished surface finish obtainable by superfinishing* ; for *truing or forming the faces of grinding wheels.*

Diamonds being small are used in the form of tips which are clamped into a tool shank and the shank is mounted in tool holder, such as micro-boring head. The diamond tips are available with cutting edge either rediused or in the form of a series of flats around 0.4 mm long. These edges are set accurately parallel to the work axis using a setting microscope. When the working cutting edge gets worn out (it cannot be sharpened), the other cutting edge is fixed for the cutting action. Alongwith high hardness, diamond tips are highly brittle also. These must, therefore, be supported rigidly in the minimum overhange and subjected to least vibrations during cutting operation.

Q. 9.18. What for lapping is used ? How much stock is left for lapping ? How does it differ from grinding ?

Ans. ● Lapping process is *used to obtain truly flat and smooth surfaces. It is also used to finish flat and round work* to tolerance of 0.01 to 0.001 mm.

- Usually about *0.001 to 0.01 mm stock is left for lapping process.*
- Lapping differs from grinding/honing process in the respect that it *uses loose abrasive instead of bonded abrasive.*

Q. 9.19. Why is it undesirable to continue running coolant on to a grinding wheel after the wheel has stopped ?

Ans. If coolant is run continuously on a grinding wheel after it has stopped, then a part of the wheel would become saturated, thereby increasing the weight of that portion and *resulting in unbalance of wheel, increasing the danger of wheel disintegration.* Further, an unbalanced wheel is *deterimental to the quality of the surface finish produced on the workpiece.*

HIGHLIGHTS

1. *Machining* is the process of cold working the metals into different shapes by using different types of machine tools.

2. Elements of machines tools are : Structure ; Slides and tool structure ; spindles and spindle bearing ; Kinematics of machine tool devices ; work holding, and tool holding elements.

3. A *lathe* is one of the oldest and perhaps most important tools ever developed.

   Types of lathe. Speed lathe, engine or central lathe, bench lathe, tool room lathe, turret and capstan lathes, automatic lathes, special purpose lathes.

   Lathe operations : Facing, plain turning, step turning, taper turning, drilling, reaming, boring, undercutting or grooving, threading, knurling and forming.

4. A *taper* may be defined as a uniform increase or decrease in diameter of a piece of work measured along its length. The half taper angle (α) can be found by using the following relationship :

$$\alpha = \tan^{-1}\left(\frac{D - d}{2L}\right)$$

Where, D = Large diameter of taper in mm ;

d = Small diameter of taper in mm, and

L = Length of tapered part in mm.

5. Turret, capstan and automatic lathes are widely used in producing goods in large quantities (they reduce or eliminate the amount of skilled labour and reduce production time).

6. The lathes that have their tools automatically fed to work and withdrawn after the cycle is completed are known as '*Automatic lathes*'.

7. A *shaper* is a reciprocating type of machine tool intended primarily to produce horizontal, vertical or inclined flat surfaces (up to 1000 mm long).

   Machining time required for shaping the surface,

$$t_m = \frac{LB\,(1 + k)}{1000\, V_c f}$$

where, L = Length of stroke (mm) = $L_j + 2c$, where j and c stand for job and clearance (mm) respectively.

B = Width of the job (mm),

k = Return stroke time/cutting stroke time,

V_c = Cutting speed (m/min), and

f = Feed (mm/stroke),

In terms of ram strokes N (per min), the time for machining the surface is given by :

$$t_m = \frac{B}{fN} \text{ min}$$

Cutting speed is determined by using the formula :

$$V_c = \frac{NL\,(1 + k)}{1000} \text{ m/min.}$$

Material removal rate (MRR) = $f.d.\ N.L\,(1 + k)$

where, d = Depth of cut.

8. Basically, the *slotting machine* is a vertical axis shaper. The stroke of ram is smaller in slotting machines than in shapers to account for the type of work that is handled in them.

9. The *planing machine (planer)* is a machine tool used in the production of flat surfaces on workpieces too large or too heavy to hold in a shaper.

   The fundamental difference between a shaper and a planer is that the latter operates with an action opposite to the shaper, *i.e.* the workpiece reciprocates past one or more stationary single point cutting tools.

10. *Drilling* is the process of making hole or enlarging a hole is an object by forcing a rotating tool called "drill".

11. A power operated machine tool, which holds the drill in its spindle rotating at high speeds and when manually actuated to move linearly simultaneously against the workpiece produces a hole is called *drilling* machine.

12. *Machining time* in a drilling machine is given by :

$$t_m = \frac{L_j + l_1 + l_2}{fN} \text{ min}$$

where, L_j = Hole length or depth, mm.

l_1 = Tool approach $\approx 0.29\,D$ (with point angle of 118°)

l_2 = Tool overtravel, 1 to 2 mm

f = Feed, mm/rev, and

N = Speed of drill, r.p.m.

Material removal rate (MRR) = $\dfrac{\pi D^2 fN}{4}$ mm³/min

where, D is the drill diameter in mm.

13. *Boring* is the process of using a single point tool to enlarge and locate a previously made hole.

14. *Jig-borers* are used as coordinate measuring machines for inspection and precision layout operations.

15. *Milling machine* is a machine tool in which metal is removed by means of a revolving cutter with many teeth, each tooth having a cutting edge which removes metal from a workpiece.

   Indexing is the rotation of a work-part by small uniform amounts, with or without the aid of change wheels, mainly in connection with milling operations.

   Common indexing methods are : (*i*) Rapid or direct indexing : (*ii*) Single or plain indexing ; (*iii*) Compound indexing ; (*iv*) Differential indexing ; (*v*) Angular indexing.

16. *Broaching* is a process of machining a surface with a special multipoint tool called a '*broach*', whose teeth remove the whole machining allowance in a single stroke. In this operation cutting speeds and feeds are low and adequate lubrication is essential.

17. *Metal sawing* is chiefly concerned with cutting bar stock to convenient length or size for machining. The blades of sawing machine are made of : (*i*) Standard carbon steel ; (*ii*) High speed steel ; (*iii*) Bimetallic high speed steel.

18. *Grinding* is a metal cutting operation performed by means of a rotating abrasive tool, called "grinding wheel". Grinding may be *classified* as :

   (*i*) External cylindrical grinding; (*ii*) Internal cylindrical grinding ; (*iii*) Surface grinding ; (*iv*) Form grinding.

19. *Centreless grinding* is the method of grinding metallic parts in which the piece to be ground is supported on a work rest, and passed between a grinding wheel running at a high speed and a controlling wheel running at a slow speed.

20. *Internal grinding* is the mechanical grinding of the internal bores of gears, bushes and a wide variety of machine parts.

21. *Surface grinding* is the method of grinding designed to carry out the removal of metal from a part or parts less expensively and with greater recision than could be achieved by machining processes with cutting tools of steel, or by hand or machine filing.

22. *Form grinding* is grinding of tools, designed for machining and other machines in such a way that they are provided with the precise form required for their work, or regrinding them, to restore the form after it has been lost as a result of service conditions.

23. The *grinding wheel* is a multi-tooth cutter made up of many hard particles known as '*abrasive*' which has been crushed to leave sharp edges which do the cutting.

24. *Honing* is a grinding or abrading process in which very little material is removed. It is used primarily to remove ι ჳ marks on the surface left by previous operations.

25. *Lapping* is a finishing process, following after grinding, and designed to produce an exceptionally high degree of surface finish as well as perfectly true surface accurate to size within extremely close limits.

26. *Superfinishing* is an abrading process, efficient is surface refining of cylindrical, flat, spherical and cone shaped parts.

27. *Burnishing* is an operation by which a bright, polished finish is produced on the surface of a metal by a rubbing action which smooths out small scratch marks.

OBJECTIVE TYPE QUESTIONS

Fill in the blanks or Say 'Yes' or "No" :

1. .... is the process of cold working the metals into different shapes by using different types of machine tools.

2. .... is defined as the ease of removing metal while maintaining dimensions and developing a satisfactory surface finish.

3. .... processes are material removing operations in which the desired shape, size and surface finish on the finished product are obtained by removing surplus material.

4. Turning is a finishing process.

5. Laser beam machining is an unconventional machining process.

6. Lapping is a finishing process.

7. The .... refers to only those processes where material removal is affected by the relative motion between tool made of harder material and the workpiece.

8. .... processes are those which use electrical, chemical and other means of material removal for shaping high strength materials and for producing complicated shapes.

9. Grinding and finishing processes are those where metal is removed by a large number of hard abrasive particles or grains which may be bonded as in grinding wheels, or be in loose form as in ....

10. .... angle is the angle between the face of the tool called the rake face and normal to the machining direction.

11. .... angle is the angle between the machined surface and underside of the tool called the flank face.

12. The cutting speed is the speed with which the tool moves through the work material.

13. .... may be defined as the small relative movement per cycle (per revolution or per stroke) of the cutting tool in a direction usually normal to the cutting speed direction.

14. The tool material must not remain harder than work material at elevated temperature.

15. The coefficient of friction at chip tool interface must remain .... for minimum wear and reasonable surface finish.

16. While selecting proper tool material the type of service to which the tool will be subjected should be given least consideration.

17. .... tool steels are characterised by the low stability of the supercooled austenite.

18. Stellites cannot be forged to shape.

19. The most important properties of cemented carbides are very high heat and wear resistance.

20. .... is a class of material used to sharpen the edges of cutting tools, and to reduce or polish metallic or other surfaces.

21. Carborundum is a trade name of .... carbide.

22. .... is defined as the time interval between the two successive grinds.

23. The continuous chips are produced while machine less ductile materials.

24. The continuous type of chip is most desirable.

25. .... chips are usually produced while cutting more brittle materials.

26. A .... is one of the oldest and perhaps the most important machine tools ever developed.

27. Facing is an operation of machining the ends of a workpiece to produce a flat surface square with the axis.

28. A .... may be defined as a uniform increase or decrease in diameter of a piece of work measured along its length.

29. .... turning means to produce a conical surface by gradual reduction in diameter from a cylindrical workpiece.

30. .... is the operation of enlarging and turning a hole produced by drilling, punching, casting or forging.
31. The tool used in a reaming process is called ....
32. Undercutting is the process of reducing the diameter of a workpiece over a very narrow surface.
33. .... is an operation of embossing a diamond shaped pattern on the surface or a workpiece.
34. .... is an operation of turning a convex, concave or any irregular shape.
35. .... is the distance the tool advances for each revolution of the work.
36. Depth of cut is the perpendicular distance measured from the machined surface to the uncut surface of work.
37. .... is the process of making hole or enlarging a hole in an object by forcing a rotating tool called 'drill'.
38. .... is an operation in which threads are cut in the existing hole.
39. .... is an operation of enlarging a drilled hole partially. That is for a specific length.
40. A .... is a reciprocating type of machine tool intended primarily to produce horizontal, vertical or inclined flat surfaces.
41. A .... is machine tool used in production of flat surfaces on workpieces too large or too heavy to hold in a shaper.
42. The .... machine is a machine tool in which metal is removed by means of a revolving cutter with many teeth, each tooth having a cutting edge which removes metal from a workpiece.
43. .... cutting is an operation in which a pair of side milling cutters is used for machining two parallel vertical faces of a workpiece simultaneously.
44. Thread cutting by machining process is done by using a ....
45. Turret lathe is classed as a .... machine tool.
46. Turret lathe has no tailstock.
47. Capstan lathe is never used on smaller work.
48. Semi-automatic machines are generally preferred in lot-production work.
49. Multi-spindle automatic screw machines seldom maintain accuracy as good as single speed machines.
50. The machine tool used for the shaping operation is called a ....
51. .... machines are the reciprocating type of machine tools in which the workpiece is held stationary and the tool reciprocates.
52. Cutting speed on shapers may be either constant or variable depending on the design of the machine tool.
53. The .... planer is used only for machining edges of large plates, used for making boiler drums or other pressure vessels.
54. The size of the shaper or planer is specified by the maximum length of stroke.
55. Basically the .... machine is a vertical axis shaper.
56. The stroke of the ram is larger in slotting machines than in shapers to account for the type of work that is handled in them.
57. .... is a process of making hole or enlarging a hole in an object by forcing a rotating tool called drill.
58. Radial drilling machine is used in tool rooms and large scale die manufacturing units.
59. Deep-hole drilling machines are operated at .... speed and .... feed.
60. The lips in a twist drill are also known as cutting edges.
61. .... is the operation of finishing a hole very smoothly and accurately in size.
62. .... is an operation of enlarging an existing hole.
63. Counterboring is an operation of forming a conical shape at the end of a drilled hole.
64. .... is an operation of enlarging a drilled hole partially, that is for a specific length.
65. .... is an operation of smoothing and squaring the surface around a hole for the seat for the nut or the head of a screw.
66. .... is an operation in which external threads are cut in the existing hole.
67. Trepanning is an operation of producing a hole by removing the metal along the circumference of a hollow cutting tool.
68. Jig boring machine is a non-precision machine.
69. .... are used as coordinate measuring machines for inspection and precision layout operations.
70. In .... process, the workpiece is fed opposite to the cutter's tangential velocity.

71. In the down milling process the workpiece is fed in the same direction as that of the cutter's tangential velocity.
72. The drum type milling operations are of continuous operation type.
73. .... milling is used to machine flat and horizontal surfaces.
74. Straddle milling is used to mill square and hexagonal surfaces.
75. .... milling is an operation of producing narrow grooves and slots on the workpiece.
76. Plain dividing head is hand operated and used for .... indexing.
77. Optical dividing head is the most precision attachment.
78. .... is the rotation of a work-part by small uniform amounts, with or without the aid of change wheels, mainly in connection with the milling operations.
79. .... is a process carried out with a grinding wheel made up of abrasive grains for removing very fine quantities of material from the workpiece surface.
80. The grinding and finishing processes are used for final finish and superfinish.
81. .... grinding is a commonly used method for removing excess material from castings, forgings and weldments etc.
82. Grinding is done on surfaces of limited shapes and materials of limited kinds.
83. .... grinding is the principal production method of cutting materials that are too hard to be machined by other conventional tools or for producing surfaces on parts to higher dimensional accuracy and finer finish as compared to other manufacturing methods.
84. External cylindrical grinding produces a straight or tapered surface on a workpiece.
85. .... cylindrical grinding produces internal cylindrical holes and tapers.
86. .... grinding produces flat surface.
87. .... grinding operation is done with specially shaped grinding wheels that grind the formed surfaces as in grinding gear teeth, threads, splined shafts, holes etc.
88. In a grinding process it is possible to achieve very accurate dimensions and smoother surface finish in a very short time.
89. .... is the only method of removing material from materials after hardening.
90. The grinding operation is intermittent in nature, and produces discontinuous chips.
91. The grinding wheel has .... character.
92. In a grinding wheel the load acting on individual cutting grains is uniform.
93. The effective rake angle of abrasive grains is highly positive.
94. The geometry of the grain is highly random and the time of contact between the chip and an abrasive grain is very small.
95. The .... size indicates the sieve number used for screening grains.
96. A higher sieve number would indicate coarser grains.
97. The grade or hardness of the wheel is designated by a letter with 'A' representing the soft end and 'Z' the hard end of the scale.
98. The .... of a wheel characterises the mean void size and the distribution of the grains.
99. Centreless grinding operations can provide surface finish in the range of 0.1 to 1.5 μm and the tolerance achievable is 0.005 to 0.03 mm.
100. When the grinding wheels lose their geometry, the original shape is restored by .... with a diamond tool.
101. The performance of a grinding wheel is usually evaluated in terms of the .... ratio.
102. The grinding ratio is defined as volume of wheel wear and volume of material removed.
103. In .... grinding process there is no need for centring and use of fixtures etc.
104. The centreless grinding process is suitable for .... volume production.
105. The centreless grinding process in not suitable for large workpiece sizes.
106. Internal grinding produces accurate results, is not expensive and gives a high degree of surface finish.
107. Disc grinding is a form of .... grinding.
108. Form grinding must be done with great accuracy.
109. A .... is a substance that is used for grinding and polishing operations.
110. Silicon carbide is employed for grinding materials of .... tensile strength.

111. Aluminium oxide is better adopted to grind materials of .... tensile strength.
112. For grinding piston rods or pins, rod shaped wheel is used.
113. .... wheels do not grind and require dressing.
114. Most of the grinding wheels possess vitrified bond.
115. Extra hard wheels can be produced with silicate bond.
116. Oxychloride bond provides a cool cutting action, but grinding is usually done dry.
117. .... is a grinding or abrading process in which little material is removed.
118. .... is primarily used to remove the marks on the surface left by previous operations.
119. A metal frame which holds the abrasive sticks during honing operation is known as a ....
120. .... is a finishing process, following after grinding, and designed to produce an exceptionally high degree of surface finish as well as perfectly true surface accurate to size within extremely close limits.
121. Soft materials are lapped with Al_2O_3 and hard materials with diamond or SiC grit.
122. .... is an abrading process, efficient in surface refining of cylindrical, flat spherical and cone shaped parts.
123. Superfinishing produces a high wear resistant surface on any object which is symmetrical.
124. A very small amount of material is removed in polishing.
125. The dimensional accuracy of the parts is not affected by polishing and buffing operations.
126. Negligible amount of material is removed in buffing.
127. .... is an operation by which a bright, polished finish is produced on the surface of metal by rubbing action which smooths out small scratch marks.
128. Burnishing is generally used to produce a decorative finish.

ANSWERS

1. Machining	2. Machinability	3. Machining	4. No	5. Yes
6. Yes	7. metal-cutting	8. Unconventional machining		9. lapping
10. Rake	11. Clearance	12. Yes	13. Feed rate	14. No
15. low	16. No	17. Carbon	18. Yes	19. Yes
20. Abrasive	21. silicon	22. Tool life	23. No	24. Yes
25. Discontinuous	26. lathe	27. Yes	28. taper	29. Taper
30. Boring	31. reamer	32. Yes	33. Knurling	34. Forming
35. Feed	36. Yes	37. Drilling	38. Tapping	39. Counterboring
40. shaper	41. planer	42. milling	43. straddle	44. lathe
45. production	46. Yes	47. No	48. Yes	49. Yes
50. shaper	51. shaping	52. Yes	53. edge	54. Yes
55. slotting	56. No	57. Drilling	58. Yes	59. high, low
60. Yes	61. Reaming	62. Boring	63. No	64. Counter boring
65. Spot facing	66. Tapping	67. Yes	68. No	69. Jig-borer
70. up milling	71. Yes	72. Yes	73. Plain	74. Yes
75. Saw	76. simple	77. Yes	78. Indexing	79. Grinding
80. Yes	81. Rough	82. No	83. Precision	84. Yes
85. Internal	86. Surface	87. Form	88. Yes	89. Grinding
90. Yes	91. self sharpening	92. No	93. No	94. Yes
95. grain	96. No	97. Yes	98. structure	99. Yes
100. truing	101. Grinding	102. No	103. Centreless	104. Large
105. Yes	106. Yes	107. Surface	108. Yes	109. abrasive
110. Low	111. High	112. Yes	113. Glazed	114. Yes
115. No	116. Yes	117. Honing	118. Honing	119. hone
120. Lapping	121. Yes	122. Superfinishing	123. Yes	124. Yes
125. Yes	126. Yes	127. Burnishing	128. Yes.	

THEORETICAL QUESTIONS

1. Define the term 'machining'.
2. Give the classification of machining processes.
3. List the major factors which influence the selection of a suitable process.
4. What is a machine tool ?
5. What are the function of a machine tool ?
6. How are machine tool classified ?
7. What are the elements of machine tools ?

Centre lath :
8. What is the working principle of a lathe ?
9. Explain briefly the parts of a lathe.
10. How is the size of a lathe specified ?
11. What are the various types of lathe ?
12. How are lathe tools classified ?
13. List the common lathe operations which can be carried out on a lathe.
14. What do you mean by the term 'Taper' ?
15. Name the methods by which taper turning can be carried out on lathes.
16. Explain briefly, with the help of a sketch, the taper turning method involving swivelling of compound rest.
17. What are the advantages of using a taper turning attachment ?
18. Explain briefly, with neat diagrams, the following lathe operations :
    (i) Threading ;   (ii) Knurling ;
    (iii) Forming ;   (iv) Reaming.
19. What do you mean by 'Lathe accessories' ?
20. Name the various accessories used on lathe.
21. Explain briefly the following lathe accessories :
    (i) Driving plate ;   (ii) Lathe centres,   (iii) Rests.
22. Enumerate the commonly used lathe attachments.
23. Explain briefly the following terms in relation to a lathe :
    (i) Cutting speed ;   (ii) Feed ;   (iii) Depth of cut.
24. What do you mean by 'Metal removal rate' ? Explain briefly.
25. How is the machining time in a lathe calculated ?
26. State the factors on which the power required at the spindle for turning depends.
27. Discuss briefly the procedure for cutting threads on lathe.
28. Write a short note on 'Eccentric turning'.
29. Explain briefly various systems which provide single-point tool specifications.
30. What factors are likely to give rise to excessive heat during a metal-cutting operation ?
31. Why should the cutting edge and top face of the tool be given a high finish ?

Turret ; Capstan and Automatic lathes :
32. What are the limitations of a centre lathe ?
33. What are the differences between a turret lathe and a centre lathe ?
34. Give the comparison between turret and capstan lathes ?
35. What are the main differences between a turret lathe and a capstan lathe ?
36. State the features of turret lathe that make it a quantity production machine.
37. Explain briefly with neat sketches the following :
    (i) Turret lathe ;   (ii) Capstan lathe.
38. Explain briefly the main parts of a turret or capstan lathe.
39. How are turret lathes classified ?
40. Explain briefly the following :
    (i) Ram type horizontal turret lathe.   (ii) Vertical turret lathe.
    (iii) Numerically controlled turret lathes.

41. How is the size of a turret lathe specified ?
42. How is a capstan lathe specified ?
43. List the common tools and attachments used on turret and capstan lathes.
44. List the various operations which can be performed on turret lathe.
45. What do you mean by the term 'Automation' ?
46. What are the advantages of automation ?
47. What is difference between 'Automatic' and 'Semi-automatic' lathes ?
48. How are automatic lathes classified ?
49. Explain briefly the following :
    (i) Automatic vertical multistation lathe.
    (ii) Automatic screw machines.
50. List briefly the primary and secondary motions of the horizontal turret, capstan and vertical lathes.
51. Explain briefly 'Bar feed mechanism' used in turret and capstan lathes.
52. Discuss briefly an 'Automatic cut-off machine'.

Shaper, Slotter and Planer :
53. What is a shaper ?
54. What is the principle of working of a shaper ?
55. State the advantages, limitations and applications of shapers.
56. How are shapers classified ?
57. Explain briefly with a neat sketch the principal parts of a shaper.
58. How is a shaper specified ?
59. Discuss briefly the crank and slotted/lever quick return motion mechanism for a shaper.
60. Explain briefly how quick return motion in the hydraulic shaper is accomplished.
61. What are the advantages and disadvantages of hydraulically driven shaper ?
62. List the various shaper work holding devices.
63. Name the various operations which can be performed on a standard shaper.
64. Discuss the working principle and operation of a shaper.
65. What is the fundamental difference between a planer and a shaper ?
66. Define cutting speed, feed and depth of cut in relation to shaper work.
67. Discuss, with suitable sketches, the various types of cutting tools used in shaper work.
68. How is material removal rate (MRR) in a shaping machine calculated ?
69. Draw the block diagram of a slotting machine and explain briefly its various parts.
70. Give the comparison between planer and shaper.
71. Enumerate the various types of planers commonly used.
72. How is the size of a planer specified ?
73. Discuss briefly the planer driving mechanisms.
74. Name the feed mechanisms used in a planer.
75. List the standard clamping devices used for holding most of the work on a planer table.
76. Name and sketch the various types of tools used on planers.
77. Enumerate the various operations which can be performed on a planer.

Drilling machines :
78. Define the terms "Drilling" and "Drill".
79. What is the function of a Drilling machine ?
80. How is a drilling machine specified ?
81. How are drilling machines classified ?
82. Explain briefly with neat sketches any two of the following drilling machines :
    (i) Sensitive drilling machine   (ii) Radial drilling machine
    (iii) Gang drilling machine   (iv) Turret drilling machine.
83. Discuss briefly the following :
    (i) Deep-hole drilling machines
    (ii) Multispindle drilling machines
    (iii) Automatic drilling machines.

84. How are 'drills' classified ?
85. List the following :
    (i) Drill holding devices   (ii) Drilling machine tools.
86. Discuss briefly 'Twist drill nomenclature' with neat sketches.
87. What are the advantages of twist drills ?
88. Enumerate the work holding devices used in drilling machines.
89. Name the various operations which can be carried out on a drilling machine.
90. Explain briefly with sketches any four of the following drilling machine operations :
    (i) Reaming   (iv) Tapping
    (ii) Boring   (v) Trepanning
    (iii) Countersinking   (vi) Spot facing.
91. What is 'reamer' ?
92. List the various types of reamers.
93. What is the difference between 'counterboring' and 'countersinking' operations ?
94. Define cutting speed, feed and depth of cut as applied to a drilling machine.
95. How are the machining time and material removal rate in a drilling machine calculated ?

Boring machines :
96. What do you understand by the term 'Boring' ?
97. Give the comparison between boring and reaming.
98. How are boring machines classified ?
99. Discuss briefly, with a neat sketch, an horizontal boring machine.
100. With the help of a neat diagram, describe briefly a double column vertical boring machine.
101. Explain briefly a jig boring machine, with a neat sketch.
102. How are horizontal and vertical boring machines specified ?

Milling machines :
103. What is a 'Milling machine' ?
104. What are the applications of milling machines ?
105. Explain briefly the following milling processes :
    (i) Up milling (or conventional) milling process
    (ii) Down (or climb) milling process.
106. State the advantages of down milling process ?
107. How are milling machines broadly classified ?
108. Draw the block diagram of a horizontal milling machine and explain briefly its various parts.
109. Explain briefly any two of the following machines :
    (i) Vertical milling machines   (ii) Universal milling machines
    (iii) Planetary milling machines   (iv) Drum type milling machines.
110. Name the various operations which can be performed on a milling machine.
111. Explain briefly, with neat sketches, any three of the follwing milling operations :
    (i) Face milling   (iv) Gang milling
    (ii) Form milling   (v) Dove-tail milling
    (iii) Straddle milling   (vi) Saw milling.

Straddle :
112. Define the following terms as applied to a milling machine :
    (i) Cutting speed ;   (ii) Feed ;   (iii) Depth of cut.
113. Define the term 'material removal rate' as applied to a milling machine. How is it calculated ?
114. Explain clearly how machining time is calculated, in case of milling machine, for plain milling and face milling operations.
115. Name the various types of attachments which are used with a standard milling machine.
116. Name the devices which are used for holding the workpiece to be machined on a milling machine.
117. Define the terms 'Indexing 'and 'Dividing head'.
118. How are the dividing heads classified ?

119. Explain briefly the following :
  (i) Plain dividing head.
  (ii) Universal dividing head.
  (iii) Optical dividing head.
120. Enumerate the common indexing methods.
121. Explain briefly any *two* of the following indexing methods :
  (i) Plain indexing   (ii) Compound indexing
  (iii) Differential indexing   (iv) Angular indexing.

Broaching and Sawing Machines :

122. Explain briefly the following terms :
  (i) Broaching;   (ii) Broach.
123. What are the advantages, limitations and applications of broaching ?
124. With the help of neat sketch, explain briefly the main elements of a broaching tool.
125. What is the difference between 'pull broaching' and 'push broaching' ?
126. How are broaching machines classified ?
127. How is the size of a broaching machine specified ?
128. What do you understand by the term "Sawing" ?
129. How are sawing machines broadly classified ?
130. Explain briefly the following saws :
  (i) Power backsaws ;   (ii) Band saws ;
  (iii) Circular saws.
131. Name the materials which are used to make blades of sawing machines.
132. With the help of neat sketches show the following :
  (i) Main parts of a band tool.   (ii) Setting of band saw teeth.
133. List the applications of sawing machines.

Grinding and Finishing processes :

134. Define the following terms :
  (i) Grinding ;   (ii) Rough grinding ;
  (iii) Precision grinding.
135. How is "grinding" classified ?
136. Discuss very briefly the following :
  (i) External cylindrical grinding   (ii) Internal cylindrical grinding
  (iii) Surface grinding   (iv) Form grinding.
137. What are the advantages of grinding process over other cutting processes ?
138. Enumerate the special features of grinding process.
139. How are grinding machines classified ?
140. Explain with a neat sketch a plain cylindrical grinder.
141. Explain the working principle of the centreless grinding operation.
142. Explain briefly the three standard methods of feeding the work in centreless grinding.
143. What are the advantages and limitations of centreless grinding ?
144. Explain the working principle of internal grinding.
145. Explain with a neat sketch 'centreless internal grinding'.
146. What is surface grinding ? Explain.
147. Explain briefly with neat sketches various types of surface grinding machines.
148. What is disc grinding ?
149. What is form grinding ? Explain.
150. What is grinding wheel ?
151. How is grinding ratio defined ?
153. What is an abrasive ? How are abrasive classified ?
153. Explain briefly the following abrasives :
  (i) Silicon carbide (SiC) ;   (ii) Aluminium oxide (Al_2O_3).

154. Discuss briefly the following :
   (i) Mounting of wheels   (ii) Wheel truing.
155. What is a 'bond' ? Name and explain principal bonds.
156. What are the purposes of finishing surfaces of metal parts ?
157. What is the purpose of honing ? Give the examples of honing work.
158. What are the advantages and disadvantages of honing ?
159. Explain briefly the 'lapping process'. Give the examples of lapping work.
160. Name the lap materials generally used.
161. How are lapping machines classified ?
162. State the advantages and disadvantages of lapping.
163. What is superfinishing ?
164. Explain briefly with a neat sketch the superfinishing operation.
165. In what respects the superfinishing operation basically differs from other abrasive finishing methods ?
166. Give the comparison among lapping, honing and superfinishing.
167. Explain briefly 'Polishing' and 'Buffing'.
168. Write a short note on 'Burnishing'.
169. Give the comparison of grinding and finishing operations.

UNSOLVED EXAMPLES

Lathe :

1. A taper pin of length 80 mm has a taper length of 48 mm. The larger diameter of taper is 83 mm and the smaller diameter is 73 mm. Determine :
   (i) Taper in mm/metre and in degrees.
   (ii) The angle to which the compound rest should be set up.
   (iii) The tail stock setting over   [**Ans.** (i) 208.33 mm/metre length, 5°, 5′ ; (ii) 5°,57′ ; (iii) 8.33 mm]
2. Determine the time taken to turn a 50 mm diameter by 70 mm long workpiece if the cutting speed is 45 m/min and the feed is 0.5 mm/revolution. One cut is to be considered.   [**Ans.** 29.33 s]
3. Determine the time taken to face a workpiece of 72 mm diameter. The spindle speed is 80 r.p.m. and cross feed is 0.3 mm/revolution.   [**Ans.** 1.5 min.]
4. A 150 mm long, 12 mm diameter rod is reduced to 11 mm diameter in a single pass straight turning. If the spindle speed is 400 r.p.m. and feed rate is 200 mm/min, determine the material removal rate and cutting time.   [**Ans.** 3768 mm^3/min, 0.75 min.]
5. A workpiece of 250 mm diameter and 500 mm length is to be turned down to 235 mm diameter for the entire length. The suggested feed is 1 mm/revolution and cutting speed is 135 m/min. The maximum allowable depth of cut is 5.0 mm.
   Calculate :
   (i) Spindle r.p.m.   (iii) Material removal rate.
   (ii) Feed speed   (iv) Cutting time.

$$\left[\begin{array}{l} \textbf{Ans.}\ (i)\ 172\ \text{r.p.m} ; (ii)\ 172\ \text{mm/min} ; \\ \qquad (iii)\ 67.5 \times 10^4\ \text{mm}^3/\text{min} ; (iv)\ 5.96\ \text{min.} \end{array} \right]$$

6. Determine the time required to machine a workpiece 170 mm long, 60 mm diameter to 165 mm length, 50 mm diameter. The workpiece rotates at 440 r.p.m., feed is 0.3 mm /rev and maximum depth of cut is 2 mm. Assume total approach and overtravel distance as 5 mm for turning operation. [**Ans** : 4.51 min]
7. From a raw material of 100 mm length and 10 mm diameter, a component having a length 100 mm and diameter 8 mm is to be produced using a cutting speed of 31.41 m/min and a feed rate of 0.7 mm/rev. How-many times we have to resharpen or regrind, if 1000 workpieces are to be produced. In the Taylor's expression, use constants as $n = 1.2$ and $C = 120$.   [**Ans.** 33]

8. For turning a carbon steel cylinder bar of length 3m and diameter 0.2 m at a feed rate of 0.5 mm/rev with an HSS tool, one of the two available cutting speeds is to be selected. These two cutting speeds are 100 m/min and 57 m/min. The tool life corresponding to the speed of 100 m/min is known to be 15 minute, with $n = 0.5$. The cost of machining time, set-up time and unproductive time together is Re.1/sec the cost of one tool re-sharpening is Rs. 20. Which of the above two cutting speeds should be selected from the point of view of the total cost of producing this part ? Prove your argument.   [**Ans.** 100 m/min]

9. The following data relate to a metal cutting test conducted on a lathe :

   Vertical force on the tool = 2000 N; Depth of cut = 3.0 mm; Feed = 6 cuts/mm; Overall efficiency of the machine = 80 per cent.

   Determine : (i) Pressure (in MN/m²) of the chip cross-sectional area on the tool.

   (ii) Power required for cutting the material.   [**Ans.** (i) 4000 MN/m²; (ii) 1250 W]

Shape and Planer

10. Calculate the machining time required for machining a surface 600 mm × 800 mm on a shaping machine. Assume cutting speed as 8 m/min. The return-to-cutting time ratio is 1 : 4 and feed is 2 mm/double stroke. The clearance at each end is 70 mm.   [**Ans.** 46.25 min]

11. A cast iron plate measuring 300 mm × 100 mm × 40 mm is to be rough shaped along its wider face. Calculate the machining time taking approach = 25 mm, overtravel = 25 mm, cutting speed = 12 m/min, return speed = 20 m/min, allowance on either side of the plate width = 5 mm and feed per cycle = 1 mm.   [**Ans.** 5.12 min]

12. Calculate the time required to machine a C.I. surface 250 mm long and 150 mm wide on a shaper with cutting-to-return ratio of 3 : 2. Use a cutting speed of 21 m/min, a feed of 2 mm/stroke and a clearance of 25 mm. The available ram strokes on the shaper are : 28, 40, 60 and 90 strokes/min. Also, determine material removal rate assuming depth of cut as 4 mm.   [**Ans.** 1.88 min; 1,60,000 mm³/min]

13. The top surface of a slab, 520 mm wide and 400 mm long is to be planed on a planer. Take cutting sped = 18.8 m/min, and return speed as 75 m/min. Take machining allowance as 10 mm. The tool approach angle is 45°. Calculate the machining time. Cross feed of tool = 3 mm/full stroke and side overtravel of tool = 3 mm.   [**Ans.** 51.7 min]

Drilling machine

14. A hole of 25 mm diameter and 62.5 mm depth is to be drilled. The suggested feed is 1.25 mm/rev. and cutting speed is 60 m/min, Assuming tool approach and tool overtravel as 5 mm, calculate.
    (i) Spindle r.p.m.
    (ii) Feed speed.
    (iii) Cutting time.
    (iv) Material removal rate.   [**Ans.** (i) 764 r.p.m. ; (ii) 955 mm/min : (ii) 0.0707 min ; (iv) 46.8974 × 10⁴ mm³/min]

Milling machine

15. Calculate the time required for cutting a 125 mm long keyway using HSS end-mill of 20 mm diameter having four cutting teeth. The depth of keyway is 4.5 mm. Feed per tooth is 0.1 mm and cutting speed is 40 m/min. Assume approach and overtravel distance as half of the diameter of the cutter and a depth of 4.5 mm can be cut in one pass.   [**Ans.** 0.53 min]

16. A steel workpiece is to be milled. Metal removal rate is 30 cm³/min. Depth of cut is 5 mm, and width of cut is 100 mm. Find the table feed.   [**Ans.** 60 mm/min]

17. Calculate the time required to mill a slot of 300 mm × 25 mm in a workpiece 300 mm length with a side and face milling cutter of 100 mm diameter and 25 mm wide and having 18 teeth. The depth of cut is 5 mm, the feed per tooth is 0.1 mm and cutting speed is 30 m/min. Assume overtravel distance of 5 mm.   [**Ans.** 1.82 min]

18. A plain surface 320 mm (long) × 100 mm (wide) is to be face-milled on vertical milling machine. The cutter has 16 teeth and the feed per tooth is 0.25 mm. The spindle speed is 125 r.p.m. Diameter of the cutter is 160 mm. Overtravel distance is 3 mm. Calculate the machining time.   [**Ans.** 0.68 min]

10

Thread Manufacturing

10.1. Screw threads—classification of threads—elements of screw threads—specifications of a screw thread—forms of threads—errors in threads. 10.2. Processes for making threads. 10.3. Using die heads. 10.4. Thread milling. 10.5. Thread rolling. 10.6. Thread grinding. 10.7. Thread tapping. 10.8. Automatic screw machines. Highlights—Objective Type Questions—Theoretical Questions.

10.1. SCREW THREADS

A screw thread is helical ridge formed on uniform section round the curved surface. The shape of the normal section of the thread depends upon the shape of the tool which produces its groove. A screw is a male threaded piece generally cylindrical in form, but sometimes, conical (or tapered), used in most cases as a temporary fastening ; less frequently used as a means of transmitting motion or power.

The screw threads are applied to many devices for various *purposes* as follows :

1. To hold parts together as in the case of fastening.

2. To transmit power.

3. To control movement as in micrometer.

4. To increase the effect of applied effort as in auto-jack.

5. To convey materials.

10.1.1. Classification of Threads

The threads may be *classified* as follows :

1. *According to the surface on which the threads are cut :*

(*i*) External threads.

(*ii*) Internal threads.

The external threads are cut into the surface of a *cylindrical bar.*

The internal threads are cut into the surface of the *cylindrical hole* of a bar or cone.

2. *According to the direction of rotation of the threaded cylinder with respect to engagement or disengagement with the other part :*

(*i*) Right handed thread.

(*ii*) Left handed thread.

A *right handed thread* is one in which the nut must be turned in a right handed direction to screw it on (Fig. 10.1 (*a*)). A left handed thread is one in which the nut would be screwed on by turning it to the left (Fig. 10.1(*b*)).

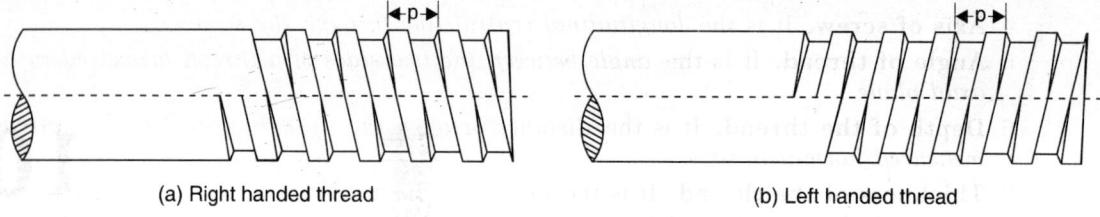

(a) Right handed thread   (b) Left handed thread

Fig. 10.1.

3. *According to number of starts* :

(*i*) Single start threads.

(*ii*) Multi-start threads.

In a piece of work it is possible to have separate and independent threads running along it. Accordingly, there are single threaded screw and multiple or multi-start threaded screw. The independent threads are called starts and we may have single start, two start, three start *etc.* (Fig. 10.2).

A *single start threaded screw* is one in which there is a movement of one thread for one complete turn round the screw or bolt. In the *multi-start threaded screw* there is a movement of more than one thread. In the case of double start thread, for one complete turn, the thread advances two times as if it were a single thread.

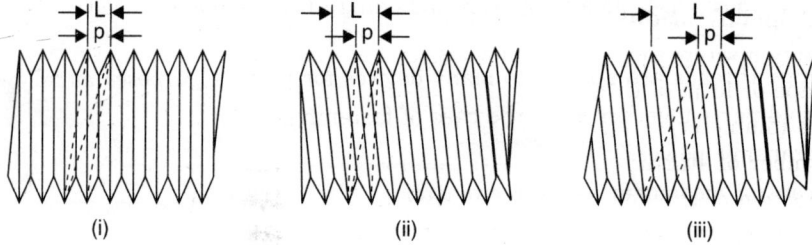

(i)   (ii)   (iii)

Fig. 10.2. (*i*) Single-start, (*ii*) Two-start, and (*iii*) Three-start threads.

Use of multi-start threads. Multi-start threads are used on those cases where *rapid movement or motion is required.*

Examples : Fountain-pen cap, screw press, bottles, tooth paste etc.

10.1.2. Elements of Screw Threads

To designate different parts of the screw threads the following *terms* are commonly employed : Refer to Fig. 10.3.

1. **Major diameter.** It is the *largest diameter of a screw thread.* It is also termed as *outside* or *crest diameter.*

2. **Minor diameter.** It is the *smallest diameter of a screw thread.* It is also known as *root* or *core diameter.*

3. **Effective or pitch diameter.** It is *an imaginary diameter in between the major and minor diameters, and is equal to the major diameter less than an amount equal to the single depth of a thread.*

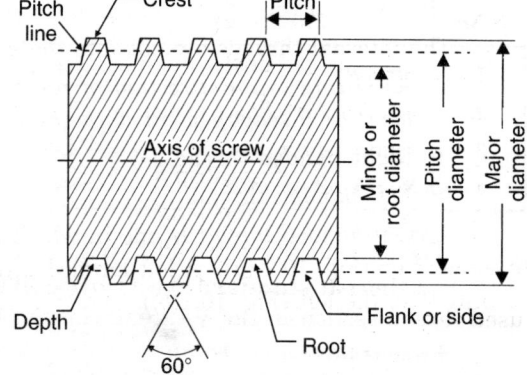

Fig. 10.3. Nomenclature of a screw thread.

4. **Axis of screw.** It is the *longitudinal central line through the screw.*

5. **Angle of thread.** It is the *angle between the two sides of a thread measured in an axial plane.*

6. **Depth of the thread.** It is the *distance between the crest and the root of a thread measured normal to axis.*

7. **Thickness of the thread.** It is the *distance between the adjacent sides of the thread measured along or parallel to the pitch line.*

8. **Side.** It is the *slant surface of the thread which connects the crest with the root.*

9. **Helix angle.** It is the *angle made by helix of the thread at the pitch or effective diameter with the plane perpendicular to the axis.*

10. **Crest.** It is top *surface joining the two sides of a thread.*

11. **Root.** It is the *bottom surface joining the two sides of a thread.*

12. **Pitch.** *It is the distance from a point on one thread to the corresponding point on the next thread measured parallel to the axis of the thread. It is denoted by 'p'.*

13. **Lead.** *It is the distance a screw thread advances axially in one turn on a single thread screw, the lead is equal to pitch and for a double threaded screw, the lead becomes two times the pitch and so on.*

Note : A screw is specified by a *nominal diameter,* it is the diameter of the cylindrical piece on which the threads are cut.

10.1.3. Specifications of a Screw Thread

To specify a screw thread the following points are given due considerations :

1. Shape or form of thread   2. Pitch
3. Size (diameter)   4. Length
5. Number of starts   6. Material
7. Direction of threads   8. Internal or external threads.

10.1.4. Forms of Threads

Many different shapes of threads are employed but they can be satisfactorily grouped into following *three* main classes :

1. V-threads
2. Square threads
3. Modification of both.

Comparison of 'V' and Square threads

S.No.	Particulars	V-threads	Square threads
1.	Strength	More strong	Less strong
2.	Frictional resistance to motion	More	Less
3.	Cutting of threads	Easy and hence cheaper	Difficult and hence costly
4.	Suitability	Suitable for fastening purposes	Suitable for power transmission

The various screw thread forms are discussed in details below :

(*i*) **British standard whitworth (B.S.W.) thread.** Refer to Fig. 10.4. It is most widely used thread section in the V-thread class in British practice.

Proportions : $H = 0.9605\ p$ H = Theoretical depth
                       $h = 0.6403\ p$ h = Actual depth

Thread angle   = 55°

$r = 0.1373\ p$ r = Radius (at the root or crest)

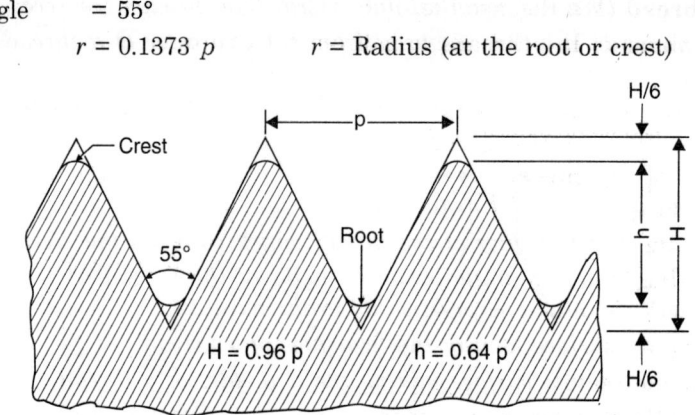

Fig. 10.4. British standard Whitworth (B.S.W.) thread.

The angle measured in an axial plane is 55°. One sixth of height of the fundamental triangle is truncated at top and bottom ; crests and roots are round, yielding thread depth of 0.64 pitch. The thread (B.S.W.) is employed in *general machine construction, where conditions favour the use of bolts, screws and other threaded pieces where quick and easy assembly of parts is required. The pitches are relatively coarse.*

(*ii*) **British standard fine (B.S.F.) thread.** *This thread is the same as the Whitworth standard thread, but the pitch for any given nominal diameter is smaller than for corresponding size of Whitworth standard thread.* Hence for a having nominal size, a bolt having a fine thread is stronger than one having a Whitworth standard thread. *In the design consideration where weight is an important factor, such as design of aircraft and automobile work, fine threads are most essential.* Thus these types of threads are largely used in *automobiles, and plant machine tool works* etc.

(*iii*) **British standard pipe thread.** These threads are of Whitworth form and have *fine pitches.* The nominal diameter is equal to the diameter of the bore of the pipe for which thread is intended.

Proportions : $H = 0.9602\ p$, $h = 0.64\ p$, Thread angle = 55° radius (at the root or crest), $r = 0.1372\ p$.

The thread is mainly used for gas, water and steam work.

(*iv*) **British association (B.A.) thread.** Refer to Fig. 10.5. The thread has rounded *V-form.* *These threads are exclusively used for small screws used in optical instruments, clocks etc.*

Proportions : $H = 1.136\ p$, $h = 0.6\ p$, $r = 2/11\ p$

Thread angle = 47½°.

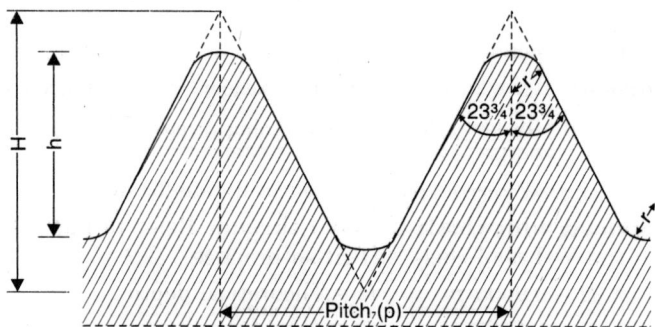

Fig. 10.5. British association (B.A.) thread.

(v) **Metric thread (M).** Refer to Fig. 10.6. These threads are in metric units. They employ 60° angle. The diameter of the threads vary from 6 to 80 mm. They are used in *motor car practice.*

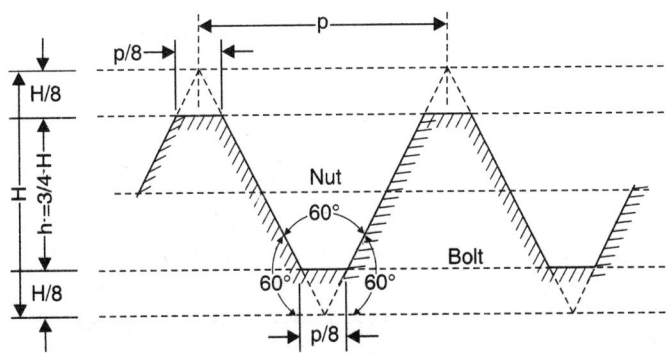

Fig. 10.6. Metric thread.

(vi) **Sellers thread.** Refer to Fig. 10.7. These are known as United States standard threads. These threads were introduced by Mr. Sellers.

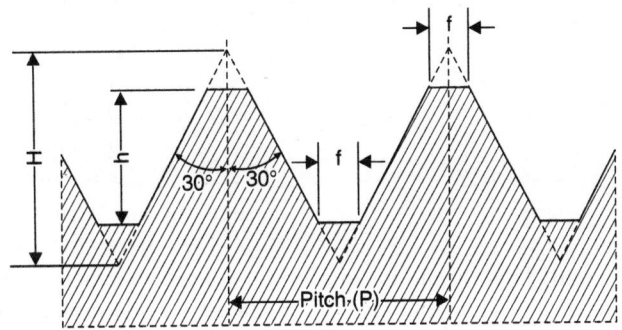

Fig. 10.7. Sellers thread.

They are employed for general use in engineering work such as for bolts, screws etc.

Proportions. $H = 0.866\ p$

$h = 0.6495\ p$

$f = 0.1252\ p$

Thread angle = 60°

(vii) **Square threads.** Refer to Fig. 10.8. This thread has its faces normal to the axis of the screw. There is less friction and less wear and most commonly used for transmission of power as in *vices, clamps* etc. They are also used for converting a rapid rotary motion into a slow linear motion *viz., screw presses, lead screw of lathe, jacks* etc. They are less stronger than *V-* threads and are more expensive to cut on a lathe.

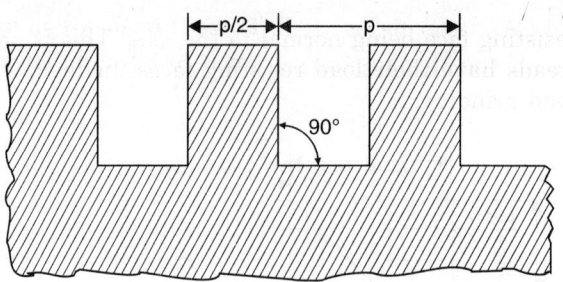

Fig. 10.8. Square thread.

(*viii*) **Acme thread.** Refer to Fig. 10.9. It is modification of a square thread having a profile of 29°. It is *stronger than the square thread and is more easily cut, milled or ground, on accounts of its sloping flanks.* An inherent advantage is that, if the tapered sides of the screw wear, the mating nut automatically comes into closer engagement, instead of allowing backlash to develop. Another point in its favour is that its shape permits the use of a disengaging or split nut. Thus *acme-threaded lead screws are very commonly used on lathes,* for it is essential to be able to engage or disengage the split nut quickly during screw cutting operation.

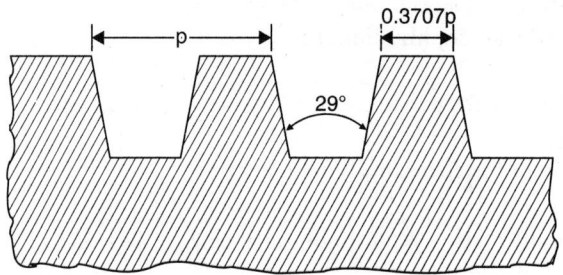

Fig. 10.9. Acme thread.

(*ix*) **Knuckle thread.** This thread is usually considered as a modification of a square thread. The cross-section of the thread is a semi-circle. The radius is one quarter of pitch, and the depth is therefore, half the pitch. In rough service the sharp outside corners of a square thread would soon be damaged, hence rounding above and below the pitch line as shown in Fig. 10.10. This rounding yields a strong thread but increases the friction, hence liberal clearance must be left between the screw and the nut. A typical application is in *railway carriage couplers.*

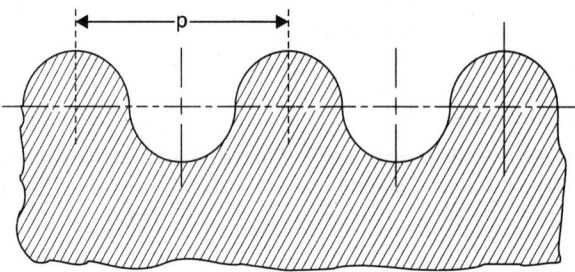

Fig. 10.10. Knuckle thread.

(*x*) **Buttress thread.** This thread resists heavy axial loads. *i.e.,* loads acting in the direction of the axis. In section this thread has various forms. Thus in common design the thread angle is 45°, the load-resisting face being normal to the axis. This is shown in Fig. 10.11. However, some buttress threads have their load resisting faces inclined, a design which facilitates thread milling and thread grinding.

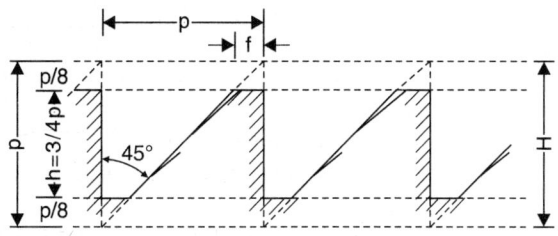

Fig. 10.11. Buttress thread.

Examples of uses of buttress thread occur in *quick-acting vices* and in breech mechanism of guns, for which reason the name *breech lock thread* is sometimes used. For power transmission the buttress thread combines some of the easy-working advantages of the square thread with the strength of Vee thread. The approximately triangular section gives a shearing strength twice that of a square thread for a given length of nut. While buttress threads are often used for transmitting motion, in some cases they are used in the reverse way, in order to prevent motion by the friction which arises on account of the sloping side of the thread. All *Vee-shaped threads absorb a large amount of power.*

Selection of a thread form. Many factors enter into selection of a thread form. Among them are :

1. Easy production of the form implying a capacity to make the tools without undue trouble and expense.
2. The measurement of the thread should be straightforward.
3. The form selected should enable a good bearing between the bolt and its nut, without unduly expensive precision cutting and gauging.
4. The thread angle should be as sharp as is consistent with the required strength.

Large thread angles cause large bursting forces on the nut. Again friction increases as thread angle increases. There is relatively little friction with square threads, hence their general use for transmitting motion and power.

10.1.5. Errors in Threads

On a screw thread there are at least five important elements, error on any one of which can cause rejection of thread. These elements are : *major, minor,* and *effective diameters, pitch,* and *angle of the thread form.* In the routine gauging of production threads, all of these features have to be checked, and the method of gauging must be such that they are covered as far as possible. *Other features, such as general form, radii at crest and root, and mutual concentricity of the various diameters, must also be controlled.*

- *Errors on the major and minor diameters will cause interference with the mating thread, weakness of the component by reduction of its root section or wall thickness, or reduction of flank contact, perhaps to the point of excessive weakness.*
- *Error on the effective diameter will cause either interference between the flanks or general slackness and possible weakening of the assembly.* The effects of errors in pitch and angle of either thread of a mating pair are not so apparent, although it will be obvious that *pitch errors are likely to cause a progressive tightening and interference on assembly. Pitch and angle errors have a special significance, and can be precisely related to effective diameter.*

10.2. PROCESSES FOR MAKING THREADS

The basic problem in manufacturing screw threads is how to produce the desired ridge on the workpiece. Various methods used are :

(*i*) Cutting ; (*ii*) Rolling ; (*iii*) Grinding ; (*iv*) Casting.

— Both external and internal threads can be *cast,* but this process is used primarily in connection with die casting or the moulding of plastics, and relatively few threads are made in this manner.

— *Rolling* also can be employed for making both external and internal threads, provided the material in reasonably *ductile*. Presently, the *majority of threads are formed by rolling.*

External threads. External threads are *produced by the following manufacturing processes* :

1. Lathe single-point turning.
2. Die and stock.
3. Automatic die head.
4. Threading machine.
5. Die casting.
6. Grinding.

Internal threads. Internal threads are *made by cutting by the following methods* :

1. On an engine lathe.
2. With a tap and holder (manual, semiautomatic, or automatic).
3. With an automatic (collapsible) tap (turret lathe, screw machine, or special threading machine).
4. By milling.

Some of these methods of cutting external and internal screw threads are discussed in the following articles.

10.3. USING DIE HEADS

Die heads are self-operating tools which are used for producing *external threads* on the work.

With the use of die heads it is possible to cut threads in comparatively less time and therefore, these are effectively used on turret and capstan types of lathes. These devices have also been successfully used on drill presses, automatic screw machines and threading machines.

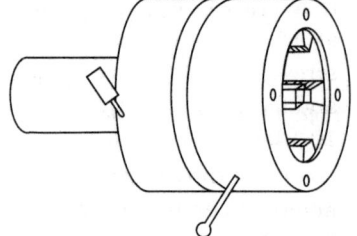

The cutting elements in a die heads are known as "*chasers*", which are of three distinct type : (*i*) *Radial,* (*ii*) *Tangential* and, (*iii*) *Circular.* Generally a die head mould accommodate four chasers (chasers are nitrided or chrome plated for better life).

Fig. 10.12. A die head.

Working. While operating the die-head, the chasers are set for the diameter of the job with the help of a closing handle. As the die head is made to move over the work the chasers come into action. When the die-head has cut the threads over the required length on the work a tripping device in the die-head throws back the closing handle into the original position. This return movement of handle opens the chasers radially, outwards. The chasers clear the outer diameter of the threads. Thus it is possible to withdraw the die-head without stopping or reversing the movement of the spindle.

Advantages :

1. Accuracy of the threads is consistent.
2. The manufacture of threads with die-heads is economical because unskilled worker can be employed.
3. The use of die head facilitates the withdrawal of any damaged chaser or the replacement of one chaser-set by the other to suit the thread.
4. Increased production rate (since the stopping and reversing of the spindle is eliminated therefore, it is possible to save considerable amount of time).

Limitations :

1. Screw threads on large work diameters cannot be cut.
2. It is difficult to cut square threads with die-heads.
3. Screw threads running upto shoulders of the work cannot be cut.

10.4. THREAD MILLING

Thread milling is an operation of producing threads both external and internal by means of thread-milling cutters either single or multi-ribbed, according to the type of thread required and the design of the thread-milling machine employed.

This process is suitable :

(i) For work upon which threads cannot be cut conveniently with die head during turning operation ;

(ii) For large diameters beyond the capacity of the die heads ;

(iii) For threads of short length running upto a shoulder ;

(iv) For internal threads which are unsuitable or inconvient for cutting with a tap.

Threads of high accuracy, particularly in larger sizes, often are cut by the thread-milling process. Either a single or a multiple-form cutter can be used.

Accurate threads of large size, both external and internal, can be cut with standard *job-type cutters.*

Long external threads. For long external threads, a threading machine similar in appearance to a lathe is used. Work is mounted either in a chuck or between centres, the milling attachment being at the rear of the machine. In cutting a long screw a single cutter is mounted in the plane of the thread angle and fed parallel to the axis of the threaded part as shown in Fig. 10.13.

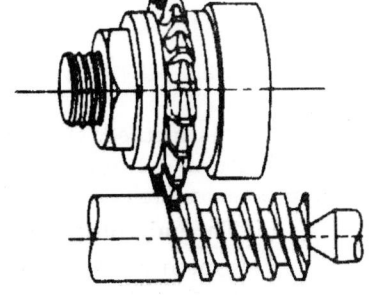

Fig. 10.13. Single thread milling cutter.

In thread milling the feed (f) is expressed as the cutter advance per tooth, or mm per cutter tooth, by the following formula :

$$f = \frac{D_{nom} \times N_j}{T_c \times N_c} \qquad ...(10.1)$$

where, D_{nom} = Nominal diameter of thread, mm,

N_j = Speed of the job, r.p.m.,

T_c = Number of teeth in cutter, and

N_c = Speed of cutter, r.p.m.

It is evident from the above expression that the cutter load per tooth, which varies directly with the feed, can be changed by varying the cutter speed, work speed or number of teeth in the cutter. This permits reducing the load on the cutter teeth so that deep threads can be cut in one pass.

Short external threads. For short external threads, a *series of single-thread cutters is placed side by side and made up as one cutter, having a width slightly more than that of the thread to be cut.* The cutter is fed radially into the work to the proper depth and, while rotating a little over one revolution, completes the milling of the thread. Proper lead is obtained by a feed mechanism that moves the cutter axially while is cutting.

- *Planetary type milling machines* are also used for mass production of short internal or external threads. The milling head carrying the *hob* is revolved eccentrically about the rigidly held work, which is rotated simultaneously on its own axis. It is advanced by means of a lead screw for a sufficient distance to produce the thread.

10.5. THREAD ROLLING

Thread rolling is the method of producing threads on screws, bolts, screw caps etc., by rolling under pressure.

More than 90% of bolts upto 51 mm in diameter have *rolled threads.* Threads can be rolled using any material that has sufficient ductility to withstand the forces of cold working without disintegration.

The parts to be threaded are caused to revolve, then compelled to make contact with the rollers, to which the required pitch and form of screw-threads have previously been given. As no metal has to be machined off to form the threads, as in machine threading, there is a great saving in material ; moreover, the pressure brings about a degree of cold working of the metallic surface of the parts, thus improving the mechanical properties.

The following *two methods* are employed in rolling threads.

(*i*) **Thread rolling with flat dies.** Refer to Fig. 10.14.

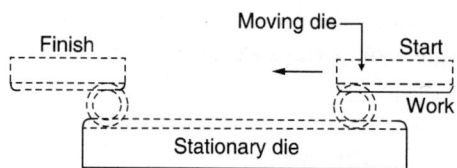

Fig. 10.14. Thread milling with flat dies.

In this method, the bolt/work is rolled between two flat dies, each being provided with parallel grooves cut to the size and shape of the thread. One die is hold stationary while the other reciprocates and rolls the blank between the two dies.

(*ii*) **Thread rolling with cylindrical dies.** Refer to Fig. 10.15.

This method employs two or three cylindrical dies.

— In the *two-die method* (Fig 10.15 (*a*)) the blank is placed on the work rest between two parallel, cylindrical rotating dies, and the right-hand die is fed into the blank until the correct size is reached, returning then to its starting position.

— The three-die machine (Fig 10.15 (*b*)) utilises cylindrical rotating dies mounted on parallel shafts driven synchronously at the desired speed. They advance radially into the blank by *cam action,* dwell for a short interval and then withdraw.

- The thread rolling dies and rolls are made of *high speed steel* or *high carbon high chromium oil and air hardening tool steels* (1.5% C, 12% Cr and small amounts of vanadium and molybdenum)

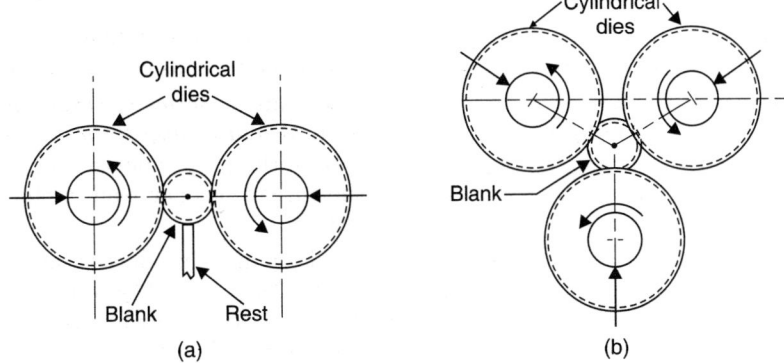

Fig. 10.15. Thread rolling two (*a*) or three (*b*) cylindrical dies.

● Thread rolling method is more economical for the following *reasons* :

(*i*) *It does not require skilled labour for operation.*

(*ii*) *It is faster as compared with other methods.*

(*iii*) The timing for threading of 50 mm long bar of 2.5 mm diameter by various methods are given below for comparison :

— *On lathe with threading tools*   ..... 20 minutes

— *On lathe with dies*   ..... 3 minutes

— *On "thread rolling machines"*   ..... 5 to 6 seconds.

(*iv*) There is no danger of the eccentricity of the workpiece.

Advantages :

1. Faster method of producing high quality threads.

2. Sharpening of dies is not required.

3. Dimensional accuracy and surface finish of the threads are quite high.

4. No wastage of material.

5. The mechanical properties of the threaded component is enhanced many times due to cold working of metal and the formation of continuous flow lines along the contour.

6. Wide variety of thread forms.

7. Cheaper materials usable because of improvement of physical properties during rolling.

8. Threads are accurate and uniform.

9. Thread rolling method can also be employed for knurling, burnishing, cutting splines and serrations and helical grooves.

Limitations :

1. It is difficult to form internal threads.

2. The blank size must be carefully decided otherwise the length of threaded components gets increased.

3. Coarse pitch and square threads cannot be made.

4. Uneconomical for low quantities.

5. Cannot roll material having a hardness exceeding Rockwell C37.

10.6. THREAD GRINDING

Thread grinding is used as either a finishing or a forming operation on many screw threads where accuracy and smooth finish are required. This process is particularly applicable for *hardened threads.*

Thread grinding, like milling, is also carried out in *two* ways :

(i) Grinding with a single wheel :

— This method is adopted for blanks of *large lengths* when a thin disc type grinding wheel is used.

— The wheel while running keeps traversing the length of the thread (Fig. 10.16 (a)). The wheel runs at a speed ranging from 2000 m p m to 3000 m p m and traverses a speed of 2 to 5 m p m. The threads are *finished in one pass of the grinding wheel.*

(ii) Plunge-cut grinding method :

— This method is adopted for threads extending over *smaller lengths and finishing at the shoulder.*

— The wheel has more than the length over which the threads are to be formed (Fig. 10.16(b)). To start the cutting action, the wheel is rotated at a high cutting speed and then *fed to entire depth of the thread.* The next step is the movement of the thread through 1.25 revolution and axial advancement of the work by a distance equal to one pitch.

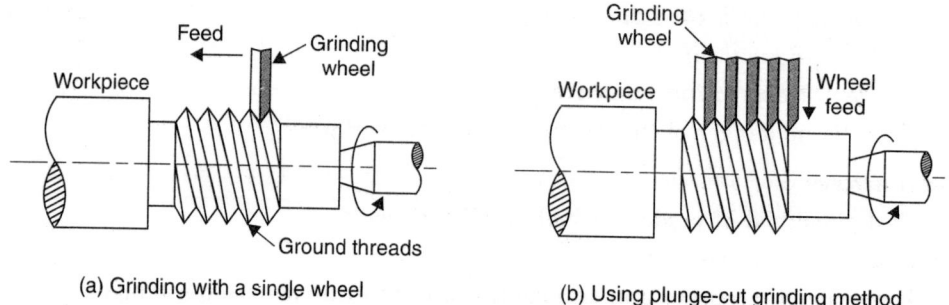

(a) Grinding with a single wheel   (b) Using plunge-cut grinding method

Fig. 10.16. Thread grinding process.

10.7. THREAD TAPPING

The cutting of an internal thread by means of a multi-point tool is called **thread tapping,** and the tool is called a *tap.* Taps are employed for the production of *internal threads.*

Fig. 10.17 illustrates a tap with the parts of the tool labelled.

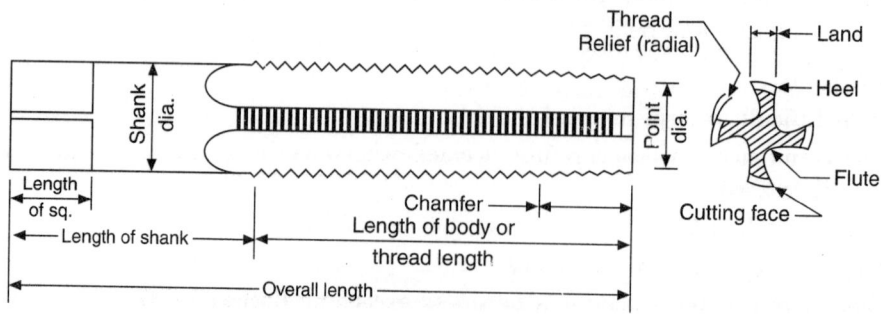

Fig. 10.17. Tap.

- For *hand tapping* these are finished in sets of three for each size. In starting the thread, the taper tap should be used because it ensures straighter turning and more gradual cutting action on the thread. If it is a through hole, no other tap is needed. For closed or blind holes with thread to the bottom, the *taper, plug,* and *bottoming* taps should all be used in the order named. Other taps are available and are named according to the kind of thread they cut.

- In small-production work on computer numerically controlled turret lathe, the tap is held by a special holder that prevents the tap from turning as the threads are cut. Near the end of the cut the turret holding the tool is stopped, and the tap holder continues to advance until it pulls away from a stop pin a sufficient distance to allow the tap to rotate with the work. The rotation of the work is then reversed and, when the tap holder is withdrawn, it is again engaged with the stop and held until the work is rotated from the tap.

- A common type of *tapping machine* has a multispindle arrangement provided with taps having *extra-long shanks*. The tap is advanced through the nut by the lead screw and, on completion of the threading, continues downward until the nut is released. The spindle then returns to its upper position with the tapped nut on its shank. When the shank has been filled with nuts, the tap is removed and the nuts are emptied.

10.8. AUTOMATIC SCREW MACHINES

Refer to Article 10.4.

HIGHLIGHTS

1. A *screw thread* is helical ridge formed on uniform section round the curved surface.
2. Multistart threads are used on those cases where rapid movement or motion is required.
3. The three main classes of threads are :
   (i) V-threads ;   (ii) Square threads ;   (iii) Modification of both.
4. Some commonly used methods for manufacture of external and internal threads are :
   (i) Using die heads ;   (ii) Thread cutting ;   (iii) Thread rolling ;
   (iv) Thread grinding ;   (v) Thread tapping ;   (vi) Automatic screw machines.

OBJECTIVE TYPE QUESTIONS

Fill in the blanks or say 'Yes' or 'No' :

1. A .... is helical ridge formed on uniform section round the curved surface.
2. A right handed thread is one in which the nut must be turned in a right handed direction to screw it on.
3. In the .... threaded screw there is a movement of more than one thread.
4. Fountain-pen cap is an example of single-start thread.
5. .... diameter is smallest-diameter of a screw thread.
6. Axis of the screw is the longitudinal central line through the screw.
7. Metric thread employs 60° angle.
8. The .... threaded lead screws are very commonly used on lathes.
9. A typical application of knuckle thread is in railway carriage complex.
10. The cutting elements in die heads are known as ....
11. Accurate threads of large size, both external and internal can be cut with standard hob-type cutters.
12. ...... is the method of producing threads on screws ; bolts, screw taps etc., by rolling under pressure.
13. In thread rolling there is no danger of the eccentricity of the workpiece.

14. Thread grinding is particularly applicable for threads.
15. Plunge-cut grinding method is adopted for threads extending over smaller lengths and finishing at the shoulder.

ANSWERS

1. screw thread	**2.** Yes	**3.** multi-start	**4.** No	**5.** Minor
6. Yes	**7.** Yes	**8.** acme	**9.** Yes	**10.** chasers
11. Yes	**12.** Thread rolling	**13.** Yes	**14.** hardened	**15.** Yes.

THEORETICAL QUESTIONS

1. Define the term 'screw thread'.
2. What purposes are served by the screw threads ?
3. How are threads classified ?
4. What is the difference between single-start and multi-start threads ?
5. Explain briefly the elements of a screw thread, with the help of a neat sketch.
6. Give the specifications of a screw thread.
7. Explain briefly various forms of threads.
8. Name the various methods used for cutting external and internal threads.
9. Discuss briefly any three of the following methods of manufacturing threads :
    (*i*) Thread rolling ;
    (*ii*) Thread grinding ;
    (*iii*) Thread milling ;
    (*iv*) Thread tapping ;
    (*v*) Using die heads.
10. Explain with neat sketches the 'thread rolling' method of making threads. State its advantages and limitations.

11

Gear Manufacturing

11.1. INTRODUCTION

*A **gear** is a wheel provided with teeth which mesh with the teeth on another wheel, or on to a rack, so as to give a positive transmission of motion from one component to another.*

- Gears constitute the most commonly used device for *power transmission* or for changing power-speed ratios in a power system. They also afford a convenient way of *changing the direction of motion*. A number of devices such as *differentials, transmission gear boxes, planetary drives etc.*, used in many construction machines employ gears as *basic components :*

- Gears are made from both metals as well as non-metals. **"Metals"** normally used for making gears include *cast iron, cast steel, structural steel, gun metal, brass*, etc. The **"non-metals"** commonly used for gear manufacture include *synthetic plastic, fibres and laminated wood.*

- With rapid industrialisation, the gears are called upon to yield noise free operation, ensure high load carrying capacity at a constant velocity ratio. The wear and fatigue strength of the gear tooth are the factors that govern its durability and reliability, and depend to a large extent on the manufacturing technology employed.

11.2. ADVANTAGES AND DISADVANTAGES OF TOOTHED GEARING

Following are the *advantages* and *disadvantages* of toothed gearing/gear drive :

Advantages :
1. High efficiency.
2. Long service life.
3. High reliability.
4. More compact.
5. Can operate at high speeds.
6. Can be used where precise timing is required.
7. Large power can be transmitted.
8. Constant speed ratio owing to absence of slipping.
9. Possibility of being applied for a wide range of torques, speeds and speed ratios.

10. The force required to hold the gears in position is much less than in an equivalent friction drive. This results in lower bearing pressure, less wear on the bearing surface and efficiency.

Disadvantages :

1. Special equipment and tools are required to manufacture the gears.
2. When one wheel gets damaged the whole set up is affected.
3. Noisy in operation at considerable speeds.

11.3. TYPES OF GEARS

The different types of gears used are enumerated and described briefly below :

1. Spur gear.
2. Helical gear.
3. Bevel gear.
4. Worm gear.
5. Rack and pinion.

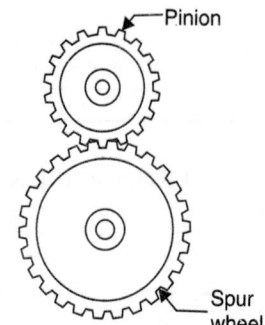

Fig. 11.1. Spur gears.

1. Spur gear. A spur gear is a gear wheel or pinion for transmitting motion between two *parallel shafts*. This is the simplest form of geared drive. The teeth are cast or machined parallel with the axis of rotation of the gear. Normally the teeth are of involute form. Fig. 11.1 illustrates a spur gear drive, consisting of a pinion and a spur wheel.

The efficiency of power transmission by these gears is very high and may be as much as 99% in case of high-speed gears with good material and workmanship of construction and good lubrication in operation. Under average conditions, efficiency of 96–98% are commonly attainable. The *disadvantages* are that they are liable to be *more noisy* in operation and may wear out and develop backlash more readily than the other types.

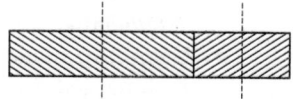

Fig. 11.2. Helical gears.

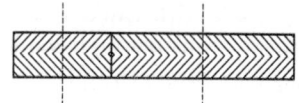

Fig. 11.3. Double helical gears.

2. Helical gear. Refer to Fig. 11.2. Helical gear is one in which teeth instead, of being parallel with shaft as in ordinary spur gears, are *inclined*. This ensures *smooth action and more accurate maintenance of velocity ratio*.

A *disadvantage* is that the inclination of the teeth *sets up a lateral thrust*. A method of neutralising this lateral or axial thrust is to use double-helical gears (also known as *Herring bone gears*) shown in Fig. 11.3.

3. Bevel gear. Refer to Fig. 11.4. A bevel gear transmits motion between two shafts which *intersect*. If the shafts are at right angles and wheels equal in size, they are called **mitre gears**. If the shafts are not at right angles, they are sometimes called angle bevel gears. Spiral toothed bevel gears are preferred to straight toothed bevels in certain applications, because they will run more smoothly and make less noise at high speeds.

4. Worm gear. Refer to Fig. 11.5. Worm gears connect two *non-parallel, non-intersecting shafts which* are usually *at right angle*. One of the gears is called the *'worm'*. It is essentially part of a screw, meshing with the teeth on a gear wheel, called the "*worm*

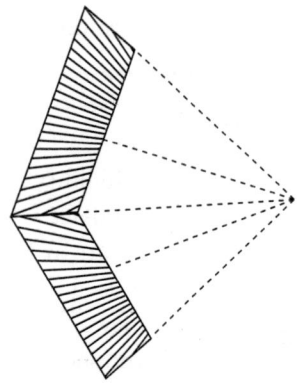

Fig. 11.4. Bevel gears.

wheel". The gear ratio is the ratio of number of teeth on the wheel to the number of threads on the worm.

One of the great advantages of worm gearing in that *high gear ratios* (*i.e.,* ratio of rotational speed of worm to that of worm wheel) are easily obtained. Worm gearing is *smooth and quiet.*

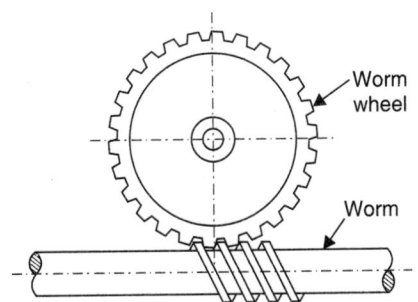

Fig. 11.5. Worm gear.

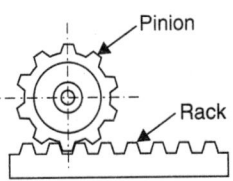

Fig. 11.6. Rack and pinion.

5. Rack and pinion. Refer to Fig. 11.6. A *rack* is a spur gear of infinite diameter, thus it assumes the shape of a straight gear. The rack is generally used with a pinion to *convert rotary motion into rectilinear motion.*

11.4. FORMS OF GEAR TEETH

The most commonly used forms of gear teeth are :

1. Involute.

2. Cycloidal.

The **involute tooth** is derived from the *trace of the point on a straight line, which rolls without slipping around a circle, which is the base circle.* It could also be defined *as the locus of a point on a piece of string which is unwounded from a stationary cylinder.*

The **cycloidal tooth** is derived *from the curve which is the locus of a point on a circle rolling on the pitch circle of the gear.* Here the addendum of the tooth is trace of the point on a circle rolling *on outside* of pitch circle and this is an *epicycloidal* curve whereas the dedendum portion of the tooth is the trace of the point on a circle rolling on *inside* of the pitch circle of the gear and is a hypocycloidal curve.

Involute gears are also called *straight tooth or spur gears* and are mainly used for general purpose in precision engineering. They possess the following **advantages** :

1. The involute teeth can be generated easily and accurately in gear-cutting machines.

2. All gears having the same pitch and pressure angle work correctly together.

3. The pressure angle is constant.

4. The face and flank of a tooth form a continuous curve.

5. An involute rack has straight teeth. This enables the complex involute form to be generated from a relatively simple cutter.

6. The gears still work currently, even when the pitch circle do not exactly touch, the velocity ratio remaining constant.

A **disadvantage** is that when pinions with a small number of teeth are used there will be interference between the points of pinion teeth and the teeth on the gear wheel with which they mesh. This is way the teeth are slightly rounded off at the tip.

- *The **cycloidal gears** are not generally used in modern engineering but used for some crude purposes where heavy and impact loads come on the machine.*

Various types of gears and their tooth traces are given below :

S. No.	Gears	Tooth traces
1.	Spur gear	Straight lines
2.	Helical gear	Straight helices
3.	Spiral gear	Curved lines
4.	Straight bevel gear	Tooth traces are straight line generators of cone. It is conical in form operating on intersecting axes usually at angles.
5.	Worm gear pair	The work and mating worm wheel have their axes non-parallel and non-intersecting.

Involute curve :

It is defined as the *curve traced by the end of a cord which, being held taut all the time, is being unwrapped from a cylinder.*

Refer to Fig. 11.7. The line (*ABCDE*) is the involute to the given circle. Tangent (*FB*) equals arc (*FA*). Tangent (*GC*) equals arc (*GA*). Tangent (*HD*) equals arc (*HA*) and so on.

It is also clear that the tangent to the involute at any point will be perpendicular to the generator at that point. This condition fulfills the requirements of laws of gearing. Further it will be noticed that the shape of the involute curve is entirely dependent upon the diameter of base circle from which the involute is generated. The curvature of the involute goes on decreasing as the base circle diameter goes on increasing and finally involute becomes straight line when base circle diameter is infinity.

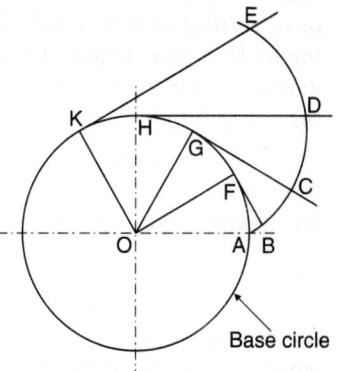

Fig. 11.7. Involute curve.

11.5. GEAR TOOTH TERMINOLOGY

Most of the terms used in connection with gears teeth are explained below :

(Refer to Fig. 11.8. (*a*) and (*b*)).

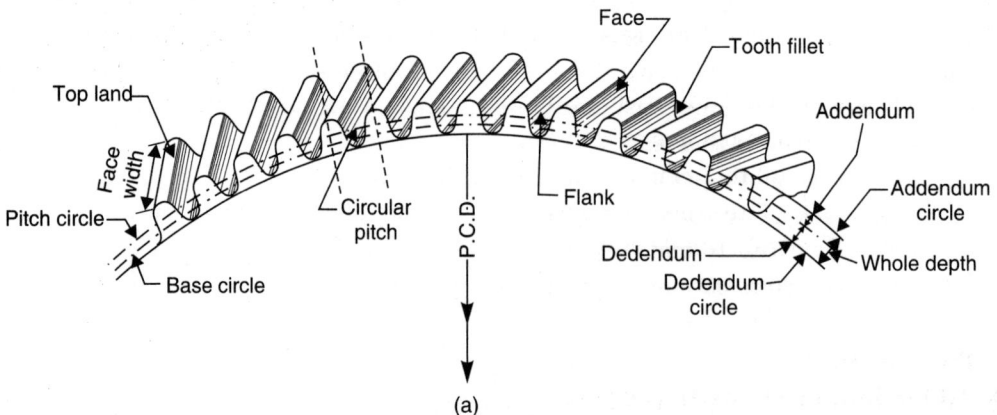

(a)

1. Pitch cylinders. Pitch cylinders of a *pair of gears in mesh are the imaginary friction cylinders which roll together without any slip and give the same velocity ratio as that of pair of gears.*

2. Pitch circle. *It is the imaginary circle most useful in calculations.* It may be noted that an infinite number of pitch circles can be chosen, each associated with its own pressure angle.

3. Base circle. *It is the circle from which involute form is generated. Only the base circle on a gear is fixed and unalterable.*

4. Pitch circle diameter (P.C.D.) or pitch diameter (D). *It is the diameter of the circle which by pure rolling action would produce the same motion as the toothed gear wheel.* In case of spur gears this is the diameter of a disc which this gear has replaced. This is the *most important diameter in gears.*

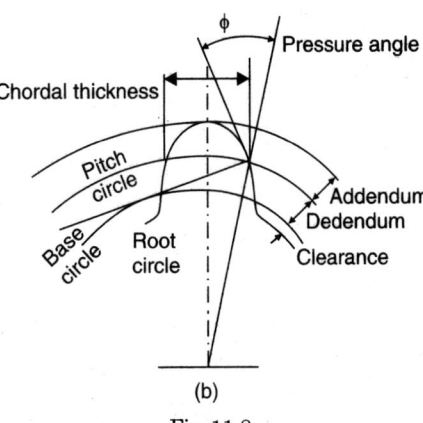

(b)

Fig. 11.8

5. Circular pitch (p_c). It is the distance measured along the circumference of pitch circle from a point on one tooth to the corresponding point on the next tooth. If T is the number of teeth on a wheel, D the pitch diameter the circular pitch (p_c) is given by

$$p_c = \frac{\pi D}{T} \qquad \qquad ...(11.1)$$

where, $\dfrac{D}{T}$ is the standard term known as **module.**

6. Diametral pitch (p_d). It is the number of tooth of the gear per mm of pitch circle diameter

$$\therefore \qquad p_d = \frac{T}{D} \qquad \qquad ...(11.2)$$

7. Module (m). It is *reverse* of the diametral pitch. Ratio between the pitch diameter and the number of teeth is known as module (m).

$$\therefore \qquad m = \frac{D}{T} \qquad \qquad ...(11.3)$$

For the purpose of standardisation, gears are available in the following standard module :

0.3 to 0.8 mm	in steps of 0.1 mm
1.0 to 5 mm	in steps of 0.25 mm
5 to 7 mm	in steps of 0.5 mm
7 to 16 mm	in steps of 1 mm
20 to 30 mm	in steps of 2 mm
30 to 45 mm	in steps of 5 mm.

8. Pitch point. *It is the point of contact of the two pitch circles of mating gears.*

9. Addendum circle or trip circle. *It is a circle concentric with the pitch circle and bounding the outer ends of the teeth.*

10. Dedendum or root circle. *It is a circle concentric with the pitch circle and bounding the bottom the teeth.* In case of a wheel having internal teeth-called *annulus*—the dedendum and dedendum circles are inside and outside the pitch circle respectively.

11. Addendum. *It is the radial distance between the pitch circle and the addendum circle.* Its standard value is *equal to one module.* This may be varied to avoid interference problems.

12. Dedendum. It *is the radial distance between the pitch circle and the dedendum circle.* Its standard value is

$$\left(1 + \frac{\pi}{20}\right) \text{ module} = 1.157 \text{ module}.$$

13. Clearance. *The difference between the dedendum and the addendum is called as clearance.* Naturally its standard value = (1.157 module − 1 module) = 0.157 module.

14. Working depth. *It is the sum of the addenda of the two mating gears.*

15. Bottom clearance. It *is the minimum distance between the tip of a tooth and the bottom of its mating space.*

16. Blank diameter. The diameter of a gear blank is equal to the pitch circle diameter plus twice addenda.

17. Tooth thickness. It *is the thickness of the tooth measured on the pitch circle.*

18. Tooth space. It *is the width of the recess between the two adjacent teeth measured along the pitch circle.*

19. Face. The *acting or working surface of the addendum is called face.*

20. Flank. It is the *working surface of the dedendum.*

21. Backlash. *The difference between tooth space and the tooth thickness is called the backlash.*

22. Pinion. The *smaller* of the two mating gears is known as *pinion* and the *larger,* the *wheel.*

23. Top land. It is the *surface of top of the gear.*

24. Bottom land. It is the *surface of the bottom of the tooth space.*

25. Whole depth. It is the *total depth of tooth space equal to sum of addendum and dedendum and is also equal to the working depth plus clearance.*

26. Tooth fillet. It is the *radius which connects the root circle to the tooth profile.*

27. Angle of obliquity or pressure angle. *Angle of obliquity or pressure angle is the angle between the tooth profile where it cuts the pitch circle and the line joining that point to the centre of the pitch circle.*

In Fig. 11.9. ∠*FPM* is the angle of obliquity (ϕ). *O* and *Q* are the centres of the gear-wheels, *A* and *B*. *PL* and *PK* are the parts of pitch circles which touch at the pitch point *P*. *PM* is the common tangent to the pitch circles through *P*. *EPF* is *"line of action"*, or *"pressure line"*. Contact between a tooth on wheel *A* and a tooth on wheel *B* takes place on this line *EPF*, and the pressure of the one tooth on the other is along this line *EPF*.

This is shown to an enlarged scale in Fig. 11.10. in which tooth *R* of wheel *A* is in contact with tooth *S* of wheel *B*, at the point *T*. *T* is always on the line *EPF*. Also, this line always cuts the line at the two tooth profiles at right angles. It is because of this that involute teeth satisfy the condition for transmitting motion from wheel *A* to wheel *B* with constant velocity ratio.

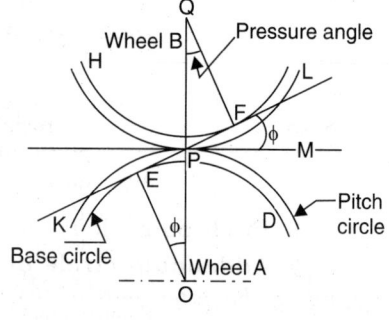

Fig. 11.9

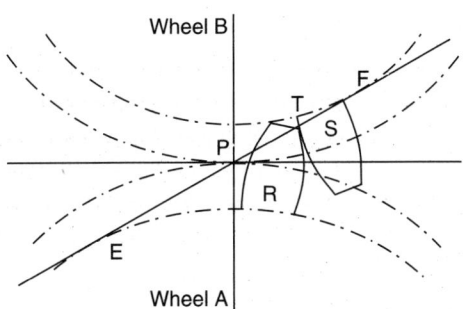

Fig. 11.10.

The *angle of obliquity determines the size of the base circles ED and FH*, on which involutes are drawn to form the tooth profiles. A value of 14½° was for a long time common value for the angle of obliquity, but an angle of 20° is now standard and widely used. *The 20° tooth is much stronger than the 14½° tooth, and undercutting in gears having a small number of teeth is not so great. At the same time its ability to withstand wear, and its quietness in running, are equal to that obtained with other tooth forms.*

In Fig. 11.10.

$$\frac{OE}{OP} = \cos \phi = \frac{R_b \text{ (radius of base circle)}}{R_p \text{ (radius of pitch circle)}} = \frac{D_b \text{ (dia. of base circle)}}{D \text{ (dia. of pitch circle)}}$$

$$D_b = D \times \cos \phi \qquad\qquad\qquad ..(11.4)$$

28. Base pitch. It is the *arc distance measured around the base circle from the origin of the involute on the tooth to the origin of a similar involute on the next tooth.*

$$\text{Base pitch} = \frac{\text{Base circumference}}{\text{Number of teeth}}$$

$$= \frac{\pi \times \text{dia. of base circle}}{T} = \frac{\pi D \cos \phi}{T}$$

$$= \phi \, m \cos \phi \qquad\qquad \left(\because \text{ module } m = \frac{D}{T} \right) \;...(11.5)$$

29. Involute function (δ). It is found from the fundamental principle of involute, that is the locus of the end of a thread (imaginary) unwound from the base circle.

Mathematically,

Involute function $\delta = \tan \phi - \phi$

where, ϕ = Pressure angle.

Basic tooth proportions for involute spur gear

S.No.	Item	Pressure angles	
		20°	**14½°**
1.	Addendum	m	m
2.	Dedendum	1.25 m	1.157 m
3.	Teeth depth	2.25 m	2.157 m
4.	Circular teeth thickness	πm/2	πm/2
5.	Fillet-radius	0.3 m	0.157 m
6.	Clearance	0.25 m	0.157 m

11.6. METHODS OF MAKING GEARS

— Most gears are produced by some *machining process.* Accurate machine work is essential for high-speed, long-wearing, quiet operating gears.

— Gears operating at slow speeds and under exposed conditions may be *sand-cast*, but such gears are not efficient in their power transmission. *Die and investment casting of gears* has proved satisfactory. The materials for such gears are limited to low-temperature melting metals and alloys ; consequently these gears do not have the wearing qualities of heat-treated steel gears.

— *Stamping*, although reasonably accurate, can be used only in making *thin gears from sheet metal.*

The various *commercial methods employed in producing gears* are as follows :

1. Casting :

 (*i*) Sand casting ;

 (*ii*) Die casting ;

(*iii*) Precision and investment casting.

2. Stamping.

3. Machining :

 A. *Formed-tooth process :*

 (*i*) Form cutter in milling machine ;

 (*ii*) Form cutter in broaching machine ;

(*iii*) Form cutter in shaper.

 B. *Template process.*

 C. *Cutter generating process :*

 (*i*) Cutter gear in shaper ;

 (*ii*) Hobbing ;

(*iii*) Rotary cutters ;

(*iv*) Reciprocating cutters simulating a rack.

4. Powder metallurgy.

5. Extruding.

6. Rolling.

7. Grinding.

8. Plastic moulding.

Sources of Errors in Manufacturing Gears :

The following two methods may be used to make gears (gear teeth) :

1. Reproducing method

2. Generating method.

1. **Reproducing method.** In this method of making gears the cutting tool is *formed involute cutter*, which forms the gear teeth profiles by reproducing the shape of the cutter itself. In this method, *each tooth space is cut independently of the other tooth spaces.* The various sources of errors in the gears made by this method are due to :

 • Incorrect profile on the cutting tool.

- Incorrect positioning of the tool in relation to the work, and
- Incorrect indexing of the blank.

2. **Generating method.** In this method the cutting tool (*hob*) forms the profiles of several teeth simultaneously during constant relative motion of the tool and blank. The sources of errors when gears are made by generating method are :

- Errors in the manufacture of the cutting tool.
- Errors in positioning the tool in relation to the work.
- Errors in the relative motion of the tool and blank during the generating operation.

11.7. GEAR MANUFACTURING BY "CASTING"

- Cast-iron gears are cast in sand moulds or permanent metal moulds. These gears are *relatively rough, weak and inaccurate, but the cost of production is very low.* This method is therefore, used for *large gears only.*

Small cast gears for light duty work are better produced by *die casting*, particularly of non-ferrous metals and alloys. These gears have sufficient accuracy and high surface finish.

- *"Plastic moulding"* is also used for producing gears of plastic materials for *very light work.*

11.8. FORMING OR FORM CUTTING

The forming process involves finish machining of gears teeth to a predetermined profile by means of form cutters or single point reciprocating form tools.

Milling machines are capable of cutting practically, every type of gear by employing an universal indexing mechanism and a form cutter. The cutter has the required tooth profile on it. This cutter may be operated on a vertical or horizontal type of milling machine.

In both cases, the cutter rotates on the spindle and work reciprocates under the cutter.

- Fig. 11.11. shows the set-up for cutting spur gear on a milling machine.

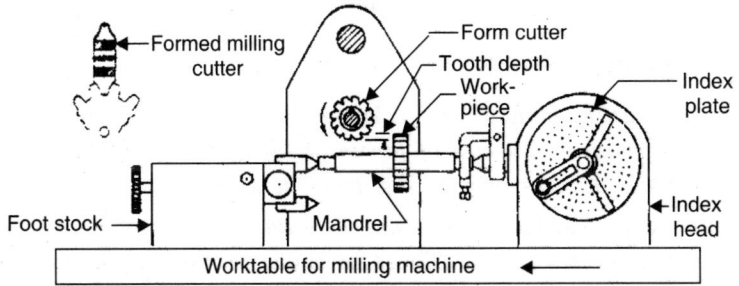
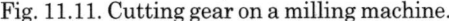
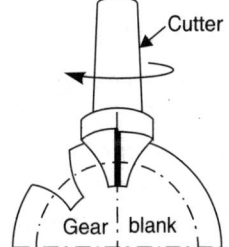

Fig. 11.11. Cutting gear on a milling machine.

Fig. 11.12. Gear teeth forming by using an end mill cutter.

- Fig. 11.12. shows gear tooth forming by using an end mill.

It should be kept in mind that two successive teeth should never be milled one after the other because the heat due to metal cutting may distort the teeth. It would be very much safer to finish the teeth on the gear blank after short intervals.

- The *forming-cutter method* is used only *when gears cut by more accurate methods cannot readily be obtained.* Only *small quantities of gears* should be cut by the forming cutter method, because this method is *relatively slow.*

Advantages :

1. All types of gears (*e.g.,* spur, helical, worm and in special circumstances bevel gears) can be produced on milling machines.

2. It can be employed for both roughing and finishing operations.

3. The gear cutting can be carried out on a conventional type of milling machine which is normally available in a modern workshop.

4. Economical and suitable for one off and small batches.

Limitations :

1. Internal teeth can not be cut.

2. The pitch accuracy is very much dependent upon the accuracy of dividing head. The tooth form is not accurate.

3. Milling is not a production process.

11.9. BROACHING

Broaching can be used to produce gear teeth and is *particularly applicable to internal teeth.*

- The process is rapid and produces *fine surface finish with high dimensional accuracy.*

- Because broaches are expensive and a separate broach is required for each size of gear, this method is *suitable mainly for high-quantity production.*

11.10. TEMPLATE METHODS

These methods are more suitable for manufacturing *large gears.* For example, a planer may be used, and a suitable template attached to the work gear may serve as a guide for manually feeding a suitable cutting tool. The operator adjusts the cutting tool to follow the outline of the template.

- *Larger straight tooth bevel gears are usually cut using a template method.*

11.11. GENERATING METHODS/PROCESSES

The term **generating** *in gear cutting stands for development of involute curve by straight cutting edges of the cutter, which produces a series of facets of the blank so as to form the involute profile.* The cutter and blank behave as two mating gears in working contact.

Generating methods for producing gear teeth *make use of certain relative motions between the work gear and the cutter during the machining.* These methods produce with one cutter theoretically correct gear-teeth profiles regardless of the number of teeth desired on the gear. A cutter for a given diametral pitch may be used to cut gears with any desired number of teeth. All teeth will have theoretically correct profiles, and *all gears will mesh interchangeably.*

Generating methods for producing gears are *faster and suited for the production of large quantities.* These methods include the use of :

(*i*) *Gear shapers and gear hobbers for spur and helical gears.*

(*ii*) *Straight-tooth bevel-gear generators for straight-tooth bevel gears.*

(*iii*) *Spiral-bevel and hypoid generators* for spiral-bevel and hypoid gears.

The common generating processes used for generating the gear teeth are as follows :

1. Gear shaper process.

2. Rack planing process.

3. Hobbing process.

11.12. GEAR SHAPER PROCESS

Fig. 11.13. shows the set up for a gear shaper.

The principal motions involved in rotary gear shaper cutting are :

(*i*) Cutting motion ;

(*ii*) Return stroke ;

(*iii*) Indexing motion ;

(*iv*) Completion of cutting operation.

— The cutter and the gear blank are connected by gears so that they will roll together as the cutter reciprocates for cutting. First the cutter must cut its way to the desired depth. The cutter and gear blank then rotate slowly together as the gear teeth are cut in the gear blank. Slightly more than one complete revolution of the work gear is required.

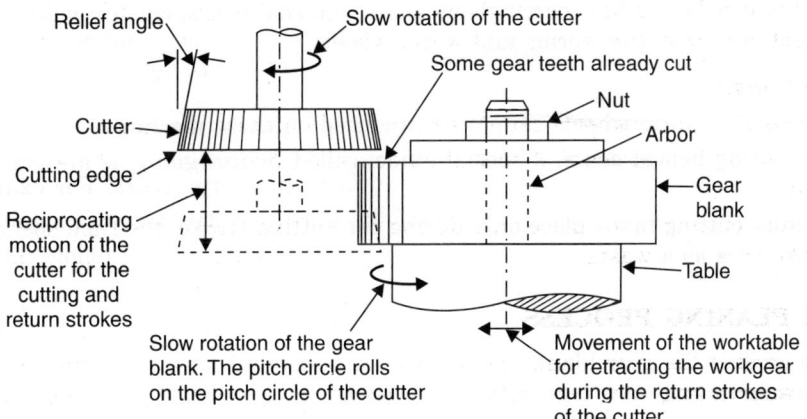

Fig. 11.13. Set up for a gear shaper.

— Often *two separate cuts* are made completely around a work gear. The first-cut is for *roughing,* and the second cut, which *increases accuracy and smoothness, is the finishing cut.*

● "Helical gears" may also be produced on gear shapers.

The cutters used should have helical teeth, and as a cutter reciprocates, a guiding cam which is installed on the machine causes the cutter to twist during its cutting strokes so that the cutter teeth will follow helical paths.

Rack shaper. On a *rack shaper*, the generating tool is a segment of a rack (Fig. 11.14), which reciprocates parallel to the axis of the gear blank. Because it is not practical to have more than 6 to 12 teeth on a rack cutter, the cutter must be disengaged at suitable intervals and returned to the starting point ; the gear blank, meanwhile, remains fixed.

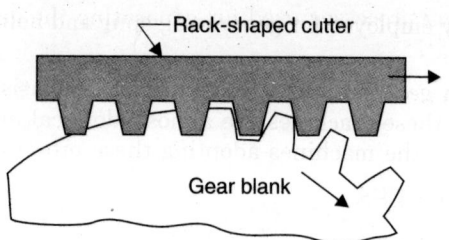

Fig. 11.14. Gear generating with a rack-shaped cutter.

Advantages and Limitations of Gear shaper processes :

Advantages :

1. The teeth cut on the gear carry a very accurate tooth profile.
2. Quite a fast process.
3. Suits well to both the medium and large size batch production.
4. The teeth can be easily cut upto quite close to a shoulder.
5. No need to change the cutter for cutting the teeth of different spur gears so long as their modules are same.
6. This process can be used to cut most of different types of gears, *viz.*, spur gears, racks, doublical helical gears, herringbone gears, internal gears, cluster gears, sprockets, etc., except of course, the worms and worm wheels.

Limitations :

1. Worms and worm wheels cannot be produced on these machines.
2. For cutting helical gears, a special guide, called *helical guide*, is always required to be used.
3. Because cutting takes place only during the cutting stroke, the time spent in the return stroke goes as a waste.

11.13. RACK PLANING PROCESS

In this process the gear blank, mounted on a horizontal axis, is rotated intermittently in mesh with a reciprocating *rack type cutter*. This results in *generation of involute teeth* on a spur gear.

The generating action of the rack is diagramatically represented in Fig. 11.15. that shows the successive position of the gear relative to rack developing involute profile.

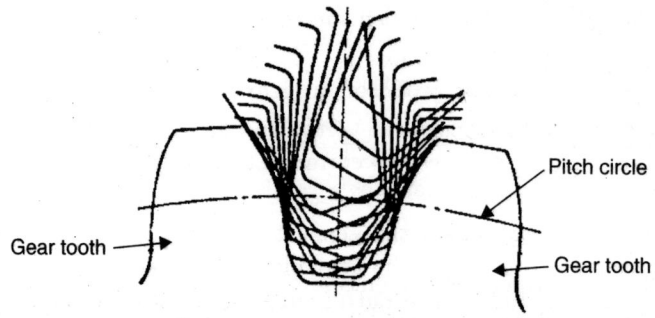

Fig. 11.15. Generating action of a rack.

— This process is mainly employed for generating spur and helical gear teeth with the help of a rack type cutter.

— The machines used in *gear planing* work either on the basis of "*Sunderland process*" or "*Maag process*". Both these machines are almost identical in principle but the constructions and operations of the machines adopting these processes differ.

11.14. HOBBING PROCESS

Gear hobbing is the method of generating gear teeth by the use of a rotating worm-shaped cutter (hob).

A majority of involute gears are produced by this method.

● Fig. 11.16 shows a gear hob.

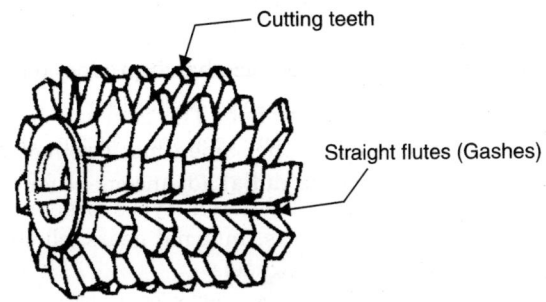

Fig. 11.16. A gear hob.

— The hobs are either *single threaded ion* or *multi-threaded*. A single threaded hob will complete one revolution for generation of each tooth whereas a double threaded will generate two teeth in its one revolution. In this way double threaded hob takes less time in finishing the gear blank. The *advantage* with the use of *single case threaded hob* lies in the fact that it is capable of giving greater accuracy.

● Fig. 11.17 illustrates the relative positions of the *hob* and the *gear blank* as the *hobbing* proceeds.

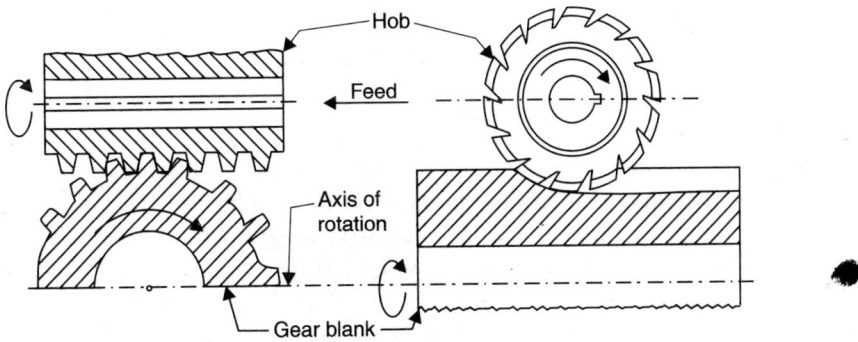

Fig. 11.17. Relative positions of hob and blank during process.

● Fig. 11.18 illustrates the arrangements of the gearing in hobbing machine. The blank is driven at a uniform speed by the index worm and gear. The bevel gears drive the hob at the correct relative speed, and the hob is fed across the face of the gear by a feed screw, which is not shown in order to simplify the diagram.

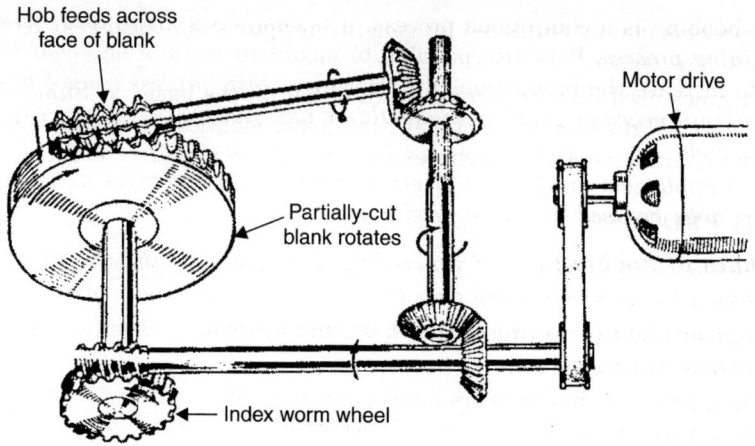

Fig. 11.18. Gear hobbing. The hob is fed across the face of the rotating blank.

- Fig. 11.19 shows the cutting action of hob. The gear cutting with hob involves *three* basic motions all of them occurring at the same time. Two motions are rotary for the hob and the work. The third motion is the radial advancement for the hob. Both, cutting and indexing takes place continuously.

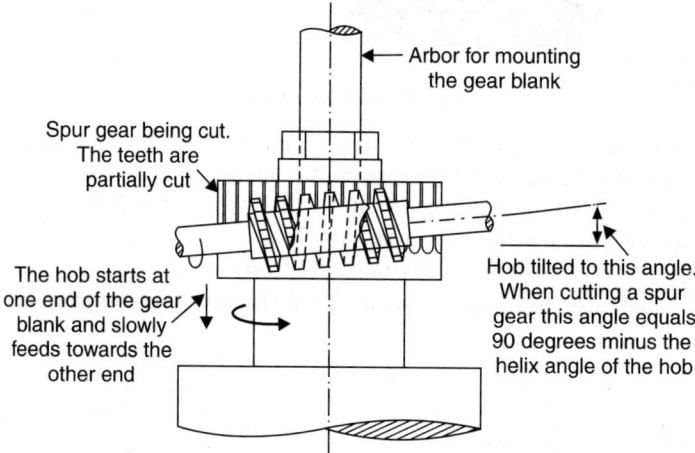

Fig. 11.19. Cutting action of a hob.

— The hob axis is set at an angle, equal to helix angle of the thread in reference to the axis of the gear blank. This brings the blank teeth in plane of the hob's teeth. This plane is termed as *'generating plane'*. The cutter finishes all the teeth in its one pass over the work. *For cutting helical gears the hob must be set over an additional amount equal to the helix angle of the gear.*

— To start the cutting, the hob is located so as to clear the blank and then moved inward until proper setting for tooth depth is obtained. The hob is then fed towards the gear blank in a direction parallel with the axis of rotation of the blank. As the gear blank rotates the teeth are generated and the feed of the hob across the face of the blank extends the teeth of the desired tooth face width.

● Since hobbing is a continuous process, it is *fast, economical and most productive gear machining process.* It is also possible to mount more than one gear blank in the work axis to increase the production rate. However, *this process cannot be used for machining internal gears or gears with shoulders and flanges because of the clearance needed for the hob.*

Hobbing is applicable to all types of gears of ferrous, non-ferrous and non-metallic materials. Carbide tips hobs be used for mass production.

Advantages of hobbing :

1. It is easier to exercise control over tooth spacing, head and tooth profile.
2. High production rate (owing to the absence of indexing, reduction in cutter approach and continuous cutting action).
3. A large number of similar gears held on a mandrel can be cut at a time ; this reduces the approach time of the hob.
4. The possibility of distortion in the gear is minimum, as the heat generated due to metal cutting in uniformly distributed over the entire work.
5. All types of spur and helical gears can be cut on metals and non metals.
6. High degree of accuracy can be maintained on the products over large period.
7. Irrespective of the number of teeth, the same hob can be used for generating teeth on different gears so long as they carry the same module.
8. *Gap type herringbone gears can be generated only through this process.*
9. The setting and operation of hobbing machines are simpler.
10. On hobbing machines, long shafts and splines can be easily accommodated.

Limitations :

1. Internal gears cannot be cut by this method.
2. Gears which have shoulders and flanges can not be cut by hobbing.

Comparison of 'Gear hobbing' and 'Gear shaping' :

● The continuous action of *"gear hobbing"* generally makes it faster and more accurate than gear shaping. Therefore *gear hobbing is normally used if either process is suitable and available and quantities be produced are sufficiently high.*

● *"Gear shaping"* can be used for some workpieces which *cannot be hobbed .* For example, *internal gears, continuous herringbone gears, and parts with gears permanently positioned too close to another gear or to a shoulder must be gear-shaped.*

Internal splines in blind holes must be machined with a gear shaper.

11.15. BEVEL GEAR CUTTING

● For cutting straight bevel gears a special gear cutting machine tool is used. This machine tool employs two *'half tooth' tools* ; one of it, cuts one face of the tooth while the other one cuts the other face of the same tooth. Both of these tools are mounted on a *cradle and have reciprocating motion, in opposite directions.* The gear cutting process on this machine tool is carried out as follows :

— At the start, the cradle carrying the tools is rolled in upward position and the blank and cradle are set for required teeth depth.

— The machine is next started when the tools start reciprocation and the cradle gradually rolls down moving the blank also along with it. This *movement accomplishes the generating action.* By the time cradle reaches the extreme downward position the tools complete the tooth profile.

— The cradle is then rolled back and the tools perform the finishing operation on the tooth profile.

— When the cradle reaches the starting position, the blank is withdrawn and indexed for next tooth.

● *Small and straight tooth bevel gears* are cut with the help of *two disc type cutters.* The cutters roll with blank and produce the teeth.

● Bevel gears with *curved tooth profile* are made on another type of generating machine by employing a circular rack type face mill cutter. The cutter axis is parallel to the axis of the roll of the cradle. For cutting the blank is fed inside towards the cutter to full depth. The cradle is in its down-most position. Both of them roll together in upward direction. The cutter engages the blank from the small end and moves towards the large end or across the width of tooth space.

11.16. CUTTING WORMS AND WORM WHEELS

● Normally, a worm is produced by the *forming process* using a rotary milling cutter on a milling machine.

● *Worm wheel* teeth are cut by *generating process* using a *hob.* The hob should be of the same form as the worm which will drive this worm wheel. The disadvantage of this process is that one hob can be used only for one design and form of worm wheel.

11.17. GEAR FINISHING

After gear cutting, it is rare that a gear is entirely free of all errors in *index, tooth profile, eccentricity, surface smoothness* and *helix angle.* The purpose of gear finishing is to eliminate or substantially reduce these possible errors, so that gears will run quietly and smoothly when transmitting large amounts of power at high rotational speeds.

Gear finishing operations include *gear shaping and gear burnishing for gears which are not to be hardened,* and gear *burnishing* for gears which are *to be hardened.* A commonly used sequence for gear finishing includes *gear shaving, surface hardening, and gear lapping.*

11.17.1. Gear Shaving

Gear shaving is the process of finishing gear teeth by running the gear at high speed in mesh with a gear shaving tool which is in the form of a rack or pinion.

● *"Rotary shaving"* gives gears an over-all accuracy. The cutter used has the shape of an accurate gear, which will mesh with work gear. Grooves or gashes, as shown in Fig. 11.20, are cut in the sides of each cutter gear tooth extending in the direction from top to bottom. The edges of these gashes are sharp so they are able to shave the tooth of a work gear. The cutter and work gear are meshed together in a gear shaving machine.

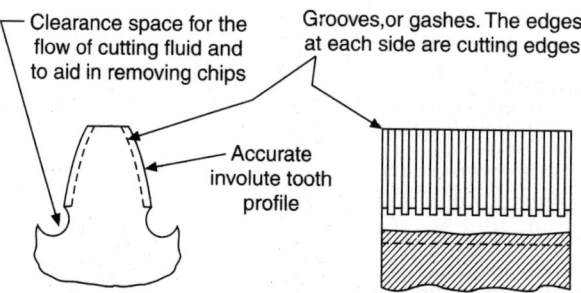

Fig. 11.20. One of the teeth of a rotary shaving cutter.

— During shaving the cutter axis is set crossed to the work axis at an angle so as to avoid burnishing. While cutting, the cutter and the gear rotate in close mesh with each other. The work also traverses back and forth longitudinally, across the shaving tool.

This process is usually applied to the *unhardened gears* and *offers a very rapid method of finishing*. The gears are corrected within 0.002 mm. It is also possible to shave as many as 9000 to 12000 gears before it becomes necessary to recondition the tool.

● When a *rack-type tool* is used, the gear is mounted on a reciprocating arbor and is brought in mesh with a horizontal rack situated under it. The rack is reciprocated longitudinally at high speed and the gear across it. The lengthwise movement of the rack type tool also rotatates the gear.

— Shaving with rack-type tool, however, *suits only small gears.*

11.17.2. Gear Burnishing

Burnishing *is a cold-working operation accomplished by rolling the gear in contact and under pressure with three hardened burnishing gears.*

Although the gears may be made accurate in tooth form, the *disadvantage* of this process is that the surface of the tooth is covered with amorphous or "smear" metal rather than metal having true crystalline structure, which is desirable from a long-life stand point.

11.17.3. Gear Grinding

Hardened gears are difficult to finish by the shaving and burnishing methods. Since the heat treatment may cause severe distortion and oxide film formation on teeth, therefore there is a necessity for removing considerable stock from the teeth. With the grinding method it is possible to finish the heat treated gears.

The application of this method is limited because it is *slower* and *more expensive*. However, it guarantees production of highest quality gears.

The grinding may be carried on generating principle or employing formed wheels (Fig. 11.21).

● The *disadvantage* of gear grinding is that considerable time is consumed in the process. Also, the surfaces of the teeth have small scratches or ridges that increase both wear and noise. To eliminate the latter defect ground gears are frequently lapped.

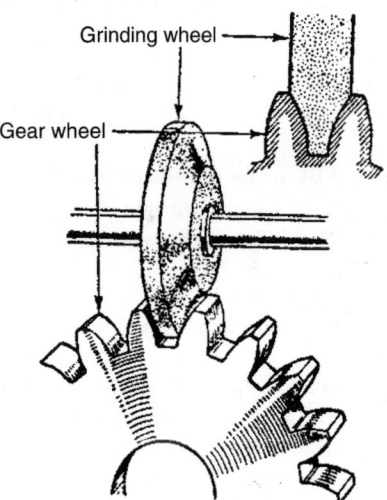

Fig. 11.21. Gear grinding.

11.17.4. Gear Lapping

Gear lapping is accomplished by having the gear in contact with one or more cast iron lap gears of true shape.

— The work is mounted between centres and is slowly driven by the *rear* lap. It in turn drives the *front* lap, and at the same time both laps are rapidly reciprocated across the gear face.

— Each lap has individual adjustment and pressure control.

— A fine abrasive is used with kerosene or light oil to assist in the cutting action.

The time consumed for average-size gears is $\frac{1}{2}$ to 2 min per side of gear teeth.

- As lapping can remove very small amount of metal, this process of finsihing can only correct small errors maximum upto 0.05 mm only.
- The results of lapping are demonstrated by *longer wearing and quieter operating gears.*

11.17.5. Gear Honing

Honing is the last operation performed on the hardened gears. It also helps in correcting gear errors.

- Honing removes minor nicks and burrs, gives a fine surface finish on the gear contacting surfaces.
- Gears so finished find applications in *gear trains where a low noise level is desirable.*

HIGHLIGHTS

1. A *gear* is a wheel provided with teeth which mesh with the teeth on another wheel, or on to a rack, so as to give a positive transmission of motion.
2. The different types of gears are :
   Spur gear ; Helical gear ; Bevel gear ; Worm gear ; Rack and pinion.
3. Methods of making gears are :
   Casting ; Stamping ; Machining (Formed tooth process, template process, Cutter generating process) ; Powder metallurgy ; Extruding ; Rolling ; Grinding ; Plastic moulding.

OBJECTIVE TYPE QUESTIONS

Fill in the blanks or say 'Yes' or 'No' :

1. .... gears are light in weight ; enable noiseless operation and are oil resistant
2. The difference between actual tooth thickness and the width of space, with which it meshes, is known as ......
3. .... can be described as the metric standard for pitch.
4. A rack is milled on a milling machine with the help of a rack milling attachment.
5. In .... process, a pinion shaped cutter is used which carries clearance on the tooth face and sides.
6. Worms and worm wheels can be produced by gear shaper process.
7. The gear teeth generated by hobbing process are very accurate.
8. .... process does not suit the generation of internal gears.
9. The shaving tools have a very short life.
10. Gear burnishing is a hot process.

ANSWERS

1. Non-metallic	2. backlash	3. Module	4. Yes	5. gear shaper
6. No	7. Yes	8. Hobbing	9. No	10. No.

THEORETICAL QUESTIONS

1. What is a gear ?
2. Name the metals and non-metals which are normally used for making gears.

3. What are the application of gears ?
4. List the advantages are disadvantages of toothed gears.
5. Explain briefly any three of this following types gears.
   (i) Spur gear   (iv) Worm gear
   (ii) Helical gears
   (iii) Bevel gear   (v) Rack and pinion.
6. Explain briefly the following forms of gear teeth :
   (i) Involute ;   (ii) Cycloidal
7. Define the following as applied to gears :
   Base circle ; Diametral pitch ; Module ; Addendum ; Dedendum ; Pressure angle ;
8. Enumerate the various methods of making gears.
9. Discuss briefly the 'sources of errors' in manufacturing gears.
10. Explain briefly the following gear manufacturing methods :
    (i) Forming or form cutting.
    (ii) Broaching.
    (iii) Template methods.
11. Discuss briefly any two of the following methods/processes for manufacturing gears :
    (i) Gear shaper process.
    (ii) Rack planing process.
    (iii) Hobbing process.
12. What are the advantages and limitation of the following gear manufacturing process ?
    (i) Hobbing process.
    (ii) Gear shaper process.
13. Give the comparison of 'Gear hobbing' and 'Gear shaping'.
14. How are Bevel gears cut ? Explain briefly
15. What is the purpose of gear finishing ?
16. Explain briefly the following methods of gear finishing.
    (i) Gear shaving   (iv) Gear lapping
    (ii) Gear burnishing   (v) Gear honing.
    (iii) Gear grinding

12

Unconventional Machining Processes

12.1. Introduction. 12.2. Classification of unconventional machining methods. 12.3. Selection of process. 12.4. Electrical Discharge Machining (EDM). 12.5. Electro-Chemical Machining (ECM). 12.6. Electro-Chemical Grinding (ECG). 12.7. Ultrasonic Machining (USM). 12.8. Electron Beam Machining (EBM). 12.9. Laser Beam Machining (LBM). 12.10. Plasma Arc Machining (PAM). 12.11. Abrasive Jet Machining (AJM). 12.12. Chemical Machining (CHM). 12.13. Comparison of Unconventional machining methods. **Question with Answers**—Highlights—Objective Type Questions—Theoretical Questions.

12.1. INTRODUCTION

In conventional machining processes the ability of the cutting tool is utilised to stress the material beyond the yield point to start the material removal process. This requires that the cutting tool material be harder than the workpiece material. The advent of harder materials for aerospace applications have made the removal process by conventional methods very difficult as well as time consuming since the material removal rate reduces with an increase in hardness of the work material. Hence *machining processes which utilise other methods such as electro-chemical processes are termed as* **Unconventional or Non-traditional machining methods.**

The main **reasons** *for using non-traditional processes are* :

(*i*) *High strength alloys.*

(*ii*) *Complex surfaces.*

(*iii*) *High accuracies and surface finish.*

12.2. CLASSIFICATION OF UNCONVENTIONAL MACHINING METHODS

The unconventional methods can be broadly *classified* on the basis of the following criteria :

1. **Type of energy :**

(*i*) Chemical

(*ii*) Electro-chemical

(*iii*) Mechanical

(*iv*) Electrothermal.

2. **Mechanism of metal removal :**

(*i*) Shear

(*ii*) Erosion

(*iii*) Chemical ablation

(*iv*) Ionic dissolution

(*v*) Spark erosion

(*vi*) Vaporisation etc.

3. **Media for energy transfer :**

(i) Physical contact

(ii) High velocity particles

(iii) Reactive atmosphere

(iv) Electrolyte

(v) Hot gases

(vi) Electrons

(vii) Radiation etc.

4. **Source of energy :**

(i) Hydraulic or pneumatic pressure

(ii) Mechanical pressure

(iii) Corrosive agent

(iv) High current

(v) High voltage

(vi) Ionized gas etc.

● **Common Unconventional Machining Methods :**

1. Electrical Discharge Machining (EDM)

2. Electro-Chemical Machining (ECM)

3. Electro-Chemical Grinding (ECG)

4. Ultrasonic Machining (USM)

5. Electron Beam Machining (EBM)

6. Laser Beam Machining (LBM)

7. Plasma Arc Machining (PAM)

8. Abrasive Jet Machining (AJM)

9. Chemical Machining (CHM).

12.3. SELECTION OF PROCESS

For selecting a particular process the following common parameters should be taken into consideration :

1. Shape and size required to be produced.

2. Physical properties of the work material.

3. Process economy.

4. Process capabilities such as expected tolerance, surface finish, rate of metal removal, power requirement etc.

5. Type of operation required (*e.g.*, cutting, hole making etc.)

Description of Unconventional Methods

12.4. ELECTRICAL DISCHARGE MACHINING (EDM)

Principle and Working : Refer to Fig. 12.1.

The **Electrical Discharge Machining (EDM)** *process involves controlled erosion of electrically conducting materials by the initiation of rapid and reptitive electrical discharge between the tool (cathode) and workpiece (anode) separated by a dielectric fluid medium.* A suitable gap between the tool and workpiece is maintained to cause the spark discharge. The gap can be varied to match the machining conditions such as metal removal rate.

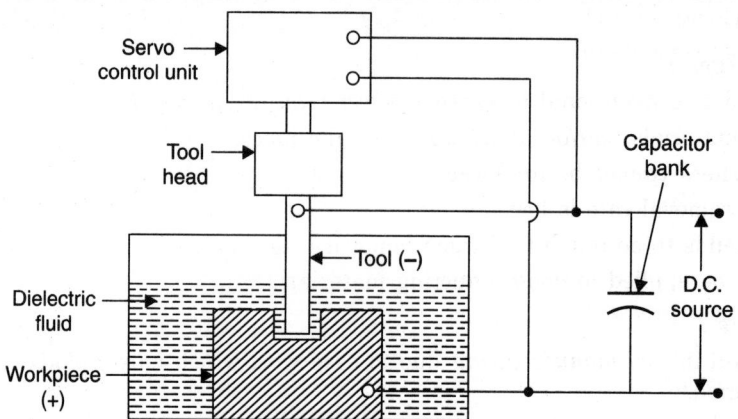

Fig. 12.1. Electric Discharge Machining (EDM).

As soon as the voltage gradient set up between the tool and the workpiece is sufficient enough to breakdown the dielectric medium, a conducting electrical path is developed for spark discharge owing to ionization of the fluid medium and thereby causes the current to flow. The temperature of the spot hit by the spark may rise upto 10000°C causing the work surface to melt and vaporize and ultimately to take the form of a sphere as it is quenched by the surrounding fluid.

If the tool is fed downwards, maintaining the predetermined gap, *the tool shape/profile will be reproduced on the workpiece.*

The spark gap, generally 0.01 to 0.1 mm, is adjusted so that the gap voltage is around 70 percent of the supply voltage for charging the capacitor bank. Higher gap although increases the discharge energy but it decreases the spark frequency due to increase in charging time of the capacitor.

The **'servocontrol unit'** is provided to *maintain the predetermined gap.* It senses the gap voltage and compares it with the preset value and the difference in voltage is then used to control the movement of servomotor to adjust the gap.

Important characteristics of EDM :
- *Tool materials* : Copper, brass and graphite.
- *Workpiece materials* : Conducting metals and alloys.
- *Process parameters* : Voltage, capacitance, spark gap and melting temperature of workpiece.
- *Material removal* : Melting and vaporisation.

Advantages :
1. Machining time is *less* than conventional machining processes.
2. *Any complicated shape* that can be made on the tool can be reproduced on the workpiece.
3. The process can be applied to *all electrically conducting metals and alloys* irrespective of their melting points, toughness, hardness or brittleness.
4. Can be employed for *extremely hardened workpiece.*
5. Fragile and slender workpieces can be machined without distortion.
6. *Considerably easier and more economical polishing* can be done on the cratering type surfaces developed by EDM.

7. *Fine holes* can be easily drilled.

8. Enables high accuracy on tools and dies, because they can be machined in '*as hard*' condition.

Disadvantages :

1. Compared to conventional processes, *power required is very high.*

2. In some materials, *surface cracking* may take place.

3. Sharp corners *cannot* be produced.

4. Material removal rate is low.

5. Surface tends to be rough for larger removal rates.

6. It cannot be applied to non-conducting materials.

Applications :

1. Very useful in *tool manufacturing* due to ease with which hard metals and alloyed can be machined.

2. Resharpening of cutting tools and broaches, trepanning of holes with straight or curved axes.

3. Machining of cavities for dies and remachining of die cavities without annealing.

- This process can be *used to preform almost all conventional machining operations.*

12.5. ELECTRO-CHEMICAL MACHINING (ECM)

Principle and Working : Refer to Figs. 12.2, 12.3.

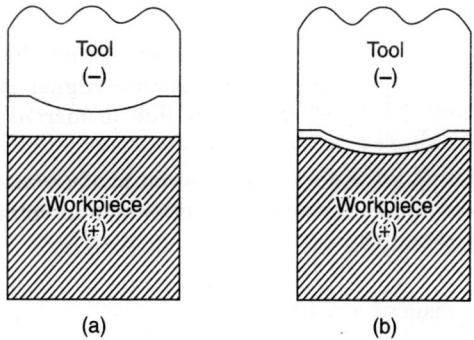

Fig. 12.2. The principle of ECM process (*a*) Shape of workpiece before machining ;
(*b*) Tool shape is reproduced on workpiece after ECM.

It is an inherently versatile process of machining because of its capability of stress free machining of various kinds of metals and alloys. *It can produce shapes and cavities which are costly and extremely difficult to machine with the conventional machining processes and a true shape of the tool (or cathode) can be made on the workpiece (or anode) by controlled dissolution of anode of an electrolytic cell.*

An electrolyte (usually a neutral salt solution such as sodium chloride, sodium nitrate, sodium chlorate) is passed through a very small gap (0.05 to 0.03 mm) created between the workpiece (or anode) and the tool (or cathode) whereas a direct current flow is made between them. When sufficient electrical energy (about 6 eV) is available, a metallic ion may be pulled out of the workpiece surface. The positive metallic ions will react with negative ions present in the electrolytic solution forming metallic hydroxides and other compounds. Hence the metal will be anodically dissolved with the formation of sludges and precipitates. *The metal removal rate is governed by Faraday classical laws of electrolysis.*

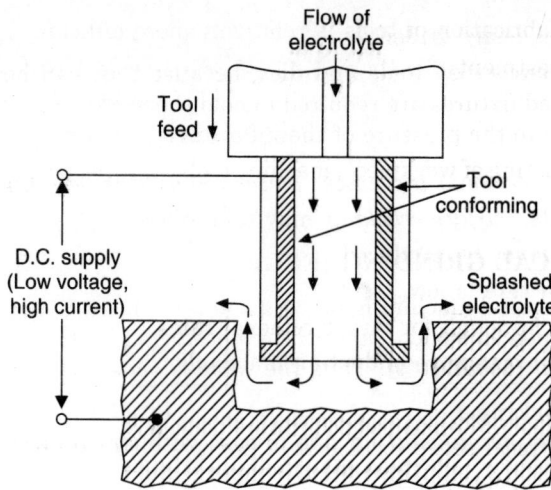

Fig. 12.3. Set up for Electro-Chemical Machining (ECM).

Characteristics of ECM :

- *Tool materials* : Copper and brass.
- *Workpiece materials* : Conducting metals and alloys.
- *Process parameters* : Current, voltage, feed rate and electrolyte.
- *Material removal* : Electrolysis.

Advantages :

1. The process is capable of machining metals and alloys irrespective of their strength and hardness.
2. Fragile parts, which are otherwise not easily machinable can be shaped by ECM.
3. Intricate and complex shapes can be machined easily through this process.
4. Metal removal rate is quite high in comparison to traditional machining, specially in respect of high tensile and high temperature resistant materials.
5. Wear on tool is insignificant or (say) almost non-existent.
6. With the application of this process, many machining operations, like grinding, milling, polishing etc. can be dispensed with.
7. No cutting forces are involved in the process.
8. The machined work surface is free of stresses.
9. High surface finish of the order of 0.1 to 0.2 microns can be obtained.
10. It is an accurate process and close tolerances of the order of 0.05 mm can be easily obtained.

Disadvantages :

1. Non-conductive materials cannot be machined.
2. The process cannot be used to machine sharp interior edges and corners less than 0.2 mm radius because of very high current densities at those points.
3. Very high power consumption.
4. Larger floor space is required.
5. A constant monitoring is required to suitably vary the tool feed rate and supply pressure of electrolyte so as to avoid formation of cavitation.

6. Designing and fabrication of tools is relatively more difficult.

7. High initial investment.

8. Specially designed fixtures are required to hold the workpiece in position, because it may be displaced due to the pressure of the inflowing electrolyte.

9. Corrosion and rusting of workpiece, machine tools, fixtures etc., by electrolyte is a constant menance.

12.6. ELECTRO-CHEMICAL GRINDING (ECG)

Based on electro-chemical machining process and refinement to it, the process of electrolytic grinding has been developed in which *metal is removed by electro-chemical deposition (about 90%) and abrasion of metal (about 10%).* Thus the wear is very less.

Principle and Working :

Fig. 12.4 shows a typical setup for Electro-Chemical Grinding (ECG).

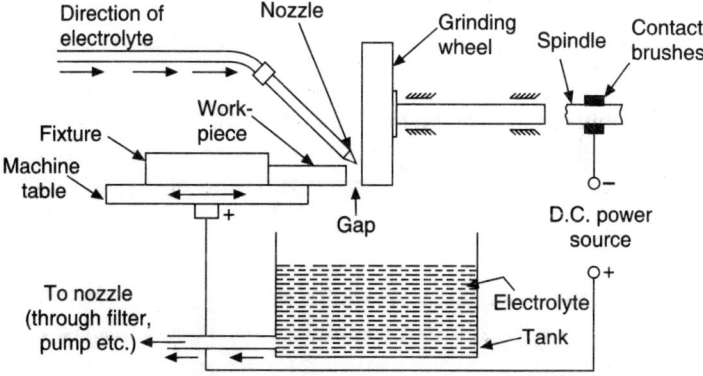

Fig. 12.4. Electro-Chemical Grinding (ECG).

The *grinding wheel* is mounted on a *spindle* which rotates inside suitable bearings. The *workpiece* is held on the *machine table* in suitable fixtures. The table can be moved forward and backward to feed the work or to withdraw it. *The grinding wheel and spindle are insulated from the rest of the machine* by using *an insulating sleeve,* as shown in the figure.

Electrolyte from the *tank* is pumped into gap between the wheel and the workpiece. Current flows from the *cathode (grinding wheel)* to the *anode (workpiece)* through the electrolyte. *This leads to an electrochemical oxidation on the work surface. The oxide film so formed is removed by the grinding wheel, producing a highly accurate and finished surface.*

Advantages :

1. Fairly good surface finish obtained (surface finish of 0.05 to 1 μm CLA is possible.

2. Accuracy of the order of 0.01 mm can be achieved by proper selection of wheel grit size and abrasive particles.

3. Negligible wear of the tool (grinding wheel).

4. Increased wheel life.

5. Considerable saving in wheel dressing time, as it is not required to be dressed very frequently.

6. As compared to conventional grinding, a very little cutting force is applied to the workpiece.

7. Since the heat is not generated in the process therefore work is free of surface cracks.

8. Work material is not subjected to any structural changes.

Disadvantages :

1. Metal removal rate is very low (of the order of 15 mm^3/s).
2. Power consumption is high.
3. Only electrically conductive materials can be machined.
4. Preventive measures are always required against corrosion by the electrolyte.
5. High initial cost.

Applications :

1. This process is *best suited for very precision grinding of hard metals like tungsten carbide tool tips* as the grinding pressure is very less and the temperature is very low due to which the defects like grinding cracks, tempering of work transformation of layers and dimensional control difficulties are eliminated.
2. Cutting thin sections of hard materials without danger of any damage or distortion.

12.7. ULTRASONIC MACHINING (USM)

Ultrasonic machining *is a process in which material is removed due to the action of abrasive grains.*

Principle and Working : Refer to Fig. 12.5.

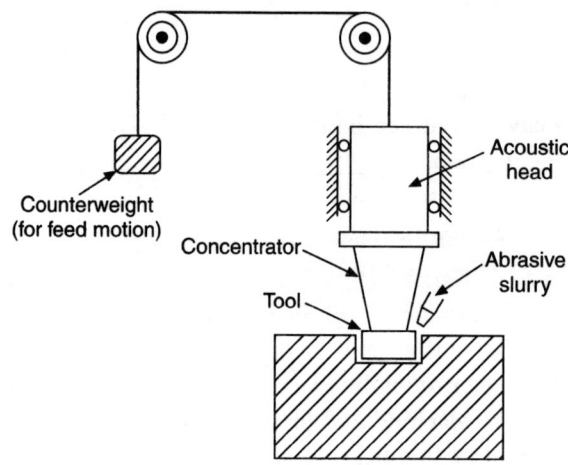

Fig. 12.5. Ultrasonic Machining (USM).

The abrasive particles are driven into the work surface by a tool oscillating normal to the work surface at high frequency. The tool is made of soft material, oscillated at frequencies of the order of 20 to 30 kHz with an amplitude of about 0.02 mm. It is pressed against the workpiece with a load of a few kg and fed downwards continuously as the cavity is cut in the work. The tool is shaped as the approximate mirror image of the configuration of the cavity desired in the work.

The *acoustic head* consists of a high frequency generator, a magnetostrictive transducer which converts mechanical motion into high vibration and transmits it to the tool. The counterweight with rope and pulley arrangement provides the feed mechanism and is designed to apply the working force during machining.

Characteristics of USM :

- *Tool materials :* Brass and mild steel.
- *Work materials :* Hard and brittle materials like semiconductors, glass and ceramics.
- *Process parameters :* Frequency, amplitude, grain size, slurry concentration and feed force.
- *Material removal :* Fracture of work material due to impact of grains.
- *Abrasive :* Aluminium oxide, silicon carbide and boron carbide.
- *Grain size :* Mesh-size 100-800.
- *Gap :* 0.2 to 0.5 mm.

Advantages :

1. Noiseless operation.
2. Low metal removal cost.
3. Extremely hard and brittle materials can be easily machined.
4. Operation of the equipment is quite safe.
5. Highly accurate profiles and good surface finish can be easily obtained.
6. Because of no heat generation in the process, the physical properties of the work materials remain unchanged.
7. The machined workpieces are free of stresses.

Disadvantages :

1. High tooling cost.
2. Low metal removal rate.
3. The size of the cavity that can be machined is limited.
4. High power consumption.
5. The initial equipment cost is higher than the conventional machine tools.
6. The process is unsuitable for heavy metal removal.
7. For maintaining an efficient cutting action the slurry may have to be replaced periodically.
8. It is difficult to machine softer materials.

Applications :

1. Several machining operations like turning, threading, grinding, milling etc.
2. Machining of hard to machine and brittle materials.
3. Dentistry work—to drill fine holes of desired shape in teeth.
4. Tool and die making, specially wire drawing and extrusion dies.

12.8. ELECTRON BEAM MACHINING (EBM)

Electron beam machining *is a process of machining materials with the use of a high velocity beam of electrons.* This process is best suited for *microcutting of material* (in mg/s) because the evaporated area is function of the beam power and the method of focusing which can be easily controlled.

Principle and Working : Refer to Fig. 12.6.

In this process the material is removed with the help of a high velocity (travelling at half the speed of light. *i.e.,* 160,000 km/s) focused stream of electrons which are focused magnetically upon a very small area. These electrons heat and raise the temperature locally above the boiling point and thus melt and vaporise the work material at the point of bombardment.

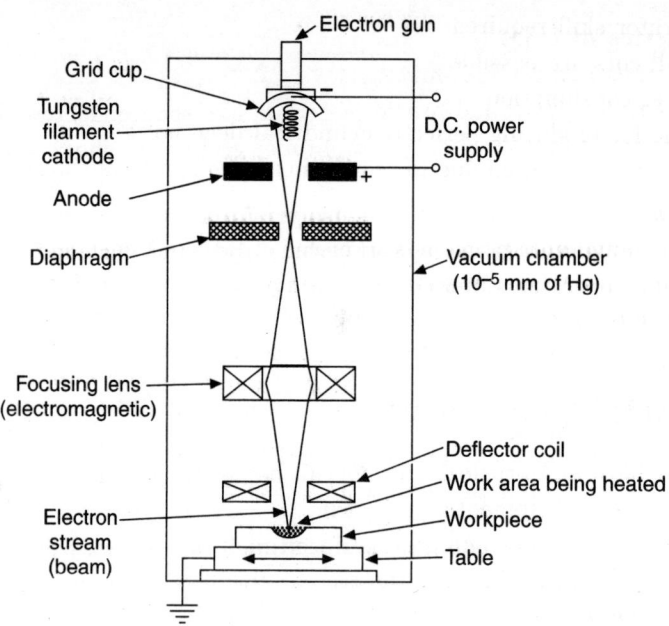

Fig. 12.6. Set up for Electron Beam Machining (EBM).

The electrons are obtained in free state by heating the cathode metal in *vacuum* to the temperature at which they attain sufficient speed for escaping to the space around the cathode. These can then be made to move under the effect of electric or magnetic field and can be accelerated greatly. The acceleration is carried out by electric field and focusing and concentration is done by controllable magnetic fields.

Characteristics of EBM :

- *Workpiece materials* : All materials.
- *Material removal :* High speed electrons impinge on surface and K.E. of electrons produces intense heating to melt or vaporise the metal.
- *Voltage :* 150 kV.
- *Power density :* 6500 billion W/mm^2
- *Medium :* Vacuum (10^{-5} mm of Hg)
- *Specific power consumption :* 500 W/mm^3 min.

Advantages :

1. It is excellent strategy for micro-machining. It can drill holes or cut slots which cannot be otherwise made.
2. It can cut any known material, metal or non-metal that would exist in vacuum.
3. No physical or metallurgical damage.
4. There is no contact between the work and tool.
5. Heat can be concentrated on a particular spot.
6. Close dimensional tolerances can be achieved because problem of tool wear is non-existent.

Disadvantages :

1. Low metal removal rate.
2. High equipment cost.

3. High operator skill required.

4. Only small cuts are possible.

5. High power consumption.

6. Unsuitable for producing perfectly cylindrical deep holes.

7. Workpiece size is limited due to requirement of vacuum in the chamber.

Applications :

1. Micro-machining operations on workpieces of thin sections.

2. Micro-drilling operations (upto 0.002 mm) for thin orifices, dies for wire drawing parts of electron microcopes, fibre spinners, injector holes for diesel engines etc.

3. Very effective for machining of metals of low heat conductivity and high melting point.

12.9. LASER BEAM MACHINING (LBM)

The word LASER stands for "Light Amplification using Stimulated Emission of Radiation". Laser provides intense and unidirectional beam of light ; this light is coherent in nature. The machining by which a laser beam removes material from the surface being worked *involves a combination of the melting and evaporation processes.* However with some materials the mechanism is purely one of evaporation.

Principle and Working :

The principle utilised in LBM is that under proper conditions light energy of a particular frequency is utilised to stimulate the electrons in an atom to emit additional light with exactly the same characteristics of the original light source.

Fig. 12.7. shows the set-up for Laser Beam Machining. The laser beam is focused with the help of the lens and the workpiece is placed near the focal point of the lens. A short pulse of laser melts and vaporises the material. The explosive escape of the vaporised metal helps in removing most of the molten metal from the hole as tiny droplets. Any of the molten metal not removed will be resolidified along the walls of the hole.

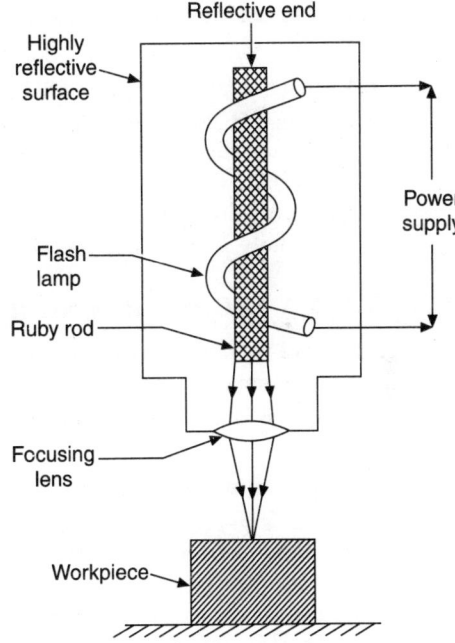

Fig. 12.7. Laser Beam Machining (LBM).

Characteristics of LBM :

- *Tool* : High powered focused laser beam.
- *Workpiece materials :* Any material.
- *Process parameters :* Power intensity of laser beam, focused diameter of laser beam and melting temperature of workpiece material.
- *Material removal :* Melting and vaporisation.
- *Medium :* Air.

Advantages :

1. No mechanical contact between the tool and the work.

2. The beam can be projected through a transparent window.

3. The laser can be used with materials sensitive to heat shock such as ceramics.

4. The workpiece is not subjected to large mechanical forces.

5. The laser operates in any transparent environment including air, inert gas, vacuum and even certain liquids.

6. Can be effectively used for welding of dissimilar metals as well.

7. Very small holes and cuts can be made with fairly high degree of accuracy.

8. Any material can be easily machined irrespective of its structure and physical and mechanical properties.

Disadvantages :

1. The laser system is quite inefficient.

2. Low production rate.

3. High capital investment required.

4. Its application is limited to only thin sections and where a very small amount of metal removal is involved.

5. Cannot be effectively used to machine highly heat conductive and reflective materials.

6. Highly skilled operators are needed.

Applications :

1. Trimming of sheet metal plastic parts and carbon resistors.

2. Cutting or engraving patterns on thin films.

3. Drilling small holes (upto 0.005 mm dia.) in hard materials like tungsten and ceramics.

4. Cutting complex profiles on thin and hard materials, viz., films for making ICs.

- Laser drilled holes exhibit a taper and also lack a high degree of roundness. Holes larger than 1.25 mm cannot be drilled because the power density decreases. Hence *laser cutting is more often used than laser drilling*. In a laser cutting operation a high velocity gas jet is used in conjunction with the laser beam. The gas jet helps to rapidly remove the metal from the hole.

12.10. PLASMA ARC MACHINING (PAM)

A **plasma** is *a high temperature ionized gas.* The plasma arc machining is done with a high speed jet of a high temperature plasma.

Principle and Working :

A plasma is generated by subjecting a flowing gas to the electron bombardment of an arc. For this, the arc is set up between the electrode and the anodic nozzle; the gas is forced to flow through this arc (Fig. 12.8). The high velocity electrons of the arc collide with the gas molecules, causing a dissociation of the diatomic molecules or atoms into ions and electrons resulting in a substantial increase in the conductivity of the gas which is now in plasma state. The free electrons subsequently accelerate and cause more ionization and heating. Afterwards a further increase in temperature takes place when the ions and free electrons recombine into atoms or when the atoms

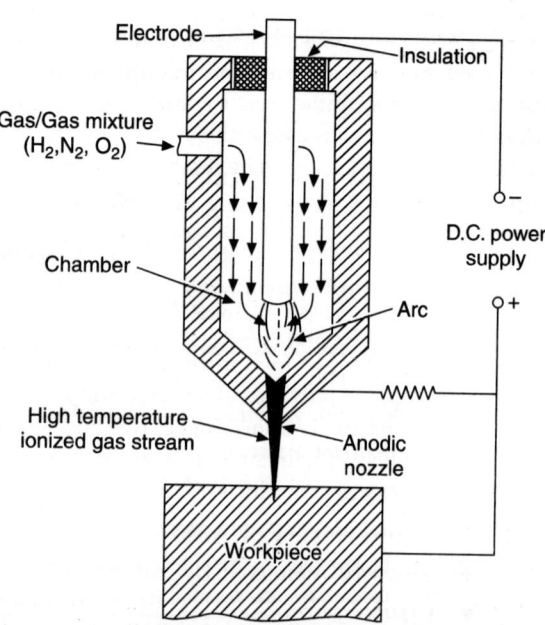

Fig. 12.8. Plasma Arc Machining (PAM).

recombine into molecules as these are exothermic processes. So a *high temperature plasma is generated which is forced through the nozzle in the form of a jet.* This jet (ionized steam of gas) is impinged on the workpiece which gets melted and eroded.

Characteristics of PAM :

- *Tool* : Plasma jet (maximum velocity = 500 m/s).
- *Workpiece materials* : All conducting materials.
- *Material removal* : Melting (maximum material removal rate = 150 cm^3/min).
- *Medium* : Plasma.
- *Maximum temperature* : 16000°C.
- *Power range* : 2 to 200 kW.
- *Critical parameter* : Voltage (30 to 250 V) ; current (upto 600 A) ; electrode gap ; nozzle dimensions ; melting temperature.

Advantages :

1. Excessively high temperatures are generated for use.
2. Can be used to cut any metal.
3. A faster process.

Disadvantages :

1. Initial cost of equipment is quite high.
2. Adequate safety precautions are always needed for the operator.
3. The work surface may undergo some metallurgical changes.

Applications :

1. Cutting of stainless steel and non-ferrous metals (such as aluminium alloys).
2. Turning and milling of 'hard to machine' materials.

12.11. ABRASIVE JET MACHINING (AJM)

Principle and Working : Refer to Fig. 12.9.

This process consists of directing a stream of fine abrasive grains, mixed with compressed air or some other gas at high pressure through a nozzle on to the surface of the workpiece to be machined. These particles impinge on the work surface at high speed and the erosion caused by their impact enables the removal of metal. *The metal removal rate depends upon the flow rate and size of abrasive particles.*

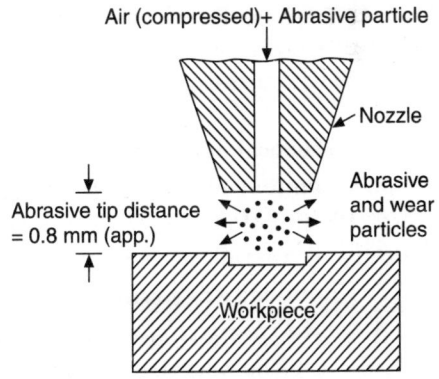

Fig. 12.9. Abrasive Jet Machining (AJM).

— The abrasives may be Al_2O_3, SiC, sodium bicarbonate, dolomite, glass beads etc. The abrasive particles should have irregular shape consisting of short edges. Round particles are useless. Their size is 10 to 50 microns. The carrier is usually air, CO_2 or N_2.

Characteristics of AJM :

- *Material removal :* By impinging abrasive grains at high speed.
- *Critical parameters :* Abrasive flow rate and velocity, nozzle tip distance, abrasive grain size.

Advantages :

1. Low capital investment required.
2. Brittle materials of thin sections can be easily machined.
3. Intricate cavities and holes of any shape can be machined in materials of any hardness.
4. There is no direct contact between the tool and workpiece.
5. Normally inaccessible portions can be machined with fairly good accuracy.

Disadvantages :

1. Low metal removal rate.
2. Unsuitable for machining of ductile materials.
3. The abrasive powder used in the process cannot be reclaimed or reused.
4. Machining accuracy is relatively poorer.
5. There is always a danger of abrasive particles getting embedded in the work material. Hence cleaning needs to be necessarily done after the operation.

Applications :

1. Machining of intricate profiles on hard and fragile materials.
2. Fine drilling and micro-welding.
3. Frosting and abrading of glass articles.
4. Aperture drilling for electronic microscopes.
5. Machining of semiconductors.

Note : This process should not be misunderstood as *sand blasting* because this process is *basically meant for metal removal with the use of small abrasive particles,* whereas the sand blasting process is a surface cleaning process which *does not involve any metal cutting.*

12.12. CHEMICAL MACHINING (CHM)

Chemical machining *is a process used to dissolve the workpiece material in chemical solutions.* Metal can be removed from selected portions or from the entire surface of the workpiece, according to requirement. If selective machining is desired, the portions required to be left unmachined are covered with a resistant material called a '*resist*' or '*maskant*' which can be stripped away after machining.

The chemical machining process can be *classified* as follows :

1. Chemical blanking. It is used for cutting out parts from thin sheet metal.

2. Chemical contour machining. It is also known as '*Chemical milling*' and is employed for selective or overall metal removal from thicker workpieces.

3. Chemical engraving.

A typical chemical operation entails the following steps :

(i) Clean the workpiece thoroughly so as to ensure that the masking material will adhere to the workpiece well to reduce any possibility of stray etching due to *maskant* debonding.

(ii) Apply a chemical resistant mask on the workpiece surface where no material is to be removed.

(iii) Dip the workpiece into the chemical solution called *etchant* and leave it for sufficient time to get the necessary depth of etching. The strength of the etching is maintained since it becomes weak by absorbing the workpiece material with time.

(iv) Remove the mask and clean the workpiece.

Advantages :

1. Very low tool cost.
2. Low design change cost.
3. Complex contours can be easily machined.
4. Tooling time is substantially reduced.
5. Both faces of the workpiece can be machined simultaneously.
6. The part produced is free of burrs.
7. Hard and brittle materials can be machined.
8. It is a flexible process from design point of view.

Disadvantages :

1. It is slow process, since the metal removal rate is low.
2. Larger floor space is required.
3. Skilled operators are needed.
4. Sharp corners cannot be produced.
5. High manufacturing cost.
6. Metal thickness that can be machined is limited.

Applications :

Chemical machining is generally used when *very small amounts of material are to be removed from the surface in any application.* In *aerospace application* a large volume of unwanted material is removed from the surface to reduce the weight, thereby increasing the strength to weight ratio which is conveniently done with chemical machining.

12.13. COMPARISON OF UNCONVENTIONAL MACHINING METHODS

The comparison of Unconventional machining methods is given (in tabular form) below :

Table 12.1. Comparison of Unconventional Machining Methods

S. No.	Process	Material removal rate (mm³/s)	Dimensional accuracy (µm)	Surface finish (µm)	Capital cost	Power consumption (kWh)
1.	Electric Discharge Machining (EDM)	10–20	15–50	0.2–2.5	Medium	2–4
2.	Electro-Chemical Machining (ECM)	200–300	15–100	0.1–2.5	Very high	100–150
3.	Ultrasonic Machining (USM)	5–10	7–15	0.2–2.5	Low	2–3
4.	Laser Beam Machining (LBM)	0.001–0.002	10–100	0.5–1.5	Low/medium	0.003–0.005
5.	Chemical Machining (CHM)	0.15–30	25–100	0.5–2.5	Medium	—

QUESTIONS WITH ANSWERS

Q. 12.1. Explain briefly 'Water Jet Machining'.

Ans. In 'Water Jet Machining' process a high velocity *water jet* is made to impinge on to the workpiece. This jet pierces the work material and performs a sort of slitting operation. Water under pressure from a hydraulic accumulator is passed through the orifice of a nozzle to increase its velocity. The nozzle orifice size (dia.) usually varies from 0.08 mm to 0.5 mm and the exit velocity of the water jet from the nozzle varies upto 920 m/s. These high velocity jets can be used to cut relatively softer and non-metallic materials like paper boards, wood, plastics asbestos, rubber, fireglass, leather etc.

- A variation of this process known as *"Hydrodynamic Jet Machining (HJM)"* has been successfully used to machine almost all types of ferrous and non-ferrous metals and alloys.

Q. 12.2. For what types of works the Electro-Chemical Grinding (ECG) is best suited ?

Ans. ECG is best suited for grinding *multiple-tooth milling cutters, carbide tipped tools and tool bits.*

- As it puts very little pressure on work it is best suited for grinding fragile and not easily supported works, like grinding slots, flats and forms in thin-walled fragile workpieces.

Q. 12.3. On what types of works the process of chemical machining is best suited and what are its advantages and limitations ?

Ans. ● Chemical machining is an excellent method of getting complex shapes on very thin and most difficult to machine tools.

- It does not require any press or punch or die. It does not distort the workpiece and no burrs are produced.
- It is a *slow process* and thus limited to machining metals upto 3 mm thickness and not used for producing large quantities.

Q. 12.4. What is 'hot machining' ?

Ans. 'Hot Machining' *is used for machining high strength, high hardness and high temperature resistant materials which are difficult to machine at room temperature.*

Machining of hard metals at elevated temperature is applied mainly to *turning and milling operations.* Since the shear strength of metal decreases at elevated temperature as compared to that at room temperature the magnitude of the cutting forces on the tool is lower. Further as the chip formation by plastic deformation in the shear plane ahead of tool becomes easier at elevated temperature and cutting forces involved are less, therefore *power requirements are low.* But as the property of tool material at elevated temperature is also changed due to its being in contact with high temperature material, therefore tool life is also affected. It is found that tool life is maximum for certain temperature of workpiece (for particular work material and tool material) at which total metal removal rate per tool grinding will be maximum irrespective of the speed.

Q. 12.5. For what purpose Electric Discharge Machines (EDM) are used ?

Ans. EDM machines are used for avoiding difficulty to machine materials through the use of electric spark.

Example : Die sinking or removal of broken tools embedded in workpieces.

Q. 12.6. Indicate the sources of energy in the following processes :
(a) EDM, (b) USM, (c) LBM, (d) ECM.

Ans.	Process	Sources of energy
(a)	EDM	Electric spark
(b)	USM	Mechanical
(c)	LBM	Radiation
(d)	ECM	Electrical

Q. 12.7. What is 'Magnetic Forming' ?

Ans. 'Magnetic Forming' is also known as 'magnetic pulse forming' or 'electromagnetic forming'.

In this process an insulated induction coil is either wrapped around or placed within the work depending upon whether the metal is to be squeezed inward or buldged outward. The high momentary current is passed through the coil which *develops intense magnetic field* causing the work to collapse, compress, shrink or expand depending on the design and placement of the coil.

The work size, that can be forced, is determined by the energy storage capacity and ability of the unit to utilize that energy. High conductivity materials can be formed if they are wrapped or coated with a high conductivity auxiliary material.

HIGHLIGHTS

1. The main reasons for using non-traditional processes are :
    (i) High strength alloys.
    (ii) Complex surfaces.
    (iii) High accuracies and surfaces finish.
2. Common " *Unconventional Machining Methods*" are :
    (i) Electrical Discharge Machining (EDM)
    (ii) Electro-Chemical Machining (ECM)
    (iii) Electro-Chemical Grinding (ECG)
    (iv) Ultrasonic Machining (USM)
    (v) Electron Beam Machining (EBM)
    (vi) Laser Beam Machining (LBM)
    (vii) Plasma Arc Machining (PAM)
    (viii) Abrasive Jet Machining (AJM)
    (ix) Chemical Machining (CHM).

OBJECTIVE TYPE QUESTIONS

A. Choose the Correct Answer :

1. Which of the following is an advantage of 'Laser beam machining' ?
    (a) Laser beam can be sent to longer distances without diffraction.
    (b) There is no contact between tool and workpiece.
    (c) Heat treated and magnetic particles can be welded without losing their properties.
    (d) None of the above.
2. For machining materials of high hardness material selected for tool in EDM should be
    (a) hard   (b) soft
    (c) any material with good electrical conductivity   (d) none of the above.

3. Surface finish produced by electrochemical grinding on 'Tungsten carbide' can be expected to be the order of .... micron.
    (a) 0.1 to 0.2   (b) 0.2 to 0.4
    (c) 0.4 to 0.8   (d) 0.8 to 0.9.

4. .... governs metal removal rate in electrochemical machining.
    (a) Fleming's rule   (b) Newton's law
    (c) Faraday's law   (d) None of the above.

5. Ultrasonic machining is based on
    (a) Uniform heating   (b) Uniform grinding
    (c) Vibratory waves of high frequency   (d) Uniform machining
    (e) None of the above.

6. In ultrasonic machining the rate of penetration is dependent on
    (a) Action of slurry   (b) Action of abrasive grains
    (c) Reduction of a chemical   (d) All of these.

7. In ultrasonic machining, the rate of penetration is dependent on
    (a) Flow path   (b) Slurry
    (c) Area of tool tip   (d) All of these.

8. In ultrasonic machining, longitudinal waves are preferred because they
    (a) Can travel at a high velocity   (b) Are easily generated
    (c) Can be propagated in solid, liquid and gases   (d) All of the above.

9. Tool tip is attached to the tool cone by
    (a) Welding   (b) Press fitting
    (c) Silver soldering.   (d) Nut and bolt
    (e) none of the above.

10. .... process is used for making a complicated contour in a carbide piece.
    (a) Laser machining   (b) Electro-chemical milling
    (c) Plasma-arc machining   (d) Electro-discharge machining
    (e) None of the above.

11. Slurry used in USM is
    (a) Alkaline only   (b) Alcohol based
    (c) Mercury based   (d) Water based
    (e) None of the above.

12. Erosion of metal in EDM is
    (a) Proportionate to the number of sparks   (b) Continuous
    (c) Either of the above   (d) None of the above.

13. AJM is used for
    (a) Plastics only   (b) Ductile materials only
    (c) Brittle materials only   (d) Any of the above.

14. In USM, slurry is fed by
    (a) A high power pump   (b) Manual system
    (c) Any of the above   (d) None of these.

15. Selection of proper tool material in EDM is influenced by which of the following parameters ?
    (a) Tolerance required   (b) Volume of material to be removed
    (c) Size of the electrode   (d) Surface finish required
    (e) None of the above.

16. Time required for machining by EDM in comparison to the conventional machining is
    (a) Less   (b) Equal
    (c) More   (d) Unpredictable.

17. In EDM, metal removal rate is proportional to
    (a) Frequency of charging
    (b) Energy delivered in each spark
    (c) Both (a) and (b)
    (d) None of these.

18. Abrasive jet machining uses a jet of
    (a) Abrasive particles suspended in oil
    (b) Fine grained abrasive particles mixed with air or some other carrier gas at high pressure
    (c) Abrasive particles suspended in water
    (d) None of the above.

19. Abrasive jet machining is used for
    (a) Cutting thin sectioned fragile components made of glass, refractories, ceramics, mica etc.
    (b) Removing flash and parting lines from injection moulded parts
    (c) Deburring and polishing plastic, nylon and teflon components
    (d) All of the above.

20. In ultrasonic machining, tool used
    (a) Oscillates at a frequency of 20 to 30 kHz
    (b) Has the shape exactly to that of a hole to be made
    (c) Is made of soft material
    (d) All of the above.

21. .... machining process needs high velocity stream of electrons for its operation.
    (a) Abrasive jet
    (b) Ultrasonic
    (c) Electro-discharge
    (d) Electron-beam.

22. In abrasive jet machining (AJM), metal removal takes place due to
    (a) Machining
    (b) Grinding
    (c) Metal erosion
    (d) All of the above
    (e) None of the above.

23. What is the principle of 'Water jet machining' (WJM) ?
    (a) Air and water mix jet is used.
    (b) Surface is dipped in the water.
    (c) A jet of water is directed on the surface at a high velocity.
    (d) None of the above.

24. In electro-discharge machining (EDM), metal removal takes place as
    (a) Chemical reaction of metal
    (b) Dissolution of metal
    (c) Erosion of metal
    (d) None of these.

25. In electrolytic grinding, metal removal takes place by
    (a) Erosion
    (b) Corrosion
    (c) Electro-chemical action
    (d) All of these.

26. In EDM, the required property of tool is
    (a) Resistivity
    (b) Dielectric strength
    (c) Conductivity
    (d) None of these.

27. LASER welding finds wide applications in
    (a) Electronic industry
    (b) Heavy industry
    (c) Structural work
    (d) None of these.

28. LASER is produced by
    (a) Aluminium
    (b) Ruby
    (c) Diamond
    (d) Graphite.

29. For converting electrical energy into mechanical energy, which of the effects form the basis of USM ?
    (a) Chemical action
    (b) Photosynthesis
    (c) Piezoelectric effect
    (d) Any of these.

ANSWERS

1. (*d*)	2. (*c*)	3. (*b*)	4. (*c*)	5. (*c*)
6. (*b*)	7. (*d*)	8. (*d*)	9. (*c*)	10. (*d*)
11. (*d*)	12. (*a*)	13. (*c*)	14. (*c*)	15. (*e*)
16. (*a*)	17. (*c*)	18. (*b*)	19. (*d*)	20. (*d*)
21. (*d*)	22. (*c*)	23. (*c*)	24. (*c*)	25. (*c*)
26. (*c*)	27. (*a*)	28. (*b*)	29. (*c*).	

B. Fill in the blanks or say 'Yes' or 'No' :

1. The .... process involves controlled erosion of electrically conducting materials by the initiation of rapid and repetitive electrical discharges between the tool and workpiece separated by a dielectric fluid medium.
2. In EDM the .... unit is provided to maintain the predetermined gap.
3. In EDM the machining time is .... than conventional machining process.
4. Fine holes cannot be drilled by EDM.
5. EDM can be employed for extremely hardened workpiece.
6. In EDM the material removing is by melting and vaporisation.
7. In .... the metal removal rate is governed by Faraday classical laws of electrolysis.
8. Tool materials in ECM are copper and brass.
9. In ECM process metal removal rate is quite low.
10. Non-conductive materials can be machined by ECM process.
11. In case of ECM the power consumption is very high.
12. In case of ECM smaller floor space is required.
13. In ECM designing is fabrication of tools is relatively more difficult.
14. ECG process is best suited for very precision grinding of hard metals like tungsten carbide tool tips.
15. In case of ECG the wear of tool is ....
16. In ECG process power consumption is ....
17. In ECG metal removal rate is very low.
18. .... machining is a process in which material is removed due to the action of abrasive grains.
19. In case of USM the tool materials are brass and mild steel.
20. USM is a noisy operation.
21. In case of USM the metal removal cost is ....
22. In case of USM tooling cost is high.
23. .... is a process of machining materials with the use of a high velocity beam of electrons.
24. In EBM the medium of working is ....
25. In EBM close dimensional tolerances can be achieved.
26. In EBM only small cuts are possible.
27. In EBM process the metal removal rate is high.
28. In LBM the material removal is by melting and vaporisation.
29. Any material can be easily machined by LBM process.
30. A .... is a high temperature ionized gas.
31. In PAM the material removal is by ....
32. PAM is a faster process.
33. In AJM the material removal is by impinging abrasive grains at high speed.
34. In AJM machining accuracy is relatively ....
35. In AJM is unsuitable for machining of ductile materials.
36. .... machining is a process used to dissolve the workpiece material in chemical solutions.
37. Chemical .... is used for cutting out parts from thin sheet metal.

38. Chemical .... is employed for selective or overall metal removal from thicker workpieces.

39. Chemical machining has a very low tooling cost.

40. Chemical machining is a .... process.

ANSWERS

1. EDM	**2.** servo-control	**3.** less	**4.** No	**5.** Yes
6. Yes	**7.** ECM	**8.** Yes	**9.** No	**10.** No
11. Yes	**12.** No	**13.** Yes	**14.** Yes	**15.** negligible
16. high	**17.** Yes	**18.** Ultrasonic	**19.** Yes	**20.** No
21. low	**22.** Yes	**23.** EBM	**24.** vacuum	**25.** Yes
26. Yes	**27.** No	**28.** Yes	**29.** Yes	**30.** plasma
31. melting	**32.** Yes	**33.** Yes	**34.** poorer	**35.** Yes
36. Chemical	**37.** blanking	**38.** milling	**39.** Yes	**40.** slow.

THEORETICAL QUESTIONS

1. What do you understand by he term "Unconventional or Non-traditional machining methods ? What is their importance ?

2. Explain why unconventional machining processes are used ?

3. How are 'Unconventional machining methods' classified ?

4. What are the main parameters to be considered while selecting a particular process and why ?

5. What is "Electrical Discharge Machining (EDM)" ? Explain its principle with the help of a suitable diagram.

6. State the advantages, disadvantages and applications of EDM.

7. List the important characteristics of EDM.

8. Explain with a neat sketch the principle and working of Electro-chemical Machining (ECM) process.

9. List the advantages, disadvantages and applications of ECM.

10. Give the important characteristics of ECM.

11. Describe briefly 'Electro-Chemical Grinding' process. State also its advantages, disadvantages and applications.

12. Explain briefly with a neat sketch the principle and working of Ultrasonic machining (USM) process/ method. List also its advantages, disadvantages and applications.

13. Give the important characteristics of USM.

14. What is the working principle of Electron Beam Machining (EBM) ? What are its advantages, disadvantages and applications ?

15. List the characteristics of EBM.

16. Describe briefly with a neat diagram the working principle of Laser Beam Machining (LBM) ? Give also its advantages, disadvantages and applications.

17. What is a 'plasma' ?

18. Explain briefly with a neat sketch the working principle of Plasma Arc Machining (PAM). State also its advantages, disadvantages and applications.

19. List the characteristics of PAM.

20. Explain clearly, with a neat diagram, Abrasive Jet Machining (AJM) method. State also its advantages, disadvantages and applications.

21. What is Chemical Machining ? State its advantages and disadvantages.

13

Machine Tools Testing

13.1. Introduction. 13.2. Machine tool tests. 13.3. Instruments required for alignment test. 13.4. Alignment tests on lathe. 13.5. Alignment test on drilling machine. 13.6. Alignment tests on milling machines. 13.7. Acceptance tests for surface grinders. 13.8. Practical tests. Objective Type Questions—Theoretical Questions.

13.1. INTRODUCTION

The accuracy of machine tools, which cut metal by removing chips or swarfs, is tested by means of the following *tests* :

1. Geometric tests
2. Practical tests.

1. Geometric tests :

These tests are *check on relationship of the various elements of the machine when it is idle and unloaded and are performed by using ordinary measuring equipment together with a few special accessories.* The various geometric checks generally made on the machine tools are :

(*i*) *Straightness*
- Straightness of a line in two planes ;
- Straightness of slideways of machine tools ;
- Straightline motion.

(*ii*) *Flatness*

(*iii*) *Parallelism, equidistance and coincidence.* The parallelism includes :
- Parallelism of lines and planes ;
- Parallel motion.

(*iv*) *Rectilinear movements or squareness of straight lines and planes. Quality of the guiding and bearing surfaces of beds, uprights and base plates are also tested.*

(*v*) *Rotations.* This includes :
- Out of round.
- Ecentricity.
- Radial throw of an axis at a given point.
- Out-of-true running (run-out).
- Camming, and
- Periodical axial slip.

Main spindle is the fundamental element of the machine and is tested for concentricity, axial slip, accuracy of axis and position relative to the other axes and surfaces.

(*vi*) *Movement of all the working components.* These are commonly referred to as 'Alignment tests'.

2. Practical tests :

These tests are intended to *check the accuracy of the work done on the machine and are performed by machining suitable pieces and measuring these.*

13.2. MACHINE TOOLS TESTS

The tests applied to machine tools, regardless of type, fall into well defined group which may be summarised as given below :

1. Tests for the level of installation of machine in horizontal and vertical planes.
2. Tests for flatness of machine bed and for straightness and parallelism of bed ways or bed surfaces.
3. Tests for perpendicularity of guideways to other guideways of bearing surfaces.
4. Tests for true running of the main spindle and its axial movements.
5. Tests for parallelism of spindle axis to guideways or bearing surfaces.
6. Tests for the line movements of various members *e.g.,* saddle and table cross-slides etc., along their ways.
7. Practical tests relate to some test pieces which are machined and their accuracy and finish are checked.

13.3. INSTRUMENTS REQUIRED FOR ALIGNMENT TEST

Generally, the following instruments may be required to conduct alignment tests :

1. Dial gauges   2. Test mandrels
3. Straightness and squares   4. Spirit levels
5. Autocollimator   6. Waviness meter.

- **Dial gauges** are widely used in alignment tests. The graduation on the dial need not be finer than 0.01 mm. The initial measuring pressure should be from 4 to 100 gm for very fine measurements ; a lower pressure as small as 20 gm is desirable.
- **Test mandrels** are used for checking the true running of the spindle. Test mandrels are hardened and ground and made to a length which varies from 100 to 300 mm.
- **Straight edges** of cast-iron or steel should be heavy, well rolled and seasoned. A square must have a wider bearing surface. The standard square should have a tolerance of ± 0.01 mm and the precision square ± 0.005 mm.
- **Spirit levels** are used for *high grade precision work.* These should have sensitivity of about 0.04 to 0.06 mm per metre for each deflected division.
- **Autocollimator** in conjuction with block, deflector and optical square is *very sensitive instrument for checking deflection of long beds in horizontal, vertical or inclined planes.*
- **Waviness meter** with 50 : 1 magnification is useful in recording and examining the surface waviness.

13.4. ALIGNMENT TESTS ON LATHE

The following alignment tests are conducted on the lathe :

1. Levelling of the machine.
2. Parallelism of spindle axis and bed.
3. True running of headstock centre.
4. True running of taper socket in main spindle.
5. Alignment of both the centres in vertical plane.

6. Cross-slide perpendicular to spindle axis.

7. Accuracy of pitch of lead screw.

8. Axial slip of lead screw.

13.4.1. Levelling of the Machine

● The set up for the testing the bed of a centre lathe is shown in Fig. 13.1. The saddle is kept approximately in the centre of the bed support feet. The spirit level is then placed at *X-X* to give the level in *longitudinal direction*. It is then traversed along the length of the bed and readings at various places noted down.

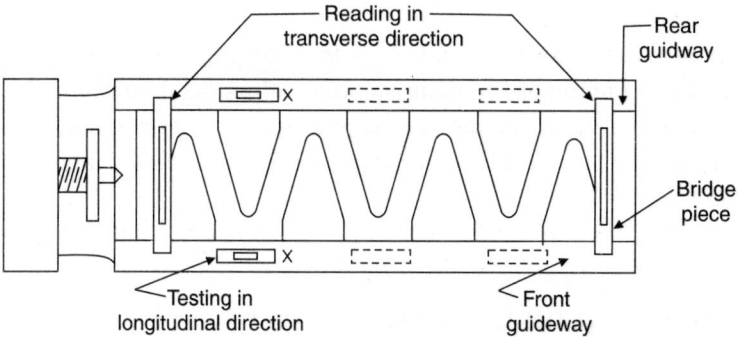

Fig. 13.1. Set-up for testing the bed of a centre lathe.

● For test in *transverse direction* the level is placed on a bridge piece to span the front and rear guideways and then reading is noted. It is preferable to take two readings in longitudinal and traverse directions simultaneously so that the effect of adjustments in one direction may also be observed in the other. The *readings in the transverse direction reveal any twist or wind in the bed*. It may be noted that two guideways may be perfectly levelled in longitudinal direction, but might not be parallel to each other. This is revealed by the transverse direction.

● The error in level may be corrected by setting wedges or shims at suitable points under the support feet or pads of the machine. The permissible errors in level are :

Lathe .... (0.02, ± 0.02) mm

Radial drilling machine .... (0.1, 0.1) mm

Milling machine .... (± 0.04, 0.04) mm.

Note : The straightness *of bed in longitudinal direction for the long beds can also be determined by other methods, e.g., using straight edges, autocollimator or by taut wire method but the test in transverse direction can be carried out only by spirit level.*

13.4.2. Parallelism of Spindle Axis and Bed.

● The equipment needed for testing parallelism of spindle axis and bed *consist of dial gauge and stand*, together with a *special mandrel*. The dial gauge should be graduated in 0.01 mm divisions and have a light measuring pressure.

● The mandrel has a parallel portion about 25 mm in diameter and 300 mm long, with a taper end to fit the headstock spindle.

● The mandrel is accurately ground on centres so that the parallel and taper parts are concentric.

● The mandrel is fitted into the spindle and the dial gauge attached to the carriage of the lathe, with its plunger touching surface of the mandrel near the ends as shown in Fig. 13.2.

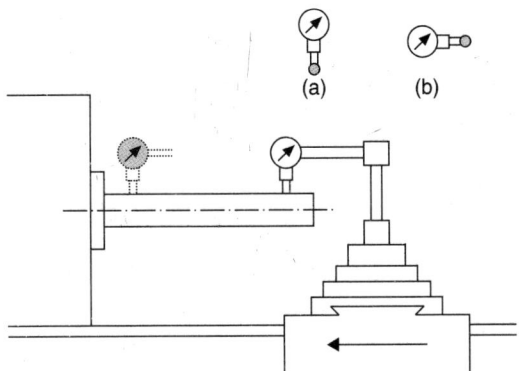

Fig. 13.2. Set-up for testing parallelism of spindle axis and bed.

- In order to eliminate the effect of any out of trueness of the mandrel or spindle taper, the spindle should be rotated slowly until the mean position is found.

- The dial gauge reading having been noted, the carriage is moved along towards the headstock and the reading taken at the other end of the parallel position. *These observations are made in both the vertical and horizontal planes.*

- The permissible inclination of the spindle axis relative to the bed is such that the free end of test mandrel must be high to allow for wear of the spindle bearing and forward to allow for horizontal forces caused by the cutting tool. The amount depends upon the size and quality of the machine. They should be less than 0.08 mm/m in case of a 300 mm swing lathe of first grade.

13.4.3. True Running of Head-stock Centre

- Headstock centre is a *live centre* and the workpiece has to rotate with this centre. If it is not true with the axis of movement of spindle, eccentricity will be caused while turning a work as the job axis would not coincide with the axis of rotation of main spindle.

- For testing this error, the feeler of dial indicator is pressed perpendicular to the taper surface of the centre (Fig. 13.3.), and spindle is rotated. The *deviation indicated by dial gauge gives trueness of the centre.*

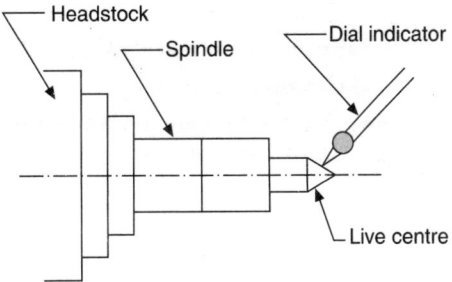

Fig. 13.3. Set-up for testing true running of headstock centre.

13.4.4. True Running of Taper Socket in Main Spindle

In case the axis of the tapered hole of the socket is not concentric with the main spindle axis, eccentric and tapered jobs will be produced.

To test it, a mandrel is fitted into the tapered hole and readings at two extremes of the mandrel are taken by means of a dial indicator as shown in Fig. 13.4.

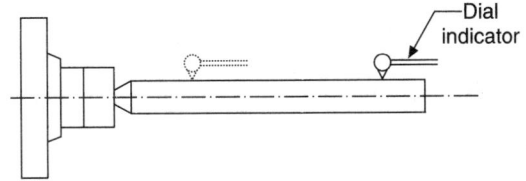

Fig. 13.4. Set-up for testing true running of taper socket in main spindle.

13.4.5. Alignment of Both the Centres in Vertical Plane

- Besides testing the parallelism of the axis individually (main spindle axis and tailstock axis) it is necessary to check the relative position of the axes also. Both the axes may be parallel to the carriage movement but they may not be coinciding. So when a job is fitted between the centres the axis of the job will not be parallel to the carriage movement.

- This test is to be carried out in *vertical plane only*. A mandrel is fitted between the two centres and dial gauge on the carriage. The feeler of dial gauge is pressed against the mandrel in vertical plane as shown in Fig. 13.5. and the carriage is moved and the error recorded.

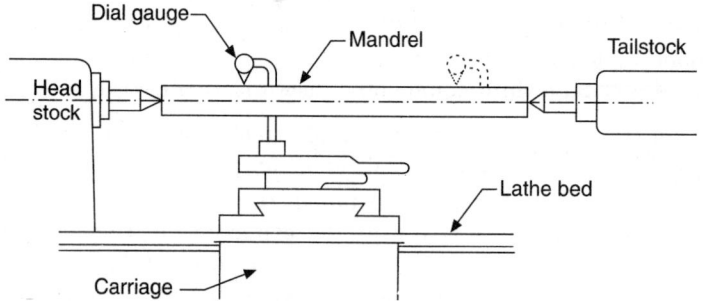

Fig. 13.5. Set-up for testing alignment of both the centres in vertical plane.

13.4.6. Cross-slide Perpendicular to Spindle Axis

This test can be performed by *two methods*. These are :

(*i*) Geometric test.

(*ii*) Practical test.

Both the above methods should be used as each one is check on the other.

(*i*) Geometric test :

- A straight edge is clamped to the cross-slide, with its edge horizontal and perpendicular to the second axis.

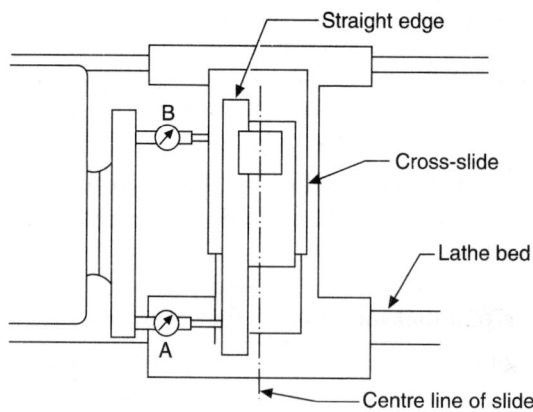

Fig. 13.6. Set-up for testing cross-slide perpendicular to spindle axis.

- A dial gauge is fixed to the edge of the face plate, or at the end of a crank bar held in the chuck, so that its feeler/plunger touches the straight edges as shown in Fig. 13.6. The straight edge is adjusted until it is accurately at right angles to the spindle axis by obtaining equal dial gauge readings when the spindle is turned through 180 degrees, that is when the gauge is in position *A* and *B* in the figure.

- The cross-slide is then moved along its guides carrying with it the straight edge across the dial plunger. If the line of movement is not parallel to the straight edges and, therefore, not perpendicular to the spindle axis, the pointer of the dial gauge will move. Readings are taken over a measured length of the straight edge in order to determine the amount by which the slide is out of square.

(ii) Practical test :

- *This method consists in taking a light finish cut on a workpiece held in chuck and then checking the straightness of a diameter across the finished surface.*
- This is done by placing a straight edge on two equal slip gauges at opposite ends of a diameter and trying a third equal slip gauge under the straight edge near the centre of the workpieces. This slip gauge should enter because the only permissible departure from straightness is slight concavity. The amount of concavity can be measured by finding the slip gauge which just fits at the centre.

The dial gauge readings in the first-method and concavity measured by slip gauges in the second method should be converted into a figure giving the departure from straightness per meter of diameter.

13.4.7. Accuracy of Pitch of Lead Screw

The accuracy with which threads are cut on the lathe depends upon the accuracy of its lead screw. Thus it is very essential that the pitch of the lead screw should be uniform throughout its length. Test for this is performed as follows :

- Positive stop is fixed on the lathe.
- Against the stop, the length bars and slip gauges can be located.
- An indicator is mounted on the carriage and first it makes contact against the calculated length of slip gauges.
- The initial loading of the dial gauge against the slip gauge is noted.
- The slip gauges are then removed and carriage is connected to the lead screw and the lead screw is disconnected from the gear train.
- An indexing arrangement is utilised for rotating the lead screw and lead screw is given some revolutions so that distance travelled by carriage is equal to the length of slip gauges. The reading of the dial indicator against the stop is noted down in this position. If it is same as before, there is no error, otherwise it can be recorded.

A suitable method for recording the progressive and periodic errors is by using a suitably divided scale, which is placed close to the line of centres. A microscope is rigidly mounted on the carriage in a convenient position to note the readings on the scale.

- *In this method care must be taken not to disturb the datum location when changing the gauges for testing different pitch lengths.*

13.4.8. Axial Slip of Lead Screw

The thrust face and collars of the lead screw (or the abutment collar and the thrust bearing of the screw) must be exactly square to the screw axis, otherwise a cyclic endwise movement is set up which is of the same nature as the axial slip in the main spindle. Thus a periodic pitch error will be additional to any true periodic errors in the pitch of the screw.

In order to *test axial slip in the lead screw,* a ball is fitted in the end of lead screw and the plunger of the dial is pressed against the ball as shown in Fig. 13.7. The lead screw is rotated and deviation, if any, is noted down.

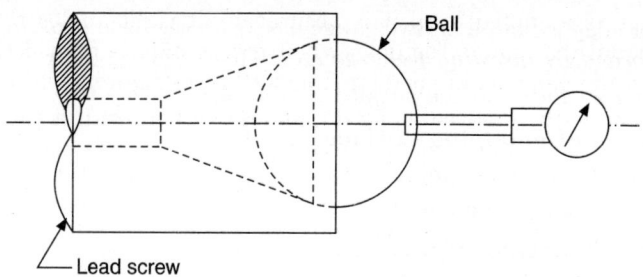

Fig. 13.7. Set-up for testing axial slip of lead screw.

13.5. ALIGNMENT TEST ON DRILLING MACHINES

13.5.1. Alignment Test on Pillar Type Drilling Machine

The various tests performed on the pillar type drilling machines are :

1. Flatness of clamping surface of base.
2. Flatness of clamping surface of table.
3. Perpendicularity of drill head guide to the base plate.
4. Perpendicularity of drill head guide with table.
5. Perpendicularity of spindle sleeve with base plate.
6. True running of spindle taper.
7. Parallelism of spindle axis with its vertical movement.
8. Squareness of clamping surface of table to its axis.
9. Squareness of spindle axis with table.
10. Tool deflection.

13.5.1.1. Flatness of clamping surface of base

The set up for checking flatness of clamping surface of base is shown in Fig. 13.8.

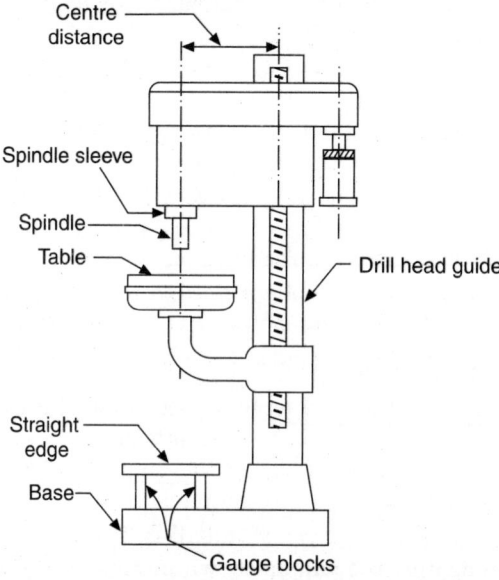

Fig. 13.8. Set-up for testing flatness of clamping surface of base.

- The straight edge is placed on two gauge blocks on the base plate in various position and *error is recorded by inserting feeler gauges.* This error should not exceed 0.1/1000 mm.
- The clamping surface and the surface should be concave only.

13.5.1.2. Flatness of clamping surface of table

The checking of flatness of clamping surface table is done the same way as given in article 13.5.1.2. The permissible error is also same.

13.5.1.3. Perpendicularity of drill head guide to the base plate

The perpendicularity/squareness of drill head guide to the base plate is tested : (a) in a vertical plane passing through the axes of both spindle and column and (b) in a plane at 90° to the plane at (a).

- The test is carried out by placing the frame level (with graduations from 0.03 to 0.05 mm/m) on guide column and base plate.
- The error is noted by noting the difference between the readings of the two levels. This error should not exceed 0.25/1000 mm guide column for (a) and the guide column should be inclined at the upper end towards the front only, and 0.15/1000 mm for (b).

(It may be noted that Fig. 13.9 shows the similar test for the squareness of drill head guide with table, the only difference being that the frame level is to be placed on the base instead of table).

13.5.1.4. Perpendicularity of drill head guide with table

Refer to Fig. 13.9. The procedure to conduct this test is exactly the same as at (b). The permissible error is same too.

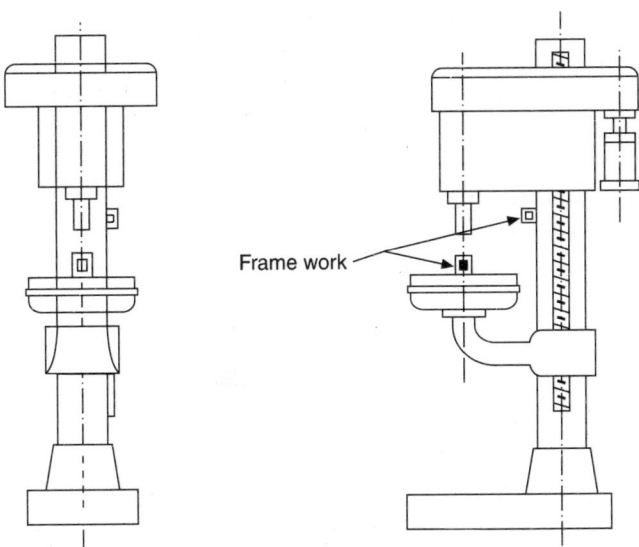

Fig. 13.9. Set-up for testing perpendicularity of drill head guide to the base plate.

13.5.1.5. Perpendicularity of spindle sleeve with base plate

The test check perpendicularity of spindle sleeve with base plate is performed in both the planes as specified in test 13.5.1.3 and in the similar manner with the difference that the frame levels are to be placed on spindle sleeve and base plate.

- The error should not exceed 0.25/1000 mm for plane (a) and the sleeve should be inclined towards column only ; and 0.15/1000 mm for plane (b).

13.5.1.6. True running of spindle taper

Fig. 13.10. shows the set up for this test.

- The test mandrel is placed in the tapered hole of spindle and a dial indicator is fixed on the table and its feeler made to scan the mandrel.
- The spindle is rotated slowly and reading of indicator recorded.
- The error should not exceed 0.3/100 mm for machines with taper upto Morse No. 2 and 0.04/300 mm for machines with taper larger than Morse No. 2.

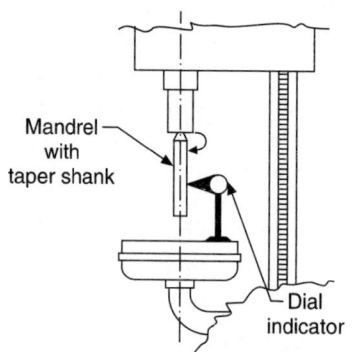

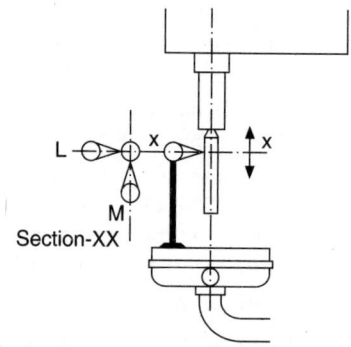

Fig. 13.10. Set-up for testing true running of spindle taper.

Fig. 13.11. Set-up for testing parallelism of spindle axis with its vertical movement.

13.5.1.7. Parallelism of spindle axis with its vertical movement

Refer to Fig. 13.11. for the set up of this test.

- This test is carried out in two planes (L) and (M) at right angles to each other.
- The test mandrel is fitted in the tapered hole of the spindle and the dial indicator is fixed on the table with its feeler touching the mandrel.
- The spindle is adjusted in the middle position of its travel.
- The reading of the dial indicator noted down when the spindle is moved in upper and lower directions of its middle position with slow vertical feed mechanism.

Permissible errors :

S.No.	Machines	Plane A	Plane B
1.	With taper upto Morse No. 2	0.03/100 mm	0.03/100 mm
2.	With taper larger than Morse No. 2	0.05/300 mm	0.05/300 mm

13.5.1.8. Squareness of clamping surface of table to its axis

- Refer to Fig. 13.12. In this test, the dial indicator is mounted in the tapered hole of the spindle and its feeler is made to touch the surface of table.
- The table is rotated slowly and the readings of the dial recorded, which should not exceed 0.05/300 mm diameter.

13.5.1.9. Squareness of spindle axis with table

Fig. 13.13 shows the set up for this test.

- A straight edge is placed in positions LL' and MM'.
- Work table is arranged in the middle position of its vertical travel.

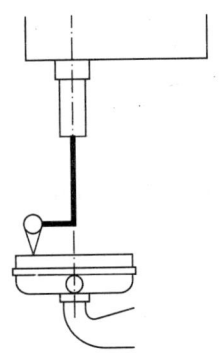

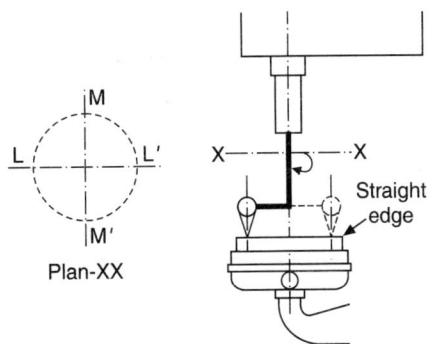

Fig. 13.12. Set-up for testing squareness of clamping surface of table to its axis.

Fig. 13.13. Set-up for testing squareness of spindle axis with table.

- The dial indicator is mounted in the spindle tapered hole and its feeler made to touch the straight edge first at L and reading noted down.
- The spindle is rotated by 180° so that the feeler touches at point L' and again reading noted down.
- The difference of two readings gives the error in squareness of spindle axis with table.
- Similar readings are noted down by placing the straight edge in position MM'.

The permissible errors are :

For set up LL' .... 0.08/300 mm

For set up MM' .... 0.05/300 mm

13.5.1.10. Total deflection

The set up for this test is shown in Fig. 13.14.

- The drill head and table are arranged in their middle position.
- Dial indicator is mounted on table with its feeler touching the lower machined surface part and spindle stock.
- Drill spindle is loaded with the dynamometer (load gauges) placed on table and the deflection of dial indicator noted down.

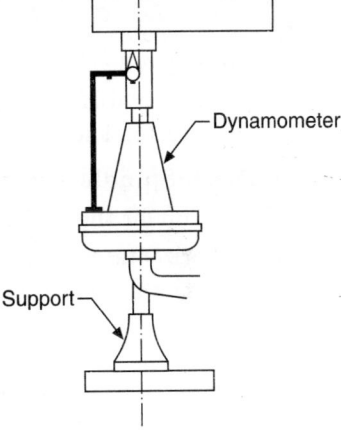

Fig. 13.14. Set-up for testing total deflection.

The drill pressure is set as follows :

Drilling diameter in open hearth steel	kg
6	100
10	200
16	350
20	550
25	750
32	950
40	1200

Permissible error	For centre distance upto and including
(i) 0.4 mm	200 mm
(ii) 0.6 mm	200-300 mm
(iii) 0.8 mm	300-400 mm
(iv) 1.0 mm	400 mm and above.

13.5.2. Alignment Tests on Radial Drilling Machines

The following alignment tests are carried out on a drilling machine :

1. Saddle and arm movements parallel to base plate.

2. Saddle and feed movement square with base plate.

13.5.2.1. Saddle and arm movements parallel to base plate

● When the saddle is moved along the arm, any deviation from parallelism with the base plate should be an inclination upwards towards the column, not exceeding 0.16 mm/m.

This is tested by fixing a dial gauge to the spindle, with the plunger bearing on the surface of the base plate, and observing the reading as the saddle is moved along the arm, local irregularities being ignored.

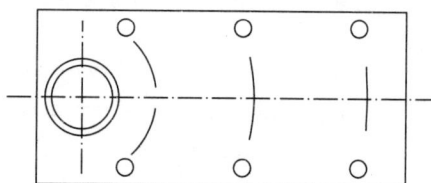

● For checking the parallelism of the arm itself as it rotates on the column, readings are taken near the edges of the base plate, with the saddle in three different positions as shown in Fig. 13.15. The tolerance is 0.16 mm/m.

Fig. 13.15

13.5.2.2. Spindle and feed movement square with base plate

The set up for this test is shown in Fig. 13.16. The test is performed as follows :

● A horizontal rod is fixed to the spindle, with a dial gauge, attached at a radius of about 300 mm as shown in Fig. 13.16.

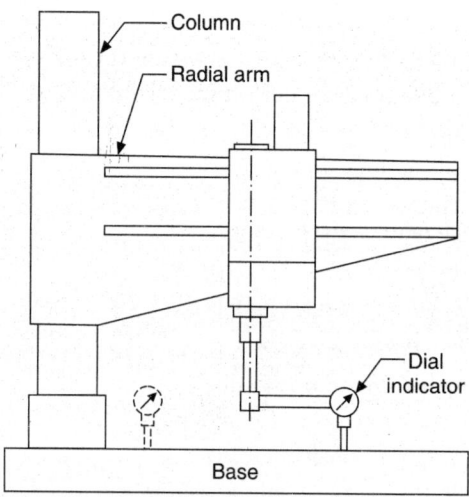

Fig. 13.16. Set-up for testing spindle and feed movement square with base plate.

- The plunger of the dial is arranged to bear on the base plate and is brought into two positions shown in the figure by rotation of the spindle. Readings taken in two positions check the squareness in the plane of the arm.

- Two readings are taken in the positions at *right angles* to those shown, in order to check the squareness in the plane perpendicular to the arm.

- Readings should be taken with the spindle in four different positions, namely *near to and remote from* the column with the arm *first low* and then *high upto the column*.

In any position, the departure from squareness should not exceed 0.16 mm/m, and the plane of the arm, must be such that the bottom of the spindle inclines towards the column.

To check the *squareness of the feed motion*, the dial gauge is held in the spindle with its plunger horizontal and bearing on the vertical edge of a true square which rests on the base plate. Any variation of the dial gauge reading, as the spindle is moved up and down, measures the error, which should not exceed 0.25 mm/m in the plane of the arm (the spindle inclined towards the column at its lower end), and 0.16 mm/m at right angles to the arm. The test should be performed with the saddle near to the column and also at the end of the arm.

13.6. ALIGNMENT TESTS ON MILLING MACHINE

The following tests are performed on the milling machine :

1. Eccentricity of external diameter.
2. True running of internal taper.
3. Work table surface parallel with arbor rising towards overarm.
4. Surface parallel with longitudinal movement.
5. Traverse movement parallel with spindle axis.
6. Central *T*-slots parallel with longitudinal movement.
7. Central *T*-slots square with arbor.
8. Tests on column.
9. Overarm parallel with spindle.
10. Alignment of the main spindle with the bore of the bracket of the overarm.

13.6.1. Eccentricity of External Diameter

Refer to Fig. 13.17.

- The feeler of the dial gauge is placed on the cylindrical surface of the shoulder. The locating shoulder is rotated, any deviation in the reading of dial gauge is recorded.

- It is due to the eccentricity of the spindle in the hole in which it fits. Due to it, vibrations are produced and the cutter will float side ways and cut over, or undersize. Force mills may dig in when leading edges stop to cut.

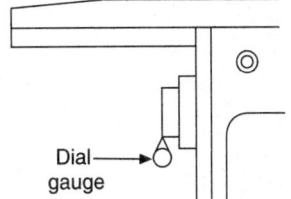

Fig. 13.17. Set-up for testing eccentricity of external diameter.

Permissible error : 0.01 mm.

13.6.2. True Running of Internal Taper

The set up for this test is shown in Fig. 13.18.

- The table is set in its main position longitudinally and the mandrel 300 mm long is fixed in the spindle taper.

- A dial gauge is set on the machine table and feeler adjusted to touch the lower surface of the mandrel.

- The mandrel is then turned and the dial readings at two points are noted *i.e.* one at the place nearest to spindle nose and other at about 300 mm from it.

- For shifting the position of dial gauge from *L* to *M* cross-slide of the machine is operated to bring the dial gauge at the bottom of the end of mandrel.

- There can be *two errors* : (*a*) Axis of the spindle and the axis of taper may not be parallel. (*b*) Eccentricity of the taper hole which, if present, should indicate same error at both the places. The error in case (*a*) will give different readings at two places. Due to this error, *cut will be shared equally between teeth of cutters,* and therefore *variations and poor finish will be obtained.*

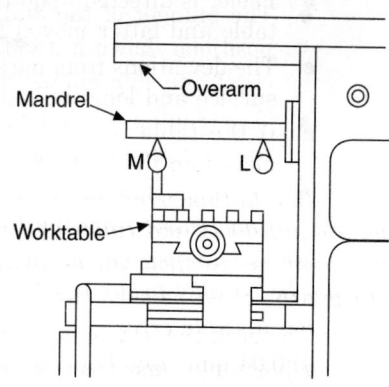

Fig. 13.18. Set-up for testing true running of internal taper.

Permissible error :

For position at *L* .... 0.01 mm

For position at *M* .... 0.025 mm

13.6.3. Work Table Surface Parallel With Arbor Rising Towards Overarm

The set up for this test is shown in Fig. 13.19.

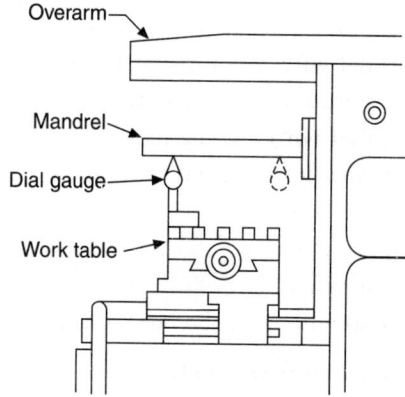

Fig. 13.19. Set-up for testing work table surface parallel with arbor towards overarm.

Procedure :

- A dial gauge is set on the machine table.

- A mandrel 300 mm long is fitted in the spindle taper.

- The feeler of dial gauge is made to touch the lower surface of the mandrel. With mandrel in position (mean) the readings at the maximum travel of the table surface are observed. The stand of the dial gauge is moved and not the table itself which remains stationary.

- Due to *this type of error the milled surface produced will not be square to the base and parallel cross ways.*

Permissible error : 0.025 mm per 300 mm.

13.6.4. Surface Parallel With Longitudinal Movement

Refer to Fig. 13.20 for the set up of this test.

- The dial gauge is fixed to the spindle.

- Feeler is directed upon the surface of the machine table and latter moved longitudinally.
- The deviations from parallelism between the table surface and longitudinal motion are noted down.
- If the table is uneven, a straight edge may be placed on the surface.

Due to this error the surface of the table will fluctuate up and down and cutter will not take equal cuts on the job which is clamped on the table and the milled surface will not be parallel to the base.

Permissible error :

(*i*) 0.04 mm *upto* 600 mm movement.

(*ii*) 0.05 mm *over* 600 mm movement.

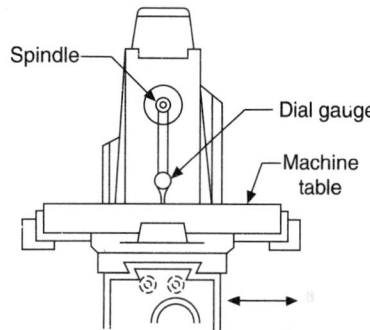

Fig. 13.20. Set-up for testing surface parallel with longitudinal movement.

13.6.5. Traverse Movement Parallel With Spindle Axis

(*i*) *in horizontal plane ;*

(*ii*) *in vertical plane.*

Refer to Fig. 13.21 for the set up of the test.

- The table is set in mean position and the dial gauge is fixed on it (table).
- The table is moved cross-wise and any deviation on the reading of dial gauge is noted with feeler on one side of the mandrel in horizontal plane and under the mandrel for error in vertical plane.
- When cross slide is moved the *depth of cut will vary due to this error.*

Permissible error :

For position at L : 0.025 mm per 300 mm

For position at M : 0.025 mm per 300 mm

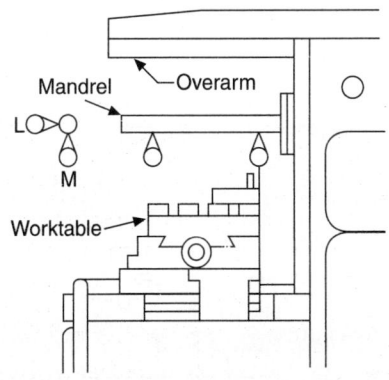

Fig. 13.21. Set-up for testing traverse movement parallel with spindle axis.

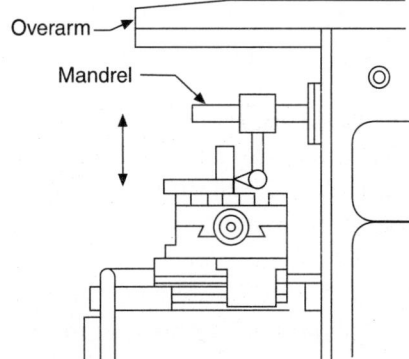

Fig. 13.22. Set-up for testing central T-slots parallel with longitudinal movement.

13.6.6. Central T-slots Parallel With Longitudinal Movement

The set up for this test is shown in Fig. 13.22.

- The general parallelism of the central slot with the longitudinal movement of the table is checked by using a bracket 150 mm long with a tenon which enters the T-slot.
- Against the upper surface of the bracket in vertical plane the feeler of the dial gauge is located.

- After fixing the dial gauge to the spindle and adjusting its feeler to the surface of the bracket the table is moved longitudinally while the tenon block is held stationary by hand and deviations from parallelism are noted from dial gauge.

- The *depth of cut*, due to this error, *will not remain constant* as the job will be inclined according to inclination of T-slots with longitudinal movement and the axis of job held between tailstock and index head will not be perpendicular to cutter.

Permissible error :

    0.04 mm upto 600 movement

    0.05 mm over 600 mm and upto 1000 mm movement.

13.6.7. Central T-slots Square With The Arbor

When the central T-slot is not perpendicular to the arbor, key way etc., cut on the machine will not be parallel to the job axis. This test is conducted as follows :

Refer to Fig. 13.23.

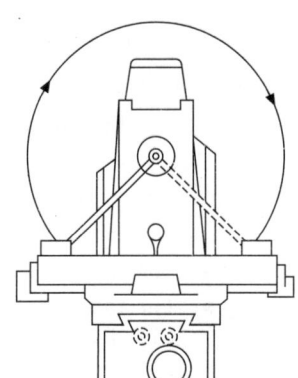

- The table is adjusted in the middle portion of its longitudinal movement and a tenon block 160 mm long inserted in the central T-slot.

- A dial gauge is fixed on the mandrel, the feeler being adjusted to touch the vertical face of the bracket.

- Observe the reading on dial gauge when the bracket or tenon block is near one end of the table. Then swing over the dial gauge and move the tenon block so that corresponding reading can be taken near the other end of the table. Generally two tenon blocks are used.

Fig. 13.23. Set-up for testing central T-slot square with the arbor.

13.6.8. Test on Column

Refer to Fig. 13.24. This test is conducted as follows :

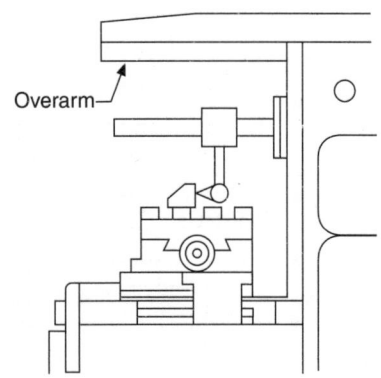

- Locate the table in its central position and fix a square with an arm about 300 mm long on the table surface and attach a dial gauge to the spindle mandrel in such a way that the feeler rests on the arm (vertical face) of the surface near the bottom edge. Note down the reading of the dial gauge.

- Move the table upwards about 300 mm and again note down the reading of the dial gauge.

The difference in readings is a direct indication of the error of perpendicularity of the table surface (front guiding surface) and knee support or side guiding support.

Fig. 13.24. Set-up for tests on column.

The above test is performed for *two positions* of the square and in the first position the dial gauge touches the square in front and in second position it faces the side of the square *i.e.* 90° to the first position.

It is worth noting that if column ways for knee are not square with the table, as the table is fed upwards in facing operation or end milling, *the surface produced will not be square* with the surface of the table.

Permissible error : 0.025 mm per 300 mm.

13.6.9. Overarm Parallel With Spindle

(i) *in horizontal plane ;*

(ii) *in vertical plane.*

The set up for this test is shown in Fig. 13.25. The procedure to conduct this test is as follows :

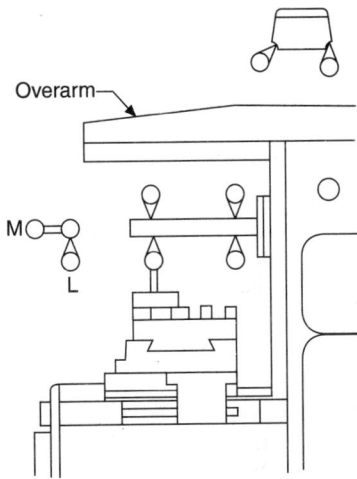

Fig. 13.25. Set-up for testing overarm parallel with spindle.

● Fix the dial gauge on the table and its feeler under the mandrel.

● Move the table crosswise and note any change in the reading.

● For error in horizontal plane, repeat above readings so that feeler is under overarm, and compare the readings on mandrel and side of overarm.

Permissible error : 0.025 mm per 300 mm.

13.6.10. Alignment of The Main Spindle With Bore of The Bracket of The Overarm

The test is conducted as follows :

● With the mandrel in the bore of the overhanging bracket and gauge holder in the mandrel fitted to the spindle taper, the feeler is adjusted so that it touches the mandrel in the bore.

● The main spindle is turned slowly and reading of the dial gauge is noted at four points (opposite ends in horizontal and vertical planes *i.e.* 90° apart).

● The difference between 180° opposite readings and other two is twice the eccentricity of the mandrel in the vertical and horizontal direction respectively.

● In case the axis of the bearing of the supporting bracket is *not co-axial with* the spindle axis, the axis of the arbor which is held in spindle and supporting bracket will not be parallel with table surface and hence the cutter *mounted on the arbor will take more cut on supporting* bracket side if the bearing axis is *somewhat lower than spindle axis and less cut if the bearing axis is above.*

13.7. ACCEPTANCE TESTS FOR SURFACE GRINDERS

For surface grinders following tests are conducted :

1. **Table top parallel to its movement** by fitting the dial gauge in the stationary spindle and dial feeler touching the table top and *traversing the table.*

2. **Slots parallel to table movement** by fitting the dial gauge in the stationary spindle and dial feeler touching the slot and *traversing the table.*

3. **Spindle axis parallel to table** by fitting dial gauge on radial arm of the spindle and its feeler touching the square resting on table, and *rotating the spindle by* 180°.

4. **Spindle axis square with the slot** by fitting dial gauge on an arm fitted to spindle and *rotating the spindle by* 180°.

5. **Vertical movement of spindle square with table top** by fitting dial gauge in spindle and its feeler touching the square resting on table.

6. Practical tests.

13.8. PRACTICAL TESTS

The practical tests form a part of the acceptance tests for machine tools. *These tests mainly reveal the alignment accuracy of the machine tools under dynamic loading. These tests also indicate whether the machine has been installed correctly on its foundation* and as per specifications. These practical tests are usually specified by the manufactures but of late there has been tendency to standardise these tests.

OBJECTIVE TYPE QUESTIONS

Fill in the blanks :

1. The accuracy of machine tools, which cut metal by removing chips or swarfs, is tested by means of ...... and ...... tests.
2. ...... tests are checked on relationship of the various elements of the machine when it is idle and unloaded and are performed by using ordinary measuring equipment together with a few special accessories.
3. ...... tests are intended to check the accuracy of the work done on the machine and are performed by machining suitable pieces and measuring these.
4. ...... gauges are widely used in alignment tests.
5. Test ...... are used for checking the true running of the spindle.
6. ...... levels are used for high grade precision work.

ANSWERS

1. Geometric, Practical	2. Geometrical	3. Practical
4. Dial	5. Mandrels	6. Spirit.

THEORETICAL QUESTIONS

1. What do you mean by 'Geometric' and 'Practical' tests ?
2. List various geometric checks generally made on the machine tools.
3. Enumerate the various tests which are applied to machine tools.
4. Name and describe briefly the instruments used for alignment test.
5. Name the various alignment tests which are conducted on lathe.
6. Explain briefly any three of the following alignment tests conducted on lathe :
   (*i*) Parallelism of spindle axis and bed   (*ii*) True running of head-stock centre
   (*iii*) Alignment of both the centres in vertical plane
   (*iv*) Accuracy of pitch of lead screw.

7. Name the various alignment tests which are conducted on the Pillar type drilling machine.
8. Explain any three of the following alignment tests performed on the Pillar type drilling machine :
   (*i*) Flatness of clamping surface of base   (*ii*) Perpendicularity of drill head guide to the base plate
   (*iii*) Perpendicularity of drill head guide with table
   (*iv*) True running of spindle taper
   (*v*) Squareness of clamping surface of table to its axis.
9. Describe briefly the alignment tests conducted on Radial drilling machine.
10. Enumerate the various alignment tests which are performed on a milling machine.
11. Explain any three of the following alignment tests conducted on a milling machine.
   (*i*) Eccentricity of external diameter   (*ii*) True running of internal taper
   (*iii*) Transverse movement parallel with spindle axis
   (*iv*) Central *T*-slot square with arbor
   (*v*) Tests on column.
12. Write a short note on "Practical tests".

Numerical Control of Machine Tools and CAD/CAM

14.1. INTRODUCTION TO MODERN MACHINE TOOLS

- Newer machine tools have been built to absorb newer machining technologies to cope with newer and tougher materials. New technologies include Ultrasonic Machining (USM), Electro-Chemical Machining (ECM), Laser Beam Machining (LBM) etc.

- Besides this the advancement in electronics and applications of computer in the machine tools have brought in a significant and revolutionary change in the machine tool control concept. This has given birth to an entirely new generation of machine tools. *"Numerically Controlled (NC) machine tools"* are *highly flexible and are economical for producing a single or a large number of parts.* **Numerical Control, NC** can be defined simply as *control by numbers.*

- A machine tool having a *dedicated computer to help prepare the program and control some or all of the operations of the machine tool* is called **Computer Numerical Control (CNC)** *machine tool.*

14.2. NC MACHINES

14.2.1. Introduction

NC machines assimilate a method of automation, where automation of medium and small volume production is done by some controls under the instructions of a program.

The definition of NC (Numerical Control) as given by EIA (Electronic Industries Association) is as under :

"A system in which actions are controlled by the direct insertion of numerical data at some point. The system must automatically interpret at least some portion of this data."

- In NC machines, the input information for controlling the machine tool motion is provided by means of *punched tapes* or *magnetic tapes in a coded language.*

14.2.2. Working of NC Machine Tool

Fig. 14.1 shows the working sequence of a NC machine tool viz-a-viz operator controlled machine tool.

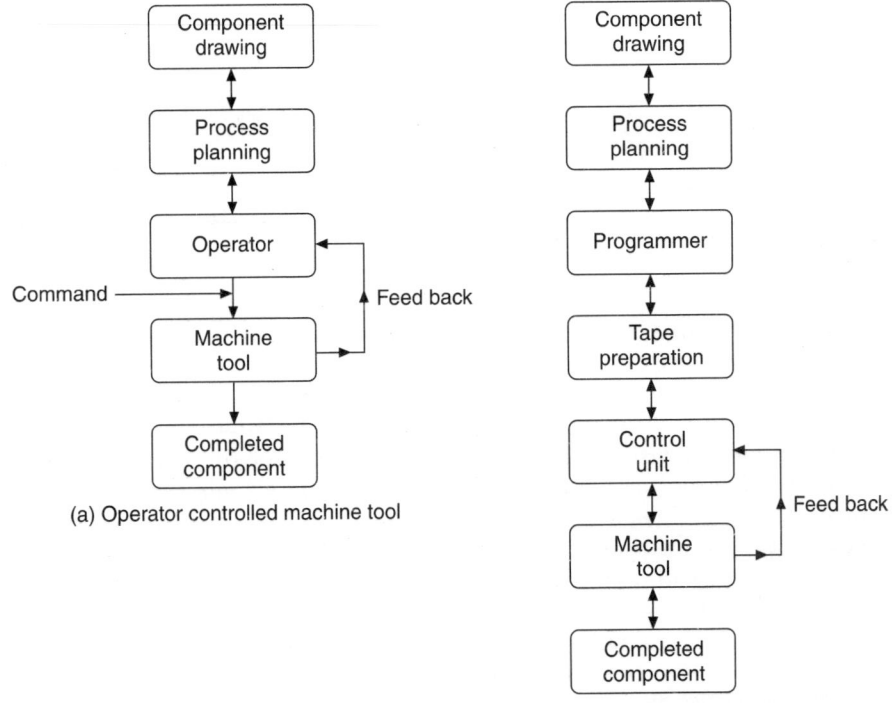

(a) Operator controlled machine tool

(b) Numerically controlled (NC) machine tool

Fig. 14.1

- The first two steps , component drawing and process planning are *similar* in both operator controlled and NC machine tools.

- In the *operator controlled* machine tools, the operator controls the cutter position during manufacture and also makes necessary adjustments and corrections to produce the desired component.

However, in **NC machine tool** the *operator is replaced by the data processing part of the system and the control unit.*

— In the **data processing unit**, the co-ordinate information regarding the component is *recorded on a tape by means of a teleprinter.*

— **Tape** is fed to the control unit which sends the *position command signals to slideway transmission elements of the machine.* At the same time, the command signal is constantly *compared* with the actual position achieved, with the help of position *feedback signal derived from automatic monitoring of the machine tool slide position.* The difference in two signals, if any, is corrected until the desired component is produced.

14.2.3. Main Elements of a NC Machine Tool

Refer to Fig. 14.2. The main elements of a NC machine tool are :

1. The control unit (also known as NC console or Director)

2. The drive units   3. The position feedback package

4. Magnetic box   5. Manual control.

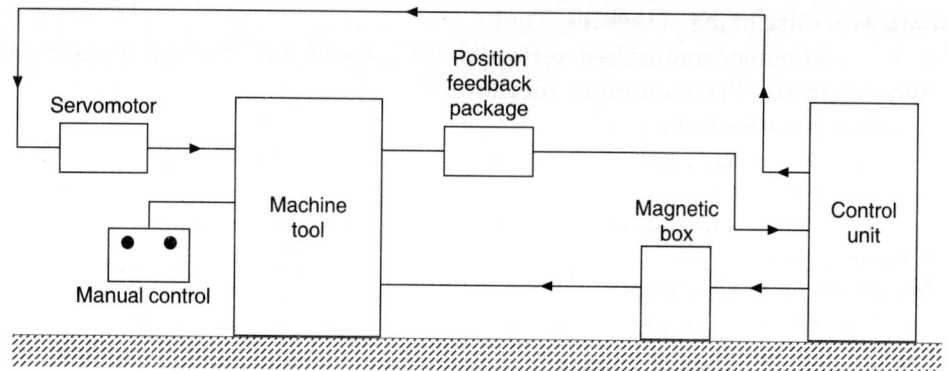

Fig. 14.2. Main elements of a NC machine.

- In the **control unit**, a tape recorder reads the instructions (written in a coded language) for manufacturing the component.
- The instructions under electronic processing and the control unit sends command signals to the **drive units** of the machine tool and also to the **Magnetic box** (Electrical control cabinet). Command signals sent to the *drive units* of the machine tool, *control the length of travel and feed rates,* while the command signals sent to the *magnetic box* control other functions such as *spindle motor starting and stopping,* selecting spindle speeds, actuation of tool change, coolant supply etc.
- A **feedback transducer** provided in the machine tool checks whether the required lengths of travel have been obtained. It sends the information of the actual position achieved to the control unit. In case there is any difference between the input command signal and the actual positions achieved, the drive unit is actuated by suitable amplifier from the error signal.
- **Manual Control** provided in the machine tool assists the operator to perform some functions manually such as motor start-stop, speed change, feed change, axes movements, coolant supply etc.

14.2.4. Classification of Numerical Control (NC) Machines

NC machines can be generally *classified* as follows :

1. **According to "Type of motion" :**

(i) Point-to-point (PTP) systems :

- In PTP motion machine tool, the cutting tool moves to a numerically defined location, the motion is stopped and the task is performed. After completing the task, the tool moves to the next point and the cycle is repeated.
- In this system, the path of movement of the tool and its velocity, when the tool moves from one point to another, is of no significance.

Example. Drilling machine operation.

(ii) Straight line system :

- It is an extension of point-to-point system.

Example. —Stepped turning on a lathe ;
              —pocket milling etc.

(iii) Contouring or continuous systems :

- In these systems there are continuous, simultaneous and coordinated motions of the tool and workpiece along different coordinate axes.
- The distinguishing features of these types of machines is their capacity for *simultaneous control of more than one axis movement of the machine tool.*

Examples. —Machining of profiles ;

—Contour and curved surfaces.

2. According to "Programming method" :

(i) Absolute positioning :

- In this method, the coordinates of the target point are defined with respect to a fixed program zero point.
- Absolute dimensions tell the control to what point the tool is to move, regardless of its current position.

(ii) Incremental positioning :

- Incremental or chain positioning tells the control by what amount the tool is to move from its present position.
- The present position of the tool acts as the program zero point.

3. According to "Control system" :

(i) Open loop system :

- There is *no 'feedback'* and no return signal to indicate whether the tool has reached the correct position at the end of operation or not.

Example. Coordinate drilling machine.

(ii) Closed loop system :

- A *feed back* is built into the system, which automatically monitors the position of the tool.
- It is *more expensive* than open loop system.

14.2.5. Applications of NC Machines

The *major applications* of NC machines are :

1. Complex parts.
2. Parts which are frequently subjected to design changes.
3. Respective and precision quality parts which are to be produced in low to medium batch quantity.
4. To cut down lead time in manufacture.
5. In situations where the investment on tooling and fixture inventory will be high if parts are made on conventional machines tools.

14.2.6. Advantages and Disadvantages of NC machines

Advantages :

Following are the *advantages* of NC machines :

1. Accuracy achieved is of high order.
2. Reduced production cost per piece.
3. Less scrap.
4. High production rates.
5. Less operator skill required.
6. Excellent reliability.
7. Tooling cost low.
8. Less cycle time and increased tool life.
9. Increased flexibility.
10. Production of complex parts.
11. Reduced set-up time.
12. Elimination of special jigs and fixtures.
13. Reduced inspection.

14. Lower labour cost.

15. Reduced floor space.

16. Easy and effective production planning.

Disadvantages :

Major *disadvantages* of NC machines are :

1. Control systems are costly.

2. Higher investment cost (in comparison to conventional machine tools).

3. If a tape or the control malfunctions there is a loss in machine flexibility.

4. Higher maintenance cost.

5. Redundancy of labour.

6. Special skill is required for programming and operating equipment.

7. NC machines are not generally considered feasible for high production.

14.2.7. Comparison Between Conventional Machines and NC Machines

The comparison between conventional and NC machines is given below :

S.No.	Aspects	Conventional machines	NC machines
1.	Rate of production	Normal	Comparatively higher ; can produce complicated profiles consistently with good accuracy.
2.	Machine hour rate	Normal	Generally higher.
3.	Maintenance cost	Less costly	Comparatively more costly.
4.	Airconditioned rooms	Generally not required	Usually required.
5.	Type of workes required	Normal skilled workers required	A high order of skill and trained personnel required to operate to produce complex contours.
6.	Cost	Normal	New NC machines cost around two to five times more than the similar capacity of conventional machines, depending upon the sophistication of the control system and the size of the machine.
7.	Type of tooling required	Normal	Require special tooling, the tools made from carbon tool steel and H.S.S. can not be used since these machines are designed to operate at very high operating conditions—NC machines are capable of operating at 50,000 r.p.m.
8.	Other features available	—	Use of automatic tool changers to change the tool automatically and simultaneous machining by using multiple tools are the features available with high-end machines. These features help to reduce down time and set-up time.

14.3. CNC MACHINES

14.3.1. General Aspects

In a CNC machine, a *minicomputer is used to control machine tool functions from stored in information or punched tape input or computer terminal input.*

The definition CNC (Computer Numerical Control) as given by EIA is as under :

"*The numerical control system where a dedicated, stored program computer is used to perform some or all of the basic numerical control functions in accordance with control programms stored in read/write memory (RAM) of the computer*".

CNC may also be defined as : "*An NC system with a microcomputer or microprocessor using software to implement control algorithms*".

Fig. 14.3 shows the control unit and panel of a CNC. The following points about CNC machines are worthnoting :

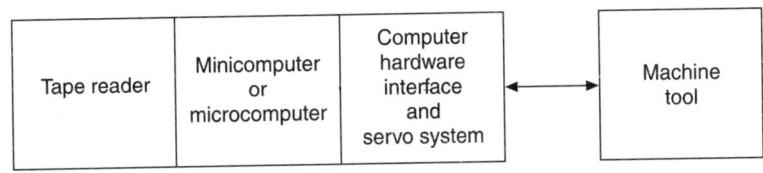

Fig. 14.3. Computer Numerical Control (CNC) system.

- The control unit and a panel of CNC differs from that of NC controls in that, it works in *ON-line mode* whereas NC works in *batch processing mode.*
- A typical CNC may need only the drawing specifications of a part to be manufactured and the computer automatically generates the part program for the loaded part.
- The part program once entered into the computer memory can be used again and again.
- The input information can be reduced to a great extent with the use of special sub-programs developed for repetitive machining sequences.
- The CNC machines have the facility for proving the part program without actually running it on the machine tool.
- CNC control unit allows compensation for any changes in the dimensions of the cutting tool.
- With CNC control systems, it is possible to obtain information on machine utilisation which is useful to the managements.

14.3.2. Functions of CNC

The principal functions of CNC are :

1. Machine tool control.

2. In-process compensation.

3. Improved programming and operating features.

4. Diagnostics.

14.3.3. Advantages and Disadvantages of CNC Machines (Over NC Machines)

Advantages :

Following are the *advantages* and *disadvantages* of CNC machines :

1. Greater flexibility.

2. Reduced data reading error.

3. CNC machine can diagnose program and can detect the machine malfunctioning even before the part is produced.

4. In highly sophisticated manufacturing system CNC machine can be integrated with DNC (Direct Numerical Control) systems.

5. Conversion of units — possible within the computer memory.

Disadvantages :

1. Higher investment cost.
2. Higher maintenance cost.
3. Costlier CNC personnel.
4. Airconditioned places are required for the installation of the machines.
5. Unsuitable for long run applications.
6. Planned support facilities.

14.3.4. Applications of CNC

CNC is being used in the following machines/areas :

- Drilling machines
- Turning machines
- Boring machines
- Milling machines
- Grinding machines
- Pipe bending machines
- Coil winding machines
- Flame cutting machines
- Welding, wire cut EDM and several other areas.

14.4. COMPARISON OF NC AND CNC MACHINES

The comparison of NC and CNC machines is given below :

S. No.	Aspects	NC machines	CNC machines
1.	Base	Purely hardware-based system.	A software-based system.
2.	Type of system	Hard-wired system (Hard-wired NC used IC digital-logic circuit packages-usually medium scale integration (MSI), which are mounted and wired in a fixed permanent arrangement on plug-in printed-circuit boards (PCBs).	Soft-wired system (the term soft-wire is applicable because the functions created to control the specific machine tool result from the *application software program* rather than from any physical wiring of a group of logic elements.
3.	Accuracy reliability	The tape is read again during the production of each part. Owing to the increased use of tape reader, the chances of inaccuracy in manufacturing are increased.	The part programme tape and tape reader are used only once to enter the program into the computer manner. This results in improved reliability.
4.	Part-programming mistakes	Common with conventional NC.	Such mistakes are not there.
5.	Control features	Cannot be easily altered to incorporate improvements in the unit.	Use of a computer as the control device provides the *flexibility* to make improvements in such features as circular interpolation when better softwares become available.
6.	Productivity	Low	Comparatively high

14.5. DIRECT NUMERICAL CONTROL (DNC)

Direct numerical control (DNC) *has been defined as a manufacturing system in which a number of machines are interconnected using a computer through direct connection in real time.*

Whereas CNC is a self contained NC system for a single machine tool, with DNC, several machine tools are *directly controlled by a central computer.* In DNC system, punched tape is no longer used. In operation, the DNC computer receives the equivalent of a punched tape information from a compiling facility and direct storage of that information in a disk or magnetic tape data storage unit.

The DNC system (Fig. 14.4) consists of the following basic four components :

(*i*) Central computer

(*ii*) Bulk memory, which stores the NC programs

(*iii*) Telecommunication lines

(*iv*) Machine tools.

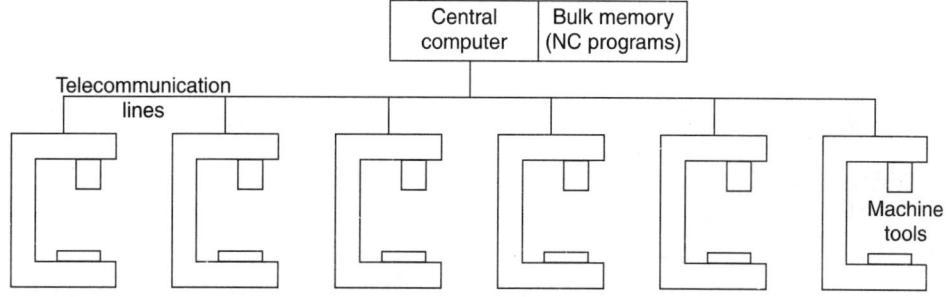

Fig. 14.4. Direct numerical control system.

— The computer calls the program instructions from bulk storage and sends them to the individual machines as the need arises. It also receives feedback data from the machines.

— The above two-way information flow occurs in real time, which mean that each machine requests for instructions that must be satisfied simultaneously. Similarly, the computer must always be ready to receive information from the machines and to respond accordingly.

● The most outstanding feature of a DNC system is that the computer is *servicing a large number of separate machines all in real time.*

Advantages :

1. NC part programs can be conveniently stored in computer files.

2. Elimination of unreliable tape reader and fragile paper tape.

3. The DNC computer is much more accessible for revision and editing or for quick and easy interaction between the programmer and the machine tool.

4. The DNC concept represents a first step in the development of production plants which will be managed by computer system.

5. The large DNC computer performs the computational and data processing functions more effectively than traditional NC.

Disadvantages :

This system is *very expensive* and a *highly skilled software knowledge is needed.* It can usually be justified only in large corporations or if a system is expanded to include process management and control (CAM and CAD).

- DNC system appeared in the market earlier than CNC system, but with the development of dedicated mini-computers, the benefits of DNC system can be realised in CNC system.

14.6. ADAPTIVE CONTROL SYSTEMS

Adaptive control system (ACS) is a logical extension of CNC system in the sense that in CNC system operating conditions are specified by the user in the form of a program. The determination of these operating conditions requires thorough knowledge of the properties of the workpiece and tool material, coolant conditions, and other factors. By contrast, in *ACS, calculation and setting of the operating conditions like speed, feed, and depth of cut are done during the machining by the control system itself.*

- In ACS it is possible to program for constraints on maximum power of the machine or optimising operating conditions for maximum tool life.

Advantages of AC machining :

1. Less intervention of the operator.
2. Increased production rate (as compared to conventional or NC machining).
3. Increased tool life (due to more efficient and uniform use of cutter ; the cutter is never severely loaded.
4. Part programming is easier (since the selection of feed etc. is left to the controller unit rather than to part programmer).
5. Part is protected against an out-of-tolerance condition and possible damage.

14.7. MACHINING CENTRE

A **machining centre** *is a numerically controlled machine tool which is capable of carrying out a range of machining functions (operations) normally performed by a number of different types of conventional machine tools.*

Operations such as milling, drilling, boring, tapping and reaming are performed automatically in accordance with instructions expressed numerically on standard punched tape.

- The machining centre has the capacity to change tools automatically under tape command. Automatic tool changing make it possible to :
  — increase output ;
  — cut down rejects ;
  — reduce the amount of inspection operations.
- *The capability of the machining centre reduces the number of set-ups, time spent in setting, transportation between sections of machines and time spent in waiting for machines to become available before the job can be started on.*
- The *major* **advantage** is that the job needs clamping on the work holding surface only once ; the machine then performs a variety of machining operations on all the job's faces except the base. Work handling time is thus decreased because there is no movement of the workpiece from one machine to another.
- A machining centre is *mainly* **used** for *batch production of main components of a product.*
  — The main components of a product are usually small (about 10 per cent) but are *expensive* (they represent about 50 per cent of the product value) because they have considerable material value and usually require a large amount of machining. For such components, the machining centre is generally most economical.

14.8. ADVANTAGES AND LIMITATIONS OF AUTOMATION AND COMPUTER CONTROLLED MACHINES

The advantages and limitations of *automation and computer controlled machines* are discussed below :

Advantages :

1. Possible to produce high quality products at the lowest price (since new manufacturing technologies give faster production rates, lesser down times and fewer rejections).

2. Overall increase in productivity, higher quality, and meeting tight schedules help to compete in the global market and customers achieve better standard of living.

3. Through new manufacturing technologies, it is possible to have fewer working hours and more leisure time for the worker as the new technologies like robots are 24 hours/day-7 days/week-52 weeks/year workers.

4. Owing to lesser physical participation by the worker, human workers are more safe.

Limitations :

1. Unemployment problems are increased (since new technologies can perform the task of many workers effectively and efficiently).

2. Although it is argued that the automation improves standard of living, in actual practice it declines the purchasing power of customer (Good quality products are available at the lowest price, but with the increased unemployment problems, purchasing power of customers will decrease).

3. Automation can even disturb the economy of a country as a whole (since with increased productivity and reduced purchasing power of customers, there will be decline in the sales of the products that will lead to a situation in which companies may have to stop production for want of customers).

14.9. INTRODUCTION TO CAD/CAM

CAD/CAM (Computer-Aided Design/Computer-Aided Manufacture) technology was initiated in the aerospace industry but presently it is spreading at a rapid pace in all industries.

It can be *defined* most simply as the *use of computers to translate a product's specific requirements into the final physical product.*

Following points are worth noting about CAD/CAM technology :

- With *this system, a product is designed, produced and inspected in one automatic process.*

- It plays a key *role in areas such as design analysis, production planning, detailing documentation, N/C part programming, tooling fabrication, assembly, jig and fixture design, quality control, and testing.*

- Whenever any deviation is noted, a programmable controller takes automatic corrective action to compensate for the deviation. Thus a *closed loop system* is formed which produces *consistent quality products, reduces wastes and improves productivity.*

- CAD/CAM system is *ideally* suited for *designing and manufacturing mechanical components of free form complex with three dimensional shapes.*

14.10. CAD

14.10.1. Definition

In the modern sense, CAD (Computer Aided Designs) is defined as :

"A design process using sophisticated computer graphics techniques, backed up with computer software packages to aid in the analytical, development, costing and ergonomic problems associated with design work".

- CAD makes use of computer systems to assist in the creation, modification, analysis or optimization of a design. The computer systems consist of the hardware and software to perform the specialized design functions required by the particular user firm.
- Modern CAD systems (also often called CAD/CAM systems) are based on *Interactive Computer Graphics* (ICG).
  - ICG denotes a user-oriented system in which the computer is employed to create, transform, and display data in the form of pictures or symbols.
  - ICG system is one component of a CAD system, the other major component is the *human designer*.
- A modern CAD system can perform the following design-related tasks :
  - (*i*) Geometric modelling ;
  - (*ii*) Engineering analysis ;
  - (*iii*) Design review and evaluation ;
  - (*iv*) Automated drafting.
- Most CAD systems are capable of generating as many as six views of the part.

14.10.2. Advantages

The following are the *advantages* of CAD :

1. Drawings can be produced at a faster rate.
2. Drawings produced by CAD systems are more accurate and neat.
3. In this system there is no repetition of the drawings.
4. CAD systems assimilate several special draughting techniques which are not available with conventional means.
5. Design calculations and analysis can be carried out quickly.
6. With CAD systems superior design forms can be produced.
7. CAD simulation and analysis techniques can drastically cut the time and money spent on prototype testing and development—often the costliest stage in the design process.
8. Using CAD systems design can be integrated with other disciplines.
9. Reduced engineering personnel requirements.
10. Improved engineering productivity.
11. Customer modifications are easier to make.
12. Fewer errors in NC part programming.
13. Lead time for process planning can be reduced.
14. Cost saving in tool design and other capital investments can be realized.
15. Quality assurance is improved.

14.11 CAM

14.11.1. General Aspects

CAM (Computer-Aided Manufacture) concerns any automatic manufacturing process which is controlled by computers.

CAM *can be defined as the use of computer systems to plan, manage and control the operations of a manufacturing plant through either direct or indirect computer interface with the plant's production resources.*

The most important elements of CAM are :

1. CNC manufacturing and programming techniques.
2. Computer controlled robotics manufacture and assembly.
3. Flexible Manufacturing Systems (FMS).
4. Computer Aided Inspection (CAI) techniques.
5. Computer Aided Testing (CAT) techniques.

- Each machine in a CAM system has the ability to select and manipulate a number of tools according to programmed instructions ; thus CAM provides a high degree of *flexibility* in performing and controlling manufacturing processes.

- CAM systems are capable of recording important around-the-clock time data such as set-up time, downtime or non-productive time, and run time.

14.11.2. Advantages of CAM

CAM entails the following *advantages :*

1. Product obtained is superior in quality.
2. The manufactured form has a greater versatility.
3. Higher production rates with lower work-forces.
4. There is less likelihood of human error.
5. As a result of increased manufacturing efficiency cost savings are materialised.
6. The production processes can be repeated via storage of data.

14.11.3. Software and Hardware for CAD/CAM

The functions of CAD/CAM system are mainly determined by the *software.*

Software usually consists of a *number of separate application packages to perform the desired function.* The size of computer depends on the number and sizes of packages and number of work stations.

Hardware *is responsible for the reliability and speed of response of the system.*

A wide range of standard software is available and generally it is not worth developing users own software. Though a system can be built up from standard software packages from different sources and standard hardware, it is often costly because of the considerable programming effort required to interface the packages to a common data base to provide user friendly software to adapt the system to the user's requirements. It is thus *advisable to adopt turn key system for turn key suppliers.*

14.11.4. Functioning of CAD/CAM System

- CAD/CAM is *an interactive computer graphic tool that enhances design and manufacturing functions to create a highly profitable product.* This technique is being applied by big industries for improving overall manufacturing performance.

- It is *not* a standard tool which can be fitted into any company but has to be tailored to suit the needs of the company. It is rather complex technology and has wide potential for immediate benefits.

- Usually this *tool consists of a dedicated computer, which is connected to a number of work-stations.* The system is used to assist in the design and manufacturing through the use of an *expandable set of linked software modules.* A designer can define dimensions and display views of 2 dimensions, $2\frac{1}{2}$ dimensions and 3 dimensions parts on modules. It is possible to generate the families of part directly by a parametric processor either by direct scaling or using a catalogue of subprograms. From the geometric definition a solid model can be constructed, to assist in a visualisation. It is possible to store complete details of designs on numerical control types for subsequent use on demand. Bench making tests are carried out to ensure system's capability.

14.11.5. Features and Characteristics of CAD/CAM System

The following are the *features and characteristics of CAD/CAM system* :

1. A major portion of the output of the engineering sector involves batch production and CAD/CAM offers immense cost and quality benefits for such requirements.

2. The work-in-progress, in batch production, is reduced considerably.

3. It is possible to produce at random all the variants and series of a product planned to be manufactured by a firm.

4. Such a system has inherent flexibility to cater to new models of the product in pipeline without major modification.

5. In such a system, several machining centres are arranged one after the other with robots and proper automatic materials handling equipment. Software is developed to integrate the machine CNC control and the handling system. Each machining centre is equipped with several tool magazines. All the tools required to complete each operation on each model of the product can be stored in the magazine.

6. All the part programs for the different models are stored in the memory. System has only to identify the model of the product presented to a machine in order to complete the machining operations. Thus it is possible to have totally random mixes of models of a product proceeding down the line at any one time.

7. The system can be conceived in multiplies of 15-20 minutes operations. If certain operations take longer, then multiples of similar machines can be installed in the line. Sometimes identical machines are introduced for each operation so that production can continue even if one machine goes down.

Procedure :

● The components are loaded on to a pallet. Means are provided to identify the exact model.

● Loaded pallets enter the line and wait at the start of the line until a signal that one of the first operation machines is vacant is obtained.

● The handling system automatically directs the pallet to the first vacant machine for first operation.

● The pallets are loaded on a fixture. The fixture is designed so that it permits access to all four sides and end faces and wherever machining operation is required. The pallets are designed to have windows where access for machining is required.

● As the pallet enters the machining area, air blast clears both the fixture and pallet locations. The fixture is then properly clamped and supported. Touch trigger probes are used to check its location in the pallet.

● Probes also identify the exact model of the component and signals from the probes active master calling program which selects the appropriate past program and sub-routines from the control memory.

● An overhead cascade coolant wash is provided to clear away swarf before the pallet is located. All coolant and swarf is carried away via underground ducts to a central separation and coolant filtration plant.

Some systems can show metal being removed dynamically.

● It is possible to store libraries of standard tools and tool holders, thus carrying out process planning.

- By calling up and manipulating standard fixtures components, like studs, stops, clamps, bushes, location devices, fixtures etc., it is possible to design a fixture for a component already designed on the CAD/CAM system.
- It also allows sheet metal development (unfolding), taking account of the material for the bends. It is also possible to layout sheet metal components on a standard sheet to reduce the waste (nesting). Factory layout process planning and robot programming have also been attempted.
- Exploded views, schematics and diagrams, 3-D colour shades like photographic views of the parts can be produced.
- Tenders and estimates can be quickly produced to high quality.

14.11.6. Application Areas for CAD/CAM

The potential application areas for CAD/CAM are :

1. Design and design analysis :

- CAD system would be best suited for *drawing offices where frequent modifications are required on drawing and several parts repeat.*
- It must be remembered that it is very easy with computer to make modifications and very fast to draw part profile once its details are fed into computer.
- Once a drawing is entered in the CAD system, later modifications can be done quickly, and detail drawings can be prepared quickly from a general arrangement drawing.
- NC tapes can be produced.
- Storing of the drawing is very convenient, easy, occupies very less space and symbols for electrical, hydraulic control and instrumentation circuits can be called up quickly and positioned on the schematic drawing.
- Standard components can be stored permanently in the data base and called up and positioned on the drawing, resulting in saving of time and enforcement of standards. It is possible to associate nongraphical information like past number, supplier, material etc. for any component assembly.
- It is very convenient to calculate properties like weight, centre of gravity, moment of inertia, etc., because 3-D models can be easily produced.
- It is also possible to carry out finite element analysis by producing meshing for analysis.

2. Manufacture :

- With CAD/CAM system the complete NC part programming process can be carried out interactively, including post processing and production of NC tape. Source programs in languages such as APT can be produced. Systems can verify tapes by producing tool centre path plots.

QUESTIONS WITH ANSWERS

Q. 14.1. Explain briefly the following categories of automation :

1. Fixed automation ;

2. Programmable automation ;

3. Flexible automation.

Ans. 1. **Fixed automation :**

- It is also known as *hard automation,* which is used to produce a standardized product such as gears, nuts and bolts, etc.

- Even though the operating conditions can be changed, *fixed automation is used for very large quantity production of one or few marginally different components.*
- Highly specialized equipment, called *special purpose machine tools,* are utilized to produce a product or a component of a product very efficiently and at high production rates such that the *unit cost are low relative to the alternative methods of production.*

2. **Programmable automation :**

- In this type of automation we can change the design of the product or even change the product by *changing the program.*
- It is *useful for the low quantity production of large number of different components.*
- The equipment used for the production is designed to be adaptable or programmable.
- The production is normally carried out in *batches.*

3. **Flexible automation :**

- This is also called *'Flexible Manufacturing System* (FMS)'.
- This *allows manufacturing different products on the same equipment in any order or mix.*

Examples :

(*i*) Numerical control—Descrete parts manufacturing.

(*ii*) Robot-programmable automation equipment that is not directly used as the production equipment. Robots are an integral part of FMS and Computer Integrated Manufacturing Systems (CIMS).

Q. 14.2. What is 'CNC retrofitting' ? Explain briefly.

Ans. The term **retrofit** *means adding accessories to a given object in order to improve the performance.*

The high initial investment required for CNC machines restricts medium and small scale industries to take benefits of this advanced technology. Especially in developing countries like India, where small scale industries form the backbone of our economy, *it is possible to modernise the existing conventional machines into CNC machines by adding accessories and making slight design modifications.* This is known as **CNC retrofitting.**

- Retrofitted machines are *not as good as CNC machines, but perform better than the conventional machines.* Even though the ultimate solution is to go in for sophisticated CNC machines, retrofitting finds a place where there is a *budgetary contraint.*

Q. 14.3. Discuss briefly features of NC machines.

Ans. The various features of NC machines are as follows :

1. **Structure :**

- The structure of NC machines should be very rigid to withstand heavy cuts and maintain precision and accuracy built-in for long period.
- Present day NC machines use good quality cast iron structure or welded structure for bed and column of these machines.

2. **Guidways :**

- In the case of NC machines, hardened and ground flat guidways or a combination of *V* and flat guidways are commonly used. Often they are in the form of replaceable steel strip.
- The slides are equipped with preloaded recirculating type roller bearings or friction reducing liners to provide smooth movement free of *stick slip.*

3. **Spindle drive :**
- The spindle drive on modern NC machines is invariably a D.C. drive in combination with three or four step gear-boxes in order to get a wide range of spindle speeds to provide full power in the entire cutting speed range. These drives are controlled by SCR (Silicon Controlled Rectifier) controllers.
- All axes are driven by powerful DC servo drives controlled by SCR controllers through amply dimensioned and rigidly mounted preloaded ball screws and nuts.

4. **Automatic tool Changer :**
- Tool selection is done at random with either tool position or with tool holder coding.
- Some NC machining centres use pocket coding with computer update facility for the location of the tools in the pockets.

5. **Thermal stabilization.** In most NC machines, suitable refrigeration system is provided as an option.

6. **Lubrication system.** All modern NC machines are equipped with centralized lubrication system with automatic supply in predetermined quantities.

HIGHLIGHTS

1. Numerical control, *NC* can be defined simply as control by numbers.
2. A machine tool having a dedicated computer to help prepare the program and control some or all of the operations of the machine tool is called Computer Numerical Control (*CNC*) machine tool.
3. The main elements of a NC machine tool are :
   The control unit ; The drive units ; The position feedback page ; Magnetic box ; Manual control.
4. The principal functions of CNC are :
   Machine tool control ; In-process compensation ; Improved programming and operating features ; Diagnostics.
5. A *machining centre* is a numerically controlled machine tool which is capable of carying out a range of machining functions (operations) usually performed by a number of different types of conventional machine tools.
6. *CAD/CAM system* is ideally suited for designing and manufacturing mechanical components of free form complex with three dimensional shapes.

OBJECTIVE TYPE QUESTIONS

Fill in the blanks or Say 'Yes' or 'No' :
1. Mechanisation means that something is done or operated by machinery and not by hand.
2. .... machine layout system is normally employed for batch production and non-repetitive kind of manufacturing activities.
3. .... control can be defined as a form of programmable automation in which the process is controlled by numbers, letters and symbols.
4. Most automated machine tools are uneconomical for small runs, custom made products or job lots.
5. In a NC machine tool, the .... is replaced by the data processing part of the system and the control unit.
6. The .... is a device for winding and reading the punched tape containing the program of instructions.
7. NC drill processes are a good example of PTP systems.
8. .... paper tape is widely used as a medium for feeding.
9. The .... of NC machines should be very rigid to withstand heavy cuts and maintain precision and accuracy built-in for long period.
10. Straight line system is an extension of point-to-point system.
11. A closed loop system is .... expensive than an open loop system.
12. Highly skilled operator is required to run NC machine.
13. The major disadvantage of NC machines are their costs.

14. In .... system, several NC machines can be controlled by a large central computer.
15. DNC system appeared in the market earlier than CNC system.
16. A machining centre is mainly used for .... production of main components of a product.
17. Machining centres have low metal removal rate capabilities.
18. The .... program is a very important software element in a NC manufacturing system.
19. CAD and CAM are separate disciplines.
20. A .... system is basically a design tool in which the computer is used to analyse various aspects of a designed product.
21. Modem CAD systems are based on Interactive Computer Graphics (ICG).
22. The typical ICG system is a combination of hardware and software.
23. .... is considered to be the ultimate in total manufacturing capability.
24. Probably the most mature of the CAM technologies is ....
25. CAM provides a high degree of flexibility in performing and controlling manufacturing processes.

ANSWERS

1. Yes	2. Conventional	3. Numerical	4. Yes	5. Operator
6. tape recorder	7. Yes	8. Punched.	9. sturcture	10. Yes
11. more	12. No	13. Yes	14. DNC	15. Yes
16. Batch	17. No	18. part	19. Yes	20. CAD
21. Yes	22. Yes	23. CAM	24. NC	25. Yes.

THEORETICAL QUESTIONS

1. Define the term 'Numerical Control (NC)'.
2. Explain briefly the working of a NC machine tool.
3. Explain with a neat diagram the main elements of a NC machine tool.
4. How are NC machines classified ? Explain briefly.
5. List the applications of NC machines.
6. What are the advantages and disadvantages of NC machines ?
7. Give the comparison between conventional machines and NC machines.
8. Define the term 'Computer Numerical Control (CNC)'.
9. List the functions of CNC.
10. What are the advantages and disadvantages of CNC machines over NC machines.
11. List the applications of CNC.
12. Give the comparison between NC and CNC machines.
13. What do you mean by "Direct Numerical Control (DNC)" ? Explain briefly. State its advantages and disadvantages also.
14. Discuss briefly 'Adoptive control systems'.
15. Write a short note on "Machining centre".
16. What are the advantages and limitations of automation and computer controlled machines ?
17. Define the term 'CAD'.
18. List the design-related tasks which a modern CAD system can perform.
19. What are the advantages of CAD ?
20. Define the term 'CAM'.
21. List the most important elements of CAM.
22. What are the advantages of CAM ?
23. Discuss briefly functioning of CAD/CAM system.
24. List the features and characteristics of CAD/CAM system.
25. What are the application areas for CAD/CAM ?

15

Automatic Machines

15.1. INTRODUCTION

Automatic machines. *"Automatic machines" are those machines in which both the workpiece handling and the metal cutting operations are performed automatically.* In these machines, operations right from feeding of the stock to clamping, machining and even inspection of the workpiece are carried out automatically.

Automation. *It means a system in which many or all of the processes in the production, movement, and inspection of parts and material are automatically performed or controlled by self-operating devices.* This term is used to denote the *continuous automatic production of a product.*

— *Partial automation* means replacement of human activities or involvement by automatic means only *partially.*

— In *full automation*, the human involvement is totally eliminated and the process is entirely carried out and controlled through automatic means alongwith a proper feedback system.

Mechanization. It means that something is done or operated *by machinery* and not by hand. Feedback is *not* provided and thus one deals with *open-loop systems.*

— The automatic screw machine with its cam actuated tools and sequencing represents a high degree of mechanization.

15.2. ADVANTAGES OF AUTOMATION

Automation entails the following *advantages :*

1. Increased productivity.

2. Reduced overall production cost.

3. Less floor area required.

4. Increased overall profits of the manufacturing concern.

5. Human fatigue is greatly minimised

6. Reduced maintenance requirements.

7. The workers avail better working conditions.

8. Owing to the use of standardised parts and assemblies the inventory requirement is minimised.

9. Uniform components are produced.

10. There is an effective control over the production process.

11. The workpiece is tidy and safe.

12. With the use of group technology a considerable saving is effected in terms of design cost, material cost, tooling cost etc.

13. There is a great improvement in the quality and reliability of the products.

14. Human safety is fully ensured.

15.3. TRANSFER MACHINES

Transfer-type production machines, frequently designated as *automated machines, complete a series of machining operations at successive stations and transfer the work from one station to the next.* They are in effect a *production line of connected machines* that are synchronized in their operation so that the workpiece, after being loaded at the first station, *progresses automatically through the various stations to its completion.*

"*Transfer machines*" perform a variety of *machining, inspecting, and quality control functions.* They drill, mill, hone, and grind, as well as control and inspect the operations.

15.3.1. Types of Transfer Machines

Transfer machines are of the following *three* types :

1. Rotary indexing table transfer machines.

2. In-line transfer machines.

3. Drum type transfer machines.

1. **Rotary indexing table transfer machines :**

In rotary transfer system the workpieces are held in fixtures on a continuous rotating table. The rotating table brings the workpieces under different machines (Fig. 15.1).

Owing to the problems of rigidity and maintenance of proper accuracy, the table size is always limited and, hence, is the number of *stations* fixed on this type of machine. Usually, such a machine carries 6 or 8 stations around it, although for smaller workpieces this number can be as high as 16. The components size and required number of work stations are the main factors which affect the determination of table size and the number of stations.

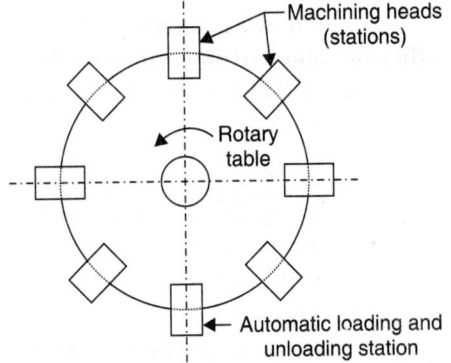

Fig. 15.1. Arrangement of rotary indexing table transfer machine.

- This method is quite compact and permits the workpiece to be loaded and unloaded at a *single location without having to interrupt the machining.*

- This type of arrangement is *best suited for automatic assembly of a product.*

2. **In-line transfer machines :**

In this arrangement the workpiece is held in a *fixture* or *special "pallet".* The fixtures are located and clamped in proper position. A schematic diagram of an In-line transfer machine is shown in Fig. 15.2.

"*Palletized*" *work holding fixtures secure the transmission during all operations.* Pallets are often carried in conveyors which are indexable.

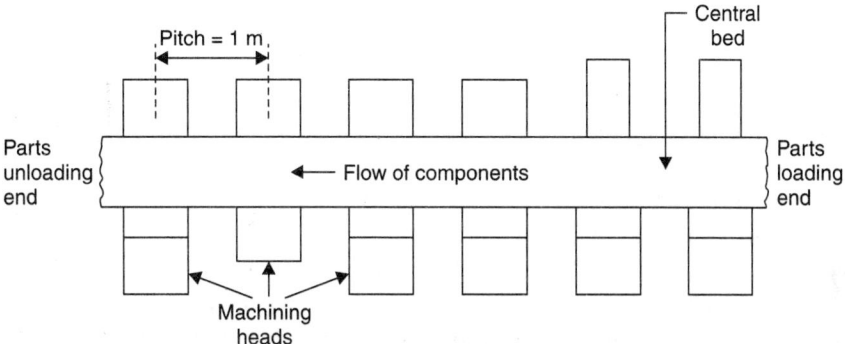

Fig. 15.2. Schematic diagram of an In-line transfer machine.

Following are the **functions** *of an in-line transfer machine :*

(*i*) Transfers workpiece to the first station and then from station to station.

(*ii*) Locates and clamps the work at each station.

(*iii*) Rapid approach of the tools to the work.

(*iv*) Feed the tools through the cutting cycle.

(*v*) Retract the tools clear of the work and guide bushes

(*vi*) Unclamp the workpiece at various stations ready for further transfer.

3. **Drum type transfer machines :**

In these types of machines work fixtures are fastened to the outside surface or periphery of the drum and work stations are positioned radially around the circular path at equal intervals.

As the work hangs from the fixture, the clamping arrangement must be fool proof and efficient. Like circular indexing arrangement, this too cannot be big in size.

● *Solution of transfer machines* depends on the following factors :

(*i*) Product size and machining requirements.

(*ii*) Handling systems used.

(*iii*) Floor space available.

● *Transfer machines range from comparatively small units having only two to three stations to long straight line machines with more than 100 stations.*

● These machines are *used primarily in the automobile industry.* Products processed by these machines include cylinder blocks, cylinder heads, refrigeration compressor bodies and similar parts.

15.3.2. Constructional Features of a "Transfer Machine"

The principal constructional features of a 'Transfer machine' consists of the following main parts and mechanisms :

1. Central bed.

2. Machining heads.

3. Automatic work holding and Transfer mechanisms.

4. Locating and clamping devices.

5. Cooling supply mechanisms.

6. Chip disposal devices.

7. Control systems.

15.3.3. Advantages and Disadvantages of Transfer Machines

Following are the *advantages* and *disadvantages* of transfer machines :

Advantages :

1. Higher production rates are achieved.
2. Higher accuracy is obtained.
3. Less floor space is required.
4. Heavy workpieces can be easily handled.
5. The quality of products is considerably improved.
6. Complex shaped components can be conveniently machined.
7. The length of production cycle is reduced.
8. Less number of operators are required.
9. Increased tool life (resulting in further reduction in production cost).

Disadvantages :

1. High initial investment.
2. A breakdown of one machine means stoppage of whole of the production line.
3. Complex control systems are required.
4. Much time is required to change over the machine to handle a different shaped components.
5. Very high overhauling and maintenance costs of transfer lines, specifically when reshuffling is required.

15.4. MACHINING CENTERS

15.4.1. General Aspects

It has been observed that each machine, regardless of how highly it is automated is designed to perform basically one type of operation ; furthermore, in manufacturing operations, most parts generally require a number of different machining operations on their various surfaces. None of the processes and machine tools described thus far could individually produce these parts.

Transfer lines are commonly used in high volume or mass production. There are situations and products, however, for which such lines are *not* feasible or economical particularly when the types of products to be machined change rapidly. An important concept (developed in late 1950) is *machining centres*.

A **machining center** is *a computer-controlled machine tool capable of performing a variety of cutting operations on different surfaces and in different orientations on a workpiece.* In general, the *workpiece is stationary*, and the *cutting tools rotate,* such as in milling and drilling. Because of the high productivity of machining centers, large amounts of chips are produced and must be collected and disposed of properly.

— In a machining center, the workpiece is placed on a *"pallet"* that can be oriented in three principal directions, as well as rotated around one or more axes on the pallet. Thus, after a particular operation is completed, the workpiece does not have to be moved to another machine for subsequent operations (say, drilling, reaming, and tapping). In other words, the tools and machine are brought to the workpiece.

— After the completion of all the cutting operations, the pallet automatically moves away, carrying the finished workpiece, and another pallet containing another workpiece to be machined is brought into position by *"automatic pallet changers"*.

— All movements are guided by computer control, and pallet-changing cycle times are of the order of 10–30s.

— Pallet stations are available with multiple pallets that serve the machining center.

- Machining centres are equipped with a *"programmable automatic tool changer"*. The cutting tools are automatically selected, with random access for the shortest route to the machine spindle. Tools (as many as 200 tools can be stored in a magazine, drum or chain) are identified by coded tags, bar codes, or memory chips applied directly to the tool holders).

- The maximum dimensions that the cutting tools can reach around a workpiece in a machining center is known as the *"work envelope"*.

15.4.2. Types of Machining Centers

The *two* basic types of designs of machine centers are :

1. Vertical-spindle machining centers.
2. Horizontal-spindle machining centers.
3. Universal machining centers.

1. **Vertical-spindle machining centers :**

- In vertical-spindle machining centers or *'vertical machining centers'* the thrust forces are directed *downward* and as such these machines have *high stiffness and produce parts with good dimensional accuracy.*

- These are generally less expensive than horizontal-spindle machines.

- These machines centers are *suitable for perfoming various machining operations on flat surfaces with deep cavities, such as in mould and die making.*

2. **Horizontal-spindle machining centers :**

- These machining centers (also simply called *'horizontal machining centers'*) are suitable for *large or tall workpieces that require machining on a number of their surfaces.* The pallet can be rotated on different axes and to various angular positions.

- Another category of these machines is *"turning centers"* which are computer-controlled lathes with several features.

3. **Universal machining centers :**

- The machining centers are equipped with both vertical and horizontal spindles.

- These are so called since they have a variety of features and are capable of machining all surfaces of a workpiece.

15.4.3. Characteristics of Machining Centers

The major characteristics of machining centers are as follows :

1. *Highly automated and relatively compact* (so that one operator can attend two or more machines at the same line).

2. Significant *reduction* in the need for a variety of machine tools and a large amount of floor space (since the machines are versatile, having as many as six axes of linear and angular movements and capability of quick change over from one type of product to another.

3. These machines *can handle a variety of part sizes and shapes* efficiently, economically, and with repetitively high dimensional tolerances (on the order of $\pm$ 0.0025 mm).

4. These machines are equipped with tool-condition monitoring devices for *detecting* tool breakage and wear, as well as probes for compensation for tool wear and for tool positioning.

5. *Improved productivity, reduced labour requirements and minimised overall cost* (owing to reduction in time required for loading and unloading workpieces, changing tools, gauging, and troubleshooting)

15.4.4. Selection of Machining Center

Some important factors on which the selection of a particular type and size of machining centre depends, are as follows :

1. Rate of production.
2. Type of products, their size, and their shape complexity.
3. Type of machining operations to be performed and type and number of tools required.
4. Dimensional accuracy of products.

- While selecting machining centers, though *vereability* is the key factor, jet *these considerations must be weighted against the high capital investment required and compared with those of manufacturing the same products by using a number of the more traditional machine tools.*

HIGHLIGHTS

1. *Automatic machines* are those machines in which both the workpiece handling and the metal cutting operations are performed automatically.
2. 'Transfer machines' perform a variety of machining, inspecting, and quality control functions. These machines are of the following types : (*i*) Rotary indexing table transfer machines, (*ii*) In-line transfer machines, and (*iii*) Drum type transfer machines.
3. A *machining center* is a computer-controlled machine tool capable of performing a variety of cutting operations on different surfaces and in different orientations on a workpiece.

OBJECTIVE TYPE QUESTIONS

Fill in the blanks or Say 'Yes' or 'No' :

1. In .... machines, operations right from feeding of the stock to clamping, machining and even inspection of the workpiece are carried out automatically.
2. .... automation means replacement of human activities or involvement by automatic means partially.
3. .... means that something is done or operated by machinery and not by hand.
4. Automation increases productivity.
5. Automation does not fully ensure the human safety.
6. Palletized workholding fixtures secure the transmission during all operations.
7. Transfer machines are used primarily in .... industry.
8. In transfer machines higher production rates are not achieved.
9. Transfer machines entail high initial investment.
10. A .... is a computer-controlled machine tool capable of performing a variety of cutting operations on different surfaces and in different orientations on a workpiece.

ANSWERS

1. automatic	2. Partial	3. Mechanization	4. Yes	5. No
6. Yes	7. automobile	8. No	9. Yes	10. machining center.

THEORETICAL QUESTIONS

1. What are 'Automatic machines' ?
2. Define the following terms :

   (*i*) Automation   (*ii*) Partial automation   (*iii*) Mechanisation.

3. What are the advantages of automation ?
4. What is a 'Transfer machine' ?
5. Name the various types of 'Transfer machines' and explain briefly any one of them.
6. Explain briefly any two of the following machines :
   (i) Rotary indexing table transfer machines.
   (ii) In-line transfer machines.
   (iii) Drum type transfer machines
7. What are the functions of an 'In-line transfer machines' ?
8. List the constructional features of a 'Transfer machine'.
9. State the advantages and disadvantages of 'Transfer machines'.
10. What is a 'Machining center' ? Explain briefly.
11. What are the various types of machining centers ? Explain any one of them briefly.
12. Describe briefly any rwo of the following machines :
   (i) Vertical-spindle machining center.
   (ii) Horizontal-spindle machining center.
   (iii) Universal machining center.
13. Mention the major characteristics of machining center.
14. List some important factors on which the selection of a particular type and size of machining center depends.

16

Jigs and Fixtures

16.1. Introduction. 16.2. Definitions and concept of jigs and fixtures. 16.3. Advantages of using jigs and fixtures. 16.4. Main components or elements of jigs and fixtures. 16.5. Principles of jigs and fixtures design. 16.6. Degrees of freedom. 16.7. Principles of location. 16.8. Locating devices. 16.9. Clamping devices—Introduction—basic requirements of clamping devices—Principles of clamping—types of clamps. 16.10. Materials used for making locating and clamping devices. 16.11. Types of jigs. 16.12. Types of fixtures. Highlights—Objective Type Questions—Theoretical Questions.

16.1. INTRODUCTION

Jigs and fixtures *are production tools used to accurately manufacture duplicate and interchangeable parts*. They are specially designed so that a large number of components can be machined or assembled identically, and to ensure interchangeability of components. They eliminate the necessity of a special set up for each individual part.

— Jigs and fixtures are *precision tools*.

— They are *expensive to produce* because they are made to fine limits from materials with good resistance to wear.

— They must be properly stored or isolated to prevent accidental damage, and they must be numbered for identification for future use.

16.2. DEFINITIONS AND CONCEPT OF JIG AND FIXTURE

Jig :

*A **jig** is a device in which a component is held and located for a specific operation in such a way that it will guide one or more cutting tools to the same zone of machining.*

The use is economically feasible, if this operation is to be performed on more than one component.

Fig. 16.1 shows a simple drilling jig. The workpiece to be drilled is held and positioned in the drilling jig. Bush(es) guide the drill(s) at desired location(s) in the workpiece.

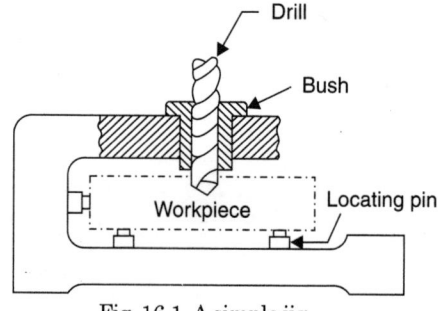

Fig. 16.1. A simple jig.

- The usual machining operations for jigs are *drilling and reaming*. Jigs are usually fitted with *hardened steel bushings* for guiding drills or cutting tools.

- The most common jigs are *drilling jigs, reaming jigs, assembly jigs* etc. When these are used, they are usually not fastened to machine tools or table but are free to be moved so as to permit the proper registering of the work and the tool.

Fixture :

A ***fixture*** *is a production tool that locates, holds and supports the work securely in a fixed orientation with respect to the tool so that the required machining operations can be performed.*

Fig. 16.2 shows a simple fixture.

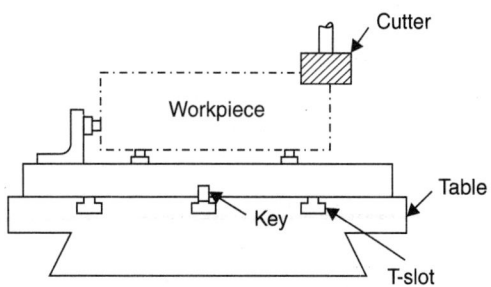

Fig. 16.2. A simple fixture.

- Fixtures vary in design from relatively simple tools to expensive complicated devices. These are most frequently attached to some machine tool or table. Consequently they are associated in name with the particular machine tool with which they are used, *e.g., milling fixtures, broaching fixtures, assembly fixtures* etc.

Differences between jigs and fixtures :

Following are the *differences* between jigs and fixtures :

1. *Essential* **difference** *between a jig and fixture* is *that the former incorporates bushes that guide the tools whilst the latter holds the component being machined with the cutters working independently, of it.*

2. **Jigs** are used on *drilling, reaming, tapping and counterboring operations,* while **fixtures** are used in connection with *turning, milling, grinding, shaping, planing and boring operations.*

3. Whereas **jigs** are *connected with operations,* **fixtures** *most commonly are related to specific machine tools.*

4. *Jigs are lighter than fixtures,* for quick handling ; *fixtures are heavier in construction and bolted rigidly on the machine table.*

16.3. ADVANTAGES OF USING JIGS AND FIXTURES

The use of *jigs* and *fixtures* in mass production of identical parts offer the following *advantages* :

1. Improved productivity.

2. Rapid production work.

3. Reduced manufacturing costs (since large number of identical and interchangeable parts are produced) using jigs and fixtures.

4. Large reduction in fatigue to the operator (since there is considerable reduction in manual handling operations).

5. Complex and heavy components can be easily machined (since such parts can be rigidly held in proper location for machining in jigs and fixtures).

6. Owing to high clamping rigidity (offered by jigs and fixtures), higher speeds, feeds and depth of cut can be used.

7. Increased machining accuracy (owing to the automatic location of the work and guidance of the tool).

8. Considerable reduction in the expenditure due to inspection and quality control of finished components (since the parts produced with the use of jigs and fixtures are very accurate).

9. Jigs and fixtures provide easy means for manufacture of interchangeable parts and, thus, facilitate easy and quick assembly.

10. Considerable saving in labour cost (since the use of jigs and fixtures facilitates deployment of less skilled labour because setting of tools and workpieces is not manual).

16.4. MAIN COMPONENTS OR ELEMENTS OF JIGS AND FIXTURES

In order to fulfil their basic functions, both jigs and fixtures should process the following components or elements :

1. Sturdy and rigid *body.*
2. *Locating elements.*
3. *Clamping elements.*
4. *Tool guiding elements* (for jigs) or *tool setting elements* (for fixtures).
5. *Positioning elements* (these elements include different types of fastening devices)
6. *Indexing elements* (not always provided).

● The complexity of a jig or fixture is determined by :

(*i*) The number of pieces that must be produced.

(*ii*) The degree of accuracy required.

(*iii*) The number and kind of machining operations that must be performed.

Classification of jigs and fixtures :

Jigs and fixtures, according to the degree of mechanization, are *classified* as follows :

1. Hand operated.
2. Power operated.
3. Semi-automatic.
4. Automatic.

16.5. PRINCIPLES OF JIG AND FIXTURE DESIGN

While designing the jigs and fixtures the following *principles* should be borne in mind :

1. *Reduction of idle time.*
2. *Rigidity.*
3. *Clearance between jig and component.*
4. *Swarf clearance.*
5. *Locating points and supports :*

● Locating and supporting surfaces should wherever possible be removable. Generally the surfaces should be of hardened material.

● Make sure that locating points are clearly defined and to hold swarf from adjacent positions.

● For easy removal of worn out locating or supporting pins, these should be fitted into through-holes and not blind holes.

6. *Easy loading and unloading of jig.*
7. *Clamping :*

● Clamping should always be arranged directly above the points supporting the work.

● Fibre pads should be riveted to clamp faces where any metallic contacts with work would cause damage.

● Arrange all clamps and adjustments on the side of the fixture nearest to the operator whilst the component is being loaded and unloaded.

- Design the clamping arrangements in such a way that they can be easily and quickly removed clear of the work and avoid the necessity for lengthy unscrewing of nuts.

- Do not rely on the clamps to hold the work against the cut, but arrange fixed stops which will take the direct thrust of the cutters.

8. *Fool-proofing :*

- Since the jigs and fixtures are mostly used by unskilled labour, care should be taken to ensure that the components can be loaded into position correctly. Pins and similar devices can be often driven into the face of the jig to present the components being loaded incorrectly.

9. *Design for safety.*

10. *Components should be ejected.*

11. *Spring location :*

- On any rough component, the number of locations should never exceed three in any one plane. The component will sit on three points without locking. Should it be necessary, however, for further supports to be provided these should be spring loaded so that after the component is on the three fixed positions, other where necessary will automatically rise to touch the component through the medium of the springs. These spring locations can then be locked in position.

12. *Jig base :*

- A jig which is not bolted to machine table must be provided with four feet instead of whole bottom surface lying flat on the machine table. In this way jig will rock if it is not standing square on the machine table due to chips under one of the feet and warn the operators.

13. *Accuracy :*

- Jig or fixture design has to determine the permissible variations in an operation. The locating and clamping system should be so designed so as to keep the variations in the desired limits.

- Variations arise from the following *causes :* (*i*) Variations in the dimensions of the workpieces ; (*ii*) Variations in material conditions ; (*iii*) Defects in tools and machines ; (*iv*) Wear ; (*v*) Deflection ; (*vi*) Thermal expansion ; (*vii*) Errors of human judgement.

14. *Jig bushes :*

- The use of bushes provide an advantage that bushes when worn out due to flowing out chips, can be replaced at intervals.

16.6. DEGREES OF FREEDOM

The complete location of a component inside a *jig* or *fixture* can be understood with the help of Fig. 16.3. The workpiece is a *cube* having perfectly flat and true faces and is located in space to act as a free body. The workpiece is free to revolve around or move parallel to any axis in either direction. To visualise this, the planes have been marked *X–X', Y–Y'* and *Z–Z'.* The directions of movement are numbered from 1 to 12.

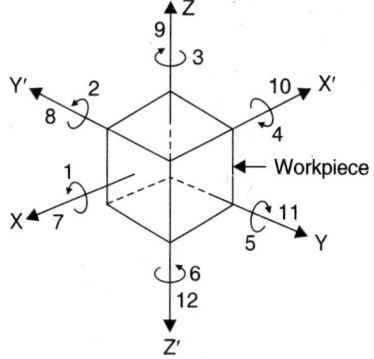

Fig. 16.3. Degrees of freedom of movements of a free body (cube).

16.7. PRINCIPLES OF LOCATION

The principles of location are described as follows :

1. 3-2-1 method of location :

● A workpiece can be positively located by means of 6 (six) pins so positioned that collectively they restrict the workpiece in nine of its degrees of freedom. This is known as *six point location method. Three* of these pins are in the first plane and *two* in the second plane perpendicular to the first and *one* in the third plane perpendicular and adjacent to the both first and second planes.

2. Principle of least point :

● In order to secure location in any one plane, points more than necessary should not be used.

● However, if more points are used such as for finished surface, the extra ones should only be inserted because they serve a useful purpose and care must be taken that they *do not impair the location.* Extra supports if needed should be made *adjustable also.*

3. The principle of mutually perpendicular planes :

● An *ideal location of a component* is achieved when it is *located on six locating points in three mutually perpendicular planes.* Other arrangements possible but not desirable; the following *disadvantages* result if the locating points do not lie on the mutually perpendicular planes :

(*i*) The displacement of locating point or a particle (chip or dirt) adhering to it introduces a correspondingly large error.

(*ii*) The adjacent locating surfaces will provide a wedging action due to which the workpiece will have a tendency to lift.

4. Location of accurate work :

● Efforts should not be made to locate from a hole or a position previously machined, on which a wide tolerance is permissible but *consider the advisability of having the tolerance tightened* so that the required results on subsequent operations may be obtained.

5. Principle of extreme position :

● On any one workpiece surface, locating points should be chosen as *far apart as possible.* Thus, for a given displacement of any locating point from another, the resulting deviation decreases as the distance between the points increases.

6. Small locating surfaces :

● The small locating surfaces (*e.g.* rest buttons, supporting pins or cylindrical locators) entails the following *advantages :*

(*i*) Reduced chance for lodgement of disturbing particles.

(*ii*) Less time for cleaning.

(*iii*) Some saving in material and labour.

(*iv*) A more realistic approach to the mean plane of a rough surface.

However, the only *disadvantage* is large rate of wear causing early replacements. Thus, these surfaces should not be too small which *also create a high specific pressure on the workpiece at such points.*

7. Swarf clearance :

● All corners that collect small chips and swarf must be avoided by relieving them.

● Undercuts should be provided to all locators where they create corners with the surface to which they are fixed.

8. **Replacement a necessity :**

● All locating points which require replacement due to wear and tear should be easily replaceable or repairable.

16.8. LOCATING DEVICES

Locating devices *place the workpiece in essentially the same position, in a jig or fixture, cycle after cycle.* In this sense, the locator provides a reference point from which all sizing or spacing way be accomplished.

The various locating devices are enumerated and described as follows :

1. Support pins and jack pins.
2. Cylindrical locators or location pins.
3. Conical locators.
4. Diamond pin locator.
5. Vee locators.

1. Support pins and jack pins :

When the work is clamped directly in the body of the jig or fixture, *i.e.*, directly on the base or on the walls, there is always a chance of error due to improper support on account of the unevenness of the work surface facing the base or the wall.

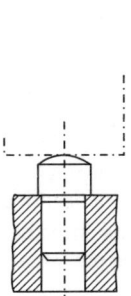

Fig. 16.4. Support pins or rest buttons (fixed type).

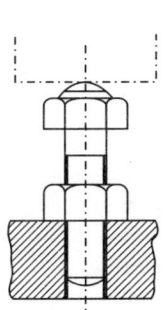

Fig. 16.5. Support pins or rest buttons (adjustable type).

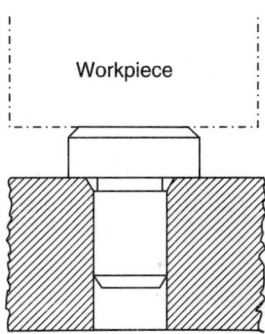

Fig. 16.6. Support pad.

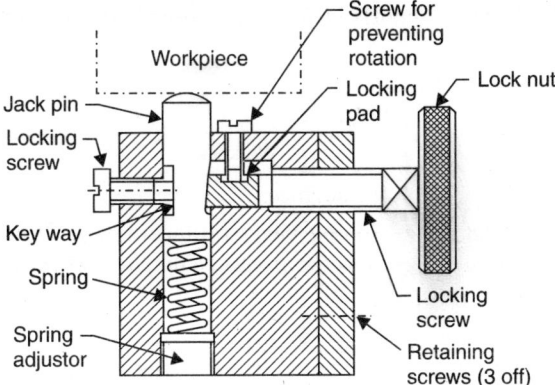

Fig. 16.7. Spring type jack pin.

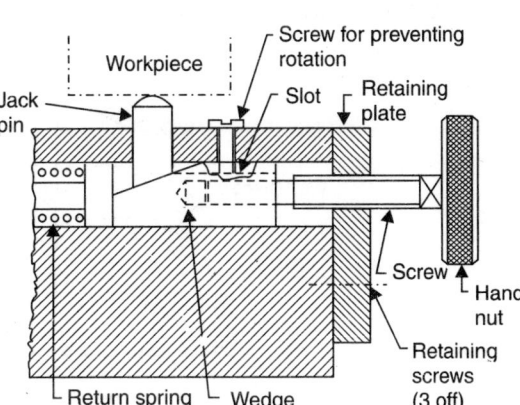

Fig. 16.8. Wedge type jack pin.

Therefore, in most cases except in case of reamed or otherwise finely finished holes, support pins or jack pins are employed to support the work on the specific points on its surfaces. These pins or jacks are of the following various types :

(i) Support pins or rest buttons (Fixed type)   ...Fig. 16.4

(ii) Support pins or rest buttons (Adjustable type)   ...Fig. 16.5

(iii) Supporting pads   ...Fig. 16.6

(iv) Jack pin-spring type   ...Fig. 16.7

(v) Jack pin-wedge type   ...Fig. 16.8

(vi) Jack screws

(vii) Stop pins.

2. Cylindrical locators or location pins :

When reamed or finally finished holes are available for positoning purpose, the following types of locating pins are generally used :

(i) Plain locating pins, for small holes   ...Fig. 16.9

(ii) Plain locating pins with flange, for small holes   ...Fig. 16.10

(iii) Locating pins for large holes.   ...Fig. 16.11

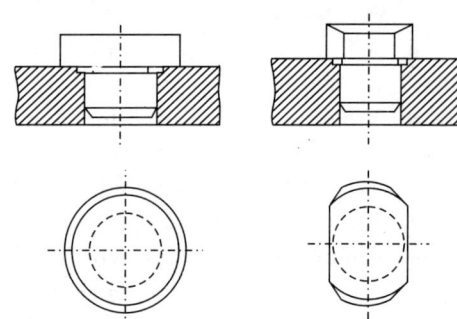
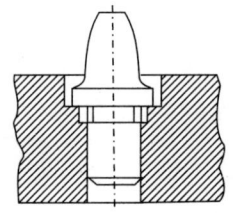

Fig. 16.9. Plain locating   Fig. 16.10. Plain locating   Fig. 16.11. Locating pin for large holes.
pins for small holes.   pin with flange for small holes.

3. Conical locators :

● Conical locating pins are used to locate a workpiece which is cylindrical and with or without a hole as shown in Fig. 16.12 (a) and (b).

● Any variation in the hole size will be easily accommodated due to the conical shape of the pin.

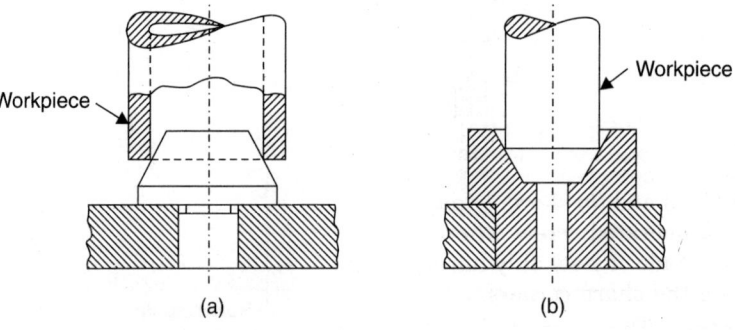

Fig. 16.12. Conical locators.

4. Diamond pin locator :

- A diamond or relieved pin is common to jigs and fixtures.
- A diamond pin when used along with a round pin (Fig. 16.13) makes it easier to locate a part than to locate the part on two round pins. *In use round pin locates the part and the diamond pin prevents the movement around the pin.*

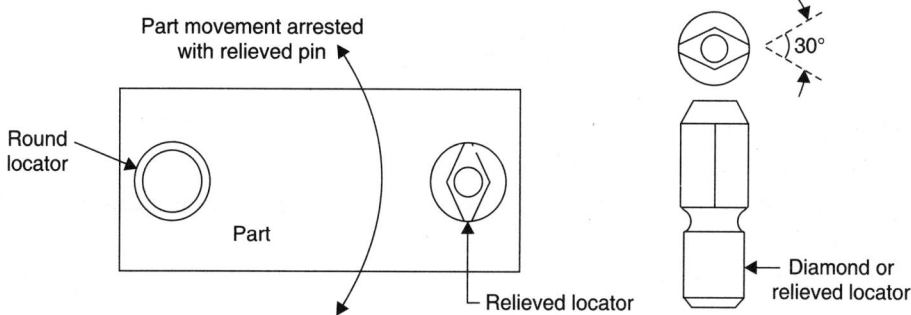

Fig. 16.13. Locating with diamond pin.

5. Vee locators :

- A V-block may be used to locate a component incorporating a circular or a semi-circular profile.
- V-blocks may be used for clamping as well locating if the faces are inclined upto 3°. V-blocks usually have an angle of 90°.

There are three types of V-locators :

(*i*) Fixed-type V-locators   (*ii*) Sliding-type V-locators

(*iii*) Cam operated sliding type V-locators.

— Fig. 16.14 shows a fixed-type V-locator.

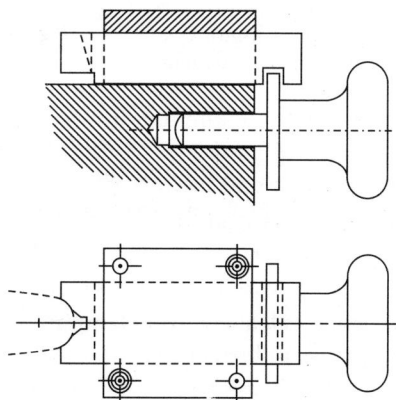

Fig. 16.14. Fixed type V-locator.

— Fig. 16.15 (*a*) and (*b*) shows two types of sliding V-locators ; former is actuated by means of a hand operated screw where as the latter is actuated by a cam. All sliding parts should be made of steel duly hardened and ground. The bottom of the V should be slotted to remove the sharp corners.

- **Ejectors.** These are employed to *remove workpiece from close-fitting locators, after the workpiece has been machined.*

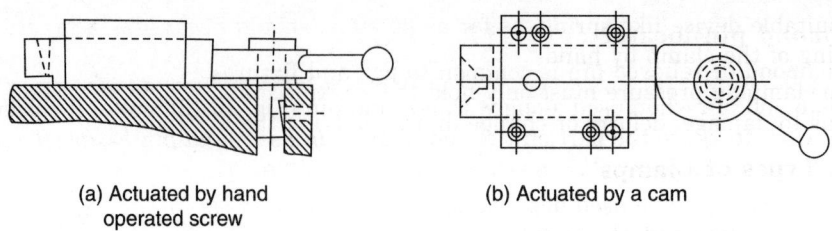

(a) Actuated by hand   (b) Actuated by a cam
    operated screw

Fig. 16.15. Sliding type V-locators.

16.9. CLAMPING DEVICES

16.9.1. Introduction

*A **clamp** is a device that holds the workpiece firmly against the locators provided, and also resists all forces generated by the cutting action of the tool on the workpiece.*

The most common example of a clamp is the bench vice, where the movable jaw of the vice exerts force on the workpiece thereby holding it in the correct position of location in the fixed jaw of the vice.

- *A **clamping device** ensures proper location and centering of the workpiece.*

16.9.2. Basic Requirements of Clamping Devices

The basic requirements of clamping devices (whether simple or complicated) are :

1. To force the workpice to remain in firm contact with the locating pins or surfaces.
2. To rigidly hold the workpiece in a jig or fixture against all the forces acting on the latter.
3. To exert just sufficient pressure on the workpiece.
4. To neither distort nor become loose under heavy pressures or vibrations.
5. To not to damage the workpiece it holds.
6. To have minimum operation (*i.e.,* releasing and reclamping) time.
7. To be wear resistant.
8. To be strong enough to resist bending.

16.9.3. Principles of Clamping

In order to have proper and adequate clamping of a workpiece, the following design and operational factors need be taken into consideration :

1. Clamping should be simple, quick and fool proof.
2. The clamps should always be arranged directly above the points supporting the work, otherwise the distortion of the work can occur.
3. The clamping pressures applied against the workpiece must counteract the tool forces.
4. The movement of clamp, while releasing and clamping should be limited and positively guided to make it quick acting and accurate.
5. The design of the clamp should be such as to deliver the required clamping force when operated by the smallest force expected.
6. The clamp position should be so arranged on the workpiece that several different operations can be performed on the workpiece without disturbing the clamp setting.
7. The clamping pressure as well as cutting forces should be directed towards the locating surfaces or pins to prevent bending and distortion of the workpiece.
8. Fibre pads should be riveted to the clamp faces so that soft and fragile workpiece do not get damaged.

9. A suitable device like spring, as far as possible, should always be incorporated to avoid lifting of the clamp by hand.

10. The clamping pressure must only hold the workpiece and should never be great enough so as to damage, deform or change dimension(s) of the workpiece.

16.9.4. Types of Clamps

The type of clamp to be used depends upon the *shape and size of the workpiece*, the *type of jig or fixture being used*, and *the work to be done.*

The tool designer should choose the clamp which is the *simplest, easiest to use and most efficient.*

Different types of **clamps** are :

1. Clamping screws.

2. Hook bolt clamps.

3. Lever type clamps.

    (*i*) Bridge clamp

    (*ii*) Heel clamps

    (*iii*) Swinging strap (latch) clamp

    (*iv*) Hinged clamps.

4. Quick acting clamps

    (*i*) C-clamps

    (*ii*) Quick acting nut

    (*iii*) Cam-operated clamp.

5. Some other clamping devices :

    (*i*) Wedge clamps

    (*ii*) Toggle clamps

    (*iii*) Power clamps

    (*iv*) Non-mechanical clamping.

1. **Clamping screws :**

The *clamping scerws* or *clamp screws* are simple in operation. They apply pressure directly on the side faces of the workpiece. Normally they carry a *floating pad* at their end to *prevent* displacement of workpiece, denting of clamping surface and deflection of screw.

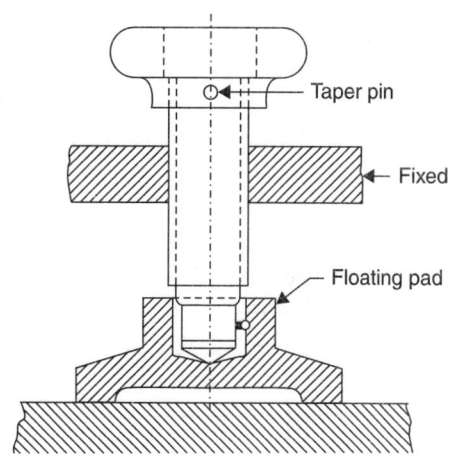

Fig. 16.16 shows a screw clamp.

● Screw clamps are the simplest and most versatile type of clamping elements.

These clamps entail the following *disadvantages.*

(*i*) Clamp force is not constant.

(*ii*) Comparatively large effort is required for operation, causing fatigue to the worker.

(*iii*) Comparatively large time is required for clamping, especially if screw clamps are to hold work on different sides.

Fig. 16.16. Screw clamp.

(*iv*) In the absence of the floating pad, the component can be displaced by the frictional force and indentation marks of the screw tip can be left on the surface clamped.

2. Hook bolt clamp :

An hook bolt clamp is a very simple clamping device and is *only suitable for light work and where the usual type of clamp is inconvenient.*

Fig. 16.17 shows a typical hook bolt clamp.

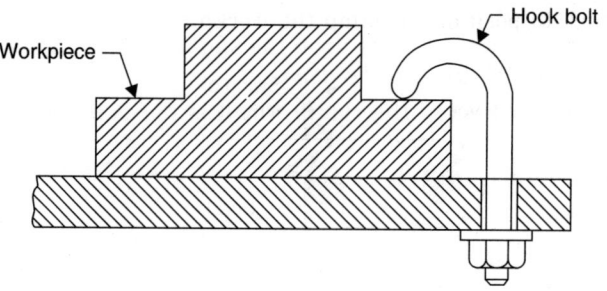

Fig. 16.17. Hook bolt clamp.

3. Lever type clamps

(i) *Bridge clamp :*

- It is a very *simple and reliable clamping device.*
- The clamping force is applied by the spring load nut as shown in Fig. 16.18 ; the distance *'a'* is less than or equal to but never greater than the distance *'b'*. The spring is fitted with the clamp for its automatic lifting when the nut is loosened to remove the workpiece from the jig or fixture. To avoid the com-

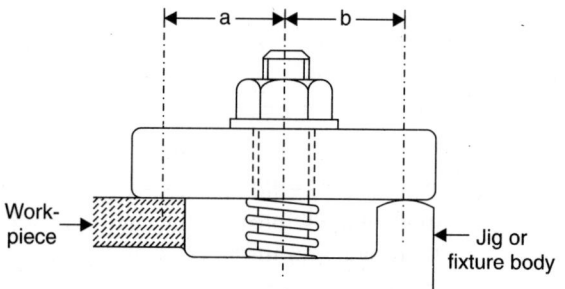

Fig. 16.18. Simple bridge clamp.

plete removal of the nut every time a workpiece is changed the clamp may be slotted to draw it back.

(ii) *Heel clamps :*

The various types of heel clamps are shown in Fig. 16.19 (*a, b, c*).

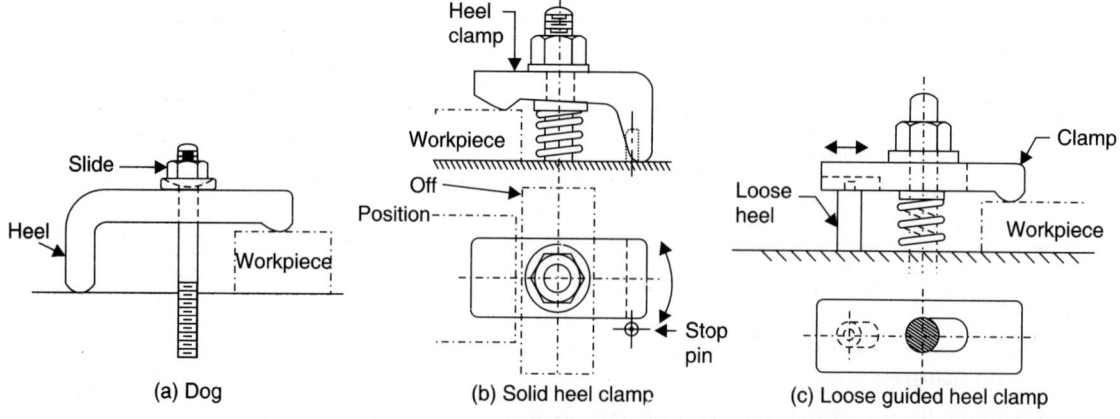

(a) Dog   (b) Solid heel clamp   (c) Loose guided heel clamp

Fig. 16.19. Heel clamps.

- These clamps consist of a *robust plate or strap, centre stud and a heel*. The strap should be strengthened at the point where the hole for stud is cut out by increasing the thickness around the hole.

(iii) *Swinging strap (latch) clamp :*

- This clamp is so named because its *strap* can be swung to a side to remove the workpiece and swing back to position to clamp the workpiece as shown in Fig. 16.20.

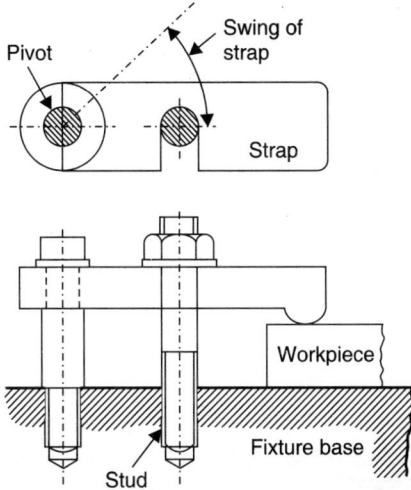

Fig. 16.20. Swinging strap (latch) clamp.

- The swinging lids or straps used in these clamps may have a number of variations in their shapes, position of slots, pivot holes etc., but the basic principle of swinging the strap out and in for releasing and clamping the work is common to all.

(iv) *Hinged clamps :*

- Several times, the requirement in a jig is that the strap (latch) should be completely lifted up for loading and unloading the workpiece. In such cases the upper strap is hinged on one side and locked by means of a bolt as shown in Fig. 16.21.

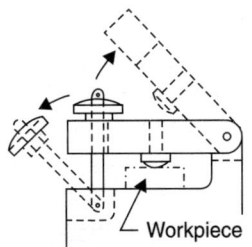

Fig. 16.21. Hinged clamp.

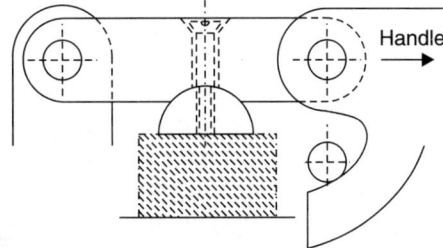

Fig. 16.22. Cam operated clamp.

- Fig. 16.22 shows a hinged clamp provided with a *hook cam*. This clamp is *much faster device* than the former and is suitable for workpiece which maintains *dimensional accuracy*.

4. **Quick acting clamps :**

(i) *C-clamps :*

C-clamps are of two types viz. *Free type* (Fig. 16.23) and *Capative type* (Fig. 16.24). Both the types are very useful devices. The essential feature in both the cases is that once the clamp has been removed or swung away, the workpiece/component can pass freely over the nut. The locking nut requires about one turn to release or lock the clamp and is therefore, quick acting.

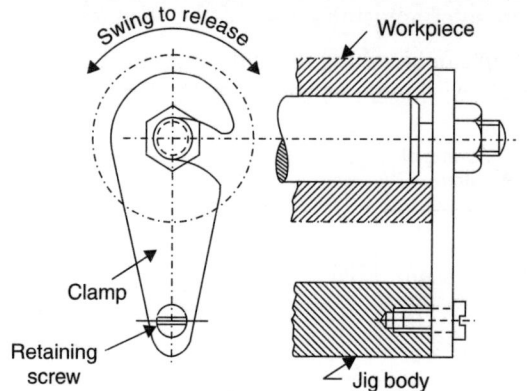

Fig. 16.23. A free type C-clamp or 'C' washer.

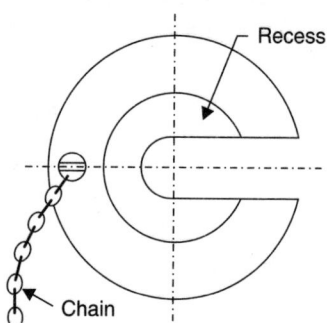

Fig. 16.24. A capative C-clamps or swing washer.

(*ii*) **Quick acting nut :** Refer to Fig. 16.25.

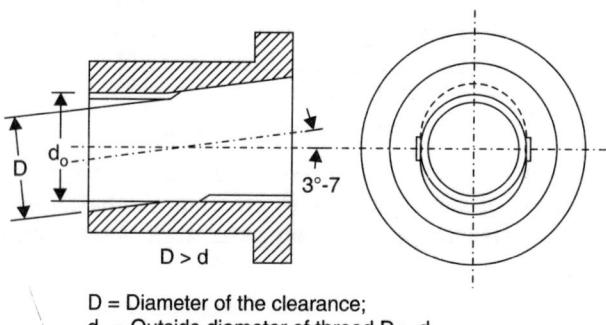

D = Diameter of the clearance;
d_o = Outside diameter of thread D > d_o.

Fig. 16.25. Quick acting nut.

- It is an effective but simple quick acting clamping device.
- The threads of the nut are not continuous but are interrupted. The length of the nut is about 2 to 3 times the thread diameter. The diameter of the clearance D is slightly bigger than the outside diameter of the thread and the axis of the hole is inclined at an angle 3° to 7° to the axis of the nut. While assembling, the nut is positioned over the stud and slightly tapped. Thus the plain portion of the nut passes over the threads of the stud. After positioning, it is slightly tapped to enable the male and female threads to engage together. At this stage, just about half a turn of the nut results in its locking in position.

(*iii*) **Cam-operated clamps :**

The *cam-operated clamps* find wide application and are *fast and positive in action.* These should *not* be used where vibrations are present or where the dimensions of the workpiece vary *e.g.* sand castings.

These clamps are of the following two types :

(*i*) Direct acting levers ;

(*ii*) Eccentrics.

● There are two variations in the 'Direct acting lever cam clamps'. In one case, Fig. 16.26(*a*), the handle is at the centre of the strap and the cam has replaced the nut of the strap clamp due to which the slow movement nut has been eliminated. In the other case, Fig. 16.26(*b*), the cam serves the purpose of adjustable heel of the strap clamp. This is also called a cam actuated strap clamp.

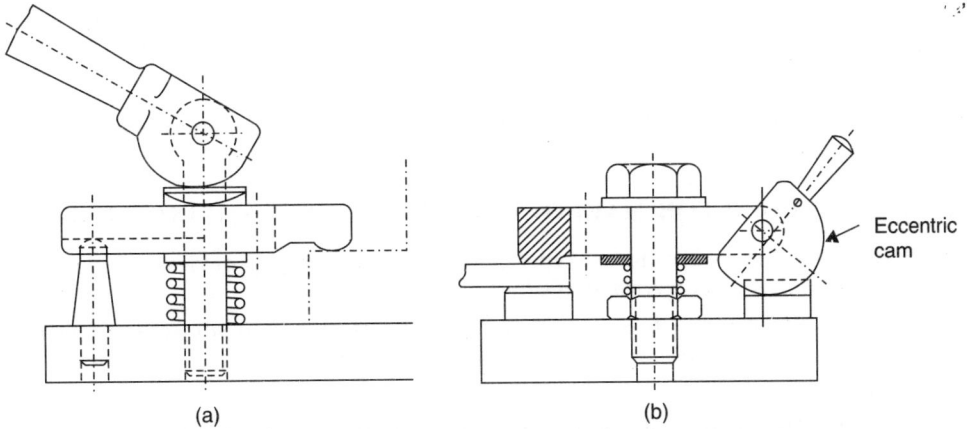

(a)   (b)

Fig. 16.26. Cam-operated clamps.

● The *'eccentric clamp' is used for universal jig.* This type of clamping *allows high pressure to be exerted and also affords an adequate range of linear movement.*

5. **Some other clamping devices :**

(*i*) **Wedge clamps :**

● These clamps utilise the *principle of inclined plane* to secure the workpiece.

● The plain wedge clamp is a crude type of clamp used in places where accuracy is not very important.

● Wedge clamps are widely used for clamping long angle bars on frames provided for the purpose. All sheets and angles used in *wagon manufacturing* are clamped to the frames by wedges only.

(*ii*) **Toggle clamps :**

These clamps use the movement of the rigid links for the clamping action. They are so designed that the movements of their links are independent of the work.

● The *working principle* of a simple toggle clamp is illustrated in Fig. 16.27. The workpiece is held securely under the clamping pad which moves downward by forward motion of piston in an air cylinder. Both clamping arm and cylinder are hinged to the body which may be part of the jig or fixture. It may also be securely bolted to another body at the work zone.

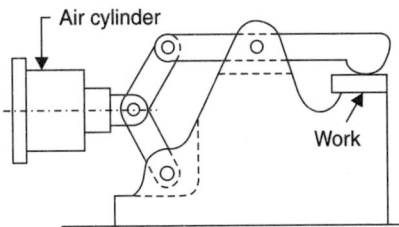

Fig. 16.27. Toggle clamp.

● These clamps are widely used *to hold sheet metal parts* in position while they are being welded or otherwise fastened.

(iii) **Power clamps :**

Power clamps are the clamps *operated by pneumatic power or hydraulic power.*

● These camps are *more efficient,* provide *higher clamping force* and are *quick acting.* However, these are *more costly* than mechanical clamps.

(iv) **Non-mechanical clamping :**

The main types of non-mechanical clamps used for production manufacturing are :

(a) Magnetic chucks *;* *(b)* Vacuum chucks.

● *Magnetic chucks* hold only ferrous metal parts.

● *Vacuum chucks* are used to clamps parts which are non-magnetic or which must be clamped uniformly.

16.10. MATERIALS USED FOR MAKING LOCATING AND CLAMPING DEVICES

● For the manufacture of different types of locating and clamping devices the most commonly used material is *"High carbon steel".*

The elements which come in contact with the workpieces or are subjected to wear, should be *hardened and* wherever necessary, the working face should be *ground.*

● For *dowel pins* and *handles,* 'silver steel' is generally used.

● For complicated shapes and when *exceptional wear is liable to occur,* good quality *case-hardened steel, tool steel or a high tensile steel may be used.*

16.11. TYPES OF JIGS

According to the operations for which they are used, *"jigs"* may be broadly *classified* as :

    I. **Drilling jigs** ...... These can be *'open or closed'* , the latter type being known as *box-type* jigs ; the former type are used when only one side of the workpiece is subjected to operation, while the *latter type* enables operations to be carried out on more than one side.

    II. **Boring jigs.**

● The **drilling jigs** are mainly used in many operations, like *drilling, reaming, counter-sinking, counterboring, tapping etc.*

● The **boring jigs** are mainly used for *boring large holes.*

Types of drilling jigs :

Main types of drilling jigs are :

1. Template jig.

2. Plate jig.

3. Channel jig.

4. Leaf jig.

5. Box jig.

6. Pot-type jig.

7. Universal jig.

8. Index jig.

9. Built-up jig.

10. Trunion jig.

11. Diameter jig.

12. Ring jig.

13. Multistation jig.

1. **Template jig :** Refer to Fig. 16.28.

● A template jig is the *simplest type of drilling jig*. It is used *more for accuracy than speed*.

● This jig fits over, on, or into the work and is not usually clamped.

● These jigs may or may not have bushings.

● This type of jig is suitable if *only a few parts are to be made*.

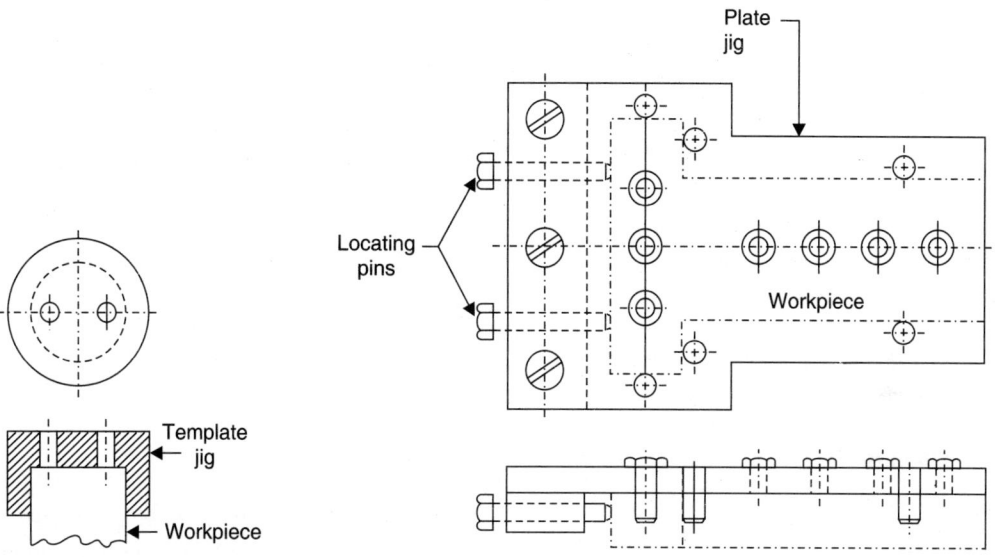

Fig. 16.28. Template jig.   Fig. 16.29. Plate jig.

2. **Plate jig :** Refer to Fig. 16.29.

The main part of such a jig is the *plate* from which it takes its name other essential parts are *drill bush* and *locating pins*. A clamping device is needed, but for some jigs this is not included in the jig proper and may consists of a *C*-clamp or machinist's set of clamps.

● Plate jigs are very common for *drilling holes in long strips or at angles*. Several strips are stacked and plate jig is positioned above them locating it by pins on mutually perpendicular sides. The clamping element is separate form the jig.

3. **Channel jig :**

● A channel jig is the simplest form of box jig. The work is placed in a channel shaped trough and clamped by means of screws (Fig. 16.30).

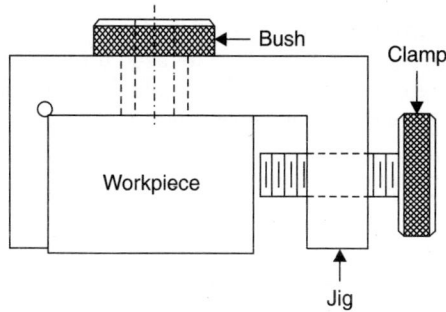

Fig. 16.30. Channel jig.

- The use of this type of jig is limited to jobs having workpieces of a *simple symmetrical shape.*

4. **Leaf jig :** Refer to Fig. 16.31.

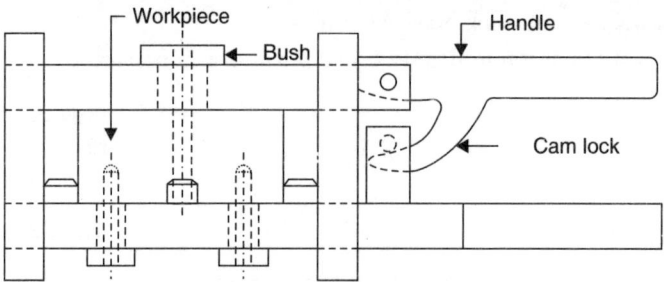

Fig. 16.31. Leaf jig.

- Leaf jigs are characterised by a *hinged cover or leaf* that can be swing open to permit the part to be inserted and then closed to clamp the part in position. A handle is provided for easier movement. Drill bushings are usually located in the leaf.
- Leaf jigs are also called *latch jigs.*
- Most jigs are easy to load, their normally open design, allows rapid and easy removal of chips. A good visibility of the workpiece is also available when the component is being machined.

5. **Box jig :**

- The box jig is so named because of its box shaped construction, closed from most of the sides.
- These jigs are normally designed and used for those components which carry an *awkward shape and need machining in more than one plane.*
- The body of this jig is made as *light as possible to reduce fatigue.*

6. **Pot type jig :**

- Circular components which have both an external diameter and an internal diameter suitable for location purposes are drilled in pot type jigs.
- The jigs essentially consists of two parts. The body is in a pot carrying the workpiece and the bush plate. The bush plate is also provided with locating spigots and can be used without the pot body.

7. **Universal jig :**

- The universal jigs are produced as basic units by a few. By further additions they can be converted into useful jigs for particular purposes.
- They have many advantages where the cost of production is low and a large production merits their higher first cost. One jig can be used for different workpieces by changing or removing the top plate or bushing. If the top plate is removable with the help of a handle the jigs are sometimes called *pump jigs,* also.

8. **Index jig :** Refer to Fig. 16.32.

This type of jig, as the name suggests, incorporates a suitable indexing mechanism, by means of which the workpiece can be indexed to required positions. Such a requirement occurs when, say, a *number of equidistant holes are to be drilled, on the circular flange of a workpiece.* By means of the indexing device, each time a hole is drilled the workpiece is indexed to the next position of the hole to come under the drill bush automatically. Thus, a series of equidistant holes can be drilled along a circle.

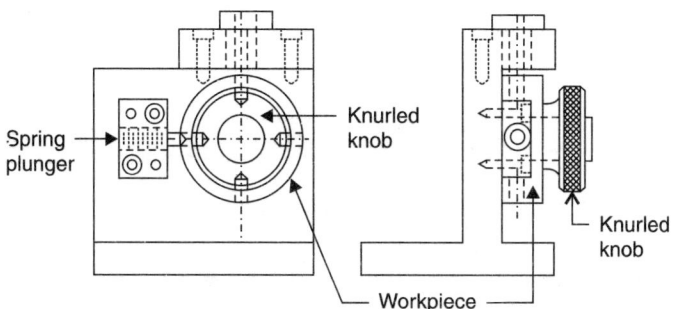

Fig. 16.32. Index jig.

9. **Built-up jig :**

● This jig can be constructed either by building up with the help of dowel pins and cap screws or fabricated by the plates and angles. Depending upon the mode of construction, it may be a 'build-up jig' or 'welded jig'.

10. **Trunion jig :**

● These jigs are used for *drilling holes on very large or odd shaped parts from different sides.* The part is first put into a box-type carrier and then loaded on the trunion.

● The pin and the locating hole for the locking of the jig should be wear resistant.

11. **Diameter jig :**

● These jigs provide a simple means of locating a drilled hole exactly on a diameter or a cylindrical or spherical piece.

12. **Ring jigs :**

● These jigs are used only for drilling round parts, such as pipe flanges.

● Adequate clamping must be provided to prevent the part from rotating in the jig.

13. **Multistation jigs :**

● On a multistation jig, several operations can go simultaneously. While one part is drilled, another can be reamed and a third counterbored. The final station is used for unloading the finished parts and loading fresh parts.

● This type of jig is commonly used on multi-spindle machines.

Types of jig bushes :

The *bushes* used in jigs serve to *locate* and *guide the cutting tools* like drills, reamers, boring bars and counterboring tools. The common types of bushes used in jigs are as follows :

1. *Fixed bushes :*

(*i*) Plain bush (Fig. 16.33) ...... These are always a force or process fit in the jig body.

(*ii*) Flanged bush (Fig. 16.34) ...... This bush enables a larger entry radius and greater bore length.

(*iii*) Screwed bush (Fig. 16.35) ...... This method of fixing the bush prevents the bush from lifting and coming out alongwith the tool, when the latter is withdrawn.

2. *Linear bushes* ...... These bushes act as hardened guides and supports for the slip and renewable bushes

3. *Slip and renewable bushes* ...... These are used in conjuction with the linear bushes. The linear bushes are first fitted in the bush plate, the slip or renewable bushes are then push-fitted into the linear bushes.

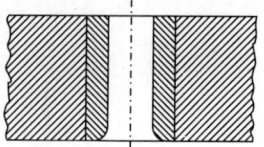

Fig. 16.33. Plain bush.

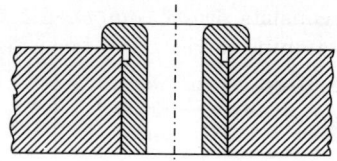

Fig. 16.34. Flanged bush.

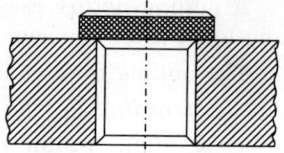

Fig. 16.35. Screwed bush.

16.12. TYPES OF FIXTURES

Fixtures are normally classified by the type of machine operations for which they are used. Some popular forms of fixtures are enumerated and described as follows :

1. Turning fixtures
2. Milling fixtures
3. Broaching fixtures
4. Grinding fixtures
5. Boring fixtures
6. Indexing fixture
7. Tapping fixture
8. Duplex fixture
9. Welding fixture
10. Profile fixture
11. Multistation fixture
12. Assembly fixtures
13. Inspection fixture.

1. **Turning fixtures :**

For a turning operation on a lathe, the simple odd shaped jobs can be held successfully by using chucks, mandrels and collets. However, workpieces having complicated shapes have to be necessarily held in position with the help of *"turning fixtures"*. These fixtures are normally mounted on the nose of the machine spindle or on a face plate and workpieces held on them.

While designing turning fixtures, the following points need be considered :

(*i*) Fixture should be free from projections likely to cause injury to the operator.

(*ii*) Fixture should be accurately balanced to avoid vibrations at high spindle speeds.

(*iii*) Grip the rotating components securely to the fixture to resist torsional forces.

(*iv*) Provide adequate support for frail sections or sections under pressure from the lathe tools.

(*v*) The fixture should be rigid and overhang should be minimum.

(*vi*) Locate the workpiece on critical surfaces from which all or major dimensional and angular tolerances are taken.

(*vii*) Provide a pilot bush for supporting tools where extreme accuracy is required in operations such as boring.

2. **Milling fixtures :**

The milling fixtures are employed on milling machines for carrying out different milling operations on workpieces. The fixture is properly located on the table of the machine and secured in position by means of bolts and nuts. The table is shifted and set in proper position, in relation to the cutter. The workpiece is located on the base of the fixture and clamped before starting the operation.

A milling fixture essentially consists of *six components* : (*i*) Base ; (*ii*) Tenon strip ; (*iii*) Setting block ; (*iv*) Tee-bolts ; (*v*) Clamping device ; (*vi*) Locating element.

The milling fixtures are **classified** as follows :

(*i*) *According to type of operation :*

- Plain milling fixtures
- Gang milling fixtures
- Straddle milling fixtures
- Slot milling fixtures
- Form milling fixtures etc.

(*ii*) *According to method of clamping :*

- Hand clamping fixtures (Mechanical clamping)
- Power clamping fixtures (Hydraulic or pneumatic)
- Automatic clamping fixtures
- Vice jaw clamping fixtures.

(*iii*) *According to method of milling :*

- Single piece milling fixtures
- String milling fixtures
- Index milling fixture
- Reciprocal milling fixture etc

3. **Broaching fixtures :**

Broaching fixtures are must for holding work owing to high forces involved and because of the manner in which the operation is performed.

Broaching fixtures are required to perform one or more of the following *functions :*

(*i*) To hold the job rigidly.

(*ii*) To locate the job in current position relative to the tool of the machine table.

(*iii*) To guide the broaching tool in relation to the job.

(*iv*) Move the job into and out of the cutting position.

(*v*) Index the job between the cuts.

- Fixtures are required for internal broaching and external broaching.
    - The fixture used for *internal broaching* consists solely of hardened bush, peg or similar device. The bush or peg is located in the machine table and the broach is either pulled or pushed. *A pull type broach is generally used instead of push type because length of push broach acts as a long column which may buckle or break.*
    - Surface broaching fixtures range from the simple hand operated type to the complex hydro-mechanical type designed integral with a hydraulically operated machine. A large variety of workpieces of numerous shapes and materials are now machined by surface broaching.

4. **Grinding fixtures :**

In grinding operation, a fixture is useful.

- A **mandrel** is a very simple type of the grinding fixture which is largely used for *external grinding* of surface on a cylindrical grinding machine. The mandrel may be plain or tapered.
- For *internal grinding operations* **chuck** is the most standard fixture. Special jaws as in case of turning may be used to hold castings or forgings. An equipment that is useful in grinding operations where job *cannot* be accurately held is a magnetic chuck.

● Surface grinding operations require fixture of the same type as used in milling. Fixtures for surface grinders are equipped with setting block and tendon strip if the fixture is to be clamped to the table.

The following *problems* are encountered by designer while designing fixtures :

(*i*) Rotating fixture or chuck generally requires dynamic balancing.

(*ii*) To make provision to ensure that coolant escape and control remain effective.

(*iii*) To ensure that the functions of coolant nozzles, spray guards, part feeders and other such devices are not hampered.

(*iv*) Mounting of wheel dressers on or close to the fixture.

5. Boring fixtures :

Boring operation is performed in the following *two* ways :

(*i*) By keeping the boring bar (tool) stationary and feeding the rotating workpiece on to the bar.

(*ii*) By keeping the workpiece stationary and feeding the rotating bar into the work.

Accordingly, therefore, the *boring fixtures* are made in *two designs :*

— One of these incorporates the principle of a drilling jig, and in this the boring bar (tool) is guided through a *pilot bush.* Such fixtures are often referred to as *"boring jigs".*

— The other design facilitates holding of the workpiece in correct position, relative to the boring bar.

  ● Boring fixtures do not normally require to be as rigidly constructed as milling fixtures, because the load imposed by the boring bar rarely approaches to that introduced by a milling cutter.

6. Indexing fixture :

  ● It is used for machining parts having evenly spaced machined surfaces (Fig. 16.36).

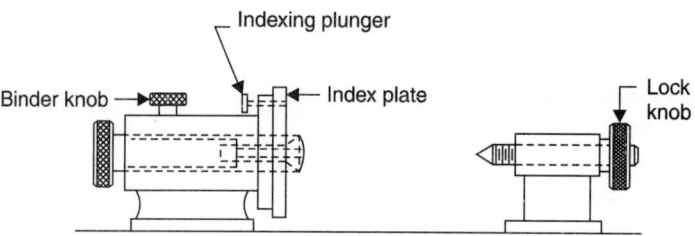

Fig. 16.36. Indexing fixture.

7. Tapping fixture :

  ● These fixtures are specially designed to position and fairly secure identical workpieces for cutting threads in drilled holes in them.

  ● Odd shaped and unbalanced components will always need the use of such fixtures, especially when the tapping operation is to be carried out repeatedly on mass scale on such components.

8. Duplex fixture :

  ● Duplex fixture is a fixture which holds two similar components simultaneously and facilitates simultaneous machining of these components at two separate stations.

9. Welding fixture :

  ● 'A welding fixture *properly locates and holds workpieces for complete welding operation.*

- A welding fixture must be designed to withstand the thermal pressures generated during the welding operations. The locators and clamps should not loose their accuracy due to the effect of heat.

10. Profile fixture :

- It is used to guide tools for machining contours which the machine cannot normally follow.

11. Multistation fixture :

- It is used for high-speed high volume production runs in which the machining cycle is continuous.
- This form of fixture allows the loading and unloading operations to be performed while the machining operations are in progress at different stations.

12. Assembly fixtures :

- The assembly fixtures *hold different components together in their proper relative positions at the time of assembling them.* These fixtures which are used for holding the components for performing mechanical operations, are known as *Mechanical assembly features.*
- Other types of fixtures are those in which the components are held for hot joining, such as welding fixtures. Thus, the welding fixtures are also assembly fixtures, but *for hot joining.*

13. Inspection fixtures :

The *"inspection fixtures" are used to check the quality of the workpieces, parts and components of machines.*

- Dimension inspection or gauging fixtures raise the efficiency of the work of human inspecters, improve their working conditions, quality of workpieces, parts and components of machines.

    — These fixtures need not be designed to withstand forces, such as shock and vibration, associated with machining or with some other fabricating and assembly processes. They are not required to resist temperatures present in welding, brazing etc.

    — Clamping forces in an inspection fixture are generally too small to affect its design, but they should not distort the workpiece.

HIGHLIGHTS

1. *Jigs and fixtures* are production tools used to accurately manufacture duplicate and interchangeable parts.
2. Jigs and fixtures may be : Hand operated, power operated, semi-automatic and automatic.
3. The various locating devices are :
   - (i) Support pins and jack pin ;
   - (ii) Cylindrical locators or location pins ;
   - (iii) Conical locators ;
   - (iv) Diamond pin locator ;
   - (v) Vee locators.
4. Different types of *clamps are :*
   - (i) Clamping screws ;
   - (ii) Hook bolt clamp ;
   - (iii) Lever type clamps ;
   - (iv) Quick acting clamps.
5. Jigs are classified as :
   - (i) Drilling jigs ;
   - (ii) Boring jigs.

OBJECTIVE TYPE QUESTIONS

Fill in the blanks or Say 'Yes' or 'No' :

1. .... is a frame or body which holds and positions the work and guides the cutting tool during the machining operation.
2. A fixture differs from a jig in the sense that the fixture does not guide the cutting tool.
3. A .... should be securely fastened to the table of the machine upon which the work is done.
4. Jigs are heavier than fixtures.
5. Jigs and fixtures are not designed for specific jobs.
6. The term .... refers to establish a definite location relationship between the workpiece and the cutting tool or jig or fixture.
7. Locating elements place the workpiece in essentially the same position cycle after cycle.
8. Adjustable supports are used when the surface is rough or uneven, such as in cast parts.
9. Solid supports are difficult to use.
10. Equalizing support is also a form of adjustable support.
11. Pin-type locators are used for smaller holes and for aligning members of the tool.
12. .... locators position the work in relation to an outside edge or the outside of a detail, such as hub or boss.
13. Ring nest is normally used for cylindrical profiles.
14. .... locators are used for parts which cannot be placed in either a nest or V-locator.
15. A .... is a device that holds the workpiece firmly against the locators provided.
16. .... are employed to remove workpiece from close-fitting locators, after the workpiece has been machined.
17. Cam action clamps provide a fast, efficient, and simple way to hold work.
18. Wedge clamps apply the basic principle of .... plane to hold work in a manner similar to a cam.
19. Boring jigs are used to bore holes that are either too large to drill or must be made an odd size.
20. Closed jigs are used for simple operations where work is done on only one side of the part.
21. A channel jig is the simplest form of .... jig.
22. .... jigs are used only for drilling round parts such as pipe flanges.
23. .... jigs provide a simple means of locating a drilled hole exactly on diameter or a cylindrical or spherical component.
24. A .... fixture is used to guide tool for machining contours which the machine cannot normally follow.
25. A duplex fixture uses only two stations.
26. .... fixtures are used for machining parts which must have machined details evenly spaced.

ANSWERS

1. Jig	2. Yes	3. fixture	4. No	5. No
6. location	7. Yes	8. Yes	9. No	10. Yes
11. Yes	12. Profile	13. Yes	14. Fixed-stop	15. clamp
16. Ejectors	17. Yes	18. inclined	19. Yes	20. No
21. box	22. Ring	23. Diamond	24. profile	25. Yes.
26. Indexing.				

THEORETICAL QUESTIONS

1. What are 'jigs and fixtures' ? Explain.
2. Define and explain the following terms :
   (*i*) Jig   (*ii*) Fixture.

3. What are the differences between jigs and fixtures ?
4. What are the advantages of using jigs and fixtures ?
5. List the main components or elements of jigs and fixtures.
6. How are jigs and fixtures classified ?
7. Explain briefly the principles of jig and fixture design.
8. What do you understand by "Degrees of freedom" ? Explain.
9. Describe briefly "Principles of location".
10. What is the main function of a locating device ?
11. Explain briefly any two of the following locating devices :

    (*i*) Support pins and jack pins   (*ii*) Cylindrical locators or location pins

    (*iii*) Conical locators   (*iv*) Diamond pin locator

    (*v*) Vee locators.

12. What is a 'clamp' ?
13. What is the function of a 'clamping device' ?
14. List the basic requirements of clamping devices.
15. What are the principles of clamping ?
16. Enumerate various types of clamps.
17. Explain briefly the following :

    (*i*) Clamping screws   (*ii*) Quick acting clamps.

18. How are jigs broadly classified ?
19. Explain briefly any three of the following types of jigs :

    (*i*) Template jig   (*ii*) Channel jig

    (*iii*) Diameter jig   (*iv*) Leaf jig

    (*v*) Ring jig   (*vi*) Universal jig.

20. Enumerate various types of fixtures.
21. Explain briefly the following fixtures :

    (*i*) Turning fixtures   (*ii*) Milling fixtures

    (*iii*) Indexing fixture   (*iv*) Grinding fixtures.

17.1. Meaning of metrology. 17.2. Objectives of metrology. 17.3. Standards of measurements—line standard—end standard—wavelength standard—classification of standards—relative characteristics of line and end standards. 17.4. Limits, fits and tolerances—General aspects—normal size and basic dimensions—definitions—basis of fit (or limit) system—systems of specifying tolerances—designation of holes, shafts and fits—commonly used holes and shaft—The Newall system—ISO system of limits and fits—types of fits—concept of interchangeability. 17.5. Measuring instruments. 17.6. Gauges—classification of gauges—description of some commonly used gauges. 17.7. Linear measurements—Engineer's steel rule—calipers—vernier calipers—vernier height gauge—vernier depth gauge—micrometers—advantages and limitations of commonly used precision instruments. 17.8. Angular and taper measurements—Angular measurements—Taper measurement. 17.9. Screw thread measurements—Introduction—classification of threads—elements of screw threads—specifications of a screw thread—forms of threads—measuring elements of a screw thread—screw thread gauges. **Questions with Answers**—Objective Type Questions—Theoretical Questions.

17.1. MEANING OF METROLOGY

Metrology, in literary sense, *means the pure science of measurements*. But for engineering purposes, it is restricted to measurements of length and angles and other quantities which are expressed in linear or angular terms.

- *Metrology is mainly concerned with :*
    - (*i*) Establishing the units of measurements, reproducing these units in the form of standards and ensuring the uniformity of measurements :
    - (*ii*) Developing methods of measurement :
    - (*iii*) Analysing the accuracy of methods of measurement, researching into the causes of measuring errors and eliminating these.

- In the broader sense, metrology is not limited to length measurement but is also concerned with *the industrial inspection* and its various techniques. Inspection is carried out with gauges and the metrologist is intimately concerned with the design, manufacturing and testing of gauges of all types.

17.2. OBJECTIVES OF METROLOGY

Although the basic objective of a measurement is to provide the *required accuracy at a minimum cost,* metrology has further *objectives* in a modern engineering plant with different shapes which are :

1. Complete evaluation of newly developed products.
2. Determination of the process capabilities and ensure that these are better than the relevant component tolerances.

3. Determination of the measuring instrument capabilities and ensure that they are quite sufficient for their respective measurements.

4. Minimising the cost of inspection by effective and efficient use of available facilities.

5. Reducing the cost of rejects and rework through application of *"Statistical quality control techniques"*.

6. To standardise the measuring methods.

7. To maintain the accuracies of measurement.

8. To prepare designs for all gauges and special inspection fixtures.

17.3. STANDARDS OF MEASUREMENTS

These days only *two standard systems of linear measurement, English* (yard) and *Metric* (metre) are in general use throughout the world.

The *metric system was originated in France* and is now being used in many countries in the world.

The British system of linear measurement is based on one arbitrarily unit known as *yard*.

For linear measurements the various standards known are :

1. Line standard.

2. End standard.

3. Wavelength standard.

17.3.1. Line Standard

A **yard** *or* **metre** *is defined as the distance between scribed lines on a bar of metal under certain conditions of temperature and support.* These are legal standards and Act of Parliament authorises their use.

- The **Metre** is defined as 1650763.73 wavelengths of the orange radiation in vacuum of krypton 86 isotope.

- The **Yard** is defined as 0.9144 metre. This is equivalent to 1509458.35 wavelengths of the same radiation.

17.3.1.1. Yard

A Yard was formerly known as the Imperial Standard Yard. Fig. 17.1 shows a diagram indicating its essential features. It consists of a bronze bar made from an alloy known as Baily's metal, consisting of 16 parts copper, $2\frac{1}{2}$ parts tin and 1 part zinc. The bar, 1 sq. in. in cross-section has an overall length of 36". Two counter bored holes, $\frac{1}{2}$" diameter by $\frac{1}{2}$" deep, at 36" centres (1" from each end of the bar) provide sighting holes for two gold plugs inserted in the holes at the base of each counter bore. The faces of the gold plugs are flush with bases of counter bores and, therefore, lie in the neutral plane of the bar where bending effects are minimised when the bar is resting on supports. These plugs are 0.01" in. diameter, and five lines are ruled on the upper polished face of each ; three lines at right angles to the length of the bar and two parallel to the bar as shown in Fig. 17.1 (b). The *length of the yard is defined as the distance between the two central transverse lines on the plugs when the temperature of the bar is constant at 62° F, and when the bar is supported on rollers, in a specified manner, to prevent flexure.*

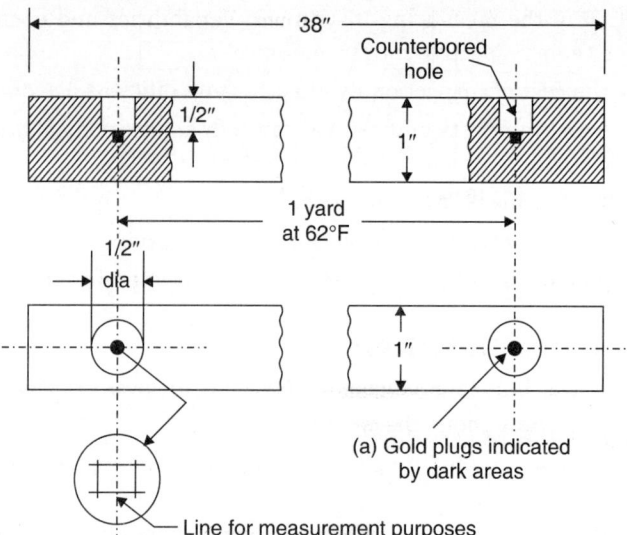

(a) Gold plugs indicated
by dark areas

(b) Enlarged view of gold plug showing
engraved lines (Actual diameter 0.10″)

Fig. 17.1. Imperial standard yard.

● The original procedure, when intercomparisons were made between the standard and its copies, was to float the bars in the mercury ; but proof that the bars could be effectively supported on rollers, while maintaining the previous accuracy, was provided by Airy. His method is shown in Fig. 17.2. The standard was directly supported on eight on equally spaced rollers in conjunction with a special frame. The distance between the rollers was

proved by Airy to be equal to $\dfrac{l}{\sqrt{n^2 - 1}}$ where, n and l represent number of rollers and length of bar respectively. This method of support was used for the purpose of inter-comparisons until 1922 when two supports at the Airy points were introduced. When a bar is supported specifically at two points symmetrically about its centre, a condition can be produced when the bar ends lie in a horizontal plane. With this condition the bar deflects at its centre, but the effective error in the length of bar is negligible. Applying Airy's formula the specific distance between the supports is equal to 0.577 l.

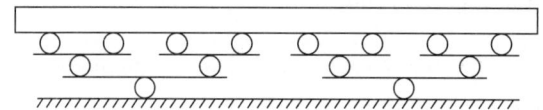

Fig. 17.2. 8-point supporting system for imperial yard till 1922.

17.3.1.2. Metre

The *length of the metre is defined as the distance, at 0°C between the centre portions of pure platinum—irridium alloy (10% irridium) of 102 cm total length—and having a cross-section as shown in Fig. 17.3.* The graduations are on the upper surface of web which coincides with the neutral axis of the section.

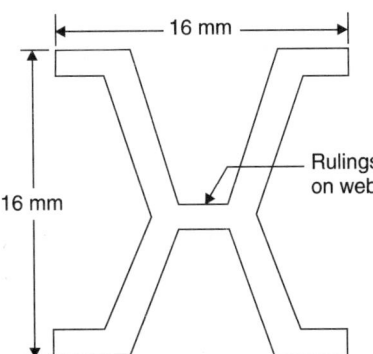

Fig. 17.3. International prototype metre (cross-section).

17.3.1.3. Sub-division of standards

The imperial standard yard and international prototype metre defined previously are considered to be perfect or master standards and cannot usually be used for general purposes. Thus depending upon the importance of accuracy required for the work, standards are subdivided into *four grades.*

1. **Primary standards.** To ensure that standard unit of length, *i.e.,* yard or metre does not change its value and it is strictly followed and precisely defined there should be one and only one material standard preserved under most careful conditions. This has no direct application to a measuring problem encountered in engineering. *These are used only at rare intervals and solely for comparison with secondary standards.*

2. **Secondary standards.** Secondary standards are made as nearly as possible to the primary standard with which they are compared at intervals. Any error existing in these bars is recorded by comparison with primary standards after long intervals. *These standards are distributed to a number of places for safe custody and used in their turn for occasional comparison with tertiary standards. These standards also act as safeguard against the loss or destruction of primary standards.*

Materials for secondary standards :

(*i*) Invar—an alloy of nickel and steel

(*ii*) Fuse silica

(*iii*) Elinvar—an alloy of nickel and chromium.

All the above materials have *usually very low coefficient of linear expansion.*

3. **Tertiary standards.** Tertiary standards are the *first standards to be used for reference purposes in laboratories and workshops.* These should also be maintained as a *reference for comparison at intervals with working standards.*

4. **Working standards.** These standards are necessary for use in metrology laboratories and similar institutions. These are derived from fundamental standards.

Sometimes standards are *classified* as :

- *Reference standards* (used for reference purposes)
- *Calibration standards* (used for calibration of inspection and working standards)
- *Inspection standards* (used by inspectors)
- *Working standards* (used by operators).

17.3.1.4. Characteristics of line standards

The characteristics of line standard are given below :

1. Accurate engraving on the scales can be done but it is difficult to take full advantage of this accuracy. For example, a steel rule can be read to about ± 0.2 mm of true dimension.

2. It is easier and quicker to use a scale over a wide range.

3. The scale markings are not subject to wear although significant wear on leading end *leads to undersizing.*

4. There is no '*built in*' datum in a scale which would allow easy scale alignment with the axis of measurement, this again *leads to undersizing.*

5. Scales are subjected to the parallax effect, a source of both positive and negative reading errors.

6. For close tolerance length measurement (except in conjunction with microscopes) scales are not convenient to be used.

17.3.2. End Standard

End standards, in the form of the bars and slip gauges, are in general use in precision engineering as well as in standard laboratories such as the N.P.L (National Physical Laboratory). Except for applications where microscopes can be used, scales are not generally convenient for the direct measurement of engineering products, whereas slip gauges are in everday use in tool-rooms, workshops, and inspection departments throughout the world.

A *modern end standard consists fundamentally of a block or bar of steel generally hardened whose end faces are lapped flat and parallel to within a few millionth of a cm.* By the process of lapping, its size too can be controlled very accurately. Although, from time to time, various types of end bar have been constructed, some having flat and some spherical faces, *the flat, parallel faced bar is firmly established as the most practical method of end measurement.*

Characteristics of end standards :

1. Highly accurate and well suited to close tolerance measurements.

2. Time-consuming in use.

3. Dimensional tolerance as small as 0.0005 mm can be obtained.

4. Subjected to wear on their measuring faces.

5. To provide a given size, the groups of blocks are "*wrung*" together. Faulty wringing leads to damage.

6. There is a "*built-in*" datum in end standards, because their measuring faces are flat and parallel and can be positively located on a datum surface.

7. As their use depends on "*feel*" they are not subject to the parallax effect.

End bars. Primary end standards usually consist of *bars of carbon steel* about 20 mm in diameter and made in sizes varying from 10 mm to 1200 mm. These are hardened at the ends only. They are used for the measurement of work of larger sizes.

Slip gauges. *Slip gauges* are used as standards of measurement in practically *every precision engineering* works in the world. These were invented by C.E. Johansom of Sweden early in the present century. These are made of *high-grade cast steel and are hardened throughout.* With the set of slip gauges, combinations of slip gauge enables measurements to be made in the range of 0.0025 to 10 mm but in combinations with end/length bars measurement range upto 1200 mm is possible.

Note : The accuracy of line and end standards is affected by temperature changes and both are originally calibrated at $20 \pm 1/2°C$. Also care is taken in manufacture to ensure that change of shape with time is reduced to negligible proportions.

17.3.3. Wavelength Standard

In 1829, Jacqnes Babinet, a French philosopher, suggested that wavelengths of monochromatic light might be used as natural and invariable units of length. It was nearly a century later that the Seventh General Conference of Weights and Measures in Paris approved the definition of a standard of length relative to the metre in terms of the wavelength of the red radiations of cadmium. Although this was not the establishment of a new legal standard of length, it set the seal on work which kept on going for a number of years.

- *Material standards are liable to destruction and their dimensions change slightly with time. But with the monochromatic light we have the advantage of constant wavelength and since the wavelength is not a physical one, it need not be preserved. This is reproducible standard of length, and the error of reproduction can be of the order of 1 part in 100 millions. It is because of this reason that International standard measures the metre in terms of wavelength of krypton 86 (Kr 86).*

- Light wavelength standard, for sometime, had to be objected because of the impossibility of producing pure monochromatic light as wavelength depends upon the amount of isotope impurity in the elements. But now with the rapid development in atomic energy industry, pure isotopes of natural elements have been produced. *Krypton 85, Mercury 198 and Cadmium 114* are possible sources of radiation of wavelength suitable as natural standard of length.

17.3.3.1. Advantages of wavelength standards

The following are the advantages of using wavelength standard as basic unit to define primary standards :

1. It is not influenced by effects of variation of environmental temperature, pressure, humidity and ageing because it is not a material standard.
2. There is no need to store it under security and thus there is no fear of its being destroyed as in the case of yard and metre.
3. It is easily available to all standardising houses, laboratories and industries.
4. It can be easily transferred to other standards.
5. This standard can be used for making comparative statement of much higher accuracy.
6. It is easily reproducible.

17.3.4. Classification of Standards

To maintain accuracy and interchangeability it is necessary that the standards be traceable to a single source, usually the National Standards of the country, which are further linked to International Standards. The accuracy of National Standards is transferred to working standards through a chain of intermediate standards in a manner given below :

National Standards

↓

National Reference Standards

↓

Working Standards

↓

Plant Laboratory Reference Standards

↓

Plant Laboratory Working Standards

↓

Shop Floor Standards

Evidently, there is degradation of accuracy in passing from the defining standards to the shop floor standards. The *accuracy of a particular standard depends on a combination of the number of times it has been compared with a standard in a higher echelon, the frequency of such comparisons, the care with which it was done, and the stability of the particular standard itself.*

17.3.5. Relative Characteristics of Line and End Standards

The relative characteristics of *line* and *end standards* are given in the tabular form below :

S. No.	Aspects	Line standard	End standard
1.	*Manufacture and cost of equipment*	Simple and low.	Complex process and high.
2.	*Accuracy in measurement*	Limited to ± 0.2 mm. In order to achieve high accuracy, scales have to be used in conjunction with microscopes.	Very accurate for measurement of close tolerances upto ± 0.001 mm.
3.	*Time of measurement*	Quick and easy.	Time consuming.
4.	*Effect of use*	Scale marking not subject to wear but the end of scale is worn. Thus it may be difficult to assume zero of scale as datum.	Measuring faces get worn out. To take care of this end pieces can be hardened, protecting type. Built-in datum is provided.
5.	*Other errors*	There can be parallax error.	Errors may get introduced due to improper wringing of slip gauges. Some errors may be caused due to change in laboratory temperature.

17.4. LIMITS, FITS AND TOLERANCE

17.4.1. General Aspects

● In the design and manufacturing of engineering products a great deal of attention has to be paid to the *mating, assembly and fitting of various components*. In the early days of mechanical engineering during the nineteenth century, the majority of such components were actually mated together, their dimensions being adjusted until the required type of fit was obtained. These methods demanded craftsmanship of a high order and a great deal of very fine work was produced.

● Present day standards of quantity production, interchangeability, and continuous assembly of many complex compounds, could not exist under such a system, neither could many of the exacting design requirements of modern machines be fulfilled without the knowledge that certain dimensions can be reproduced with precision on any number of components.

● *Modern mechanical production engineering is based on a system of limits and fits, which, while not only itself ensuring the necessary accuracies of manufacture, forms a schedule or specifications to which manufacturers can adhere.*

In order that a system of limits and fits may be successful, following *conditions* must be fulfilled :

(*i*) The range of sizes covered by the system must be sufficient for most purposes.

(*ii*) It must be based on some standards so that everybody understands alike and a given dimension has the same meaning at all places.

(*iii*) For any basic size it must be possible to select from a carefully designed range of fit the most suitable one for a given application.

(*iv*) Each basic size of hole and shaft must have a range of tolerance values for each of the different fits.

(v) The system must provide for both unilateral and bilateral methods of applying the tolerance.

(vi) It must be possible for a manufacturer to use the system to apply either a hole-based or a shaft-based system as his manufacturing requirements may need.

(vii) The system should cover work from high class tool and gauge work where very wide limits of sizes are permissible.

17.4.2. Nominal Size and Basic Dimensions

Nominal size. A *'nominal size' is the size which is used for purpose of general identification.* Thus the nominal size of a hole and shaft assembly is 60 mm, even though the basic size of the hole may be 60 mm and the basic size of the shaft 59.5 mm.

Basic dimension. A *'basic dimension' is the dimension, as worked out by purely design considerations.* Since the ideal conditions of producing basic dimension, do not exist, the basic dimensions can be treated as theoretical or nominal size, and it has only to be approximated. A study of function of machine part would reveal that it is unnecessary to attain perfection because some variations in dimension, however small, can be tolerated on size of various parts. It is, thus, *general practice to specify a basic dimension and indicate by tolerances as to how much variation in the basic dimension can be tolerated without affecting the functioning of the assembly into which this part will be used.*

17.4.3. Definitions

The definitions given below are based on those given in IS : 919 *Recommendation for Limits and Fits for Engineering,* which is in line with the ISO recommendation.

Shaft. The term *shaft* refers not only to diameter of a circular shaft but to *any external dimension on a component.*

Hole. This term refers *not only* to the diameter of a circular hole but to *any internal dimension on a component.*

Actual size of the shaft. This is the measured dimension of the part.

Basic size. Refer to Fig. 17.4. The *basic size is the standard size for the part and is the same for both the hole and its shaft.* Example : A 60 mm diameter hole and shaft.

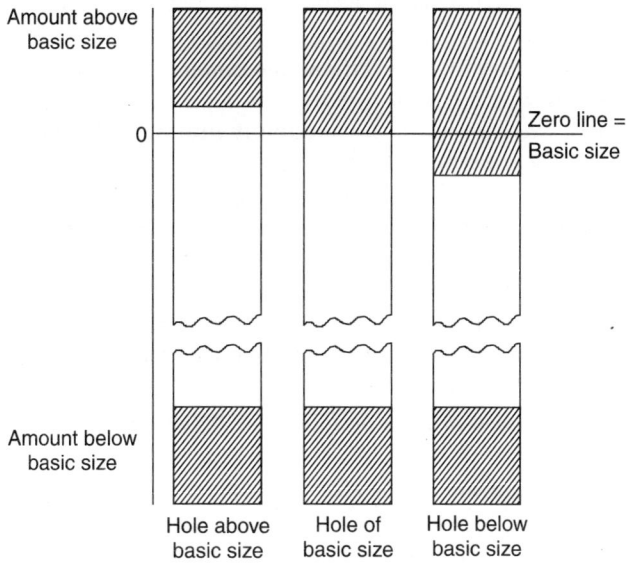

Fig. 17.4.

Zero line. Refer to Fig. 17.4. *This is the line which represents the basic size so that the deviation from the basic size is zero.*

Limits of size. These are *maximum and minimum permissible sizes of the part.*

Minimum limit of size. Refer to Fig. 17.5. The *minimum size permitted for the part.*

Maximum limit of size. Refer to Fig. 17.5. The *maximum size permitted for the part.*

Tolerance. Refer to Fig. 17.5. *The difference between the maximum and minimum limits of size.*

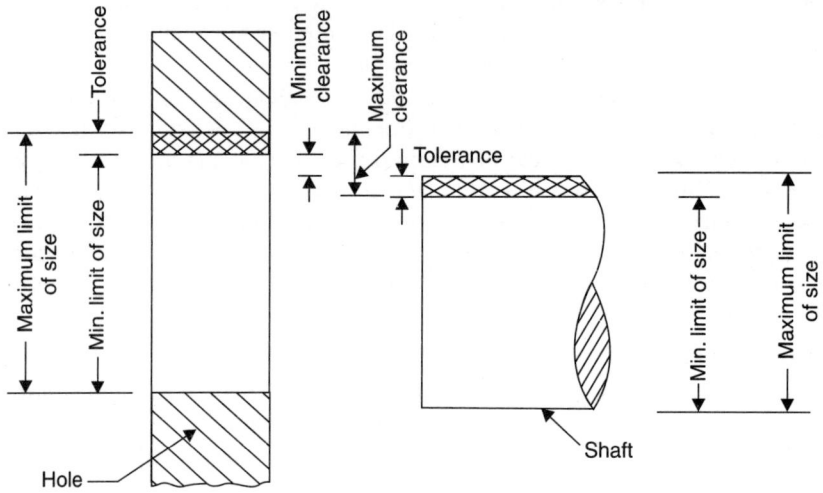

Fig. 17.5.

Tolerance size. This is the difference between the two limits of size.

Grade of tolerance. The tolerance grade *is an indication of the degree of accuracy of manufacture.* It is designated by the letter IT followed by a number. Tolerance grades are ITOI, ITO, upto IT16. The *larger the number the larger the tolerance.*

Standard tolerance unit. This is the unit used to calculate the various grades of tolerance for a given basic size.

Upper deviation. Refer to Fig. 17.6. This is the *amount from the basic zero or zero line, on the maximum limit of size for either a hole or a shaft.* It is designated ES for a *hole* and *es* for a *shaft.* Upper deviation is a *positive quantity* when the maximum limit of size is *greater* than the basic size and *negative quantity* when the maximum limit of size is less than the basic size.

Lower deviation. Refer to Fig. 17.6. This is the *amount from basic size, or zero line,* to the *minimum limit of size.* It is designated EI for a *hole* and *ei* for a *shaft.* Lower deviation is a *positive quantity* when the minimum limits of *sizes* is *greater* than the basic size and a *negative quantity* when the minimum limit of size is *less* than the basic size.

Fundamental deviation. Refer to Fig. 17.6. This is the deviation, either the upper or lower deviation, which is the *nearest one to the zero line* for either a hole or a shaft. It fixes the position of the tolerance zone in relation to zero line.

Fit. Fit *means a degree of tightness or looseness between two mating parts to perform a definite function.*

Clearance. The *difference between the sizes of a hole and a shaft* which are to be assembled together when the *shaft is smaller than the hole.*

Interference. The *difference between the sizes of a hole and a shaft* which are to be assembled together when the *shaft is larger than the hole.*

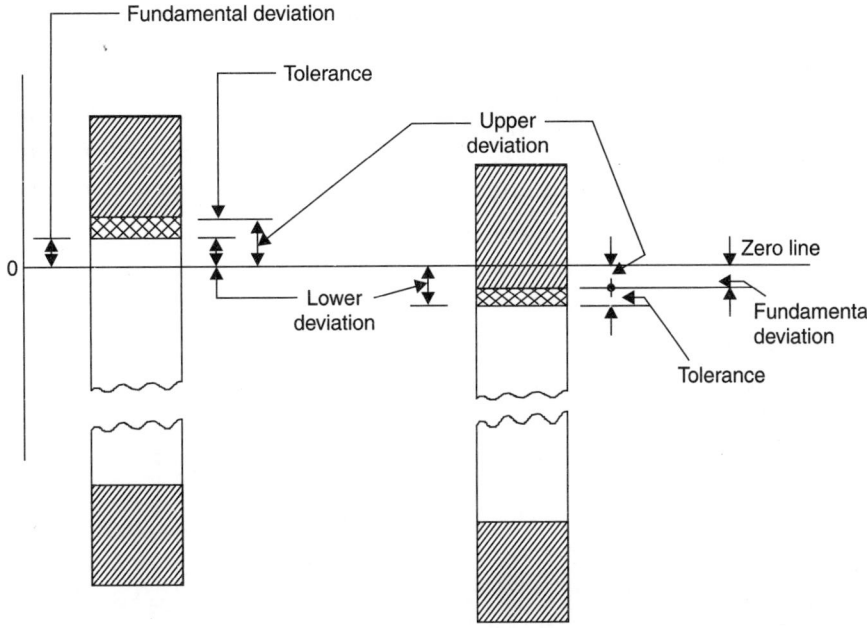

Fig. 17.6.

Classes of fit. Refer to Fig. 17.7.

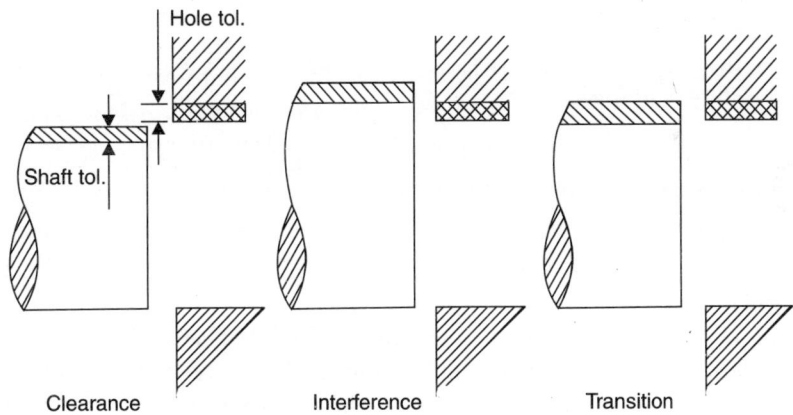

Fig. 17.7. Classes of fit.

1. Clearance fit. A *clearance fit could be obtained by making the lower limit on the hole equal to or larger than the upper limit on the shaft.* Any hole and any shaft made to these tolerances would assemble with a clearance fit with certainty.

2. Interference fit. An *interference fit would be obtained with equal certainty by making the lower limit on the shaft equal to or larger than the upper limit on the hole.*

3. Transition fit. Between these two conditions lies a *range of fits known as transition fit.* These are obtained when the upper limit on the shaft is larger than the lower limit on the hole, and the lower limit on the shaft is smaller than the upper limit on the hole. It must be realised that transition fits exist only as a class ; any actual hole and shaft must assemble with either a clearance or interference fit.

Allowance. The *difference between the maximum shaft and minimum hole is known as allowance.* In a *clearance fit,* this is the *minimum clearance* and is a *positive allowance.* In an *interference fit,* it is the *maximum interference and is a negative allowance.*

17.4.4. Basis of Fit (or Limit) System

A fit or limit system consists of a series of tolerances arranged to suit a specific range of sizes and functions, so that limits of size may be selected and given to mating components to ensure specific classes of fit. This system may be arranged on the following basis :

1. Hole basis system
2. Shaft basis system.

1. **Hole basis system.** Refer to Fig. 17.8.

'Hole basis system' is one in which the limits on the hole are kept constant and the variations necessary to obtain the classes of fit are arranged by varying those on the shaft.

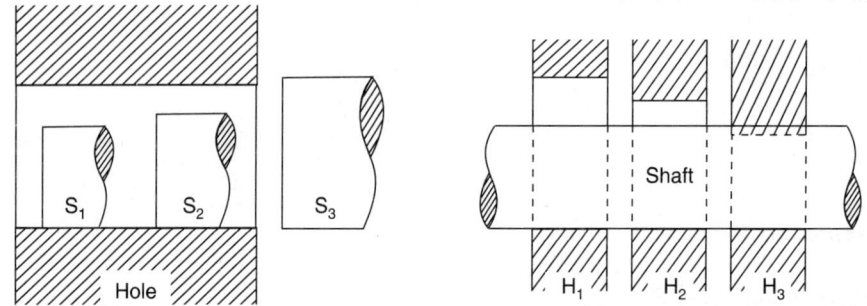

'S' denotes shaft to give various fits with hole
Hole Basis System

'H' denotes hole to give various fits with shaft
Shaft Basis System

Fig. 17.8. Hole and shaft basis systems.

2. **Shaft basis system.** Refer to Fig. 17.8.

'Shaft basis system' is one in which the limits on the shaft are kept constant and the variations necessary to obtain the classes of fit are arranged by varying the limits on the holes.

In present day industrial practice *hole basis system is used because a great many holes are produced by standard tooling,* for example, reamers drills, etc., whose size is not adjustable. Subsequently the shaft sizes are more readily variable about the basic size by means of turning or grinding operations. Thus the *hole basis system results in considerable reduction in reamers and other precision tools as compared to a shaft basis system because in shaft basis system due to non-adjustable nature of reamers, drills etc. great variety (of sizes) of these tools are required for producing different classes of holes for one class of shaft for obtaining different fits.*

17.4.5. Systems of Specifying Tolerances

The tolerance or the error permitted in manufacturing a particular dimension may be allowed to vary either on *one side* of the basic size or on *either side* of the basic size. Accordingly two systems of specifying tolerances exit.

Refer to Fig. 17.9.

1. Unilateral system.
2. Bilateral system.

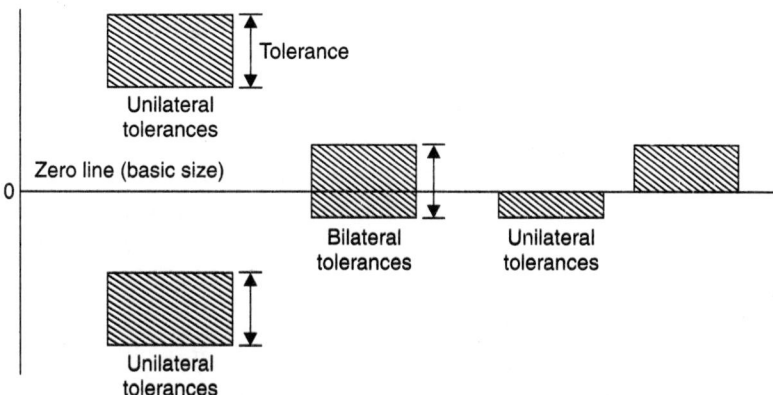

Fig. 17.9. Unilateral and bilateral tolerance.

In the **unilateral system,** tolerance is applied only in *one direction*,

$$\text{Example}: 40.0 \begin{array}{c} +\,0.04 \\ +\,0.02 \end{array} \quad \text{or} \quad 40.0 \begin{array}{c} -\,0.02 \\ -\,0.04 \end{array}$$

In the **bilateral system** of writing tolerances, a dimension is permitted to vary in *two directions.*

$$\text{Example}: 40.0 \begin{array}{c} +\,0.02 \\ -\,0.04 \end{array}$$

● *Unilateral system* is more satisfactorily and realistically applied to certain machining processes where it is common knowledge that dimensions will most likely deviate in one direction. Further, in this system the tolerance can be revised without affecting the allowance or clearance conditions between mating parts *i.e., without changing the type of fit.* This system is *most commonly used in interchangeable manufacture especially where precision fits are required.*

● It is not possible, in *bilateral system*, to retain the same fit when tolerance is varied. The basic size dimension of one or both of the mating parts will also have to be changed. This system clearly points out the theoretically desired size and indicates the possible and probable deviations that can be expected on each side of basic size. *Bilateral tolerances help in machine setting and are used in large scale manufacture.*

17.4.6. Designation of Holes, Shafts and Fits

A *hole or shaft is completely described if the basic size, followed by the appropriate letter and by the number of the tolerance grade, is given.*

Example : ● A 25 mm H-hole with the tolerance grade IT8 is given as :
25 mm H8 or simply 25 H8.

● A 25 mm *f*-shaft with the tolerance grade IT7 is given as :
25 mm *f*7 or simply 25 *f*7.

A *'fit'* is indicated by combining the designations for both the hole and shaft with the hole designation written first, regardless of system (*i.e.,* hole-basis or shaft-basis)

Example : 26 H8—*f* 7 or

$$25 \text{ H8} - f\,7 \quad \text{or} \quad 25\frac{\text{H8}}{f\,7}$$

17.4.7. Commonly Used Holes and Shafts

● In several engineering applications the fits required can be met by a quite small selection from the full range available in the standards. The holes and shafts commonly used are as follows :

Holes (*commonly used*) : H6, H7, H8, H9, H11.

Shafts (*commonly used*) : c11, d10, e9, f7, g6, h6, k6, n6, p6, s6.

IS : 919 gives the most commonly used holes and shafts upto 500 mm for the purpose of general engineering.

17.4.8. The Newall System

● The Newall system is the first standard evolved in Great Britain to standardise limits and fits and is still used to a certain extent although all the fits provided by this system can be obtained with approximately the same values by selection from 1916. This system provides a range of clearance, transition and interference fits for size upto 12″. *It is a hole basis system,* which stipulates two grades of holes, specified with bilateral tolerances, together with 6 grades of shaft tolerances.

● This system is extremely simple and is earliest of all the systems. It specifies too few fits and those listed do not enforce to modern ideas as regards their basic deviations. Though this served a useful purpose in the past but is *not considered suitable for modern production.*

● Since it is based on hole basis system, therefore, *in this system provision is made in the size of the hole for error in workmanship, and the variation to obtain the quality of fit required is allowed for on the size of the shaft which has to enter the hole.*

17.4.9. ISO System of Limits and Fits

ISO system has presently been universally adopted and as a matter of fact IS : 919 is almost in line with this system. While ISO specifies 28 classes of holes designated A, B, C, CD, E, EF, F, FG, G, H, J, JS, K, M, N, P, R, S, T, U, V, X, Y, Z, ZA, ZB, ZC and 18 grades of tolerance exactly matching with those of ISO systems. Similarly ISO has 28 classes of shafts while IS : 919 specifies only 25 classes of shaft. Other characteristics such as fundamental deviation and tolerance unit etc. are same in both the systems.

17.4.10. Types of Fits

Some important types of fits are discussed below :

1. Selective fit. A *selective fit* may be a *transition* or an *interference fit.* This type of fit is required where the object is to make a shaft and hole with a finite and not a permissible range on it. It is customarily used for tight or interference fits where it is desired to avoid the extremes of maximum tightness or looseness. The ideal selective fit for the tightest class of fit would stress the hole just to its elastic limit, thereby giving the maximum holding power without overstressing or distorting the grain structure.

2. Push fit. A *push fit* is a *transition fit.* It is also known as 'sung fit' and represents the closest fit that permits assembling parts by hand. With a push fit, there should be no perceptible play between the mating parts.

3. Driving fit. A *driving fit* is an *interference fit.* When a plug or shaft is made slightly larger than the hole into which it is to be inserted and the allowance is such that the parts can be assembled by driving, this is known as a driving fit. *Such fits are employed when the parts are to remain in a fixed position relative to each other.* Before assembling parts with a driving fit, the bearing surfaces should be oiled. A hydraulic press is usually preferable for assembling.

4. Forced or pressed fit. A forced or pressed fit is an *interference fit.* It is the term used when a pin, shaft or other cylindrical part is forced into a hole of slightly smaller diameter, ordinarily by the use of hydraulic press or some other type of press capable of exerting a considerable pressure.

A *force fit has a larger allowance than a driving fit, and therefore requires greater pressure for assembling*. Forced fits are restricted to parts of *small* and *medium size e.g. crankpins, car wheel axles,* and *similar parts* (which must be held very securely).

5. Shrinkage fit. A *shrinkage fit* is an *interference fit*. It is obtained by making the internal member slightly larger than the hole in the external member. In this type of fit, the pressure is not required for assembly but instead the external member is heated and expanded sufficiently to permit inserting the internal member easily. Then as the external part cools or is cooled by applying water or dry ice, it shrinks tightly around the internal part. In general practice, a smaller allowance for shrinkage fit is favoured.

6. Freeze fit. In a *freeze fit*, instead of heating the female member, the male member may be *contracted by cooling and subsequently allowed to expand into the female*. This process uses an industrial refrigerator giving a temperature of about $-50°C$, or to obtain lower temperatures the component is cooled in liquid air (boiling point $-190°C$). Examples of this process are in the insertion of exhaust valve seats inserts in engine cylinder heads or blocks, or in the insertion of brass bushes in various assemblies.

This process is convenient only for small parts as otherwise the size of the refrigerator equipment is prohibitive or consumption of liquid air is excessive. However, on suitable parts the process is very convenient, since the temperature is controlled and is unlikely to damage the structure of the material in any way. A combination of freezing and heating is also used as this enables reasonable maximum temperatures to be used. A *convenient method of reaching $-50/-60°C$ without expensive equipment, and suitable for occasional use is to cool the part in alcohol to which solid frozen dioxide (known as 'dry ice' or 'dry cold') is added*.

17.4.11. Concept of Interchangeability

'Interchangeability' refers to assembling a number of unit components taken at random from stock so as to build up a complete assembly without fitting or adjustment.

A modern motor-car, for example, consists of many hundreds of separate components each of which is manufactured in large numbers. For complete interchangeability it should be possible simply to collect at random the constituent parts then to assemble the whole without the use of any cutting tools and for the assembly to function satisfactorily.

The contacts between the various parts constitute what are termed *fits*.

For correct functioning of parts the fits must be good within certain limits of accuracy. It would be possible to choose these limits as to ensure absolute interchangeability, and this should always be done in the case of less-important fits such as bolts-fitting in bolt-holes and so on. Experience shows, however, that to do this in the case where the tolerance on the fit is very small may call for such fine limits that the cost is excessive.

In cases like this a process known as *selective assembly* is used. Thus if we have a shaft required to run in a close fitting bearing we can arrange, during inspection, to sort the shafts into say, three grades, those near the upper limit, those near the middle and those nearly at the bottom limit. The same selection is made and those nearly at the bottom limit. The same selection is made with the bearings. By arranging to mate the top-limit shafts with the top-limit holes, for example, we shall ensure a much better assembly than if the parts were chosen at random. We can, infact, increase the limits on the components and thereby very much reduce the cost of production.

One of the objects of interchangeability is to make it possible to replace a work part, such as a complete ball bearing, without making any adjustment to the old or new parts. Here, of course, selective assembly is difficult except by actual manufacturers of the components, and in such cases it is necessary that absolute interchangeability should be possible.

17.5. MEASURING INSTRUMENTS

I. Linear Measurement :

A. *Direct reading*

1. Rule
2. Combination set
3. Depth gauge
4. Vernier caliper
5. Micrometer
6. Measuring machine
   (*a*) Mechanical
   (*b*) Optical.

B. *Instruments for transferring measurements*

1. Calipers and dividers
2. Telescopic gauges.

II. Angular measurement :

1. Protractors
2. Sine bar
3. Combination set
4. Angle gauge blocks
5. Dividing head.

III. Plan surface measurement :

1. Level
2. Combination set
3. Surface gauge
4. Profilometer
5. Optimal flat.

IV. All-purpose special measurement :

1. Pneumatic
2. Electric
3. Electronics
4. Lasers.

17.6. GAUGES

A **gauge** *is tool or instrument used to measure or compare a component.* It is here employed in the sense of an instrument which, having fixed dimension, is used to determine whether the size of some component exceeds or is less than the size of the gauge itself.

The true value of a gauge is measured by its accuracy and service life which, in turn, depends on the workmanship and materials used in its manufacture. Since all gauges are continually subjected to abrasive wear while in use the selection of the proper material is of great importance. *High carbon and alloy tool steels* have been the *principal materials* used for many years. *These materials can be accurately machined to shape,* and they *respond readily to heat treating operations which increase their hardness and abrasive resistance. Steel gauges* are subjected to *some distortion* because of the heat-treating operation and their *surface hardness is limited.* These objections are largely overcome by the use of *chrome plating or converted carbides as the surface material.* Chrome plating permits the use of steels having inert qualities, since wear resistance is obtained by the hard chromium surface. This process is also widely *used in the reclaiming of the worn gauges. Cemented carbides, applied on metal shanks by powder-metallurgy technique, provide the hardest wearing surface obtainable.*

17.6.1. Classification of Gauges

The following gauges represent those most commonly used in *production work*. The classification is principally according to the shape or purpose for which each is used. A complete classification would require more sub-divisions under the various headings.

1. Snap gauges
2. Plug gauges
3. Ring gauges
4. Length gauges
5. Form comparison gauges
6. Thickness gauges
7. Indicating gauges
8. Pneumatic gauges
9. Electric gauges
10. Electronic gauges
11. Projecting gauges
12. Multiple dimension gauges.

17.6.2. Description of Some Commonly Used Gauges

Snap gauges. A snap gauge, used in the measurement of *plain external dimensions*, consists of a U-shaped frame having jaws equipped with suitable gauging surfaces. A plain gauge has two parallel jaws or anvils which are made to some standard size and cannot be adjusted. This type of gauge is largely being replaced by adjustable gauges which provide means of changing tolerance settings or adjusting to wear. Most gauges are provided with the *"Go"* and *"Not Go"* feature in a single jaw, such a design being both satisfactory and rapid. The general design shown in Fig. 17.10 has been selected because it incorporates most of the advantages of similar gauges. It is *light in weight, sufficiently rigid, easy to adjust, provided with suitable locking means and is designed to permit interchangeability of as many of the parts as possible.*

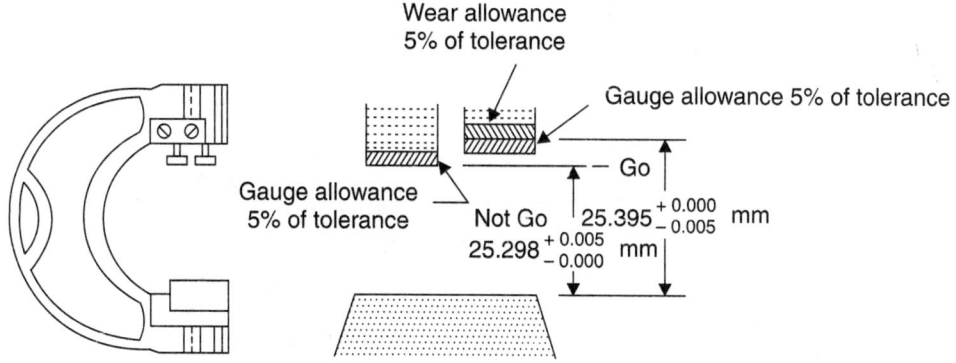

Fig. 17.10. Snap gauge.

The tolerance for settings on such gauges as shown in Fig. 17.10 must account for the total *gauge allowance* which is customarily taken as 10% of the tolerance of the part ; that is, 5% of the part tolerance for each button and the *wear allowance* which is 5% of the part tolerance. The allowance for wear is usually made only for 'Go' gauges, since 'Not Go' gauges have little wear.

The usual practice is to allocate both gauge tolerance and wear allowance entirely within the tolerance limits of the part to be inspected. An occasional part may be rejected though it has been made within its tolerance.

Fig. 17.10 indicates the appropriate size for the "Go" and "Not Go" dimensions when the gauge is set to measure $25.40^{+0.00}_{-0.10}$ mm.

Other designs of snap gauges are shown in Figs. 17.11, 17.12, and 17.13.

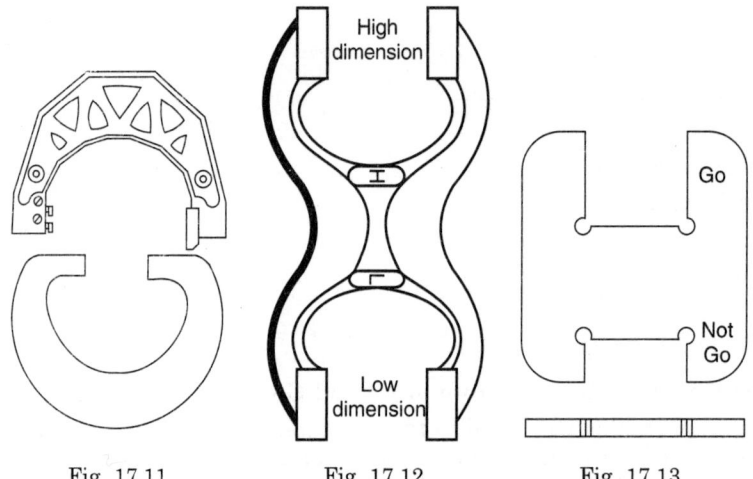

Fig. 17.11.   Fig. 17.12.   Fig. 17.13.

Plug gauges. A plain plug gauge is an accurate cylinder used as an internal gauge for size *control of holes*. It is provided with a suitable handle for holding and is made in a variety of styles. These gauges may be either single or double ended. Double ended, plain gauges have *"Go"* and *"Not Go"* members assembled on *opposite ends,* whereas *progressive gauges have both gauging sections combined on one end.*

As for snap gauges, the allowance for manufacturing the gauge, as well as the allowance of the part to be inspected, must be considered in the design. Fig. 17.14 shows a "Go" and "Not Go" gauge with the approximate dimensions for checking a hole $19.05^{+0.00}_{+0.10}$ mm.

$18.961^{+0.005}_{-0.000}$ mm

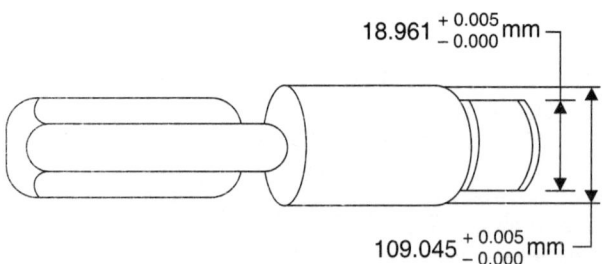

$109.045^{+0.005}_{-0.000}$ mm

Fig. 17.14.

Other types of gauges include *ring, taper, thread,* and *thickness.*

● **Ring gauges** (Fig. 17.15) for outside diameters, are used in pairs, a *"Go"* and *"Not Go"*.

● **Taper gauges are** shown in Figs. 17.16 and 17.17. Taper gauges are not dimensional gauges but rather a means of checking in terms of degrees. Their use is a matter more of fitting rather than measuring.

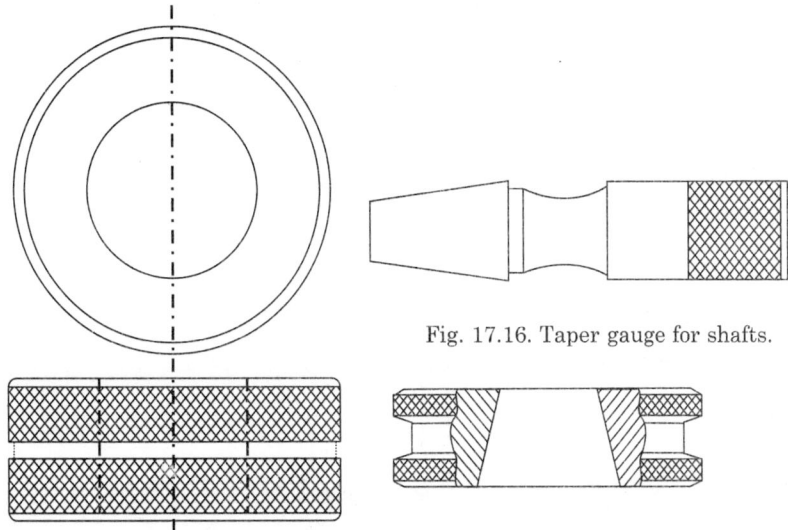

Fig. 17.16. Taper gauge for shafts.

Fig. 17.15. Ring gauge.   Fig. 17.17. Taper gauge for holes.

A *thickness or feeler gauge* (Fig. 17.18) consists of a number of thin blades and is used in *checking clearances and for gauging in narrow places.*

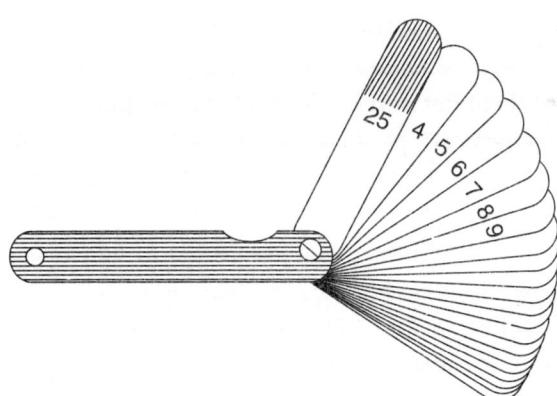

Fig. 17.18. Thickness or feeler gauge.

Dial gauges. Dial gauges or dial test indicators are used for checking flatness of surfaces and parallelism of bars and rods. They are also used for testing the machine tools. They can also be used for measurement of linear dimensions of jobs which require easy readability and moderate precision. A dial gauge is shown in Fig. 17.19. It has two pointer arms which are actuated by a rack and pinion arrangement which acts as a mechanical amplifier. The rack is cut in a spindle. The spindle is made to come in contact with the workpiece. The linear displacement is converted into rotary movement of the pointers. The dial is divided into 100 equal divisions, each division represents a spindle movement of 0.01 mm. For 1 mm movement the bigger arm turns through one complete revolution. The smaller arm registers the number of full turns made by the bigger arm.

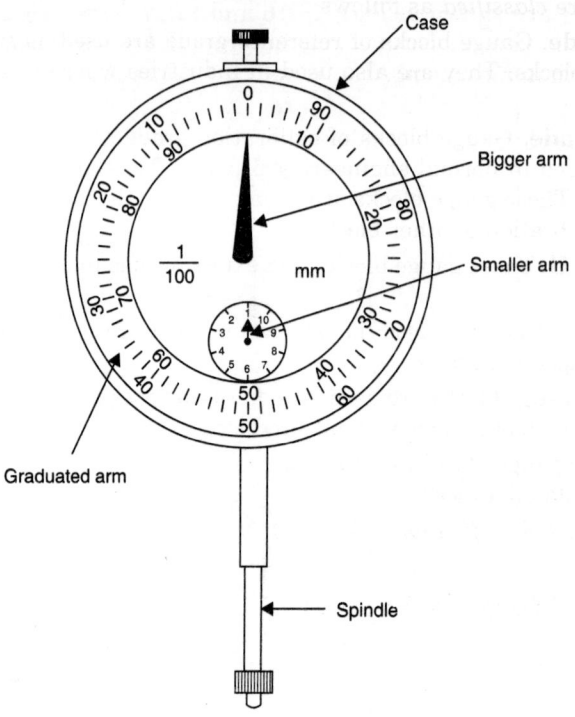

Fig. 17.19. Dial gauge.

Slip gauges. These are also known as *gauge blocks*. A gauge block is a *rectangular block made up of hardened steel with two opposite faces separated by a defined distance*. The faces are ground and lapped to make them flat and parallel within fine limits.

- The blocks (slip gauges) are produced in sets with which it is possible to build a stack of any required height. The flatness and finish of the surfaces of the block is such that they are made to adhere to each other strongly sufficient enough to keep the stack together during normal use. *The process of making the blocks adhere is called 'wringing'.*

'Wringing' is done by first very *carefully cleaning* the faces, placing them in contact as in Fig. 17.20, *pressing them together lightly and twisting* as indicated by the arrows. Two true faces will adhere with very considerable force. It is not a question of air pressure, but a *molecular attraction; in fact if left too long in contact they may be damaged in separating them.* It is quite easy to wring ten or more slip gauges together and handle them as if they were one piece.

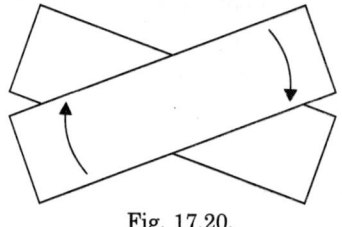

Fig. 17.20.

- The rectangular blocks are 9 mm × 30 mm in size upto 10 mm length and 9 mm × 35 mm in size greater than 10 mm length. The lengths range upto 100 mm and the gauge blocks are graded depending upon their minimum accuracy of flatness, parallelism and face distance.

The gauge blocks are *classified* as follows :

1. **Reference grade.** Gauge blocks of reference grade are used as standards by manufactures of gauge blocks. They are also used in industries where work of similar nature is produced.

2. **Calibration grade.** Gauge blocks of calibration grade are used where highest level of accuracy is desired in normal engineering practice. They are used where tolerances are less than 2 mm. These gauge blocks are primarily used, in conjunction with suitable comparators, for calibration of other blocks.

3. **Grade 0, I and II.** These gauge blocks find extensive use in general engineering applications.

   Grade 0. Gauge blocks are used for jobs where high precision is required. These gauge blocks are of inspection grade.

   Grade I. These gauge blocks are used as general purpose manufacturing gauges in applications like gauge, tool and component production.

   Grade II. These gauge blocks are used for rough setting purposes and checking of components having wide tolerances.

The *Indian Standard Specifications No.* 2984–1966 lists *two sets of gauge blocks.* They are given in the table below.

Table. Details of Gauge Blocks according to Indian Standard Specifications

Normal Set		
Size (mm)	*Increment (mm)*	*Number of pieces*
1.001—1.009	0.001	9
1.01—1.09	0.01	9
1.1—1.9	0.1	9
1—9	1	9
10—90	10	9
	Total	45
Special Set		
1.001—1.009	0.001	9
1.01—14.9	0.01	49
0.5—9.5	0.5	19
10—90	10	9
	Total	86

17.7. LINEAR MEASUREMENTS

17.7.1. Engineer's Steel Rule

An engineer's steel rule is also known as '*Scale*' and is a *line measuring device.* It is a precision measuring instrument and must be treated as such, and kept in a nicely polished condition.

It works on the basic measuring technique of *comparing an unknown length to the one previously calibrated.*

It consists of *strip of hardened steel* having line graduations etched or engraved at interval of fraction of a standard unit of length. Depending upon the interval at which the graduations are made, the scale can be manufactured in different sizes and styles. The scale is available in 150 mm, 300 mm, 600 mm and 1000 mm lengths.

Some scales are provided with some attachments and special features to make their use versatile *e.g.*, very small scales may be provided with a handle to use it conveniently. They may be made in folded form so that they can be kept in pocket also. *Shrink rules* are the scales (used in foundry and pattern making shops) which take into account the shrinkage of materials after cooling.

Following are the *desirable qualities of a steel rule* :

(*i*) Good quality spring steel   (*ii*) Clearly engraved lines

(*iii*) Reputed make   (*iv*) Metric on two edges

(*v*) Thickness should be minimum.

Use of scale. To get good results it is necessary that certain techniques must be followed while using a scale.

- The end of the scale must never be set with the edge of the part to be measured, because generally the scale is worn out at the ends and also it is very difficult to line up the end of the scale accurately with the part of the edge to be measured.

- The scale should never be laid flat on the part to be measured because it is difficult to read the correct dimension.

The principle of *common datums* should be employed while using a scale or rule. The principle is shown in Fig. 17.21 (*a*), the set up indicating the correct method of measuring the length of a component. A surface plate is used as a datum face and its purpose is to provide a common location or position from which the measurement can be made. It may be noted that both the rule and key are at right angles to the working surface of the surface plate and the use of an angle plate simplifies the set-up.

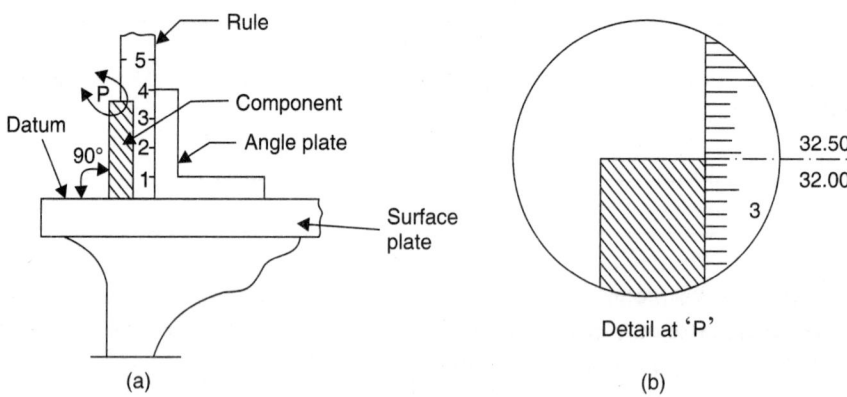

Fig. 17.21. Correct use of scale.

The degree of accuracy which can be obtained while making measurements with a steel rule or scale depends upon :

(*i*) Quality of the rule, and

(*ii*) Skill of the user.

The correct technique of reading the scale is simply illustrated in Fig. 17.21 (*b*).

It is important when making measurements with engineer's rule *to have the eye directly opposite and at 90° to the mark on the work,* otherwise there will be an error—known as *'parallax'*—which is the result of any sideways positioning of the direction of sighting. In Fig. 17.22 the point A represents the mark on the work whose position is required to be measured by means of a rule laid alongside it. The graduations of measurement are on the upper face of the scale or steel rule. If the eye is placed along the sighting line P–A, which is at 90° to the work surface, a true *reading* will be obtained at '*p*', for it is then directly opposite 'A'. If however, the eye is not on this sighting line, but displaced to the right, as at 'Q' the division '*q*' on the graduated scale will appear to be opposite 'A' and an incorrect reading will be obtained. Similarly if the eye is displaced to the left, as at 'R', an incorrect reading on the opposite side as at '*r*' will result.

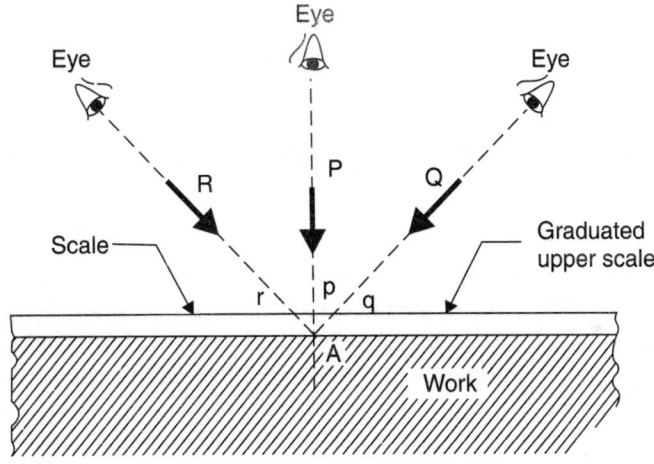

Fig. 17.22. Reading of scale.

Care of the scale or steel rule. A good scale or steel rule should be looked after carefully to prevent damage to its ends as these provide the datums from which measurements are taken. It should never be used as scraper or a screw driver, and it should never be used to remove swarf from the machine-tool table 'Tee' slots. After use the rule should be wiped clean and lightly oiled to prevent rusting.

17.7.2. Calipers

Caliper is an instrument used for measuring distance between or over surfaces, or for comparing dimensions of workpieces with such standards as plug gauges, graduated rules etc. In modern precision engineering they are not employed on finishing operations where high accuracy is essential, but in skilled hands they remain extremely useful. No one can prevent the spring of the legs affecting the measurement, and adjustment of the *firm-joint* type can be made only by tapping a leg or the head. Thus results obtained by using calipers depends very largely on the degree to which the user has developed a sense of touch.

Some firm-joint calipers (Fig. 17.23) have an adjusting screw which enables finer and more controlled adjustment than is possible by tapping methods. Thus at (*e*) in Fig. 17.23 is shown the blacksmith's caliper made with firm joints and a long handle, the latter enabling the measurement of hot forgings without discomfort. The long arm is used for the greater, and the small arm for the smaller or furnished *size*.

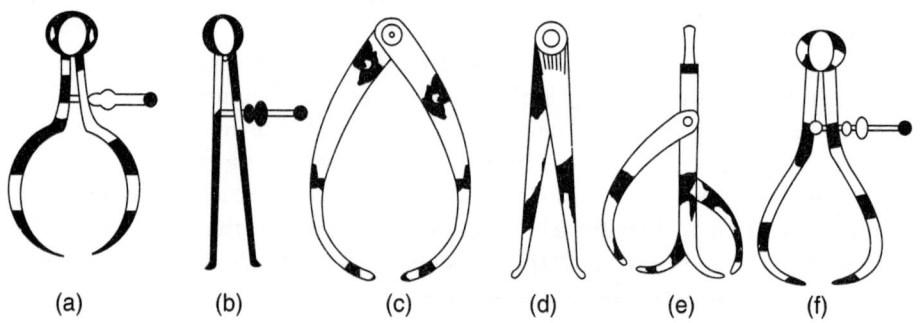

(a)   (b)   (c)   (d)   (e)   (f)

Fig. 17.23. Firm-joint calipers.

At (*f*) is shown a *wide jawed caliper* used in rough measurement of the major diameters of threaded places. For measuring minor diameters a caliper with specially thinned points is sometimes used.

It is *unwise to use calipers on work revolving in a lathe.* If one contact point of the caliper touches revolving work, the other is likely to be sprung and drawn over it by friction.

17.7.3. Vernier Calipers

17.7.3.1. Introduction

The vernier instruments generally used in workshop and engineering metrology have comparatively *low accuracy*. The line of measurement of such instruments does not coincide with the line of scale. The accuracy therefore depends upon the straightness of the beam and the squareness of the sliding jaw with respect to the beam. To ensure the squareness, the sliding jaw must be clamped before taking the reading. The zero error must also be taken into consideration. Instruments are now available with a measuring range upto one metre with a scale value of 0.1 or 0.2 mm. They are made of *alloy steel*, hardened and tempered (to about 58 Rockwell C), and the contact surfaces are lap-finished. In some cases stainless steel is used.

17.7.3.2. The vernier principle

The principle of vernier is that when two scales or division slightly different in size are used, the difference between them can be utilised to enhance the accuracy of measurement.

Principle of 0.1 mm vernier. In the Fig. 17.24 is shown the *principle of 0.1 mm vernier.* The *main scale* is accurately graduated in 1 mm steps, and terminates in the form of a caliper jaw. There is a second scale which is movable, and is also fixed to the caliper jaw. The movable scale is equally divided into 10 parts but *its length* is only 9 mm ; therefore one division on this scale is equivalent to 9/10 = 0.9 mm. This means the *difference between one graduation on the main scale and one graduation on the sliding or vernier scale* is 1.0 − 0.9 = 0.1 mm. Hence if the vernier caliper is initially closed and then opened so that the *first graduation on the sliding scale corresponds to the first graduation on the main scale a distance equal to 0.1 mm* has moved as shown in Fig. 17.25. Such a vernier scale is of limited use because measurements of greater accuracy are normally required in precision engineering work.

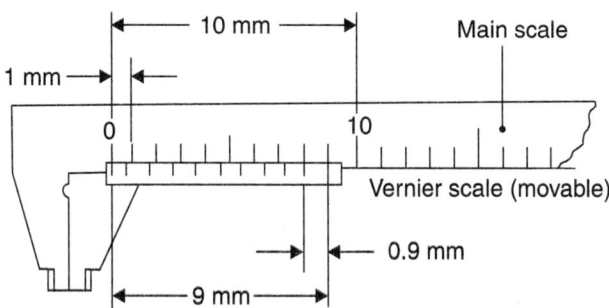

Fig. 17.24. Principle of 0.1 mm vernier.

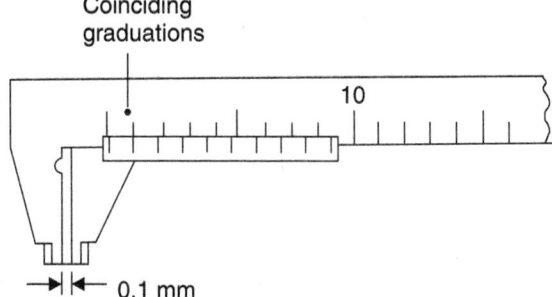

Fig. 17.25.

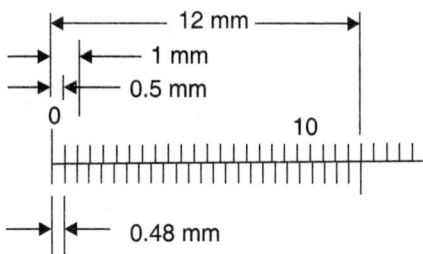

Fig. 17.26. Principle of 0.02 mm vernier.

Principle of 0.02 mm vernier. In Fig. 17.26 is shown the principle of a *0.02 mm vernier*. The vernier scale has main scale graduation of 0.5 mm, whilst the vernier scale has 25 graduations equally spaced over 24 main scale graduations, or 12 mm. Hence each division on the vernier scale = 12/25 = 0.48 mm. *The difference between one division on the main scale and one division on the vernier scale*

$$= 0.5 - 0.48 = 0.02 \text{ mm.}$$

This type of vernier is read follows :

● Note the number of millimetres and half millimetres on the main scale that are coincident with the zero on the vernier scale.

● Find the graduation on the vernier scale that coincides with a graduation on the main scale. This figure must be multiplied by 0.02 to give the reading in millimetres.

● Obtain the total reading by *adding the main scale reading to the vernier scale reading.*

Example. An example of a 0.02 mm vernier reading is given in Fig. 17.27.

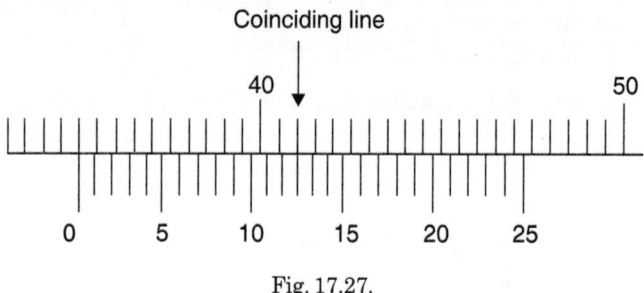

Fig. 17.27.

Reading on the main scale upto zero of the vernier scale = 34.5 mm

The number of graduation that coincides with the graduation on the main scale = 13th

This represents a distance of: 13 × 0.02 = 0.26 mm

∴   Total reading = 34.5 + 0.26 = 34.76 mm.

Note : While taking measurements using vernier calipers it is important to *set the caliper faces parallel to the surface across which measurements are to be made. Incorrect* reading will result if it is not done.

17.7.3.3. Types of vernier calipers

According to Indian Standard IS : 3651–1974, three types of vernier calipers have been specified to make external and internal measurements and are shown in Figs. 17.28, 17.29 and 17.30 respectively. All the three types are made with one scale on the front of the beam for direct reading.

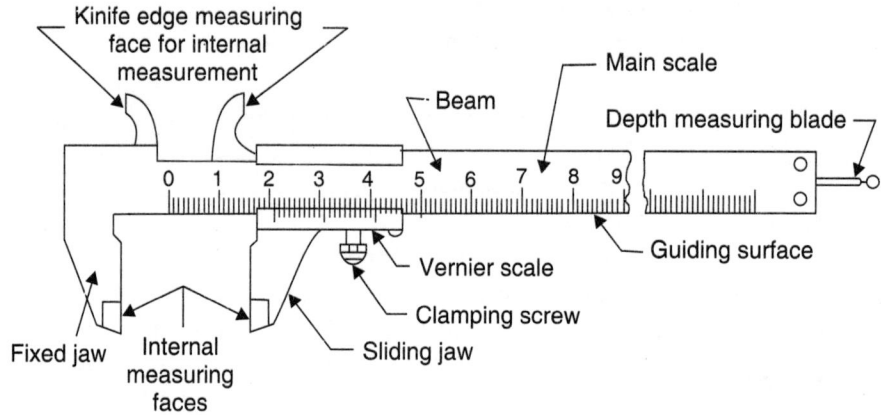

Fig. 17.28. Vernier caliper—Type A.

- **Type A** (Fig. 17.28) vernier has jaws on both sides for *external and internal measurements, and a blade for depth measurement.*
- **Type B** (Fig. 17.29) is provided with jaws on one side for *external and internal measurements.*
- **Type C** (Fig. 17.30) has jaws on both sides for making the *measurements and for marking operations.*

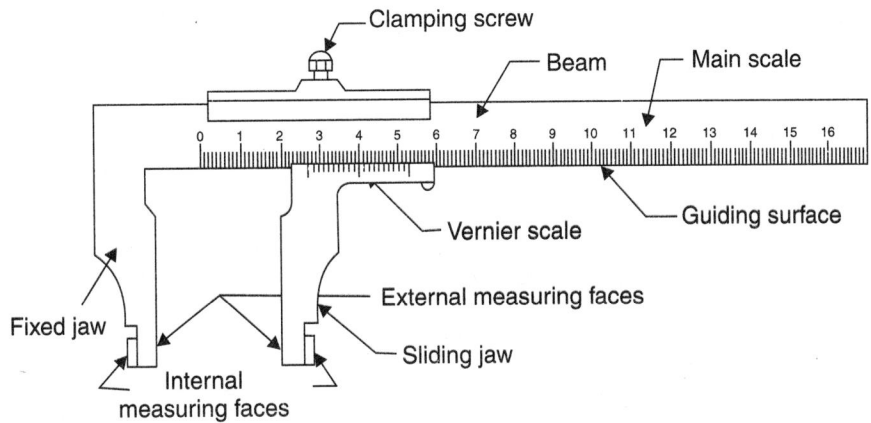

Fig. 17.29. Vernier caliper—Type B.

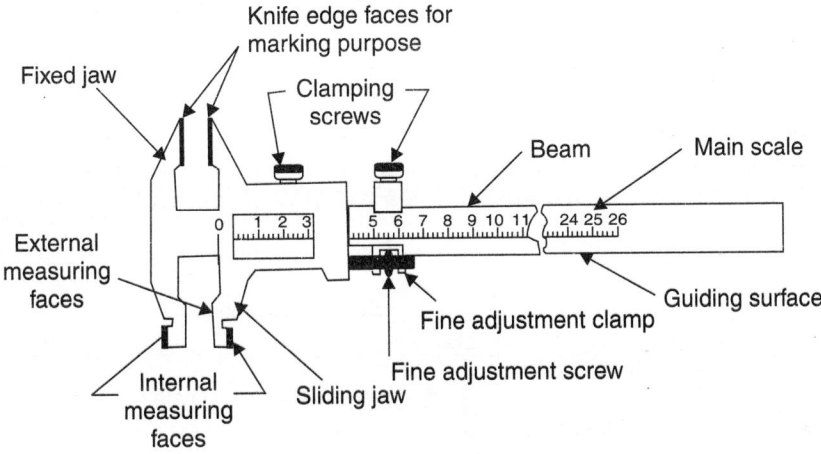

Fig. 17.30. Vernier caliper—Type C.

All parts of the vernier calipers should be of good quality steel and the measuring faces should possess a minimum hardness of 650 HV. The recommended measuring ranges (nominal size) of vernier calipers as per IS 3651-1974 are :

0—125, 0—200, 0—250, 0—300, 0—500, 0—750, 0—1000, 750—1500 and 750—2000 mm.

17.7.3.4. Errors in calipers

The degree of accuracy obtained in measurement greatly depends upon the condition of the jaws of the calipers and a special attention is needed before proceeding for the measurement. The accuracy and natural wear, and warping of vernier caliper jaws should be tested frequently by closing them together tightly and setting them to 0–0 point of the main and vernier scales. In this position, the caliper is held against a *light source.*

- If there is wear, spring or warp, a knock-kneed condition as shown in Fig. 17.31 (*a*) will be observed. If the measurement error on this account is expected to be greater than 0.005 mm the instrument should not be used and sent for repair.

- When the sliding jaw frame has become worn or warped so that it does not slide squarely and snugly on the main caliper beam, then jaws would appear as shown in Fig. 17.31(*b*).

● Where a vernier caliper is used mostly for measuring inside diameters, the jaws may become bow legged as in Fig. 17.31(c) and its outside edges worn down as in Fig. 17.31(d).

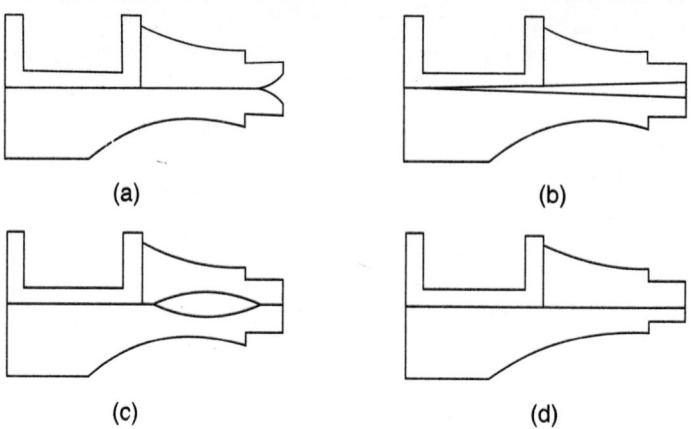

Fig. 17.31.

17.7.4. Vernier Height Gauge

Refer to Fig. 17.32. The *vernier height gauge is mainly used in the inspection of parts and layout work*. It may by *used to measure and mark vertical distances above a reference surface, or an outside caliper*.

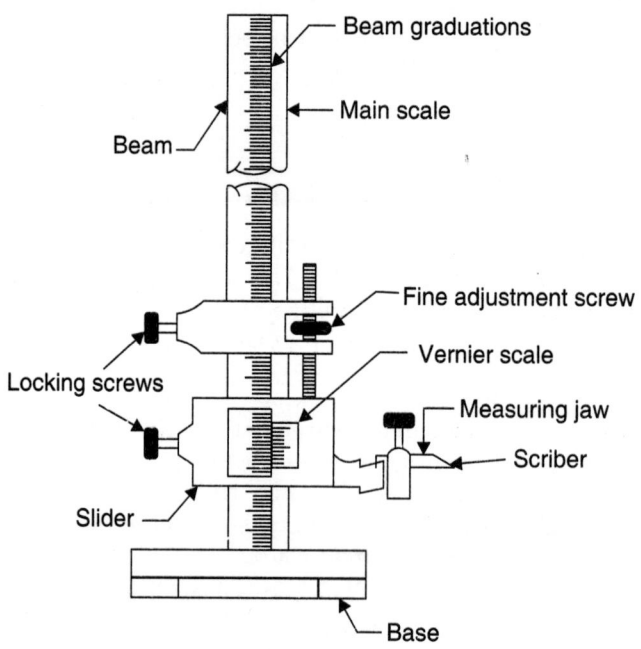

Fig. 17.32. Height gauge.

It consists of the following *parts* :

1. Base

2. Beam

3. Measuring jaw and scriber

4. Graduations

5. Slider.

Base. It is made quite robust to ensure rigidity and stability of the instrument. The underside of the base is relieved leaving a surface round the outside edge of a least 7 mm width and an air gap is provided across the surface to connect the relieved part with the outside. The base is ground and lapped to an accuracy of 0.005 mm as measured over the total span of the surface considered.

Beam. The section of the beam is so chosen as to ensure rigidity during the use. The guiding edge of the beam should be perfectly flat within the tolerances of 0.02, 0.04, 0.06, 0.08 mm for measuring range of 250, 500, 750, 1000 mm respectively. The faces of the beam should also be flat within the tolerances of 0.04, 0.06, 0.10, 0.12 mm for vernier measuring heights of 250, 500, 750, 1000 mm respectively.

Measuring jaw and scriber. The clear projection of the measuring jaw from the edge of the beam should be atleast equal to the projection of the beam from the base. For all position of the slider, the upper and lower gauging surfaces of the measuring jaw should be flat and parallel to the base to within 0.008 mm. The measuring faces of the scriber should be flat and parallel to within 0.005 mm. The projection of the scriber beyond the jaw should be at least 25 mm. Vernier height gauges may also have an offset scriber and the scale on the beam is so positioned that when the scriber is co-planar with the base, the vernier is at zero position.

Graduations. The following *requirements* should be fulfilled in respect of graduations on scales :

● All graduations on the scale and vernier should be clearly engraved and the thickness of graduation both on scale and vernier should be identical and should be in between 0.05 mm and 0.1 mm.

● The visible length of the shortest graduation should be about 2 to 3 times the width of the interval between the adjacent lines.

● The perpendicular distance between the graduations on scale and the graduations on vernier should in no case be more than 0.01 mm.

● For easy reading, it is recommended that the surfaces of the beam and vernier should have dull finish and the graduations lines blackened in. Sometimes a magnifying lens is also provided to facilitate taking the readings.

Slider. The slider has a good sliding fit along the full working length of the beam. A suitable fitting is incorporated to give a fine adjustment of the slider and a suitable clamp provided so that the slider could be effectively clamped to the beam after the fine adjustment has been made.

An important feature of the height gauge is that a special attachment can be fitted to the part to which the scriber is normally fitted, to convert it, in effect, into a depth gauge.

17.7.5. Vernier Depth Gauge

A vernier depth gauge is used to *measure the depth of holes, recesses and distances from a plane surface to a projection.*

In Fig. 17.33 is shown a vernier depth gauge in use. The vernier scale is fixed to the main body of depth gauge, and is read in the same way as vernier calipers. Running through the depth gauge body is the main scale the end of which provides the datum surface from which the measurements are taken.

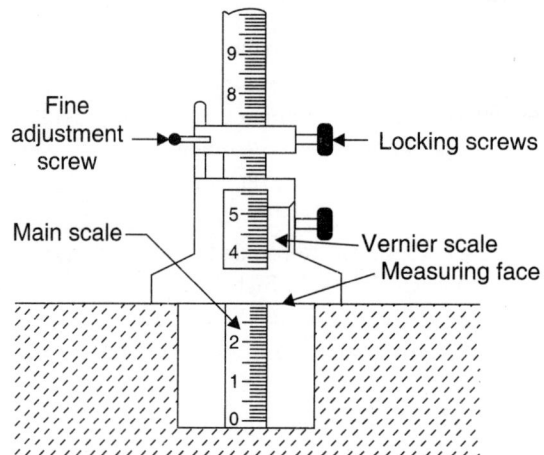

Fig. 17.33. Vernier depth gauge in use.

The depth gauge is carefully made so that the beam is perpendicular to the base in both directions. The end of the beam is square and flat, like the end of a steel rule and the base is flat and true, free from curves or waviness.

Use of vernier depth gauge :

● While using the vernier depth gauge, first of all, make sure that the reference surface, on which the depth gauge base is rested, is satisfactorily true, flat and square. Measuring depth is a little like measuring an inside diameter. The gauge itself is true and square but can be imperceptibly tipped or wanted, because of the reference surface perhaps and offer erroneous reading.

● In using a depth gauge, press the base or anvil firmly on the reference surface and keep several kilogrammes hand pressure on it. Then, in manipulating the gauge beam to measure depth, be sure to apply only standard light, measuring pressure one to two kg—like making a light dot on paper with a pencil.

17.7.6. Micrometers

Micrometers are designed on the principle of 'Screw and Nut'.

17.7.6.1. Description of a micrometer

Fig. 17.34 shows a 0–25 mm micrometer which is used for *quick, accurate measurements* to the *two-thousandths of a millimetre*. It consists of the following parts :

1. Frame
2. Anvil
3. Spindle
4. Thimble
5. Ratchet
6. Locknut.

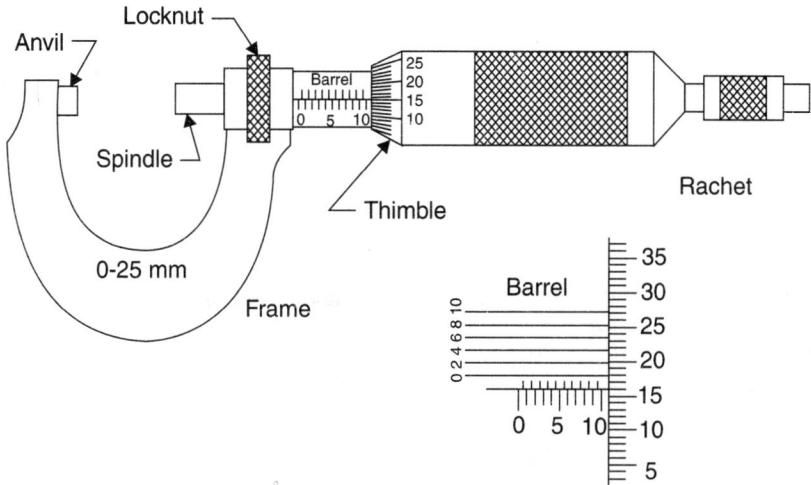

Fig. 17.34. 0—25 mm micrometer.

The micrometer requires the use of an *accurate screw thread* as a means of obtaining a measurement. The screw is attached to a *spindle* and is turned by the movement of a *thimble or ratchet* at the end. The barrel, which is attached to the *frame*, acts as a nut to engage the screw threads, which are accurately made with a *pitch* of *0.05 mm*. Each revolution of the thimble advances the screw 0.05 mm. On the barrel a datum line is graduated with two sets of division marks. The set below the datum line reads in millimetres, and the set above the line reads in half millimetres. The thimble scale is marked in 50 equal divisions, figured in fives, so that each small division on the thimble represents 1/50 of 1/2 mm which is 1/100 mm on 0.01 mm.

- To *read the metric micrometer to 0.01 mm,* examine Fig. 17.34 and first note the whole number of major divisions on the barrel, then observe whether there is a half millimetre visible on the top of the datum line, and last read the thimble for hundredths. The *thimble reading is the line coinciding with the datum line.* The reading for Fig. 17.34 is as follows :

    Major divisions = 10 × 1.00 mm = 10.00 mm
    Minor divisions = 1 × 0.50 mm   = 0.50 mm
    Thimble divisions = 16 × 0.01 mm = 0.16 mm
    _____
                      Reading = 10.66 mm

Since a micrometer reads only over a 25 mm range, to *cover a wide range of dimensions, several sizes of micrometers are necessary.*

The micrometer principle of measurement is also applied to inside measurements and depth reading and to the measurements of screw threads.

- To read the metric micrometer to 0.002 mm, vernier on the barrel is next considered. The vernier, shown rolled out in Fig. 17.34 has each vernier graduation represent two thousandths of a millimetre (0.002 mm), and each graduation is marked with a number 0, 2, 4, 6, 8 and 0 to help in the reading. To *read a metric vernier micrometer note the major, minor and thimble divisions. Next observe which vernier line coincides with a graduated line on the thimble.* This gives the number of two thousandths of a millimetre to be added to the hundredth's reading. For the cut out in Fig. 17.34, the reading is as follows :

$$\text{Major divisions} = 10 \times 1.00 \text{ mm} \quad = 10.00 \text{ mm}$$
$$\text{Minor divisions} = 1 \times 0.50 \text{ mm} \quad = 0.50 \text{ mm}$$
$$\text{Thimble divisions} = 16 \times 0.01 \text{ mm} \quad = 0.16 \text{ mm}$$
$$\underline{\text{Vernier divisions} = 3 \times 0.002 \text{ mm} \quad = 0.006 \text{ mm}}$$
$$\text{Reading} = 10.666 \text{ mm}$$

If the vernier *line coincident with the datum line is 0,* no thousandths of a millimetre are added to the reading.

Note : For shop measurements to 0.001 mm, a mechanical bench micrometer may be used. This machine is set to correct size by precision gauge blocks, and readings may be made directly from a dial on the head-stock. Constant pressure is maintained on all objects being measured and comparative measurements to 0.0005 mm are possible. Precision measuring machines utilizing a combination of electronic and mechanical principles are capable of an accuracy of 0.000 001 m.

17.7.6.2. Sources of errors in micrometers

Some possible *sources of errors* which may result in incorrect functioning of the instrument are :

(*i*) The anvils may not be truly flat.

(*ii*) Lack of parallelism and squareness of anvils at some, or all, parts of the scale.

(*iii*) Setting of zero reading may be inaccurate.

(*iv*) Inaccurate readings following the zero position.

(*v*) Inaccurate readings shown by fractional divisions on the thimble.

The *parallelism* is checked by measuring the diameter of a standard accurate ball across at least three different points on the anvil faces. The *squareness* of the anvils to the measuring axis is checked by using two standard balls whose diameters differ by an odd multiple of half a pitch which calls for turning the movable anvil at 180° with respect to fixed one. *Flatness* of the anvils is tested by the interference method using optical flats. The face must not show more than one complete interference band, *i.e.,* must be within 0.25 μm.

Whenever tested at 20°C, the total error should not exceed the following values :

$$\text{for grade 1, total error} = \left(4 + \frac{L}{100} \right) \mu m$$

$$\text{for grade 2, total error} = \left(10 + \frac{L}{100} \right) \mu m$$

where, L = Upper limit of the measuring range in mm.

The micrometer must be so adjusted that the cumulative error at the lower and upper limits of the measuring range *does not exceed half the total error.*

17.7.6.3. Precautions in using the micrometer

The following *precautions* should be observed while using a micrometer :

1. Micrometer should be cleaned of any dust and spindle should move freely.

2. The *part* whose dimension is to be measured must be held in *left hand and* the *micrometer in right hand.*

The way for holding the micrometer is to place the small finger and adjoining finger in the U-shaped frame. The forefinger and thumb are placed near the thimble to rotate it and the middle finger supports the micrometer holding it firmly. Then the micrometer dimension is set slightly larger than the size of the part and the part is slid over the contact surfaces of micrometer gently. After it, the thimble is turned till the measuring tip first touches the part and the final movement given by ratchet so that uniform measuring pressure is applied. In case of circular parts, the micrometer must be moved carefully over representative arc so as to note maximum dimension only.

   3. Error in readings may occur due to lack of flatness of anvils, lack of parallelism of the anvils as part of the scale or throughout, inaccurate setting of zero reading etc. Various tests to ensure these conditions should be carried out from time to time.

   4. The micrometers are available in various sizes and ranges, and the corresponding micrometer should be chosen depending upon the dimension.

17.7.6.4. Types of micrometers

Different types of micrometers are described below :

   1. **Depth micrometer.** It is also known as '*micrometer depth gauge*'. Fig. 17.35 illustrates a depth micrometer. The measurement is made between the end face of a measuring rod and a measuring face. Because the measurement increases as the measuring rod extends from the face, the readings on the barrel are *reversed* from the normal ; *the start at a maximum* (when *the measuring rod is fully extended from the measuring face*) and *finish at zero* (when *end of the measuring rod is flush with the face*).

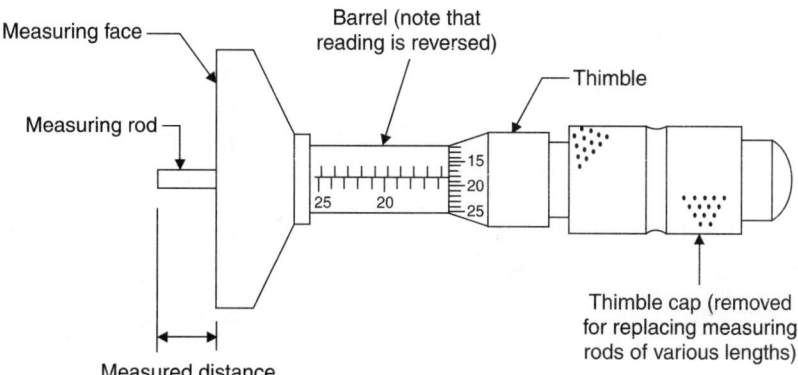

Fig. 17.35. Depth micrometer.

For example, the measurement on the depth micrometer as shown in Fig. 17.35 is :
16 + (18 × 0.01) mm

$$= 16 + 0.18 \text{ mm} = 16.18 \text{ mm}$$

Measuring rods in steps of 25 mm can be interchanged to give a wide measuring range. The thimble cap is unscrewed from the thimble which allows the rod to be withdrawn. The desired rod is then inserted and thimble cap replaced, so holding the rod firmly against a rigid face.

Fig. 17.36 shows the applications of a depth micrometer.

A depth micrometer is *tested for accuracy as* follows :

   ● In order to check the accuracy of a depth micrometer *unscrew the spindle and set the base of the micrometer on a flat surface like a surface plate or tool maker's flat.*

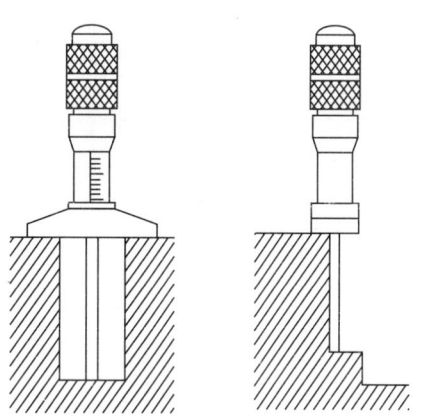

Fig. 17.36. Applications of a depth micrometer.

- Holding the base down firmly turn the thimble or screw in, or down, and when the tip of the micrometer depth stem contacts the flat firmly, with not more than one kg gauging pressure, read the barrel. *If the micrometer is accurate it should read zero.*
- Then rest the micrometer on a 25 mm slip gauge and screw the stem all the way down to contact with the flat. There it should register 25 mm.

2. **Height micrometer.** Fig. 17.37 shows a height micrometer. The same idea as discussed under depth micrometer is applied to the height micrometer.

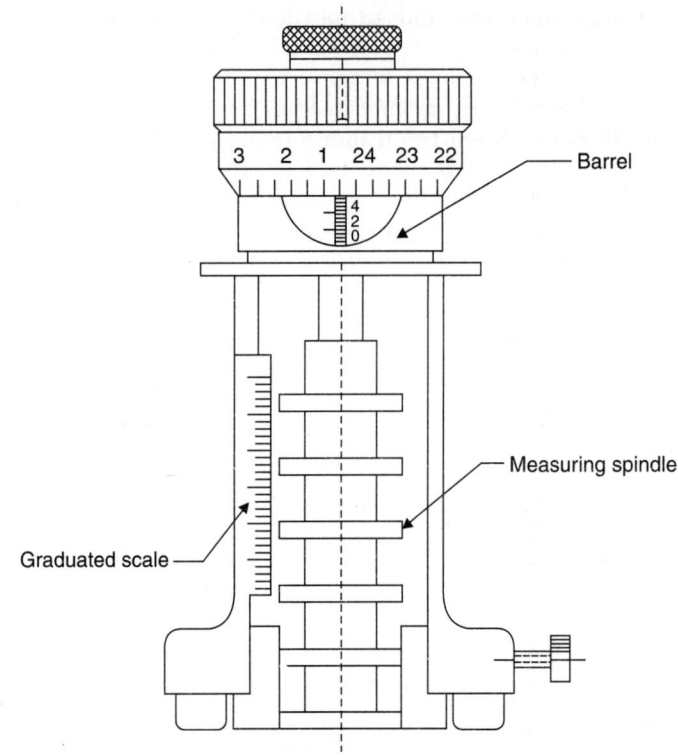

Fig. 17.37. Height micrometer.

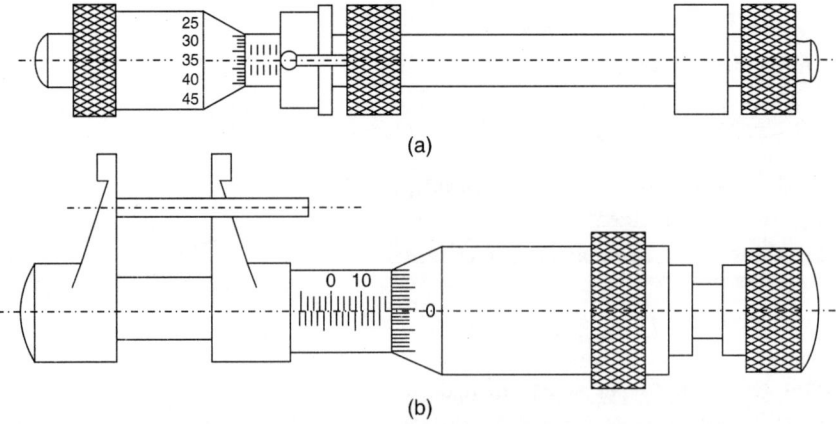

(a)

(b)

Fig. 17.38. Internal micrometer.

3. **Internal micrometers.** These micrometers are used for measuring internal dimensions. The micrometer can be a rod provided with spherical anvils as shown in Fig. 17.38 (*a*). The measuring range of this micrometer is from 25 to 37.5 mm *i.e*, 12.5 mm. By means of exchangeable anvil rods, the measuring capacity can be increased in steps of 12.5 mm upto 1000 mm. Another type of internal micrometer is that shown in Fig. 17.38 (*b*), in which the measuring anvils are inverted cantilevers. The measuring range of this micrometer is from 5 to 30 mm *i.e.* 25 mm.

4. **Differential micrometer.** Refer to Fig. 17.39. This type of micrometer is used to *increase the accuracy of the micrometers*. The right hand screws of different pitches p_1 (= 1.05 mm) and p_2 (= 1 mm) are arranged such that due to rotation of the thimble, the thimble will move relative to the graduated barrel in one direction while the movable anvil, which is not fixed to the thimble but slides inside the barrel, moves in the other direction. The net result is that the movable anvil receives a total movement in one direction given by 1.05 – 1.0 = 0.05 *i.e.,* 1/20 mm per one revolution of the thimble. When the thimble scale is divided to 50 equal divisions, the scale value of the differential manometer will be $\dfrac{1}{20} \times \dfrac{1}{50} = 0.01$ mm. If a vernier scale is provided on the barrel, the micrometer would have a scale value of 0.1 μm. The measuring range, however, is comparatively small (about 5 mm).

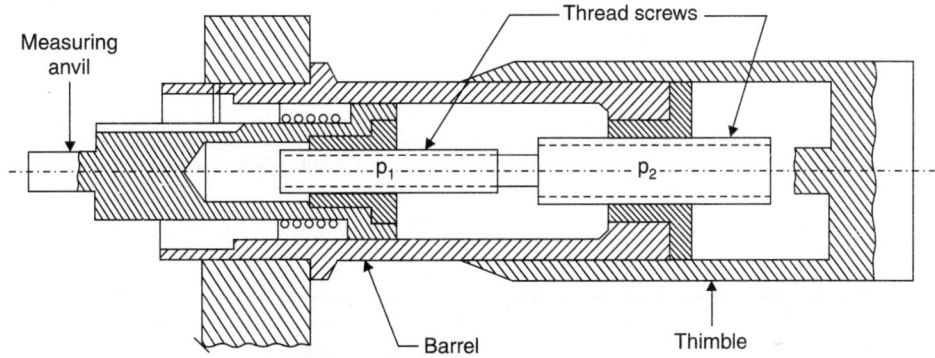

Fig. 17.39. Differential micrometer.

5. **Micrometer with dial gauge.** Refer to Fig. 17.40. In order to enhance the accuracy of micrometers, different types are designed in which the fixed anvil is not merely a fixed one but

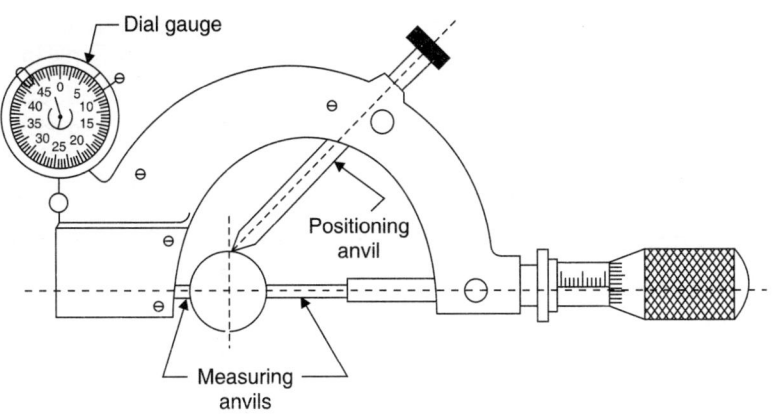

Fig. 17.40. Micrometer with dial-gauge.

moves axially to actuate a dial gauge through a lever mechanism. The micrometer can be used with the dial gauge anvil clamped as an ordinary micrometer for external measurement. Using the dial gauge, the micrometer works as comparator for checking similar components. The micrometer can be provided with a third anvil to improve and facilitate the mounting of the workpiece. Such a micrometer is called *snap dial gauge or snap dial micrometer.*

17.7.7. Advantages and Limitations of Commonly used Precision Instruments

Some of the advantages and limitations of verniers, micrometers and dial indicators are given below :

S. No.	Instruments	Advantages	Limitations
1.	Vernier calipers	Large measuring range on one instrument, upto 2000 mm, will measure external and internal dimensions.	Accuracy 0.02 mm. Point of measuring contact not in line with adjusting nut. Jaws can spring. Lack of 'feel'. Length of jaws limits measurement to short distance from end of component. No adjustment for wear.
2.	Vernier height gauge	Large range on one instrument, upto 1000 mm.	Accuracy 0.02 mm. Lack of 'feel'. No adjustment for wear.
3.	Vernier depth gauge	Large range on one instrument, upto 600 mm.	Accuracy 0.02 mm. Lack of 'feel'. No adjustment for wear.
4.	External micrometer	Accuracy 0.01 mm or, with vernier, 0.002 mm. Adjustable for wear. Ratchet or friction thimble available to aid constant 'feel'.	Micrometer head limited to 25 mm range. Separate instruments required in steps of 25 mm or by using interchangeable anvils.
5.	Internal micrometer	Accuracy 0.01 mm. Adjustable to wear. Can be used at various points along length of bore.	Micrometer head limited to 5 mm or 10 mm range. Extension rods and spacing collars required to extend range to 300 mm. Difficulty in obtaining feel.
6.	Depth micrometer	Accuracy 0.01 mm. Adjustable for wear. Ratchet or friction thimble available to aid constant 'feel'.	Micrometer head limited to 25 mm range. Interchangeable rods required to extend range to 300 mm.
7.	Dial indicator	Accuracy can be as high as 0.001 mm. Operating range upto 100 mm. Mechanism ensures constant 'feel'. Easy to read. Quick in use if only comparison is required.	Does not measure but will only indicate differences in size. Must be used with gauge blocks to determine measurement. Easily damaged if mishandled.

17.8. ANGULAR AND TAPER MEASUREMENTS

17.8.1. Angular Measurements

17.8.1.1. Introduction

The **angle** *is defined as the opening between two lines which meet at a point.* If one line is moved around a point in an arc, a complete circle can be formed and it is from this circle that units of angular measurement are derived. If a circle is divided into 360 parts, each part is called a degree (°). *So angle is such a thing which be just generated very easily requiring no absolute standard,* and it is the precision with which circle can be divided to get the correct measure of angle. Each degree is further divided into sixty parts called minutes ('), and each minute is further subdivided into 60 parts called seconds (").

As in linear measurement, where two kinds of standards (*viz.* end standard and line standard) are used ; so also in angular measurement we have two types of tools. They are *angle gauges* corresponding to slip gauges and *dividing scales* corresponding to line standard.

The *difference in the linear and angular measurement is that no absolute standard is required for angular measurement.*

17.8.1.2. Instruments for angular measurement

The following instruments are used for angular measurement :

1. Protractors
    (*i*) Vernier bevel protractor
    (*ii*) Dial bevel protractor
    (*iii*) Optical bevel protractor.
2. Sine bars
3. Sine tables
4. Sine centre
5. Angle gauges
6. Spirit level
7. Clinometers
8. Plain index centre
9. Optical instruments for angular measurement such as Auto-collimators.

1. Protractors :

(*i*) **Vernier bevel protractor**. The angle between two faces of a component can be simply measured by means of a protractor. As shown in Fig. 17.41 the protractor has *two blades* (*viz.* fixed measuring blade and rotatable or movable blade) which can be set along the faces containing the angle. Its body contains a circular scale which is extended to form one of the blades. The second blade (rotatable blade) is sliding and can be locked in any position along its length to a *rotating turret* mounted on the body. Either the body or the turret carries the divided circular scale, while the other member carries a vernier or an index mark.

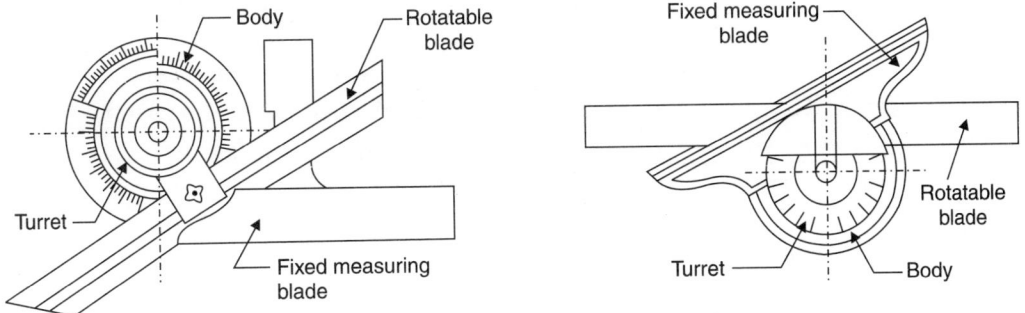

Fig. 17.41. Vernier bevel protractor.

Fig.17.41 shows an ordinary bevel protractor which is a workshop instrument having scale value of 5' or 2.5'.

(*ii*) **Dial bevel protractor.** A dial bevel protractor is shown in Fig. 17.42. With this protractor the *angles can be measured to within 5′.*

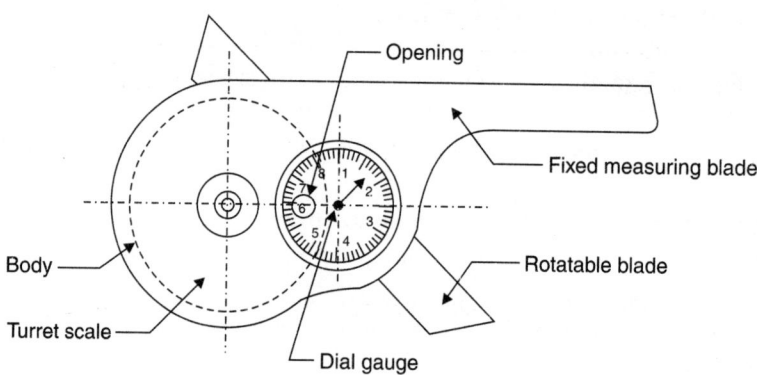

Fig. 17.42. Dial bevel protractor.

In this type of protractor, the turret rotates a circular scale that serves as a gear actuating the dial gauge pointer. The magnification ratio is made so that the pointer rotates through one complete revolution of every 10° of the turret scale, *i.e., for* every 10° of blade angle movement. The scale of the dial gauge is divided into 10 equal divisions, each is subdivided into 12 so that a scale value of $\dfrac{1°}{12}$ or 5′ is obtained. The scale of the dial gauge has an *opening* through which the angle in degrees is read against a fixed mark on the dial.

(*iii*) **Optical bevel protractor.** Refer to Fig. 17.43. A recent development of the vernier protractor is optical bevel protracor. In this device, a glass circle divided at 10 mm intervals throughout the circle is fitted inside the body. A small microscope is fitted throughout which the circle graduations can be viewed. The adjustable blade is clamped to a rotating member which carries its microscope. With the aid of microscope it is possible to read by estimation to about 2 min.

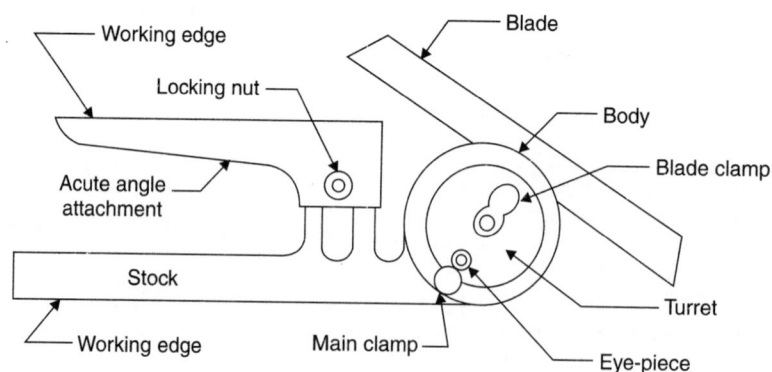

Fig. 17.43. Optical bevel protractor.

Fig. 17.44 shows some applications of the optical protractor which can also be done by any other type of protractor.

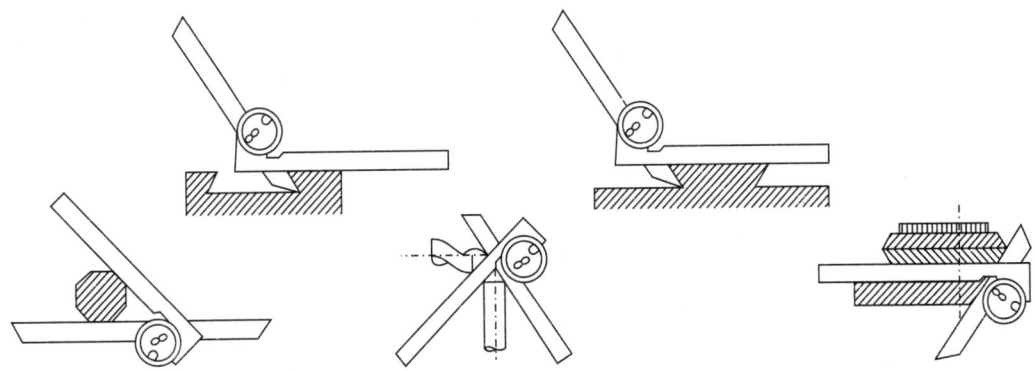

Fig. 17.44. Some applications of optical protractor.

2. Sine bars :

'Sine bar' is a tool used for accurate setting out of angles by arranging to convert angular measurements to linear ones.

Fig. 17.45 shows the *principle of sine bar*. If we wish to set out any angle θ, we could do so (Fig. 17.45) by setting off a horizontal line AB, next scribing an arc CD with a measured radius R and then setting out a vertical dimension EF, made so that $EF = R \sin \theta$. This is quite easily done by means of the sine bar.

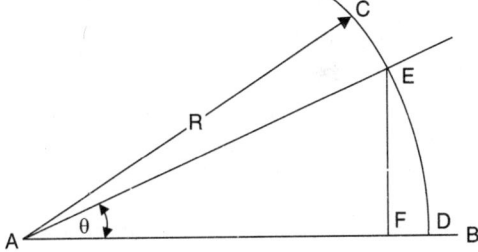

Fig. 17.45. Principle of sine bar.

- *Sine bars used in conjunction with slip gauges constitute a very good device for the precise measurement of angles. Sine bars are used either to measure angles very accurately or for locating any work* to a given angle within very close limits.

- *Sine bars are made from high carbon, high chromium, corrosion resistant steel, hardened, ground and stabilised.* Two cylinders of equal diameter are attached at the ends. The axes of these two cylinders are mutually parallel to each other and also parallel to and at exact distance from the upper surface of the sine bar. The distance between the axes of the two cylinders is exactly 5 inches or 10 inches in British system, and 100 and 200 and 300 mm in metric system. The above recommendations are met and maintained by taking due care in the manufacture of all parts. The various parts are hardened and stabilised before grinding and lapping. All the working surfaces and the cylindrical surfaces of the rollers are finished to surface finish of 0.2 mm or better.

- Depending upon accuracy of the centre distance, sine bars are graded as of *A grade or B grade. B grade of sine bars are guaranteed accuracy upto 0.02 mm/m of length and A grade sine bars are more accurate upto 0.01 mm/m length.*

Types of sine bars :

Different forms of sine bars are discussed below :

- Fig. 17.46 shows a sine bar of simplest type. It consists of a lapped steel bar, at each end of which is attached an accurate cylinder, the axis of the cylinder being mutually parallel, and

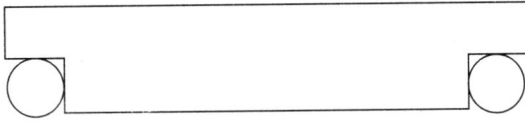

Fig. 17.46. Sine bar of simplest type.

parallel to the upper surface of the bar. The axes are separated by a nominal distance, usually 100 mm or 300 mm for metric bars and 5 or 10 inches, for those in inch units. *The distance between the rollers forms the basis of the designation of sine-bars.* Thus a sine bar designated as 100 mm sine bar will have a distance of 100 mm between its rollers.

- The form of the sine bar shown in Fig. 17.47 is most commonly used. Some holes are drilled in the body of the bar to *reduce the weight and to facilitate handling.*

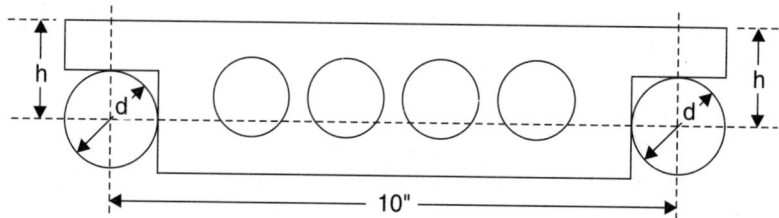

Fig. 17.47. Commonly used sine bar.

- From the point of view of ease in manufacturing to ensure an exact distance between the cylinders, the form shown in Fig. 17.48 is preferable. Also this type of sine bar can be set to a steep angle without the slip gauges fouling the underside of the bar. But this point is immaterial as the accuracy of setting appreciably decreases with steep angles and from the point of stability also sine bars are not generally used for steep angles.

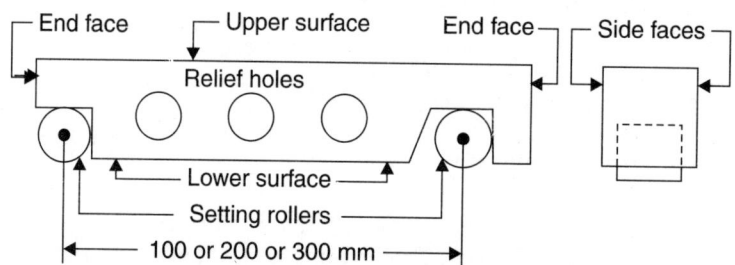

Fig. 17.48.

- In the Fig. 17.49 is shown a most commonly used form of sine bar in which the rollers are so arranged that their outer surfaces on one side are level with the plane to surface of the sine bar.

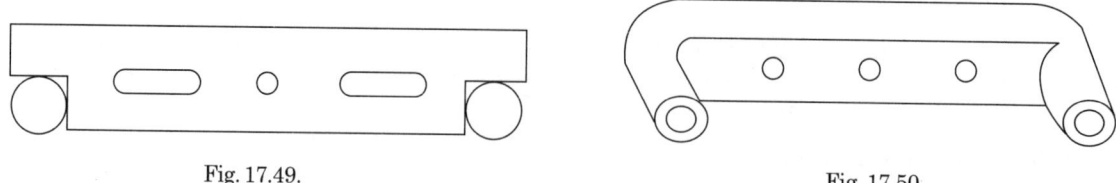

Fig. 17.49.   Fig. 17.50.

- In Fig. 17.50 is shown the form of sine bar which has hollow rollers whose *outside diameter is equal to the width of the sine bar* and is *used where the width* of the *sine bar enters into calculation of work height.*
- Fig. 17.51 shows the form of the bar which is used where ordinary type cannot be used on the top surface due to interrupt.

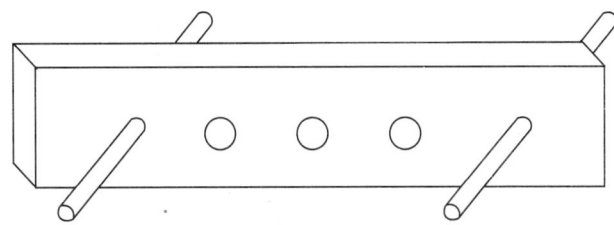

Fig. 17.51.

To get accurate results, in the use of a sine bar, it is essential that :

1. The *contact rollers must be of equal diameter and true geometric cylinders.*
2. The *distance between the roller axes must be precise and known, and these axes must be mutually parallel.*
3. The *upper surface of the bar must be flat and parallel with the roller axes, and equidistance from each other.*

Use of sine bar. A sine bar may be used in the following cases :

1. *Measuring known angles or locating any work to a given angle.*
2. *Checking of unknown angles.*
3. *Checking of unknown angles of heavy components.*

● Fig. 17.52 shows sine bar set up on gauge blocks for measuring an angle on workpiece. In this case :

$$\sin \theta = \frac{h_1 - h_2}{L}$$

where L = Known distance, such as 100 mm or 200 mm, depending on the size of bar used,

$h_1 - h_2$ = Heights built up to correct amounts with precision gauge blocks, and

θ = Angle being checked.

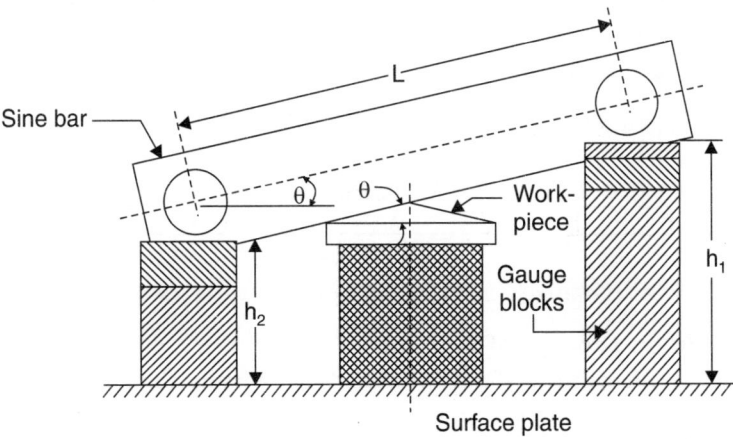

Fig. 17.52. Sine bars set up for measuring an angle of workpiece.

● Fig. 17.53 shows how the sine bar is used to check small components that may be mounted upon it.

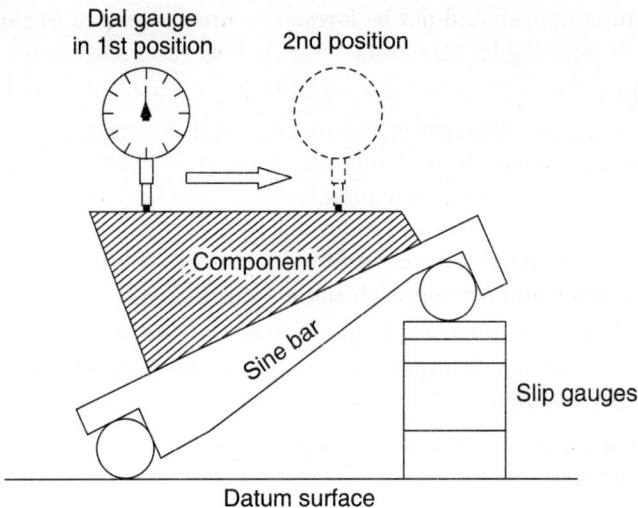

Fig. 17.53. Set up of sine bars for checking small components.

- In such cases where components are heavy and cannot be mounted on the sine bar, then sine bar is mounted on the component as shown in Fig. 17.54. In the example shown, the height over the rollers is measured by means of a vernier height gauge, using a dial test gauge. This is achieved by adjusting the height gauge until the dial gauge shows the same zero reading each time. The difference of the two readings being the rise of the sine bar as shown.

$$\sin \theta = \frac{R_1 - R_2}{L}$$

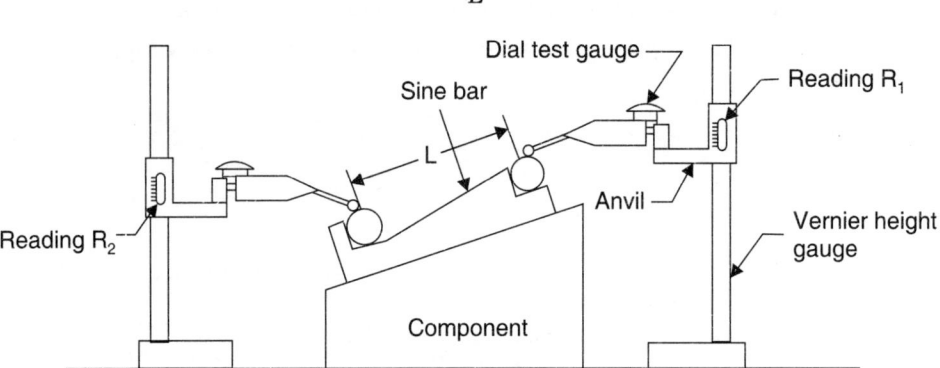

Fig. 17.54.

Precautions in the use of sine bars :

The following *precautions* should be taken while using the sine bars :

1. As far as possible longer sine bar should be used since many errors are reduced by using longer sine bars.

2. The sine *bar should not be used for angle greater that 60°* because any possible error in construction is accentuated at this limit.

3. Accuracy of sine bar should be ensured.

4. A compound angle should not be formed by misaligning of workpiece with the sine bar. This can be avoided by attaching the sine bar and work against an angle plate.

3. Sine table :

A sine table is the *development* of *the sine bar* and the *procedure of setting* it at any angle is *same for sine bars*. It has one hinged roller and the slots to facilitate work holding. A further development of this is the *compound sine table* having their axes of tilt set at right angles to each other are provided. These two tables are mounted on a common base and the table can be set at compound angle by revolving this compound angle into its individual angles in two planes at right angles to each other and setting each table accordingly.

Since a sine table is suitable for large work of greater height, so it is made a robust one. *Usually* the *sine tables are used to measure angles in two planes* (*i.e.*, compound angles) *but these are also used for linear as well as radial measurement.*

4. Sine centre :

Refer to Fig. 17.55. A *sine centre* is basically a sine bar with block holding centres which can be adjusted and rigidly clamped in any position. The *principle of setting is same as of sine table.*

The *sine centres are used for inspection of conical objects* (having male and female centres) *between centres.* These *are used upto inclination of 60°.* Rollers are clamped firmly to the body without any play. This is a very useful device for *testing the conical work centre at each end.*

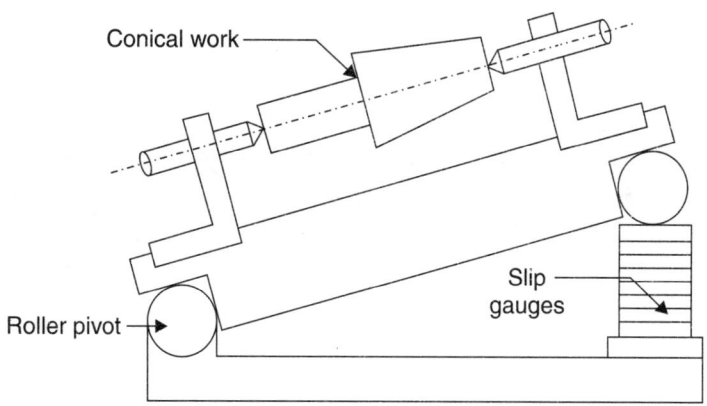

Fig. 17.55. Sine centre.

5. Angle gauges :

An *angle gauge* (Fig. 17.56) *is a hardened steel block approximately 75 mm long 16 mm wide which has two lapped flat working faces lying at a very precise angle to each other.* These are supplied in set, and can be *wrung together to form desired angles.* The angles of the gauges in thirteen piece set, which also includes a square block, are shown in the Table 17.1. The arrangement allows angles in 3 seconds step to

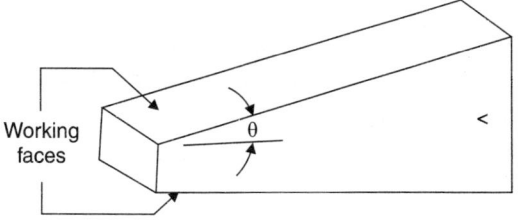

Fig. 17.56. Angle gauge.

be obtained, thus no required angle expressed in whole seconds can be more than 1.5 seconds in nominal error.

Table 17.1. Nominal angles of combination angle gauges

Degrees (°)	1	3	9	27	41
Minutes (')	1	3	9	27	
Fraction of minutes or seconds (")	0.05 (3)	0.1 (6)	0.3 (18)	0.5 (30)	

An additional block having an angle of 9° is also available so that a 90° angle can be built up. Two different grades *A* and *B* are available. The only difference between these two types is that angle gauge of 0.05 minutes is not included in the set *B* and thus consists of only twelve gauges. For the visual inspection of angles, each set of angle gauges is provided with a straight-edge having parallel faces.

Method of using angle gauges :

Fig. 17.57. shows the correct use of angle gauges for making angles. The angles made are : 30° 9′ 18″ and 26° 50′ 30″.

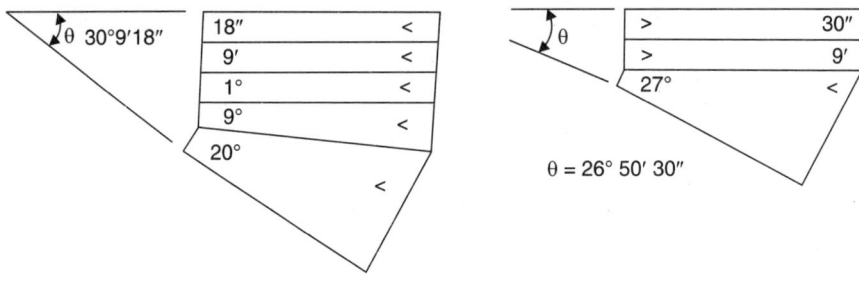

Fig. 17.57.

It may be noted that each angle gauge is marked with engraved *V* which indicates the direction of included angle. *When the angles of individual gauges are to be added up then the* Vs *of all should be* in line and when any angle is to be *subtracted, its engraved V should be in other direction.*

Practical applications of angle gauges :

- In engineering industries the angle gauges have been widely used for the quick measurement of angles between the two surfaces. A frequent use of these gauges is to check whether the component is of angle tolerance.

- Where the angle to be measured between the two surfaces exceeds 90 degrees, the use of precision becomes essential. Figure 17.58 shows a set up to test the angle of V-gauge whose include angle is 102°.

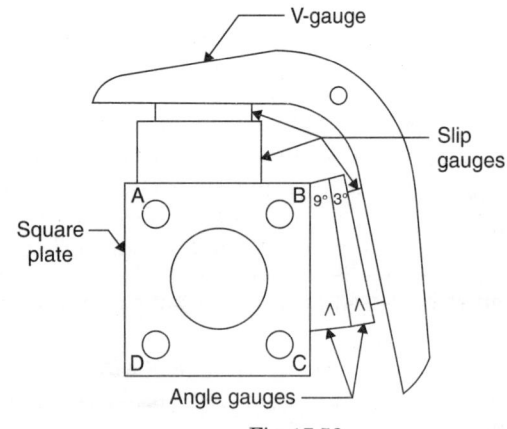

Fig. 17.58.

6. Spirit level :

The spirit level is a simple form of mechanical measuring device. This instrument has been used in engineering for many years, mainly for static levelling of machinery and other equipment, yet a calibrated level can be, in fact, an angle measuring instrument of high precision with a wide variety of applications.

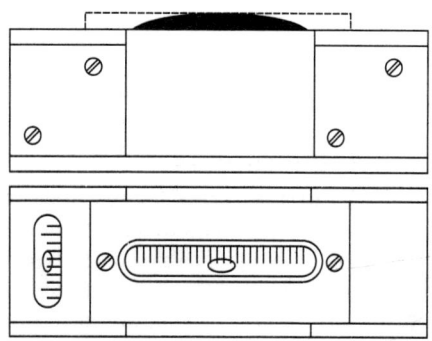

Refer to Fig. 17.59 (Block spirit level). The sensitive spirit level consists of a curved glass tube of uniform bore, nearly filled with a liquid so that an air bubble is left. The glass tube is fitted to a metal block in such a manner that when the lower surface of the block is horizontal, the air bubble will centre itself at the highest point of the tube, *i.e.*, in its centre. In order that the movement of the air bubble be proportional to the angle

Fig. 17.59. Block spirit level.

of tilt, it is necessary that the radius of curvature of the glass tube is made constant over its whole length. The internal cross-section of the tube should be uniform.

The sensitivity of a spirit level is *expressed as the angle of tilt in seconds for which bubble will move by one division on the tube. One* division is generally about 2.5 mm is length.

$$\text{Thus,} \qquad \text{Sensitivity} = \frac{\text{Angle in seconds}}{\text{1 division on the tube}}$$

Fig. 17.60 shows the *principle of spirit level.*

If, R = Radius of curvature of glass tube,

S = Scale division *i.e.*, the length of one division, and

α = Angle of tilt corresponding to 1 division movement of bubble.

$$\text{Then,} \qquad \alpha = \frac{S}{R}$$

Generally the graduations are at 2.5 mm intervals and these represent a tilt of 10 seconds of arc, *i.e.*, sensitivity of level desired is 10 sec. per 2.5 mm movement of the bubble.

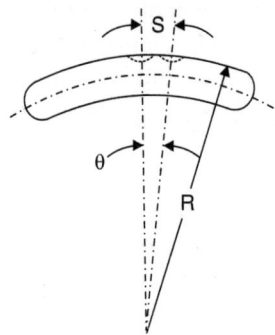

$$\text{Then,} \qquad 10 \text{ sec.} = 0.0000485 \text{ radian} = \frac{2.5 \text{ mm}}{R}$$

$$\text{or,} \qquad R = \frac{2.5}{0.0000485} = 51500 \text{ mm}$$

$$\text{or,} \qquad R = 51.5 \text{ m approximately.}$$

Thus for above sensitivity, radius of curvature of the tube must be about 51.5 m and it is obvious now that *sensitivity of the spirit level is governed solely by the radius of the tube and the base length of its mount.*

Fig. 17.60. Principle of spirit level.

If the base length of any spirit level be about 250 mm, then the height h by which one end must be raised for 2.5 mm bubble movement is given by :

$$0.0000485 = \frac{h}{250}$$

$$\text{or,} \qquad h = 0.0000485 \times 250 = 0.0121 \text{ mm.}$$

If the base length be reduced to 125 mm, then sensitivity is increased twice, and in this case each graduation represents 0.006 mm.

The sensitivity of a spirit level can also be expressed by tangent of the angle of tilt in mm/m.

The *accuracy of a spirit level depends upon the setting of the tube relative to the base.*

7. Clinometers :

A *"clinometer" is an instrument used for measuring angle relative to the horizontal plane.* By a simple calculation it is possible to check the angles between two parts of a job even if separated by considerable distances or on different levels. There are various makes and types of clinometers obtainable, but they are all designed on one of the two basic principles : One involves the *use of a spirit level* and the other makes use of the *plumb-bob (or pendulum) principle* and works from the vertical.

Clinometers are *used for checking angular faces, and relief angles on large cutting tools and milling cutter inserts.* These can also be *used for setting inclined table on jig boring machines and angular work on grinding machines etc.*

The different types of clinometers available are :

1. Vernier clinometer
2. Micrometer clinometer
3. Dial clinometer
4. Pendulum clinometer
5. Optical clinometer.

The most commonly used clinometer is of Hilger and Watt type. The circular glass scale is totally enclosed and is divided from 0° to 360° at 10′ intervals. Sub-division of 10′ is possible by the use of an optical micrometer. A coarse scale figured every 10 degrees is provided outside the body for coarse work and approximate angular reading. In some instruments worm and quadrant arrangement is provided so that reading upto 1′ is possible.

In some clinometers, there is no bubble but a graduated circle is supported on accurate ball bearings and it is so designed that when released, it always takes up the position relative to the true vertical. The reading is taken against the circle to an accuracy of one second with the aid of vernier.

8. Plain index centre :

Fundamentally index centres were meant for use on milling machines but they are now being used for inspection work also. This is specially suited to those problems which involve the measurement of a large number of angular dimensions about a common centre. The work is set between the centres and correct angular dimensions are set directly from the indexing plate.

To obtain wide range of angles the following types of indexing methods may be employed :

1. Single indexing
2. Differential indexing
3. Compound indexing.

17.8.2. Taper Measurement

17.8.2.1. Gauges for tapers

- A taper is tested by using taper plug and ring gauges. The important thing in testing a tapered job is to check the diameter at bigger end and the change of diameter per unit length. For testing the correctness of a taper, three light lines are drawn with persian blue about equidistant along the length on the (male portion) plug gauge or spindle to be tested and it is fitted in the gauge (female) and rotated once or twice. If persian blue *marks do not rub off evenly*, the *tapper is incorrect* and setting must be adjusted until persian blue marks are rubbed equally all along its length.

- *Taper holes* can be checked by a *'Go' and 'Not Go'* taper plug limit gauge as shown in Fig. 17.61. At the large end of the gauge where the large diameter of the taper hole should theoretically lie, a flat surface is machined on the gauge on which two

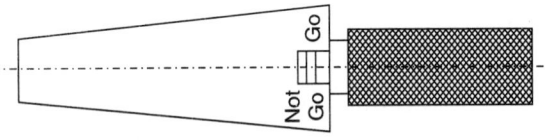

Fig. 17.61.

lines are engraved, the distance between them is equal to the tolerance on the base distance. The line nearer to the small end represents the *'Go'* limit and the other near the large end represents the *'Not Go'* limit of the taper hole. The *dimensional accuracy of the taper hole will be accepted when the taper gauge fits in the hole such that the "Not Go" limit remains outside the hole.* This *does not provide a positive check* of the type of contact between the mating parts along the whole length of the fits. These geometrical inaccuracies can be checked by marking light lines with persian blue as explained earlier.

- *Taper shafts* can be checked by the reversible procedure using taper hole gauges as shown in Fig. 17.62 (*a, b*). The *solid taper sleeve gauge* [Fig. 17.62 (*a*)] has cut away portion A, on which the *"Go" and "Not Go"* limits are engraved. By inserting the taper shaft in the hole gauge, the small end of the taper shaft should lie between these two limits. The geometrical accuracy of the shaft can be checked by the persian blue procedure.

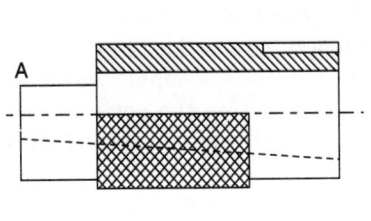

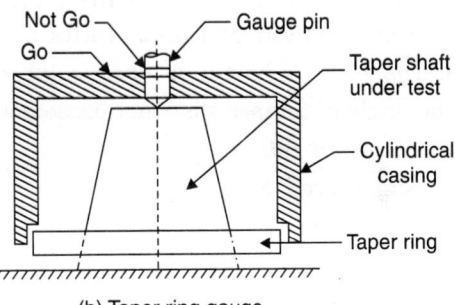

(a) Solid taper sleeve gauge.   (b) Taper ring gauge.

Fig. 17.62.

- The *taper ring gauge* [Fig. 17.62 (*b*)] consists of a taper ring that can be placed on the taper shaft under test. A cylindrical casing is mounted on the taper ring and the position of the taper ring relative to the small end of the shaft can be checked by a gauge pin on which *"GO" and "Not Go"* limits are marked. To check the geometrical accuracy of the taper shaft, more than one taper ring should be used.

A positive check of the dimensional and geometrical accuracies of taper shafts can be achieved by using *"May" taper gauge* [Fig. 17.63]. It consists of two similar side members separated by a standard central taper gauge. The taper gauge is provided with a taper work support on which the taper shaft has to rest. The height of the work support is made such that the two side members make contact with the taper shaft at two enveloping lines diametrically opposite. The two side members and the taper gauge are made separate for the universal use of the gauge and they can be screwed together. On the top surface of one of the side members the two *"Go" and*

"Not Go" limits are provided within which the large diameter of taper shaft should lie. Geometrical errors can be observed against an illuminated background. Error as small as 0.01 mm. or even less can be detected by this gauge.

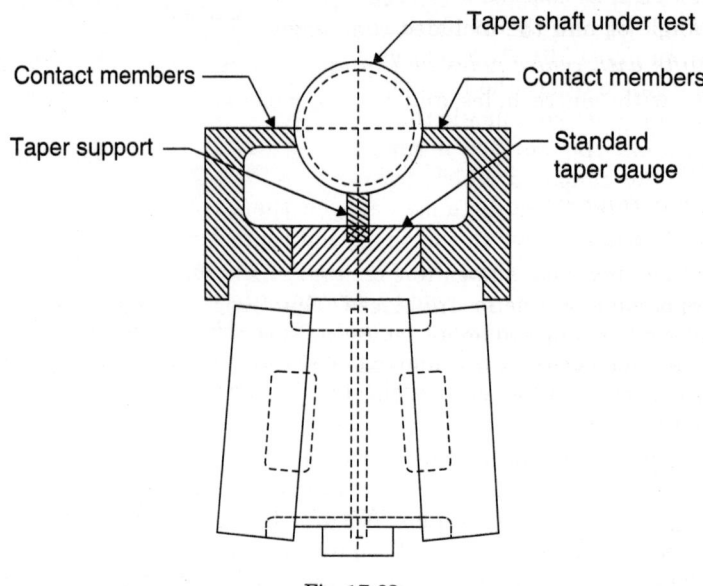

Fig. 17.63.

17.8.2.2. Taper measuring instruments
1. Measurement of taper shafts :

A. *Taper shafts having a taper length not exceeding 100 mm :*

The angle of a taper shaft not exceeding 100 mm length can be measured by means of :

(*i*) *Standard rollers,*

(*ii*) *Slip gauges,* and

(*iii*) *Micrometer.*

The taper plug gauge is to be placed with its small end on a surface plate as shown in Fig. 17.64. Two precise rollers of equal diameters are then placed on the surface plate to contact the surface of the taper plug of two points diametrically opposite. The distance L_1 across the rollers can be measured by the micrometer. The same rollers are placed on two equal columns of slip gauges of height h and the dimension L_2 across them is also measured.

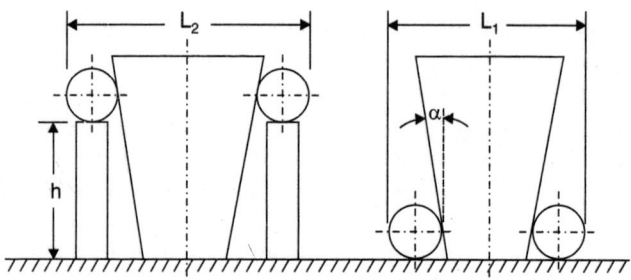

Fig. 17.64.

Then, $L_2 - L_1 = 2h.\tan \alpha,$

or, $\tan \alpha = \dfrac{L_2 - L_1}{2h}$

where, α = Taper angle or half the included cone angle.

B. *Taper shafts with centre holes and having a taper length more than 100 mm :*

Taper shafts with centre holes and having a taper length more than 100 mm can be measured on :

    (*i*) *Sine table,*

    (*ii*) *Taper comparator or,*

    (*iii*) *Eccentricity tester.*

Sine table. The *sine table* shown in Fig. 17.65, is provided with locating angular grooves in which two rollers of equal diameter are placed. The taper is mounted between the two centres on the table. The sine table is placed on a surface plate and adjusted by means of slip gauges until the top surface of the taper shaft is made parallel to the surface plate. This can be checked by a dial gauge mounted on the surface plate. The difference between the height of the slip gauges column under each roller divided by the given distance between the rollers, L gives the sine of the taper angle α as,

$$\sin \alpha = \frac{h_1 - h_2}{L}$$

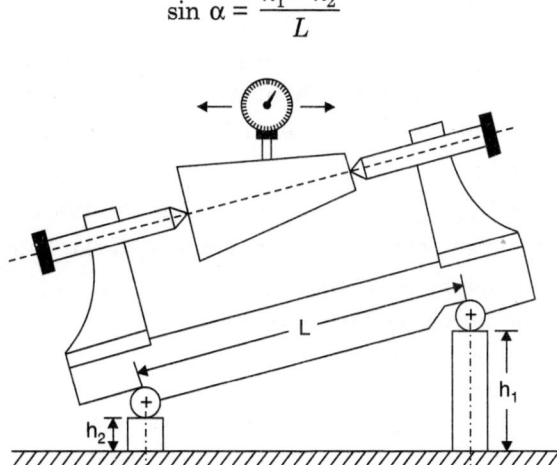

Fig. 17.65.

Taper comparator :

● The *taper comparator* is shown in Fig. 17.66. It is provided with one concentric and one eccentric centres so that they can be exactly aligned. By mounting a mandrel between them, the centres can be adjusted until their alignment is achieved when the two dial gauges 1 and 2 of the comparator show the *same reading*. The taper shaft under test is then mounted between the two centres and the readings of the two dial gauge are observed. The taper angle is then calculated as :

$$\tan \alpha = \frac{G_1 - G_2}{L}$$

where, G_1, G_2 = Readings of dial gauges, and

        L = Distance between the measuring plungers of the dial gauges.

Eccentricity tester's description is out of scope of this book.

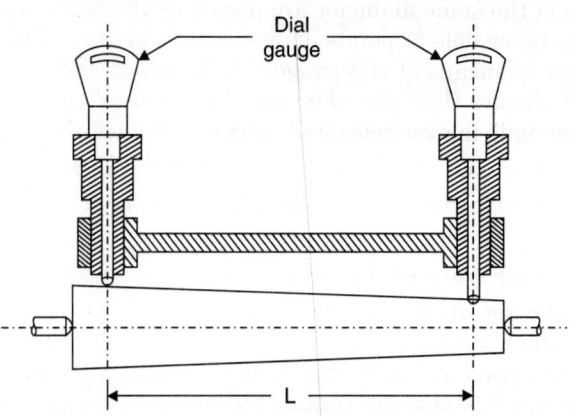

Fig. 17.66. Taper comparator.

2. Measurement of taper holes :

A. *For long and small taper holes below 25 mm* :

In order to measure internal tapers, a suitable accurate ball of radius r_1 is placed inside the taper hole, as shown in Fig. 17.67, and the height h_1 above the ball is measured by a depth micrometer. A large ball of radius r_2 is then placed in the hole and the height h_2 above the ball is also measured. The taper angle α of the hole under test can be calculated as :

$$\sin \alpha = \frac{r_2 - r_1}{(h_1 - h_2) - (r_2 - r_1)}$$

Every care should be taken to avoid wedging the balls, either by dropping them inside the hole or by applying high pressures during the measurement.

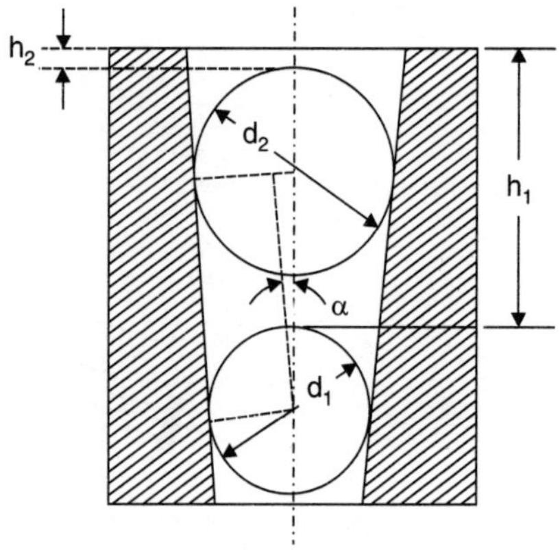

Fig. 17.67.

B. *For medium and large size taper holes* :

For medium and large size taper holes, a more precise method can be applied using *balls and slip gauge*, as shown in Fig. 17.68. The large end of the component is placed downwards on

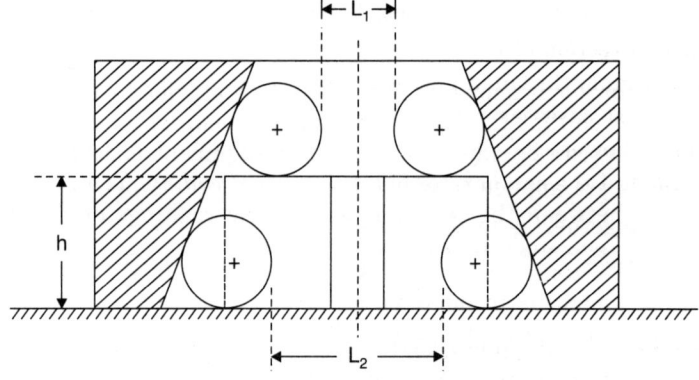

Fig. 17.68. Use of balls and slip gauges.

a surface plate. Two balls of the same diameter are placed on the face plate and make contact with the internal surface of the taper hole at points diametrically opposite. The distance L_2 between the two balls can be measured by means of slip gauges blocks which should be inserted without force to avoid errors. The two balls are then placed on equal columns of slip gauges of height h and the distance L_1 separating the balls is also measured with slip gauges. The taper angle then be calculated as follows :

$$\tan \alpha = \frac{L_2 - L_1}{2h}$$

C. Measuring a taper hole on the sine table using internal adaptor :

Fig. 17.69 shows the set-up for measuring a taper hole on the sine table using internal adaptor. The internal adaptor used consists of a hollow cylinder with suitable external diameter and a lever, hinged in the cylinder, whose magnification ratio is 1. The taper ring as placed on one of the centres and adapter is made to contact the bottom surface of the taper hole. The sine table is adjusted by slip gauges until the lower surface of the taper hole is set parallel to the surface plate. This can be ensured when the reading of the dial gauge, in contact with the adaptor, remains constant during the axial movement of the adaptor inside the taper hole. In this position, the angle made by the sine table is equal to the taper angle, angle of the component being tested.

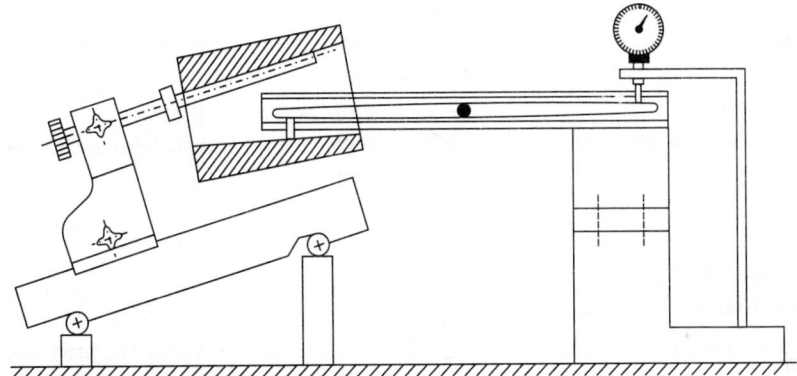

Fig. 17.69. Set-up measuring a taper hole on the sine table using internal adaptor.

17.9. SCREW THREAD MEASUREMENTS

17.9.1. Introduction

A screw thread is helical ridge formed on uniform section round the curved surface. The shape of the normal section of the thread depends upon the shape of the tool which produces its groove. A screw is a male threaded piece generally cylindrical in form, but sometimes, conical (or tapered), used in most cases as a temporary fastening ; less frequently used as a means of transmitting motion or power.

The screw threads are applied to many devices for various *purposes* as follows :

1. To hold parts together as in the case of fastening.

2. To transmit power.

3. To control movement as in micrometer.

4. To increase the effect of applied effort as in auto jack.

5. To convey materials as in the case of fastening.

17.9.2. Classification of Threads

The threads may be *classified* as follows :

1. *According to the surface on which the threads are cut :*

    (*i*) External threads

    (*ii*) Internal threads.

The external threads are cut into the surface of a *cylindrical bar.*

The internal threads are cut into the surface of the *cylindrical hole* of a bar or cone.

2. *According to the direction of rotation of the threaded cylinder with respect to engagement or disengagement with the other part :*

    (*i*) Right handed thread

    (*ii*) Left handed thread.

A *right handed thread* is one in which the nut must be turned in a right handed direction to screw it on (Fig. 17.70 (*a*)). A left handed thread is one in which the nut would be screwed on by turning in to the left (Fig. 17.70 (*b*)).

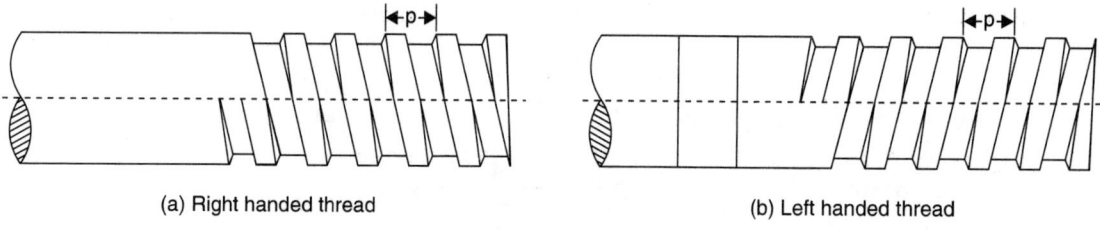

(a) Right handed thread   (b) Left handed thread

Fig. 17.70.

3. *According to number of starts :*

    (*i*) Single start threads

    (*iii*) Multi-start threads.

In a piece of work it is possible to have separate and independent threads running along it. Accordingly there are single threaded screw and multiple or multi-start threaded screw. The independent threads are called starts and we may have single start, two start, three start etc. (Fig. 17.71).

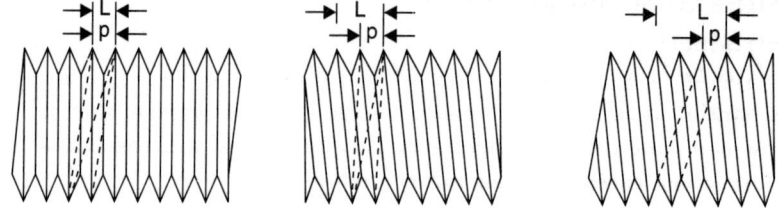

Fig. 17.71. (*i*) Single-start, (*ii*) Two-start, and (*iii*) Three-start threads.

A *single start threaded screw* is one in which there is a movement of one thread for one complete turn round the screw or bolt. In the *multi-start threaded screw* there is a movement of more than one thread. In the case of double start thread, for one complete turn, the thread advances two times as if were a single thread.

Use of multi-start threads. Multi-start threads are used on those cases where *rapid movement or motion is required.*

Examples : Fountain-pen cap, screw press, bottles, tooth paste etc.

17.9.3. Elements of Screw Threads

To designate different parts of the screw threads the following *terms* are commonly employed :

Refer to Fig. 17.72.

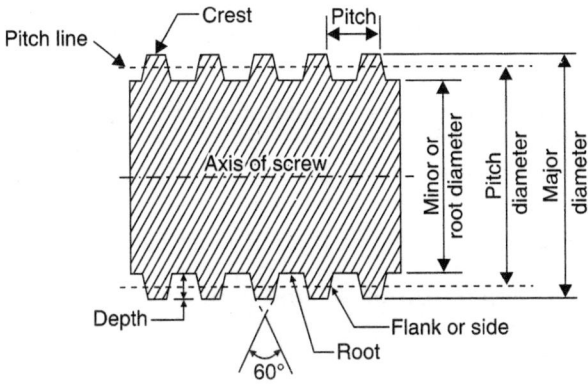

Fig. 17.72. Nomenclature of a screw thread.

1. **Major diameter.** It is the *largest diameter of a screw thread.* It is also termed as *outside or crest diameter.*

2. **Minor diameter.** It is the *smallest diameter of a screw thread.* It is also known as *root* or *core diameter.*

3. **Effective or pitch diameter.** It is *an imaginary diameter in between the major and minor diameters, and is equal to the major diameter less than an amount equal to the single depth of a thread.*

4. **Axis of screw.** It is the *longitudinal central line through the screw.*

5. **Angle of thread.** It is the *angle between the two sides of a thread measured in an axial plane.*

6. **Depth of the thread.** It is the *angle between the crest and the root of a thread measured normal to axis.*

7. **Thickness of the thread.** It is the *distance between the adjacent sides of the thread measured along or parallel to the pitch line.*

8. **Side.** It is the *slant surface of the thread which connects the crest with the root.*

9. **Helix angle.** It is the *angle made by helix of the thread at the pitch or effective diameter with the plane perpendicular to the axis.*

10. **Crest.** It is top *surface joining the two sides of a thread.*

11. **Root.** It is the *bottom surface joining the two sides of a thread.*

12. **Pitch.** *It is the distance from a point on one thread to the corresponding point on the next thread measured parallel to the axis of the thread.* It is denoted by *'p'.*

13. **Lead.** *It is the distance a screw thread advances axially in one turn on a single thread screw, the lead is equal to pitch and for a double threaded screw, the lead becomes two times the pitch and so on.*

Note : A screw is specified by a nominal diameter, it is the diameter of the cylindrical piece on which the threads are cut.

17.9.4. Specifications of Screw Thread

To specify a screw thread the following points are given due considerations :

1. Shape or form of thread   2. Pitch

3. Size (diameter)   4. Length

5. Number of starts   6. Material

7. Direction of threads   8. Internal or external threads.

17.9.5. Forms of Threads

Many different shapes of threads are employed but they can be satisfactorily grouped into following three main classes :

1. V-threads

2. Square threads

3. Modification of both.

Comparison of 'V' and Square threads

S. No.	Particulars	V-threads	Square threads
1.	Strength	More strong	Less strong
2.	Frictional resistance to motion	More	Less
3.	Cutting of threads	Easy and hence cheaper	Difficult and hence costly
4.	Suitability	Suitable for fastening purposes	Suitable for power transmission

17.9.6. Measuring Elements of a Screw-Thread

To determine the accuracy of a screw thread, it will be necessary to measure all the following :

1. Major diameter

2. Minor diameter

3. Effective or pitch diameter

4. Pitch

5. Thread angle form.

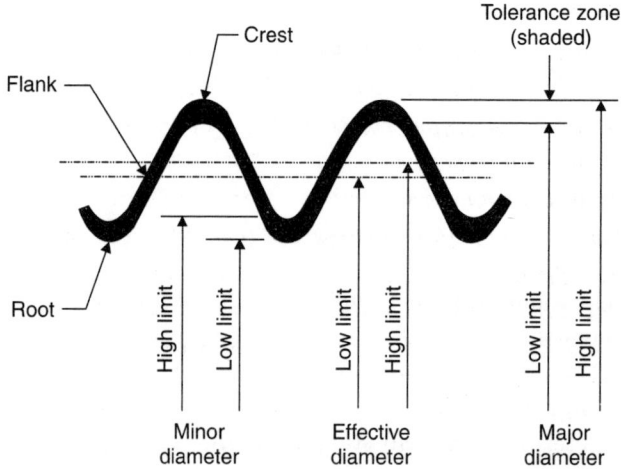

Fig. 17.73. Tolerances for minor, major and effective diameters.

The first four are all linear dimensions and must be subject to tolerances as shown in the Fig. 17.73, according to the class of fit (fine, medium, coarse) required, the *object of all thread measurement is to ensure that these tolerances are within the limits laid down.*

17.9.6.1. External screw thread measurements

1. Measurement of major diameter (D_{major}) :

The major diameter of a screw thread can be determined by using an ordinary micrometer or, preferably, by using a bench micrometer.

Hand micrometer. For most applications a good hand micrometer is quite suitable for measurement of the major diameter of external thread. However *extreme care is necessary to ensure that only light pressure is applied because the anvils of the micrometer make contact only at the points of the screw.* Excessive pressure may lead to *elastic deformation* due to elastic compression need to be applied. It is, however, also desirable to check the micrometer reading on a cylindrical standard (setting gauge or standard) of approximately the same size, so that the zero error etc. might be eliminated.

Bench micrometer. *Greater accuracy is probably obtainable if the major diameter is measured by a 'bench micrometer'* (Fig. 17.74). A bench micrometer uses constant measuring pressure and with this machine the *error due to pitch error in the micrometer thread is avoided.* In order that all measurements be made at the same pressure, a fiducial indicator is used in place of the fixed anvil. In this machine *there is no provision for mounting the workpiece between the centres and* it is to be held in hand. This is so, because, generally the *centres of the workpiece are not true with its diameter. This machine is used as comparator in order to avoid any pitch error of micrometers, zero error setting etc.* A calibrated setting cylinder is used as the setting standard or gauge.

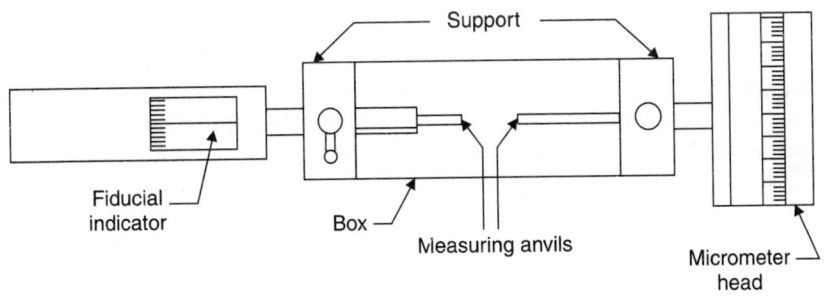

Fig. 17.74. Bench micrometer.

The *advantage of using cylinder as setting standard and not slip gauges etc., is that it gives greater similarity of contact at the anvils. The diameter of the setting cylinder must be nearly same as the major diameter.*

The *procedure of measuring major diameter* is as follows :

- The cylindrical standard (of known size) is held and reading of the micrometer is noted down [Fig. 17.75(a)]
- The cylindrical standard is then replaced by threaded workpiece and again micrometer reading is noted [Fig. 17.75(b)]

Let, M_1 = Micrometer reading over cylindrical standard,

M_2 = Micrometer reading over threaded workpiece major diameter,

D_{cs} = Diameter of cylindrical standard, and

D_{major} = Major diameter of thread.

Then, $D_{major} = D_{cs} \pm$ the difference between M_1 and M_2

The (+) or (–) is determined by whether the standard is a smaller or larger diameter than the thread major diameter.

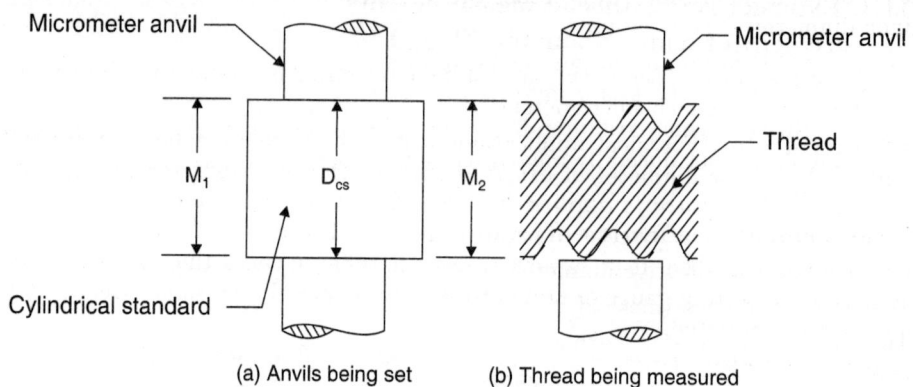

(a) Anvils being set   (b) Thread being measured

Fig. 17.75. Measurement of major diameter.

Note : The setting guage used for the initial reading should always be of similar geometrical form to that of the part to be measured. In this instance, a cylindrical form, such as a plug gauge, or roller, or other cylindrical standard, should be used.

2. Measurement of Minor Diameter (D_{minor}) :

The use of external hand micrometer is rarely made to measure the minor diameter screw threads because of *difficulty of ensuring that the measurement takes place at 90° to the thread axis.*

For more accuracy and convenience the *bench micrometer* is used for this measurement. Here small Vee pieces which make contact with the root of the thread are used as shown in the Fig. 17.76. These Vee pieces (made of hardened steel) are made in several sizes, having suitable radii at the edges, and with included angles *less than the angle of the thread to be checked.* It is not necessary to know the dimension of the Vee pieces, since they are interposed between the micrometer faces and the cylindrical standard when the standard reading is taken [Fig. 17.76 (a)]. The thread being measured is shown in Fig. 17.76 (b).

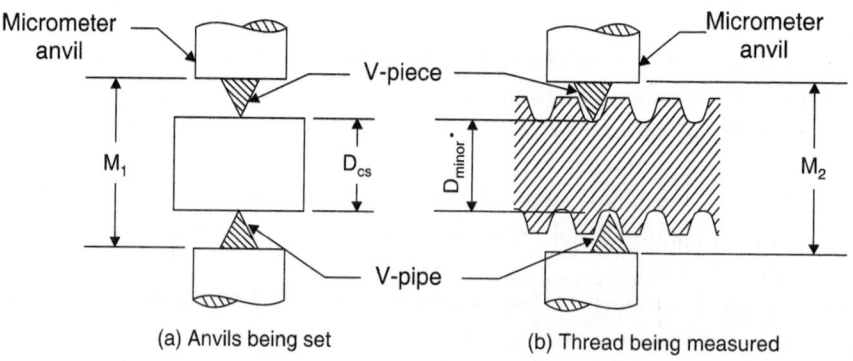

(a) Anvils being set   (b) Thread being measured

Fig. 17.76. Measurement of minor diameter.

Refer to Fig. 17.76.

Let M_1 = Micrometer reading over cylindrical and V-pieces.

       M_2 = Micrometer reading over minor diameter and V-pieces,

D_{cs} = Diameter of the cylindrical standard, and

$D_{(minor)}$ = Minor diameter of thread.

Then, $D_{(minor)} = D_{cs}$ difference between M_1 and M_2.

The (+) or (–) is determined by whether the standard is a small or larger diameter than the thread minor diameter.

This method of minor diameter measurement is *only applicable to threads such as Whitworth and B.A., where there is a definite radius at the root of the thread.*

Gauges for threads which are intended to make close contact at the minor diameter, which is not closely limited, are not normally measured by this method, but are checked visually on the projector or microscope.

3. Measurement of effective diameter (D_e) :

For measuring the effective diameter, one of the following methods can be used :

1. Three and two wires method.

2. The thread micrometer method

3. Optical measurement on tool-maker microscope

4. Axial section method.

Three and two wires method :

Effective diameter is measured by placing small cylinders or *wires of known diameter* in the Vees of the thread, and measuring over the tops of the wires with a micrometer or special machine. These *wires are made of hardened steel, and are lapped to sizes suitable for various pitches.* For each pitch of thread there is a "best size" wire ; this is of such diameter that it makes contact with the flanks of the thread on the effective diameter or pitch line. Effective diameter may be measured with any diameter of wire which makes contact on the true flak of the thread, but values so obtained will differ from those obtained with *"best size"* wires if there is any error in angle or form of thread.

Three-wire method. Fig. 17.77 shows diagrammatically how three wires may be used for measuring effective diameter. If a hand micrometer is used, three wires must be used, so that the micrometer faces can be aligned parallel to the thread axis. With small diameter threads it is possible to hold gauge, with wires in place, with one hand, and the micrometer in the other, although one sometimes wishes for a third hand. Other effective methods of securing the wires include the use of grease in the threads, or sticking the ends of the wires into Plasticine or wax. In any case the wires must not be constrained but must be free to adjust themselves under the micrometer pressure.

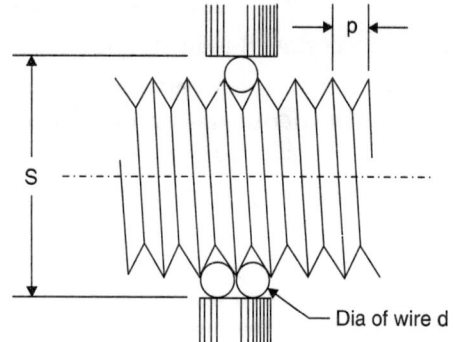

Fig. 17.77. Three-wire method of measuring effective diameter.

For screw gauge measurement, it is better to use a specially designed Diameter Measuring Machine incorporating a bench micrometer which maintains the axis of the micrometer spindle square to thread axis.

Two-wire method. The two-wire method of measuring the effective diameter of a screw thread is illustrated in Fig. 17.78.

- Two wires/cylinders having been selected, a cylindrical standard is used and a micrometer reading is taken over the cylindrical standard, and the two wires.

- The cylindrical standard is then removed and is replaced by the screw, with the wires inserted in opposite grooves of the thread. The micrometer reading is then noted.

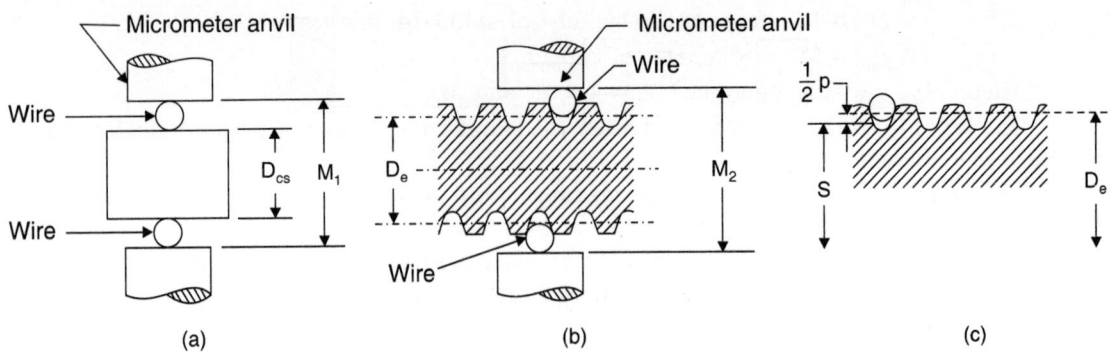

Fig. 17.78. Two-wire method of measuring effective diameter.

Let, M_1 = Micrometer reading over cylindrical standard and wires,

M_2 = Micrometer reading over screw thread and wires,

D_{cs} = Diameter of the cylindrical standard,

S_1 = Diameter under wires when the latter are in position on threads,

D_c = Effective diameter of thread, and

P = The difference between the effective diameter and the diameter under the wires.

The value of P is dependent on the diameter of the wires and form of the thread to be measured. Its value may be found as follows :

Let, p = Pitch of thread, and

d = Mean diameter of wires used.

For Whitworth threads (55°)

$$P = (0.96049\ p) - (1.16568\ d)$$

For B.A. threads (47°),

$$P = (1.13634\ p) - (1.48239\ d)$$

For Metric and Unified threads (60°),

$$P = (0.86603\ p) - d$$

Now, $D_e = D_{cs} - M_1 + M_2 + P$

When the effective diameters of screw plug gauges are being measured it is necessary to subtract the "rake correction" C and add the "compression correction" e to the micrometer reading to obtain accurate measurements. Both corrections are small, and they result from the rake angle of the thread and the compression of the wires respectively.

Thread micrometer method :

When the *threads are accurate the effective diameter of the thread will be equal to its pitch diameter.* In such cases, a specially designed thread micrometer is found to be most convenient way of measuring the effective diameter. *The thread micrometer* is shown in Fig. 17.79. The end of the spindle is pointed to a 60° cone for ISO thread and an accurate 60° Vee is grooved in the anvil. The anvil is free to rotate so as to adjust itself to the helix angle of the thread being measured. The sharp tip of the spindle point is ground off. Likewise, flats are ground on the peaks of the vee and the foot of the vee is ground out or cleared to ensure that only the pitch diameter is measured by the thread micrometer and not the root diameter or major diameter of the screw thread. Before using a thread micrometer, screw cone point of its spindle down to contact with the vee-anvil and check the micrometer thimble's zero reading.

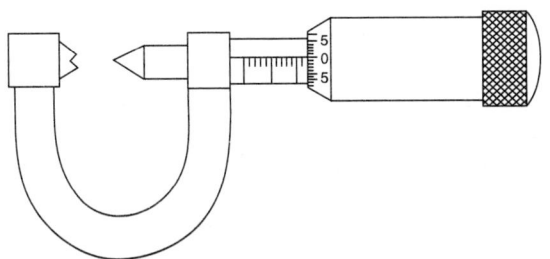

Fig. 17.79. Thread micrometer.

A thread micrometer is designed to measure threads within a certain range of thread pitches. Since any given thread micrometer is required to measure a range of threads of different pitches, each of which may cause a slight variation of anvil position on the thread, small errors in measurement are sometimes introduced. For this the best pitch diameter to follow in using the thread micrometer is to *first measure the pitch diameter of a standard thread plug gauge of the same size as the threads to be measured and then compensate for error.*

17.9.7. Screw Thread Gauges

Although it is possible to measure threads by methods earlier discussed, the *normal method* of checking threads in production is by the use of screw thread gauges. In broad principle, the gauges take the form of the mating thread, and are assembled with the thread to be checked. By suitably designing the gauge it is possible to provide a limiting gauge system which will control the complex dimensions of the thread within the limits which are laid down in the specifications.

The various types of thread gauges are described below :

Plug screw gauges :

(*For internal screw threads*) : In order to gauge nuts or internal threads it is obvious that a full form plug screw gauge, made accurately to the minimum dimensions of the internal thread, will ensure that all dimensions are not less than that minimum, if it will assemble with thread. Major, minor, and effective diameters will be checked, and the gauge will ensure that pitch, angle or form errors are not reducing the virtual effective diameter below the minimum.

The plug screw gauges are usually made *double ended,* either solid with the "*Go*" and "*Not Go*" threads ground on the ends, or as a built up gauge with a metal or plastic handle having a ground gauging unit of each end, fitting into a taper socket or otherwise attached. The "*GO*" gauge has a length of some *dozen pitches* or more but "*Not Go*" need be only about three threads long. In the use, the "*Not Go*" gauge may enter the work by one or two turns and still be regarded as "*Not Go*", since the work near the upper limit is usually slightly bell-mouthed, a condition which can normally be permitted.

Fig. 17.80 shows various types of plug-screws gauges.

- *One-sided thread* plug gauges (Fig. 17.80 (*a*)) are used for nominal diameters between 100 and 150 mm.
- Smaller thread plug gauges can be *double-sided* combining the "*Go*" and "*Not Go*" gauge as shown in Fig. 17.80 (*b*).
- Fig. 17.80 (*c*) shows a "*Go*" thread gauge and the "*Go*" plain cylindrical gauge.
- Fig. 17.80 (*d*) shows a "*Not Go*" thread gauge and the "*Not Go*" plain cylindrical gauge.
- Fig. 17.80 (*e*) shows external threaded ring gauges which are used to check internal threads having nominal diameters above 150 mm.

The threads on all types of thread limit gauges should be removed to the extent of at least one-half a pitch, so that the thread can start at its full form to minimise the risk of damage.

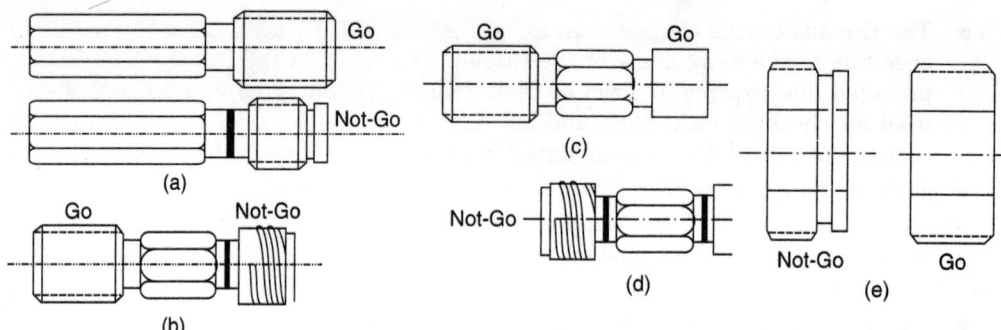

Fig. 17.80. Plug screw gauges.

Ring screw gauges :

(For external screw threads) : For the production gauging of bolts, the equivalent mating
gauges are known as *ring screw gauges*. As in the
plug screw gauges, a limit gauge system can be
provided by a full form *"Go"* and a *"Not Go"* effective" ring gauge. The *"Not Go"* ring is truncated
on its minor diameter, and is cleared on its major
diameter. All the factors involved are exact counterparts of the plug gauging of internal threads.

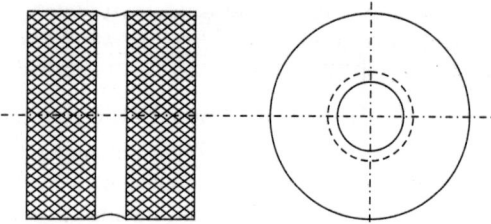

Fig. 17.81. *"Go"* thread ring gauge.

The *"Go"* ring gauge has a full form of thread, the *"Not Go"* gauge being truncated on the minor
diameter and cleared on the major diameter at the root of the thread.

A *"Go"* thread ring gauge is shown in Fig. 17.81.

Thread snap gauges :

Thread snap gauges are far widely used as compared to thread ring gauges which have a
limited field of application. These have the *advantage* of being a combination of the *"Go"* and *"Not
Go"* gauges required for checking the upper and lower limits of the tolerance in one operation,
thus reducing the gauging time, which is particularly important for mass production.

In the Figs. 17.82 and 17.83 respectively are shown the *caliper thread snap gauge and roll
thread snap gauge.*

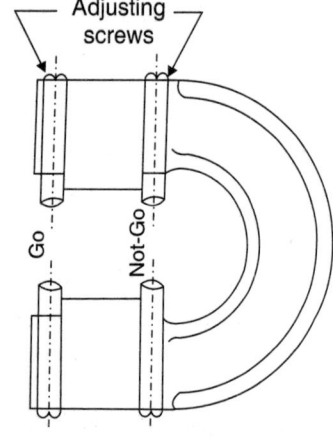

Fig. 17.82. Caliper thread snap gauge.

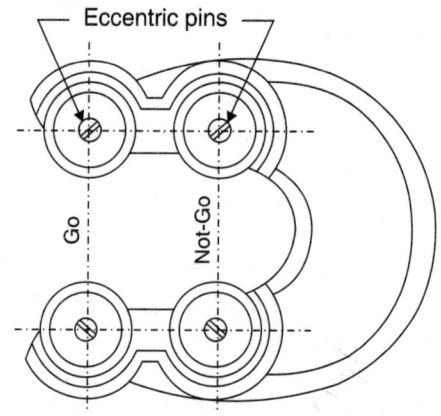

Fig. 17.83. Roll thread snap gauge.

- The threads on the caliper snap gauges are provided with a relief to prevent interference due to the helix angle of the thread. The relief on the roll thread snap gauges is provided due to circular form of their rollers. Thread snap gauges can, therefore, be used for checking right-hand and left hand threads, Moreover they allow checking the components when they are mounted between the centres without difficulty.

- Thread snap gauges are adjustable and they can be easily and accurately checked.

The *adjustment* of the *caliper snap gauge is done by screws* while the roll *snap gauge* can be adjusted by means of eccentric pins on which two rollers are mounted as shown in Fig. 17.82 and 17.83.

- Fig. 17.84 shows wide range *adjustable thread snap gauges*. These are used for checking threads of the same pitch that have different nominal diameters. The dimensions of the "Go" and the "Not Go" sides can be adjusted with slip gauge blocks.

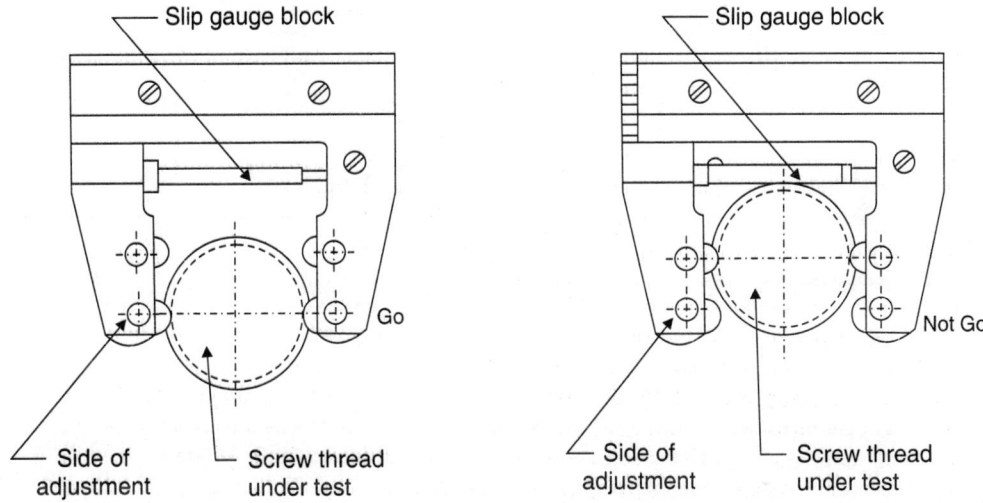

Fig. 17.84. Adjustable thread snap gauge.

Since a thread caliper gauge is adjustable, it is *necessary to provide a method of setting with speed and certainty*. This is done with a setting plug, which is a double-ended plug screw gauge whose sizes are made are the upper and lower limits of the work. The "Go" end of this plug is large than the "Not Go" end and the thread forms of both ends are cleared to make contact on the flanks only.

QUESTIONS WITH ANSWERS

Q. 17.1. Although aluminium is light yet it is not used for micrometer frames, why ?

Ans. Micrometer frames are made of *steel forgings because it has the same coefficient of thermal expansion as most of the parts to be measured*. With aluminium frame, the differential in the rate of thermal expansion would *result in measuring errors*.

Q. 17.2. What are the important points in wringing slip gauges together ?

Ans. The most important points to be remembered in the method of wringing of two slip gauges together is to ensure that any moisture, oil, dust, or similar foreign matter will be pushed away from gauging surfaces instead of being trapped. Therefore, first the surfaces should be slided transversly and then twisted together to form the wrung pair.

Q. 17.3. What do you understand by line and end measurements ?

Ans. Line measurement involves measurement of distance between two lines and end measurement involves measurement of distance between two surfaces.

Q. 17.4. State Taylor's principle in design of limit gauges.

Ans. According to Taylor's principle for limit gauges, the *Go gauge* should be designed to check *all or as many maximum* metal limits of the part as possible in single operation. In other words *GO* gauge should be a full form gauge. If it is not done then hole of non-circular shape and not perfectly straight could be accepted. Further, *No GO gauges* should be designed to check the minimum metal limit of only one *dimension at a time*. If *NO GO* gauge is made of full form, then elliptical holes could be accepted. On the other hand, *NO GO* gauge of bar type would detect major axis of elliptical hole and reject it.

OBJECTIVE TYPE QUESTIONS

Fill in the blanks :

1. The .... gauge should be used only in air-conditioned rooms free from dust and maintained at constant temperature.
2. The standard method of .... the slip gauges together is first to bring the faces of the gauges into contact at right angles to one another and then turn them through 90°.
3. The .... gauges should never be left wrung together for an unnecessary length of time.
4. .... gauge is used for controlling the reference gauge and can be found only in standard room.
5. .... gauge is used for checking check gauge and for arbitration purposes in case of doubt.
6. .... gauge is used for testing the accuracy of workshop gauges.
7. .... gauges are used for general checking of component parts.
8. .... is the commonest material for the manufacture of gauges.
9. .... plating has proved most useful method in reclaiming the worn gauges.
10. .... gauges have the advantage of eliminating the corrosive effects due to perspiration from hands.
11. .... scraping is principally used on cast-iron machine beds and surface plates.
12. .... grinding is a common method of producing fairly accurate flat surface of moderate dimensions.
13. .... grinding process is used to produce very accurate flat surfaces.
14. .... is essentially a cutting and not a polishing action.
15. The process of .... is applied to the gauges to prevent movement of the gauge—a phenomenon known as creep.
16. .... means the pure science.
17. Most important parameter in metrology is ....
18. .... metrology is that part of metrology which treats units of measurement, methods of measurement and the measuring instruments, in relation to the technical and legal requirements.
19. .... metrology is the technique of measuring small variations of a continuous nature.
20. The basic objective of a measurement is to provide the required accuracy at a .... cost.
21. .... measurement is concerned with the precise determination of the linear, angular and non-linear functions of the machined surfaces of tools and devices used to produce engineering components.
22. The term .... manufacture implies that the parts which go into an assembly may be selected at random from large number of parts.
23. The metric system was originated in ....
24. The British system of linear measurement is based on an arbitrary unit known as ....
25. A .... or .... is defined as the distance between scribed lines on a bar of metal under certain conditions of temperature and support.
26. A modern .... standard consists of fundamentally of a block or bar of steel, generally hardened, whose end faces are lapped flat and parallel to within a ten millionth of a cm.
27. .... gauges are used as standards of measurement in practically every precision engineering works in the world.

28. A '............ size' is the size which is used for purpose of general identification.
29. A '............ dimension' is the dimension as worked out by purely design considerations.
30. .... is the difference between the maximum and minimum limits of size.
31. .... means a degree of tightness or looseness between two mating parts to perform a definite function.
32. .... is the difference between the sizes of a hole and shaft which are to be assembled together when the shaft is smaller than the hole.
33. .... is the difference between the sizes of a hole and a shaft which are to be assembled together when the shaft is larger than the hole.
34. The three classes of fit are :
     (i) .... fit,   (ii) .... fit and   (iii) .... fit.
35. .... is the difference between the maximum shaft and minimum hole.
36. '............ basis system' is one in which the limits on the hole are kept constant and the variations necessary to obtain the classes of fit are arranged by varying those on the shaft.
37. '............ basis system' is one in which the limits on the shaft are kept constant and the variations necessary to obtain the classes of fit are arranged by varying the limits on the holes.
38. In the .... system, tolerance is applied only in one direction.
39. In the .... system of writing tolerances, a dimension is permitted to vary in two directions.
40. .... system is most commonly in interchangeable manufacture especially where precision fits are required.
41. In .... system, it is not possible to retain the same fit when tolerance is varied.
42. A push fit is .... fit.
43. A driving fit is an .... fit.
44. Forced or pressed fit is an .... fit.
45. Shrinkage fit is an .... fit.
46. .... refers to assembling a number of unit components taken at random from stock so as to build up a complete assembly without fitting or adjustment.
47. .... is defined as the repeatability of a measuring process.
48. .... is the agreement of the result of a measurement with the true value of the measured quantity.
49. Controllable errors are also known as .... errors.
50. A .... instrument is any device that may be used to obtain a dimensional or angular measurement.
51. International system of units (SI) was established in the year ....
52. International Organisation of Weights and Measures was established in the year ....
53. A .... is a tool or instrument used to measure or compare a component.
54. .... is the numerical evaluation of a dimension or property by comparison with standards of the same kind.
55. .... is to judge if similar components are alike, by comparing them together,
56. .... is to be certain that exact value of the dimension or property is within the allowable tolerance.
57. .... is to set up a serial number of abstract and concrete things.
58. .... is to range the components in groups according to a certain order.
59. .... gauges are gauges by means of which a certain dimension or a certain form can be checked for which gauges are designed or adjusted.
60. In a .... measuring device pneumatic means are used in the measuring process.
61. .... value is the smallest possible value that can be read on the scale.
62. .... division is the distance in mm between two successive graduations on the scale.
63. .... range is the product of the total number of scale dimenions and the scale value.
64. .... value is the smallest change in the dimension or property being measured that can just be indicated on the measuring device.
65. .... is the ratio between the displacement of the pointer or mark on the scale, measured in scale division or in mm, and the corresponding change in the measured value.
66. .... is the exactness of the indicated readings with the actual values of the measured dimensions or property.
67. A .... may be regarded as an infinite number of prismatic units of constantly changing angles which decrease from the edges to the centre.

68. A .... lens is thinner at the centre than at the edges and diverges parallel light which then appears to come from a focal point behind the lens.

69. Optical .... consist of solid piece of highly transparent homogenous glass with three or more polished plane faces.

70. The mechanical-electrical and electrical-electrical energy transducers make use of the principle of .... induction.

71. The .... transducers work on the principle of mechanical-electrical energy transformation.

72. In a .... cell the light energy can be transformed to electrical energy.

73. .... strain gauges are used where a very high gauge factor and a small envelope are required.

74. An Engineer's steel rule is also known as ....

75. The scale is a .... measuring device.

76. The .... works on the basic measuring technique of comparing an unknown length to the one previously calibrated.

77. The principle of .... datum should be employed while using a scale or rule.

78. .... is an instrument used for measuring distances between or over surfaces, or for comparing dimensions of workpieces with such standards as plug gauges, graduated rules etc.

79. The vernier instruments generally used in workshop and engineering metrology have comparatively .... accuracy.

80. While taking measurements using vernier calipers it is important to set the caliper faces .... to the surface across which measurements are to be made.

81. The vernier .... gauge is mainly used in the inspection of parts and layout work.

82. The vernier .... gauge is used to measure the depth of holes, recesses and distances from a plane surface to a projection.

83. Micrometers are designed on the principle of .... and ....

84. .... micrometers are used for measuring internal dimensions.

85. .... micrometer is used to increase the accuracy of micrometers.

86. The .... is defined as the opening between two lines which meet at a point.

87. With a .... bevel protractor the angles can be measured to within 5'.

88. .... is a tool used for accurate setting out of angles by arranging to convert angular measurements to linear ones.

89. The sine bars should not be used for angle greater than .... degrees.

90. A sine table is the development of the ....

91. The .... centres are used for inspection of conical objects between centres.

92. A sine centre is basically a .... with block holding centres which can be adjusted and rigidly clamped in any position.

93. An .... gauge is a hardened steel block approximately 75 mm long and 16 mm wide which has two lapped flat working faces lying at a very precise angle to each other.

94. .... gauge can be wrung together to form desired angles.

95. The .... level is a simple form of mechanical measuring device.

96. The sensitivity of the .... level is governed solely by the radius of the tube and the base length of its mount.

97. The accuracy of .... level depends upon the setting of the tube relative to the base.

98. A .... is an instrument used for measuring angle relative to the horizontal plane.

99. An .... is an instrument desired to measure small angular deflections.

100. An .... is essentially an infinity telescope and a collimator combined into one instrument.

101. An .... is based on the principle that a collimating lens can project and receive a parallel beam of light and that the reflected beam of light will change its direction by changing the angle of the surface reflecting the light.

102. A .... is tested by using taper plug and ring gauges.

103. A positive check of the dimensional and geometrical accuracies of taper shafts can be achieved by using .... taper gauge.

104. The angle of a taper shaft not exceeding .... mm length can be measured by means of standard rollers, slip gauges and micrometer.

105. A .... is a helical ridge formed on a uniform section round the curved surface.
106. A .... handed thread is one in which the nuts must be turned in a right handed direction to screw it on.
107. A .... handed thread is one in which the nuts would be screwed on by turning it to the left.
108. A .... start threaded screw is one in which there is a movement of one thread for one complete turn round the screw or belt.
109. In the .... threaded screw there is a movement of more than one thread.
110. In a .... start thread, for one complete turn, the thread advances two times as if it were a single thread.
111. .... start threads are used on those cases where rapid movement or motion is required.
112. .... diameter is the largest diameter of a screw. It is also known as .... or .... diameter.
113. .... diameter is the smallest diameter of a screw thread. It is also known as .... or .... diameter.
114. .... diameter is an imaginary diameter in between the major and minor diameter, and is equal to the major diameter less than an amount equal to the single depth of a thread.
115. .... of screw is the longitudinal central line through the screw.
116. .... of thread is the angle between the two sides of a thread measured in an axial plane.
117. .... of thread is the distance between the crest and the root of a thread measured normal to axis.
118. .... of thread is the distance between the adjacent sides of the measured along or parallel to the pitch line.
119. .... is the slant surface of the thread which connects the crest with the root.
120. .... angle is the angle made by the helix of the thread at the pitch or effective diameter with a plane perpendicular to the axis.
121. .... is the top surface joining the two sides of a thread.
122. .... is the bottom surface joining the two sides of a thread.
123. .... is the distance form a point on one thread to the corresponding point on the next thread measured parallel to the axis of the thread.
124. .... is the distance a screw thread advances axially in one turn on a single thread screw.
125. In a double threaded screw, the lead is two times the ....
126. V-threads are .... strong than square threads.
127. .... threads are suitable for power transmission.
128. .... threads are suitable for fastening purposes.
129. .... thread are exclusively used for small screws used in optical instruments, clocks etc.
130. .... thread are used in motor car practice.
131. .... threaded lead screws are very commonly used on lathes.
132. .... threads are used in railway carriage couplers.
133. All .... shaped threads absorb a large amount of power.
134. .... thread angles cause large bursting forces in the nut.
135. The major diameter of a screw thread can be determined by using an .... micrometer or preferably, by using a .... micrometer.
136. .... diameter can be measured by two or three wires method.

ANSWERS

1. Slip	2. Wringing	3. Slip	4. Master	5. Reference
6. Check	7. Workshop	8. Steel	9. Chromium	10. Glass
11. Hand	12. Surface	13. Spot	14. Lapping	15. Stabilisation
16. Metrology	17. Length	18. Legal	19. Dynamic	20. Minimum
21. Precision	22. Interchangeable	23. France	24. Yard	25. Yard or metre
26. End	27. Slip	28. Nominal	29. Basic	30. Tolerance
31. Fit	32. Clearance	33. Inteference	34. Clearance, Interference, Transition	

35. Allowance	**36.** Hole		**37.** Shaft	**38.** Unilateral		**39.** Bilateral	
40. Unilateral system			**41.** Bilateral	**42.** Transition		**43.** Interference	
44. Interference	**45.** Interference fit		**46.** Interchangeability			**47.** Precision	
48. Accuracy	**49.** Systematic		**50.** Measuring instrument			**51.** 1980	
52. 1875	**53.** Gauge		**54.** Measure	**55.** Compare		**56.** Test.	
57. Count	**58.** Classify		**59.** Limit	**60.** Pneumatic		**61.** Scale	
62. Scale	**63.** Graduation		**64.** Threshold	**65.** Sensitivity		**66.** Accuracy	
67. Lens	**68.** Concave		**69.** Prism	**70.** Magnetic		**71.** Piezo-electric	
72. Photo	**73.** Semiconductor		**74.** Scale	**75.** Line		**76.** Scale	
77. Common	**78.** Caliper		**79.** Low	**80.** Parallel		**81.** Height	
82. Depth	**83.** Screw, Nut		**84.** Internal	**85.** Differential		**86.** Angle	
87. Dial	**88.** Sine bar		**89.** 60	**90.** Sine bar		**91.** Sine	
92. Sine bar	**93.** Angle		**94.** Angle	**95.** Spirit		**96.** Spirit	
97. Spirit	**98.** Clinometer		**99.** Auto-collimator			**100.** Auto-collimator	
101. Auto-collimator	**102.** Taper		**103.** "May"	**104.** 100		**105.** Screw thread	
106. Right	**107.** Left		**108.** Single	**109.** Multistart		**110.** Double	
111. Multi	**112.** Major, outside, crest			**113.** Minor, root, core		**114.** Effective	
115. Axis	**116.** Angle		**117.** Depth	**118.** Thickness		**119.** Side	
120. Helix	**121.** Crest		**122.** Root	**123.** Pitch		**124.** Lead	
125. Pitch	**126.** More		**127.** Square	**128.** Vee-thread		**129.** B.A.	
130. Metric	**131.** Acme		**132.** Knuckl	**133.** Vee		**134.** Large	
135. Ordinary, bench			**136.** Effective.				

THEORETICAL QUESTIONS

1. Define the following :
   (*i*) Metrology   (*ii*) Legal metrology
   (*iii*) Dynamic metrology.
2. Enumerate the objectives of metrology.
3. What is the necessity and importance of metrology ?
4. State the need of precision measurement.
5. Discuss 'Metrology' as a means to achieve quality control.
6. What is actually understood by measurement ? What are the two standard systems of linear measurements ?
7. Explain briefly the following :
   (*i*) Line standard   (*ii*) End standard
   (*iii*) Wave length standard.
8. Discus briefly the following :
   (*i*) Primary standards   (*ii*) Secondary standards
   (*iii*) Tertiary standards   (*iv*) Working standards.
9. State the characteristics of line standard.
10. What are the advantages of wave length standards ?
11. Give classification of standards.
12. Give the relative characteristics of 'line' and 'end' standards.
13. Define the following :
    Nominal size, Basic dimension, Tolerance, Upper deviations, Fit, Allowance.
14. How do you classify fits ?
15. Explain briefly the basis of Fit (or limit) system.

16. What are the systems of specifying tolerances ; which system is used most and why ?
17. How are holes ; shafts and fits designated ? Give examples.
18. Write a short note on 'Newell system'.
19. Explain briefly the following :
    (i) Selective fit
    (iii) Driving fit
    (v) Shrinkage fit
    (ii) Push fit
    (iv) Forced or pressed fit
    (vi) Freeze fit.
20. Write a short note on "Concept of Interchangeability".
21. Define the term 'Precision' and 'Accuracy' and describe the methods to achieve them.
22. What are sources of errors ? Explain them briefly.
23. Distinguish between 'Controllable errors' and 'Random errors'.
24. State the classification of measuring instruments.
25. Write a short note on 'Selection of Instrument'.
26. What steps should be taken to care and handle equipment/apparatus ?
27. Write a short not on 'Standardising organisations'.
28. Explain any two of the following gauges :
    (i) Snap gauges.
    (iii) Dial gauges.
    (ii) Plug gauges.
29. What is an Engineer's steel rule (or, scale) ?
30. What are the desirable qualities of a steel rule ?
31. State the techniques which must be used to get good results.
32. Write a short note on 'Steel rule'.
33. How should a steel rule or scale be taken 'care' of ?
34. What is the vernier principle ? Explain the following vernier principles :
    (i) Principle of 0.1 mm vernier
    (ii) Principle of 0.02 mm vernier.
35. Name the different types of vernier calipers and draw their neat sketches.
36. State the probable errors in calipers.
37. Write down the precautions which should be taken while using a vernier caliper.
38. Explain the construction and working of a vernier height gauge with the help of a neat sketch.
39. State the field of application of a vernier height gauge.
40. What precautions should be taken while using a vernier height gauge ?
41. Draw a neat sketch of a vernier depth gauge and explain its construction and working.
42. State the 'principle' on which micrometers are designed.
43. Give the description of an outside micrometer. How is it read ?
44. Enumerate the sources of errors in micrometers.
45. What precautions should be observed while using the micrometer ?
46. Enumerate different types of micrometers and explain with neat sketches any two of them.
47. Explain briefly the construction and working of a depth micrometer.
48. Name different types of anvils, draw their sketches and state their uses.
49. State the advantages and limitations of following precision instruments :
    (i) Vernier caliper
    (iii) Vernier depth gauge
    (v) Internal micrometer
    (ii) Vernier height gauge
    (iv) External micrometer
    (vi) Depth micrometer.
50. Define an 'angle.'
51. Enumerate the instruments used for angular measurement.
52. Explain any two of the following instruments :
    (i) Vernier bevel protractor.
    (iii) Optical bevel protractor.
    (ii) Dial bevel protractor.
53. Differentiate between the following instruments :
    (i) Sine bar
    (iii) Sine centre.
    (ii) Sine table

54. What is a sine bar ? Name the materials of which it is made up of.
55. Explain with the help of a diagram the principle of a sine bar.
56. Sketch various forms of sine bars and explain any one of them in detail.
57. Enumerate the uses of sine bars.
58. Write down the essential requirements in the use of a sine bar to get accurate results.
59. What precautions should be taken while using sine bars ?
60. What is sine table ?
61. What is a sine centre ? Is its principle of setting same as that of sine table ?
62. What are angle gauges ? How are they used ?
63. State the practical applications of angle gauges.
64. Explain with a neat diagram block spirit level and its principle of working.
65. Wht is a clinometer ? Name different types of commonly used clinometers.
66. Write a short note on 'Plain index centre'.
67. Define a screw thread.
68. Enumerate the purposes for which screw threads are used.
69. How are threads classified ?
70. Explain any four of the following elements of a screw threads :

    (i) Major diameter   (ii) Minor diameter
   (iii) Effective material   (iv) Helix angle
    (v) Thickness of thread   (vi) Axis of the screw
   (vii) Crest.

71. Enumerate the points which should be given due consideration while giving specifications of a screw thread.
72. Give the comparison between Vee and square threads.
73. Explain two and three wires methods of measuring effective diameter of a screw thread.
74. Explain briefly 'Thread micrometer method' of measuring effective diameter of a screw thread.
75. Explain how can the pitch of a screw thread be measured on a pitch measuring machine.
76. Describe briefly the method of measuring thread angle of a screw thread.
77. Explain briefly how the measurements of the following elements of internal screw threads can be done :

    (i) Major diameter   (ii) Minor diameter
   (iii) Effective diameter   (iv) Pitch
    (v) Thread angle and form.

78. Explain any two of the following :

    (i) Plug screw gauges   (ii) Ring screw gauges
   (iii) Thread snap gauges.

Economics of Machining and Manufacturing

18.1. Economics of Machining. 18.2. Economics of manufacturing. Highlights—Theoretical Questions.

18.1. ECONOMICS OF MACHINING

The primary concern of the manufacturing engineer is to produce the objects at the most economical cost. In order to achieve this he should be able to analyse the machining process for all possible costs, so that he would be able to optimise the process to get the minimum possible costs satisfying all the requirements.

The most important *parameter* associated with the economics of machining are :

(i) *Minimum cost per part ;*

(ii) *Maximum production rate.*

The **total cost per piece (C_{tp})** consists of the following four items :

(i) *Cost of machining (C_m) ;*

(ii) *Cost of setting up for machining (C_s)* such as mounting the cutter and fixtures and preparing the machine for the particular operation ;

(iii) *Cost of loading, unloading, and machine loading (C_l) ;*

(iv) *Cost of tooling (C_t),* which includes tool changing, regrinding, and depreciation of the cutter.

Thus, $\qquad C_{tp} = C_m + C_s + C_l + C_t$ $\hfill$ ...(18.1)

Machining cost (C_m). The machining cost is given by :

$$C_m = T_{mp} \, (C_{labour} + O_m) \hspace{3cm} ...(18.2)$$

where, $\qquad T_{mp}$ = Machining time per piece,

$\qquad C_{labour}$ = Labour cost of production operator per hour, and

O_m = Overhead charge (or burden rate) of the machine, including depreciation, maintenance, indirect labour and the like.

Set-up cost (C_s). Set up cost is the fixed figure in rupees per piece.

Loading, unloading, and machine handling cost (C_l). This cost is given by :

$$C_l = T_l \, (C_{labour} + O_m) \hspace{3cm} ...(18.3)$$

where, T_l = Time involved in loading and unloading the part, changing speeds, changing feed rates, and so on.

Tooling cost (C_t) : The tooling cost is expressed as :

$$C_t = \frac{1}{N_p} \, [T_c \, (C_{labour} + O_m) + T_g \, (C_{labour\text{-}tg} + O_{tg}) + D_t] \hspace{2cm} ...(18.4)$$

where,　　N_p = Number of parts machined *per tool grind,*

T_c = Time required to change the tool,

T_g = Time required to grind the tool,

$C_{labour-tg}$ = Labour cost of tool grinder operator per hour,

O_{tg} = Burden rate of tool grinder per hour, and

D_{tool} = Depreciation of the tool in rupees per grind.

Total time needed to produce *one part* T_{tp} is given by :

$$T_{tp} = T_l + T_m + \frac{T_c}{N_p} \qquad \qquad ...(18.5)$$

where T_m has to be calculated for each particular operation.

Let us consider a **"turning operation"** :

The machining time (T_m) is given by :

$$T_m = \frac{L}{fN} = \frac{\pi LD}{fV} \qquad \qquad ...(18.6)$$

$$\left(\because \quad V = \pi DN, \text{ and } N = \frac{V}{\pi D} \right)$$

where,　　L = Length of cut,

f = Feed,

N = Speed of the workpiece in r.p.m ;

D = Diameter of the workpiece, and

V = Cutting speed.

It may be noted that consistency of units must be maintained in these equations.

From Taylor's tool life equation, we have

$$VT^n = C$$

or,　　　　$$T = \left(\frac{C}{V} \right)^{\frac{1}{n}} \qquad \qquad ...(18.7)$$

where,　　T = Time in minutes, required to reach a flank wear of certain dimension, after which the tool *has to be reground or changed.*

Thus, the number of pieces per tool grind,

$$N_p = \frac{T}{T_m} \qquad \qquad ...(18.8)$$

By combining equations (18.6), (18.7) and (18.8), we have

$$N_p = \frac{(C/V)^{1/n}}{\pi LD/(fV)}$$

or,　　　　$$N_p = \frac{fC^{1/n}}{\pi LD V^{\left(\frac{1}{n} - 1 \right)}} \qquad \qquad ...(18.9)$$

The total cost per piece (C_{tp}) in equation (18.1) can now be defined in terms of several variables.

● *Optimum cutting speed (V_{opt}) and optimum tool life (T_{opt}) for* **minimum cost** :

In order to find the optimum cutting speed and the *optimum tool life* for *"minimum cost"* let us differentiate C_{tp} with respect to V and set it to zero. Thus, we have

$$\frac{\partial C_{tp}}{\partial V} = 0$$

By doing so, we find that, the optimum cutting speed, V_{opt}, is

$$V_{opt} = \frac{C(C_{labour} + O_m)^n}{\left[\left\{\left(\frac{1}{n} - 1\right)\right\}\{T_c(C_{labour} + O_m) + T_g(C_{labour\text{-}tg} + O_{tg}) + D_{tool}\}\right]^n} \qquad ...(18.10)$$

and the *optimum tool life, T_{opt}*, is

$$T_{opt} = \left[\left(\frac{1}{n} - 1\right)\left\{\frac{T_c(C_{labour} + O_m) + T_g(C_{labour\text{-}tg} + O_{tg}) + D_{tool}}{C_{labour} + O_m}\right\}\right] \qquad ...(18.11)$$

● *Optimum cutting speed and optimum tool life for* **maximum production :**

In order to find optimum cutting speed and the optimum tool life for "*maximum production*", let us differentiate T_{tp} with respect to V and set the result to zero. Thus, we have

$$\frac{\partial T_{tp}}{\partial V} = 0 \qquad ...(18.12)$$

$$V_{opt} = \frac{C}{\left[\left\{\frac{1}{n} - 1\right\}T_c\right]^n} \qquad ...(18.13)$$

$$T_{opt} = \left(\frac{1}{n} - 1\right)T_c \qquad ...(18.14)$$

Fig. 18.1 shows the graphs of cost per piece *v/s* cutting speed and time per piece *v/s* cutting speed respectively, in a machining process.

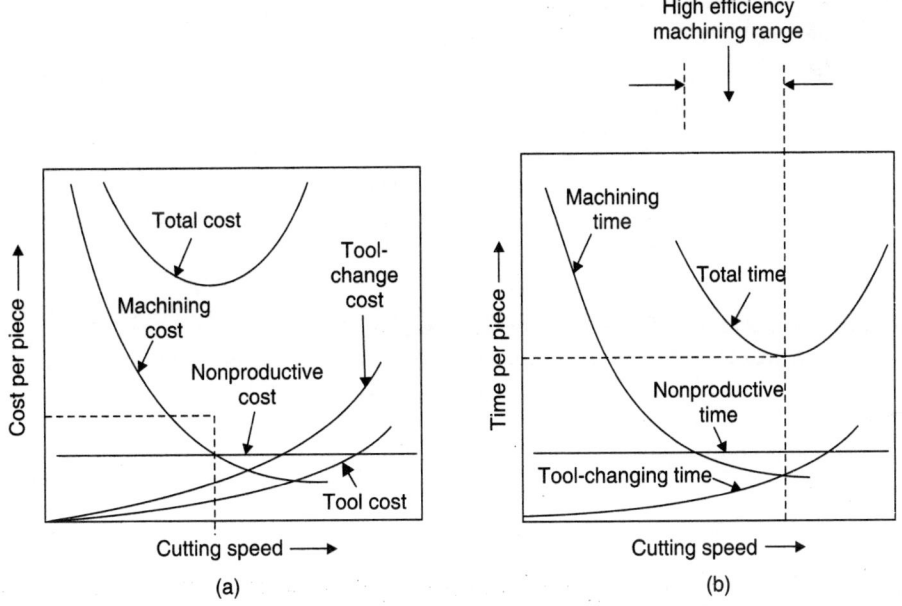

Fig. 18.1. Graphs showing cost per piece and time per piece in machining.

— The cost of a machined surface also depends on the degree of finish required ; the *machining cost increases rapidly with finer surface finish.*

18.2. ECONOMICS OF MANUFACTURING

It is imperative to have a through knowledge of *economics of manufacturing* if one wants to succeed in the market and to have leading edge over competitors. In order to earn more profits, it is essential that *production cost* should be *less* because selling price cannot be increased in a competitive market. This can be achieved only if the *right quality of a product is produced using the appropriate manufacturing method.*

In a competitive market, manufacturers usually do not have complete control over the volume of sales and selling prices where as it is *possible to cut or at least control the manufacturing cost to increase the profit.*

The cost of producing a part can broadly be *classified* as follow :

  1. Fixed cost ;   2. Variable cost

  1. **Fixed cost** (C_f) :

● It is the cost required to *set up the facilities* to produce a product.

● It does *not* depend on the quality of the product produced ; rather it *tends to be fixed for a given period of time and for a given level of installed capacity.*

Examples. *Salaries and wages of permanent employees, rent, land, building, machinery etc.*

  2. **Variable cost** (C_v) :

● It *varies with the quantity or number of parts produced.*

Example. *Cost of raw materials, fuel, power etc.*

The total cost (C_t) for producing a quantity x will be the sum of fixed cost and total variable cost, that is

$$C_t = C_f + x\,C_v \qquad ...(18.15)$$

By using above equation we can determine the cost of production of one piece by dividing total cost by the quantity produced.

Break-even analysis :

● Let us consider two manufacturing processes, 1 and 2 used to produce x parts each. Further, let for these processes,

$$C_{f_1},\ C_{f_2} = \text{Fixed costs respectively, and}$$
$$C_{v1},\ C_{v2} = \text{Variable costs respectively}$$

Then the total costs for these process, 1 and 2, respectively will be :

$$C_{t1} = C_{f1} + xC_{v1} \qquad ...(18.16)$$
$$C_{t2} = C_{f2} + xC_{v2} \qquad ...(18.17)$$

The total costs C_{t1} and C_{t2} of the two processes given by the equations (18.16) and (18.17) can be plotted with the cost on one axis and quantity on the other, as shown in Fig. 18.2. This plot is known as *break-even chart*. The quantity corresponding to break-even point is known as *break-even quantity.*

At break-even point,

$$C_{t1} = C_{t2} \qquad ...(18.18)$$

or,
$$C_{f1} + x_{\text{BEP}} \times C_{v1} = C_{f2} + x_{\text{BEP}} \times C_{v2}$$

or,
$$x_{\text{BEP}} = \frac{C_{f2} - C_{f1}}{C_{v1} - C_{v2}} \qquad ...(18.19)$$

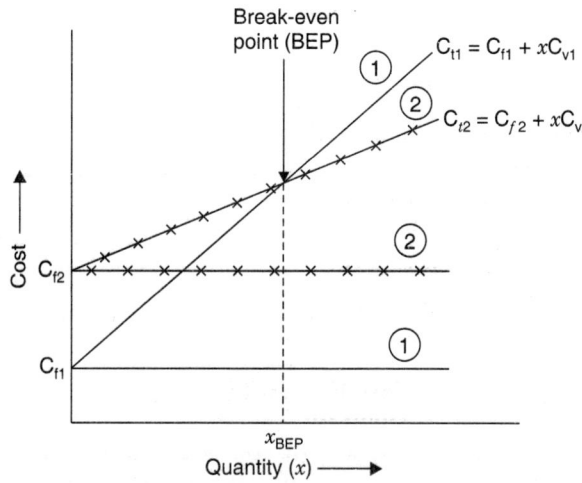

Fig. 18.2. Break-even chart for composing two manufacturing processes.

Equation (18.19) can be used to determine break-even quantity. Thus using break-even analysis we can determine which process shall be used for production of a given quantity in the most economical manner. This analysis can be easily extended for comparing more than two processes at a time.

- The concept of break-even analysis can also be employed for the following purposes :

  (i) To analyse the *profitability* of a *"single product"* :

  (ii) To determine the *minimum quantity of production of the product.*

The total cost of the product (C_t) is given by equation (18.15).

$$C_t = C_f + xC_v \qquad \qquad ...(1)$$

Let, P_s = Selling price per piece (or per unit), and

I_t = Total income for the sale of x units

Then, $I_t = x \times P_s$ $\qquad \qquad ...(2)$

The plot of equations (1) and (2) is shown in Fig. 18.3. The point of intersection of total cost line with total income line gives the break-even point, corresponding to sales volume x_{BEP}. At the point of intersection

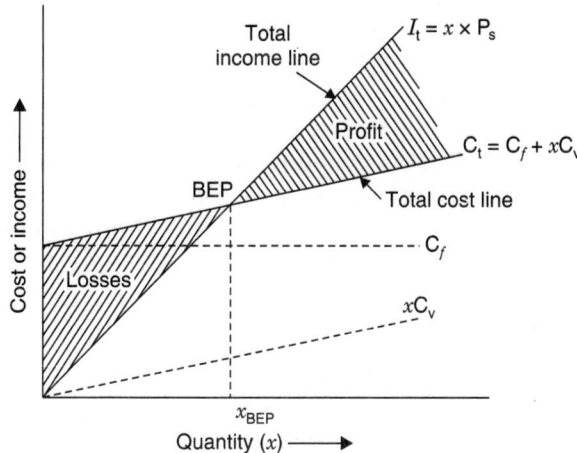

Fig. 18.3. Break-even chart for a single product.

$$C_f + x\, C_v = x \times P_s \qquad\qquad\qquad\qquad ...(18.20)$$

or
$$x_{BEP} = \frac{C_t}{P_s - C_v} \qquad\qquad\qquad\qquad ...(18.21)$$

The term $P_s - C_v$ indicates the *contribution to profit*.

● The value of BEP should be *as small as possible* for a healthy organisation ; the BEP can be *reduced* by the following methods :

    (*i*) By *reducing* the fixed cost (C_f) ;

    (*ii*) By *reducing* the variable cost (C_v) ;

    (*iii*) By *increasing* the selling piece (P_s).

HIGHLIGHTS

1. The most important parameters associated with the economics of machining are :
       (*i*) Minimum cost per part ;   (*ii*) Maximum production rate.
2. The value of BEP should be as small as possible.

THEORETICAL QUESTIONS

1. List the most important parameters associated with the economics of machining ?
2. List the items of which the total cost per piece is made up of.
3. Explain briefly the following items of cost :
       (*i*) Machining cost ;   (*ii*) Set up cost ;
       (*iii*) Loading, unloading and machine handling costs ;
       (*iv*) Tooling cost.
4. For a turning operation, derive relations for optimum cutting speed and 'optimum tool life' for the following cases.
       (*i*) For minimum cost ;   (*ii*) For maximum production.
5. Write a short note on 'Economics of manufacturing'.
6. How can the value of BEP reduced in case of a single product ?

RECENT TRENDS IN MANUFACTURING

19

Need for Integration of Design and Manufacturing

19.1. INTRODUCTION

Now-a-days the manufacturing is facing a formidable challenge in the national and international market place. The need for manufacturing high quality, low cost products is demanding major changes in the manufacturing function resulting in *facility modernisation, high level of automation and reduction in inventory and lead times.* The designer must understand and accommodate these changes to ensure that a design can be efficiently produced.

After the revolution which took place in computer technology *Computer Integrated Manufacture (CIM) is the term used to describe the complete and whole automation of the factory with all processes functioning under computer control and only digital information integrating them together.*

— CIM is the evolutionary outcome of Computer Aided Design (CAD) and drafting (CADD) and Computer Aided Manufacturing (CAM).

— CIM is required and desirable because it reduces the human component of manufacturing and thereby relieve the process of its most expensive and error prone ingredient.

19.2. INTEGRATION OF DESIGN AND MANUFACTURING

Traditionally the product design and the procedures by which the products are manufactured have been performed separately by different groups of individuals. Now-a-days, the design engineers are well versed in manufacturing procedures and processes.

Concurrent engineer attempts to integrate the product design and its related manufacturing process design.

— For each product, teams are formed that consist of design, manufacturing, quality assurance, and purchasing personnel.

— The design of the product is performing concurrently with the selection of processes, equipment, and steps that will determine how it is to be built.

— Inputs by the *quality assurance* members of the team assure that the finished product can be manufactured to predetermined levels of quality. Inputs by the *purchasing members* of the team ensure that the product will be built with parts that are available and economically procurable.

19.3. PROCESS SELECTION

Owing to the developing in *'manufacturing automation'* the number of process alternatives available to manufacturing planners have increased. *The selection of the most cost effective process is dependent upon* :

— Part geometry ;

— Tolerance requirements ;

— Lot size ;

— Other factors.

Following are some important *'alternatives' available to process planners* :

1. Numerical Control (NC). In recent decades Numerical Control (NC) has revolutionised machining processes.

— Numerical controlled machine tools have the ability to machine *geometric contours* not achievable with conventional equipment. There is the additional advantages of *increased repeatability* from part to part, without variations induced by a human operator.

— Numerical control *reduces tooling cost* by eliminating the costs of sophisticated jigs and fixtures.

— Set-up costs involved with the consecutive machining of different parts are usually small.

— Numerical control is not a high-volume production process and *may be considered a form of programmable automation.*

— The purchase *costs* of NC machine tools are *higher* than conventional, manual counterparts.

— The advantages of NC machining are partially *offset* by some *overhead costs* that are created by this form of automation. In addition to higher equipment costs, other costs are incurred for the programming, check out, and storage of tapes containing the programmes for parts to be manufactured.

2. Flexible manufacturing. *"Flexible manufacturing" involves the interconnection of groups of automated machine tools by an integrated and automated material handling system.*

● *"Flexible Manufacturing Systems (FMSs)"* are used *to complete sophisticated sequences of machining operations on families of manufactured parts.* These systems often contain provisions for automated part inspection. *Hierarchical computer control* is used to schedule and route parts through the system. Such control facilitates the tracking of machine tool usage and parts produced. This information is used to schedule periodic maintenance during which cutting tools and fluids may be changed.

● FMS implementation involves significant initial costs. Product volume and projected labour savings must therefore be sufficiently high to justify these extremely high initial investment costs.

3. Robotics. During the last two decades 'Industrial robots' have been successfully utilized in a number of manufacturing applications. Typical applications include :

— Spray painting ;

— Application of coating ;

— Spot welding ;

— Arc welding ;

— Part transport during heat-treating applications ;

— Palletizing operations ;

— Machine loading and unloading.

● Robots eliminate repetitive, manual labour operations. They have successfully removed human workers from hazardous working environments and from physically stressful labour tasks.

● Robots, like NC machine tools, have the requirements for the generation, checkout and storage of various movement programs.

— Languages used are often unique to a particular robot. This information has impeded the integration of robots into automatic manufacturing cells and systems.

● Robots have achieved only *limited use in complex assembly operations* (due to limitations in gripper design, tactile sensing, and vision).

4. Fixed versus programmable automation. Decisions to use fixed as opposed to programmable automation alternatives are usually based on the production quality to be built.

● *"Fixed automation facilities"* are usually dedicated to the production of a single product, usually in *extremely high volumes. Fixed automation is characterised by extremely high costs associated with equipment procurement.* Any change in the product being manufactured results in a significant line change-over cost.

● Examples of *"programmable automated equipment"* include both *NC machine tools* and *industrial robots.* These machines are capable of producing a variety of different parts, with comparatively modest set up and programming cost. *Production output is much lower than dedicated, fixed automation equipment.*

19.4. PROCESS ROUTING

The **process routing** *specifies that sequence of manufacturing operations that is required to produce a product or sub-assembly.*

The process routing is divided into *separate steps* or *operations.* Each operation specifies the factory location or work centre in which a set of manufacturing operation is to take place. The routing contains a description of these operations along with set of machines or manufacturing facilities that are to be used.

— *Inspection operations* are also often defined as *separated routing operations.*

19.5. CAPACITY PLANNING

Capacity planning *is a method by which the master schedule is adjusted to balance the due dates of jobs or orders against the capacity of the plant and its individual work cells and facilities.*

Vollman, Berry, and Whybark describe three separate types of planning methods. The methods *differ in the amount of production data used to afford increasing levels of detail in assessing workload levels.* These methods are discussed as follows :

1. Capacity planning using overall factors (CPOF). It is a relatively simple approach that results in a "rough cut" capacity plan.

— The inputs come from the master schedule rather than from MRP tables associated with individual parts in the bill of materials.

— The workload levels are derived from performance standards or historical data for end products only.

● This method does not consider the time study associated with the lead times for all component parts in the end item.

2. Capacity bills :

● This method provides a more direct linkage between different end products being produced and respective capacities required by these different end items in various work centres. This method is *responsive to changes in product mix of the end items produced.*

● In order to use this approach, additional data are required. *Lot sizes* for each end product and their respective components *must be known. Set-up and run times for each lot must be defined* for each work centre in which processing is required.

3. Resource profiles. This approach further *refines* the capacity bills procedure.

— It considers the lead time requirements associated with each mode in the parts explosion diagram.

— All data for previous method are used, but are defined to occur in the specific period during which the work on a specific part or subassembly is scheduled to take place.

19.6. INTRAWORK CELL SCHEDULING METHODS

For the scheduling of jobs or orders within a given work cell a variety of methods exist. For *most rules* a notation of the form $n/m/C$ applies :

Here, n = The number of *jobs* or *orders* that are to be scheduled ;

m = The number of *machines* within the work cell, and

C = The objective or criterion addressed by the developed schedule.

The *common scheduling 'objectives'* are to deliver or complete the orders by the due dates required by the customer. This is accomplished by minimizing the average or maximum lateness for a sequence of jobs or orders.

● Most early work in analyzing scheduling methodologies focused on work cells consisting of only one or two machines and observed the following rules/procedures :

— Shortest processing time rule ;

— Due date rule ;

— Slack time rule ;

— Multiple machine rules ;

— First come, first served and random scheduling ;

— The RAND Simulations ;

— Scheduling in FMS environments.

19.7. DISPATCHING

Dispatching, an important facet of the production control process, *involves the movement of parts, components, sub-assemblies, and end items so that they arrive at the appropriate work centre exactly at the time they are needed in the production process :*

In this process the following three types of materials are usually *moved.*

(*i*) The movement of a partially completed part or subassembly to the appropriate work centre.

(*ii*) The movement of raw materials or components that are to be added at a particular process operation.

(*iii*) The movement of tooling, fixtures, gauges, and inspection equipment to the work centre.

19.8. BASIC TOOLS FOR INTEGRATION

— The *"computer"* has been the *core of the information age.* Continued technological developments of the past two to three decades have propelled society into an era whereby the computer is an integral part of daily life.

— One of the most interesting developments of the information age revolution is that the computer is no longer the sale domain of the computer expert. As more users gain familiarity with the knowledge and productivity gain made possible by using computer as a tool, applications continue to grow.

— A knowledge of computer basics facilitates productivity for the computer user. For the computer systems purchaser, an understanding of computer fundamentals is essential to ensure the procurement of a unit containing appropriate hardware, software, and expansion capabilities for current and future use as well as interfaces for existing and potential data sharing environments.

— There is a central difference between computer field and other industrial products whose growth has stabilized after reaching maturity. Rather than an end product is itself, the *computer is also a* **"tool"** *which has spurred development of new products and applications. It has become the core of a new revolution of innovations, products, and applications which will be emerging for years to come.*

  • *Computers are increasingly used for doing engineering drawings and graphics work because computers allow the graphics engineer or the draughtsman to easily change the contents format, colour, size etc. of a drawing.*

  • A computer is an electronic machine for processing information. The processing and storage of information inside a computer is using binary digits (0 and 1) only, and computers are called **digital computers.** Computers of other types (analogy computers) are no longer in common use.

19.9. COMPUTERS AND MICROPROCESSORS

19.9.1. History and Development of Computers

  • Charles Babbage (an English Mathematician) was responsible for conceiving the concept of the Modern computer, and is called "Father of Computers".

  • He designed the early computer called *"Difference Engine"* in the year 1822, with which reliable tables could be produced. In 1833 he improved upon the machine and put forth new of idea of *"Analytical Engine"* which could perform the basic arithmetic functions automatically. In this machine punched cards were used as input/output devices for basic input and output.

The concept of use of punched cards was developed further by Horman Hollerith in the year 1889.

  • Leonards Torres demonstrated a *digital calculating machine* in Paris in 1920.

  • In 1944 Prof. Howard Aiken (Howard University) developed Electromechanical calculators known as Mark-I. This machine could handle about a sequence of 5 arithmetic operations by usign memory for previous results.

  • On the basis of research done for U.S. army during the World War-II in 1946, the first electronic computer, ENIAC (Electronic Numerical Integer and Computer) was designed in 1946. This computer was about 15 metres long and 2 metres high and weighed about 50 tons. It consumed about 200 kW power. This machine did not have any facility for storing programs.

- In 1949, the *concept of stored program was adopted.*
- In 1951, was introduced the commercial version of stored program computer UNIVAC-1 (Universal Automatic Computer)—the first digital computer.

Generation of Computers :

First Generation ...... Developed during the year 1951-1959.

- These computers are *"based"* on *"Vacuum Tubes".*
- Very slow in operation (10^3 operations/sec).
- Big in size and unreliable.
- Short span of life.
- Frequent breakdowns.
- High power consumption and great amount of heat generation.
- Small primitive memories and no auxiliary storage.
- Limited programming capabilities.

Examples. UNIVAC-1 and IMB 650.

Second generation ...... Developed during the years 1960-1965

- These computers are *based on "transistors".*
- Faster in operation, comparatively (10^6 operations/sec).
- Smaller in size.
- More reliable.
- Consume less power.
- Generate less heat than vacuum tubes.
- Auxiliary memory in the form of magnetic tape was introduced.

Examples. UNIVAC 1107, IBM 7090, CDC 1604, Honeywell 800 etc.

Third generation ....Introduced during 1965-1970, also being used presently.

- These computers are based on *"Integrated circuits",* based on silicon technology.
- Much more smaller in size.
- More reliable.
- Faster in operation (10^9 operation/sec).
- Less expensive.
- Employ higher capacity internal storage.
- Wide range of peripherals used.
- Make use of new concepts like *multi-programming, multi-processing, high level languages.*

Examples. IBM-360/370, Honeywell 6000.

Fourth generation ...... Introduced in 70s.

- These computers are based on VLSI (Very large scale integration) chips and microprocessors chips.
- Possess high processing power.
- Low maintenance.
- Faster in operation.

- High reliability.
- Very low power consumption.
- Less expensive.
- Small size.

This generation also includes the following :

— Microcomputers ;

— Office automation systems ;

— Distributed processing systems.

Fifth generation ...... Introduced during the 1990's

- These computers use optic fibre technology to handle *Artificial Intelligence, Expert Systems, Robotics etc.*
- Possess very high processing speeds.
- More reliable.

19.9.2. Definition of a Computer

A **computer** is a *machine that processes data according to set of instructions stored within the machine.*

- It *receives data as input, processes the data, i.e., performs arithmetic and logical operations on the same and produces output in the desired form on output device as per the instructions coded in the program.*
- The process function of the computer is *directed by the stored program, a set of codes instructions stored in the memory unit, which guides the sequence of steps to be followed during processing.*

19.9.3. Characteristics of a Computer

The following are the *characteristics* which make a computer an indispensable unit :

1. Speed
2. Consistency
3. Accuracy
4. Flexibility
5. Reliability
6. Large storage capacity
7. Automatic operation
8. Diligent
9. No emotional ego and psychological problems.

Limitations of a computer :

A computer entails the following *limitations :*

1. It does not work on itself, a set of instructions is required for its operation.
2. It cannot take decision on its own, it has to be programmed as per requirements.
3. It is not intelligent, it has to be instructed in detail for the performance of each and every task.
4. It cannot learn by experience, as human beings do.

19.9.4. Classification of Computers

The computers may be *classified* as follows :

1. On the basis of the type of data :

(*i*) *Analog computers* (These computers process the data in analog form).

(*ii*) *Digital computers* (These computers process the data in digital form).

2. On the basis of the size and capacity :

(*i*) *Microcomputers.*

(*ii*) *Minicomputers.*

(*iii*) *Main frame.*

(*iv*) *Supercomputers.*

3. On the basis of the type of application :

(*i*) *Special purpose computers.*

(*ii*) *General purpose computers.*

4. On the basis of the number of users :

(*i*) *Single user computers.*

(*ii*) *Multi-user computers.*

5. On the basis of the number of processors :

(*i*) *Single processor computers.*

(*ii*) *Multiprocessor computers.*

6. On the basis of the type of instructions set :

(*i*) *Complex Instruction Set Computers (CISC).*

(*ii*) *Reduced Instruction Set Computers (RISC).*

19.9.5. Analog Computers

● The *principle of operation of analog* computers is to *create a physical analog of mathematical problems.*

● Measure physical variables *continuously.*

● Use signals as input (which may be supplied by devices like barometers, speedometers, thermometers etc.)

● The *result* given by an analog computer is *not very precise, accurate and consistent.*

● These computers find limited applications.

Examples. Speedometer of a vehicle (here speed varies continuously).

19.9.6. Digital Computers

● The digital computers *accept digits and alphabets as input.*

● Receive data in the form of discrete signals representing ON (high) or OFF (low) voltage.

● The data input can be represented as sets of O's and I's representing low and high reprectively.

● The digital computers convert data into discrete form before operating on it.

● The *most important characteristic* of a digital computer is that it is general purpose device capable of being used in a *number of different* applications. By changing the stored program, the same machine can be used to implement totally different tasks.

Example. Digital watches.

Digital computers may be further classified based upon : (*i*) Purpose of use (*e.g.,* General purpose, special purpose) ; (*ii*) Size and capabilities.

On the basis of *size and capabilities,* the digital computers are classified as :

1. Super computers.
2. Mainframe computers.
3. Medium sized computers.
4. Minicomputers.
5. Microcomputers.

1. Super computers :

- These computers are the fastest (speed of calculations upto 1.2 billion instructions per second) and have very high processing speeds.
- They are very large in size and most powerful and costliest.
- Their fields of applications include processing weather data, geological data, genetic engineering etc.
- Word length : 64 bits and more.
- These computers can receive input from more than 1000 individual work stations.

Example. PARAM (a super computer developed in India).

2. Main frame computers :

- These are large scale general purpose computer systems.
- Possess large storage capacities in several million words.
- Secondary storage directly accessible—of the order of several billion words.
- Can support a large number of terminals (upto 100 or more).
- Faster in operation (100 million instructions/sec. approx).
- *Accept all types of high level languages.*
- Word length—16 or 32 or 64 bits.

3. Medium sized computers :

- Mini verisons of mainframe computers.
- They have smaller power than mainframes.
- Processing speeds relatively high with support for about 200 remote systems.

4. Minicomputers :

- These are general computer systems.
- Reduced storage capacity and performance (as compared to main frame).
- CPU speed—few million instructions/sec.
- Word length—16 or 32 bits.
- Can accept all types of high level languages.
- Can support upto about 20 terminals.

Note : In view of fast development in electronics it is difficult to draw a line of demarcation between small main frame computers and large minicomputers.

5. Microcomputers :

- These are small size computers *utilising microprocessors.* These are popularly known as personal computer (P.C).
- CPU is usually contained on one chip.
- Possess low storage capacity (maximum being 256 K words).
- Slow in operation (10^5 instructions/sec).

- Usually provided with *video display unit, floppy drive and printer*. Some microcomputers can support hard disc also.

- Maximum word length is 16 bits ; however most of these use 8-bit words.

- Commonly used language—BASIC. However these computers can also accept other high level languages, *viz.* PASCAL, FORTRAN etc.

Note : *A single chip microcomputer* consists of a single chip on which the central processing unit, input/output and memory units are integrated. This is used for *industrial applications* and also in *product calculators*.

* *Its advantage is the reduction in cost and size, increase in performance and reliability.*

19.9.7. Differences between Analog and Digital Computers

The differences between analog and digital computers are given in the Table 19.1.

Table 19.1. Differences between Analog and Digital Computers

S.No.	Digital computer	Analog computer
1.	It performs calculations by counting and thus counts directly. It is the most *versatile machine.*	It processes work electronically by *analogy.* It does not produce number but produces its results *in the form of graph.* It is *more efficient in continuous calculations.*
2.	It operates on inputs which are on-off type (being digits 0 or 1) and its output is in the form of off signals.	It accepts variable electrical signals (analog values) as inputs, and its output is also in the form of analog electrical signals.
3.	It is based on counting operation.	It operates by measuring analog signals.

These days *digital computers are being widely used.*

A hybrid computer is combination of both analog and digital computers. It is used for *simulation applications.*

19.9.8. Block Diagram of a Digital Computer

Fig. 19.1. shows a block diagram of a typical digital computer.

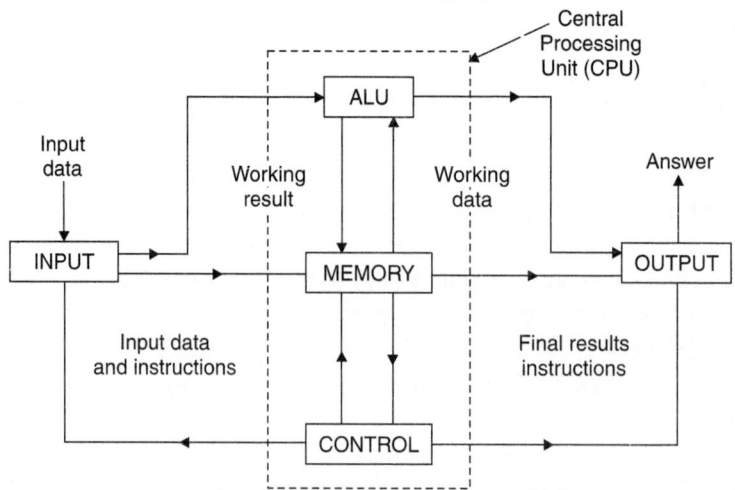

Fig. 19.1. Block diagram of a digital computer.

The following are the *five basic elements* of a computer system :

1. Input :
- The data and instructions are first recorded on a machine readable medium, like punched card, and then fed into the computer via a device that codes them in a manner which is suited to conversion into electrical pulses before entering memory.
- The input supplies data to the computer in digital (binary) form.

2. Memory :
- The memory section within the computer is where data are stored or memorized.
- Problems to be solved, inputs for the problem, a program of instruction, working data intermediate results and final results are *types of memory data.*
- The memory section holds data between high speed computer operation and slower input and output devices.

3. Arithmetic Logic Unit (ALU) :
- ALU performs necessary arithmetical operations on the data and ensures that instructions are obeyed.
- It also performs *logical operations.*
- The ALU combined with control unit is called *central processing unit.*

4. Control Unit :
- It fetches instructions from main memory, interprets them and issues the necessary signals to the components making up the system.
- It issues commands for all hardware operations necessary in obeying instructions.

5. Output :
- The output is the path for data out of the computer and may include devices for reading out answers.

19.9.9. Rating of Chips

Chips are rated in terms of their *capacity* and *speed.*
- *Capacity* of a chip refers to the amount of *kilo-bites it can store.*
- *Chip speed* refers to the rate at which the microprocessor can write to the chip. It is usually measured in nano-seconds (ns). As the chip speed increases, its cost goes up.

19.9.10. Computer Peripherals

A **peripheral** *is any device commonly used with a CPU of a computer for input or output of information or for memory functionally separate from the CPU and electronically detachable.*

Input Devices :

1. Keyboard :
- It is the most common and simplest input device.
- It is merely a collection of momentary switches. The outputs of the key switches are fed to electronic circuitry known as *keyboard encodes* which convert them into binary coded values. The values are then fed into the computer which interprets the key which was pressed. Thus the function of the key changes with the type of work we are doing.

2. Mouse :
- It is a pointing device and its size is about the size of palm.
- It is hand-held device that *controls a pointer on the screen.*
- It rolls on a small ball. A mouse has one or more buttons on the top. When the user moves the mouse over a flat surface, the screen cursor moves in the direction of the mouse movement.

3. Digitizer (or Graphic tablet) :

- It is similar to light pen.
- It consists of a glass plate on which digitizing tablet is moved.
- It is used for fine drawing works and for image manipulation application such as Autocad.

4. Optical Mark Reader (OMR) :

- OMR is being used for reading the answer sheet by means of light. It can read upto 150 documents per minute ; when on-line with respect to the computer system, can read upto 2000 documents per minute.
- OMR can also be used for such applications as *order writing payroll, inventory control, insurance, questionnaires etc.*

5. Magnetic Ink Character Reader (MICR) :

- MICR uses a special ink to print character. These character can be decoded by special magnetic devices.
- This system is employed *by banks for processing cheques.*

6. Scanner :

- It is used for getting existing graphical images (like photographs, mats, etc.) into computer.
- Once the graphical image is scanned and brought into the computer user can include them into documents or can edit them.

7. Light pen :

- It consists of a pen like device and photoelectric cell.
- It is used to draw pictures on the screen.
- When light pen is in contact with screen, the electron beam activates the photoelectric cell which in turn sends signals into the computer ultimately a mark is made on the screen where light pen contacted the screen.

8. Joy-stick :

- It is screen-pointing device.
- A stick is present with a button at the top. It can be held in the hand and bent in any one of the four directions. As the stick is moved, the action on the screen changes in the appropriate direction.
- A joy-stick is *widely used for playing computer games.*

9. Touch screen :

- The touch screen technique involves beam and ultrasonic waves.
- By using touch screen we can issue command to the computer by touching the screen.
- Limited amount of data can be entered via a terminal or a microcomputer that has a touch screen.

10. Compact Disk Read Only Memory (CDROM) :

- It is a 120 mm diameter disc with a polycarbonate subtrate, a reflective metalised layer on one side, with a protective lacquer finish.
- Here a laser beam is used to burn a small hole or pit which represent binary '1'. The absence of pit represents '0'. In this way digital information is stored on the disc in large quantities (in Giga Bytes).

11. Voice Recognition System or Voice Synthesizer :

- Voice recognition techniques, along with several other techniques, are used to convert the voice signals to appropriate words and devoice the correct meanings of words. There has been a limited success in this area and these days devices are available commercially to recognize and interpret human voices.

Output Devices :

1. Printer :

- A printer is *device that produces copies of text and graphics on paper.*
- The printers are classified/categorised as follows :

A. *Impact printers :*

   (*i*) Solid Font   (*ii*) Dot Matrix.

B. *Non-impact printers :*

   (*i*) Thermal printer   (*ii*) Inkjet printer

   (*iii*) Laser printer   (*iv*) Electrographic printer

   (*v*) Electrostatic printer.

2. Plotters :

- Plotters are those devices which *reproduce drawings using pens that are attached to movable arms.*
- Plotting in different colours is possible.

3. Monitors or Visual Display Unit (VDU) :

- A monitor is a television like device, which is used to display information, output and input data.
- It consists of a cathode ray tube (CRT), on which the information is displayed. When the user processes any key on the keyboard, the keyboard encoder generates code of that key which is depressed. This code is then fed to the computer ; from there VDU system takes that code and displays it on the screen.

19.9.11. Storage Devices

The memory devices in a memory unit (which stores the data, instructions and intermediate results) may be of the following types :

1. Internal storage device ...... also known as *main or primary* storage device.

The primary storage devices currently in use in computers are :

   (*i*) Magnetic core memory device.   (*ii*) Thin film memory device.

(*iii*) Thin rod memory device.   (*iv*) Plated wire memory device.

2. Auxiliary storage device :

The popular *secondary memory devices* are :

   (*i*) Magnetic tape drive.   (*ii*) Magnetic disk drive.

(*iii*) Magnetic drum.   (*iv*) Floppy disk.

   (*v*) Winchester disk.

Method of Input to Backing Stores :

The following methods are generally used :

   (*i*) Key-to-tape   (*ii*) Key-to-cassette/cartridge

(*iii*) Key-to-disk/diskette.

Memory. The memory is used to store information/data so that it can be retrieved whenever required. There are mainly two types of memories :

1. Primary memory.

2. Secondary memory.

1. Primary memory :

● It is also known as core memory, main memory, RAM (Random Access Memory).

● It is constructed using purely semiconductor devices, data is stored in the form of voltages.

● It is a volatile memory where as ROM (Read Only Memories) are non-volatile memories.

2. Secondary memory :

● Secondary memory, also known as *auxiliary memory,* is used to store large volumes of data.

● Data is stored in the form of magnetic energy and can be stored (in the secondary memory) for large periods.

Difference between Read Only Memory (ROM) and Random Access Memory (RAM)

ROM :

● As the name implies ROM is a memory unit that performs the read operation only ; it *does not have a write capability.* This implies that the *binary information stored in a ROM is made permanent during the hardware production of the unit and cannot be altered by writing different words into it.* Whereas a *RAM is a general-purpose device whose contents can be altered during the computational process.*

● ROM is a type of memory chip that we can read only and we cannot write on it.

● ROM provides permanent storage for program instructions.

● The most important ROM chip in any computer is ROM BIOS (Basic Input/Output System).

● ROM is most oftenly used in microprocessors that always execute the same program such as BOOT STRAP LOADER.

Disadvantages of ROM :

(*i*) A ROM is prepared by the manufacturer and cannot be altered once the chip has been made.

(*ii*) It is slow.

The *ROM memory* may be **classified** as follows :

(*i*) *Programmable Read Only Memory (PROM).* Here, the information can be altered, but *not as easily as in the ordinary memory.* Once the operations to be performed have been written into a PROM chip, they are permanent and cannot be changed.

(*ii*) *Erasable Programmable Read Only Memory (EPROM).* This type of ROM can be erased and programmed with the help of special equipment. It has a window at its top, which if exposed to ultraviolet light, allows data to be erased.

(*iii*) *Electrically Erasable Programmable ROM (EEPROM).* In order to erase and reprogramme this type of ROM, it is required to be removed from the socket.

(*iv*) *Flash EPROM.* It is the latest type of ROM. A manufacturer can make changes to the flash EPROM while it remain in the PC, by running a special program.

RAM :

● This memory is so named since memory registers can be accessed for information transfer as required.

● RAM chip is made with Metal Oxide Semiconductor (MOS).

● RAM chips may be classified as :

   (*i*) *Dynamic RAM :*

      It provides volatile storage (*i.e.,* the data stored is lost in the event of a power failure).

(ii) Static RAM :

These chips are more complicated and take up more space for a given storage capacity than dynamic RMA chips. These chips are also volatile in nature but as long as they are supplied with power, they need not require special regenerator circuits to retain the stored data.

- *Static RAM chips are thus used in specialised applications while Dynamic RAM chips are used in the primary locations.*
- Owing to the volatile nature of these storage elements, a back up *Uninterrupted Power Systems (UPS) is often installed along with larger computer systems.*

19.9.12. Hardware, Software and Liveware

Hardware :

The set of *physical components, modules and peripherals comprising a computer system is* called *Hardware.*

Apart from wires and nut bolts, the major hardware components are :

(*i*) Input-output devices

(*ii*) Control unit

(*iii*) Memory

(*iv*) ALU.

Software :

The software is a *set of programs required for data processing activities of the computer.* In other words, the program written in any one of the computer languages, is called *software.*

System software includes the following :

(*i*) Operating systems.

(*ii*) Language processors (assemblers, compilers, interpreters).

(*iii*) Utility program.

(*iv*) Subroutine program.

Liveware :

All persons concerned with computers, *i.e.,* compiler, programmer, etc. are called *liveware.*

19.9.13. Translators

A **translator** is *a software program which converts statements written in one language into another, e.g.,* converting assembly language to machine code etc. The assembly language program is called *'source program'* and the machine code program is called *'object program'.*

There are three types of *translators :*

1. Assembler.

2. Compiler.

3. Interpreter.

19.9.14. Computer Languages

1. Machine language. It is a programming language in which the instructions are in a form which allows the computer to perform them immediately, without any further translation. Instructions in machine language are *in the form of a binary code,* also called machine code and are known as machine *instructions.*

2. Low level language. Low level languages are machine oriented languages in which each instruction corresponds or resembles a machine instruction. The low level language must be translated into machine language before use.

3. High level language. The development of high level language was intended to overcome main limitations of low level language. The high level languages have an extensive vocabulary of word, symbols and sentences.

Different types of high level languages are :

(*i*) *Commercial language* ...... The most well known commercial language is COBOL (Commercial Business Oriented Languages).

(*ii*) *Scientific language* ...... The most well-known languages among this group are :

      (*a*) ALGOL (Arithmetic Oriented Language)

      (*b*) FORTRAN (Formula Translation)

      (*c*) BASIC (Beginner All Purpose Symbolic Instruction Code).

(*iii*) *Special purpose language.*

(*iv*) *Command language.*

(*v*) *Multipurpose language.*

19.9.15. Computer Programming Process for Writing Programs

The complete computer programming process followed by programmer for writing comprises the following **steps :**

1. Analysis	2. Flow charting
3. Coding	4. Debugging
5. Documentation	6. Production.

19.9.16. Computer Elements of Analog Computers

1. *Attenuators* ...... are used to multiply a variable quantity by a constant.

2. *Summing amplifiers* ...... are used to add or subtract variables as required.

3. *Servo multipliers* ...... are used to multiply two variables.

4. *Function generators* ...... are used to simulate the arbitrary behaviour of variables.

5. *Integrating amplifiers* ...... are used to integrate a variable with respect to time.

19.9.17. Microprocessors

19.9.17.1. General aspects

- The **microprocessor** *is a semiconductor device consisting of electronic logic circuits manufactured by using either a Large-Scale Integration (LSI) or Very-Large Scale Integration (VLSI) technique.*

- The microprocessor is *capable of performing computing functions and making decisions to change the sequence of program execution.* In large computers, the Central Processing Unit (CPU) performs these computing functions and it is implemented on one or more circuit boards.

- The microprocessor is in many ways *similar* to the CPU, however, the microprocessor includes all the *logic circuitry* (including the control unit) on one chip.

The microprocessor consists of the following three *segments* (see Fig. 19.2)

1. Arithmetic/Logic Unit (ALU). In this area of the microprocessor, computing functions are performed on data. The ALU performs arithmetic operations such as addition and subtraction, and logic operations such as AND, OR and exclusive OR. Results are stored either in registers or in memory or sent to output devices.

2. Register Unit. This is of the microprocessor consists of various registers. The registers are used primarily to store data temporarily during the execution of a program. Some of the registers are accessible to the user through instructions.

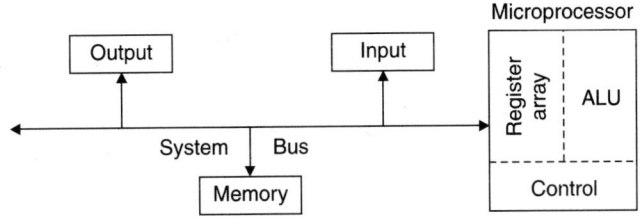

Fig. 19.2. Block diagram of microcomputer.

3. Control Unit. The control unit provides the necessary timing and control signals to all the operations in the microcomputer. It controls the flow of data between the microprocessor and peripherals including memory.

In short a microprocessor performs the following *functions :*

- Communicates with all peripherals (memory and I/O) using system.
- Controls timing of information flow.
- Performs the computing tasks specified in a programme.

19.9.17.2. Characteristics of microprocessor

In nearly every type of design, with any complexity at all, microprocessors have potential for *drastically reducing component count and shortening design time.* In fact a microprocessor is considered to represent long-awaited next generation of digital building blocks, and that microprocessor will provide the best single approach to the system-level digital integrated circuit.

Some of the characteristics of a microprocessor are listed below :

1. It *handles shorter words than other computers,* usually from 4 to as many as 16 bits.
2. It consists of *integrated circuits* from 1 to 30 in number.
3. It contains arithmetic logic unit (ALU), registers, control, random access memory (RAM), data bases and read only memory (ROM) with programmes.

19.9.17.3. Important features

The *important features* of the microprocessors are :

1. Low cost
2. Small size
3. Low power consumption
4. Versatile (The versatility of a microprocessor results from its 'stored programme' mode of operation).
5. Extremely reliable.

Note : Probably the term 'micro' in the name of the device can be contributed to its *low cost, small size and low power consumption.* The processing capability of a microprocessor should not, however, be underestimated. Currently available 32-bit microprocessors have a processing power similar to that of the mainframe computer of a few years ago. Even the early 8-bit microprocessors are powerful enough to perform several applications.

19.9.17.4. Uses of microprocessors in instrumentation

The processing power of the 8-bit microprocessors is more than adequate to satisfy the requirements of most of the *instrumentation applications.* By making an instrument microprocessor based, it can be made *intelligent by incorporating new features like programmability,* which cannot be provided in its hard-wired counterpart.

Some important *uses* of microprocessors in instrumentation area are listed below :

1. Frequency meters.
2. Function generators.

3. Frequency synthesizers.

4. Spectrum synthesizers.

5. **Intelligent instruments CRT terminals**

6. Digital millimeters

7. Oscilloscopes

8. Caunters

9. **Process control**

— Instrumentation

— Monitoring and control

— Data acquasition

— Logging and processing

10. **Medical Electronics**

— Patient-monitoring in intensive care unit

— Pathological analysis

— Measurement of parameters like blood pressure and temperature.

Under this heading the following instruments/machines are included :

(i) Microprocessor based medical instrument

(ii) Microprocessor based ECG machines

(iii) Microprocessor based EEG machines etc.

Other Applications of Microprocessors :

(i) High level language computers

(ii) Replacing hard-wired logic by a microprocessor

(iii) Control of automation and continuous processes

(iv) Computer peripheral controllers

(v) Home entertainment and games.

19.9.18. Computer Terms

Abort. To terminate the execution ⌐f a program and to control the operating system.

ALU (Arithmetic and Logic Unit). The portion of the CPU that performs arithmetic and logical operations.

Access. To locate desired data.

Accummulator. A register, or set of registers in the central processor used for temporarily storing the numerical result on an operation performed by the ALU.

Adder. A *logic device* that performs the arithmetic addition of two binary numbers.

ALGOL (Algorithmic language). Arithmetic language by which numerical procedures may be presented to the computer in a blended form.

Assembly. The process of translating a program written in symbolic code into its equivalent machine code ; the time during which this process occurs is called *assembly time.*

ASCII. An eight level (7 bits + 1 parity bit) code from American Standard Code for Information. In it, the letters, numbers and symbols are coded as 7 binary characters, 8th bit being used for parity check. $2^7 = 128$ character can be represented by this code.

Bar Code. A pattern of printed lines in binary coding that can be read into the computer by light pen scanning.

BASIC. Beginner's All Purpose Symbolic Instruction Code—a programming language that is easy to learn and widely used as first programming language taught in schools and as the principal language in many minicomputers and microcomputers. Although it is simple to use, it contains many advanced features for handling mathematical formulae and character strings.

BCD (Binary Code Decimal) Numbers. It is a code in which notation is preserved and each decimal digit is coded in binary, form, using 4 bits (called a nibble) for each successive digit.

Binary. A numbering system using only the digits 0 and 1. Also called "base-2".

Bit. An acronym for Binary Digit. It is the simplest possible information element. It is an entity which may have one of the two states, *i.e.*, on or off represented by 1 or 0. It is the *smallest* unit of information in the binary numbering system.

Boolean Algebra. An algebra defining the rules for manipulating variables in symbolic logic Boolean algebra was developed as a method for expressing logical concepts in a mathematical form and uses such logical operators as AND, OR, NOR and IF-THEN.

Bootstrap. When power supply to a computer using main memory as semiconductor memory fails, all its memory is washed off. In order to restart, *i.e.*, enable it to work, it has to be programmed to accept instructions. This process is called *bootstrap.*

Bubble memory. Latest art in a memory device. When an external field is applied to a ferromagnetic specimen, the domains in which magnetisations are antiparallel get converted into cylindrical domains known as *bubble*. This size of the *bubble* is of the order of 1 to 100 microns.

Bug. Refers to fault resulting from a programming error. Sometimes it also refers to faults resulting from hardware design or construction errors.

Bus :
- It is a digital highway or an electrical channel along which data can be sent and received.
- It interconnects various elements of a computer and conveys data, addresses, instructions and control signals between the registers, arithmetic and logic unit (ALU), control unit and memory.
- There may be separate buses for data and instruction or a common bus. These can be unidirectional or bi-directional.

Byte :
- A group of consecutive bits forming a unit of storage in the computer and used to represent one alphanumeric character.
- It usually consists of 8-bits but may contain more or fewer bits depending on the model of computer.

CAD/CAM :
- Acronym for Computer-Aided Design/Computer-Aided Manufacturing.
- A computer system used in engineering for such projects as designing parts and machinery, precisely calculating parts specifications and generating complex wiring diagrams.

Call. A transfer of program control to a subroutine.

Capacity. The amount of information that all or a part of a computer system, such as main memory or a disk, pack, can store. For example ; the capacity of a computer's main memory could be 512 K information (524, 288 characters)

Character. An alphabetic letter, digit or special symbol.

Chip :
- It is a tiny piece of semiconductor material on which microscopic electronic components (*e.g.*, resistors, capacitors, diodes etc.) are all created by photoetching at the same time in one chip of silicon to form one or more circuits.
- It is usually a few millimeter square in size and is encapsuled in rectangular plastic or ceramic package, usually 20 mm wide 400 mm long.
- After connection leads and a core are added to the chip, it is called an IC (Integrated Circuit).

CMOS (Complementary Metal Oxide Semiconductor)

- This is an integrated circuit family, having high threshold logic and a technology which consumes very low power compared to other semiconductor technologies.
- It has moderate speed and high integrated device density.

COBOL :

- Acronym for Common Business Oriented Language.
- A high level programming language capable of performing all the necessary calculations most often used in *business*.

Compiler :

- A program that translates a source program written in a high level language into its equivalent machine language.
- The output program from a compiler is called an *Object Program*.

Computer :

- A machine capable of receiving, storing, manipulating and yielding information such as numbers, words, pictures.
- Unless qualified, the word computer means electronic digital computer.

Computer Graphics. The use of a computer to produce pictorial representations or relationships such as charts and two-or-three dimensional images, by means of dots, lines, curves etc.

Computer Program. A series of instructions or statements, in a form acceptable to a computer prepared in order to achieve a certain result.

Control Unit. It generates control signals (switching signals to control the sequencing of data flows and ALU operations).

Controller. A device (*e.g.,* a register) used to represent the number of occurrences of an event.

CPU :

- Abbreviation for Central Processing Unit the portion of a computer composed of the ALU and the Control Unit.
- It is where instructions are fetched, decoded and executed and the overall activity of the computer is controlled.

Crash. A term used when the computer breaks down at the time of programming.

Data :

- Characters grouped together in specific patterns, to which meaning is assigned.
- Commonly used to designate the numbers, facts, concepts, or the like to be processed by a program although any information input to a computer system is considered data.

Data base. A collection of logically related data elements that may be structured in various ways to meet the multiple processing and retrieval needs of individuals/organisation.

Debug. To trace and correct errors in programming code of hardware malfunctions in a computer system.

Decode. To interpret a code.

Documentation. A collection of written description and procedures that provide information and distance about a program or about all or part of a computer system so that it can be properly used and maintained.

DOS. Acronym for Disk Operating System.

DP. Abbreviation for Data Processing.

Encode. To convert data into a code.

Feedback. Data produced as output by a program and used as input to another phase in the same program so as to modify or correct the factors that have produced the output.

File :
- A collection of logically related records dealt with as a unit.
- It is usually referenced by a symbolic name.

Floppy Disk. Auxiliary memory storage device consisting of magnetic film coated on the flat plastic substrate.

Flow Chart :
- A graphical representation of the processing steps performed or sequence of logic operations implemented in hardware, software, firmware or manual procedures.
- It is chart illustrating the logic sequence of events that must be performed to attain a predetermined aim.

Format. The arrangement and location of data items within a large unit of storage.

FORTRAN. Acronym for FORmula TRANslation, a scientific programming language used to perform mathematical computation.

Gate :
- A circuit that has one or more input signals and produces a signal output of binary 1 or 0, depending on the type of logic built into the circuit.
- The relationship of input and output logic gets is generally described in a *"truth table"*.

Hardware. The physical equipment and components in a computer system.

Hybrid computer :
- The computer that is a combination of an analog and digital computer linked together by an interface system for converting analog data or vice-versa.
- Used in scientific research and other such specialized applications.

Input. Data fed to computer and process of feeding it.

Inverter :
- A gate with only one input and one output.
- The output is always the complement of the input.
- Also known as a NOT gate.

Karnaugh Map. A graphical display of the fundamental products in a truth table.

Language. A means of conveying information (data) between people and machines.

LIFO. Acronym for Last In-First Out.

Latch. The simplest type of flip-flop, consisting of two cross coupled NAND or NOR latches.

LED (Light Emitting Diode) :
- A semiconductor diode, the junction of which emits light when energised [passing a current in the forward (junction ON) direction].
- Used in the construction of display indicators.

Logic Circuit. A circuit whose input and output signals are two states, either high or low voltage.

Loop. A series of instructions which are executed interactively.

Machine Language. The language with which a computer works directly.

Master file :
- A file containing relatively permanent data.
- This file is often updated by records in a transaction file.

Microcomputer :

- A small, low cost computer containing a microprocessor.
- Used for a wide variety of purposes, as in a small department within large businesses, and in home, as for household management, video games etc.

Microprocessor :

- A chip that contains the ALU, SCRATCH PAD MEMORY, and CONTROL UNIT in a microcomputer.

Minicomputer. A computer, size wise, in between a micro and mainframe types.

Modular Programming. Technique of working programs in modules.

MOSFET. Metal Oxide Semiconductor Field Effect Transistor.

Parity. The concept of parity is a check on the accuracy of data.

PASCAL. A popular high-level language that facilitates the use of structured programming techniques.

Personal Computer (P.C). A relatively low-cost portable microcomputer, generally sold with software packages and useful and word processing, maintaining a budget, storing mailing lists, playing computer games etc.

Program. A list of instructions defining the sequential activities or operations to be performed by a computer to solve a problem.

Programming. Giving instructions to a computer before it begins to work.

RAM (Random Access Memory). A type of memory chip that can be read but cannot be written on or altered.

Word Processor :

- An automated, computerized system incorporating variously an electronic type writer, CRT terminal, memory, printer and the like.
- It is used to prepare, edit, store, transmit, or duplicate letters, reports, records etc., ..., as for business some programs now have spelling and syllabification verifiers.

19.10. COMPUTER NETWORKS FOR MANUFACTURING

19.10.1. Introduction

With the increase of technological developments the use and capabilities of computers, the need likewise occurred for communications from *peripherals* to the computer and between computers themselves. Advances in the telecommunications industry have kept pace and interfaced with technology enhancements in the computer field, making the two industries very compatible. Early computer networking involved connecting terminals to nearby host computers. The deployment of modern and other communication devices which enabled the conversion of digital-based computer codes to analog signals carried over telephone lines brought computer capabilities closer to the user. Networks linking computers and peripherals over great distances are referred to as *wide-area networks*.

19.10.2. Computer Network

- **A Computer Network** *is a number of computers interconnected by one or more transmission paths.* The transmission path is usually the *telephone line.*
- The *main aim of* network is *"the transfer and exchange of data between the computers and terminals".*

Advantages of networks :

Following are the *advantages of networks.*

1. It is possible to manufacture high quality products at relatively low cost.

2. The networking of computers permits the sharing of resources.

3. Network systems make it possible for companies to accomplish the synthesis, analysis, evaluation and documentation of the design in much less time than with manual methods.

4. The use of networking allows a 'very flexible working environment'.

5. Now-a-days, modern organisations are widely dispersed with offices located in diverse parts of a country and the world. Many of the computers and terminals at the sites need to exchange information and data often daily. *A network provides means to exchange data among these computers and to make programs and data available to the people of the enterprise.*

19.10.3. Local Area Network (LAN)

19.10.3.1. General aspects

Local Area Networking (LAN) *provides short range interconnectivity between the devices in the network, normally within one site.* Data transmission on a local area network is *typically in digital form and does not require conversion.*

*In a **factory** the computers communicate with each other by means of LAN.* A local area network is a non public communication system which permits the various devices connected to the network to communicate with each other over distances from several feet to several metres. The factory devices that can be attached to the network include :

— Computers ;
— Programmable controller ;
— CNC machine ;
— Robots ;
— Data collection devices ;
— Bar code readers ;
— Vision systems etc.

- • Major companies with branches usually have the need for both wide area and local networking. Networking has enabled companies to choose between centralised and decentralised computer systems. A decentralised system, known as a "*Distributed Data Processing* (DDP) system, utilises several linked computers to facilitate data sharing.

— In a *DDP system,* multiple computers perform specialized tasks and communicate with other CPUs and/or devices as needed.

— In a *centralised system,* user terminals connects to the main frame at the headquarters.

19.10.3.2. Network topologies

A *network configuration is called a* **network topology.** A network topology is the *shape or the physical connectivity of the network.*

The network design has the following three major goals when establishing the topology of a network :

(*i*) To give the end user the best possible response time and throughout.

(*ii*) To provide maximum possible reliability to assume proper receipt of the all traffic (alternate routing).

(*iii*) To route the traffic across the least cost path within the network between the sending and receiving DETS (Data Terminal Equipments).

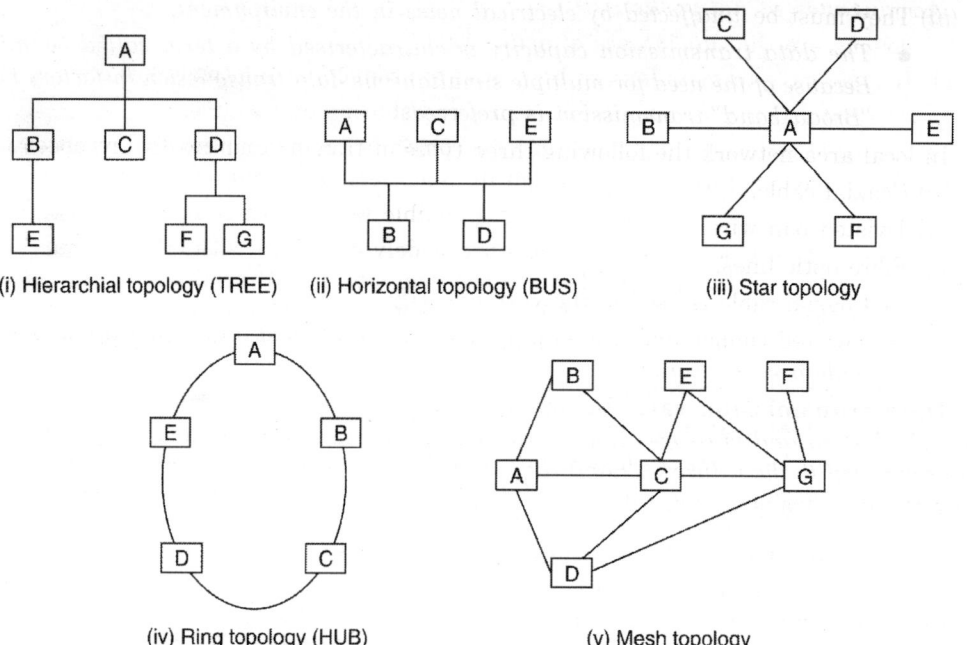

Fig. 19.3. Network topologies.

The more common network topologies, shown in Fig. 19.3 are :

(*i*) Hierachial topology (TREE).

(*ii*) Horizontal topology (BUS).

(*iii*) Star topology.

(*iv*) Ring topology (HUB).

(*v*) Mesh topology.

For a *factory local area network the "bus network" is the most appropriate* due to the following *reasons :*

(*i*) The bus network is generally easier to maintain and repair than the star or ring configuration.

(*ii*) The main transmission line of the bus network can be laid out in a pattern that corresponds closely to the layout of machine in the factory, thus facilitating installations of the communication systems.

Example. The product flow line layout.

(*iii*) Machines and other devices in the plant are often being rearranged to match the changing production requirements. On bus network each device can be connected to the main transmissions bus without major disruptions to the rest of the network.

● The major differences between local area and wide-area networks are the communication medium and subsequent need to use or not to use the signal conversion equipments.

Transmission line. The *"transmission line" is the message and data carrying medium that constitutes the physical distribution elements of the network.* Following are the requirements of the transmission data for factory networks :

(*i*) They must be *inexpensive* to install service and alter.

(*ii*) They must be capable of a high data transmission capacity.

(*iii*) They must be *unaffected by electrical noise in the environment.*

- *The data transmission capacity is characterised by a term called "bandwidth". Because of the need for multiple simultaneous data transmission in factory networks, "Broad band" transmission is preferred.*

In local area network the following three types of transmission media are used :

(*i*) Coaxial cable.

(*ii*) Twisted pan wire.

(*iii*) Fibre optic lines.

— Coaxial cable is used in the vast *majority of* LASNs ;

— Twisted copper line, fibre-optic, microwave, and satellite connections are used in *wide area networks.*

Data transmission rate. In communication network, the *"transmission rate" is the rate at which data and massages can be transferred among computers and computers controlled devices connected to the network.* "Band rate" and "bit rate" are the two units of measure used to indicate the data transmission rate.

19.10.4. Manufacturing Automation Protocol (MAP)

In implementing the networks, one of the major problems encountered is that, the various computers and computer based devices in the structure are not always compatible in terms of their ability to communicate with each other. The procedure used to *rectify* the above mentioned problem is known as *"communication protocols".* **MAP** *is a set of protocol standard designed for use in a factory local area network.*

Automation protocol uses :

— A bus network configuration ;

— Broad band transmission ;

— A token passing access scheme ;

— A data transmission rate of 10 megabits per second.

- **MAP** is based on specification defined by the International Standards Organisation (ISD) called the Open System Interconnected (OSI) reference model.

19.10.5. Computer Software

The collection of programs, together with the data used by the programs and documentation describing them, is known as **Computer software.** Data on the computer are organised into files which consist of various records, typically organised.

Software is typically categorised into two basic types. *"Systems software"* and *"Applications software".*

Systems software. *It is the collection of programs that coordinate the various components of the computer system.* The basic language of the computer is machine language, where the coding is binary digits. Since machine language is complex, intermediate assembler language and high-level programming languages such as COBOL, FORTRAN, PASCAL, BASIC, C, LISP, and host of other have been developed. *Programs are characterised by a definite set of procedures which instruct the computer how to proceed.*

To enable the computer to process a program written in *higher-level language, it must be converted to machine language.*

Operating systems. It is the main program that allocates the system resources and manages and controls the computer's activities.

The operating system *schedules and performs input/output, allocates space and resources, provides monitoring and security functions, and governs the execution and operation for the various system programs and applications.*

- The system's overall performance depends on the operating system's ability to manage various functions efficiently, as well as on the hardware's capabilities.

Software tools. Another type of software is a group of programs referred to as *software tools.*

The most frequently utilised software tools are :

— Database mangers ;
— Spreadsheet packages ;
— Graphics packages ;
— Word-processing systems.

Applications programs. Applications programs are the *programs which are written in a programming language to solve a specific problem.* Although hardware and systems software provide the major access point to problem solving by computer, neither would be of much use without applications programs. The capabilities of modern computers have created a lost of new applications for virtually every profession and industry. *Typically used industrial engineering applications are :*

— Computer-Aided Design (CAD) ;
— Computer-Aided Manufacturing (CAM) ;
— Computer-Aided Process Planning (CAPP) ;
— Cost estimating ;
— Routing and scheduling ;
— Time standards ;
— Machine processes ;
— Material Resource Planning (MRP) ;
— Plant layout ;
— Quality control.

- Computer-Integrated Manufacturing (CIM) is the interfacing of all the various manufacturing modules into a unified system, networked to the company's other major systems (finance, distribution etc.). A fully operational CIM system is extremely complex because of its interfaces, and requires a very large investment.
- The *computer has also greatly assisted companies in the implementation of Flexible Manufacturing Systems (FMS) and Just-In-Time (JIT) inventory systems.*

Artificial Intelligence (AI). Artificial intelligence concentrates on emulating human reasoning in order to solve problems. AI also thus analyses the methodology of how a human being solves a problem and translates the thought process to the computer. The computer then approaches the human reasoning process to solve the problem in contrast to executing an ordered set of instructions.

Expert systems, neural networks and fuzzy logic are alternative within the artificial intelligence field.

19.11. INFORMATION TECHNOLOGY

19.11.1. General Aspects

Information technology *is defined as the system which collects and processes data and disseminates and spreads information.* It may also be called as a set of organised procedures that when executed, provides information to support the organisation.

Communication, which is a part of information technology, transmits messages through selected channels involving sender and receiver. Information technology is very useful to society as without it no section of society can flourish and prosper.

The *basic functions of information technology are* depicted in Fig. 19.4.

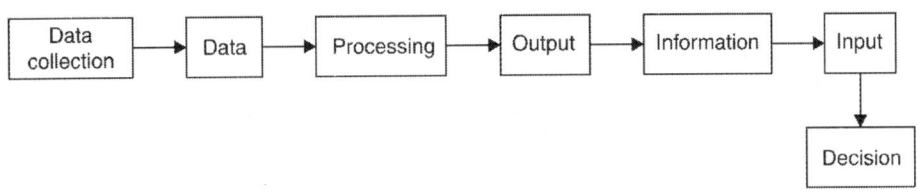

Fig. 19.4. The basic functions of information technology.

19.11.2. Data and Information

Data *is defined as facts which are generally recorded and filed.* It is the raw material which serves as the starting point.

Information *is defined as the processed data which is found handy and useful to arrive at making decisions.* It consists of data that have been retrieved, processed, or otherwise used for informative or inference purposes, argument or as a basis for forecasting and decision analysis.

19.11.3. Information System

The system which collects and processes data and disseminates or spreads information is called **Information system.** It may also be called a set of organised procedures that, when executed, provides informative to support the organisation. The basic functions of an information system are shown in Fig. 19.4.

An information system serves an individual with certain cognitive (faculty of knowing) style faced with a particular decision problem in some organisational setting.

Fig. 19.5 summarises all the variables that *influence the interpretation of an information.*

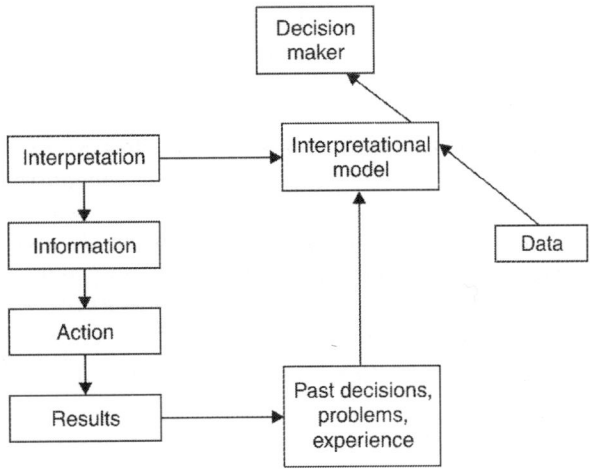

Fig. 19.5. Summary of variables that influence the interpretation of an information.

QUESTIONS WITH ANSWERS

Q. 19.1. Explain briefly the 'Hierarchy of computers in manufacturing'.

Ans. Computers and computer driven devices (*e.g.,* CNC machine tools, robots), in the manufacturing firm tend to form a *"pyramidal control structure"* as shown in Fig. 19.6. The structure represents the common structure that exists between the computers and computer driven devices in the factory.

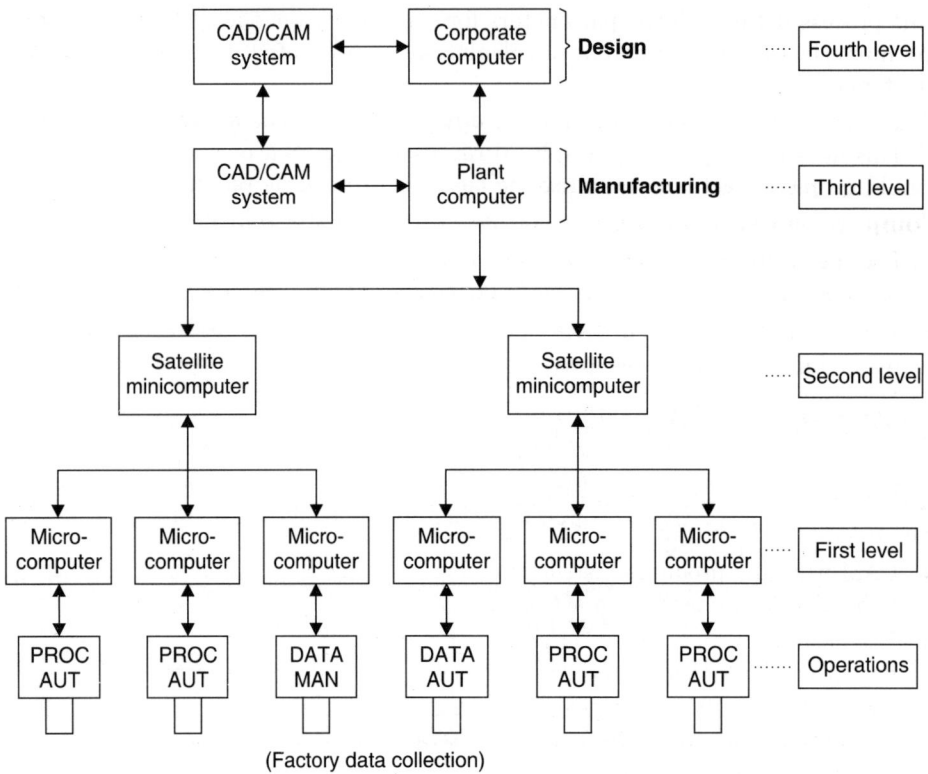

Fig. 19.6. Hierarchy of computers in manufacturing.

The various computers in the hierarchy are fixed together by communication links to a distributed computer system. The individual links provide a computer communications network to forward data and information through the various level from the manufacturing operations all the way to the corporate computer level. In the opposite direction, product design, process plans, production schedules, machine commands and so on are passed down to the individual production cell. *The hierarchy allows for design engineering and manufacturing engineering function to be included within the computer network.*

The following are the *benefits of the hierarchial approach to CIM* (Computer Integrated manufacture) :

(*i*) In the hierarchial configuration, the software development can be managed more easily since computers are separated with pyramidal arrangement, programming for each project can be handled separately.

(*ii*) The hierarchial computer can be installed gradually rather than all at once.

(*iii*) This structure contains redundancy. In the event of a computer breakdown, other computers in the system are programmed to assume the critical task of the broken-down computer.

Q. 19.2. What is the difference between Computer-Aided-Designed (CAD) and Computer-Aided-Drifting (CAD) ?

Ans. Computer-aided-design. It is used for the *initial design of the parts, sub-systems and the total system.*

This process defines all the parameters for construction of the system. Once this design is made, the numerical figures found out of calculations is presented in the form of a rough sketch to the draftsman.

Example : In the design of a beam, the dimensions are to be found out by doing a 'strength analysis'. This does not require any graphical representation. But once design is made, to convey it to the field people for actually building the beam, a *drawing* is required.

Computer-aided drafting :

● It comes into picture *only after the design is completed.* It is *basically used to generate the working drawing to be given to the field or the shop floor to help them in production.*

● Computer-aided-drafting simply provides an *electronic drafting board* to the draftsmen and a lot of electronic tools for creation, modification, storage, retrieval, and outputting of the drawings.

Q. 19.3. What is 'Computer Aided Manufacturing (CAM)' ? What are its applications ?

Ans. Computer Aided Manufacturing (CAM). *"Computer aided manufacturing is the use of computer to help in the manufacturing or production after the design process.*

● CAM basically produces instructions for the controller, a cutting or processing machine, to automatically perform a set of operations to produce the final product.

● The design outputs of a CAD application, can directly become the input to a CAM system and from the design, directly such a machine control program-called otherwise as *"Numerical Control Code"* can be generated.

— Different machine controllers are available with every type of NC numerical control machine. Each of these controllers support one or more of these languages ; ISO, APT, UNIAPT. The output of the CAM system can directly be input into these machines to perform the machining or processing required.

Applications of CAM. Following are the *applications* of CAM :

1. *Computer monitoring and control.* These are the *direct applications in which the computer is connected directly to the manufacturing process for the purpose of monitoring or controlling the process.*

2. *Manufacturing support applications.* These are the *indirect applications* in which computer is used in support of the *product operations* in the plant, but there is no direct interface between the computer and the manufacturing process.

● Computer monitoring and control can be separated into monitoring applications and control applications. *"Computer process monitoring" involves a direct computer interface with the manufacturing process and associated equipment and collecting data from process.* The computer is not used to control the operation directly. The *control* of the process remains in the hands of *human operators,* who may be guided by the information compiled by the computer.

Q. 19.4. What is CAD workstation ?

Ans. *The CAD workstation is the system interface with the outside world.* It represents a significant factor in determining how corvenient and efficient it is for a design to use the CAD system. The workstation must accomplish the following *five functions :*

1. To interface with the central processing unit.
2. To generate a steady graphic image for the user.
3. To provide digital descriptions of graphic image.
4. To translate computer commands into operating functions.
5. To facilitate communications between the user and the system.

Q. 19.5. List the various input and output devices used for graphics.

Ans. *Input devices for graphics :*

- Cursor pad or cursors
- Thumb wheels
- Light pens
- Mouse
- Digitizing tablet and table
- Track ball
- Control disc
- Function keys.

Output devices :

The output devices are essentially termed as peripherals associated with the computer. Some of the output devices do not require CRT (Cathode Ray Tube) or interactive display devices. The hard copy output is printed on a paper, magnetic tape, or slide film which can be read and visualised and transported from one place to another without any distortion. Graphics hard-copy devices may be categorised in the three major types namely, *pen plotters, electrostatic printers, plotters,* and *computer-output microfilm units.* There are some other types of graphics output equipments such as *photographic devices, ink-jet* and *laser plotters, and computer-controlled copies.*

Q. 19.6. Discuss briefly the steps in the 'Numerical Control (NC)' in manufacturing.

Ans. Following are the steps which must be accomplished to utilise NC in manufacturing :

1. Process planning.
2. Part programming.
3. Tape preparation.
4. Tape verification.
5. Production.

1. Processing planning :

- The engineering drawing of the work part must be interpreted in terms of the manufacturing processes to be used. This step is referred to as process planning and it is concerned with the preparation of a route sheet.
- The *route sheet* is a listing of the sequence of operations which must be performed on the work-part.

2. Part programming :

- A part programmer plans the process for the portions of the job to be accomplished by NC.

- There are two ways to program for NC :
  - (*i*) Manual part-programming.
  - (*ii*) Computer-assisted part programming.
    - — In *manual part programming,* the machining instructions are prepared on a form called a part *program manuscript.*
    - — In computer-assisted part programming much of the tedious computational work required in the manual part programming is *transferred to the computer.*

3. Tape preparation :

- A punched tape is prepared from the part programmer's NC process plan.
  - — In *manual part programming* the punched tape is prepared directly from the part program manuscript or a type writer like device equipped with tape punching capability.
  - — In *computer assisted part programming,* the computer interprets the list of part programming instructions, performs the necessary calculations to convert this into detailed set of machine tool motion commands, and then controls a tape punch device to prepare the tape for the specific NC machine.

4. Tape verification :

- After the prunched tape has been prepared, a method is usually provided for *checking the accuracy of the tape.*
  - — Sometimes the tape is checked by running it through a computer program which plots the various tool movements (or table movements) on paper. In this way, major error in the tape can be discovered.
  - — The "acid test" of the tape involves trying it out on the machine tool to make the part. A foam or plastic material is sometimes used for this layout.

Programming errors are not uncommon, and it may require about three attempts before the tape is correct and ready to use.

5. Production :

- The final step in the NC procedure is to use the NC tape in production. This involves ordering the raw workparts, specifying and preparing the tooling and any special fixturing that may be required, and setting up the NC machine tool for the job.
- The machine tool operator's function during production is to load the raw workpart in the machine and establish the starting position of the cutting tool relative to the workpiece. The *NC system* then takes over and *machines the part according to the instructions on the tape.*

Q. 19.7. Discuss briefly "Applications of Numerical Controls".

Ans. Now-a-days the NC systems are widely used in industry especially in the *metal working* industry. By far the most common applications of NC is for metal cutting machine tools. Within this category, NC equipment has been built to perform virtually the entire range of material removal processes including :

- — Milling,
- — Drilling and related processes,
- — Boring,
- — Turning,

— Grinding, and

— Sawing.

Following are the *general characteristics of production jobs in metal machining for which numerical control would be most appropriate* :

(*i*) The part geometry is complex.

(*ii*) The parts are processed frequently and in small lot sizes.

(*iii*) Many operations must be performed on the part in its processing.

(*iv*) Much metal needs to be removed.

(*v*) Engineering design changes are likely.

(*vi*) Close tolerances must be held on the workpart.

(*vii*) The parts require 100 percent inspections.

(*viii*) Expensive part where mistakes in processing would be costly.

Q. 19.8. Discuss briefly the steps involved in a computer aided engineering design problem.

Ans. CAD/CAM technology was initiated in the area space industry but is now widely used in all industries. It can be defined most simply as *use of computer to translate a product's specific requirements into the final physical products.* With this system a *product is designed, produced and inspected in one automatic process.*

The various steps involved in computer aided engineering design problems are shown in Fig. 19.7.

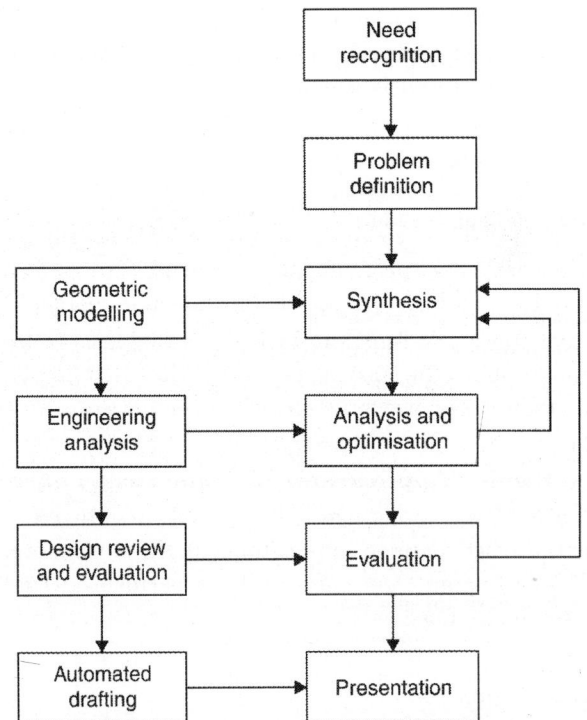

Fig. 19.7. Steps involved in computer aided engineering problems.

1. *Computer Integrated Manufacture (CIM)* is the term used to describe the complete and whole animation of the factory with all processes functioning under computer control and only digital information integrating them together.
2. *"Flexible manufacturing"* involves the interconnection of groups of automated machine tools by an integrated and automated material handling system.
3. *Capacity planning* is the method by which the master schedule is adjusted to balance the due dates of jobs or orders against the capacity of the plant and its individual work cells and facilities.
4. *Despatching* involves the movement of the parts, components, sub-assemblies, and end items so that they arrive at the appropriate work centre exactly at the time they are needed in the production process.
5. A *computer* is a machine that processes data according to set of instructions stored within the machine.
6. A *hybrid computer* is a combination of both analog and digital computers. It is used for simulation applications.
7. A *peripheral* is any device commonly used with a CPU of a computer for input or output of information or for memory functionally separate from the CPU and electronically detachable.
8. The *microprocessor* is a semiconductor device comping of electronic logic circuits manufactured by using either a large scale integration (LSI) or very-large scale integration (VLSI) technique.
9. A *computer network* is a number of computers interconnected by one or more transmission paths.
10. *Local Area Network (LAN)* provides short range interconnectivity between the devices in the network, normally within one site.
11. A network configuration is called network *topology*.
12. The *transmission line* is the message and data carrying medium that constitutes the physical distribution element of the network.

OBJECTIVE TYPE QUESTIONS

Fill in the blanks or Say 'Yes' or 'No'.

1. Numerical control reduces tooling cost by eliminating the costs of sophisticated jigs and fixtures.
2. .... manufacturing involves the interconnection of groups of automated machine tools by an integrated and automated material handling system.
3. Flexible manufacturing systems, (FMSs) are used to complete .... sequences of machining operations on families of manufactured parts.
4. .... eliminate repetitive, manual labour operations.
5. Fixed automation is characterised by extremely low costs associated with equipment procurement.
6. The process .... specifies that sequence of manufacturing operations that is required to produce a product or sub assembly.
7. The .... planning is the method by which the master schedule is adjusted to balance the due dates of jobs or orders against the capacity of the plant and its individual work cells and facilities.
8. The computer has been the .... of the information age.
9. A .... is an electronic machine for processing information.
10. .... are popularly known as personal computer (P.C.)
11. .... Computer is based on counting operation.
12. A hybrid computer is used for simulation applications.
13. The ALU combined with control unit is called .... processing unit.
14. Chips are rated in terms of their .... and ....
15. Keyboard is the most common and simplest .... device.
16. .... is a pointing device and its size is about the size of palm.

17. Digitizer is similar to light pen.
18. A .... is widely used for playing computer games.
19. .... is also known as core memory, main memory, RAM.
20. .... is a type of memory chip that one can read only and we cannot write on it.
21. ROM is slow.
22. The set of physical components, modules, and peripherals comprising a computer system is called ....
23. The .... is a set of programs required for data processing activities of the computer.
24. All persons concerned with computers, *i.e.*, compiler, programmer, etc. are called. ....
25. Attenuators are used to multiply a variable quantity by a constant.
26. The microprocessor is in many ways similar to the ....
27. A computer network is a number of computers interconnected by one or more transmission paths.
28. LAN provides .... range interconnectivity between the devices in the network, normally within one site.
29. A network configuration is called network ....
30. Coaxial cable is used in vast majority of LANs.

ANSWERS

1. Yes	2. Flexible	3. sophisticated	4. Robots	5. No
6. routing	7. Capacity	8. core	9. Computer	10. Microcomputer
11. Digital	12. Yes	13. central	14. capacity, speed	15. input
16. mouse	17. Yes	18. Joy-stick	19. Primary	20. ROM
21. Yes	22. hardware	23. software	24. liveware	25. Yes
26. CPU	27. Yes	28. short	29. topology	30. Yes.

THEORETICAL QUESTIONS

1. What do you mean by the term 'Computer Integrated Manufacture (CIM)' ?
2. Discuss briefly "Integration of design and manufacturing".
3. What do you understand by "Process selection" ? Explain
4. What do you mean by "Flexible manufacturing" ? Explain
5. Write short note on "Flexible Manufacturing Systems (FMSs)".
6. Explain briefly the following :
   (*i*) Process routing   (*ii*) Capacity planning
   (*iii*) Intra work cell scheduling methods.
7. Discuss briefly "Basic tools for integration".
8. Briefly discuss "History and development of computers".
9. What is a computer ? What are its characteristics ?
10. What are the limitations of a computer ?
11. How are computers classified ?
12. State the differences between 'Analog and Digital' computers.
13. Explain briefly the basis elements of digital computer with the help of a block diagram.
14. How are the chips rated ? Explain.
15. Discuss briefly 'Computer peripherals'.
16. Give the difference between ROM and RAM.
17. Explain briefly the terms : Hardware, Software and Liveware.

18. Discuss briefly the 'computer languages'.
19. Briefly discuss the 'computer programming process for writing programs'.
20. List the computing elements of analog computers.
21. What is a 'microprocessor' ? Where are its characteristics ?
22. What are the uses of microprocessors in instrumentation ?
23. What is computer network ?
24. What are the advantages of network ?
25. What is 'Local area network (LAN)' ?
26. Discuss briefly the more common network topologies.
27. What is Manufacturing Automation Protocol (MAP) ? Explain.
28. Explain briefly the following :
    (*i*) Systems software
    (*ii*) Applications software
    (*iii*) Artificial Intelligence (AI).
29. Explain briefly the term "Information technology".
30. What is an 'Information system' ? Explain.

20

Elements of Integration

20.1. Computer integrated manufacturing. 20.2. Mechanisation and automation. 20.3. Automation and production. 20.4. Mechatronics and concurrent engineering. 20.5. Industrial robots—Introduction—objectives of using industrial robots—advantages of employing robots—robot components—robot classification—control systems and components—types of industrial robots—terms related to the construction and operation of robots—robot programming and languages—intelligent robots—applications of robots. 20.6. Automated guided vehicle system (AGVS)—General aspects—AGVs equipment—applications of AGVs. 20.7. Automated storage systems—General aspects—objectives of automated storage systems—automatic storage/retrieval systems . **Questions with Answers**—Highlights—Objective Type Questions—Theoretical Questions.

20.1. COMPUTER INTEGRATED MANUFACTURING

The *Computer Integrated Manufacturing* (*CIM*) is the best context in which *industrial controls and sensors* can be understood. This concept, in some quarters, was formerly called *hierarchical control*. This concept is *applicable in both "manufacturing" and the "process" industries*. The structure is very similar, although different terms are used at certain levels, as discussed below.

Sensors and actuators. These devices provide the interface between the *controller and machine or process being controlled*. *"Sensors" translate some variables on the machine or process to a signal* (usually electrical) *that can be subsequently processed by the control equipment to obtain the current value of the variable of interest.*

Sensors can be either continuous or discrete.

Examples : 'Continuous'—Measurement devices for temperatures, pressures, flows etc.

'Discrete'—Proximity switches, limit switches, on/off level switches etc.

"Actuators" accept a signal from the control system and in some fashion impose this signal on the machine or process being controlled.

Examples : Continuous'—Variable speed drives, pneumatic positioning valves etc.

'Discrete'—On/off motors, solenoid valves etc.

Machine or regulatory control. Functions at this level provide the basic regulation of the machine or process being controlled.

● For a *machine,* regulatory control is provided by *controllers* such as *numerical controllers or programmable logic controllers.*

● For *processes*, regulatory control is provided by *single-loop or multi-loop microprocessor based controllers.*

Cell control or supervisory control :

- In *"manufacturing facilities"* several machines are often grouped to form a **cell**.
- In *"process plants"*, several unit operations are *interconnected* to form the process.

Plant control. At this level reside the various functions required for efficient operation of the entire plant. *Functions* at this level include :

— Keeping track of the raw materials and products ;

— Routing production activities throughout the plant ;

— Monitoring equipment utilisation etc.

- The computers at this level communicate with the cell or supervisory computer in a plant Local Area Network (LAN).

Corporate :

- Normally implemented as part of the Management Information System (MIS) capabilities, functions at this level are directed to the long term or strategic issues.
- An essential part of this function is the collection, analysis and presentation of information to corporate-level managerial personnel.

20.2. MECHANISATION AND AUTOMATION

The **mechanisation** (*the precursor to automation*) *can be defined in its simplest sense as the transfer of skills and manual activities to machine operations.*

The primary difference between mechanisation and automation includes *feedback for controlling an automated system.*

"Microautomation" differs from "macroautomation" in that the latter pertains to the low-level control system commonly envisioned for industrial automation.

1. "Microautomation" is primarily concerned with logical control focused on *individual machines* and the logical linkage between machines and devices. In particular microautomation pertains to *servomechanism, hydraulic and pneumatic devices* and associated low-level software used for physical movement of parts through a production system which typically results in "islands" of automation.

2. "Macroautomation" as the term indicates, has a larger scope and *deals with the coordination and supervision of various smaller scooped islands of automation.* This has led to various models and landings for communications and control of large-scale manufacturing (automated) system. Pertinent issues include refractive models and corresponding architectures for implementation of large scale systems.

20.3. AUTOMATION AND PRODUCTION

While classifying automated production systems, production volume and product variety need be considered. Such systems may be *classified* into three basic types :

1. *Fixed automation :*

- The fixed automation is characterised by having the sequence of operations necessary to manufacture or assemble a product fixed by the equipment configuration. As such, there is typical equipment which is inflexible to product changes.
- Its advantage is *high production rates.*

2. *Programmable automation :*

- Programmable automation equipment is highly reprogrammable to accommodate high product variety but has low production rates relative to fixed automation.

- Parts are typically loaded into programmable automated production system in batches. Each batch consists of a different part ; and the machines comprising the systems to manufacture the part are reprogrammed for each batch. Change-over from one batch to the next also requires a change in physical set-up of the machines tools, that is their fixing and tooling. Such change-over results in a loss of production time.

Example. Numerically Controlled (NC) machine tools and industrial robots.

3. *Flexible automation :*

- Conceptually, a flexible automated system has the *capability of producing a variety of parts with minimal changeover time from one part to the next.* The ability to change part programs and to change the physical set-up of the production system with little or no loss of production time is the primary difference between flexible automation and programmable automation.

- To accomplish this, one strategy has been to formulate flexible manufacturing class based on group technology principles. The overall objective is to increase productivity by properly designing a flexible system that offers advantages similar to fixed automation schemes of mass production but has capability for handling part variety.

20.4. MECHATRONICS AND CONCURRENT ENGINEERING

The *concept of "Mechatronics"* (originated by Japanese in the early 1970s) *is clearly related to the issues of robotics and flexible automation.*

Mechatronics has been described as *"the union of mechanical and electrical engineering needed in producing the next generation of machines, robots and smart mechanisms for applications such as manufacturing, large-scale construction and work in hazardous environment."* It is a multi-disciplinary approach using integrated terms of designers, manufacturing engineers, purchasing, marketing, and sales personnel to design both the product and the system to manufacture it.

Complementary to the term mechatronics would be **concurrent** (*simultaneous*) **engineering,** which is a term more prominent in U.S. industries.

The focus is being given on four key areas in regard to concurrent engineering. These areas are, in many ways, fundamental to the existing discipline of industrial engineering and its future development include :

— Intelligent computer-aided design ;

— Intelligent manufacturing systems ;

— Human-machine interface ;

— Maintenance.

The *function of "machatronics"* or in more general terms, *"concurrent engineering",* is *to provide a national framework for the integration and implementation of the various sub-systems which comprise the disciplines and organisational functions of an engineering and manufacturing facility.*

20.5. INDUSTRIAL ROBOTS

20.5.1. Introduction

According to the Robotics Industries Association (RIA) an industrial robot is *defined* as under :

"An **industrial robot** *is a reprogrammable, multi-functional manipulator designed to move materials, part tools, or special devices through variable programmed motions for the performance of a variety of tasks."*

This definition is compatible with the classification of programmable automation, but robots are also used in flexible automation and fixed automation systems, *Example*. An automation line using many robots to perform spot welds in which robot programs are downloaded from a computer to a specific robot controller could be considered a high-production flexible automation system.

20.5.2. Objectives of Using Industrial Robots

The use of *Industrial Robots* is increasing day-by-day with a view to achieve the following *main objectives* :

1. To reduce production time.
2. To minimise the labour requirement.
3. To raise the quality level of products.
4. To increase productivity.
5. To improve existing manufacturing processes.
6. To enhance the life of production machines.
7. To minimise the loss of man-hours on account of accidents and diseases.
8. To make the reliability and applicability of new high-speed production processes and their related machinery possible.
9. To take advantage of fatigue-free continuous deployment of robots, because the human beings are always bound to experience fatigue when put to continuous working.

20.5.3. Advantages of Employing Robots

Followings are the *advantages of employing robots* :

1. Lifting and moving heavy objects.
2. Working in hostile environments.
3. Providing repeatability and consistency.
4. Working during unfavourable hours.
5. Performing dull or monotonous jobs.

20.5.4. Robot Components

The *main components of a robot* are as follows :

1. Base :
- The base may be fixed on mobile.

2. Manipulator arm :
- This arm is provided with a number or degrees of freedom of movement.

3. Gripper or end effector :
- It is used for holding a piece or a tool, depending upon the application of the robots.

4. Drives :
- The drives are also known as "*actuators*".
- They move the manipulator arm and end deflector to the required position in space.

5. Controller :
- It delivers commands to the actuators with the help of hardware and software support.

6. Sensors :
- They perform the following *functions* :
  - (*i*) To act as feedback devices to direct further actions of the manipulator arm and the end effector (gripper), and
  - (*ii*) To interact with the robot's working environment.

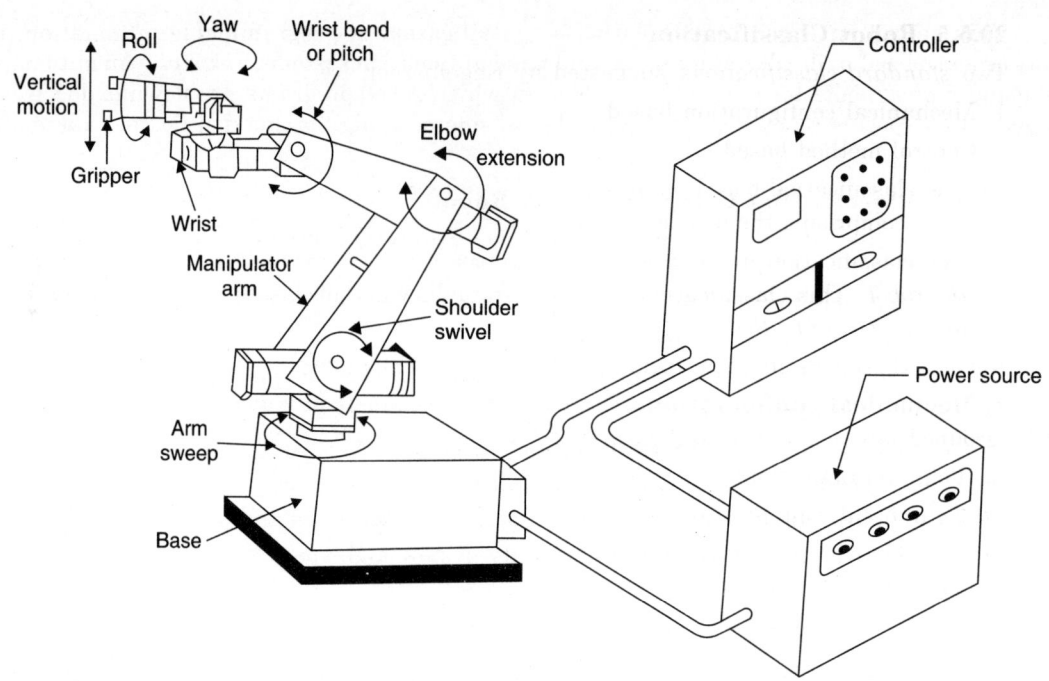

Fig. 20.1. Main components of a robot and the basic motions.

Power source :

The power source is the source of energy used to move and *regulate the robot's drive mechanisms*. The energy comes from three sources : *Electric, hydraulic* and *pneumatic*.

- *Electric drives* are *clean and quite* with a *high degree of accuracy* and repeatability. They also offer a *wide range of payload capacity,* accompanied by an equally wide *range of costs*.

- *Hydraulic drives* today's most popular, have *high payload capacities* and are relatively easy to maintain. They are, however, rather *expensive* and *not as accurate as either the electric or pneumatic drives*.

- *Pneumatic drives* although limited to *smaller payloads,* are *relatively inexpensive, fast and reliable*.

Basic Motions :

The *six basic motions* or degrees of freedom are as follows (see Fig. 20.1).

1. *Vertical motion.* The entire manipulator arm can be moved up and down vertically either by means of the *shoulder swivel, i.e.,* turning it about a horizontal axis, or by sliding it in a vertical slide.

2. *Radial motion.* Radial movement, *i.e., in* and *out* movements, to the manipulator arm is provided by *Elbow extension* by extending it and drawing back.

3. *Rotational motion.* Clockwise or anticlockwise rotation about the vertical axis to the manipulator arm is provided through *arm sweep.*

4. *Pitch motion.* It enables up and down movement of the wrist and involves rotational movement as well. It is also known as *wrist bend.*

5. *Roll motion.* Also known as *wrist swivel,* it enables rotation of wrist.

6. *Yaw.* Also called *wrist yaw,* it facilitates rightward or leftward swivelling movement of the wrist.

20.5.5. Robot Classification

Two *standard classifications* suggested by Engelberger are :

1. Mechanical configuration based.

2. Control method based.

— The classification based on *'mechanical' representation* considers the *various joints and links* comprising the physical structure of the robot and their relationship to each other.

— The classification by *'control'* pertains to the type of *technique implemented to control the robot.* This classification considers the following subclasses : *Non-servo controlled and servo-controlled.*

Each of these classification techniques is discussed as under :

1. Mechanical configuration. The majority of commercially available industrial robots can be grouped into *four basic configurations :*

(*i*) Polar configuration.

(*ii*) Cylindrical configuration.

(*iii*) Cartesian coordinate configuration.

(*iv*) Jointed-arm configuration.

(*i*) *Polar configuration.* Refer to Fig. 20.2.

— This configuration has a telescopic arm which pivots about a horizontal axis and also rotates about a vertical axis.

— The work volume, a term referring to the space within which the robot can manipulate the wrist end, thus defines a spherical volume in which the robot can perform its task.

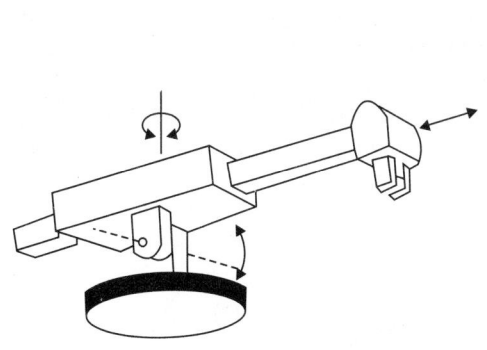
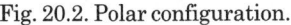
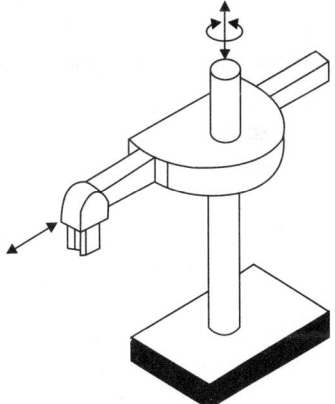

Fig. 20.2. Polar configuration. Fig. 20.3. Cylindrical configuration.

(*ii*) *Cylindrical configuration.* Refer to Fig. 20.3.

— Cylindrical configured robots use a vertical column with the robot arm attached to a side which can move up and down the column.

— Simultaneously, the arm can move radially with respect to the column.

(*iii*) *Certesian coordinate configuration.* Refer to Fig. 20.4.

— The cartesian or rectilinear robot also termed as *gentry robot*, has three mutually perpendicular axes which define a rectangular work volume.

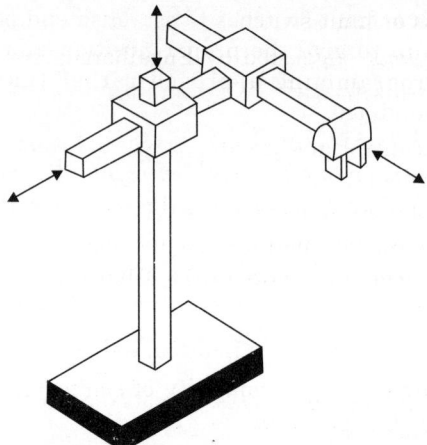

Fig. 20.4. Cartesian configuration.

(iv) Jointed-arm configuration. Refer to Fig. 20.5.

— The jointed-arm robot most resembles an human arm and consists of a series of links connected by rotary joints which when referenced from the base are referred to as the shoulder, arm and wrist joints.

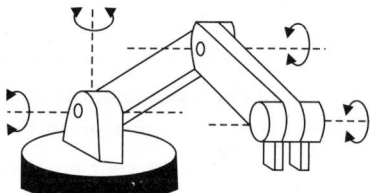

— Regardless of configuration, the function of the robot arm configuration is to position a wrist assembly which orients an end effector to the proper position and orientation (jointly referred to as the POSE) dictated by the task at hand.

Fig. 20.5. Jointed-arm configuration.

Fig. 20.6 shows the degree of freedom associated with a robot wrist.

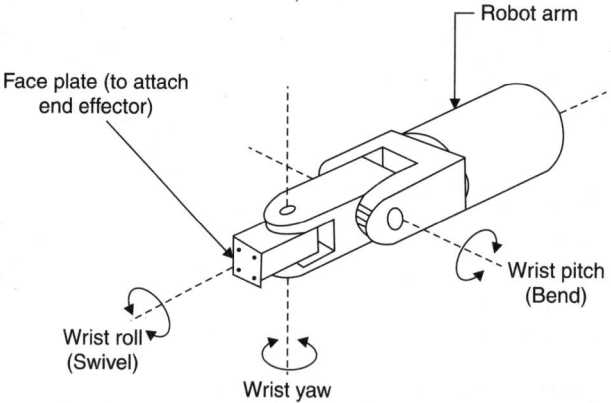

Fig. 20.6. Degrees of freedom associated with a robot wrist.

2. Control method. Several techniques have been developed to control the various axes of a robot simultaneously.

● The simplest type of control is the ***"non-servo controlled"*** or *"limited-sequence robot"*.

   — Non-servo-controlled robots are also called *limited-sequence robots, end-point robots, pick and place robots* or *bang-bang robots*. Such a robot is controlled by setting mechanical stops or limit switches to establish end points of travel of each joint. The mechanical set-up to give the proper position and sequence of stops serves as a rudimentary programming approach rather than a computer-intensive robot programming language.

- For more complex control and greater flexibility, current industrial practice employs *servo-controlled* robots.
  — The servo control of an industrial robot is accomplished by comparing feedback information to the command input such that a desired trajectory will be followed, that is, a *closed-loop system*. Feedback information on position, velocity, or other physical variables is provided by continuously monitoring the variables of interest. Several sophisticated foot arm control techniques and algorithms have been developed for robot controllers.
  — Servo-controlled robots can be *classified* as playback robots with *Point-To-Point* (PTP) control or playback robots with *Continuous Path* (CP) control.

Mechanical inaccuracies result from the various errors that can occur in a robots's construction or operation. These include *gear backlash, stretching of pulley cords, leakage of fluids* or *material and structural imperfections*. Such errors degrade the overall accuracy of the robot and are magnified when the arm is fully extended. Mechanical inaccuracies are principally responsible for repeatability errors. A robot manufacturer typically specifies repeatability as the radius of an idealised sphere and expresses the specification as a plus or minus value. The majority of robots have extremely high repeatability (can be ± 0.0025 mm or less) but many of these robots can have comparatively greater inaccuracy values.

20.5.6. Control System and Components

A brief overview of the control structure actuation devices, grippers and types of sensors that may be used with an industrial robot is given below :

1. Controller :

- Its function is to control the manipulator as programmed by the user to perform the prescribed task.
- A robot controller not only can be used to control the robot itself but through interfacing with various other equipment can function as a work-cell controller.

2. Actuators :

- Actuators which provide the power to move the joints of a robot are typically *pneumatic, hydraulic or electrical devices.*
  — *D.C. servomotors provide excellent controllability with minimum maintenance requirements.* Torque control is possible by controlling the voltage or current, respectively, applied to the motor. D.C. motors using permanent magnets to generate the magnetic field are a common method of activating a robot joint. *Advantages* offered by such motors include *high stall torque, a small frame size and light-weight and a linear-speed curve which reduces computational effort.*

3. End effectors :

The end effector is considered as special-purpose tooling and is typically customs engineered. A wide variety of end effectors exist to perform different work functions. The various types can be classified into two major groups :

  (*i*) Grippers.

  (*ii*) Tools.

Grippers are used to *grasp and hold objects* whereas *tools* can be used to *perform work on a part rather than merely grasp it.*

- *"Grippers"* consist of mechanical devices, magnets, suction devices, adhesives, or other devices to grasp and hold objects. These can be classified as *single* or multiple grippers depending on the number of grouping members attached.

Examples :

— Standard hand ;

— Cam operated hand with inside and outside jaws ;

— Special hand for cartons ;

— Double hand.

4. Sensors :

- Sensors are fundamental in the use of robots and other automated systems. The primary use of sensors can be categorised as follows :

1. Part inspection for quality control.

2. Safety monitoring.

3. Work-cell control interlocks.

4. Position and related information on parts in a work cell.

- Sensor use in robotics and automated systems includes a wide range of devices. These devices can be divided into the following categories :

(*i*) Proximity and range sensors.

(*ii*) Tactile sensors.

(*iii*) Machine vision.

(*iv*) Miscellaneous sensors and sensor-based systems.

20.5.7. Types of Industrial Robots

Industrial robots can be broadly *classified* into two main groups as follows :

1. General purpose robots.

2. Special purpose robots.

1. *General purpose robots :*

- These robots carry standard designs and parts and are readily available.

- They can be easily adapted to the users' requirements by attaching suitable end effectors or fingers to them according to the requirement of the work, such as a part picking operation, welding operation, spray painting, etc.

- Since such robots are mass produced, they are cheaper.

2. *Special purpose robots :*

- These robots are tailor made to specific job requirements. The ultimate user has to feed his requirements and, based on them, these robots are specially designed and built to cater to such specific needs. Obviously, their designing and manufacturing consumes a lot of time. As such, they cannot be readily available in market.

- Since they cannot be manufactured on mass scale, their *prices* are bound to be *higher*.

20.5.8. Terms Related to the Construction and Operation of Robots

The following important terms are commonly related to the construction and operation of robots :

1. Work envelope.

2. Speed of movement.

3. Load carrying capacity.

4. Precision of movement.

5. Drive systems.

1. Work envelope :

- The work envelope is also known as *"work volume"* of a robot. It represents the volume of space around the manipulator arm of a robot, within which it can operate. Since the arm movements in different robot configurations are different ; the *work volumes* or *work envelopes* of different coordinate systems are also different.

Example : The work envelopes of cartesian, polar, cylindrical and revolute coordinate system robots respectively will be rectangular, partially spherical, cylindrical and non-uniform or irregular.

2. Speed of movement :

- *"Speed of movement" is the speed at which a robot is capable of manipulating its end effector.* It is governed by the distance to be moved, weight of the part to be moved and the accuracy required in placement of the part in position.

- Heavier parts and higher placement accuracy demands slower movement while the higher parts can be moved at faster speeds.

3. Load carrying capacity :

- The robot's capacity to carry load (weight) varies according to its structure.

- While this capacity for very light models can be as low as 1.5 kg (including the weight of end effector), the heavier class of robots can have their capacities as high as 1000 kg.

4. Precision of movement :

- *"Precision of movement"* is defined *as the degree of precision with which a robot is capable of moving the end point of the wrist of its manipulator arm.*

5. Drive systems :

In majority of robots, *pneumatic, hydraulic* and *electrical drive systems* are used.

- *"Pneumatic drives"* are generally used for *lighter class and simpler type of robots,* *"hydraulic drives,"* for heavier class of robots and *electrical drives* for medium and heavier varieties in which better accuracy and repeatability are used.

20.5.9. Robot programming and languages

In order to programme the work cycle of a robot the following *four methods* are used :

1. Manual programming method.

2. Walk through programming method.

3. Teach pendant method.

4. Off-line programming method.

1. *Manual method :*

- Various controlling components of a robot, like cams, stops, switches, control relays etc., are set through this method.

- This method can be employed only for such robots (with relatively short work cycles) which have to perform simple tasks like *pick and place, loading and unloading* etc.

2. *Walk through programming method :*

- In this method, the arm and hand of the robot are initially moved manually and these movements are stored in computer memory for being followed driving further operations.

- This method suites well for robots used in *arc welding* and *spray painting operations.*

3. *Teach pendant method :*

- It is also known as *'Lead through method'.*
- The 'Teach pendant' is a small control device held in hand. It is used to switch-on power drive for various robot movements in predetermined sequence of motions. The initial robot motion in this case (just like 'walk through method') is also recorded in computer memory and played back during further operations.
- It is *quite an easy and convenient method to use.*

4. *Off-line programming method :*

- Such a programme is separately prepared and fed into the computer memory. Then it is readily available for use whenever the operation is to be performed. Since it is separately prepared, it can be prepared simultaneously while the robot is operating on some other task and, therefore, a lot of time can be saved and the robot can be utilised more.
- This type of robot can be connected to a central CAD/CAM data base system.

Programming languages :

- For first three types of the above methods, no specific programming languages are required.
- Several different languages have been developed through which different robot motions can be effectively controlled. Some of these languages are modifications of the existing common computer languages, while many others are completely new.

Some commonly used computer languages are :

> AL ; AML ; VAL ; RAIL ; RPL ; MCL ; HELP.

20.5.10. Intelligent Robots

- A continuous effort is being made to develop control algorithms for reference model hierarchies. Even more intensive efforts have been applied to the development of intelligent controls at the lower levels of the hierarchy, particularly the equipment level. This has lead to concept of intelligent machines or robots. Fundamental issues that need to be addressed include not *only the control structure* but also *sensor strategies, multi-sensor integration or fusion, learning and decision making, programming, mobility and navigation, trajectory planning and overall systems integration.*
- Intelligent robots, as with intelligent factory automation, will provide an abundance of problems and opportunities for industrial engineers. It is anticipated that there will be heightened activity in both theoretical and practical developments in the issues pertaining to the technology and management of such technologies with automated systems.

20.5.11. Applications of Robots

Some common areas of applications of robots are the following :

1. Welding :

- Mostly *spot welding* and *arc welding* in automobile industries.

2. Spray painting :

Robots are used for *spray painting* of automobile bodies. Use of robots for this provides the following *advantages :*

- Higher productivity.
- Substantial saving in consumption of paint.
- Saves human operators from likely health hazards due to toxic fumes and mist, noise, fire etc.
- Consistency of paint layer over the entire surface.
- Saving in energy consumed.

3. Other processing operations

Besides welding and spray paintings, applications of robots are found in number of other processing operations like :

— Polishing ;

— Wire brushing ;

— Riveting ;

— Heat treatment etc.

4. Machine loading and unloading :

Robots are commonly used for loading of stock parts and unloading of finished parts on :

— Die casting machines ;	— CNC machine tools ;
— Injection and transfer plastic moulding machines ;	
— Forging presses and hammers ;	— Stamping and punch presses ;
— Transfer machines ;	— Testing machines, etc.

5. Material handling and transfer :

Robots are commonly used for shifting an object from one location to the other. The prominent *areas of application* for this purpose are :

● Transfer of parts from one conveyor to the other.

● Palletising and depalletising.

● Transfer of blanks from an incoming conveyor to the machine tool for further processing.

6. Assembly operations :

Robots find applications in assembly areas involving :

— Screwing of studs and screws in threaded holes ;

— Screwing and unscrewing of nuts ;

— Insertion of shafts in holes ;

— Insertion of electronic components in electronic assemblies ;

— Assemblies of small electric motors, plugs, switches, etc.

● So far maximum potential for robot application is observed only in *back type assembly work.*

7. Sorting of parts :

Robots are used to sort out correct parts from a lot of finished parts and place them in proper locations in respective bin.

8. Inspection of parts :

Robots are finding applications, on a fairly good scale, for inspection of finished workpieces and subassemblies, specially of electronic components and devices.

20.6. AUTOMATED GUIDED VEHICLE SYSTEM (AGVS)

20.6.1. General Aspects

● In a Flexible Manufacturing System (FMS), Automated Guided Vehicles (AGVs) is one of the widely used type of material handling device.

● AGVs *are battery-powered vehicles that can move and transfer materials by following prescribed paths around the manufacturing floor.* They are neither physically tied to the production line nor driven by an operator like a fork lift.

● Such vehicles have *on-board controllers* that can be programmed for complicated and varying routes as well as load and unload operations. The computer for the material handling system or the central computer provides overall control functions such as dispatching, routing, traffic control and collision avoidance.

- AGVs usually *complement* in automated production line consisting of conveyor or transfer systems *by providing the flexibility of complex and programmable movements around the manufacturing shop.*

- The first automated guided vehicle was developed in 1954 by A.M. Barrett.

20.6.2. AGVs Equipment

Several types of equipment, *e.g.,* conveyors, rollers, self-powered monorails, carts, forklift trucks and various mechanical, electromagnetic, pneumatic, and hydraulic-devices and *manipulators* can be used to move materials.

- *"Manipulators"* are designed to be controlled directly by the operator, or they are automated for repeated operations, such as loading and unloading parts from machine tools, presses and furnaces. The manipulators are capable of *gripping and moving heavy parts and operating them as required between manufacturing and assembly operations.*

- Machines are often used in a sequence where workpieces are transferred directly from machine to machine.

20.6.3. Applications of AGVs

- In Flexible Manufacturing System (FMS), industrial robots, specially designed pallets and *Automated Guided* Vehicles (AGVs) are extensively used.

- *AVGs,* which are the latest development in material movement in plants, *operate automatically along pathways with in-flow wiring or tapes for optical scanning and without any operator intervention.*

  — This transport system has *great flexibility* and is capable of *random delivery to different workstations.*

  — It *optimises the movement of material and parts in case of cogestion around workstations, machine breakdown (downtime) or failure of part of the system.*

- The movements of the AGVs are planned so that they interface with *Automated Storage/Retrieval* Systems (AS/RS), which utilise workhouse spaces efficiently and reduce labour costs.

20.7. AUTOMATED STORAGE SYSTEMS

20.7.1. General Aspects

Industrial business firms are presently using traditional (non-automated) systems according to the personnel managing and budgetary considerations available to them. These storage systems include :

    (*i*) Bulk storage system.   (*ii*) Rack storage system.

   (*iii*) Shelves and bins.   (*iv*) Drawer storage system etc.

- The above systems require human labour to access for storage (which is *static*).

- In *highly 'automated system' loads are entered and retrieved by computer controls without any human participation except for input or control.*

- Automated systems have one *disadvantage* that they require *large investment and change in style of working.*

20.7.2. Objectives of Automated Storage Systems

The automated storage systems have the following *objectives* to serve :

1. To improve stock rotation.

2. To improve security.

3. To increase capacity to store.

4. To increase storage density.

5. To recover factory or warehouse floor space which is used for storing work in process.

6. To reduce pilferage and provide safety to stores.

7. To reduce labour costs.

8. To improve customer service.

20.7.3. Automatic Storage/Retrieval Systems

- This type of storage system allows the user to do store and retrieval operations with *speed and accuracy.*

- An AS/RS system consists of one or more aisles and each is serviced by storage retrieval machine. The aisles have storage racks for holding the stored materials.

  — The retrieval machine can be crane or any other type.

  — Each aisle of AS/RS is provided with one or more input/output stations (also referred as pick up/delivery station) where material is delivered for storage and picked up. These input/output stations are computer controlled or manually operated or can be interfaced with any automated handling system.

Types of AS/R systems. Various types of AS/R system are enumerated and discussed as under :

(*i*) Unit load AS/R system   (*ii*) Deeplan AS/R system

(*iii*) Miniload AS/R system   (*iv*) Man on board AS/R system

(*v*) Automated item retrieval system   (*vi*) Vertical lift storage/retrieval system.

(*i*) **Unit load AS/R system :**

- It is typically a large automated system to handle unit load stored on pallets or in containers.

- The system is computer controlled.

(*ii*) **Deeplan AS/R system :**

- It is high density unit load storage system which is used when large quantities of stock is to be stored but number of separate stock racks is small.

- This system stores ten or more loads in a single rack, are behind the next. Each rack is designed for flow with input on one side and output on another side.

(*iii*) **Miniload AS/R system :**

- In this storage system small loads are handled that are contained in bins or drawers.

- The S/R is specially designed to retrieve the bin and deliver it to output station at the end of aisles, so that individual item can be withdrawn from the bins. The bin or drawer must be retrieved back to its position or location.

(*iv*) **Man on board AS/R system :**

- In this system, a human operator, rides on the carriage of the machine (S/R) while the miniload system delivers an entire bin to the end of aisles pick station. The bin or container is then returned to its position in compartment.

- The man on board permits individual items to be picked directly at their storage locations.

(*v*) **Automated item retrieval system :**

- In this system, items are stored in lanes than bin or drawers.

- Any item to be retrieved is pushed from its lane and dropped into a conveyor for delivery.

(vi) Vertical lift storage/retrieved system :
- In this system a centre aisle is used to access loads in vertical position.
- Vertical lift, storage modules with heights of 10 m or more are capable of holding large inventories and help in saving floor space.

Applications of AS/R system :

Following are the applications of AS/R system :

1. Warehouse and distribution systems.

2. Storage of raw materials.

3. Work-in-process manufacturing :
   — Computer control and tracking of materials ;
   — Buffer storage in production ;
   — Support in JIT ;
   — Knitting of parts for assembly.

QUESTIONS WITH ANSWERS

Q. 20.1. In what ways a robot differ from a NC machine tool ?

Ans. A robot differs from a machine tools in the following respects :

1. As compared to a NC machine tool, a *robot is a lighter and more portable equipment.*

2. *Programming* of a robot is different from the *tool programming* used in NC machine tools.

3. The applications of robots are more general in nature as compared to a NC machine tool.

Q. 20.2. Give the companion between 'Automated machine' and 'Robot'.

Ans. The comparison between 'Automated machine' and 'Robot' is as follows :

S.No.	Aspects	Automated machine	Robot
1.	Type of job	It performs a particular job only in the designed sequence.	In can be made to do different jobs at different times and in different sequences.
2.	Programming	It cannot be programmed.	It can be programmed to change the sequence of tasks.
3.	Knowledge base, intelligence	It has neither a knowledge base nor intelligence.	It is possible to impart intelligence base by suitable programming technique.

Q. 20.3. Discuss briefly the 'motion systems' used in industrial robots.

Ans. The following two types of *motion systems* are used in *industrial robots :*

1. Point-to-point system.

2. Continuous path system.

1. Point-to-Point system :

- In this system, only the movement of robot from one point location to other location in space is controlled with no regard to the path followed by it in doing so.
- Majority of the robots employed in production activities, like *pick and place, loading and unloading of parts* etc., work on Point-to-point motion system.

2. Continuous path system :

- In this system a robot follows a predetermined *path* or *contour* in reaching from one point to the other.
- These robots need higher level of memory and control as compared to Point-to-point robots.
- Such robots are widely required in several operations like *continuous welding, spray painting* etc.

Q. 20.4. Discuss briefly "Robot sensing and sensors".

Ans. In order to impart more and more artificial intelligence to a robot, in order to bring its operation nearer to that of a human being, efforts continue to be made for improving its *sensing abilities, i.e.,* its vision, hearing, feeling by touching other objects, vision and coordination between its hand and eye. These objectives are achieved by using different types of *sensors*. These sensors may be of :

— Digital type or analog type ;

— Contact type or non-contact type ;

— Visual type or non-visual type, etc.

Sensor may be *Internal State Sensors* and *External State Sensors*.

- ***Internal State Sensors.*** These are used for *measuring position, velocity and acceleration of the end effector or the joints of a robot.* This class of sensors includes devices like :

— Linear inductive scales ;

— Synchronous ;

— Resolvers ;

— Potentiometers ;

— LVDTs, RVDTs ;

— Optical encoders ;

— Tachometers etc.

- ***External State Sensors.*** These sensors *determine the relationship of the robot with its environment and the objects handled* by it. Some of the devices used for this purpose include :

— Proximity devices ;

— Electromagnetic sensors ;

— Strain gauges ;

— Ultraviolet sensors ;

— Pressure transducers, etc.

All these are *non-visual* sensors and enable sensing of force, torque, obstruction, distance from an object, touch, slip, etc.

The *visual external state sensors includes* a video camera with a programmed computer and a source of light. With the help of this sensing unit a robot is able to sense the presence, position and orientation of an object in the line of its vision within its work volume.

Q. 20.5. Discuss briefly "Workcell and Interlocks".

Ans. Workcell :

- The term ***workcell*** represents a collection of automated equipment and controls that *acts as a means to coordinate all the activities within the robot workstation.* It is required because the robot system by itself has a very limited utility. In order to utilise it for performing some really useful tasks, it has to be integrated with many other things like work parts, part feeders, conveyors, control devices, process machinery, tools, machine tools etc.

- A work cell need not essentially cover a single robot only, but may contain several robots. In a robot *"work station"*, some of the activities take place in a sequence and some others occur simultaneously. A controlling device, called a *"workcell controller" or "workstation controller"*, is, therefore suitably incorporated, which coordinates and regulates these different activities and ensures that they are performed in a proper sequence.

Interlocks :

- In *workcell control,* there has to be some means to check the continuation of sequence of the work cycle whenever needed. The means used to apply this check is known as an *"Interlock".*

- The workcell carries *two types facilitates of interlocks* – one *"outgoing"* and the other *"incoming"*. The provision of interlocks facilitates an effective control over the workcycle by permitting the sequential activities to take place only when they should and preventing them when they should not.

Q. 20.6. What are the advantages of using 'Automated Guided Vehicle System (AGVS)' in 'Flexible Manufacturing System (FMS)' ?

Ans.   Following are the *advantages* of using AGVS in FMS :

1. Flexibility.

2. Real time monitoring and control.

3. Safety.

1. Flexibility :

- The route of AGVs can be easily altered and modified, simply by changing the guide path of the vehicles. This is more *cost effective* than modifying the fixed conveyor line or rail guided vehicles.

- It provides a *direct access* material handling system for loading and unloading FMS cells and accessing the automated storage and retrieval system.

2. Real time monitoring and control :

— AGVs can be monitored in real time because of the computer control. If the FMS centre system decides to change the schedule, the vehicles can be re-routed and urgent requests can be served.

— *AGVs are usually controlled through wires implanted on the factory floor using a variable frequency approach.*

3. Safety :

- AGVs can travel at a slow speed but typically operate in the range 10 to 70 m/min. They have *on-board microprocessor control* to communicate with the local zone controllers which direct the traffic and prevent the collision between vehicles as well as the vehicle and other objects.

- AGVs may also incorporate :

— Warning lights ;

— Fire safety interlocks ;

— Controls for safety in shops.

Q. 20.7. Explain briefly how you can achieve vehicle guidance, routing and traffic control in a manufacturing plant employing AGVs.

Ans. The conventional conveyor systems are sometimes inadequate to fulfil the requirements of plants where production is carried out through interconnected work cells and where flexibility and rapid change-over times are of chief importance. Automated Guided Vehicles (AGVs) offer a variable solution for such needs.

- **AGVs** *possess intrinsic flexibility and capability of synchronous operations and easiness of integration with other automatic devices like robots, CNC, automatic storage/retrieval systems.*

- AGVs are designed for *automatic, efficient and flexible execution of transport tasks* in which a variety of loads (workpieces, fixtures, empty pallets, tool dispensers) have to be transported at regular intervals with varying frequencies and from a fairly large number of locations (input/output station set-up areas, machining centres and auxiliary machines, wait stations and storage areas).

HIGHLIGHTS

1. *Sensors* translate some variable on the machine or process to a signal (usually electrical) that can be subsequently processed by the control equipment to obtain the current value of the variable of interest.
2. *Actuators* accept a signal from the control system and in some fashion impose this signal on the machine or process being controlled.
3. The *mechanisation* is the transfer of skills and manual activities to machine operations.
4. *Mechatronics* has been described as "the union of mechanical and electronic engineering needed in producing the next generation of machines, robots and smart mechanisms for applications such as manufacturing, large-scale construction and work in hazardous equipment".
5. An "*Industrial robot*" is a reprogrammable multi-functional manipulator designed to move materials, part tools, or special devices through variable programmed motions for the performance of a variety of tasks.
6. Industrial robots are of two types :
   (*i*) General purpose robots.
   (*ii*) Special purpose robots.
7. The following four methods are used to programme the work cycle of a robot :
   (*i*) Manual programming method.
   (*ii*) Walk through programming method.
   (*iii*) Teach pendant method.
   (*iv*) Off-line programming method.
8. *Automated Guided Vehicles (AGVs)* are battery-powered vehicles that can move and transfer materials by following prescribed paths around the manufacturing floor.

OBJECTIVE TYPE QUESTIONS

Fill in the blanks, say 'Yes' or 'No' :

1. It is said that CIM has .... percent computer technology and .... percent is the business process and people.
2. Automated production systems can be put into three groups : .... automation, programmable automation, flexible automation.
3. .... automation is most suitable for mass/flow production.
4. In programmable automation there is a high flexibility to deal the changes in product configuration provided the differences are not dramatic.

5. .... automation is basically an improvement upon programmable automation.

6. Flexible automation is not suitable for planned work-load.

7. Flexible automation can produce various combinations and schedules of products and does not require batch production.

8. .... programming is concerned with the planning and documentation of the sequence of processing steps to be performed on a numerical control machine.

9. Automatically Programmed Tools (APT) is a programming language which allows geometrical data to be specified together with tool motion statements for any NC material processing machine.

10. An industrial .... may be defined as 'a reprogrammable multi-functional manipulator designed to move material, parts, tools or other specialised devices through variable programmed motion for the performance of a variety of task'.

11. Basically the robot needs .... axes of motion (or degrees of freedom) to reach a point with a specific 'orientation' in the space.

12. An industrial robot has three types of components : Physical parts or anatomy, built in instruction or instinct (placed there by the manufacturer) and learned behaviour or task programmes (on-the-job training).

13. The .... is the part of the robot that physically performs the task.

14. The .... is the brain of the robot.

15. Robot .... is concerned with the physical construction of the body, arm and wrist of the machine.

16. The polar configuration robot uses a telescopic arm that can be raised or lowered about a horizontal pivot.

17. The .... configuration robot uses a vertical column and slide that can be moved up or down along the column.

18. The cartesian coordinate robot uses three perpendicular slides to construct the x, y and z axes.

19. The cartesian configuration robot can reach out only in front of itself.

20. The joints used in the design of industrial robots can be classified in two categories : Linear joints and .... joints.

21. Work envelope refers to the space within which the robot can manipulate its wrist end.

22. The .... system of a robot is used to power a robot and moves its body, arm and wrist.

23. Limited-sequence robots make use of servo-controlled system to indicate relative positions of the joints.

24. Playback robots with point-to-point control use a more sophisticated system where in a series of positions or motions are 'taught' to the robot, recorded into system memory, and repeated by the robot under its control.

25. Robots with continuous path capability are used for works such as spray painting, arc welding etc.

26. .... robots are generally recommended for simple applications such as pick-and-place operations.

27. .... robots constitute a growing class of industrial robots that possesses the capability to interact with the working environment in a way that seems intelligent.

28. The speed of response and .... are two important characteristics related with the dynamic performance of the robot.

29. One of the important aspects of the robot's performance is the .... of its movements.

30. .... of a robot refers to its ability to position its wrist end at a desired target position within the work volume.

31. .... defines Robot's capability to position its wrist or end effector at a point in space that had previously been taught to the robot.

32. The robot cannot be used in production while it is being programmed.

33. Lead through programming is compatible with modern CAD/CAM and other information processing systems.

34. Lead through programming is difficult for complex robot movements.

35. Most of the robot languages in use today are a combination of textual programming and .... programming.

36. The first generation robot languages use a combination of compound statements and teaching box procedures for developing robot programmes.

37. The .... language is an example of first generation robot programming language.
38. The second generation robot languages are also called "Structured Programming Languages" because they are similar to computer languages.
39. .... technology is an approach to the organisation of work in which organisational units are relatively independent groups and each group of machines or facilities is responsible for production of a given family of product(s).
40. *Part-family* is defined as 'Collection of parts which are similar in terms of geometric shape, size and similar processing steps required in manufacturing, so that flow of materials through the plant improves'.
41. Factory Flow Analysis (FFA) is the first level of planning in progressive production flow analysis.
42. Live analysis is the final level of planning in Production Flow Analysis (PFA).
43. Numerical Control (NC) is the least sophisticated form of automatic control of machine tools.
44. *Part-family programming* is an NC programme system that groups common or similar programmes for machining of part families into a major computer programme which is a permanent base from which an NC tape can be prepared for any part in the part-family.
45. Travel heart handles a large quantity of material handling data in a concise and rapid way.
46. .... analysis is based upon the sequence of operations for parts and products.
47. Systematic layout planning is analysing the relative relationships between various workcentres.
48. Machine loading and product mix decisions are not important in group technology.
49. The problems of distributing the work-load amongst the facilities/work-centre is referred to as *machine loading problem*.
50. .... *manufacturing system* (FMS) is a manufacturing system based on multi-operation machine tools, incorporating integrated computer control with associated support function and material handling.

ANSWERS

1. 20, 80	2. Fixed	3. Fixed	4. Yes	5. Flexible
6. No	7. Yes	8. NC	9. Yes	10. Robot
11. Six	12. Yes	13. Manipulation	14. Controller	15. Anatomy
16. Yes	17. Cylindrical	18. Yes	19. Yes	20. Rotational
21. Yes	22. Drive	23. No	24. Yes	25. Yes
26. Limited-sequence	27. Intelligent	28. Stability	29. Precision	30. Accuracy
31. Repeatability	32. Yes	33. No	34. Yes	35. Teaching box
36. Yes	37. VAL	38. Yes	39. Group	40. Yes
41. Yes	42. Yes	43. No	44. Yes	45. Yes
46. Sequence	47. Yes	48. No	49. Yes	50. Flexible.

THEORETICAL QUESTIONS

1. Discuss briefly "Mechanisation and automation".
2. What is the difference between "Microautomation" and "Macroautomation" ?
3. How "Automated production systems" are classified ?
4. Explain briefly the following :
   (i) Fixed automation   (ii) Programmable automation
   (iii) Flexible automation.
5. How is "Industrial Robot" defined ?
6. What are the objectives of using industrial robots ?
7. What are the advantages of employing robots ?
8. Explain briefly with a neat sketch the main components of a robot.

9. Explain briefly any four of the following robot components :
    (i) Base   (ii) Manipulator arm
    (iii) Gripper or effector   (iv) Drives
    (v) Controller   (vi) Sensors.

10. Discuss briefly the six basic motions or degrees of freedom of a robot.

11. Explain briefly the classification of 'robot'.

12. Discuss briefly the following basic configurations of industrial robots :
    (i) Polar configuration   (ii) Cylindrical configuration
    (iii) Cartesian coordinate configuration   (iv) Jointed-arm configuration.

13. Explain briefly the following is relation to an industrial robot :
    (i) Controller   (ii) Actuators
    (iii) End effectors   (iv) Sensors.

14. Discuss briefly the following types of industrial robots :
    (i) General purpose robots   (ii) Special purpose robots.

15. Explain briefly the following terms related to construction and operation of robots :
    (i) Work envelope   (ii) Speed of movement
    (iii) Load carrying capacity   (iv) Precision of movement
    (v) Drive systems.

16. Discuss briefly the following methods used to programme the work cycle of a robot.
    (i) Manual programming method   (ii) Walk through programming method
    (iii) Teach pendant method   (iv) Off-line programming method.

17. Write a short note on "Industrial Robots".

18. What are the applications of robots ? Explain.

19. What do you mean by AGVS ? Explain.

20. Discuss briefly AGVS equipment.

21. State the applications of Automated Guided Vehicles (AGVs).

22. What do you mean by "Automated storage systems" ? Explain.

23. What are the objectives of Automated storage systems ?

24. Explain briefly the various types of Automatic storage/retrieval systems.

25. List the applications of Automatic storage/retrieval systems.

21

Product and Process Design for Integration

21.1. PRODUCT ANALYSIS

The product which is to be manufactured is first analysed primarily from a technological point of view to determine the processes required.

Assembly charts :

● For visualising the material flow and the relationship of parts, *schematic and graphic models* are commonly developed.

● The *"assembly chart"* can be useful in making preliminary plans regarding subassemblies, where purchased parts are used in the assembly sequence, and appropriate general methods of manufacture. The assembly chart is often called a *"Goznito chart"* for the words, "goes into".

Operation process chart :

● Presuming that the product is already engineered, we have complete drawings and specifications of the parts, their dimensions and tolerances, and materials to be used.

The engineering drawings specify locations, sizes, and tolerances for holes to be drilled and surfaces to be finished for each part.

With this information, the most economical equipment, processes, and sequences of processes can be specified.

● An *operation process chart is a summary of all required operations and inspections.* It is a *general plan for manufacture.*

Although the focus of such charts is on the technological processing required, it is obvious that the jobs to be performed by humans have also been specified. Some discretion in the make up of jobs still exists, especially, in the assembly phase ; however, it is clear that technology is in the saddle.

21.2. CONCEPTS OF PROCESS AND JOB

The technology imposes constraints that limit the possible arrangement of processes and jobs ; the job satisfaction and social system needs impose also certain constraints.

— The circle marked "Technological constraints" in Fig. 21.1 indicates that all job designs within the circle represent feasible solutions from a technological point of view, and all points outside the circle are infeasible.

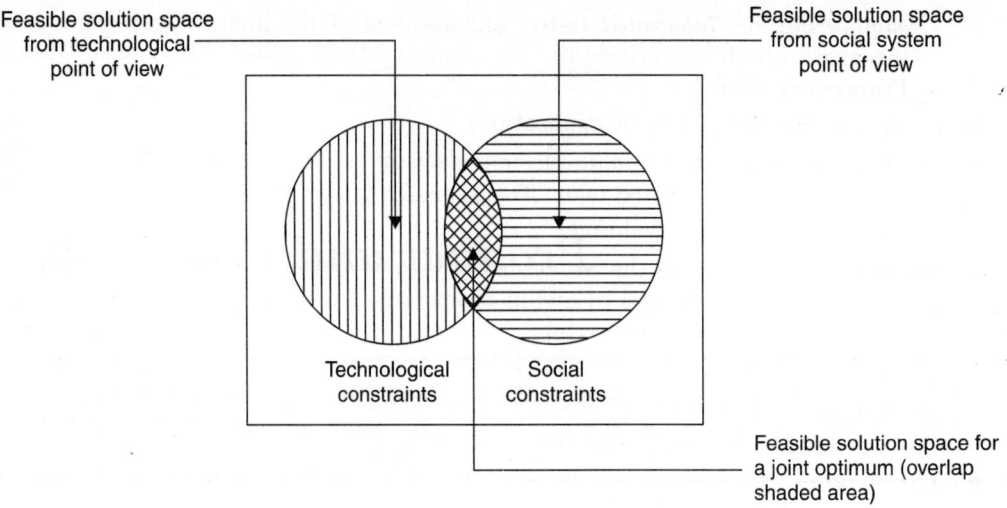

Fig. 21.1. Feasible solution spaces for technological and
social systems, and joint solution space.

— Similarly, the circle marked "Social system constraints" indicates that all designs within that circle represent feasible solutions from the socio-job satisfaction point of view.

— Within the shaded area of overlap between the two circles, we have a solution space that *meets the constraints of both technology and the social system, The shaded area defines the only solutions that can be considered as feasible in joint terms.* Our objective, then is to consider jointly the economic and social system variables, and find the *best solution.*

● *Because optimisation is an unclear process in job design, we seek solutions that are acceptable.*

21.3. STRATEGIES FOR INTEGRATED PRODUCT AND PROCESS DESIGN

In view of extreme competitive pressure, the general aims in the entire process of manufacturing are to reduce the cycle time from product concept to market by fifty percent, reducing costs by similar amounts while increasing quality—but particularly to give early visibility to many downstream options *at the product concept level,* thus reducing risk.

To accomplish these aims, *a team is assembled and* applied on the next new or redesigned product that the company is planning. The *'advantage'* of this method is that it can be implemented instantly with your present people without the necessity of investing a single piece of new technology (computers, FMS etc.). The investment is in people, that is, *changing their awareness by training.* Teams are then formed with designers, manufacturers, suppliers, and users. The *'limitation'* with this technique is that it presents many institutional-social problems to management ;

— how individual team members resolved their carrier issues ;

— what happens to them when they leave the team and return to their old organisations ;

— the use of outside "facilitators" ;

— what types of training are needed for staff, for management and so on.

● This larger strategy attempts to apply all the typical downstream activities at the *product concept level.* It is called *"Concurrent Engineering (CE)"* or *"Concurrent Design (CD)"* or *"Integrated product and process design"* or *"Simultaneous engineering"* or *"Integrated system management"*. On a higher level, which seeks to integrate all the activities of the company, it is called "Enterprise Integrated Framework (EIF)".

Following are the *three levels* of applications :

- **First level.** It involves a *management decision* to integrate all the normal serial operations into an integrated team. The approach is very effective. However, there is no long-term competitive advantage.

- **Second level.** It involves the *use of new tools and methods* (a number of them computer based) to aid this process by giving quantitative visibility to options, and constraints due to economic-technological issues.

  It is generally the approach used by companies after they have tried a few projects using the first method and realise that new tools and methods may be helpful to alleviate some of the problems associated with the integrated team only concept.

- **Third level.** This level represents the *integration of these various tools into a feature-based design system* involving :

  — solid models ;

  — extensive databases with analysis tools for process planning ;

  — scheduling ;

  — economic modelling ;

  — automatic system design.

21.4. TOOLS AND METHODS

- A number of tools, methods, and techniques have been developed *to help assembly or manufacturing system designers.*

- Some of the *"new tools"* available are :

  — Part-mating science ;

  — Liaison sequence analysis ;

  — System synthesizing techniques ;

  — Programs capable of automatically generating process plans and task-resource matrix needed to automate the generation of the system synthesizing technique.

These tools were developed primarily to support assembly system design but their potential use is much broader.

Part-mating science :

- This is an analysis technique for specifying the engineering requirements for physically mating piece parts, both rigid and spring parts.

Assembly System Design Program (ADSP) :

- This is a *computer program based on dynamic programming* with heuristics capable of synthesizing manufacturing systems based on economic-technological issues created from a process plan, technical resources including people and constraints.

- It does not simultate a system that you have given it, it creates completely new systems based on a detail description of the problem, resources and constraints.

Fig. 21.2 shows a simplified block diagram of the system.

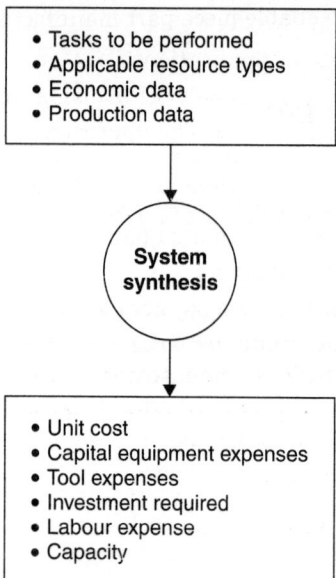

Fig. 21.2. Simplified block diagram of an assembly
system design program (ASDP).

SPM (Assembly sequence, Assembly process plan and Assembly task resource matrix) :

- This is a method for *automatically generating the necessary process plan and task-resource matrix* needed by ASDP starting from a simple bounding-box description of the individual piece-part and data on the part mates.

Liaison Sequence Analysis (LSA) :

- This is a method for *identifying all possible ways to assembly a product* starting from liaisons between parts, where a liaison may be a process, a test, or the physical mating of two parts.

- A typical industrial engineering study might identify one or at most two ways of assembling something ; LSA will show *all possible ways* by which something can be assembled. The results of LSA can then be correlated with reliability modelling and economic-technological analysis to obtain the best product quality for a certain capital investment.

Integration of tools :

- The tools listed can be used individually, as a group, or as an integrated set. They have been applied individually on studies ranging from precision mechanisms to an automation strategy for shipyards.

- *Basically the tools are generic and thus independent of the product domain.*

Feature-Based Design (FBD) Systems :

- *Feature-based design captures design intent* (assembly topology, product-function, manufacturing, or field use) *while creating part and product geometry.*

- Design for assembly, as used here, extends existing ideas about critiquing part shapes and part count to include assembly process planning, assembly sequence generation, assembly fixturing assessments and assembly process costs. As used here, features are not restricted to the set required to accomplish the machining of an individual part, they may describe attributes to enable piece-part manufacture, or describe a process, or define some form of testing.

21.5. CONCURRENT ENGINEERING

Concurrent engineering (also known as *'Simultaneous engineering'*) *is an approach used in product development in which the functions of design engineering, manufacturing engineering and other functions are integrated in order to reduce the required time to bring a new product in the market.*

Earlier approach of launching a new product, with two functions, *"Design"* and *"Manufacturing"* tend to be separated. The design development team develops new design sometime without much consideration given to the manufacturing capabilities of the company. There is little opportunity for manufacturing engineers to offer advice how the design might be altered to make it more manufacturable. *In contrast* a company practising concurrent engineering, *manufacturing engineering department also becomes involved in the product development cycle,* providing necessary advice how the product and its components can be designed to facilitate manufacture and assembly. It also offers the early stages of manufacturing planning of the product.

- Concurrent engineering, besides manufacturing, involves other functions of product development cycle such as :
  - — Quality ;
  - — Field service ;
  - — Vendors supplying critical components, etc.
- Concurrent engineering includes several *elements* such as :
  - — Design for manufacturing and assembly ;
  - — Design for quality ;
  - — Design for life cycle ;
  - — Design for cost.

QUESTIONS WITH ANSWERS

Q. 21.1. Discuss briefly "Technological view of process planning and job design".

Ans. The technological view of process planning and job design is briefly discussed as under :

- **"Process planning"** takes as *input the drawings or other specifications* that indicate what is to be made and the forecasts, orders, or contracts that indicate the *quantity* to be made.
- The *drawings* are then analysed to determine the overall scope of the project. If it is a complex assembled product, considerable effort may go into "exploding" the final product into implied component parts and subassemblies. This overall planning may take the form of special drawings that show the relationship of parts, cutaway models, and assembly drawings.
- Preliminary decisions are made concerning some assembly groupings to determine which parts to make and buy as well as the general level of tooling expenditures. Then for *each part, a detailed routing through the system is developed.* Technical knowledge is required concerning processes, machines and their capabilities, cost and production economics. Ordinarily, a range of processing alternatives is available. The selection may be influenced strongly by the projected volume and stabling of product design.

Q. 21.2. Discuss briefly "Process-job design in relation to layout".

Ans. Following are the two basic types of physical systems :

(*i*) Process focused system.

(*ii*) Product focused system.

Although many actual systems are combination of these two extremes, the two types illustrate the nature of jobs that result.

— At one extreme, where *product volumes are low and variety flexibility and quality are the prime strategies,* the economical production system will be **functional layout.** Such a solution to the layout of physical facilities results in relatively good utilisation of equipment by time sharing it for various jobs requiring a certain production technology.

— At the other extreme, for products that are *standardised and produced in high volume,* there is little if any difference in process requirements, from unit to unit. Specialisation in the form of **linear or product layout** is *economically justified.*

Between the extreme, we find mixed types of layout.

● The appropriateness of a given layout in supporting the competitive strategy depends on *economics of alternate solutions.*

Examples. If plant, equipment, and other resources were *costless,* we would setup separate product-focused facilities for each product. Each product would have its own production line. Because these factors of production are *not free, we must have sufficient utilisation of facilities* to justify "single-product" lines.

— For an *"oil refinery"* or a steel mill, a *very high utilisation* is important because of the immense capital requirements,

— For a *"fast-food operations",* perhaps *fifty percent utilisation is sufficient.*

Q. 21.3. Explain briefly "Functional layout for process-focused systems".

Ans. *In functional layout,* the processing units are organised by functions on the assumption that certain skills and expertise are available in that serve facility.

In practice, many examples of process-focussed systems can be found, for instance :

— Manufacturing ;

— Hospital and medical clinics ;

— Large offices ;

— Municipal services ;

— Libraries.

In every situation, the work is organised according to the *function performed.*

● The *machine shop* is one of the most common examples, and the name and much of our knowledge of process-focussed system results from the study of such manufacturing system.

Q. 21.4. What is CIM ? List some of the technologies of CIM.

Ans.

● **Computer Integrated Manufacturing (CIM)** *is the automation of the entire manufacturing process with computers.*

— Designing a CIM system means applying systems thinking to the manufacturing enterprise ; in brief, it means viewing the organisation as a unit with certain inputs and certain desired output and designing systems that are both *computer-based and people-integrated to cause the inputs to be transformed into the outputs.*

- Some of the technologies of CIM include :

  (*i*) Computer-aided design.

  (*ii*) Numerical control.

  (*iii*) Robotics.

  (*iv*) Expert systems.

  (*v*) Automatic guided vehicles.

  (*vi*) Computer-aided testing.

  (*vii*) Automatic assembly.

  (*viii*) Communication networks.

  (*ix*) Remote sensing.

  (*x*) Digital instrumentation.

Q. 21.5. Discuss briefly the components of "Computer Integrated Manufacturing (CIM)".

Ans. The components of CIM are discussed below :

1. Computer-aided design and drafting (CADD) :

- CADD has evolved from drafting automation. Automated drafting systems simplifying many aspects of its creation.
- The engineering portion of a CIM organisation must use computer-aided *design,* not computer-aided drafting and must produce three-dimensional geometric data.
  - The three-dimensional geometric data can take three common forms : *Wire frame, surfaces* and *solids,* in order of progressively diminishing abstraction.

2. Computer-aided engineering (CAE) :

- It is the term used to describe computer-based engineering analysis of *digital models.*
- The most popular form of CAE for *mechanical engineers is finite-element analysis.* In *electronics,* there are *various forms of circuit analysis* that are applied in this manner.
- For CIM to be efficient, the model created in the computer-aided design process should be able to be digitally tested, through the design verification process that follows initial design.

3. Computer aided manufacturing (CAM) :

- It usually refers to *numerically controlled machine tools,* the means of controlling them and the associated computer-based systems within the factory.
- This category also includes other physical devices such as :
  - Flexible machining system (FMS) ;
  - Robots ;
  - Automatic guided vehicles (AGVs) ; and
  - Data control systems, such as Manufacturing Requirements Planning (MRP) software.

4. Data base management systems (DBMS) :

- CIM, is in a sense, *data based manufacturing.*
- A robust CIM data base must be equally efficient with graphical and non-graphical data. Recent developments in data base management technology have made the implementation of CIM easier than before.

5. **Networks :**

● Networking is the most important technology for CIM : The digital connectivity that allows the data to flow from design through production to field support and back is key to the "integrated" aspect of CIM.

● Salient features of networks include :

— Topology ;

— Physical media ;

— Protocols ;

— Bandwidth ;

— Cost.

● Popular networking schemes include :

— *Token ring and collision sensing* protocols ;

— Ring and star topologies ;

— *Coal and twisted-pair physical media.*

6. **Computational hierarchies :**

● In addition to the engineering workstations and servers, the engineering data management network and accounting computers, many other computers serve the CIM organisation. Most of them are *programmable logic controllers,* or PLCs, simple computers designed for integration into industrial systems and therefore offering a particularly a wide range of input-output options, as well as a variety of robust installation possibilities.

Q. 21.6. List the points in regard to design for economy in mass production.

Ans. The head of the design organisation must be part of or have direct access to company's top management. The design organisation must have direct links with those responsible for production, purchasing and marketing functions.

Following points need be considered in regard to *design for economy in mass production :*

1. Before taking up the design, the designer must be given precise and comprehensive terms of reference. Some formal procedure should be there to avoid wastage of time upon ill-conceived ideas.

2. Adequate time must be given to designer to arrive at optimum solution for a given problem. This can mean considerable cost savings later on.

3. Designer must have fairly good idea of production techniques. The design of product and selection of manufacturing process must proceed in parallel.

4. Process selection is of paramount importance.

5. Good detail design is of vital importance and may also be critical to the cost of manufacture, quality and reliability of the product.

6. Simplicity should always be aimed for.

7. If design can't be produced at a cost which will permit it to be sold at a profit, then it is as much a failure as if it did not function correctly.

8. Great savings in manufacturing, stocking and distribution can be made by standardisation and variety reduction.

9. The quality and reliability of the design must be right.

10. Simplification, specialisation and standardisation can bring handsome rewards.

Q. 21.7. Discuss briefly 'Generic CIM architecture'.

Ans. CIM involves a great many functions within an organisation. The complexity of this relationship is illustrated in Fig. 21.3.

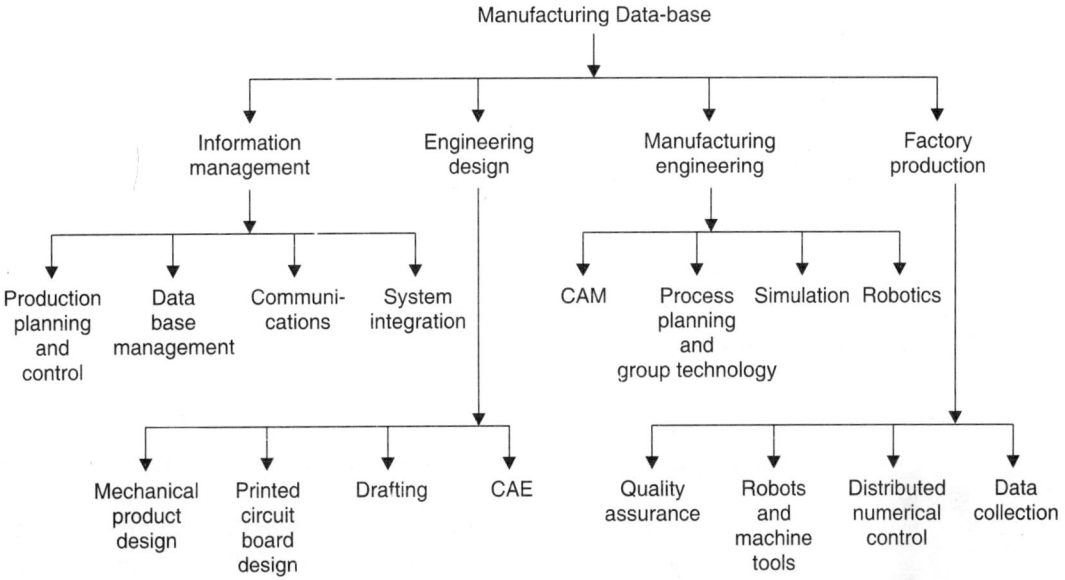

Fig. 21.3. Generic CIM architecture.

HIGHLIGHTS

1. *Part-mating science* is an analysis technique for specifying the engineering requirements for physically mating piece parts, with rigid and spring parts.
2. *ADSP (Assembly System Design Program)* is a computer program based on dynamic programming.
3. *Feature-based design* captures design intent while creating part and product geometry.
4. *Concurrent engineering* is an approach used in product development in which the functions of design engineering, manufacturing engineering and other functions are integrated in order to reduce the required time to bring a new product in the market.

OBJECTIVE TYPE QUESTIONS

Fill in the Blanks or Say 'Yes' or 'No' :

1. EIF stands for enterprise Integrated Framework.
2. CD means .... design.
3. .... science is an analysis technique for specifying the engineering requirements for physically mating piece parts, both rigid and springs parts.
4. ASDP is a computer program based on .... programming.
5. .... is a method for identifying all possible ways to assemble a product starting from liaison between parts, where a liaison may be a process, a test, or the physical mating of two parts.
6. Feature-based design captures design intent while creating part and the .... geometry.
7. .... is the automation of the entire manufacturing process with computers.
8. Designing a CIM means applying systems thinking to the manufacturing enterprise.

9. A well-designed CIM-based factory can have a breakdown point of about 70 percent of its operating capacity.
10. Computer-aided design in one of the technologies of CIM.

ANSWERS

1. Yes	**2.** concurrent	**3.** Part-making	**4.** dynamic	**5.** LSA
6. product	**7.** CIM	**8.** Yes	**9.** No	**10.** Yes.

THEORETICAL QUESTIONS

1. Explain briefly the following :
    (*i*) Assembly charts.
    (*ii*) Operation process chart.
2. Discuss briefly " Concepts of process and job".
3. Discuss briefly the strategies for integrated product and process design.
4. Discuss briefly the following tools :
    (*i*) Assembly System Design Program (ADSP).
    (*ii*) Liaison Sequence Analysis (LSA).
    (*iii*) Feature-based design systems.
5. Write a short note on "Concurrent Engineering".

22

Computer Aided Process Planning

22.1. Introduction. 22.2. Analysis and planning data required. 22.3. Analysis and planning steps and considerations. 22.4. Process format and content. 22.5. Computer Aided Process Planning (CAPP) systems. **Questions with Answers**—Highlights—Objective Type Questions—Theoretical Questions.

22.1. INTRODUCTION

- The manufacturing activities necessary to accomplish the production of an end product must be processed or arranged in an orderly, worktable sequence. *This analysis and planning is the bridge between design engineering and product manufacturing.* It encompasses every phase of industrial and manufacturing engineering by establishing a manufacturing plan which is economical and supplies a quality product.

- *Process analysis and operation planning are required for all manufactured products regardless of their size, material make up, or type of construction.*

- An adequately defined operation plan can improve any task which adds value to a product with materials, labour power, or equipment. The detail and complexity of process plan will vary greatly from simple handwritten notes to detailed CAD/CAM-created points and computer-generated procedures specifying parts, tools and equipment, gauges and their readings, and even which hand is employed for each movement.

- The personnel performing process analysis and operation planning are in a pivotal position between product engineering and manufacturing operations. They receive product information from design engineering, interpret the data into terms of process requirements, and then create operation plans with work standards and tooling requirements for manufacturing to use.

22.2. ANALYSIS AND PLANNING DATA REQUIRED

The amount of detailed desired in process analysis and operation plan will determine the volume and scope of data necessary. The following are the examples of the type of *data required for effective operations plans :*

1. Part drawings and bills of material.
2. Expected product life and manufactured quantities.
3. Equipment list.
4. Work centre load forecasts.
5. Material specifications.
6. Speed and feed data.
7. Tooling and gauging standards and inventory.
8. Work measurement and standard data.

9. Abbreviation listing.

10. Scrap and rework history.

11. Cutting fluid applications.

22.3. ANALYSIS AND PLANNING STEPS AND CONSIDERATIONS

The sequence of events necessary to establish an effective manufacturing process varies from company to company and depends upon plant size and the product manufactured. Companies with large and specialised staff normally require more steps and operate with greater detail.

The following examples contain the *most normally used sequences* :

1. *Preproduction drawings* :

— The first release of drawings must be considered as preliminary or preproduction.

2. *Manufacturing feasibility review* :

All preproduction drawings should be analysed for the following :

— Are dimensioning and datum surfaces compatible with accepted manufacturing practices ?

— Are bills of material and casting or forging drawings available ?

— Are sufficient stock allowances provided on castings, forgings and stampings to allow for anticipated mismatch or distortion in heat treatment ?

— Are sufficient clearances and access allowed for proper assembly of all components ?

— Are maximum allowable tolerances, applied to non-functional characteristics ?

— Are tolerances and surface finishes on functional characteristic realistic, and is statistical tolerancing used where possible ?

— Are adequate clamping and locating surfaces provided for manufacturing ?

— Are value analysis suggestions for lower cost applicable ?

The information on the above items is collected in the form of comments and is reviewed with responsible design engineer. Acceptable suggestions or trade-offs are agreed upon and incorporated into each applicable drawings, these revised drawings become the *production drawings*.

3. *Make or buy decisions* :

— Proper decisions (make or buy) are based upon true cost comparisons, in-plant workloads, lead time comparisons, and inplant *versus* vendor capability.

4. *In-plant production considerations* :

— The two points to consider when preparing to establish process routing are the *anticipated product life cycle* and the *required production quantities*.

5. *Developing the process* :

The method of developing a process remains the same regardless of the variation in product nature, level of production, lead time allowed and the like. The formal *steps* in this procedure are :

(*i*) Create a general statement of the manufacturing operations to be performed.

(*ii*) Establish a provisional process.

(*iii*) Develope alternative processes.

(*iv*) Select a production process.

(*v*) Communicate the selected process to other affected activities.

(*vi*) Perform detailed processing.

6. *Pilot production run* :

— The first production run of a part can be considered a pilot run, as any new process may require some modifications.

7. *Process review and updates :*

Any process change after the first production run must consider the following :

(*i*) Parts or assemblies in process and material on hand.

(*ii*) Cost of change, including effect on tooling, material, and delivery schedule.

(*iii*) Anticipated savings or added cost.

Although cost reduction is a constant philosophy of manufacturing, changes should be made with caution.

22.4. PROCESS FORMAT AND CONTENT

Process format can vary greatly but tends to follow similar patterns in like industries. The document received by the production operator will depend on methods of duplication, distribution and presentation.

Types of formats :

— In some job shops, drawings and handwritten notes serve as format. Still pictures of elaborate operation set-ups or part assemblies are often utilised either to aid workers or even to be their only process document. At the other end of spectrum are part files kept at each machine consisting of detailed process prints showing specific surfaces, dimensions, finishes and tolerances to be machined for that defined operation step accompanied by computerized procedures detailing every movement made.

— Currently the approach is to use a computer to store and reproduce the required process documentation.

Process content :

For all but very small plants, the following *items are normally mandatory for inclusion :*

- Part identification.
- Process revision information.
- Drawing identification.
- Bill of material information.
- Material identification.
- Operation number.
- Workstation identity.
- Operation description.
- Production time standard.
- Tooling and gauges required.
- Speeds and feeds to use.
- Tool layouts.
- Workpiece layout.
- Process operation drawing.
- Handling equipment.
- Special protection and conditions.

22.5. COMPUTER AIDED PROCESS PLANNING (CAPP) SYSTEMS

— **Process planning** *involves the preparation and documentation of the plans for manufacturing the products. Computer Aided Process Planning (CAPP) represents the link between design and manufacturing in a CAD/CAM system.*

— The process planning is concerned with determining the sequence of processing and assembly steps that must be accomplished to make the product. The processing sequence is documented on a form called a *"route sheet"*. The route sheet typically lists the production operations, machine cells or work-stations where each operation is performed, fixtures and tooling required, and the standard time for each task. During the last decade or so, there has been much interest in automatising the task of process planning by means of CAPP systems.

Following are the *two approaches around which the computer aided process planning systems are designed* :

1. Retrieval CAPP systems.

2. Generative CAPP systems.

1. *Retrieval CAPP systems* :

● These types of systems (also known as *"variant CAPP systems"*) are based on the principles of *"Group technology and parts classification and coding"*.

● With these systems, a *standard process plan (route sheet) is stored in computer files and each part code number.* The standard route sheets are based on current part routings in use in the factory, or an ideal plan that is prepared for each family.

Fig. 22.1 shows the *general procedure for using the retrieval computer aided process planning system.*

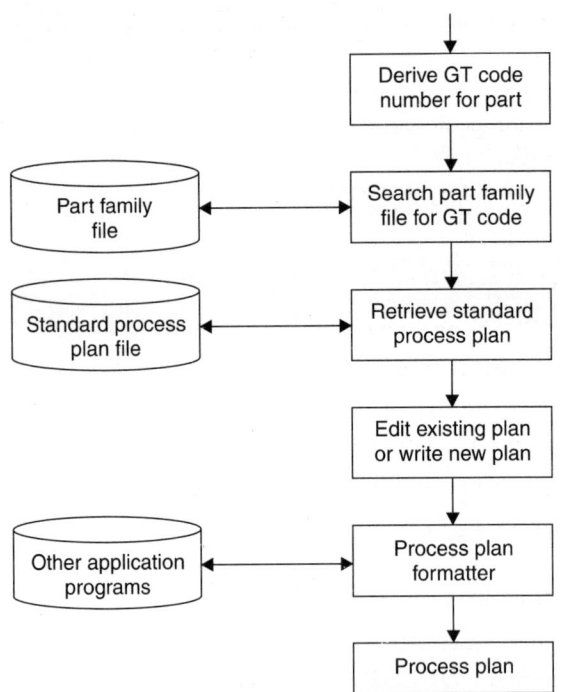

Fig. 22.1. General procedure for using one of the retrieval computer
aided process planning systems.

● A large number of GT based variant process plans are available, such as :

— NIPLA ;

— MICLASS ;

— CAPP etc.

2. *Generative CAPP systems :*

These systems *utilise an automatic computerised system consisting of decision logic, formulae, technology algorithms, and geometry based data to uniquely determine the many processing decisions for converting a part from a rough to a finished state.*

In this approach, unlike the variant approach, *no standard manufacturing plans are predefined or stored.* Instead, the computer automatically *generates a unique operation sheet for a part every time the part is ordered and released for manufacturing.*

This system consists of the following two major components :

(*i*) *Geometry based coding system :*

— This component *translates physical features and engineering drawing specifications into computer interpretable data.*

— The coding defines all geometric features, feature sizes and locations, and feature tolerances for all process-related surfaces. The coding scheme not only describes both the rough and finished states, but must be defined in similar terms for the part in each operation. Consequently, the coding scheme for a *generative system requires for greater detail than is required in the variant methods.*

(*ii*) *Process knowledge :*

— This component is *in the form of decision logic and data in order to compare the part geometry requirements to manufacturing capabilities.* This logic is used to automatically determine the appropriate sequence, selecting the machine for each operation, determining cut planning or any other operation details subject to available tooling and fixturing, and calculating the setup time and cycle times for each operation.

  ● An additional element of the system is the *software for printing the operation sheet* and/or procedure sheet for instructing the operator according to the methods selected by the computer.

  ● Although a truly universal general system has yet to be developed, a number of generative systems are available today for specific manufacturing processes or types of parts.

— CAPPGENPLAN packages are normally used.

Advantages of Computer Aided Process Planning (CAPP) :

Following are the *advantage of CAPP :*

1. Increased productivity of process planners.

2. Process rationalisation and standardisation.

3. Incorporation of other application programs.

4. Improved legibility.

5. Reduced lead time for process planning.

QUESTIONS WITH ANSWERS

Q. 22.1. What is process planning ?

Ans.

● **Process planning** *means the preparation of work detail plans.* It involves the detailed planning of the process of production for the product of a specified quality and quality at a minimum cost with available resources.

● This activity is not confined only to the time when a new product is introduced but it is reviewed continuously with a view to achieve increased production of higher quality products at lower manufacturing costs.

Q. 22.2. What are the advantages of "Process planning" ?

Ans. Following are the *advantages of process planning :*

1. The net result of process planning is more effective and efficient operation.

2. It facilitates coordination among various agencies.

3. It helps in process improvement.

4. Supervisors are free from detailed task of process planning so that they can concentrate on more productive activities.

Q. 22.3. What information is required to do process planning ?

Ans. To carry out process planning, the following information is required :

1. Quantity of work to be done along with product specifications.

2. Quality of work to be completed.

3. Availability of equipments, tools and personnels.

4. Sequence in which operation will be performed on the raw material.

5. Names of equipments on which the operations will be performed.

6. Standard time for each operation.

7. When the operations will be performed ?

Q. 22.4. What is the procedure of process planning ?

Ans. Following *steps* are involved in the procedure of process planning :

1. *Selection of process.* The selection of process depends upon :
   — Current production equipment ;
   — Delivery date ;
   — Quantity to be produced ;
   — Quality standards.

2. *Selection of materials :*
   — The materials should be of right quality and chemical composition as per the product specifications.
   — The shape and size of materials should restrict scrap.

3. *Selection of jigs, fixtures, and other special attachments :*
   — These supporting devices are necessary to give higher production rate and to reduce cost of production per piece.

4. *Selection of cutting tools and inspection gauges :*
   — They are necessary to reduce production time and inspect accurately and at a faster rate.

5. *Making the layout :*
   — To make the process layout indicating every operation and the sequence in which each operation is to be carried out.

6. *Set-up time and standard time :*
   — To find set-up time and standard time for each operation.

7. *Manifest process planning :*
   — To manifest planning by documents such as operation and route sheets, which summarise the operations required, the preferred sequence of operations, auxiliary tools required, estimated operation times etc.

Q. 22.5. What are the basic factors which influence process planning ?

Ans. The process planning activity begins with the receipt of product specifications and ends with the plans for manufacture of product. As such the first step is a careful review of product-design and specifications, leading to revised product specifications which form the basis for process planning. Thus, the principal initial data required are :

(i) Assembly and detail drawings of product.

(ii) Specifications for acceptance of the finished products and service function.

(iii) Volume and rate of output.

(iv) Information regarding plant facilities.

Q. 22.6. Discuss briefly "Process planning activities".

Ans. The task of determining a process plan that defines how to produce a part involves the following distinct *steps* :

(i) To analyse part requirements.

(ii) To determine operating sequence.

(iii) To select machines.

(iv) To calculate processing times.

(v) To process documentation.

(vi) To communicate process knowledge.

Q. 22.7. List the notable objectives for process planning systems.

Ans. In the light of the magnitude and complexity of the data involved, it is essential to have clearly defined objectives both for the development and maintenance of a process planning system. Notable objectives are :

(i) Consistency.

(ii) Accuracy.

(iii) Ease of application.

(iv) Completeness.

(v) Low maintenance efforts.

Q. 22.8. Discuss briefly "The evaluation and selection of the best process planning system".

Ans. The evaluation and selection of the best process planning system for a given company encompasses numerous engineering management decisions. The process involves identifying, weighing and comparing various interrelated factors. The *major factors to consider* are :

(i) General environment in which process planning is conducted.

(ii) The organisational structure of the company.

(iii) The technical expertise available to the company.

(iv) The needs and objectives of the company regarding the generation of manufacturing information and process plans.

Q. 22.9. List the 'aids' and 'techniques' relating to process planning.

Ans. Following is the *list of aids and techniques relating to process planning :*

1. Group technology.

2. Computer aided process planning.

3. Outside processing assistance.

4. Tolerance charts.

5. Line balancing.

6. Simulation.

7. Open-loop control.

8. Closed-loop control.

Q. 22.10. What do you understand by "Discrete process control" ?

Ans. Discrete process control *deals with systems that discrete information and parts.* Generally, the information is binary ; however, this is not a requirement.

Example. An example of discrete process control is *a conveyor and packaging system for crayons.* The machine dispenses the crayons, one of each colour at the same time, into a box. Sensors detect if the crayons were damaged for the box was not filled. A conveyor carries the good boxes to a bundling unit and reject the bad boxes for inspection and packaging.

Q. 22.11. What are "Programmable Logic Controllers (PLC)" ?

Ans.

- Programmable Logic Controllers (PLC) are in widespread use for both continuous and discrete process control. Recent developments have provided PLC with increased speed, programming and communication capabilities.

- Modules are now available which provide connection to other factory networks using MAP (Manufacturing Automation Protocol). Most PLCs are designed in module fashion :
  — Input and output modules.
  — Communication modules.
  — Special purpose modules.

- PLC programs offer some important advantages over relay circuits. The most important is that relays and other output devices can have an unlimited number of contacts. In contrast, most control relays are limited to a few sets of a contacts and circuits must be designed with this limitation in mind. Generally, this results in few rungs with more complex logic. PLCs, on the other hand, use an electronic bit to represent each input and output. This single bit can be referenced as many times as required in the program without adversely affecting scans of other rungs because there is not electrical interaction between the elements represented in the rungs.

HIGHLIGHTS

1. *Process analysis* and *operation planning* are required for all manufactured products regardless of their size, material make up, or type of construction.

2. *Process planning* involves the preparation and documentation of the plans for manufacturing the products.

3. Following are the two approaches around which the computer aided process planning systems are designed :
   (*i*) Retrieval CAPP systems.
   (*ii*) Generative CAPP systems.

OBJECTIVE TYPE QUESTIONS

Fill in the Blanks or Say 'Yes' or 'No' :

1. .... analysis and .... planning are required for all the manufactured products regardless of their, size, material make-up, or type of construction.

2. The first release of drawings in process planning must be considered as preliminary or preproduction.

3. The first production run of a part can be considered as a .... run.

4. Process format can vary greatly but tends to follow similar patterns in like industries.

5. .... planning involves the preparation and documentation of the plans for manufacturing the products.
6. In .... CAPP systems a standard process plan is stored in computer files for each part code number.
7. In .... CAPP systems no standard manufacturing plans are predefined or stored.
8. Geometry based coding system translates physical features and engineering drawing specifications into computer interpretable data.

ANSWERS

1. Process, operation   2. Yes   3. pilot   4. Yes
5. Process   6. retrieval   7. Generative   8. Yes.

THEORETICAL QUESTIONS

1. List the data required for effective operation plans.
2. What are the normally used sequences to establish an effective manufacturing process ?
3. Discuss briefly 'Process format and content.
4. What is "Process planning" ? Explain.
5. Explain briefly the following two approaches around which the computer aided process planning systems are designed :
   (i) Retrieval CAPP systems.
   (ii) Generative CAPP systems.
6. What are the advantages of Computer Aided Process Planning (CAPP) ?

23

Group Technology

23.1. General aspects. 23.2. Advantages and limitations of group technology. 23.3. Part families. 23.4. Formation and establishment of component family. 23.5. Collection of production data. 23.6. Classification and codification—Introduction—basic requirements of classification and coding system—advantages of well-designed classification and coding system—classification and coding for group technology. **Questions with Answers**—Highlights—Objective Type Questions—Theoretical Questions.

23.1. GENERAL ASPECTS

*GROUP TECHNOLOGY (GT) *is an approach in which similar parts are identified and grouped together in order to take advantage of their similarities in design and production.* The group technology **aims** at *high productivity at a low cost on short run production.*

- *"Group technology"* in manufacturing, is the replacing of traditional jobbing shop manufacture *by analysis and grouping of work into families and the information of groups of machines to manufacture these families on a flow-line principle with the object of minimizing setting times and throughout times.*
  - The traditional approach of any production unit is to use **"line layout"** where possible and **"functional layout"** in all other cases. In *"line layout"* generally the machines are laid out in a line in their sequence of usage. It is mainly used in simple process industries where all components made on the line use the same machines in the same sequence. In *functional layout* **"batch production"** is used and is based on process specialisation. In this type of factory the workers are divided into organisational units each of which specialises in a particular process or part of a process. This type of layout result in *low machine utilisation and high wastage of time. The high cost of set-ups for small batches results in high manufacturing costs.*
- *Group technology is based on* **"product specialisation".** *In this case each group of workers specialises in the production of a particular list or family of production and is equipped with all machines and equipment needed to complete these products.*
- Group technology is a new name for production system and many have used different names *viz.* **"Family grouping"**, **"Family manufacturing"** etc. to describe this system.
  - While examining the meaning of the 'Group technology' one finds that *"Group"* is a number of things classed together, and *"Technology"* is the science of industrial arts. Thus, *"group technology is the science of the industrial arts to a number of things classed together".*

**Group technology was originated in Russia and was used during World War II.

- In 'group technology' similar parts are arranged into 'part families'. For example, a plant producing 10000 different part numbers may be able to group the vast majority of these parts into 40 or 50 distinct families. Each family would possess similar design and manufacturing characteristics. Hence the processing of each member of a given family would be similar and this *results in manufacturing efficiencies. These efficiencies are achieved by arranging the production equipment into machine groups, or cells, to facilitate workflow.*

  — In product design, there are also advantages obtained by grouping parts into families. *These advantages lie in the classification and coding of parts.*

23.2. ADVANTAGE AND LIMITATIONS OF GROUP TECHNOLOGY (GT)

Following are the *advantage and disadvantages of group technology :*

Advantages :

The benefits of group technology are typically realised in the following areas :

(*i*) **Product design benefits :**

- In the area of product design, the principle benefit is derived from the use of parts classification and coding system.
- Group technology promotes *design standardisation.*

(*ii*) **Tooling and set-ups.** Group technology tends to promote standardisation of several areas of manufacturing. Two of these are tooling and *set-ups.*

- In tooling, an effort is made to design jigs and fixtures that will accommodate every member of a part's family.
- The machine tools in a GT cell do not require drastic changeovers in set-up because of the similarity in the workparts processed on them. Hence, *set-up time is saved* and it becomes feasible to try to process parts in another so as to achieve a bare minimum of sets changeovers.

(*iii*) **Materials handling.** The group technology machine layout lend themselves to *efficient flow of materials* through the shop. The contrast is sharpest when the flow line cell design is compared to the conventional process-type layout.

(*iv*) **Production and inventory control.** Production scheduling is *simplified* with group technology.

  — Grouping of machines into cells reduces the number of production centres that must be scheduled.

  — Grouping of parts into families reduces the complexity, and size of the parts scheduling, more attention can be devoted to the control of these parts.

Owing to the reduced set-ups and more efficient materials handling within machine cells, *manufacturing lead time and work-in-process are reduced.*

(*v*) **Process planning.** Proper parts classification and coding can lead to an automated process planning system.

  — Even without an automated process planning system, reductions in the time and cost of process planning can still be accomplished. This is done through standardisation. New part designs are identified by their code as belonging to a certain parts family, for which the general process routing is already known.

(*vi*) **Employee satisfaction :**

  — The machine cell often allows parts to be processed from raw material to finished state by small group of workers. The workers are able to visualise their contributors to the firm more clearly. This tends to cultivate an improved worker attitude and a higher level of job satisfaction.

— Another employee-related benefit of group technology is that more attention tends to be given to product quality. Workpart quality is more easily traced to a particular machine cell in group technology. Consequertly, workers are more responsible for the quality of work they accomplish.

Disadvantages/Limitations :

(i) The problem of identifying part families among the many components produced by a plant.

(ii) The expense of parts classification and coding.

(iii) Rearranging the machines in the plant into the appropriate machine cells.

(iv) The general resistance that is commonly encountered when changeover to a new system is contemplated.

23.3. PART FAMILIES

A **part family** or **group** *is a collection of a parts, which either because of geometric shape and size or because of similar processing steps are required in their manufacture.* The parts within a family are different, but their similarities are close enough to merit their identification as members of the part family.

The biggest single obstacle in changing over to group technology from a traditional production shop is the *problem of grouping parts into families*. There are three general methods for solving this problems.

(i) Visual inspection.

(ii) Classification and coding by examination of design and production data.

(iii) Production Flow Analysis (PFA).

(i) *Visual Inspection :*

— This method is the *least sophisticated* and *least expensive.*

— It involves the classification of parts into families by looking at either their physical parts or their photographs and arranging them into similar groupings.

(ii) *Classification and coding by examination of design and production data :*

— This method *involves classifying the parts into families by examining the individual design and/or manufacturing attributes of each part.* The classification results in a *code number that uniquely identifies the parts attributes.* This classification and coding may be carried out on the entire list of active parts of the firm or some sort of sampling procedure may be used to establish the part families. *Example :* The parts produced in the shop during a certain given time period could be examined to identify part family categories. The trouble with any sampling procedure is the risk that sample may be unrepresentative of the entire population.

— The method of parts classification and coding *seems to be the most commonly used method today.*

(iii) *Production Flow Analysis (PFA) :*

— This method *makes use of information contained on route sheets rather than part drawings.*

— *Worksparts with identical or similar routings are classified into part families.*

• All the above three methods are *time consuming and involves the analysis of much data by trained personnel.*

23.4. FORMATION AND ESTABLISHMENT OF COMPONENT FAMILY

Formation of component family :

— Initially, a *pilot study* be carried out by examining simple components first, and gradually progressing to the more complex parts. However, the component family analysis should *preferably be based on the complete range of components from the products manufactured by the company.* The drawings and associated production data of the selected components are collected together, classified and sorted into code number order.

— The investigation should not be limited to the formation of potential component families, but also to access the necessary diversity in company operation. This will include the preparation of component family paper work and scheduling and control of the components into the machine group. While this investigation is in progress, the remainder of the drawing, new designs and obsolete drawings still liable to be called forward for spares replacement, should be coded.

— It is *only one-time exercise and effort is fully justified particularly, for variety control and computerization of production planning and control.*

Establishment of component family :

— The coding provides the first stage in sorting and makes it possible to gather the components into families. If tabulated lists are visually examined, the naturally occurring families can be easily determined. These families are normally of the type *'identical in shape and function' and 'identical in shape but different in function'* and appear as blocks of near identical blocks of code number of listings. These will be the most obvious component families to begin to develop and establish machine groups.

— Finally, recheck the component and tabulations and revise where necessary.

23.5. COLLECTION OF PRODUCTION DATA

● The number and sequence of machining operations, setting times and numbers of each component within a definite period of time need to be collected. By analysis of the *'machining operations and sequence',* it is possible to *derive the types of machines required to form the machine group.* From the 'machining and setting times' and the numbers of each component, the potential load on the machine group may be established. Line balancing, however, cannot be adequately established until the component tooling requirements have been assessed. This is the *first stage* in *'family formation'.* The drawing of each component within the family is examined and the type and number of tools necessary to produce the component are determined. With the tooling analysis complete, the family is established and group layout balanced, *i.e.,* data should now include :

(*i*) Geometric shape.

(*ii*) Maximum and minimum size.

(*iii*) Material type.

(*iv*) Form and method of holding.

(*v*) Tools-type and holding.

(*vi*) Machine tools-type and capacity.

From the above information the *profile and parameters of the component family are constructed against which the acceptance or non-acceptance of new components into the family can be used.*

● Once a component family has been formed and integrated into a group layout, it does not necessarily have to remain static. Some components will become obsolete while new components will appear. The more the flexibility that can be built into the system, the more one can expect to get out of it.

23.6. CLASSIFICATION AND CODIFICATION

23.6.1. Introduction

Classification *involves arranging items into groups according to some system whereby like things are brought together by virtue of their similarities and are then separated according to a specific difference.*

A **code** *can be a system used in information processing in which numbers or letters or a combination there of are given a certain meaning.*

23.6.2. Basic Requirements of Classification and Coding Systems

The classification and coding system should meet the following *basic requirements :*
1. To be based upon permanent characteristics.
2. To be mutually exclusive.
3. To be all embracing and offer company wide applications.
4. To be adaptable to computer processing.
5. To be specific to user needs.
6. To be adaptable to future changes.

23.6.3. Advantages of Well-designed Classification and Coding System

The major benefits of a well-designed classification and coding system for group technology have been summarised as followed by Ham :
1. It facilitates the formation of part families and machine cells.
2. It permits quick retrieval of design, drawings, and process plan.
3. It reduces design duplication.
4. It provides reliable workpiece statistics.
5. It facilitates accurate estimation of machine tool requirements and logical machine loading.
6. It permits rationalisation of tooling set-ups, reduces set-up time, and reduces production throughout time.
7. It allows rationalisation of improvement in tool design.
8. It aids production planning and scheduling procedures.
9. It improves cost estimation and facilitates cost accounting procedures.
10. It provides for better machine utilization and better use of tools, fixtures, and manpower.
11. It facilities NC part programming.

23.6.4. Classification and Coding for Group Technology

In group technology, parts are identified and grouped into families for convenience by classification and coding systems, abbreviated as (C/C). This process is a complex and critical first step in group technology. This process is done according to :

(*i*) **Design attributes.** These relate to *similarities in geometric features of the part.*

(*ii*) **Manufacturing attributes.** These relate to *similarities in process involved* in manufacturing of the part.

- Since the above attributes consist of various situations it is time consuming for coding them. So, coding can be done by viewing the shapes of the parts in generic way and then classifying the parts accordingly.
- Parts can also be classified by studying their production flow during the manufacturing cycle ; this approach is known as *production flow analysis.*

Coding :

A company can do *coding* on its own method or any of the following methods, depending upon the situation :

● The code structure for part families typically consists of *numbers, of letters* or of a *combination of the two*. Each specific component is assigned a code (generally less than 12 digits). This code may belong to design attributes only or to manufacturing attributes only.

Three levels of coding (varying in degree and complexity) are as follows :

(*i*) Monocodes hierarchial coding.

(*ii*) Polycodes.

(*iii*) Decision-three coding.

(*i*) *Monocodes hierarchial coding :*

— The interpretation of each succeeding digit depends on the value of the preceding digit.

— Each symbol amplifies the information contained in the preceding digit, so a digit in the code cannot be interpreted alone.

● The *advantage* of this system is that *a short code can contain a large amount of information*. But this method is *difficult to apply in a computerised system*.

(*ii*) *Polycodes :*

— In this code (also known as *chain type*) each digit has its own interpretation, which *does not depend on the preceeding digit*.

— This structure looks to be *lengthy* but it allows the identification of specific parts attributes and it is *suited to computerised system*.

(*iii*) *Decision-tree coding :*

— This system is also called *"hybrid codes"*.

— It is the *most advanced and it combines both design and manufacturing attributes*.

Coding systems :

Some of the important classification and coding systems include :

● Brisch Systems   ● CODE

● CUTPLA   ● D CLASS

● Multiclass   ● Part Analog System

Here we shall discuss *Optiz* and *Multiclass* systems only.

Optiz system. Prof. H. Optiz (University of Aachen in west Germany) and his co-workers developed this classification and coding system. It represents one of the pioneering efforts in group technology area and is probably the best known of the classification and coding systems.

The Optiz coding system uses the following digit sequence :

12345   6789   ABCD

● The basic code consists of nine digits, which can be extended by adding four more digits. The *first nine digits are intended to convey both design and manufacturing data*. The general interpretation of the nine digits is indicated in Fig. 23.1.

— The first five digits, 12345, are called '**form code**' and describe the *primary design attributes of the part*.

— The next four digits, 6789, constitute the '**supplementary code**' which indicates some of the attributes that would be of use to manufacturing (dimension, work material, starting raw workpiece shape and accuracy).

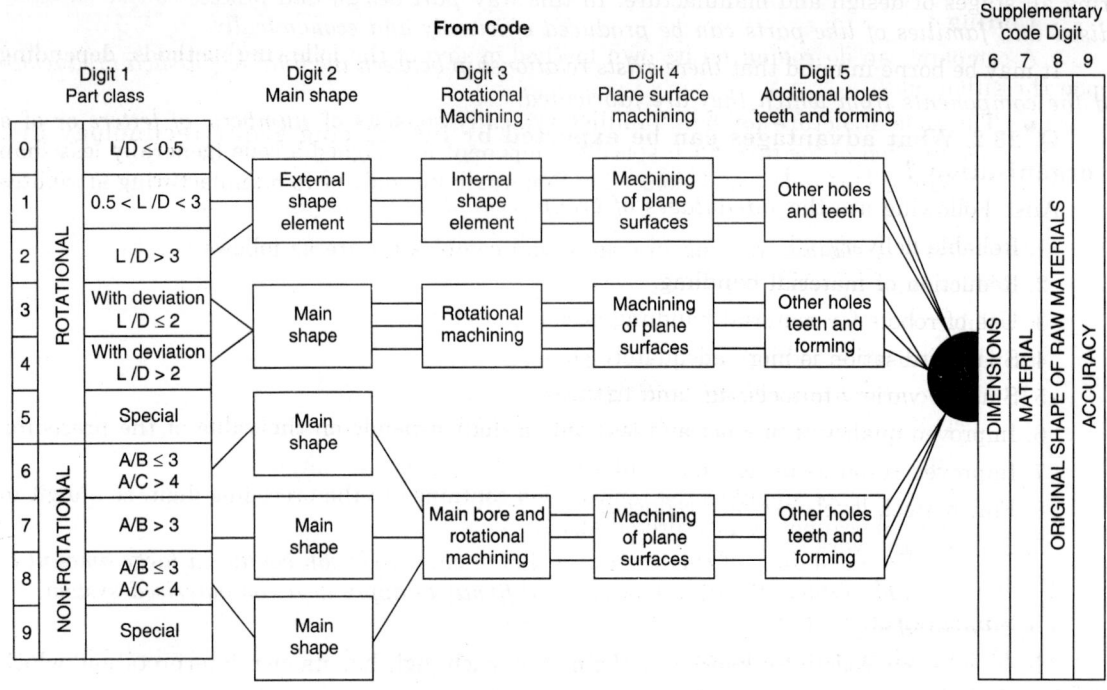

Fig. 23.1. Basic structure of Optiz system.

— The extra four digits, ABCD, are referred to as the **'secondary code'** and are intended to identify the production type and sequence.

Multiclass system. This coding system is developed by the Organisation for Industrial Research.

- The system is *relatively flexible,* allowing the user company to customize the classification and coding scheme to a large extent to fit its own products and application.

- It can be used for a variety of different types of manufactured product, including machined and sheet metal parts, tooling, electronics, purchased parts, assemblies and sub-assemblies, machine tools and other elements. Upto nine different types of components can be included within a single multiclass software structure.

- This coding systems uses hierarchial or decision-tree coding structure in which the succeeding digits depend on values of the previous digits.

- In the application of the system, a series of menus, pick lists, tables, and other interactive prompting routines are used to code the part. This *helps to organise and provide discipline to the coding procedure.*

- The coding structure consists of upto 30 digits. These digits are divided into two regions, one provided by OIR and the second designed by the user to identify the type of part.

QUESTIONS WITH ANSWERS

Q. 23.1. What is the basic philosophy of group technology ?

Ans. Basic philosophy of group technology is to identify and bring together related or similar parts and processes, to take advantage of the *similarities* which exist between them during all stages of design and manufacture. In this way *part design and process can be standardised and families of like parts can be produced efficiently and economically.*

It may be borne in mind that *there exists relationship between the final or finished products and the components from which they are fabricated.*

Q. 23.2. What advantages can be expected by introducing group technology in an organisation ?

Ans. Following are the *advantages of group technology in manufacturing :*

1. Reliable delivery.
2. Reduction of material handling.
3. Use of robots for material handling becomes useful.
4. Space utilisation is more adequately possible.
5. Smaller variety to tools, jigs and fixtures are required.
6. Improved quality of product and less amount of wastage is possible.
7. Improved resource utilisation results in greater amount of output.
8. Finish stock levels are also reduced.
9. Satisfactory work program.
10. Job satisfaction.
11. Standardisation of part design and minimisation of design duplication.

Q. 23.3. Explain the concept of group technology and U-line in the modern production system.

Ans. Group technology :

- Group technology is a concept of grouping machines in 'Group technology cells' so that parts having similar geometrical and processing features are processed in the same set or group of machines.
- Group technology layout is different from process type layout in which machines are grouped according to thier function. Flow of parts through a factory with a group technology layout is *more streamlined* than the flow in a process layout when (quantity to be produced/part variety) or (Q/P) ratio is more.
  - Very small Q/P ratio suggests the use of process layout and very large Q/P ratio support the use of a flow shop.
  - *Moderate Q/P ratio demands a group technology layout.*

U-line layout. Refer to Fig. 23.2.

- In flow shop type situations, various arrangements of machines are possible. The common ones are I flow or line flow ; L flow when space is a problem, U flow which has feeding to and ejection from the line at the same end.

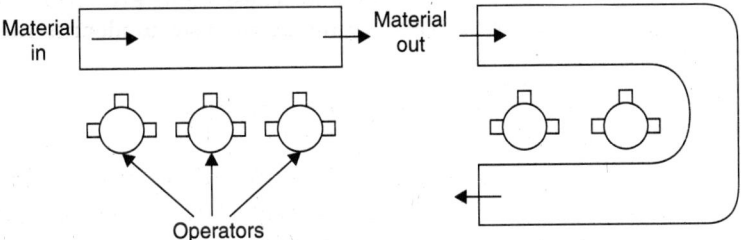

Fig. 23.2.

● A U-line is *easier to supervise.* A U-line can also adopt to different design and give different output rates through the same number of operators. When the flow rate is less, one operator can look after two machines with ease ; the machine in front and the machine behind him.

Q. 23.4. What is group technology ? State its area of application.

Ans. Group technology is the realisation that many problems are similar, and that by grouping similar problems, a single solution can be found to a set of problems, thus saving time and efforts.

Group technology is a manufacturing philosophy or principle whose basic concept is to *identify and bring together related or similar parts and processes, to take advantage of similarities which exist, during all stages of design and manufacturing.* It is the replacing of traditional jobbing shop manufacture by the analysis and grouping of work into families and the formation of group of machines to manufacture these families on a flow-line principle with the object of minimising setting lines and throughout times.

● Group technology has become an increasingly popular concept of manufacturing which can be applied in any industry (such as *"machining" "welding", "foundry", "presswork", "forging", "plastic moulding"* etc.) that is designed to take advantage of *mass production layout and techniques, in smaller batch-production systems.*

Features of group technology are shown in Fig. 23.3.

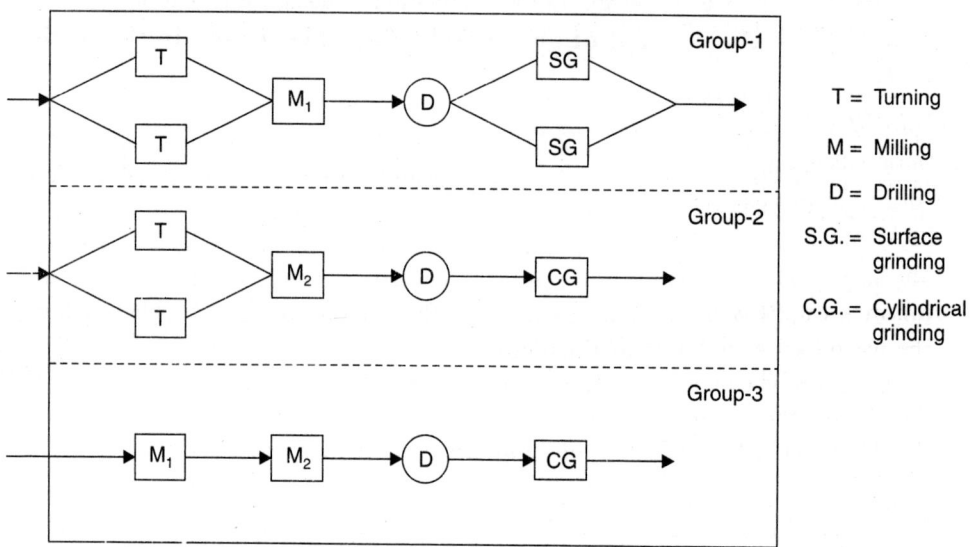

Fig. 23.3. Group layouts.

Q. 23.5. Give the reasons for adopting group technology.

Ans. Modern manufacturing industries are facing a lot of challenges caused by growing international competition and fast changing market demands. These challenges which are examplified in the following list, have and can be successfully met by group technology :

1. There is an industrial trend toward the low-volume production (small lot size) of a wider variety of products in order to meet the rising demand for specifically ordered products in today's affluent societies. In other words the share of the batch-type production in industry is grouping day after day, and it is anticipated that 75 percent of all manufactured parts will be in small lot sizes.

2. As a result of the first factor, the conventional shop organisation (*i.e.*, departmentalization by process) is becoming very inefficient and obsolete because of the wasteful routing paths of the products between the various machine tool departments.

3. There is no need to integrate the design and manufacturing phases in order to cut short the lead time, thus winning a competitive situation in the international market.

Q. 23.6. What do you understand by 'Production flow analysis' ? Explain.

Ans. Production Flow Analysis (PFA) is a *method for identifying part families and associated grouping of machine tools.* It neither uses a classification and coding system, nor uses part drawings to identity families. Instead PFA is *used to analyse the operation sequence and machine routing for parts produced in the given shop. It groups parts with identical or similar routings together.* These groups can then be used *to form logical machine cells in a group technology layout.*

Since PFA uses manufacturing data rather the design data to identify part families, it can overcome the following two possible anomalies :

— *First,* parts whose geometries are *quite different* may nevertheless require similar or indentical process routings.

— *Second,* parts whose geometries are *similar* may nevertheless require process routings that are quite different.

  ● However, the *'disadvantage'* of using production flow analysis is that it *provides no mechanism for rationalising the manufacturing routings.* It takes the route sheets the way they, are, with no consideration being given to whether the routings are optimal or consistent or even logical.

The *procedure in production flow analysis* can be organised into the following "steps" :

(*i*) Data collection.

(*ii*) Sorting of process routings.

(*iii*) PFA chart.

(*iv*) Analysis.

Comments on PFA :

— The *weakness* of PFA is that the data used in the analysis are derived from production route sheets. The routings may contain processing steps that are non-optimal, illogical, and unnecessary. Consequently the final machine groupings that result from the analysis may be suboptimal. Notwithstanding this weakness, *PFA has the virtue of requiring less time to perform than a complete parts classification and coding procedure.* It therefore provides a technique that is attractive to many firms for making the changeover to a group technology machine layout.

Q. 23.7. What is 'cluster analysis" ? Explain.

Ans. "Cluster analysis" *is employed for the analysis of production flow analysis chart to determine feasible group of processes and their respective packs of parts.* It makes use of algorithms for the study of similarities between objects in a quantitative manner as compared to the classification techniques, which appears to be descriptive.

"Clustering" may be defined as the *science of the classification of objects based on their possession or lack of defined characteristics.* This technique shows an approach to study the similarities between a diverse population of objects in a quantitative manner.

Clustering consists of the following *three stages :*

1. ***Preparing a post-operation matrix.*** This shows whether certain features (like a keyway on shaft) are present or absent.

2. *Computing a similarity coefficient matrix.* The bases of this is the extent to which the parts share common characteristics. In this case, coefficient would have a value one (1) when parts are identical and ten (10) when they have no common entity.

3. *Performing a clustering analysis.* In this case, the similarity between each pair of objects is examined and group of objects formed such that within each group, the objects are similar to each other according to set of rules which have been formulated previously.

Q. 23.8. Discuss briefly "cellular and flexible manufacturing".

Ans. For the development of effective quality systems and automation, low volume is a major obstacle. When only a few parts will be produced, little capital is made available for purchasing automated systems and up-to-date test equipment. In addition, other benefits resulting from learning curves and quality feedback may not be possible. Long lead times, high levels of inventory, and poor levels of quality usually result. *Cellular manufacturing* using the techniques of group technology and *flexible manufacturing systems* was developed to help solve these problems. **Cellular manufacturing** *increases the apparent production volumes by grouping similar parts.*

- The **Flexible Manufacturing System** (FMS) is a different approach to development of a specialised production line suitable for many different parts *using sophisticated computer control and a material handling system.*
    - **FMS** *consists of a group of manufacturing workstations connected together by an automated workpart handling system.* The system is capable of simultaneous processing of a variety of different part types at the various workstations under program control.
    - FMS system, however, is a *very expensive bridge for the gap between high-production transfer lines and lower production rates.*

Q. 23.9. Explain briefly "Group technology machine cells".

Ans. The traditional view of GT includes *the concept of "GT Machine cells"—groups of machines arranged to produce similar part families.* This cellular arrangement of production equipment is *designed to achieve an efficient work flow within the cell.* It also results in *labour and machine specialisation* for the particular part families produced by the cell. This presumably raises the productivity of the cell.

Although these advantages exist in GT machine cell, it is a matter of considerable inconvenience and disruption for the shop to make the conversion from a conventional process type layout to the GT cell layout. Today many practitioners argue that it is possible to achieve a good share of the benefits of GT without physically rearranging the machining into cells.

The organisation of machines into cells can follow one of three general patterns :

1. **Single machine cell.** The single machine approach can be used for work parts, whose attributes allow them to be made on *basically one type of process,* such as turning or milling.

2. **Group machine layout :**
    - The group machine layout is a cell design in which several machines are used together, with no provision for conveyorised parts movement between the machines.
    - The cell contains the machines needed to produce a certain family of parts.

3. **Flow line design :**
    - The flow line cell design is a *group of machines connected by a conveyor system.*
    - This design *approaches the efficiency of an automated transfer line.*

HIGHLIGHTS

1. *Group technology* is an approach in which similar parts are identified and grouped together in order to take advantages of their similarities in design and production.

2. Group technology is based on *"product specialisation"*.

3. A *part family or group* is a collection of parts, which either because of geometric shape and size or because of similar processing steps are required in their manufacture.

4. *Classification* involves arranging items into groups according to some system whereby like things are brought together by virtue of their similarities and are then separated according to a specific difference.

   A *code* can be a system used in information processing in which numbers or letters or a combination there of are given a certain meaning.

5. Three *levels* of coding are :

   (*i*) Monocodes hierarchial coding.

   (*ii*) Polycodes.

   (*iii*) decision-tree coding.

OBJECTIVE TYPE QUESTIONS

Fill in the blanks or say 'Yes' or 'No' :

1. Group technology is based on .... specialisation.

2. Group technology was originated in U.S.A.

3. Group technology is a new name for production system.

4. In group technology similar parts are arranged into .... families.

5. Group technology does not promote design standardisation.

6. Production scheduling is .... with group technology.

7. A .... is a collection of parts, which either because of geometric shape and size or because of similar processing steps are required in their manufacture.

8. Visual inspection method is the least sophisticated and least expensive.

9. Production flow analysis method makes use of information contained on .... sheets than part drawings.

10. A .... can be a system used in information processing in which numbers or letters or a combination there of are given a certain meaning.

ANSWERS

1. product	2. No	3. Yes	4. part	5. No
6. simplified	7. part family	8. Yes	9. route	10. code.

THEORETICAL QUESTIONS

1. What do you mean by 'Group technology' ? Explain.

2. List the advantages and limitations of group technology.

3. Discuss briefly "part families".

4. What is 'production flow analysis'?

5. Briefly discuss 'Formation and establishment of component family'.

6. Write short note on 'Collection of production data'.

7. What do you understand by 'Classification and codification'? Explain.

8. What are the basic requirements of classification and coding systems?

9. List the advantages of well-designed classification and coding system.

10. Explain briefly the following three levels of coding :
   (*i*) Monocodes hierarchial coding
   (*ii*) Polycodes
   (*iii*) Decision-tree coding.

11. Explain briefly the following coding systems :
   (*i*) Optiz system
   (*ii*) Multiclass system.

ADDITIONAL TOPICS

24

Inspection and Quality Control

24.1. Inspection—Introduction—aims of inspection—inspection standards—types of inspection—inspection devices. 24.2. Quality control—definition and scope—advantages of quality control—objectives of quality control—essential or principles of quality control—statistical quality control—control charts—sampling—advantages of sampling—acceptance sampling—single sampling—double sampling—standards and specifications—quality assurance concepts—functions of quality assurance. Highlights—Objective Type Questions—Theoretical Questions.

24.1. INSPECTION

24.1.1. Introduction

Inspection may be defined as the function by which the control of quality is maintained. It is one of the most important functions of production control. Just as design and production form the main stream of any production line, inspection acts as the control valve which regulates the flow of production along the main stream according to design and specifications. In an engineering factory, the design department sets the standards of accuracy which production must comply with and inspection department must enforce. Similarly in a chemical factory the technical department sets the standards for production to comply with and quality control section keeps the check. Not only are the standards of the end products to be maintained but in order to make this possible and easy to detect, inspection exercises quality control at different stages in the process. To do this the inspection must know the critical points at which a stricter quality control is called for. *The responsibility for detection of faulty work and returning it for correction or scrapping lies with the inspection department.*

24.1.2. Aims of Inspection

The main *principles or aims* of inspection are :

1. To determine that the material in process is of uniform quality.

2. To ensure that product is to the desired standard.

3. To initiate means to determine variations during manufacture.

4. To provide means to discover inefficiency during manufacture.

24.1.3. Inspection Standards

Inspection standards are ascertained by design department as per the requirement of goods and are set with an idea to check goods by physical means (such as size, weight, tensile strength and hardness etc.). The inspection standards by which all materials or products are checked are : (*i*) Physical properties, (*ii*) Finish and, (*iii*) Dimensions. Physical properties are variables and depend upon the materials or products concerned. These are determined by various means, the process used depending upon the property to be determined.

24.1.4. Types of Inspection

The inspection may be of the following *types :*

1. Process inspection.
2. Hundred percent inspection.
3. Sampling inspection.
4. Centralised inspection.
5. Floor inspection.
6. Cage inspection.
7. Final inspection.

1. *Process inspection.* This type of inspection is mainly concerned with the inspecting goods in process. The type of process inspection employed depends largely on the skill of work force, equipment and the tolerances specified. *First piece inspection* is very common type of process inspection. In a machine shop, for example, after a machine has been set up to turn out a part, one or two of the parts are run on the machine and are inspected before the entire lot is produced. The first piece inspection might indicate that adjustments should be made in the machine or it might show that first piece product is of desired specifications and that operator can go a head with the job. Where several successive steps are required to make a product, and the accuracy of each step is dependent upon the accuracy of prior operations, many companies run several pieces through all the steps required in the manufacture of the part. Thus, the acceptability of entire operation is assured prior to processing all the parts through any one step. This type of inspection is usually an extension of first piece inspection and is some times called *pilot-piece* inspection (because a pilot or trial run is made of all the equipment to be used).

Probable causes of defective work affecting the usefulness of the process inspection in whatever form should be given consideration. These are indicated below :

(*i*) Failure of operator to carry out instructions due to (*i*) Unsatisfactory or insufficient training, and (*ii*) Inability of operator to perform the job.

(*ii*) Bad setting.

(*iii*) Wear on tools (which results in the setting becoming incorrect).

(*iv*) Bad setting.

2. *Hundred percent inspection.* Hundred percent or *cent percent* inspection is quite common when the number of parts to be inspected is relatively small and also where a fault in one major part is likely to affect the operation of a finished product or where there is difficulty in manufacture of the parts and it is liable to errors. Here every part is examined as per the specifications or standard established and acceptance or rejection of the parts depend on the examination.

3. *Sampling inspection.* The use of sampling inspection is made when it is not practicable or too costly to inspect each piece. A random sample from a batch is inspected and the batch is accepted if the sample is satisfactory. If the sample is not to the desired specification then either entire batch may be inspected piece by piece or rejected as a whole. Statistical methods are employed to determine the portion of total quantity of batch which will serve as a reliable sample.

4. *Centralised inspection.* Centralised inspection is carried out in a specially designed inspection area which is separate from production area. The product is brought into this area where supervised inspection procedures are carried out. Highly centralised inspection is an impracticable proposition where larger parts are involved as in the manufacture of steam turbine. These are properly inspected on the floor or at the machine. Central inspection becomes a necessity in the case of small repetitive job tending towards mass production.

Advantages :

The centralised inspection has the following *advantages :*

1. With inspectors in one location, closer supervision over them can be established. With this closer supervision, less skilled inspectors can be employed and in addition, the close supervision makes possible the achievement of a more nearly uniform level of quality.

2. The worker and the inspector are separated, thereby practically eliminatiʌg any collusive action concerning whether or not a doubtful product should pass inspection because work is almost always stacked up and waiting for them.

3. Since the inspection work is concentrated in an area, very effective procedures and specialized equipment can be setup.

4. Production control is facilitated.

Disadvantages :

The centralised inspection is associated with the following *disadvantages :*

1. Defective work may be discovered too late to stop a large amount of spoilage.

2. There is more material handling and tie-up of product.

3. There is the need for greater work effort in co-ordinating the flow of materials through production and inspection.

5. *Floor inspection.* In floor inspection the parts are checked at the point of manufacture. It is usually more effective, and more desirable because time is lost in transporting materials to and from the centralized inspection cribs. This system keeps a constant check on production. In fact, to prevent the manufacture of bad parts many companies provide the machine operators with simple gauges to check the product quality, and the *patrolling inspectors* are only called where questions arise or where the products are not meeting the specifications. In such cases, the inspector usually helps determine and remedy the difficulty in the manufacturing process. Because of the independent nature of their work, floor inspectors are necessarily more highly skilled than those in centralised inspection.

6. *Cage inspection.* Cage inspection revolves round the idea that inspection should keep in step with production. Here the machines are placed round a fenced portion of works in which inspection staff is accommodated. To carry out the inspection the inspectors are provided with benches and gauges. Work performed in each machine is allowed to pass in convenient batches to the cage suitable for inspector to handle. The parts which come upto the standard specifications or standard desired are retained by the inspector, while the defective parts are sent back to the operator for rectification.

Cage inspection entails the following *advantages :*

1. Flow of work more even.

2. Adequate saving of labour.

3. Scrap reduction to a minimum.

4. Less movement of work.

5. Easy control of quantities.

7. *Final inspection.* The final inspection of a product usually takes place close to the point where the work is finished. It ensures that all components comprising the mechanism are in order. Some plants use only a visual inspection where as others give the product rigid tests.

24.1.5. Inspection Devices

Depending on the established level of quality, various techniques and devices are used to determine if a product meets the standards and specifications established by the engineering organisation. An effective inspection procedure for many products consists simply of a visual inspection. This is true in textiles, where colour and flows are determined usually. On more complex products where measurements are critical, various devices are used. Some of the commonly used devices are : (*i*) Inside and outside callipers ; (*ii*) Vernier ; (*iii*) Straight edge ; (*iv*) Try-squares ; (*v*) Flexible steel rule ; (*vi*) Protector ; (*vii*) Trammel steel beam ; (*viii*) Inside micrometer and outside micrometer ; (*ix*) "Go" and "Not Go" gauges of the following varieties, Plug, Ring, Gap and Screw ; (*x*) Fillet and radius gauge ; (*xi*) Dial gauge and (*xii*) Sine bar.

24.2. QUALITY CONTROL

24.2.1. Definition and Scope

Quality control *is an effective system for co-ordinating the quality maintenance and quality improvement efforts of the various groups in an organisation so as to enable production at the most economical levels which allow for further customer satisfaction.* Since this activity is one of the major responsibilities of management, quality control must be classified as a "management tool", along with similar tools such as production control and budget control. From the administrative point of view, quality control enters into all phases of industrial production process. It starts with the customer's specification, goes on to engineering, laboratory work, and materials purchasing through factory methods, job planning, manufacturing and mechanical inspection and electrical test to packaging and shipping and then back to the customer, whose needs must be satisfied with a quality product.

Effective human relations is basic to quality control. A major feature of this activity is its positive effect in building up operator responsibility for and interest in product quality. In the final analysis it is a pair of human hands which performs the important operations affecting product quality. It is of utmost importance to successful quality-control work that these hands be guided in a skilled, conscientious and quality minded fashion.

24.2.2. Advantages of Quality Control

1. Improvement in product quality.
2. Improvement in product design.
3. Reduction in operating costs.
4. Reduction in operating losses.
5. Reduction of production line bottlenecks.
6. Improvement in employee moral.

24.2.3. Objectives of Quality Control

1. To determine size, material, design, appearance, workmanship, finish and other relevant properties.

2. To ensure that products of lower quality may not go into hands of customer.

3. To carefully observe and analyse the deviation from the set standard of quality during manufacture and to investigate the causes leading to such deviation.

4. To apply corrective measures to achieve the real mission of quality control.

24.2.4. Essentials or Principles of Quality Control

1. A competent personnel should shoulder the responsibility for the quality of products.

2. For determining the variations in quality, clear-cut standards should be setup before hand.

3. To ensure that standards of measurement are uniformly applied, an efficient routine should be prepared.

24.2.5. Statistical Quality Control (S.Q.C.)

Statistical quality control *is a special type of inspection which employs mathematical techniques and probability.* In many instances it may be used in place of ordinary inspection procedures. It has been defined *"a method of applying statistical techniques to the collection and analysing of inspection and other data in order to achieve and maintain maximum economy in manufacturing process"*. Statistical quality control is based on the statistical theories and methods of probability to sample testing. Many of the efforts to ensure proper quality have always been done on sampling basis, that is, a relatively few of the entirely are inspected. However, with statistical quality control the risk involved, in assuming the sample has the same characteristics as a lot, is known and better quality control with minimum inspection cost can be achieved. The risk is not eliminated, but the probability of the reliability of the samples is expressed in numerical items.

Why use statistical quality control ?

One important reason is because it can help prevent defects from being made. In operation, accurate measurements of the parts at the machine are taken, compared to predetermined standards, and the decision reached whether the operation should continue. When and where to look for sources of trouble are revealed. Costly errors can be located and corrected before large scrap work and rework losses, due to which quality deficiency occurs. Another important reason for using statistical quality control is to supply an audit of quality regarding the producer's products. An universally understood measurement is supplied. In addition, the reasonableness of the quality standards established are checked. Also, savings are enjoyed in that losses due to operations giving non-acceptable quality are minimized.

24.2.6. Control Charts

Controlling the quality of materials, batches, parts and assemblies during the course of their actual manufacture is probably the most popularly recognized quality-control activity. The statistical tool most generally recommended for this work is the *control chart*. Its most prominent pioneer has been Dr. Walter A. Shewhart of the Bell Telephone Laboratories.

A **control chart** *may be defined as "A chronological (hour by hour, day by day) graphical comparison of actual product-quality characteristics with limits reflecting the ability to produce as shown by past experience on the product characteristics*. This comparison is usually made by selecting the measuring samples rather than by examination of each piece produced. The control chart method is a device for carrying out, on a factual basis the shopman's separation of variation into "usual" and "unusual" components. It compares actual production variation of manufactured parts with the control limits that have been set up for those parts. When these limits have been computed and then judged acceptable for use in production, the control chart takes up its major role-aiding the control of quality of materials, batches, parts and assemblies during their actual manufacture.

There are four main control charts : (*i*) $\overline{X}$-chart gives the mean measurement in the sample ; (*ii*) R-chart gives the range of measurement in the sample ; (*iii*) P-chart gives defectives in the sample, and (*iv*) C-chart gives the number of defects in the sample.

24.2.7. Sampling

Statistical quality control is based on sampling, probability and statistical interference, that is judging an entire lot by the characteristics of sample. Strange as it may seem, sampling has usually been found to be more nearly accurate than 100 percent inspection. The reason for this is that humans get tired of doing the same monotonous work over and over, and errors due to "inspection fatigue" creep in. Another and perhaps more easily understood reason for employing sampling rather than 100 percent inspection is that it is much less expensive where relatively large quantities are involved. This is true because sample sizes do not have to be large.

24.2.8. Advantages of Sampling

1. Time, money and labour are saved.

2. Less damage to the lot during inspection.

3. Number of inspectors required is less.

4. If entire lot on inspection is rejected, it is a pressure for quality improvement.

5. It can be applied to obtain a desired assurance of lot quality even in cases where the tests are destructive in nature.

24.2.9. Acceptance Sampling

A sample is a true representation of the entire lot. The inspection of a sample gives information regarding the goodness or badness of the lot. The sample being a representative of the variables present in the process gives exact idea as to whether the process is doing good job or bad at the time it produces. From this conclusion it is certain that sample gives an idea about the uninspected pieces because the uninspected pieces are the neighbours of the inspected pieces. Now acceptance sampling is valid because uninspected pieces have come and we have labelled as good.

24.2.10. Single Sampling

Acceptance or rejection of lot is based upon the units in one sample drawn from the lot. Select a sample at random from the lot of M say m pieces. Then the sample of m pieces is inspected and if it contains say c or less defectives, the lot is accepted. If it contains more than c defectives the lot is rejected. This procedure is called a *single sampling*.

In short, in single sampling if,

M = Entire lot,

m = Sample,

d = Number of defectives found in the sample,

c = Allowable defectives in the sample,

Then,   Accept the lot if $d \leq c$

Reject the lot if $d \geq c$

24.2.11. Double Sampling

Sometimes from the first sample if the number of defective articles exceed the number of allowable articles, then it is desirable to give the lot a second chance. This leads us to *double sampling*. Double sampling is illustrated diagrammatically in the Fig. 24.1.

The advantage of a double sampling plan over a single sampling plan is that a smaller first sample is taken which often provides the evidence necessary for the acceptance or rejection of the lot. Or if additional samples are required, the average size of the total sample is usually smaller. Thus inspection time and effort are saved.

In *multiple* or *sequential* sampling plans the first samples are even smaller than those of the double sampling plan with same types of advantages.

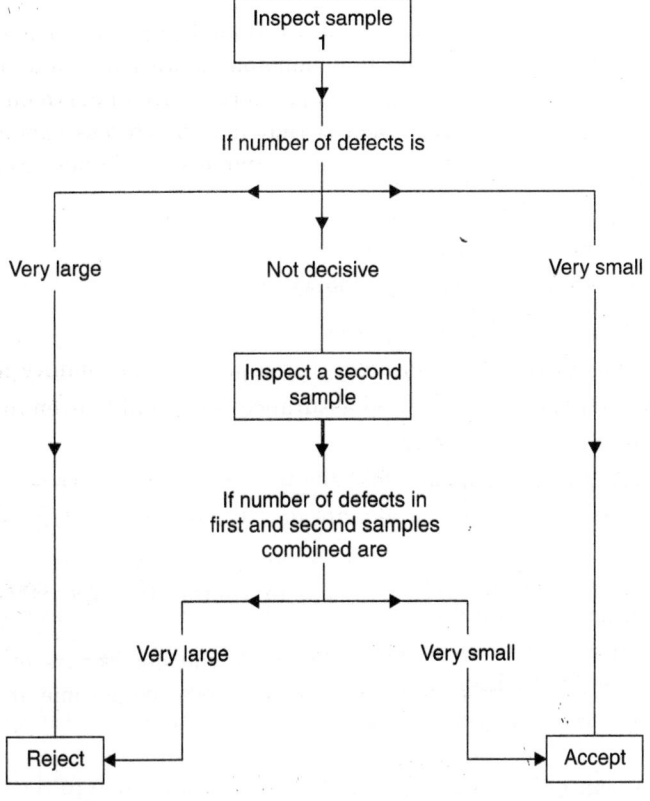

Fig. 24.1.

Techniques of quality control :

The principles and techniques by which the variables in a manufacturing process are controlled to achieve the degree of quality desired are as follows :

1. Standards and specifications.

2. Inspection (of materials, parts and products).

3. Statistical technique (including, sampling, analysis and charting).

4. Inspection devices.

24.2.12. Standards and Specifications

It is seen that the imperfections which come up in a manufacturing operation are caused by the variables. To reduce the imperfections the variables can be minimised through proper control. Generally the standards of engineering, design process, and materials are determined by those standards which establish the quality of products. Quality standards should be (i) measurable, (ii) reasonable, (ii) available, and (iv) understandable. These are generally set by product engineering department in cooperation with the sales, production control, cost and inspection departments. *The standards should in no case be established by inspection department.* The main function of this department is to interpret the standards and to ensure that they are rightly observed.

Inspection. Under quality control programme the two main functions performed by inspections are (i) To segregate defective goods and to ensure that only goods of approved quality pass into the hands of customers. (ii) To determine flaws in the raw material or in the processing of that material which may crop up trouble at subsequent operations.

The inspection practices usually recognized as the main steps towards contributing to the control of quality through inspection are : (*i*) Control raw materials and purchased parts ; (*ii*) Locate inspection strategically ; (*iii*) Plan the inspection operations ; (*iv*) Inspect for the defects promptly ; (*v*) Control inspection output and accuracy ; (*vi*) Set-up a procedure for handling border line material and (*vii*) Make use of inspection records.

24.2.13. Quality Assurance Concepts

Assurance *means insurance.* **Quality assurance** *means quality insurance. The main purpose of quality assurance is to assure the product is fit for use.* The concept of quality assurance has much in common with the concept of financial assurance. In the quality assurance concept qualified independent auditors examine the quality of the product and issue a certificate that the product is fit to use. *A quality assurance system is an effective methods of attaining and maintaining the desired quality standards.* This concept is based on the fact that quality is the responsibility of all functions.

Importance of quality assurance :

Importance of quality assurance is felt due to the following reasons :

- *It separates the defective components and parts from other components of required quality.*
- Quality assurance *finds out defects in raw materials and tools,* only good quality materials, proper tools and materials should be used in the production.
- Quality assurance *prevents further work being spoiled* which have been already detected to be defective during inspection.
- Quality assurance of finished product will ensure that the *operation of the product will be safe.*
- Quality assurance detects sources of weakness and trouble in the finished products and thus checks the work of designer.
- Quality assurance *ensures that quality of goods* supplied to the consumers is up to the mark.
- Inspected products sent to the market are of quality and this builds the reputation of the industry and earns goodwill of the consumers, because of quality assurance concept.

24.2.14. Functions of Quality Assurance

Right from the raw materials to the finished product, quality assurances are required to be carried out. They are as under :

- Quality assurance of raw materials.
- Process quality assurance during manufacture.
- Metallurgical and metallographic quality assurance.
- Purchased parts quality assurance.
- Finished goods quality assurance.
- Tool quality assurance.

HIGHLIGHTS

1. *Inspection* may be defined as the function by which the control of quality is maintained.
2. *Quality control* is an effective system for coordinating the quality maintenance and quality improvement efforts of the various groups in an organisation so as to enable production at the most economical levels which allow for further customer satisfaction.

3. *Statistical Quality Control* is a method of applying statistical techniques to the collection and analyzing of inspection and other data in order to achieve and maintain maximum economy in manufacturing process.

4. A 'control chart' may be defined as a chronological (hour by hour, day by day) graphical comparison of actual product-quality characteristics with limits reflecting the ability to produce as shown by past experience on the product characteristics.

5. *Quality assurance* means quality insurance.

OBJECTIVE TYPE QUESTIONS

Fill in the blanks or Say 'Yes' or 'No' :

1. The responsibility for detection of faulty work and returning it for correction or scrapping lies with the .... department.

2. .... may be defined as the function by which the control of quality is maintained.

3. .... inspection is mainly concerned with the inspecting goods in process.

4. The use of .... inspection is made when it is not practicable or too costly to inspect each piece.

5. .... inspection is carried out in a specially designed inspection area which is separate from production area.

6. In floor inspection the parts are checked at the point of manufacture.

7. .... inspection revolves round the idea that inspection should keep in contact with production.

8. In cage inspection the flow of work is less even.

9. Effective human relations are basic to quality control.

10. One of the objectives of quality control is to ensure that products of lower quality may not go into hands of customer.

11. Statistical quality control is a special type of inspection which employs mathematical techniques and probability.

12. Statistical quality control is based on sampling probability and statistical interference.

13. A .... is true representation of the entire lot.

14. The standards in no case be established by inspection department.

15. Quality assurance means quality ....

ANSWERS

1. inspection	2. Inspection	3. Process	4. sampling	5. Centralised
6. Yes	7. Cage	8. No	9. Yes	10. Yes
11. Yes	12. Yes	13. sample	14. Yes	15. insurance.

THEORETICAL QUESTIONS

1. Define the term 'Inspection'.
2. What are the aims of inspection ?
3. What do you understand by 'Inspection Standards' ?
4. Enumerate various types of inspection and explain briefly any two of them.
5. Describe briefly any two of the following types of inspection :
   (*i*) Process inspection.
   (*ii*) Hundred percent inspection.
   (*iii*) Centralised inspection.
   (*iv*) Cage inspection.

6. Explain briefly the term 'Quality Control'.
7. What are the advantages of 'Quality Control' ?
8. List the objectives of 'Quality Control' ?
9. What are the essentials or principles of 'Quality Control' ?
10. What do you mean by 'Statistical Quality Control'.
11. What are control charts ? Explain ?
12. What is sampling ?
13. What are the advantages of sampling ?
14. Explain briefly the following :
   (i) Single sampling.
   (ii) Double sampling.
15. Write a short noted on 'Standards and Specifications'.
16. What do you mean by 'Quality Assurance' ?
17. What are the functions of 'Quality Assurance' ?

25

Maintenance Management, Work Study and MIS

25.1. MAINTENANCE MANAGEMENT

Types of maintenance :

It is very difficult to classify the types of maintenance as it is an integral part of smooth functioning of a plant as a whole. Fig. 25.1 shows types of maintenance. However, the following types of maintenance are important from subject point of view :

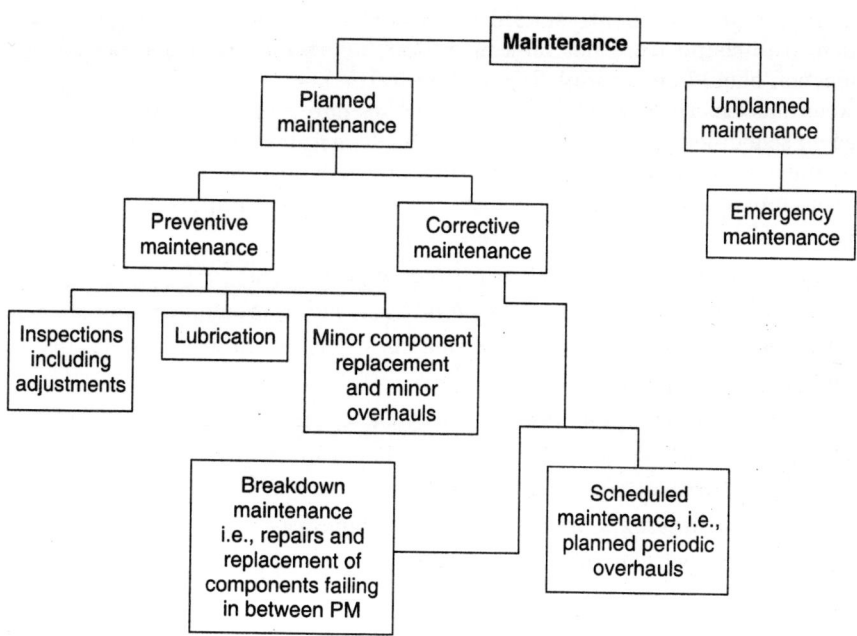

Fig. 25.1. Types of maintenance.

873

1. Preplanned maintenance.
2. Breakdown or Corrective maintenance.
3. Preventive maintenance.
4. Schedule maintenance.

1. Preplanned maintenance :

- The chief feature of organizing proper maintenance is the adoption of orderly and systematic methods. A regular programme should be drawn up in advance for the routine inspections, adjustments and lubrication of machines. When a machine is well cleaned and serviced, both the maintenance man and the operator feel proud of it and try to maintain it still better. *Cleanliness directly helps in tracing cracks and other defects.* Preplanned maintenance is required due to the following *reasons* :
- Increased mechanisation.
- To keep the production failure at minimum rate.
- To reduce the quantity of spare parts.
- To equip machines upto date so as to make maximum use of the available man power and to minimise the disturbance of operation.

2. Breakdown maintenance :

- *Without prior warning or without giving notice sometimes the production machine would have breakdown suddenly.* Priority will be given to emergency breakdown in order to minimise interruption of current productive operation. In all such cases, the causes of the breakdown should be noted and steps are to be taken to see that similar breakdowns do not occur in future.

3. Preventive maintenance :

- *Prevention is better than cure.* It is more efficient than remedial maintenance. The machines and equipments should be inspected periodically to avoid risks and losses which are arising from not making repair before damages had occurred. Preventive inspection should be scheduled according to the needs of given situation. The aims are, to keep machinery in good running order, to maintain continuity of production and to keep off scheduled preventive inspection the department makes timely preventive repairs and adjustments to avoid serious machine failure which would disturb production. Highly skilled employees can help for a good maintenance of equipment in industries.

(a) Objectives of preventive maintenance :

Following are the *objectives of 'Preventive maintenance'* :

- To get *maximum availability of the plant* by avoiding breakdown and shut-down period.
- To keep the *machine in proper condition* in order to maintain the quality of product.
- To ensure *safety of workers.*
- To maintain the maximum *production efficiency* of the plant.
- To keep the *maintenance economical* and at optimum cost.

(b) Functions of preventive maintenance :

Following are the *functions of Preventive maintenance :*

- Proper and timely *inspection.*
- Servicing includes *cleaning, lubrication* etc.
- Planning and *scheduling.*
- Records and *analysis.*
- *Training the maintenance staff.*
- *Storage of spare* parts.

(c) *Advantages of preventive maintenance :*

Following are the *advantages of preventive maintenance :*

- *Reduction* in production *down time.*
- *Less standby* equipment.
- *Less overtime* pay for maintenance staff.
- *Less expenditure* on repairs.
- *Storage of less spare parts.*
- *Greater safety* of employees.
- *Increased* equipment *life.*
- Better product quality and lesser product repairs.

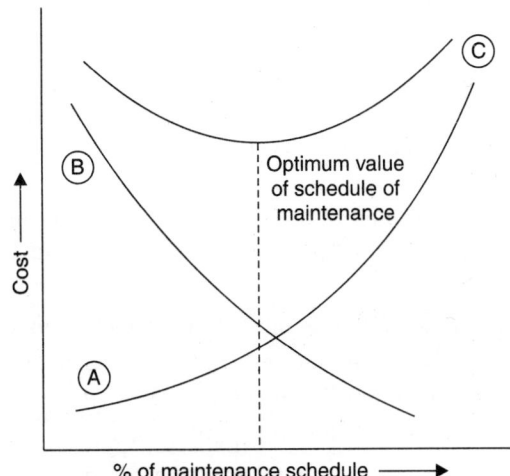

A—Breakdown maintenance,
B—Schedule maintenance,
C—Breakdown cost and schedule maintenance cost.

Fig. 25.2. Maintenance strategies.

4. Schedule maintenance :

The aim of this maintenance is to avoid breakdown. It includes inspection, lubrication, repair and overhauling of equipment, in a predetermined schedule. It is generally followed for overhauling of machines, cleaning of water and other leaks, whitewashing of building etc.

25.2. WORK STUDY

25.2.1. Definitions

As defined by British Standard Institution, *work study* is a generic term for those techniques particularly *'Method study'* and *'Work measurement'* which are used in the examination of human work in all its contexts and which lead systematically to the investigation of all the factors which affect the efficiency of the situation being reviewed in order to seek improvements.

Standard time. The time taken by a normal worker for a specific task or job under moderate conditions of working.

Standard time = Average time × rating factor + other allowances.

Rating factor. It is also known as *'Levelling factor'*. Time study engineer multiplies actual time with rating factor to get the average time which a normal worker would take. This is expressed as a percentage of the efficiency of representative operator, which indicates how efficient, a worker is as compared to average fellow workers. Rating factor is generally assumed to be 90 to 120 percent.

Performance rating. Comparison of performance of the operator under observation with normal performance.

$$\text{Performance rating} = \frac{\text{Observed performance}}{\text{Normal performance}} \times 100$$

25.2.2. Symbols used in Work Study

Symbols used in process charting are as follows :

| Operation | Storage | Inspection | Delay | Transport |

25.2.3. Steps Involved in Method Study

1. Select .... the work and area to be studied
2. Define .... the problem
3. Record .... all relevant files
4. Examine .... all relevant facts critically
5. Develop .... a new most economical and effective method
6. Sell .... the new method and find out discripencies
7. Install .... the new method as standard practice
8. Maintain .... the new method by regular checks.

25.2.4. Recording Techniques used in Method Study

(*i*) The Operation Process Chart   (*ii*) The Outline Process Chart

(*iii*) The Flow Process Chart (Material)   (*iv*) The Flow Process Chart (Man)

(*v*) The Multiple Activity Chart   (*vi*) Two Handed Process Chart

(*vii*) The simultaneous Motion Cycle Chart (SIMO Chart)

(*viii*) The Flow Diagram   (*ix*) The String Diagram

(*x*) The Travel Chart.

25.2.5. Time Study

Definition. *Time study is the analysis of a job for the purpose of determining the time that it should take a qualified person working at a normal pace, to do a job using a definite and prescribed method. This time is called standard time for the operation.*

Objectives. The following are the purposes or objectives of time study :

1. To serve as basis for determination a standard time during which an operation may be performed efficiently.

2. A basis for establishing a standard time-data for preparing a fair incentive wage plan.

3. An aid in bringing improvement in methods.

4. Production planning and control purposes.

5. Cost control purposes.

6. To be of great help to the motion study of job.

7. To achieve a uniform flow of work and thus to be helpful in layout of a plant on a scientific basis so that machine capacity may not be unbalanced.

8. To strive for improvement in operating efficiency.

Methods of time study. The three commonly used methods of conducting time study are :

1. Stop watch method.

2. Time recording machine.

3. Motion picture camera.

Time study allowances. The normal time for an operation does not include any allowances ; it is simply the time that a qualified operator would need for the performance of the job if he works at a normal pace. It is quite obvious that it cannot be expected that an operator will continue working all day without some interruptions. A little time will be consumed by the worker for his personal needs, for rest and for reasons beyond his control. Such interruptions necessitate allowance which may be classified in the following way :

1. Personal allowance.

2. Fatigue allowance.

3. Delay allowance.

25.2.6. Motion Study

Definition. *Motion study is a management technique linking motion to each other in such a scientific way that bodily and mental fatigue may be eliminated ; working conditions, machines and materials best suited to men may be provided resulting in the production of best product.* Frank Gilbreth, the leading exponent of this novel management technique defined motion study as *"the science of eliminating wastefulness resulting from unnecessary, ill directed and in-efficient motions".*

Advantages :

Although motion study is a time consuming and costly process yet it has the following *advantages :*

(*i*) Unnecessary, wasteful and tiresome motions are eliminated.

(*ii*) Minor changes in method and in equipment may be devised.

(*iii*) It makes possible the effective distribution of work, arrangement of work place and tools.

(*iv*) It exerts a solutary influence upon the general morale of an organisation when the savings made are shared with the employees.

(*v*) Various methods of performing operations may be changed and newer and more effective ones found.

(*vi*) It enhances productivity and results in the reduced cost of production.

(*vii*) Motion study promotes operation planning in a scientific way and establishes the time standard accordingly.

(*viii*) Data are always secured from which a series of job specifications may be developed.

Operation charts. The operation chart or the left hand and right hand chart is a very simple and effective aid for analyzing an operation. Here no timing device is required and on most kinds of work the analyst is able to construct such a chart from observations of the operator at work. The main objective of such a chart is to assist in finding a better way of performing the task, but this chart also has definite value in training operators.

While preparing operation charts two symbols are commonly used. The small circle represents transportation, such as moving the hand to hold an article and the large circle indicates such actions as holding, positioning, using or releasing the article. In signing a letter with a fountain pen the left hand holds the paper while the right hand performs the various movement as indicated in the Fig. 25.3.

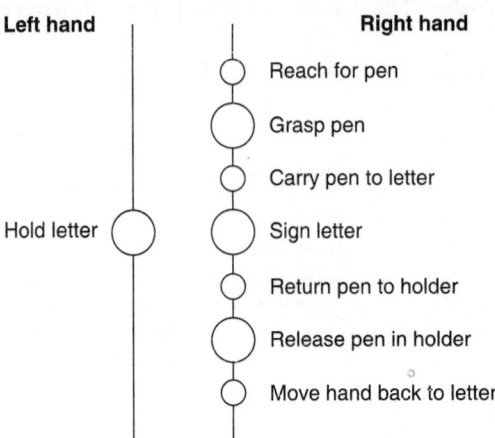

Fig. 25.3. Operation chart showing the movements
of the two hands in signing a letter.

An operation chart involves the following *steps* in its preparation :

(*i*) Draw a sketch of work place, indicating the contents of the bins and location of tools and materials.

(*ii*) Watch the operator and movement of his hands, observing one hand at a time.

(*iii*) Record the motions or elements for the left hand on the left hand side of a sheet of paper, and in the similar way record the motions for the right hand on the right hand side of the sheet. Usually a necessity is felt to redraw the chart as it is rarely possible to obtain the motion of the two hands in proper relationship on the first draft.

Flow process chart. It *is a graphic representation of various activities occurring on the plant floor.* It is an elaboration of an operation process chart and assimilates transportation, storage and delay. A flow process chart traces the parallel flow of two or more components through their respective fabricating operations and their subassembly and final assembly.

The flow process chart accumulates and classifies the complete information necessary for the analysis and improvement of plant operations as a whole or of one phase. An improved flow process chart provides an important basis for revising an existing plant layout. This chart is also used to check and verify the efficiency of a proposed floor plan for a new plant.

To prepare a flow process chart (See Fig. 25.4) the engineer must visit the plant to study and record the plant activities step-by-step. When the chart is completed the engineer will have an intimate knowledge of the present process and layout. These charts can be prepared with little difficulty for a plant that is manufacturing standardized products. In case of job lot plants, however, it is usually felt necessary to prepare a number of generalised process charts, which will account for large bulk of output. It necessitates the accumulation of data found on the route cards of past orders. The production orders are studied and classified into categories that require the same or similar production processes. When such data are ready, the engineer will follow step-by-step and record on the chart all the activities necessary in the production of each important category of customer's orders.

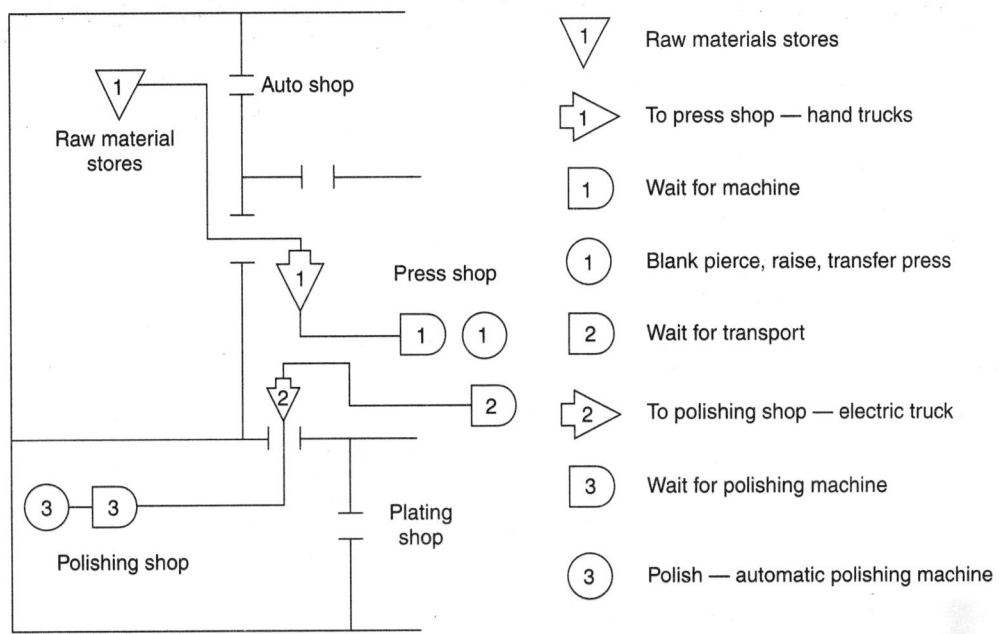

Fig. 25.4. Flow process chart.

Flow diagram. *A flow diagram is an aid to the visualisation of the movement of the materials on an existing floor layout.* This type of diagram is prepared on the floor-plan drawing or on an onion sheet superimposed on the floor plan drawing or blue prints. Use of colour lines is made to trace the flow of work through the machine stations and fabrication and assembly operations required for the output of different products. In case of job-lot plants, the flow of typical jobs or general flow of groups of jobs that require more or less similar production operations can be represented by a number of colour lines. A flow diagram analysis indicates where long lines of handling, back tracking, criss crossing, bottlenecks and confusion exist in the present arrangement and where production operations and service activities are located. In short, the flow diagram checks the effectiveness of overall arrangement of plant activities for materials handling and suggests where revisions can be made. Such an analysis enables the engineer to determine which machine stations, assembly areas, store rooms, office space, locker rooms etc. should be relocated to attain a greater economy in handling Fig. 25.4 shows a flow diagram and a flow process chart.

Multiactivity charts. A multiacitivity chart is used whenever it is necessary to consider the activities of a subject in relation to one or more others on the same document. By means of separate bars placed against a common time scale to represent the activities of individual worker or machine during a process, this chart indicates clearly periods of ineffective time with the process. The construction of this chart also helps in a way that the most important subject from the aspects of costs receives the major emphasis. Furthermore it is particularly useful for enabling maintenance and similar work to be organized in order that the time expensive equipment out of commission is reduced to a minimum. It is also a useful means when organising team work deciding the number of machine workers and for such like purposes. It assists in recording complex processes in a simple way for study at leisure.

Chart's construction. The activities in respect of workers and machines are usually recorded by shading the respective bars. The timings which can be built up from previous measurements or by direct timing, need only be adequately accurate to ensure that the chart will be as effective as possible. The timings determined by clock or wrist watch are sometimes sufficient ; however, frequently it will be necessary to ascertain times by one of the techniques of work measurement. The activities are then plotted in sequence against time scale within their own particular bar on the chart Fig. 25.5 shows a multiple activity chart for the job writing a letter using a short hand typist.

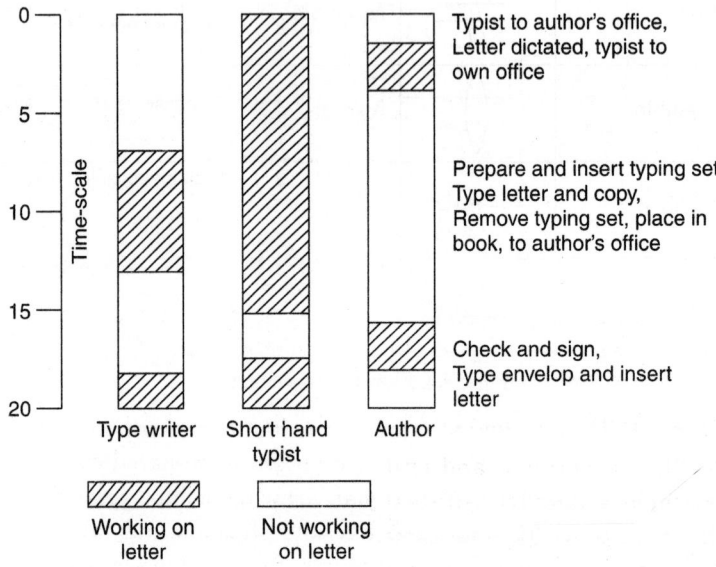

Fig. 25.5.

Man machine charts. A man-machine chart is a record of the simultaneous activities of men and machines. It is similar to the multi-activity chart except that the work done by a machine is also listed in a sequence column. Its purpose is to facilitate the study of work performed so as to arrive at a better distribution of work between the man and the machine. By analysing these charts we find, for example that one man can easily run three machines or a man machine study may show that we need two men to operate one machine efficiently. Man machine charts can be used to study any combination of men and machines—two men and six machines, one man and two machines and so on to develop a better method of performing the work, thereby gaining for management a greater control of labour costs.

Micromotion study. Micromotion study is the study of very small elements of motions (called *therbligs*) and their analysis for the work cycle frequently with the help of a special camera. Frank B. Gilbreth was the leading protagnist of this study. The study makes use of a microchronometer ('wink' clock) in the back grounds which records the therbligs, *i.e., ultra small elements* and measures the time by taking pictures at a constant speed. *Its application is economical only in those concerns where identical tasks are being performed by a number of highly skilled operators.*

Therbligs. Glibreth introduced certain symbols called *'therbligs'* to indicate human movement such as follows :

	search		find		select
	grasp		hold		transport (loaded)
	position		assemble		use
	disassemble		inspect		pre-position
	release load		transport empty		rest
	unavoidable delay		avoidable delay		plan

25.3. JOB ANALYSIS, JOB EVALUATION AND MERIT RATING

Job analysis. It is a detailed and systematic study of job to find out the nature and qualifications of the people required for efficient performance of job. Job analysis reveals the tasks which constitute the job, the skills, and knowledge required for the successful performance of various tasks.

Job description. It is a formal and organised statement of the contents of a job.

Job specification. It states the minimum human qualities required to perform a job efficiently. It lays the requirements which the person selected for the job should satisfy.

Job standardisation. It involves the establishment of uniform mechanical facilities and methods and combines the specifications of human standards, so that efficiency can be maintained.

Methods of job evaluation. The important methods of job evaluation can be *classified* as under :

A. *Non-qualitative methods :*

    (*i*) Ranking or Grading methods

    (*ii*) Classification method.

B. *Quantitative methods :*

    (*i*) Factor comparison method

    (*ii*) The point rating method.

Merit rating or performance. It is defined as *"the process of evaluating the employee performance on the job in terms of requirements of the job".*

25.4. WAGES AND INCENTIVES

Nominal wages. The amount of money paid to a worker for his efforts is known as *nominal or monetary wages.*

Real wages. It is money value of all the facilities like free accommodation, free medical aid etc.

Living wages. The money that can meet some requirements of workers like food, cloth, education for children, medical care insurance etc. are known as *living wages.*

Fair wages. The wages that are sufficient to meet the basic necessities of life, are known as *fair wages.*

Classification of wage payment plans. A wages payment plan is one which satisfies the workers and at the same time brings profits to management.

Wage payment plans can be *classified* under two groups :

1. Non-incentive plans like Time or Day rate system, and

2. Incentive wage plans like piece rate and other schemes.

Wage plans not based on time study. The following are the wage payment plans or systems not based on time study.

1. Day rate or time study.

2. Piece rate system.

3. Combinations of time rate and piece work systems.

4. Halsey premium wage plan.

5. Rowan plan.

1. Day rate or time rate system. This system of payment involves the time as the basis of payment. It is the oldest of wage payment systems. Under this system the employees are paid at the rate so much per day or per hour of work done irrespective of the quantity of the work performed. *Presently this system of remuneration is commonly used as the basis of calculating the amount payable to indirect workers such as formen, supervisors, time-keepers cleaners, engine men, gate men etc. as the nature of their nature of work is such that 'time' above can be considered as the basis on which to remunerate them.* Furthermore 'Time Rates' are essential in certain industries where the work cannot be standardised or classified into suitable grades for piece rating. This system can be made more effective by careful classification of workers into certain grades according to their skill, capacity and willingness to work, and by fixing different scales of day wage in direct relation to their usefulness.

Advantages :

1. It is simple and easy to understand.

2. It proves quite satisfactory if the rates indicate the value of time spent on work.

3. The quality of work is improved as the workers are in no hurry to enhance the output.

4. It requires less administrative attention.

5. Trade unions favour this system as it involves simpler calculations.

6. It does not involve any physical overpressure because here it is not essential for the workers to overstrain themselves for boosting up the production.

7. It provides the worker some security against sudden reductions in his income as a result of unavoidable accident or sickness or fatigue from outside activities.

Disadvantages :

1. It provides no incentive to an efficient person.

2. It does not provide any measure to reward a good worker and punish the loafer.

3. "The day work method of payment" says Franklin, "permits many a man to work at a task for which he has neither taste nor ability, when he might make his mark in some other ....".

4. The herding together of men into classes regardless of personal character and performance leads to employer-employee trouble.

5. As it provides no encouragement to work hard ; to keep the workers working foremen and supervisors have to keep a strict watch on them.

2. Piece rate system. Under this plan workmen working in given conditions with given machinery are paid exactly in proportion to their physical output .... a workman is paid, from the stand point of the moment in direct proportion to his output, the actual amount of the pay per unit of service being approximately equal to the (marginal) value of his services in assisting the machinery to make his output. In this system employee makes all the gain or loss of his time. If he puts in more labour he will get more remuneration and if he waste his time his remuneration may fall below time wages. *This system is liked by the employers as they get more profit with the increase in production.*

Advantages :

(*i*) The efficient workers are benefited whilst the inefficient are punished in terms of decreased remuneration.

(*ii*) The workers are contented as wage payment is in due proportion to their efforts.

(*iii*) It enables easy computation of cost.

(*iv*) Estimating of jobs is facilitated as the piece work rates are known.

(*v*) The quality is maintained as the worker is paid only when his work has passed inspection.

(*vi*) The total costs are reduced, for while direct labour costs remain constant at all speed of performance, the increase in the amount of work done per hour decreases the hourly charge on account of plant or management.

(*vii*) This system not only increases the output and wages but brings improvement in methods of production as the workers demand materials free from defects and machinery in proper working order.

Disadvantages :

(*i*) When some workers earn too much money, the employers tend to exercise a 'cut' in the rate which creates grounds for discontentment of the workers and consequently friction between employers and employees develops.

(*ii*) Too much hard work done by the workers to earn more and more tells upon their health.

(*iii*) While payment based on speed provides an incentive for greater volume, it discourages quality.

(*iv*) Straight piece work does not guarantee day wages, so that a worker may at times earn below the subsistence level. Such fluctuations in earnings burden the workers with constant worry and annoyance.

(*v*) The whole of the benefit of extra wages earned goes to the workmen.

3. Combination of time rate and piece rate system. Under this system, a workman receives a fixed minimum weekly wage irrespective of the work done by him, provided he has worked for the full week. In case he works for lessers number of hours, his weekly wages would abate proportionately. For example, if the minimum weekly (assuming the week to consist of 48 hours) wage is fixed Rs. 24/- ; if the workman works for 40 hours only, he will get Rs. 20/-. So far the system of payment is based on time rate. Let us further assume that each piece of work allotted to the worker is priced at Rs. 3/-. The worker will have to prepare 8 pieces to earn Rs. 24/-.

If he completes 9 pieces within 48 hours he will earn Rs. 27/-. In case he is able to complete only 6 pieces, he will still be paid his minimum weekly wage of Rs. 24/- but as he earned Rs. 18/- on the piece rate basis he will have to make good the excess Rs. 6/- paid to him out of his subsequent wages.

The main disadvantage to the employer is that benefit of extra wages goes to the workers. This has resulted in the development of several Premium and Bonus methods of payment which provide for a portion of savings in wages to pass to the employer, and thus serves to unify the apparently opposite interests of employer and the employed.

Incentives. Incentives may be defined as *type of motivation that influences or arouses interest in the people to work.* When a number of persons are working on similar types of jobs ; it is but natural that productivity capacity of individual will vary. The worker whose productivity is higher than the normal, naturally expects some kind of incentive from the management. It is obligatory on the part of the management to give some reward to the efficient workers.

The incentives may be broadly *classified* as :

1. Positive incentives.

2. Negative incentives.

1. *Positive incentives :*

Positive incentives are further *classified* as :

(*i*) Financial.

(*ii*) Non-financial.

(*i*) *Financial incentives.* These incentives may be (*a*) short range and (*b*) long range.

Short range incentives include (*i*) Time rate, (*ii*) Piece rate, (*iii*) Other incentive plans, such as Hasley plan, Rowan plan, Gantt plan, Taylor plan etc.

Long range incentives includes profit sharing and partnership.

(*ii*) *Non-financial incentives.* The different non-financial incentives are :

1. Good working conditions.

2. Less hours of work.

3. Security of job.

4. Pride of belonging to an organisation.

5. Good designation.

6. Fairness in dealings.

7. Proper welfare and security etc.

8. Counselling facilities.

9. Recreational facilities.

10. Full opportunities for training and advancement.

2. *Negative incentives :*

1. Fear of punishment.

2. Fear of dismissal.

3. Fear of demotions.

4. Halsey premium wage plan. This was one of the first incentive plans that deviated from some form of straight piece work. It was devised by F.A. Halsey and was one of the first, if not the first, to use a guaranteed base and express standards in terms of time rather than money. It sets a standard time, usually by determining the average previous time during which the job can be completed, and offers the workmen an agreed percentages of wages of any portion of this time that he may save, in addition to his hourly or daily rate for the time consumed on the job, it will not be reduced, despite the fact the conditions may not have been standardized or jobs studied.

Under this system *day rate is guaranteed.* The system is liberal with the time allowance rather than with the premium percentage. *Here* $33\frac{1}{3}$ *to 50 percent of the time saved by the worker on the standard time for a particular job is credited to the worker.* So the worker's earnings will amount to wages on actual hours worked plus $\frac{1}{3}$ to $\frac{1}{2}$ of time saved on the basis of his time rate.

Advantages :

1. It is easy to introduce as no preliminary study is necessary except the circulation of previous average times.

2. By distributing the profit of saved time between management and men, it makes for the permanence of the bonus rate, as both parties benefit by it.

3. The psychology of the plan is adroit : an employee is satisfied with what he gains although part of the saved time by is shared by the employer.

Disadvantages :

The main *disadvantage* of the plan is that it possesses the weakness of the straight piece rate of taking an unscientifically determined standard time for its job. It depends upon the past performances instead of making new standards. From the point of view of administration, the policy is one of drift as in this plan the worker is left alone to decide whether or not produce more after the standard is reached.

5. Rowan plan. It is the modified form of Halsey system. This system like that of Halsey leaves previous conditions of operation and management undisturbed. Standard times are based on experience. *A time wage is guaranteed to those who fail to reach the standard.* Like the Halsey system the chief aim of Rowan plan is to ensure the performance of the premium rate, by limiting the earnings a workman can make by unusual saving in time. The plan differs from the Halsey plan in the method of bonus determination. Briefly stated, the rule of remuneration under this plan is that the *wage of time taken shall be increased by the same percentage as that by which the time set for the job has been reduced.*

Rowan system is more liberal than the Halsey system upto $\frac{2}{3}$ *time economy, but after that it is less liberal. Moreover, the maximum a worker can earn under Rowan plan is double the guaranteed wage, which is humanly impossible. Like Halsey plan, the Rowan plan can also be used for transitional purposes. The Rowan plan, however is fairly difficult for employees to understand.*

Wage plan based on time study. The following wage plans are based on carefully established times :

1. Taylor's different piece rates.

2. Gantt task and bonus wage plan.

3. Emerson plan.

4. Bedeaux or point system.

5. 100 percent bonus plan.

1. Taylor's differential piece rates. In this system of payment, the task is set at a very high level. Here two piece work rates are used, the higher rate being applicable to a workman who completes the task in the standard time or less, and, the lower one for the worker who falls short of such standard. In this method of wage payment *day wage is not guaranteed.* The standard of efficiency is arrived at as a result of close analytical time study of each job. In order to facilitate highest efficiency, plant conditions are highly standardised and every possible care is taken by the management to eliminate all waste of time such as waiting for tools, material etc.

This system is suited to those establishments in which indirect expenditure is very heavy in comparison with the cost of labour.

Advantages :

1. Sincere and good workers are always able to earn wages at a higher rate than the scheduled time rate.

2. If the task level is reached by the worker, his earnings will rise sharply. This acts as a great incentive for him to do his best.

3. Management is also benefitted whether a worker turns out one, two or three pieces within a stated time, the "fixed overhead for that period will remain unaltered". The overhead charges per piece will be reduced, and the production of the articles will become more economical.

Disadvantages :

1. The penalty inflicted on the slow workers is rather harsh.

2. Since the standard is set at very high level, the workers have to exert themselves tremendously ; consequently they lose their health with the passage of time.

2. Gantt task and bonus wage plan. Under this plan, not only does the workman receive a reward which is large enough to make him want to make standard, but also he is guaranteed his hourly rate if he fails to reach the goal. If he completes the work, he is paid at his regular hourly rate for the time allowed for the task, plus a percentage of that time. The piece work rate in this case is usually lower than the higher rate offered under the Taylor system, as the slow workers are not penalised but receive their usual time rate. The time and bonus for each job are fixed, the bonus being a fixed percentage on the time taken.

Advantages :

1. By guaranteeing a minimum wage rate, the system does not penalise those workers who fail to achieve the standard of bonus.

2. As foreman also get the bonus, they put in hard labour to bring the work of their workers to the bonus standard.

Disadvantages :

1. As a time rate is guaranteed to every worker, many workers do not exert themselves to reach the set standard and usually feel satisfied with the guaranteed wages. This increases the direct labour and overhead charges per piece.

2. As the piece work rate is usually lower than the highest rate offered by Taylor system, therefore efficient workers do not like this system in comparison to Taylor System.

3. Emerson plan. This plan of wage payment sets also a high task level similar to those of Taylor and Gantt. The system guarantees a daily wage rate and fixes a certain standard output which represents 100 percent efficiency. The workers which show upto $66\frac{2}{3}$ percent of the efficiency standard are paid only wage rate and no bonus. If the efficiency exceeds this percentage bonus is paid on a graded scale in a fixed ratio to the increased output. Thus when the efficiency is 90 percent, the bonus payable is 10 percent ; when 100 percent, the bonus is 20 percent, and further, with every 1 percent rise in efficiency the bonus increases by 1 percent.

Its chief disadvantage is the majority of workers excepting a few highly ambitious ones, feel satisfied with the guaranteed day wage plus low bonus and thus do not show keen interest in increasing the efficiency.

4. "Bedeaux or Point system". Bedeaux system of wage payment makes use of a special unit of measurement of human effort called a 'point' or 'B' which represents a certain amount of useful work plus time allowances for unavoidable delay and rest, the total of which can be accomplished by an average worker working at a normal speed in one minute. The ratio of actual work to rest or delay in a 'point' will of course vary in different types of work, but in all cases, the standard task would be measured by "60 point hour", *i.e.,* the achievement of 60 units or *B's* in one hour. The number *B's* allowed for a given piece work is the point standard for a job. The value of the number *B's* produced by an individual worker in an hour excess of 60 *B's* per hour is shared by direct and indirect labour. The worker usually gets 75 percent and the remaining goes as bonus to foremen, supervisors and indirect labour.

This plan has the limitations in the cost of time studies, amount of inspection needed and the extensive use of clerical work.

5. 100 percent bonus plan. The main feature of the plan is that the worker gets 100 percent of the bonus earned. In this way it compares with 'Straight Piece' work wage payment system, and differs from Holsey plan in which the worker gets a portion of the time saved. It deviates from straight piece rate in respects that the standards are expressed in time per unit of production rather in money. The time saved is multiplied by the full value of the hourly rate, with result that a rapid worker is paid a guaranteed hourly wage, plus the additional amount that a slower, standard worker would have eventually received upon completing the same quantity of work. This plan, in effect, is a straight piece-rate together with a guaranteed rate per hour, which is paid regardless of speed. After the standard speed has been attained, the total pay is exactly the same as if a straight piece rate were used.

Special forms of wage payments :

1. Group bonus plans.
2. Profit-sharing plans.

25.5. PURCHASING

- **Purchasing** may be defined as, *"Business activity directed to secure the materials, supplies and equipment required in the operation of an organisation".*
- The following techniques are used for carrying out purchasing :
  - (*i*) Single tender basis
  - (*ii*) Spot quotations
  - (*iii*) Limited tender basis
  - (*iv*) Open tender basis.
- The following procedure is adopted for processing of tenders and issue of supply orders :
  (*i*) After opening the tenders, comparative statement is prepared to compare the relative price and data.
  (*ii*) Normally, lowest rates are accepted if it conforms strictly to specifications.
  (*iii*) When rates other than lowest are to be accepted, a note mentioning reasons for accepting next higher should be recorded in the comparative statement.
  (*iv*) Finally, supply order is placed and follow up action taken for timely supply of materials.
  - In **Centralised purchasing,** the entire purchasing is carried under the responsibility of a single person.
  - Generally *small units apply centralised purchasing* techniques *while large units having multiple activities use decentralised purchasing.*

25.6. STORES AND STORE-KEEPING

- Store-keeping is a *function of receiving, storing and issuing of materials.*
- *Centralised stores* are used in small industries.
- Decentralised stores are used in large industries.
- The principle of *'first in first out'* should be followed by a store-keeper.
- A number of records are maintained by a store-keeper for keeping record of items.

These are :

(*i*) Bin card   (*ii*) Inward registers

(*iii*) Outward register   (*iv*) Stock registers

(*v*) Railway receipt register   (*vi*) Issue register

(*vii*) Surplus stock register.

- As soon as material is received in an industry, it must be subjected to proper verification and inspection.
- Store verification is necessary :

(*i*) To check accuracy of stores,   (*ii*) To verify physical count in case of doubt,

(*iii*) To be ready for internal audit,   (*iv*) To prevent theft/pilferage of costly items.

25.7. INVENTORY CONTROL

Inventory :

- *Inventory* includes not only materials and supplies but also machinery spares or other items which are subject to yearly depreciation charges. These latter must be stored ; they must also be separated and carefully accounted for. Inventory costs money. This cost has been rightly called *cost of possession*. It involves several items all important. It is to control the cost of possession of every business that employs an inventory control system. Some firms exercise too much control, and cost of control, eats them up. Others do not employ enough control, inventory costs that is cost of possession are not excessive. In others, the system inventory control is economical and adequate at the same time.

- *Inventory control* refers to the physical verification of the stock in the stores, intended to determine the state of affairs of the store organisation. It brings out to light if goods are being issued and maintained properly.

Objectives of inventory Control. An inventory control routine logically established and applied, employing competent personnel and utilizing where applicable the forms and techniques should accomplish the following *objectives* :

1. Maintain a supply of materials adequate to meet the production requirement in both quantity and quality.

2. Reduce the investment in these materials to a minimum.

3. Assure that materials received are in accordance with the specifications set forth on the purchase order.

4. Safeguard all the materials by proper storage against theft, breakage and deterioration.

5. Supply the producing department with the materials required at the times and places designated and prevent the misuse and diversion of the materials to improper destinations.

6. Maintain inventory records showing receipt, disposition and use of materials issued and quantities and kind of materials in stock.

Inventory control must be fluid, must adjust rapidly to external forces. Some of these external forces are not merely business variables, they have social implications—war, threat of war, storm, persistent strike in supplier industry, crop failure etc.

Pre-requisities of an inventory control system :

1. An adequate, enclosed, well arranged store room with (*i*) definite location of material (*ii*) definite identification of material.

2. Provision for eliminating slow moving and obsolete items.

3. Centralisation of authority and responsibility with an adequate staff to man the function, particularly a responsible, preferably bonded store-keeper.

4. A system for count and inspection of material upon receipt.

5. Absolute control of issuance of materials.

6. Periodic physical inventory.

7. Checks to ascertain enforcements of routine.

Organisation. Inventory control, in continuous manufacturing, is generally part of Production control because of the necessity for maintaining a flow of materials needed for the efficient and continuous operation of the production line. In intermittent manufacturing, as the need for a steady flow of materials is not so urgent, inventory control may be the responsibility of the plant manager, production superintendent, purchasing agent or a similar person, depending on the plant size and organisation. In smaller plants, purchasing agent is usually responsible for purchasing as well as maintaining an adequate supply of material.

Inventory quantity standards :

There are four quantity standards :

1. The reorder point.

2. The standard order.

3. The maximum.

4. The minimum.

Each of these quantity standards are illustrated by Fig. 25.6.

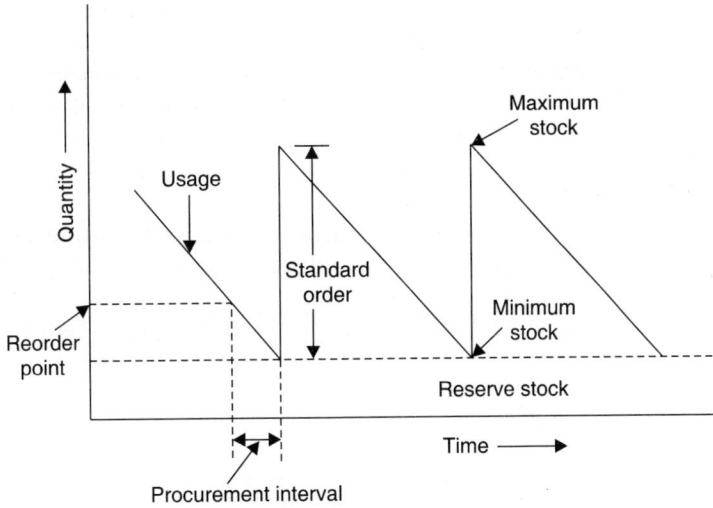

Fig. 25.6. The four quantity standards.

1. *The reorder point.* It indicates when to order and prevents the exhaustion of material. Thus if a recorder point of say 130 has been set, a new order will be issued when the quantity on hand reaches 130, or when the quantity available reaches 130 if the apportionment feature is used.

2. *The standard order.* It shows how much to order. It indicates the most economical purchase lot size. A purchase requisition is issued for the standard amount when the reorder point is reached.

3. *The maximum.* It represents the upper limit of the inventory. Ordinarily, no quantity of materials will be purchased that will cause the stock condition to rise above the maximum quantity. The purpose of establishing a maximum inventory is to prevent acquiring too much material and thereby needlessly tying up storage space and capital. A maximum inventory also assists in determining the storage space to be allocated for materials.

4. *The minimum.* It indicates the lower limit of the inventory. The minimum acts as a margin of safety.

Economic Ordering Quantity. The most economic ordering quantity is calculated by procurement and inventory (storage) carrying costs.

Purchasing cost consist of expenses for calling quotations, tenders, placing orders, inspection and other incidental charges. Inventory cost includes taxes, insurance, storage, handling and spoilage of the material. The theoretical economic ordering quantity is decided as allows :

Let, A = Annual usage in units,

S = Set-up cost per order,

I = Annual inventory carrying cost per unit, and

Q = Economic ordering quantity.

Procurement cost per year

= Number of orders placed per year × buying cost per order

$$= \frac{\text{Annual usage in Units}}{\text{Economic Ordering Quantity}} \times \text{Buying cost per order}$$

$$= \frac{A}{Q} \times S = \frac{AS}{Q}$$

Inventory carrying cost per year

= Average quantity of stock in inventory per year

× annual inventory carrying cost per unit

$$= \frac{Q}{2} \times I = \frac{IQ}{2}$$

Total cost per year = $\dfrac{AS}{Q} + \dfrac{IQ}{2}$

This will be minimum when the procurement cost equals the inventorying carry cost

i.e.,

$$\frac{AS}{Q} = \frac{IQ}{2}$$

∴ $Q^2 = \dfrac{2AS}{I}$ ∴ $Q = \sqrt{\dfrac{2AS}{I}}$

Economic ordering Quantity = $\sqrt{\dfrac{2AS}{I}}$.

Example 25.1. *The manager of a factory is purchasing forgings of outer-ring Rs. 10000 annually. Records reveal that cost of an order is Rs. 15 ; cost of inventory carrying is 9% of the average inventory value and unit price is Rs. 65. Determine :*

(i) Economic order quantity (Q_{opt}).

(ii) Optimum number of orders per year.

Solution. Set up cost (S) = Rs. 15 per order

Inventory carrying cost (I) = 0.09 × 65

I = 5.85 per unit per year

Annual demand A = Rs. 10000 per year

(i)
$$\text{EOQ} = Q_{opt} = \sqrt{\frac{2AS}{I}}$$

$$= \sqrt{\frac{2 \times 10000 \times 15}{5.85}} = \textbf{226.45 (Ans.)}$$

(ii) Optimum number of orders per year

$$= \frac{\text{Demand}}{Q_{opt}} = \frac{10000}{226.45} = 44.15 = \textbf{45 (Ans.)}$$

ABC Analysis (Value classification) :

- It is meant for relative inventory control in which maximum attention is given to those items which consumes more money and a moderate attention is given to those items which are of medium value, while the attention for low value items is reduced to routine procedure only.

- The *first category*, small numbers of high consumption cost items are called **"A-items"**. *Second category* of medium consumption value items are called **B-items**. *Third category* of large number of items with small annual consumption value are classed as **C-items.**

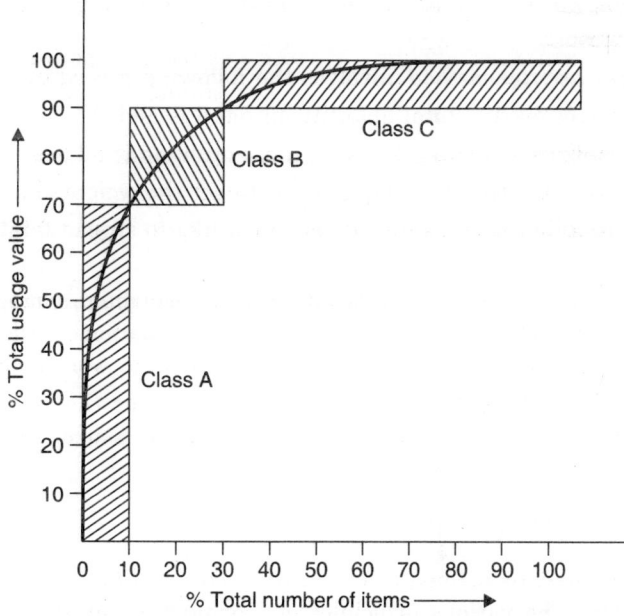

Fig. 25.7. Usage value classification of stock items.

The general principles of *ABC* Analysis can be applied to almost any activity. For example, a business may divide sales into *ABC* categories. Its best lines of sales and the most lucrative ones would be *A* items whereas miscellaneous items which do not bring in much profit would be *C* items, the rest being categorized as *B* items. Similarly, production programmes could be so divided depending on the importance of various jobs to be performed. The same applies to office routine. All successful organizations, in fact, adopt an *ABC* analysis system, perhaps, without actually knowing that they are doing so. *Although ABC Analysis is basically a matter of common sense, its formal application through a well laid out procedure, can attain maximum efficiency.*

In the field of materials management we are concerned mostly with purchasing, stores-inventory control and expenditing and follow up of purchase orders/requisitions. *ABC* analysis is eminently suited for use in this field.

Methods of pricing issue from stores. The method of pricing the material to be issued from the stores are described below :

1. *First-in first-out method.* This method is also known as the *original cost method.* Items first received into the stores are the first to be issued until exhausted. Materials are priced at the cost at which these items were placed in stock. This method is suitable for those industries where stocked items do not move very fast and have a high unit cost.

2. *Fixed prices method.* A fixed price is charged for each article issued but an adjustment account is opened for necessary adjustment of price of an article. This system may prove useful when market trend exhibits steady level and is used in connection with standard costing procedure.

3. *Average price method.* This method is further classified as :

   (*i*) Simple average method.

   (*ii*) Weighted average method.

   (*iii*) Moving average method.

   - In *simple average method,* the simple average of unit price is calculated by dividing the total of all unit pieces by the number of invoices, the quantity on each invoice being ignored.

   - The *weight average method* involves the following procedure :

   (*a*) Total quantity received to total quantity on hand is added.

   (*b*) Cost of materials received to the cost of those on hand is added.

   (*c*) Total values are divided by total quantities to get the weighted average.

This system is used by those organisations which like to spread total costs uniformly over all goods on hand.

   - The *moving average method* is a variation of the weighted average method. It is most suitable for application when there is a frequent fluctuations in the price of raw material and is considered desirable to bring uniform charges to working-process.

4. *Last-in first-out method.* This method is also known as the *replacement cost method.* Its basis are that the last items purchased are the first to be issued, the stock on hand being charged at the cost of the earliest purchases.

25.8. MATERIALS HANDLING

Material handling may be defined *as an art and science of the moving, packing and storing of substance in any from.* The various principles of economic materials handling are :

   1. Reduction in line.

   2. Reduction in handling.

   3. Equipment design.

Types of material handling equipment :

The material handling equipments may be classified as follows :

   1. *Horizontal fixed path equipment :*

      (*i*) Band and belt conveyors.

      (*ii*) Roller conveyors.

(iii) Slat and cross bar conveyors.

(iv) Vibrating conveyors.

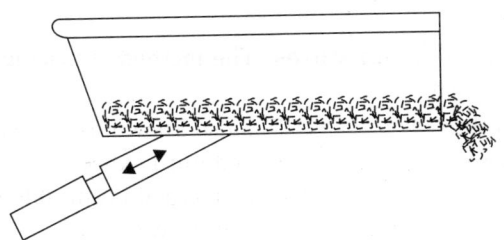

Fig. 25.8. Vibrating conveyor.

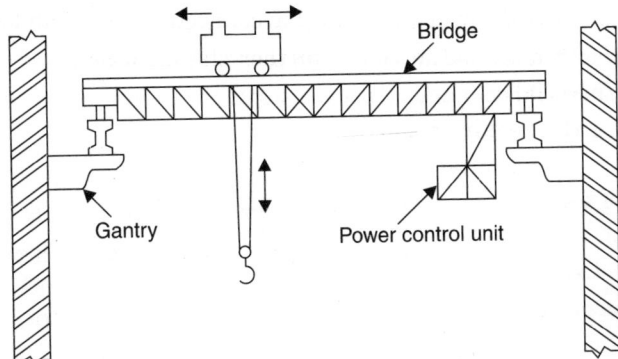

Fig. 25.9. Overhead crane.

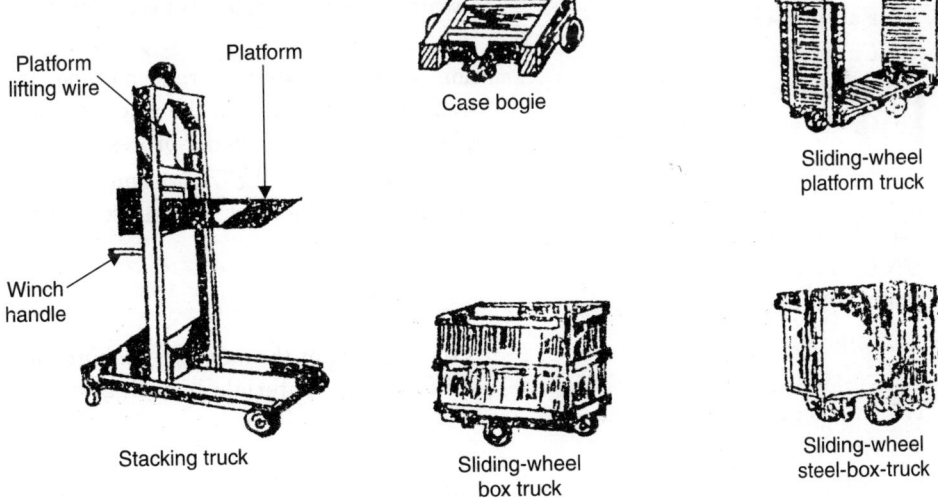

Fig. 25.10.

2. *Vertical movement :*

    (i) Bucket elevators.

    (ii) Roller spiral conveyor.

    (iii) Lifts.

3. *Overhead movement :*
 (*i*) Chain conveyors
(*ii*) Cranes.

4. *Combined vertical and horizontal movement :*
 (*i*) Flight conveyors.
(*ii*) Pneumatic conveyors.

5. *Horizontal non-fixed path equipment :*
 (*i*) Hand trucks
(*ii*) Power trucks.

25.3. FINANCIAL MANAGEMENT AND BUDGETING

- *Finance* can be said as an activity concerned with planning, raising, controlling and administering of funds used in the business.
- The scope of financial management consists of the following activities :
    (*i*) Estimating the requirements.
    (*ii*) Determining the capital structure.
    (*iii*) Sources of funds.
    (*iv*) Utilisation of funds.
     (*v*) Disposal of surplus.
    (*vi*) Management of cash.
    (*vii*) Financial controls.
- *Financial planning* refers to planning the requirements of finance, the sources from which it is to be raised and the application of funds. It involves decisions like :
    (*i*) Capitalisation.
    (*ii*) Capital structure.
    (*iii*) Capital budgeting.
    (*iv*) Divided policy.
     (*v*) Credit and collection.
- A *sound financial plan* should satisfy the following requirements :
    (*i*) Simplicity.
    (*ii*) Flexibility.
    (*iii*) Economy.
    (*iv*) Availability of adequate funds.
     (*v*) Liquidity.
    (*vi*) Solvency.
    (*vii*) Provision of contingencies.
    (*viii*) Optimum use of funds.
- While formulating financial plans the main factors that need consideration are :
    (*i*) Objectives.
    (*ii*) Nature and size of business.
    (*iii*) Status of enterprise.
    (*iv*) Growth and expansion plans.
     (*v*) Attitudes of management.

    (*vi*) Capital market trends.

    (*vii*) Goverment regulations.

- Some common sources of raising finances are :

    (*i*) Industrial banks.

    (*ii*) Unit trust of India.

    (*iii*) Industrial Finance Corporation of India.

    (*iv*) Life Insurance Corporation of India.

    (*v*) Industrial Development Bank of India.

    (*vi*) Shares.

    (*vii*) Debentures.

    (*viii*) Mutual funds.

"Budgetary control" is the process of defining desired performance through the preparation of budgets, measuring and comparing actual results with the corresponding budget data for taking of appropriate actions to correct deviations if any.

- *Budgets may be classified* as follows :

    A. 1. Fixed budget   2. Variable budget

    B. 1. Main budget   2. Master budget   3. Subsidiary budget.

- Sales budget, production budget, capital expenditure budget, materials and purchase budget, labour budget, selling and distribution cost budget, cash budget etc.

25.10. NETWORK ANALYSIS

Network analysis is a system which plans projects both large and small by analysing the project activities. It helps designing, planning, co-ordinating, controlling and in decision-making in order to accomplish the project economically in the minimum available time with the limited available sources.

Programme Evaluation and Review Technique (PERT). *It is a time event network analysis technique designed to watch how the parts of a programme fit together during the passage of time and events. This technique involves the application of network theory of scheduling problems.*

Project planning network. It is another name of PERT, which depicts the sequence of activities necessary to complete a project.

Event. It is a specific instant of time which marks the start and end of an activity.

Activity. It is an element of a project. It may be a process, material handling or material procurement cycle.

Critical path. It is the sequence of activities which decides the total project duration.

Total time. It is the time taken to complete a project. It is found from the sequence of critical activities.

Network diagram. It represent all the events and activities in sequence along with their interrelationship and interdependencies.

Bar chart. *It deals with complex activities.* It was developed by Henry Gantt in 1900.

Gantt progress chart. *It is used to make comparison between the actual and planned performance.*

Critical Path Method (CPM). It is highly similar to PERT. It is essentially an *arrow diagramming technique employing the concept of critical path and integrating all the usual factors of production, land, labour and capital.*

PERT uses *event-oriented network diagram* while **CPM** *uses activity oriented network diagram.*

25.11. OPERATION RESEARCH (OR)

- *Operation research* is the application of scientific methods, techniques and tools to problems involving the operations of a system sc as to provide those in control of the system with optimum solutions to problems.
- *Operations research* is carried out to solve a large number of problems.
- The procedure involved in operation research involves the following *seven phases :*
  - (*i*) Formulating the problem.
  - (*ii*) Collecting the data.
  - (*iii*) Constructing a mathematical model to represent system under study.
  - (*iv*) Deriving a solution from the model.
  - (*v*) Testing the model and the solution so derived.
  - (*vi*) Establishing controls over the solution with any degree of satisfaction.
  - (*vii*) Putting the solution to work, *i.e.,* implementation.
- Applications of 'operation research' are innumerous. It can be applied to almost all engineering problems.
- The various methods used in 'operation research' can be divided into the following two categories :
  - (*i*) Deterministic models   (*ii*) Probalistic models.
- **Linear programming.** *It is a classical operation research technique* that was primarily used for military applications during World War II.
- The linear programming techniques can be classified as follows :
  - (*i*) Graphical linear programming.
  - (*ii*) Transportation method : (*a*) Vogel's approximate method, (*b*) North-west corner method.
  - (*iii*) Simplex method.
- **Waiting line or queuing theory** *is used to examine the problem of waiting and minimising the waiting period.*
- *Simulation and Monte Carlo Technique* is a quantitative technique used for evaluating alternative courses of action based on facts and assumptions, with a mathematical model representing actual decision-making under conditions of uncertainty.

25.12. MANPOWER PLANNING AND CONTROL

- Manpower training, according to Vetter, is defined as *"the process by which the management determines how the organisation should move from its current manpower position to its desired manpower position".*
- Manpower planning should be scientific, need based and must be done well in advance.
- The various factors that need consideration in manpower planning are hours of working, number of shifts, type of production, product mix and performance rate.
- Manpower planning is an important activity as it provides plans for the recruitment, selection, placement and promotion of employees.
- *Span of control* is said as the number and range of direct habitual communication contacts between chief executive of an enterprise and his principal fellow officer. Its main objective is as to how many persons should a supervisor in an industry supervise.

- **Recruitment** refers to the discovery and development of the sources of required personnel so that sufficient number of candidates will always be available for employment in the organisation.
- **Training** of employees for specific job is necessary to bring the employee to the standards where he can carry out the industrial activity successfully.

25.13. FORECASTING

Forecasting means estimation of type, quantity and quality of future work.

Purpose or need of forecasting :

Sales forecasting is *essential* because ;

(*i*) It determines the volume of production and the production rate.

(*ii*) It forms basis for production budget, labour budget, material budget etc.

(*iii*) It suggests the need for plant expansion.

(*iv*) It emphasizes the need for product research development.

(*v*) It suggests the need for changes in production methods.

(*vi*) It helps establishing pricing policies.

(*vii*) It helps deciding the extent of advertising, product distribution, etc.

Forecasting techniques :

Following *techniques are used for forecasting :*

(*i*) Historical estimate   (*ii*) Estimation by salesmen

(*iii*) Statistical analysis   (*iv*) Moving average data method

(*v*) The exponential smoothing method   (*vi*) Market research by suitable questionnaire

(*vii*) Survey or buyer's views   (*viii*) Collective opinion.

25.14. MANAGEMENT INFORMATION SYSTEMS (M.I.S.)

- Management information system aims at providing reliable information to the management to take timely, sound and accurate decisions. The system should be flexible and capable of updating quickly.
- A few important steps involved in the design of M.I.S. are as follows :

1. Analysis :

(*i*) Problem recognition   (*ii*) Problem identification.

2. Synthesis of problem :

(*i*) Preparation of flow chart.   (*ii*) Examination of information documents.

(*iii*) Working out quantities.   (*iv*) Establishing inputs and outputs.

(*v*) Assigning the tasks and responsibilities.

(*vi*) Running in parallel.

- M.I.S. is necessary to take decisions. Decisions are taken at every stage but what is to be ensured is that decisions should be timely and practical.

Desirable Characteristics of Management Information System (MIS) :

Following are the desirable characteristics of MIS :

- It *should be able to generate information* for identifying alternatives and then selecting an alternative based on laid down criteria.
- It *must help in planning* of end results and must throw up as to what specific activities are required to achieve these.

- It should have an *in-built system* of measuring own and subordinate's performance, pinpoint responsib```y and indicate the corrective action.
- It should be *comprehensive* and cover all aspects of organizational activities.
- It should help in *achievement of objectives of organisation.*
- It should concentrate on developing information rather than facts. Only relevant information should be provided while irrelevant information is omitted.
- The information flow should follow organisation structure and should keep in view the delegation of various authorities.
- It must be an *integrated system* from which qualitative information for higher-level of management can be easily extracted.
- It *should identify the differential needs of information for planning and control* and develop both the types together.
- Data generated must be timely, cover past, present and future anticipation.
- It is preferable to use common data to the fullest extent on a simple format which satisfies maximum number of personnel.
- It should dovetail high level human communication which is indispensable for any information system.
- Information collection and flow should be designed keeping the managerial styles in view.
- It *must achieve appropriate and full utilisation of computer.*
- The system should provide for continuous reviewing.

HIGHLIGHTS

1. Maintanance can be classified as follows :
   (*i*) Preplanned maintenance ;   (*ii*) Breakdown or corrective maintenance ;
   (*iii*) Preventive maintenance ;   (*iv*) Schedule maintenance.
2. *Time study* is the analysis of a job for the purpose of determining the time it should take a qualified person working at a normal face, to do a job using a definite and prescribed method. This time is called *standard* time for the operation.
3. *Motion study* is a management technique linking motion to each other in such scientific way that bodies and mental fatigue may be eliminated.
4. *Merit rating* is the process of evaluating the employee performance or the job in terms of requirements of the job.
5. *Inventory control* refers to the physical verification of the stock in the stores, intended to determine the state of affairs of the store organisation.

OBJECTIVE TYPE QUESTIONS

Fill in the blanks or say 'Yes' or 'No' :
1. Cleanliness directly helps in tracing cracks and other defects.
2. The aim of .... maintenance is to avoid breakdown.
3. .... factor is also known as 'Levelling factor'.
4. .... study is defined as the science of eliminating wastefulness resulting from unnecessary, ill directed and inefficient motions.
5. A .... chart is an aid to the visualisation of the movement of the materials on an existing floor layout.
6. A .... chart is a record of the simultaneous activities of men and machines.
7. Gillbreth introduced certain elements called .... to indicate human movement.

8. .... is the process of evaluating the employee performance on the job in terms of requirements of the job.

9. The wages that are sufficient to meet the basic necessities of life, are known as .... wages.

10. Centralised stores are used in large industries.

11. .... includes not only those materials and supplies but also machinery spares or other items which are subject to yearly depreciation charges.

12. Inventory control must be fluid, must adjust rapidly to external forces.

13. The .... order shows how much to order.

14. .... can be said as an activity concerned with planning, raising, controlling and administering of funds used in the business.

15. Network analysis is a system which plans projects both large and small by analysing the project activities.

ANSWERS

1. Yes	2. schedule	3. Rating	4. Motion	5. flow
6. man-machine	7. therbligs	8. Merit rating	9. fair	10. No
11. Inventory	12. Yes	13. standard	14. Finance	15. Yes

THEORETICAL QUESTIONS

1. Explain briefly any two of the following types of maintenance :
   (i) Preplanned maintenance   (ii) Breakdown or corrective maintenance
   (iii) Preventive maintenance   (iv) Schedule maintenance.

2. What do you mean by 'Work study' ? Explain.

3. List the steps involved in 'Method study'.

4. What is 'time study' ? What are its objectives ?

5. Describe brief 'Motion study'.

6. What is micromotion study ? Explain.

7. What are 'Therbligs' ?

8. Explain briefly the following terms :
   (i) Job analysis   (ii) Job evaluation
   (iii) Merit rating.

9. Give the classification of wage payment plans.

10. Describe briefly the following wage payment plans :
    (i) Time rate system.
    (ii) Piece rate system.

11. List five non-financial incentives.

12. What do you mean by 'Contralised purchasing' ?

13. What is 'Inventory control' ? What are its objectives ?

14. What is 'Economic Ordering Quantity' ?

15. Explain briefly 'ABC analysis'.

16. Write a short note on 'Materials handling'.

17. What is 'Network analysis' ? Explain.

18. Define the following terms :
    Event ; Activity ; Critical path ; Total time.

19. Write a short note on Operation Research (OR).

20. Explain briefly 'Management Information Systems' (MIS).

21. State the desirable characteristics of MIS.

Section:
SHORT ANSWER QUESTIONS

SHORT ANSWER QUESTIONS

Q. 1. *What are the features of a production system?*

Ans. The *features of a production system* are:

1. A production system is goal oriented; goal being production of goods and services.
2. The goal is achieved through technological transformation of raw material, using energy.
3. Technological transformation is carried on by a suitable choice of technique, an optimal combination of capital (machinery) and labour from the broad spectrum of techniques ranging from complete automation (all capital, no labour) to completely manual labour (no capital, all labour).
4. Irrespective of choice of technique, specialisation or breaking down total quantum of transformation needed for goal to smaller simplified packets, so that the labour and machine repetitively perform them with reflexive (involuntary) and automatic efficiency.

Q. 2. *Discuss with examples the following types of manufacturing systems.*

   (i) Discrete manufacturing

   (ii) Continuous manufacturing.

Ans. *(i)* **Discrete manufacturing:**

   "Discrete manufacturing system" is typified by the intermittent or interrupted flow of *material through the plant.*

   • It makes use of general purpose machines and produces components different in nature and in small quantities.

   Machine-shops, repair and maintenance shops, welding shops, etc. are some of the *examples of intermittent production.*

   • Intermittent production or discrete manufacturing can be *classified* as:

   (*a*) Batch production or manufacturing.

   (*b*) Job manufacturing.

   (ii) **Continuous manufacturing:**

   "Continuous manufacturing" involves a continuous or almost continuous physical flow of material.

   • It makes use of special purpose machines and produces standardised items in large quantities.

   • Chemical processing, Cigarette manufacturing and Cement manufacturing are some of the industries engaged in continuous production or manufacturing.

CASTING PROCESSES

Q. 3. *What are the requirements of a good moulding sand?*

Ans. The *requirements of a good moulding sand* are:

1. It must allow the free passage of air and gases generated when in contact with molten metal. This is the *"permeability"* of the sand.
2. When rammed it must retain the shape given to it and resist the pressure of the molten metal. This is known as its *"cohesive"* quality.
3. It must be able to withstand high temperature without fusing. This is called *"refractory"* quality.

4. It should easily come away from the cold casting, and leave a clean, smooth surfaces. This is known as it *"stripping "* quality.

Q. 4. *What is "Stop-off" ? Explain briefly.*

Ans. ***Stop-off*** *(Fig.1) is the portion of the pattern which is added for its strength only if the pattern is fragile.*

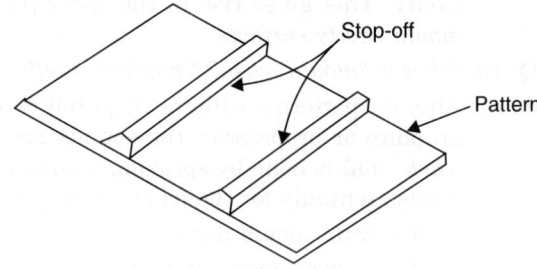

Fig. 1

- It forms a cavity in the mould when the moulder withdraws the pattern, which is *refilled* with sand before pouring.
- Stop-offs are wooden pieces used to reinforce some portions of the pattern which are structurally weak, especially from the stand point of repeated handling. They have no connection with the completed casting.

Q. 5. *What are "Mould surface coatings"? Explain.*

Ans. After the pattern is drawn, the mould surfaces are coated with certain materials possessing high refractioness. It eliminates the possibility of burn-on and enables castings with smooth surface to be obtained. However, the permeability of the mould gets reduced. Therefore, the *coatings should not contain gas forming materials.*

- Mould surface coatings which are also known as *facing, dressings, washes, blackings* or *whitening,* may be applied dry (by dusting) or wet in the form of thin cream.
- The various mould surface coating materials include: Coal dust, pitch, graphite, China clay, Zircon flour or French chalk (calcium oxide).

Q. 6. *Write short note on "Moulding sand for non-ferrous castings".*

Ans. The moulding sands for non-ferrous castings may be *less refractory and permeable* since the melting point of non-ferrous metals is much lower than that of ferrous metals. Further, a smooth surface is desirable on non-ferrous castings. In view of this, the moulding sand for non-ferrous castings contains a *considerable amount of clay and are fine grained.*

Q. 7. *What is "Stack moulding"?*

Ans. In an upright stack moulding process from 10 to 12 flask sections are arranged one above, having a common sprue through which all the moulds are poured.

- An advantage of this process is that it *requires much less floor space in the foundry.*
- Stack moulding is used to make *small castings.*

Q. 8. *(a) Define the terms "Casting" and "Foundry".*

(b) What are the types of foundries?

Ans. *(a)* **Casting** *is the solidified piece of metal which is taken out of the mould.*

Foundry *is a plant where the castings are made.*

(b) All the foundries are basically of the following two types:

(i) ***Jobbing foundries:***

- These foundries are mostly independently owned.
- They produce castings on contract, within their capacity.

(ii) ***Captive foundries:***

- The foundries are usually a department of a big manufacturing company. They produce casting exclusively for the parent company.
- Some captive foundries which achieve high production, sell a part of their output.

Q. 9. *What are "Cushion materials"? Explain.*

Ans. The **cushion materials** (*e.g., wood flour, cereals, cellulose, etc.*) when added to the moulding sand burn and form gases when the molten metal is poured into the mould cavity. This gives rise to the *space for accommodating the expansion of the sand at the mould cavity surfaces.*

Q. 10. *What is "mould wash" ? Explain briefly.*

Ans. After the pattern is withdrawn, purely carbonaceous materials such as sea coal, finely powdered graphite or proprietary compounds are applied to the mould cavity. This is called *"mould wash"* and is done by spraying, swabbing or painting in the form of a wet paste. These are used essentially for the following *reasons:*

   (*i*) To avoid mould-metal interaction and prevent sand fusion.

   (*ii*) To prevent metal penetration into the sand grains and thus ensure a good casting finish.

   • To deposit the mould wash, either water or alcohol can be used as a carrier. But because of the problem of getting the water out of mould, alcohol is preferred as a carrier. The proprietary washes are available in powder, paste or liquid form. The powder needs to be first prepared and applied whereas the paste and liquid can be applied straight away.

Q. 11. *How can the casting defect 'Hot tears' be eliminated?*

Ans. In a casting, **hot tears** result from temperature gradients, establishing different rates of contraction during solidification and thereby inducing stresses due to resistance of the sand of a magnitude sufficient to cause fracture. These can be *minimised by adopting good design i.e., avoiding abrupt changes in section, sharp angles and non-uniform webs connected to flanges.*

Q. 12. *Explain briefly "CO_2 moulding process".*

Ans. **CO_2 moulding process** is a sand moulding process in which sodium silicate ($Na_2O.x\ SiO_2$), that is, water glass is used as a binder, 2 to 6 per cent, rather than clay. After the mould is prepared, CO_2 gas is made to flow through the mould and the sand mixture hardens through a very rapid reaction (of duration about 1 minute, which is very less than several hours needed to produce a dry sand mould) yielding a stiff gel.

   • CO_2 moulds can be *used for producing very smooth and intricate castings, because the sand mix has a very high flowability to fill up corners and intricate contours.*

   Advantages:

   (*i*) CO_2 moulds can be made without flasks.

   (*ii*) Sands are free flowing, therefore, ramming is eliminated or reduced.

   (*iii*) Tensile strength of moulds are higher than those of conventional moulds. This permits reduction of mould weight and easier handling of large moulds.

   Disadvantage:

   Sand must be used *immediately*.

Q. 13. *What are composite moulds? Explain briefly.*

Ans. The **composite moulds** are made of two or more different materials, such as shells, plaster, sand with binder and graphite. These moulds combine the advantages of each material.

   • These moulds are used in shell moulding and other casting process, generally for casting *complex shapes,* such as turbine impellers.

   Advantages:

   (*i*) increased mould strength.

   (*ii*) Improved dimensional accuracy, and surface finish of castings.

   (*iii*) Reduced overall costs and processing times.

Q. 14. *What is 'dry sand moulding'?*

Ans. Invariably the *"sand mould"* is *dried* after pattern is removed in order to *increase the strength* of moulding sand so that is can withstand the higher static pressures of the liquid in case of *big castings*.

- Sometimes clay-bonded dried mould castings are used to obtain *greater dimensional tolerances* and obtain *better surface finish*.

Q. 15. *Explain briefly CO_2 process of making cores.*

Ans. CO_2 process of making *cores is* briefly discussed below:

- Clean and dry sand is mixed with a solution of sodium silicate and rammed or blown into the core box.
- The mixture is then gassed with CO_2 gas for several seconds, and CO_2 forms a silica gel which binds the sand grains into a strong solid form. If required, additional hardening of resulting core can be carried out by baking.
- — Sodium silicate and sand mixture gel cannot be reclaimed after use.

Q. 16. *List the various moulding defects.*

Ans. Various **moulding defects** are:

- Sand spots.
- Internal air pockets.
- Sand holes.
- Blow holes.
- Honey combing or sponginess.
- Oversize castings.
- Poured short.
- Swelling.
- Mismatch, lifts and shifts.

Q. 17. *List some important design rules which must be considered in sand casting.*

Ans. Following are some *important design rules* which need be considered in sand casting to ensure maximum dispersal of stress, and minimum stress concentration:

1. In order to increase resilience of ductile metals to fatigue rupture external corners should be *rounded with* radii of 10–20 per cent of the section thickness.
2. Complex sections like *X, Y, V* and *K* should be staggard to staggard *T*.
3. In joining sections of dissimilar size or at *L* or *T* junction radii equal to the thickness of small section should be provided.
   - — In case of *L* junctions, largest possible radii should be used.
4. The load bearing capability with complex loads is increased by adopting tubular or reinforced *C* sections rather than *I, H* or channel sections.

Q. 18. *Explain briefly "Casting yield".*

Ans. All the metal that is used while pouring does not end up as a casting. There are some losses in the melting. Also there is a possibility that some castings may be rejected because of the presence of various defects. On completion of the casting process, the gating system used is removed from the solidified casting and remelted to be used again as raw material. Hence, the ***casting yield*** *is the proportion of actual casting mass.*

- Casting yield depends to a great extent on the *costing materials and complexity of the shape*. Generally those materials which shrink heavily have lower casting yields. Also massive and simple shapes have higher casting yield compared to small and complex parts.

- *Higher the casting yield higher is the economics of the foundry practice.* It is therefore desirable to give consideration to *maximising the casting yield, at the design stage itself.*

Q. 19. *How are gases formed in casting and how can these be eliminated?*

Ans. The gases in casting may appear as *gas holes* and *pin holes*. These defects can be avoided by proper riser design and adequate venting of permeable moulds.

- Another source of gases is from the dissolved gases in the liquid metal at high temperature, which on cooling are given off. The gases in melts can be reduced by *vacuum melting and vacuum degassing* (placing liquid metal in low pressure chamber to remove dissolved gases).

Q. 20. *Write short note on "Residual stresses in castings".*

Ans. In case of non-uniform cross-section casting different sections solidify at different rates, depending on their cross-sectional areas. This results in varying amount of contraction in different parts, producing high internal stresses which may cause *tearing or cracking of casting.* High residual internal stresses can be avoided by *placing chills over large cross-sectional areas so that whole of casting cools at uniform rate.*

Such stresses can also be controlled by taking out casting at an average temperature of around 750°C and putting it in an *insulated pit and allowing to cool at 5.5°C/hour.*

Following points are worth mentioning:

- Any temperature gradient above 540°C does not give rise to elastic strain because same is relaxed to plastic strain due to high rate of creep.
- In case high residual stresses exist in casting it *has to be relieved by a suitable heat treatment or by other methods of stress relieving.*

Q. 21. *What is chill casting? Explain briefly.*

Ans. **Chill casting process** is nearly similar to sand casting and is *used where very hard outer surfaces and wear resistant castings are required.*

- Moulds are mode of sand or cast iron and for the purpose of *chilling the cast iron, steel blocks are used.*
- Metallic chills are used at outer surfaces so that the rate of cooling increases and hence hardness increases. Where hardness is of ample importance, *metallic moulds* are used as in the case of railway brake shoe. The excessive chilling effect is reduced by preheating the moulds.
- In case of bushes and bearings, the inner surfaces of the holes should be hard and wear resistant and to meet this requirement *core chills* are used in the moulds. Extensive chills are used to reduce the possibility of the defect called *'hot-tear'.*
- The cooling rate has a considerable effect upon the hardness of the surface; the greater the rate of cooling the lesser amount of carbon will come out in graphite state. In other words, carbon will be *in-combined form* and hence casting will be hard.

Q. 22. *What do you mean by the term "Fluidity"? Explain briefly.*

Ans. The term **fluidity** is normally used in a foundry to *designate the casting material's* ability to fill the *mould cavity.*

Fluidity depends on the *casting material* as well as the *mould.*

The *casting material's properties* which affect the fluidity to a great extent are:

- Viscosity of the melt.
- Heat content of the melt.
- Surface tension.
- Freezing range.
- Specific weight of the liquid metal.

The *mould properties* that affect the fluidity are:
— Thermal characteristics.
— Permeability.
— The mould cavity surface.
• The *most commonly used fluid test* is the **spiral fluidity test.**

Q. 23. *What is "System sand"?*

Ans. In mechanised foundaries, where machine moulding is employed so called **system sand** is used to fill the whole flask. Since the whole mould is made up of this system sand, the strength, the permeability and refractoriness of the sand must be higher that of backing sand.

• Whereas facing sand is always used to make dry sand moulds, the *system sand is frequently used for green sand moulding.*

Q. 24. *Why risers are not used in die casting?*

Ans. Die castings are made by forcing molten metal at high pressure into a split steel die cavity. Within a fraction of second, the fluid alloy fills the entire die. The die is water cooled; therefore low temperature is being maintained. *Because of the low temperature of the die, the casting solidifies quickly.* Therefore *risers are not required in die casting.*

Q. 25. *What purpose is served by risers in sand casting?*

Ans. **Riser** *is a hole cut or moulded in the cope to permit the molten metal to rise above the highest point in the casting.*

• Riser serves as feeder to feed the molten metal into the main casting cavity to *compensate for shrinage.* The design of the riser should be such that it establishes temperature gradients within casting so that the casting solidifies directionally toward the riser. *It also helps in easy ejection of steam, gas and air from the mould cavity while filling the mould with the molten metal.*

Risers act as reservoir and heat gradient regulator and provide the necessary fluid metal to *compensate solidification.*

• *In case no riser is provided during casting the solidification will start from walls and liquid metal in the centre will be surrounded by a solidified shell and the contracting liquid will produce the voids towards the centre of casting.*

Q. 26. *What is a centrifugal casting ? For what type of jobs would you recommend this casting process?*

Ans. The **centrifugal casting** process is carried out in a permanent mould which is *rotated during the solidification of the casting.*

For producing a *"hollow part"* the axis of rotation is placed at the centre of the desired casting. The speed of rotation produces a centripetal acceleration which segregates less dense non-metallic inclusions near the centre of rotation.

"Solid parts " can be made by a variation of this process by placing the entire mould cavity *on outside of the axis of rotation.*

• The castings produced by this method are *very dense and are used for such critical parts as cylinder liners, etc.*

Q. 27. *Describe the need of investment casting. Explain the investment casting process.*

Ans. **Need of investment casting:**

• Investment casting is used *when intricate shapes, good dimensional accuracy and a very good surface finish are required.*

• Investment casting is suitable for *high melting point alloys* as well as difficult to machine metals. It is also suitable for *processing small size castings having intricate shapes.*

Investment casting process:

The various *steps* in investment casting are as under:

1. Preparation of heat-disposable pattern, together with its gating system is done by injecting wax or thermoplastic into the die cavity.

2. A *'tree'* is prepared from number of such pattern fixed to a wax or plastic runner bar with a suitable ceramic cup to act as pouring basin.

3. The tree is then dipped into a ceramic slurry (containing silica flour in ethyl acetate). Sufficient fine silica sand is sprinkled on the tree dipped in ceramic slurry. This enables the formation of a self-supporting ceramic shell mould to be formed around the wax assembly.

4. The ceramic shell mould is then baked so that the wax melts and flows out leaving a precise mould cavity.

5. The shell is fired between 850°C and 1000°C to eliminate all the wax and give more strength to the mould.

6. The molten metal is poured into the mould white it is still hot and cluster of castings is obtained.

- These days, *lost wax process* is used for manufacturing larger objects like *cylinder heads, crankshafts,* etc. In these processes *'styrofoam'* is used *instead of wax.*

Q. 28. *Explain briefly 'slush casting'.*

Ans. *Slush casting* generally *involves the process of low-temperature-melting alloys.*

- In this process hollow castings can be produced without the use of cores. The *principle involves pouring the molten metal into a permanent mould. After the skin has frozen the mould is turned upside down or slung to remove the metal still liquid. A thin-walled casting results, the thickness depending on the chilling effect from the mould and the time of operation. Toys and ornaments* are made by this process from zinc, lead or tin alloys.

Q. 29. *What is a continuous casting?*

Ans. The *continuous casting process* consists of *continuously pouring molten metal into a mould that has the facilities for rapidly chilling the metal to the point of solidification and then withdrawing it from the mould.*

- It has been proved through research and experimental work that there are many opportunities for cost economies in the continuous casting having a degree of soundness and uniformity not possessed by other methods of producing bars and billets.

Q. 30. *What is 'Electro-magnetic casting'? Explain.*

Ans. This casting process is a variant of conventional continuous casting method. The *molten metal is contained and solidified in an electro-magnetic field instead of in a conventional mould. There is no sliding contact of the molten metal with the mould walls, and the metal can be solidified by direct water impingement.*

The surface finish of the casting is very good and the process can be automated. The method is being used for *obtaining continuous strand of metals.*

Q. 31. *Enumerate the methods used for repairing and salvaging efective castings.*

Ans. The methods used for *repairing* minor defects of the castings are:

1. Cold welding.
2. Hot welding.
3. Liquid metal welding.
4. Metal spraying.
5. Luting and impregnation—In these methods, the defects in the castings are repaired with the help of sealing agents.

Q. 32. *What is 'Chvorinov's rule'?*

Ans. **Chvorinov's rule:**

Chvorinov's rule for metal casting states the postulation that *total freezing (solidification) time for a casting is a function of the ratio of volume to surface area.*

$$\text{Solidification time,} \quad t = C \left[\frac{\text{Volume}\,(V)}{\text{Surface area}\,(A)} \right]^2 \quad i.e., \quad t = C \left(\frac{V}{A} \right)^2$$

where, C = a constant, that reflects mould material, metal properties like latent heat, and temperature.

Best riser is one whose $\left(\dfrac{V}{A} \right)^2$ is *10 to 15% larger than that of the casting.*

Since V and A for the casting are known, $\left(\dfrac{V}{A} \right)_{\text{riser}}$ can be determined. Assuming the height to diameter ratio for the cylindrical riser, the riser size can be determined. *This rule helps in determining the solidification time of casting.* Accordingly we can select the method of casting.

- Chvorinov's rule/formula is not *very accurate,* since it does not take into account the solidification contraction or shrinkage. This method is *valid for calculating proper riser size for short freezing range alloys such as steel and pure metals.*
— There is no *satisfactory relationship for determining riser size for non-ferrous alloys.*

METAL FORMING PROCESSES

Q. 33. *What is work (or strain) hardening?*

Ans. **Work (or strain) hardening** is a *phenomenon* which results in an *increase in hardening and strength of a metal* (specimen) subjected to plastic deformation at temperature lower than the recrystallization range *(cold working).*

- When a material is subject to plastic deformation, a certain amount of work done on it is *stored internally as stain energy. This additional energy in a crystal results in a strengthening or work hardening of solids.*

  Thus *work (or strain) hardening may be defined as increased hardness accompanying plastic deformation.* This increase in hardening is accompanied by an *increase in both tensile and yield strength.*

- Work hardening *reduces ductility and plasticity.*

- Work hardening is used in many manufacturing processes such as *rolling of bars and drawing of tubes.* It is also used to improve the elastic strength in the manufacture of many parts such as: (*i*) Prestretching of hoisting chains and cables, (*ii*) Initial pressurisation of pressure vessels, cylinders of hydraulic press and guns.

Q. 34. *What do you mean by the following terms?*

(*i*) *Coining;* (*ii*) *High-energy rate forming;* (*iii*) *Progressive piercing.*

Ans. (*i*) **Coining.** The operation of coining is performed in dies that confine the metal and restrict its flow in a lateral direction.

(*ii*) **High-energy rate forming.** It includes a number of processes in which parts are formed at a rapid rate *by extremely high pressures.*

(*iii*) **Progressive piercing.** It is the method frequently employed on upset forging machines for producing parts such as *artillery shells and radial engine cylinder forgings.*

Q. 35. *What is 'warm working'?*

Ans. **Warm working** is deformation under the conditions of *transition, i.e.,* a working temperature between 0.3 and 0.6 times the melting point".

Q. 36. *Explain briefly stretch forming process.*

Ans. In order to form large sheets of thin metal involving symmetrical shapes or double curve bends, a **metal stretch press** can be used effectively. In a simpler hydraulically operated press a single die mounted on a ram is placed between two slides that grip the metal sheet. The die moves in a vertical direction and the slides move horizontal. Large forces of 0.5 to 1.3 MN are provided for the dies and slides. The process is a stretching one and cause the sheet to be stressed above its elastic limit while confirming to the die shape. This is *accompanied by a slight thinning of the sheet, and the action is such that there is little spring back to the metal once it is formed.*

- The *process can be used with many hard-to-form alloys, there is little severe localised cold working, and the problem of unequal metal thin out is minimized.*
- *Scrap loss is fairly high* because material must be left at the ends and sides for trimming and there is a *limitation to the shapes that can be formed.*

Q. 37. *What is 'electrohydraulic forming'?*

Ans. **Electrohydraulic forming,** also known as *electrospark forming*, is a process whereby *electrical energy is directly converted into work.*

This process is *safe to operate and has low die and equipment cost. The energy rates can also be closely controlled.*

Q. 38. *What is 'magnetic forming'?*

Ans. **Magnetic forming** is another example of the direct conversion of electrical energy into useful work. At first it served primarily for swaging-type operations such as fastening fittings on the ends of tubes and crimping terminal ends of cables. More recent applications are embossing, blanking, focussing and drawing, all using the same power source but efferently designed work coils.

The process has the *limitations:* Complex shapes may be impossible to form, pressure cannot be varied over the workpiece and present units are limited to 400 MPs pressure.

Q. 39. *What is "forgeability"? Explain.*

Ans. **Forgeability** *may be defined as the tolerance of a metal or alloy for deformation without failure, regardless of forging-pressure requirements.*

Forgeability of a metal or alloy is influenced by the following *mechanical factors:*

(*i*) Strain rate,

(*ii*) Strain distribution.

Metals exhibiting low ductility at cold working temperatures show reduced forgeability at increasing strain rates and metals exhibiting high ductility at cold working temperatures are not noticeably affected by increasing strain rates.

Q. 40. *What is "Presintering"?*

Ans **Presintering** *means heating the green compact to a temperature below the sintering temperature.* It is done to increase strength of green compact and remove the lubricants and binders added during blending.

- Some metals like tungsten carbide are easily machined in presintered state as they become too hard after sintering. However, if machining is not required, this operation can be avoided for them.

Q. 41. *What are the objects of 'Sintering'?*

Ans. The *main objects of sintering* are:

1. To achieve good bonding of powder particles.
2. To produce a dense and compact structure.
3. To achieve high strength.
4. To produce parts free of oxides,
5. To obtain desired structure and improved mechanical properties.

Q. 42. *What do you understand by 'Spark discharge sintering' ? Explain briefly.*

Ans. **Spark discharge sintering:**

- In this type of sintering, a high energy electric discharge spark is produced during compaction which results in instantaneous diffusion bonding. This method thus combines compacting and sintering the metal powders to a dense metal part in 12 to 15 seconds.

   Advantages:

   (*i*) Dimensional accuracy of the parts is maintained.

   (*ii*) No separate sintering furnace is required for the process.

Q. 43. *Explain briefly "Self lubricating bearings".*

Ans. **Self lubricating or porous metallic bearing** *is that bearing in which the metal is porous, the pores containing the oil necessary for lubrication.* These bearings are produced by *powder metallurgy.* Powders of copper, tin and graphite are sintered, and are then passed through a sizing die to the correct dimensions. The metal thus produced is of a porous nature, capable of holding upto one third of its volume of oil. A pressure on the bearing surface, or a temperature rise will cause the oil to exude, so that lubrication is ensured continuously and automatically where it is needed.

- These bearings *help eliminating much routine oiling.* Dirt cannot enter the bearings since there are no oil holes or grooves. There is no leakage of oil hence no contamination of any products which are being handled, a point of vital importance in machinery for dealing with foodstuffs or delicate textiles.

PROCESSING OF PLASTICS

Q. 44. *What is "slush moulding"? Explain briefly.*

Ans. *Slush moulding* resembles casting in that no pressure is applied.

The process consists of preparing a slurry of thermoplastic resin and then pouring the same into a preheated mould. On account of the heat of the mould a uniform layer of resin sets all along the surface of the mould cavity. Excess slurry, if any, is then poured out. Additional heat is provided for curing the resin set in the mould, followed by hardening of the product by cutting and removing the same from the mould.

- Slush moulding is used to produce *flexible toys, artificial flowers, etc.*

Q. 45. *What do you understand by "Rigidized vacuum forming"? Explain briefly.*

Ans. *Rigidized vacuum forming* *is an intermediate process between basic hand layup and costly matched die moulding.*

A part is first made by thermoforming and then it is reinforced and rigidized by glass reinforcement. The thermoformed parts are placed on a holding fixture to ensure that during the curing of the reinforcing laminate the parts will not change dimensionally.

This process is employed to produce:

— Bathroom tubs and sinks and shower stalls;

— Automotive body and recreational vertical panels;

— Motor shrouds;

— Other items where production and uniformity are desired.

Q. 46. *Explain briefly "Pultrusion".*

Ans. **Pultrusion,** as the name implies, *is like pulling an extrusion.* Constant cross-sectional shapes are produced by pulling resin-impregnated, reinforcing material through a heated steel shaping die in such a way as to promote adequate polymerisation of the resin.

— Pultrusion consists of pulling continuous-strand woven roving, surfacing mat, woven fabrics, and reinforcing mat through a resin bath to wet out the fibrres and then through a heated steel die. The entrance section of the die is water cooled to prevent premature curing or binding. The pultrusion moves through the die at speeds ranging from 50 to 1540 mm/min, depending on the cure time, which in turn depends on thickness of the section.

• Pultrusion is used mainly with thermosetting plastics, but it may also be used with thermoplastics.

• The main use of pultrusion has been for electrical, recreational, and construction purposes and where corrosion resistance is necessary.

Q. 47. *What do you understand by "Filament winding"? Explain briefly.*

Ans. **Filament winding** *is a process of wrapping reinforcements consisting of resin-impregnated, tensioned, continuous filaments around a form called a mandrel.* After the resin mix is cured, the mandrel may be removed, leaving a hollow, monolithic shell.

The most common filament used is a continuous glass filament, but graphite is also used. The matrix resin is generally epoxy or polyester.

• The outstanding property of filament-wound structures is their *high strength-to-weight ratio,* which is better than that of steel or even titanium.

***Applications*:**

— Rail road freight car;

— Solid rocket cases;

— Metal-lined compressed gas cylinder, etc.

Q. 48. *Give the comparison between "Addition polymerisation" and "Condensation polymerisation".*

Ans. Comparison between Addition and Condensation polymerisation

S.No.	Addition polymerisation	Condensation polymerisation
1.	It requires molecules which are unsaturated.	It requires two unlike molecules.
2.	It does not yield a by product	It yields a by product.
3.	Reaction is very fast and may take 10^{-2} to 10^{-6} sec.	Reaction normally takes hours and days to complete.
4.	It is kinetic chain reaction.	It involves inter-molecular reaction.
5.	Polymer formed is a thermoplastic type.	Polymer formed is thermosetting type.

Q. 49. *What are 'laminated plastics'? Explain briefly.*

Ans. **Laminated plastics** are also called *plastic laminates* and are formed by impregnating sheets of fibrous materials such as paper, linen, canvas or silk with a synthetic resin and then compressing the sheets together with application of heat. The synthetic resins may be phenolic resin, urea formaldehyde or a vinyl resin. The resin is usually dissolved in alcohol.

The material in roll form is immersed in the resin solution at atmospheric pressure at room temperature and then run through a drier at 150°C. The rolls are next cut into sheets of given size, which are arranged into stacks. These stacks finally are compressed in a hydraulic press at about 170°C under a pressure of 200 bar. The sheets are thus bonded to one another.

Q. 50. *What are 'laminates'? Give examples.*

Ans. **Laminates** *(or laminar composites) are those structures which have alternate layers of materials bonded together in some manner.*

Common *examples* of laminar composite:

    (*i*) Plywood   (*ii*) Bimetallic strips

  (*iii*) Safety glass   (*iv*) Sandwich material

    (*v*) Roll cladding (bonding) and explosive cladding (welding)

  (*vi*) Laminated plastic sheet   (*vii*) Tufnol

 (*viii*) Laminated carbides   (*ix*) Laminated wood.

WELDING AND ALLIED PROCESSES

Q. 51. *Define the term 'Welding'?*

Ans. **Welding:** It *is method of joining metals by applications of heat, without the use of solder or any other metal or alloy having a lower melting point than the metal being joined.*

Or

Welding is defined as *"a localised coalescence of metals, wherein coalescence is obtained by heating to suitable temperature, with or without the applications of pressure and with or without the use of filler metal".* The filler metal has a melting point approximately the same as the base metal.

— The large bulk of materials that are welded are metals and their alloys, although the term welding is also applied to the joining of other materials such as thermoplastics. Welding joins different metals/alloys with the help of a number of processes in which heat is supplied either electrically or by means of a gas torch. In order to join two or more pieces of metals together by one of the welding processes, the *most essential requirement is heat. Pressure may also be employed, but this is not, in many processes essential.*

— A good welded joint is as strong as the parent metal. The product is known as *"weldment".*

Q. 52. *Explain the terms 'Soldering' and 'Brazing'.*

Ans. **Soldering:** *It is a process of joining two pieces of metal with a different fusible metal applied in a molten state. The fusible metal is called 'solder'.*

Or

It is a process of joining two metals with low melting point metal.

Brazing: *It is a process of joining two metal pieces in which a non-ferrous alloy is introduced in the liquid state between the pieces to be joined and allowed to solidify.*

• Soldering and brazing are two common *solid/liquid-state bonding processes.* These are *different from welding* as bonding here requires *capillary action* and that some degree *of alloying action* between the filler and the base metal always occurs. Also the *composition of filler metal is significantly different and its strength and melting point are substantially lower than that of the base metal.*

— In *'soldering'* (very similar to brazing) the filler material is usually a *lead-tin based alloy* which has much lower strength and melting temperature (about 250°C). Also, less alloying action between the base and filler metals gives *lower joint strength*. Since in this process much lower temperatures are involved, it is usually carried out with *electric resistance heating*.

— In *'brazing'* the joint is made by heating the base metal red hot and filling the gap with molten filler metal whose melting temperature is above 427°C but below the temperature of base metal. The filler metals, generally used for brazing are copper alloys. This process is usually carried out with a *gas flame*.

Q. 53. *Differentiate between 'pressure welding' and 'fusion welding'.*

Ans.   **Pressure welding.** Pressure (or solid state) welding may be carried out at room temperature or at elevated temperature. The joint is formed *owing to plastic flow at the joint due to applied pressure*. In pressure welding at high temperature, heat is generated due to resistance at the joint due to flow of current being high. Other methods of achieving high temperature include friction, induction heating, impact energy as in the case of explosive welding. Seam welding, projection welding, butt welding are well known resistance welding techniques.

Fusion welding. In fusion (or liquid state) welding the material around the joint is melted in both the parts to be joined. If necessary a molten filler metal is added from a filler rod (or otherwise). The important zones in fusion welding are: (*i*) Fusion zone; (*ii*) Heat affected unmelted zone around the fusion zone; (*iii*) the unaffected original part.

Important factors affecting fusion welding process are:

(*i*) Nature of weld pool.

(*ii*) Chemical reaction in the fusion zone.

(*iii*) Characteristics of heat source.

(*iv*) Contraction, residual stresses and metallurgical changes.

(*v*) Heat flow from the joint.

Q. 54. *State the effects of current and voltage on the quality of weld.*

Ans.   *The effects of current and voltage on the quality of weld* are as follows:

(*i*) *Too high current*	...	Gives deeper crater and penetration, flat bead, much spatter, electrode becomes red hot, etc.
(*ii*) *Too low current*	...	Imparts poor penetration, shallow crater, weld overlapping on the plate, etc.
(*iii*) *Too high voltage*	...	Produces a fierce wandering and noisy hissing arc, bead is often porous and flat, spattering of metal.
(*iv*) *Too low voltage*	...	Causes sticking of electrode with work and arc becomes difficult to maintain, weld is deposited in blobs with no penetration, etc.

Q. 55. *Explain briefly "Principle of resistance welding".*

Ans.   In **resistance welding** the heat required for welding is produced by means of *electrical resistance* between the two members to be joined. The heat produced is given by the expression.

$$H = I^2\, Rt\, K \qquad\qquad\qquad ...(1)$$

where, H = The total heat generated in the work, J,

I = Electric current, A,

R = The resistance of the joints, Ω (ohms),

t = Time for which electric current is flowing through the joint, s, and

K = A factor that represents the energy losses through radiation and conduction; its value is *less than unit*.

The resistance of the joint, R is composed of:

(*i*) The resistance of the electrodes;

(*ii*) The electrode-workpiece contact resistance.

(*iii*) The resistance of the individual parts to be welded.

(*iv*) The workpiece-workpiece contact resistances.

- In resistance-welding operations, the magnitude of the current may be as high as 100,000 A, although the voltage is typically only 0.5 V to 10 V.

- The actual temperature rise at the joint depends on the specific heat and thermal conductivity of the materials to be joined; consequently, because they have high thermal conductivity, *metals such as aluminium and copper require high heat concentrations.*

Q. 56. *Explain briefly the following resistance welding processes:*

(*i*) *Metal fibre welding.*   (*ii*) *Metal foil welding.*   (*iii*) *Spike welding.*

Ans. (*i*) **Metal fibre welding:**

- In this process (may be described as an extension of resistance spot welding), the equipment used is same as in spot welding.

- In metal fibre welding a small piece of *metal fibre sheet* is introduced between the workpieces to be joined. The workpieces are held as usual between the electrodes. The electrodes pressure employed in *very low* and, therefore, the indentation on the work surface is also very small.

- It is mainly used for making lap joints.

(*ii*) **Metal foil welding:**

- In this process the metal sheets to be joined are placed between electrode wheels (as for seam welding) and the abutting edges are *covered at top and bottom with very thin metal foils,* the foil thickness is approximately 0.25 mm and the same is made of the *same material as the workpiece metal.*

- A distinct feature of the weld formed is that is has a *raised heat.*

(*iii*) **Spike welding:**

- In this resistance welding process a *large amount of electricity is stored up in capacitors and then the same is released rapidly through the electrodes and the metal pieces to be joined.* This current flow is allowed for a very small interval and its time is controlled electronically. The welding is, therefore, *very rapid and consequently chances of warpage of workpieces and of contamination are altogether eliminated.*

Applications. Successfully used for welding of almost all metals and alloys, metals of different thicknesses together and dissimilar metals as well.

Q. 57. *Explain briefly the following variations on the friction welding process:*

(*i*) *Inertia friction welding;*   (*ii*) *Linear friction welding;*

(*iii*) *Friction stir welding.*

Ans. (*i*) **Inertia friction welding:**

- The energy required in this process is supplied *through the K.E. of a flywheel.* The flywheel is accelerated to the proper speed, the two members are brought into contact, and an axial force is applied; as friction at the interface begins to slow down the flywheel, the axial force is increased. The weld is completed when the flywheel comes to rest.

(*ii*) **Linear friction welding:**

- In this process, one part is moved across the face of the other part, using a balanced reciprocating mechanism.
- This welding process is capable of joining square or rectangular components, as well as round parts, made of metals or plastics.

(*iii*) **Friction stir welding:**

- Whereas in conventional friction welding heating of interfaces is achieved through friction by rubbing two contacting surfaces, in the *friction stir-welding* process, a third body is rubbed against the two surfaces to be joined. The rotating tool is a small (5 mm to 6 mm in diameter, 5 mm in height) rotating member that is plunged into the joint. The contact pressures cause frictional heating, raising the temperature to the range 230°C to 260°C. The probe at the tip of the rotating tool forces heating and mixing or stirring of the material in the joint.
- The welding equipment can be a conventional, vertical spindle milling machine, and the process is relatively easy to implement.

Q. 58. *Why is laser welding used only for microwelding applications?*

Ans. Laser is used only for microwelding, *i.e.,* welding very small wires to electronic devices as laser for generating high energy is *very costly* but *microwelding applications can be precisely controlled by laser.*

Q. 59. *What is "Arc cutting"? Explain briefly.*

Ans. *In* **arc cutting,** *the metal is simply melted by the intense heat of the arc and is then blown away by the force of arc itself or by other gases such as air or shielding gases.* Depending on the source of heat input, the various processes of arc cutting are:

— Carbon arc cutting;

— Air carbon arc cutting;

— Oxygen arc cutting;

— Shielded metal arc cutting;

— Gas metal arc cutting;

— Gas tungsten arc cutting;

— Plasma arc cutting.

In all these processes, the equipment used is similar to that used for the corresponding welding process with the exception of the *torch which is different. The torch holds the electrodes as also has the provision for the supply of high pressure gas wherever needed.*

- Any metal can easily be melted and blown away by shielded gas cutting processes.

Q. 60. *Why is it difficult to start A.C. arc? How is it simplified in practice?*

Ans. It is difficult to start A.C. arc *because of alternating current flow.* This difficulty is overcome by having hot start circuit which provides an extra flow of very high frequency current at the time of striking the arc. In some machines, capacitors are employed in arc (secondary) circuit to give high current surges for the arc striking.

Q. 61. *What are the advantages of A.C. arc welding?*

Ans. The major advantage of A.C. arc welding is *complete absence of magnetic arc blow and thus quality welds are produced.*

- Once the arc is started, it is easy to control and maintain it.
- It is usually faster because large electrodes and more current can be used due to minimum magnetic blow conditions.

- It is very well suited to weld aluminium and is very popular for welding on heavy gauge steel.

Q. 62. *Describe briefly the relative applications of A.C. and D.C. welding.*

Ans. While **D.C. welding** is best suited for thinner steel metal (below 6 mm) and also for welding non-ferrous metals, **A.C. welding** is used for most manual welding of 6 mm and thicker steel. As steel is the largest used structural material, A.C. welding finds maximum use, though D.C. welding has a greater variety of welding processes like GTAW and GMAW, straight polarity and reverse polarity, etc.

- Direct current straight polarity and reverse polarity welding can be used for overhead and vertical welds but A.C. welding is used for welding steel in the flat or horizontal position.

Q. 63. *Name the commercially used gas welding and cutting flames.*

Ans. The commercially used gas welding and cutting flames are:

- Oxygen — acetylene
- Oxygen — hydrogen
- Oxygen — natural gas or artificial gas
- Oxygen — liquefied petroleum gas.

Q. 64. *Give the comparison between TIG and MIG welding processes.*

Ans. The comparison between TIG and MIG welding processes is given in tabular form below:

S.No.	Aspects	TIG welding	MIG welding
1.	*Name of the process*	Tungsten inert-gas welding.	Metal inert-gas welding.
2.	*Type of electrode used*	Non-consumable tungsten electrode	Consumable metallic electrode
3.	*Electrode feed*	Electrode feed not required	Electrode need to be fed at a constant speed from a wire reel
4.	*Electrode holder*	It is called welding torch and has got a cap filled on the back to cover the tungsten electrode. It has also got connections for shielding gas, cooling water and control cable. It may be air-cooled also	It is called welding gun or torch. It has facility to continuously feed wire electrodes; shielding inert-gas, cooling water and control table
5.	*Welding current*	Both A.C. and D.C. can be used	D.C. with reverse polarity is used
6.	*Feed metal*	Filler metal may or may not be used	Filler metal in the form of fine wire is used
7.	*Bases metal thickness*	Metal thickness which can be welded is limited to about 5 mm	Thickness limited to about 40 mm
8.	*Welding speed*	Slow	Fast

Q. 65. *Name the common welding troubles.*

Ans. The common welding troubles are:

(*i*) Warping

(*iii*) Poor penetration

(*v*) Undercutting

(*vii*) Magnetic blow

(*ix*) Spatter

(*xi*) Poor fusion.

(*ii*) Porous welds

(*iv*) Distortion

(*vi*) Cracked welds

(*viii*) Brittle welds

(*x*) Poor appearance

Q. 66. *Explain briefly 'soft soldering' and 'hard soldering'.*

Ans. **Soft soldering.** It is used extensively in *sheet-metal work* for joining parts that are *not exposed to the action of high temperature and are not subjected to excessive loads and forces.* It is also employed for *joining wires and small parts.*

- The solder, which is mostly composed of lead and tin has a melting range of 150°C to 350°C. A suitable *flux* is always used in soft soldering. *Its function is to prevent oxidation of the surfaces to be soldered or to dissolve oxides that settled on the metal surfaces during the heating process.* Although corrosive, *zinc chloride* is the most common soldering flux. Rosin is non-corrosive, but it does not have the cleaning properties of zinc chloride.

- A blow torch or soldering iron constitutes the equipment for heating the base metals and melting the solder and the flux.

Hard soldering. It employs solders which *melt at higher temperatures* and are *stronger* than those used in soft soldering.

- *Silver soldering is a hard soldering method, and silver alloyed with tin is used as solder.* The temperatures of the various hard solders vary from 600°C to 900°C. *The fluxes are mostly in paste form and are applied to the joint with a brush before heating.*

Q. 67. *What is "braze welding"? Explain briefly.*

Ans. **Braze welding** is similar to brazing in that the base metal is not melted but joined by an alloy of lower melting point. *The main difference is that in braze welding the alloy is not drawn into the joint by capillary action.*

A braze welded joint is prepared very much like a joint as prepared for welding except that an effort should be made to *avoid sharp corners,* because they are easily overheated and may also be points of stress concentration.

Applications:

- Braze welding is extensively *used* for *repair work,* as well as some *fabrication* on such metals as cast iron, malleable iron, wrought iron, and steel.

- It is used, but to a lesser extent, on copper, nickel, and high-melting point brasses.

METAL CUTTING

Q. 68. *What do you mean by term 'Metal cutting'? Explain.*

Ans. The **metal cutting** (**machining,** a generic term, refers to all material removal processes) *refers to only those processes where material removal is affected by the relative motion between tool made of harder material and the workpiece.* The tool would be *single-point cutting tool* as used in operations like turning or shaping, or a *multi-point tool* as used in milling or drilling operation.

- The *conditions* which have an important influence on metal cutting are:

(*i*) Work material;

(*ii*) Cutting tool material;

(*iii*) Cutting tool geometry;

(*iv*) Cutting speed;

      (*v*) Feed rate;

      (*vi*) Depth of cut;

      (*vii*) Cutting fluid used.

- Metal cutting processes are performed on metal cutting machines, more commonly termed as *"Machine tools"* by means of various types of *"cutting tools"*.
- One major drawback of metal cutting or machining process is the loss of material in the form of *chips*.

Q. 69. *What are the factors due to which tool failure occurs?*

Ans. The tool failure may occur due to the following/factors:

    1. Excessive temperature.

    2. Excessive stress.

    3. Flank wear.

    4. Crater wear.

Q. 70. *Why can relief or clearance angles never be zero or negative?*

Ans. Relief (or clearance) angles are provided to prevent the end which is parallel to work and the side, which is at the cutting edge, from rubbing on the work. If these are made zero or negative or even very small, then these will *wear down and rubbing starts. This will lead to heating up of tool, chatter marks and marking up of smeared surfaces on the work.* Thus relief or clearance angle can *never be made zero or negative.*

Q. 71. *Why a 'built-up edge' on a tool is undesirable?*

Ans. **A *built up edge*** on a tool *increases the frictional resistance* to chip flow across the face of the tool. It *result in increased heat at chip-tool interface, absorption of more power* and poor surface finish on the workpiece.

Q. 72. *What are the factors influencing in the selection of cutting speeds 'and feeds for a machining operation?*

Ans. *Important factors influencing speeds and feeds for metal cutting are:*

    1. Machinability of workpiece material.

    2. Material of cutting tool.

    3. Objective criteria, *i.e.*, whether cost or time or surface finish/tolerance are of priority.

    4. Cutting fluid used.

Q. 73. *What is the effect of cutting speed, depth of cut and feed rate on the force on cutting tool?*

Ans. Forces on the cutting tool *increases only slightly with increase in speed though the* friction is less at higher speeds. With negative rake tools and carbides, the forces *actually* decrease at very high speeds.

Both depth of cut and feed rate also lead to *increase in forces on cutting tool,* but forces due to *increase in feed rate are less than due to increase in depth.*

Q. 74. *Explain the process of crater formation on cutting tools. Why crater is formed at some distance above the tool tip?*

Ans. ***Crater formation*** *in metal cutting occurs due to rubbing action of continuous chip with the tool material at chip-tool interface.*

The *crater is formed ahead of the cutting edge because cratering effect is temperature dependent.* Maximum interface temperature is mid-way of chip-tool contact length. Each mode, *i.e.*, adhesion wear, diffusion wear, abrasion wear, chemical wear is temperature dependent.

Q. 75. *Why are 'chip breakers' used?*

Ans. A continuous type of chip from a long cut is usually quite troublesome. Such chips foul the tools, clutter up the machine and workpiece, besides being extremely difficult to remove from the swarf travy. They should be broken into comparatively small pieces for ease of handling and to prevent it from being a work hazard. Hence *chip breakers* are used *to reduce the swarf* *into small pieces as they are formed.* The fact that the metal is already work hardened helps the chip breaker to perform effectively. Various types of chip breakers are made, but all of them consist mainly of a *step or groove ground into the leading edge of the tool or a piece of cutting tool material clamped on top of the cutting tool.*

Q. 76. *What should be done to remove maximum material per minute with the same tool life and at the same time keeping good finish?*

Ans. The best method to increase material removal rate is *to increase depth of cut, as tool life is least affected by increase in depth of cut.* If depth of cut is increased, speed needs to be decreased for same tool life. However, depth of cut is also *restricted by the strength of the workpiece and amount of stock to be removed. Increase in feed rate has smallest decrease in tool life in relation to the increased metal removal rate.* Increasing feed rate beyond finish requirements is not possible. *Thus increase in feed is best method, of course, within limits of allowable finish.*

Q. 77. *Explain why chatter and vibration are considerably reduced when a helix cutter is used instead of a plain milling cutter during milling.*

Ans. A helical cutter operates more smoothly than a plain milling cutter. It is due to the fact that if the cutter teeth are made parallel to the axis, they strike the work simultaneously across the entire width. This *result in hammering action by the tool* on the work in a regular frequency and ultimately *results in chattering of the entire set-up.* This causes a poorer surface finish and shorter tool life. A *cutter having helical teeth engages with work progressively* and the cutting action is *continuous.* This eliminates chattering and smooth cutting action takes place.

Q. 78. *State the conditions under which positive and negative rake angles are recommended.*

Ans. Following are the *conditions* under which positive and negative rake angles are recommended:

Positive rake angles:
1. When cutting at low cutting speeds.
2. When machining long shafts of small diameters.
3. When machining low strength ferrous and non-ferrous materials and work-hardening materials.
4. When using low power machines.
5. When the set up lacks strength and rigidity.

Negative rake angles:
1. For rigid set-ups and when cutting at high speeds.
2. When machining high strength alloys.
3. When there are heavy impact loads such as in interrupted machining.

Q. 79. *Describe briefly "Coated tools".*

Ans. Since long, new alloys and engineered materials are being developed continuously. These materials have high strength and toughness, *but are generally abrasive and highly chemically reactive with tool materials.* The difficulty of machining these materials efficiently and the need to improve the performance in machining the more common engineering materials have led to important development in **coated tools.**

- The *coated tools,* because of their *unique properties can be used at high cutting speeds thus reducing the time required for machining operations and, hence, costs.*
- Commonly used *coating materials* include:
  — Titanium nitride;
  — Titanium carbide;
  — Titanium carbonitride;
  — Aluminium oxide.

Q. 80. *What do you understand by 'boundary or extreme pressure (EP) lubrication mechanism' in metal cutting?*

Ans. The boundary lubrication conditions can be achieved by adding some fluid additive (known as extreme-pressure or *EP* additives) which react chemically with metal to *form compounds on the metal surface.* These compounds may form layered structures which are easily sheared in sliding thus *reducing friction; or these may inhibit the welding which would occur with base metal surface in contact.*

- The use of *EP* additives results in *improvement of surface finish, reduction in cutting forces and tool wear.*
- *Compound of sulphur and chlorine* (like chlorinated paraffins, elemental sulphur, or sulphurised fats) are often used as *EP* additives in many commercial cutting fluids used for more severe cutting operation.

Q. 81. *What is "Cryogenic machining"? Explain briefly.*

Ans. In order to reduce or eliminate the adverse environmental impact of using metal working fluids, a recent technology is the use of *liquid nitrogen as a coolant* in machining, as well as in grinding.

- The liquid nitrogen at around 200°C is injected with small-diameter nozzles into the tool-workpiece interface, thus reducing its temperature. As a result, tool hardness and, hence, *tool life is enhanced, thus allowing for higher cutting speeds; further more, the chips become more brittle and, thus, easier to remove from the cutting zone.*
- Since no fluids are involved and the liquid nitrogen simply *evaporates,* the chips can be *recycled more easily,* thus also improving the economics of machining operations, without any adverse effects on the environment.

Q. 82. *State the means by which 'chip welding to the tool' can be prevented.*

Ans. The chip welding can be *prevented by* the following means:

1. *Preventing metal-to-metal contact* by the use of a high pressure lubricant between chip and tool interface.
2. *Reduce friction* by increasing the rake angle of the cutting tool and by using a lubricant between the rake face and the chip. The rake face may also be polished.
3. *Reduce temperature* by reducing friction and by reducing cutting speed.
4. *Reduce pressure* between the chip and the tool by increasing the rake angle, reducing the feed rate and using oblique instead of orthogonal cutting.

Q. 83. *What is the most practical measure of machinability?*

Ans. The most practical measure of machinability is the comparison of the possible rate of metal removal (cm³/min.) for different materials consistent with cost, tolerance and finish.

Q. 84. *Explain briefly the effect of cutting speed, feed and depth of cut on the finish obtainable.*

Ans. In general, *increase in cutting speed tends to improve the finish.*

- With *carbide tools* particularly, *slow speed is not at all desirable* since it means wastage of time and money and *tools wear out faster.*
- The *increase in depth of cut* influences the finish *slightly,* but greater depth makes the finish poor, of course, slightly.
- As the *feed rate* increases, finish gets poorest because the tool marks show on the work. However, its effect is modified by the nose radius of the tool bit.

Q. 85. *Why should the cutting edge and top face of the tool be given a high finish?*

Ans. If the cutting edge and top face of the tool are *not* provided with high quality surface finish, then the following will happen:

- Friction between tool and chip would be high and more heat would be generated in the chip-tool contact zone.

- More power would be absorbed during the cutting process, and correspondingly greater forces would tend to dislodge the carbide tip from its shank. Under these conditions, the built-up edge is more likely to be formed.
- Roughness of the tool's cutting edge could result in a concentration of stresses which may cause surface cracks and eventual chipping of the tool.

Q. 86. *What factors are likely to give rise to excessive heat during a metal-cutting operation?*

Ans. The main factors likely to give rise to excessive heat during a metal cutting operation are:

1. Formation of built-up edge on cutting face of tool.
2. Too high cutting speed.
3. Inadequate supply of cutting speed.
4. Poor surface finish on cutting face of tool.
5. Worn or incorrectly ground cutting face of tool.

Q. 87. *Why are rake and clearance angles provided on cutting tools and on what factors the values of these angles depend?*

Ans. In order to remove a chip from the workpiece, it is essential to force a wedge type tool into the work material. When this tool is forced into the workpiece, the material of the workpiece is compressed and at certain load it fractures. With continuous cutting, chips are parted off from the workpiece and the leading tip of the tool keeps on cleaning the workpiece with a scraping action.

For getting better finish, it is also essential that only the leading tip of the tool should contact the workpiece and accordingly *clearance angle* of order of 5–10° is provided on the tool. *Rake angle* is essential to provide wedge shape. It would be noted that the *sum of rake angle, wedge angle and clearance angle is equal to 90°.*

The *value of rake angle* depends upon the following *factors:*

(*i*) Tensile strength of material (rake angle being lower for more tensile materials and may be even negative for difficult-to-cut materials);

(*ii*) The desired tool life (being shorter for higher rake angle);

(*iii*) The surface finish (being poorer for greater rake angle).

Q. 88. *What is grinding 'wheel' and why better surface finish is obtained when this is used as a tool? Explain.*

Ans. A ***grinding wheel*** *is a multi-tooth cutter made up of many hard particles called as abrasive which is crushed to leave sharp edges, these edges perform the cutting.* The abrasive grains are mixed with a suitable bond. This bond acts as a matrix or holder when the wheel is in use. The wheel may consist of one piece or of segment of abrasive blocks built up into a solid wheel.

By ***bond*** *we mean here an adhesive substance that is employed to hold abrasive grains together in the form of sharpening stones or grinding wheels.*

Q. 89. *What is 'surface grinding'? When do you recommend the use of this process?*

Ans. ***Surface grinding*** *produces flat or plane surfaces.* The work may be grounded by either the periphery or by the end face of the grinding wheel. The workpiece is reciprocated at a constant speed below or on the end face of the grinding wheel.

Surface grinding is done by two machines: (*i*) Planer type with a reciprocating table for work; (*ii*) Machines with rotating table for continuous rapid grinding.

Modern surface grinding machine is provided with hydraulic control of table movements and wheel cross feed. Machine of this type is *adapted to reconditioning dies, grinding tool ways and other long surfaces.*

Q. 90. *List the factors on which the performance of grinding wheel depends.*

Ans. The performance of a grinding wheel depends upon the following *factors:*

1. *Type of abrasive*—SiC is more hard and brittle, so it is chosen for C.I., brass, hard alloy etc. Al_2O_3 is more suitable for grinding of steels and bronzes.

2. *The grade of the wheel*—The grade designates the force holding the grains. The grade is also called the hardness of the wheel. The grade of a wheel depends on the kind of bond, structure of wheel and amount of abrasive grains. It is designated by a letter, with 'A' representing the soft end and 'Z' the hard end of the scale.

3. *Wheel structure*—Structure relates to the spacing of abrasive grains. Open structures are used for high stock removal and consequently produce a rough finish. Dense structures are used for precision forms and profile grinding.

4. *Grain size.*

5. *The properties of the grains.*

6. *The geometry of the cutting edges of grains* (rake angles and cutting edge radius compared to depth of cut).

7. The process parameters (speeds, feeds, cutting fluids) and type of grinding (cylindrical, surface).

Q. 91. *What is a lap? Explain.*

Ans.   A *lap* is a soft material disc, ring, plate or cylinder, charged with abrasive power or compound, and used for producing extremely accurate and finished surfaces.

A "lap" is a cutting tool that is made by "charging" a metal body of lead, copper, soft cast iron or any other suitable soft material *with a fine abrasive*. Charging here means that the abrasive powder is embedded in the lap by rubbing or rolling.

- In general, copper and soft steel laps cut faster than cast iron laps but *cast iron laps retain their form better. Cast iron laps are widely used.* The faces of the cast iron laps are serrated, this makes it easier to remove the work after lapping and provides a storage space for oil and abrasives.

- A *'hand lap'* for finishing flat surfaces is *in the form* of a *flat plate.* The *'lapping machine'* usually utilizes circular disc like laps.

Q. 92. *What is tumbling? Explain.*

Ans.   *Tumbling (or liquid honing)* is the process of revolving small workpiece in a barrel with abrasive and water for the purpose of producing a high lustre or removing burrs.

The following are the *purposes of tumbling:*

- To debur the parts.
- To produce a high finish.
- To improve micro finish.
- To finish gears or threaded parts without damage.
- To remove paint or plating from parts, or descale the parts.
- To form uniform radii around the workpieces, *i.e.,* generating controlled radii.
- To finish high precision work to a high lustre.

Q. 93. *What is 'diamond machining'? Explain.*

Ans.   Diamond is the hardest cutting tool material. It has *low friction, high thermal conductivity and low coefficient of expansion.* It finds applications for special jobs like *turning outside surface of aluminium alloy pistons, boring of white metal bearing liner to produce highly polished surface finish obtainable by superfinishing;* for *truing or forming the faces* of grinding wheels.

Diamonds being small are used in the form of tips which are clamped into a tool shank and the shank is mounted in tool holder, such as micro-boring head. The diamond tips are available with cutting edge either reduced or in the form of a series of flats around 0.4 mm long. These edges are set accurately parallel to the work axis using a setting microscope. When the working cutting edge gets worn out (it cannot be sharpened), the other cutting edge is fixed for the cutting action. Alongwith high hardness, diamond tips are highly brittle also. These must, therefore, be supported rigidly in the minimum overhang and subjected to least vibrations during cutting operation.

Q. 94. *What for 'lapping' is used? How much stock is left for lapping? How does it differ from grinding?*

Ans. • *Lapping process* is *used to obtain truly flat and smooth surfaces. It is also used to finish flat and round work* to tolerance of 0.01 to 0.001 mm.

• Usually about *0.001 to 0.01 mm stock is left for lapping process.*

• Lapping differs from grinding/honing process in the respect that it *uses loose abrasive instead of bonded abrasive.*

Q. 95. *Why is it undesirable to continue running coolant on to a grinding wheel after the wheel has stopped?*

Ans. If coolant is run continuously on a grinding wheel after it has stopped, then a part of the wheel would become saturated, thereby increasing the weight of that portion and *resulting in unbalance of wheel, increasing the danger of wheel disintegration.* Further, an unbalanced wheel is *deterimental to the quality of the surface finish produced on the workpiece.*

Q. 96. *What is 'Centreless grinding'? Explain.*

Ans. **Centreless grinding.** Refer to Fig. 2. *It is the method of grinding metallic parts in which the piece to be ground is supported on a **work-rest**, and passed between a **grinding wheel** running at a high speed and a **controlling wheel** running at a slow speed.* The controlling wheel causes the work, which must be cylinder, to revolve in the opposite direction to that of the grinding wheel. The work-rest includes a number of guides that feed the work to the revolving wheels, and remove it as soon as the operation is over. The *pressure exerted by the grinding wheel drives the work into contact with controlling wheel and the work-rest.* It will be realised from this that a point on the surface of the work where it comes into contact with the grinding wheel revolves in the same direction as, but at a slower speed than a corresponding point on the grinding wheel. This has the *effect of producing a more accurately cylindrical surface.*

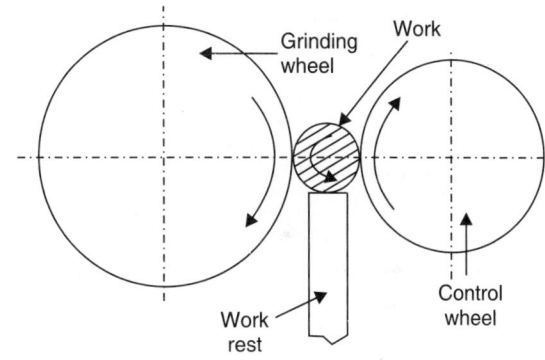

Fig. 2. Centreless grinding.

UNCONVENTIONAL MACHINING PROCESSES

Q. 97. *Explain briefly 'Water Jet Machining'.*

Ans. In *Water Jet Machining* process a high velocity *water jet* is made to impinge on to the workpiece. This jet pierces the work material and performs a sort of slitting operation. Water under pressure from a hydraulic accumulator is passed through the orifice of a nozzle to increase its velocity. The nozzle orifice size (dia.) usually varies from 0.08 mm to 0.5 mm and the exit velocity of the water jet from the nozzle varies upto 920 m/s. These high velocity jets can be used to cut relatively softer and non-metallic materials like paper boards, plastics asbestos, rubber, fireglass, leather etc.

• A variation of this process known as *"Hydrodynamic Jet Machining (HJM)"* has been successfully used to machine almost all types of ferrous and non-ferrous metals and alloys.

Q. 98. *For what types of works the Electro-Chemical Grinding (ECG) is best suited?*

Ans.   ***ECG*** is best suited for grinding *multiple-tooth milling cutters, carbide tipped tools and tool bits.*

- As it puts very little pressure on work it is best suited for grinding fragile and not easily supported works, like grinding slots, flats and forms in thin-walled fragile workpieces.

Q. 99. *On what types of works the process of chemical machining is best suited and what are its advantages and limitations?*

Ans.   **Chemical machining** *is an excellent method of getting complex shapes on very thin and most difficult to machine tools.*

- It does not require any press or punch or die. It does not distort the workpiece and no burrs are produced.
- It is a *slow process* and thus limited to machining metals upto 3 mm thickness and not used for producing large quantities.

Q. 100. *What is 'hot machining'?*

Ans.   **Hot machining** *is used for machining high strength, high hardness and high temperature resistant materials which are difficult to machine at room temperature.*

Machining of hard metals at elevated temperature is applied mainly to *turning and milling operations.* Since the shear strength of metal decreases at elevated temperature as compared to that at room temperature the magnitude of the cutting forces on the tool is lower. Further as the chip formation by plastic deformation in the shear plane ahead of tool becomes easier at elevated temperature and cutting forces involved are less, therefore *power requirements are low.* But as the property of tool material at elevated temperature is also changed due to its being in contact with high temperature material, therefore tool life is also affected. It is found that tool life is maximum for certain temperature of workpiece (for particular work material and tool material) at which total metal removal rate per tool grinding will be maximum irrespective of the speed.

Q. 101. *For what purpose Electric Discharge Machines (EDM) are used?*

Ans.   ***EDM machines*** are used for avoiding difficulty to machine materials through the use of electric spark.

Example: Die sinking or removal of broken tools embedded in workpieces.

Q. 102. *Indicate the sources of energy in the following processes:*

(a) EDM,	*(b) USM,*
(c) LBM,	*(d) ECM.*

Ans.

Process	*Sources of energy*
(*a*) EDM	*Electric spark*
(*b*) USM	*Mechanical*
(*c*) LBM	*Radiation*
(*d*) ECM	*Electrical*

Q. 103. *What is 'Magnetic forming'?*

Ans.   ***Magnetic forming*** is also known as 'magnetic pulse forming' or 'electromagnetic forming'. In this process an insulated induction coil is either wrapped around or placed within the work depending upon whether the metal is to be squeezed inward or buldged outward. The high momentary current is passed through the coil which *develops intense magnetic field* causing the work to collapse, compress, shrink or expand depending on the design and placement of the coil.

The work size, that can be forced, is determined by the energy storage capacity and ability of the unit to utilize that energy. High conductivity materials can be formed if they are wrapped or coated with a high conductivity auxiliary material.

JIGS AND FIXTURES

Q. 104. *What are 'Jigs and Fixtures'?*

Ans. *Jigs and fixtures are production tools used to accurately manufacture duplicate and interchangeable parts.* They are specially designed so that a large number of components can be machined or assembled identically, and to ensure interchangeability of components. They eliminate the necessity of a special set up for each individual part.

— Jigs and fixtures are *precision tools.*

— They are *expensive to produce* because they are made to fine limits from materials with good resistance to wear.

— They must be properly stored or isolated to prevent accident damage, and they must be numbered for identification for future use.

Q. 105. *What are the differences between jigs and fixtures?*

Ans. Following are the *differences* between jigs and fixtures:

1. *Essential **difference** between a jig and fixture is that the former incorporates bushes that guide the tools whilst the latter holds the component being machined with the cutters working independently, of it.*

2. **Jigs** are used on *drilling, reaming, tapping and counter-boring operations,* while **fixtures** are used in connection with *turning, milling, grinding, snapping, planing and boring operations.*

3. Whereas **jigs** are *connected with operations,* **fixtures** *most commonly are related to specific machine tools.*

4. *Jigs are lighter than fixtures,* for quick handling; *fixtures are heavier in construction and bolted rigidly on the machine table.*

Q. 106. *What are the advantages of using jigs and fixtures?*

Ans. The use of *jigs* and *fixtures* in mass production of identical parts offer the following *advantages:*

1. Improved productivity.

2. Rapid production work.

3. Reduced manufacturing costs (since large number of identical and interchangeable parts are produced) using jigs and fixtures.

4. Large reduction in fatigue to the operator (since there is considerable reduction in manual handling operations).

5. Complex and heavy components can be easily machined (since such parts can be rigidly held in proper location for machining in jigs and fixtures).

6. Owing to high clamping rigidity (offered by jigs and fixtures), higher speeds, feeds and depth of cut can be used.

7. Increased machining accuracy (owing to the automatic location of the work and guidance of the tool).

8. Considerable reduction in the expenditure due to inspection and quality control of finished components (since the parts produced with the use of jigs and fixtures are very accurate).

CNC MACHINES AND CAD/CAM

Q. 107. *What are 'Numerical Control (NC)' and 'Computer Numerical Control (CNC)' machine tools?*

Ans. "Numerically Controlled (NC) machine tools" are *highly flexible and are economical for producing a single or a large number of parts.* **Numerical Control, NC** can be defined simply as *control by numbers.* A machine tool having a *dedicated computer to* *help prepare the program and control some or all of the operations of the machine tool is called* **Computer Numerical Control (CNC)** *machine tool.*

Q. 108. *What are NC machines?*

Ans.   **NC machines** assimilate a method of automation, where automation of medium and small volume production is done by some controls under the instructions of a program. The definition of NC (Numerical Control) as given by EIA (Electronic Industries Association) is as under:

"A system in which actions are controlled by the direct insertion of numerical data at some point. The system must automatically interpret at least some portion of this data." In NC machines, the input information for controlling the machine tool motion is provided by means of *punched tapes* or *magnetic tapes in a coded language.*

Q. 109. *What is a CNC machine? Explain.*

Ans.   In a **CNC machine,** a *minicomputer is used to control machine tool functions from stored in information or punched tope input or computer terminal input.*

The definition CNC (Computer Numerical Control) as given by EIA is as under:

"The numerical control system where a dedicated, stored program computer is used to perform some or all of the basic numerical control functions in accordance with control programms stored in read/write memory (RAM) of the computer".

CNC may also be defined as: *"An NC system with a microcomputer or microprocessor using software to implement control algorithms".*

Fig. 3 shows the control unit and panel of a CNC. The following points about CNC machines are worthnoting:

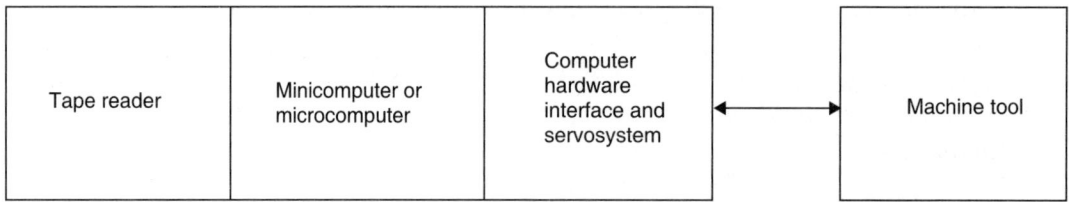

Fig. 3. Computer numerical system (CNC).

- The control unit and a panel of CNC differs from that of NC controls in that, it works in *ON-line mode* whereas NC works in *batch processing mode.*
- A typical CNC may need only the drawing specifications of a part to be manufactured and the computer automatically generates the part program for the loaded part.
- The part program once entered into the computer memory can be used again and again.
- The input information can be reduced to a great extent with the use of special sub-programs developed for repetitive machining sequences.
- The CNC machines have the facility for proving the part program without actually running it on the machine tool.
- CNC control unit allows compensation for any changes in the dimensions of the cutting tool.
- With CNC control systems, it is possible to obtain information on machine utilisation which is useful to the managements.

Q. 110. *What are the function of CNC?*

Ans.   The *principal functions of CNC* are:

    1. Machine tool control.

    2. In-process compensation.

3. Improved programming and operating features.

4. Diagnostics.

Q. 111. *What are the advantages, disadvantages and applications of CNC machines?*

Ans. **Advantages of CNC machines:**

Following are the *advantages* of CNC machines (over NC machines):

1. Greater flexibility.

2. Reduced data reading error.

3. CNC machine can diagnose program and can detect the machine malfunctioning even before the part is produced.

4. In highly sophisticated manufacturing systems CNC machine can be integrated with DNC (Direct Numerical Control) systems.

5. Conversion of units— possible within the computer memory.

Disadvantages of CNC machines:

1. Higher investment cost.

2. Higher maintenance cost.

3. Costlier CNC personnel.

4. Air-conditioned places are required for the installation of the machines.

5. Unsuitable for long run applications.

6. Planned support facilities.

Q. 112. *What is 'CAD? What are its advantages?*

Ans. In the modern sense, **CAD** (Computer Aided Designs) is defined as:

"A design process using sophisticated computer graphics techniques, backed up with computer software packages to aid in the analytical, development, costing and ergonomic problems associated with design work".

Advantages:

The following are the *advantages of CAD:*

1. Drawings can be produced at a faster rate.

2. Drawings produced by CAD systems are more accurate and neat.

3. In this system there is no repetition of the drawings.

4. CAD systems assimilate several special draughting techniques which are not available with conventional means.

5. Design calculations and analysis can be carried out quickly.

6. With CAD systems superior design forms can be produced.

7. CAD simulation and analysis techniques can drastically cut the time and money spent on prototype testing and development — often the costliest stage in the design process.

8. Using CAD systems design can be integrated with other disciplines.

Q. 113. *What is 'CAM? What are its advantages?*

Ans. **CAM** *(Computer-Aided Manufacture) concerns any automatic manufacturing process which is controlled by computers.*

The most important elements of CAM are:

1. CNC manufacturing and programming techniques.

2. Computer controlled robotics manufacture and assembly.

3. Flexible Manufacturing Systems (FMS).

4. Computer Aided Inspection (CAI) techniques.

5. Computer Aided Testing (CAT) techniques.

Advantages:

CAM entails the following *advantages:*

1. Product obtained is superior in quality.
2. The manufactured form has a greater versatility.
3. Higher production rates with lower workforces.
4. There is less likelyhood of human error.
5. As a result of increased manufacturing efficiency, cost savings are materialised.
6. The production processes can be repeated via storage of data.

ELEMENTS OF INTEGRATION

Q. 114. *In what ways a robot differ from a NC machine tool?*

Ans.   A *robot* differs from a *NC machine tools* in the following *respects:*

1. As compared to a NC machine tool, a *robot is a lighter and more portable equipment.*
2. *Programming* of a robot is different from the *tool programming* used in NC machine tools.
3. The applications of robots are more general in nature as compared to a NC machine tool.

Q. 115. *Give the comparison between 'Automated machine 'and 'Robot'.*

Ans.   The comparison between 'Automated machine' and 'Robot' is as follows:

S.No.	Aspects	Automated machine	Robot
1.	*Type of job*	It performs a particular job only in the designed sequence.	In can be made to do different jobs at different times and in different sequences.
2.	*Programming*	It cannot be programmed.	It can be programmed to change the sequence of tasks.
3.	*Knowledge base, intelligence*	It has neither a knowledge base nor intelligence.	It is possible to impart intelligence base by suitable programming technique.

Q. 116. *Discuss briefly the 'motion systems' used in industrial robots.*

Ans.   The following *two types of **motion systems*** are used in *'industrial robots':*

1. Point-to-point system.
2. Continuous path system.

1. Point-to-point system:

- In this system, only the movement of robot from one point location to other location in space is controlled with no regard to the path followed by it in doing so.
- Majority of the robots employed in production activities, like *pick and place, loading and unloading of parts* etc., work on point-to-point motion system.

2. Continuous path system:

- In this system a robot follows a predetermined *path* or *contour* in reaching from one point to the other.
- These robots need higher level of memory and control as compared to point-to-point robots.
- Such robots are widely required in several operations like *continuous welding, spray painting* etc.

Q. 117. *Discuss briefly "Workcell and Interlocks".*

Ans.   **Workcell:**

- The term **workcell** *represents a collection of automated equipment and controls that acts as a means to coordinate all the activities within the robot work station.* It is required because the robot system by itself has a very limited utility. In order to utilise it for performing some really useful tasks, it has to be integrated with many other things like work parts, part feeders, conveyors, control devices, process machinery, tools, machine tools etc.

- A *"work cell"* need not essentially cover a single robot only, but may contain several robots. In a robot *"workstation"*, some of the activities take place in a sequence and some others occur simultaneously. A controlling device, called a *"workcell controller " or "workstation controller"*, is, therefore suitably incorporated, which coordinates and regulates these different activities and ensures that they are performed in a proper sequence.

Interlocks:

- In *workcell control,* there has to the some means to check the continuation of sequence of the work cycle whenever needed. The means used to apply this check is known as an *"Interlock".*

- The workcell carries *two types facilitates of interlocks* - one *"outgoing"* and the other *"incoming"*. The provision of interlocks facilitates an effective control over the workcycle by permitting the sequential activities to take place only when they should and preventing them when they should not.

Q. 118. *Explain briefly the following categories of automation:*

1. Fixed automation;

2. Programmable automation;

3. Flexible automation.

Ans.   1. **Fixed automation:**

- It is also known as *hard automation,* which is used to produce a standardized product such as gears, nuts and bolts, etc.

- Even though the operating conditions can be changed, *fixed automation is used for very large quantity production of one or few marginally different components.*

- Highly specialized equipment, called *special purpose machine tools,* are utilized to produce a product or a component of a product very efficiently and at high production rates such that the *unit cost are low relative to the alternative methods of production.*

2. **Programmable automation:**

- In this type of automation we can change the design of the product or even change the product by *changing the program.*

- It is *useful for the low quantity production of large number of different components.*

- The equipment used for the production is designed to be adaptable or programmable.

- The production is normally carried out in *batches.*

3. **Flexible automation:**

- This is also called *'Flexible Manufacturing System* (FMS)'.

- This *allows manufacturing different products on the same equipment in any order or mix.*

Examples:

(*i*) Numerical control—Descrete parts manufacturing.

(*ii*) Robot-programmable automation equipment that is not directly used as the production equipment. Robots are an integral part of FMS and Computer Integrated Manufacturing Systems (CIMS).

Q. 119. *What is group technology? State its area of application.*

Ans.   **Group technology** *is the realisation that many problems are similar, and that by grouping similar problems, a single solution can be found to a set of problems, thus saving time and efforts.*

Group technology is a manufacturing philosophy or principle whose basic concept is to *identify and bring together related or similar parts and processes, to take advantage of similarities which exist, during all stages of design and manufacturing.* It is the replacing of traditional jobbing shop manufacture by the analysis and grouping of work into families and the formation of group of machines to manufacture these families on a flow-line principle with the object of minimising setting times and throughout times.

  • Group technology has become an increasingly popular concept of manufacturing which can be applied in any industry (such as *"machining" "welding", "foundry", "presswork", "forging ", "plastic moulding"* etc.) that is designed to take advantage of *mass production layout and techniques, in smaller batch-production systems.*

Q. 120. *What is 'cluster analysis'? Explain.*

Ans.   **Cluster analysis** *is employed for the analysis of production flow analysis chart to determine feasible group of processes and their respective packs of parts.* It makes use of algorithms for the study of similarities between objects in a quantitative manner as compared to the classification techniques, which appears to be descriptive.

Clustering may be defined as the *science of the classification of objects based on their possession or lack of defined characteristics.* This technique shows an approach to study the similarities between a diverse population of objects in a quantitative manner.

Clustering consists of the following *three* stages.

  1. *Preparing a post-operation matrix.* This shows whether certain features (like a keyway on shaft) are present or absent.

  2. *Computing a similarity coefficient matrix.* The bases of this is the extent to which the parts share common characteristics. In this case, coefficient would have a value one (1) when parts are identical and ten (10) when they have no common entity.

  3. *Performing a clustering analysis.* In this case, the similarity between each pair of objects is examined and group of objects formed such that within each group, the objects are similar to each other according to set of rules which have been formulated previously.

Q. 121. *Discuss briefly "Cellular and flexible manufacturing".*

Ans.   For the development of effective quality systems and automation, low volume is a major obstacle. When only a few parts will be produced little capital is made available for purchasing automated systems and up-to-date test equipment. In addition, other benefits resulting from learning curves and quality feedback may not be possible. Long lead times, high levels of inventory, and poor levels of quality usually result. **Cellular manufacturing** using the techniques of group technology and *flexible manufacturing systems* was developed to help solve these problems. *"Cellular manufacturing" increases the apparent production volumes by grouping similar parts.*

  • The **Flexible manufacturing system** (FMS) is a different approach to development of a specialised production line suitable for many different parts *using sophisticated computer control and a material handling system.*

  — *FMS consists of a group of manufacturing work stations connected together by an automated workpart handling system.* The system is capable of simultaneous processing of a variety of different part types at the various work stations under program control.

  — FMS system however, is a *very expensive bridge for the gap between high production transfer lines and lower production rates.*

ADDITIONAL OBJECTIVE TYPE QUESTIONS

(Selected from IES, IAS, GATE etc. Examinations)

ADDITIONAL OBJECTIVE TYPE QUESTIONS
(Selected from IES, IAS, GATE etc. Examinations)

A. Choose the Correct Answer :

1. Two plates of the same metal having equal thickness are to be butt welded with electric arc. When the plate thickness changes, welding is achieved by
   (a) adjusting the current
   (b) adjusting the duration of current
   (c) changing the electrode size
   (d) changing the electrode coating.

2. Which of the following arc welding processes does not use consumable electrodes :
   (a) SAW
   (b) GMAW
   (c) GTAW
   (d) none of these.

3. The temperature of a carburising flame in gas welding is ...... that of a neutral or oxidising flame.
   (a) lower than.
   (b) higher than
   (c) equal to
   (d) unrelated to.

4. In resistance welding, the greatest resistance, when the power is switched on, is :
   (a) at the point of contact between the electrode and the job.
   (b) at the point of contact of workpieces to be joined
   (c) at the surface
   (d) none of the above.

5. The maximum temperature produced by oxy-hydrogen flame is about :
   (a) 3300°C
   (b) 2500°C
   (c) 2000°C
   (d) 1800°C.

6. Consider the following statements :
   Fibre Reinforced Plastics are :
   1. made of thermosetting resins and glass fibre
   2. made of thermoplastic resins and glass fibre
   3. anisotropic
   4. isotropic. Of these statements
   (a) 1 and 4 are correct
   (b) 1 and 3 are correct
   (c) 2 and 3 are correct
   (d) 2 and 4 are correct.

7. A lead-screw with half nuts in a lathe, free to rotate in both directions has
   (a) V-threads
   (b) Whitworth threads
   (c) Buttress threads
   (d) ACME threads.

8. The time taken to drill hole through a 25 mm thick plate with the drill rotating at 300 rpm and moving at a feed rate of 0.25 mm/rev is
   (a) 10 s
   (b) 20 s
   (c) 100 s
   (d) 200 s.

9. If the melting ratio of a cupola is 10 : 1, then the coke requirement for one ton melt will be
   (a) 0.1 ton
   (b) 10 tons
   (c) 1 ton
   (d) 11 tons.

10. Which one of the following are the requirements of an ideal gating system ?
    1. The molten metal should enter the mould cavity with as high a velocity as possible
    2. It should facilitate complete filling of the mould cavity
    3. It should be able to prevent the absorption of air or gases from the surroundings of the molten metal while flowing through it

Select the correct answer using the codes given below :

(*a*) 1, 2 and 3   (*b*) 1 and 2   (*c*) 2 and 3   (*d*) 1 and 3.

11. Consider the following statements :

the strength of a single point cutting tool depends upon

1. Rake angle   2. Clearance angle   3. Lip angle

Which of these statements are correct ?

(*a*) 1 and 3   (*b*) 2 and 3   (*c*) 1 and 2   (*d*) 1, 2 and 3.

12. The hardness of a grinding wheel is determined by the

(*a*) hardness of abrasive grains   (*b*) ability of the bond to retain abrasives

(*c*) hardness of the bond

(*d*) ability of grinding wheel to penetrate the work piece.

13. A built-up-edge is formed while machining

(*a*) ductile material at high speed   (*b*) ductile material at low speed

(*c*) brittle material at high speed   (*d*) brittle material at low speed.

14. Trepanning is performed for

(*a*) finishing a drilled hole   (*b*) producing a large hole without drilling

(*c*) truing a hole for alignment   (*d*) enlarging a drilled hole.

15. The compositions of some of the alloy steels are as under :

1. 18 W 4 Cr 1 V   2. 12 Mo 1 W 4 Cr 1 V

3. 6 Mo 6W 4 Cr 1 V   4. 18 W 8 Cr 1 V

The compositions of commonly used high speed steels would include :

(*a*) 1 and 2   (*b*) 2 and 3   (*c*) 1 and 4   (*d*) 1 and 3.

16. The straigh grades of cemented carbide cuttingtool materials contain

(*a*) Tungsten carbide only   (*b*) tungsten carbide and titanium carbide

(*c*) tungsten carbide and cobalt.   (*d*) tungsten carbide and cobalt carbide.

17. Selection of electrolyte for ECM is as follows :

(*a*) non-passivating electrolyte for stock removal and passivating electrolyte for finish control

(*b*) passivating electrolyte for stock removal and non-passivating electrolyte for finish control

(*c*) selection of electrolyte is dependent on current density

(*d*) electrolyte is based on tool-work electrodes.

18. Inter electrode gap in ECG is controlled by

(*a*) controlling the pressure of electrolyte flow

(*b*) Controlling the applied static load

(*c*) Controlling the size of diamond particle in the wheel

(*d*) Controlling the texture of the workpiece.

19. Climb milling is chosen while machining because

(*a*) the chip thickness increases gradually

(*b*) it enables the cutter to dig in and start the cut

(*c*) the specific power consumption is reduced

(*d*) better surface finish can be obtained.

20. In centreless grinding, the workpiece centre will be

(a) above the line joining the two wheel centres

(b) below the line joining the two wheel centres

(c) on the line joining the two wheel centres

(d) at the intersection of the line joining the wheel centres with the work plate/plane.

21. In a milling operation, two side milling cutters are mounted with a desired distance between them so that both sides of a workpiece can be milled simultaneously. This set up is called

(a) gang milling   (b) straddle milling   (c) side milling.

22. In a mechanical shaper, the length of stroke is increased by

(a) increasing the centre distance of bull gear and crank pin

(b) decreasing the centre distance of bull gear and crank pin

(c) increasing the length of ram.

(d) decreasing the length of slot in the slotted lever.

23. In EDM process, the workpiece is connected to :

(a) cathode   (b) anode   (c) earth   (d) any of these.

24. USM is best suited for which materials ?

(a) soft and ductile materials   (b) hard and brittle materials

(c) Both of these materials.

25. With increase in the frequency of tool oscillation, the MRR in USM will

(a) first increase and then remain constant   (b) first decrease and then remain constant

(c) increase   (d) decrease.

26. Consider the following pairs of plastics and their distinct characteristics :

1. Acrylics ...... very good transparency to light

2. Polycarbonate ...... Poor impact resistance

3. PTFE ...... low co-efficient of friction

4. Polypropylene ...... excellent fatigue strength.

Which of these pairs are correctly matched ?

(a) 2 and 3   (b) 1 and 3   (c) 1 and 4   (d) 2 and 4.

27. In Electro-Discharge Machining (EDM), the tool is made of

(a) copper   (b) high speed steel

(c) cast iron   (d) plain carbon steel.

28. Deep hole drilling of small diameter, say 0.2 mm is done with EDM by selecting the tool material as

(a) copper wire   (b) tungsten wire

(c) brass wire   (d) tungsten Carbide.

29. The simultaneous compacting and sintering is achieved by which method :

(a) cold isostatic pressing   (b) hot isostatic pressing

(c) P/M forging   (d) none of the above.

30. Which of the following are the rules of programming NC machine tools in APT language ?

1. Only capital letters are used

2. A period is placed at the end of each statement

3. Insertion of space does not affect the APT word

Select the correct answer using the codes given below :

(a) 1 and 2   (b) 2 and 3   (c) 1 and 3   (d) alone.

31. Consider the following characteristics of production jobs :

1. Processing of parts frequently in small lots
2. Need to accommodate design changes of products
3. Low rate of metal removal
4. Need for holding close tolerances.

The characteristics which favour the choice of NC machines would include

(a) 1, 2 and 3   (b) 2 and 3, 4   (c) 1, 3 and 4   (d) 1, 2 and 4.

32. Which of the following joining processes are best suited for manufacturing pipes to carry gas products ?

1. riveting   2. welding   3. nuts and bolts

Select the correct answer using the codes given below :

Codes :

(a) 1 and 2   (b) 1 and 3   (c) 2 alone   (d) 1, 2 and 3.

33. The correct sequence of the given materials in descending order of their weldability is

(a) MS, Cu, C.I., Al   (b) C.I., M.S., Al, Cu

(c) Cu, C.I., MS, Al   (d) Al, Cu, C.I., MS.

34. Which gas from the following will form a hard constituent when it combines with molten steel and remains after solidification has taken place :

(a) O_2 (b) N_2 (c) H_2 (d) He.

35. The purpose of preheating low-alloy steel pipes, before they are electric arc welded is to :

(a) Refine grain structure   (b) Relieve internal stresses

(c) Retard rapid cooling   (d) Regulate excessive expansion.

36. Hard-Zone cracking in low-alloy steel due to welding is the result of an absorption of :

(a) N_2 (b) O_2 (c) H_2 (d) C.

37. In sheet metal work, the cutting force on the tool can be reduced by

(a) grinding the cutting edges sharp   (b) increasing the hardness of tool

(c) providing shear angle on tool   (d) increasing the hardness of die.

38. Tandem drawing of wires and tubes is necessary because

(a) it is not possible to reduce in one stage
(b) annealing is needed between stages
(c) accuracy in dimensions is not possible otherwise
(d) surface finish improves after every drawing stage.

39. In submerged arc welding, the arc is struck between

(a) Consumable coated electrode and job   (b) Non-consumable electrode and job

(c) Consumable bare electrode and job   (d) Two tungsten electrodes and the job.

Note : For MIG, the answer will be c

TIG, the answer will be b

SMAW, the answer will be a.

40. Resistance spot welding is performed on two plates of 1.5 mm thickness with 6 mm diameter electrode, using 15000 A current for a time duration of 0.25 s. Assuming the interface resistance to be 0.0001 ohm, the heat generated to form the weld is

(a) 5625 W-sec   (b) 8437 W-sec   (c) 22500 W-sec   (d) 33750 W-sec.

41. During heat treatment of steel, the hardness of various structures in the increasing order is

(a) martensite, fine pearlite, coarse pearlite, spheroidite

(b) fine pearlite, coarse pearlite, spheroidite, martensite

(c) martensite, coarse pearlite, fine pearlite, spheroidite

(d) spheroidite, coarse pearlite, fine pearlite, martensite.

42. Hardness of green sand mould increases with

(a) increase in moisture content beyond 6 percent

(b) increase in permeability   (c) decrease in permeability

(d) increase in both moisture content and permeability.

43. Consider the following statements regarding grinding of high carbon steel

1. Grinding at high speeds results in the reduction of chip thickness and cutting forces per grit.

2. Aluminium oxide wheels are employed

3. The grinding wheel has to be of open structure. Of these statements

(a) 1, 2 and 3 are correct   (b) 1 and 2 are correct

(c) 1 and 3 are correct   (d) 2 and 3 are correct.

44. Decreasing grain size in a polycrystalline material

(a) increases yield strength and corrosion resistance

(b) decreases yield strength and corrosion resistance

(c) decreases yield strength but increases corrosion resistance

(d) increases yield strength but decreases corrosion resistance.

45. Carburised machine components have higher endurance limit, because carburization

(a) raises the yield point of the material

(b) produces a better surface finish

(c) introduces a compressive layer on the surface

(d) suppresses any stress concentration produced in the component.

46. The Iron-Carbon diagram and the TTT curves are determined under :

(a) equilibrium and non-equilibrium conditions respectively

(b) non-equilibrium and equilibrium conditions respectively

(c) equilibrium conditions for both

(d) non-equilibrium conditions for both.

47. On completion of heat treatment, the following structure will have retained austenite if

(a) rate of cooling is greater than the critical cooling rate

(b) rate of cooling is less than the critical cooling rate

(c) martensite formation starting temperature is above the room temperature

(d) martensite formation finish temperature is below the room temperature.

48. Consider the following statements :

For precision machining of non-ferrous alloys, diamond is preferred because it has

1. low co-efficient of thermal expansion

2. high wear resistance

3. high compression strength

4. low fracture toughness

Which of the following statements are correct ?

(a) 1 and 2   (b) 1 and 4   (c) 2 and 3   (d) 3 and 4.

49. Which one of the following processes results in the best accuracy of the hole made ?

(a) Drilling   (b) Reaming   (c) Broaching   (d) Boring.

50. For butt-welding 40 mm thick steel plates when the expected quantity of such jobs is 5000 per month over a period of 10 years, choose the best suitable welding process out of the following available alternatives

(a) Submerged arc welding   (b) Oxy-acetylene gas welding

(c) Electron beam welding   (d) MIG welding.

51. Electron-beam welding can be carried out in

(a) open air   (b) a shielding gas environment

(c) a pressurized inert gas chamber   (d) vacuum.

52. In gas welding of mild steel using an oxy-acetylene flame, the total amount of acetylene used was 10 litre. The oxygen consumption from the cylinder is

(a) 5 litre   (b) 10 litre   (c) 15 litre   (d) 20 litre.

[**Hint** : For gas welding of M.S. neutral flame is used].

53. Mass production of cooking utensils is usually done by

(a) metal spinning   (b) deep drawing   (c) coining   (d) embossing.

54. The equipment generally used for metal spinning is

(a) mechanical press   (b) hydraulic press   (c) drop hammer   (d) speed lathe.

55. Collapsible tubes are made by

(a) direct extrusion   (b) indirect extrusion   (c) cold impact extrusion   (d) piercing.

56. In d.c. welding the straight polarity (electrode negative) results in

(a) lower penetration   (b) lower depositing rate

(c) less heating of work piece   (d) smaller weld pod.

57. The electrodes used in arc welding are coated electrodes. The coating is not expected to

(a) provide protective atmosphere to weld   (b) stabilize the arc

(c) add alloying elements   (d) prevent electrode from contamination.

58. A riser

(a) acts as a reservoir for the molten metal

(b) delivers molten metal into the mould cavity

(c) delivers molten metal from the pouring basin to runner

(d) feeds the molten metal to the casting in order to compensate for solidification shrinkage.

59. A spherical drop of molten metal of radius 2 mm was found to solidify in 10 seconds. A similar drop of radius 4 mm will solidify in

(a) 14.14 s   (b) 20 s   (c) 28.30 s   (d) 40 s.

60. In solidification of metal during casting, compensation for solid contraction is

(a) provided by the oversize pattern   (b) achieved by properly placed risers

(c) obtained by promoting directional solidification

(d) made by providing chills.

61. The purpose of a gas-welding flux is to :

(a) lower the melting point of the metal   (b) lower the melting point of the oxide

(c) remove oxides from the surface of the metal

(d) remove elements from parent metal.

62. The flux coating of an electric arc electrode has a melting point :
    (a) higher than the metallic core   (b) lower than the metallic core
    (c) the same as the metallic core   (d) the same as the metal being welded.

63. Majority of the oxy-acetylene welding is done with :
    (a) neutral flame   (b) reducing flame   (c) oxidising flame.

64. Brasses and Bronzes are welded by :
    (a) neutral flame   (b) reducing flame   (c) oxidising flame

65. Fluxes are used while welding :
    (a) to increase the rate of welding   (b) to clean the joint
    (c) to prevent oxidation of metal during welding
    (d) all the above.

66. Consider the following statements related to piercing and blanking :
    1. Shear on the punch reduces the maximum cutting force.
    2. Shear increases the capacity of the press needed.
    3. Shear increases the life of the punch.
    4. The total energy needed to make the cut remains unaltered due to provision of shear.
    Which of the above statements are correct ?
    (a) 1 and 2   (b) 1 and 4   (c) 2 and 3   (d) 3 and 4.

67. Which of the following pairs are correctly matched ?
    1. CNC machine ...... Post processor   2. Machining Centre ...... Tool Magazine
    3. DNC ...... FMS
    (a) 1, 2 and 3   (b) 1 and 2   (c) 1 and 3   (d) 2 and 3.

68. Which of the following is/are the advantage (s) of numerical control of machine tools ?
    1. Reduced lead time   2. Consistently good quality
    3. Elaborate fixtures are not required
    (a) 2 and 3   (b) 1 and 2   (c) 1 alone   (d) 1 and 3.

69. The mode of deformation of the metal during spinning is
    (a) bending   (b) stretching
    (c) rolling and stretching   (d) bending and stretching.

70. In drop forging, forging is done by dropping
    (a) the work piece at high velocity   (b) the hammer at high velocity
    (c) the die with hammer at high velocity
    (d) a weight on hammer to produce the requisite impact.

71. Consider the following statements :
    The strength of the fibre reinforced plastic product
    1. depends upon the strength of the fibre alone
    2. depends upon the fibre and the plastic
    3. is isotropic
    4. is anisotropic
    Which of these statements are correct ?
    (a) 1 and 3   (b) 1 and 4   (c) 2 and 3   (d) 2 and 4.

72. In reaming process
    (a) metal removal rate is high   (b) high surface finish is obtained
    (c) high form accuracy is obtained   (d) high dimensional accuracy is obtained.

73. Cubic boron nitride is used
   (a) as lining material in induction furnaces   (b) for making optical quality glass
   (c) for heat treatment   (d) for none of the above.

74. Match the following components with the appropriate machining process :

Component	Process
(a) Square hole in a high strength alloy	(p) Milling
(b) Square hole in a ceramic component	(q) Drilling
(c) Blind hole in a die	(r) ECM
(d) Turbine blade profile on high strength alloy	(s) Jig boring
	(t) EDM
	(u) USM.

75. With reference to NC machines which of the following statements is wrong :
   (a) both closed-loop and open-loop control systems are used
   (b) paper tapes, floppy tapes and cassettes are used for data storage
   (c) digitizers may be used as interactive input devices
   (d) post processor is an item of hardware.

76. CNC machines are more accurate than conventional machines because they have a high resolution encoder and digital readouts for positioning. (Yes/No)

77. Which of the following fibre materials are used for reinforcement in composite materials :
   1. Glass   2. Boron Carbide   3. Graphite.
   Select the correct answer using the codes given below :
   Codes :
   (a) 1 and 2   (b) 1 and 3   (c) 2 and 3   (d) 1, 2 and 3.

78. In sheet metal blanking, shear is provided on punches and dies so that
   (a) press load is reduced   (b) good cut edge is obtained
   (c) warping of sheet is minimised   (d) cut blanks are straight.

79. Which of the following pairs of process and draft are correctly matched ?
   1. Rolling 2   2. Extrusion 50   3. Forging 4
   Select the correct answer using the codes given below :
   Codes :
   (a) 1, 2 and 3   (b) 1 and 2   (c) 1 and 3   (d) 2 and 3.

80. In metal cutting operation, the approximate ratio of heat distributed among chip, tool and work, in that order is
   (a) 80 : 10 : 10   (b) 33 : 33 : 33   (c) 20 : 60 : 10   (d) 10 : 10 : 80.

81. Which one of the following sets of forces are encountered by a lathe parting tool while groove cutting :
   (a) Tangential, radial and axial   (b) Tangential and radial
   (c) Tangential and axial   (d) Radial and axial.

82. In a single-point turning operation of steel with cemented carbide tool, Taylor's tool life exponent is 0.25. If the cutting speed is halved, the tool life will increase by
   (a) two times   (b) four times   (c) eight times   (d) sixteen times.

83. The type of quick return mechanism employed mostly in shaping machines is :
   (a) D.C reversible motor   (b) Fast and loose pulleys
   (c) Whit worth motion   (d) Slotted link mechanism.

84. A 400 mm long shaft has 100 mm tapered step at the middle with 4° included angle. The tail stock offset required to produce this taper on a lathe would be

(a) 400 sin 4°   (b) 400 sin 2°   (c) 100 sin 4°   (d) 100 sin 2°.

85. In Power Metallurgy, the strength of the green compact is achieved by

(a) Tempering   (b) Compressed tempering

(c) Sintering   (d) None of the above.

86. The operation of "Pre-Sintering" is done with

(a) Soft and ductile material   (b) Hard and strong materials

(c) Blended powders   (d) None of the above.

87. Consider the following statements :

Thermosetting plastics are

1. Formed by addition polymerisation

2. Formed by condensation polymerisation

3. Softened on heating and hardened on cooling for any number of times

4. Moulded by heating and cooling. Of the statements

(a) 1 and 3 are correct   (b) 2 and 4 are correct

(c) 1 and 4 are correct   (d) 2 and 3 are correct.

88. Light impurities in the molten metal are prevented from reaching the mould cavity by providing a

(a) Strainer   (b) Bottom well   (c) Skim bob   (d) all of the above.

89. Chills are used in moulds to

(a) achieve directional solidification   (b) reduce the possibility of blow holes

(c) reduce freezing time   (d) smoothen metal flow for reducing splatter.

90. The two main criteria for selecting the electrolyte in Electrochemical Machining (ECM) is that the electrolyte should

(a) be chemically stable   (b) not allow dissolution of cathode material

(c) not allow dissolution of anode material   (d) have high electrical conductivity.

91. In ultrasonic machining process, the material removal rate will be higher for materials with

(a) higher toughness   (b) higher ductility

(c) lower toughness   (d) higher fracture strain.

92. Consider the following statements :

In up-milling process

1. The cutter starts the cut from the machine surface and proceeds upwards

2. The cutter starts the cut from the top surface and proceeds downwards

3. The job is fed in a direction opposite to that of cutter rotation

4. The job is fed in the same direction as that of cutter rotation of these statements :

(a) 1 and 3 are correct   (b) 1 and 4 are correct

(c) 2 and 3 are correct   (d) 2 and 4 are correct.

93. The correct sequence of the given processes in manufacturing by powder metallurgy is

(a) blending, compacting, sintering, sizing   (b) blending, compacting, sizing, sintering

(c) compacting, sizing, blending, sintering   (d) compacting, blending, sizing and sintering.

94. The operation of filling the pores of a sintered component with molten metal is known as :
(a) Sizing   (b) Coining   (c) Impregnation   (d) Infiltration.

95. In thermit welding, heat is generated
(a) From combustion of gas   (b) By an arc
(c) By chemical reaction between aluminium and iron oxide
(d) None of the above.

96. For resistance welding
(a) voltage is high, current is low   (b) voltage is low, current is high
(c) both voltage and current are low   (d) both voltage and current are high.

97. A block of information in N.C. machine program means :
(a) one row on tape   (b) a word comprising several rows on tape
(c) one complete instruction   (d) one complete program for a job.

98. Feed drives in CNC milling machines are provided by
(a) synchronous motors   (b) induction motors
(c) stepper motors   (d) servo-motors.

99. The completion of transformation of austenite into ferrite and pearlite is represented by
(a) A_1 line   (b) A_3 line
(c) Acm line   (d) None of the above.

100. The limit of Carbon solubility in austenite is represented by
(a) A_1 line   (b) A_3 line
(c) Acm line   (d) None of the above.

101. Fine grains of austenite
(a) decrease hardenability   (b) increase hardenability
(c) first decrease, then increase hardenability (d) first increase, then decrease hardenability.

102. As tool and work are not in contact in EDM process
(a) no relative motion Occurs between them   (b) no wear of tool occurs
(c) no power is consumed during metal cutting (d) no force between tool and work occurs.

103. Hardness of steel greatly improves with
(a) Annealing   (b) Cyaniding   (c) Normalising   (d) Tempering.

104. With a solidification factor of 0.97×10^6 s/m^2, the solidification time (in seconds) for a spherical casting of 200 mm diameter is
(a) 539   (b) 1078   (c) 4311   (d) 3233.
Hint : In Chvorinov's equation

$$t = C \cdot \left(\frac{V}{A_{surface}} \right)^2, \ C = 0.97 \times 10^6 \ \text{s/m}^2$$

105. Consider the following reason :
1. Grinding wheel is soft   2. RPM of grinding wheel is too low
3. Cut is very fine   4. An important cutting fluid is used.
A grinding wheel may become loaded due to reasons stated at
(a) 1 and 4   (b) 1 and 3   (c) 2 and 4   (d) 2 and 3.

106. Consider the following operations :
1. Under cutting   2. Plain turning   3. Taper turning   4. Thread cutting.

The correct sequence of these operations in machining a product is

(a) 2, 3, 4, 1   (b) 3, 2, 4, 1   (c) 2, 3, 1, 4   (d) 3, 2, 1, 4.

107. Which one of the following materials is used as the bonding material for grinding wheels ?

(a) Silicon carbide   (b) Sodium silicate

(c) Boron carbide   (d) Aluminium oxide.

108. The rate of production of a powder metallurgy part depends on

(a) flow rate of powder   (b) green strength of powder

(c) apparent density of compact   (d) Compressibility of powder.

109. In a machining process, the percentage of heat carried away by the chips is typically

(a) 5%   (b) 25%   (c) 50%   (d) 75%.

110. T.T.T. diagram indicates time, and temperature transformation of

(a) Cementite   (b) Pearlite   (c) Ferrite   (d) Austenite.

111. The correct composition of austenitic stainless steel used for domestic utensils is :

(a) 0.08% C, 18% Cr, 8% Ni, 2% Mn, 1% Si   (b) 0.08% C, 24% Cr, 12% Ni, 2% Mn, 1% Si

(c) 0.15% C, 12% Cr, 0.5% Ni, 1% Mn, 1% Si (d) 0.30% C, 12% Cr, 0.4% Ni, 1% Mn, 1% Si

112. Consider the following steps involved in hammer forging a connecting rod from bar stock :

1. Blocking   2. Trimming

3. Finishing   4. Fullering

5. Edging

Which of the following is the correct sequence of operations ?

(a) 1, 4, 3, 2 and 5   (b) 4, 5, 1, 3 and 2   (c) 5, 4, 3, 2 and 2   (d) 5, 1, 4, 2 and 3.

113. Size of a shaper is given by

(a) stroke length   (b) motor power   (c) weight of the machine (d) rate size.

114. Consider the following tool materials :

1. Carbide   2. Cermet   3. Ceramic   4. Borazon

Correct sequence of these tool materials in increasing order of their ability to retain their hot hardness is

(a) 1, 2, 3, 4   (b) 1, 2, 4, 3   (c) 2, 1, 3, 4   (d) 2, 1, 4, 3.

115. Enlarging an existing circular hole with a rotating single point tool is called

(a) boring   (b) drilling

(c) reaming   (d) internal turning

116. The limit to the maximum hardness of a work material which can be machined with HSS tools even at low speeds is set by which of the following tool failure mechanism ?

(a) Attrition   (b) Abrasion

(c) Diffusion   (d) Plastic deformation under compression.

117. A machanist desires to turn a round steel block of outside diameter 100 mm at 1000 rpm. The material has tensile strength of 75 kg/mm^2. The depth of cut chosen is 3 mm at a feed rate of 0.3 mm/rev. Which one of the following tool materials will be suitable for machining the component under the specified cutting conditions ?

(a) Sintered Carbides (b) Ceramic   (c) HSS   (d) Diamond.

[**Hint** : Find cutting speed]

118. The molten metal is poured from the pouring basin to the gate with the help of a

(a) Riser   (b) Sprue   (c) Runner   (d) Core.

119. In cold chamber die-casting process, only non-ferrous alloys with ...... (high melting point/ low melting point) are cast.

120. In the casting of large pipes by true centrifugal casting
 (a) Core is of sand   (b) Core is of metal
 (c) no core is used.

121. Consider the following ingredients used in moulding :
 1. Dry silica sand 2. Clay 3. Phenol formaldehyde 4. Sodium silicate
 Those used for shell mould casting include :
 (a) 1, 2 and 4   (b) 2, 3 and 4   (c) 1 and 3   (d) 1, 2, 3, 4.

122. Consider the following statements :
 MIG welding process uses
 1. consumable electrode   2. non-consumable electrode
 3. D.C. Power supply   4. A.C. Power supply
 Of these statements
 (a) 2 and 4 are correct   (b) 2 and 3 are correct
 (c) 1 and 4 are correct   (d) 1 and 3 are correct.

123. The process of producing thin sheets by squeezing a thermo-plastic material between revolving cylinders, is known as :
 (a) Transfer moulding   (b) Injection moulding
 (c) Blow moulding   (d) Calendering.

124. The process of producing plastic components in moulds without the application of pressure, is known as
 (a) Moulding   (b) Laminating   (c) Calendering   (d) Casting.

125. Consider the following parameters :
 1. Grinding wheel diameter   2. Regulating wheel diameter
 3. Speed of grinding wheel   4. Speed of the regulating wheel
 5. Angle between the axes of grinding
    and regulating wheels
 Among these parameters, those which influence the axial feed rate in centreless grinding would include :
 (a) 2, 4 and 5   (b) 1, 2 and 3   (c) 1, 4 and 5   (d) 3, 4 and 5.

126. It is required to cut screw threads of 2 mm pitch on a lathe. The lead screw has a pitch of 6 mm. If the spindle speed is 60 rpm, then the speed of the lead screw will be
 (a) 10 rpm   (b) 20 rpm   (c) 120 rpm   (d) 180 rpm.

127. Which one of the following is the correct temperature range for hot extrusion of aluminium ?
 (a) 300 – 340°C   (b) 350 – 400°C   (c) 430 – 480°C   (d) 550 – 650°C.

128. Major operations in the manufacture of steel balls used for Ball bearings are given below :
 1. Oil lapping 2. Cold heading 3. Annealing 4. Hardening 5. Rough grinding
 The correct sequence of these operations is
 (a) 3, 2, 4, 1, 5   (b) 3, 2, 1, 4, 5   (c) 2, 3, 4, 5, 1   (d) 2, 3, 5, 4, 1.

129. The bottles from thermo-plastic materials are made by
 (a) Compression moulding   (b) Extrusion
 (c) Injection moulding   (d) Blow moulding.

130. The long plastic rods and tubes are produced by
     (a) Compression moulding
     (b) Extrusion
     (c) Injection moulding
     (d) Blow moulding.

131. Which of the following components can be manufactured by powder metallurgy methods ?
     1. Carbide tool tips   2. Bearings   3. Filters   4. Brake linings
     Select the correct answer using the codes given below :
     (a) 1, 3 and 4   (b) 2 and 3   (c) 1, 2 and 4   (d) 1, 2, 3 and 4.

132. In Powder metallurgy, the operation carried out to improve the bearing property of a bush is called
     (a) Infiltration   (b) Impregnation   (c) Plating   (d) Heat treatment.

133. Which of the following materials can be used for making patterns ?
     1. Aluminium   2. Wax   3. Mercury   4. Lead
     Select the correct answer using the codes given below :
     Codes :
     (a) 1, 3 and 4   (b) 2, 3 and 4   (c) 1, 2 and 4   (d) 1, 2 and 3.

134. Which of the following materials will require the largest size of riser for the same size of casting ?
     (a) Aluminium   (b) cast iron   (c) steel   (d) copper.

135. Dissolved alloying elements in steel
     (a) decrease hardenability
     (b) increase hardenability
     (c) has no effect on hardenability
     (d) first decrease, then increase hardenability.

136. The temperature at which the first new grain appears is known as
     (a) Melting temperature
     (b) Critical temperature
     (c) Boiling temperature
     (d) Recrystallisation temperature.

137. Which of the following pairs are correctly matched ?
     1. Pit moulding ...... for large jobs
     2. Investment moulding ...... Lost wax process
     3. Plaster moulding ...... Mould prepared in gypsum
     (a) 1, 2 and 3   (b) 1 and 2   (c) 1 and 3   (d) 2 and 3.

138. Which one of the following pairs is not correctly matched ?
     (a) Aluminium alloy Piston ...... Pressure die casting
     (b) Jewellery ...... Lost wax process   (c) Large pipes ...... Centrifugal casting
     (d) Large bells ..... Loam moulding.

139. When a steel is heated to above its upper critical temperature, the structure produced is one of :
     (a) Martensite   (b) Austenite   (c) Pearlite   (d) Sorbite.

140. The predominant structure of a hypereutectoid steel that has been quenched at above its upper critical temperature will be :
     (a) Austenite   (b) Martensite   (c) Troostite   (d) Sorbite.

141. Small amount of which one of the following elements/pairs of elements is added to steel to increase machinability ?
     (a) Nickle
     (b) Sulphur and phosphorous
     (c) Silicon
     (d) Manganese and copper.

142. Which of the following pairs regarding the effect of alloying elements in steel are correctly matched ?

    1. Molybdenum : Forms abrasion resisting particles

    2. Phosphorus : Improves machinability in free cutting steels

    3. Cobalt : Contributes to red hardness by hardening ferrite

    4. Silicon : Reduces oxidation resistance

Select the correct answer using the codes given below :

    (a) 2, 3, 4   (b) 1, 3 and 4   (c) 1, 2 and 4   (d) 1, 2 and 3.

143. Which of the following materials are used in grinding wheels ?

    1. Aluminium oxide   2. Cubic boron nitride

    3. Silicon carbide

Select the correct answer using the codes given below :

Codes :

    (a) 1, 2 and 3   (b) 1 and 2   (c) 2 and 3   (d) 1 and 3.

144. Metal extrusion process is generally used for producing

    (a) uniform solid sections   (b) uniform hollow sections

    (c) uniform solid and hollow sections   (d) varying solid and hollow sections

145. Which one of the following is an advantage of forging ?

    (a) Good surface finish   (b) Low tooling cost

    (c) close tolerance   (d) Improved physical property.

146. In wire drawing process, the bright shining surface on the wire is obtained if one

    (a) does not use a lubricant   (b) uses solid powdery lubricant

    (c) uses thick paste lubricant   (d) uses thin fluid lubricant.

147. Which one of the following processes does not cause tool wear ?

    (a) USM   (b) ECM

    (c) EDM   (d) Anode mechanical machining.

148. In ECM, the material removal is due to

    (a) Corrosion   (b) Erosion

    (c) Fusion   (d) Ion displacement.

149. With increase in the abrasive slurry concentration, the MRR in USM will

    (a) increase   (b) decrease

    (c) first increase and then remain constant   (d) first decrease and then remain constant.

150. Holes in Nylon buttons are made by :

    (a) EDM   (b) CHM   (c) USM   (d) LBM.

151. Integrated circuits and printed circuits are produced by

    (a) EDM   (b) CHM   (c) USM   (d) LBM.

152. The upper critical temperature for steel

    (a) is constant   (b) depends upon the rate of heating

    (c) varies according to the carbon content in steel

    (d) none of the above.

153. Which of the following pairs are correctly matched ?

1. Silicon steels ...... Transformer stampings

2. Duralumin ...... Cooking utensils

3. Gun metal ...... Bearings

Select the correct answer using the codes given below :

Codes :

(a) 1, 2 and 3   (b) 1 and 2   (c) 1 and 3   (d) 2 and 3.

154. Which of the following methods are used for obtaining directional solidification for riser design ?

1. Suitable placement of chills 2. Suitable placement of chaplets 3. Employing padding

Select the correct answer :

(a) 1 and 2   (b) 1 and 3   (c) 2 and 3   (d) 1, 2 and 3.

155. Misrun is a casting defect which occurs due to

(a) very high pouring temperature of the metal

(b) insufficient fluidity of the molten metal

(c) absorption of gases by the liquid metal

(d) improper alignment of the mould flasks.

156. The primary purpose of sprue in a casting mould is to

(a) Feed the casting at a rate consistent with the rate of solidification

(b) Act as a reservoir for molten metal

(c) Feed molten metal from the pouring basin to the gate

(d) Help feed the casting until all solidification takes place.

157. In the grinding wheel of A60G7B23, B stands for

(a) resinoid bond   (b) rubber bond   (c) shellac bond   (d) silicate bond.

158. Soft materials cannot be economically ground due to

(a) the high temperature involved   (b) frequent wheel clogging

(c) rapid wheel wear   (d) low work piece stiffness.

159. In turning of slender rods, it is necessary to keep the transverse force minimum mainly to

(a) improve the surface finish   (b) increase productivity

(c) improve cutting efficiency   (d) reduce vibrations and chatter.

160. A grinding wheel of 150 mm diameter is rotating at 3000 rpm. The grinding speed is

(a) 7.5 π m/s   (b) 15 π m/s   (c) 45 π m/s   (d) 450 π m/s.

161. Failure of a bead weld between a heavy steel section and a thin section is mainly due to the formation of

(a) spheriodite   (b) bainite

(c) carbon free zone due to burning of carbon at high temperature

(d) martensite.

162. The open circuit voltage (OCV) in arc welding ranges from :

(a) 40 to 80 V   (b) 100 to 150 V   (c) 200 to 220 V   (d) 400 to 440 V.

163. In machining using abrasive material, increasing abrasive grain size

(a) increases the material removal rate   (b) decreases the material removal rate

(c) first decreases and then increases the material removal rate

(d) first increases and then decreases the material removal rate.

164. Abrasive material used in grinding wheel selected for grinding ferrous alloys is

(a) Silicon carbide   (b) diamond

(c) aluminium oxide   (d) boron carbide.

165. Which of the following are the advantages of a hydraulic shaper over a mechanically driven shaper ?

1. More strokes can be obtained per minute at a given cutting speed

2. The cutting stroke has a definite stopping point

3. It is simpler in construction

4. Cutting speed is constant throughout most of the cutting stroke.

Select the correct answer using the codes given below :

(a) 1 and 2   (b) 1 and 4   (c) 2 and 4   (d) 1, 3 and 4.

166. Which one of the following pairs of parameters and effects is not correctly matched ?

(a) Large wheel diameter ...... Reduced wheel wear

(b) Large depth of cut ...... Increased wheel wear

(c) Large work diameter ...... Increased wheel wear

(d) Large wheel speed ...... Reduced wheel wear.

167. The recrystallisation temperature depends upon :

(a) grain size   (b) type of metal

(c) extent of cold deformation   (d) annealing time

(e) purity of metal   (f) all of the above.

168 .The recrystallisation temperature for pure metals is

(a) $0.2\ T_m$ (b) $0.3\ T_m$ (c) $0.5\ T_m$ (d) $0.8\ T_m$

Where T_m = melting temperature.

169. The lower critical temperature for steel

(a) is constant   (b) depends upon the rate of heating

(c) varies according to the carbon content in steel

(d) None of the above.

170. The beginning of separation of ferrite from solid solution of Austenite is represented by

(a) A_1 line   (b) A_3 line

(c) Acm line   (d) None of the above.

171. Disk-shaped components are cast by

(a) True centrifugal casting   (b) Semi-centrifugal casting

(c) Centifuge casting

172. Consider the following ingredients used in moulding :

1. Dry silica sand 2. Clay 3. Ethyl silicate 4. Phenol formaldehyde

Those used for Lost Wax casting method include :

(a) 1, 2 and 4   (b) 2, 3 and 4   (c) 1 and 3   (d) 1, 2, 3 and 4.

173. Heating the Hypoeutectoid steels 30° above the upper critical temperature line, soaking at that temperature and then cooling slowly to room temperature to form a pearlite and ferrite structure is known as

(a) hardening   (b) normalising   (c) tempering   (d) annealing.

174. In a eutectic system, two elements are completely

(a) insoluble in solid and liquid states   (b) soluble in liquid state

(c) soluble in solid state   (d) insoluble in liquid state.

175. A steel with 0.8% C is called
  (a) Hypo-eutectoid steel
  (b) Hyper-eutectoid steel
  (c) Eutectoid steel
  (d) None of these.

176. Alloy steel which is work hardenable and which is used to make the blades of bulldozers bucket wheels excavators and other earth moving equipment contains iron, carbon and
  (a) Mn
  (b) Si
  (c) Cr
  (d) Mg.

177. Guide ways of lathe beds are hardened by
  (a) Carburising
  (b) Cyaniding
  (c) Nitriding
  (d) Flame hardening.

178. A given steel test specimen is studied under metallurgical microscope. Magnification used is 100 X. In that different phases are observed. One of them is Fe_3C.
  The observed phase Fe_3C is also known as
  (a) ferrite
  (c) cementite
  (c) austenite
  (d) martensite.

179. A steel with 0.8% C contain :
  (a) 100% pearlite
  (b) 100% austenite
  (c) ferrite and pearlite
  (d) pearlite and cementite.

180. The lower critical temperature for all steels is :
  (a) 700°C
  (b) 723°C
  (c) 650°C
  (d) 910°C.

181. Machine tool guideways are usually hardened by
  (a) Vacuum hardening
  (b) Martempering
  (c) Induction hardening
  (d) Flame hardening.

182. 18/8 stainless steel contains :
  (a) 18% stainless, 8% Cr
  (b) 18% Cr, 8% Ni
  (c) 18% Cr, 8% W
  (d) 18% W, 8% Cr.

183. Tin base white metals are used where the bearings are subjected to :
  (a) large surface wear
  (b) elevated temperatures
  (c) high load and pressure
  (d) high pressure and load.

184. A steel with 0.8% C has
  (a) One critical point
  (b) Two critical points
  (c) No critical point.

185. Cementite consists of :
  (a) 13% ferrite and 87% pearlite
  (b) 6.67% C and 93.33% Iron
  (c) 13% C and 87% ferrite.

186. Consider the following statements :
  Addition of Silicon to cast iron
  1. promotes graphite nodule formation
  2. promotes graphite flake formation
  3. increases fluidity of the molten metal
  4. improves the ductility of cast iron.
  Of these statements :
  (a) 1 and 4 are correct
  (b) 2 and 3 are correct
  (c) 1 and 3 are correct
  (d) 3 and 4 are correct.

187. Eutectic reaction for iron-carbon system occurs at
    (a) 600° C   (b) 723°C   (c) 1147°C   (d) 1493°C

188. The blade of a power saw is made of
    (a) boron steel   (b) high speed steel
    (c) stainless steel   (d) malleable cast iron.

189. Quartz is a
    (a) ferroelectric material   (b) ferromagnetic material
    (c) piezo electric material   (d) diamagnetic material.

190. Which of the following pairs are correctly matched ?
    1. Lead screw nut ...... Phosphor bronze
    2. Piston ...... Cast iron
    3. Cam ...... EN 31 steel
    4. Lead screw ...... Wrought Iron
    Select the correct answer using the codes given below :–
    Codes :
    (a) 2, 3 and 4   (b) 1, 3 and 4   (c) 1, 2 and 4   (d) 1, 2 and 3.

191. Consider the following treatments :
    1. Normalising   2. Hardening   3. Martempering   4. Cold working
    Hardness and tensile strength in austenitic steel can be increased by
    (a) 1, 2 and 3   (b) 1 and 3   (c) 2 and 4   (d) 4 alone.

192. In oxy-acetylene gas welding, for complete combustion, the volume of oxygen required per unit ton of acetylene is
    (a) 1   (b) 1.5   (c) 2   (d) 2.5.

193. The voltage-current characteristics of a dc generator for arc welding is a straight line between an open circuit voltage of 80 V and short circuit current of 300 A. The generator setting for maximum arc power will be
    (a) 80 V and 150 A   (b) 40 V and 300 A
    (c) 40 V and 150 A   (d) 80 V and 300 A.

194. In oxyacetylene gas welding, the temperature at the inner cone of the flame is around
    (a) 3500°C   (b) 3200°C   (c) 2900°C   (d) 2550°C.

195. Cold working of steel is defined as working
    (a) at its recrystallisation temperature   (b) above its recrystallisation temperature
    (c) below its recrystallisation temperature
    (d) at two thirds of the melting temperature of the metal.

196. In metals subjected to cold working, strain hardening effect is due to
    (a) slip mechanism   (b) twining mechanism
    (c) distocation mechanism   (d) fracture mechanism.

197. Which one of the following processes is most commonly used for the forging of bolt heads of hexagonal shape ?
    (a) closed die drop forging   (b) open die upset forging
    (c) closed die press forging   (d) open die progressive forging.

198. In blanking operation, the clearance provided is
    (a) 50% on punch and 50% on die   (b) on die   (c) on punch
    (d) on die or punch depending upon designer's choice.

199. Consider the following states of stress :

1. compressive stress in flange   2. tensile stress in the wall

3. tensile stress on the bottom part.

During drawing operation, the states of stress in cup would include

(a) 1 and 2   (b) 1 and 3   (c) 2 and 3   (d) 1, 2, and 3.

200. Blanking and Punching operations can be performed simultaneously on

(a) combination die   (b) compound die   (c) progressive die   (d) simple die.

201. Cutting and forming operations can be performed simultaneously on

(a) combination die   (b) compound die   (c) progressive die   (d) simple die.

202. Cold working of metal increases

(a) tensile strength   (b) hardness   (c) yield strength   (d) all of these.

203. The cutting force in punching and blanking operations mainly depends on

(a) the modulus of elasticity of metal   (b) the shear strength of metal

(c) the bulk modulus of the metal   (d) the yield strength of metal.

204. Cutting power consumption in turning can be significantly reduced by

(a) increasing rake angle of the tool   (b) increasing the cutting angle of the tool

(c) widening the nose radius of the tool   (d) increasing the clearance angle.

205. Plain milling of mild steel plate produces

(a) irregular shaped discontinuous chips   (b) regular shaped discontinuous chips

(c) continuous chips without built up edge   (d) jointed chips.

206. In cold working of metals, the working temperature is

(a) room temperature   (b) below the recrystallisation temperature

(c) above the recrystallisation temperature   (d) less than the room temperature.

207. In hot working of metals, the working temperature is

(a) below the recrystallisation temperature   (b) above the recrystallisation temperature

(c) equal to the melting point of the metal   (d) 150°C.

208. Which mechanical property a metal should possess to enable it to be mechanically formed ?

(a) ductility   (b) malleability   (c) elasticity   (d) machinability.

209. Duralumin alloy contains aluminium and copper in the ratio of

	% Al	% Cu
(a)	94	4
(b)	90	8
(c)	98	10
(d)	86	12

210. Hot rolling of mild steel is carried out

(a) at recrystallisation temperature   (b) Below 100°C and 150°C

(c) Above recrystallisation temperature   (d) Below recrystallisation temperature.

211. When a steel undergoes a cold working process, it becomes progressively

(a) Softer   (b) Harder   (c) Ductile   (d) Malleable.

212. Which of the following methods can be used for manufacturing 2 metre long seamless metallic tubes ?

1. Drawing   2. Extrusion   3. Rolling   4. Spinning

Select the correct answer using the codes given below :

Codes :

(a) 1 and 3   (b) 2 and 3   (c) 1, 3 and 4   (d) 2, 3 and 4.

213. A side and face cutter 125 mm diameter has 10 teeth. It operates at a cutting speed of 14 m/min. with a table traverse 100 mm/min. The feed per tooth of the cutter is

(a) 10 mm   (b) 2.86 mm   (c) 0.286 mm   (d) 0.8 mm.

214. Which one is not a method of reducing cutting forces to prevent the overloading of press ?

(a) providing shear on die   (b) providing shear on punch

(c) increasing die clearance   (d) stepping punches.

215. Crater wear on tools always starts at some distance from the tool tip because at that point

(a) Cutting fluid does not penerate   (b) normal stress on rake face in maximum

(c) temperature is maximum   (d) tool strength is minimum.

216. A 31.8 mm H.S.S drill is used to drill a hole in a cast iron block 100 mm thick at a cutting speed of 20 m/min and feed 0.3 mm/rev. If the overtravel of the drill is 4 mm and approach 9 mm, the time required to drill the hole is

(a) 1 min 40 s   (b) 1 min 44 s   (c) 1 min 49 s   (d) 1 min 52 s.

217. Pearlite consists of :

(a) 87% ferrite and 13% cementite   (b) 87% cementite and 13% ferrite

(c) 6.67% C and 93.33% iron.

218. In the austempering heat treatment process, austenite decomposes into :

(a) Sorbite   (b) Troostite   (c) Bainite   (d) Martensite.

219. Which of the following are fabricated using engineering plastics ?

1. Surface plate   2. Gears

3. Guideways for machine tools   4. Foundry patterns

Select the correct answer using the codes given below

Codes :

(a) 1, 2, and 3   (b) 1   (c) 2, 3 and 4   (d) 1, 2, 3 and 4.

220. Which one of the following is the hardest cutting tool material next only to diamond ?

(a) Cemented Carbides   (b) Ceramics

(c) Silicon   (d) Cubic boron nitride.

221. Core prints are provided on patterns

(a) to support the core   (b) to locate the core in the mould

(c) to support as well as locate the core in the mould.

222. Contraction allowance is provided on a pattern :

(a) for machining of castings   (b) for contraction in metal on cooling

(c) for making a good casting.

223. When the molten metal flows into the cavity of a metallic mould by gravity, the method of casting is known as :

(a) Die-casting   (b) Centrifugal casting

(c) Permanent mould casting   (d) Plaster mould casting.

224. The toys and ornaments of non-ferrous alloys are made by :

(a) Die-casting   (b) Lost-wax method

(c) Permanent mould casting   (d) Slush casting

225. A milling cutter having 8 teeth is rotating at 150 rpm. If the feed per tooth is 0.1 mm, the value of the table speed in mm/min is

    (a) 120   (b) 187   (c) 125   (d) 70.

226. To get good surface finish on a turned job, one should use a sharp tool with a ...... feed and ...... speed of rotation of the job.

227. Among the conventional, machining process, maximum specific energy is consumed in

    (a) Turning   (b) Drilling   (c) Planing   (d) Grinding.

228. In a CNC machine tool, encoder is used to sense and control

    (a) table position   (b) table velocity   (c) spindle speed   (d) coolant flow.

229. In rolling a strip between two rolls, the position of the neutral point in the arc of contact does not depend on

    (a) amount of reduction   (b) diameter of the rolls

    (c) co-efficient of friction   (d) material of the rolls.

230. Babbit lining is used on brass/bronze bearings to

    (a) increase bearing resistance   (b) increase compressive strength

    (c) provide any friction properties   (d) increase wear resistance.

231. In low carbon steels, presence of small quantities of sulphur improves :

    (a) Weldability   (b) formability   (c) machinability   (d) hardenability.

232. A standard dividing head is equipped with the following index plates

    1. Plate with 15, 16, 17, 18, 19, 20 holes circles

    2. Plate with 21, 23, 27, 29, 31, 33 holes circles

    3. Plate with 37, 39, 41, 43, 47, 49 holes circles

For obtaining 24 divisions on a workpiece by simple indexing

    (a) hole plate 2 alone can be used   (b) hole plates 1 and 2 can be used

    (c) hole plates 1 and 3 can be used   (d) any of the three hole plates can be used.

233. A hole of 30 mm diameter is to be produced by reaming. The minimum diameter permisible is 30.00 mm while the maximum diameter permissible is 30.05 mm. In this regard, consider the following statements about the reamer size :

    1. The minimum diameter of the reamer can be less than 30 mm

    2. The minimum diameter of the reamer can not be less than 30 mm

    3. The maximum diameter of the reamer can be more than 30.05 mm

    4. The maximum diameter of the reamer must be less than 30.05 mm

       of these statements

    (a) 1 and 4 correct   (b) 1 and 3 are correct

    (c) 2 and 3 correct   (d) 2 and 4 are correct.

234. When a cold worked metal is heated upto its recrystallization temperature, it becomes

    (a) Harder   (b) Softer

    (c) Stays unchanged in hardness   (d) Stays unchanged in softness.

235. Which of the following statements are true of annealing of steels ?

    1. Steels are heated to 500 to 700°C

    2. Cooling is done slowly and steadily

    3. Internal stresses are relieved

    4. Ductility of steel is increased

Select the correct answer using the codes given below :

Codes :

(a) 2, 3 and 4   (b) 1, 3 and 4   (c) 1, 2 and 4   (d) 1, 2 and 3.

236. Which one of the following constituents is expected in equilibrium cooling of a hypereutectoid steel from austenitic state ?

(a) Ferrite and Pearlite   (b) Cementite and Pearlite

(c) Ferrite and bainite   (d) Cementite and Martensite.

237. Addition of Magnesium to cast iron increases its

(a) hardness   (b) ductility and strength in tension

(c) corrosion resistance   (d) creep strength.

238. The selection of welding electrodes depend upon :

(a) the thickness of the metal to be joined   (b) the type of metal to be joined

(c) position of welding   (d) strength of joint

(e) all the above.

239. Low pressure OAW is the name given to the welding process in which

(a) oxygen is supplied at low pressure   (b) acetylene is supplied at low pressure

(c) both oxygen and acetylene are supplied at low pressure.

240. In arc welding, the electric arc is produced between the job and the electrode due to

(a) Voltage   (b) flow of current   (c) contact resistance   (d) All of the above.

241. Consider the following components

1. A dedicated computer   2. Bulk memory

3. Telecommunication lines

Which of these components are required for a DNC system ?

(a) 2 and 3   (b) 1 and 2   (c) 1, 2 and 3   (d) 1 and 3.

242. In a point to point type of NC system

(a) Control of position and velocity of tool is essential

(b) Control of only position of the tool is sufficient

(c) Control of only velocity of the tool is sufficient

(d) neither position nor velocity need be controlled.

243. Shrinkage allowance on pattern is provided to compensate for shrinkage when

(a) the temperature of liquid metal drops from pouring to freezing temperature

(b) the metal changes from liquid to solid state at freezing temperature

(c) the temperature of solid phase drops from freezing to room temperature

(d) the temperature of metal drops from pouring to room temperature.

244. A gate

(a) acts as a reservoir for the molten metal

(b) delivers molten metal into the mould cavity

(c) delivers molten metal from the pouring basin to runner

(d) all of the above.

245. The recrystallisation temperature for alloys is approximately :

(a) $0.2\ T_m$ (b) $0.3\ T_m$ (c) $0.5\ T_m$ (d) $0.8\ T_m$.

246. Carbon occurs in steel in the combined state with iron to form the component :

(a) Ferrite   (b) Cementite   (c) Pearlite   (d) Bainite.

247. For producing self lubricated bearings by powder metallurgy process, the secondary operation carried out is called :

(a) Infiltration   (b) Impregnation   (c) Sintering   (d) Coining.

248. The method of Powder Metallurgy is used for :

(a) Mass production   (b) Small lot production

(c) Intricate shaped components   (d) None of the above.

249. A dynamometer is a device used for the measurement of

(a) chip thickness ratio   (b) forces during metal cutting

(c) wear of the cutting tool   (d) deflection of the cutting tool.

250. The main purpose of boring operation, as compared to drilling is to :

(a) drill a hole   (b) finish the drilled hole

(c) correct the hole   (d) enlarge the existing hole.

251. For machine a complex contour on Tungsten Carbide work-piece, which process will be used ?

(a) ECM   (b) EDM   (c) USM   (d) EBM.

252. The current density used in ECM process is of the order of :

(a) 100,00 A/cm^2 (b) 1000 A/cm^2 (c) 200 A/cm^2 (d) 10 A/cm^2.

253. In ECM process, the electrolyte used is

(a) kerosene   (b) water   (c) air   (d) Brine solution.

254. A single start thread of pitch 2 mm is to be produced on a lathe having a lead screw with a double start thread of pitch 4 mm. The ratio of speeds between the spindle and lead screw for this operation is

(a) 1 : 2   (b) 2 : 1   (c) 1 : 4   (d) 4 : 1.

255. The thickness of the blank needed to produce, by power spinning a missile cone of thickness 1.5 mm and half cone angle 30°, is

(a) 3.00 mm   (b) 2.5 mm   (c) 2.0 mm   (d) 1.55 mm

256. Which of the following statements are correct ?

   1. A boring machine is suitable for job shop.

   2. A jig boring machine is designed specially for doing more accurate work when compared to a vertical milling machine.

   3. A vertical precision boring machine is suitable for boring holes in cylindrical blocks and lines

(a) 1, 2 and 3   (b) 1 and 2   (c) 2 and 3   (d) 1 and 3.

257. The function of a core is :

(a) to improve mould surface   (b) to form internal cavities in the casting

(c) to form a part of a green sand mould (d) None of the above

(e) All of the above.

258. For resistance spot welding of 1.5 mm thick steel sheets, the current required is of the order of

(a) 10 A   (b) 100 A   (c) 1000 A   (d) 10,000 A.

259. In metal working processes, the stresses induced in the metal are

(a) less than the yield strength of the metal   (b) greater than the yield strength of the metal

(c) less than the breaking strength of the metal

(d) greater than the breaking strength of the metal.

260. Killed steels
    (*a*) have minimum impurity level
    (*b*) have almost zero percentage of phosphorus and sulphur
    (*c*) are produced by LD process   (*d*) are free from O_2.

261. Cast steel crankshaft surface is hardened by
    (*a*) nitriding   (*b*) normalising
    (*c*) carburising   (*d*) induction heating.

262. The alloying element mainly used to improve the endurance strength of steel materials is :
    (*a*) Nickle   (*b*) Vanadium   (*c*) Molybdenum   (*d*) Tungsten.

263. Which one of the following structures is predominant in a normalized steel :
    (*a*) Troostite   (*b*) Bainite   (*c*) Sorbite   (*d*) Martensite.

264. When a steel is heated in a furnace and then cooled in air at ordinary temperature, the process is one of
    (*a*) Annealing   (*b*) Hardening   (*c*) Normalizing   (*d*) Tempering.

265. Which one of the following materials is used for car tyres as a standard material ?
    (*a*) Styrene-Butadiene Rubber (SBR)   (*b*) Butyl rubber
    (*c*) Nitrile rubber   (*d*) Any of the above depending upon the need.

266. Which one of the following refractory materials is recommended for steel furnaces containing CaO slags ?
    (*a*) Alumina   (*b*) Silica   (*c*) Magnesia   (*d*) Fireclay.

267. The rake angle in a drill
    (*a*) increases from centre to periphery   (*b*) decreases from centre to periphery
    (*c*) remains constant   (*d*) is irrelevant to the drilling operation.

268. Helix angle of a fast helix drill is normally
    (*a*) 35°   (*b*) 60°   (*c*) 90°   (*d*) 5°.

269. In centrifugal casting, the impurities are
    (*a*) Uniformly distributed   (*b*) Forced towards the outer surface
    (*c*) trapped near the mean radius of the casting
    (*d*) collected at the centre of casting.

270. In order to get uniform thickness of plate by rolling process, one provides
    (*a*) Camber on the rolls   (*b*) offset on the rolls
    (*c*) hardening of the rolls   (*d*) antifriction bearings.

271. A moving mandrel is used in
    (*a*) wire drawing   (*b*) tube drawing   (*c*) metal cutting   (*d*) forging.

272. Centrifugally cast products have
    (*a*) large grain structure with high porosity (*b*) fine grain structure with high density
    (*c*) fine grain structure with low density
    (*d*) segregation of slag towards the outer skin of the casting.

273 The bending force required for V-bending, U-bending and Edge bending will be in the ratio of
    (*a*) 1 : 2 : 0.5   (*b*) 2 : 1 : 0.5   (*c*) 1 : 2 : 1   (*d*) 1 : 1 : 1.

274. Preheating before welding is done to
(a) make the steel softer
(b) burn away oil, grease etc. from the plate surface
(c) prevent cold cracks
(d) prevent plate distortion.

275. The following operations are performed while preparing the billets for extrusion process :
1. Alkaline cleaning
2. Phosphate coating
3. Pickling
4. Lubrication with reactive soap.
The correct sequence of these operations is
(a) 3, 1, 4, 2
(b) 1, 3, 2, 4
(c) 1, 3, 4, 2
(d) 3, 1, 2, 4

276. In a green-sand moulding process, uniform ramming leads to
(a) less change of gas porosity
(b) uniform flow of molten metal into the mould cavity
(c) greater dimensional stability of the casting
(d) less sand expansion type of casting defects.

277. The effect of rake angle on the mean friction angle in machining can be explained by
(a) sliding (Coulomb) model of friction
(b) sticking and then sliding model of friction
(c) sticking friction
(d) sliding and then sticking model of friction.

278. Directional solidification in castings can be improved by
(a) Chills and chaplets
(b) Chills and padding
(c) Chaplets and padding
(d) Chills, chaplets and padding.

279 In which one of the following welding techniques is vacuum environment required ?
(a) Ultrasonic welding
(b) Laser beam welding
(c) Plasma arc welding
(d) Electron beam welding.

280. Consider the following operations :
1 . Cutting keyways on shafts
2. Cutting external screw threads
3. Cutting teeth of spur gears
4. Cutting external splines.
Those which can be performed with milling cutters would include :
(a) 1 and 2
(b) 2, 3 and 4
(c) 1 and 3
(d) 1, 2, 3 and 4.

B. Match List I with List II :

281. Match List I (Machine tools) with List II (Features) and select the correct answer using the codes given below the list :

List I (Machine Tool)	List II (Features)
A. Lathe	1. Push or Pull tool
B. Drilling machine	2. Ratchet and pawl mechanism
C. Shaper	3. Dividing head
D. Broaching machine	4. Hollow tapered spindle
	5. Face plate

Codes :

	A	B	C	D
(a)	2	4	5	1
(b)	5	3	2	4
(c)	2	3	5	4
(d)	5	4	2	1

282. Match List I (NC machine tool systems) with List II (Features) and select the correct answer using the codes given below the lists :

List I	List II
(NC machine tool systems)	(Features)
A. NC system	1. It has integrated automatic tool changing unit and a component indexing device.
B. CNC system	2. A number of machine tools are controlled by a computer. No tape reader, the part programme is transmitted directly to the machine tool from the computer memory.
C. DNC system	3. The controller consists of soft-wired computer and hard-wired logic system. Graphic display of tool path is also possible.
D. Machining centre	4. The instructions on tape are prepared in binary decimal form and operated by a series of coded instructions.

Code :

	A	B	C	D
(a)	4	2	3	1
(b)	1	3	2	4
(c)	4	3	2	1
(d)	1	2	3	4

283. Match List I with List II and select the correct answer using codes given below the lists :

List I	List II
(Heat treatment)	(Effect on properties)
A. Annealing	1. Refines grain structure
B. Nitriding	2. Improves hardness of the whole mass
C. Martempering	3. Increases surface hardness
D. Normalising	4. Improves ductility

Codes :

	A	B	C	D
(a)	4	3	2	1
(b)	1	3	4	2
(c)	4	2	1	3
(d)	2	1	3	4

284. Match List I with List II and select the correct answer using the codes given below the list :

List I	List II
(Materials)	(Applications)
A. Tungsten Carbide	1. Abrasive wheels
B. Silicon nitride	2. Heating elements
C. Aluminium Oxide	3. Pipes for conveying liquid metals
D. Silicon Carbide	4. Drawing dies

Codes :

	A	B	C	D
(a)	3	4	1	2
(b)	4	3	2	1
(c)	3	4	2	1
(d)	4	3	1	2

285. Match List I (Ingredients) with List II (Welding functions) and select the correct answer using the codes given below the lists :

List I (Ingredients)	List II (Welding functions)
A. Silica	1. Arc stabilizer
B. Potassium silicate	2. De-oxidizer
C. Ferro-silicon	3. Fluxing agent
D. Cellulose	4. Gas forming material

Codes :

	A	B	C	D
(a)	3	4	2	1
(b)	2	1	3	4
(c)	3	1	2	4
(d)	2	4	3	1

286. Match List I with List II and select the correct answer using the codes given below the lists :

List I (Equipments)	List II (Functions)
A Hot chamber machine	1. Cleaning
B. Muller	2. Core making
C. Dielectric baker	3. Die casting
D. Sand blasting	4. Annealing
	5. Mixing

Codes :

	A	B	C	D
(a)	3	5	2	1
(b)	4	2	5	3
(c)	4	2	3	1
(d)	3	5	1	2

287. Match List I with List II and select the correct answer using the codes given below the lists :

List I (Materials)	List II (Applications)
A. Engineering Ceramics	1. Bearings
B. Fibre reinforced plastic	2. Control rods in nuclear reactors
C. Synthetic carbon	3. Aerospace industry
D. Boron	4. Electrical insulator

Codes :

	A	B	C	D
(a)	1	2	3	4
(b)	1	4	3	2
(c)	2	3	1	4
(d)	4	3	1	2

288. Match List I with List II and select the correct answer using the codes given below the lists :

List I	List II
(Metal forming process)	*(A similar process)*
A. Blanking	1. Wire drawing
B. Coining	2. Piercing
C. Extrusion	3. Embossing
D. Cup drawing	4. Rolling
	5. Bending

Codes :

	A	B	C	D
(a)	2	3	4	1
(b)	2	3	1	4
(c)	3	2	1	5
(d)	2	3	1	5

289. Match List I with List II and select the correct answer using the codes given below the lists :

List I	List II
(Material/Part)	*(Techniques)*
A. Ductile iron	1. Inoculation
B. Malleable iron	2. Chilled
C. Rail steel joints	3. Annealing
D. White cast iron	4. Thermit welding
	5. Isothermal annealing

Codes :

	A	B	C	D
(a)	1	3	4	2
(b)	5	3	2	1
(c)	2	1	4	5
(d)	1	4	2	3

290. Match the following :

Work material	Type of joining
P. Aluminium	1. Submerged Arc Welding
Q. Die Steel	2. Soldering
R. Copper wire	3. Thermit welding
S. Titanium sheet	4. Atomic Hydrogen Welding
	5. Gas Tungsten Arc Welding
	6. Laser Beam Welding

	A	B	C	D
	P – 2	P – 6	P – 4	P – 5
	Q – 5	Q – 3	Q – 1	Q – 4
	R – 1	R – 4	R – 6	R – 2
	S – 3	S – 4	S – 2	S – 6

291. Match List I with List II and select the correct answer

List I	*List II*
(*Metal forming process*)	(*Associated force*)
A. Wire drawing	1. Shear force
B. Extrusion	2. Tensile stress (force)
C. Blanking	3. Compressive force
A. Bending	4. Spring back force

Codes :

	A	B	C	D
(a)	4	2	1	3
(b)	2	1	3	4
(c)	2	3	1	4
(d)	4	3	2	1

292. Match the terms used in connection with heat-treatment of steel with the microstructural physical characteristics :

Terms	*Characteristics*
(A) Pearlite	(P) Extremely hard and brittle phase
(B) Martensite	(Q) Cementite is finely dispersed in ferrite
(C) Austenite	(R) Alternate layers of cementite and ferrite
(D) Eutectoid	(S) Can exist only above 723°C
	(T) Pertaining to state of equilibrium between three solid phases
	(U) Pertaining to state of equilibrium between one liquid and two solid phases

293. Match List I (Machine tools) with List II (Machine tool parts) and select the correct answer using codes given below the lists :

List I	*List II*
A. Lathe	I. Lead screw
B. Milling machine	2. Rocker arm
C. Shaper	3. Universal indexing
D. Drilling machine	4. Flute

Codes :

	A	B	C	D
(a)	4	2	3	1
(b)	1	3	2	4
(c)	4	3	2	1
(d)	1	2	3	4

294. Match List I with List II and select the correct answer using the codes given below the lists :

List I	List II
A. Drawing	1. Soap solution
B. Rolling	2. Camber
C. Wire drawing	3. Pilots
D. Sheet metal operations using progressive dies	4. Crater
	5. Ironing

Codes :

	A	B	C	D
(a)	2	5	1	4
(b)	4	1	5	3
(c)	5	2	3	4
(d)	5	2	1	3

295. Match List I with List II and select the correct answer using codes given below the lists :

List I (Steel type)	List II (Product)
A. Mild steel	1. Screw driver
B. Toll steel	2. Commercial beams
C. Medium Carbon steel	3. Crane hooks
D. High carbon steel	4. Blanking die

Codes :

	A	B	C	D
(a)	1	4	3	2
(b)	2	4	1	3
(c)	1	3	4	2
(d)	2	4	3	1

296. Match List I with List II and select the correct answer using the codes given below the Lists :

List I	List II
A. Neoprene	1. Electric switches
B. Bakelite	2. Adhesive
C. Foamed polyurethane	3. Thermal insulator
D. Araldite	3. Oil seal

Codes :

	A	B	C	D
(a)	4	1	2	3
(b)	1	4	2	3
(c)	4	1	3	2
(d)	1	4	3	2

297. Match List I with List II and select the correct answer taking the help of codes given below the lists :

List I (*Products*)	*List II* (*Process of Manufacture*)
A. Automobile piston in aluminium alloy	1. Pressure die casting
B. Engine crankshaft in spheroidal graphite iron	2. Gravity die-casting
C. Carburettor housing in aluminium alloy	3. Sand casting
D. Cast titanium blades	4. Precision investment casting
	5. Shell moulding

Codes :

	A	B	C	D
(a)	2	3	1	5
(b)	3	2	1	5
(c)	2	1	3	4
(d)	4	1	2	3

298. Match List-I (Drill bits) with List-II (Applications) and select the correct answer using the codes given below the lists :

List I	*List II*
A. Core drill	1. To enlarge a hole to a certain depth so as to accommodate the bolt head of a screw
B. Reamer	2. To drill and enlarge an already existing hole in a casting
C. Counter bore drill	3. To drill a hole before making internal threads
D. Tap drill	4. To improve the surface finish and dimensional accuracy of the already drilled hole

Codes :

	A	B	C	D
(a)	1	3	2	4
(b)	2	3	1	4
(c)	2	4	1	3
(d)	3	2	4	1

299. Match List I (Material) with List II (Application) and select the correct answer using the codes given below the lists :

List I (*Material*)	*List II* (*Application*)
A. Ceramics	1. Construction of Chemical plants
B. Refractories	2. Columns and pillars
C. Stones	3. Lining of furnaces
D. High silica glass	4. Tiles

Codes :

	A	B	C	D
(a)	4	3	2	1
(b)	2	1	4	3
(c)	4	1	2	3
(d)	2	3	4	1

300. Match 4 correct pairs

List I	List II
(A) ECM	(1) Plastic Shear
(B) EDM	(2) Erosion/Brittle fracture
(C) USM	(3) Corrosive reaction
(D) LBM	(4) Melting and vapourization
	(5) Ion displacement
	(6) Plastic shear and ion displacement

301. Match List I (materials) with List II (applications) and select the correct answer using the codes given below the lists :

List I	List II
A. Engineering ceramics	1. Bearings
B. Fibre reinforced plastics	2. Control rods in nuclear reactors
C. Synthetic carbon	3. Aerospace industry
D. Boron	4. Electrical insulator

Codes :

	A	B	C	D
(a)	1	2	3	4
(b)	1	4	3	2
(c)	2	3	1	4
(d)	4	3	1	2

302. Match List I (Parts) with List II (Manufacturing processes) and select the correct answer using the codes given below the lists :

List I	List II
(Parts)	(Manufacturing Process)
A. Seamless tubes	1. Roll forming
B. Accurate and smooth tubes	2. Shot peening
C. Surface having higher hardness and fatigue strength.	3. Forging
	4. Cold forming

Codes :

	A	B	C
(a)	1	4	2
(b)	2	3	1
(c)	1	3	2
(d)	2	4	1

303. Match list I with list-II and select the correct answer using the codes given bellow the lists :

List I	*List II*
(A) Aluminium brake shoe	(1) Deep drawing
(B) Plastic water bottle	(2) Blow moulding
(C) Stainless steel cups	(3) sand casting
(D) Soft drink can (aluminium)	(4) Centrifugal casting
	(5) Impact extrusion
	(6) Upset forging

304. Match List I (Alloys) with List II (Applications) and select the correct answer using the codes given below the lists :

List I	*List II*
A Chromel	1. Journal bearing
B Babbit alloy	2. Milling cutter
C Nimonic alloy	3. Thermocouple wire
D High speed steel	4. Gas turbine blades

Codes :

	A	B	C	D
(a)	3	1	4	2
(b)	3	4	1	2
(c)	2	4	1	3
(d)	2	1	4	3

305. Match List I with List II and select the correct answer

List I (Material)	*List II* (Nature of product)
A. Polyethylene	1. Adhesive
B. Polyurathane	2. Film
C. Cyano-acrylate	3. Wire
D. Nylon	4. Foam

Codes :

	A	B	C	D
(a)	2	4	3	1
(b)	4	2	3	1
(c)	2	4	1	3
(d)	4	2	1	3

306. Match List I with List II and select the correct answer using the codes given below the lists :

List I (Mech. properties)	*List II* (Related to)
(A) Malleability	1. Wire drawing
(B) Hardness	2. Impact loads
(C) Resilience	3. Cold rolling
(D) Isotropy	4. Indentation
	5. Direction

Codes :

	A	B	C	D
(a)	4	2	1	3
(b)	3	4	2	5
(c)	5	4	2	3
(d)	3	2	1	5

307. Match the following moulding casting processes with the product :

Moulding/Casting processes	*Product*
(A) Slush casting	(P) Turbine blades
(B) Shell moulding	(Q) Machine tool bed
(C) Dry sand moulding	(R) Cylinder block
(D) Centrifugal casting	(S) Hollow castings like lamp shades
	(T) Rain water pipe
	(U) Cast iron shoe brake

308. Match List I with List II and select the correct answer using the codes given below the lists :

List I	*List II*
(Name of Material)	*(% Carbon Range)*
A. Hypo-eutectoid steel	1. 4.3 – 6.67
B. Hyper-eutectoid steel	2. 2.0 – 4.3
C. Hypo-eutectic cast iron	3. 0.8 – 2.0
D. Hyper-eutectic cast iron	4. 0.008 – 0.8

Codes :

	A	B	C	D
(a)	4	3	2	1
(b)	1	3	2	4
(c)	4	1	2	3
(d)	1	2	3	4

309. Match List I (Machining Process) with List II (Associated Medium) and select the correct answer using the codes given below the lists :

List I	*List II*
A. USM	1. Kerosene
B. EDM	2. Abrasive slurry
C. ECM	3. Vacuum
D. EBM	4. Salt solution

Codes :

	A	B	C	D
(a)	2	3	4	1
(b)	2	1	4	3
(c)	4	1	2	3
(d)	4	3	2	1

310. List I gives a number of processes and List II gives a number of products. Match correct pairs :

List I	List II
(A) Investment casting	1. Turbine rotors
(B) Centrifugal casting	2. Turbine blades
(C) Die-casting	3. Connecting rods
(D) Drop forging	4. Galvanized iron pipe
(E) Extrusion	5. Cast iron pipes
(F) Shell moulding	6. Carburettor body

311. Match List I with List II and select the correct answer using the codes given below the lists

List I	List II
A. Die-sinking	1. Abrasive jet machining
B. Deburring	2. Laser beam machining
C. Fine hole drilling (thin material)	3. EDM
D. Cutting/hardening sharp materials	4. Ultrasonic machining
	5. Electrochemical grinding

Codes :

	A	B	C	D
(a)	3	5	4	1
(b)	2	4	1	3
(c)	3	1	2	5
(d)	4	5	1	3

312. Match List I (Alloying elements in steel) with List II (Property conferred on steel by the element) and select the correct answer using the codes given below the lists :

List I	List II
A. Nickel	1. Corrosion resistance
B. Chromium	2. Magnetic permeability
C. Tungsten	3. Heat resistance
D. Silicon	4. Hardenability

Codes :

	A	B	C	D
(a)	4	1	3	2
(b)	4	1	2	3
(c)	1	4	3	2
(d)	1	4	2	3

313. Match List I (ISO classification of carbide tools) with List II (applications) and select the correct answer using the codes given below the lists :

List I	List II
A. P – 10	1. Non-ferrous roughing cut
B. P – 50	2. Non-ferrous, finishing cut
C. K – 10	3. Ferrous materials, roughing cut
D. K – 50	4. Ferrous materials, finishing cut

Codes :

	A	B	C	D
(a)	4	3	1	2
(b)	3	4	2	1
(c)	4	3	2	1
(d)	3	4	1	2

314. Match 4 correct pairs between List I and List II

List I	List II
(A) Sand casting	(1) Symmetrical and circular shapes only
(B) Plaster mould casting	(2) Parts have hardened skin and soft interior
(C) Shell mould casting	(3) Minimum post casting processing
(D) Investment casting	(4) Parts have a tendency to warp
	(5) Parts have soft skin and hard interior
	(6) Suitable only for non-ferrous metals.

315. Match List I with List II and select the correct answer using the codes given below the lists

List I	List II
A. Reaming	1. Smoothing and squaring surface around the hole for proper seating
B. Counter-boring	2. Sizing and finishing the hole
C. Counter-sinking	3. Enlarging the end of the hole
D. Spot facing	4. Making a conical enlargement at the end of the hole

Codes :

	A	B	C	D
(a)	3	2	4	1
(b)	2	3	1	4
(c)	3	2	1	4
(d)	2	3	4	1

316. Match List I with List II and select the correct answer using the codes given below the lists :

List I (Filler)	List II (Joining process)
A. Cu, Zn, Ag alloy	1. Braze welding
B. Cu, Zn, alloy	2. Brazing
C. Pb, Sn alloy	3. Soldering
D. Iron oxide and aluminium powder	4. TIG welding of Al

Codes :

	A	B	C	D
(a)	2	1	3	–
(b)	1	2	4	–
(c)	2	1	3	4
(d)	2	–	3	4

317. Match 4 correct pairs between List I and List II

List I	*List II*
(A) Rivets for air craft body	1. Forging
(B) Carburettor body	2. Cold heading
(C) Crank shafts	3. Aluminium-based alloy
(D) Nails	4. Pressure die casting
	5. Investment casting.

318. Match 4 correct pairs between List I and List II

List I	*List II*
(A) Welding of aluminium alloy	1. Submerged arc welding
(B) Ship building	2. Electron beam welding
(C) Joining of HSS drill bit to carbon steel shank	3. TIG welding
(D) Deep penetration precision welds	4. Gas welding.

319. Match List I with List II and select the correct answer using the codes given below the lists :

List I (Cutting tool Material)	*List II* (Major characteristic constituent)
A. High Speed Steel	1. Carbon
B. Stellite	2. Molybdenum
C. Diamond	3. Nitride
D. Coated carbide tool	4. Columbuim
	5. Cobalt

Codes :

	A	B	C	D
(a)	2	1	3	5
(b)	2	5	1	3
(c)	5	2	4	3
(d)	5	4	2	3

320. Match 4 correct pairs

List I (Manufacturing processes)	*List II* (Conditions)
(A) Finish turning.	1. Backlash eliminator
(B) Forming	2. Zero rake
(C) Thread cutting	3. Nose radius
(D) Down milling	4. Low speed

ANSWERS

A. Choose the Correct Answer :

1. (a, c)	**2.** (c)	**3.** (a)	**4.** (b)	**5.** (b)
6. (b)	**7.** (d)	**8.** (b)	**9.** (b)	**10.** (c)
11. (d)	**12.** (b)	**13.** (b)	**14.** (b)	**15.** (c)

16. (c)	**17.** (d)	**18.** (c)	**19.** (d)	**20.** (a)
21. (b)	**22.** (a)	**23.** (b)	**24.** (b)	**25.** (a)
26. (c)	**27.** (a)	**28.** (d)	**29.** (b)	**30.** (c)
31. (d)	**32.** (c)	**33.** (a)	**34.** (b)	**35.** (c)
36. (c)	**37.** (c)	**38.** (a)	**39.** (c)	**40.** (a)
41. (d)	**42.** (c)	**43.** (b)	**44.** (a)	**45.** (c).
46. (a)	**47.** (c)	**48.** (a)	**49.** (d)	**50.** (b)
51. (d)	**52.** (b)	**53.** (b)	**54.** (d)	**55.** (c)
56. (b)	**57.** (d)	**58.** (d)	**59.** (d)	**60.** (a)
61. (c)	**62.** (a)	**63.** (a)	**64.** (c)	**65.** (d)
66. (b)	**67.** (a)	**68.** (a)	**69.** (d)	**70.** (c)
71. (b)	**72.** (d)	**73.** (d)	**74.** A–T, B–U, C–P, D–R	
75. (d)	**76.** Yes	**77.** (d)	**78.** (a)	**79.** (a)
80. (a)	**81.** (a)	**82.** (d)	**83.** (d)	**84.** (b)
85. (c)	**86.** (b)	**87.** (b)	**88.** (c)	**89.** (a)
90. (a, d)	**91.** (c)	**92.** (a)	**93.** (a)	**94.** (d)
95. (c)	**96.** (b)	**97.** (c)	**98.** (d)	**99.** (a)
100. (c)	**101.** (a)	**102.** (d)	**103.** (b)	**104.** (b)
105. (b)	**106.** (c)	**107.** (b)	**108.** (a)	**109.** (d)
110. (d)	**111.** (a)	**112.** (b)	**113.** (a)	**114.** (a)
115. (a)	**116.** (a)	**117.** (b)	**118.** (b)	
119. High melting point		**120.** (c)	**121.** (c)	**122.** (d)
123. (d)	**124.** (d)	**125.** (a)	**126.** (b)	**127.** (b)
128. (c)	**129.** (d)	**130.** (b)	**131.** (d)	**132.** (b)
133. (d)	**134.** (a)	**135.** (b)	**136.** (d)	**137.** (a)
138. (d)	**139.** (b)	**140.** (b)	**141.** (b)	**142.** (d)
143. (a)	**144.** (c)	**145.** (d)	**146.** (d)	**147.** (b)
148. (d)	**149.** (c)	**150.** (d)	**151.** (b)	**152.** (c)
153. (a)	**154.** (b)	**155.** (b)	**156.** (c)	**157.** (a)
158. (b)	**159.** (d)	**160.** (a)	**161.** (c)	**162.** (a)
163. (d)	**164.** (c)	**165.** (b)	**166.** (d)	**167.** (f)
168. (b)	**169.** (a)	**170.** (b)	**171.** (b)	**172.** (c)
173. (d)	**174.** (b)	**175.** (c)	**176.** (a)	**177.** (d)
178. (b)	**179.** (a)	**180.** (b)	**181.** (d)	**182.** (b)
183. (d)	**184.** (a)	**185.** (b)	**186.** (b)	**187.** (c)
188. (b)	**189.** (c)	**190.** (d)	**191.** (d)	**192.** (d)
193. (c)	**194.** (a)	**195.** (c)	**196.** (a)	**197.** (d)
198. (c)	**199.** (a)	**200.** (b)	**201.** (a)	**202.** (d)
203. (b)	**204.** (a)	**205.** (c)	**206.** (b)	**207.** (b)
208. (a)	**209.** (a)	**210.** (c)	**211.** (b)	**212.** (b)
213. (c)	**214.** (c)	**215.** (c)	**216.** (d)	**217.** (a)
218. (c)	**219.** (c)	**220.** (d)	**221.** (c)	**222.** (b)
223. (c)	**224.** (d)	**225.** (a)	**226.** min, max	**227.** (d)
228. (a)	**229.** (d)	**230.** (d)	**231.** (c)	**232.** (d)
233. (a)	**234.** (b)	**235.** (a)	**236.** (b)	**237.** (b)
238. (e)	**239.** (b)	**240.** (c)	**241.** (c)	**242.** (b)
243. (c)	**244.** (b)	**245.** (c)	**246.** (b)	**247.** (b)

248. (*a*)	**249.** (*b*)	**250.** (*d*)	**251.** (*b*)	**252.** (*c*)
253. (*d*)	**254.** (*d*)	**255.** (*a*)	**256.** (*c*)	**257.** (*e*)
258. (*d*)	**259.** (*b, c*)	**260.** (*d*)	**261.** (*d*)	**262.** (*b*)
263. (*c*)	**264.** (*c*)	**265.** (*a*)	**266.** (*b*)	**267.** (*d*)
268. (*a*)	**269.** (*a*)	**270.** (*a*)	**271.** (*b*)	**272.** (*b*)
273. (*a*)	**274.** (*c*)	**275.** (*d*)	**276.** (*c*)	**277.** (*d*)
278. (*b*)	**279.** (*d*)	**280.** (*c*).		

B. Match List I with List II :

281. (*d*)	**282.** (*c*)	**283.** (*a*)	**284.** (*d*)	**285.** (*c*)
286. (*a*)	**287.** (*d*)	**288.** (*b*)	**289.** (*a*)	**290.** (*d*)
291. (*c*)	**292.** A–R, B–P, C–S, D–T		**293.** (*b*)	**294.** (*d*)
295. (*d*)	**296.** (*c*)	**297.** (*a*)	**298.** (*c*)	**299.** (*a*)
300. A–5, B–1, C–2, D–4		**301.** (*d*)	**302.** (*a*)	
303. A–3, B–2, C–1, D–5		**304.** (*a*)	**305.** (*c*)	**306.** (*b*)
307. A–S, B–P, C–R, D–T		**308.** (*a*)	**309.** (*b*)	
310. A–2, B–5, C–6, D–3, E–4, F–1		**311.** (*a*)	**312.** (*a*)	**313.** (*d*)
314. A–2, B–6, C–1, D–3		**315.** (*d*)	**316.** (*a*)	
317. A–3, B–4, C–1, D–2		**318.** A–3, B–1, C–4, D–2		**319.** (*b*)
320. A–2, B–3, C–4, D–1				

Index